D0782187

SEDIMENTARY ROCKS

SEDIMENTARY ROCKS

THIRD EDITION

F. J. PETTIJOHN

The Johns Hopkins University, Baltimore

HARPER & ROW, PUBLISHERS

New York, Evanston, San Francisco, and London

To my wife, Dorothy,
without whose patience and
understanding this book would
never have been written

Sponsoring Editor: Ronald K. Taylor
Project Editor: Brenda Goldberg
Designer: Jared Pratt
Production Supervisor: Stefania J. Taflinska

SEDIMENTARY ROCKS, Third Edition

Library of Congress Cataloging in Publication Data

Pettijohn, Francis John, 1904-
Sedimentary rocks.

Includes bibliographies.
1. Rocks, Sedimentary. 2. Sedimentation and deposition. I. Title.
QE471.P46 1975 552'.5 74-12043
ISBN 0-06-045191-2

CONTENTS

5
GEOMETRY OF SEDIMENTARY BODIES 140

6
GRAVELS, CONGLOMERATES AND BRECCIAS 154

7
SANDS AND SANDSTONES 195

11 NONCLASTIC SEDIMENTS (EXCLUDING LIMESTONES) 392

12
CONCRETIONS, NODULES, AND OTHER DIAGENETIC SEGREGATIONS 462

13
PROVENANCE 483

14
PALEOCURRENTS AND PALEOGEOGRAPHY 506

15
ENVIRONMENTAL ANALYSIS 530

16
SEDIMENTATION AND TECTONICS 571

17
SEDIMENTS AND EARTH HISTORY 588

PREFACE

The second edition of *Sedimentary Rocks* was written over eighteen years ago. Since then there has been a vast expansion of our knowledge of limestones and modern carbonate sediments, a renewed interest in sedimentary structures and in paleocurrent analysis, important advances in our understanding of chemical and physical sedimentation, and a recognition of the importance of the vertical profile in environmental analysis. The study of modern sediments has moved from an epidermal or two-dimensional approach to a three-dimensional one. A number of monographic works on sandstones, limestones, sedimentary structures, and sedimentary processes have been published, and we are now being forced to reexamine our concepts of sedimentation and tectonics on a global scale. All this and more has made revision of *Sedimentary Rocks* impossible; it has made rewriting imperative.

But the basic approach established in the first two editions and embodied in the third edition remains unchanged. My aim, as it was in the earlier editions, is to tell the user something about *sedimentary rocks* rather than sedimentation, and hence this edition, like the earlier ones, is *rock*-oriented rather than *process*-oriented. Its basic aim is still to show the student—confronted by an outcrop, hand specimen, or thin section—how to read rock history and how to make an interpretative analysis of what he sees. What can be said about the origin of the rock itself and what can be inferred about the environment of deposition? As in all of geology, the outcrop is the final court of appeal where all concepts or theories must be tested. Theoretical and experimental studies may suggest possible mechanisms, but it is in the field where the question of origin must be finally answered.

Like the earlier editions, most of the third edition is devoted to the task of describing and classifying the various families of sedimentary rocks and to an understanding of their textures and structures. Beyond this description there is a brief statement on origin, insofar as it can be deduced from the relevant field and laboratory observations. No attempt is made to explore the physics or chemistry of the formative processes. We leave that to such specialized works as Berner's *Principles of Chemical Sedimentology* and Allen's *Physical Processes of Sedimentation.*

In general the emphasis has moved away from thin-section analysis of sedimentary rocks (though this remains an important topic) to a closer look at the outcrop and the utilization of both sedimentary structures and the vertical profile or sequence in sedimentary analysis.

In addition to updating the various chapters on sedimentary rocks, we have included several interpretative chapters. Many of our somewhat poorly organized concepts and vague goals have become, as a result of new approaches, more clearly defined and better formulated. It is these interpretative principles and concepts that make up the last third of this edition. It is my hope that here I have given the student some insight and an elementary understanding of the contributions of sedimentology to environmental and paleocurrent analysis, to the problems of provenance, and to the general subject of global evolution and development. I would place the most emphasis on the identification of the agent and/or environment of deposition in which the sediment accumulated. How does one go about reconstruction of the environment? This is seldom made explicit in existing textbooks on stratigraphy and sedimentology. To some extent it is still difficult to make the principles explicit; understanding may best come from examples, and hence several are given for the reader's benefit.

In these chapters we have encroached, to

some extent, on what traditionally has been taught in some stratigraphy courses. We do not, however, touch on the core problems of stratigraphy—the age and correlation problems.

In rewriting the third edition and attempting to integrate the study of sedimentary rocks with the larger problems, it soon became apparent that some policy had to be adopted to deal with the vast literature. An exhaustive treatment is impossible. The references to the literature, therefore, are very selective. An effort was made to include all the larger monographic works on special topics such as sedimentary structures, carbonate rocks, other special classes of sediments, paleocurrents, and the like. I have attempted also to cite at least one modern reference on each topic and also to provide an "illustrative reference" as an example of study of each of the several sedimentary rock groups or classes. For the reader's convenience, the references are collected at the end of each chapter.

As usual I have been helped immeasurably in preparing this edition by many persons, some of whom have read portions of the text and made helpful suggestions, others who have contributed photographs or given permission to use other illustrative materials. Credit for use of photographic and other materials has been acknowledged at the appropriate places in the text. In particular I want to acknowledge my debt and thanks to my colleagues S. M. Stanley and O. P. Bricker for critical reading of parts of the text, to C. H. Weber for help with photographic materials, to Ernst Cloos for his advice and assistance in taking many of the microphotographs, and last but not least to Kathleen Shannon for her typing of the entire manuscript. I am indebted also to many others who have contributed a picture or two or have given permission to use previously published materials. Appropriate acknowledgment has been made to these persons at the relevant places in the text.

Baltimore, Maryland
February 28, 1974 *F. J. Pettijohn*

1

INTRODUCTION

DEFINITIONS

A sedimentary deposit is that body of solid materials accumulated at or near the surface of the earth under the low temperatures and pressures which normally characterize this environment. The sediment is generally, but not always, deposited from a fluid in which it was contained either in a state of suspension or solution. This definition encompasses most of the materials considered sediments (or sedimentary rocks), although some accumulations, such as the fragmental materials expelled from volcanoes, commonly airborne and deposited in a solid condition, may be formed at higher temperatures and others, such as the deposits made on the deep-sea floor, collect under pressures much greater than normal.

Sedimentary petrology is that branch of petrology which deals with the sedimentary deposits, and their ancient equivalents—the sedimentary rocks. As petrography is the description of rocks, *sedimentary petrography* is the description of the sedimentary rocks. In the United States the term *sedimentation* has been generally used to designate the study of sedimentary deposits. In a strict sense, however, sedimentation is the process of sediment accumulation and is primarily applied (in areas such as physics and engineering) to the settling of solid particles from a fluid. Because sedimentary petrology is often thought to be only the microscopic investigation of the sediments, the term *sedimentology* has been suggested (by Wadell in 1932) as the proper name for the science of sedimentary deposits. Sedimentology is considered by many to be a broader field than sedimentary petrology. The latter has often been applied only to the laboratory study—mainly of thin sec-

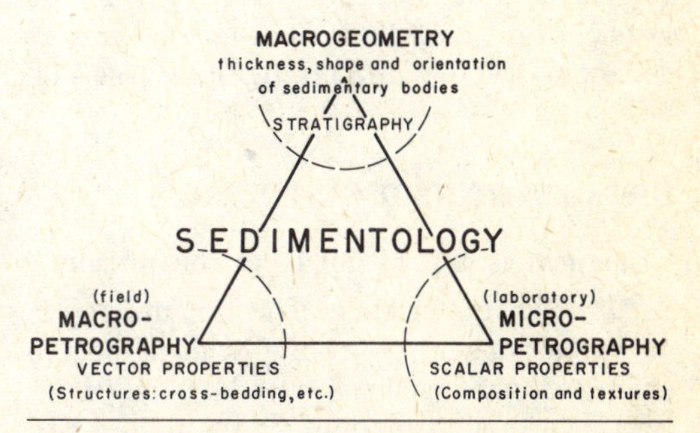

FIG. 1-1. Relations of sedimentology to stratigraphy and sedimentary petrology.

tions—of sedimentary rocks, whereas sedimentology combines the laboratory observations with those made in the field (Vatan, 1954, pp. 3–8). The term has found favor among European workers and was given status at the time the International Association of Sedimentology was organized.

The line between sedimentology and stratigraphy has not been clearly drawn. *Stratigraphy* in the broadest sense is the science dealing with strata and could be construed to cover all aspects—including textures, structures, and composition. In practice, however, stratigraphers are mainly concerned with the stratigraphic order and the construction of the geologic column. Hence the central problems of stratigraphy are temporal and involve the local succession of beds (order of superposition), the correlation of local sections, and the formulation of a column of worldwide validity. Although these are the objectives of stratigraphy, the measurement of thickness and description of gross lithology are commonly considered a part of the stratigrapher's task. Much of our knowledge of the many peculiarities of sedimentary deposits—bedding, cross-bedding, and other features readily

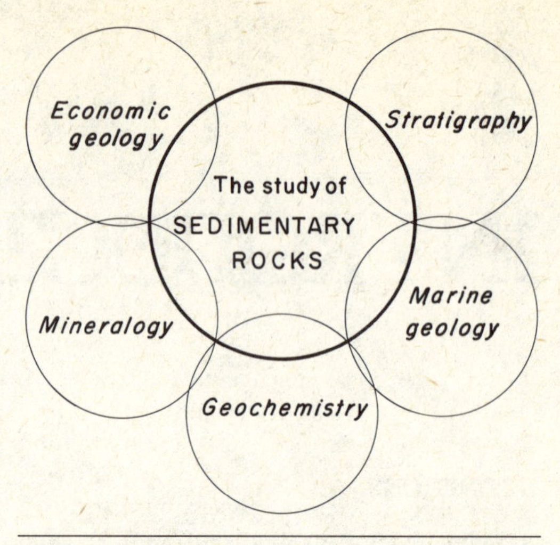

FIG. 1-2. Relation of sedimentology to the other geological sciences.

seen in the outcrop—is attributable to the close observation of stratigraphers.

The study of the sedimentary rocks cannot be pursued independently of the other geologic disciplines. Many of these, such as mineralogy, geochemistry, and marine geology, contribute to the understanding of the sedimentary deposits. The study of the latter in turn contributes to the problems of economic geology and stratigraphy (see Fig. 1-2).

HISTORY OF SEDIMENTOLOGY

"The past is but prologue to the present."

Although the science of sedimentology is of recent origin, man's knowledge of the sediments goes far back into the past. Primitive man knew something of the nature and manner of occurrence of flint, from which he fashioned knives and spear and arrow points, something about the clays from which he made his pottery, and something of the ochers which he used for pigments. No doubt he coined many of the terms which we still use, such as cobble, pebble, and flint. These terms inherited from a prescientific era are still used; others such as hornstone, flagstone, or puddingstone have become obsolete.

The first recorded speculations seem to be those of the ancient Greeks, some of whom made surprisingly astute observations on natural processes of sedimentation (Krynine, 1960, p. 1721). Despite this early interest no real science emerged.

The first truly significant work on sedimentary rocks was largely stratigraphic. Field studies were directed toward the gross geometry of the sedimentary bodies—a determination of their thickness and lateral extent. A significant milestone in the study of the sedimentary deposits was the 1815 publication of the geologic map of England by William ("Strata") Smith, a self-taught engineer and land surveyor. Smith's map, the result of many years of work and more than 11,000 miles on horseback, was one of the first successful attempts to portray the distribution and ordered arrangement of the sedimentary rocks of a region. Smith's special contribution was his discovery of the usefulness of fossils for correlation. Sedimentology, therefore, had its beginnings in the science of stratigraphy to which it is indeed closely related. The early stratigraphers contributed much to our knowledge of sedimentary rocks and many of their observations were embodied in their scientific reports and papers. Grabau's *Principles of Stratigraphy* (1913) contains an extended summary of such knowledge. The *Treatise on Sedimentation* by W. H. Twenhofel, first published in 1928, is in many ways a "lineal descendent" of Grabau's work.

The study of sediments as a discipline separate from stratigraphy may be said to have begun with the presidential address that Henry Clifton Sorby delivered before the Geological Society of London in 1879. Although Sorby's interest in sediments is clearly shown by his first paper published in 1850, his address "On the Structure and Origin of Limestones" and his paper "On the Structure and Origin of the Non-Calcareous Stratified Rocks" published the following year marks a turning point in the study of sedimentary rocks. Sorby inaugurated the thin-section study of sedimentary rocks, a technique destined to play a highly important role in the study of rocks in general. But Sorby's interest in sedimentary rocks was more than an interest in them as parts of a stratigraphic column. He was concerned with these rocks as entities in themselves worthy of study and understanding. Sorby was indeed a pioneer and is generally acknowledged as the "father of petrology." He was in many ways far ahead of his time. His 1859 paper on cross-lamination, for example, discloses an understanding of the usefulness of primary sedimentary structures in paleogeographic reconstruction —a view which only recently gained wide recognition. His last paper, published in 1908 just before his death at the age of 82, on "On the Application of Quantitative Methods to the Study of the Structure and History of Rocks," foreshadowed developments destined to come several decades later. Sorby's role in the development of petrology has been summarized by Folk (1965).

The technique of thin-section study of rocks was avidly seized upon by the students of the igneous rocks, notably by German petrographers Rosenbusch and Zirkel, among others. On the whole it was neglected by the students of the sedimentary rocks, primarily, perhaps, because these investigators were trained in stratigraphy and not petrography. The most notable exception was Lucien Cayeux, whose monographic publications on the petrography of the sedimentary rocks of France have not yet been surpassed. Cayeux's first paper appeared in 1899, his last in 1931. They include the well-known, beautifully illustrated, monographs on the sedimentary ironstones, phosphates, and siliceous and calcareous rocks of France. The thin-section study of sedimentary rocks, although long neglected, is now commonplace perhaps, because of the growth of sedimentology as a geological science independent of stratigraphy actively pursued by a new generation trained in petrographic methods. Noteworthy contributions include the extended series of longer papers by Hadding on the sedimentary rocks of Sweden, the pioneer work of Marcus Goldman and the more recent studies of Krynine and his students in the United States, Folk's work on limestones, Hallimond and Taylor's work on the British ironstones, similar work by Deverin in Switzerland, the studies of microfacies by Cuvier, the publications of Shvetzov, Ruhkin, Strakhov, and other Soviet sedimentologists, the work of Correns and his institute at Göttingen, and so forth.

In the main and until recently, however, the sedimentary petrologists—Cayeux excepted—devoted their energies mainly to the study of the mineralogy of sediments, most especially the mineralogy of the sands and to the so-called "heavy minerals" in particular (see Boswell 1933 for resumé of the history of this branch of sedimentary petrology). The study of the heavy minerals (SG $>$ 2.85) was especially prosecuted by investigators of continental Europe and Great Britain. Among the early workers in this field were Artini in Italy, Thoulet in France, Retgers in Germany, and Thomas and others in Britain. The observation of Illing (1916) that each sedimentary unit in a particular basin tended to have a unique assemblage of detrital minerals led to the so-called "heavy-mineral correlation." The usefulness of heavy minerals as a stratigraphic tool and the application of such knowledge to subsurface correlations in the exploration for petroleum greatly increased the interest in the subject. This phase of the science culminated in Milner's *Principles of Sedimentary Petrography,* which first appeared in 1922, and which is essentially a handbook for the study of the detrital minerals of sands. Edelman and others of the Dutch school have been most active in recent years. In general, however, the interest in the subject has declined, in part because of the doubts cast on the stratigraphic value of heavy minerals by Sindowski, Weyl, and others, but mainly because of the greater usefulness of microfossils and, more recently, of geophysical logging techniques for subsurface correlation. Unfortunately the use of heavy minerals as a crutch for stratigraphers in the absence of fossils has tended to obscure the value of these minerals in other respects and has tended also to make the term *sedimentary petrography* synonomous with heavy-mineral correlation.

In 1919 C. K. Wentworth's master's thesis, "A Field and Laboratory Study of Cobble Abrasion," was published in the *Journal of Geology*. Wentworth, then a graduate student at the University of Iowa, initiated a new approach to the study of sedimentary materials. In essence he was able to frame a definition of roundness such that this property, exhibited by detrital materials, could be readily *measured*. The replacement of subjective judgments by objective measurements led to the collection of a large volume of numerical data and also made possible experimental studies in the laboratory of such processes as cobble abrasion. Thus Wentworth inaugurated an epoch of measurement and controlled experimentation. It is true that there had been earlier experimenters, the most noted being Daubrée, and some earlier quantitative studies (the size analyses of detrital sediments, for example), but by and large most work on sediments had been of a qualitative and somewhat subjective nature.

Wentworth's first paper was followed by others which further demonstrated the fruitfulness of this approach and led to a great expansion of quantitative studies so that during the next two decades quantitative methods were applied to many other sedimentary properties. Following the collection of numerical data came the need for summarizing and characterizing such data. The improvised methods used at first were soon supplanted by more generally accepted statistical procedures. Although Wentworth himself attempted to use statistical methods for grain size data, Parker Trask appears to be the first to have devised statistical procedures that found ready acceptance.

The measurement of grain size of clastic sedimentary materials ("mechanical analyses") had been achieved in various disciplines, notably the soil sciences. But the work of Udden, a professor at Augustana College in Rock Island, Illinois, on windblown sediments (1899) and his larger paper (1914) are among the first studies in which size analyses were used for the elucidation of the geologic history of a sedimentary deposit. (For a history of the study of grain size see Krumbein, 1932.) Krumbein and others have since refined the methods of analysis and applied statistical techniques in a more sophisticated manner to particle size and other attributes.

The rise of *geochemistry* as a separate geological science has led to new methods and new observations of great interest to the sedimentologist. The early work of the geochemists was largely devoted to quantitative studies of the distribution of the chemical elements in nature, including their distribution in the sedimentary deposits. Such data led gradually to an understanding of the "geochemical cycles" or the laws which govern this distribution and the processes which bring it about. More recently, nuclear chemistry has contributed a "clock" and a "thermometer" which have opened up new lines of investigation. The radioactive elements, especially ^{14}C and ^{40}K, have made possible direct-age determinations of some sediments. The ^{14}C method of Libby is applicable to Recent and near-Recent deposits; the $^{40}K/^{40}A$ method of age determination has been applied with some success to glauconite and to a lesser extent to authigenic potash feldspars, clay minerals, and sylvite in the older deposits. Isotopic analysis also makes possible the determination of paleotemperatures. The method of Urey, involving the temperature-dependent $^{16}O/^{18}O$ ratio, has been applied to estimation of temperature of formation of shells in ancient marine deposits. Although the "clock" and the "thermometer" are faulty in many respects, they have made significant contributions to the study of the sedimentary record.

Van't Hoff first utilized the principles of the phase rule to study the crystallization of brines and the formation of salt deposits. Experimental study of crystallizing mixtures was, however, largely confined to high-temperature silicate systems of interest to the students of igneous and metamorphic rocks. Only recently have significant advances been made in the investigation of systems of interest to sedimentologists. These developments have been reviewed and summarized by Eugster (1971). It has been shown, for example, that this approach is applicable to the nonmarine salt deposits and to those minerals which characterize the Green River Formation (Eocene) of Wyoming and Colorado (Milton and Eugster, 1959). Zen (1959) has demonstrated the applicability of Gibbs's phase rule to the clay mineral–carbonate relations in sedimentary rocks, an approach applied by Peterson (1962) to the carbonate rocks of eastern Tennessee.

Some theoretical and experimental work

on the stabilities of minerals at various oxidation–reduction potentials (*Eh*) and at various values of *p*H has been done by Garrels and others (Garrels and Christ, 1965). The developments in sedimentary geochemistry promise to increase our understanding greatly of the sedimentary deposits. Indicative of the expanded interest in this subject are Degens's *Geochemistry of Sediments* (1965) and Berner's *Principles of Chemical Sedimentology* (1971).

That the study of Recent sediments is essential to the understanding of the ancient sediments is a corollary which follows from the uniformitarianism theory of James Hutton. With the exception of Walther, Thoulet, and a few others, sedimentologists have been prone to neglect, until recent years, this aspect of their science. Our knowledge of Recent sediments, especially of the marine sediments, has been largely the product of oceanographic surveys; one of the first, and perhaps the most famous, was that of the *Challenger*. The publication of the epoch-making *Challenger Reports* in 1891 ensured the establishment of oceanography as a discipline in its own right. Contained in these reports are many data on the distribution and nature of the sediments on the sea floor, especially in the deep oceans. Subsequent expeditions by the *Gazelle*, the *Meteor*, the *Blake*, and other pioneer oceanographic vessels have contributed greatly to our knowledge of marine sediments. In more recent years geologists have assumed a greater role in the study of these deposits. Stetson at Woods Hole and Shepard at Scripps were among those who have contributed much and who are responsible for reawakened interest in the marine deposits. Deltaic and littoral sediments have also been intensively studied in later years, in particular by Fisk in the United States, by van Straaten and others of the Dutch school, and by the group at Senckenberg. *Recent Marine Sediments*, edited by Parker Trask (1939), is tangible proof of the increased interest in the subject. The American Petroleum Institute project on the sediments of the Gulf of Mexico and the studies of van Straaten on the tidal flats of Holland, of van Andel on the Rhine and the Orinoco, of Kruit and van Andel on the Rhone delta, and of Ginsburg on the carbonate deposits of Florida and the Bahamas are examples of the interest of geologists in modern sediments.

With notable exceptions, too often the studies of the modern deposits have been made without reference to the geologic record so that such studies have failed to yield the information needed to understand the outcrops with which the field geologist is concerned. This has been primarily because samples were collected from the present sediment–fluid interface, and only the textural and mineralogical attributes were studied. The more fruitful studies of Holocene sediments have been three-dimensional, involving borings which reveal the *geometry* of the deposit, the *vertical sequence* of beds, and the primary *sedimentary structures*.

The three-dimensional approach to the study of Recent sediments has led to a closer look at the *geometry* and the vertical profile of both Recent and ancient sediments. Much of the interest in sedimentary geometry has centered on the shapes and dimensions of sand bodies, a topic that was the theme of a symposium in 1960 (Peterson and Osmond, 1961). A comparable interest has been shown in the morphology of both ancient and modern reefs (see, for example, the Reef Issue of the *Bulletin of the American Association of Petroleum Geologists*, v. 34, no. 2, 1950).

Historically, stratigraphy has been an observational science with little attention paid to the genesis of stratigraphic sequences. Walther's Law of Facies states that where there are no time breaks in a stratigraphic section, those sediments which are areally adjacent must succeed each other vertically. As a result of the study of Recent sediments, this concept has been used to construct sedimentary or *facies* models related to a sedimentary process such as transgression or regression. The process permits not only description of the vertical profile but also an understanding of the mechanism of its formation. The facies model concept is, perhaps, the single most important advance in environmental analysis in recent years. An early summary of the utility of the vertical profile in environmental reconstruction is outlined by Visher (1965), and a good elementary treatment is given by Selley (1970). Excellent examples of this approach to stratigraphy are the papers by de Raaf, Reading, and Walker (1965) on the Carboniferous strata of the north coast of Devon and the landmark studies of Allen (1962) on the Lower Old Red Sandstone of the Welsh borderlands and similar sequences elsewhere (Allen, 1970b).

Studies of the vertical sequences involve not only lithology and facies fossils but also *sedimentary structures.* The renewed interest in sedimentary structures has led to a plethora of papers on their classification, their environmental significance, their utility in paleocurrent analyses, and, of course, their origin—an aspect that involves both fluid and soil mechanics. As McKee (1971) has noted, ". . . the long-known but frequently ignored principle that only the sedimentary structures—not the composition or texture—unequivocally indicate the depositional environment." Hence many of the studies of Recent sediments have been directed to identification of the particular structure or constellation of structures which are diagnostic of each particular environment. Perhaps just as important is the sequence of structures seen in the vertical profile.

The study of sedimentary structures is not only of interest in relation to environmental analysis but also as a guide to the current system prevailing during the accumulation of the sediment. The paleocurrent system can be reconstructed by measurement and mapping of the primary current structures, as Sorby foresaw over 100 years ago. Although such current structures—cross-bedding, ripple marks, and the like—have long been known, the systematic recording of current azimuths in a "businesslike manner," as Sorby expressed it, is a comparatively new development. Partial and incomplete studies are those of Ruedemann (who plotted the orientation of fossil debris in the Utica Shale in New York in 1897), of Hyde (who mapped Berea ripple marks in Ohio in 1911), and of Rubey and Bass (who mapped the cross-bedding in channel sandstones in the Dakota Formation of Kansas in 1925). Integrated paleocurrent studies date from the work of Hans Cloos and his students in 1938. Since 1950 such studies have become commonplace. The renewed interest in sedimentary structures is reflected in the publication of several monographic compilations on the subject (see Chapter 4, Sedimentary Structures).

Interest in the primary sedimentary structures has led naturally to closer inquiry into the mode of their formation. Inasmuch as most are currently produced, the study of the hydrodynamics of the process has received increasing attention. This interest is reflected in the publication of a symposium volume, *Primary Sedimentary Structures and Their Hydrodynamic Interpretation* in 1964 (Middleton, 1965), of a series of significant papers by Allen (1969, 1970a, 1971), and of others.

The impact of the renewed interest in sedimentary geometry, in vertical sequences, and in sedimentary structures has wrought a great change in the study of sedimentary rocks. It has led to deemphasis on the study of sedimentary textures and mineralogy—essentially a laboratory enterprise—and a reaffirmation of the study of sedimentary structures, geometry, and vertical sequences. This fusion of sedimentary petrology and stratigraphy may be properly called *sedimentology* (Doeglas, 1951). It has led also to a return to and revitalization of field studies. It is here we leave the subject in the hands of the future.

As the scope of one's interest broadens, one turns from the environmental analysis of a particular vertical sequence to consideration of the basin as a whole. Basin analysis relates tectonics and sedimentation. The study of the latter involves paleocurrent systems, facies mapping, and paleogeographic reconstructions. A growing concept holds that a clastic dispersal system produces interrelated scalar and directional properties that can be used to reconstruct the configuration of the basin and the original conditions of sedimentation and paleogeography. This concept thus brings together most of the earlier methods and concepts of sedimentary petrology and the newer field studies and leads to the formulation of a basin model. Such a model makes possible a better understanding of the fill-

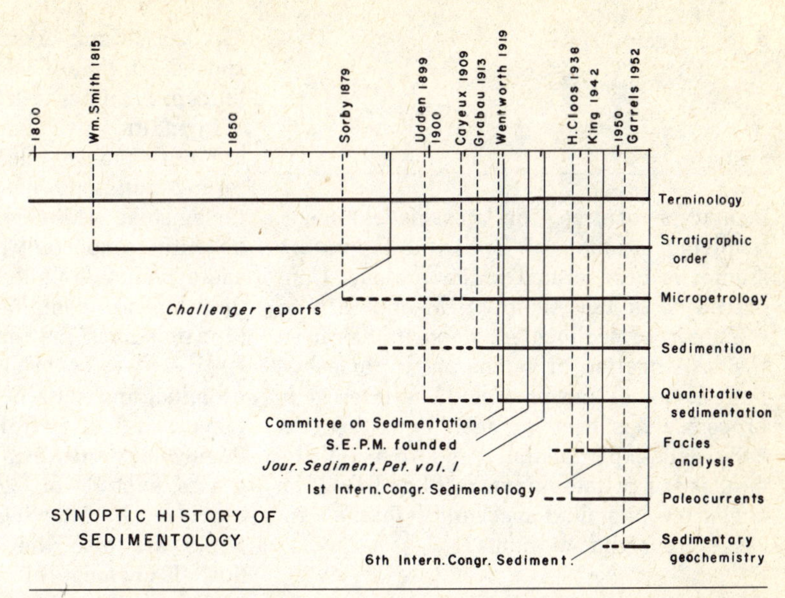

FIG. 1-3. Graphic "historogram" of the development of the various aspects of sedimentology.

ing of sedimentary basins and enables one to make downdip predictions of the distribution and character of sediments as yet unseen.

Correlative with basinwide paleocurrent analyses is the quantification and mapping of facies—largely a post-World War II development. Earlier crude attempts were made to depict facies, but the construction of adequate facies maps had to wait until a vast amount of subsurface data—both lithologic and geophysical—became available. The areas that had been extensively drilled for oil and gas provided the needed information, and the relations between oil occurrence and facies gave further impetus to the study of this subject. The symposium on facies sponsored by the Geological Society of America in 1948 is an expression of the rapid post-war rise in interest in facies. A synoptic atlas of North American Phanerozoic facies maps by Sloss, Dapples, and Krumbein appeared in 1960.

The study of sedimentary basins and an investigation of their fill and history lead one to the problem of continental evolution. The relation between tectonics and sedimentation, between cratonic and geosynclinal sediments, and between sedimentation and plate tectonics are some of the larger problems being investigated. The interest in the sedimentation and tectonics goes back to early papers by Bertrand (1897) and Tercier (1939), but the present interest began with papers by Krynine (1942, 1951) and Pettijohn (1943). Related problems are the questions of mass balance and transfer of materials, subjects being pursued especially by Ronov, Migdisov, and Barskaya (1969) and Garrels and Mackenzie (1971). Such studies have a strong historical aspect and involve earth history from the earliest days. Such problems involve all of geology —all of its subdisciplines—but the sedimentologic aspects are among the most important, as they carry the record of past events.

It can thus be seen that sedimentology has passed through four stages: (1) the study of sedimentary deposits as stratigraphic entities; (2) the collection of data on the sedimentary rocks and the formulations of tentative interpretations, exemplified by the work of Sorby, Murray and Renard, Grabau, and Twenhofel; (3) the development of sedimentary petrography as a separate discipline through the efforts of Cayeux, Hadding, Milner, Krynine, and others with special emphasis on thin-section studies of the ancient rocks and laboratory analyses of the textural and mineral attributes of the unconsolidated sediments; and finally (4) the three-dimensional study of sedimentary deposits, both modern and ancient, and environmental analysis based on geometry, the vertical profile, and sedimentary structures. The latest development, involving a union of field and laboratory studies is best termed *sedimentology*. Concurrent with this evolution in approach to the study of sedimentary *deposits* has been the investigation of sedimentary *processes*, both chemical and physical. The interest in processes has been pursued both theoretically and experimentally and involves the application of thermodynamics and solution chemistry and fluid mechanics to the processes of sedimentation.

ECONOMIC VALUE OF SEDIMENTS

"According to available statistical data, about 85–90 percent of the annual yield of mineral products come from sedimentary mineral and ore deposits . . ." (Goldschmidt, 1937, p. 664). This being so, it is manifestly impossible to enumerate and discuss all the mineral products of sedimentary origin which are useful to man. A brief review of the economic value of sands has been made elsewhere (Boswell, 1919, pp. 4–11; Pettijohn, Potter, and Siever, 1972, pp. 12–13). Special works too numerous to mention deal with clays, coal, petroleum, and other economic products of sedimentary origin.

Of sedimentary origin are the mineral fuels—natural gas, petroleum, coal, and oil shale. The first two occupy pores in the sediments, whereas the other two are sedimentary rocks. Sedimentary deposits are the raw materials for the ceramic and portland cement industries. Other nonmetallic resources of sedimentary origin include sand and gravel, lime, building stone, molding sand, Fuller's earth, and diatomite. Of sedimentary origin also are the mineral fertilizers—phosphate, potash salts, and some nitrates. Even the ores of many metals are won from sedimentary deposits. These include most of the ores of iron and aluminum and some of the ores of manganese, copper, uranium, and magnesium. From sedimentary placers tin, tungsten, gold, platinum, various gem stones, and certain rare elements such as zirconium and thorium are recovered. The search and exploitation of placers is a special art, that of alluvial prospecting (Raeburn and Milner, 1927; Smirnov, 1965).

In addition to the uses to which sediments are put, and in addition to the ores and rarer constituents which are recovered from them, some sediments (notably the sands) constitute important reservoirs for the storage of valuable fluids. The large volumes of fluid that are contained in the pore systems of sandstones are important reservoirs of fresh water; of brines from which iodine, bromine, and salts of various kinds are obtained; and of petroleum and natural gas. Some limestones also are oil-bearing. From others, such as oil shale, hydrocarbons are recovered by retorting and distillation. Sand strata are also the conduits for artesian flow. Knowledge of the shape and attitude of these reservoirs and their porosity and permeability is necessary for extraction of their fluids. Because fluids may also be injected into sands, sands thus may be utilized for natural gas storage and for recharge with fresh waters for future use or injected with water to drive out the contained oil.

The economic aspects of sediments are not confined to their value as raw materials and their various uses. Sediment production, movement, and deposition are of concern to engineering geologists and geomorphologists, especially those concerned with shore erosion, harbor development, floodplain management, and soil erosion.

In short, to be an economic geologist one must first be a sedimentologist.

The growth of sedimentology as a separate geological discipline has been accompanied by the establishment of specialist societies and journals. The growth of the petroleum industry and the employment of geologists led to an upsurge in interest in the sedimentary deposits. In 1920, in response to this interest, the Committee on Sedimentation of the National Research Council (USA) was organized under the chairmanship of W. H. Twenhofel. The committee undertook the compilation and publication of the *Treatise on Sedimentation*, first appearing in 1928 and revised and reissued in 1932. *Recent Marine Sediments* was also prepared and published under the auspices of the Committee, as was *Applied Sedimentation* (1950). The Society of Economic Paleontologists and Mineralogists was organized as a division of the American

Association of Petroleum Geologists in 1927. It has become the principal society of stratigraphers (micropaleontologists) and sedimentologists in the United States. The *Journal of Sedimentary Petrology,* published since 1930, is one of the society's journals. The International Association of Sedimentology was established in 1946; it held its eighth international meeting in Heidelberg in 1971. Its official journal is *Sedimentology,* now in its twentieth volume. Still another journal, *Sedimentary Geology,* devoted to the study of sediments, appeared in 1967.

Whereas a decade or two ago the student had few major works on sediments and sedimentary processes, he is now confronted with a proliferation of books on these subjects. An annotated list is appended to this chapter.

References

Allen, J. R. L., 1962, Petrology, origin and deposition of the highest Old Red Sandstone of Shropshire, England: Jour. Sed. Petrology, v. 32, pp. 657–697.

———, 1969, Some recent advances in the physics of sedimentation: Proc. Geol. Assoc. London, v. 80 pp. 1–42.

———, 1970a, Physical processes of sedimentation: London, Allen and Unwin, 248 pp.

———, 1970b, Studies in fluviatile sedimentation: a comparison of fining-upwards cyclothems, with special reference to coarse-member composition and interpretations: Jour. Sed. Petrology, v. 40, pp. 298–323.

———, 1971, Bed forms due to mass transfer in turbulent flows: a kaleidoscope of phenomena: Jour. Fluid Mechanics, v. 79, pp. 49–63.

Berner, R. A., 1971, Principles of chemical sedimentology: New York, McGraw-Hill, 240 pp.

Bertrand, M., 1897, Structure des Alpes françaises et recurrence de certain facies sédimentaires: Proc. 16th Int. Géol. Congr., 1894, pp. 161–177.

Boswell, P. G. H., 1919, Sands: considered geologically and industrially, under war conditions: Inaugural Lecture, 1917, Univ. Liverpool, Univ. Press of Liverpool.

———, 1933, On the mineralogy of the sedimentary rocks: London, Murby, 393 pp.

Cloos, H., 1938, Primäre Richtungen in Sedimenten der rheinischen Geosynkline: Geol. Rundschau, v. 29, pp. 357–367.

Degens, E. T., 1965, Geochemistry of sediments: Englewood Cliffs, N.J., Prentice-Hall, 342 pp.

de Raaf, J. F. M., Reading, H. G., and Walker, R. G., 1965, Cyclic sedimentation in the Lower Westphalian of North Devon, England: Sedimentology, v. 4, pp. 1–52.

Doeglas, D. J., 1951, From sedimentary petrology to sedimentology: Proc. 3rd Int. Congr. Sedimentology, pp. 15–22.

Eugster, H. P., 1971, The beginnings of experimental petrology: Science, v. 173, pp. 481–489.

Folk, R. L., 1965, Henry Clifton Sorby (1826–1908), the founder of petrography: Jour. Geol. Educ., v. 13, pp. 43–47.

Garrels, R. M., and Christ, C. L., 1965, Solutions, minerals, and equilibria: New York, Harper & Row, 450 pp.

Garrels, R. M., and Mackenzie, F. T., 1971, Evolution of sedimentary rocks: New York, Norton, 397 pp.

Goldschmidt, V. M., 1937, The principles of distribution of chemical elements in minerals and rocks: Jour. Chem. Soc. for 1937, pp. 655–673.

Grabau, A. W., 1913, Principles of stratigraphy: v. 1, pp. 1–581, v. 2, pp. 582–1185, New York, Dover (reprinted 1960).

Hyde, J. E., 1911, The ripples of the Bedford and Berea formations of central Ohio with notes on the paleogeography of that epoch: Jour. Geol., v. 19, pp. 257–269.

Illing, V. C., 1916, The oilfields of Trinidad: Proc. Geol. Assoc., v. 27, p. 115.

Krumbein, W. C., 1932, A history of the principles and methods of mechanical analysis: Jour. Sed. Petrology, v. 2, pp. 89–124.

Krynine, P. D., 1942, Differential sedimentation and its products during one complete geosynclinal cycle: Proc. 1st Pan-American Congr. Min. Eng. Geol., v. 2, part 1, pp. 537–561.

———, 1951, A critique of geotectonic elements: Trans. Amer. Geophys. Union, v. 32, pp. 743–748.

———, 1960, On the antiquity of "sedimentation" and hydrology: Bull. Geol. Soc. Amer., v. 71, pp. 1721–1726.

McKee, E. D., 1971, Book review: Geotimes, v. 16, p. 38.

Middleton, G. V., ed., 1965, Primary sedimentary structures and their hydrodynamic interpretation: Soc. Econ. Paleont. Min. Spec. Publ. 12, 265 pp.

Milner, H. B., 1922, An introduction to sedimentary petrography, 1st ed.: London, Murby, 125 pp.

Milton, C., and Eugster, H. P., 1959, Mineral assemblages of the Green River Formation, *in*

Researches in geochemistry (Abelson, P. H., ed.): New York, Wiley, pp. 118–150.

Peterson, J. A., and Osmond, J. C., eds., 1961, Geometry of sandstone bodies: Bull. Amer. Assoc. Petrol. Geol., 240 pp.

Peterson, M. N. A., 1962, The mineralogy and petrology of Upper Mississippian carbonate rocks of the Cumberland Plateau in Tennessee: Jour. Geol., v. 70, pp. 1–31.

Pettijohn, F. J., 1943, Archean sedimentation: Bull. Geol. Soc. Amer., v. 54, pp. 925–972.

Pettijohn, F. J., Potter, P. E., and Siever, R., 1972, Sand and sandstone: New York, Springer, 618 pp.

Raeburn, C., and Milner, H. B., 1927, Alluvial prospecting: London, Murby, 478 pp.

Ronov, A. B., Migdisov, A. A., and Barskaya, N. V., 1969, Tectonic cycles and regularities in the development of sedimentary rocks and paleogeographic environments of sedimentation of the Russian platform (an approach to a quantitative study): Sedimentology, v. 13, pp. 179–212.

Rubey, W. W., and Bass, N. W., 1925, The geology of Russell County, Kansas, pt. I: Bull. Kansas Geol. Survey 10, 104 pp.

Ruedemann, R., 1897, Evidence of current action in the Ordovician of New York: Amer. Geol., v. 19, pp. 367–391.

Selley, R. C., 1970, Ancient sedimentary environments: Ithaca, N.Y., Cornell Univ. Press, 237 pp.

Sloss, L. L., Dapples, E. C., and Krumbein, W. C., 1960, Lithofacies maps: New York, Wiley, 108 pp.

Smirnov, V. I., 1965, Geologiia rossypei (Geology of placer deposits): Div. Earth Sci., Sci. Council Ore-formation, Acad. Sci., Moscow, USSR, 400 pp.

Sorby, H. C., 1859, On the structures produced by the current present during the deposition of stratified rocks: The Geologist, v. 2, pp. 137–147.

———, 1879, On the structure and origin of limestones (presidential address, 1879): Proc. Geol. Soc. London, v. 35, pp. 56–95.

———, 1880, On the structure and origin of non-calcareous stratified rocks (presidential address, 1879): Proc. Geol. Soc. London, v. 36, pp. 46–92.

———, 1908, On the application of quantitative methods to the study of the structure and history of rocks: Quart. Jour. Geol. Soc. London, v. 64, pp. 171–232.

Tercier, J., 1939, Dépôts marins recents et séries géologiques: Eclogae Géol. Helvetiae, v. 32, pp. 47–100.

Trask, P. D., ed., 1939, Recent marine sediments, a symposium: Tulsa, Okla., Soc. Econ. Paleont. Min., 726 pp. (reprinted 1955).

Twenhofel, W. H. (1928), 1932, Treatise on sedimentation, 2nd ed.: Baltimore, Williams and Wilkins, 926 pp.

Udden, J. A., 1899, Mechanical composition of wind deposits: Augustana Library Publ. No. 1, 69 pp.

———, 1914, Mechanical composition of clastic sediments: Bull. Geol. Soc. Amer., v. 25, pp. 655–744.

Vatan, A., 1954, Pétrographie sédimentaire: Paris, Editions Technip, 279 pp.

Visher, G. S., 1965, Use of vertical profile in environmental reconstruction: Bull. Amer. Assoc. Petrol. Geol., v. 49, pp. 41–61.

Wadell, H., 1932, Sedimentation and sedimentology: Science, n.s., v. 75, p. 20.

Wentworth, C. K., 1919, A laboratory and field study of cobble abrasion: Jour. Geol., v. 27, pp. 507–521.

Zen, E-an, 1959, Clay–carbonate relations in sedimentary rocks: Amer. Jour. Sci., v. 257, pp. 29–43.

TEXTBOOKS AND GENERAL REFERENCES

Bathurst, R. G. C., 1971, Carbonate sediments and their diagenesis: Amsterdam, Elsevier, 620 pp.

Blatt, H., Middleton, G., and Murray, R., 1972, Origin of sedimentary rocks: Englewood Cliffs, N.J., Prentice-Hall, 634 pp.

Chilingar, G. V., Bissell, H. J., and Fairbridge, R. W., 1967, Carbonate rocks, v. 1, 471 pp.; v. 2, 413 pp.: Amsterdam, Elsevier.

Garrels, R. M., and Mackenzie, F. T., 1971, Evolution of sedimentary rocks: New York, Norton, 397 pp.

Pettijohn, F. J., Potter, P. E., and Siever, Raymond, 1972, Sand and sandstone: New York, Springer, 618 pp.

Ruchin, L. B. (Schüller, A., transl.), 1958, Grundzüge der Lithologie: Berlin, Akademie-Verlag, 806 pp.

LABORATORY AND FIELD MANUALS

Bouma, A. H., 1969, Methods for the study of sedimentary structures: New York, Wiley-Interscience, 458 pp.

Brajnikov, B., et al., 1943, Techniques d'étude des sédiments, pt. II: Paris, Hermann, 110 pp.

Carver, R. E., ed., 1971, Procedures in sedimentary petrology: New York, Wiley-Interscience, 458 pp.

Cayeux, L., 1931, Introduction à l'étude pétrographique des roches sédimentaires: Paris, Imp. Nat., 524 pp.

Duplaix, S., 1948, Détermination microscopique de minéraux des sablé: Paris–Liège, Librairie Polytech. Ch. Beranger, 80 pp.

Folk, R. L., 1968, Petrology of sedimentary rocks: Austin, Texas, Hemphill's Book Store, 170 pp.

Griffiths, J. C., 1967, Scientific method in analysis of sediments: New York, McGraw-Hill. 508 pp.

Jones, M. P., and Fleming, M. G., 1965, Identification of mineral grains: Amsterdam, Elsevier, 102 pp.

Köster, E., 1964, Granulometrische und morphometrische Messmethoden an Mineralkörnern, Steinen, und sondstigen Stoffen: Stuttgart, F. Enke, 336 pp.

Krumbein, W. C., and Pettijohn, F. J., 1938, Manual of sedimentary petrography: New York, Plenum, 549 pp.

Milner, H. B., 1962, Sedimentary petrography, v. 1, Methods in sedimentary petrography, 643 pp.; v. 2, Principles and applications, 715 pp.: New York, Macmillan.

Müller, G., 1967, Methods in sedimentary petrology: Stuttgart, Schweizerbart'sche Verlagsbuchhandlung, 283 pp.

Plas, Leedert van der, 1966, The identification of detrital feldspars: Amsterdam, Elsevier, 305 pp.

Russell, R. D., 1942, Tables for the determination of detrital minerals, *in* Report of the Committee Sedimentation, 1940–1941: Div. Geol. Geogr., Nat. Res. Coun., pp. 6–8 (reprints available from Dept. Earth and Planetary Sci., The Johns Hopkins Univ., Baltimore, Md. 21218, $0.50.)

Strakhov, N. M., 1957, Méthode d'étude des roches sédimentaires: Service Inf. Geol., Ann., v. 1, 2, no. 35, 1007 pp.

Tickell, F. G., 1947, The examination of fragmental rocks: Palo Alto, Calif., Stanford Univ. Press, 127 pp.

———, 1965, The techniques of sedimentary mineralogy: Amsterdam, Elsevier, 220 pp.

Twenhofel, W. H., and Tyler, S. A., 1941, Methods of study of sediments: New York, McGraw-Hill, 183 pp.

SEDIMENTARY PETROGRAPHY

Carozzi, A., 1960, Microscopic sedimentary petrography: New York, Wiley, 485 pp.

Cayeux, L. (Carozzi, A. V., transl.) 1970, Carbonate rocks: Riverside, N.J., Hafner, 472 pp.

———, 1929, Les roches sédimentaires de France, v. 1, Roches siliceuses: Paris, Masson, 696 pp.

Füchtbauer, H., and Müller, G., 1970, Sedimente und Sedimentgesteine, pt. II: Stuttgart, E. Schweizerbart'sche Verlagsbuchhandlung, 696 pp.

Hatch, F. H., and Rastall, R. H. (rev. Greensmith, J. T.), 1965, Petrology of the sedimentary rocks, 4th ed.: London, Murby, 408 pp.

Horowitz, A. S., and Potter, P. E., 1971, Introductory petrography of fossils: New York, Springer, 320 pp.

Schürmann, H. M. E. (ed.), International sedimentary petrographical series (14 volumes to 1971): Leiden, Brill.

Vatan, A., 1954, Pétrographie sédimentaire: Paris, Editions Technip., 279 pp.

SEDIMENTATION AND STRATIGRAPHY

Dunbar, C. O., and Rodgers, John, 1957, Principles of stratigraphy: New York, Wiley, 356 pp.

Grabau, A. W., 1913, Principles of stratigraphy: New York, Seiler, 1185 pp. (reprinted 1960, Dover, New York).

Krumbein, W. C., and Sloss, L. L., 1963, Stratigraphy and sedimentation, 2nd ed.: San Francisco, Freeman, 660 pp.

Twenhofel, W. H., 1950, Principles of sedimentation, 2nd ed.: New York, McGraw-Hill, 673 pp.

Twenhofel, W. H., et al., 1932, Treatise on sedimentation, 2nd ed.: Baltimore, Williams and Wilkins, 926 pp.

PROCESSES, PHYSICAL AND CHEMICAL

Allen, J. R. L., 1970, Physical processes of sedimentation: New York, American Elsevier, 248 pp.

Berner, R. A., 1971, Principles of chemical sedimentology: New York, McGraw-Hill, 240 pp.

Degens, E. T., 1965, Geochemistry of sediments: Englewood Cliffs, N.J., Prentice-Hall, 342 pp.

SEDIMENTARY ENVIRONMENTS, ANCIENT AND MODERN

Kukal, Z., 1970, Geology of Recent sediments: Prague, Czech. Acad. Sci., 490 pp.

Morgan, J. P., ed., 1970, Deltaic sedimentation, modern and ancient: Soc. Econ. Paleont. Min. Spec. Publ. 15, 312 pp.

Reineck, H.-E., and Singh, I. B., 1973, Depositional sedimentary environments: New York, Springer, 439 pp.

Rigby, J. K., and Hamblin, W. K., eds., 1970, Recognition of ancient sedimentary environments: Soc. Econ. Paleont. Min.: Spec. Publ. 16, 340 pp.

Selley, R. C., 1971, Ancient sedimentary environments: Ithaca, N.Y., Cornell Univ. Press, 254 pp.

Trask, P. D., ed., 1939, Recent marine sediments: Tulsa, Okla., Amer. Assoc. Petrol. Geol., 736 pp.

SEDIMENTARY STRUCTURES

Conybeare, C. E. B., and Crook, K. A. W., 1968, Manual of sedimentary structures: Australian Dept. Nat. Devel., Bull. Bur. Min. Res. Geol. Geophys. 102, 327 pp.

Gubler, Y., Bugnicourt, D., Faber, J., Kubler, B., and Nyssen, R., 1966, Essai de nomenclature et caractèrization des principales structures sédimentaires: Paris, Editions Technip, 291 pp.

Khabakov, A. V., ed., 1962, Atlas tekstur: struktur osadochyhk gornykh porod (An atlas of textures and structures of sedimentary rocks, pt. 1, Clastic and argillaceous rocks): Moscow, VSEGEI, 578 pp. (Russian with French translation of plate captions).

Pettijohn, F. J., and Potter, P. E., 1964, Atlas and glossary of primary sedimentary structures: New York, Springer, 370 pp.

Potter, P. E., and Pettijohn, F. J., 1963, Paleocurrents and basin analysis: New York, Springer, 296 pp.

Ricci Lucchi, F., 1970, Sedimentografia: Bologna, Zanichelli, 288 pp. (Italian with Italian-English glossary).

Shrock, R. R., 1948, Sequence in layered rocks: New York, McGraw-Hill, 507 pp.

PERIODICALS AND SERIALS

Developments in Sedimentology (14 vols. to 1971): Amsterdam, Elsevier.

Journal of Sedimentary Petrology: Tulsa, Okla., Soc. Econ. Paleont. Min., since 1930.

Maritime Sediments: Halifax, N.S., and Fredericton, N.B., Canada.

Sedimentary Geology: Amsterdam, Elsevier, since 1969.

Sedimentology (official journal Int. Assoc. Sedimentologists): Amsterdam, Elsevier, since 1962; Oxford, Blackwell, since 1973.

2

NATURE AND ORIGIN OF SEDIMENTARY ROCKS

INTRODUCTION

As noted in Chapter 1, a sedimentary deposit is a body of solid materials accumulated at or near the surface of the earth under the low temperatures and pressures normally characteristic of this environment. The deposits are, traditionally, largely the result of breakdown of preexisting rocks whose weathering products have been redistributed by waves and currents or precipitated chemically or biochemically from solution. In general these are the sedimentary rocks (as opposed to the igneous rocks, which are not a product of weathering but are of magmatic or volcanic origin). This way of looking at sediments is geologically meaningful, because it relates the rocks formed to two differing geological processes —*weathering* on the one hand, *volcanism* on the other. Some sediments, however, are not derived from any preexisting rock. These deposits, generally small in volume and not so common, include coal (principally an organic residue derived from plant matter and hence derived from the atmosphere) and volcanogenic sediments (the stratified deposits of ash and other products of volcanic action). Still rarer and of almost no importance are the deposits containing meteoric matter of cosmic origin.

Grabau (1904) chose to look at rocks differently and to group them according to their physical or chemical processes of origin. He therefore divided rocks into two classes, namely exogenetic and endogenetic (Fig. 2-1). *Exogenetic rocks* are fragmental or clastic rocks. The constituents of these rocks are emplaced in the rock fabric as solid particles which have been formed by the fragmentation of preexisting materials. In other words they are mechanically deposited; they are a product of physical sedimentation. The great bulk of the sediments (by volume) belong in this category. To this group also belong the pyroclastic igneous rocks, which are similar to the clastic rocks in most essential details of texture and structure and therefore in their dynamics of accumulation.

In marked contrast are *endogenetic rocks,* which are both amorphous and crystalline precipitates from solution. Many sediments, including the saline deposits—rock salt, anhydrite, and the like—belong here, as do the bulk of the igneous rocks. These igneous rocks, like the chemical sediments, are precipitates from solution. Phase-rule chemistry governs the formation of each. There is no basic difference in principle between the crystallization of a brine and a magma.

If we then look at sediments in terms of *process* of accumulation, Grabau's classification is more meaningful than the traditional division into the sedimentary and igneous categories. In terms of process the formation of a rock salt deposit is more akin to a diabase than it is to a shale or limestone with which it is associated. On the other hand, a sandstone is closely related to a tuff. Aerodynamic or hydrodynamic principles govern the accumulation of each. These two major categories of rocks each have their own textures and structures. Exogenetic rocks have a framework-pore fabric and may be cross-bedded. Endogenetic rocks have an interlocking, crystalline granular fabric.

We can also look at sedimentary rocks in another way, namely in terms of *provenance,* which again divides these rocks into two major groups: *Intrabasinal* deposits are those formed in the basin in which they accumulate; *extrabasinal* rocks are those generated outside the basin and carried into it by waves or currents. The first category includes chemical and biochemical sediments extracted from the basin waters; the second includes both the terrigeneous sediments and the pyroclastic deposits.

The origin and accumulation of sedimentary rocks might at first seem relatively simple. Sands and muds are seen to form and to be carried by the rivers from the continents into the seas. The origin of sedimentary rocks, unlike that of many igneous and all metamorphic rocks, is apparently open to inspection and study. Unfortunately the matter is not so simple. Not all of the formative processes can be seen. Diagenetic changes in particular, which include intrastratal solution, cementation, formation of concretions, and so forth, cannot readily be observed. Neither can we observe the turbidity currents responsible for the transport, deposition, and structures of many marine sediments. The formation of many chemical sediments has never been observed. And so, as in the case of most other rocks, the origin of the rock must be reconstructed from the geologic record—the effects produced by processes no longer operative. The "effects" are primarily the textures, the structures, and the minerals of the deposit in question. This then is the proper task of sedimentary petrology: to go to the record and there read and unravel the natural history of the rock.

	SEDIMENTARY ROCKS	IGNEOUS ROCKS
EXOGENETIC ROCKS (*allogenic*)	Epiclastic (*epiclasts*)	Pyroclastic (*pyroclasts*)
ENDOGENETIC ROCKS (*authigenic*)	Biogenic and chemical (*bioliths and evaporites*)	Pyrogenic (*pyroliths*)

FIG. 2-1. Two contrasting ways of classifying rocks, the conventional and that of Grabau (1904). Either category can be remade by recrystallization or replacement at either low temperature and pressure (diagenetic and epigenetic rocks) or at high temperature and pressure (metamorphic rocks).

Even the common sediments are not so clearly either chemical (endogenetic) or mechanical (exogenetic) in origin. Most are both, and the sediment is therefore a hybrid or polygenetic deposit. As seen in Fig. 2-2, the constituents may be derived by mechanical abrasion of the parent rock, by washing or winnowing of the weathered residuum, or by chemical or biochemical additions from the seawater which fills the basin in which the sediment collects. Circulating groundwater may later deposit a large quantity of mineral matter in the pore system of the sediment.

The kind of sedimentary deposit depends on the relative importance of the several lines of supply. Those formed predominantly of the solid debris of preexisting rocks are *clastic* sediments—gravels, sands, and silts, and their indurated equivalents (conglomerate, sandstone, and siltstone). Those containing an appreciable quantity of clay minerals produced by weathering are muds (or shales and argillites if indurated). Those accumulations derived largely or exclusively from the materials in solution in seawater are the *chemical* (or biochemical) sediments. Here belong limestones and

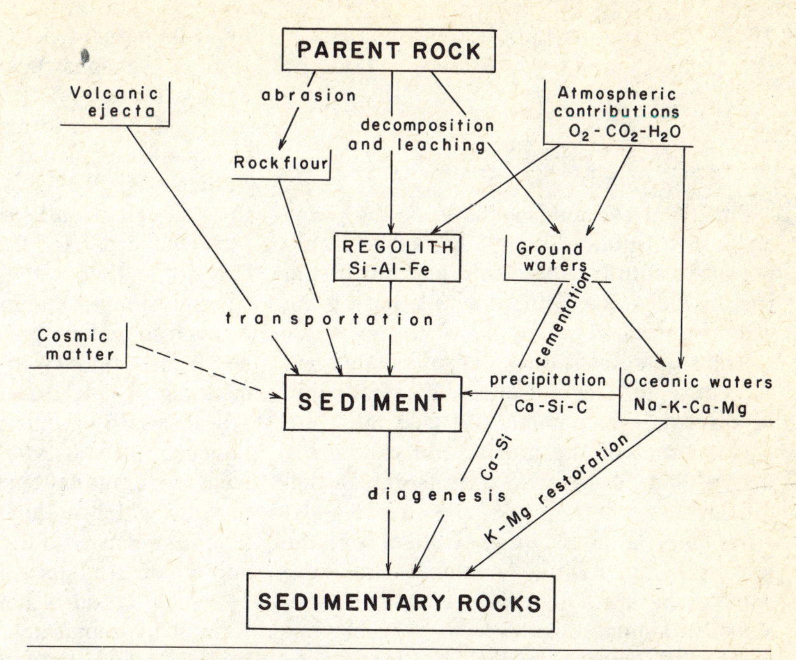

FIG. 2-2. Diagram of the complexity of origin of a sedimentary rock.

dolomites, evaporites (the products of brine evaporation), the ironstones, phosphorites, and cherts and related siliceous rocks.

FABRIC

In metamorphic and igneous rocks the constituent minerals are in continuous tightly interlocking contact; most sedimentary rocks, on the other hand, have a distinctive framework, whose elements display tangential or point contacts with one another. Because of this geometry, sediments, unlike other rocks, have high porosity and permeability. Hence they have the ability to both store and allow the passage of fluids. They are important reservoirs of natural gas, petroleum, artesian waters, and various brines. The initial porosity may be high—35 to 45 percent of the bulk volume—but it is commonly much reduced by precipitation of mineral water in the pores.

The peculiar microgeometry of sediments is caused by their manner of formation. Each framework element—sand grain, pebble, shell fragment—is formed elsewhere and is mechanically placed in its present position in the fabric. Although a few igneous rocks (notably the stratiform pyroclastics) display sedimentary geometry, most are crystalline aggregates which formed in place. The sedimentary fabric described above characterizes the clastic sediments—gravels and sands and their indurated equivalents, conglomerates and sandstones—but it also characterizes many, if not most, limestones. These limestones are merely carbonate sands and silts; their fabric differs in no essential way from ordinary sands and silts.

Because the individual grains are not and cannot be in continuous contact with one another, an inhomogeneity in the distribution of pressures exists within the sedimentary rock. The weight of overlying rocks is carried on the relatively small contact area of individual grains, whereas in the pore system the contained fluid is only under pressure equal to a column of water extending to the earth's surface (assuming a pore system freely connected to the surface). Under these conditions of unequal pressures one might expect solution of the solid materials at points of contact and precipitation of the dissolved materials in the pore spaces, a process which, if carried to completion, would eliminate the pore space and lead to equalization of the pressure field.

The solutions that fill the pore system constitute a medium in which reactions between these solutions and the framework elements can take place.

Finally, if the fluid contents of the pores are also in motion, materials can be transported in solution, both into and out of the rock, which affects the composition of the sedimentary body.

The student of sedimentary deposits must therefore consider not only the composition of the solid components but also take into account that of the mobile and chemically active fluid phases which constitute a significant part of the rock. The nearer a detrital rock is to its original condition, the greater is the relative volume of the nonsolid phase; the higher the grade of diagenesis, or metamorphism, such a rock has reached, the more does the solid framework predominate and the more does the sedimentary rock resemble a rock of metamorphic or even of igneous origin.

Some sedimentary rocks, however, do not have normal sedimentary frameworks and pore systems. These include aqueous precipitates, such as rock salt and gypsum, *in situ* accumulations such as coal, and sediments which when formed had a normal fabric but which have since been transformed by recrystallization and replacements (such as dolomites) or which have been converted to crystalline mosaics by grain growth ("secondary enlargement") such as crystalline limestones or "sedimentary marbles."

COMPOSITION

Sedimentary rocks differ from the igneous rocks in that they display a wider range of composition; but many have singularly restricted compositions. The greatest concentrations of many chemical elements known in the earth's crust are in sediments. Such concentrations are the results of repeated washings and millings of the weathered residues of older rocks; the glass sands which attain a silica content in excess of 99 percent are an example. Others are formed by remarkably selective chemical or biochemical processes which, if carried out under optimum conditions, produce such end-products as beds of high-calcium limestone (over 99 percent $CaCO_3$), rock salt, and gypsum. No igneous rocks of comparable character are known.

The minerals formed elsewhere and mechanically emplaced in the fabric of the sediment at the time of accumulation are the so-called *allogenic* minerals; those formed *in situ* are the *authigenic* minerals. It is not sufficient, therefore, merely to identify the mineral components of a sediment or even to make a quantitative modal analysis. It is necessary to place the minerals in meaningful categories—to determine which are allogenic or detrital and which are authigenic (or post-depositional). One may wish to pursue the matter further and to determine which authigenic minerals are contemporaneous with the accumulation process and which are later; to decide which formed by precipitation in a void space and which formed by replacement; and to distinguish those formed by weathering from those formed in the sedimentary and in the subterranean environments. All these decisions about the time of emplacement and the process involved, whether diagenetic, metamorphic, or weathering, require astute observations of the textural relations between the component grains. To do this requires a thin-section analysis.

Unlike the minerals of igneous and metamorphic rocks, the minerals of the clastic sedimentary rocks are not equilibrium assemblages. They are not precipitated in equilibrium with one another or with the fluid from which they settled. But although not in equilibrium, the chemical changes required to attain equilibrium generally do not take place because of the inert character of the solid materials or of the low temperatures which prevail. The mineral composition may be profoundly altered if the temperature is raised and the barriers to reaction overcome; this happens during metamorphism in the deeper levels of the earth's crust. Some reactions do occur, however, at the lower temperatures and pressures. These *diagenetic* reactions are mainly between the detrital framework components and the contained pore fluids. In those sediments which are formed by precipitation from solutions or are biochemical accumulations—many of which are metastable—diagenetic transformations are more prone to occur, and the resulting changes are more profound.

In some, perhaps most, nonclastic sediments the mineral assemblage is an equilibrium assemblage. Zen (1959) has shown that such equilibrium appears to have been attained in the sediments of the Peruvian Trench by diagenetic reactions on the sea floor. Similar equilibrium relations were established in some carbonate rocks of the Cumberland Plateau in Tennessee (Peterson, 1962). Less ambiguous examples are the deposits from evaporating brines.

CLASSIFICATION

The classification of the sedimentary rocks is a problem on which much thought has been expended and one for which no mutually satisfactory or complete solution has yet been found. It is appropriate, therefore, to inquire into the objectives of rock classifications and to continue to examine more closely the principles involved in their construction. Many authors fail to make these objectives and principles explicit, although there has been some discussion of the philosophy underlying the classification of sedimentary rocks (Grabau, 1904; Wadell, 1938; Krynine, 1948; Pettijohn, 1948; Lombard, 1949; Rodgers, 1950; Middleton, 1973).

As noted by Rodgers, the problem of classification inevitably entails the problem of terminology or nomenclature. Scientific names denote a group or class of objects and hence imply a classification. And classification is basically an attempt to group objects of concern into classes or categories to which names can be given. The first purpose of a classification, then, is to provide such groups and appropriate names which can substitute for a description of the objects so classified. Only by the aid of such names or categories can individuals communicate effectively. To be successful, therefore, a system of nomenclature and classification must have general agreement of those who have need of a system.

But, as noted by Grabau, precision in classification leads to precision in thought and so is of great value as a mental discipline. A classification embodies in shorthand form our knowledge about a subject. Thus the construction of a classification is an attempt to organize our knowledge. In short, its second objective is the schematic representation of ideas or concepts.

The definition of a class requires the choice of limiting parameters. The choice of parameters may be governed by convention or usage or simply by agreement among the interested parties. But, as genesis is the ultimate aim in any study of rocks, the parameter chosen should be genetically meaningful or significant. In biology the basis of all sound taxomonic work must be the selection of significant characters for classificatory purposes and avoidance of irrelevant peculiarities. (Not all organisms with wings, for example, should be grouped together.) Difficulties in classifying sedimentary rocks stem in large part from failure to recognize the basic differences between clastic (exogenetic) and chemical (endogenetic) rocks. The significant properties of one group are not the significant properties of the other. So to apply the same textural terms to all carbonate rocks, which are in fact polygenetic, obscures rather than elucidates their natural history.

A rock is, after all, a complex of properties of which only two or three will be chosen for purposes of classification. It is not possible to construct a classification based on all known or knowable properties. A workable classification will consider two or three and ignore all others. Such choice requires that the defining parameters be not only genetically significant but that they be the most significant. All properties no doubt have some meaning, but many are much less relevant than others. The magnetic susceptibility of a sandstone is surely genetically less important than its grain size. In general, too, the properties chosen should be readily observable and not ones that require long and complex methods of detection or measurement. Chemical analyses are important and useful but are hardly usable for a working classification of sedimentary rocks.

What are the significant properties? Those chosen as such will be the ones believed meaningful in the light of our present concepts and knowledge. New ideas and discoveries may render our choice obsolete and hence require a revision of our system of classification and nomenclature. Such instability of nomenclature is disturbing, but it is the penalty for progress and is only another proof that classifications are a codification of ideas or concepts which are subject to constant revision.

Most classifications of sedimentary rocks are traditional, and, like Topsy, they "just growed." There have been several efforts to standardize the nomenclature—to redefine terms, set quantitative limits, and weed out poor and obsolete terms (see, for example, the various reports of the Committee on Sedimentation, including Wentworth and Williams, 1932, on the pyroclastic sediments; Wentworth, 1935, on the coarse clastics; Allen, 1936, on the medium-grained clastics; Twenhofel, 1937, on the fine-grained clastics; and Tarr, 1938, on the siliceous sediments). Few efforts have been made, however, to review the whole problem and find a single integrated solution such as those proposed for the classification of the igneous rocks.

Space does not permit a detailed review of all the classifications or partial classifications proposed for sedimentary rocks. This has been treated in some measure by Lombard (1949). Inasmuch as the classification of the several subgroups or families of sediments is treated in detail in their respective sections of this book, there is no need to review these here. One can say in conclusion that any attempt to impose a single classification scheme on all sediments encounters difficulties simply because of the polygenetic nature of sedimentary materials. If a scheme is to be based on genetically significant properties, it follows that those properties significant for one group of sediments may be inappropriate to another group of different origin. The maturity concept, for example, is fundamental, but it can be applied only to those sediments which are residues derived by weathering from a metastable source rock. It has no meaning if applied to pyroclastic materials. Similarly the concepts of provenance basic to the understanding of the texture and composition of the clastic sediments have little or no meaning if applied to the chemical sediments. Hence it is difficult to build an all-embracing classification.

It is possible, however, to build partial classifications—appropriate for certain classes of sediments—which are closely correlated with fundamental concepts of origin. Much effort, for example, has gone into the problems of the classification of sandstones and of limestones. (These classifications are discussed fully in the chapters dealing with these sediments.) On the other hand, the classification of the argillaceous sediments is rather unsatisfactory at present.

In this book the sedimentary rocks are grouped according to prevailing usage (Fig. 2-3). This is done for convenience only. A chapter each is devoted to coarse clastics (conglomerates and sedimentary breccias), to sandstones, and to shales and siltstones. Volcanogenic sediments are given separate treatment, although in terms of textures and structures they are akin to the other clastic deposits. But because their constituents are of volcanic origin, the concepts of maturity and provenance applied to the weathering residues have no meaning for this group of sediments. Inclusion of all limestones (and dolomites) in a single chapter, despite their polygenetic nature, is again justified by prevailing usage. Moreover, despite the clastic textures and structures shown by many limestones, they are intrabasinal in provenance so the concept of compositional maturity as applied to the other clastic rocks has no meaning when applied to the carbonate rocks.

The remaining sediments are entirely nonclastic and have been placed together in a single chapter because of the relatively minor volume of such sediments in the crust of the earth. Some—the cherts, the iron-bearing sediments, and the phosphatic deposits—have common problems of origin. Considerable difficulty is encountered in deciding whether these rocks are original precipitates or are diagenetic (or epigenetic) replacements. Their origin seems, too, to be governed by the acidity or alkalinity of the solution and its oxidation–reduction potential. On the other hand, the evaporates and the carbonaceous sediments are classes quite apart, although they too have been included in this chapter.

The reader is referred to these several chapters for further discussion of the classification problems.

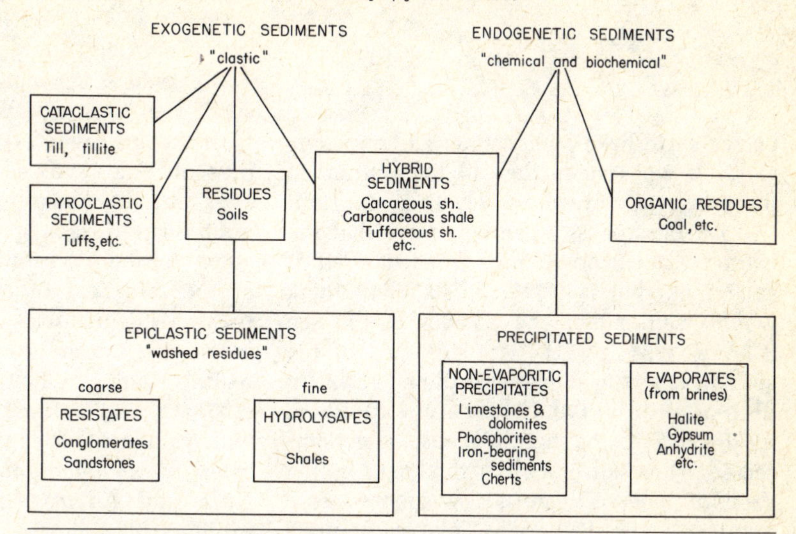

FIG. 2-3. Simplified classification of sedimentary rocks.

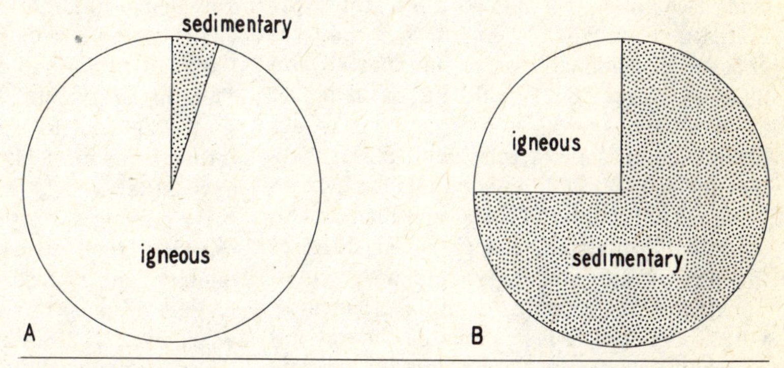

FIG. 2-4. Relative abundance of igneous and sedimentary rocks in the "crust" of the earth. A: by volume; B: by area. (Based on data by Clarke, 1924, p. 34.)

TOTAL VOLUME AND MASS OF SEDIMENT

By volume the sedimentary (and metasedimentary) rocks are estimated to constitute only 5 percent of the known lithosphere (the 10-mile-thick outer shell of the earth), whereas the igneous (and meta-igneous) rocks form 95 percent (Clarke, 1924, p. 34). On the other hand, the area of exposure of the sediments is 75 percent of the total land area, whereas the igneous rocks crop out over only 25 percent of the total (Fig. 2-4). It is evident, therefore, that the sediments must form only a thin superficial layer. Although their thickness ranges from 0 up to about 8 miles (13 km), they would average, according to Leith and Mead (1915, p. 73), only 1.39 miles (2.2 km) in thickness over the continental areas, a figure not greatly different from the most recent estimate of 6000 feet (1.8 km) made by Blatt (1970). The ocean floor is covered by sediment from shore to shore. The thickness of this cover is highly uncertain. It has been variously estimated from as little as 800 feet or 0.2 km (Blatt, 1970, p. 259) to over 3.0 km (Kuenen, 1941, p. 189) but is now thought to average about 1 km in thickness (Garrels and Mackenzie, 1971, p. 199). How have these various estimates been obtained?

The total volume and mass of sediments

on the earth have been estimated in various ways. If we assume that all the sodium in the ocean is derived by leaching of primitive igneous rocks, it can be shown that the complete decomposition of an earth shell of such rock that is 0.33 mile (0.5 km) thick would yield all of the sodium in the sea (Clarke, 1924, p. 31). Because some of the sodium is retained in the sediments and in deep-basin brines and is thus not all stored in the ocean, the figure needs to be corrected. Thus corrected the thickness of the shell of igneous materials weathered is a scant 0.5 mile (0.8 km). This shell, upon changing to a sediment, is increased in volume by oxidation, carbonation, and hydration. Assuming, as Clarke does, that this increase is roughly 10 percent, then the total volume of sediment corresponding to the igneous source rock would be about 9.3×10^7 cubic miles or 4.4×10^8 km³. This would be enough to form a rock shell about 2447 feet (735 m) thick enveloping the whole earth. If this material were confined to the continental platform (roughly one-third of the area of the globe), its thickness would be about 7300 feet (2200 m). Goldschmidt (1933, p. 866) likewise calculated the amount of weathered igneous rock and the volume of sediments formed therefrom based on the sodium content of the sea; his estimate was 3.0×10^8 km³. Kuenen (1941) applied various corrections to Clarke's data and obtained a figure of 8×10^8 km³ to which he added an estimated 5×10^8 km³ of disintegrated but undecomposed material (tuffs, graywacke, etc.) for a total of 13.0×10^8 km³ (Table 2-1).

TABLE 2-1. Estimated volume of sedimentary rocks

Reference	Cubic kilometers
Clarke (1924, p. 32)	3.7×10^8
Goldschmidt (1933)	3.0×10^8
Kuenen (1941, p. 188)	13.0×10^8
Wickman (1954)	$4.1 \pm 0.6 \times 10^8$
Poldervaart (1955, pp. 126–130)	6.3×10^8
Horn and Adams (1966, p. 282)	10.8×10^8
Ronov (1968, p. 29)	9.0×10^8
Blatt (1970, p. 259)	4.8×10^8

Others have approached the problem from a different point of view and have attempted to make an inventory of what actually is present. This involves estimating areas covered and thickness of the accumulations. Poldervaart (1955, pp. 124 passim), for example, based his estimates on the estimates of the thickness of sediments in the continental-shield areas, the younger folded belts, the ocean basins, and the continental shelves. Poldervaart used Kay's (1951, p. 92) estimates for the first two and estimates based on geophysics and rates of sedimentation for the last two. Horn and Adams (1966) approached the problem in somewhat the same manner as Poldervaart, but using somewhat different data, they arrived at a figure of 10.8×10^8 km³ instead of the 6.3×10^8 km³ of Poldervaart. Using a somewhat similar approach, Blatt (1970) derived a total of 4.8×10^8 km³. This is equivalent to an earth-enveloping shell 2690 feet (810 m) thick, not much different from the 2447 feet (735 m) estimated by Clarke.

By making appropriate assumptions about mineral density and porosity, or a bulk rock density, one can convert these estimates to weight or mass instead of volume. Poldervaart (1955), for example, estimates the mass as 1702×10^{15} metric tons; Garrels and Mackenzie (1971, p. 247) give 3200×10^{15} tons.

The question arises, Is the volume or mass fixed or does it slowly increase with time as new sediment is formed? Or, in other words, do we have a steady state where the sediment formed is balanced by that destroyed by granitization? This problem has been treated by Garrels and Mackenzie (1971, p. 259) and will be discussed in Chapter 17.

RELATIVE ABUNDANCE OF THE COMMON SEDIMENTS

Of the many kinds of sediment known, only a few are common. Three principal types, sandstone, shale, and limestone, form 95 or more percent of all sediments. But these three types are not equally abundant. Vari-

TABLE 2-2. Percentage of sedimentary rocks as measured

	Leith and Mead (1915)[a]	Schuchert (1931)[b]	Kuenen (1941)[c]	Krynine (1948)[d]	Horn and Adams (1966)[e]		Ronov (1968)[f]	
					Continent shield	Mobile belt shelf	Platform	Geosyncline
Shale	46	44	57	42	53	59	49	39
Sandstone	32	37	14	40	28	36	24	19
Limestone	22	19	29	18	19[g]	5	21	16

[a] Based on North American section totaling 520,000 feet.
[b] Based on North American Paleozoic maximum of 259,000 feet.
[c] Based on measurements in East Indies.
[d] Basis of estimate not given.
[e] Horn and Adams use data from several sources. They do not explain how their estimates were made.
[f] Ronov (1968), p. 30. Volcanic sediments form an additional 25 percent of the geosynclinal fill.
[g] Also includes evaporites.

TABLE 2-3. Computed proportions of sedimentary rocks

	Mead (1907)[a]	Clarke (1924)[b]	Holmes (1913)[c]	Wickman (1954)[d]
Shale	82	80	70	83
Sandstone	12	15	16	8
Limestone	6	5	14	9

[a] Mead calculated the proportions of average shale, sandstone, and limestone, which combined would be as nearly like the average igneous rock as possible. Based on chemical analyses.

[b] Clarke (p. 34) assigned all free quartz of the average crystalline rock to sandstone and one-half of the calcium to the formation of limestone.

[c] Holmes's (1913, p. 81) estimates are the proportions now contributed by the large rivers of the world.

[d] Wickman used the same procedure as did Mead. Percentage values were calculated from Wickman's data.

ous workers have attempted to estimate their relative proportions.

An estimate of relative abundance can be made in several ways, either by actual measurement of many stratigraphic sections (Table 2-2) or by calculation based on some geochemical considerations (Table 2-3). The second approach attempts to determine the proportions in which the average analyses of shale, sandstone, and limestone must be combined to yield an average analysis as nearly like that of the average igneous rock (from which they must all be derived). As early as 1907, Mead made the calculation and estimated the proportions to be 80, 11, and 9 for shale, sandstone, and limestone, respectively. A recent estimate (Garrels and Mackenzie, 1971, p. 243), based on better data, is 81, 11, and 8, figures not appreciably different from those of Mead.

The results obtained by measurement and by calculation are somewhat different. In general the inventory obtained by stratigraphic measurements shows a higher proportion of sandstone and limestone and a lesser amount of shale than is estimated by calculations. This discrepancy may in part be attributable to loss of the finer materials to the deep sea so that shales are underrepresented in the stratigraphic record on the continents.

Have the proportions of the common sedimentary rocks changed through geologic time? This question has been discussed by Ronov (1964, 1968) and by Garrels and Mackenzie (1971, pp. 258–259). We will review the subject in Chapter 17.

References

Allen, V. T., 1936, Terminology of medium-grained sediments: Rept. Comm. Sed. 1935–1936, Nat. Res. Coun., pp. 18–47 (mimeographed).

Blatt, H., 1970, Determination of mean sediment thickness in the crust: a sedimentologic model: Bull. Geol. Soc. Amer., v. 81, pp. 255–262.

Clarke, F. W., 1924, The data of geochemistry: Bull. U.S. Geol. Surv., no. 770, 841 pp.

Garrels, R. M., and Mackenzie, F. T., 1971, Evolution of sedimentary rocks: New York, Norton, 397 pp.

Goldschmidt, V. M., 1933, Grundlagen der quantitativen Geochemie: Fortschr. Min. Krist., Petrogr., v. 17, pp. 112–156.

Grabau, A. W., 1904, On the classification of sedimentary rocks: Amer. Geol., v. 33, pp. 228–247.

Holmes, A., 1913, The age of the earth: London and New York, Harper & Row, 195 pp.

Horn, M. K., and Adams, J. A. S., 1966, Computer-derived geochemical balances and element abundances: Geochim. Cosmochim. Acta, v. 30, pp. 279–290.

Kay, M., 1951, North American geosynclines: Geol. Soc. Amer. Mem. 48, 143 pp.

Krynine, P. D., 1948, The megascopic study and field classification of the sedimentary rocks: Jour. Geol., v. 56, pp. 130–165.

Kuenen, Ph. H., 1941, Geochemical calculations concerning the total mass of sediments in the earth: Amer. Jour. Sci., v. 239, pp. 161–190.

Leith, C. K., and Mead, W. J., 1915, Metamorphic geology: New York, Holt, Rinehart and Winston, 337 pp.

Lombard, A., 1949, Critères descriptifs et critères génétiques dans l'étude des roches sédimentaires: Bull. Soc. Belge Géol., v. 58, pp. 214–271.

Mead, W. J., 1907, Redistribution of elements in the formation of sedimentary rocks: Jour. Geol., v. 15, pp. 238–256.

Middleton, G. V., 1973, Basic concepts used in classifying sedimentary rocks: Symposium on classification of soils and sedimentary rocks, Univ. of Guelph, Guelph, Ontario, Canada.

Peterson, M. N. A., 1962, The mineralogy and petrology of Upper Mississippian carbonate rocks of the Cumberland Plateau in Tennessee: Jour. Geol., v. 70, pp. 1–31.

Pettijohn, F. J., 1948, A preface to the classification of the sedimentary rocks: Jour. Geol., v. 56, pp. 112–118.

Poldervaart, A., 1955, Chemistry of the earth's crust, *in* Crust of the earth—a symposium (Poldervaart, A., ed.): Geol. Soc. Amer. Spec. Paper 62, pp. 119–144.

Rodgers, J., 1950, The nomenclature and classification of sedimentary rocks: Amer. Jour. Sci., v. 248, pp. 297–311.

Ronov, A. B., 1964, Common tendencies in the chemical evolution of the earth's crust, ocean, and atmosphere: Geokhimiya, v. 8, pp. 715–743.

———, 1968, Probable changes in the composition of sea water during the course of geologic time: Sedimentology, v. 10, pp. 25–43.

Schuchert, C., 1931, Geochronology or the age of the earth on the basis of sediments and life, *in* The age of the earth: Bull. Nat. Res. Coun., v. 80, pp. 10–64.

Tarr, W. A., 1938, Terminology of the chemical siliceous rocks: Rept. Comm. Sed. 1937–1938, Nat. Res. Coun., pp. 8–27 (mimeographed).

Twenhofel, W. H., 1937, Terminology of the fine-grained mechanical sediments: Rept. Comm. Sed. 1936–1937, Nat. Res. Coun., pp. 81–104 (mimeographed).

Wadell, H., 1938, Proper names, nomenclature and classification: Jour. Geol., v. 46, pp. 546–568.

Wentworth, C. K., 1935, The terminology of coarse sediments: Bull. Nat. Res. Coun. 98, pp. 225–246.

Wentworth, C. K., and Williams, H., 1932, The classification and terminology of the pyroclastic rocks: Bull. Nat. Res. Coun. 89, pp. 19–53.

Wickman, F. E., 1954, The "total" amount of sediment and the composition of the "average igneous rock": Geochim. Cosmochim. Acta, v. 5, pp. 97–110.

Zen, E-an, 1959, Clay mineral–carbonate relations in sedimentary rocks: Amer. Jour. Sci., v. 257, pp. 29–43.

3

THE TEXTURE OF SEDIMENTS

Texture deals with the size, shape, and arrangement of the component minerals of a rock. It is essentially the microgeometry of the rock. Terms such as *coarse-grained, angular,* or *imbricated* describe textural attributes. The geologist may require a more precise description and ask how coarse, how angular, and at what angle and direction the components are imbricated. It is necessary, therefore, to formulate a clear definition of the property, to devise a method of measurement, to make a statistical summation for the sediment as a whole, and, most important, to understand the geological significance of the textural attribute in question.

Some textural properties are not simple attributes but are complex and are dependent on the more fundamental grain characteristics. Porosity, for example, is dependent on packing, on grain shape, and on sorting.

Unlike texture, *structure* deals with the larger features of the rock and is best seen in the field. Whereas texture has to do with the grain-to-grain relations, structure takes account of such features as bedding, ripple-marking, and the like. Texture is studied best in the thin section or by analysis of a small sample. Structure, on the other hand, is usually studied in the outcrop and less commonly in the hand specimen.

Most sediments—as initially deposited—differ from igneous and other crystalline rocks in consisting of a framework of grains, a framework stable in the earth's gravitational field. Unlike the grains of igneous and metamorphic rocks, which are in continuous contact with their neighbors, the grains in a sediment are generally in tangential contact only and thus form an open, three-dimensional network. The grains of most sediments, preformed, are emplaced as solid particles in the fabric of the rock by the movement of fluids under the influence of gravity. They did not form *in situ*. Rocks thus formed may be said to have a *hydrodynamic texture.*

Newly formed sediments have a great deal of intergranular space, i.e., a high pore volume. Porosity of sands may be 35 to 40 percent of the bulk volume of the rock; that of freshly deposited silts or clays may be as high as 80 percent. Sediments thus contrast markedly with igneous and metamorphic rocks which have very little or no porosity. With passage of time, however, the fluid-filled pores are the loci of deposition from solution of new minerals—the cementing minerals. These gradually and ultimately reduce the porosity to zero. The textures exhibited by these chemical precipitates and by alteration of the framework elements may be collectively referred to as *diagenetic textures.* They are, for the most part, crystalline. They may be so pervasive that the original depositional fabric is obscured or even obliterated, although in a great many cases the original fabric persists as a ghost or relict feature.

Nearly all sediments, therefore, exhibit two fabrics—hydrodynamic and diagenetic. This is not only true of sandstones but also of many, if not most, limestones which differ more in the composition of the framework elements than in the fabrics seen under the microscope. A quartz sand may be cemented by quartz in optical continuity and converted to a crystalline quartz mosaic (a quartzite). Likewise a crinoidal sand may be cemented by calcite in optical continuity with the framework elements and be converted to a "sedimentary marble." Many diagenetic fabrics are microcrystalline either because the original components were microcrystalline or because of the degradation of the larger framework elements. Partial degradation of some sands (the process of "graywackisation") yields a fine-grained matrix. Devitrification of glass particles or shards does likewise. Micritization affects oolites and skeletal debris in limestones. Carbonates, however, are more susceptible at ordinary temperatures to recrystallization, which leads to a coarser crystalline fabric. Unfortunately the texture of many sediments—the shales in particular—is so fine-grained that it can be studied only with great difficulty under the microscope. In these cases one cannot usually distinguish between primary depositional fabrics and those caused by diagenesis. Far less is known about the fabrics of such rocks than about the textural patterns of sandstone and limestone.

Some textures are neither hydrodynamic nor diagenetic. They may be *biogenic* (if produced by organisms) or *colloform* (if formed by production and coagulation of a gel).

In this chapter we will deal first with depositional fabrics, mainly hydrodynamic —fabrics which characterize sands and silts, whether they be siliceous or calcareous—and second with the crystalline fabrics of diagenetic and related origins.

PARTICLE SIZE OF DETRITAL ROCKS

The grain size of a detrital sediment is of considerable importance. The size of the fragments comprising a sediment is, in part, the basis of the subdivision into conglomerates, sandstones, and shales. The size and uniformity of size or sorting are measures of the competence and efficiency of the transporting agent. In normal terriginous water-deposited materials, size is in some measure a guide to the proximity of the source area. Deposits of great coarseness usually have not moved far. The several agents and modes of transport differ materially in their sorting and transporting ability. Clearly grain size and sorting is of first-rate importance to a sedimentologist.

A fuller understanding of the geological significance can come only from a clearer concept of what is meant by "grain size" and from a knowledge of the characteristics of grain-size distributions, the processes responsible for them, and the relation between size and the distance and direction of transport.

CONCEPT OF SIZE

If the particles composing a sediment were all spheres, no special difficulty would arise in defining or in determining their size. A statement of their diameters would suffice. But the pebbles in a gravel, for example, are nonspherical and are commonly highly irregular; they defy ordinary shape classification. Yet a gravel is said to be composed of pebbles of a certain "diameter." What is the "diameter" of a nonregular solid?

Direct measurements of the particle diameter are commonly made, although the irregular shape of the fragment creates difficulties. Some investigators report the *length, breadth,* and *thickness* of a fragment without clearly defining these terms. The long, intermediate, and short diameters of a triaxial ellipsoid are easily recognized, but even casual inspection of a collection of pebbles will demonstrate the difficulty of defining such terms for irregular solids. Must the "diameters" or intercepts be at right angles? Must they pass through a common point? How should they be combined to give an "average" diameter? Or will the intermediate diameter alone suffice as a measure of size? Krumbein (1941a) has reviewed these questions and framed objective operational definitions of these terms (Fig. 3-1). Somewhat different definitions were given by Humbert (1968, p. 11).

In practice the term *diameter* varies widely in meaning in accordance with the way in which it was measured. All methods of measurement are based on the premise that the constituent particles are spheres, or nearly so, or that the measurements can be expressed as diameters of equivalent spheres in some manner. Insofar as these conditions are not fulfilled, the reported sizes are incorrect or inaccurate. Very commonly the diameter itself is not measured. Instead some other property is measured and converted, on the basis of certain assumptions, to diameter terms. One could, for example, measure the volume of a pebble and then *calculate* the diameter of a sphere having the same volume. This is the *nominal diameter* of Wadell (1932) and is a measure independent of either the shape or density of the pebble. On the other hand, the settling velocity, commonly used to determine grain size, is valid only if both the density and shape are constant. Settling velocity measurements are reduced to a diameter or radius on the assumption of spherical form and a density of 2.65 (quartz). We will not review nor evaluate all of these methods here (Fig. 3-2). The subject has

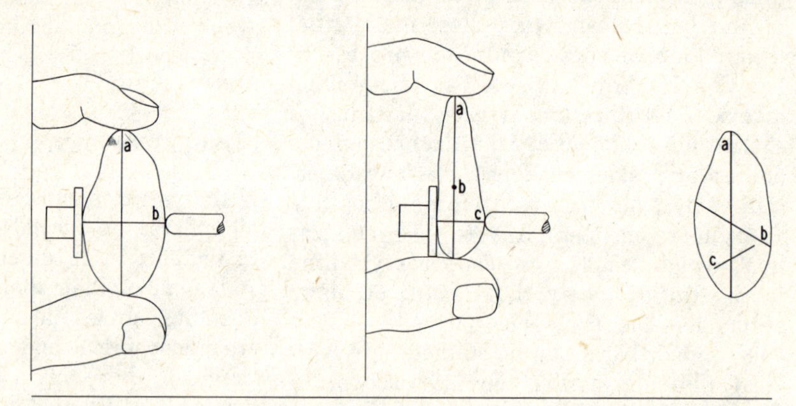

FIG. 3-1. Concept and measurement of pebble diameter. Left, the *b*-axis in position; center, the *c*-axis in position; right, the pebble in perspective. (From Krumbein, 1941a.)

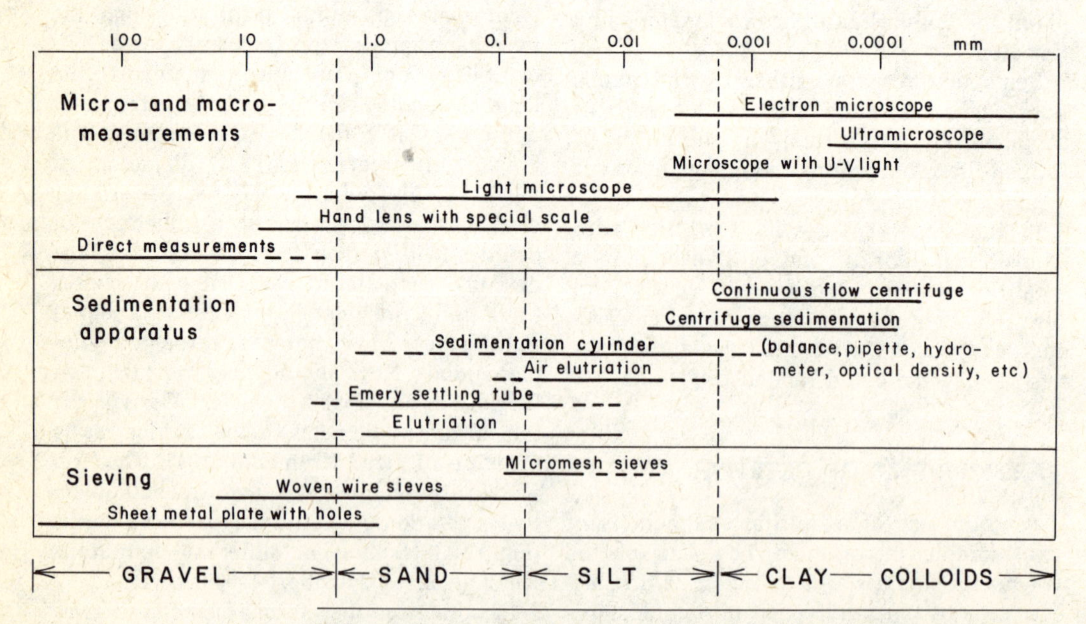

FIG. 3-2. Range of grain size and methods of size analyses. (Modified from Mueller, 1967, Methods in sedimentary petrology; Fig. 16, top, by permission of E. Schweizbart'sche Verlagsbuchhandlung.)

been treated at length in various comprehensive laboratory manuals (Krumbein and Pettijohn, 1938; Dalla Valle, 1943; Irani and Callis, 1963; Köster, 1964, pp. 43–147; Müller, 1967, pp. 52–96; Folk, 1966; Allen, 1968; and Carver, 1971, pp. 47–128). A summary is presented in Table 3-1. But a clear concept of size should be kept in mind when interpreting the results of size analyses which, because of the limitations of the analytical method, yield only approximations.

SIZE TERMS

Various terms in common use have been adopted by the geologist. Some writers have replaced these common language terms with others less familiar. Several sets of such size terms (and their adjectival modifications) are given in Table 3-2.

The terms *psephite, psammite*, and *pelite*, derived from the Greek and Grabau's (1904) more or less equivalent Latin-derived terms (*rudite, arenite*, and *lutite*), were proposed to supplant the common terms *gravel, sand*, and *clay*. The latter carry some implication of composition or other characters and are not, therefore, strictly size terms. If a mineralogical restriction is to be placed on such a term as *clay*, then no size term exists for material of fine grain. Even if the term were to be used with a double meaning, that is, both as a size term and as a particular sediment type, it is doubtful whether geologists would be satisfied to call a pure lime mud a clay and the consolidated rock a claystone. The term *lutite*, however, is acceptable for such materials and the fine-grained limestones could be called *calcilutite*, whereas the argillaceous equivalent would be *argillutite*. Likewise a pure carbonate sand, if lithified, would more probably be called a limestone rather than a sandstone. Tyrrell (1921, p. 501) would restrict the

TABLE 3-1. Concepts of size and related diameters

Measured property	Method of measurement	Diameter or equivalent
"Dimensions" of actual pebble or cobble	Macroscopic; calipers or gauge	"Length," "breadth," "thickness"; arithmetic, geometric, or log mean of same
Dimensions of projected or magnified image of grains	Suitable scale; micrometer ocular	Nominal projection diameter (diameter of circle of same area as projected image)[a]
Dimensions of cross-sectional image of small grains	As above from enlarged thin-sectional image	Thin-sectional diameter (diameter of circle having same area as cross section)[b]
Minimal or least area of cross section	Sieves	"Sieve diameter" (width of minimum square aperture through which the particle will pass)
Weight	Balance	Diameter of sphere of same weight and density as particle
Volume	Volumeter	Nominal diameter (diameter of sphere of same volume)
Surface area	Gas adsorption apparatus to estimate m^2/g	—
Settling velocity	Sedimentation cylinders, elutriators, sedimentation balance, and so on	"Equivalent radius"; sedimentation diameter (diameter of sphere having same settling velocity and density as particle)

[a] More often the short dimension of the projected image is reported.
[b] More often a random intercept of the cross-sectional image is reported.

Latin-derived terms to sedimentary rocks and the Greek-derived terms for their metamorphic derivatives—pelitic schist, for example.

Regardless of the choice, the terms used are likely to mean different things to different people. The meaning attached to the word *sand* as a size term will vary greatly (Fig. 3-3). Hence it is desirable that usage be standardized. Unfortunately there are several standards. The engineers, the soil scientists, and the geologists all use different standards. Even among sedimentologists there is no universal agreement; European usage is different from that prevalent in North America.

The generally accepted standard used by sedimentologists in North America stems from the work of Udden (1898, 1914). Udden devised a geometrical scale of size classes and redefined the common terms *gravel, sand, silt,* and *clay.* In 1922, Wentworth modified the definitions of Udden to the then prevailing opinion among research workers as indicated by a questionnaire. In 1947, a committee of geologists and hydrologists recommended adoption of the Wentworth-Udden scale and size terms, omitting only the granule class (Lane et al., 1947). This scale is now universal among North American investigators and since the use of the phi notation, introduced by Krumbein in 1938, has become widely used elsewhere.

The Committee on Sedimentation of the National Research Council issued a series of reports on the nomenclature of sediments including redefinitions of the common size terms (Wentworth and Williams, 1932; Wentworth, 1935; Allen, 1936; Twenhofel, 1937). The recommendations of the Committee are outlined in Table 3-3.

Boulder was defined as "a detached rock mass, somewhat rounded or otherwise modified by abrasion in transport, and larger than a cobble" with a minimum size of 256 mm (about 10 inches). For those objects produced by weathering *in situ* such terms as *boulders of disintegration* or *boulders of exfoliation* were recommended. The term *block* was reserved for "a large angular fragment showing little or no modification by transporting agencies."

A *cobble* is defined in the same manner as a boulder, but it is restricted in size from 64 to 256 mm. In like manner there may be *cobbles of exfoliation.*

A *pebble* is a "rock fragment larger than a coarse sand grain or granule and smaller than a cobble, which has been rounded or otherwise abraded by the action of water, wind, or glacial ice." It is, therefore, between 4 and 64 mm in diameter.

The unconsolidated accumulation of pebbles, cobbles, and boulders is *gravel,* which accordingly may be designated pebble-gravel, cobble-gravel, and so forth. The consolidated equivalent is *conglomerate,* likewise designated pebble-conglomerate, and so forth. *Rubble* is an unconsolidated accumulation of angular rock fragments

TABLE 3-2. Descriptive size terms

Texture	Common terms	Greek-derived terms	Latin-derived terms[a]	
			Clastics	Nonclastic constructional[b]
Coarse	Gravel (gravelly)	Psephite (psephitic)	Rudite (rudaceous)	Spherite
Medium	Sand (sandy)	Psammite (psammitic)	Arenite (arenaceous)	Granulite
Fine	Clay (clayey)	Pelite (pelitic)	Lutite (lutaceous)	Pulverite

[a] Grabau (1904) spelled the terms rudyte, arenyte, spheryte, and so on.
[b] Now generally obsolete.

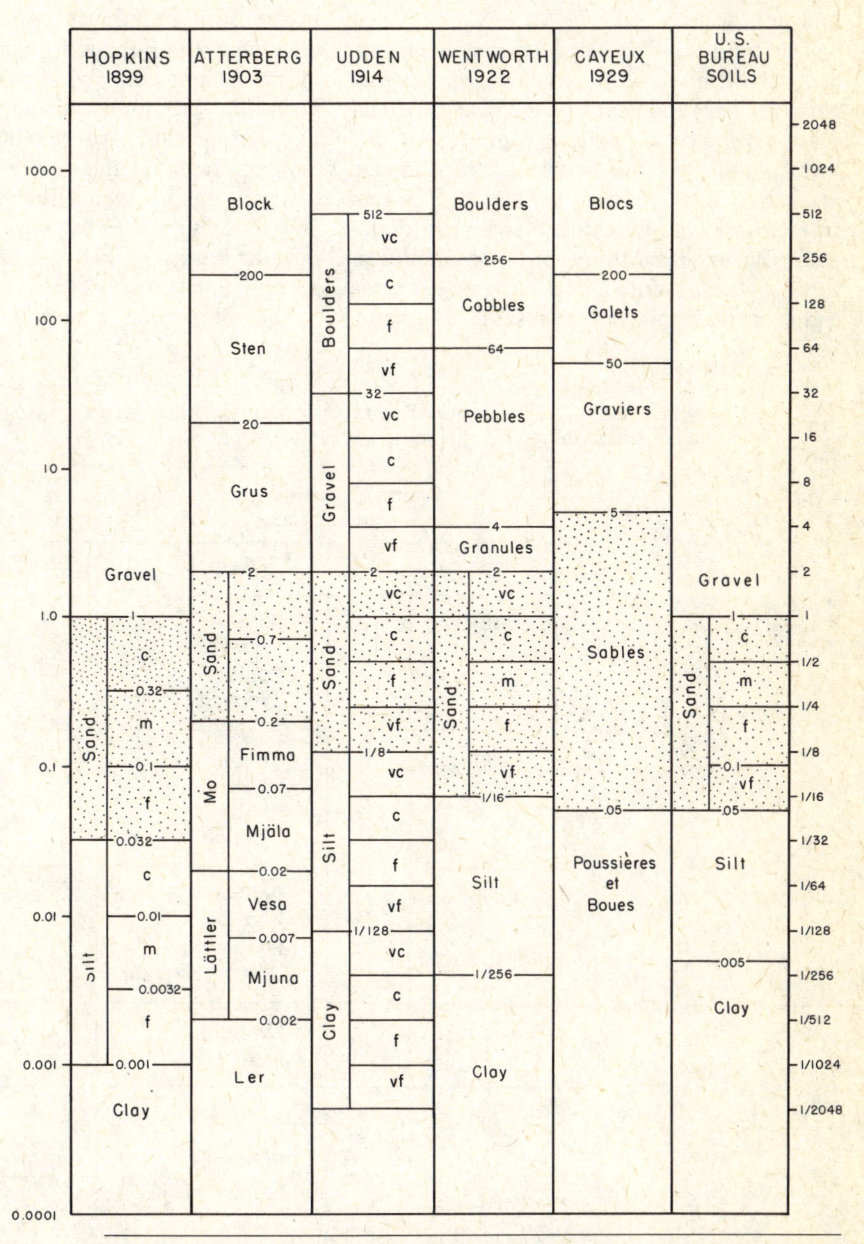

FIG. 3-3. Three types of representative grade scales. The scales of Hopkins and Atterberg are geometric, decimal, and cyclic. Udden's scale, later adopted by Wentworth and since accepted by the Lane Committee, is geometric but noncyclic and nondecimal. The scales of the Bureau of Soils (U.S. Dept. of Agriculture) and of Cayeux are nonregular. The diagram illustrates the diverse meanings of the size terms and need for standardization. Note the variations in the limits of sand (stippled).

coarser than sand. Its consolidated equivalent is *breccia*.

The term *sand* is used to denote an aggregate of mineral or rock grains greater than 1/16 mm and less than 2 mm in diameter. Wentworth (1922c) proposed the term *granule* for the material 2 to 4 mm in size. *Silt* was defined as 1/16 to 1/256 mm in size and *clay* less than 1/256 mm in diameter. Closer description requires modification of these terms, such as very coarse sand, coarse sand, medium sand, fine sand, and so forth. Indurated equivalents are sandstone and siltstone.

The above definitions as size terms are defective in several respects. Several concepts other than size have been inadvertently introduced into definitions of what are supposed to be size or grade terms. The injection of roundness, a particular process of shape modification (abrasion) and the agents responsible (wind, water, ice), is undesirable. The terms are not purely descriptive terms; they are genetically defined. It may, in fact, be impossible or undesirable to try to define them otherwise. Were sand grains defined strictly on size, one would not discriminate between the grains of a granite and those of a sandstone.

The Committee's terminology is incom-

TABLE 3-3. Size limits of common grade and rock terms of sedimentary or epiclastic rocks

Size	Rounded, subrounded, subangular			Angular	
	Fragment		Aggregate	Fragment	Aggregate
	Boulder		Boulder gravel Boulder conglomerate	Block	
256 mm					
	Cobble	"Roundstone"	Cobble gravel Cobble conglomerate	—	Rubble
64 mm					
	Pebble		Pebble gravel Pebble conglomerate	—	Breccia
4 mm					
	Granule		Granule gravel	—	
2 mm					
	Sand		Sand Sandstone	—	— —1 mm— Grit —½ mm—
1/16 mm					
	Silt		Silt Siltstone	—	—
1/256 mm					
	Clay		Clay Shale	—	

plete in some respects. It is unfortunate, for example, that the term *block* had previously been defined in a restricted and special way for certain fragments of pyroclastic origin. Moreover, there is no provision for terms analogous to *block* for fragments of less than boulder size—a defect which was overlooked despite the fact that Woodford (1925, p. 183) extended the term to designate any more or less equidimensional angular fragment over 4 mm in size. The term *slab* was proposed by Woodford for flat fragments with a maximum dimension over 64 mm; the terms *chip* and *flake* were likewise used for angular flat fragments the maximum dimensions of which were 64 and 4 mm, respectively. Note that these definitions involve two attributes: size and shape. They are thus not strictly size terms.

Both prior to and following the publication of the Committee reports, there have been other contributions to the problem of size nomenclature. Fernald (1929), for example, proposed the term *roundstone* to designate collectively the largest sizes—boulders, cobbles, and pebbles. Shrock (1948b) suggested the analogous term *sharpstone* to designate the clastic elements of rubble. Hence the term *sharpstone conglomerate* would be an appropriate term for a *sedimentary* breccia and *roundstone conglomerates* would designate ordinary conglomerates. Here again we find two concepts united in one term: size and rounding.

The term *granule*, as defined by Wentworth, is also ambiguous. It has been used to describe precipitated bodies, notably of certain iron silicates such as the greenalite and glauconite granules. It has been generally abandoned as a size term. The Lane Committee assigned the 2 to 4 mm materials to the gravel class.

Other reviews of the nomenclatural problems or compendia of size and aggregate terms will be found in Bonorino and Teruggi (1952) and in the various larger works such as Köster (1964, pp. 7–20).

The limits placed on the various size terms are mainly arbitrary and are to be considered "correct" only insofar as they are generally agreed upon and adhered to by students of sediments. Wentworth (1933) has, however, argued that there is also a "natural" basis for the divisions generally made. He believed that the several major classes of materials correlate closely with the several fundamental modes of transport by running water and with the several modes of rock disintegration (see p. 42). Bagnold (1941, p. 6) likewise utilized a dynamic behavior in defining sand. The lower limit of "sand" was that at which the terminal fall velocity is less than the upward eddy currents, and the upper limit is that size such that a grain resting on the surface ceases to be movable either by direct pressure of the fluid or by impact of other moving grains. This sort of behavioral definition depends on the nature of the moving fluid and is valid for the "average" conditions of flow. Bagnold further pointed out that sand has one characteristic not shared by coarser or finer materials—namely, the power of self-accumulation—of utilizing the energy of the transporting medium to collect the scattered grains together into definite heaps leaving the intervening surface free of grains.

CLASSIFICATION OF SEDIMENTARY AGGREGATES

Although some agreement has been reached with respect to the terms to be applied to individual grains or fragments, no such agreement has been attained concerning the names to be applied to an aggregate of such fragments. Because natural aggregates are rarely composed of fragments of the same size, the problem is one of the nomenclature of mixed sizes. A pebble, for example, has been closely defined, but the term *gravel* or *conglomerate* has not been so circumscribed. Various suggestions have been made. Perhaps to warrant the term *gravel*, the mean size should fall within the prescribed range, or perhaps 50 percent (or some other specified amount) or more of the material must lie in the gravel range.

These or other methods for naming an aggregate are not all equivalent nor wholly satisfactory. A poorly sorted sediment, for example, which is a mixture of coarse gravel and sand might be classified as a coarse sand if the mean fell within this size grade, whereas only 10 or 20 percent of the

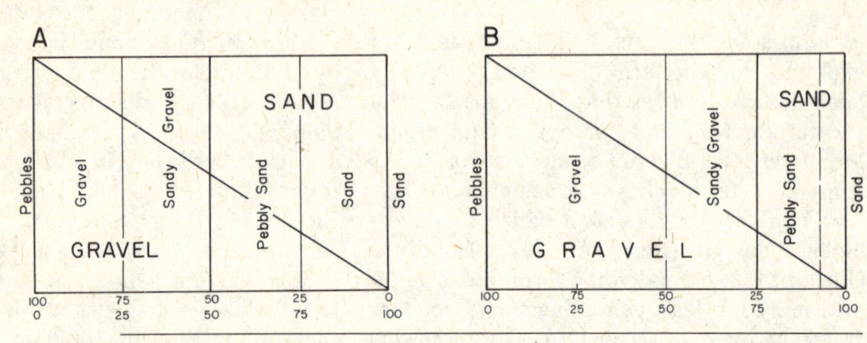

FIG. 3-4. Nomenclature of sand and gravel mixtures. A: idealized "symmetrical" classification; B: classification based on field usage. (Willman, 1942.)

material might actually be in the coarse sand grade, with the remainder divided between the gravel and finer sand grades. In extremely poorly sorted mixtures of gravel, sand, and silt or clay, less than 50 percent might fall in the range of any one of these materials. What name should be applied to this sediment? Various special names have been proposed (Flint, Sanders, and Rodgers, 1960a, 1960b; Schermerhorn, 1966).

There are numerous attempts at solution of the problem of naming sedimentary mixtures. There have been two approaches to this problem. One is to attempt to standardize prevalent usage. In this case existing practice is ascertained and the aggregate limits and terms are redefined to correspond. The other is to set more or less arbitrary limits to the various mixtures and to name them according to some systematic plan. The first approach leads to irregular and seemingly illogical boundaries and definitions; the second leads to trouble with one's co-workers. These two approaches are illustrated by the problem of naming mixtures of sand and gravel. This mixture constitutes a simple binary system consisting of the two end-members, sand and gravel, which could be subdivided and named as shown in Fig. 3-4A. Willman (1942, p. 343), however, has noted that a large number of deposits which are called "gravel" and commercially worked as such, contain about 50 percent sand, many containing from 50 to 75 percent sand. He therefore proposed the classification (Fig. 3-4B) in which *sand* has at least 75 percent of material in the sand range, *pebbly sand* if it contains a conspicuous number of pebbles but less than 25 percent; *sandy gravel* contains 50 to 75 percent sand and 25 to 50 percent gravel. According to this scheme a deposit with as little as 25 percent in the gravel range would be called *gravel,* and almost certainly, if it were consolidated, the field geologist would call it a conglomerate.

Three-component mixtures (ternary systems) such as sand, silt, and clay, though uncommon, do exist. Some attempted solutions of the naming of such mixtures is shown in Fig. 3-5. As can be seen, three-component aggregates can be represented by an equilateral triangular diagram in which a single point is a graphic representation of the proportions of the three components. The perpendiculars from that point to the three sides of the triangle are proportional to the abundances of the several components. The triangle may be divided into fields and appropriate names given to the several mixtures represented.

As seen in Fig. 3-5, there is no general agreement among geologists, oceanographers, pedologists, and engineers. The term *clay* as applied to aggregates, for example, may contain as little as 50 percent clay-sized material (A), or it may be defined as containing no less than 80 percent of such materials (D).

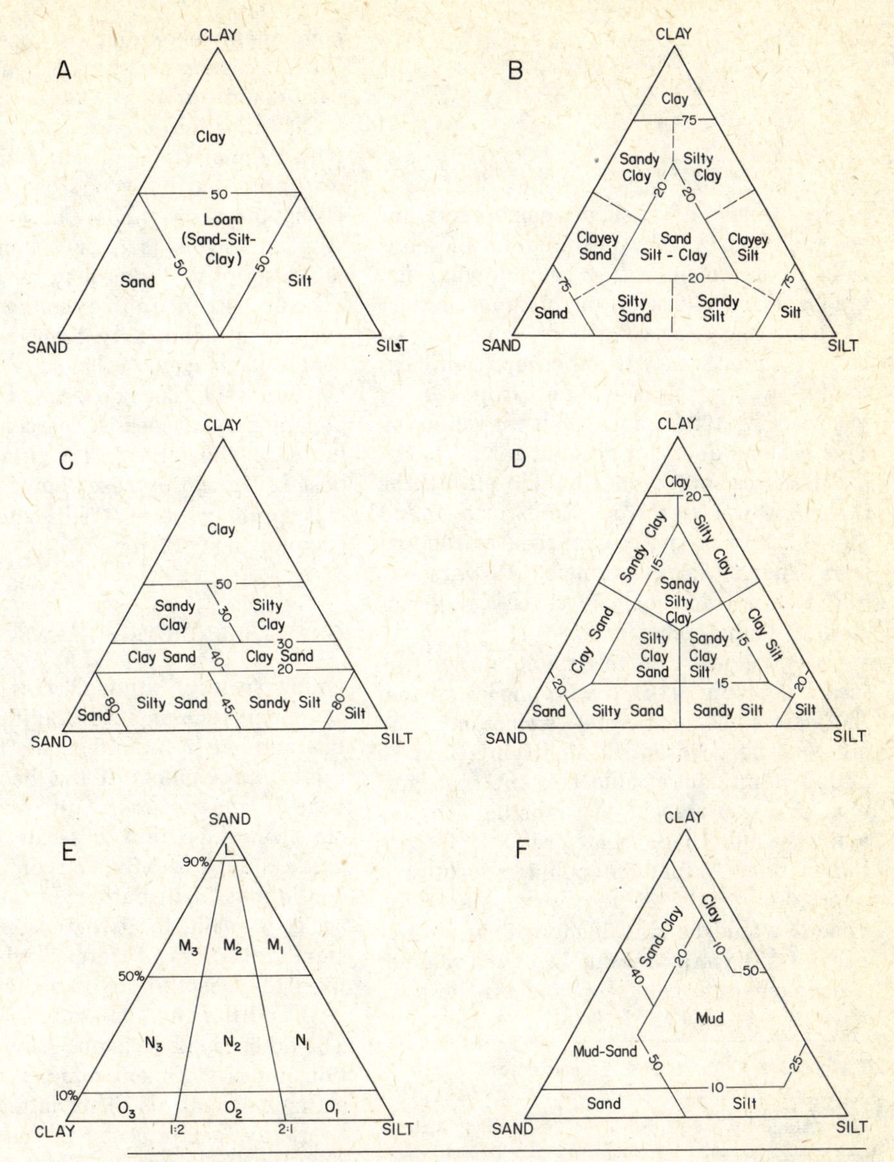

FIG. 3-5. Nomenclature of sand, silt, and clay mixtures. A: after Robinson (1949); B: after Shepard (1954); C: after U.S. Army Corps of Engineers; D: after Trefethen (1950); E: after Folk (1954); F: after A.P.I. Project 51 (Shepard, 1954).

Instead of a ternary system based on the proportions of three components, other schemes for classifying sediments based on two parameters have been proposed. Baker (1920) based his scheme on the "equivalent grade" (mean size) and the "grading factor" (a sorting coefficient). Niggli (1934) proposed a classification which, like Baker's, requires a knowledge of the entire size distribution. It is based on the ratio of two critical values on the size distribution curve.

It is clear that not only is there no general agreement on nomenclature, but it is also evident that most of the proposed systems of nomenclature could not be applied without a complete size analysis—an analysis impossible to make, or at best very difficult and tedious to secure, on well-indurated sediments. These schemes have, therefore, limited value to the students of ancient sediments.

Although a clastic sediment could be a mixture of three (or even four) components, in *practice* such is rarely the case. Most deposits consist of a single component and are modified only by introduction of materials from the preceding or following size grades. For that reason the complex ternary or quaternary classification schemes are largely unnecessary. It is perhaps sufficient to follow the classification proposed by Wentworth (1922c) and given in Table 3-4. This scheme does not account for all possible mixtures; but it does handle all but the rarest natural deposits. Wentworth found that, of 50 size analyses chosen at random from Udden's compilation (1914), only one (till) was not amenable to classification according to his scheme.

Essentially the same principle was utilized by Krynine (1948), who suggested that the terms *conglomerate, sandstone,* and *siltstone* be used and a modifying term be added if an appreciable content of a foreign size was present. A sandstone, for example, would be *conglomeratic* if it contained over 20 percent pebbles, *pebbly* if over 10 but under 20 percent pebbles. Likewise it would be *silty* if more than 20 percent of silt were present. In a like manner a conglomerate would be sandy if the sand component exceeded 20 percent and so forth. Presumably the foreign size in each case could not exceed 50 percent.

TABLE 3-4. Class terms for sediments

Percentage by grade	Class term
Gravel > 80	Gravel
Gravel > sand > 10 others < 10	Sandy gravel
Sand > gravel > 10 others < 10	Gravelly sand
Sand > 80	Sand
Sand > silt > 10 others < 10	Silty sand
Silt > sand > 10 others < 10	Sandy silt
Silt > 80	Silt
Silt > clay > 10 others < 10	Clayey silt
Clay > silt > 10 others < 10	Silty clay
Clay > 80	Clay

Source: After Wentworth (1922).

Classification of breccias or aggregates with angular fragments in a similar manner was proposed by Woodford (1925, p. 183). Using the term *rubble* for an aggregate of angular fragments over 2 mm, the following descriptive terms may be used: *breccia,* over 80 percent rubble; *sandy breccia,* over 10 percent sand; *silty breccia,* over 10 percent silt; *clayey breccia,* over 10 percent clay. In each case no second foreign component may exceed 10 percent. If such be the case, the term *earthy breccia* was proposed. The latter case, however, is a special problem (see "Till and tilloid" in Chapter 6, p. 171).

GRAIN SIZE DISTRIBUTIONS

Grade scales Although the particles of a sediment, such as a sand, differ in size from one another by infinitesimals from the largest to the smallest, it has been found convenient and necessary to divide the particle size range into a series of classes or grades. Such subdivision of an essentially continuous distribution of sizes, a *grade scale,* is made for two reasons. One is the standardization of terms, which systematizes the description of sedimentary materials and thus avoids confusion of meaning. The other is subdivision of the size distribution into a sufficient number of classes for statistical analysis. This latter requirement is greatly facilitated by the use of a regular scale in which the subdivisions bear some simple relation to each other.

The range of sizes to be subdivided is very great. An extreme case is boulder clay or till in which a boulder, 1 m in diameter, is a million times as large as a 1-micron clay particle. For such a range an ordinary linear scale is unsuitable, for if 1 mm were taken as the class unit, almost all of the material known as sand, silt, and clay would fall into one class, and the coarse sand and gravel fraction would be divided between 999 classes or grades! Clearly a graduated or geometric scale is required for subdividing such a range of values. In this scale the larger classes are used for the larger sizes and smaller classes for the smaller sizes. As Bagnold (1941, p. 2) puts it, linear scales are seldom acceptable to Nature. Nature, if

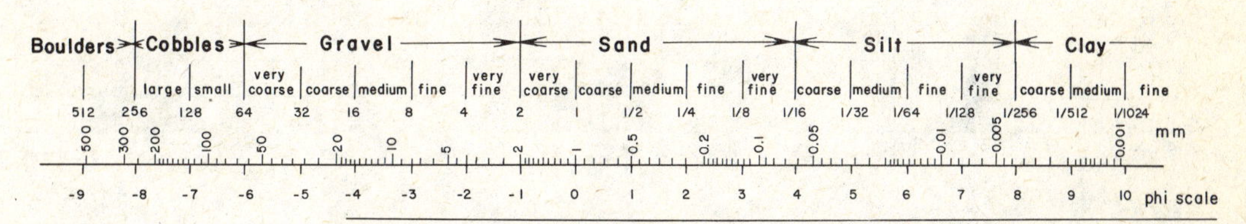

FIG. 3-6. Relation between Wentworth-Lane class limits and phi scale.

she has any preference, probably takes more interest in the ratios between quantities; she is rarely concerned with size for the sake of size. A millimeter difference between the diameters of two boulders is insignificant, but a millimeter difference between one sand grain and the next is a large and important inequality.

The natural scale for size classification, therefore, is geometric. Udden, in the United States, recognized this as early as 1898. He chose 1 mm as the starting point and used the ratio ½ (or 2, depending on the sense of direction) and obtained the diameter limits to his size classes of 1, ½, ¼, and so forth, or 1, 2, 4, 8, and so forth in the other direction (Fig. 3-3). His scale has remained in use to the present day and was adopted by Wentworth in 1922 and by the Lane Committee of the National Research Council in 1947 (Lane et al., 1947). See Table 3-3.

The Udden scale has some disadvantages. It is not suited to the analysis of very well-sorted sediments such as dune sand, because the number of classes into which a sediment will be divided is too small for statistical analysis. The scale, therefore, must be subdivided by splitting each class into two or, in some cases, into four subclasses. Such subdivision, however, gives rise to a series of irrational class limits, numbers difficult to recall. Moreover, the midpoints (geometric means) of the various classes of these and the unmodified Udden scale required for statistical computation all have irrational values.

To avoid both irrational class limits and midpoints and also to simplify statistical computations, Krumbein (1934) proposed his phi scale. This scale is based on the observation that the class limits of Udden's scale can be expressed as powers of 2. Four millimeters is 2^2, 8 is 2^3, 1 is 2^0, ½ is 2^{-1} and so forth. He therefore proposed to use the exponent (logarithm to the base 2 of the diameter) instead of the diameter itself. To avoid negative numbers in the various sand grades and finer materials, the log was multiplied by -1, or, in other words, phi $= -\log_2$ diam (mm). See Fig. 3-6.

Many other grade scales have been proposed or used; these fall into several categories (Fig. 3-3). Some, like that of Udden, are regular geometric progressions. Some, like that of Atterberg (1905), are also geometrical but differ from Udden's in that they are also decimal and cyclical. In a decimal scale, the size limits are cyclical and are regularly repeated with only a change in the decimal point. The Atterberg scale, for example, starts with 2 mm and the major subdivisions are 2, 20, 200, and so on in the upward direction and 0.2, 0.02, 0.002, and so on in the downward direction. This scale does not provide a sufficient number of classes for analytical purposes, so subdivision is necessary. If the subdivision is to follow the logarithmic rule, the divisions will be the square root of the product of class limits (geometric mean). These values are also irrational and difficult to remember unless they are rounded off. Atterberg's scale has been widely used by European soil scientists and geologists.

Some scales are neither wholly geometric nor linear. Such a nonregular scale is used by the U.S. Department of Agriculture and is commonly taken as a standard by the students of soils in the United States. The scale is satisfactory for descriptive purposes only for the materials of medium and fine grain but is not adequate for statistical analysis nor suited for coarse material.

Although many grade scales have been

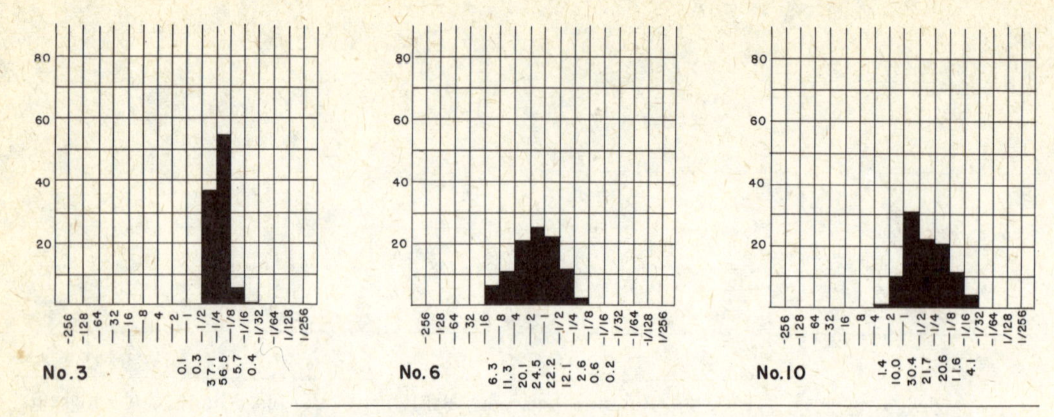

FIG. 3-7. Histogram of size distribution in sand analyzed (Table 3-5).

constructed, 20 appearing on the chart compiled by Truesdell and Varnes (1950), there is no universally accepted scale which meets the needs of pedologists, civil engineers, oceanographers, and geologists. For sedimentologists a standard scale should be geometrical to provide a sufficient number of classes for size analysis and statistical summarization. Though neither decimal nor cyclic, the Udden scale, and the phi scale based on it, meets this need. It has been widely used and is the basis for the limits placed on the various grade terms used in this book.

Representation of size frequency distributions If arranged according to size, the detrital elements in a mechanically deposited sediment (sand grains, pebbles, and the like) would be found to differ in size from one another by infinitesimals. The *size frequency distribution* is said to be continuous. As we have seen, it is convenient to subdivide this continuous distribution into a series of classes or grades. This facilitates the comparison of the size distribution of one sediment with another and also statistical analysis of the distribution.

Although a size frequency distribution of a sediment so subdivided into classes can be summarized (Table 3-5), it may also be presented graphically. Such graphical representation is easier to grasp than columns of figures in a table. It facilitates comparison of several different analyses.

The usual graphical representations are a species of bar diagram, the *histogram* (Fig. 3-7) and the *cumulative curve* (Fig.

TABLE 3-5. Representative mechanical analyses

Size grade (mm)	No. 6		No. 10		No. 3	
	%	Cum. %[a]	%	Cum. %	%	Cum %
16–8	6.3	6.3	—	—	—	—
8–4	11.3	17.6	—	—	—	—
4–2	20.1	37.7	1.4	1.4	—	—
2–1	24.5	62.2	10.0	11.4	0.1	0.1
1–½	22.2	84.4	30.4	41.8	0.3	0.4
½–¼	12.2	96.6	21.7	63.5	37.1	37.5
¼–⅛	2.6	99.2	20.9	84.4	56.5	94.0
⅛–¹⁄₁₆	0.6	99.8	11.6	96.0	5.7	99.7
Under ¹⁄₁₆	0.2	100.0	4.1	100.1	0.4	100.1
Total	100.0		100.1		100.1	

[a] Cumulative percent.

Samples are Wisconsin glacial outwash (Pleistocene), Dundee, Illinois. M. A. Rosenfeld, analyst.

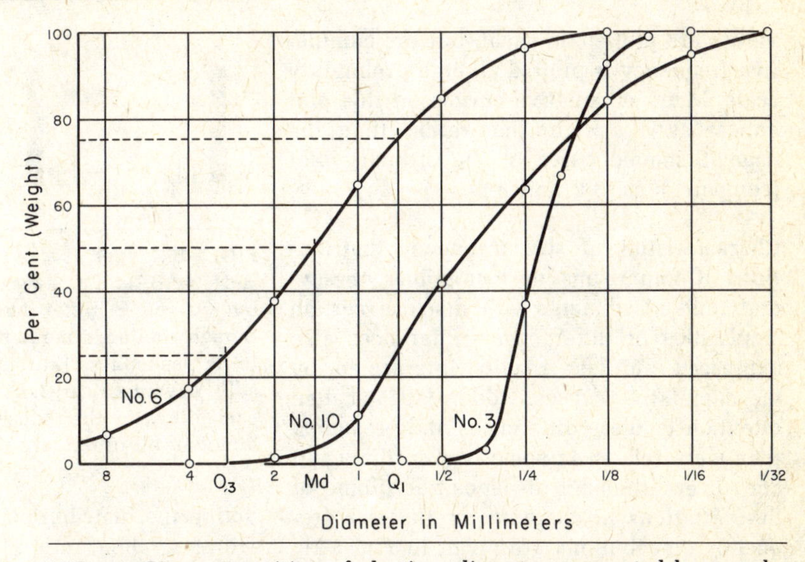

FIG. 3-8. Size composition of clastic sediments represented by cumulative curves. The curves are based on the analyses of Table 3-5. Sands nos. 6 and 10 have about the same sorting but differ markedly in average size, whereas nos. 10 and 3 have about the same average size but differ notably in sorting. The significant points for determination of the median size and coefficient of sorting are shown for no. 6 Md—median, Q_1—first quartile, and Q_3—third quartile.

3-8), both well-known devices for presenting frequency distributions of any kind. As used by the sedimentologists, however, these devices depart somewhat from standard practice. As used for size analyses, they commonly show the percent in each class based on *weight* of the material in that class rather than the *number* or a percentage based on count. Moreover, the size values plotted on the x-axis are in fact the logarithms of the "diameter" rather than the sizes themselves. The widths of the bars of the histogram, for example, are equal even though the classes represented are not. Similarly the size scale (X-scale) of the cumulative frequency curve is in fact a logarithmic scale. In recent years it has become common practice to plot the log—the phi (ϕ) value—instead of the diameter itself. This makes for easier interpolation on the cumulative curve plots. Moreover, the sense of the scales in both these diagrams has been reversed from the more conventional usage. Values decrease (instead of

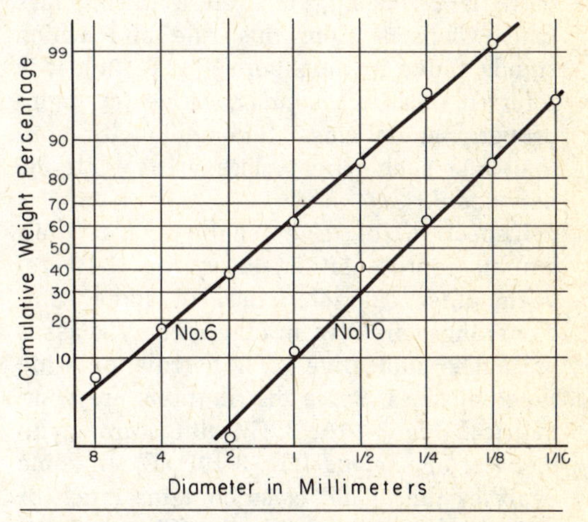

FIG. 3-9. Size composition of sands shown by cumulative weight percentage plotted on logarithmic probability paper. (Data from Table 3-5.)

increase) to the right. It has become an increasingly common practice to plot the cumulative percentage curve on log probability paper (Fig. 3-9). The diameter or phi values are plotted as usual, but the cumulative frequency is plotted along a probability scale. Many cumulative curves on this plot appear as a straight line rather than the usual S-shaped curve of the ordinary plot (compare Figs. 3-8 and 3-9).

Characteristics of size frequency distributions Comparisons of histograms of several unlike sediments will disclose certain similarities or differences—differences also expressed but less readily interpreted by the cumulative curves. Udden (1914) called attention to these differences and presumed they were related in some way to the agent and/or environment of deposition. Some of the variations in character of the size frequency distributions are given in Fig. 3-10.

The significant attributes of frequency distributions are several. As noted in Fig. 3-10, in all cases there is one size class more prominent or larger than any other. This is the *modal class.* The quantity in the other grades or classes tend to fall off away from this class in a regular manner. Exceptionally there is a second, or even a third, class which departs from this rule and which stands above its neighbors (Fig. 3-10F). It is referred to as a *secondary mode* (or, more properly, modal class). Sediments with more than one such modal class are said to be *polymodal.*

Inspection of size analyses, or their graphic representation, as in Fig. 3-10, reveals other characteristics. In some sediments there are few, in others many, classes or grades indicative of a narrow or wide *range* of sizes. Some distributions are symmetrical (Fig. 3-10A, B, C); others are asymmetrical or *skewed* (Fig. 3-10D, E). In some analyses the modal class contains much or most of the material (Fig. 3-10C); in others, having the same range or number of classes, a lesser quantity is found in the modal class (Fig. 3-10B). In other words, there is a difference in the peakedness, or *kurtosis,* of the two sediments. These various attributes of a frequency distribution can be expressed as simple numerical parameters which describe the distribution. The fundamental attributes, then, are (1) "average" size or central tendency of the distribution (*mean, median, mode*); (2) "sorting" or dispersion of the values about the mean (mean or *standard deviation*); (3) symmetry (*skewness*); and (4) peakedness (*kurtosis*). These attributes, which can be used to describe any frequency distribution, are more closely defined in any elementary text on statistics.

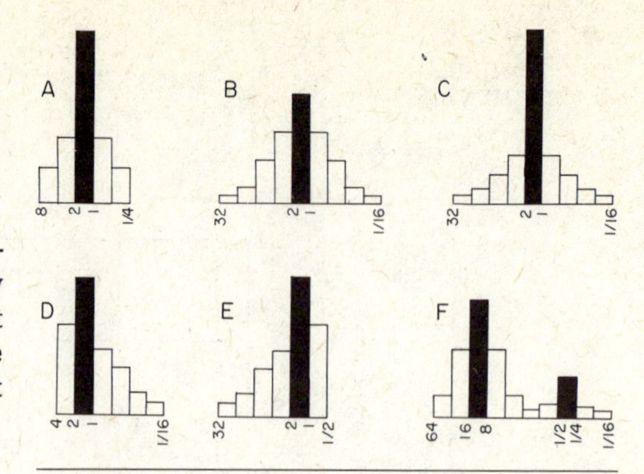

FIG. 3-10. Types of size frequency distributions. Here A, B, and C have similar modal classes, but A and B differ in their sorting; and C, though similar in mode and sorting to B, differs in its peakedness (kurtosis). D and E are markedly asymmetrical (skewed) and differ from each other in their direction of skewing. F is bimodal; all others are unimodal.

The advantages of single-number representation of these properties are obvious. Such summarization enables sedimentologists not only to say a sediment is well sorted or better sorted (having a smaller standard deviation) than some other deposit, but also to say how much better sorted such a sediment may be. Such values also enable him to plot average size (or other characteristic) against distance and to express quantitatively the relation of size change to distance of travel and the like, or to plot the values of the median or other size parameters on a map, each value at the corresponding sample point, and then to contour such a map and thereby deduce the direction of current flow and other data.

The parameters of the frequency distribution may be read or calculated from certain critical points on the cumulative curve. Or comparable parameters may be calculated from the analysis itself (the so-called "moment measures"). A great deal of effort

TABLE 3-6. Graphic parameters of size frequency distributions

Attribute	Technical measure	Graphical measures based on diameters	Graphical measures based on $-\log_2$ diameter (ϕ): After Krumbein and Pettijohn (1938, p. 230)	Graphical measures based on $-\log_2$ diameter (ϕ): After Inman (1952)
Central tendency or "average" size	Median (Md)	$Md = P_{50}$	$Md_\phi = \phi_{50}$	(same)
	Mean (M)	—	—	$M_\phi = \frac{1}{2}(\phi_{16} + \phi_{54})$ (phi mean diameter)
	Mode (Mo)	Mo = midpoint of most abundant class	$Mo_\phi = \phi$ midpoint of most abundant class	(same)
"Sorting"	Dispersion	$So = \sqrt{Q_3/Q_1}$ (coefficient of sorting)	$\sigma_\phi = \frac{1}{2}(Q_{3\phi} - Q_{1\phi})$ (phi quartile deviation)	$\sigma_\phi = \frac{1}{2}(\phi_{84} - \phi_{16})$ (phi deviation measure)
Symmetry	Skewness	$Sk = \sqrt{Q_1 Q_3/Md^2}$ (coefficient of skewness)	$Sk_\phi = \frac{1}{2}[(Q_{1\phi} + Q_{3\phi}) - 2Md_\phi]$ (phi quartile skewness)	$\alpha_{2\phi} = \dfrac{\frac{1}{2}(\phi_5 + \phi_{95}) - Md_\phi}{\sigma_\phi}$ (phi skewness measure)
Peakedness	Kurtosis	$K = \dfrac{(Q_3 - Q_1)}{2(P_{90} - P_{10})}$ (Kelley's quartile kurtosis)	$K_\phi = \dfrac{(Q_{3\phi} - Q_{1\phi})}{2(P_{90\phi} - P_{10\phi})}$ (phi quartile kurtosis)	$\beta\phi = \dfrac{\frac{1}{2}(\phi_{95} - \phi_5) - \sigma_\phi}{\sigma_\phi}$ (phi kurtosis measure)

Abbreviations: P = percentile; Q_1 = first quartile; Q_3 = third quartile.

Note: $Q_3 > Q_1$ and $P_{90} > P_{10}$.

has been expended, and diverse ways of estimating the size parameters have been proposed or used; these have been summarized by Folk (1966). Refer also to other papers (Inman, 1952) and the larger works in which the various methods and calculations are presented (Krumbein and Pettijohn, 1938, pp. 228–267; McBride, 1971, pp. 109–127). It is not possible to discuss and evaluate here all the possible options available. In general the tendency has been to utilize the phi value (the negative $\log_2$ of the diameter) rather than the diameter itself in the various schemes for characterizing size frequency distributions and to calculate the size parameters from certain points on the cumulative curve, either the quartile values (the 25, 50, and 75 percentiles) together with the 10 and 90 percentiles, or the 5, 16, 50, 84, and 95 percentiles. Table 3-6 summarizes some of the formulae for expressing the size parameters.

Mathematical nature of size frequency distributions Udden observed (1914), as others have since, that a geometrical grade scale tends to symmetrize the frequency curve (or histogram). This is just another way of saying that the size distribution is more or less symmetrical on a log basis, that is, if the frequency is plotted against the log size and not size itself. This observation has stimulated interest in the nature of the size frequency distribution and has led investigators to determine the kind of function, to express this function as an equation, and to seek out the physical causes that underlie it.

Krumbein (1938) concluded that the size distribution of many clastic sediments was log normal, and he expressed the distribution as a Gaussian function in which log size was substituted for actual size. Krumbein applied the tests for normality and found that the requirements were reasonably well satisfied by many sediments. The log normal character of a size distribution can be quickly assessed by use of a modified probability paper (Otto, 1939). The size frequency, expressed as weight percentage, is cumulated in the usual manner and plotted against the log of the size (Fig. 3-9). Many sediments deviate only a little from a straight line; some depart not at all. Other sediments, however, appear not to have a log normal size distribution.

Bagnold (1941, p. 116) believed that the distribution is not log normal but is some other probability function. Roller (1937, 1941) has called attention to the theoretical and actual failures of the Gaussian probability law for both the coarsest and finest grades of many sediments. In some cases it is probable that the size distribution more closely approximates that observed in crushed products—products produced by random breakage. This size distribution, observed in powdered coal, was expressed in equation form by Rosin and Rammler (1934). Some coarse pyroclastic deposits, glacial boulder clay or till, and residual weathering products have been shown by Krumbein and Tisdel (1940) to have size distributions that suggest an origin by random breakage, an observation confirmed by Kittleman (1964). See Fig. 3-11. Even the size distributions in some ordinary sediments (such as both arkosic and quartzose sandstones) approximate Rosin's Law (Dapples, Krumbein, and Sloss, 1953, Fig. 4). Roller (1937, 1941), however, has pointed out that Rosin's Law also has theoretical and practical drawbacks.

Some evidence suggests that many, if not most, size frequency distributions are composites of two or more discrete distributions in many natural sediments. Each distribution is a separate population, perhaps log normal, and the combination of these results in a distribution that tends to be markedly skewed or even in some cases bimodal (or polymodal). Several efforts have been made to "dissect" the cumulative curve and separate the component populations (Tanner, 1959; Spencer, 1963; Visher, 1969, for example).

GRAIN SIZE DISTRIBUTIONS AND CAUSAL FACTORS

In general the interpretation of grain size analyses has followed three paths. One path relates the characteristics of the grading curve to hydrodynamics (to the depositional process itself). This view was advanced by Udden (1914) to account for the

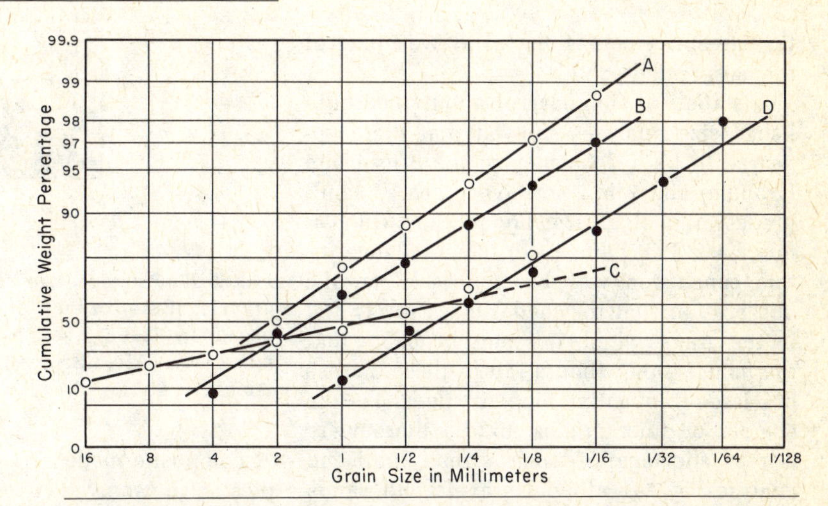

FIG. 3-11. Cumulative weight–percentage curves plotted on Rosin's Law paper. A: artificially crushed quartz (after Krumbein and Tisdel, 1940); B: disintegrated igneous rock (after Krumbein and Tisdel, 1940); C: tuff (after Moore, 1934); D: detritus from weathered gneiss (after Wentworth, 1931a).

bimodal distribution of many coarse river sediments, the coarser mode being a product of traction transport, the lesser and finer mode being the result of saltation transport. The interpretation of grading curves in hydrodynamic terms has been advanced by Inman (1949), Moss (1962, 1963), Friedman (1967), and Visher (1969) in particular. A second approach considers the grain size distribution largely a product of the sediment generative processes. In this case the distribution is attributed to the source materials and the size distributions generated by their disintegration. The breakage theories of Rosin and Rammler (1934), Tanner (1959), and Kolmogorov (1941) and the observations of Rogers, Krueger, and Krog (1963) and Smalley (1966) are illustrative of this approach. A third approach is to make an empirical study of grading characteristics of sediments from various natural geomorphic environments to see what relation, if any, exists between them. This approach was initiated by Udden (1914), expanded by Wentworth (1931a), and in more recent years promulgated by Sindowski (1957), Friedman (1961, 1962), Moiola and Weiser (1968), and many others.

In the following section we will examine each of these approaches to the study of grain size of sediments.

Grain size and provenance Certain particle sizes appear to be underrepresented in the sedimentary system. Wentworth (1933) called attention to this subject and argued that this observation provided a natural basis for defining the larger classes in the grade scale. He attributed the paucity of certain sizes and the overrepresentation of others to both the nature of the processes of particle production and to certain hydrodynamic factors (Table 3-7).

What evidence is there that such a paucity of certain sizes exists? Einstein, Anderson, and Johnson (1940) cite Nesper's work on the Rhine in Switzerland, in which the bed material ranges from 5 mm in diameter to "boulders" 100 mm or larger. Between the boulders and in protected pools, sand 1 mm in diameter and finer is found, but particles ranging from 1 to 5 mm are not present. These authors conclude that such particles "occur rarely for certain geologic or hydraulic reasons." In the Rhine the coarse material was part of the bed load of

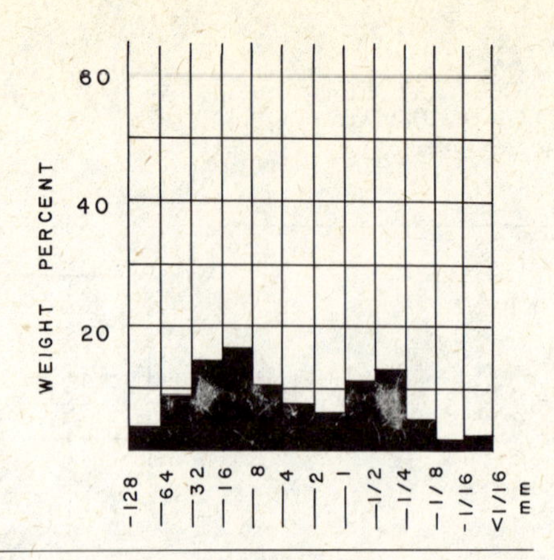

FIG. 3-12. Average size composition of upland gravels (Brandywine) of southern Maryland, based on 72 analyses. Note deficiency in 2–1 mm class. (After Schlee, 1957, Bull. Geol. Soc. Amer., v. 68, Fig. 5.)

the stream, whereas the sand was part of the suspended load.

A statistical summary of about 1000 published size analyses showed that there appeared to be a deficiency in the 2 to 4 mm (granule) and 2 to 1 mm (very coarse sand) grades, and also probably in the 1/8 to 1/16 mm class (Pettijohn, 1940). The evidence for this conclusion was based largely on the observation that the modal class rarely fell in the grades cited. It seems probable that the modal group should fall in these classes no less frequently than it does in the coarser or finer grades, unless there were a real shortage of such sizes. Published analyses of 241 alluvial gravels and sands of southern California (Conkling, Eckis, and Gross, 1934) show that the chief mode fell into the 2 to 4 mm class only three times, in contrast to 63 times in the 1/2 to 1/4 mm grade and 41 times in the 64 to 32 mm class. These observations were confirmed by Schlee (1957, p. 1379) in his study of the alluvial upland gravels of southern Maryland. No mode fell in either the 1 to 2 or 2 to 4 mm grades in 72 channel samples. A composite made by averaging all 72 samples also showed a deficiency in these grades (Fig 3-12).

That fluvial sediments are not unique in this peculiarity is shown by data secured by Hough (1942, p. 25) from shore and bottom sediments of Buzzard's and Cape Cod bays. Hough noted that the *medians* of several hundred samples rarely fell in the 2 to 4 mm class or in the 1/16 to 1/32 mm grade. Similarly, a composite from Massachusetts Bay (an average of 64 samples analyzed by

TABLE 3-7. Modes of transport and chief natural aggregates

Mode of transport	Usual source	Name of aggregate
Traction	All available hard rocks	Gravel
Inertia suspension[a]	Chiefly monomineral grains of phanerites	Sand
Viscous suspension[b]	Chiefly monomineral grains of any rocks	Silt
Colloidal suspension	Molecularly decomposable materials	Clay

[a] Traction and inertia suspension take place approximately in accordance with the so-called Sixth-Power Law, which postulates a complete transfer of kinetic energy from the water to a particle and which makes no allowance for the subsidiary effect of viscous drag.

[b] Viscous suspension accounts for the transport of finer particles in which the surface effect is greater relative to the mass. The size–velocity relationship in this range is defined by the well-known Stokes Law. Still smaller particles are kept in suspension, chiefly by the kinetic effects found in dispersed systems, i.e., colloid.

Source: After Wentworth (1933).

Trowbridge and Shepard, 1932, p. 29) disclosed a low frequency in the 1 to 2 mm grade. This low frequency was explained as the gap between two sediment loads, one moved by storm waves and the other moved by the more gentle waves. It should be noted that in general the sediments, both offshore and beaches, are not generally bimodal as are the coarser fluvial sediments, and the deficiency of certain grades is apparent only when all the available analyses are taken together. Not all workers are convinced that there is an underrepresentation of the 1 to 2 and 2 to 4 mm grades. Russell (1968) has called attention to the concentration of very coarse sand and fine gravel on certain beaches where these materials are present in more-than-ordinary abundance. He concluded that these grades are hydrodynamically unstable in rivers and tend to be sorted out and rapidly transported seaward to accumulate on beaches.

Wind-blown sediments seem to show a deficiency in the ⅛ to 1/16 mm grade. This peculiarity was noted by Udden (1914, p. 741). Like fine-grained stream deposits, eolian sediments are rarely bimodal, but only rarely does the modal class fall into the ⅛ to 1/16 mm class. Udden did not account for this peculiarity but suggested that it might not be general and that other wind deposits might be found in which the mode would fall in the apparently deficient grade. That there is a gap between silt and sand has also been noted by Rogers, Krueger, and Krog (1963, p. 631) and by Tanner (1958). The matter has been looked into further by Wolff (1964). A composite of 930 size analyses showed a deficiency in the coarse silt grade. Wolff thought this deficiency might be related to the use of an analytical technique for the silt and finer materials different from that used for the sands. If the apparent gap is not an artifact thus produced, perhaps it is due to sample inadequacies and would disappear if other sediments were included in the composite.

There are several possible explanations for the apparent scarcity of certain size grades or at least for the scarcity of sediments with their mode in these grades. One may assume that materials within these classes were produced by weathering and were never deposited as a modal class for certain hydrodynamic reasons, or they disappeared during transportation because of mechanical instability. Or we may suppose that there is a primary deficiency of certain size classes. It may be that all sizes are not produced in equal quantities by the breakdown of source rocks. It is difficult to determine, under any of the hypotheses accounting for the apparent deficiencies of certain sizes. If these sizes were produced by weathering or abrasion, what has become of them? Hydraulic considerations might prevent their deposition in some particular place or with certain other sizes but can hardly prevent their deposition everywhere. Perhaps they are segregated, as Russell thought, and separately accumulated. Otherwise we must suppose that they have been produced by rock disintegration but are mechanically unstable and destroyed, or that they were never formed in appreciable volume in the first place. The first hypothesis has been invoked to explain the apparent deficiency of particles in the 2 to 4 mm range (Hough, 1942, p. 26 fn). Such particles do appear to form by disintegration (but not decomposition) of plutonic rocks. The mineral grains of which they are composed are relatively large compared with the whole fragment; hence it might be supposed that such granules are structurally weak and unable to survive vigorous stream action.

On the other hand, it seems more probable that the processes of rock disintegration would produce more of certain sizes and less of others and thus create an initial deficiency of certain sizes. Three separate classes of particles might be expected from the breakdown of rocks (we are concerned with crystalline source rocks; in the case of clastic sedimentary rocks, breakdown will simply lead to the liberation of particles from an earlier phase of primary production). Some rocks characteristically yield *blocks* on breakdown, whereas others undergo granular disintegration and yield *grains* of sand size. The first type is illustrated by quartzite; the latter by many coarse-grained acid igneous rocks and gneisses. Products of disintegration of intermediate size may be relatively rare.

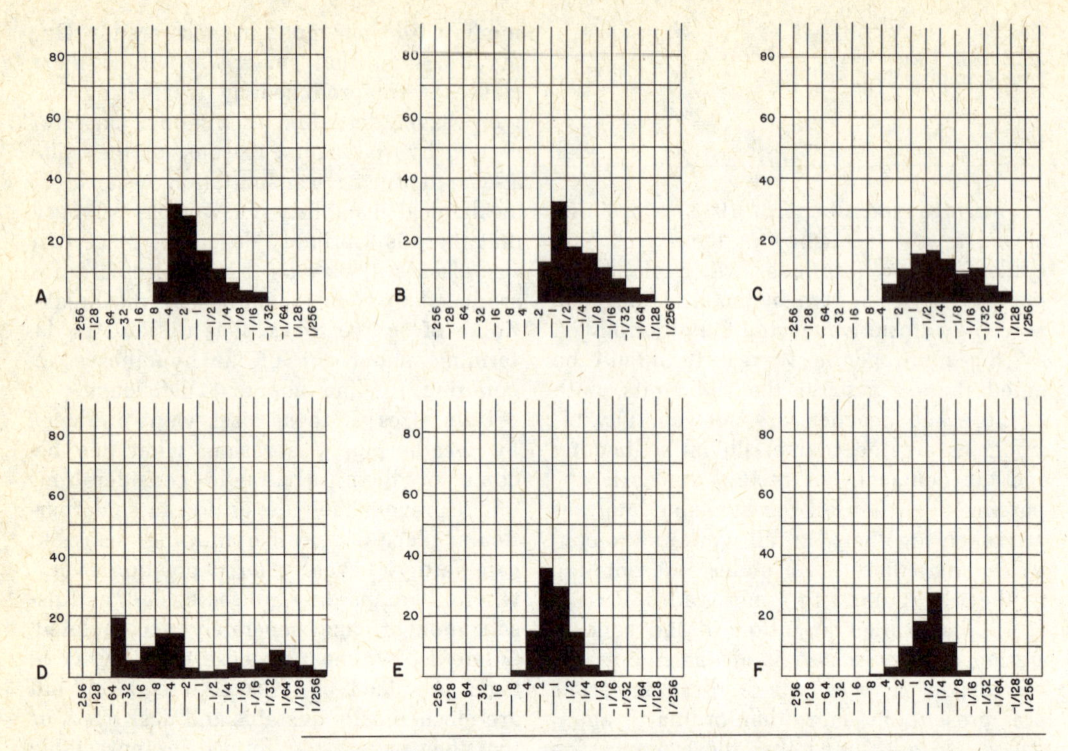

FIG. 3-13. Mechanical analyses of residual materials and related deposits. A: average of five samples of disintegrated granitic and gneissic boulders from glacial outwash (Krumbein and Tisdel, 1940); B: detritus from weathered gneiss, District of Columbia (Wentworth, 1931a); C: average of two samples of residual soil over gneiss, North Carolina (Krumbein and Tisdel, 1940); analyses incomplete; about 6 percent under 1/128 mm; D: rain-washed slope detritus (Wentworth, 1931); E: Sand from dry wash, El Centro, California (essentially a disintegrated granite); F: dry wash, Superior, Arizona.

Present data, however, are inconclusive; five samples of disintegrated (but not decayed) rocks of granitic composition analyzed by Krumbein and Tisdel (1940) showed the largest quantity in the 2 to 4 mm class (Fig. 3-13). Nevertheless, as pointed out by Dake (1921, p. 102) and by Smalley (1966), the size distribution of the quartz grains must be closely restricted by the size distribution of the quartz in the phaneritic crystalline rocks. Grains larger than 1 mm are rare. The blocks generated become pebbles whose abrasion produces fine silt or clay-sized material—not sand. Moreover, further breakage in general does not occur unless exceptional forces come into operation.

The decomposition products are of clay size; hence it would appear that there should be a deficiency in the silt range. Silt, however, is relatively common, and its production is a problem. Rogers, Krueger, and Krog (1963) presume that silt is produced by the chipping off of particles of silt size from larger quartz grains. This view was also advanced by Smalley and Vita-Finzi (1968), who thought the process most effective during wind transport in deserts. Kuenen's experimental work on eolian transport (1969) failed to substantiate this hypothesis. Kuenen attributed silt to the weathering of fine-grained quartzose rocks. Vita-Finzi and Smalley (1970) later concluded that glacial grinding was responsible for the bulk of the silt in the geologic record. The close association of loess—primarily a silt—with continental glaciation lends some credence to this view.

To what extent does the size composition of the larger population affect the grading curves of specific sediments? The peculiar bimodality shown by coarse fluvial sediments has been attributed to a primary deficiency in those grades which separate the modes (Fig. 6-4). Would this bimodal character of the materials supplied be retained

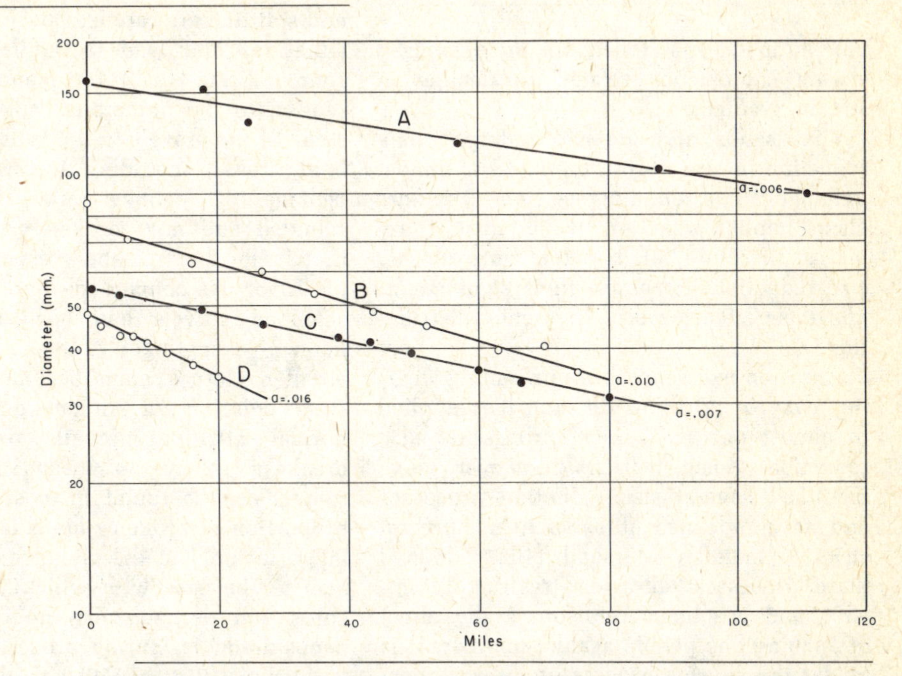

FIG. 3-14. Relation of diameter of pebbles to distance of travel. A: largest cobbles of Rhine River (after Sternberg and Barrell, 1925); B: River Mur (after Grabau, 1913, p. 246, from Hochenburger, 1886); C: limestone test piece, T-33, in abrasion mill (after Wentworth, 1919); D: limestone in abrasion mill (after Krumbein, 1941). a is a coefficient of size reduction.

in the deposits made by streams? This seems to be true in the case of ice-deposited materials—boulder clay or till. Analyses of till commonly show one or more secondary modes which seem not to be random (Fig. 6-17). These lesser modes probably express "loading" of the ice with some particular material which had a previous history of sorting and deposition. Ice moving over sandy outwash might pick up much of this material, and the size analysis thus might reveal a small mode in the sand class. Whether analogous loading occurs in water-laid materials is less certain, although some observers have believed this to be the case. Swenson (1942) thought the bank sediment of the Mississippi was significantly modified by input from a major tributary, the Maquoketa River. Curray (1960) believed that the grading curves of many bottom sediments in the Gulf of Mexico were determined by the proportions of several unlike sediments deposited concurrently.

Grain size and transportation To what extent and in what way are size and size distributions modified by the transportation processes? The effects of transportation are only incompletely understood. The prevailing concepts are based mainly on deductive reasoning and supported by very little experimental or field data. Although most attention has been given to the effects of transport on the size of the materials moved, we are not yet certain of the cause of the observed effects. In general the gravels carried by streams appear to decrease in size downstream (Fig. 3-14). And because the corners and surfaces of the larger materials are rounded and smoothed, it has been presumed that abrasion is an active process during transport and that the downcurrent decline in size, therefore, is caused by such wear. This is true in part, but, as pointed out elsewhere (Chapter 14), the size decline in some cases is very probably not caused solely by abrasive action but is instead a reflection of decreased stream competence related to decline in stream gradient.

That sands and gravels undergo some reduction in size during transport is almost axiomatic. The rounding seen on all mature clasts implies wear and weight loss. It remains, therefore, to examine more closely size reduction processes, their significance, and their effect on size frequency distribution.

Abrasion is a general term meaning wearing away or attrition. As such it is applied to almost any mechanical process of size reduction. Some workers, however, have identified several size reduction processes and have redefined abrasion in a more restrictive manner. Marshall (1927) defined three processes: abrasion (restricted), impact, and grinding. *Abrasion* is the effect of rubbing one pebble against another. It is by far the slowest process of wear. *Impact* is the effect of definite blows of relatively larger fragments on others of a smaller size and hence it is important only when there is an appreciable disparity in the size of the largest and smallest fragments. If such a disparity does exist, and if the larger sizes greatly predominate, the smallest sizes undergo heavy losses in a short period of time. *Grinding* is the crushing of small grains by continued contact and pressure of the pebbles of somewhat larger size. It was found to be more rapid in its action than impact. Sand mixed with gravel was reduced to silt and clay in a few hours in an abrasion mill.

Wadell (1932) recognized four size reduction processes: solution, attrition, chipping, and splitting. The differences are primarily a matter of the ratio of the size of the particles removed to the size of the original fragment. The mode of action is not considered. If the particles removed are of suboptical size, *solution* is implied. Solution may be ionic or colloidal. If the particles removed are visible but less than $\frac{1}{150}$ that of the fragments being worn, the process is termed *attrition*. If the material removed is still larger, such as results from flaking off the corners, the term *chipping* is applied. If the destruction process produces two subequal fragments, the term *splitting* may be employed.

Normal attrition of gravels produces silt- and clay-sized debris, not sand. Chipping and splitting are rare except in high-velocity situations and lead to *spalls* and *broken rounds*. Bretz (1929), for example, called attention to the numerous broken rounds of some of the gravels in the Washington scabland areas. According to Bretz, the percentage of pebbles and cobbles once rounded and now broken greatly exceeds that in the bars of the present-day Columbia River. He concluded, therefore, that the scabland gravels had been moved by a flood of exceptional violence. Yet the gravels of modern streams become rounded despite episodes of violent flow. Although normal attrition prevails over splitting, breakage is by no means rare; broken rounds can be found in most gravels. The proportion of broken rounds is probably related not only to the violence of action but also to the ease with which the rock type splits and perhaps also upon some post-depositional fracturing processes.

Kuenen (1956, p. 350) has also attempted to analyze the abrading process. He recognized seven size reduction processes: splitting, crushing, chipping, cracking, grinding, solution, and sand blasting. *Cracking* is the process responsible for the crescentic percussion marks on a pebble's surface; *sand blasting* is the wear caused by sand that is swept past the pebble, which itself remains at rest.

No complete or comprehensive study has been made of the effects of the size reduction processes on the size parameters, in part because of the concomitant sorting action which takes place in the natural environment and which makes it difficult to separate the effects of size reduction from those of sorting. As noted by Marshall, under some conditions certain sizes are subject to more rapid size reduction than others. This greatly alters the composition of the original "mix" in an abrasion mill. If the finer products are removed by sorting, the result is an increase in mean size of the residue and an improvement in sorting (decrease in standard deviation).

There is a great deal of interest in the *rate* of size reduction and the factors which control it. Something about rate can be learned in abrasion mills and related experiments, but it is difficult to apply these results to natural situations in which the size decline is only partially, perhaps only in a small way, related to wear. Most of the observed downcurrent decline in size is attributable to sorting.

Controlled laboratory studies offer another approach to the problem. Experimenters beginning with Daubrée (1879) have done important work on this problem. Such experimental work, however, has several shortcomings. It is oversimplified; one cannot be certain that conditions in a mill even approximate those in a stream or on a beach. Experimental work may be extrapolated beyond reasonable limits; because most abrasion mill studies have dealt with pebble-sized fragments, conclusions drawn from such work have been erroneously applied to sand-sized materials. Since Daubrée's early work, there have been many experimental studies of the wear of gravels, including those of Wentworth (1919, 1931b), Marshall (1927), Schoklitsch (1933), Krumbein (1941b), Raleigh (1943, 1944), and Potter (1955, p. 22). See Fig. 3-14. These studies utilized an abrasion mill or tumbling barrel. More recent work (Kuenen, 1955, 1956, 1964; Bradley, Fahnestock, and Rowehamp, 1972) employed a circular moat, a device believed to be more nearly comparable with natural streams.

These experimental studies have shown that the reduction in size of gravels by abrasion and related processes is a function of size of the materials, nature (durability), nature and violence of action (rigor), size and proportions of the associated materials, nature of bed materials over which the gravel moves (sand or gravel), and duration or distance involved in the abrasive action.

The effect of *size* is most marked; gravels are rapidly abraded and rounded, whereas sand is abraded only with extreme slowness. Even within the pebble size range the percent lost for a given distance of travel is larger for larger sizes (Kuenen, 1956, Fig. 7). But the result is complicated when a mixture instead of a single size is used. In mixtures the smallest sizes undergo the greatest loss, perhaps because of the action of the larger sizes on them. The *durability* of the materials is of obvious importance, as all investigators have realized. In general chert, quartzites, and vein quartz are the most resistant to wear, metamorphics less so, and limestones and friable sandstones least (Plumley, 1948; Kuenen, 1956, pp. 349–350). The violence of action or *rigor* is also an important factor. The rate of weight loss appears to increase with increasing rigor. Experiments of Kuenen (1956, p. 344) showed that the abrasion was proportional to the square of the velocity. Whether there are critical velocities for certain minerals or rocks above which chipping or splitting dominate over normal abrasion (as Krynine suggests, 1942) is not known. The *nature of the bed surface* over which the gravel is moved has been shown by Kuenen (1956, p. 350) to be important. Losses are lessened if the bed is sand and notably greater, perhaps by as much as five times, if the bed is gravel. The effect of the original *shape* of the fragments seems to be a very minor factor, but with increasing roundness the rate of wear declines. The influence of the geologic *agent* is less significant in the wearing of gravels than it is for sands. Experimental studies of surf action, however, are few (Kuenen, 1964). Apparently the wear of gravels by surf action proceeds rapidly.

All experimental studies have shown that the rate of size reduction is greatest in the early stages of the process and tends to decline exponentially with time or distance (Krumbein, 1941b; Schoklitsch, 1933).

Experimental studies of sand abrasion include those of Daubrée (1879), Anderson (1926), Thiel (1940), and more recently of Kuenen (1959, 1960a, 1960b) and Berthois and Portier (1957). All show that, in the absence of coarse materials, the abrasion of sand is a much slower process. Daubrée (1879, p. 256), for example, showed that a sand grain lost only 0.01 percent per kilometer of travel. Using a circular moat instead of a revolving mill, Kuenen (1958, p. 50) found quartz grains of about 0.5 mm in diameter to lose 0.0001 percent per kilometer—a loss so small that 10,000 km of transport would fail to show any visible rounding of the quartz. Because the average river travel is 1,000 km or less, 10 cycles of mechanical fluvial abrasion would cause less than a 1 percent weight loss. Other experimenters, using abrasion mills (Berthois and Portier, 1957; Thiel, 1940) have reported larger losses, but it still seems improbable that much size reduction of quartz sand is accomplished by river action. Even to reduce a cube 1 mm on a side to a sphere 1 mm in diameter requires a removal of 47.5 percent of the original volume. Even the higher losses reported would not significantly modify the original grain size or shape and at best would produce only minor rounding of the grains. Thiel's photographs of quartz sand, before and after a presumed equivalent of 5,000 miles of travel, confirm this conclusion. Kuenen's photographs (1958, Pl. 1) show no visible effects on crushed quartz sand that traveled 248 km. The effects of eolian action appear to be several orders of magnitude greater than that of water action (Kuenen, 1960b), an observation that led Kuenen to the view that rounding of sand must be attributed to wind action. Grains less than 0.05 mm in diameter are wholly unabraded.

In the light of these experimental studies, what is the role of natural abrasion in reducing or altering the size of the individual clasts or in modifying the size parameters of a grain population? Clearly the process modifies shape, rounding, and surface textures—effects which are discussed in other sections of this chapter. The marked downcurrent size decline of gravels seen in many streams, however, is probably caused only in small part by wear. It is more largely the result of decline in stream gradient and competence (see Chapter 14). It is probable that there is little or no size diminution of quartz sands by abrasive processes in most natural environments. In short, the size distribution is a product of hydraulic rather than abrasive action, and in general actual size is inherited from the parent rock or is a product of rock disintegration and not the result of the transportation agent or process.

Grain size and depositional processes The views that most size frequency distributions are mixtures of two or three grain populations related to differing modes of sediment transport and that the interpretation of grading curves was a problem in identifying and relating these subpopulations to specific hydrodynamic processes have gradually gained ground (Fig. 3-15).

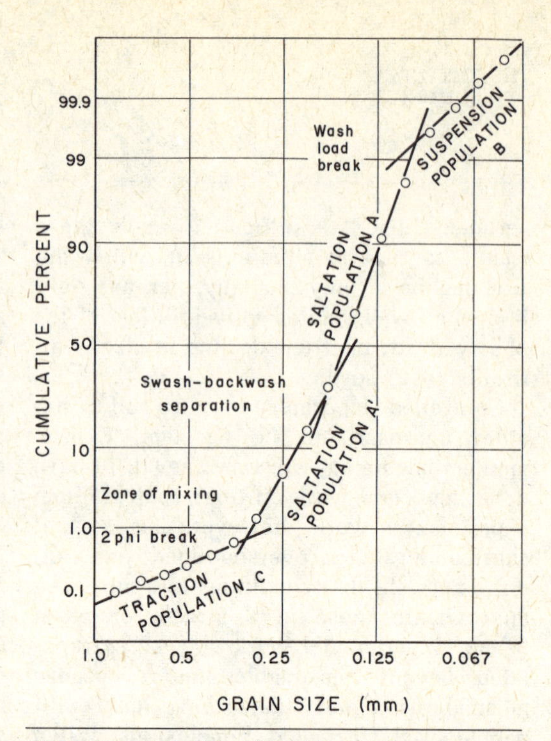

FIG. 3-15. Relation of sediment transport dynamics to populations and truncation points in a grain size distribution. (After Visher, 1969, Jour. Sed. Petrology, v. 39, Fig. 4.)

This approach was first applied to the coarse deposits of streams. Bimodality is especially common in these deposits. Sands, on the other hand, tend to have a single mode. Of the several hundred published analyses of California alluvial gravels, for example, 92 percent had more than one mode (Conkling, Eckis, and Gross, 1934). Only 42 percent of the associated sands were so characterized. In general the bimodal gravel had its chief mode in some gravel class and its secondary or lesser mode in the sand grades (Fig. 6-4). The modes are 4 to 5 grades apart on the average. The chief ingredient, therefore, has a diameter 16 to 32 times that of the material in the secondary mode. Other alluvial gravels are also characteristically bimodal. Krumbein (1940, 1942a) found 85 percent of the flood gravels of the San Gabriel and Arroyo Seco to be bimodal. Twenty of 23 samples of Black Hills terrace gravels were bimodal (Plumley, 1948).

These observations have been explained as indicating two populations related to two modes of transport. Udden (1914, p. 737) noted that "... a transporting medium... tend[s] to carry and to deposit more of two certain sizes of material than of other sizes. The principal deposit it makes will have an excess of another considerably coarser ingredient which it can roll, smaller in quantity." He presumed the principal mode would be in the finer sizes, the secondary mode in the gravel. The reverse is the more common.

Fraser (1935) thought that simultaneous deposition of cobbles and finer sands was improbable and pointed out that the velocity of a current carrying 10-inch (25-cm) cobbles would have to be decreased 60 percent before the 1-mm size could be deposited. Such violent changes in velocity were thought to be unlikely, and Fraser believed that at any given instant a river usually deposits only a very limited range of material and that the finer sizes in gravels were the result of later infiltration. This view was also expressed by Dal Cin (1967, p. 154), based on his study of gravels of the River Piave in Italy. Plumley (1948) also interpreted the fines, a secondary mode in most cases, as entrapped materials filling the voids in the primary gravel framework. To support this view he noted that if one assumed two sizes of spheres—the smallest small enough to be contained in the spaces between the larger—that the smaller fraction would form 22 to 32 percent by weight of the whole deposit, depending on the tightness of the packing. Because natural gravels contained, on the average, 20 percent of the secondary mode and closely related grades, it seems probable that these grades indeed were trapped fines. Considering the departure of pebbles from spherical forms, the nonuniform size of the two fractions involved, and the chaotic nature of the packing, the agreement between theory and observation is surprisingly good.

It is possible, of course, that these bimodal distributions could result from the sampling procedure in which the material collected was from two distinct beds, each with its own population (Bagnold, 1941, pp. 118–124). But that many bimodal distributions are not artifacts produced by the sampling technique is shown by the bimodal character of some single layers. They seem to be two populations deposited in the same episode of deposition.

The bimodal distribution is only a special case of the mixing of two populations —populations in this case separated widely enough to show two modes. Where separation is slight, only a single mode is present. But the composite population is markedly skewed, and the size frequency distribution departs markedly from log normality. The increasing awareness that many distributions are composites of two or three subpopulations has led to attempts to relate these subpopulations to differing modes of sedimentary transport—to move from the special case of bimodality to the more general case involving many, if not most, sediments.

Doeglas (1946), as well as Harris (1958a, 1958b), recognized the composite nature of many size distributions and attempted to relate these to varying transport conditions. A major advance in this direction was that of Moss (1962, 1963, 1972). Moss identified three subpopulations associated with different processes of sedimentation. These populations can be identified from the cumulative curve. The main part of the distribution lies between percentiles 20 and 80. This is the principal saltatory load. The coarse "tail" of the distribution is the traction load; the fine "tail" of the distribution is formed of suspension materials trapped in the interstices of the main framework of the deposit. As noted by Moss (1962), in some coarse river deposits the traction load becomes the main part of the deposit. These are the bimodal sediments described above.

The principal study in recent years involving the subpopulations and hydrodynamics is that of Visher (1969). He presumes that grading curves of all clastic sediments are composites of three basic populations—populations identified with traction, saltation, and suspension transport —and that each population has a log normal size distribution and plots, therefore, as a straight line segment on a probability scale (using the log of the diameter or phi notation). See Fig. 3-15. These relations seem to prevail in some 2000 size analyses from many differing environments.

What is the relation between the shape of the grading curve—the relative abundance and character of its components—and the depositional environments defined in the usual geomorphic terms? Visher believed that the characteristics of individual grain size distribution curves do provide a basis for environmental identification but noted that "any attempt to define precise limits for the slope, truncation points, and percentages of the three basic populations for individual environments is impossible."

In summary we can say that the several approaches to search for meaning of the size grading curves of clastic sediments have some validity. The hydrodynamic factors may to some extent correlate with geomorphic environments. A particular process may prevail in a particular environment but be subordinate in another.

GRAIN SIZE DISTRIBUTIONS AND ENVIRONMENTAL ANALYSIS

Udden believed that the size composition of a sediment was controlled by hydrodynamic conditions prevailing during deposition of a clastic sediment. It therefore follows that, if ancient sediments were deposited under conditions similar to those now forming, study of modern sediments would reveal the grading characteristics of each type, which could then be used in turn to decipher the origin of ancient deposits. Udden accordingly made many "mechanical analyses" of sediments, especially wind deposits, and published (1914) over 350 such analyses together with a summary of certain "laws" thought by him to govern the mechanical composition of clastic sediments. Wentworth (1931a) published more than 800 such size analyses in graphic form in an effort to extend Udden's initial approach. Inspection of Wentworth's histograms shows that the graphic patterns of sediments from different environments are different. Glacial till and beach sand, for example, are strikingly unlike. On the other hand, some unlike sediments are quite similar in their mechanical composition, such as, for example, beach and dune sands.

Our inability to discriminate between closely related environments or agents of deposition by means of the size analysis or their graphic representation has not discouraged search for other, more subtle differences. Keller (1945) used the ratio of the quantities in the two classes proximal to the modal class—his F : C ratio—to distinguish between wind and beach sands; his ratio is a rough measure of skewness. Since Keller, more sophisticated efforts utilizing one or several size parameters have been used to discriminate between river, beach, and dune sands. Friedman (1961, 1962, 1967) attempted to distinguish between beach and dune sands by plotting skewness against the mean and between river and beach sand by a plot of skewness against standard deviation (Fig. 3-16). A similar approach by Moiola and Weiser (1968) substantiated Friedman's conclusions. Others have also used scatter diagrams involving two variables. Passega (1957, 1964), for example, plotted the 1 percentile, *C* (essentially the coarsest size), against the mean size, *M*, as did Bull (1962). It was presumed that the particular *CM* pattern was indicative of the depositional agent or process. Passega, for example, thought the *CM* diagram differentiated between turbidity and ordinary current action.

The use of size parameters in various combinations as environmental indicators has also been tried by Mason and Folk (1958), Gees (1965), Schlee, Uchupi, and Trumbull (1965), Kolduk (1968), Doeglas (1968), Solohub and Klovan (1970), and Buller and McManus (1972). Not all of these investigators achieved a successful separation of environments of sand deposition by this approach. In some cases no success was achieved; in others considerable overlap appeared on the scatter diagrams, so that the results were ambiguous.

Other more sophisticated techniques have utilized more than two variables at a time. This technique of discriminant analysis was tested by Sahu (1964a, 1964b). Klovan (1966) used a factor analysis on sediments from Barataria Bay, Louisiana, to isolate factors related to causal processes: surf, bottom currents, and quiet-water settling.

In summary, we can say that the efforts

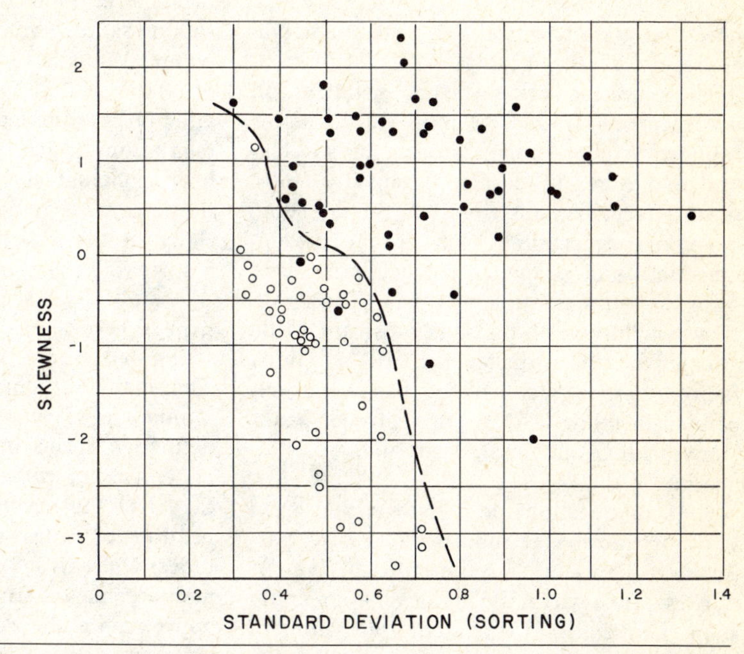

FIG. 3-16. Plot of third moment (skewness) and standard deviation, using phi scale for beach and river sands. Solid circles represent river sand. (After Friedman, 1961, Jour. Sed. Petrology, v. 31, Fig. 4.)

to relate the size distribution of a sediment to its environment of deposition have had limited success. In part the negative results may be based on the premise that the grading curve is wholly hydrodynamic in origin and that each environment is characterized by a different hydrodynamic regimen. Neither assumption may be wholly valid. The effect of provenance (that is, the size characteristics of the sediment put into the environment) may not have been fully assessed. And the same hydrodynamic process may operate in more than one environment. To put it differently the hydrodynamic environment and the environment as usually defined in geomorphic terms may not coincide (Solohub and Klovan, 1970).

Summary 1. Various measurements may be made on individual fragments (such as weight and volume) or on procedures which are applicable to the aggregate itself (such as sieving, elutriation, and gas absorption). All of these are converted to grain "diameter" based on some simplifying assumptions, many of which are at best only crude approximations.

2. The grain size distribution thus determined is expressed in terms of the percentages of the several size classes of the whole sample, based on the measured or calculated weight of material in each class or on the number of grains in the respective classes. The distribution curves, by weight and by number, differ materially from one another.

3. The standard analytical procedures are applied with a certain success to unconsolidated sediments, mostly to modern sediments. They are largely inapplicable to well-indurated ancient sediments and are therefore severely limited in their usefulness. They are limited also in the applicability of the conclusions drawn from analyses of modern sediments, such as the distinction between the several environments based on textural parameters.

4. A further serious restriction is im-

posed on grain size studies by the diagenetic alteration of the original size distribution. Such alteration arising from pelletization by organisms, from degradation of the larger framework elements, or from recrystallization and other processes, drastically alters the primary size distribution. So do some analytical procedures such as the dissaggregation and dispersion procedures used prior to sedimentation analysis. The grain size distribution of shales is most apt to be affected by these diagenetic and technique factors. Hence most effort has been expended in study of the grain size of sandstones, which are less susceptible to modification.

5. It is questionable whether the grain size distribution is indicative of any particular agent and/or environment. Even if it were, it is virtually impossible to apply such knowledge to ancient well-lithified rocks.

6. In short, despite the voluminous literature and extended efforts made to define grain size, measure it, and calculate the parameters of grain size distribution, the net input toward the solution of geological questions has been disappointing and disproportionately small relative to the effort expended.

7. There remains, however, some meaningful grain size observations that can be made even on thoroughly indurated rocks. We include here measurement and mapping of maximum pebble size in the conglomerates, recognition of phenoclasts (outsize fragments indicative of rafting), and, of course, observations on the framework elements which, although not strictly size attributes, are geologically significant. We include here shape, roundness, and surface texture and, of course, composition.

SHAPE AND ROUNDNESS

Shape and roundness of pebbles and sand grains have long been used to decipher the history of a deposit of which they are a part. The distinctive shapes of ice-faceted pebbles and wind-whetted ventifacts are known even to beginners. The effects of other agents are less clear and are the subject of much controversy. Are beach pebbles flatter than river pebbles? Does wind round sand grains more effectively than water? What is the lower size limit, if any, of water rounding? Can quartz grains be effectively rounded in one cycle of sedimentation? Such questions have not yet been conclusively answered. It is obvious that a conclusive answer would aid us materially in the interpretation of the geologic history of a sediment.

SHAPE (FORM)

The shapes of objects may be classified in a number of ways. The geometrician has defined a series of regular shapes such as cube, prism, sphere, cylinder, and cone. Likewise the crystallographer has a classification of solids bounded by plane surfaces. Neither system is adequate for sedimentary clasts. At best pebble shapes only approximate the regular solids of the geometrician. Terms to indicate similarity, such as prismoidal, bipyramidal, pyramidal, wedge-shaped, or parallel-tabular, may be used (Wentworth, 1936a). But such a classification is only a qualitative description and does not, as a rule, bear any relation to the dynamic behavior of these objects during transportation. Instead, a single-number index of shape is required, which is amenable to mathematical or graphic analysis and by means of which a shape distribution or frequency curve can be constructed.

Certain distinctive shapes, however, are not described in a simple numerical manner. Such are the crystal euhedra displayed by some detrital heavy minerals and the diagnostic curving forms exhibited by volcanic glass fragments (the *shards* of tuffs and tuffaceous sediments). The distinctive shapes of wind-faceted stones—the familiar einkanter and dreikanter (Bryan, 1931; Whitney and Dietrich, 1973), the broken rounds of some torrential gravels (Bretz, 1929), and the flat-iron form of the glacial cobble (Von Engelen, 1930) likewise cannot be reduced to a single number. Nevertheless, for reasons stated, a quantitative or numerical index of shape is useful, and many efforts have been made to find such a form index.

One approach is to choose a standard of reference, a sphere being such a standard. Not only is the limiting shape assumed by many rock and mineral fragments upon prolonged abrasion that of a sphere, or approximately so, but the sphere has certain unique properties which make it an acceptable standard of reference. Of all possible shapes, the sphere has the least surface area for a given volume. As a consequence of this property, the sphere has the greatest settling velocity in a fluid of any possible shape, volume and density being fixed (Krumbein, 1942b). See Fig. 3-17. Hence under conditions of suspension transport, more spherical particles tend to become separated from others of the same size and density but of less spherical form.

Ideally the property of *sphericity* might be defined as s/S, where s is the surface area of a sphere of the same volume as the fragment in question and S is the actual surface area of the object. For a sphere the ratio has a value of 1.0; for all other solids the value is less than 1.0. Because of the difficulty of measuring the surface area of an irregular solid, the sphericity may be approximated by d_n/D_s, where d_n is the diameter of a sphere with the same volume as the object and D_s is the diameter of a circumscribing sphere—generally the long diameter (Wadell, 1935).

In a sample of sand or gravel each particle or fragment will have its own sphericity value. Some, however, will be disk-shaped or notably flat and elongated in two directions and shortened in the third. Others will be elongated in one direction only and will be prolate or roller-shaped. Both shapes yield a low sphericity value. The sphericity index as defined above fails to discriminate between the two. This discrimination is important for some investigations as, for example, for the fabric of a gravel.

Thus various other form indices have been proposed. These all involve defining and measuring the several "diameters" of

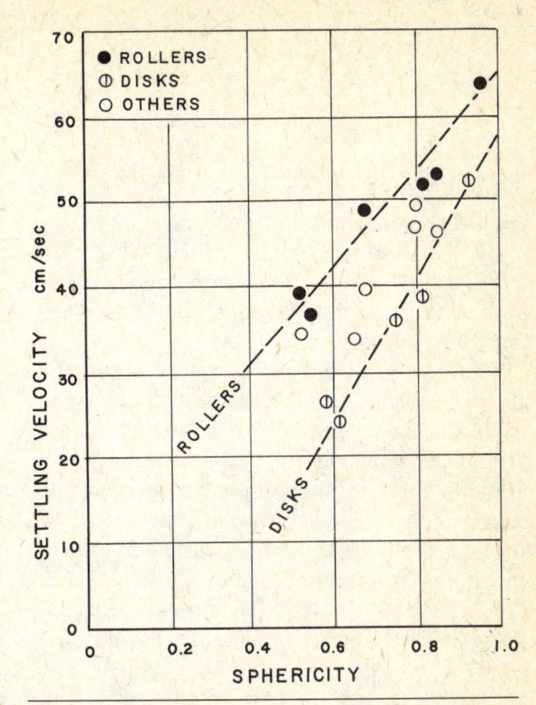

FIG. 3-17. Shape-settling velocity relations. (From Krumbein, 1942.)

the fragment and the choice of one or more ratios to express the shape. Zingg (1935) used the ratios b/a and c/b (where a, b, and c are length, breadth, and thickness, respectively) to define four shape classes (Fig. 3-18 and Table 3-8). These classes—oblate, prolate, triaxial, and equiaxial—and the relation to the Wadell sphericity index are shown in Fig. 3-19.

Other measures, including indices of flatness and of elongation, have been proposed. Most of these have been reviewed by Konzewitsch (1961), Köster (1964, pp. 138–147, 161–176), Flemming (1965), Humbert (1968), and Carver (1971). Sneed and Folk (1958) proposed a modification of the Zingg-Wadell approach and defined a maximum projection sphericity index $(c^2a^{-1}b^{-1})^{1/3}$, which they found to correlate better with the observed settling velocity than with the operational sphericity of Wadell.

Practical difficulties arise with all the methods of measuring and expressing form or sphericity in that they all involve measurements that can be made easily only on matrix-free pebbles and are difficult or impossible to apply to sand grains or to well-indurated gravels and sands. Nonetheless, it is necessary to study the processes of pebble shaping and the geological meaning of this attribute. Pebbles can be extracted from many ancient sediments and no approach to understanding should be overlooked.

What actually can we say now about the geological significance of the shape of a pebble or sand grain? The quartz grains of sands are variable in shape. In the main they tend to be subspherical. Even in the most mature sands, however, they tend to show a slight elongation, a ratio of long to short axis of 1.0 to 2.5, generally nearer 1.5. Wayland (1939) noted the tendency for the elongation of detrital quartz to be greatest in the direction of the *c*-axis and attributed this to unequal abrasion caused by slight differences in hardness in the several crystallographic directions. Ingerson and Ramisch (1942), however, observed that the quartz grains of igneous and metamorphic source rocks, even those of granites, tended to be elongated parallel to the *c*-axis (Fig. 3-20). The final end-shape was therefore determined in large part by the initial shape. Bloss (1957) and Moss (1966) showed ex-

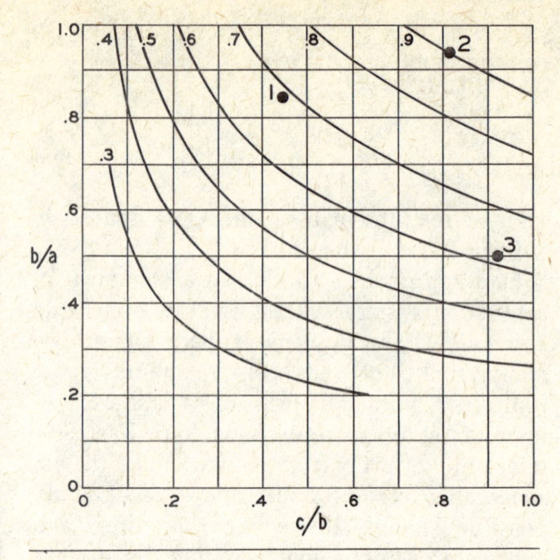

FIG. 3-19. Relations between sphericity and Zingg shape ratios. The curves represent lines of equal sphericity. Solids 1 and 3 fall in the same sphericity class (0.6–0.7), but 1 is oblate and 3 is prolate.

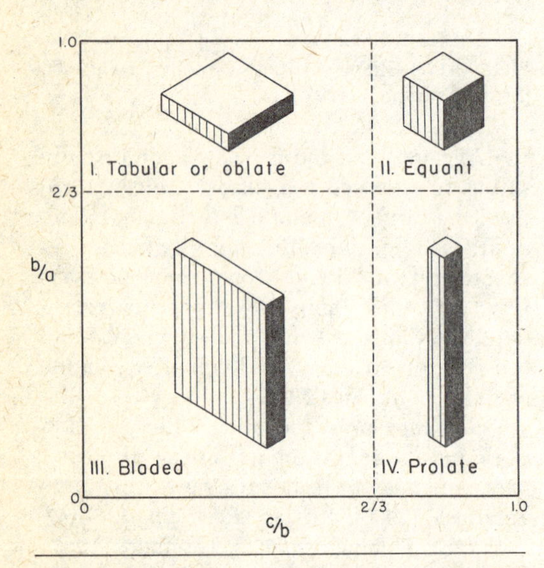

FIG. 3-18. Zingg's classification of pebble shapes. Note that the representative solids shown (rectangular parallelopipeds) have the same roundness (0) but that they have different shapes.

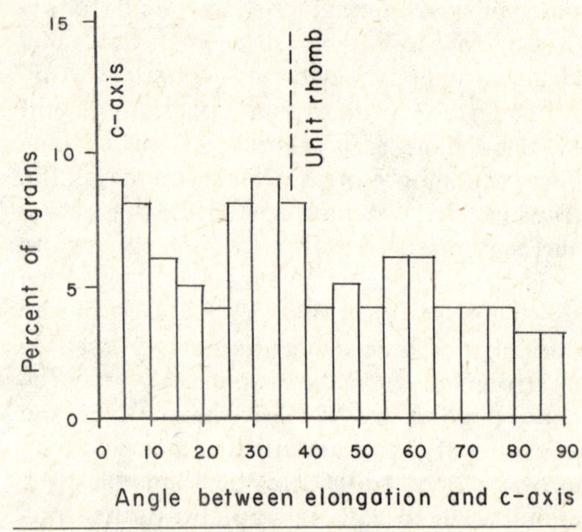

FIG. 3-20. Histogram of relation of elongation and crystallographic directions in original quartz grains from a chlorite schist. (After Ingerson and Ramisch, 1974, Amer. Mineral., v. 27, Fig. 10.)

TABLE 3-8. Zingg shape classes[a]

Class number	b/a	c/b	Shape
I	> ⅔	< ⅔	Oblate (discoidal)
II	> ⅔	> ⅔	Equiaxial (spherical)
III	< ⅔	< ⅔	Triaxial (bladed)
IV	< ⅔	> ⅔	Prolate (rod-shaped)

[a] *a*, length; *b*, breadth; *c*, thickness.

perimentally that quartz has both a weak prismatic and rhombohedral cleavage, so that the grains produced by fracture tend to be elongated parallel to the *c*-axis or at a fixed angle to it, an observation made earlier by Turnau-Morawska (1955). The shape of detrital quartz is therefore largely attributable to original growth or fracture. It has been presumed that the quartz of metamorphic rocks had an initial more elongate form than that from igneous rocks (Krynine, 1946) and that this difference would enable one to discriminate between detrital quartz from these two sources (Bokman, 1952). Later investigations (Blatt and Christie, 1963) seem not to support this notion.

In general, also, the shapes of pebbles are believed to be largely determined by the original shape of the fragment, which may in some cases be controlled by the structure of the rock. It cannot be denied, however, that some geologic agents do modify the shapes of pebbles and leave their own imprint as in the case of eolian sandblast and glacial action. Does this notion extend to beaches where, according to some, the swash tends to produce flatter pebbles than does stream action? This view has been supported by the observations of Landon (1930), Cailleux (1945), Lenk-Chevitch (1959), and Dobkins and Folk (1970) but disputed by Gregory (1915), Wentworth (1922b), and Kuenen (1964) and by the observation of Grogan (1945). See Figs 3-21 and 3-22. From experimental results and field observations, others concluded that mechanical wear on the beach has very little influence on flatness (Kuenen, 1964, p. 37). It could be, however, that gravel is sorted in such a manner that flatter pebbles tend to accumulate on beaches, a view early advanced by Landon. To a certain extent this view finds support in the work of Humbert (1968, p. 36), who found the flatter pebbles to migrate downbeach and the more spherical ones to lag behind. This does not mean that sphericity is in no way modified by abrasion. But most of the published studies show this modification to be small, and perhaps many of the downcurrent changes observed are caused by shape selection rather than shape modification. Included here are such studies as those of Russell and Taylor (1937a), Plumley (1948), Sneed and Folk (1958), Humbert (1968), Unrug (1957), and Dal Cin (1967). However, Dobkins and Folk (1970), who studied and measured sphericity and roundness of a large number of basaltic pebbles of rivers and beaches of Tahiti, found the beach pebbles to have the highest roundness and lowest sphericity and to be distinctly oblate in contrast to the

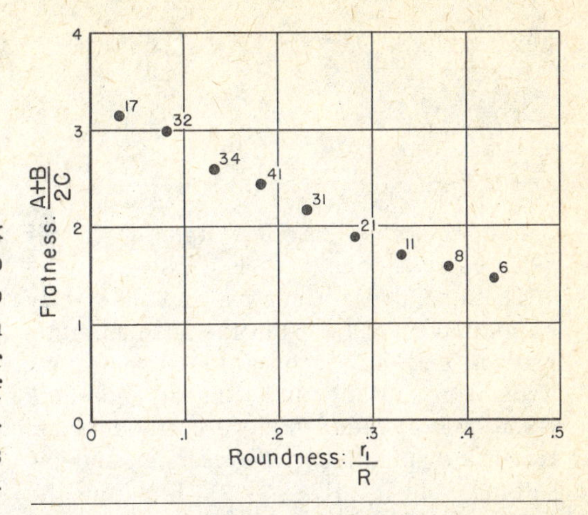

FIG. 3-21. Relation of roundness to flatness of 201 pebbles, Nantasket Beach, Massachusetts. For sake of simplicity, the pebbles were arranged in subgroups, and each dot gives the mean position of the pebbles in a particular group. The associated figure is the number of pebbles within the group. It is apparent that, as the roundness increases, the flatness decreases. (After Wentworth, 1922.)

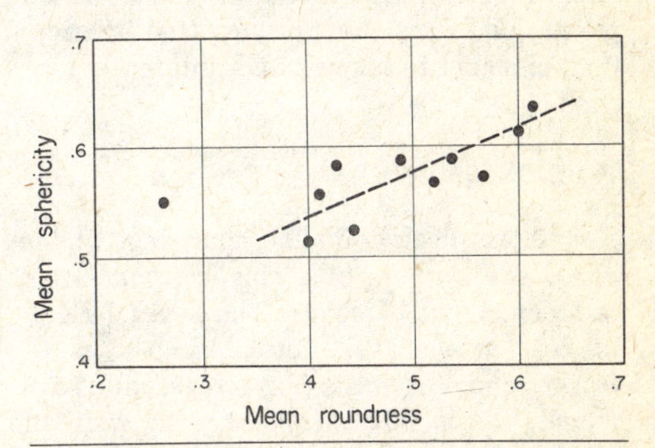

FIG. 3-22. Relation of sphericity and roundness of rhyolite pebbles (24 to 75 mm) from Lake Superior beach. Because the better rounded pebbles are also the more spherical, it follows that prolonged abrasion of a beach tends to make the pebbles more spherical and hence less flat. (After Grogan, 1945.)

gravel of rivers carrying material of the same composition.

These conclusions suggest that shape may be an important factor in the sedimentation process and in response to current flow. Krumbein (1942b) and Sneed and Folk (1958, Fig. 4) found a close correlation between the sphericity or form index and the settling velocity (Fig. 3–17). Experimental work by Briggs, McCulloch, and Moser (1962) showed the shape factor to be as important as the specific gravity in affecting the fall velocity of various species of heavy minerals. Certainly, also, the responses of sand grains or pebbles to current flow is markedly different for different shapes. An equidimensional grain cannot, by definition, have a preferred orientation; those which are flat disks assume a pronounced imbrication; elongate forms respond differently (see page 69).

ROUNDNESS

Roundness deals with the sharpness of the edges and corners of a clastic fragment; it is independent of shape. The several right-angled geometrical forms, cube, plate, prism, and the like (Fig. 3-18), all have equally sharp edges in that their radii of curvature are zero. Yet they differ from one another in shape (and therefore sphericity). The term *roundness*, however, has been used or misused in the literature and in many cases has been used interchangeably with shape (Russell and Taylor, 1937a). The distinction between the two is fundamental and should be kept clearly in mind. Roundness was first clearly defined by Wentworth (1919) as r_i/R, where r_i is the radius of curvature of the sharpest edge, and R is one-half the longest diameter. Wadell (1932) defined roundness as the ratio of the average radius of curvature of the several corners or edges to the radius of curvature of the maximum inscribed sphere. Because this is difficult to apply, it is more convenient to work with a two-dimensional figure, a sectional or projected image of the fragment or particle in question rather than the three-dimensional object itself. In this case the roundness is defined as the average radius of curvature of the corners of the grain image divided by the radius of the maximum inscribed circle (Fig. 3-23). This is expressed as:

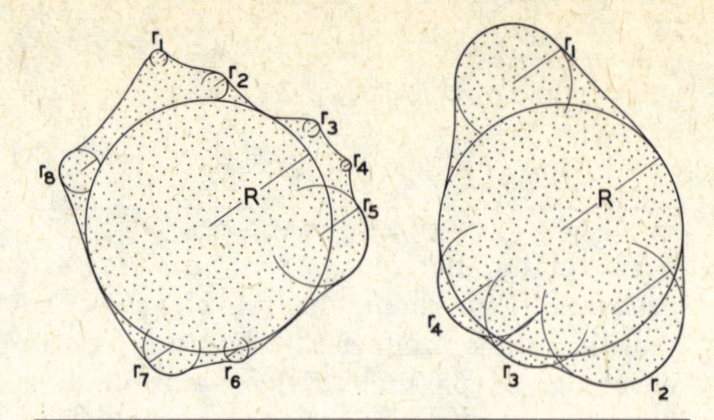

FIG. 3-23. Geometrical nature of roundness of pebbles. (From Krumbein, 1940.)

$$P\ (\text{rho}) = \frac{\sum\left(\frac{r_i}{R}\right)}{N}$$

where r_i are the individual radii of the corners, N is the number of corners, and R is the radius of the maximum inscribed circle. By such a definition a sphere has both a roundness of 1.0 and a sphericity of 1.0. Other objects, however, which are nonspherical may also have a roundness of 1.0, such as a capsule-shaped body which is essentially a cylinder capped by two hemispheres. Various modifications of this definition of roundness have been proposed by other workers. These are reviewed by Köster (1964, pp. 147–161), Humbert (1968, pp. 11–15), and Pryor (1971, pp. 138–142).

As pointed out, the term *roundness* has been used carelessly. Loosely used also are such terms *rounded, subrounded, subangular,* and *angular.* In order that these terms have a more precise meaning, they have been redefined in quantitative terms in a manner analogous to the more exact redefinitions of the common size terms. Most such efforts utilize the Wadell roundness values (Russell and Taylor, 1937b; Folk, 1955). The classes so defined (Table 3-9) are not equal; the inequality arises from the observation that it is difficult to distinguish slight differences in roundness if the roundness values are high but similar differences can be readily detected at the lower end of the scale. Pettijohn has therefore redefined the class

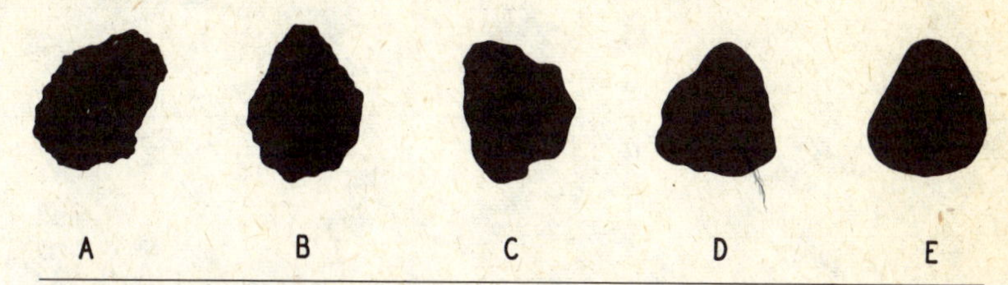

FIG. 3-24. Roundness classes. A: angular; B: subangular; C: subrounded; D: rounded; E: well rounded.

limits in such a way that the midpoints of the classes form an approximate geometric progression. Powers (1953) named and defined six roundness grades (instead of five) in such a way that the class limits closely approximate a $\sqrt{2}$ geometric scale. These were given rho values by Folk (1955) in the same way that phi values were assigned by Krumbein (1938) to the diameter values expressing grain size.

Pettijohn's roundness grades (Table 3-9 and Fig. 3-24) are as follows:

Angular (0–0.15): very little or no evidence of wear; edges and corners sharp; secondary corners (the minor convexities of the grain profile, not the principal interfacial angles) numerous and sharp.

Subangular (0.15–0.25): definite wear; edges and corners rounded off to some extent; secondary corners numerous (10–20), although less so than in the angular class.

Subrounded (0.25–0.40): considerable wear; edges and corners rounded; secondary corners much rounded and reduced in numbers (5–10). Area of the original faces reduced, original interfacial angles, though rounded, still distinct.

Rounded (0.40–0.60): original faces almost completely destroyed; some comparatively flat surfaces may be present. May be broad re-entrant angles between remnant faces; all original edges and corners smoothed off to rather broad curves; secondary corners greatly subdued and few (0–5). At roundness 0.60 all secondary corners disappear. Original shape still apparent.

Well-rounded (0.60–1.00): no original faces, edges, or corners left; entire surface consists of broad convexities; flat areas absent; no secondary corners present. Original shape is suggested by the present form of the grain.

What is the geological significance of rounding, and what is its usefulness in as-

TABLE 3-9. Roundness grades

	Russell and Taylor		This text	
Grade terms	Class limits	Midpoint[a]	Class limits	Midpoint[b]
Angular	0–0.15	0.075	0–0.15	0.125
Subangular	0.15–0.30	0.225	0.15–0.25	0.200
Subrounded	0.30–0.50	0.400	0.25–0.40	0.315
Rounded	0.50–0.70	0.600	0.40–0.60	0.500
Well-rounded	0.70–1.00	0.850	0.60–1.00	0.800

[a] Arithmetic midpoints.
[b] Geometric (approximately) except for angular class. Most particles, even freshly broken, have a finite roundness, rarely less than 0.10. Hence lower limit of the angular class is not, in practice, zero. The midpoint of the fragments in the group, therefore, is probably near 0.125.

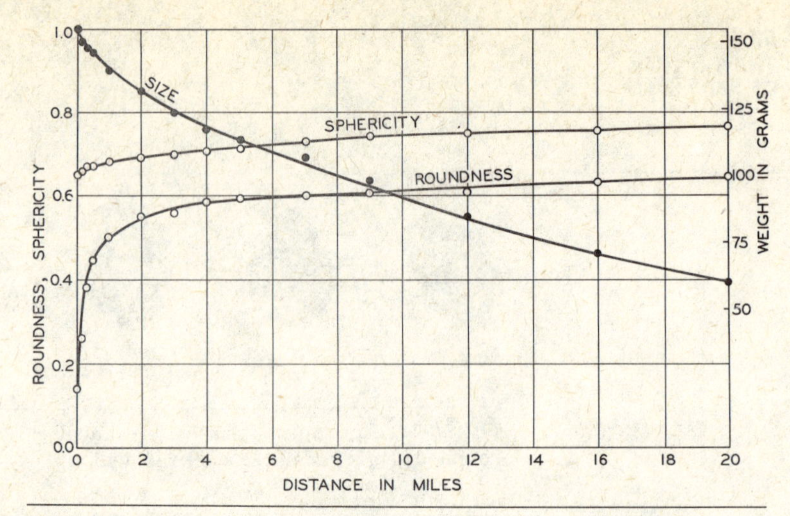

FIG. 3-25. Size (weight), roundness, and sphericity of limestone fragments as functions of distance during experimental abrasion. (After Krumbein, 1941.)

certaining distance, direction, and velocity of movement of the sedimentary particles? Beginning with Daubrée (1879), many workers have attempted to answer these questions by both field and laboratory studies. All such studies have shown that roundness increases with distance of travel most rapidly at first and then more slowly (Figs. 3-25 and 3-26). This observation, though noted by Daubrée, was first given a clear quantitative demonstration by Wentworth (1919, 1922a, 1922b). It remained for Krumbein (1941b) to attempt formulation of the relations in mathematical terms. He observed that the rate of change of roundness was a function of the difference between the roundness at any point and a certain limiting roundness, a value dependent in some measure on the material involved and the particular stream or beach regime. This relation may be expressed as:

$$P = P_L(1 - e^{kx})$$

where P is the roundness at any point, P_L is the limiting roundness, x is the distance, and k is the coefficient of rounding. This equation seemed to fit both experimental and field data (Krumbein, 1940, 1942a). See Fig. 3-27. Later experimental work by Krumbein (1941b, p. 482) and Plumley's work on the gravels of Black Hills streams (1948, p. 566) cast doubt on the validity of Krumbein's equation. The rounding process is more complex. Plumley concluded that the change of roundness with distance is proportional not only to the difference between the roundness at any point and a limiting roundness but also to some power of the distance traveled.

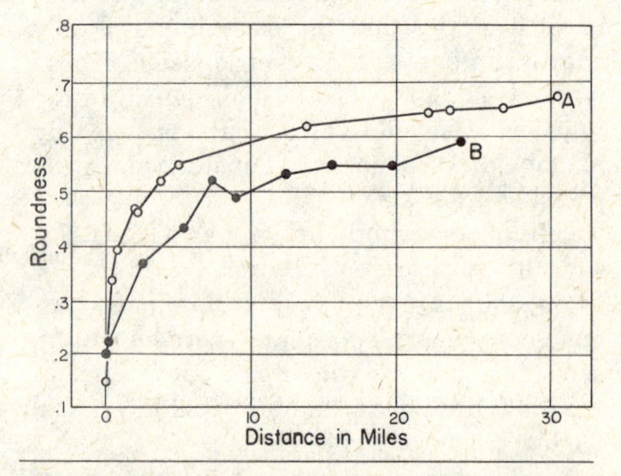

FIG. 3-26. Relation of roundness of limestone pebbles to distance of transport in Black Hills, South Dakota, streams. A: Rapid Creek; B: Battle Creek. (Data from Plumley, 1948.)

Whatever the precise meaning of the rounding equation, whether derived from abrasion mill experiments or from natural streams, the rounding does increase with time (distance), most rapidly at first and then more slowly. There does seem to be a limiting roundness, in part related to the lithology of the materials—lower, for example, for chert than for quartz or limestone (Sneed and Folk, 1958). Rounding of gravel, moreover, is rapid. How far must a pebble be transported to become well rounded (0.60)? Experimental and field studies do not

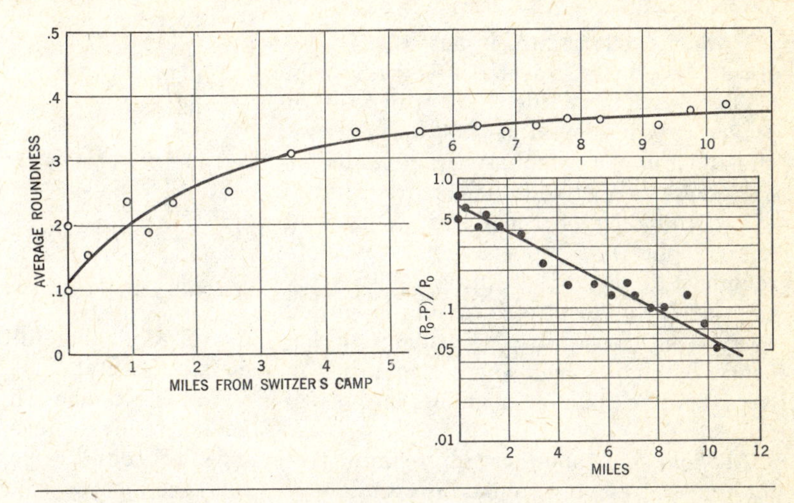

FIG. 3-27. Variation in pebble roundness along Arroyo Seco, California. Lowe granodiorite, 16 to 32 mm class. Inset shows plot of $(P_0-P)/P_0$ to distance, where P is roundness at any point and P_0 is the limiting roundness. (After Krumbein, 1942.)

provide a precise answer but do suggest distances of probably the right order of magnitude. Reduction of a cube to a sphere with a diameter equal to one edge of the cube involves removal of 47.5 percent of the original volume or weight. It may be supposed, therefore, that a weight loss of one-third to one-half would lead to a maximum roundness and that further reduction in size would not be accompanied by any increase in roundness. As seen from Krumbein's data (1941b), a loss of about one-third is associated with a roundness of about 0.60 (well rounded). His data also show that further weight loss did not result in much change in roundness. His limestone reached this roundness in 7 miles. Using Daubrée's figure of 0.001 to 0.004 loss in weight per kilometer of travel of granite pebbles, to become well rounded (that is, to lose one-third of the original weight) requires a transport of between 50 and 200 miles (84 to 333 km). The calculation, though very crude, is probably of the correct magnitude. Kuenen (1956b), who experimented with gravel movement in a circular flume, found limestone to become well rounded in about 31 miles (50 km) of travel; gabbro had a weight loss of 35 to 40 percent at about 87 miles (140 km). Vein quartz showed a 0.001 loss in weight per kilometer of travel and hence should be well rounded at about 186 miles (300 km).

Plumley (1948) found that limestone pebbles in two Black Hills streams became well rounded (0.60) at about 11 and 23 miles (18 and 37 km), respectively (Fig. 3-26). The quartzite pebbles in the Brandywine upland gravels of Maryland have a roundness of 0.59 (Schlee, 1957). The nearest ledges from which they could have come are about 45 miles (72 km) distant. Quartz in the gravels of the Colorado River of Texas became well rounded in less than 100 miles (161 km); the limestone pebbles had already attained their maximum roundness by the time they reached the main stream (Sneed and Folk, 1958). These results are confirmed by Unrug (1957), who found the granite pebbles to reach a maximum in about 77.5 miles (125 km) in the River Dunajec in Poland, and by Dal Cin (1967), who noted rapid rounding of both limestone and quartz followed by little or no rounding further downstream of the River Piave in Italy.

Knowing that most of the observed rounding is acquired in the first few miles of transport, it is evident that angular or subangular gravel cannot have been moved

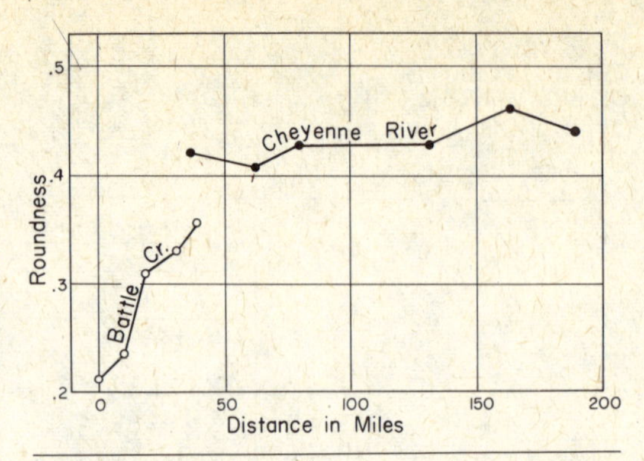

FIG. 3-28. Relation of roundness to distance of transport of quartz sand, 1.0 to 1.414 mm grade, in South Dakota streams. (After Plumley, 1948.)

more than a mile or two, at most 10–15 miles (16–24 km), by a stream. Moreover, except in the proximal portions of a deposit of gravel, roundness will show little or no regional variation, which severely limits its usefulness as a guide to paleocurrent flow.

The rounding of beach gravels is well known, but it is more difficult to relate rounding to distance of travel (Fig. 3-22). One can only say that as gravel is shifted away from the sources of supply, it becomes better rounded. There also seems to be a limiting roundness on beaches as in rivers.

All field and laboratory data show that the rounding of sand, unlike the rounding of gravel, is a very slow process. Daubrée (1879) found sand grains to lose only 0.0001 part of their weight per kilometer of travel. Thiel's (1940) abrasion experiments with quartz sand showed a 22 percent loss in 100 hours in an abrasion mill, presumed to be equivalent to about 5000 miles of travel. This is an average of less than 0.0001 part per mile of travel. Marshall (1927) showed a loss of 0.0005 part per mile of travel for grains 2 to 3 mm in diameter. Kuenen (1960a), using a flume instead of a tumbling barrel, found even smaller losses. Quartz lost only 1.0 percent of its weight in 10,000 km of transport (1958), a loss so small that the rounding achieved is virtually undetectable. Because most stream transport is under 1,000 km, a stream will not round sand if these experimental data are valid.

Eolian action was shown by experimental studies of Kuenen (1960b) to be a much more effective agent in the rounding of sand; the loss for quartz was 100 to 1,000 times greater than for aqueous transport over the same distance. Cubes of quartz were worn down to perfect spheres. The outcome of Kuenen's experiments suggests that fluviatile transport is wholly ineffective in rounding quartz or feldspar. Beach action is perhaps more effective but is not thought to have much influence on the average of all sands. Eolian action is a potent abrading mechanism of sand down to 0.1 mm in diameter; action is nil below 0.05 mm. Rounded quartz sand, therefore, is taken to indicate eolian action at some time in its history.

The effectiveness of beach action has not yet been fully evaluated. Folk (1960), observing interbedding of poorly rounded and well-rounded quartz sands in the Tuscarora Quartzite (Silurian) of West Virginia, ascribed the better rounded materials to surf action. Swett, Klein, and Smith (1971, p. 412) have estimated distances of sand transport in tidal estuaries to be of a magnitude sufficient to round quartz even at the low rate estimated by Kuenen.

Field observations tend to support the experimental results. The classic work of Russell and Taylor (1937b) showed that sand carried by the Mississippi River between Cairo, Illinois, and the Gulf of Mexico, a distance of about 1100 miles (1770 km), apparently decreased in roundness downcurrent. They concluded, therefore, that streams do not round sand but that the decrease observed was caused by progressive fracturing. The observed decline was from 0.24 to 0.18 or 23.5 percent in 1100 miles (1770 km). On the other hand, Plumley (1948) showed that coarse sand (1.0 to 1.414 mm grade) of Battle Creek, in the Black Hills of South Dakota, increased 71 percent in roundness, from 0.21 to 0.36, in a distance of only 40 miles (64 km). The same sand grade in the Cheyenne River of South Dakota, however, was found to increase in roundness from 0.42 to 0.44, barely 5 percent, in 150 miles of transport (Fig. 3-28). The quartz (.088–.250 mm) in the sands of the Rio Grande, Argentina, showed no discernible change in rounding in over 60 miles (100 km) of travel (Mazzoni and Spalletti, 1972, Figs. 16 and 17). Sands shifted along the beach of Lake

Erie, like those of the Mississippi, decreased in roundness in the downcurrent direction (Pettijohn and Lundahl, 1943), a decrease apparently related to a sorting action. Because sphericity and roundness are generally positively correlated, a downcurrent decline in sphericity would carry with it a downcurrent decline in rounding. Apparently in large, slow-flowing streams, such as the Mississippi, this sorting action prevails so that any downstream increase in rounding is masked by the downcurrent sorting action. It is hardly conceivable that the steep-gradient, gravel-carrying streams of the Black Hills would round sand and that the Mississippi River would actually reduce rounding by progressive fracturing.

The role of solution in the rounding of quartz likewise has not been fully assessed. Kuenen (1960b, p. 448) believed it negligible, for if it were an important agent it should affect the smallest grains most—those grains which are observed to be the least rounded. Under some circumstances solution *in situ* of quartz does occur, especially in some soils. Crook (1968) in particular has called attention to this action.

It is noteworthy that rounding, especially of quartz sand, once acquired is not likely to be lost. Moreover, quartz sand is commonly recycled, so the rounding observed in any particular deposit may be inherited from an earlier episode of transport. This is also the case in the pebbles of quartzite and vein quartz.

Efforts to use roundness of sand grains to make environmental discriminations have had limited success. Beal and Shepard (1956) and Waskom (1958) found only relatively small differences in roundness in present-day sands in the several subenvironments of the shore zone in the Gulf Coastal region.

SURFACE TEXTURES

The minor or *microrelief* features of the grain surface, which are independent of size, shape, or roundness, are termed *surface textures*. Polish, frosting, striations, and the like belong in this category. Some are visible to the unaided eye; others are discernible only under the microscope, or in some cases the electron microscope. Many such features are believed to be of genetic significance (Krinsley, 1973). The striations on glacial cobbles illustrate the point. The frosting on sand grains has been attributed to eolian action.

But just as a sand grain or pebble may inherit its shape and roundness from an earlier deposit of different origin, so, too, a particle or fragment may inherit the surface markings which it bears. However, less abrasion and far less transport are required to modify these details than are needed to change roundness, shape, or size. Surface markings are easily erased or imparted to a grain or fragment. Wentworth (1922a, p. 114), for example, determined experimentally that about 0.35 miles (560 m) of travel would remove glacial striations from limestone pebbles without significant modification of their shape. Bond (1954) has noted that the frosting of the sands of the Kalahari Desert is lost in less than 40 miles (64 km) of travel in the Zambezi River. The surface textures, therefore, are most likely to record the last cycle of transportation. But, as with other characteristics, a sand of mixed parentage will contain grains with a diversity of surface textures. It has even been presumed that surface textures acquired in one cycle can be overprinted by those of a later cycle (Krinsley and Funnell, 1965), so that a single grain may record several episodes in its history.

Surface textures are diverse, but they may be grouped into two categories. One class deals with the dullness or the polish of the fragment. The other concerns the markings on the surface—the microrelief features, such as striations, percussion scars, and the like.

POLISH VERSUS FROST

Polish, or gloss, referring to surface luster, is a quality related to the regularity of light reflection. Scattering or diffusion of light produces a *dull* or *matte* surface. Polish is indicated by the presence of highlights. The cause of polish or lack of it is not fully understood. Very probably the causes are several. Polish may be mechanically produced by gentle attrition or wear, particularly if the abrasive agent is of fine grain. Such is thought to be the cause of the wind polish on some quartzite outcrops and fragments (ventifacts). It may also be produced by the deposition of a vitreous film or glaze such as that known as *desert varnish*. Although the origin of this feature is not certain, prevailing opinion (Laudermilk, 1931) holds that desert varnish seems to have been produced by water, which is present in the rock and is drawn to the surface and evaporated by the heat of the sun, depositing the substances dissolved in a relatively insoluble form, as thin hard coatings consisting of silica, iron oxide, and some manganese oxides. Some geologists attribute the high polish to subsequent sandblasting. Laudermilk thought that certain lichens played a role as accumulators of iron and manganese compounds. The lichen growth was terminated by the accumulation of these products, which were then spread over the pebble surface, by acid generated upon the death of the lichens, onto adjacent rocks that themselves had no manganese content. Dehydration and oxidation under the desert sun leaves the varnishlike residue. Hunt (1954), however, while noting that desert varnish is most conspicuous in arid regions, thought that it may form also in humid regions and that much of the desert varnish seen in deserts is a product of an earlier, more humid climate.

A most enigmatic polish appears on some clay-embedded pebbles, such as the *gastroliths*, or "stomach stones," of ancient plesiosauran reptiles. Most famous of these are the pebbles found in Cretaceous marine shales (Hares, 1917; Stauffer, 1945). Although much has been written about these objects, no agreement has been reached concerning the origin of their polish, which has been attributed to wind action, to abrasion in the stomach of the animal, and to compaction movements of the shale matrix.

Polish, or certainly a high polish or gloss, is exceptional. Most pebbles have a dull surface. Rarely do quartz grains have a high polish. Some sand grains, on the other hand, have a striking surface character variously described as "matte" or "frosted." This surface is seen, for example, in the highly quartzose, well-rounded grains of the St. Peter Sandstone (Ordovician) of the Upper Mississippi Valley region. Frosting has been attributed to eolian abrasion and has even been mapped in the European Pleistocene deposits by Cailleux (1942), who considered the feature a criterion of periglacial wind action. The superficial similarity of the frosted surface to that produced on glass by sandblast gives credence to this theory. However, recent work by Kuenen and Perdok (1961, 1962) and Ricci Lucchi and Casa (1970) has shown that chemical corrosion is the more likely cause of this feature. It can be produced on quartz grains by etching with a very dilute solution of hydrofluoric acid for a very short time. Quartz grains in calcareous sands are slightly corroded or replaced by the carbonate cement. Such chemically attacked grains have a frosted surface (Walker, 1957), which suggests a post-depositional origin of this texture. But Roth (1932) believed that frosting was the result of incipient enlargement rather than abrasion or solution.

As noted by Kuenen and Perdok (1962), the microrelief responsible for the scattering of light, and the frosted appearance which results, may be caused by several processes. The coarser-textured features may be attributable to abrasion, but the fine-textured microrelief, of the order of 2 microns or less (which is chiefly responsible for frosting), is chemically produced by alternation of wetting and drying related to dew formation and evaporation, and to correlative solution and precipitation. This "chemical frost" affects the entire grain, even the recesses in the grain surface. The coarser-textured frost caused by abrasion action affects only the protruding parts and not the protected reentrants.

MICRORELIEF

Well-known microrelief features on pebbles and cobbles are readily seen with the unaided eye and include striations, scratches, percussion marks, and indentations or pits. Striations are scratches mainly the product of ice action, generally glacial ice, that mark the surface of some pebbles. Wentworth

(1932, 1936b) has called attention to the work of subarctic streams in producing striated cobbles. The percentage of striated cobbles in some streams is quite high; in many cases a majority of all those above a certain size are so marked. The percentage may equal or exceed that shown in true glacial deposits, but those striated by river ice lack the characteristic glacial facets. Striated stones are not abundant even in glacial sediments. Wentworth (1936a) studied several Wisconsin morainal deposits that are noted for the excellence of their striated stones. Over 600 pebbles or cobbles were examined, of which 40 percent showed no striations at all, 50 percent had only faint striations or clear striations on one side, and as few as 10 percent displayed conspicuous markings (Fig. 6-20). Striations were most prevalent and most conspicuous on limestone cobbles, whereas the cobbles of siliceous and of coarse-grained igneous rocks were virtually unmarked. It is not surprising, therefore, that ancient, well-indurated tillites, from which extraction of entire pebbles or cobbles is almost impossible, show few or no striated stones.

Striations are ideally narrow, straight, or nearly straight scratches clearly cut into the surface on which they appear. Related to striations are *bruises*, which are cruder, shorter, and broader than striations and are commonly found in *en echelon* patterns. *Nailhead* scratches are striations which have a definite head or point of origin. Such scratches tend to narrow or taper slightly from this point and come to an indefinite or ill-defined end. If the cobbles are embedded in the matrix, their striations tend to be aligned with the direction of the ice flow. They are therefore in general parallel to the long axes of the cobbles (see p. 69 on fabric).

Four chief striation patterns can be defined: parallel, subparallel, scatter or random, and grid. The grid is marked by the crossing of two or more parallel systems. Subparallel and random patterns are most common on glacial cobbles. Parallel and subparallel striations tend to run lengthwise of the cobble. Wentworth (1936b) states that the grid pattern, especially those wide-space striations, as well as curved striations, are much more abundant, relatively, on ice-jam cobbles of rivers than on glacial cobbles.

Striations (and slickensides) also form during deformation of a rock under pressure. Pebbles or cobbles embedded in a rather fine-grained matrix may show such markings. Striations produced by such movements are generally *microstriations*, the largest of which are just visible to the unaided eye (Judson and Barks, 1961; Clifton, 1965). Microstriations are generally parallel rather than random and pebbles so marked commonly show a "tectonic polish." These characters plus microfaulting distinguish these surface textures from those imparted to the pebble prior to deposition.

Crescentic impact scars or *percussion marks* are common on some pebbles, particularly the cherts and dense quartzites. These small marks are caused by blows on the surface of the cobble or pebble and are indicative, perhaps, of high-velocity flow. They have been attributed to fluvial rather than beach action (Klein, 1963).

Many pebbles have surface indentations or pits. These may be produced by etching and differential solution of inhomogeneities of the rock. The coarse-grained igneous rocks are characteristically pitted and pockmarked, whereas the fine-grained rocks, such as cherts, quartzites, and many limestones, are smooth. Under many conditions of abrasion even the coarsest-grained rocks may be smooth. More generally the term *pitted pebbles* is applied to those pebbles or cobbles which have concavities not related to the texture of the rock or to differential weathering. Such depressions are common at the contacts between adjacent pebbles. They vary in size but may be as much as a few centimeters across and a centimeter deep (Fig. 3-29). Normally the pits are smooth and sharply cut as if scooped out with a small spoon. Kuenen has reviewed the literature on such pitted pebbles and the problem of their formation (1942). They have been explained as mutual indenture by pressure, a hypothesis readily shown to be untenable, and by solution induced by pressure at points of contact (Sorby, 1863; Kuenen, 1942).

Pitted pebbles are not to be confused with "cupped pebbles," which have been subject to solution on their upper side and so corroded that some are mere shells (Scott, 1947).

The microrelief of pebbles is readily seen with the unaided eye. That of sand grains, however, can only be seen under the microscope. Hence only recently has the microrelief of sand grains been rigorously investigated, primarily by the use of the electron or scanning-electron microscope (Krinsley and Takahashi, 1962a, 1962b, 1962c; Porter, 1962; Wolfe, 1967; Krinsley and Donahue, 1968; Margolis, 1968; Stieglitz, 1969; Krinsley and Margolis, 1969; Ricci Lucchi and Casa, 1970; Fitzpatrick and Summerson, 1971). These investigations have revealed a myriad of markings of varied size and form on the surface of quartz grains. Much effort has been expended to relate particular microrelief patterns to specific depositional environments. Patterns displayed by grains from the littoral, eolian, and glacial environments have received special attention. The approach has been to sample these environments in order to establish the characteristic surface textures of each. Unfortunately, in some environments sampled, the sand grains were found to have had a complex history, having been previously transported by ice, and perhaps running water, before accumulating on a present-day beach or dune. That surface textures, presumed to have formed in differing environments, can be superimposed was recognized (Krinsley and Funnell, 1965, pp. 453–454), although it was once thought that early patterns were quickly erased. Not knowing which features may have been formed by which agent, or whether some features may have been produced by more than one agent and the lack of any common means of measurement or description of the features displayed, has reduced the value of surface textures of sand grains as a criterion of the agent and/or environment of deposition. The usefulness of this approach for ancient sandstones is almost wholly unknown; diagenesis no doubt drastically alters the grain surface so that, even if environmental criteria based on nonsubjective and reproducible data were available, they might be difficult to apply to the older rocks, especially those sandstones so well indurated that they can only be studied in thin section.

FIG. 3-29. Pitted pebble of Precambrian age, Great Slave Lake, Northwest Territories, Canada. (Photograph by C. Weber.)

FABRIC AND FRAMEWORK GEOMETRY

FABRIC

Geologists have long been interested in the fabric of sediments, especially of clastic sediments. Jamieson (1860, p. 349) observed the imbrication of stones in creek beds in Scotland, but systematic study of fabrics did not begin until the publication of *Gefügekunde der Gesteine* in 1936 by Bruno Sander. Although Sander's book dealt mainly with metamorphic fabrics, it provided a systematic methodology and developed principles readily transferrable to the fabric of sedimentary rocks. The literature on sedimentary fabrics has grown much in recent years. It has been most thoroughly summarized by Potter and Pettijohn (1963, pp. 23–61) and by Johansson (1965a).

The principal object of most studies of the primary fabrics of clastic sediments has been the reconstruction of the current direction that prevailed at the time of the deposition of the sediment. Only recently have fabrics been used to elucidate the transport process itself. Fabric studies have chiefly been made on sands and gravels and on tills of both Pleistocene and pre-Pleistocene age.

In addition, fabrics have an important bearing on the physical properties of rocks such as thermal, electrical, fluid, and sonic conductivity.

The study of the fabrics of dolomites and limestones has received less attention than it deserves. The potential usefulness of this field of study is well shown by the monographic work of Sander on Triassic limestones and dolomites of Austria (Sander, 1936, transl. Knopf, 1951).

DEFINITION AND CONCEPTS

Fabric, as the term is generally used by sedimentologists, refers to the spatial arrangement and orientation of the fabric elements. As used, the term is more restrictive than *Gefüge*, a term used by Sander (1936) to include such properties as grain size, sorting, porosity, and so forth, which are usually considered textural attributes. A *fabric element* of a sedimentary rock may be a single crystal, a pebble or sand grain, a shell, or any other component.

Packing is the spacing or "density" of the fabric elements. Even in a rock composed of wholly uniformly sized spherical elements, there are several ways in which these spheres can be arranged or packed. When the sizes and shapes are more varied, the manner of packing becomes more complex. Although fabric and packing are closely related, they are not the same thing.

Any nonspherical element (such as a pebble) has an orientation. When, out of all possible orientations, a significant number of like elements has assumed a certain orientation or orientations in preference to all others, these objects (or the gravel containing such oriented pebbles, for example) are said to show a preferred orientation or to have an anisotropic fabric pattern. Such a pattern may be expressed by the alignment of the long axes of the pebbles—the subparallelism of graptolites in a shale, the uniform convex-upward arrangements of the valves of mollusks, and the like. Such a fabric is *dimensional*, because the actual dimensions of the elements govern the alignment observed. If the fabric is shown by the alignment of the crystallographic directions (c-axes of the quartz grains, for example), it is termed *crystallographic*. The dimensional and crystallographic fabrics may or may not be closely related; in the case of rock fragments or fossils, of course, no crystallographic fabric is possible.

Genetically there are two kinds of fabric, namely, deformation and apposition. A *deformation fabric* is produced by external stress on the rock and results from a rotation or movement of the constituent elements under stress or the growth of new elements in common orientation in the stress field. This type of fabric is essentially that exhibited by metamorphic rocks. An *apposition fabric* is formed at the time of deposition of the material and is a "primary" fabric. For the most part, the fabrics of sedimentary rocks are apposition fabrics, although compaction of sedimentary rocks, accompanied by decrease in porosity, is in part a deformation phenomenon which modifies the primary fabric. Such deformation may be arrested by early cementation and the several stages of the process may be recorded in some concretions (Oertal and Curtis, 1972). An apposition or primary fabric records a response of the linear elements (such as long axes of pebbles) to a force field, such as the earth's gravitational or magnetic fields. Most nonspherical objects tend to lie in their most stable position with their longer dimensions parallel to the surface of deposition, a response to gravitational forces. The positions of these elements, however, may be modified by fluid flow and they may reorient in response to this movement.

Not all apposition or primary fabrics are of this kind. Some are *growth fabrics*, orientations resulting from crystal growth and often related to a free surface. The growth of crystals normal to such surfaces, as in geodes, veins, and the like, are primary fabrics of this type. These fabrics are discussed in the section on diagenetic textures.

FABRIC ELEMENTS AND ANALYSIS

Only those fabric elements having dimensional inequalities respond to fluid flow and have a preferred orientation. A sphere having no such inequalities can show no response. A triaxial ellipsoid, on the other hand, is oriented in space, and its space position can be determined. The orientation of the long axis, if it is prolate, or the short axis, if it is oblate, is commonly determined.

Almost any detrital component is a potential fabric element, although those with the greatest dimensional inequalities are most useful. Pebbles and sand grains are most commonly studied for the space position of their most atypical axes. Mica flakes, even the clay micas, and much organic debris (plant stems, strap-shaped leaf fragments) are usable fabric elements, especially in the argillaceous sediments. Skeletal materials, especially orthocerids, *Tentaculites*, bivalve shells, and high-spired gastropods, are readily oriented and are also useful fabric elements.

The orientation of a fabric element, a pebble for example, may be described in terms of two angles. One is the direction ("strike") or azimuth angle between some axis of the pebble and the meridian; the other is the inclination ("plunge" or "dip") of this axis, the angle between the axis in question and the horizontal. The long axes of pebbles may show a preferred orientation, but for some pebbles, such as those approaching a disk, a preferred orientation of the long axes may be lacking, or at most very feeble. In this case the orientation is controlled by the large flat faces of the object. The position of these faces (the *a*–*b* plane) is closely approximated by giving the azimuth and plunge of the normal ("face pole") to that surface. This is essentially the orientation of the shortest diameter (*c*-axis) of the pebble.

If a pebble which can be extracted from its matrix can be suitably marked in the outcrop so that it can be reoriented in the same position with reference to the meridian and zenith in the laboratory as it had in the field, it is possible, by use of a goniometer, to measure the azimuth and angle of dip of the long axis and the normal to the maximum projection area (Karlstrom, 1952). If, of course, the strata have been tilted, the effects of such tilt must be properly taken into account. Those interested in the techniques of sampling and measurement procedures should consult appropriate reference works (Potter and Pettijohn, 1963, p. 28, 40; Bonham and Spotts, 1971, pp. 258–312).

The observations thus made on 100 or more pebbles can be graphically summarized in several ways. It may be necessary, for example, to find the direction of ice flow from the many observed azimuths of the long axes of till stones. The azimuth readings may be collected into classes (using an appropriate interval, such as 20 degrees), and the modal class observed, or the arithmetic mean azimuth computed. The methods of computation involving such "circular" frequency distributions have been reviewed and summarized by Jizba (1971, pp. 313–333). Or it may be more honest to present the data as a circular histogram. The inclinations can be similarly treated.

A diagram that shows *both* the azimuth and the inclination of the long axis of a fabric element is the so-called "petrofabric diagram" (Knopf and Ingerson, 1938, pp. 226–262). The position of each long axis measured is represented by a point or dot on polar coordinate paper, by a point on a Lambert equiarea polar net, or on the so-called Schmidt net (Fig. 3-30A). The clustering of points, or lack of it, shows whether any preferred orientation is present or not. Such diagrams are easily visualized if one imagines that each pebble is placed in turn at the center of a hollow sphere in precisely the position that it had in the undisturbed outcrop. The long axis (or any other axis) of the pebble is extended until that axis intersects the surface of the sphere. The piercing point on the lower half of the sphere ("southern hemisphere") is then plotted on the "polar" map of that hemisphere.

The orientation of a line in space (pebble axis and so forth) is thus represented on the diagram by a point. A plane may also be represented by a point which is the piercing point of the normal or perpendicular to the plane and the surface of the sphere. One might wish, for example, to show the orientation of a cross-bedding lamination in this manner. By so doing, the orientation of many laminations in a cross-bedded deposit might be shown on a single diagram.

If the linear elements are randomly oriented, points representing these lines are scattered in a haphazard manner over the diagram. If they are preferentially oriented in one direction, there will be a cluster of points on the diagram. In order to show the clustering or density of points, appropriate contours are drawn. On a population density map, contours are drawn, and appropriate shading used, to show the number of persons per square mile. So on the petrofabric diagrams, contours are drawn to show the number of points per unit area (Fig. 3-30B). Relative numbers (percent) rather than actual numbers are usually shown. The unit area is commonly 1 percent of the area of the diagram.

The points representing axes of the fabric elements or the normals to specified planes (cross-bedding laminations, for example) may show centers (termed *poles*) or zones or belts of greater concentration (termed *girdles*).

Although the concept of dimensional fabrics applies equally to all clastic sediments, including some limestones, their measurement and representation as described above are difficult to make in the case of well-indurated rocks. Imbrication of flatter pebbles of a conglomerate can readily be seen in favorably oriented surfaces, but in general the inability to mark, remove, and reorient pebbles makes a full-fabric analysis impossible. Bedding-plane surfaces may be informative, and on these can be seen the orientation of elongate pebbles, conical or elongated fossils, and plant debris.

The dimensional orientation of sand grains in a sandstone cannot be easily determined. Thin sections perpendicular to the bedding usually show the longest grain dimension parallel to the bedding or, in some cases, imbricated. Sections cut parallel to the bedding commonly show a preferred alignment of elongate grains. Several techniques have been developed to cope with the study of the fabrics of sandstones (Martinez, 1963; Nanz, 1955; Bonham and Spotts, 1971).

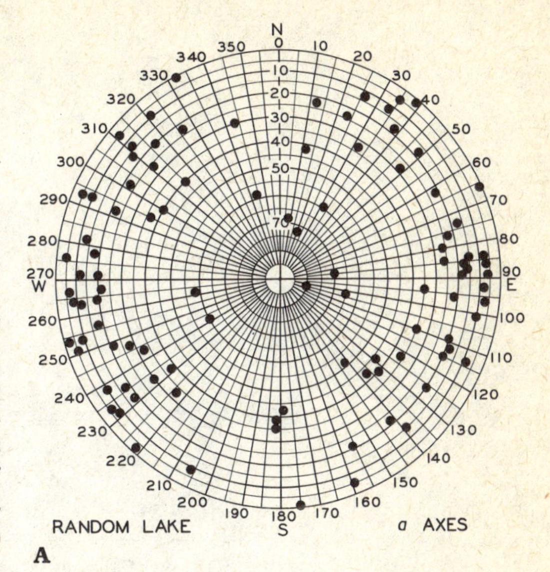

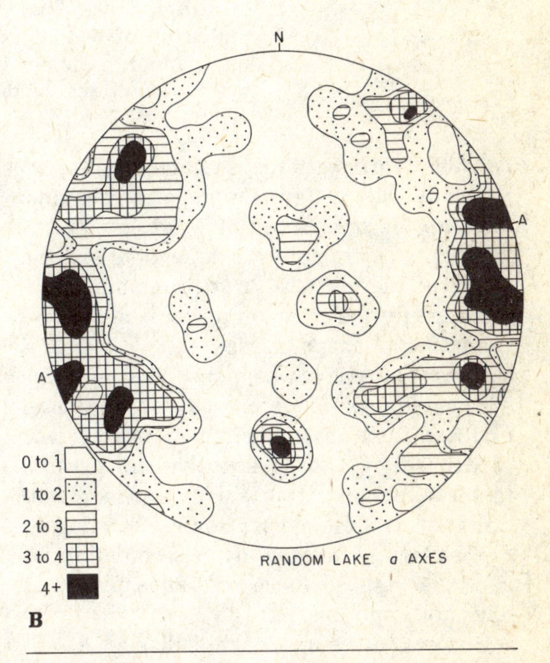

FIG. 3-30. Till fabric. A: distribution of long (*a*) axes of Lake Wisconsin till on equiareal polar coordinate paper; B: same contoured as a "petrofabric" diagram. Black: 4 percent maxima. (From Krumbein, 1939.)

SYMMETRY CONCEPTS AND FABRIC TYPES

When orientation of the fabric elements is random, the fabric is *isotropic;* when a preferred orientation is present, the fabric is *anisotropic.* Although there are many con-

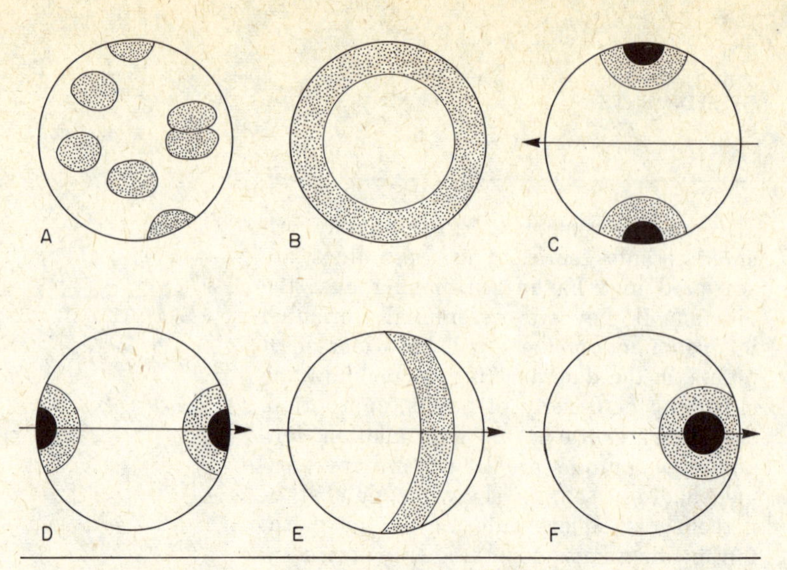

FIG. 3-31. Schematic diagram showing types of depositional fabric. A: isotropic or random; B: girdle in plane of bedding; C: transverse pole in plane of bedding normal to current (arrow); D: longitudinal pole in plane of bedding; E: girdle dipping upcurrent; F: pole dipping upcurrent. Various combinations of these types are also possible.

ceivable fabric patterns, sedimentary deposits show only a few simple arrangements. Because the orientation of a fabric element (pebble, for example) is dependent on the shape of the element, it is profitable to review briefly the common patterns shown by the principal shape classes.

These patterns are best described in terms of two planes of reference: a horizontal plane, closely approximated by the surface of deposition, and a vertical plane parallel to the direction of current flow. The orientation of a fabric element or the relation of the pattern generated by a population of such elements to these reference planes is of major importance.

Spheres, of course, cannot display a fabric pattern. Prolate forms are described in terms of the orientation of their long axes. These may be random or isotropic (Fig. 3-31A). They may be confined to the horizontal plane but random *within* that plane and their poles form a girdle, a simple gravity fabric (Fig. 3-31B). Or they may exhibit a current fabric caused by realignment in the horizontal plane with poles either transverse (Fig. 3-31C) or parallel (Fig. 3-31D) to the prevailing current. Other arrangements of prolate forms are possible but very unusual, such as a single pole in the center of the diagram (the vertical long axis of some dropstones).

The orientation of oblate or disk-shaped elements may be described in terms of the fabric pattern of the short axes—essentially the normals to the disk. The disks may lie in the bedding plane with short axis pole perpendicular to the bed, a simple gravity fabric (Fig. 3-31E), or they may be rearranged by a current and acquire an imbrication with upstream dip, in which case the short-axis pole is displaced and is "off-center" (Fig. 3-31F).

FABRIC OF SEDIMENTS

Gravel fabric The preferred orientation of pebbles in some gravel fabrics has long been noted. The overlapping shingling effect of flat pebbles in certain gravels and conglomerates has been described as an "imbricate structure" (Becker, 1893, pp. 53–54). See Figs. 3-32 and 6-7. Cailleux (1945) studied the inclination of about 4000 pebbles in formations ranging from Paleozoic to Recent in age. Imbrication was very common and in marine formations was somewhat variable in direction, whereas in river deposits the inclination was notably uniform. The mean upstream inclination in fluvial deposits was 15 to 30 degrees; marine deposits showed inclinations of only 2 to 15 degrees. In general, flat pebbles had a lower

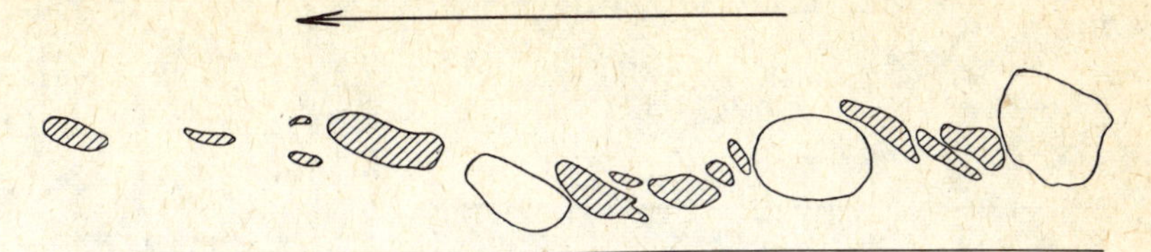

FIG. 3-32. Sketch showing the imbricate arrangement of pebbles, Archean conglomerate, Little Vermilion Lake, western Ontario, Canada. See also Fig 6-7. (From Pettijohn, 1930.)

inclination than less flat ones; the larger sizes were better oriented than the smaller, and pebbles in contact were better oriented than those wholly isolated. According to Unrug (1957, Fig. 21) the angle of inclination tends to decrease downstream attributed, by him, to "poorer sorting of gravel." Johansson (1965b), who has made the most comprehensive study of the subject since Cailleux, noted that the imbrication was the most reliable guide to direction of current flow in modern streams. Inclination varied between 10 and 30 degrees, these variations in some way related both to oblateness and "hydrodynamic conditions." Overly steep inclinations, near 40 degrees found by White (1952) in a Keweenawan conglomerate (Precambrian), were attributed to concentrations of flat pebbles on the flanks of scour pits and thus are a measure of the angle of repose. If so, the difference in inclination of the obtuse bisectrix of the angle between the crater walls and the normal to the bedding plane should be the initial dip of the bed.

The orientation of the long axes of the prolate pebbles has been less well understood. Even the facts seem to be controversial. Alignment parallel to the current has been described by many authors including Krumbein (1940, 1942a), Schlee (1957), and Dumitriu and Dumitriu (1961). However, Twenhofel (1932, p. 36), Unrug (1957), Doeglas (1962), the Sedimentary Petrology Seminar (1965), and Rust (1972) have reported a transverse orientation, an orientation confirmed experimentally by Sarkisian and Klimova (1955) and Kelling and Williams (1967). These conflicting observations are probably the results of various factors. Johansson noted that those pebbles transported in contact with a frictional substratum tend to come to rest transverse to the current flow; those immersed in the transporting medium, such as glacial ice, mud flows, and the like, tend to be aligned parallel to the direction of movement because of the shearing stress of the moving medium. As noted by Rust (1972), the transverse orientation was most pronounced when the pebbles were isolated on a sand bed. With increasing pebble concentration the transverse orientation was lost, and orientation more nearly parallel to current flow was acquired. Flow velocity seemed also to be a factor; torrential flow led to parallel rather than transverse orientation.

Till fabric The preferred orientation of till stones has been used as a criterion of the direction of ice flow (Richter, 1932, 1936; Krumbein, 1939; Holmes, 1941; Karlstrom, 1952; Harrison, 1957; Virkkala, 1960; Seifert, 1954; West and Donner, 1956; Kauranne, 1960, and many others). Prolate stones in ground morainal till tend to be parallel to the direction of ice movement, as inferred from glacial striations, chatter marks, and other criteria of ice motion. In some cases there is a subordinate transverse orientation (Fig. 6-21). In other morainal tills the fabric may be more complex. Till fabrics have in general proved a useful tool in determining the direction of ice flow, particularly in "fossil" tills where other criteria are lacking (Lindsey, 1966; Hälbach, 1962).

Sand fabric The fabric of sands and sandstones is less well known than that of gravels primarily because of the difficulties of study of finer-grained materials. Attempts have been made to measure the position of the actual long axes of grains (Schwarzacher, 1951), the apparent long axes (Griffiths, 1949; Griffiths and Rosenfeld, 1950, 1953), and the orientation of the crystallographic *c*-axes (Rowland, 1946) on the premise that the dimensional long axis and the *c*-crystallographic axis are closely related.

Wayland (1939) noted that the long axis

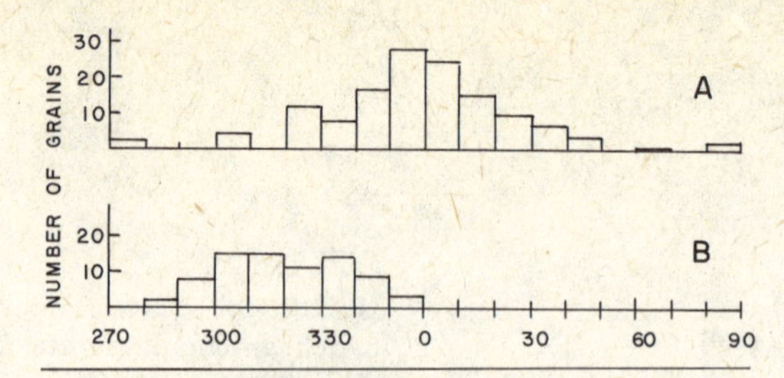

FIG. 3-33. Orientation of long axes of sand grains in sample of Tanner Graywacke. A: section perpendicular to bedding (note tendency to lie parallel to bedding); B: section cut parallel to bedding. Note clustering at about 320° azimuth. (After Helmbold, 1952, Heidelberger Beitr. Min. Pet., v. 3, Fig. 9.) By permission of Springer-Verlag.

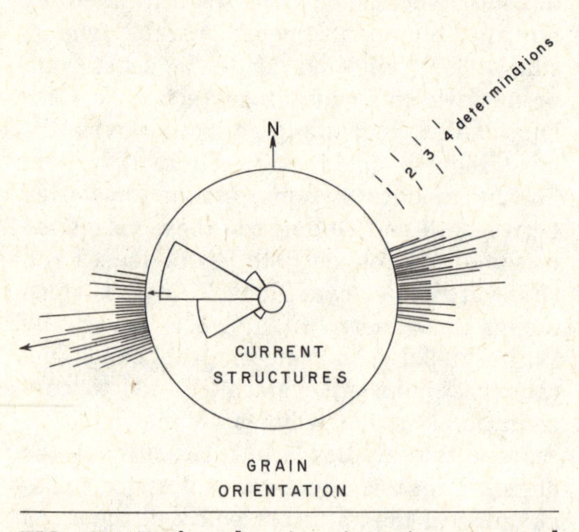

FIG. 3-34. Preferred grain orientation compared with associated current structure azimuths for 57 "Portage" (Devonian) sandstones, central Appalachians. (From McIver, 1961, Ph.D. thesis, Johns Hopkins Univ., Fig. 55.)

and the *c*-crystallographic axis of clastic quartz grains tend to be the same. Ingerson and Ramisch (1942) confirmed Wayland's observation. Accordingly it might be expected that, if nonspherical quartz grains tend to be preferentially oriented by bottom currents at the time of deposition, sandstones would therefore display a crystallographic fabric. Wayland's petrofabric analysis of the St. Peter Sandstone (Ordovician) showed that the optic or *c*-axes of quartz did indeed show such an orientation. Rowland (1946) attempted to explore further the relations between dimensional and crystallographic directions in clastic quartz, but his results were somewhat inconclusive. The difficulty appears to arise in part from the fact that quartz has a rhombohedral cleavage, and, even though very imperfect, it does tend to yield elongate fragments in which the *c*-axis is inclined markedly in the direction of elongation (Bloss, 1957; Bonham, 1957, Fig. 6; Zimmerle and Bonham, 1962). Nevertheless the relation between dimensional elongation and the crystallographic orientation makes it possible to determine the former by use of a photometer on thin sections cut parallel to the bedding (Martinez, 1958, 1963).

In general, dimensional quartz fabrics that relate to current flow are those seen in sections cut parallel to the bedding, particularly of those sandstones with undisturbed horizontal bedding (Fig. 3-33). A plot of the apparent long axes of grains seen in such sections generally shows the mean direction of these axes to be parallel or nearly so to the current direction, as shown by sole marks (Sestini and Pranzini, 1965). A similar agreement between grain fabric and dielectric anisotropy was noted by McIver (1970, Fig. 13). See Fig. 3-34. Notable exceptions, however, have been reported (Onions and Middleton, 1968; Parkash and Middleton, 1970).

These relations were confirmed by experimental studies (Dapples and Rominger, 1945) that showed the grain long axes to be parallel to the current; the larger ends of the asymmetric grains tended to be upcurrent. Studies of modern beach, river, and dune sediments show that clear dimensional bedding-plane fabrics are present (Nanz, 1955; Curray, 1956).

Thin sections of sandstones cut perpendicular to the bedding and parallel to the current flow commonly show a sand grain imbrication generally, though not always, dipping upcurrent (Sestini and Pranzini, 1965).

Sand fabrics have been found to be closely related to vectorial permeability (Griffiths, 1949; Griffiths and Rosenfeld, 1950, 1953).

Fabrics of clays and shales It has been shown that particles of clays, especially clay minerals, are micaceous in habit and that clastic fragments are commonly platelike (Marshall, 1941; Bates, 1958). Even if these were randomly deposited, gravitational pressure and resultant compaction would rotate them into the same plane so that they would have a parallel or subparallel orientation. Such orientation reduces porosity and gives the shale or clay an anisotropic fabric and a fissility. This is well demonstrated by x-ray analysis of the kaolinite fabrics of a sequence of samples taken from a sideritic concretion from the center outward (Oertal and Curtis, 1972). It was apparent that the concretion recorded the compactional history of the enclosing shale. The concretion began before an appreciable compaction and continued until compaction was nearly complete. The kaolinite fabric showed a progressive change from near random at the center to highly oriented at the surface. For an extended discussion of the chemical and mechanical factors that control clay fabrics, refer to Meade (1964).

Inspection of thin sections of shales cut perpendicular to the bedding show "mass extinction" effects under crossed nicols, indicating a marked parallelism of the clay mineral platelets. Keller (1946), however, showed that some fire clays have randomly oriented mineral plates; he believed this to be the result of growth of these plates in a clay gel after deposition. Such clays have a conchoidal to irregular fracture.

Fabrics of limestones and dolomites Primary fabrics of limestones and dolomites have been investigated by Sander (1936) and Hohlt (1948). Well-defined crystallographic fabrics were reported by Hohlt. The patterns described by Sander are largely growth fabrics in pores and other openings produced by druselike implantations of crystals on the walls of such cavities. It seems unlikely that pronounced crystallographic fabrics will be found in unstressed limestones or dolomites.

Dimensional fabrics, however, are common. These are related to preferred orientation of various flat (or elongate) or concavoconvex skeletal elements (Dunham, 1962, Pl. II). These fabrics are discussed more fully below; diagenetic fabrics are discussed in the last part of this chapter and also in the chapter on limestones.

Fossil orientation Organic structures also respond to current flow. Detached valves of concavoconvex form may lie with either the concave or the convex side upward; but if moved by a current the orientation tends to become uniform, in this case with the convex side upward. The preferred orientation of such shells is therefore an index both of current velocity and of upper and lower surfaces of steeply tilted or overturned strata (Shrock, 1948a, p. 314). It has been noted, however, that in some presumed turbidite deposits the single valves have a contrary orientation, namely, with the concave side upward (Crowell et al., 1966, p. 30). Such an orientation can indeed be produced by a turbidity current (Middleton, 1967).

Oriented fossils may also be indices of current *direction*. As noted long ago, *Tentaculites*, graptolite colonies (Ruedemann, 1897; Moors, 1969), and the like tend to show a preferred orientation in the plane of the bedding. Chenowith (1952, pp. 556–559) showed that orthoceracone cephalopods and high-spired gastropods were well oriented in the Trenton of New York State. They tended to have their longest dimension either parallel to or perpendicular to the pararipples of the same beds. Chenowith believed those normal to the ripples and parallel to current flow became so oriented because of their displaced center of gravity. This view was supported by plotting the position of the long axes and noting the direction of the apical end of the forms studied (Fig. 3-35). According to Seilacher (1960, p. 59), the equally opposing modes of a current rose ("bow-tie" configuration) represents orientation of detrital shells perpendicular to current direction, whereas unequally developed opposing modes indicate orientation parallel to the current. The larger mode points upcurrent (Fig. 3-36).

One of the most common paleocurrent criteria is "charcoal lineation"—a lineation expressed by parallelism of carbonized plant debris. Orientation both perpendicular (Pelletier, 1958, p. 1051) and parallel (Colton and DeWitt, 1959, p. 1759) to current direction, as inferred from other sedimentary structures, has been observed. The normal orientation is probably parallel to the current, but, as in the case of some elongate sand grains (Ingerson, 1940) and many elongate fossil forms (Seilacher, 1960, p. 95), the alignment can be controlled by rippled bed forms; the elongation then becomes parallel to the ripple trough.

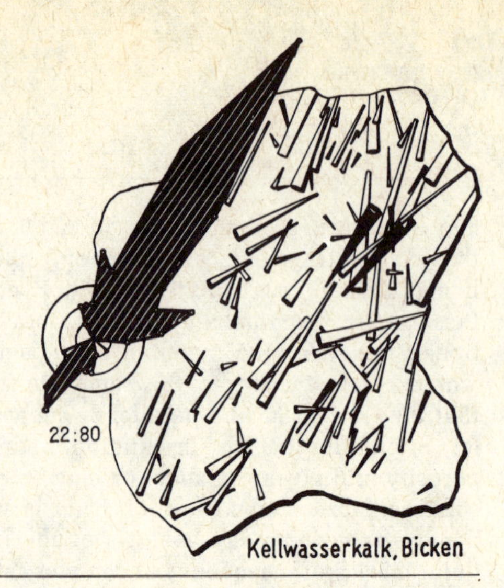

FIG. 3-35. Orientation of orthocone cephalopods in Kellwasserkalk (Upper Devonian), near Bicken, Germany. (After Seilacher, 1960, Notizbl. Hessischen Landesanst. Bodenforschung, v. 88, Fig. 9.)

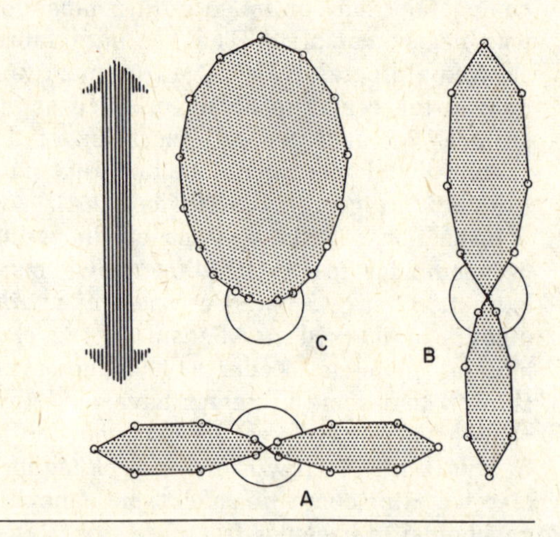

FIG. 3-36. Ideal forms of fossil orientation. A: transverse orientation of elongate shell forms; B: longitudinal orientation of elongate asymmetrical forms; C: orientation caused by response of living organism to current. *a* and *b* are caused by mechanical accumulation only. (After Seilacher, 1960, Notizbl. Hessischen Landesanst. Bodenforschung, v. 88, Fig. 6.)

EVALUATION OF SEDIMENTARY FABRICS

Like the study of grain size, a good deal of effort has been expended in the study of grain fabrics. And, like the study of grain size, the results have not been commensurate with the efforts made. As in the case of grain size, this has been caused in part by the fact that techniques readily applied to modern uncemented gravels and sands cannot be used on the well-lithified rocks. Moreover, dimensional fabrics of sands tend to be disturbed or erased by slump, rippling, and bioturbation. Tectonic movements obscure the primary fabric and may wholly replace it with a deformation fabric. The study of fabric has been directed largely toward determining the direction of flow of the depositing current. Other criteria of current flow—cross-bedding, ripple marks, and sole marks—are more readily seen and measured, so one tends to undertake more tedious fabric analysis only when no other criteria are available.

The greatest value of fabrics, especially sand fabric, is in determining the orientation of sand bodies encountered in drilling. If there is a correlation between fabric and sand body shape, and if the fabric of an oriented core can be ascertained, the value in predicting sand body trend from a single hole is obvious. An understanding of fabric also contributes to our understanding of geophysical properties related to anisotropy of the sand body.

FRAMEWORK GEOMETRY OF DETRITAL SEDIMENTS

Packing Packing has to do with the manner of arrangement of the framework elements in which each element is supported and held in place in the earth's gravitational field by a tangential or point contact with its neighbors (Graton and Fraser, 1935, p. 790).

The study of packing is important for several reasons. Close packing reduces both volume of pore space and size of the pores, and hence significantly alters both porosity and permeability. "Open" or "loose" packing has the converse effect. The question of what processes and agents are responsible for variations in packing has intrigued the students of beaches, some of which show loose packing, whereas others do not (Kindle, 1936). Although the initial grain contacts are mostly tangential, these may be modified by intrastratal solution in such a manner that the character of contact is altered, and the grains brought into closer proximity with one another. A study of grain-to-grain relations may shed light on the extent and nature of these post-depositional diagenetic changes.

The study of packing requires a closer definition of packing, the development of a suitable measure of "closeness" of packing, and an assessment of modifications of packing in the post-depositional period. The study can proceed by theoretical and experimental analysis of the packing of spheres, both uniform and nonuniform in size, and the investigation of packing of natural aggregates both experimentally and in nature. A number of papers deal with one or several of these approaches to the problem, notably the monographic studies of Graton and Fraser (1935; Fraser, 1935) and the more recent review of the problem by Kahn (1956a, 1956b, 1959).

The framework elements of coarser clastic sediments—gravels and sands—are the pebbles and sand grains of which these deposits are composed. These clastic elements are nonspherical and are nonuniform in size. Nevertheless, an understanding of the phenomenon of packing and its effects on porosity and permeability is furthered by assuming the solid units to be spheres (in most cases the constituents of these coarser sediments are nearly spherical, with an average sphericity of 0.80 or more in many sands). Consideration must first be given to an aggregate of uniform-sized spheres and then to nonuniform mixtures.

The packing of spheres of uniform size may be either disorderly or constantly repetitive and geometrically systematic. Consideration of the problem shows that although there are six fundamentally different ways of systematic packing, one case, that of rhombohedral packing (Slichter, 1899, p. 305), is the "tightest," that is, has minimum porosity and is the most compact possible arrangement of solid spheres. Because this is also the most stable, most natural closely sized aggregates approach the rhombohedral arrangement. A considerable degree of disorder occurs in most deposits, although within any given deposit may be "colonies" or regions in which the closest packing prevails. Rhombohedral packing is characterized by a unit cell of six planes passed through eight sphere centers situated at the corners of a regular rhombohedron, each edge of which measures $2R$ (Fig. 3-37). Rhombohedral packing stands in marked contrast with cubic packing (the "loosest" possible systematic packing), in which the unit cell is a cube, the eight corners of which are the centers of the spheres involved (Fig. 3-37). In rhombohedral packing the porosity is 25.95 percent; for cubic packing it is 47.64 percent.

Any plane passed at random through systematically packed spheres reveals alternating areas of solid materials and voids. The area of voids, however, is not a true measure of the total area available for transfer of fluid, because part of the voids are closed by other grains which block the passageway. If, however, a plane is passed through the centers of the spheres in one of the rhombohedral layers of tightest packing, that is, a "throat plane," the void area in this plane is a true measure of the minimal cross-sectional area of the channelways or what might be called "useful porosity." In the tightest rhombohedral packing, with total porosity of 25.95 percent, the effective porosity is 9.30 percent. These differences do not affect capacity of the pore system for storage, but they do relate to the movement of fluids through the rock, that is, their permeability.

If a large number of spheres of equal diameter is arranged in any strictly systematic manner, there is a fixed diameter ratio for the smallest sphere which can pass between the throats of the larger spheres and into the interstices. For tightest packing, the critical diameter is $0.154D$ (where D is the diameter of the larger spheres). Similarly there is a critical ratio for the diameter of a sphere which, although too large to pass into the void from without, can exist in such a void and therefore must have entered that

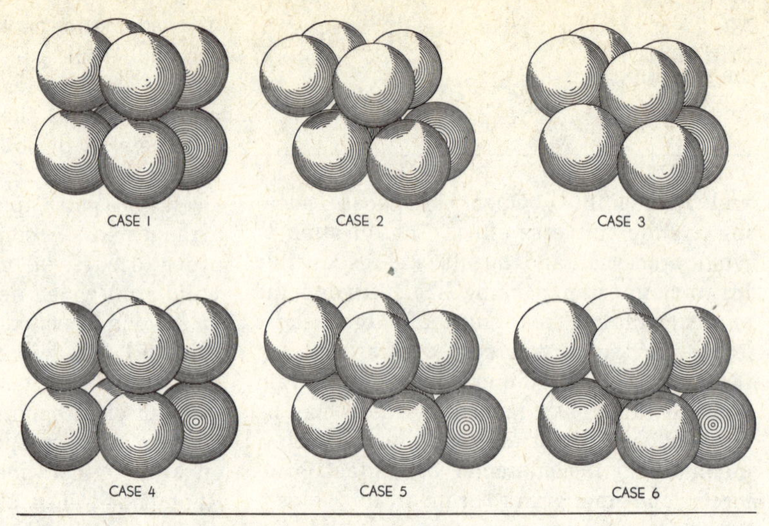

FIG. 3-37. Spheres may be packed in six possible arrangements. Case 1 is most "open" or cubic packing; case 6 is the "closest" or rhombohedral packing. (From Graton and Fraser, 1935, p. 796.)

position at the time of deposition. The critical ratios of occupation are 0.414*D* and 0.225*D* for the rhombohedral arrangement (two types and sizes of voids are present in this arrangement). These theoretical concepts cannot be applied too literally to natural deposits, because the latter neither are composed of spheres nor are they packed in a wholly systematic manner. Nevertheless, if the materials filling the interstices of a gravel are mostly larger than 0.154 times the diameter of the pebbles, then these "fines" have not been washed in but have been laid down contemporaneously. These observations have some bearing on the bimodal size distributions displayed by some gravels (see p. 49).

In the initial deposit the contacts between grains are basically tangential or point contacts. Hence a random section through such a matrix seldom passes through these points of contact. In such sections, therefore, many grains appear not to be in contact at all (Fig. 3-38). If, however, the contacts are altered in such a way to increase the area of contact, the random slice will pass through more grain contacts, and the number of contacts per grain in a given field of view will increase (Table 3-10). As the tangential contacts are modified, they become long, concavoconvex, or sutured (Fig. 3-39). Jane Taylor (1950) studied the grain contacts in sandstones from various depths in deep wells in Wyoming. Normal sands were

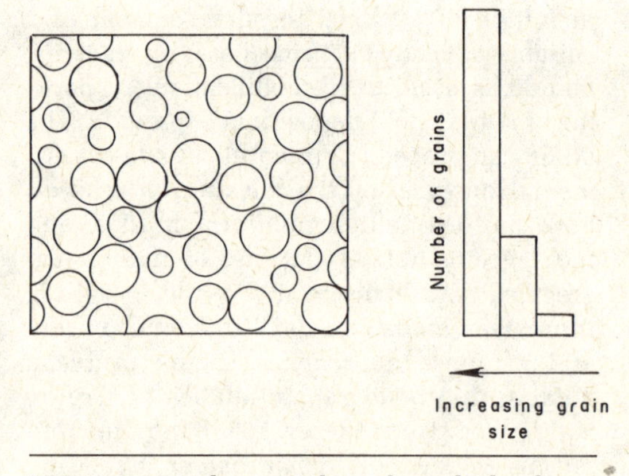

FIG. 3-38. Random cut-through packed spheres of equal size and their appearance in plane of section. Note *apparent* absence of contact between many spheres and *apparent* variations in size. (After Graton and Fraser, 1935, Fig. 14.)

found to have 1.6 contacts per grain (more likely, 0.85 contacts per grain, according to Gaither, 1953). At 2885 feet (900 m) the sands had 2.5 contacts per grain and at 8343 feet (2570 m) the contacts per grain were 5.2. Seemingly, therefore, sandstones do undergo a "condensation" process which tends to bring the grains into closer and more extended contact with one another. Taylor attributed this change to intrastratal solution and precipitation and to solid flow of the quartz grains. Taylor cited various evidence of pressure, such as bent micas and fractured quartz grains. But solid flow itself is difficult to prove, and the concavo-convex contacts seen by Taylor may be, as Waldschmidt (1943) supposed, a solution effect. Some grains other than quartz may be ductile; their deformation may lead to some loss of porosity. Rittenhouse (1971) has made estimates of the effects of such mechanical compaction.

A number of attempts to measure the packing have been made. The number of contacts per grain is one such measure. Kahn (1956a) proposed two measures: *packing proximity*, which is essentially the number of contacts per grain (ratio of number of grain-to-grain contacts to the total number of grains counted along a traverse) and *packing density*, which is the sum of all grain intercepts to the total length of a traverse (essentially the complement of the porosity on a matrix or cement-free basis). Other indices of packing have been proposed by Smalley (1964a, 1964b), Allen (1962, p. 678), Emery and Griffiths (1954, p. 71), and Mellon (1964).

Unfortunately the study of grain contacts and the measurement of packing are still

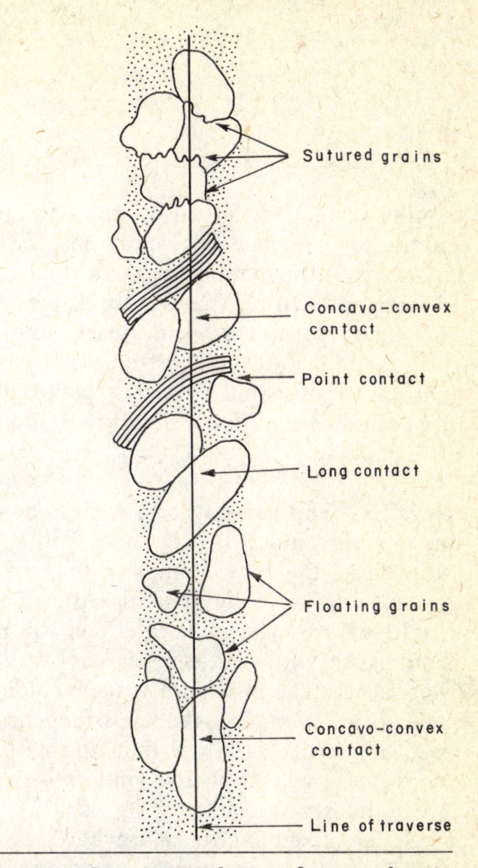

FIG. 3-39. Fabric terminology and types of grain contacts. (After Pettijohn, Potter, and Siever, 1972, Sand and sandstone, Fig. 3–10. By permission of Springer-Verlag.)

TABLE 3–10. Contacts per grain in diverse sands and sandstones

Type	Number of contacts: Range	Number of contacts: Average	Reference
Uniform spheres	—	0.63	Graton and Fraser (1935)
Artificial sand	1.00–1.52	1.28	Gaither (1953)
Artificial sand	—	1.6	Taylor (1949)
Calcareous oolite	0.70–1.3	0.90	—
Orthoquartzites[a]	1.2–3.1	(2.2)	Manry (1949)
Orthoquartzites[b]	2.3–3.5	2.7	—
Sandstone	2.1–3.6	(2.8)	Beaudry (1949)
Subgraywackes	1.7–2.2	(1.9)	Hays (1951)
Subgraywackes	2.3–5.2	4.1	Taylor (1949)

[a] Six samples.
[b] Three samples.
Numbers in parentheses are average of extreme values.

rather subjective, in part because of inaccurate or inadequate observations. Original quartz grain boundaries in some sandstones disappear partially or wholly because of secondary overgrowths of quartz and can be seen only with catholuminescence. Others are obscured by matrix materials so that one has real doubt about any grain-to-grain contact.

Porosity The porosity of a rock is defined as that percentage of pore space in the total volume of the rock, that is, the space not occupied by the solid mineral matter. Porosity thus expressed is the *total* pore space as contrasted with the *effective* or available pore space. The total pore space includes all interstices or voids, whether connecting or not, and hence is larger than the effective pore space, which includes only those voids freely interconnected.

Unlike crystalline rocks in which porosity is nil, clastic sediments are moderately to highly porous. This porosity is attributable to the fact that clastic components cannot, at the time of deposition, be in continuous contact with one another; they are instead in tangential contact only. The pore system constitutes the channelways for the passage of fluids through the rock as well as for storage of fluids. Hence the volume of such spaces and the resulting storage capacity of the rock and its transmissability are of great importance in the study of oil and gas, brines, and ordinary groundwaters. For this reason, also, a great deal of effort has been expended in devising ways to measure porosity and to make porosity determinations. Refer to the various laboratory manuals in which such methods are described (Müller, 1967, pp. 235–244; Curtis, 1971). For an extended treatment on all aspects of porosity, refer to von Engelhardt (1960).

The porous nature of detrital sediments is a major cause and condition for diagenetic reorganization. Porosity, for example, leads to inhomogeneities in the distribution of pressure because of the weight of superincumbent beds; this weight is carried on the relatively small areas of contact between the grains. This leads to solution at points of contact and precipitation in the voids. Moreover, the fluid that occupies the pores constitutes a medium in which chemical reactions occur or which itself reacts with the solid framework of the sediment. As a result of solution and precipitation, infilling, and other diagenetic changes, the porosity of a sediment is diminished with time and hence with depth (Füchtbauer and Reineck, 1963, Fig. 7; Füchtbauer, 1967). The higher the grade of diagenesis, the closer the similarity of the rock to those of metamorphic or igneous origin.

Porosity may be considered as either original or secondary (Fraser, 1935). Original porosity is an inherent characteristic and was determined at the time the rock was formed. Secondary porosity results from later changes that may increase the original porosity. Carbonate sediments are most prone to develop secondary porosity, although some sandstones acquire a secondary porosity following leaching of their carbonate cement.

The original porosity of a sediment is affected by uniformity of grain size, shape of the grains, method of deposition and packing of the sediment, and compaction during and after deposition.

Theoretically the actual size of grain has no influence on porosity. In fact, however, the finer-grained sediments have a higher porosity than the coarse-grained sediments (Table 3-11). This observation, however, does not establish a cause-and-effect relation, because size may be closely correlated with other properties, such as shape, which may be the primary cause of the porosity differences noted.

Whether the size is uniform or nonuniform is of fundamental importance (Rogers and Head, 1961). The highest porosity is commonly obtained when the grains are all of the same size. To such an assemblage the

TABLE 3-11. Relation between porosity and grain size

Size of material	Porosity (percent)
Coarse sand	39 to 41
Medium sand	41 to 48
Fine sandy loam	50 to 54
Fine sand	44 to 49

Source: After Lee (1919, p. 121).

addition of other sand grains, either larger or smaller, tends to lower the porosity; and this lowering, within certain limits, is directly proportional to the amount added (Gaither, 1953, Fig. 2) until the mixture consists of nearly equal parts of the sizes involved. On the other hand, the addition of clay increases the porosity (Füchtbauer and Reineck, 1963, Fig. 4). No simple relation, however, exists between the size distribution and porosity. Fraser (1935) and others have shown that quite different mixtures can have the same porosity.

The effect of shape on porosity is little understood. In general, grains of high sphericity tend to pack with minimum pore space. Fraser found, for example, that uniformly sized beach and dune sands that had been experimentally compacted had porosities of about 38 and 39 percent, respectively, whereas crushed quartz had a porosity of about 44 percent. Because the sphericity of crushed quartz is about 0.60 to 0.65 and that of beach sand probably near 0.82 to 0.84, it is clear that grain shape has a small but noticeable effect on the porosity. Fraser found the effect of shape to be most marked in the case of very flat pebbles. Certain detrital limestones, such as coquina, are very porous, displaying a "potato-chip" fabric. Such deposits may have a porosity as high as 80 percent (Dunham, 1962, p. 108). Likewise freshly deposited clays have very high porosities, even up to 85 percent.

The method of deposition and packing have a marked influence on porosity. For uniform spheres calculations show porosities ranging from 26 to 48 percent for the closest packed to the most open arrangements, respectively. A given sand, experimentally packed, has been found to vary from 28 to 36 percent. In nature, however, the tightest packing with minimum voids tends to prevail, so in general the importance of packing tends to be minimized.

The effect of compaction on porosity has been reviewed elsewhere (Chap. 8, p. 276 and Chap. 12, p. 468). The effect on clay and shale is very great; porosity seems to be a function of depth of burial according to the expression:

$$P = p(e^{-bx})$$

where P is the porosity, p is the average porosity of surface clays, b is a constant, and x is the depth of burial (Athy, 1930). Clays are reduced from initial porosities of much in excess of 50 percent to less than 10 percent in many cases. The compaction of sands, however, is negligible. The initial porosity of a sand (35 to 40 percent) may be reduced to a very small figure by solution and reprecipitation or by infilling with introduced cement. The average sandstone has a porosity of 15 to 20 percent. The high porosity of some sandstones, such as Devonian Oriskany of the central Appalachian area, has been attributed to the leaching of a preexisting carbonate cement (Krynine, 1941).

Permeability Permeability is the property of a rock which allows the passage of fluids without impairment of its structure or displacement of its parts. A rock is said to be permeable if it permits an appreciable quantity of fluid to pass through it in a given time and impermeable if the rate of passage is negligible. Obviously the rate of discharge through a given cross section depends not only on the rock but also on the nature of the fluid and the hydraulic head or pressure.

The permeability of a porous medium can be expressed as the quantity of fluid Q (cm^3/sec) passing through a given cross section C (cm^2) and a given length L (cm). This quantity is directionally proportional to the pressure difference, P (in atmospheres), at the two ends of the system and inversely proportional to the viscosity of the fluid, V (in centipoises):

$$Q = K\frac{C\ P}{V\ L}$$

The proportionality factor, K, is the permeability, a factor characteristic of the rock under consideration. This permeability coefficient has been called a *darcy*. A sand is said to have a permeability of 1 darcy when it yields 1 cm^3 of fluid (viscosity 1 centipoise) per second through a cross section of 1 cm^2 under a pressure gradient of 1 atmosphere per cm of length. Modern sands have permeabilities of 10 to 100 or more darcys (Fig. 3-40). But in most consolidated sandstones the permeability is likely to be considerably less than 1 or 2 darcys. Hence permeability is usually reported in millidarcys.

PERMEABILITY (darcys)

10^5 10^4 10^3 10^2 10^1 10^0 10^{-1} 10^{-2} 10^{-3} 10^{-4} 10^{-5}

Material	Clean gravel	Clean sands, mixtures of sand and gravel	Very fine sand, silt, mixtures of sand, silt, and clay; glacial till, stratified clays, etc.	Unweathered clays
Flow characteristics	Good aquifers		Poor aquifers	Impervious

FIG. 3-40. Magnitude of permeability for various uncemented sediments. (After Todd, 1960, Ground water hydrology, Fig. 304. By permission of John Wiley and Sons.)

The importance of permeability in the study of oil sands and aquifers cannot be exaggerated. Much effort has been expended, therefore, to measure permeability and to determine the geologic factors which control it. The techniques of permeability measurement have been detailed in various papers and laboratory manuals (Curtis, 1971, pp. 335–364; Müller, 1967, pp. 244–249).

The coefficient of permeability, K, of an unconsolidated sand is affected by grain size, grain sorting, grain shape, and packing. The effects of size and uniformity of size have been studied experimentally. Krumbein and Monk (1942), for example, used glacial outwash sand which was sieved and recombined into mixtures of desired composition. Because many natural sands have log normal size distributions, prepared mixtures were made log normal. They were made into sets with either a common mean size but variable standard deviation (sorting), or a common standard deviation but variable mean size. Thus the effects of size and sorting could be studied separately. Krumbein and Monk found the coefficient of permeability to vary as the square of the diameter and inversely as the log of the standard deviation (Fig. 3-41). In actual sandstones the permeability shows a close correlation with grain size, increasing as size increases (Fig. 3-42).

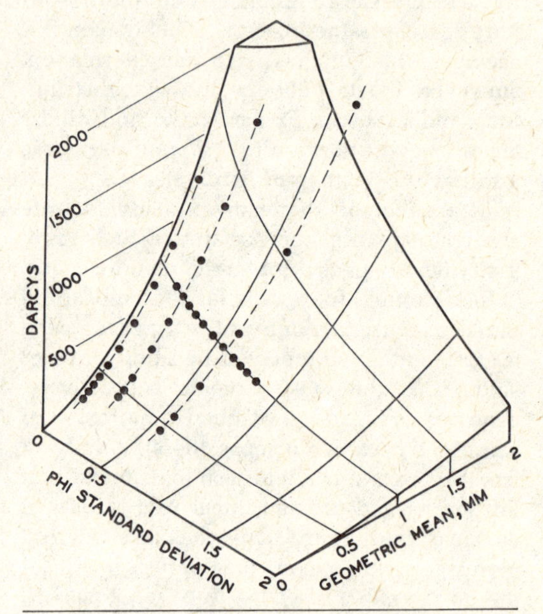

FIG. 3-41. Permeability surface. The mutual control of permeability by mean size and the standard deviation is shown. The vertical axis represents the permeability in darcys, and the horizontal axes are the geometric mean diameter in millimeters and the phi standard deviation. The nature of the surface is shown by the grid lines, which are parabolas parallel to the darcy-size plane, and negative exponentials parallel to the darcy-standard deviation plane. (Krumbein and Monk, 1942, Fig. 5.)

The shapes of constituent grains, expressed by their sphericity, in some way affect permeability, perhaps because sands with lower sphericities tend to have high porosity and looser packing and hence higher permeability.

Permeability is also dependent on the packing arrangement, because, as we have seen for a given size of spheres, the pore dimensions (which govern permeability) are dependent on the style of packing. Hence any change in packing that increases the porosity will increase the permeability, a conclusion verified experimentally by von Engelhardt and Pitter (1951).

Theoretically permeability is independent of porosity, although obviously a nonporous rock is also nonpermeable. On the other hand, a highly porous rock is not necessarily permeable. Fine-grained rocks even though highly porous are only slightly permeable.

The relations between porosity, permeability, and grain size have been studied both experimentally and theoretically by von Engelhardt and Pitter (1951) and theoretically by Scheidegger (1957, pp. 100–108) and others. As a first approximation permeability is proportional to the first power of the porosity and inversely proportional to the second power of the specific surface (cm^2/cm^3). Hence the finer the grain (and therefore the larger the specific surface), the smaller the permeability. In sandstones, porosity and permeability show a rough correlation. Permeability shows a much wider range in magnitude than does porosity (Fig. 3-43).

In stratified sediments, permeability has been found to be greater parallel to the bedding than perpendicular to it, and in some sands it has been found to vary in value in diverse directions parallel to the bedding, presumably because of some kind of anisotropic grain fabric (Mast and Potter, 1963; Potter and Pettijohn, 1963, p. 50).

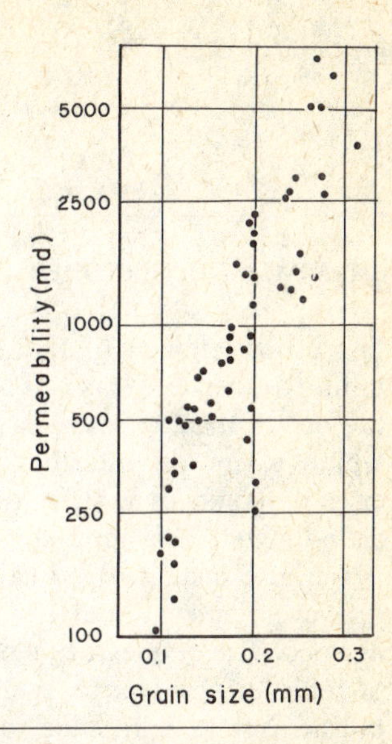

FIG. 3-42. Dependence of permeability on grain size in the Benheimer Sandstone of the Scherrhorn oil field near Longen, Germany. (Based on random selection of data of von Engelhardt, 1960, Der Porenräum der Sedimente, Fig. 49. By permission of Springer-Verlag.)

CRYSTALLINE AND OTHER ENDOGENETIC TEXTURES

Endogenetic textures, those shown by precipitates from solution or produced by recrystallization or alteration of preexisting materials, are distinctive and notably different from exogenetic textures—the textures of clastic deposits. In the former the minerals involved are formed on the spot where they now are; in the latter they are emplaced in the framework of the rock as solid particles.

We have reviewed what is known about primary textures of clastic sediments; we will now summarize the salient facts of the textures created by chemical action. It is not to be supposed that the two are unrelated. Many sedimentary rocks display both. A sandstone, for example, has an exogene or clastic framework, but it may also have an endogenetic cement which displays a crystalline fabric. Many limestones likewise display both fabrics; in some cases carbonate sands mechanically accumulated are bound together by a precipitated crystalline cement.

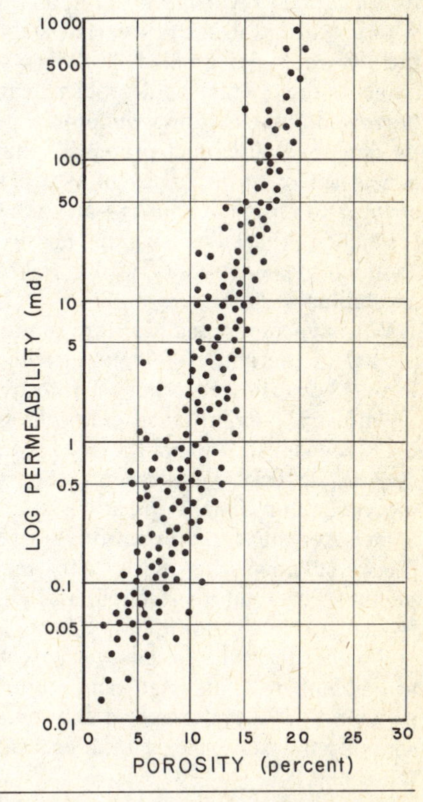

FIG. 3-43. Plot of permeability against porosity, Upper Carboniferous sandstone. (Füchtbauer, 1967, Proc. 7th World Petrol. Congr., Fig. 10.)

CRYSTALLINE TEXTURES

In a strict sense virtually all rocks are crystalline—even the clays—but the term *crystalline* is usually reserved for those rocks which show an interlocking aggregate of crystals (such as rock salt). Such rocks are called *crystalline granular* or *saccharoidal.* An arkose cemented by calcite is a *crystalline aggregate,* although it is not described as such. The framework elements are grains of feldspar and quartz, each obviously crystalline, but in deference to the detrital nature of these grains the rock is said to have a clastic rather than a crystalline texture. The carbonate cement, however, has a strictly crystalline fabric.

The terminology to be applied to crystalline fabrics and textures of sedimentary rocks has not been standardized. The question has been reviewed by Friedman (1965), who noted that usage has varied widely and the terminologies applied to the crystalline igneous and metamorphic rocks have been borrowed and used for sediments, a practice he deplored. Friedman proposed, therefore, a new set of terms, many of which are restricted to carbonate rocks—an unfortunate restriction, because similar textures are found in gypsum and anhydrite and in other crystalline sediments.

We have here used the terminology employed in the study of metamorphic rocks, in part because we do not wish to overload an already overburdened terminology, but mainly because we hold that diagenesis—recrystallization, replacement, and internal reorganization (*neomorphism*)—is after all a metamorphic transformation. There is no break between diagenesis and metamorphism as customarily defined, and all transformations in the solid state are considered metamorphic in the broadest sense whether at ambient or at elevated temperatures and pressures. The textures and fabrics are essentially the same because the processes are the same.

Crystal elements The basic components of a crystalline fabric are its individual crystal elements. If large, the texture is *macrocrystalline;* if small, the texture is *microcrystalline.* If in between, the texture is *mesocrystalline.* Crystalline textures difficult to resolve under the microscope are *cryptocrystalline.* Several attempts to quantify these terms have been made (Table 3-12). Several terms, such as *micrite, microspar,* and *sparry,* have been used to describe the crystallinity of limestones. These are defined in the chapter on limestones.

We have to consider not only the question of crystal size but also the question of uniformity of size. When uniform, the term *equigranular* may be applied; when unequal sizes are involved, the term *inequigranular* may be used. In some cases crystal size displays a discontinuity; that is, large crystals, perhaps an order of magnitude larger than those of their matrix, stand out. Such crys-

TABLE 3-12. Crystalline textures and fabric

Macrocrystalline ("granoblastic"; over 0.75 mm)
Equigranular (grain boundaries sutured or mosaic)
- Xenotopic (most crystals anhedral)
- Hypidiotopic (most crystals subhedral)
- Idiotopic (most crystals euhedral)

Inequigranular
- Xenotopic
  - Porphyrotopic ("porphyroblastic")
  - Poikilotopic ("poikiloblastic")
- Hypidiotopic
  - Porphyrotopic
  - Poikilotopic

Mesocrystalline (0.20–0.75 mm)
Equigranular
Inequigranular
- "Porphyroblasts" dominant (in contact, forming a loose net), groundmass micro- or cryptocrystalline)
- "Porphyroblasts" subordinate (scattered or isolated), groundmass micro- or cryptocrystalline or pelitomorphic

Microcrystalline (crystalline nature recognizable only under the microscope; 0.01–0.20 mm)
Equigranular
Inequigranular
- Cryptocrystalline material subordinate
- Cryptocrystalline material dominant

Cryptocrystalline (less than 0.01 mm)

tals are the homologues of the *porphyroblasts* of garnets or staurolite in schists and may be so termed (porphyrotopes of Friedman; porphyrocrystallic fabric of Phemister, 1956). Some anhydrite beds, for example, have large conspicuous crystals of gypsum in a microcrystalline matrix of anhydrite.

When a large crystal encloses smaller crystals of another mineral, the texture is *poikiloblastic* (poikilotopic of Friedman; poikilocrystallic of Phemister). An analogous fabric is seen in some sandstones when the calcite cement has a uniform crystallographic orientation over a large area and encloses many sand grains (sand crystals). Large barite crystals may similarly enclose sand grains (barite rosettes).

The *shape* of the crystal elements may be described in terms of the extent to which they display external crystal faces and symmetry. Those which show no crystal faces are *anhedral;* those which are bounded by well-formed crystal faces are *euhedral. Subhedral* may be applied to those showing a few faces or incomplete crystal outlines.

Of importance also is the nature of the boundaries between crystals in a crystalline aggregate. The boundaries between the crystal elements may be *straight, curved, embayed, scalloped (cuspate),* or *sutured.* The question and significance of grain boundaries have been explored in detail by Spry (1969, pp. 19–46) and discussed in respect to the crystalline fabrics of the carbonates by Bathurst (1971, pp. 421–425, 462–465) and by Folk (1965a, pp. 17–20). The character of the boundary may be a clue to the relative ages of the minerals or a criterion of mutual solution (microstylolitic boundaries), of replacement and corrosion (embayment), and the like.

The fabric of cements The fabric of cements is the fabric of pore filling. In this respect the crystalline fabric of sediments differs from that of metamorphic rocks. Two situations prevail. In one the framework is inert and does not react with the cement or solutions from which the cement was precipitated. In the second case the framework

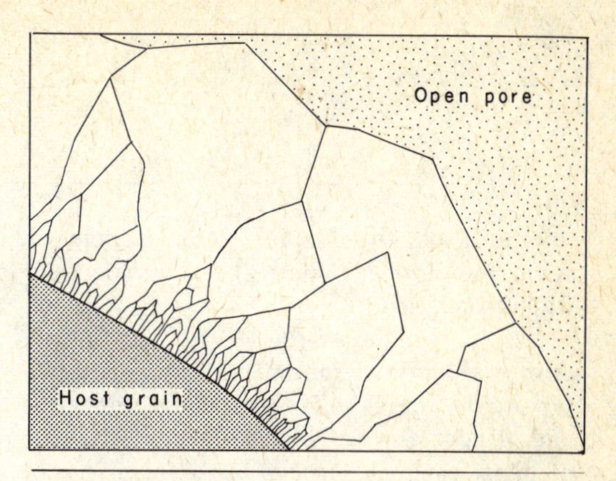

FIG. 3-44. Diagram to show crystal fabric of pore filling. (After Pettijohn, Potter, and Siever, 1972, Sand and sandstone, Fig. 10–6. By permission of Springer-Verlag.)

and cement interact, and the framework itself is altered. With an inert framework, the problem is one of mineral precipitation on a surface with free growth into a pore. In general the precipitated materials form a drusy implantation of crystals on the pore walls. These tend to grow outward into the pore; some crystals become dominant and suppress the growth of others (Fig. 3-44). Ultimately the crystals growing from various surfaces meet, and the pore is filled. In other cases the cementing mineral (calcite or barite) forms large crystal units unrelated to the pore system, and the cement acquires a uniform crystallographic and optical orientation over a large area and which encloses many detrital grains (a poikiloblastic texture).

When the framework grains are reactive, a different cementing fabric prevails. In some cases the framework grains grow or are added to by precipitation from the pore solutions. In essence the framework grains are "seed crystals" and form the nuclei for growing crystals. This is the case with "secondary enlargement" of quartz and feldspar and even calcite in some crinoidal sands. The cement is then in optical and crystallographic continuity with the framework grain (Fig. 7-20, 10-30). The end-result is a crystalline granular texture. In other cases the framework grains are corroded by the cement and are partially replaced. In some extreme cases the unstable framework grains are degraded or even decomposed and form a microcrystalline aggregate. Carbonate detritus in many limestones display micritic rims; the unstable rock particles in some sandstones yield the *epimatrix* of Dickinson (1970).

Recrystallization fabrics Many sedimentary rocks recrystallize at normal pressures and temperatures. This is particularly true of many carbonate rocks but is also true of gypsum, anhydrite, and even chert. The aragonitic shells and skeletal debris and cements recrystallize to calcite. The conversion of anhydrite to gypsum or the reverse is a good example of this kind of recrystallization, as is the microcrystalline texture of chert presumed to be the result of neocrystallization of a silica gel. In some cases the change is truly a recrystallization (aragonite to calcite); in other cases it involves hydration or dehydration (opal to chalcedony or gypsum to anhydrite). In some cases new material is added (calcite to dolomite).

Crystallization or recrystallization in the solid state produces textures essentially "metamorphic" or crystalloblastic in character. Crystals grown in this way are apt to be full of inclusions which may be concentrated in the center of the new crystal or zonally distributed. As might be expected, rocks which owe their crystallization to diagenetic changes would carry, like their metamorphic counterparts, relict textures and structures. These are traces of original textures and structures of the rock which have not been wholly destroyed by postdepositional reorganization. Bedding laminations, oolites, fossils, and even clastic textures may persist and be recognizable.

Recrystallization may be selective (involving only some rock components), or it may be pervasive (involving all). Conversion of aragonite shells to calcite illustrates the first; wholesale dolomitization is an example of the second. Recrystallization may decrease the grain size; more commonly it leads to a coarsening of the rock texture.

The sedimentary petrologist is confronted with some troublesome questions such as discrimination between matrix-like materials (epimatrix) formed by degradation of framework elements and microcrystalline matrix formed by recrystallization of interstitial muds (the *orthomatrix* of Dickinson). The coarsely crystalline cements of some limestones are products of precipitation in a pore system, but a similar product may be the result of recrystallization of an interstitial carbonate mud. These questions have been reviewed by Dickinson (1970) in the case of the noncarbonate sands and by Folk (1965a) and by Bathurst (1971, p. 475) in the case of the carbonate rocks. These topics are also discussed in the chapters on sandstones (Chapter 7) and limestones (Chapter 10).

In some rocks new crystal growth is incomplete; new material occurs in the form of large porphyroblasts. In other cases new mineral growth takes the form of minute spherulites (Fig. 10-3).

Textures of replacement and paragenesis Minerals which are chemically precipitated and exhibit a crystalline texture or fabric may be formed at the time of accumulation of the sediment; or they may be formed later, perhaps precipitated in the interstices of the sediment, or perhaps in part replacing the preexisting minerals, either detrital or chemical. It is evident, therefore, that to understand the history of a sedimentary rock it is necessary (1) to discriminate between detrital minerals and those chemically precipitated; (2) to determine the relative ages of the several precipitated minerals; and (3) to ascertain the manner of emplacement of the precipitated materials, whether formed in openings or formed by replacement. To make these discriminations, the student will need criteria and will need to evaluate those criteria in reading the history of the rock (Grout, 1932, pp. 1–3). These criteria are in large part textural, having to do with the shape of the crystal, the nature of the grain boundaries, and similar considerations.

For a full understanding of the history of the rock, it is imperative that the relative ages or paragenesis of the several minerals present be determined. The problem of mineral paragenesis and the related problem of replacement of one mineral by another have occupied the attention of petrographers and the students of ore deposits for many years. The students of ores in particular have been astute observers of textures of ores and rocks and have formulated many criteria for attacking the problems of relative age and replacement (Bastin et al., 1931; Bastin, 1950; Edwards, 1947). Those papers are invaluable when dealing with the needed criteria and proper evaluation of mineral paragenesis. Refer also to the works of Grout, Bastin, and others in which such criteria are collected, summarized, and evaluated. These criteria are best understood with reference to specific cases. Many such cases are cited elsewhere and are illustrated by appropriate sketches or photomicrographs.

Most microtextures and mineral contact relations found in other rocks and ores are also present in sediments. The relative age of two minerals in contact is in part determined by the fabric of the rock. The minerals of the detrital framework are obviously earlier than those which fill the voids within the framework. Some have contended, however, that the cementing minerals are penecontemporaneous with the detrital minerals which they bind together (Krynine, 1941). Minerals filling vugs or fractures as well as voids are obviously later than the host rock. When several minerals fill the same cavity, their relative ages are determined by mutual contact relations. In general the younger mineral is molded about the older. Because the early formed minerals grew into empty or fluid-filled cavities, they will be euhedral; later minerals will conform to the remaining spaces between early euhedra and will themselves be anhedral. Unfortunately euhedralism is not an infallible criterion of early formation. If the euhedral crystal was formed by replacement, it may be later than the mineral which surrounds it. Quartz euhedra found in some limestones are excellent illustrations of this principle (Fig. 10-4). It is necessary, therefore, to discriminate between euhedra formed by growth in a fluid medium and euhedra formed by replacement within a solid matrix.

The criteria of replacement are many and include automorphic crystals which transect earlier structures such as bedding, fossils, and oolites (Fig. 10-4). Minerals formed by replacement contain inclusions of the material replaced. These unreplaced residuals may have a common crystallographic orientation, or they may be distributed in a relict or ghost pattern inherited from the structure replaced (Fig. 11-8). Embayed contacts, as well as the residuals isolated by extreme embayment, are indicative of replacement relations (Fig. 7-21). The best-known criterion of replacement, perhaps, is pseudomorphism. Pseudomorphic replacement of organic structures (fossil wood, shells, and the like) and crystal pseudomorphs (silica after dolomite, for example) are common and conclusive evidences of replacement (Fig. 11-7). The alert observer will discover and use other criteria of relative age and replacement. It is essential that these relations be carefully worked out.

Fabric of veins Veins display a great variety of textures and structures. Those of quartz veins have been described by Adams (1920), those of carbonate veins by Grout (1946).

In some veins, crystalline filling is equigranular; in some are unfilled vugs or druses. Cross-fiber veins of quartz, calcite, and gypsum are common and are produced by fibrous crystals normal to the walls of the vein, and some are modified by deformation. Comb structure is an analogous feature in which the deposited crystals are prismatic but not fibrous. In places these are flamboyant; that is, they show an expansion of elongate crystals from their point of origin. This structure passes into a true radial pattern of crystal growth.

OOLITES, SPHERULITES, AND PELOIDS

Many sediments contain small spherical bodies of diverse mineral composition and various internal organization. Included here are oolites (also termed ooids, ooliths, ovulites); pisolites, pseudo-oolites or false oolites; and peloids, spastoliths, and spherulites.

Oolites and pisolites A rock may be said to possess an oolitic texture if it consists largely of *oolites*. These are small spherical or subspherical, accretionary bodies, 0.25 to 2.00 mm in diameter; most commonly they are 0.5 to 1.0 mm in size. If they are over 2.00 mm in diameter, they are termed pisolites. Although spherical form is the rule, some oolites are oblate ellipsoidal. Within any given rock, however, they are highly uniform in both size and shape. Oolites are found in rocks of all ages.

Oolites have long fascinated petrographers, and there is much literature on the subject. Refer to papers by Rothpletz (1892), Barbour and Torrey (1890), Linck (1903), Brown (1914), Bucher (1918), Carozzi (1957, 1961a, 1961b, 1963), and Monaghan and Lytle (1956), which deal with the topic in general terms, and to the innumerable papers dealing with particular modern or ancient oolitic deposits, such as those in the Bahamas (Newell, Purdy, and Imbrie, 1960), and with the problems of terminology (DeFord and Waldschmidt, 1946; Flügel and Kirchmayer, 1962).

The term *oolite* has been applied both to the minute concretionary bodies described above and to the rock composed of such objects. To avoid confusion some writers apply the term *oolite* only to the rock and use some other term such as *oolith* for the spherites of which the rock is made (DeFord and Waldschmidt, 1946). The suffix *-lith,* however, has been used by some petrographers to denote a rock (as in the terms *biolith* or *calclithite*). Hence the term *oolith* is also ambiguous. The terms *ooid,* as well as *ooide* (Kalkowsky, 1908, p. 72) and *ovulite* (Deverin, 1945), have also been used. Twenhofel (1950, p. 625) avoids ambiguities by restricting the term *oolite* to the microconcretionary bodies and would use the adjective *oolitic* for the rocks: Oolitic chert, oolitic limestone, and so forth, are certainly clear enough.

Carozzi (1957) distinguishes between oolites and *superficial oolites* (a mineral or skeletal grain surrounded by *one* concentric layer). If *two* or more layers are present, the object is a true oolite. Superficial oolites may be confused with those calcareous grains which display a dense micritic rim, a product in some cases of peripheral micritization of the original grain and not an added precipitated layer.

False oolites or *pseudo-oolites* are those

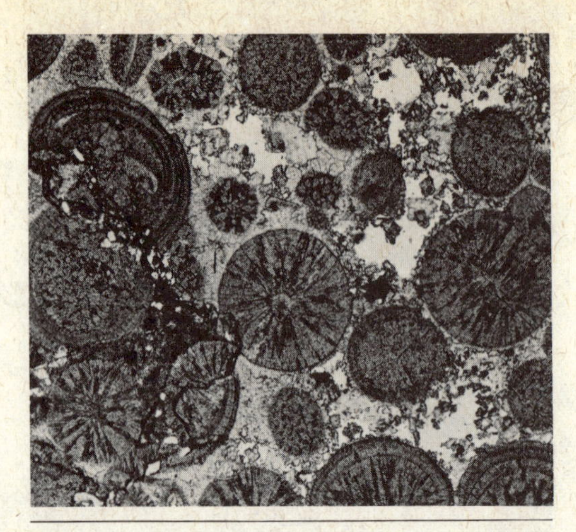

FIG. 3-45. Warrior (Cambrian) Limestone, State College, Pennsylvania. Ordinary light, ×22. Ooids with radial and concentric structure cemented by carbonate. Note stylolitic seam along which ferruginous clay (black) and bits of detrital quartz (white) have been concentrated. The amount of solution involved in the formation of the stylolite is indicated by the concentration of the quartz silt, which is very sparsely distributed throughout the rock.

bodies lacking internal structure. They may be fecal pellets or degraded oolites which have lost their primary structure by micritization. Some may even be well-worn intraclasts of micritic limestone. To these ambiguous forms McKee and Gutschick (1969) applied the term *peloid.* This seems like a good choice when the origin of these bodies is in doubt.

In cross section, oolites appear concentric or radial or both (Figs. 3-45 and 10-25). Oolites appear to have grown outward from a center. In many the growth has taken place around a nucleus, such as a quartz sand grain or a small fragment of a shell. In others no nucleus can be seen, either because the section did not pass through the nucleus or because there was none. In some ooids the regular concentric structure is interrupted, apparently by breakage or erosion of the original body and then followed by regeneration or renewed growth (Carozzi, 1961a). The result is an "unconformity" between the outer concentric shells and those of the internal core. In other ooids are shells or layers of cryptocrystalline unoriented aragonitic material. A few oolites are composite and contain several centers of growth.

Spastoliths are distorted ooids (Rastall and Hemingway, 1941, p. 362 and Fig. 4). Some ooids, especially those of chamosite, are closely appressed, twisted, or misshapen (Figs. 11-21 and 11-22). The majority are thick in their central portion, flattened and pointed at the ends (*delphinformig*, Berg, 1944), or are more rarely compressed in the middle with relatively thick ends (*knockenformig*, Berg, 1944). Distortion has been attributed to the soft condition of the ooid at the time of its burial (Taylor, 1949, p. 32) and therefore believed to establish the primary nature of these chamositic bodies. Even some calcareous ooids show effects of synsedimentary deformation (Cayeux, 1935, p. 232; Carozzi, 1961b). Ooids, of course, may be subject to flattening and elongation by the tectonic forces which deform the rock and all structures which it contains (Cloos, 1947). Strangest of all distorted ooids are those which display arcuate apophyses and those joined together by such apophyses (Cayeux, 1935, p. 233; Carozzi, 1961b). See Fig. 11-22. The origin of these chains of linked ooids is obscure. They seem to be more typical of noncalcareous ooids, although they are reported also from some limestones.

Diagenesis leads to partial to complete obliteration of the ooid structure. Recrystallization may occur, leading to a coarse granoblastic texture which may contain faint inclusions marking the original concentric structure. This is common when the original ooid was aragonite which has recrystallized to calcite. In rare cases the interior of the ooid has become one single calcite crystal occupying the whole of the interior. In other cases the ooid is converted to a dense microcrystalline carbonate (micrite) with near total loss of concentric structure. Such ooids may be confused with pellets and micritic intraclasts. One interesting diagenetic modification is that of partial to complete solution of the ooid, which leaves a void that may later be filled by inward growth of crystalline material. In rare cases "half-moon" ooids are formed, divided into two parts, with a lower dense micritic sediment and an upper coarse sparry mosaic (Carozzi, 1963). Other diagenetic modifications involve replacement of the original ooid with other material such as silica in calcareous ooids. Dolomitization is most common. Dolomite first appears as rhombic euhedra—veritable metacrysts—in the ooid. Such rhombs lie athwart the concentric structure and may even lie across the boundary of the ooid itself.

Ooids formed at the present time seemed to be largely aragonitic, although some invert to calcite while still in the environment of deposition (Eardley, 1938). The inversion, as noted above, may lead to loss of structure and conversion of the ooid to a dense micritic body or, in other cases, to a coarse calcitic mosaic. Most commonly it is altered to a radial calcitic structure—a structure which characterizes a great many older oolitic limestones (Fig. 3-45). Under crossed nicols both the aragonitic and calcitic ooids display a dark cross, a pseudointerference figure of the uniaxial type caused, in the case of the aragonitic ooids, by the tangential orientation of the aragonite needles and by radial fibers of calcite in the calcitic ooids.

"Fossil" oolites may also be siliceous, dolomitic, hematitic, pyritic, and so forth. Some of these may be replacements of a calcareous oolite; others may be original precipitates. As mentioned above, dolomite is a common replacement. Silica may also replace a calcareous oolite, in some cases before dolomitization and in some cases afterward. Evidence of the secondary nature of siliceous oolite includes chert pseudomorphs of dolomite rhombs (Fig. 11-7), secondary enlargement of the detrital quartz nucleus in some oolites with a thin ring of carbonate inclusions at the interface between the detrital grains, and quartz overlay indicating entrapment of the calcareous matrix at the time of overgrowth (Fig. 10-4). Additional evidence includes (Henbest, 1968) the clear encroachment of reconstructed quartz crystals on the structure of the ooid (Fig. 3-46), and "hybrid" oolites, which are in part carbonate and in part chert with the chert–carbonate boundary (Fig. 3-47) crossing the original concentric structure. The diagenetic history of some siliceous oolites is very complex (Choquette, 1955).

Not all noncalcareous ooids are replace-

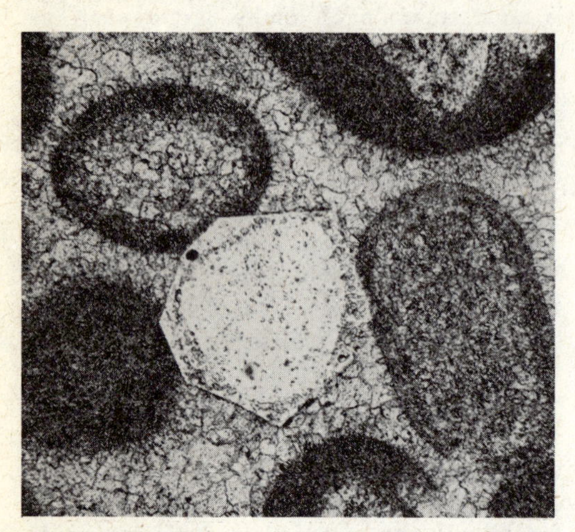

FIG. 3-46. (Arbuckle Limestone, Carter County, Oklahoma. Ordinary light, ×80.) Reconstructed quartz crystal with ringlike arrangement of inclusions along border of detrital nucleus. Carbonate was engulfed on enlargement of quartz.

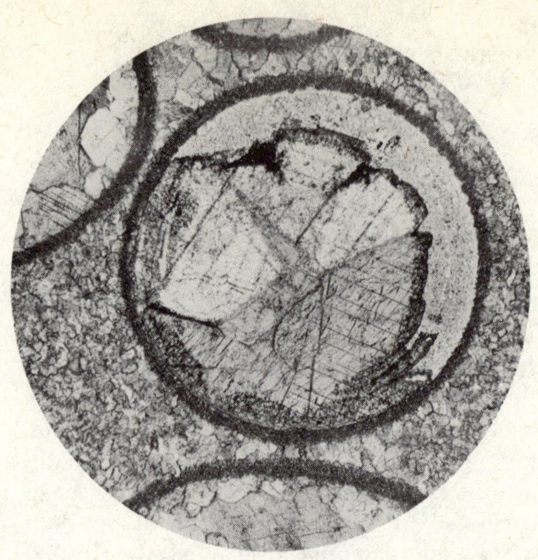

FIG. 3-47. (Cambrian, Tyrone, Pennsylvania. Crossed nicols, ×40.) Incompletely silicified ooid in a dolomitic matrix. Original calcareous (probably aragonite) ooid was replaced by mosaic of coarsely crystalline calcite. Note cleavage cracks of latter, which pass through concentrically arranged inclusions—vestigial remnants of original concentric structure. Silica (chert) now eating into and replacing calcite mosaic. Matrix has been dolomitized. This illustrates a complex sequence of replacements.

ments. Although their formation has not been observed, petrologic data indicate a primary origin for both the oolitic phosphates and the oolitic ironstones, especially the chamositic ironstones. These are discussed in Chapter 11.

Many theories have been advanced to explain the formation of oolites and pisolites. Some theories demand direct or indirect intervention of organisms, especially algae (Rothpletz, 1892); others require the formation of a gel medium (Bucher, 1918); some theories presume that pisolites form by replacement of nonoolitic detritus. Some of the larger concentric-structured bodies are certainly "algal pisolites," a type of oncolite (Fig. 3-48); some, such as those found in pisolitic bauxite and some ferruginous laterites, may be products of replacement of nonoolitic materials. Most calcareous oolites, however, and many noncalcareous oolites as well, seem to be the product of direct precipitation of dissolved materials on nuclei in a "free-rolling" environment. The close association of oolites and detrital quartz grains (Fig. 10-31), the cross-bedded character of many oolitic limestones, and the sorting of such deposits in-

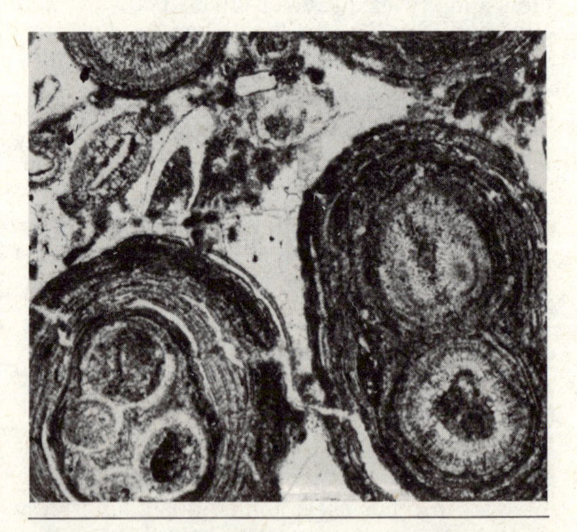

FIG. 3-48. (Silurian, Malvern, England. Ordinary light, ×32.) Calcareous pisolite, probably algal. Note composite character of several of the pisolites. Structure is chiefly concentric.

dicate accumulation in a turbulent medium. Illing (1954), who has given a good review of the problem of the calcareous oolites, concluded that the oolitic sands of the Bahamas are forming today only where the sediment is subjected to strong tidal current action and that oolites form where cooler oceanic waters moving onto shallow banks are sufficiently warmed to become appreciably supersaturated with calcium carbonate. Neither algae nor any other organisms play any part in their formation, but some boring algae may play a role in their destruction if they are left exposed in a "dead" environment, one where they are no longer in motion (Illing, 1954, p. 44). Illing's observations and conclusions have been further substantiated (Newell and Rigby, 1957; Newell, Purdy, and Imbrie, 1960).

Despite a general consensus on the origin of ooids as precipitation products in a shallow turbulent environment, there are some exceptions and contradictory observations. Freeman (1962), for example, described accretionary pellets, to which he applies the name *oolite*, that display off-center growth layers. These somewhat irregular bodies, lacking the highly spherical form and polish characteristic of normal ooids and possessing an asymmetric growth, are found in shallow-water areas of the Laguna Madre, Texas. Some oolitic limestones display ooids in a micritic matrix—a matrix hardly compatible with a turbulent, high-energy environment presumably required for oolitic growth.

Spherulites The term *spherulite* has been applied to any spherical body with a radial structure. Certain concretionary bodies are spherulitic. Many oolites are also spherulitic, but, as mentioned above, it seems probable that radial structure is a secondary development (Eardley, 1938, p. 1380). As used here the term *spherulite* is reserved for those minute subspherical bodies with a radial structure which have formed *in situ*. They are somewhat similar to the spherulitic bodies which form by the devitrification of a glass and which characterize certain spherulitic lavas. Spherulites of chalcedony, for example, form in some limestones (Fig. 10-3), as do spherulites of calcite (Muir and Walton, 1957, p. 162). Spherulites of this type, unlike oolites, have a somewhat irregular surface. Moreover, if the centers of growth are too closely spaced, mutual interference of these bodies may take place. If mutual interference is common, the growing spherulites become closely packed polyhedra. Oolites become broken, and the broken parts become nuclei for new oolites; this feature is obviously impossible in a spherulitic formation. Some of the so-called "polyhedral pisolites" may be in-place spherulitic bodies (Shrock, 1930).

BIOGENIC FABRICS

Many sedimentary rocks contain *fossils* as an integral part of the rock. They may be a minor component of the rock, or, as in the case of some limestones, they may be the dominant constituent. Our task here is to inquire neither into the biological aspects of the fossil form (morphology and taxonomy) nor into its stratigraphic significance but rather to consider fossils as component parts of the rock. The petrologist should be able to recognize fossils in thin section and should, from their composition, preservation, and manner of occurrence, secure further pertinent data concerning the origin of the rock in question.

We are not concerned here with the larger structures of organic origin. Reefs, for example, are bodies, of sedimentary rock and are discussed in Chapter 5; we exclude also various tracks, trails, burrows, such modification of bedding caused by bioturbation, and even those forms of growth bedding (stromatolites) attributable to organisms. These biogenic sedimentary *structures* are described in Chapter 4. We are concerned here with fossils and fossil detritus—with recognition of the biogenic skeletal components of the framework of the coarser sediments of which they are a part. We would like to know the origin of the skeletal fragments, that is, the organisms responsible for them, whether algal, foraminiferal, coralline, or the like.

The upsurge of interest in the petrology of limestones during the past two decades has stimulated interest in these questions. Although Sorby (1879) was the first to describe the petrographic nature of fossil debris, little interest was shown in the topic until recently. Presented here is the barest outline of the subject; refer to the more modern monographic works of Majewske (1969) and of Horowitz and Potter (1971) for detailed information.

COMPOSITION AND MODE OF PRESERVATION

Fossils are any evidence of past life. They may be buried as relatively unaltered remains, such as the more resistant organic structure—bones, teeth, shells. Most shells and similar structures are originally calcium carbonate, but even these show a varied composition in terms of magnesium content and minor elements (see Table 10-5). Others are phosphatic, siliceous, or chitinoid.

Organic remains may be altered in varying degrees. Some carbonate shells are leached; others have been recrystallized with loss of internal structure; bones and similar materials may be enriched in fluorine. Organic materials such as cellulose may be much degraded and in the oldest rocks consist only of a carbonaceous film. This is true of plant tissues, some chitinous materials, and even soft-bodied animal matter may be so altered. Such fossils are little more than carbon films, left after the loss of the volatile constituents of the original organic materials which were driven off under anaerobic conditions. Carbonized plant fossils are abundant in shales associated with many coal beds. Numerous bits of carbonized wood are also found in shales and sandstones. In a few cases, as in "mineral charcoal" (fusain) and in "coal balls" (Stopes and Watson, 1909), cell walls are carbonized, but the cells themselves are filled with mineral matter, usually calcite (Fig. 11-43). Some of the most exquisitely preserved plant remains known are so preserved. Graptolite fossils of some Ordovician slates and black shales are preserved as carbon films.

Organic structures may be completely replaced so that the present composition of the fossils bears no relation to its original nature. Such replacements (petrifications) are in truth segregations of minor mineral constituents of the rock and have therefore the same geochemical significance as concretions, nodules, and the like. Silica, carbonate, and iron sulfides are among the replacing materials. Such replacements may take place with preservation of considerable detail.

In many cases neither the original organic structure nor its replacement is present. Instead are found cavities, called *molds,* formed by removal in solution of the original shell. Molds show the form and external ornamentation of the original object. If a mold is filled with foreign matter, the filling is properly termed a *cast.* It, also, will show the external features (but none of the internal structures) of the original shell. The filling of the interior body cavity of a bivalve is often called a "cast of the interior." In a strict sense it is an internal mold.

Fecal pellets or excreta (mainly of invertebrates) are present in modern marine deposits and in some sedimentary rocks where they have been lithified (Moore, 1939; Dapples, 1942). The pellets are, for the most part, the product of mud-eating forms and hence consist of silt and mud particles bound by organic matter. Many are soft and subject to disintegration; others become mineralized and an integral component of the sediment. In some cases a large part of the deposit, 30 to 50 percent, is composed of pellets. The pellets may be transformed into glauconite or become pyritized, or they may serve as centers for the accumulation of phosphate. Most abundant are those of simple ovoid form and of a small size—1 mm or less in length. These have been reported by many workers from ancient sediments, especially limestone, but they have not always been identified as being of fecal origin. Rarer are the rod-shaped bodies with longitudinal or transverse sculpturing or both. Most fecal pellets are devoid of internal structures and in many cases have been overlooked or assigned to an inorganic origin.

Coprolites are larger objects with an origin similar to that of fecal pellets. The earliest recognition of their fecal nature seems to have been that of Buckland, who described their occurrence in the Lias of England in 1835 (Folk, 1965b). The literature on coprolites has been thoroughly reviewed by Amstutz (1958).

Coprolites are generally light to dark brown or black, ovoid to elongate structures, 1 to 15 cm in length, with surfaces marked by annular convolutions. Longitudinal striations or grooves are rarely present. The brown vitrous material of many coprolites is phosphatic, optically isotropic with refractive index near that of collophane (1.58–1.62). Bradley (1946) has shown that the coprolites of the Bridger Formation (Eocene) of Wyoming are probably a carbonate-apatite (francolite). Those described by Amstutz were limonitic but with a sideritic core from which the limonite was presumably derived. Obviously the siderite replaced the original coprolitic material. As a rule, coprolites are a relatively rare constituent of sedimentary rocks; exceptionally they are more common in nonmarine Tertiary deposits containing abundant mammalian remains.

PETROLOGY OF FOSSILS

The petrographer encounters, especially in the limestones, granular components of the sediment which are invertebrate skeletal fragments. Some rocks consist of little else. It is important to be able to identify these as shell fragments and, if possible, to determine the class of organisms to which they belong. Because this problem is of the most importance in the case of limestones, the details of the petrology of fossils is included in the chapter on limestone.

References

Adams, S. F., 1920, A microscopic study of vein quartz: Econ. Geol., v. 15, pp. 623–664.

Allen, J. R. L., 1962, Petrology, origin, and deposition of the highest Lower Old Red Sandstone of Shropshire, England: Jour. Sed. Petrology, v. 32, pp. 657–697.

Allen, Terence, 1968, Particle size measurement: London, Chapman and Hall, 248 pp.

Allen, V. T., 1936, Terminology of medium-grained sediments: Rept. Comm. Sedimentation 1935–1936, Nat. Res. Coun., pp. 18–47 (mimeographed).

Amstutz, G. C., 1958, Coprolites: a review of the literature and a study of specimens from southern Washington: Jour. Sed. Petrology, v. 28, pp. 498–508.

Anderson, G. E., 1926, Experiments on the rate of wear of sand grains: Jour. Geol., v. 34, pp. 144–158.

Athy, L. F., 1930, Density, porosity, and compaction of sedimentary rocks: Bull. Amer. Assoc. Petrol. Geol., v. 14, pp. 1–24.

Atterberg, A., 1905, Die rationelle Klassifikation der Sande und Kiese: Chem. Zeitschr., v. 29, pp. 195–198.

Bagnold, R. A., 1941, The physics of blown sand and desert dunes: London, Methuen, 265 pp.

Baker, H. A., 1920, On the investigation of the mechanical constitution of loose arenaceous sediments, etc.: Geol. Mag., v. 57, pp. 321–332, 363–370, 411–420, 463–470.

Barbour, E. H., and Torrey, J., Jr., 1890, Notes on the microscopic structure of oolites, with analyses: Amer. Jour. Sci., ser. 3, v. 40, pp. 246–249.

Barrell, J., 1925, Marine and terrestrial conglomerates: Bull. Geol. Soc. Amer., v. 36, pp. 279–342.

Bastin, E. S., 1950, Interpretation of ore textures: Geol. Soc. Amer. Mem. 45, 101 pp.

Bastin, E. S., *et al.*, 1931, Criteria of age relations of minerals with special reference to polished sections of ores: Econ. Geol., v. 26, pp. 561–610.

Bates, T. F., 1958, Selected electron micrographs of clays and other fine-grained minerals: Pennsylvania State Univ., Min. Ind. Exp. Sta. Circ. 51, 61 pp.

Bathurst, R. G. C., 1971, Carbonate sediments and their diagenesis: Amsterdam: Elsevier, 620 pp.

Beal, M. A., and Shepard, F. P., 1956, A use of roundness to determine depositional environments: Jour. Sed. Petrology, v. 26, pp. 49–60.

Becker, G. F., 1893, Finite homogeneous strain, flow, and rupture of rock: Bull. Geol. Soc. Amer. v. 4, pp. 13–90.

Berg, G., 1944, Vergleichende Petrographie oolithischer Eisenerze: Archiv. Lagerstättenforschung, v. 76, 128 pp.

Berthois, L., and Portier, J., 1957, Reserches expérimentales sur le façonnement des grains des sables quartzeux: C. R. Acad. Sci., v. 245, pp. 1152–1154.

Blatt, Harvey, and Christie, J. M., 1963, Undulatory extinction in quartz of igneous and metamorphic rocks and its significance in provenance studies of sedimentary rocks: Jour. Sed. Petrology, v. 33, pp. 559–579.

Bloss, F. D., 1957, Anisotropy of fracture in quartz: Amer. Jour. Sci., v. 255, pp. 214–225.

Bokman, J., 1952, Clastic quartz particles as indices of provenance: Jour. Sed. Petrology, v. 22, pp. 17–24.
Bond, G., 1954, Surface textures of sand grains from the Victoria Falls area: Jour. Sed. Petrology, v. 24, pp. 191–195.
Bonham, L. C., 1957, Structural petrology of the Pico Anticline, Los Angeles County, California: Jour. Sed. Petrology, v. 32, pp. 251–264.
Bonham, L. C., and Spotts, J. H., 1971, Measurement of grain orientation, *in* Procedures in sedimentary petrology (Carver, R. E., ed.): New York, Wiley-Interscience, pp. 285–312.
Bonorino, F. G., and Teruggi, M. E., 1952, Lexico sedimentológico: Inst. Nac. Invest. Ciencias Nat., publ. no. 6, 164 pp.
Bradley, W. C., Fahnestock, R. K., and Rowehamp, E. T., 1972, Coarse sediment transport by flood flows on Knik River, Alaska: Bull. Geol. Soc. Amer., v. 83, pp. 1261–1284.
Bradley, W. H., 1946, Coprolites from the Bridger Formation of Wyoming: their composition and microorganisms: Amer. Jour. Sci., v. 244, pp. 215–239.
Bretz, J Harlen, 1929, Valley deposits immediately east of the channeled scablands of Washington II: Jour. Geol., v. 37, pp. 505–541.
Briggs, L. I., McCulloch, D. S., and Moser, F., 1962, The hydraulic shape of sand particles: Jour. Sed. Petrology, v. 32, pp. 645–646.
Brown, T. C., 1914, Origin of oolites and the oolitic texture in rocks: Bull. Geol. Soc. Amer. v. 25, pp. 745–780.
Bryan, K., 1931, Wind-worn stones or ventifacts—a discussion and bibliography: Rept. Comm. Sedimentation, 1929–1930, Nat. Res. Coun., reprint and circular ser., no. 98, 29–50.
Bucher, W. H., 1918, On oolites and spherulites: Jour. Geol., v. 26, pp. 593–609.
Bull, W. B., 1962, Relation of textural (CM) patterns to depositional environment of alluvial-fan deposits: Jour. Sed. Petrology, v. 32, pp. 211–217.
Buller, A. T., and McManus, J., 1972, Simple metric sedimentary statistics used to recognize different environments: Sedimentology, v. 18, pp. 1–21.
Cailleux, A., 1942, Les actions eoliennes periglacial en Europe: Bull. Soc. Geol. France, ser. 5, v. 6, pp. 495–505.
———, 1945, Distinction des galets marins et fluviatiles: Bull. Soc. Geol. France, ser. 5, v. 15, pp. 375–404.
Carozzi, A., 1957, Contribution a l'étude des propriétés géomètrique des oolithes—l'exemple du Grand Lac Sale, Utah, U.S.A.: Bull. Inst. Nat. Genevois, v. 58, pp. 3–52.
———, 1961a, Oolithes remainées, brisées et regenerées dans le Mississippien des chaines frontales, Alberta central, Canada: Arch. des Sciences, Soc. Phys. Hist. Nat. de Geneve, vol. 14, pp. 281–296.
———, 1961b, Distorted oolites and pseudoolites: Jour. Sed. Petrology, v. 31, pp. 262–274.
———, 1963, Half-moon oolites: Jour. Sed. Petrology, v. 33, pp. 633–645.
Carver, R. E., ed., 1971, Procedures in sedimentary petrology: New York, Wiley-Interscience, 653 pp.
Cayeux, L., 1935, Les roches sédimentaires de France—roches carbonatées: Paris, Masson, 447 pp.
Chenowith, P. A., 1952, Statistical methods applied to Trentonian stratigraphy in New York: Bull. Geol. Soc. Amer., v. 63, pp. 521–560.
Choquette, P. W., 1955, A petrographic study of the "State College" siliceous oolite: Jour. Geol., v. 63, pp. 337–347.
Clifton, H. E., 1965, Tectonic polish of pebbles: Jour. Sed. Petrology, v. 35, pp. 867–873.
Cloos, E., 1947, Oolite deformation in the South Mountain fold, Maryland: Bull. Geol. Soc. Amer., v. 68, pp. 843–918.
Colton, G. W., and DeWitt, W., Jr., 1959, Current oriented structures in some Upper Devonian rocks in western New York (abstract): Bull. Geol. Soc. Amer., v. 70, pp. 1759–1760.
Conkling, H., Eckis, R., and Gross, P. L. K., 1934, Groundwater storage capacity of valley fill: California Div. Water Resources, Bull. 45, 279 pp.
Crook, K. A. W., 1968, Weathering and roundness of quartz sand grains: Sedimentology, v. 11, pp. 171–182.
Crowell, J. R., Hope, R. A., Kahle, J. E., Ovenshine, A. T., and Sams, R. H., 1966, Deepwater sedimentary structures Pliocene Pico Formation, Santa Paula Creek, Ventura Basin, California: California Div. Mines Geol., Spec. Rept. 89, 40 pp.
Curray, J. R., 1956, Dimensional grain orientation studies of Recent coastal sands: Bull. Amer. Assoc. Petrol. Geol., v. 40, pp. 2440–2456.
———, 1960, Tracing sediment masses by grain size modes: Repts. 21st Int. Geol. Congr. Norden, pp. 119–130.
Curtis, B. F., 1971, Measurement of porosity and permeability, *in* Procedures in sedimentary petrology (Carver, R. E., ed.): New York, Wiley-Interscience, pp. 335–364.
Dake, C. L., 1921, The problem of the St. Peter Sandstone: Bull. Univ. Missouri Sch. Mines Metall., Tech. Ser., v. 6, 158 pp.
Dal Cin, R., 1967, Le Ghiaie del Piave: Mem. Mus. Tridentino Sci. Nat., Ann. Rept. 29–30, 1966–1967, v. 16, no. 3, pp. 1–177.

Dalla Valle, J. M., 1943, Micromeritics, the technology of fine particles, 2nd ed.: New York, Pitman, 555 pp.

Dapples, E. C., 1942, The effect of macro-organisms upon near-shore sediments: Jour. Sed. Petrology, v. 12, pp. 118–126.

Dapples, E. C., Krumbein, W. C., and Sloss, L. L., 1953, Petrographic and lithologic attributes of sandstones: Jour. Geol. v. 61, pp. 291–317.

Dapples, E. C., and Rominger, J. F., 1945, Orientation analysis of fine-grained clastic sediments—a report of progress: Jour. Geol., v. 53, pp. 246–261.

Daubrée, A., 1879, Études synthétiques de géologies expérimentale: Paris, Dunod, 828 pp.

DeFord, R. K., and Waldschmidt, W. A., 1946, Oolite and oolith: Bull. Amer. Assoc. Petrol. Geol., v. 30, pp. 1587–1588.

Deverin, L., 1945, Étude pétrographique de minerais de fer oolithique du Dogger des Alpes suisses: Beitr. Geol. Schweiz, Lieferung 13, v. 2, 115 pp.

Dickinson, W. R., 1970, Interpreting detrital modes of graywacke and arkose: Jour. Sed. Petrology, v. 40, pp. 695–707.

Dobkins, J. E., Jr., and Folk, R. L., 1970, Shape development on Tahiti-Jui: Jour. Sed. Petrology, v. 40, pp. 1167–1203.

Doeglas, D. J., 1946, Interpretation of results of mechanical analyses: Jour. Sed. Petrology, v. 16, pp. 19–40.

———, 1962, The structure of sedimentary deposits of braided rivers: Sedimentology, v. 1, pp. 167–190.

———, 1968, Grain-size indices, classification and environment: Jour. Sed. Petrology, v. 10, pp. 83–100.

Dumitriu, M., and Dumitriu, C., 1961, Analize de orientare texturala si directii de transport ale sedimentelor (French summ.): Acad. Republ. Pop. Romine, Inst. Geol. Geogr., Stud. Cercetari Geol., v. 6, pp. 709–717.

Dunham, R. J., 1962, Classification of carbonate rocks according to depositional texture, *in* Classification of carbonate rocks (Ham, W. E., ed.): Tulsa, Okla., Amer. Assoc. Petrol. Geol., Mem. 1, pp. 108–121.

Eardley, A. J., 1938, Sediments of Great Salt Lake, Utah: Bull. Amer. Assoc. Petrol. Geol., v. 22, pp. 1359–1387.

Edwards, A. B., 1947, Textures of the ore minerals and their significance: Australian Inst. Min. Metall., Melbourne, 185 pp.

Einstein, H. A., Anderson, A. G., and Johnson, J. W., 1940, A distinction between bed load and suspended load in natural streams: Trans. Amer. Geophys. Union, 21st Ann. Mtg., pt. II, pp. 628–633.

Emery, J. R., and Griffiths, J. C., 1954, Reconnaissance investigation into relationships between behavior and petrographic properties of some Mississippian sediments: Pennsylvania State Univ., Min. Ind. Expt. Sta. Bull. 62, pp. 67–80.

von Engelhardt, W., 1960, Der Porenraum der Sedimente: New York, Springer, 207 pp.

von Engelhardt, W., and Pitter, H., 1951, Über die Zusammenhänge zwischen Porosität, Permeabilität, und Korngrösse bei Sand und Sandsteinen: Heidelberger Beitr. Min. Pet., v. 2, pp. 477–491.

Fernald, F. A., 1929, Roundstone, a new geologic term: Science, v. 70, p. 240.

Fitzpatrick, K. T., and Summerson, C. H., 1971, Some observations on electron micrographs of quartz sand grains: Ohio Jour. Sci., v. 71, pp. 106–119.

Flemming, N. C., 1965, Form and function of sedimentary particles: Jour. Sed. Petrology, v. 35, pp. 381–391.

Flint, R. F., Sanders, J., and Rodgers, J., 1960a, Symmictite, a name for nonsorted terrigenous sedimentary rocks that contain a wide range of particle sizes: Bull. Geol. Soc. Amer., v. 71, pp. 507–510.

———, 1960b, Diamictite, a substitute term for symmictite: Bull. Geol. Soc. Amer., v. 71, p. 1809.

Flügel, E., and Kirchmayer, M., 1962, Zur Terminologie der Ooide, Onkoide und Pseudooide: Neues Jahrb. Geol. Paläont. Mh., v. 3, pp. 113–123.

Folk, R. L., 1954, The distinction between grain size and mineral composition in sedimentary rock nomenclature: Jour. Geol., v. 62, pp. 344–359.

———, 1955, Student operator error in determination of roundness, sphericity, and grain size: Jour. Sed. Petrology, v. 25, pp. 297–301.

———, 1960, Petrography and origin of the Tuscarora, Rose Hill, and Keefer formations, Lower and Middle Silurian of eastern West Virginia: Jour. Sed. Petrology, v. 30, pp. 1–58.

———, 1965a, Some aspects of recrystallization in ancient limestones, *in* Dolomitization and limestone diagenesis: a symposium (Pray, L. C., and Murray, R. C., eds.): Soc. Econ. Paleont. Min. Spec. Publ. 13, pp. 14–48

———, 1965b, On the earliest recognition of coprolites: Jour. Sed. Petrology, v. 35, pp. 272–273.

———, 1966, A review of grain size parameters: Sedimentology, v. 6, pp. 73–93.

Fraser, H. J., 1935, Experimental study of the porosity and permeability of clastic sediments: Jour. Geol., v. 43, pp. 910–1010.

Freeman, T., 1962, Quiet water oolites from Laguna Madre, Texas: Jour. Sed. Petrology, v. 32, pp. 475–483.

Friedman, G. M., 1961, Distribution between dune, beach, and river sands from their textural characteristics: Jour. Sed. Petrology, v. 31, pp. 514–529.

———, 1962, On sorting, sorting coefficients, and the log-normality of the grain-size distribution of clastic sandstones: Jour. Geol., v. 70, pp. 737–753.

———, 1965, Terminology of *crystallization textures* and *fabrics* in sedimentary rocks: Jour. Sed. Petrology, v. 35, pp. 643–655.

———, 1967, Dynamic processes and statistical parameters compared for size frequency distribution of beach and river sands: Jour. Sed. Petrology, v. 37, pp. 327–354.

Füchtbauer, H., 1967, Influence of different types of diagenesis on sandstone porosity: Proc. 7th, World Petrol. Congr., Mexico, v. 2, pp. 353–369.

Füchtbauer, H., and Reineck, H.-E., 1963, Porosität und Verdichtung rezenter, mariner Sedimente: Sedimentology, v. 2, pp. 294–306.

Gaither, A., 1953, A study of porosity and grain relationships in experimental sands: Jour. Sed. Petrology, v. 23, pp. 180–195.

Gees, R. A., 1965, Moment measures in relation to the depositional environments of sands: Eclogae Géol. Helvetiae, v. 56, pp. 209–213.

Grabau, A. W., 1904, On the classification of sedimentary rocks: Amer. Geol., v. 33, pp. 228–247.

———, 1960, Principles of stratigraphy, reprint ed.: New York, Dover, 1185 pp.

Graton, L. C., and Fraser, H. J., 1935, Systematic packing of spheres—with particular relation to porosity and permeability: Jour. Geol., v. 43, pp. 785–909.

Gregory, H. E., 1915, The formation and distribution of fluviatile and marine gravels: Amer. Jour. Sci., ser. 4, v. 39, pp. 487–508.

Griffiths, J. C., 1949, Directional permeability and dimensional orientation in Bradford sand: Bull. Pennsylvania State Coll., Min. Ind. Expt. Sta., v. 54, pp. 138–163.

Griffiths, J. C., and Rosenfeld, M. A., 1950, Progress in measuring grain orientation in Bradford sand: Bull. Pennsylvania State Coll., Min. Ind. Expt. Sta., v. 56, pp. 202–236.

———, 1953, A further test of dimensional orientation of quartz grains in Bradford Sand: Amer. Jour. Sci., v. 251, pp. 192–214.

Grogan, R., 1945, Shape variation of some Lake Superior beach pebbles: Jour. Sed. Petrology, v. 15, pp. 3–10.

Grout, F. F., 1932, Petrography and petrology: New York, McGraw-Hill, 522 pp.

———, 1946, Microscopic characters of vein carbonates: Econ. Geol., v. 41, pp. 475–502.

Hälbach, I. W., 1962, On the morphology of the Dwyka Series in the vicinity of Loeriesfontein, Cape Province: Ann. Univ. Stell., v. 37, ser. A, no. 2, 162 pp.

Hares, C. J., 1917 Gastroliths in the Cloverly Formation: Jour. Washington Acad. Sci., v. 7, p. 429.

Harris, S. A., 1958a, Differentiation of various Egyptian aeolian microenvironments by mechanical composition: Jour. Sed. Petrology, v. 28, pp. 164–174.

———, 1958b, Probability curves and the recognition of adjustment to depositional environment: Jour. Sed. Petrology, v. 28, pp. 151–163.

Harrison, P. W., 1957, A clay-till fabric—its character and origin: Jour. Geol., v. 65, pp. 275–308.

Hatch, F. H., Rastall, R. H., and Black, M., 1938, The petrology of the sedimentary rocks, 3rd ed.: London, Murby, 383 pp.

Helmbold, R., 1952, Beitrag zur Petrographie der Tanner Grauwacken: Heidelberger Beitr. Min. Pet., v. 3, pp. 253–288.

Henbest, L. G., 1968, Diagenesis in oolitic limestones of Morrow (Early Pennsylvanian) age in northwestern Arkansas and adjacent Oklahoma: U.S. Geol. Surv. Prof. Paper 594-H, 22 pp.

Hirschwald, J., 1912, Handbuch der bautechnischen Gesteinsprüfung: Berlin, Gebrüder Borntraeger, pp. 511–512.

Hohlt, R. B., 1948, The nature and origin of limestone porosity: Colorado School Mines Quart., v. 43, no. 4, 51 pp.

Holmes, C. D., 1941, Till fabric: Bull. Geol. Soc. Amer., v. 52, pp. 1299–1354.

Horowitz, A. S., and Potter, P. E., 1971, Introductory petrography of fossils: New York, Springer, 302 pp.

Hough, J. L., 1942, Sediments of Cape Cod Bay, Massachusetts: Jour. Sed. Petrology, v. 12, pp. 10–30.

Howell, J. V., Notes on the pre-Permian Paleozoics of the Wichita Mountains area: Bull. Amer. Assoc. Petrol. Geol., v. 6, pp. 413–425.

Humbert, F. L., 1968, Selection and wear of pebbles on gravel beaches: Ph.D. dissertation, Groningen, 144 pp.

Hunt, C. B., 1954, Desert varnish: Science, v. 120, pp. 183–184.

Illing, L. V., 1954, Bahaman calcareous sands: Bull. Amer. Assoc. Petrol. Geol., v. 38, pp. 1–95.

Ingerson, E., 1940, Fabric criteria for distinguishing pseudo-ripple marks from ripple marks: Bull. Geol. Soc. Amer., v. 51, pp. 557–570.

Ingerson, E., and Ramisch, J. L., 1942, Origin of

shapes of quartz sand grains: Amer. Mineral., v. 27, pp. 595–606.

Inman, D. L., 1949, Sorting of sediments in the light of fluid mechanics: Jour. Sed. Petrology, v. 19, pp. 51–70.

———, 1952, Measures for describing the size distribution of sediments: Jour. Sed. Petrology, v. 22, pp. 125–145.

Irani, R. R., and Callis, C. F., 1963, Particle size: measurement, interpretation, and application: New York, Wiley, 165 pp.

Jamieson, T. F., 1860, On drift and rolled gravel of the north of Scotland: Quart. Jour. Geol. Soc. London, v. 16, pp. 347–371.

Jizba, Z. V., 1971, Mathematical analysis of grain orientation, *in* Procedures in sedimentary petrology (Carver, R. E., ed.): New York, Wiley-Interscience, pp. 313–334.

Johansson, C. E., 1965a, Structural studies of sedimentary deposits (orientation analysis, literature digest, and field investigations): Lund Stud. Geogr., ser. A, Phys. Geogr., no. 32, 61 pp.

———, 1965b, Orientation of pebbles in running water: Geol. Fören. Stockholm Förh., v. 87, pp. 3–61.

Judson, S., and Barks, R. E., 1961, Microstriations on polished pebbles: Amer. Jour. Sci., v. 259, pp. 371–381.

Kahn, J. S., 1956a, The analysis and distribution of the properties of packing in sand-size sediments. 1. On the measurement of packing in sandstones: Jour. Geol., v. 64, pp. 385–395.

———, 1956b, Analysis and distribution of packing properties in sand-size sediments. 2. The distribution of the packing measurements and an example of packing analysis: Jour. Geol., v. 64, pp. 578–606.

———, 1959, Anisotropic sedimentary parameters: Trans. New York Acad. Sci., ser. 2, v. 21, pp. 373–386.

Kalkowsky, E., 1908, Oolith und Stromatolith im norddeutschen Bundsandstein: Zeitschr. Deutsche. Geol. Ges., v. 60, pp. 68–125.

Karlstrom, T. N. V., 1952, Improved equipment and techniques for orientation studies of large particles in sediments: Jour. Geol., v. 60, pp. 489–493.

Kauranne, L. K., 1960, A statistical study of stone orientation in glacial till: Bull. Comm. Geol. Finlande, no. 188, pp. 87–97.

Keller, W. D., 1945, Size distribution of sand in some dunes, beaches, and sandstones: Bull. Amer. Assoc. Petrol. Geol., v. 29, pp. 215–221.

———, 1946, Evidence of texture on the origin of the Cheltenham fireclay of Missouri and associated shales: Jour. Sed. Petrology, v. 16, pp. 63–71.

Kelling, G., and Williams, P. F., 1967, Flume studies of the reorientation of pebbles and shells: Jour. Geol., v. 75, pp. 243–267.

Kindle, E. M., 1936, Dominant factors in the formation of firm and soft sand beaches: Jour. Sed. Petrology, v. 6, pp. 16–22.

Kittleman, L. R., Jr., 1964, Application of Rosin's distribution to size-frequency analysis of clastic rocks: Jour. Sed. Petrology, v. 34, pp. 483–502.

Klein, G. deV., 1963, Boulder surface markings on Quaco Formation (Upper Triassic), St. Martin's, New Brunswick, Canada: Jour. Sed. Petrology, v. 33, pp. 49–52.

Klovan, J. E., 1966, The use of factor analysis in determining depositional environments from grain-size distributions: Jour. Sed. Petrology, v. 36, pp. 115–125.

Knopf, E. B., and Ingerson, E., 1938, Structural petrology: Geol. Soc. Amer. Mem. 6, 270 pp.

Kolduk, W. S., 1968, On environment-sensitive grain-size parameters: Sedimentology, v. 10, pp. 57–69.

Kolmogorov, A. N., 1941, Über das logarithmische Verteilungsgesetz der Teichen bei Zerstückelung: Dokl. Akad. Nauk. SSSR, v. 31, pp. 99–101.

Konzewitsch, N., 1961, La Forma de los clastos: Serv. Hidrografia Naval Publ., v. 626, 113 pp. (with English summary).

Köster, E., 1964, Granulometrische und morphometrische Messmethoden an Mineralkörnen, Steinen, und sonstigen Stoffen: Stuttgart, Enke, 336 pp.

Krinsley, D. H., 1973, Atlas of sand surface textures: Cambridge Earth Sci. Series, 91 pp.

Krinsley, D., and Donahue, J., 1968, Environmental interpretation of sand grain surface texture by electron microscopy: Bull. Geol. Soc. Amer., v. 79, pp. 743–748.

Krinsley, D., and Funnell, B., 1965, Environmental history of sand grains from the Lower and Middle Pleistocene of Norfolk, England: Quart. Jour. Geol. Soc. London, v. 121, pp. 435–461.

Krinsley, D., and Margolis, S., 1969, A study of quartz sand grain surface textures with the scanning electron microscope: Trans. New York Acad. Sci., ser. II, v. 31, pp. 457–477.

Krinsley, D., and Takahashi, T., 1962a, The surface textures of sand grains, an application of electron microscopy: Science, v. 138, pp. 923–925.

———, 1962b, The surface textures of sand grains, an application of electron microscopy. Glaciation: Science, v. 138, pp. 1262–1264.

———, 1962c, Applications of electron microscopy to geology: Trans. New York Acad. Sci., ser. II, v. 25, pp. 3–22.

Krumbein, W. C., 1934, Size frequency distributions of sediments: Jour. Sed. Petrology, v. 4, pp. 65–77.
———, 1938, Size frequency distributions of sediments and the normal phi curve: Jour. Sed. Petrology, v. 8, pp. 84–90.
———, 1939, Preferred orientation of pebbles in sedimentary deposits: Jour. Geol., v. 47, pp. 673–706.
———, 1940, Flood gravel of San Gabriel Canyon, California: Bull. Geol. Soc. Amer., v. 51, pp. 636–676.
———, 1941a, Measurement and geologic significance of shape and roundness of sedimentary particles: Jour. Sed. Petrology, v. 11, pp. 64–72.
———, 1941b, The effects of abrasion on the size, shape, and roundness of rock fragments: Jour. Geol., v. 49, pp. 482–520.
———, 1942a, Flood deposits of Arroyo Seco, Los Angeles County, California: Bull. Geol. Soc. Amer., v. 53, pp. 1355–1402.
———, 1942b, Settling velocity and flume-behaviour of non-spherical particles: Trans. Amer. Geophys. Union, 1942, pp. 621–633.
Krumbein, W. C., and Monk, G. D., 1942, Permeability as a function of the size parameters of unconsolidated sand: Amer. Inst. Min. Metall. Eng., Tech. Publ. 1492, 11 pp.
Krumbein, W. C., and Pettijohn, F. J., 1938, Manual of sedimentary petrography: New York, Plenum, 549 pp.
Krumbein, W. C., and Tisdel, F. W., 1940, Size distributions of source rocks of sediments: Amer. Jour. Sci., v. 238, pp. 296–305.
Krynine, P. D., 1941, Petrographic studies of variations in cementing material in the Oriskany sand: Bull. Pennsylvania State Coll., 33, pp. 108–116.
———, 1942, Critical velocity as a controlling factor in sedimentation (abstract): Bull. Geol. Soc. Amer., v. 52, p. 1805.
———, 1946, Microscopic morphology of quartz types: Proc. 2nd Pan Am. Congr. Min. Eng. Geol., v. 3, 2nd Comm., pp. 35–49.
———, 1948, The megascopic study and field classification of sedimentary rocks: Jour. Geol., v. 56, pp. 130–165.
Kuenen, Ph. H., 1942, Pitted pebbles: Leidse Geol. Meded., v. 13, pp. 189–201.
———, 1955, Experimental abrasion of pebbles. 1. Wet sandblasting: Leidse Geol. Meded., v. 20, pp. 131–137.
———, 1956, Experimental abrasion of pebbles. 2. Rolling by current: Jour. Geol., v. 64, pp. 336–368.
———, 1958, Some experiments on fluviatile rounding: Proc. Koninkl. Nederl. Akad. Wetensch., ser. B., v. 61, pp. 47–53.
———, 1959, Experimental abrasion. 3. Fluviatile action on sand: Amer. Jour. Sci., v. 257, p. 192–190.
———, 1960a, Experimental abrasion of sand grains: 21st Sess. Int. Geol. Congr., Norden, pt. 10, pp. 50–53.
———, 1960b, Experimental abrasion. 4. Eolian action: Jour. Geol., v. 68, pp. 427–449.
———, 1964, Experimental abrasion. 6. Surf action: Sedimentology, v. 3, pp. 29–43.
———, 1969, Origin of quartz silt: Jour. Sed. Petrology, v. 39, pp. 1631–1633.
Kuenen, Ph. H., and Perdok, W. G., 1961, Frosting on quartz grains: Proc. Koninkl. Nederl. Akad. Wetensch., ser. B., pp. 343–345.
———, 1962, Experimental abrasion. 5. Frosting and defrosting of quartz grains: Jour. Geol., v. 70, pp. 648–658.
Landon, R. E., 1930, An analysis of beach pebble abrasion and transportation: Jour. Geol., v. 38, pp. 437–446.
Lane, E. W., et al., 1947, Report of the subcommittee on sediment terminology: Trans. Amer. Geophys. Union, v. 28, pp. 936–938.
Laudermilk, J. D., 1931, On the origin of desert varnish: Amer. Jour. Sci., ser. 5, v. 21, pp. 51–66.
Lenk-Chevitch, P., 1959, Beach and stream pebbles: Jour. Geol., v. 67, pp. 103–108.
Linck, G., 1903, Die Bildung der Oolithe und Rogensteine: Neues Jahrb. Min., B.B.: v. 16, pp. 495–513.
Lindsey, D. A., 1966, Sediment transport in a Precambrian ice age—the Huronian Gowganda Formation: Science, v. 154, pp. 1442–1443.
McBride, E. F., 1971, Mathematical treatment of size distribution data, *in* Procedures in sedimentary petrology (Carver, R. E., ed.): New York, Wiley-Interscience, pp. 109–127.
McIver, N. L., 1970, Appalachian turbidites, *in* Studies of Appalachian geology: central and southern (Fisher, G. W., Pettijohn, F. J., Reed, J. C., and Weaver, K. N., eds.): New York, Wiley-Interscience, pp. 69–81.
McKee, E. D., and Gutschick, R. C., 1969, History of Redwall Limestone of northern Arizona: Geol. Soc. Amer. Mem. 114, pp. 1–726.
Majewske, O. P., 1969, Recognition of invertebrate fossil fragments in rocks and thin section: (Int. Sed. Petrog. Ser., v. 13) Leiden, Brill, 102 pp.
Margolis, S. V., 1968, Electron microscopy of chemical solution and mechanical abrasion features on quartz sand grains: Sed. Geol., v. 2, pp. 243–256.

Marshall, C. E., 1941, Studies in the degree of dispersion of clays. IV. The shapes of clay particles: Jour. Phys. Chem., v. 41, pp. 81–93.

Marshall, P. E., 1927, The wearing of beach gravels: Proc. New Zealand Inst., v. 58, pp. 507–532.

Martinez, J. D., 1958, Photometer method for studying quartz grain orientation: Bull. Amer. Assoc. Petrol. Geol., v. 42, pp. 588–608.

———, 1963, Discussion: rapid methods for dimensional grain orientation measurements (W. Zimmerle and L. C. Bonham): Jour. Sed. Petrology, v. 33, pp. 483–484.

Mason, C. C., and Folk, R. L., 1958, Differentiation of beach, dune, and aeolian flat environments by size analyses, Mustang Island, Texas: Jour. Sed. Petrology, v. 28, pp. 211–226.

Mast, R. F., and Potter, P. E., 1963, Sedimentary structures, sand shape fabrics, and permeability, pt. 2: Jour. Geol., v. 71, p. 548–565.

Mazzoni, M. M., and Spalletti, L. A., 1972, Sedimentológia de las arenas del Rio Grande de Jujuy: Rev. Mus. La Plata (ms.), Sec. Geol., v. 8, pp. 35–117.

Meade, R. H., 1964, Removal of water and rearrangement of particles during compaction of clayey sediments—a review: U.S. Geol. Surv. Prof. Paper 497-B, 23 pp.

Mellon, G. B., 1964, Discriminatory analysis of calcite- and silicate-cemented phases of the Mountain Park Sandstone: Jour. Geol., v. 72, pp. 786–809.

Middleton, G. V., 1967, The orientation of concavo-convex particles deposited from experimental turbidity currents: Jour. Sed. Petrology, v. 37, pp. 229–232.

Moiola, R. J., and Weiser, D., 1968, Textural parameters: an evaluation: Jour. Sed. Petrology, v. 38, pp. 45–53.

———, 1969, Environmental analysis of ancient sandstone bodies by discriminant analysis (abstr.): Bull. Amer. Assoc. Petrol. Geol., v. 53, p. 733.

Monaghan, P. H., and Lytle, M. A., 1956, The origin of calcareous ooliths: Jour. Sed. Petrology, v. 26, pp. 111–118.

Moore, B. N., 1934, Deposits of possible *nuée ardente* origin in the Crater Lake Region, Oregon: Jour. Geol., v. 42, pp. 358–375.

Moore, H. B., 1939, Faecal pellets in relation to marine deposits, *in* Recent marine sediments (Trask, P., ed.): Tulsa, Okla., Amer. Assoc. Petrol. Geol., pp. 516–524.

Moors, H. T., 1969, The position of graptolites in turbidites: Sed. Geol., v. 3, no. 4, pp. 241–261.

———, 1970, Current orientation of graptolites: its significance and interpretation: Sed. Geol., v. 4, no. 2, pp. 117–134.

Moss, A. J., 1962, The physical nature of common sandy and pebbly deposits, pt. 1: Amer. Jour. Sci., v. 260, pp. 337–373.

———, 1963, The physical nature of common sandy and pebbly deposits, pt. 2: Amer. Jour. Sci., v. 261, pp. 297–343.

———, 1966, Origin, shaping, and significance of quartz sand grains: Jour. Geol. Soc. Australia, v. 13, pp. 97–136.

———, 1972, Bed-load sediments: Sedimentology, v. 18, pp. 157–219.

Muir, R. O., and Walton, E. K., 1957, The East Kirton limestone: Trans. Geol. Soc. Glasgow, v. 22, pp. 157–168.

Müller, G., 1967, Methods in sedimentary petrology: Stuttgart, E. Schweizerbart'sche Verlagsbuchhandlung, 283 pp.

Nanz, R. H., 1955, Grain orientation in beach sands—a possible means for predicting reservoir trend (abstr.): Jour. Sed. Petrology, v. 25, p. 130.

Newell, N. D., Purdy, E. G., and Imbrie, J., 1960, Bahamian oolitic sand: Jour. Geol., v. 68, pp. 481–497.

Newell, N. D., and Rigby, J. K., 1957, Geological studies on the Great Bahama Bank, *in* Regional aspects of carbonate deposition: Soc. Econ. Paleont. Min. Spec. Pbl. 5, pp. 15–72.

Niggli, P., 1934, Die Charakterisierung der klastischen Sedimente nach Kornsusammensetzung: Schweiz, Min. Pet. Mitt., v. 15, pp. 31–38.

Oertal, G., and Curtis, C. D., 1972, Clay-ironstone concretions preserving fabrics due to progressive compaction: Bull. Geol. Soc. Amer., v. 83, pp. 2597–2606.

Onions, D., and Middleton, G. V., 1968, Dimensional grain orientation of Ordovician turbidite graywackes: Jour. Sed. Petrology, v. 38, pp. 164–174.

Otto, G. H., 1939, A modified logarithmic probability graph for interpretation of mechanical analyses of sediments: Jour. Sed. Petrology, v. 9, pp. 62–76.

Parkash, B., and Middleton, G. V., 1970, Downcurrent textural changes in Ordovician turbidite graywackes: Sedimentology, v. 14, pp. 259–293.

Passega, R., 1957, Texture as a characteristic of clastic deposition: Bull. Amer. Assoc. Petrol. Geol., v. 41, pp. 1952–1984.

———, 1964, Grain size representation by CM patterns as a geological tool: Jour. Sed. Petrology, v. 34, pp. 830–847.

Pelletier, B. R., 1958, Pocono paleocurrents in

Pennsylvania and Maryland: Bull. Geol. Soc. Amer., v. 69, pp. 1033–1064.

Pettijohn, F. J., 1930, Imbricate arrangement of pebbles in a pre-Cambrian conglomerate: Jour. Geol., v. 38, pp. 568–573.

———, 1940, Relative abundance of size grades of clastic sediments (abstr.): Program Soc. Econ. Paleont. Min. 1940 meeting.

Pettijohn, F. J., and Lundahl, A. C., 1943, Shape and roundness of Lake Erie beach sands: Jour. Sed. Petrology, v. 13, pp. 69–78.

Pettijohn, F. J., Potter, P. E., and Siever, R., 1972, Sand and sandstone: New York, Springer, 618 pp.

Phemister, J., 1956, Petrography, *in* The limestones of Scotland: Mem. Geol. Surv., Spec. Repts. Min. Res. Great Britain, v. 37, pp. 66–74.

Plumley, W. J., 1948, Black Hills terrace gravels: a study in sediment transport: Jour. Geol., v. 56, pp. 526–577.

Porter, J. J., 1962, Electron microscopy of sand surface texture: Jour. Sed. Petrology, v. 32, pp. 124–135.

Potter, P. E., 1955, Petrology and origin of the Lafayette gravel: Jour. Geol., v. 63, pp. 1–38.

Potter, P. E., and Pettijohn, F. J., 1963, Paleocurrents and basin analysis: New York, Springer, 295 pp.

Powers, M. C., 1953, A new roundness scale for sedimentary particles: Jour. Sed. Petrology, v. 23, pp. 117–119.

Pryor, W. A., 1971, Grain shape, *in* Procedures in sedimentary petrology (Carver, R. E., ed.): New York, Wiley-Interscience, pp. 131–150.

Raleigh, L., 1943, The ultimate shape of pebbles, natural and artificial: Proc. Roy. Soc. London, v. 181, pp. 107–118.

———, 1944, Pebbles, natural and artificial. Their shape under various conditions of abrasion: Proc. Roy. Soc. London, v. 182, pp. 321–335.

Rastall, R. H., and Hemingway, J. E., 1940, The Yorkshire Dogger, I. The coastal region: Geol. Mag., v. 67, pp. 177–197.

———, 1941, The Yorkshire Dogger, II. Lower Eskdale: Geol. Mag., v. 78, pp. 351–370.

Ricci Lucchi, F., and Casa, G. D., 1970, Surface textures of desert quartz grains. A new attempt to explain the origin of desert frosting: Ann. Museo. Geol. Bologna, ser. 2a, v. 36, pp. 751–766.

Richter, K., 1932, Die Bewegungsrichtung des Inlandeises, rekonstruiert aus dem Kritzen und Längsachsen der Geschiebe: Zeitsch. Geschieberforsch., v. 8, pp. 62–66.

———, 1936, Ergebnisse und Aussichten der Gefügeforschung im pommereschen Diluvium: Geol. Rundschau, v. 27, pp. 196–206.

Rittenhouse, G. R., 1943, Sedimentation near junction of Maquoketa and Mississippi Rivers —a discussion: Jour. Sed. Petrology, v. 13, pp. 40–42.

———, 1949, Petrology and paleogeography of Greenbriar Formation: Bull. Amer. Assoc. Petrol. Geol., v. 33, pp. 1704–1730.

———, 1971, Mechanical compaction of sands containing different percentages of ductile grains: a theoretical approach: Bull. Amer. Assoc. Petrol. Geol., v. 55, pp. 92–96.

Robinson, G. W., 1949, Soils, their origin constitution and classification, 3rd ed.: London, Murby, 573 pp.

Rogers, J. J. W., and Head, W. B., 1961, Relationships between porosity, median size, and sorting coefficients of synthetic sands: Jour. Sed. Petrology, v. 31, pp. 467–470.

Rogers, J. J. W., Krueger, W. C., and Krog, M., 1963, Sizes of naturally abraded materials: Jour. Sed. Petrology, v. 33, pp. 628–632.

Roller, P. S., 1937, Law of size distribution and statistical description of particulate materials: Jour. Franklin Inst., v. 223, pp. 609–633.

———, 1941, Statistical analysis of size distribution of particulate materials with special reference to bimodal frequency distributions: Jour. Phys. Chem., v. 45, pp. 241–281.

Rosin, P. O., and Rammler, E., 1934, Die Kornzusammensetzung des Mahlgutes im Lichte der Wahrscheinlichkeitslehre: Kolloid Zeitschr., v. 67, pp. 16–26.

Roth, R., 1932, Evidence indicating the limits of Triassic in Kansas, Oklahoma, and Texas: Jour. Geol., v. 40, pp. 718–719.

Rothpletz, A., 1892, On the formation of oolite: Bot. Centralbl., no. 35 (Crasin, F. F., trans.), in Amer. Geol., v. 10, pp. 279–282.

Rowland, R. A., 1946, Grain-shape fabrics of clastic quartz: Bull. Geol. Soc. Amer., v. 57, pp. 547–564.

Ruedemann, R., 1897, Evidence of current action in the Ordovician of New York: Amer. Geol., v. 19, pp. 367–391.

Russell, R. D., and Taylor, R. E., 1937a, Bibliography on roundness and shape of sedimentary rock particles: Rept. Comm. Sedimentation 1936–1937, Nat. Res. Coun., pp. 65–80.

———, 1937b, Roundness and shape of Mississippi River sands: Jour. Geol., v. 45, pp. 225–267.

Russell, R. J., 1968, Where most grains of very coarse sand and gravel are deposited: Sedimentology, v. 11, pp. 31–38.

Rust, B. R., 1972, Pebble orientation in fluvial sediments: Jour. Sed. Petrology, v. 42, pp. 384–388.

Sahu, B. K., 1964a, Depositional mechanisms from the size analysis of clastic sediments: Jour. Sed. Petrology, v. 34, pp. 73–83.

———, 1964b, Significance of the size-distribution statistics in the interpretation of depositional environments: Res. Bull. Panjab Univ. (n.s.), v. 15, pts. 3–4, pp. 213–219.

Sander, Bruno, 1936, Beiträge zur Kenntniss der Anlagersgefüge (Rhythmische Kalke und Dolomite aus der Trias): Min. Pet. Mitt., v. 48, pp. 27–139; Contributions to the study of depositional fabrics (Rhythmically deposited Triassic limestones and dolomites), Tulsa, Okla., Amer. Assoc. Petrol. Geol., 207 pp. (Knopf, E. B., trans., 1951).

Sarkisian, S. G., and Klimova, L. T., 1955, Orientation of pebbles and methods of studying them for paleogeographic construction: Akad. Nauk. SSSR, Isvestia, 164 pp.

Scheidegger, A. E., 1957, The physics of flow through porous media: New York, Macmillan, Inc., 236 pp.

Schermerhorn, L. J. G., 1966, Terminology of mixed coarse-fine sediments: Jour. Sed. Petrology, v. 36, pp. 831–835.

Schlee, J., 1957, Upland gravels of southern Maryland: Bull. Geol. Soc. Amer., v. 68, pp. 1371–1410.

Schlee, J., Uchupi, E., and Trumbull, J. V. A., 1965, Statistical parameters of Cape Cod beach and eolian sands: U.S. Geol. Surv. Prof. Paper 501-D, pp. 118–122.

Schoklitsch, A., 1933, Ueber die Verkleinung der Geschiebe in Flussläufen: Sitzber. Akad. Wiss. Wien. Math.-Natur. Kl., sec. II2, v. 142, no. 8, pp. 343–366.

Schwarzacher, W., 1951, Grain orientation in sands and sandstones: Jour. Sed. Petrology, v. 21, pp. 162–172.

Scott, H. W., 1947, Solution sculpturing in limestone pebbles: Bull. Geol. Soc. Amer., v. 58, pp. 141–152.

Scott, K. M., and Gravlee, G. C., Jr., 1968, Flood surge on the Rubicon River, California—hydrology, hydraulics and boulder transport: U.S. Geol. Surv. Prof. Paper, 422-M, 40 pp.

Sedimentary Petrology Seminar, 1965, Gravel fabric in Wolf Run: Sedimentology, v. 4, pp. 273–283.

Seifert, G. V., 1954, Das mikroskopische Korngefüge des Eisabbaues in Fehmarn, Ost-Wagrien und dem dänischen Wohld: Meyniana, v. 2, pp. 126–189.

Seilacher, A., 1960, Strömungsanzeichen im Hunsrückschiefer: Notizbl. Hessischen Landesanst. Bodenforsch., Wiesbaden, v. 88, pp. 88–106.

Sestini, G., and Pranzini, G., 1965, Correlation of sedimentary fabric and sole marks as current indicators in turbidites: Jour. Sed. Petrology, v. 35, pp. 100–108.

Shepard, F. P., 1954, Nomenclature based on sand–silt–clay ratios: Jour. Sed. Petrology, v. 24, pp. 151–158.

Shrock, R. R., 1930, Polyhedral pisolites: Amer. Jour. Sci., ser. 5, v. 19, pp. 368–372.

———, 1948a, Sequence in layered rocks: New York, McGraw-Hill, 507 pp.

———, 1948b, A classification of sedimentary rocks: Jour. Geol., v. 56, pp. 118–129.

Sindowski, K. -H., 1957, Die synoptische Methode des Kornkurben-Vergleiches zur Ausdeutung fossiler Sedimentationsräume: Geol. Jahrb., v. 73, pp. 235–275.

Slichter, C. S., 1899, Theoretical investigation of the motion of ground water: U.S. Geol. Surv., 19th Ann. Rept., pt. II, p. 305.

Smalley, I. J., 1964a, Representation of packing in a clastic sediment: Amer. Jour. Sci., v. 262, pp. 242–248.

———, 1964b, A method for describing the packing texture of clastic sediments: Nature, v. 203, pp. 281–284.

———, 1966, Origin of quartz sand: Nature, v. 211, pp. 476–479.

Smalley, I. J., and Vita-Finzi, C., 1968, The formation of fine particles in sandy deserts and the nature of "desert" loess: Jour. Sed. Petrology, v. 38, pp. 766–774.

Sneed, E. D., and Folk, R. L., 1958, Pebbles in the lower Colorado River, Texas—a study in particle morphogenesis: Jour. Geol., v. 66, pp. 114–150.

Solohub, J. T., and Klovan, J. E., 1970, Evaluation of grain-size parameters in lacustrine environments: Jour. Sed. Petrology, v. 40, pp. 81–101.

Sorby, H. C., 1863, Ueber Kalkstein-Geschiebe mit Eindrücke: Neues Jahrb. Min., pp. 801–807.

———, 1879, The structure and origin of limestones: Proc. Geol. Soc. London, v. 35, pp. 56–95.

Spencer, D. W., 1963, The interpretation of grain size distribution curves of clastic sediments: Jour. Sed. Petrology, v. 33, pp. 180–190.

Spry, A., 1969, Metamorphic textures: London, Pergamon, 350 pp.

Stauffer, C. R., 1945, Gastroliths from Minnesota: Amer. Jour. Sci., v. 243, pp. 336–340.

Stieglitz, R. D., 1969, Surface textures of quartz and heavy mineral grains from fresh-water environments: an application of scanning electron microscopy: Bull. Geol. Soc. Amer., v. 80, pp. 2091–2094.

Stopes, M. C., and Watson, D. M. S., 1909, On the present distribution and origin of the calcareous concretions in coal-seams, known as coal balls: Trans. Roy. Phil. Soc., London, ser. B, v. 200, pp. 167–218.

Swensen, F. A., 1942, Sedimentation near junction of Maquoketa and Mississippi Rivers: Jour. Sed. Petrology, v. 12, pp. 3–9.

Swett, Keene, Klein, G. deV., and Smith, D. M., 1971, A Cambrian tidal sand body—the Eriboll sandstone of northwest Scotland: an ancient-recent analog: Jour. Geol., v. 79, pp. 400–415.

Tanner, W. F., 1958, The zig-zag nature of Type I and Type IV curves: Jour. Sed. Petrology, v. 28, pp. 372–375.

———, 1959, Sample components obtained by the method of differences: Jour. Sed. Petrology, v. 29, pp. 408–411.

Taylor, J. H., 1949, Petrology of the Northampton Sand ironstone formation: Mem. Geol. Surv. Great Britain, 111 pp.

Taylor, J. M., 1950, Pore-space reduction in sandstones: Bull. Amer. Assoc. Petrol. Geol., v. 34, pp. 701–716.

Thiel, G. A., 1940, The relative resistance to abrasion of mineral grains of sand size: Jour. Sed. Petrology, v. 10, pp. 102–124.

Todd, D. K., 1960, Ground water hydrology: New York, Wiley, 336 pp.

Trefethen, J. M., 1950, Classification of sediments: Amer. Jour. Sci., v. 248, pp. 55–62.

Trowbridge, A. C., and Shepard, F. J., 1932, Sedimentation in Massachusetts Bay: Jour. Sed. Petrology, v. 2, pp. 3–37.

Truesdell, P. E., and Varnes, D. J., 1950, Chart correlating various grain-size definitions of sedimentary materials: U.S. Geol. Surv.

Turnau-Morawska, M., 1955, Optical orientation of elongated quartz sand grains: Arch. Min., v. 18, pp. 293–302 (Polish with English summary).

Twenhofel, W. H., 1932, Treatise on sedimentation, 2nd ed.: Baltimore, Williams and Wilkens, 926 pp.

———, 1937, Terminology of the fine-grained mechanical sediments: Rept. Comm. Sedimentation 1936–1937, Nat. Res. Coun., pp. 81–104 (mimeographed).

———, 1950, Principles of sedimentation: New York, McGraw-Hill, 673 pp.

Tyrrell, G. W., 1921, Some points in petrographic nomenclature: Geol. Mag., v. 58, pp. 501–502.

Udden, J. A., 1898, Mechanical composition of wind deposits: Augustana Library Publ. 1, pp. 1–69.

———, 1914, The mechanical composition of clastic sediments: Bull. Geol. Soc. Amer., v. 25, pp. 655–744.

Unrug, R., 1956, Preferred orientation of pebbles in Recent gravels of the Dunajec River in the Western Carpathians: Bull. Acad. Polonaise Sci., Ch. 3, v. 14, pp. 469–473.

———, 1957, Recent transport and sedimentation of gravels in the Dunajec Valley (western Carpathians): Acta Geol. Polonica, v. 7, pp. 217–257 (Polish with English summary).

Virkkala, K., 1960, On the striations and glacier movements in the Tampere region, southern Finland: Bull. Comm. Geol. Finlande 188, pp. 161–176.

Visher, G. S., 1969, Grain size distributions and depositional processes: Jour. Sed. Petrology, v. 39, pp. 1074–1106.

Vita-Finzi, C., and Smalley, I. J., 1970, Origin of quartz silt: comments on a note by Ph. H. Kuenen: Jour. Sed. Petrology, v. 40, pp. 1367–1368.

Von Engelen, O. D., 1930, Type form of faceted and striated glacial pebbles: Amer. Jour. Sci., ser. 5, v. 19, pp. 9–16.

Wadell, H., 1932, Volume, shape, and roundness of rock particles: Jour. Geol., v. 40, pp. 443–451.

———, 1935, Volume, shape, and roundness of quartz particles: Jour. Geol., v. 43, pp. 250–280.

Waldschmidt, W. A., 1943, Cementing materials in sandstones and their influence on the migration of oil: Bull. Amer. Assoc. Petrol. Geol., v. 25, pp. 1839–1879.

Walker, T. R., 1957, Frosting of quartz grains by carbonate replacement: Bull. Geol. Soc. Amer., v. 68, pp. 267–268.

Waskom, J. D., 1958, Roundness as an indicator of environment along the coast of panhandle Florida: Jour. Sed. Petrology, v. 28, pp. 351–360.

Wayland, R. G., 1939, Optical orientation in elongate clastic quartz: Amer. Jour. Sci., v. 237, pp. 99–109.

Wentworth, C. K., 1919, A laboratory and field study of cobble abrasion: Jour. Geol., v. 27, pp. 507–521.

———, 1922a, A field study of the shapes of river pebbles: Bull. U.S. Geol. Surv., no. 730-C, p. 114.

———, 1922b, The shapes of beach pebbles: U.S. Geol. Surv. Prof. Paper 131-C, pp. 75–83.

———, 1922c, A scale of grade and class terms for clastic sediments: Jour. Geol., v. 30, pp. 377–392.
———, 1931a, The mechanical composition of sediments in graphic form: Univ. Iowa Studies Nat. Hist., v. 14, no. 3, 127 pp.
———, 1931b, Pebble wear on Jarvis Island beach: Washington Univ. Studies, Sci. and Tech., n.s., no. 5, pp. 11–37.
———, 1932, The geologic work of ice jams in subarctic rivers: Washington Univ. Studies, Sci. and Tech., no. 7, pp. 49–80.
———, 1933, Fundamental limits to the sizes of clastic grains: Science, v. 77, pp. 633–634.
———, 1935, The terminology of coarse sediments: Bull. Nat. Res. Coun., v. 98, pp. 225–246.
———, 1936a, An analysis of the shapes of glacial cobbles: Jour. Sed. Petrology, v. 6, pp. 85–96.
———, 1936b, The shapes of glacial and ice jam cobbles: Jour. Sed. Petrology, v. 6, pp. 97–108.
Wentworth, C. K., and Williams, H., 1932, The classification and terminology of the pyroclastic rocks: Bull. Nat. Res. Coun., v. 89, pp. 19–53.
West, R. C., and Donner, J. J., 1956, The glaciation of East Anglia and the East Midlands: a differentiation based on stone orientation measurement of tills: Quart. Jour. Geol. Soc. London, v. 112, pp. 69–91.
White, W. S., 1952, Imbrication and initial dip in a Keweenawan conglomerate bed: Jour. Sed. Petrology, v. 22, pp. 189–199.
Whitney, M. L., and Dietrich, R. V., 1973, Ventifact sculpture by windblown dust: Bull. Geol. Soc. Amer., v. 84, pp. 2561–2581.
Willman, H. B., 1942, Geology and mineral resources of the Marseilles, Ottawa, and Streater quadrangles: Bull. Illinois Geol. Surv. 66, pp. 343–344.
Wolfe, M. J., 1967, An electron microscope study of the surface texture of sand grains from a basal conglomerate: Sedimentology, v. 8, pp. 239–247.
Wolff, R. G., 1964, The dearth of certain sizes of materials in sediments: Jour. Sed. Petrology, v. 34, pp. 320–327.
Woodford, A. O., 1925, The San Onofre Breccia: Bull. Univ. California Dept. Geol. Sci., v. 15, pp. 159–280.
Zimmerle, W., and Bonham, L. C., 1962, Rapid methods for dimensional grain orientation measurements: Jour. Sed. Petrology, v. 32, pp. 751–763.
Zingg, Th., 1935, Beiträge zur Schotteranalyse: Min. Petrog. Mitt. Schweiz., v. 15, pp. 39–140.

4

INTERNAL ORGANIZATION AND STRUCTURE OF SEDIMENTARY ROCKS

INTRODUCTION AND CLASSIFICATION

The structures of sediments are those features that, unlike textures, are seen or studied best in the outcrop rather than in the hand specimen or thin section. Texture deals with grain-to-grain relations, best seen under the microscope; structure deals with larger organizational units and is most clearly seen in the field. Study of structures, therefore, is as old as geology itself. Structures such as cross-bedding, ripple marks, and mud cracks were described in the earliest writings of geologists.

Primary sedimentary structures have been used as guides to the agent and/or environment of deposition. Such structures as graded and cross-bedding have been used to ascertain stratigraphic sequence in vertical or overturned strata (Shrock, 1948). In recent years directional structures have been utilized to map paleocurrents and to determine paleoslope and sedimentary strike (Pettijohn, 1962; Potter and Pettijohn, 1963). Special attention has been given also to biogenic structures (ichnofossils) as guides to depositional environment. Unlike body fossils these structures are not susceptible to reworking or transport (Seilacher, 1964a). Most recently there has been a renewed interest in current structures and the flow conditions that produced them (Middleton, 1965).

Renewed interest in sedimentary structures growing out of studies of modern sediments and paleocurrent mapping of ancient sediments has led to the publication of a number of monographs on various aspects of the subject. Included here are such works as the atlases of Khabakov (1962), of Potter and Pettijohn (1964), and of Ricci Lucchi (1970); the manual on sedimentary structures by Conybeare and Crook (1968); and the monograph of Gubler and associates (1966). There are also a number of larger works dealing with specialized classes of structures, especially sole marks (Vassoevich, 1953; Dzulynski and Sanders, 1962; Dzulynski, 1963; Dzulynski and Walton, 1965) and biogenic structures or ichnofossils (Abel, 1935; Lessertisseur, 1955; Seilacher, 1964b). Of note also are the proceedings of a symposium on primary sedimentary structures and their hydrodynamic interpretation (Middleton, 1965). In addition to these larger works is a rash of lesser papers on particular structures, as well as a number of longer papers dealing with bedding itself, the most universal structure characterizing sedimentary deposits (Andrée, 1915; Zhemchuzhnikov, 1940; Bruns, 1954; Birkenmajer, 1959; Botvinkina, 1959, 1962). There is even a manual on methods for studying sedimentary structures, especially those of modern marine sediments (Bouma, 1969).

A result of the greatly expanded interest in sedimentary structures have been various attempts to straighten out the classification and nomenclature of these features. Two approaches to the problem of classification are possible. One is morphologic, the other genetic. In the first, an effort is made to group sedimentary structures on the basis of their form or geometry and on their place of occurrence—on the base of the bed, for example. Genetic classification groups structures according to the process involved in their formation, such as biogenic, hydrodynamic, or rheologic (Nagtegaal, 1965; Elliott, 1966). A genetic classification presupposes that we are certain of the origin, an assumption not always warranted. Moreover, some structures are complex, and several processes are involved in their formation. Such is the case with flute casts and sand ripples subjected to load deformation concurrent with their formation by currents. But purely morphological classification is

TABLE 4-1. Classification of structures of sedimentary rocks

Inorganic structures		
Mechanical ("primary")	Chemical ("secondary")	Organic Structures
A. Bedding: geometry 1. Laminations 2. Wavy bedding	A. Solution structures 1. Stylolites 2. Corrosion zones 3. Vugs, oolicasts, and so on	A. Petrifactions
B. Bedding internal structures 1. Cross-bedding 2. Ripple-bedding 3. Graded bedding 4. Growth bedding	B. Accretionary structures 1. Nodules 2. Concretions 3. Crystal aggregates (spherulites and rosettes) 4. Veinlets 5. Color banding	B. Bedding (*weedia* and other stromatolites)
C. Bedding-plane markings (on sole) 1. Scour or current marks (flutes) 2. Tool marks (grooves, and so on)	C. Composite structures 1. Geodes 2. Septaria 3. Cone-in-cone	C. Miscellaneous 1. Borings 2. Tracks and trails 3. Casts and molds 4. Fecal pellets and coprolites
D. Bedding-plane markings (on surface) 1. Wave and swash marks 2. Pits and prints (rain, and so on) 3. Parting lineation		
E. Deformed bedding 1. Load and founder structures 2. Synsedimentary folds and breccias 3. Sandstone dikes and sills		

not without its difficulties. Ripple structures can be treated as a surface feature of a bed, but they also lead to formation of a microcross lamination, an internal structure. Even mud cracks can be a surface feature, an internal structure, or a sole mark (as mudcrack casts on the base of a sandstone). A purely morphologic classification (Conybeare and Crook, 1968) is rather artificial and leads to grouping of wholly unrelated structures. Such a scheme may be a suitable key to identification, but it does not contribute to our understanding.

As is the case with our systems of rock classification, it is more meaningful to use a system involving both genetic and morphologic groupings. It is convenient to make our largest groupings genetic and the lesser ones morphologic. Hence sedimentary structures are classed as physical, chemical, and biologic. Physical (mechanical) structures are essentially primary features formed at the time of sediment accumulation. They may be either hydrodynamic (current produced) or rheologic, that is, formed by hydroplastic synsedimentary deformation. Chemical structures are diagenetic in origin and may be formed in the long post-depositional interval. Organic or biogenic structures are formed during the depositional process by organisms. (Because diagenetic structures are post-depositional and are largely unrelated to sedimentation processes, they are treated separately in Chapter 12 on diagenetic differentiation and segregations.)

In general, sedimentary structures are independent of rock composition or lithology. Cross-bedding is characteristic of any noncohesive granular material, whether it is a quartz or a carbonate sand. Graded bedding and sole marks characterize some limestones as well as certain classes of sandstone, but there are some exceptions. Cross-bedding and ripples occur only in noncohesive granular materials; mud cracks characterize only cohesive muds. Stromatolites, with rare exceptions, occur only in carbonate sediments. Preservation of sedimentary structures is, however, very much dependent on rock composition. Those formed on argillaceous mud surfaces are apt to be preserved only as casts on the base of overlying sandstone beds. If the mud is calcareous, however, and if it becomes a well-lithified limestone, the features will be retained and visible on the topside of the bed. For this reason most published photographs of modern mud cracks are from argillaceous muds, whereas those of ancient equivalents are of limestones.

BEDDING

The most nearly universal primary structure of sedimentary rocks is their bedding or stratification. In fact, the expression "stratified rocks" is nearly synonomous with "sedimentary rocks," although a few rare sediments, such as tillite, are without internal stratification, and some igneous rocks, the surface flows, are bedded.

Bedding or stratification is expressed by rock units of general tabular or lenticular form that have some lithologic or structural unity and are thus set off from other strata with which they are interleaved. Payne (1942) has used the term *stratum* for a layer "greater than 1 centimeter in thickness... visually separable from other layers above and below, the separation being determined by a discrete change in lithology, a sharp physical break, or both." The term *lamination* is restricted to similar units under 1 cm in thickness. The difference, therefore, is wholly one of degree. Payne further redefined the common terms applied to the strata, namely *fissile, shaly, flaggy,* and *massive,* and assigned specific thickness limits to each. McKee and Weir (1953) have attempted to discriminate between terms applied to *thickness* of strata and those which describe *splitting* properties. Like Payne, McKee and Weir call all units less than 1 on thick *laminations* and those thicker than 1 cm *beds*. Beds 1 to 5 cm thick are *very thin,* 5 to 60 cm layers are *thin;* those from 60 to 120 cm are *thick,* and if over 120 cm the term *thick-bedded* is applied. If the beds split into units of the same order of thickness, they are termed *flaggy, slabby, blocky,* and *massive,* respectively. The thinner strata are *laminated*, or if less than 2 mm thick, *thinly laminated*.

Otto (1938) attempted to define two genetically significant units, namely, the *sedimentation unit* and the *lamination*. The sedimentation unit was defined as "that thickness of sediment which was deposited under essentially constant physical conditions." Current flow in nature is never absolutely uniform; hence no sediment, for example, is composed of particles of uniform size. Actually there is some prevailing current that deposits some prevailing size. This prevailing current has a mean velocity and deposits some mean size for a considerable period of time. The sedimentation unit is made during this time period. When the current is radically changed and a new set of conditions is established at another time, a new unit will be formed. There are momentary fluctuations in the current velocity, of course, which accounts for laminations or *phases* (Apfel, 1938), differing slightly from each other. A cross-bedded layer of sand, for example, is a sedimentation unit. It was deposited under essentially uniform conditions. The depositing current maintained a more or less uniform direction and velocity of flow. The cross-laminations, however, record local and short-lived fluctuations in the velocity of the depositing current. A second cross-bedded unit above the first, either with the same or differently oriented cross-laminations, is a separate and distinct sedimentation unit and records a new and different episode of deposition.

The distinction between a sedimentation unit and the lamination is not, according to Otto, a matter of thickness. The annual layers or varves of Pleistocene proglacial lakes, though commonly over 1 cm in thickness may also be, in part, less than 1 cm thick. It does not seem reasonable to class some as beds or strata and others as laminations. All are, instead, thin sedimentation units. Because silty and sandy portions of some thicker varves are commonly laminated, it does seem necessary to discriminate between laminations and varves proper and therefore between beds and laminations on a basis other than a predetermined arbitrary thickness.

Although the sedimentation unit is a useful concept, it is difficult to apply to some types of rocks and in many situations. It is more appropriate for coarser clastic deposits than those of purely chemical or biologic origin.

Considerable attention has been given to external form and geometry of bedding units and to character and significance of the bedding planes which separate these units. Beds are described as *planar* if the bounding surfaces are parallel within the limits of the outcrop, and *lenticular* if they converge. Bounding surfaces may be irregular. Terms such as *wavy* or even *lumpy* or *nodular* are applied to beds that pinch and swell or even disintegrate into discrete lenses or nodular bodies. The regularity of a bedded sequence can be described in terms of uniformity in thickness from bed to bed and lateral continuity and uniformity of thickness of individual beds. Four classes can be recognized: (1) beds *equal* or subequal in thickness, laterally uniform in thickness and continuous; (2) beds *unequal* in thickness, laterally uniform and continuous; (3) beds *unequal* in thickness, laterally *variable* in thickness but still continuous; and (4) beds *unequal* in thickness, laterally *variable* and *discontinuous*.

Bedding is susceptible to measurement, and hence its grosser aspects can be in some degree quantified. It has been shown that thicknesses of individual beds in many sequences, especially turbidites and ash falls, are logarithmically normal (Schwarzacher, 1953; Fiege, 1937). See Figs. 4-1 and 4-2. In general, though by no means universal, coarseness of grain and thickness of a bedding unit are related (Fig. 4-3). This has been observed even in cross-bedded strata (Schwarzacher, 1953), as well as in sandstones of turbidite sequences (Fiege, 1937; Potter and Scheidegger, 1966). In the case of turbidite sands and ash falls, both coarseness of grain and bed thickness decline downcurrent (Scheidegger and Potter, 1971). Bedding geometry, therefore, is clearly an important means of distinguishing between proximal and distal facies in these deposits. Whether log normal or not, bedding thicknesses are strongly skewed toward thinner beds. Bokman (1957) proposed a geometric *theta* scale, which tends to normalize originally skewed thickness distributions in much the same way as the phi scale does for size distributions.

It has been long known that bedding planes probably record an interval of nondeposition, in some cases, perhaps, erosion. Such breaks have been termed *diastems* (Barrell, 1917, p. 794) and probably record a longer period of time than do beds themselves.

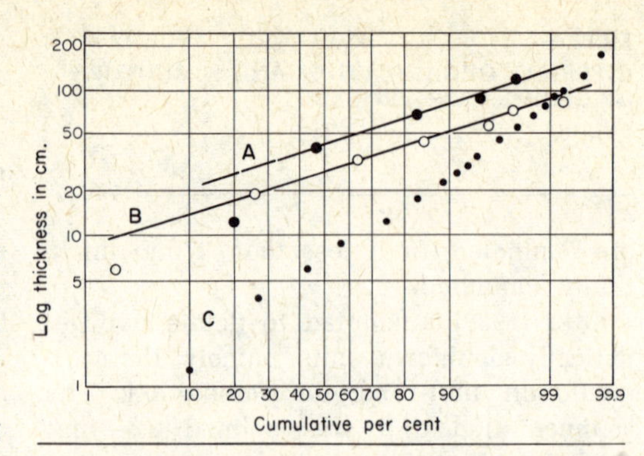

FIG. 4-1. Bedding thickness plotted on logarithmic probability paper. A: Archean graded beds, Minnitaki Lake, western Ontario, Canada; B: Pleistocene varved sands and silts, Patagonia (data from Caldenius, 1932); C: Archean graded beds, near Tampere, Finland (Simonen and Kuovo, 1951).

INTERNAL ORGANIZATION AND STRUCTURE OF BEDS

MASSIVE BEDDING

Beds are rarely without some sort of internal fabric or structure. Those which are apparently structureless have been termed *massive*. X-ray photographs of such seemingly homogeneous beds have in many cases revealed internal lamination (Hamblin, 1965). Truly massive beds are, therefore, probably very rare.

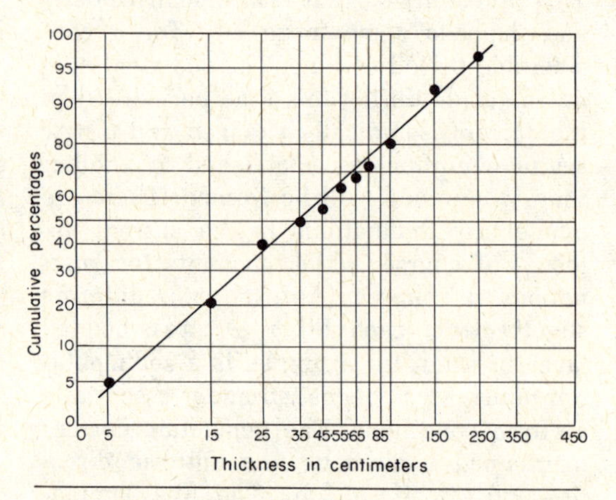

FIG. 4-2. Bedding thickness plotted on logarithmic probability paper. Cross-bedded sedimentation units of Cretaceous sandstones, England. (Schwarzacher, 1953.)

LAMINATIONS

Many beds show an internal lamination of some sort. In many these laminations are parallel to the bounding surfaces of the bed. In others laminations are inclined to these boundaries, in some cases at low angles (1 to 10 degrees), in others at a higher angle (10 to 35 degrees and even more). The latter is termed cross-bedding and is characteristic of sands. Laminations in these materials are merely records of transitory phases or minor chance fluctuations in velocity of the depositing current.

Laminations are most characteristic of the finer-grained sediments, notably siltstones and shales. They appear as more or less distinct alternations of material which differ one from the other in grain size or composition. Commonly they are 0.5 to 1.0 mm thick. They may be continuous and distinct or discontinuous and obscure. Ex-

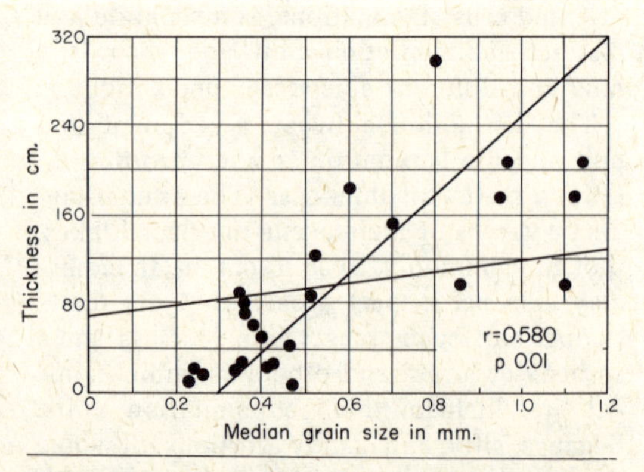

FIG. 4-3. Thickness and median grain size of cross-bedded sedimentation units of Cretaceous sandstones, England. (Schwarzacher, 1953.)

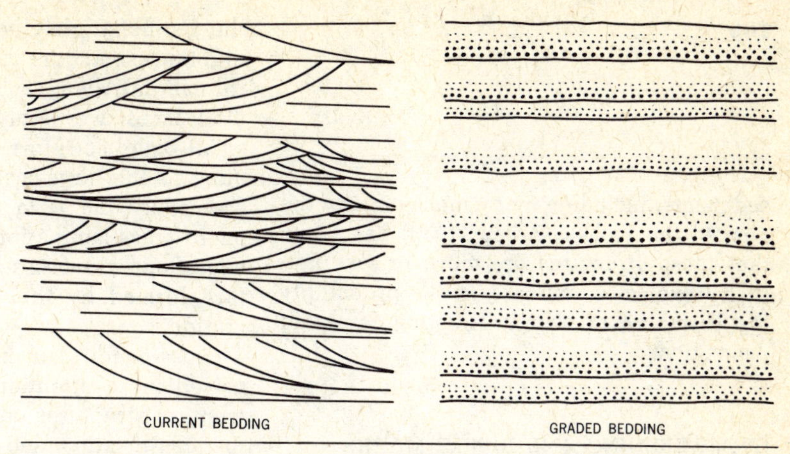

FIG. 4-4. Current and graded bedding. (From Bailey, 1936, Fig. 1.)

amples of laminations are those formed by alternations of coarse and fine particles—silt and even fine sand and clay—(Fig. 8-3), dark- and light-colored silt layers caused by differences in content of organic matter (Fig. 8-4), and by alternations of calcium carbonate and silt.

The causes of such laminations are variations in the rate of supply or deposition of the different materials. Such variations have been attributed to the fortuitous shift in the depositing current, to climatic causes (especially cyclical changes related to diurnal or annual rhythms), and also to aperiodic storms or floods. For a discussion of the conditions needed for formation and preservation of annual layers and the criteria for their recognitions, refer to the extended discussions of Bradley (1929) and Rubey (1930).

Some shales are noted for the excellence of their laminations; others are conspicuous because they lack this structure. The most perfect examples of laminated shales are commonly lacustrine. Least perfect are those found in the blocky mudstones and other terrestrial sediments. The deposits of some carbonate tidal flats are also well laminated. Such indurated sediments have been called *laminites.*

The distinctiveness and the degree of preservation of laminations is in part a rough measure of the quietness of the waters in which the sediments accumulated. Even slight bottom currents would destroy any previously formed laminations. Hence laminations often record deposition below wave base. The distinctness of laminations in clays may also be related to salinity of the water. Certain electrolytes, of which sodium chloride is the most common, induce flocculation or *symmixis,* which results in mixing of silt and clay particles and leads to a nearly homogeneous rather than laminated clay. It is also probable that stratification of a sediment may be destroyed by organisms that feed on organic matter contained in bottom muds. Repeated ingestion of muds results in a thorough working over of sediment and partial or complete destruction of laminations. Because this is nearly universal, the preservation of laminations, therefore, indicates either very rapid deposition or toxic bottom conditions and suppression of benthonic fauna. Under the latter conditions individual paper-thin laminations may persist and be traceable over distances of several kilometers (Anderson et al., 1972, Fig. 9).

In general, the thinner the laminations, the slower the rate of accumulation. This is obviously true of paired laminations that were formed during equal time intervals, such as a year.

INTERNAL ORGANIZATION AND STRUCTURE

Next to the grosser dimensions—thickness and lateral extent—the internal structure of a bed is its most important property (Fig. 4-4). There are two principal types of internal structure: cross-bedded and graded. Although these structures are most characteristic of sandstone beds, they may be present in both coarser and finer clastic sediments, including many mechanically deposited limestones. Bailey (1930) has noted that these structures are more or less mutually exclusive and are in all probability the earmarks of two contrasting facies of deposition.

CROSS-BEDDING AND RIPPLE MARKS

Cross-bedding and ripple marks are commonly treated as two unrelated phenomena. Cross-bedding is considered to be an internal feature of a bed, whereas ripple marks are treated as a surface or bedding-plane structure. They are, in fact, closely related and are two aspects of the same thing. Cross-bedding is the product of the migration of a megaripple or sand wave; ripples produce a small-scale cross-bedding (ripple bedding) on migration.

In general ripple mark is a small-scale structure. The wave length is only a few centimeters, and the height is measured in millimeters. In certain environments, however, giant ripples are common. These megaripples have a wave length of a meter or more—tens of meters in some cases—and an amplitude of several tens of centimeters. Ripples of these dimensions have been described from tidal channels (van Straaten, 1950; Off, 1963) and rivers (Sundborg, 1956, p. 270). There is some question whether the largest structures are indeed ripples. They are frequently called dunes or sand waves (Carey and Keller, 1957, p. 17). They have a very gentle upstream slope, only a degree or two, and are themselves frequently surmounted by normal small-scale current ripples.

Even though there is some kinship between small-scale ripples and larger sand waves or dunes in their gross morphology and the formation of cross-bedding by their migration, they are here treated separately. This is in part because the ripple form of the small-scale structure is commonly seen on bedding planes in the geologic record; that of the larger sand waves is not. Also, the distinction may be related to some basic differences in the physical processes of formation (Allen, 1963). To emphasize these differences we apply the term *ripple* to small-scale bed forms and *sand wave* or *dune* to the large-scale feature and in general apply the term *ripple bedding* to the micro-cross-lamination produced by ripple migration and *cross-bedding* to the structure formed by the larger sand wave migration.

Cross-bedding or cross-lamination has received more attention than all other sedimentary structures combined. It lends itself to quantification and is particularly useful in paleocurrent analyses.

Cross-bedding is a structure characteristic of sands—noncohesive granular materials no matter what their composition. Known also as cross-lamination, current bedding, diagonal bedding, and false bedding, it is a difficult structure to define. For some it means only inclined bedding—bedding with a high initial dip. It is here restricted, however, to internal bedding, called foreset bedding, inclined to the principal surface of accumulation *within a single sedimentation unit.* This definition, which makes cross-bedding the internal structure of a bed, excludes inclined bedding of other origins, such as beach bedding, delta foresets, and lateral accretion bedding. The definition given is independent of scale. A cross-bedded layer may vary in thickness from 3 mm to over 30 m.

The above definition is widely used and would apply to most of what is called cross-bedding. McKee and Weir (1953, p. 382) define the foreset bed as a "cross-stratum" and the cross-bedded unit as a "set of cross-strata." They distinguish between cross-bedding, which has foreset layers greater than 1 cm in thickness, and "cross-lamination," with foresets under 1 cm in thickness.

The classification of cross-bedding is difficult, in part because of its great variability in size and form and in part from inadequacies of exposure which make complex schemes difficult or impossible to apply. There are two general types of cross-bedding (Fig. 4-5). One is a simple tabular set, with foresets approximating planes. The other is a trough-shaped set of cross-strata which are usually curved surfaces. Even the distinction between these two contrasting

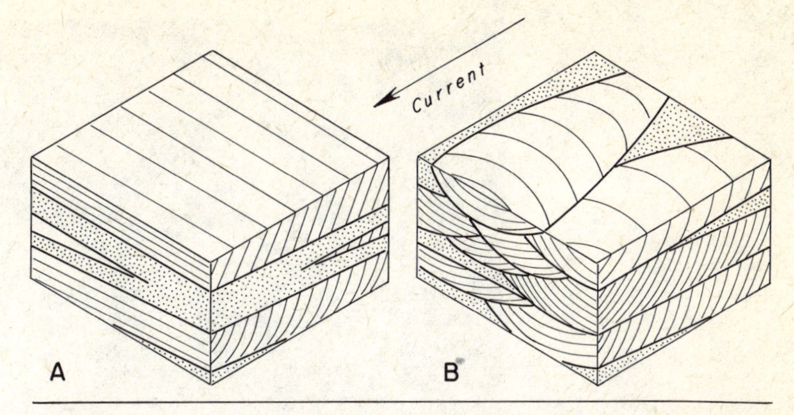

FIG. 4-5. Two principal types of cross-bedding. A: Planar tabular; B: trough (or festoon). Arrow indicates direction of current flow. (Redrawn from Potter and Pettijohn, 1963, Paleocurrents and basin analysis, Fig. 4-2. By permission of Springer-Verlag.)

FIG. 4-6. Large-scale cross-bedding of Navajo Sandstone (Jurassic), Zion National Park, Utah. (Photograph by W. C. Hamblin.)

types is difficult in small or incomplete and unfavorably oriented sections. The distinction is most easily made on a bedding-plane exposure. The traces of the foreset in the first case are straight lines; in the second case they are markedly curved and concave downcurrent. The bisectrix is the direction of current flow. McKee and Weir (1953) and Allen (1963) have identified and defined many variations of these basic cross-bedding patterns.

A simple tabular cross-stratified layer exhibits scale, inclination, and azimuth. *Scale* deals with the thickness of the cross-bedded unit and as noted may vary from 1 or 2 cms to many m (Figs. 4-6 and 4-7). Most cross-sets are less than 1 m thick. *Inclination* refers to the dip of the foreset laminations. It is usually taken to be the solid or dihedral angle between the foreset planes, or a plane tangent to the foreset at the place of its greatest inclination to the true bedding. The latter is presumed to have been horizontal at the time of deposition, an assumption nearly but not precisely correct. Inclination is usually equated with the "angle of repose," which it does indeed approximate. Although the angle of repose is commonly said to be 33 to 34 degrees, the *average* inclination of cross-sets is more likely to be between 15 and 20 degrees. In some cases the inclination is oversteepened and exceptionally even overturned (Fig. 4-8). Clearly this is the result of post-

FIG. 4-7. Small-scale cross-bedding in siltstone of Martinsburg Formation (Ordovician), from Middletown, New York. (Photograph by E. F. McBride, 1962, Jour. Sed. Petrology, v. 32, Fig. 9.) Natural size.

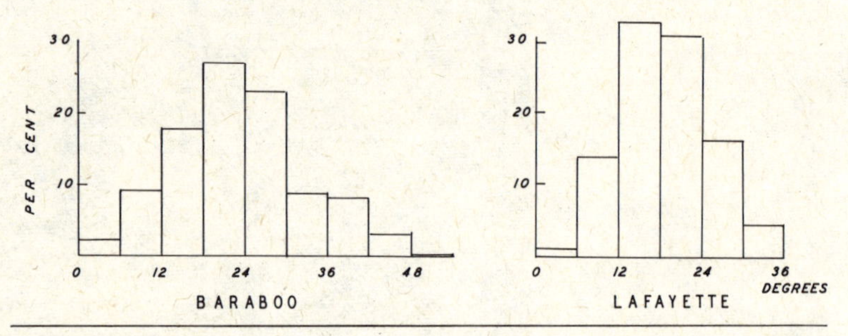

FIG. 4-8. Frequency distribution of cross-bedding inclinations in (left) Baraboo Quartzite (Precambrian) and (right) Lafayette Formation (Pliocene or Pleistocene). (Left, after Brett, 1955; right, after Potter, 1955.)

depositional deformation. The *azimuth* is the horizontal angle between the meridian and the horizontal projection of the dip line of the foreset. It is, in short, the downcurrent direction. When the cross-bedded unit is not a simple planar tabular body, these definitions need modification. The geometry of the trough cross-set is best described in terms of *width* and *depth* of the trough. The width–depth ratio tends to be fixed even for a wide range of values (Allen, 1963, p. 215). See Fig. 4-9. Troughs vary from a few centimeters to over 30 meters in breadth and from a fraction of a centimeter to 10 or more meters in depth. Horizontal traces of foresets are markedly curved, concave downcurrent (Fig. 4-5).

The bedding-plane pattern shown by very small-scale trough cross-stratification has been called "rib-and-furrow" by Stokes (1953, p. 17), "micro-cross-lamination" by Hamblin (1961), and "*Schrägschichtungsbögen*" by Gürich (1933). See Fig. 4-10.

Although many foresets approximate planes and meet both upper and lower surfaces of the tabular cross-bedded unit at the same angle, others are curved surfaces and are tangent to the base of the bed.

Although the terms *topset* and *bottomset* are often applied to the strata respectively overlying and underlying tabular cross-strata, they are inappropriate. Normally foreset layers do not pass into either. Cross-strata are not a product of microdelta growth, as the terminology implies. Foreset layers are described as "truncated," implying beveling by erosion—a concept probably also erroneous.

Cross-bedding has been explained in various ways. Here defined, however, it seems to be clearly the product of the migration of a sand wave whose size determines the scale of the cross-bedding. Dune migration produces large-scale cross-strata, ripple migration forms small-scale cross-stratification. The origin of planar-tabular cross-bedding is easily understood by reference to Fig. 4-11. Here the cross-stratified unit itself has an initial dip upcurrent; the foresets dip downcurrent. The initial dip in the former case is small, usually a degree or two and not evident in a single outcrop. Whereas the scale of cross-bedding is determined by dune height, the morphology of the cross-bedded unit is determined by the morphology of the ripple if small scale, or dune (sand wave) if large scale (Allen, 1963). The regular, linear ripples or sand waves produce the simple planar-tabular cross-stratification. The linguloid wave forms produce the trough cross-stratification.

The significance of cross-bedding has been long debated. Cross-bedding is not random but tends to show a strong preferred orientation in a given formation (Fig. 4-12). In alluvial deposits the mean direction is downslope. In the marine environment its significance is less clear, although a preferred orientation is the rule. Opposed azimuths, recording ebb-and-flood tidal currents, are common in some formations and appear in some outcrops as a "herringbone" structure. Eolian cross-beds reflect the prevailing or most effective surface winds, not necessarily the planetary circulation. So far no particular style or scale of cross-bedding is characteristic of any particular agent and/or environment, but very large scale is probably either eolian or marine rather than fluvial.

The scale of the cross-bedding (and the sand wave responsible for it) in subaqueous deposits seems to be related to the water depth (Allen, 1963). See Fig. 4-13. As noted by Carey and Keller (1957), the size of dunes or sand waves in the Mississippi River in-

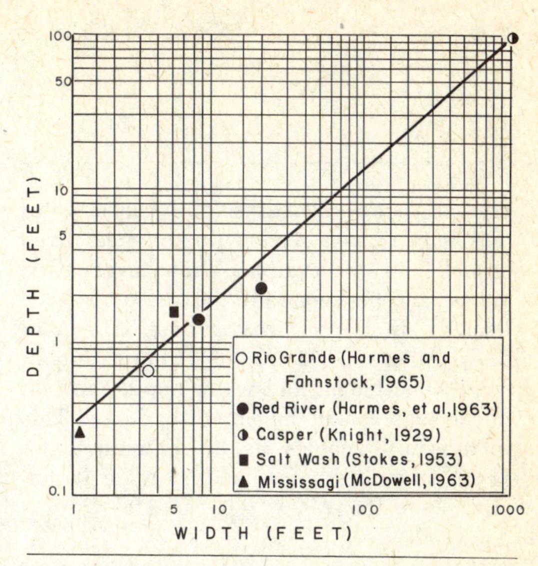

FIG. 4-9. Relation of width to depth of trough cross-stratified sets.

FIG. 4-10. Rib-and-furrow markings indicate direction of current movement. Salt Wash Member, Morrison Formation (Jurassic), Carrizo Mountains, Arizona. (Stokes, 1953, Fig. 6.)

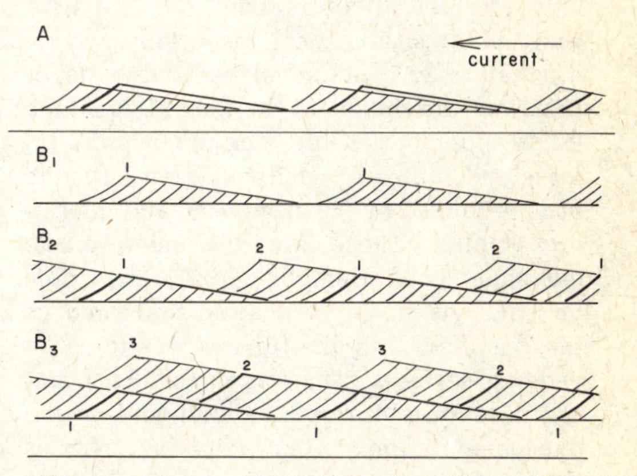

FIG. 4-11. Origin of cross-bedding. (After Shotten, 1937, Geol. Mag., v. 74, Fig. 3.)

creases as the discharge (and hence water depth) is greater during flood times. Even casual field observations show larger sand waves, and hence greater thickness of cross-bedding, in larger streams than in smaller ones. Allen's (1963, p. 198) compilation of sand wave height and water depth shows a linear increase in height with depth. This relationship enables us to estimate water depth from the cross-bedding scale in ancient deposits (Allen, 1963, p. 212).

Ripple marks A very common feature of both modern sand flats and of bedding planes of many ancient sandstones, *ripple marks* have excited the curiosity not only of geologists but also of physicists interested in wave phenomena. Consequently there is a very large literature on the subject.

Much attention has been focused on ripples as an interface phenomenon. When a current flowing over a bed of sand reaches a certain velocity, sand particles begin to move, and a rippling appears on the surface of the sand. Much of the earlier work was directed toward the study of this process and the resulting rippled pattern. Among the earlier more comprehensive geologic works dealing with the subject are the papers of Bucher (1919) and Kindle (1917), who were also concerned with rippled sandstones of the past.

Two aspects of the study of ripple marks have received attention. One is their paleogeographic significance, especially ripple orientation as exemplified by Hyde's paper (1911) on the ripples of the Berea Sandstone (Mississippian) in Ohio. The other is the internal structures of sandstone and siltstone beds produced by superimposed and migrating ripples. These are the micro-cross-laminations of Hamblin (1961) or ripple bedding, visible in vertical sections and as the familiar "rib-and-furrow" pattern on bedding surfaces. There is considerable recent literature on ripple bedding and on the phenomena known as *climbing ripples* (Walker, 1963, 1969; Allen, 1963; McKee, 1966). The most comprehensive recent works treating ripples from all points of view is Allen (1963, 1969). When a current flowing over a bed of sand reaches a certain velocity, sand grains begin to move, and very promptly a rippling appears on the surface of the sand. These current ripples consist of numerous long, essentially parallel, more or less equidistant ridges, trending in straight or gently curved lines at right angles to the current (Fig. 4-14). Under some

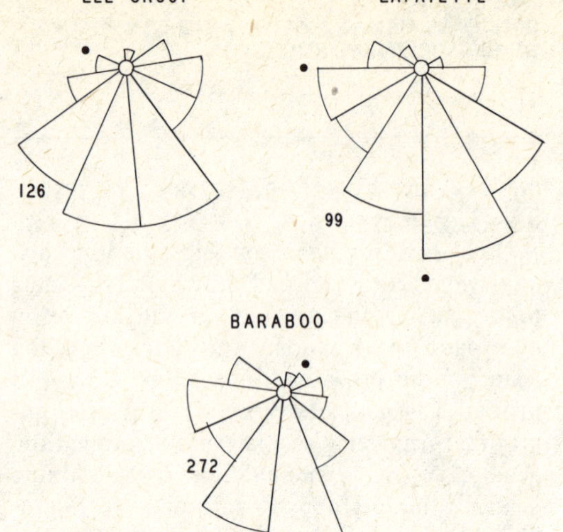

FIG. 4-12. Azimuthal distribution of cross-bedding in Baraboo Quartzite (Precambrian); Lee Group (Pennsylvanian); and Lafayette Formation (Pliocene or Pleistocene). Baraboo, after Brett, 1955; Lafayette, after Potter, 1955.

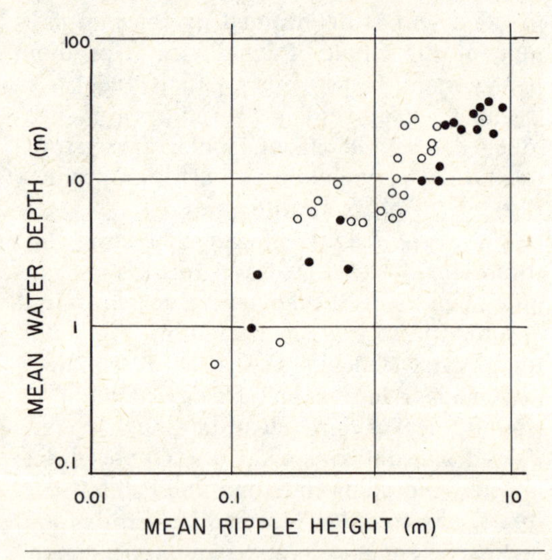

FIG. 4-13. Relation between water depth and ripple height; the latter determines the scale of the cross-bedding. Solid dots, seas and estuaries; open dots, rivers. (After Allen, 1963, Liverpool and Manchester Geol. Jour., v. 3, Fig. 6.)

flow conditions, the current ripple pattern becomes less regular, and ultimately the ripple crests break up into laterally compressed U-shaped crescentic structures. Some are barchanoid or *lunate* if the form is convex upstream; others are tongue-shaped or *linguloid* if the convexity points downstream (Fig. 4-15). The latter seem to be much more common. With a continued increase in velocity the rippling disappears, and a smooth bed is formed over which sand is swept.

If sand is being moved across a nonsandy bottom (a firm mud surface, for example), and the supply of sand is insufficient to form a continuous layer, the sand will accumulate in isolated ripple ridges. These are the *starved ripples* of some authors and appear in cross-section as planoconvex lenses of sand embedded in a mudstone. The term *lenticular bedding* has been applied to this structure (Reineck and Wunderlich, 1968) as has also *flaser bedding* (Conybeare and Crook, 1968, p. 98B).

Shallow sandy bottoms of standing water bodies are commonly covered by oscillation ripple marks generated by the to-and-fro motion of the water agitated by waves. The ground plan of oscillation ripples is much like that of current ripples and is perhaps even more regular. In profile the oscillation ripples, unlike the current ripples, are symmetrical, which, together with the sharper crests and broadly rounded troughs within which a minor medial ridge is commonly present, distinguishes these ripples from the current type. The contrasts between the original and the cast makes these structures a valuable criterion for distinguishing the top and bottom of steeply inclined beds (Cox and Dake, 1916; Shrock, 1948, p. 114).

The terminology applied to oscillation ripples is given in Fig. 4-16. *Length* is the distance between two corresponding points on two consecutive ripples. *Height* (called *amplitude* in some of the older literature) is the vertical distance between crest and trough. *Ripple index* is the ratio of length to height. The same nomenclature can be used for current ripples, but its application becomes difficult in the less regular lunate or

FIG. 4-14. Ripple-marked Baraboo Quartzite (Precambrian), Devils Lake, Wisconsin.

linguloid forms. Current ripple differs, moreover, because it is asymmetrical in cross section. It has a gentle stoss or upcurrent slope and a steep downcurrent or lee slope and is thus a good criterion of direction of current flow.

Other ripple patterns are also known. Two superimposed ripple patterns produce "interference ripples" known also as "tadpole nests." A rather peculiar pattern in mud, which has been termed ripple mark and which has been described by van Straaten, consists of fairly regularly spaced, more or less symmetrical, continuous ridges which are *parallel* to the current flow. Another peculiar pattern is the so-called *rhomboid ripple* (Hoyt and Henry, 1963), a form seemingly restricted to the swash face of beaches.

Various ripple patterns may be complex because of the combined action of waves and currents. These various hybrid forms have been described by van Straaten.

Ripple bedding A most important aspect of sand ripples is their internal structure and the small-scale (and often complex) cross-lamination which they generate on migration. In cross section, the migration of a ripple produces a small-scale cross-bedded layer—the micro-cross-lamination of Hamblin (1961). The simplest form generated is an uncomplicated cross-bedded layer 1 cm more or less in thickness. But if rippling persists through time, superposition of one such rippled bed follows another, and in many cases a very complex composite bed is the result (Fig. 4-17). Andersen (1931, p. 175) called attention to these complex forms (which he called "rolling

FIG. 4-15. Modern current ripples on sand bar, Vermilion River, Indiana. Current from upper left to lower right. (Photograph by Potter. Potter and Pettijohn, 1963, Paleocurrents and basin analysis, Pl. 10A. By permission of Springer-Verlag.)

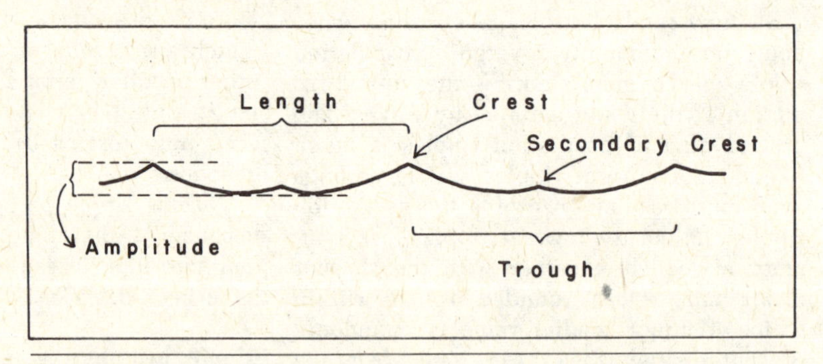

FIG. 4-16. Ripple-mark terminology.

strata") as they occur in the fluvioglacial sediments of Denmark. McKee (1938, 1939) described similar complex ripple bedding in the flood deposits of the Colorado River of the Grand Canyon region. Several possibilities for superposition exist. Ripples may be superimposed *in phase,* so that the ripple appears not to have migrated at all but has persisted in one place as deposition progressed. A more common relation is superposition with a progressive migration of the crests, so that the ripples "drift" and appear to "climb," each on the stoss side of the underlying ripple. A less regular structure is produced by superposition of several sets of ripples completely out of phase with each other. The result is a confused bedding with lens-like structures, sometimes called flaser bedding (Fig. 4-18).

Climbing ripples and their cross-laminated deposits, variously termed *climbing ripple structure* (McKee, 1966), *ripple-drift lamination* or *ripple-drift cross-lamination* (Walker, 1963, 1969) display a variety of forms (Fig. 4-17). In some is a linkage between the laminations of one ripple form and the next.

FIG. 4-17. Complex ripple-bedding and climbing ripples, Colorado River alluvium. (Photograph by E. D. McKee, 1938, Jour. Sed. Petrology, v. 8, Fig. 4-D.)

FIG. 4-18. Wavy bedding and flaser structure. (After Reinecke and Wunderlich, 1968, Sedimentology, v. 11, Figs. 2, 3, 4, and 5. By permission of Elsevier Publishing Co.)

In others the cross-laminations are sharply bounded by backset bedding planes. In the first case stoss-side laminations are preserved, although they are thinner than the lee-side laminations. In the second, the stoss-side laminations are not present. Either they were not preserved, or they have been eroded. A special case of the first type is marked by the accumulation of mud in the trough and silt and sand on the stoss slopes. This segregation of materials forms a series of alternating silt and mud layers with a steep upcurrent dip simulating larger-scale cross-bedding and which on superficial observation might be mistaken for such. This type seems to be most characteristic of turbidite sequences (Walker, 1963). Hydraulic factors governing type and angle of climb of climbing ripples have been analyzed by Allen (1970).

Less regular stacking of ripple bedding produces a confused internal lamination. The structure is clearly that produced by rippling, but no well-organized or persistent stacking pattern is evident. The term *wavy bedding* has been applied to such sand- or siltstone beds.

When mud is present, the form of the ripple-bedded unit becomes more evident. Intercalated mud may be in the form of lenticles or flasers as a result of isolated accumulations of mud in ripple troughs; this is flaser bedding. If the mud lenticles coalesce, the result is wavy bedding. If mud dominates, the ripple-bedded units are isolated and enclosed in a mud matrix; this is lenticular bedding or starved ripples (Reineck and Wunderlich, 1968). See Fig. 4-18.

Another aspect of internal structures of sandstones, presumed to be related to ripple bedding, is the "rib-and-furrow" structure of Stokes (1953) (see p. 108). The structure was described by Gürich (1933) in the flagstones of the Maulbronn monastery in central Germany and termed "*Schrägschichtungsbögen*" by him. The structure as seen on a bedding plane consists of small, transverse crescentic markings, which occur in sets confined to relative long narrow furrows separated from one another by very narrow and not altogether continuous ribs (Fig. 4-10). The longitudinal furrows are essentially parallel to one another and to the direction of current flow. They are a few centimeters wide and up to a meter in length. The small transverse markings are arcuate, the convex side is upcurrent, and the bisectrix is parallel to the current flow. These transverse markings are the eroded edges of an imbricate structure—upturned arcuate laminations.

The rib-and-furrow structure seems to be the bedding-plane expression of a species of small-scale trough cross-bedding produced by the migration of a linguloid ripple system. It has been described by Stokes from the Moenkopi Formation (Triassic) and the Saltwash Sandstone Member of the Morrison Formation (Jurassic) of Utah. It has also been observed in the Devonian flagstones of Pennsylvania. Ripple bedding is subject to synsedimentary deformation. This deformation is most commonly expressed by oversteepening of the ripple laminations. By degrees this deformation becomes more acute and overturning occurs. Closely related to, and perhaps an exaggerated form of, ripple deformation is convolute bedding. When ripple accumulations are isolated on a mud surface, they may trigger load deformation and sink or burrow down into the underlying mud. These are "load-casted" ripples (Dzulynski, 1962).

Ripple marks, like cross-bedding, have proved useful in determining stratigraphic order, in giving evidence of the direction of current flow, and in indicating flow conditions. They have proved less useful in defining the environment of deposition, because they form under many conditions

FIG. 4-19. Graded bedding, Archean turbidite, Minnitaki Lake, western Ontario, Canada. (Photograph by R. G. Walker. Pettijohn, Potter, and Siever, 1972, Sand and sandstone, Fig. 4–8. By permission of Springer-Verlag.)

and under water of almost any depth whenever a current moves across a sand surface. Wave-induced ripples differ from those of a unidirectional current, and wind ripples differ markedly from aqueous ripples; but unfortunately the former are seldom if ever seen in the geologic record. Ripples have proved useful in regional paleogeographic analysis.

Graded bedding Graded bedding, a common feature of some sedimentary sequences, has been called to attention by field geologists who have found it exceptionally useful in determining the order of superposition in isoclinally folded and overturned strata. The geologic significance of graded bedding and the recognition of graded bedding and cross-bedding as distinguishing attributes of two contrasting facies of sand deposition were first clearly pointed out by Bailey (1930, 1936). It is now generally conceded that graded bedding is perhaps the most characteristic feature of turbidite deposition generally in waters of considerable depth.

Graded beds are sedimentation units marked by a gradation in grain size, from coarse to fine, upward from the base to the top of the unit (Fig. 4-19). Graded beds are deposited from a waning current and may range in thickness from a centimeter or less to one or more meters. Graded materials may be silt, sand, or, in rare cases, gravels. Usually most graded beds are sandstones (commonly graywackes in the older sequences) and range from a few centimeters to a meter in thickness. In general, the thicker the graded unit, the coarser the materials at the base of the bed (Potter and Scheidegger, 1966). Graded beds display a log normal thickness distribution (Fig. 4-2).

Grading is of several types. Some graded beds are composites apparently formed by a second surge which arrived before the first current had completed its deposition or alternately formed by truncation of an earlier deposit before deposition of a new graded unit.

Despite the variation in styles of grading shown in the field, it is clear that there is an ideal or normal sequence of structures found in the most completely graded unit. This ideal cycle has come to be called the *Bouma cycle,* as it was first most explicitly described by Bouma (1962, p. 48). His ideal sequence (Fig. 4-20) consists of five subdivisions or "intervals." The lowest, the graded interval (a), displays the marked grading and is normally the thickest part of the bed. In some cases the grading is indistinct or even absent if the sand available was exceptionally well-sorted. The graded interval is followed in some cases by an interval of laminated sand (b), which is commonly followed by an interval displaying ripple cross-lamination (c). According to Bouma, this in turn is followed by an upper interval of indistinctly laminated sandy or silty pelitic material (d), an interval generally poorly displayed and not often observable. The top interval (e) is the pelitic interval (shale or slate), which terminates the Bouma sequence.

As noted by Bouma, beds displaying the entire sequence are not common. Many display top truncation, that is, incomplete cycles starting with the graded interval but lacking one or more of the overlying members. More common perhaps is the bottom "truncation," that is, beds beginning with a higher interval. But, as noted by Bouma, where truncation has occurred, the remaining intervals present are in their proper sequence.

Incomplete sequences may be attributable

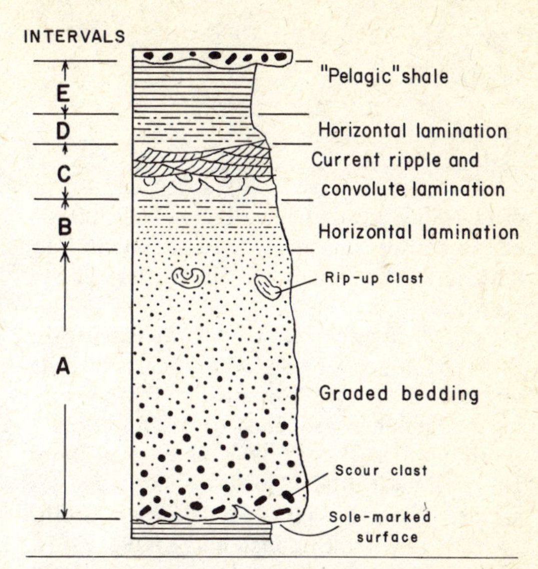

FIG. 4-20. Diagram showing ideal sequence of structures (Bouma cycle) in a graded bed. (After Stanley, 1963, Jour. Sed. Petrology, v. 33, Fig. 2.)

to the weakening of the formative current as it deploys over the floor of the basin. As the coarser sand supply is dropped out and the current wanes, the graded interval will be missing and deposition will begin with the lower laminated sands. As the current further weakens, the first deposit is the ripple-laminated interval.

This lateral change in character of graded beds is accompanied by a decline in thickness of the bed involved and explains the correlation of thickness and grain size. The thickness and grain size decline in the downcurrent direction of the ideal graded bed is a negative exponential (Scheidegger and Potter, 1971). The successive dropping-out of lower intervals provides us with a clue to the "proximality" of the deposit. Proximal beds, close to their source, display the full sequence. Those most distal are apt to show bottom truncation. From these relations Walker (1967) has calculated a *proximality index, P,* which was defined as $P = A + \frac{1}{2}B$, where A and B are the percentage of beds in a sequence beginning with Bouma divisions *a* and *b*, respectively.

As noted above, graded bedding and cross-bedding are the earmarks of two contrasting facies of sand deposition. They are, therefore, mutually exclusive features and do not occur in the same sedimentary sequence. But, as noted also, *small*-scale cross-bedding, or ripple cross-lamination, is an integral part of the ideal graded bed. *Large*-scale cross-bedding, involving the whole sedimentation unit, is conspicuously absent in graded sequences.

Graded bedding is widespread both in time and place. It is a common feature of virtually all early Precambrian (Temiskaming) sequences of the Canadian Shield (Pettijohn, 1943; Walker and Pettijohn, 1971). Similar graded bedding has been described from the early Precambrian (Bothnian) of Finland (Simonen and Kuovo, 1951).

The structure has also been reported and described from the Archean of South Africa and Australia (Dunbar and McCall, 1971). But graded bedding is not solely an Archean feature. It is well displayed by later rocks. Excellent descriptions of this structure have been published on the Silurian of Aberystwyth area of Wales (Rich, 1950; Kuenen, 1953b; Wood and Smith, 1959), on the Miocene of the Apennines (Kuenen and Migliorini, 1950; ten Haaf, 1959), on the Cambrian of the Harlech Dome of Wales (Kopstein, 1954), on the Pliocene of the Santa Paula Creek section of California (Natland and Kuenen, 1951), on the Carboniferous Kulm of central Germany (Kuenen and Sanders, 1956), on the Carpathian flysch (Dzulynski, Ksiazkiewicz, and Kuenen, 1959), on the Ordovician Martinsburg of the central Appalachians (McBride, 1962), and on the upper marine Devonian of the same area (McIver, 1970). It is present also in the Late Paleozoic Stanley-Jackfork beds of the Ouachita Mountains of Arkansas and Oklahoma (Cline, 1966) and in the Cretaceous of the Sacramento Valley of California (Ojakangas, 1968), as well as in many other sequences of diverse ages. Graded bedding probably characterizes all thick geosynclinal accumulations of graywackes interbedded with shales or slates. It is seen also in many of the cores taken of modern deep-sea sands (Nesteroff, 1961; Kuenen, 1964, p. 10).

Graded bedding is a feature primarily of sandstone, principally graywackes in the Paleozoic and older sequences. It is, however, not confined to this class of sands. It is even known from some limestones which, however, were in fact deposited as sands at the time of their formation. These are the *allodapic* limestones of Meischner (1964, p. 156). An occasional graded bed is seen in quartzites and other sands, both ancient and modern, which are atypical because they are not deep-water deposits characterized by a complete or partial Bouma cycles. In these cases the graded bed is generally a solitary or sporadic occurrence.

Graded beds have been explained in several ways. Bailey (1930) ascribed them to earthquakes which served as "intermittent distributors of sand and mud." He presumed that graded beds were "the products of settling through comparatively still bottom water, which allows the sand and mud to accumulate in one and the same locality, though with a lag on the part of the mud determined by its finer texture." According to Bailey, "the sand and the mud, which formed unstable accumulations on the borders of the geosyncline, are periodically dislodged by submarine temblors and thrown into suspension to settle out in comparatively deep and quiet water."

Kuenen and Migliorini (1950) first expressed the view that turbidity currents are the probable cause of most graded bedding. Kuenen (1953a) has presented a detailed review of the evidence for the turbidity origin of graded beds. The most compelling evidence lies in the structure of the bed itself, that is, the grading, a feature which can be experimentally reproduced by turbidity flows (Kuenen and Migliorini, 1950; Kuenen and Menard, 1952). Also significant is the uniformity of thickness, even of the coarsest graded units (normal swift currents would produce lenticular cross-bedded units), absence of cross-bedding, evidence of deep-water origin (deep-water microfauna of the associated shale interbeds), and deposition of coarse debris on underlying mud without disturbance of the mud surface (delicate worm trails preserved as casts on the underlying surface of the overlying sand bed). Obviously each graded bed records a single short-lived episode and is a product of deep-water sedimentation beyond the reach of normal bottom currents and waves. The accumulated evidence now almost certainly indicates deposition from a dense turbidity flow, which may be the product of submarine slump, triggered perhaps by earthquakes. Despite a general consensus on the subject, some opposition to the turbidity concept persists (van der Lingen, 1969; Hubert, 1966 p. 696). Refer to these papers and to a discussion of their views by Kuenen (1967; 1970).

Graded beds conceivably might originate in other ways. The close resemblance of the thinner graded beds to the varved silts and sands of the Pleistocene proglacial lakes led to the view that a seasonal influx of sediment controlled by seasonal melting of a glacier, was responsible for the graded bed. This explanation was invoked to explain the graded beds of the Sudbury Series of Ontario, Canada (Coleman, 1926, p. 234), the graded beds of the Archean of Tampere, Finland (Simonen and Kuovo, 1951), and the Archean graded beds of Minnitaki Lake, Ontario (Pettijohn, 1936). These explanations are most certainly incorrect and were advanced before the concept of turbidity current emplacement was developed. If these graded beds had been seasonal, their thickness would imply an unreasonably high rate of deposition. Although some Pleistocene lake sediments are known to possess thick sandy varves, it is improbable that the older graded beds were so deposited. For one thing, older graded beds lack *dropstones*, a most telling feature of glaciolacustrine or glaciomarine deposits.

If the time usually estimated for the deposition of a graded sequence is divided by the number of graded beds, one must conclude that the graded beds record events widely spaced in time. Kuenen (1953) thus estimated that several hundred to several thousand years elapsed between the formation of one graded bed and the next, a view also expressed by Sujkowski (1957, p. 550). The graded beds record very short-lived events. Intervening pelitic layers are the indigenous sediment of the basin and accumulated very slowly.

Although some isolated or sporadic graded beds can be produced by volcanic eruptions, heavy floods, or hurricanelike storms, most repetitive marine graded beds are almost certainly the product of turbidity flows. Graded beds arising from other causes apparently are relatively rare, are likely to be solitary, and are different in structure or associated features so that confusion with grading caused by turbidity flow is unlikely. Possible exceptions are the much thinner, evenly bedded, fine-grained siltstones. Discrimination between these deposits and truly seasonal sediments may be less easy.

The origin of graded beds is inextricably bound up with the subject of turbidites. Hence for a more extended discussion of graded bedding, refer to the comprehensive works on turbidites (Bouma, 1962; Bouma and Brouwer, 1964; Walker, 1970).

Growth bedding The term *growth bedding* is here applied to stratification produced by the *in situ* activity of organisms or by chemical precipitation on surfaces of accumulation. It differs from bedding previously described, in which the component grains were emplaced in the rock fabric by current action. Growth bedding, thus set in opposition to current bedding, is particularly characteristic of some classes of limestones and many travertine and tufa deposits.

Most important perhaps is *stromatolitic bedding*, a form of growth bedding very prominent in many early Paleozoic and Precambrian limestones. Because this type of bedding is related to the formation and properties of an algal mat, it partakes somewhat of the character of both a sedimentary structure and a fossil. Like burrows, trails, and tracks, it is neither. It is therefore discussed more fully in the section on biogenic structures.

Many precipitated materials—travertine, onyx, tufa of various sorts, and caliche—show a banding or stratification, some of which closely mimics stromatolitic bedding (Westphal, 1957). Bedding of this type is commonly closely related to the crystal fabric of the rock and to some types of diagenetic bedding, particularly certain caliches (Multer and Hoffmeister, 1968). These are discussed in Chapter 12 on diagenetic segregations.

BEDDING-PLANE MARKINGS AND STRUCTURES

When beds separate readily along bedding planes, the surfaces produced commonly display various markings or structures. These features form on the surface of the accumulating sediment, but many, if not most, are preserved as casts on the underside or *sole* of the overlying bed. This is particularly true when the underlying mate-

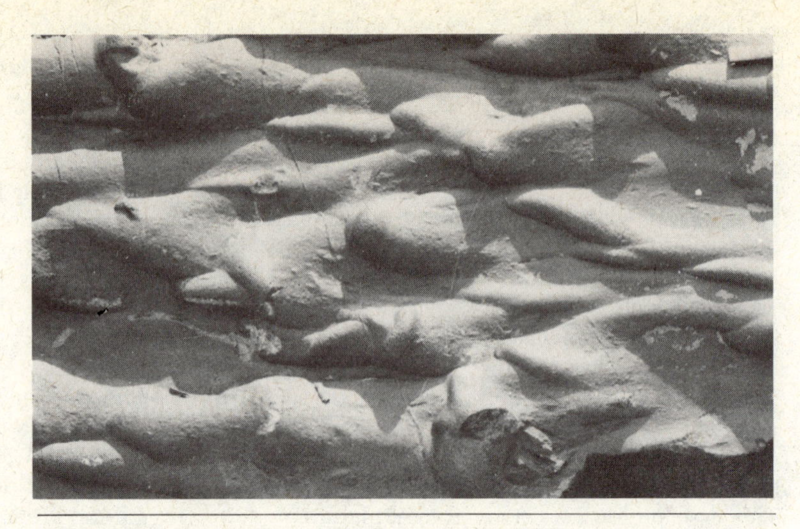

FIG. 4-21. Underside of sandstone slab showing swarm of flute casts, Denbigh Grits (Silurion), Wales. Length of specimen about 22 inches (55 cm).

rial was mud and the overlying sediment is a sand. Rain prints, mud cracks, flutes and grooves, and so forth are preserved as "casts" on the underside of the overlying sand. On the other hand, some features appear either on the underside or the topside of a bed. Ripple marks, for example, appear both as casts and as normal ripple structure on the bedding planes of sandstone. So also parting lineation appears on either the upper or lower surface of a sandstone flag. Those features which characteristically form on mud surfaces, however, generally appear only as *sole marks* and hence are here given separate treatment. The structures which develop on sandy surfaces and which can appear either on bottom or top of a bed are separately discussed as "surface markings."

SOLE MARKS

Sole markings are bedding-plane features which characterize the undersurfaces of some sandstone beds and, less commonly, some limestone beds that overlie shales. They are raised structures formed by the filling of depressions in the mud surface over which the sand was deposited. Although known for many years (see Hall, 1843), their origin was little understood. They were indeed hieroglyphs and only recently have been deciphered (Vassoevich, 1953; Kuenen, 1957; Dzulynski, 1963; Dzulynski and Sanders, 1962; Dzulynski and Walton, 1965). Much of the earlier work was directed toward describing and classifying these markings and toward their utility as paleocurrent indicators. Efforts to understand their origin have led to experimental studies (Dzulynski, 1966; Dzulynski and Walton, 1963; Allen, 1971).

Sole markings originate by current action, by deformation caused mainly by loading, and by the action of organisms (Table 4-1). We are concerned here mainly with those produced by currents. These belong to two classes, namely, those attributable to current scour and those caused by the action of debris carried by the current. The latter are referred to as *tool marks.*

Scour and tool marks Current scour produces flutes which, upon becoming filled with sand and welded to the overlying sand bed, are termed *flute casts* (Fig. 4-21). Flute casts appear, therefore, as raised structures on the base of the overlying sand bed. They vary in shape, size, and arrangement. The cast is a slightly elevated, elongate, moundlike form with a bulbous upcurrent nose, the downcurrent end of the flute flaring and merging with the bedding plane. Flute casts vary from a centimeter or two to over a meter in length and a few millimeters to a few centimeters in height. Some are much elongated; others are more triangular or deltoid. The prominent upcurrent nose or beak displays a twist or hook in some cases. Flute casts occur in swarms; solitary flutes are rare. They may be widely spaced or they may cover most of the sole and even overlap; in more rare cases they are arranged in diagonal rows or patterns (Kuenen, 1957).

The filling of the flute is sand, in many cases coarser than most of the bed of which the flute casts are a part. The less regular flutes may resemble load casts, but they differ in that the flute is cut or eroded and transects the laminations in the underlying material. In fact, a few are sculptured or terraced, expressing differential erosion of the laminations in the underlying materials. The laminations surrounding load casts or pockets are deformed and are not transected by the structure.

Flutes seem to have formed by eddy scour. When flow conditions are right, a swarm of eddies develops and scours the underlying mud surface. Their size is probably dependent on the flow conditions. The flute size seems to bear some relation to the coarseness of the material and hence to the current strength. The resulting flutes are very useful in paleocurrent studies. Although they may form in several environments, they seem to be most prominent on the underside of turbidite sandstones (and limestones) and hence are one of the earmarks of the flysch facies.

Another structure produced by current scour (and hence related to flutes) is the *current crescent*, a horseshoe-shaped depression (German, *Hufeisenwülste*). This structure is the result of scour around an obstacle, such as a pebble, lying on a sand surface. The scour or moat is greatest on the upcurrent side and extends downcurrent on each side of the obstruction involved. In many cases the obstruction to current flow was a shale chip or fragment, which later weathers out and leaves a hole partially surrounded by a crescentic raised ridge-like cast of the moat.

Currents also move various objects—sand grains, shells, mud chips and the like. These travel across the underlying mud surface, rolling or intermittently impinging on that surface and leaving marks which become preserved as weak positive features on the base of the overlying sandstone. These features are collectively called *tool marks*.

FIG. 4-22. Lower side of flagstone from Hatch Formation, Naples Group ("Portage"), Upper Devonian, Conesus Lake, New York. A: simple groove cast; B: multiple or complex groove cast; C: flute casts of low relief; D: prod cast. Current moved from upper to lower part of specimen. Base of specimen about 15 inches (38 cm).

Most striking are *groove casts* which appear as raised, rectilinear, rounded to sharp-crested features on the base of some sandstone beds (Fig. 4-22). Some are multiple and show a second-order set of microgrooves or ridges. A few are "feathered" or characterized by smaller slightly divergent grooves arranged in a symmetrical manner on either side of the principal groove cast. These features are presumed to originate by the filling of corresponding furrows engraved in a firm mud surface by various moving objects. They have also been called "drag marks" and "drag casts" (Kuenen, 1957, p. 243).

Groove casts occur generally in sets. More than one set is commonly seen on the same surface, the second set intersecting the first at an acute angle. One set is usually partially erased by the second. Within a given set there is little or no deviation in azimuths. Groove casts seldom occur with flute casts; these two seem to be mutually exclusive. The individual groove cast exhibits a relief of only 1 or 2 mm, is remarkably straight, and in most exposures shows neither a beginning nor an end. Hence seldom does one find the tool responsible for the structure itself.

Groove casts should be distinguished from

slide marks or casts formed by the movement of a large object or mass, such as a shale raft. Such sliding masses tend to pivot or rotate so that the marks produced all curve in a similar manner, reflecting this rotation. Groove casts display no such coordinated behavior; they are associated with other tool marks such as prod and skip casts but (as noted above) seldom with flute casts. Like flute casts, however, they are most characteristic of the soles of turbidite beds. They are perhaps the most common sole mark of the flysch facies.

The origin of groove casts has long been an enigma. They are a current-produced structure, and their orientation is closely correlated with the current directions deduced from other structures. Moreover, that they are a tool mark is further proven by the very rare cases in which a sand grain or shell fragment has been found at the downcurrent end of a groove. The exact dynamics of formation is, however, not entirely clear. Most objects moved by currents proceed by rolling or by saltatory leaps, as the various impact marks testify. A groove, however, requires continuous contact, even pressure, and, as the ornamented grooves show, nonrotational motion of the tool. Eddy motion produces flutes, not grooves. The mechanism of groove formation is, therefore, not well understood.

Intersecting sets of grooves also present a problem. Presumably grooves were formed by turbidity currents moving as dense flows down a subaqueous slope. But if one set of grooves records the downslope motion, the other does not. Were the sets formed by the same or different currents?

Because groove casts are common, they are one of the most useful guides to paleocurrent flow. They must be used, however, in conjunction with other structures; they themselves give only the azimuth of the flow and not its direction.

In addition to grooves, there exists also a diverse group of tool marks. Some of these are produced by objects which touch bottom intermittently; others roll along the bottom and leave a characteristic trail or signature. The first group includes bounce, brush, and prod casts. *Bounce casts,* also *skip casts,* are records of impacts made by an object pursuing a saltatory path. They appear as small, slightly raised structures spaced at somewhat regular intervals. *Brush marks* or cast is the term applied to similar features when impact is accidental or casual and not regularly repeated. Brush casts are also characterized by a slight mound of material pushed up in advance of the impinging object. The *prod cast* is characterized by penetration of the mud bottom by an object, such as a waterlogged stick which, after penetration, is rotated forward and lifted free of the bottom. Prod casts, therefore, appear as a very short groove casts with an abrupt more prominent downcurrent termination (Fig. 4-22).

Roll marks are records of rolling objects. Common in some flysch sequences are those produced by planar coiled shells which apparently were propelled in the fashion of a wheel or hoop and, like the tread on an automobile tire, left a characteristic imprint or signature (Seilacher, 1963).

Mud crack casts Another type of sole mark, unrelated to current action, is the *mud crack cast.* Mud cracks develop in cohesive materials, such as mud, as a result of drying and shrinkage. A polygonal crack system is formed; the cracks are widest at the surface and taper downward. When such a surface is abruptly flooded and buried by sand, the sand filling of the crack system becomes welded to the superjacent sand bed and on lithification and weathering away of the underlying shale appears as a polygonal network of sharp-crested ridges on the sole of the sandstone bed itself.

Load structures Soft-sediment deformation produces a variety of structures, some of large scale. Among the smaller structures are sole marks caused by unequal loading or unstable density stratification. These have been called *load casts* and are discussed here because of their close association with sole marks of other origins. The larger topic of deformed bedding and the resultant structures are covered in the next section.

Load casts, more properly called *load pockets,* are somewhat irregular bulbous or mammillary features on the base of a sandstone bed that overlies shale (Fig. 4-23).

FIG. 4-23. Load casts on base of sandstone bed, Aux Vases (Mississippian), Illinois. (Photograph by C. Weber.)

FIG. 4-24. Parting lineation in Salt Wash Member, Morrison Formation (Jurassic) of Arizona. Current flow parallel to scale. (Stokes, 1953, Fig. 7.)

They resemble flute casts in size and relief but differ in their irregularity and lack of symmetry and orientation. They are not "casts" in any sense, because the downward protusion of sand is achieved by deformation of the laminations of the underlying mud and not by the filling of a scour. Apparently these structures are a product of unequal loading of the underlying hydroplastic mud and owe their origin to a vertical readjustment, with downward motion of the sand and compensatory upward movement of mud. These structures may, in extreme cases, become saclike, attached to the mother bed by a constricted neck; even in rare cases they may be detached from the overlying bed and sink downward into the underlying materials. These are *load pouches* and, if detached, *load balls*.

Sometimes load structures were initiated by unequal loading related to the sedimentation process. If the properties of the underlying mud are correct, flutes and grooves may themselves subside and assume some of the characteristics of a load cast. Even "starved" or isolated sand ripples may constitute an unequal load and under the right conditions "burrow" down into the underlying mud (Dzulynski, 1962). In the case of the ripples there is a pattern of arrangement and an internal structure inherited from the parent structure.

Load casts may form in any environment where sand is deposited on a water-saturated hydroplastic mud. They are common in turbidite sequences, but even here some layers show much deformation by loading, whereas others do not. Perhaps where one turbidite flow follows on the heels of another, the underlying mud did not have time to dewater; hence loading effects are striking. If the time lapse between flows is sufficient, natural compaction will render load deformation unlikely.

SURFACE MARKS

Included here are various rill and current markings and other features, most of which form on sandy surfaces. They may be seen either as a normal structure of the top side of a bed or as the "negative" or "cast" on the underside. Ripple marks, the most prominent and common structure, were discussed above; as were the various biogenic structures. Somewhat illogically we include mud cracks, which characterize muds.

Parting lineation A common but less well-known feature is a faint but distinct structure found on the bedding planes of some thin-bedded sandstones (Fig. 4-24). It is especially prominent on those sandstones from which flagstones are split and is well displayed on the flagstone walks all over the world. This feature was termed *primary current lineation* by Stokes (1947) but had been earlier described and figured by Hans Cloos (1938), who pointed out that it was parallel to the direction of the depositing current. Because it is prominent on the cleaved surfaces, it was called *parting lineation* by Crowell (1955, p. 1316).

The structure appears as subparallel, very faint linear grooves and ridges of very slight relief on plane-parting surfaces. In other cases the parting is less perfect, and somewhat irregular, plasterlike remnants of an adjacent lamination cling to the split surface. The term *parting-step lineation* has been applied by McBride and Yeakel (1963) to such cases to distinguish them from *parting-plane lineations* seen on the smoother surfaces. These authors have shown that the mean direction of grain elongation is parallel to the lineation. Stokes (1953) presumed that the structure indicates "formation in a fluvial environment or at least under shallow sheets of flowing water." It is, however, also known from deep-water turbidite sandstones.

Rill, swash, and related marks Sandy surfaces may exhibit a host of markings many of which are rarely preserved. Among these are *rill marks*, small dendritic, bifurcating-upstream rivulets commonly found in the swash zone on beaches but also in sand bars and sandflats. They appear to be formed by the flow of a thin sheet of water. *Swash marks* are the thin wavy lines on beaches left by the upper limit of the swash of waves (Shrock, 1948, fig. 89). *Rhomboid "ripple" marks* are a low-relief, reticulate pattern (Hoyt and Henry, 1963; Otvos, 1965) apparently a product of backwash on beaches. In general, preservation of rill, swash, and rhomboid patterns is very rare in ancient sediments.

Rain, hail, and spray pits Rain, drip, and spray impressions are small circular to elliptical pits formed in wet muds by those agents. Rain prints have been reported from ancient strata, commonly as casts on the base of a sandstone or siltstone bed. Like desiccation cracks they indicate subaerial exposure and are most likely to be preserved in continental beds. *Bubble impressions* may resemble rain prints and be mistaken for them.

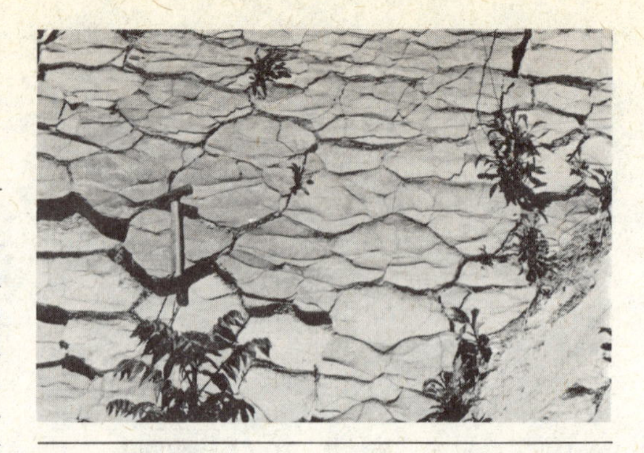

FIG. 4-25. Mud cracks in argillaceous limestone of Wills Creek Formation (Silurian), near Roundtop, Maryland. (Photograph by Warren White.)

Mud cracks Some bedding planes are marked by an irregular checkered polygonal pattern of fractures which are now filled with sand or silt (Fig. 4-25). The host rock was originally a mud, and the crack system developed as a result of shrinkage—shrinkage in most cases caused by loss of water by drying, which implies exposure. The cracks have therefore been termed *desiccation cracks* or *sun cracks*. Not all mud-cracked sediments are argillaceous. Mud cracks are known from some micritic limestones and may be filled with lime silt or even dolomitic silt or sand. Those found in argillaceous muds are most likely to be seen as casts on the underside of sandstones; those formed in lime muds may appear on the top side of the now lithified rock.

The size of the polygons, the width of the cracks, and their depth of penetration vary widely. The polygons vary in width from a few millimeters to over 30 cm and the cracks from under 1 mm to 3 to 5 cm wide. They may penetrate to a depth of 1 or 2 cm or for several tens of centimeters. The coarseness of the network probably is related in some way to the thickness of the desiccating bed.

Normally the cracks taper downward and ultimately pinch out. They are filled by sand or coarser materials. If the drying seam is relatively thin (a few millimeters), the cracks may extend through the clay bed to the next underlying sand layer. The polygons thus formed may become detached, slightly displaced, rotated, or even overturned and picked up by the next sand-depositing torrent and incorporated in the next overlying sand bed. Thus originate some of the flat shale-pebble conglomerates with a sand matrix.

In many cases the tapered sand filling is seen in cross section to be deformed, to be "crumpled" or contorted (Fig. 4-26). The wider upper end of the filling may even appear to have punched upward into the overlying stratum. This crumpling is the result of the noncompactable filling trying to accommodate itself to compaction and thickness reduction of the host material. The crumpling can even be used to make a quantitative estimate of the compaction (Shelton, 1962).

Because mud cracks are formed by shrinkage, they cannot be formed in pure sand; the latter undergoes no volume decrease on drying. Mud cracks cannot truly be preserved but are represented only by their fillings ("casts"). The cracked clayey rock generally crumbles, but in the overlying sandstone bed the entire crack system is preserved as a polygonal network of elevated sharp-crested ridges on the underside of the overlying sandstone bed.

The polygonal crack system is presumed to be attributable to shrinkage upon loss of water. Generally this implies subaerial drying. Some shrinkage crack systems, however, have been attributed to a spontaneous dehydration of gel-like materials, even in an aqueous environment. This origin has been invoked to explain the crack system in some septaria and chert nodules (Taliaferro, 1934). These have been termed *synaeresis cracks*. Synaeresis has even been called on to explain crack patterns in mudstones, especially in those of unusual composition, such as dolomitic mudstone. In general such cracks are thought to be characteristic of gel-like materials (White, 1961, p. 566; Burst, 1965). The criteria for distinguishing between a normal desiccation crack network and synaeresis cracks are obscure.

FIG. 4-26. Crumpled mud-crack fillings, Buffalo Springs Member, Conococheague Formation (Cambrian), Morgantown, Pennsylvania. (Photograph by C. Weber.)

Clearly the three-dimensional radial system in nodular bodies are of different origin than the sand-filled polygonal networks of ordinary mudstones.

The environments most favorable to normal mud cracking are intertidal zone, ephemeral playa lakes, and overbank mud flats of the floodplain. Barrell (1906, p. 550) believed that opportunities for preservation in the tidal flats were poor and hence "... mud cracks form one of the surest indications of the continental origin of argillaceous sediments."

DEFORMED AND DISTURBED BEDDING

Gravitational displacements may occur during sedimentation or shortly thereafter, which deform or alter the depositional structures. Bedding in particular may be disturbed and even destroyed. Many of these effects are caused by some instability which triggers movement under gravitational forces. Three situations prevail. In one, movement is largely vertical, a convectivelike transfer of material initiated by an unstable density stratification of materials such as that, for example, created by deposition of a bed of sand over a less dense, water-saturated mud or silt. If underlying material undergoes a thixotropic transformation with loss of strength, a series of convective cells may be set up with downward movement of the sand and a compensatory upward movement of the clay or silt (Artyushkov, 1960a, 1960b; Anketell, Cegla, and Dzulynski, 1970). The motion may be slow or rapid and catastrophic.

In another situation instability is present on an oversteepened depositional slope. Movement generated by oversteepening has a large lateral component and hence results in a near-horizontal transfer of material. Such displacement, if slow, is termed *creep;* if rapid, it is a *slide* or *slump*. It may be either subaqueous or subaerial.

The third process usually involves only sand. Under some conditions this material is rendered "quick" and is capable of injection as a dike or sill into adjacent beds. In these cases the sand affected loses all primary structures. It is of interest to note that only sand—not clay—becomes mobile. Shale or clay may be disrupted, but it tends to form fragments engulfed in the more mobile sand.

LOAD CASTS AND BALL-AND-PILLOW STRUCTURES

Small-scale vertical readjustment leads to the formation of *load casts,* described on page 120, and in extreme cases the formation of *load pouches* and detached pouches or *load balls.* The tongues of shale which penetrate the overlying sand constitute, in cross-sectional views, *flame structure.* In some cases these "flames" of shale show overturning in one direction or even a spiral configuration as though acted upon by a lateral stress.

Some sandstones, like some subaqueous lava flows, display a pillow structure (Fig. 4-27). The sand now appears as numerous, closely packed ball- or pillowlike forms. They have also been called "pseudonodules" (Macar, 1948), and "hassocks." They have also been designated "flow-rolls" (Sorauf, 1965). They are not a primary depositional feature but are, instead, a product of deformation which took place before the deposition of the overlying strata. Ball-and-pillow structure characterizes certain sandstones, but the same structure is found in some limestones—limestones which were in fact sands at the time of their accumulation.

The structure characterizes mainly the lower part of the bed affected. The individual balls or pillows range from a few centimeters to over a meter in diameter. They are generally oblate spheroids or ellipsoids. Not uncommonly they are kidney-shaped or even resemble a large inverted mushroom with recurved rims. The cuplike or basinal structure is convex downward, concave upward, and in many cases slightly tilted but not overturned. Laminations within the balls or pillows are deformed and conform to the lower half of the ball or pillow. The pillows are separated and partially or wholly surrounded by the finer shale or silt derived from the subjacent bed.

FIG. 4-27. Ball-and-pillow structure ("flow rolls"), Devonian, near Port Treverton, Pennsylvania.

The pillows are clearly not concretions nor the product of spheroidal weathering, both of which are found in sandstones. Nor are they a product of slumping as commonly stated. Their symmetry and orientation imply a downward and not a lateral movement. That such saucer- or kidney-shaped structures can be produced by the foundering of unconsolidated sand into a quasiliquid substrate was demonstrated by the experiments of Kuenen (1958, p. 18). Recent field studies of ball-and-pillow structures in the New York Devonian (Sorauf, 1965) and elsewhere (Howard and Lohrengel, 1969) confirm the concept of origin by foundering rather than by slump. Perhaps this action was sudden or catastrophic.

SYNSEDIMENTARY FOLDS AND BRECCIAS

As noted, unconsolidated sediments may also be deformed by gravity-induced movements with a large lateral component. We are concerned here only with deformation which took place while the sediment was still in the environment of deposition, thus excluding tectonic and other later deformation. The resulting slide or slump produces folds, faults, and breccias in the affected materials. Inasmuch as these structures are also produced by tectonic deformation and perhaps by other synsedimentary processes, it becomes necessary to consider the criteria for discriminating "soft-sediment" deformation from true tectonic deformation. This discrimination is generally not difficult, but there are some ambiguous situations (Miller, 1922). The preconsolidation structures are usually confined to particular beds, in some cases to beds only 1 or 2 cm thick. Unlike drag folds, they bear no relation to the larger structures or to the tectonic pattern of the region. Notable is the absence of vein fillings either in the microfaults or in the interstices of breccias. In most cases the folds produced are of small scale and are commonly truncated or beveled by a bedding plane, showing that they were formed and partially eroded before deposition of the overlying bed. All preconsolidation structures are presumed to be caused by a gravitational force directed downslope. If so, they become a criterion of slope direction and accordingly should be carefully observed and mapped. The paleogeographic significance of these features has been pointed out by Kuenen (1952) and others (Murphy and Schlanger, 1962; Marschalko, 1963; Scott, 1966; Hubert, 1966).

There are, in addition, other ways in which deformed bedding can be produced. Some soft-sediment folds have been attributed to grounded icebergs, shove of shore ice, and so on. Although soft-sediment deformation is common in glaciolacustrine deposits, such features also occur in deposits where ice action is highly improbable. Gravitational action is adequate and common enough to produce most all soft-sediment structures.

Soft-sediment folding is common in many sediments. It is conspicuous in some thin-bedded sand-shale sequences, but slump folds and breccias are prominent in some limestone sequences, especially in the vicinity of reefs.

As noted by Rich (1950), synsedimentary folds are of several types. One variety is confined to a single thin sandstone or siltstone layer, siliceous or calcareous silt. In such folds the stratum itself is not involved; only the internal laminations are contorted. This structure, termed *convolute folding*, is of uncertain origin, is probably not caused by slumping, and hence is discussed elsewhere.

Slump folding proper usually involves more than one bed, in contrast to the convolute structure. Folding of this type, described and well illustrated by Hadding (1931), affects many interbedded layers and seems to be the result of a mass flowage of these materials which if long continued leads to partial to complete disruption of the bedding and to the production of a breccia or pseudoconglomerate. When movement is distributed throughout the sliding mass, the thin, more competent beds are broken into irregular, small to large, slablike fragments. In general, shale or mudstone layers form fragments, and the matrix is apt to be sand. In some cases the fragments show only slight separation and no rotation (*pull-aparts*). In other cases the fragments are rotated and twisted into hook-shaped forms, termed *slump overfolds* by Crowell (1957, p. 998). These slump overfolds and the spiral slump balls or "snowball structures" of Hadding (1931, p. 386) may give a clue to the direction of the slide movement. The result is a chaotic mixture, which with a higher water content may acquire greater mobility and pass into a mud flow and lead to the formation of a "pebbly mudstone" (Crowell, 1957) or *tilloid*. These deposits are discussed in Chapter 6.

In other cases the slump leads to tight folding of the superincumbent beds over a detachment surface. This décollement type of movement over a sole generates a nappe-like structure accompanied by attenuation or even a hiatus in the upslope detachment area. Structures of this type are common in the varved clays of Pleistocene glacial lakes (van Straaten, 1949; Fairbridge, 1947).

Slump deposits may be both thick and widespread. Ksiazkiewicz (1958, p. 135) described deposits 55 m thick. Crowell has described equally impressive slump beds in the Cretaceous of California. Some slump sheets are thick enough to be mapped (Jones, 1937); they may extend over many square kilometers. Most of those in the geologic record appear to have been submarine.

Slumping in calcareous sediments differs in no significant way from that in clastic sequences. Slide structures, varying from small-scale crumpling to large-scale folds with amplitudes of 10 to 15 m and coarse breccias 10 to 15 m thick extending over many square miles, have been reported in the Permian limestones near the Guadalupe Reef complex in New Mexico (Newell et al., 1953, pp. 69, 86; Rigby, 1958). Limestone breccias in the Alps associated with graded limestones, the "allodapic limestones" of Meischner, were attributed by Kuenen and Carozzi (1953, p. 369) to slumps or slides along a reef front.

SANDSTONE DIKES AND SILLS

It is not uncommon to find sand-filled dikelets which extend across the bedding for a few centimeters. These are essentially sand-filled mud cracks. They may become "welded" to the overlying sand bed and appear, after the underlying shale has weathered away, as a polygonal system of mud crack casts (p. 120). Such a feature is a minor sedimentary structure. But if the dikes are several meters thick and traceable for hundreds or even thousands of meters, they are substantial rock bodies. These and the related sandstone sills are treated at length in Chapter 5 on the geometry of sandstone bodies.

CONVOLUTE BEDDING

Convolute bedding, also known as convolute lamination or slip bedding, is an enigmatic deformational structure (Fig. 4-28). Rich (1950) termed it *intrastratal contortions,* and this is perhaps a most appropriate description of the phenomenon. Convolute bedding is indeed intrastratal and involves laminations *within* a bed but *not* the bed itself.

FIG. 4-28. Convolute folding of siltstone bed in Martinsburg Formation (Ordovician), New Jersey. Length of specimen about 5 inches (13 cm). (Photograph by W. S. Starks. Van Houten, 1954, Pl. 2.)

Convolute folding seems to be characteristic of coarse silt or fine sand beds 2 to 25 cm thick. Within such beds, which may be either siliceous or calcareous, is a set of complex folds. Individual laminations are traceable from fold to fold, although micro-unconformities are common. In general, the synclines tend to be broad and U-shaped, whereas the intervening anticlines are tight and peaked. Convolute folds tend to die out toward both top and bottom of the bed. In some cases the anticlines appear to be truncated by erosion.

The distortions are not ordinary linear folds, because the bedding-plane pattern shows no continuity of fold crests. Instead the structures are a series of sharp domes and basins. This pattern suggests a complex system of vertical motion rather than lateral displacement. The geometry of the structure, along with its confinement to a single bed and to materials restricted in size (coarse silt or fine sand), seems to indicate an origin by some sort of internal readjustment of material in a quick or near-quick condition.

Many theories have been advanced to explain the structure (see Potter and Pettijohn, 1963, p. 154, for a summary), and none seem wholly satisfactory. Convolute bedding is commonly associated with ripple-bedded silts and sands, the ripple bedding of which is itself oversteepened and even overturned. Part of the problem may also be confusion between true convolute bedding and some other load deformation structures.

STROMATOLITES AND OTHER BIOGENIC STRUCTURES

STROMATOLITES

The term *stromatolite,* apparently derived from the German *Stromatolith* (first used by Kalkowsky, 1908, p. 68), has come to mean a laminated structure composed of particulate sand-, silt-, and clay-sized sediment which has been formed by the trapping and binding of detrital sedimentary particles by an algal mat. Perhaps the term *algal stromatolite* would be more appropriate. In general, the particulate matter is calcareous and only rarely otherwise (Davis, 1968). The structure varies from flat laminations, which require close inspection to distinguish between these and ordinary sedimentary laminations to small moundlike forms of varying degrees of convexity and size, to columnar structures not unlike an inverted stack of soup bowls, to various digitate and branching forms. In addition to these fixed or attached forms, there are free-rolling or mobile *oncolites,* concentrically structured bodies superficially resembling concretions.

There are also structures that have an external form and size of a typical hemispherical stromatolite which lack the characteristic internal laminations. These are the *thrombolites,* so named because of their clotted internal appearance (Aitken, 1967).

It is not possible to review and summarize the whole field of stromatology. Refer to the earlier literature very thoroughly summarized by Hofmann (1969). But sedimentologists should be familiar with stromatolitic bedding and also the various pseudostromatolitic growth forms generated by inorganic processes.

The classification and nomenclature of stromatolites have grown increasingly complex. Early workers considered these structures to be fossils and applied generic and specific names to them. It was presumed that the structure was secreted and formed by a specific organism. This view has been challenged and the concept grown that the algal mat responsible for stromatolites was probably a complex of several kinds of filamentous and unicellular green and blue-green algae. The particular form and size are dependent on environmental rather than genetic factors. Thus various generic names applied would be invalid, because names refer only to the various forms assumed by the accumulated entrapped sediment and may not be directly related to specific organisms. They are not fossil algae. The distinction between fossil algae and stromatolites (Rezak, 1957, p. 129) is that fossil algae have recognizable skeletal structures such as cell walls and reproductive organs, whereas algal stromatolites have only fine laminated and fragmental textures.

A number of papers attempt to classify and name the various growth forms (Hofmann, 1969; Logan, Rezak, and Ginsburg, 1964; Maslov, 1953; Aitken, 1967). The algal stromatolites vary from small pseudopisolitic forms and crusts to biscuit- and cabbagelike heads of considerable size (Figs. 4-29, 4-30, and 4-31). The pseudoconcretionary forms attributed to algae range from small subspherical bodies 0.5 to 1.0 cm in diameter to larger and somewhat flatter *oncolites.* The larger bodies have thicker and more irregular outer coatings. Commonly the growth is not equal in all directions except in the earliest stages. Later growth is most active on the upper surface, and if by chance the oncolite was rolled over, the new growth may be on the side opposite to that which earlier received the greatest deposit. Some pisolites are composite, that is, are formed of several smaller bodies grown together and enveloped by later growth. The nuclei of these structures may be a bit of foreign substance or in some cases a piece of algal material.

The algal *crusts* are simple laminated and commonly crinkled crusts which grade into irregular nodular masses having mammillary surfaces. The crusts may be nearly flat, essentially parallel to the bedding ("*Weedia*" type of stromatolite); slightly arched, several centimeters in diameter and only a centimeter in height; or more hemispherical to cabbagelike in form with a height equal to or greater than the equatorial diameter. Some of the larger hemispherical forms, measured in decimeters or

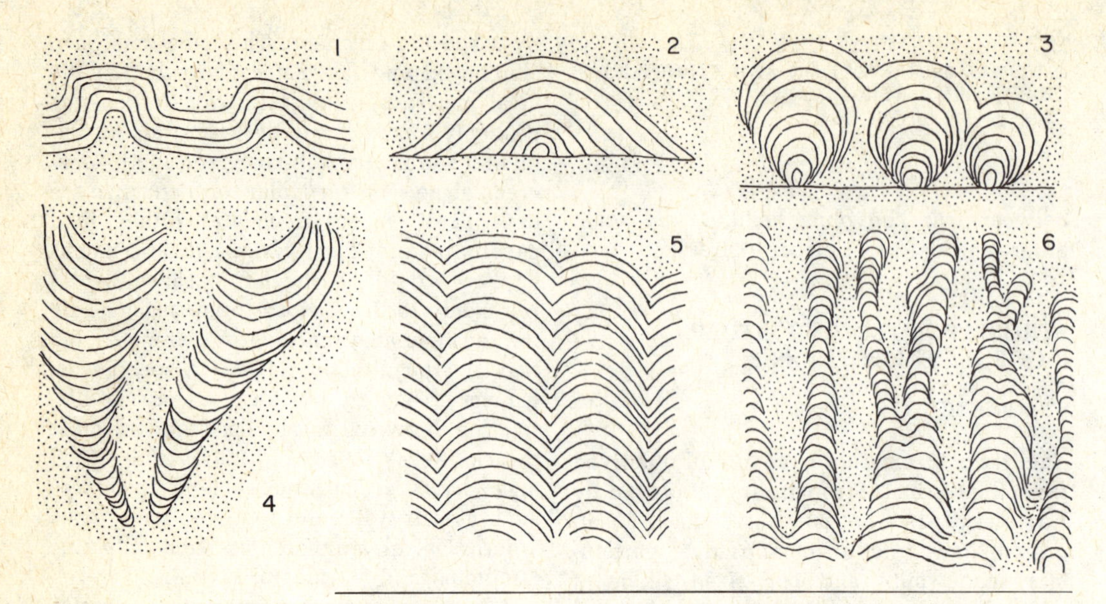

FIG. 4-29. Diversity of stromatolitic forms. (1) *Weedia* Walcott; (2) *Collenia* Walcott; (3) *Cryptozoon* Hall; (4) *Cryptozoon boreale* Dawson; (5) *Archaeozoon* Matthew; (6) *Gymnosolen* Steinmann. Not shown are oncolites or algal concretions. (After Pia, 1927, Handbuch der Paläobotanik, v. 1, p. 37.)

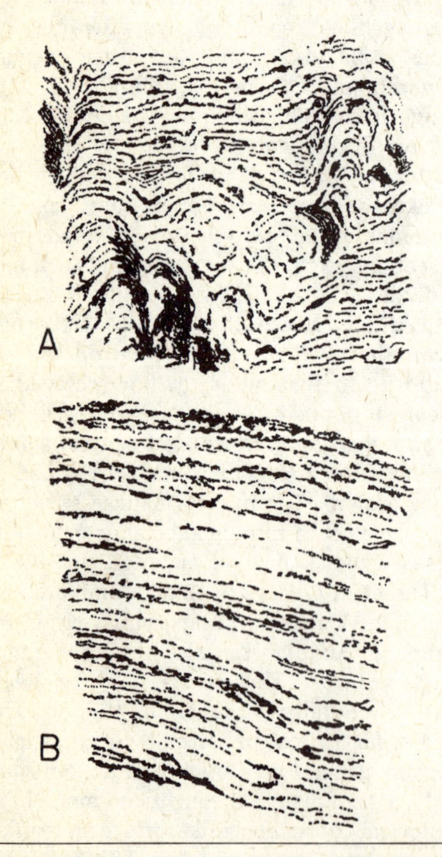

FIG. 4-30. Two types of stromatolitic bedding. A: Upper Paleozoic, Nova Scotia, Canada. B: Brighton Limestone, Australia. About three-fourths natural size.

FIG. 4-31. Stromatolitic limestone, Precambrian, East Arm, Great Slave Lake, Northwest Territories, Canada.

even meters, are complex in that the smaller mammillary outgrowths occur on the surface of the hemispherical head. Another common type resembles a series of stacked inverted thimbles or, if large, of similarly arranged soup bowls. These form vertical columns, usually several in number, which reach upward from a few centimeters to several meters. In some cases the columns expand upward to form a clublike structure. In other cases the columns or fingers split into two or more branches in the upward direction.

Some of the algal structures show asymmetrical growth. The stromatolitic heads are elliptical instead of circular; the elongation is parallel to the prevailing current system (Hoffman, 1967). The drapeover laminations also reflect asymmetrical growth and appear thickest on the side that faces upcurrent.

Some of the complex stromatolitic forms attain a large size. Individual columnar stromatolites may be several meters or more in height. It is unlikely, however, that any had more than a meter of relief at the time of their formation. Their height is attained by upward growth of the structure during the sedimentation process. Substantial algal bioherms up to 18 m thick and 60 m across have been described in Precambrian limestones (Hoffman, 1969, p. 448).

The relations of one stromatolitic head to another and to the surrounding sediment are varied. In some cases the internal laminations of one stromatolite are traceable through the host rock and appear to be linked to the next adjacent stromatolitic column. In other cases there is no linkage and the interstromatolitic material is fragmental carbonate sand. Stromatolitic heads are rarely solitary. In general, they occur closely packed together in a given bed characterized by a particular kind of stromatolite.

The term *thrombolite* has been proposed by Aitken (1967, p. 1164) for cryptalgal structures closely related to stromatolites but lacking laminations and characterized by a macroscopic clotted fabric. In external form and size thrombolites resemble stromatolites.

Stromatolites are in essence modified bedding—bedding modified by the activity of algal mats which under different conditions assume different forms. Under the microscope the only structure visible is a lamination parallel to the surface of the stromatolite. Laminations are generally thin, a millimeter or less in thickness, and are marked by a greater or lesser concentration of carbonate and other debris. Even grains of quartz silt may be trapped in the laminations.

Stromatolites and related structures are found in limestones ranging in age from Precambrian to Recent. They are best displayed and are by far most abundant in the older rocks, especially the Precambrian and early Paleozoic. Their scarcity in the later Phanerozoic strata has been attributed to the destruction of the algal mats by grazing forms, particularly snails, and destruction of algal laminations by burrowing organisms (Garrett, 1970). It is presumed that such organisms were not present in the Precambrian and absent only in later times when salinity or other environmental factors restricted or extinguished the usual biota.

Only recently has the algal origin of stromatolites and related structures been firmly established. Black's work in the Bahamas (1933) first placed the present interpretation of ancient stromatolites as an organosedimentary structure on a firm footing. The discovery of a superb display of present-day lithified stromatolites in Shark Bay, Western Australia, resolved any lingering doubt about their algal origin (Logan, 1961). Recent work on present-day stromatolitic forms in Bermuda and in the Bahamas has elucidated details of mat development and sediment entrapment (Gebelein, 1969). Observations of both modern and ancient stromatolites demonstrate the very-shallow-water origin of these structures. Because the crinkling shown by the algal laminations has been attributed to drying, the water must have been exceedingly shallow, probably intertidal. Algae seem not to be restricted either by salinity or by water temperature. The close association of mud-cracked limestones, flat-pebble conglomerates, and oolites further suggests very shallow waters. The asymmetry displayed by some stromatolites has proved a useful paleocurrent indicator. The upward convexity of the stromatolitic laminations has also proved useful as a criterion to determine stratigraphic order in vertical or overturned beds.

OTHER BIOGENIC STRUCTURES

Introduction The sedimentologist should be alert for other structures caused by the activity of organisms, such as tracks, trails, and burrows. These features are very common in some sedimentary sequences. They appear as markings on bedding planes (both upper and lower surfaces of beds) and also as features seen in sections perpendicular to the bedding.

Although long known, their study has only recently been pursued systematically. As with stromatolites many earlier workers described them as fossils and gave them generic and specific names, some forms being misinterpreted as plant remains. Recent work has uncovered their true nature and shown that their form, both geometry and detailed ornamentation, is a record of the activity of various unrelated organisms. We owe much present-day knowledge about various tracks, trails, and burrows to studies of modern sedimentary environments. Important early work of this sort includes that of J. Walther at a marine station on the Bay of Naples and especially to Rudolph Richter at the Senckenberg-am-Meer station in the North Sea.

We can present here only the briefest account of the subject. For a more extended treatment of trace fossils or *ichnofossils,* as they are called, refer to the monographic and prolific studies of Abel (1935), Krejci-Graf (1932), Lessertisseur (1955), Häntzschel (1962), Seilacher (1953, 1964a, 1964b), and Crimes and Harper (1970).

Biogenic structures, fossils in the broadest sense of the term, differ from body fossils in that they are incapable of reworking or redeposition. Although they record a particular activity of an animal such as the dwelling habit or manner of feeding, they are particularly useful in determining the environment in which the organism lived. The assemblage of such "trace fossils" has proved to be a very good index of the sedimentary facies and water depth (Fig. 4-32).

Trace fossils also yield information on the rate of sedimentation and are a guide to toxicity of bottom waters. They have also proved helpful in determining the stratigraphic order in steeply inclined or overturned beds.

Classification Trace fossils may be classified in several ways. Seilacher (1964a), for example, recognized five functional classes, classes defined in terms of animal behavior (1) resting marks (*Ruhrspuren: Cubichnia*) or shallow marks made by mobile animals resting on the bottom; (2) crawling trails (*Kreichspuren: Repichnia*), trails made by mobile animals while moving over the sediment surface; (3) residence or shelter structures (*Wohnbauten: Domichnia*), essentially permanent structures, usually burrows made either by mobile or semi-attached organisms, primarily to protect the animal from predators or from bottom scour; (4) feeding structures (*Fressbauten: Fodinchnia*), or burrows made by sessile bottom feeders, in general having a radial pattern; and (5) grazing trails (*Weidespuren: Pasichnia*), generally sinuous trails or burrows of mud-eating organisms at or below the sediment–water interface.

One can also classify these biogenic features on the basis of their relation to the bedding and on their geometrical configuration and ornamentation or internal structure. Some are confined to a bedding plane; this is especially true of various tracks and trails. The form and pattern of these vary from the small resting marks of free-swimming forms to footprints of dinosaurs. It includes also the continuous though meandering trails made by crawling forms. Many resting marks exhibit bilateral symmetry; many trails also show a bilateral character because of the bilateral symmetry of the animal which made them. Some are complicated by marks of appendages and tails.

Grazing trails are also bedding-plane features displaying a variety of patterns. Some are irregular sinuous markings; others show a remarkable regularity; some are spiral, others have a regular systematic sinuosity (Fig. 4-32), and some even displaying a polygonal network (*Paleodycton*). In general these grazing trails form only on mud surfaces and are preserved, therefore, only as casts on the underside of siltstone or fine sandstone beds.

Other biogenic structures are better seen in sections perpendicular to the bedding than on the bedding plane itself. Some are simple tubes, such as *Skolithus,* whereas others are more complex; many are U-shaped tubes. Burrows may be simple or

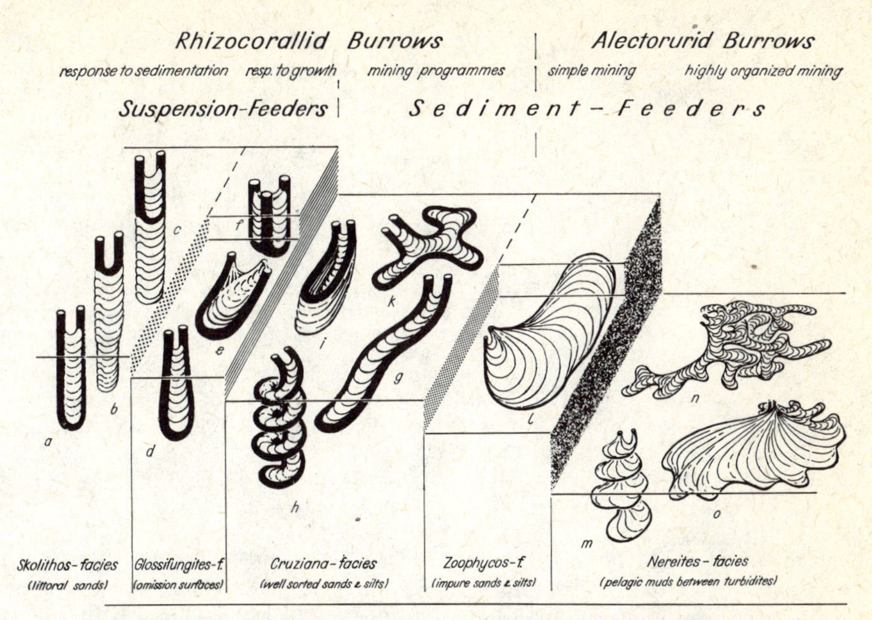

FIG. 4-32. Bathymetric zonation of trace fossils. Suspension feeders prevail in shallow waters and elaborate sediment feeders in deeper waters. (After Seilacher, 1967, Marine Geology, v. 5, Fig. 3.) By permission of Elsevier Pub. Co.

branched. The burrow filling commonly differs in texture from the host rock, and in some the filling is sequentially and intermittently deposited. Burrows do, of course, reach the sediment–fluid interface, and in the feeding structures are bedding-plane markings related to the burrow, usually radiating from it so that the structure has both a bedding-plane and a transectional aspect.

Burrows may also be largely in a horizontal plane at the interface between beds or even within a given bed. Some extend downward a long distance (20 cm or more) from the surface of the sediment; others are shallow.

Burrows are recognized in sectional view by the contrast in the texture of their fill and by that of the host rock, especially by destruction of the bedding that they transect. If burrows are abundant enough, only vestiges of the original bedding remain (Moore and Scruton, 1957). The rock may be much "churned" or "plowed over" by the organisms. *Bioturbation* is the term applied to this action, and *bioturbite* is the name given to a rock so affected (Fig. 4-33).

Geological significance In review, these various biogenic structures are useful in places for determining stratigraphic order in a vertical or overturned sequence of beds (Shrock, 1948, p. 173). Many are preserved only as casts on the *under*side of the sand beds.

They may also shed light on the rate of sedimentation. Seilacher (1962) was able to show that the sandstone beds in a flysch sequence were essentially instantaneous deposits. If they were not, burrows would be initiated at different levels within the bed and not just from the top downward. The sandstones in some of the Devonian "Portage" sequences in Pennsylvania are delicately laminated; the associated shales are bioturbated. The undisturbed laminated sands were deposited very quickly (in a few days at most), whereas the muds were exposed to burrowing activity over many years, perhaps centuries.

The general absence of burrows and the delicate preservation of laminations do not always mean rapid sedimentation. It may only mean suppression of the benthos because of toxicity from free H_2S or lack of oxygen. The trace-fossil assemblage may also correlate with salinity (Seilacher, 1963).

The most useful aspect of the trace-fossil assemblage is their guide to facies. Seilacher (1964a), for example, defined four major

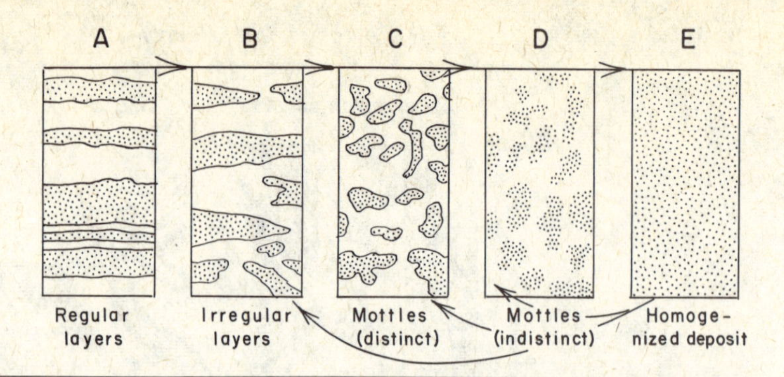

FIG. 4-33. Progressive alteration of sediment by burrowing organisms. (After Moore and Scruton, 1957, Bull. Amer. Assoc. Petrol. Geol., v. 41, Fig. 12.)

facies, each characterized by its own assemblage of ichnofossils. One facies, the *Nereites* facies, is characteristic of the flysch or turbidite basin; the *Zoophycus* facies is a shallow but tranquil-water facies. The *Cruziana* facies occupies a shallow shelf; and the *Scolithus* facies is essentially a rough-water shore facies. The deep-water turbidite environment (*Nereites* facies) is characterized mainly by grazing trails in contrast to the turbulent shore zone, where both dwelling and feeding burrows dominate. Clearly the trace-fossil morphology reflects an organism response and adaptation to the conditions which prevail in a particular environment.

In summary, trace fossils are a valuable aid to the sedimentologist. Like other aspects of a sedimentary rock, they can be mapped and used to define major facies belts (Farrow, 1966, Fig. 11) and to help establish progressive changes in water depth (Seilacher, 1967).

DIAGENETIC STRUCTURES

There are a host of structures—concretions, nodules, and other bodies—formed by post-depositional solution and precipitation. These epigenetic structures are described in detail in Chapter 13.

References

Aario, R., 1971, Syndepositional deformation in the Kurkiselkä esker, Kiiminki, Finland: Bull. Geol. Soc. Finland, v. 43, pt. 2, pp. 163–170.

Abel, O., 1935, Vorzeitliche Lebenspuren: Jena: Gustav Fischer, 644 pp.

Aitken, J. D., 1967, Classification and environmental significance of cryptalgal limestones and dolomites, with illustrations from the Cambrian and Ordovician of southwestern Alberta: Jour. Sed. Petrology, v. 37, pp. 1163–1178.

Allen, J. R. L., 1962, Asymmetrical ripple marks and the origin of cross-stratification: Nature, v. 194, pp. 167–169.

———, 1963a, Asymmetrical ripple marks and the origin of water-laid cosets of cross-strata: Liverpool and Manchester Geol. Jour., v. 3, pt. 2, pp. 187–236.

———, 1963b, The classification of cross-stratified units with notes on their origin: Sedimentology, v. 2, pp. 93–114.

———, 1964, Primary current lineation in the Lower Old Red Sandstone (Devonian), Anglo-Welsh Basin: Sedimentology, v. 3, pp. 89–108.

———, 1969, Current ripples: Amsterdam: North Holland, 433 pp.

———, 1970, A quantitative model of climbing ripples and their cross-laminated deposits: Sedimentology, v. 14, pp. 5–26.

———, 1971, Some techniques in experimental geology: Jour. Sed. Petrology, v. 41, pp. 695–702.

Andersen, S. A., 1931, Om Aase og Terrasser inden for Susaas Vandomraade og deres Vidnesbyrd om Isafsmeltningens Forlöb (The eskers and terraces in the basin of the River Susaa and their evidences of the process of the ice-waning): Danmarks geol. Undersögelse Raekke II, no. 5, 201 pp. (Danish with English summary).

Anderson, R. Y., Dean, W. E., Jr., Kirkland, D. W., and Snider, H. I., 1972, Permian Castile varved evaporite sequence, west Texas and New Mexico: Bull. Geol. Soc. Amer., v. 83, pp. 59–86.

Andrée, K., 1915, Ursachen und Arten der Schichtung: Geol. Rundschau, v. 6, pp. 351–397.

Anketell, J. M., Cegla, J., and Dzulynski, S., 1970, On the deformational structures in systems with reversed density gradients: Ann. Soc. Geol. Pologne, v. 40, pp. 3–30.

Apfel, E. T., 1938, Phase sampling of sediments: Jour. Sed. Petrology, v. 8, pp. 67–68.

Artyushkov, Ye. V., 1960a, Possibility of convective instability in sedimentary rocks and the general laws of its development: Dokl. Akad. Nauk. SSSR, Geol. Ser., v. 153, pp. 26–28.

———, 1960b, Principal forms of convective structures in sedimentary rocks: Dokl. Akad. Nauk. SSSR, v. 153, pp. 43–45.

Bailey, E. B., 1930, New light on sedimentation and tectonics: Geol. Mag., v. 67, pp. 77–92.

———, 1936, Sedimentation in relation to tectonics: Bull. Geol. Soc. Amer., v. 47, pp. 1713–1726.

Bailey, R. J., 1964, Ludlovian facies boundary in south central Wales: Geol. Jour., v. 4, pt. 1, pp. 1–20.

Bajard, J., 1966, Figures et structures sédimentaires dans la zone intertidal de la partie de la Baie du Mont-Saint-Michel: Rev. Géog. Phys. Géol. Dyn., v. 8, pp. 39–111.

Barrell, J., 1906, Relative geological importance of continental, littoral, and marine sedimentation: Jour. Geol., v. 14, pp. 316–356.

———, 1917, Rhythms and measurement of geologic time: Bull. Geol. Soc. Amer., v. 28, pp. 745–904.

Birkenmajer, K., 1959, Systematyka warstwawan w ufworach fliszowych i podobynch (Classification of bedding in flysch and similar graded deposits): Stud. Geol. Polonica, v. 3, 133 pp. (Polish with English summary).

Black, M., 1933, The algal sediments of Andros Island, Bahamas: Trans. Roy. Phil. Soc. London, ser. B, v. 222, pp. 165–192.

Bokman, J., 1957, Suggested use of bed-thickness measurements in stratigraphic descriptions: Jour. Sed. Petrology, v. 27, pp. 333–335.

Borrello, A. V., 1962, Sobre los diques clasticos: Rev. Mus. La Plata (n.s.), v. 5, pp. 155–191.

Botvinkina, L. N., 1959, Morfologicheskaia klassificatsiia sloistosti osadochaykh porod (Morphological classification of bedding in sedimentary rocks): Dokl. Akad. Nauk. SSSR, Ser. Geol., no. 6, pp. 16–33 (Amer. Geol. Inst. translation issued 1961, pp. 13–30).

———, 1962, Slosistot osadochaykh porod (Bedding of sedimentary rocks): Akad. Nauk. SSSR, Trudy Geol. Inst., no. 59, 542 pp.

———, 1965, Metodicheskoe rukovodstvo po izucheniiu stoistostiv (Manual on the methods of studying bedding): Akad. Nauk. SSSR, Trans. Geol. Inst., v. 119, 253 pp.

Bouma, A. H., 1962, Sedimentology of some flysch deposits: Amsterdam, Elsevier, 169 pp.

———, 1969, Methods for the study of sedimentary structures: New York, Wiley, 458 pp.

Bouma, A. H., and Brouwer, A., eds., 1964, Turbidites: Developments in Sedimentology, v. 3: Amsterdam, Elsevier, 264 pp.

Bradley, W. H., 1929, The varves and climate of the Green River epoch: U.S. Geol. Surv. Prof. Paper 158E, pp. 87–110.

Brinkmann, R., 1933, Uber Kreuzchichtung in deutschen Buntsandsteinbecken: Göttingen Nachr., Math. Physik, Kl. IV, no. 32, pp. 1–12.

Bruns, Ye. P., 1954, Observations of the peculiarities of bedding in deposits: VSEGEI, Gosgeoltekhzdat. (Manual of methods used in geological surveying and prospecting). In Russian.

Bucher, W. H., 1919, On ripples and related sedimentary surface forms and their paleogeographic interpretations: Amer. Jour. Sci., ser. 4, v. 47, pp. 149–210, 241–269.

Burst, J. F., 1965, Subaqueously formed shrinkage cracks in clay: Jour. Sed. Petrology, v. 35, pp. 348–353.

Campbell, C. V., 1967, Lamina, laminaset, bed, and bedset: Sedimentology, v. 8, pp. 7–26.

Carey, W. C., and Keller, M. D., 1957, Systematic changes in the beds of alluvial rivers: Jour. Hydraulics Div., Proc. Amer. Soc. Civil Eng., hy. 4, paper 1331, 24 pp.

Castellarin, A., 1966, Filon sedimentari nel Giurese di Loppio: Giornale di Geologia, ser. 2, vol. 33, pp. 527–546.

Chamberlain, C. K., 1971, Morphology and ethology of trace fossils from the Ouachita Mountains, southeastern Oklahoma: Jour. Paleont., v. 45, pp. 212–246.

Cline, L. M., 1966, Late Paleozoic rocks of Ouachita Mountains, a flysch facies: Kansas Geol. Surv., Guidebook to Ouachita Mountains, pp. 91–111.

Cloos, H., 1938, Primäre Richtungen in Sedimenten der rheinischen Geosynkline: Geol. Rundschau, v. 29, pp. 357–367.

Coleman, A. P., 1926, Ice ages, recent and ancient: New York: Macmillan, Inc., 296 pp.

Collins, W. H., 1925, North shore of Lake Huron: Geol. Surv. Canada Mem. 143, 160 pp.

Conybeare, C. E. B., and Crook, K. A. W., 1968, Manual of sedimentary structures: Bull. Australian Dept. Nat. Dev., Bur. Min. Res. Geol. Geophys. 102, 327 pp.

Cox, G. H., and Dake, C. L., 1916, Geological criteria for determining the structural position of sedimentary beds: Bull. Missouri Univ. School Mines, v. 2, pp. 1–59.

Crimes, T. P., and Harper, J. C., eds., 1970, Trace fossils: Geol. Jour., spec. issue, v. 3, 547 pp.

Crowell, J. C., 1955, Directional current structures from the pre-Alpine flysch, Switzerland: Bull. Geol. Soc. Amer., v. 66, pp. 1351–1384.
———, 1957, Origin of pebbly mudstones: Bull. Geol. Soc. Amer., v. 68, pp. 993–1010.
———, 1958, Sole markings of graded graywacke beds: a discussion: Jour. Geol., v. 66, p. 333.
Davis, R. A., Jr., 1968, Algal stromatolites composed of quartz sandstone: Jour. Sed. Petrology, v. 38, pp. 953–955.
Diller, J. S., 1890, Sandstone dikes: Bull. Geol. Soc. Amer., v. 1, pp. 411–442.
Dionne, J. C., 1971, Contorted structures in unconsolidated Quaternary deposits, Lake Saint-Jean and Saguenay regions, Quebec: Rev. Geogr. Montreal, v. 25, no. 1, pp. 5–33.
Dorr, J. A., Jr., and Kauffman, E. G., 1963, Rippled toroids from the Napoleon Sandstone Member (Mississippian) of southern Michigan: Jour. Sed. Petrology, v. 33, pp. 751–758.
Dott, R. H., Jr., 1961, Squantum "tillite," Massachusetts—evidence of glaciation or subaqueous mass movement?: Bull. Geol. Soc. Amer., v. 72, 1289–1306.
Dott, R. H., Jr., and Howard, J. K., 1962, Convolute lamination in non-graded sequences: Jour. Geol., v. 70, pp. 114–121.
Dunbar, G. J., and McCall, G. J. H., 1971, Archean turbidites and banded ironstones of the Mt. Belches area (Western Australia): Sed. Geol., v. 5, pp. 93–133.
Dzulynski, S., 1962, On load-casted ripples: Ann. Soc. Geol. Pologne, v. 32, pp. 147–160.
———, 1963, Directional structures in flysch: Stud. Geol. Polonica, v. 12, 136 pp.
———, 1966a, Influence of bottom irregularities and transported tools upon experimental scour markings: Ann. Soc. Geol. Pologne, v. 36, pp. 285–294.
———, 1966b, Sedimentary structures resulting from convection-like pattern of motion: Ann. Soc. Geol. Pologne, v. 36, pp. 3–21.
Dzulynski, S., Ksiazkiewicz, M., and Kuenen, Ph. H., 1959, Turbidites in flysch of the Polish Carpathian Mountains: Bull. Geol. Soc. Amer., v. 70, pp. 1089–1118.
Dzulynski, S., and Radomski, A., 1956, Clastic dikes in the Carpathian Flysch: Ann. Soc. Geol. Pologne, v. 26, pp. 225–264.
———, 1966, Experiments on bedding disturbances produced by the impact of heavy suspensions upon horizontal sedimentary layers: Bull. Acad. Sci. Polonaise, v. 14, no. 4, pp. 227–230.
Dzulynski, S., and Sanders, J. E., 1962, Current marks on firm mud bottoms: Trans. Connecticut Acad. Arts Sci., v. 42, pp. 57–96.
Dzulynski, S., and Smith, A. J., 1963, Convolute lamination, its origin, preservation, and directional significance: Jour. Sed. Petrology, v. 33, pp. 617–627.
Dzulynski, S., and Walton, E. K., 1963, Experimental production of sole markings: Trans. Edinburgh Geol. Soc., v. 19, pp. 279–305.
———, 1965, Sedimentary features of flysch and greywackes: Amsterdam, Elsevier, 300 pp.
Eisbacher, G. H., 1970, Contemporaneous faulting and clastic intrusions in the Quirke Lake Group, Elliot Lake, Ontario: Canad. Jour. Earth Sci., v. 7, pp. 215–225.
Elliott, R. E., 1965, A classification of subaqueous sedimentary structures based on rheological and kinematical parameters: Sedimentology, v. 5, pp. 193–209.
Fairbridge, R. W., 1946, Submarine slumping and the location of oil bodies: Bull. Amer. Assoc. Petrol. Geol., v. 30, pp. 84–92.
———, 1947, Possible causes of intraformational disturbances in the Carboniferous rocks of Australia: Jour. Proc. Roy. Soc. New South Wales, v. 81, pp. 99–121.
Farrow, G. E., 1966, Bathymetric zonation of Jurassic trace fossils from the coast of Yorkshire: Paleogeogr. Paleoclim. Paleoecol., v. 2, pp. 103–151.
Feige, K., 1937, Untersuchungen über zyklische Sedimentation geosynklinaler und epikontinentaler Räume: Abhandl. Preuss. Geol. Landesanst., n.s., v. 177, 218 pp.
Franks, P. C., 1969, Synaeresis features and genesis of siderite concretions, Kiowa Formation (early Cretaceous), north-central Kansas: Jour. Sed. Petrology, v. 39, pp. 799–803.
Garrett, P., 1970, Phanerozoic stromatolites: noncompetitive ecologic restriction by grazing and burrowing animals: Science, v. 169, pp. 171–173.
Gebelein, C. D., 1969, Distribution, morphology, and accretion rate of Recent subtidal algal stromatolites, Bermuda: Jour. Sed. Petrology, v. 39, pp. 49–69.
Ginsburg, R. N., 1967, Stromatolites: Science, v. 157, pp. 339–340.
Glaessner, M. F., 1958, Sedimentary flow structures on bedding planes: Jour. Geol., v. 66, pp. 1–7.
Glennie, K. W., 1963, An interpretation of turbidites whose sole markings show multiple directional trends: Jour. Geol., v. 71, pp. 525–527.
Gottis, Ch., 1953, Les filons clastique "intra formationnels" du flysch numidien tunisien: Bull. Soc. Géol. France, ser. 6, v. 3, pp. 775–784.
Gubler, Y., Bugnicourt, D., Faber, J., Kubler, B.,

and Nyssen, R., 1966, Essai de nomenclature et caractérisation des principales structures sédimentaires: Paris, Editions Technip, 291 pp.

Gürich, G., 1933, Schragschichtungsbögen und zapfenformige Fliesswülste im "Flagstone" von Pretoria und öhnliche Vorkommnisse im Quarzit von Kubis, S.W.A., dem Schilfstein von Maulbrunn, u.a.: Zeitschr. Deutsch Geol. Ges., v. 85, pp. 652–654.

ten Haaf, E., 1959, Graded beds of the northern Apennines: Ph.D. thesis, Rijks Univ., Groningen, 102 pp.

———, 1965, Significance of convolute laminations: Geol. en Mijn. v. 18, pp. 188–194.

Hadding, A., 1931, On subaqueous slides: Fören. Geol. Stockholm Förh., v. 53, pp. 378–393.

Hall, J., 1843a, Geology of New York: pt. IV, Survey of the Fourth Geol. Dist.: Albany, 683 pp.

———, 1843b, Remarks upon casts of mud furrows, wave lines, and other markings upon rocks of the New York System: Assoc. Amer. Geol. Nat., Rept., pp. 422–432.

Hamblin, W. K., 1961, Micro-cross-lamination in Upper Keweenawan sediments of northern Michigan: Jour. Sed. Petrology, v. 31, pp. 390–401.

———, 1965, Internal structures of "homogeneous" sandstones: Bull. Kansas Geol. Survey 175, pp. 569–582.

Häntzschel, W., 1962, Trace fossils and problematica, *in* Treatise on invertebrate paleontology (Moore, R. C., ed.). Pt. W, Miscellanea: Boulder, Colo., Geol. Soc. Amer. and Univ. Kansas, pp. W177–W245.

Harms, J. C., Mackenzie, D. B., and McCubbin, D. G., 1963, Stratification in modern sands of the Red River, Louisiana: Jour. Geol., v. 71, pp. 566–580.

Heim, A., 1908, Uber rezente und fossile subaquatische Rutschungen und deren lithologische Bedeutung: Neues. Jahrb. Min. Geol. Paläont., v. 1, pp. 136–157.

Hoffman, P. F., 1967, Algal stromatolites: use in stratigraphic correlation and paleocurrent determination: Science, v. 157, pp. 1043–1045.

———, 1969, Proterozoic paleocurrents and depositional history of East Arm fold belt, Great Slave Lake, Northwest Territories: Canadian Jour., Earth Sci., v. 6, pp. 441–462.

Hoffman, P. F., Logan, B. W., and Gebelein, C. D., 1969, Biological versus environmental factors governing the morphology and internal structure of recent algal stromatolites in Shark Bay, Western Australia: Geol. Soc. Amer., N.E. Section; 4th Ann. Mtg. Abstr., pt. 1, pp. 28–29.

Hofmann, H. J., 1964, Mudcracks in the Ordovician Maysville group: Jour. Geol., v. 72, pp. 638–640.

———, 1969, Attributes of stromatolites: Geol. Surv. Canada, Paper 69–39, 58 pp.

Holland, C. H., 1959, On convolute bedding in the Lower Ludlow rocks of North-East Radnorshire: Geol. Mag., v. 96, pp. 230–236.

———, 1960, Load-cast terminology and origin of convolute bedding: some comments: Bull. Geol. Soc. Amer., v. 71, p. 633.

———, 1963, Convolute lamination: a discussion: Jour. Geol., v. 71, p. 658.

Howard, J. D., and Lohrengel, C. F. II, 1969, Large nontectonic deformational structures from Upper Cretaceous rocks of Utah: Jour. Sed. Petrology, v. 39, pp. 1032–1039.

Hoyt, J. H., and Henry, V. J., Jr., 1963, Rhomboid ripple mark, indicator of current direction and environment: Jour. Sed. Petrology, v. 33, pp. 604–608.

Hubert, J. F., 1966, Sedimentary history of Upper Ordovician geosynclinal rocks, Girvan, Scotland: Jour. Sed. Petrology, v. 36, pp. 677–699.

Hyde, J. E., 1911, The ripples of the Bedford and Berea formations of central Ohio, with notes on the paleogeography of that epoch: Jour. Geol., v. 19, pp. 257–269.

Jones, O. T., 1937, On the sliding or slumping of submarine sediments in Denbighshire, north Wales, during the Ludlow period: Quart. Jour. Geol. Soc. London, v. 95, pp. 335–382.

Jordan, G. F., 1962, Large submarine sand waves: Science, v. 136, pp. 839–847.

Kalkowsky, E., 1908, Oolith und Stromatolith im norddeutschen Buntsandstein: Zeitschr. Deutsch Geol. Ges., v. 60, pp. 68–125.

Kelling, G., and Walton, E. K., 1961, Flow structure in sedimentary rocks: a discussion: Jour. Geol., v. 69, p. 224.

Khabakov, A. V., ed., 1962, Atlas tekstur i struktur osadochnykh gornykh porod (An atlas of textures and structures of sedimentary rocks. Pt. 1, Clastic and argillaceous rocks): Moscow, VSEGEI, 578 pp., 268 pls. (Russian, with French translation of plate captions).

Kindle, E. M., 1917, Recent and fossil ripple marks: Bull. Mus. Geol. Surv. Canada, 25, pp. 1–56.

Klein, G. de V., 1967, Paleocurrent analysis in relation to modern marine sediment dispersal patterns: Bull. Amer. Assoc. Petrol. Geol., v. 51, pp. 366–382.

Kopstein, F. P. H. W., 1954, Graded bedding of the Harlech Dome: Ph.D. thesis, Rijks Univ., Groningen, 97 pp.

Krejci-Graf, K., 1932, Definition der Begriffe Marken, Spuren, Fahrten, Bauten, Hieroglyfen und Fucoiden: Senkenbergiana, v. 4, pp. 19–39.

Ksiazkiewicz, M., 1954, Graded and laminated bedding in the Carpathian flysch: Ann. Soc. Geol. Pologne, v. 22, pp. 399–449 (Polish with English summary).

———, 1958, Submarine slumping in the Carpathian flysch: Ann. Soc. Geol. Pologne, v. 28, pp. 123–150.

Kuenen, Ph. H., 1952, Paleogeographic significance of graded bedding and associated features: Proc. Koninkl. Nederl. Akad. Wetensch. ser. B, v. 55, pp. 28–36.

———, 1953a, Significant features of graded bedding: Bull. Amer. Assoc. Petrol. Geol., v. 37, pp. 1044–1066.

———, 1953b, Graded bedding with observations on Lower Paleozoic rocks of Britain: Verhandl. Koninkl. Nederl. Akad. Wetensch., Afd. Nat., v. 20, no. 3, pp. 1–47.

———, 1957, Sole markings of graded graywacke beds: Jour. Geol., v. 65, pp. 231–258.

———, 1958, Experiments in geology: Trans. Glasgow Geol. Soc., v. 23, pp. 1–28.

———, 1964, Deep-sea sands and ancient turbidites, *in* Turbidites (Bouma, A. H., and Brouwer, A., eds.) Amsterdam: Elsevier, pp. 3–33.

———, 1967, Emplacement of flysch-type sand beds: Sedimentology, v. 9, pp. 203–243.

———, 1970, The turbidite problem: some comments: New Zealand Jour. Geol. Geophys. v. 13, pp. 852–857.

Kuenen, Ph. H., and Carozzi, A., 1953, Turbidity currents and sliding in geosynclinal basins in the Alps: Jour. Geol., v. 61, pp. 363–372.

Kuenen, Ph. H., and Menard, H. W., 1952, Turbidity currents, graded and non-graded deposits: Jour. Sed. Petrology, v. 22, pp. 83–96.

Kuenen, Ph. H., and Migliorini, C. I., 1950, Turbidity currents as a cause of graded bedding: Jour. Geol., v. 58, pp. 91–127.

Kuenen, Ph. H., and Sanders, J. E., 1956, Sedimentation phenomena in Kulm and Flözleeres graywackes, Sauerland and Oberharz, Germany: Amer. Jour. Sci., v. 254, pp. 649–671.

Lanteaume, M., Beaudoin, B., and Campredon, R., 1967, Figures sédimentaires du flysch "Gres D'Annot" du synclinal de Peira-Cava.: Paris, Editions du Cent. Nat. Rech. Sci., 99 pp.

Lessertisseur, J., 1955, Traces fossiles d'activité animale et leur signification paléobiologique: Soc. Geol. France Mem. 74, 142 pp.

Logan, B. W., 1961, *Cryptozoon* and associate stromatolites from the Recent, Shark Bay, Western Australia. Jour. Geol., v. 69, pp. 517–533.

Logan, B, W., Rezak, R., and Ginsburg, R. N., 1964, Classification and environmental significance of algal stromatolites: Jour. Geol., v. 72, pp. 68–83.

Lowenstam, H. A., 1950, Niagaran reefs of the Great Lakes area: Jour. Geol., v. 58, pp. 430–487.

McBride, E. F., 1962, Flysch and associated beds of the Martinsburg Formation (Ordovician), central Appalachians: Jour. Sed. Petrology, v. 32, pp. 39–91.

McBride, E. F., and Yeakel, L. S., 1963, Relationship between parting lineation and rock fabric: Jour. Sed. Petrology, v. 33, pp. 779–782.

McGowen, J. H., and Garner, L. E., 1970, Physiographic features and stratification types of coarse-grained point bars; modern and ancient examples: Sedimentology, v. 14, pp. 77–112.

McIver, N. L., 1970, Appalachian turbidites, *in* Studies of Appalachian geology: central and southern (Fisher, G. W., Pettijohn, F. J., Reed, J. C., Jr., and Weaver, K. N., eds.): New York, Wiley-Interscience, pp. 69–82.

McKee, E. D., 1938, Original structures in Colorado River flood deposits of Grand Canyon: Jour. Sed. Petrology, v. 8, pp. 77–83.

———, 1939, Some types of bedding in the Colorado River delta: Jour. Geol., v. 47, pp. 64–81.

———, 1966a, Dune structures: Sedimentology, spec. ed., v. 7, pp. 1–69.

———, 1966b, Significance of climbing ripple structure: U.S. Geol. Surv., Prof. Paper 550-D, pp. 94–103.

McKee, E. D., and Weir, G. W., 1953, Terminology of stratification and cross-stratification: Bull. Geol. Soc. Amer., v. 64, pp. 381–390.

Macar, P., 1948, Les pseudonodules du Famenien et leur origine: Ann. Soc. Geol. Belgique, v. 72, pp. 47–74.

Marschalko, R., 1961, Sedimentologic investigation of marginal lithofacies in flysch of central Carpathians: Geol. Prace (Bratislava), v. 60, pp. 197–230.

———, 1963, Sedimentary slump folds and the deposition slope (flysch of central Carpathians): Geol. Prace (Bratislava), v. 28, pp. 161–168.

Maslov, V. P., 1953, Printzipy nomenclatury i sistematiki stromatolitov (Principles of nomenclature and systematics of stromatolites): Akad. Nauk. SSSR, Izvestia, Ser. Geol., pp. 105–112.

Meischner, K.-D., 1964, Allodapische Kalke, Turbidite in Riff-nahen Sedimentations-Becken, *in* Turbidites (Bouma, A. H., and Brouwer, A., eds.): Amsterdam, Elsevier, pp. 156–191.

Middleton, G. R., ed., 1965, Primary sedimentary structures and their hydrodynamic interpretation: Soc. Econ. Paleont. Min. Spec. Pub. 12, 265 pp.

Miller, W. J., 1922, Intraformational corrugated rocks: Jour. Geol., v. 30, pp. 587–610.

Moore, D. G., and Scruton, P. C., 1957, Minor internal structures of some recent unconsolidated sediments: Bull. Amer. Assoc. Petrol. Geol., v. 41, pp. 2723–2751.

Multer, H. G., and Hoffmeister, J. E., 1968, Subaerial laminated crusts of the Florida Keys: Bull. Geol. Soc. Amer., v. 79, pp. 183–192.

Murphy, M. A., and Schlanger, S. O., 1962, Sedimentary structures in Ilhas and São Sabastiao formations (Cretaceous), Reconcavo Basin, Brazil: Bull. Amer. Assoc. Petrol. Geol., v. 46, pp. 457–477.

Nagtegaal, P. J. C., 1965, An approximation to the genetic classification of non-organic sedimentary structures: Geol. Mijn., v. 44, pp. 347–352.

Nagahama, H., Hirokawa, O., and Enda, T., 1968, History of researches on paleocurrents in reference to sedimentary structures, with paleocurrent maps and photographs of sedimentary structures: Bull. Japan Geol. Surv. 19, pp. 1–9.

Natland, M. L., and Kuenen, Ph. H., 1951, Sedimentary history of the Ventura Basin, California, and the action of turbidity currents: Soc. Econ. Paleont. Min. Spec. Pub. 2, pp. 76–107.

Nesteroff, W. D., 1961, La "sequence type" dans les turbidities terrigènes moderns: Rev. Geograph. Phys. Geol. Dyn., v. 4, pp. 263–268.

Newell, N. D., Rigby, J. K., Fischer, A. G., Whiteman, A. G., Hickox, J. E., and Bradley, J. S., 1953, The Permian reef complex of the Guadalupe Mountains region, Texas and New Mexico: San Francisco: Freeman, 236 pp.

Niehoff, W., 1958, Die Primär gerichteten Sedimentstrukturen inbesondere die Schrägschichtung im Koblenzquarzit am Mittelrhein: Geol. Rundschau, v. 47, pp. 252–321.

Off, T., 1963, Rhythmic linear sand bodies caused by tidal currents: Bull. Amer. Assoc. Petrol. Geol., v. 47, pp. 324–341.

Ojakangas, R. W., 1968, Cretaceous sedimentation, Sacramento Valley, California: Bull. Geol. Soc. Amer., v. 79, pp. 973–1008.

Otto, G. H., 1938, The sedimentation unit and its use in field sampling: Jour. Geol., v. 46, pp. 569–582.

Otvos, E. G., Jr., 1965, Types of rhomboid beach surface patterns: Amer. Jour. Sci., v. 263, pp. 271–276.

Pavoni, N., 1959, Rollmarken von Fischwirbeln aus der oligozänen Fischschiefern von Engi-Matt (K. Glarus): Eclogae Géol. Helvetiae, v. 52, pp. 941–949.

Payne, T. G., 1942, Stratigraphical analysis and environmental reconstruction: Bull. Amer. Assoc. Petrol. Geol., v. 26, pp. 1697–1770.

Pettijohn, F. J., 1936, Early Precambrian varved slate in northwestern Ontario: Bull. Geol. Soc. Amer., v. 47, pp. 621–628.

———, 1943, Archean sedimentation: Bull. Geol. Soc. Amer., v. 54, pp. 925–972.

———, 1962, Paleocurrents and paleogeography: Bull. Amer. Assoc. Petrol. Geol., v. 46, pp. 1468–1493.

Pettijohn, F. J., and Potter, P. E., 1964, Atlas and glossary of primary sedimentary structures: New York, Springer, 370 pp.

Potter, P. E., and Pettijohn, F. J., 1963, Paleocurrents and basin analysis: New York, Springer, 296 pp.

Potter, P. E., and Scheidegger, A. E., 1966, Bed thickness and grain size: graded beds: Sedimentology, v. 7, pp. 233–240.

Prentice, J. E., 1960, Flow structures in sedimentary rocks: Jour. Geol., v. 68, pp. 217–224.

Quirke, T. T., 1917, Espanola district, Ontario: Geol. Surv. Canada Mem. 102, 92 pp.

Reineck, H.-E., 1955, Marken, Spuren und Fährten in den Waderner Schichten (ro) bei Martinstein/Nahe.: Neues Jahrb. Geol. Paläont. Abh. 101, pp. 75–90.

———, 1963, Sedimentgefüge im Bereich der südlichen Nordsee: Senckenberg. Naturf. Gesell. Abh., v. 505, 138 pp.

Reineck, H.-E., and Wunderlich, F., 1968, Classification and origin of flaser and lenticular bedding: Sedimentology, v. 11, pp. 99–105.

Rettger, R. E., 1935, Experiments on soft-rock deformation: Bull. Amer. Assoc. Petrol. Geol., 19, 271–292.

Rezak, R., 1957, Stromatolites of the Belt Series in Glacier National Park and vicinity, Montana: U.S. Geol. Surv. Prof. Paper 294-D, pp. 127–154.

Ricci Lucchi, F., 1970, Sedimentografia: Bologna, Zonichelli, 288 pp.

Rich, J. L., 1950, Flow markings, groovings, and intrastratal crumplings, etc.: Bull. Amer. Assoc. Petrol. Geol., v. 34, pp. 717–741.

Rigby, J. K., 1958, Mass movements in Permian rocks of Trans-Pecos Texas: Jour. Sed. Petrology, v. 28, pp. 298–315.

Rubey, W. W., 1931, Lithologic studies of fine-grained Upper Cretaceous rocks of the Black Hills region: U.S. Geol. Surv. Prof. Paper 165-A, pp. 1–54.

Sanders, J. E., 1960, Origin of convoluted laminae: Geol. Mag., v. 97, pp. 409–421.

Schafer, W., 1956, Wirkungen der Benthosorganismen auf den jungen Schichtverband: Senckenbergiana, 37, pp. 183–263.

———, 1962, Aktuo-Paläontologie: Frankfurt-am-Main, Kramer, 666 pp.

Scheidegger, A. E., and Potter, P. E., 1971, Downcurrent decline of grain size and thickness of single turbidite beds: a semi-quantitative analysis: Sedimentology, v. 17, pp. 41–49.

Schleiger, N. W., 1964, Primary scalar bedding features of the Siluro-Devonian sediments of the Seymour district, Victoria: Jour. Geol. Soc. Australia, v. 11, pt. 1, pp. 1–31.

Schwarzacher, W., 1953, Cross-bedding and grain size in the Lower Cretaceous sands of East Anglia: Geol. Mag., v. 90, pp. 322–330.

Scott, K. M., 1966, Sedimentology and dispersal pattern of a Cretaceous flysch sequence, Patagonian Andes, southern Chile: Bull. Amer. Assoc. Petrol. Geol., v. 50, pp. 72–107.

Seilacher, A., 1953, Über die Methoden der Palichnologie: Neues Jahrb. Geol. Paläont. Abh. 96, pp. 421–452.

———, 1962, Paleontological studies on turbidite sedimentation and erosion: Jour. Geol., v. 70, pp. 227–234.

———, 1963, Umlagerung und Rolltransport von Cephalopoden-Gehäusen: Neues Jahrb. Geol. Paläont. Mh. 11, pp. 593–615.

———, 1964a, Biogenic sedimentary structures, *in* Approaches to Paleoecology (Imbrie, J., and Newell, N., eds.): New York, Wiley, pp. 296–316.

———, 1964b, Sedimentological classification and nomenclature of trace fossils: Sedimentology, v. 3, pp. 253–256.

———, 1967, Bathymetry of trace fossils: Marine Geol., v. 5, pp. 413–428.

Selley, R. C., 1964, The penecontemporaneous deformation of heavy mineral bands in the Torridonian Sandstone of northwest Scotland, *in* Deltaic and shallow marine deposits: Amsterdam, Elsevier, pp. 362–367.

Selley, R. C., Shearman, D. J., Sutton, J., and Watson, J., 1963, Some underwater disturbances in the Torridonian of Skye and Raasay: Geol. Mag., v. 100, pp. 224–243.

Sharp, R. P., 1963, Wind ripples: Jour. Geol., v. 71, pp. 617–636.

Shelton, J. W., 1962, Shale compaction in a section of Cretaceous Dakota Sandstone, northwestern North Dakota: Jour. Sed. Petrology, v. 32, pp. 873–877.

Shrock, R. R., 1948, Sequence in layered rocks: New York, McGraw-Hill, 507 pp.

Simonen, A., and Kuovo, O., 1951, Archean varved schists north of Tampere in Finland: Soc. Geol. Finlande, Compte rendus, v. 24, pp. 93–117.

Sorauf, J. E., 1965, Flow rolls of Upper Devonian rocks of south-central New York State: Jour. Sed. Petrology, v. 35, pp. 553–563.

Stewart, J. H., 1961, Origin of cross-strata in fluvial sandstone layers in the Chinle Formation (Upper Triassic) on the Colorado Plateau: U.S. Geol. Surv. Prof. Paper 424-B, art. 54, pp. 127–129.

Stokes, W. L., 1947, Primary lineation in fluvial sediments: a criterion of current direction: Jour. Geol., v. 45, pp. 52–54.

———, 1953, Primary sedimentary trend indicators as applied to ore finding in the Carrizo Mountains, Arizona and New Mexico: U.S. Atomic Energy Comm. RME-3043, pt. 1.

van Straaten, L. M. J. U., 1949, Occurrence in Finland of structures due to subaqueous sliding of sediments: Bull. Geol. Comm. Finlande 144, pp. 7–18.

———, 1950, Giant ripples in tidal channels: Koninkl. Nederl. Aardr., Genootsch., v. 67, pp. 76–81.

Sujkowski, Zb. L., 1957, Flysch sedimentation: Bull. Geol. Soc. Amer., v. 68, pp. 543–554.

Sullwold, H. H., Jr., 1959, Nomenclature of load deformation in turbidites: Bull. Geol. Soc. Amer., v. 70, pp. 1247–1248.

Sundborg, A., 1956, The River Klarälven: a study of fluvial processes: Geografiska Annaler, v. 38, pp. 127–316.

Taliaferro, N. L., 1934, Contraction phenomena in cherts: Bull. Geol. Soc. Amer., v. 45, pp. 189–232.

van der Linden, W. J. M., 1963, Sedimentary structures and facies interpretation of some molasse deposits: Geol. Ultraiectina, Med. Geol. Inst. Utrecht, no. 12, 42 pp.

van der Lingen, G. J., 1969, The turbidite problem: New Zealand Jour. Geol. Geophys., v. 12, pp. 7–50.

Vassoevich, N. B., 1953, O nekotorykh flishevykh tekturakh (Znakakh) (On some flysch textures) Trudy Lvovs: Geol. Obsh. Univ. Ivan Franko, geol. ser. 3, pp. 17–85.

Vitanage, P. W., 1954, Sandstone dikes in the South Platte area, Colorado: Jour. Geol., v. 62, pp. 493–500.

Walker, R. G., 1963, Distinctive types of ripple-drift cross-lamination: Sedimentology, v. 2, pp. 173–188.

———, 1967, Turbidite sedimentary structures and their relationship to proximal and distal depositional environments: Jour. Sed. Petrology, v. 37, pp. 25–43.

———, 1969, Geometrical analysis of ripple-drift cross-lamination: Canadian Jour. Earth Sci., v. 6, no. 3, pp. 383–391.

———, 1970, Review of the geometry and facies

organization of turbidites and turbidite-bearing basins, *in* Flysch sedimentology in North America (Lajoie, J., ed.): Geol. Assoc. Canada, spec. paper no. 7, pp. 219–252.
Walker, R. G., and Pettijohn, F. J., 1971, Archean sedimentation: Analysis of the Minnitaki Basin, northwestern Ontario, Canada: Bull. Geol. Soc. Amer., v. 82, pp. 2099–2130.
Weimer, R. J., and Hoyt, J. H., 1964, Burrows of *Callianassa major* Say, geologic indicators of littoral and shallow neritic environments: Jour. Paleont., v. 38, pp. 761–767.
Westphal, F., 1957, Synsedimentär gequollene Gesteine in alttertiären Sedimenten des Oberrheintalgrabens: Ber. Naturf. Ges. Freiburg im Br., v. 47, pp. 103–114.
White, W. A., 1961, Colloid phenomena in sedimentation of argillaceous rocks: Jour. Sed. Petrology, v. 31, pp. 560–570.
Williams, E., 1960, Intra-stratal flow and convolute folding: Geol. Mag., v. 97, pp. 208–214.
Wood, A., and Smith, A. J., 1959, The sedimentation and sedimentary history of the Aberystwyth Grits (Upper Llandoverian): Quart. Jour. Geol. Soc. London, v. 114, pp. 163–195.
Wurster, P., 1958, Geometrie und Geologie von Kreuzschichtungs-Körpern: Geol. Rundschau, v. 58, pp. 322–359.
Zhemchuzhnikov, Yu. A., 1940, An attempt at a morphological classification of bedding in sedimentary rocks: Gornyatskaya Pravda, Nauchno-tekhnicheskiy listok.

5

GEOMETRY OF SEDIMENTARY BODIES

INTRODUCTION

Just as igneous rocks occur in various forms—some as extrusive flows and domes, others as intrusive sheets or stocks—so also sedimentary rocks exhibit a variety of forms. Some are sheetlike deposits, produced in some cases by fallout of sediment from suspension and in other cases by lateral accretion or sedimentation. Some form somewhat circumscribed flows (mud-flows, for example) or deposits of grain and turbidity flows. These deposits are not necessarily coextensive with the basin in which they accumulate. Some may be localized by proximity to the source of sediment supply and constitute fans, either subaerial or subaqueous, or form composite fans or wedgelike aprons. Others are restricted for other reasons and accumulate at certain loci, as in the case of shoestring deposits related to river channels or shorelines.

Still other sedimentary bodies are accumulates, either mechanical or organic, which bear only an indirect relation (if any) to the geomorphic situation. Such organic accumulates are the reefs and banks, whose locations and shapes are the result of the interactions of climate, hydrology, and topography on organic communities. The size and shape of mechanical accumulates such as dunes are likewise controlled by subtle factors difficult to identify or measure.

A few small and rare sedimentary deposits are analogous to the intrusive igneous rocks and are in fact dikes, sills, or diapirs. Most striking of these, perhaps, are salt "domes" or stocks. More common, but very small in volume, are deposits in large and small cavities. Included here are deposits in caves and sinks.

Sandstone and limestone are more prone to accumulate as localized well-defined masses; shales tend to form wherever tranquil waters prevail, but we have no shale analogues of shoestring sands or limestone reefs. An exception, perhaps, are carbonate mud mounds. Shales frequently form the "matrix" in which the sand bodies are embedded (Fig. 5-5).

A geomorphic entity may coincide with or delimit a sedimentary body as in the case of a reef or dune. On the other hand, some geomorphic entities are complex and contain many discrete sedimentary bodies. Such is the case with a delta, which may include various types of sand bodies of several different origins.

Interest in the geometry of sedimentary bodies was greatly intensified by exploration for oil, in part because some oil accumulations are contained in or related to certain sedimentary bodies (such as reefs and shoestring sands) and also because closely spaced drilling made it possible to delineate in considerable detail sedimentary geometry. By geometry we mean gross forms and dimensions of the sedimentary body and not its internal organization, although these two are related. The latter is treated in the chapter on environmental analysis in which both external form and the internal structures of a sedimentary body are of major significance.

One of the earliest papers on the geometry of sandstone bodies was that of Rich (1938), who attempted to formulate a rational classification of sand bodies and to summarize their salient features. Interest in the subject culminated in the publication of a symposium volume, *Geometry of Sandstone Bodies,* in 1961, containing some 14 papers on this subject. More recent contributions include those of Potter (1963) on the shape and distribution patterns of Late Paleozoic sand bodies in the Illinois Basin, of LeBlanc (1972) on the geometry of sandstone reservoir bodies, of Shelton and others (1972). There are, of course, many papers on specific sandstone bodies, some of which are cited here; others are included in a reprint volume on sandstone reservoirs (Weimer, 1973).

The geometry of carbonate bodies received an important impetus when oil was found to be associated with some ancient limestone reefs. This discovery led to a great resurgence of interest in ancient and modern reefs. And a great deal of this interest dealt with reef geometry—form and size. It was soon learned that not all bioherms, a term applied by Cumings and Shrock (1928) to a locally restricted carbonate buildup, were reefs. Some were calcilutitic bodies, essentially mud mounds. Although considerable attention has been paid to the size, shape, and orientation of these carbonate buildups, the geometry of carbonate bodies is less well understood than that of sands.

On the other hand, the geometry of salt "domes" or stocks and related structures has long been known, again the result of the association of petroleum with these structures. Current knowledge of these features has been summarized in various text and reference books such as Levorson's *Geology of Petroleum* (1967, pp. 356–379) and the memoir on diapirism and diapirs (Braunstein and O'Brien, 1968). Unlike sand bodies or reefs, salt domes are secondary or postdepositional bodies, so they have no modern counterpart. In like manner, the geometry of sandstone dikes and sills is known only from a study of ancient examples.

In general, the geometry of sedimentary bodies can be ascertained from careful mapping of exhumed and dissected reefs, sand bodies, and the like, or from study of geophysical logs and cores or cuttings in a closely drilled situation.

SANDSTONE BODIES

Initially sandstones were described either as *sheet sands,* extended in two dimensions and restricted in the third, or *shoestring sands,* extended in only one dimension and restricted in the other two. Krynine (1948, p. 146) expanded the terminology to include four types: blanket or sheet, tabular, prism, and shoestring (Fig. 5-1). Each of these was defined in terms of specific ratios of width to thickness. Others have classified sand bodies in genetic terms (Rittenhouse, 1961; LeBlanc, 1972), designating them as alluvial fans, dunes, beaches, and so forth. Although this approach is most geologically meaningful, it tends to confuse the question of geometry (shape and size) with the whole sedimentary model concept, which involves many other attributes. Moreover, as noted by Potter (1963, p. 15), genetic classifications are subject to error, because establishing the environment of deposition of many ancient sand bodies is difficult. Furthermore, one type of sand body may grade into another. And subsurface data, commonly geophysical, do not readily permit a genetic assignment. Hence Potter's classification is strictly descriptive and geometrical. He recognized two classes: widespread sheet sands and linear or elongate (in some cases discontinuous) sand bodies. In Illinois Late Paleozoic sheet sands tend to be thin (under 20 feet or 6 m), fine-grained, and rippled, and contain marine fossils. They constitute conformable sheets. Elongate bodies are thicker (up to 125 feet or 38 m), and generally somewhat coarser, in some cases with shale intraclasts and small quartz pebbles. They are cross-bedded and contain reworked marine fossils and plant debris. With erosional and disconformable bases, elongate sand bodies show one or another of four distribution patterns—pods, ribbons, dendroids, and belts (Fig. 5-2).

For our purposes we recognize simple linear or shoestring sands, complex and

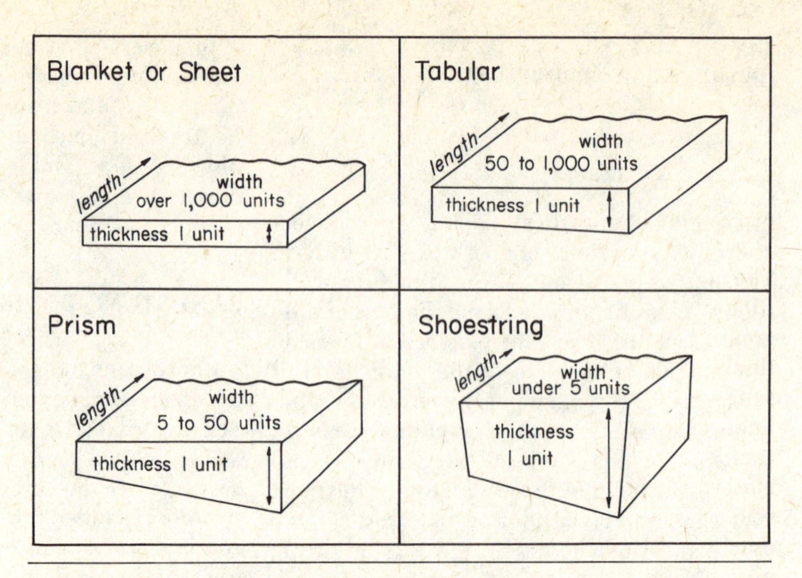

FIG. 5-1. Geometry of sedimentary bodies. (After Krynine, 1948, Fig. 9.)

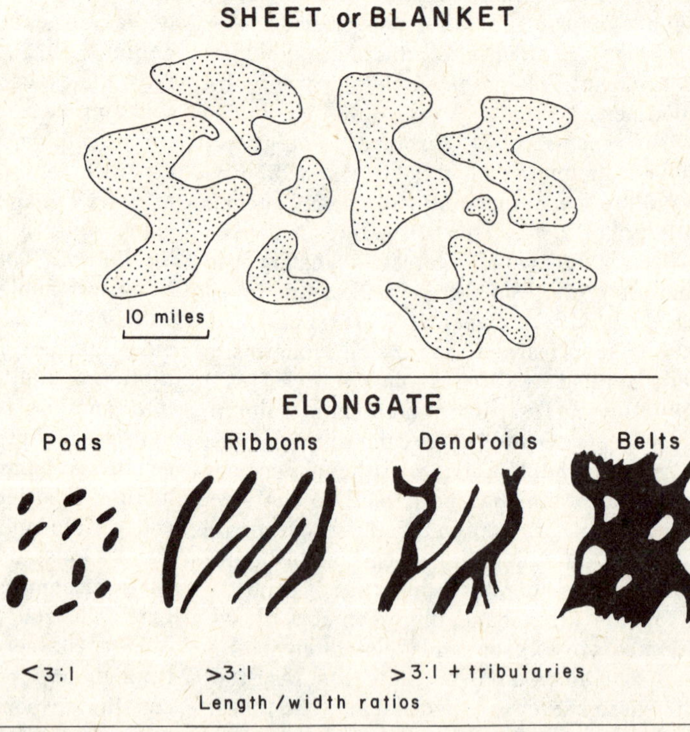

FIG. 5-2. Patterns of sheet-sand bodies and four types of elongate sand bodies—pods, ribbons, dendroids, and belts. (From Potter, 1963, Fig. 7.)

bifurcating linear sands, wedge-shaped sand bodies, and sheet sands. Sandstone dikes and sills constitute a separate class because they are secondary and not a primary depositional accumulation.

Sand bodies, of course, need not be siliceous sands. Carbonate sands also form discrete accumulations (Ball, 1967), but, for reasons pointed out elsewhere, they are more difficult to identify in the geologic record than are siliceous sand bodies encased in shale.

SHOESTRING SANDS

The term *shoestring sand,* introduced by Rich (1923), is applied to those sand bodies greatly elongated in proportion to their thickness and width. They are the result of sand accumulation in a relatively narrow linear belt. Their origin and prediction of trend have long been of interest; the now classic paper by Bass (1934) on the Bartlesville shoestring sands of Kansas generated much interest in the problem of these deposits.

Shoestring sands vary from small bodies, commonly showing their whole cross section in a single outcrop, to large bodies several hundred feet thick, with widths varying from a half mile (0.8 km) or less to several miles (3 km) wide and up to a hundred or more miles (160 km) in length. The Bethel Sandstone (Mississippian) of western Kentucky and south-central Indiana, for example, has been traced nearly 200 miles (320 km) (Reynolds and Vincent, 1967). See Fig. 5-3. Near Fort Knox it is 150 to 200 feet (46 to 61 m) thick and 0.5 to 0.8 miles (0.8 to 1.3 km) wide (Sedimentation Seminar, 1969). The Caseyville channel (Pennsylvanian) in Kentucky is traceable for over 100 miles (161 km) and is 100 to 200 feet (30 to 60 m) thick, and up to 4 to 6 miles (6.4 to 9.9 km) wide (Fig. 5-4). A similar but smaller example is the Anvil Rock Sandstone (Pennsylvanian) of Illinois (Fig. 5-5). The Bartlesville shoestring sands (Pennsylvanian) of Kansas and Oklahoma are 50 to 150 feet (15.2 to 45.7 m) thick, 0.5 to 2 miles (0.8 to 3.2 km) wide, and individual sand bodies are 2 to 6 miles (3.2 to 9.6 km) long. They are commonly arranged in "trends" or belts over 50 miles (80 km) long (Fig. 5-6).

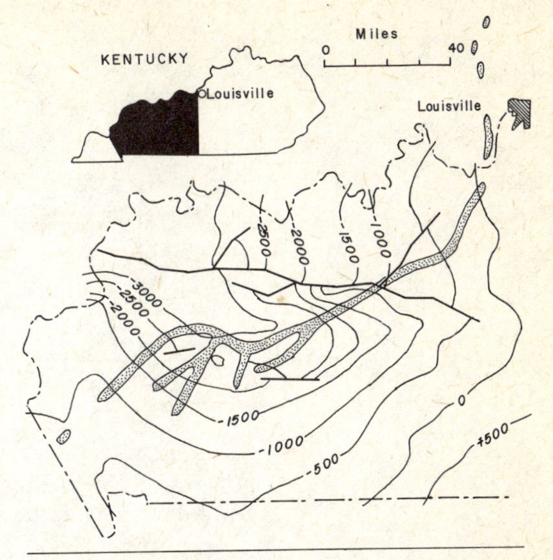

FIG. 5-3. Bethel (Mississippian) channel and structure on top of New Albany Shale, western Kentucky. (From Sedimentation Seminar, 1969, Fig. 1, as modified from Reynolds and Vincent, 1967.)

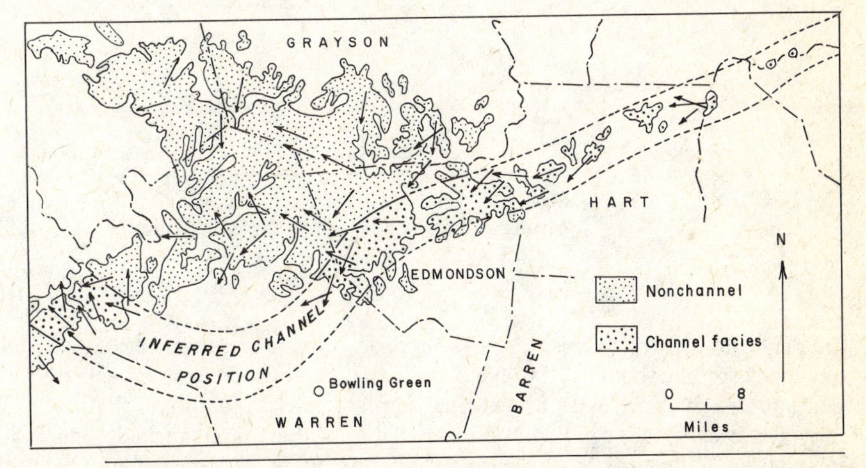

FIG. 5-4. Semidiagrammatic map of the pre-Pennsylvanian channel in Edmonson County, Kentucky. (After Potter and Siever, 1956, Jour. Geol., v. 64, Fig. 6.)

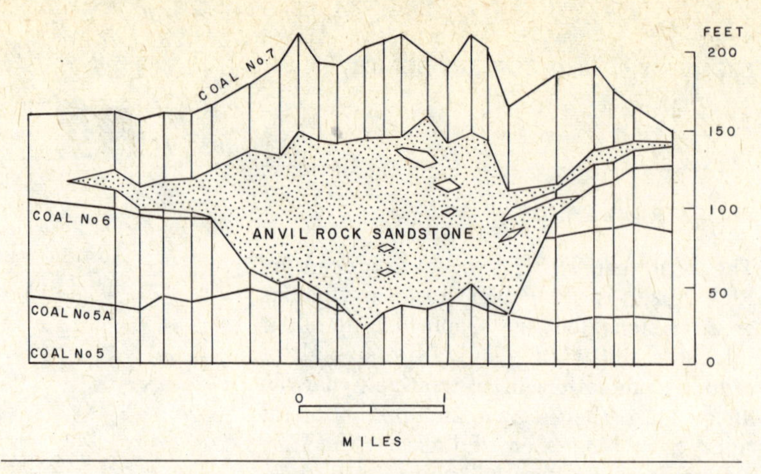

FIG. 5-5. Cross section of Anvil Rock Sandstone (Pennsylvanian) in Edwards County, Illinois. (From Potter, 1962, Fig. 7.)

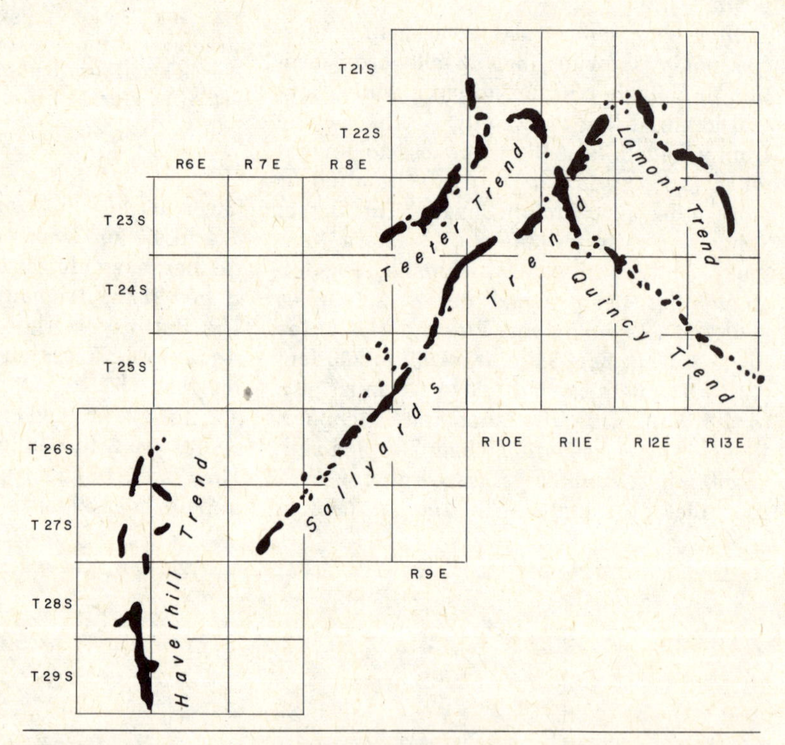

FIG. 5-6. Map of Bartlesville shoestring sands in Greenwood-Butler county area, Kansas. (After Bass, 1934, Bull. Amer. Assoc. Petrol. Geol., v. 18, Fig. 2.)

Some shoestring sands occupy incised channels; others have a flat base and appear not to be channels. Shoestring sands may be interrupted or discontinuous. This intermittent pattern is in part brought about by erosion that leaves isolated remnants. In other cases it is depositional; the original sand formed discrete bodies that may have been arranged, as noted above, in a belt or trend. Some shoestring sands display a fairly simple, slightly sinuous pattern; others are markedly meandering. Some are more complex deposits with either divergent or convergent branches. These are discussed below.

Shoestring sands have multiple origins.

Some seem to be fluvial channels (such as the Caseyville in Kentucky); others seem to occupy submarine channels cut in a carbonate shelf (Bethel); still others were barrier islands (Bartlesville).

COMPLEX SAND BODIES

We include here those sand bodies which show complex bifurcation patterns, divergent or convergent branches, and anastomosing patterns.

One complexity arises when more or less simple linear sand bodies are superposed on one another. Such superposition produces an inordinate sand thickness. These superposed sands may be termed *multistory sands* (Fig. 5-7).

A more common complexity is attributable to multiple bifurcation of the sand body such as one might expect in a distributary region on a delta. In some cases the sand bodies show an anastomosing pattern. Examples of such complex sands include the "Frio" (Oligocene) sands of the Seeligson Field, the pattern of which is interpreted by Nanz (1954) as a distributary system in the upper portions of a deltaic plain (Fig. 5-8). The "Jackpile" Sandstone (Jurassic) in New Mexico has similar aspects and is presumed to be of the same origin (Schlee and Moench, 1961). One of the best known, a somewhat more widely divergent branching sand body, is the Booch Sandstone (Pennsylvanian) in eastern Oklahoma (Busch, 1959). It is regarded as a deltaic distributary sand (Fig. 5-9).

WEDGE-SHAPED (FAN) SAND BODIES

Some sedimentary bodies, mainly sands and gravels, are notably wedge-shaped in cross section. In plan these tend to be divergent from a thicker apical area, and their cross-bedding tends to radiate from this apex. One of the largest such deposits described is the Salt Wash Member of the Morrison Formation (Jurassic) in Utah and Colorado. This unit is some 600 feet (183 m) thick at its apex and extends north, northeast, and

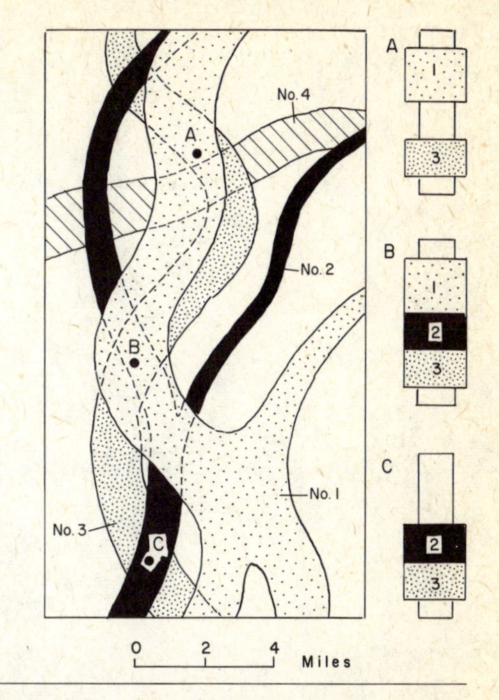

FIG. 5-7. The origin of multistory sand bodies. A: separate sand bodies; B: a three-story sand body; C: a two-story sand body. The position of an underlying sand body (diagonal ruling) was not controlled by the same factors that localized sand bodies 1, 2, and 3. (From Potter, 1963, Fig. 50.)

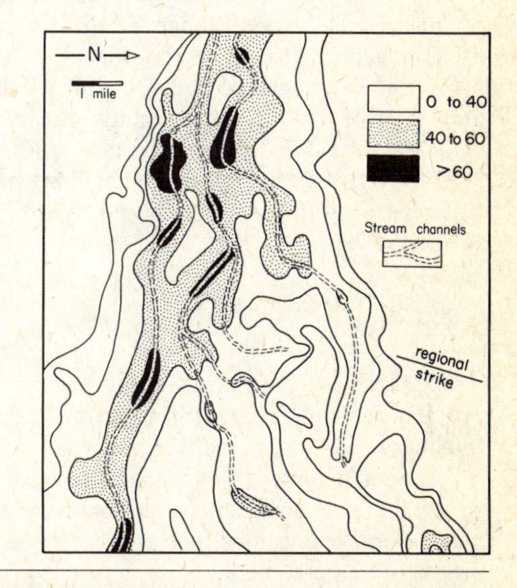

FIG. 5-8. Map of complex shoestring sand, "Frio" sand body (Oligocene). Distributary channels of a shoal-water delta. (After Nanz, 1954, p. 110.)

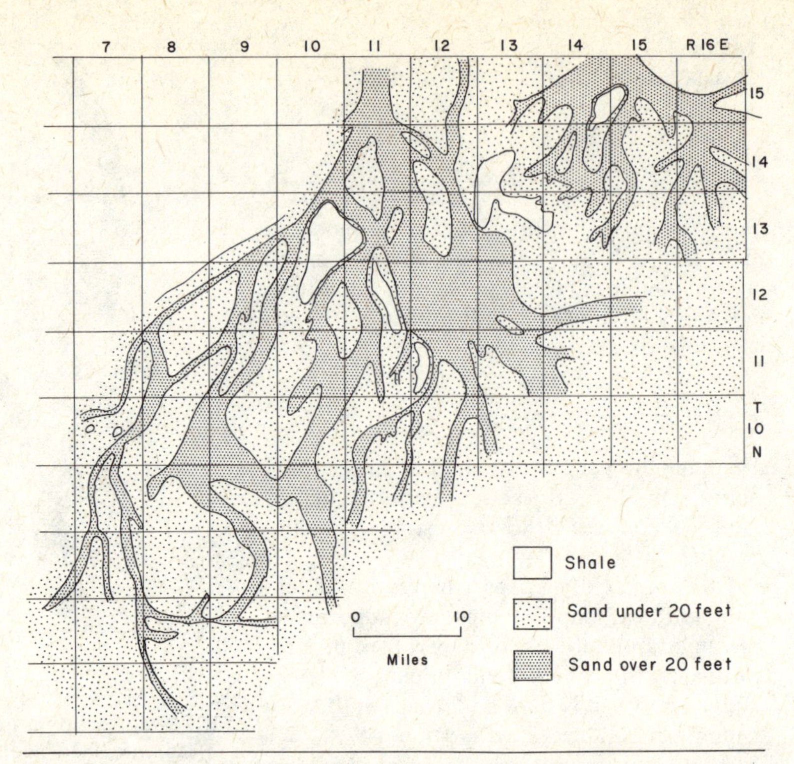

FIG. 5-9. Booch Sandstone (Pennsylvanian), Greater Seminole District, eastern Oklahoma. (After Busch, 1961, Geometry of sandstone bodies, Fig. 12. By permission of Amer. Assoc. Petrol. Geol.)

east for nearly 200 miles (322 km) (Fig. 5-10). In some cases fan-shaped deposits are numerous enough to merge and form a composite wedge-shaped apron.

Sedimentary bodies of similar shape and form are known to accumulate at the toes of steep subaqueous slopes at the mouths of submarine canyons. Such turbidite fans have been recognized only recently. Examples of such deposits include the Miocene Tarzana fan (Sullwold, 1960) and the Miocene Capistrano fan at Dana Point near Newport, California (Piper and Normark, 1971).

SHEET SANDS

Sheet sands are, as the name implies, sands of great horizontal extent in relation to their thickness. A great many, perhaps most, sandstones belong to this class. These sands cover thousands of square kilometers and may be but a few tens, perhaps a few hundreds, of meters thick.

The origin of such sheets has long posed a problem, inasmuch as modern sands are generally associated with rivers and beaches, both linear features. How is sand dispersed and deposited to form an extensive blanket deposit? The prevailing view is that such sheets form by "lateral sedimentation." They are not, therefore, of the same age everywhere. Their boundaries transect, at a low angle, time planes. One can consider that such a sheet sand is formed by the coalescence of a series of linear sand bodies each of which, if it were isolated, would be a shoestring sand. Such a composite sheet sand is illustrated in the Mesaverde Group (Cretaceous) in the San Juan Basin of Colorado, especially Point Lookout Sandstone (Hollenshead and Pritchard, 1961) which is a product of a migrating shoreline (Fig. 15-15).

SAND BODY ORIENTATION

The orientation of a sand body relative to the depositional or sedimentary strike is of considerable interest. Some bodies, in fact, are termed *strike sands* if they are parallel to the depositional slope; others are *dip*

FIG. 5-10. Relations between cross-bedding directions (arrows) and isopach lines in Salt Wash Member, Morrison Formation (Jurassic), southeastern Utah and adjacent states. (Data from Craig, et al., 1955.)

sands if they extend down the paleoslope. Those linear sand bodies that originate as barrier islands are parallel to the shoreline. Others, particularly those related to stream channels, are normal to the strand. Some are irregular and show no clear relation to the paleoslope.

If a sand body occurs above an unconformity developed on tilted beds, sand accumulation may be controlled by buried topography, resulting in a *strike-valley sand body* (Busch, 1959, p. 2829).

A problem related to sand body orientation is the relation between the internal textures and structures of a sand body and its external form. Presumably both are a response to the paleocurrent system from which they were deposited. Hence it might be presumed that grain orientation and cross-bedding, for example, might bear systematic relations to the long axis of the sand body; such indeed seems to be the case. This question is more fully explored by Potter and Pettijohn (1963, pp. 173–190).

SANDSTONE DIKES, SILLS, AND AUTOINTRUSIONS

Injection or intrusion of sand into sedimentary and other rocks is a characteristic feature of certain sedimentary assemblages. Sandstone dikes, sills, and related bodies, ranging in age from Precambrian to Pleistocene, have been described many times.

Sandstone dikes range from 2 cm to over 10 m in thickness (Fig. 5-11). Some Precambrian quartzite dikes that cut the Espanola on the north shore of Lake Huron are 30 feet (9 m) thick (Quirke, 1917, p. 37, Pl. 11B; Collins, 1925, p. 53, Pl. 7; Eisbacher, 1970). These dikes are unusual in that they contain large granite pebbles up to 6 inches

(15 cm) in diameter, which are commonly concentrated in the center of the dike (Fig. 5-11). Many dikes are traceable no further than across the outcrop, but Vitanage (1954) mapped sandstone dikes in Colorado granite for a distance of about 8 miles (12.9 km). And Diller (1890) traced the "Great Dike," which cuts Cretaceous sediments in the Sacramento Valley of California, for a distance of about 9 miles (14.5 km).

Many dikes are sharp-walled, vertical, tabular bodies, but in some cases, such as those in the Carpathians of Poland (Dzulynski and Radomski, 1956) and in the Paleogene of Tunisia (Gottis, 1953), dikes are very irregular and sinuous bodies and in places are an interrupted series of pod-shaped sand masses. The irregular sinuosities have been attributed to reduction in thickness of the host shales as a result of compaction. The same phenomenon produces "crumpled mud-crack casts" (p. 123). If this concept is correct, injection of the dikes took place at an early precompaction stage. It is even possible to make a quantitative estimate of compaction since injection. On the other hand many dikes, such as those described by Diller (1890), are not deformed. Because they are vertical and the host strata tilted, Diller believed them to be injected after tilting and hence at a late rather than at an early stage.

In general the dike filling is massive sand, although, as noted above, pebbles (not wall rock inclusions) may be present. Diller (1890, p. 425) observed a faint alignment of mica flakes parallel to the walls. A similar arrangement of micas and long axes of quartz grains was noted by Vitanage (1954, p. 498). These alignments, parallel to the walls rather than transverse to the dike, support the view that the dike is emplaced by injection and is not the grain-by-grain filling of an open fissure.

Sandstone sills are virtual proof of the injection process. Because they are tabular bodies, parallel to the bedding, they resemble and are commonly mistaken for beds. Unlike associated beds, they are not graded or cross-stratified and lack sole marks characteristic of the facies in which they occur. If traced far enough these sills commonly cross the beds and continue at a slightly higher or lower stratigraphic level.

FIG. 5-11. Sandstone dike cutting Espanola Formation (Huronian) near Espanola, Ontario, Canada. The dike, essentially a graywacke, cuts the Espanola Formation, an impure, thin-bedded carbonate silt, the weathered surface of which is etched into relief and is harshly corrugated.

Why sandstone dikes are common in some sedimentary terranes (especially in the flysch facies) is not clear. Fissuring has been attributed to earthquakes, the injection to movement under hydrostatic pressure of water-saturated sands that are "quick" or capable of mobilization. Fairbridge (1946) and Dzulynski and Radomski (1956, p. 258) have pointed out the association of sandstone dikes and slumping. Fairbridge believed dikes and associated structures to be indicative of a particular facies and environment of sedimentation, namely, tectonically active unstable "foredeep areas of geosynclinal belts." This concept may apply to many dikes, but others, such as the multiple dikes in Colorado granite, owe their origin to other causes.

CARBONATE ROCK GEOMETRY

The geometry of carbonate rock bodies is, in general, much less well known than that of ordinary sands. This is true in part because it is easier to distinguish in geophysical logs between a terrigenous sand and a shale than to distinguish a carbonate sand from a micritic limestone. In other words, geophysical logs are inadequate to separate a particular limestone from associated limestones. (Most sandstones are treated as single entities so that one need only to separate them from the enclosing shales to outline the form or shape of the sandstone body.) Consequently our knowledge has been too meager to develop a classification of various carbonate bodies. Two exceptions are Ball's (1967) classification of modern carbonate sand bodies in the Bahama region, and the study of carbonate "buildups" or bioherms, including both true reefs and the carbonate mud mounds or "knolls."

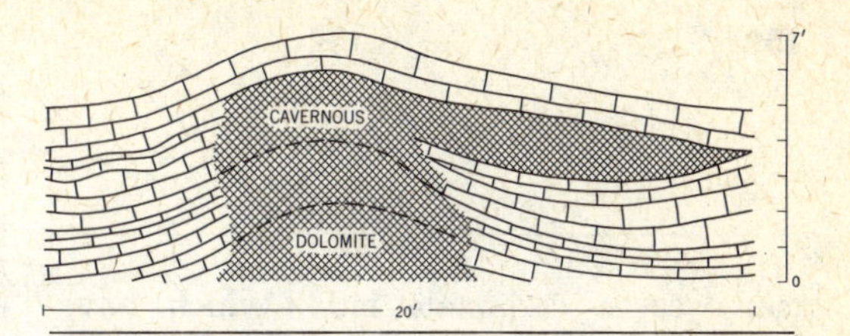

FIG. 5-12. Limestone reef (Silurian), Wisconsin. This is a relatively small reef body. (From Shrock, 1939, Fig. 3.)

REEFS

Reefs are bioherms that today appear as structureless masses of unevenly textured, porous rock (commonly dolomite), which interrupts the uniform bedding of regional stratigraphic sequence. These bodies were islands of intense vital activity and hence were sites of rapid accumulation of carbonate; once initiated, they formed mounds rising above the surrounding bottom and ultimately were built up to sea level and even became partially emergent. If true reefs, they were wave resistant; when they reached the surf zone, they shed debris to the surrounding area, the coarser portion of which formed an apron of waste flanking the reef core itself. Such steeply dipping flank beds form a part of the reef complex (Fig. 5-12).

Bioherms are usually the product of a community of organisms; the most important are the frame builders, which bind the sediment together and create the reef body. Among these the algae play a leading role; the oldest bioherms seem to have been solely algal.

Reef bodies vary in size from small structures, such as the serpulid mounds and stromatolitic heads 1 m or less in size, to great reef complexes whose dimensions are measured in hundreds of meters or more. Precambrian stromatolitic bioherms up to 60 feet (18 m) thick and 200 feet (61 m) across have been reported from Great Slave Lake in the Northwest Territories (Hoffman, 1969, p. 448). Younger reef bodies are considerably more diverse in makeup and are generally much larger. One of the most impressive is the Devonian Attendorn reef complex in central Germany, which is over 900 m thick and occupies an area of over 100 km^2. This is an atoll-like structure (Krebs, 1971). A similar, though somewhat smaller, reef is the Iberg-Winterberg complex in the Harz Mountains (Franke, 1971). The Virgilian reefs of Pennsylvanian age in the Sacramento Mountains of New Mexico range from a few hundred feet to 1 mile (1.6 km) in diameter and up to 200 feet (61 m) in thickness (Plumley and Graves, 1953). These bodies are thick planoconvex lenses of massive carbonate in thin-bedded limestone.

The reef body normally has a flat base and a marked convex top and, as noted, is surrounded by steeply dipping flank beds. In some cases the reef structure is surrounded by a weak peripheral syncline, suggesting settlement of the reef body into its substrate.

In plan many reef bodies are roughly circular, although some are elongate. Others show some asymmetry, presumed to be a response to the prevailing wind direction (Lowenstam, 1957, p. 223; Ingels, 1963).

Most reefs occur in groups and form a belt or trend. They may form on the edge of a platform or may be isolated and located on a volcanic rise, as is presumed to be the case of the Iberg reef in Germany (Franke, 1971).

MOUNDS AND BANKS

Not all bioherms are true reefs, that is, have a wave-resistant framework. Some are, instead, banks or mounds of carbonate, in many cases carbonate mud, which have been localized for some reason. It has been thought that some modern mud banks have formed as a result of the baffle effect of sea grass meadows. Other organisms, such as bryozoans, may have played a similar role in the past. Pray (1958) describes Mississippian bioherms 25 to 350 feet (7.6 to 107 m) thick, having a flat base and convex top, which he attributes to this process. A similar explanation has been advanced for the Waulsortian (Carboniferous) "reefs" of west-central Ireland (Lees, 1964). These form "knolls" primarily of massive calcitic mudstone, fewer than 100 to more than 1000 feet (304 m) in diameter and of variable thickness of the order of some tens of feet. Their flanks have dips up to 50 degrees.

For references to reef and mound literature, refer to the discussion of reef facies in the chapter on limestones and dolomites.

OTHER CARBONATE BODIES

In addition to true reefs and mud mounds or "knolls" are other circumscribed carbonate bioherms, which belong to neither of the above categories. Notable are the restricted accumulations of crinoidal limestone, such as that which forms the reservoir rock of the "crinoidal pool" in the Todd Oil Field in Texas (Imbt and McCollum, 1950). This Pennsylvanian crinoidal body is 2 miles (3.2 km) in diameter and grades to a maximum of 400 feet (122 m) in thickness; it is surrounded on all sides by black and green shales. Although referred to as a "reef," it probably was formed by a thickly populated crinoid colony which persisted in time. Very probably it was never much elevated above the surrounding sea floor at any time. Similar large disconnected and irregularly distributed masses of crinoidal limestone enclosed in clastic strata have been reported from the Bordon Group (Mississippian) of Indiana (Stockdale, 1931). The largest of these is 2 miles (3.2 km) in diameter and has a maximum thickness of about 70 feet (21.3 m). See Fig. 5-13.

Other carbonate bodies include oolitic limestones such as the "McClosky Sand" in the Passport Oil Pool, Clay County, Illinois. These "sands," in the Ste. Geneviève Formation (Mississippian), are essentially lenses 1 mile wide, about 1.5 to 2 miles long, and not much over 12 to 14 feet (3.7 to 4.3 m) thick. They pinch out in all directions. These lenses have been interpreted as oolite shoals formed in a shallow marine environment (Carr, 1973). They may be related in origin to some of the oolitic shoals forming today on the Bahama Banks (Rich, 1948).

SALT DOMES, STOCKS, AND ANTICLINES

The salt dome, an unusual sedimentary body, is a product of flow and injection of salt from a deeper layer into overlying strata. Like sandstone dikes and sills it is a secondary or post-depositional structure.

Salt domes are near-vertical cylindrical stocks of salt, from 0.5 mile to 1 or 2 miles (0.8 to 3.2 km) in diameter, and in places their walls overhang (Fig. 5-14). The strata through which the salt is injected are deformed, are generally upturned against the stock, and approach verticality in places. They thin notably as they approach the salt plug. The salt column is surmounted by a "cap rock," generally of gypsum and limestone. Overlying strata show a gentle arching and may be complexly faulted. Salt domes show a complex set of minor folds and other internal structures generated by the flow of the salt during intrusion (Balk, 1949, 1953).

Salt domes are presumed to form because of a gravitational instability produced by the superposition of sediments with a greater density than that of the subjacent salt. The plastic salt rises as a kind of convective column. The cap rock is thought to be the accumulation of insoluble components of the salt solution-concentrated at the top of the rising salt column.

Salt domes, like reefs, are prone to occur in clusters. Notable examples are found in the Gulf Coast of Louisiana and Texas (both on land and offshore), in the Zechstein of Germany, in Iran, and elsewhere. Their relation to petroleum accumulation has made them of special interest (Moore, 1926). A good summary of this question is given by Levorsen (1967, pp. 356–379).

Salt anticlines are structures in which the

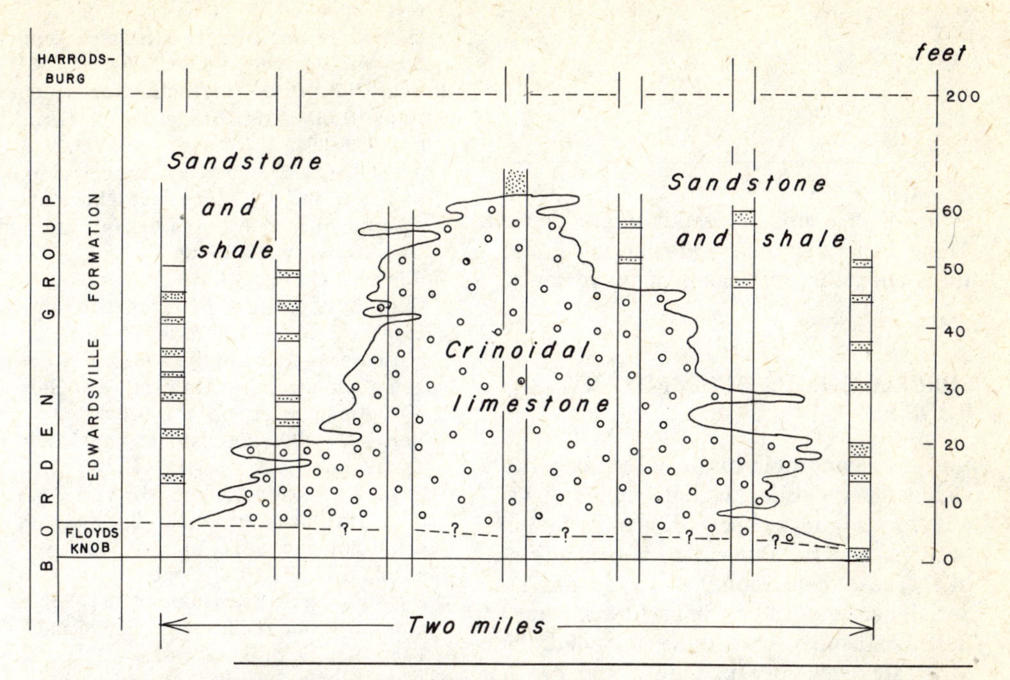

FIG. 5-13. Crinoidal bioherm, Bordon Group (Mississippian), Indiana. (After Stockdale, 1931, Fig. 2.)

salt forms the core of the anticline. Examples occur in the Paradox Basin of Utah and Colorado (Prommel and Crum, 1927).

FILLED CAVITIES AND SINKS

Some sedimentary bodies, generally rather small, are very restricted, because they were deposited in cavities or very small basins such as limestone sinks. A great many small cavities, such as those in reefs, show internal mechanical sedimentation. Fine sediment sifts down in or is trapped in these openings. Fillings may be complex and include precipitated material as well as silt and mud.

For the most part these internal deposits are more largely a minor sedimentary structure *within* a larger sedimentary body. In some unusual cases, however, they are large enough to warrant the appelation *sedimentary body*. Even the largest of these, cavern fillings, are rare in the older geologic record and are of interest mainly in Holocene history.

Sink fillings are somewhat different and attain appreciable size. They include deposits introduced from above into sinks formed by solution and collapse of limestone caverns. Some such fillings, which include shale, sandstone, and coal, have been formed by gradual subsidence of overlying strata which kept pace with solutional removal of subjacent calcareous rock (Bretz, 1940, 1950). These are, in effect, downward intrusions of superjacent strata and are, like salt domes and sandstone dikes, a postdepositional deformational phenomenon

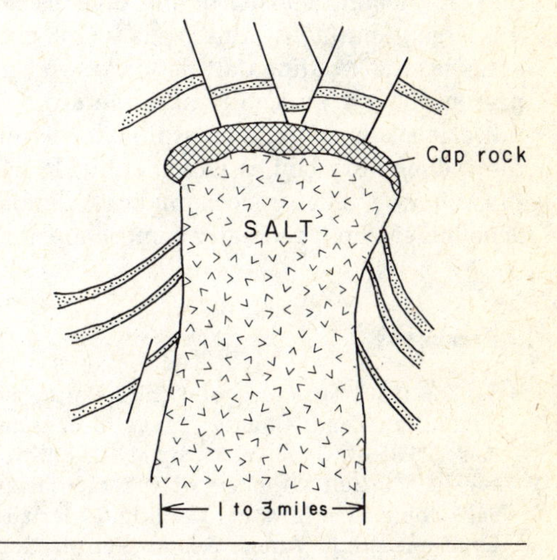

FIG. 5-14. Schematic diagram of salt dome. (Modified from Fig. 8-26, Geology of petroleum, 2nd ed., by A. I. Levorsen. W. H. Freeman and Co. Copyright © 1967.)

rather than a primary sedimentary deposit. They display various deformational structures and other evidences of movement.

MISCELLANEOUS SEDIMENTARY BODIES

Various geomorphic forms, which constitute sedimentary entities, include dunes of various forms and a host of glacial features such as eskers, kames, drumlins, and the like. These constructional forms are subaerial, but some subaqueous forms, some of dunelike aspect, should be included here. Most are transitory in nature and are not likely to be buried or preserved intact. Dunes, for example, tend to coalesce or merge and lose their identity as discrete bodies. Glacial deposits of the older record are largely marine, so the forms which characterize the continental deposits are not present. The general absence, therefore, of these miscellaneous geomorphic forms and the sedimentary bodies which coincide with them in the older record make a detailed account of their geometry superfluous.

References

Balk, R., 1949, Structure of Grand Saline salt dome, Van Zandt County, Texas: Bull. Amer. Assoc. Petrol. Geol., v. 33, pp. 1791–1829.

———, 1953, Salt structure of Jefferson Island salt dome, Iberia and Vermilion parishes, Louisiana: Bull. Amer. Assoc. Petrol. Geol., v. 37, pp. 2455–2474.

Ball, M. M., 1967, Carbonate sand bodies of Florida and the Bahamas: Jour. Sed. Petrology, v. 37, pp. 556–591.

Bass, N. W., 1934, Origin of Bartlesville shoestring sand, Greenwood and Butler counties, Kansas: Bull. Amer. Assoc. Petrol. Geol., v. 18, pp. 1313–1345.

Braunstein, J., and O'Brien, G. D., eds., 1968, Diapirism and diapirs: Amer. Assoc. Petrol. Geol. Mem. 8, 444 pp.

Bretz, J H., 1940, Solution cavities in the Joliet limestone of northeastern Illinois: Jour. Geol., v. 48, pp. 337–384.

———, 1950, Origin of the filled sink-structures and circle deposits of Missouri: Bull. Geol. Soc. Amer., v. 61, pp. 789–834.

Busch, D. A., 1959, Prospecting for stratigraphic traps: Bull. Amer. Assoc. Petrol. Geol., v. 43, pp. 2829–2843.

Carr, D. D., 1973, Geometry and origin of oolitic bodies in the Ste. Genevieve Limestone (Mississippian) in the Illinois Basin: Bull. Indiana Geol. Surv., v. 48, 81 pp.

Collins, W. H., 1925, North shore of Lake Huron: Geol. Surv. Canada Mem. 143, 160 pp.

Cumings, E. R., and Shrock, R. R., 1928, Niagaran coral reefs of Indiana and adjacent states and their stratigraphic relations: Bull. Geol. Soc. Amer., v. 39, pp. 579–620.

Diller, J. S., 1890, Sandstone dikes: Bull. Geol. Soc. Amer., v. 1, pp. 411–442.

Dzulynski, S., and Radomski, A., 1956, Clastic dikes in the Carpathian flysch: Ann. Soc. Geol. Pologne, v. 26, pp. 225–264.

Eisbacher, G. H., 1970, Contemporaneous faulting and clastic intrusions in the Quirke Lake Group, Elliot Lake, Ontario: Canad. Jour. Earth Sci., v. 7, pp. 215–225.

Fairbridge, R. W., 1946, Submarine slumping and the location of oil bodies: Bull. Amer. Assoc. Petrol. Geol., v. 30, pp. 84–92.

Franke, W., 1971, Structure and development of the Iberg-Winterberg reef (Devonian to Lower Carboniferous), Harz, West Germany, *in* Sedimentology in parts of central Europe (Müller, G., ed.): Frankfurt am Main, Waldemar Kromer, pp. 83–90.

Gottis, Ch., 1953, Les filons clastique "intra formationnels" du flysch numidien tunisien: Bull. Soc. Géol. France, ser. 6, v. 3, pp. 775–784.

Hoffman, P. F., 1969, Proterozic paleocurrents and depositional history of the East Arm fold belt, Great Slave Lake, Northwest Territories: Canad. Jour. Earth Sci., v. 6, pp. 441–462.

Hollenshead, C. T., and Pritchard, R. L., 1961, Geometry of producing Mesaverde sandstones, San Juan Basin, *in* Geometry of sandstone bodies (Peterson, J. A., and Osmond, J. C., eds.): Tulsa, Okla., Amer. Assoc. Petrol. Geol., pp. 98–118.

Imbt, R. F., and McCollum, S. V., 1950, Todd deep field, Crockett County, Texas: Bull. Amer. Assoc. Petrol. Geol., v. 34, pp. 239–262.

Ingels, J. J. C., 1963, Geometry, paleontology, and petrography of Thornton reef complex, Silurian of northeastern Illinois: Bull. Amer. Assoc. Petrol. Geol., v. 47, pp. 405–440.

Krebs, W., 1971, Devonian reef limestones in eastern Rhenisch Schiefergebirge, *in* Sedimentology of parts of central Europe (Müller, G., ed.): Frankfurt am Main, Waldemar Kramer, pp. 45–81.

Krynine, P. D., 1948, The megascopic study and field classification of the sedimentary rocks: Jour. Geol., v. 56, pp. 130–165.

LeBlanc, R. J., 1972, Geometry of sandstone bodies: Amer. Assoc. Petrol. Geol. Mem. 18, pp. 133–190.
Lees, A., 1964, The structure and origin of the Waulsortian (Lower Carboniferous) "reefs" of west-central Eire: Trans. Roy. Phil. Soc. London, ser. B, v. 247, pp. 483–531.
Levorson, A. I., 1967, Geology of petroleum, 2nd ed.: San Francisco, Freeman, 724 pp.
Lowenstam, H. A., 1957, Niagaran reefs in the Great Lakes area, *in* Treatise on marine ecology and paleoecology, vol. 2 (Ladd, H. S., ed.): Geol. Soc. Amer. Mem. 67, pp. 215–248.
Moore, R. C., ed., 1926, Geology of salt dome oil fields: Tulsa, Okla., Amer. Assoc. Petrol. Geol., 797 pp.
Nanz, R. H., Jr., 1954, Genesis of Oligocene sandstone reservoir, Seeligson Field, Jim Wells and Kleberg counties, Texas: Bull. Amer. Assoc. Petrol. Geol., v. 38, pp. 96–117.
Peterson, J. A., and Osmond, J. C., eds., 1961, Geometry of sandstone bodies: Tulsa, Okla., Amer. Assoc. Petrol. Geol., 240 pp.
Piper, D. J. W., and Normark, W. R., 1971, Reexamination of a Miocene deep-sea fan and fan-valley, southern California: Bull. Geol. Soc. Amer., v. 82, pp. 1823–1830.
Plumley, W. J., and Graves, R. W., Jr., 1953, Virgilian reefs of the Sacramento Mountains, New Mexico: Jour. Geol., v. 61, pp. 1–16.
Potter, P. E., 1963, Late Paleozoic sandstones of the Illinois Basin: Illinois Geol. Surv., Rept. Inv. 217, 92 pp.
Potter, P. E., and Pettijohn, F. J., 1963, Paleocurrents and basin analysis: New York, Springer, 296 pp.
Pray, L. C., 1958, Fenestrate bryozoan core facies, Mississippian bioherms, southwestern United States: Jour. Sed. Petrology, v. 28, pp. 261–273.
Prommel, H. W. C., and Crum, H. E., 1927, Salt domes of Permian and Pennsylvanian age in southeastern Utah and their influence on oil accumulation: Bull. Amer. Assoc. Petrol. Geol., v. 11, pp. 373–393.
Quirke, T. T., 1917, Espanola district, Ontario: Geol. Surv. Canada Mem. 102, 92 pp.
Reynolds, D. W., and Vincent, J. K., 1967, Western Kentucky's Bethel channel—the largest reservoir in the Illinois Basin: Kentucky Geol. Surv., ser. 10, Spec. Pub. 14, pp. 19–30.
Rich, J. L., 1923, Shoestring sands of eastern Kansas: Bull. Amer. Assoc. Petrol. Geol., v. 7, pp. 103–113.
———, 1938, Shorelines and lenticular sands as factors in oil accumulation, *in* The science of petroleum, v. 1 (Dunstan, A. E., ed.): London, Oxford Univ. Press, pp. 230–239.
———, 1948, Submarine sedimentary features on Bahama Banks and their bearing on distribution patterns of lenticular oil sands: Bull. Amer. Assoc. Petrol. Geol., v. 32, pp. 767–779.
Rittenhouse, G., 1961, Problems and principles of sandstone-body classification, *in* Geometry of sandstone bodies (Peterson, J. A., and Osmond, J. C., eds.): Tulsa, Okla., Amer. Assoc. Petrol. Geol., pp. 3–12.
Schlee, J. S., and Moench, R. H., 1961, Properties and genesis of "Jackpile" sandstone, Laguna, New Mexico, *in* Geometry of sandstone bodies (Peterson, J. A., and Osmond, J. C., eds.): Tulsa, Okla., Amer. Assoc. Petrol. Geol., pp. 134–150.
Sedimentation Seminar, 1969, Bethel Sandstone (Mississippian) of western Kentucky and south-central Indiana, a submarine channel fill: Kentucky Geol. Surv., Ser. 10, Rept. Inv. 11, 24 pp.
Shelton, J. W., Terrell, D. W., and Karvelut, M. D., eds., 1972, A guidebook to the genesis and geometry of sandstones: Oklahoma City, Geol. Soc., 66 pp.
Shrock, R. R., 1939, Wisconsin Silurian bioherms (organic reefs): Bull. Geol. Soc. Amer., v. 50, pp. 529–562.
Stockdale, P. B., 1931, Bioherms in the Borden group of Indiana: Bull. Geol. Soc. Amer., v. 42, pp. 707–718.
Sullwold, H. H., Jr., 1960, Tarzana fan, deep submarine fan of late Miocene age, Los Angeles County, California: Bull. Amer. Assoc. Petrol. Geol., v. 44, pp. 433–457.
Vitanage, P. W., 1954, Sandstone dikes in the South Platte area, Colorado: Jour. Geol., v. 62, pp. 493–500.
Weimer, R. J., 1973, Sandstone reservoirs and stratigraphic concepts: Amer. Assoc. Petrol. Repr. Ser., no. 7, 212 pp.; no. 8, 216 pp.

6

GRAVELS, CONGLOMERATES, AND BRECCIAS

INTRODUCTION

A *gravel* (German, *Schotter* for coarse gravel, *Kies* for fine gravel; French, *gravier*) is an unconsolidated accumulation of rounded fragments larger than sand. The lower size limit is variously set at generally 2 mm (Wentworth, 1922a; 1935) or 5 mm (Cayeux, 1929). Material in the 2 to 4 mm range has been termed *granule gravel* (Wentworth) or *very fine gravel* (Lane et al., 1947). There is no general agreement on the percentage of gravel-sized fragments that must be present before the aggregate warrants the term *gravel*. Actual analysis shows that the field geologist is prone to call a deposit gravel even if pebbles and like sizes form less than one-half of the whole. Some rocks, such as tillite or indurated boulder clay containing less than 10 percent of gravel-sized fragments, are nonetheless designated as conglomerate. Willman (1942) suggested the following definitions to correspond to prevailing field usage: *gravel* contains 50 to 100 percent pebbles; *sandy gravel* contains 25 to 50 percent pebbles and 50 to 75 percent sand; *pebbly sand* contains a conspicuous number of pebbles but less than 25 percent; *sand* must contain 75 to 100 percent sand (Fig. 6-1). Folk (1954) would apply the term *gravel* only to mixtures containing 30 or more percent gravel size and would apply the adjective *gravelly* to sands and muds containing 5 to 30 percent pebbles or larger fragments. Other proposals for naming mixtures include those of Wentworth (1922a) and Krynine (1948).

The term *conglomerate* (German, *Konglomerat;* French, *conglomerat*) is applied to indurated gravels. As in the case of gravel, the terms *pebble, cobble* or *boulder* may be prefixed according to the dominant fragment size.

The term *rubble* has been applied to an accumulation of angular fragments coarser than sand; the term *scree* may be applied if the blocks are large. *Breccia* (German, *Bresche;* French, *breche*) is the consolidated equivalent of rubble. Breccia applies to nonsedimentary rocks also (such as fault breccia, volcanic breccia). Various terms have been proposed for the several sizes of fragments of which rubble and breccias are composed (Woodford, 1925, p. 183).

To what extent must the fragments be rounded before the terms *gravel* or *conglomerate* rather than *rubble* or *breccia* are used? Usage is variable, but most workers would apply the term *conglomerate* to subangular and better-rounded materials and *breccia* only to those rocks whose fragments are angular. The terms *roundstone* (Fernald, 1929) and the term *sharpstone* have been used to distinguish between rounded and angular fragments; hence the term *roundstone conglomerate* and *sharpstone conglomerate* have been proposed as substitutes for sedimentary conglomerates and breccias, respectively.

The term *conglomerite* (Willard, 1930) has been suggested for conglomerates which have reached the same state of induration as a quartzite. It has, however, been little used; a deformed conglomerate or one

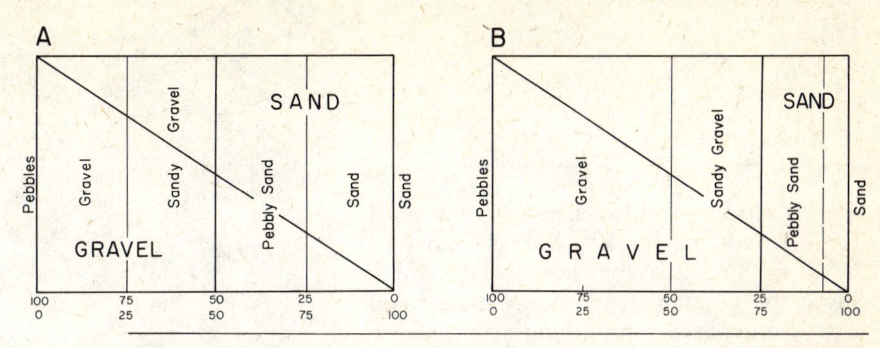

FIG. 6-1. Nomenclature of sand and gravel mixtures. A: idealized "symmetrical" classification; B: classification based on field usage. (After Willman, 1942.)

otherwise altered by metamorphic changes is more commonly called a *metaconglomerate.*

MODERN GRAVELS

Gravels may be accumulated and laid down as subaqueous deposits beneath permanent water bodies as the result of wave action on rocky shores. They may also form on land as a result of weathering and the flow of water when relief is sufficient and rainfall adequate to provide the moving power. Gravels, the coarsest product of erosion, are moved shorter distances from the places of their origin and are deposited in more restricted areas than are sand, clay, or matter in solution.

According to place of accumulation, whether above or below sea surface, gravels can be considered terrestrial or subaqueous. The latter, in addition to normal shoreline gravels, include such special types as ice-rafted deposits with their dropstones and resedimented gravels, which are those shifted from the shore zone to deeper waters by subaqueous slumps or slides and the stronger turbidity currents which they induce. Some gravels, rather small in volume, also occur in those bodies of water with less effective wave action such as estuaries and small lakes. The gravel of large lakes is essentially similar in all important respects to those of the sea.

Terrestrial gravels include some local deposits which have undergone little or no transport (such as talus) that have moved only a short distance, such as the coarse rock glaciers or scree and the more common but generally superficial solifluction deposits. Included here also are the gravelly deposits of continental glaciers—morainal materials, and gravels of eskers and kames, and most especially gravel fans laid down by the larger meltwater streams. But by far the most important terrestrial gravels are those of streams.

Excepting, therefore, local deposits of limited extent and those attributable to climatic accidents, we are here concerned mainly with marine and fluviatile gravels, and ancient conglomerates of marine or fluvial origin.

Shoreline marine gravels range, according to Barrell (1925, p. 305), from about 1 fathom above high-water level to 4 or 5 fathoms below. They are kept against the coast and seldom extend more than 1 to 3 miles from shore. Exceptionally concentrated undertow may sweep out gravel to depths of 20 or 30 fathoms and up to 10 miles from land. Modern gravels found at greater distances offshore may be relict gravels left behind by the rapid rise of sea level during post-glacial times. Or they may be lag deposits caused by submarine erosion of till and other pebbly deposits—a pavement of cobbles too large to be shifted landward (Hough, 1932). In general the sea "rejects" or "repels" the debris of continents (certainly the coarse debris). On flat profiles the gravel migrates shoreward, whereas the fines drift seaward. Hence the gravel tends to hug the shore and accumulate in a zone seldom extending far from shore.

Beach gravels are local accumulations, especially common in pocket beaches where gravel is trapped between two rocky headlands whose erosion supplies the detritus. In some instances the gravel shifts longshore and accumulates in beach ridges and bars. On exposed coasts the gravel tends to be very coarse. It is generally very well sorted and highly rounded; the rounding proceeds rapidly as the gravel migrates away from the source areas (Grogan, 1945; Wentworth, 1922b).

Comprehensive sedimentological studies of modern beach gravels include that of Krumbein and Griffith (1938) on the limestone gravels of Little Sister Bay of Lake Michigan, of Bluck (1967) on the gravels of south Wales, of Humbert (1968, p. 89) on the shingle complex of Bridgewater Bay on the coast of Somerset (England), and of Emery (1955) on modern marine gravels.

Shoreline gravels are of small volume and are confined to narrow linear belts; with a changing sea level, however, the loci of gravel accumulation shift. With rising sea level and transgression of the sea, shoreline gravel deposits are extended so that the result is a thin sheet of "basal" gravel superimposed unconformably on older rocks. With falling sea level, gravel beach ridges may be abandoned and appear as low topographic forms far inland and now elevated above sea level. They are apt to be eroded or destroyed by subaerial processes.

Rivers, on the other hand, carry gravels many tens of miles (even hundreds of miles) from the places of gravel formation, an order of magnitude or more greater than transport under marine conditions. Fluvial gravels are widespread, forming great piedmont fans and flooring the valleys of large rivers in regions of high relief. The thickness of these accumulations is many times greater than that formed along the strandline.

Alluvial fans are of vast extent and particularly characterize arid regions of bold relief. Probably more than half the area of Nevada and large parts of Utah, New Mexico, Arizona, California, and Mexico are mantled by alluvial fan accumulations. The deposits of fans are among the coarsest and poorest sorted alluvial deposits. These deposits are situated at the places where streams emerge from a mountainous terrane and deploy into an open basin. The stream divides into many distributaries and drops much or all of its load; the coarsest occurs near the apex, and the size decreases rapidly away from the apex of the fan (the decrease in size in many cases is exponential). The composition of gravels changes little down the fan, but pebbles become appreciably better rounded in the same direction. Bedding varies from large scale (15 to 20 feet; 5 to 6 m) in the coarse gravels to a few centimeters in the interbedded sands. Imbrication is common. Fans of arid or semiarid regions are characterized by interbedding of mudflow deposits with more normal gravels. Fan gravels interfinger downcurrent with ordinary alluvium and lacustrine sediments. These gravels become cemented in arid regions by caliche; the lithified rock has been termed *fanglomerate* (Lawson, 1925). Refer to the papers of Blissenbach (1952, 1954), Bluck (1964), and Denny (1965) for study of the deposits of modern alluvial fans and to the paper of Lawson (1925) on the role of such deposits in the present world and in the geologic record.

FIG. 6-2. Lake Michigan beach gravel, Little Sister Bay, Wisconsin. Note absence of interstitial sand (compare with outwash of Fig. 6-3). Size analysis of these gravels is in Fig. 6-4E. (Krumbein and Griffith, 1938, Pl. 1.)

Gravel-carrying streams alluviate large areas and form extensive gravel deposits. We are not concerned with gravels found in the upper reaches of high-gradient streams. These gravels are in transit—only temporarily lodged in gravel bars or exposed in terrace remnants. Our concern is with the interior basins, where alluviation dominates and gravel accumulations reach hundreds or even thousands of feet in thickness. Such basins are found in arid regions of high relief, as is the Great Basin, and in downfaulted tracts, as in the Rhine Graben. These alluvial gravels are commonly coarse and interbedded with coarse and fine sands and show a progressive decrease in size downstream, accompanied by a notable increase in roundness. The composition is modified downcurrent by selective abrasion and removal of the least resistant clasts. Imbrication is well displayed in sections parallel to stream flow. Bedding and sorting are generally better than that of alluvial fans; mudflows that characterize fans are absent or rare. Many studies of modern alluvial gravels include those of Krumbein (1940, 1942) on the deposits of exceptional floods, of Plumley (1948) on the terrace gravels of several small streams that emerge from the Black Hills of South Dakota, of Unrug (1957) on the gravels of the river Dunajec arising in the Polish Tatra Mountains, of Dal Cin (1967) on the River Piave in northern Italy, and of Teruggi et al. (1971) on the Rio Sarmiento of Argentina. The paper by Conkling, Eckis, and Gross (1934) and the classic paper by Udden (1914) contain a large number of size analyses of modern alluvial gravels. Sedimentological studies of unconsolidated upland gravels of alluvial origin include those of Potter (1955) and Schlee (1957).

We can say, in summary, that there are doubtless more gravel and boulders strewn over the surface at present than at any time in the earth's history. Youthful and mature landscapes are common; old age forms are rare. Vigorous waves and rivers supply gravel in maximum amounts, and extensive coarse deposits of the Pleistocene glaciers are still in place. Of the gravel now being deposited, by far the greater amount is furnished by continental agents—rivers and glaciers. Marine erosion yields relatively little (Gregory, 1915; Barrell, 1925). At present the ratio of the volume of sediment produced (and hence the volume of gravel) by rivers is of the order of 50 times that produced by marine erosion. The proportional efficiency of these agents was probably not essentially different during any previous geological period.

FIG. 6-3. Glacial outwash gravel (Pleistocene), Cary, Illinois. Note tendency toward imbricate structure. Current flowed from right to left.

FABRIC AND COMPOSITION OF GRAVELS

Most gravels consist of a *framework* and *voids*. The framework is composed of gravel-size materials (*phenoclasts: pebbles, cobbles, boulders*); the voids are the openings between these framework elements. Normally framework components touch one another and form a structure stable in the gravitational field. Voids are rarely empty; they are generally filled with detritus, sand, or smaller sizes, with introduced precipitated cements. Gravels with unfilled voids have been termed *openwork* gravels, attributed to "vortex action on the downstream face of a gravel bar or delta" (Cary, 1951). In most gravels the matrix material completely fills the space between the pebbles or cobbles; it forms, therefore, about one-third of the whole rock volume. In some conglomerates, especially those with a clayey matrix, the matrix greatly exceeds this figure, so a framework of clasts in tangential contacts no longer exists. The pebbles and the like are, instead, isolated and scattered throughout the matrix.

TEXTURE OF GRAVELS AND CONGLOMERATES

The size frequency distribution in gravels varies greatly. Openwork gravels, such as some coarse beach gravels, consist solely of pebbles or cobbles without fines; these are unimodal (Fig. 6-2). Gravels with a sandy

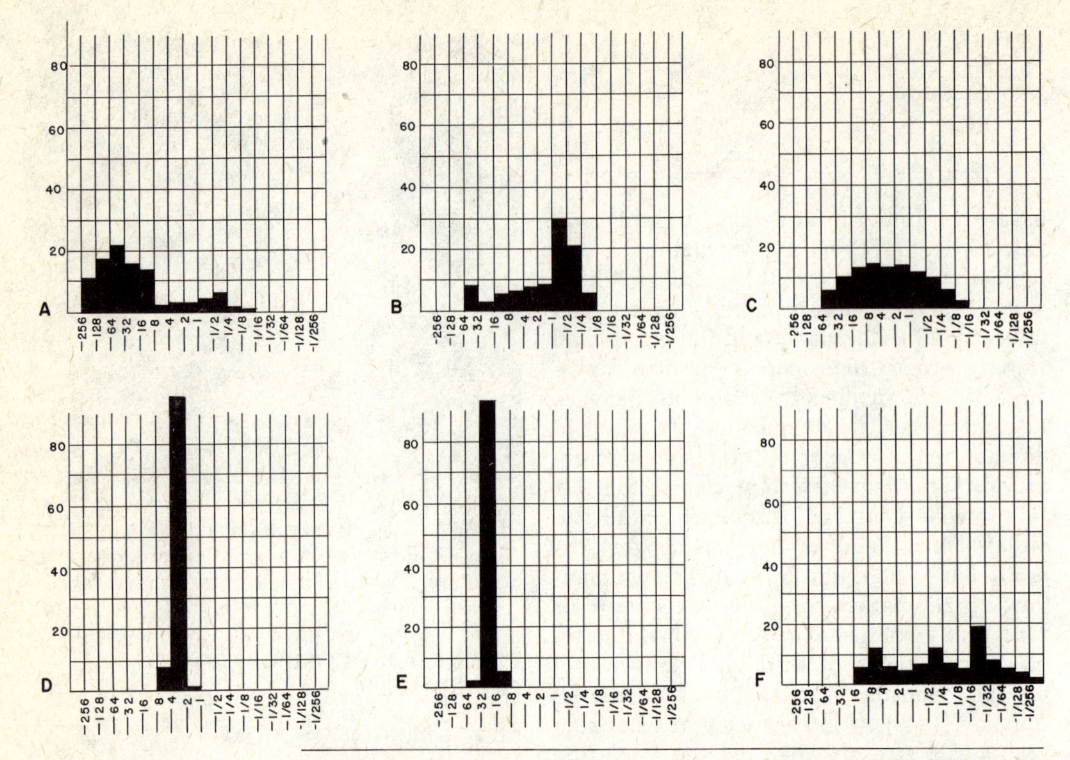

FIG. 6-4. Mechanical composition of gravels. A: Glacial outwash gravel (Pleistocene), Cary, Illinois, see Fig. 6-3. B: Mississippi River gravel, LaGrange, Missouri, from commercial dredge (Lugn, 1927). C: Flood gravel, Arroyo Seco, California. Represents deposit near maximum current velocity and turbulence (Krumbein, 1942). D: Beach gravel, Lake Michigan, Kenilworth, Illinois. E: Lake Michigan beach gravel, Little Sister Bay, Wisconsin (Krumbein and Griffith, 1938); see Fig. 6-2. Note lack of interstitial material in this gravel, compare with A. F: Pleistocene till, composite, Cary, Illinois.

matrix tend to be bimodal (Fig. 6-3). Alluvial or river gravels are commonly bimodal even when great care is taken to secure samples of single beds. Bimodal gravels have their chief mode in the gravel class and their secondary mode in the sand grades (Fig. 6-4). These modes are 4 to 5 grades apart on the average. The chief mode, therefore, has a diameter 16 to 32 times that of the material in the secondary mode. In alluvial gravels the quantity of material in the modal class is small. Such gravels in California (Conkling, Eckis, and Gross, 1934), 92 percent of which are bimodal, had only 15 to 25 percent in this class, in contrast to the unimodal gravels of beaches, whose modal class may contain as much as 90 or more percent of the whole sample (Krumbein and Griffith, 1938, Table 2). The secondary mode, usually a sand grade, contains even less material—half of that of the principal modal class, or roughly 5 to 10 percent. A large size range, 9 or 10 or even 12 grades with 1 percent or more is common. The modern gravels of the San Gabriel River and the Arroyo Seco of California, for example, fall into 9 to 11 size grades; the modal class has 15 to 35 percent (average 20). Eighty-five percent of the samples (35) have more than one mode (Krumbein, 1940; 1942). Glacial outwash gravels, even those of single layers, are characterized by 7 to 12 or more size classes, most of which are bimodal (Kurk, 1941). The quantity in the chief mode ranges from 14 to 35 percent (average 28 percent). All but one of 37 samples of Lafayette Gravel (?Pliocene) from western Kentucky are bimodal (Potter, 1955). In 23 samples the principal mode is in the gravel fraction. The number of grades (with over 1 percent) ranges from 7 to 11; the quantity in the modal class varies from 19 to 40 percent (average 26). The Brandywine upland gravels of southern Maryland are essentially the same (Schlee, 1957).

Present-day beach gravels, like beach sands, are characterized by their good sorting. They are generally better sorted than fluvial gravels of comparable coarseness (Emery, 1955). See Table 6-1. Unlike river gravels they are almost all unimodal. They have 2 to 9 grades that contain one or more percent of the material. The average number of such grades is 4 or 5, although 2 or 3 classes may contain over 90 percent of the distribution. Fifty to 60 percent of the whole distribution usually falls within the modal class and in some cases 90 percent of the total falls in this grade.

It is difficult to make meaningful size analyses of well-cemented gravels. It is convenient, therefore, to substitute the more easily determined "maximum size" for the mean size in studies of ancient conglomerates. In general, conglomeratic layers are prone to outcrop; also, the thickest layers are generally the coarsest. As a result it is not difficult to select the bed most likely to yield the largest fragments. To secure a more stable "maximum size," the average of the ten largest pebbles or cobbles is used. That the largest size bears a direct relation to the mean size has been established by analyses of modern, unconsolidated gravels (Kurk, 1941; Schlee, 1957, p. 1385). See Fig. 6-5. Both maximum and average size of gravels carried by streams decrease markedly downstream, a decline in alluviating

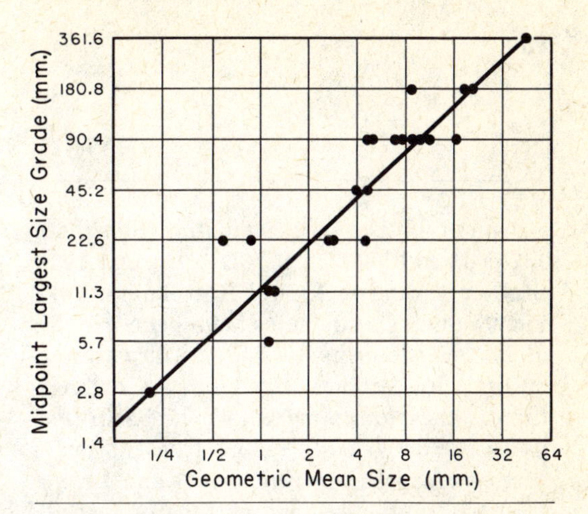

FIG. 6-5. Relation of mean size to maximum size of fluvioglacial gravels. (Unpublished data from Kurk, 1941.)

TABLE 6-1. Size characteristics of modern gravels

Gravel type	Number of samples	Number of size classes > 1%		% Bimodal	% Modal class	
		Range	Average		Range	Average
Beach[a]	26	2–9	4 to 5	3	35–95	55
Beach[b]	38	2–3	2½	none	54–96	79
Flood[c]	35	9–11	10	85	14–35	20
Glacial outwash[d]	19	7–12	9	52	14–35	28
Alluvial[e]	37	7–11	9	97	19–40	27
Terrace[f]	23	9–11	10	87	14–33	23
Alluvial[g]	72	8–12	10½	97	14–52	23

[a] Miscellaneous sea and lake beaches (Udden, 1914, and others).
[b] Lake Michigan beach gravel, Wisconsin (Krumbein and Griffith, 1938).
[c] Flood gravels of San Gabriel Canyon and Arroyo Seco, California (Krumbein, 1940, 1942).
[d] Pleistocene outwash, Illinois (Kurk, 1941). *Single layers* only.
[e] Lafayette Formation (?Pliocene), western Kentucky and adjacent states (Potter, 1955).
[f] Pleistocene terrace gravels, Black Hills, South Dakota (Plumley, 1948).
[g] (?) Pliocene Brandywine gravels, Maryland (Schlee, 1957). Channel samples.

streams found to be exponential (Sternberg, 1875; Barrell, 1925, p. 328; Unrug, 1957; Bradley, Fahnestock, and Rowehamp, 1972, Fig. 9). Such size declines have been mapped in ancient deposits to deduce direction of current flow (Bluck, 1965) and to estimate the distance to the margin of the basin (Pelletier, 1958; Yeakel, 1962). This matter is further discussed in Chapters 3 and 14.

Shape, roundness, and surface textures of the gravel clasts may aid in determination of the agent responsible for transport and deposition of the gravel. Distinctive pebble shapes and markings include faceted and snubbed ice-shaped cobbles, wind-faceted ein- and dreikanter, striations and scars produced by ice action, chink facets of some beach pebbles, percussion marks and spalls of the pebbles of high-velocity streams, and so forth (see p. 63). In highly lithified conglomerates, however, it is difficult to remove the pebble from its matrix, so these features are seldom seen.

The shape of a pebble is more dependent on the shape of the original fragment than on agent or transport history. The shape of the original fragment is a function of the bedding, jointing, and cleavage of the source rock. Hence flatness is largely a function of lithology (Cailleux, 1945); slates and thin-bedded rocks yield flat pebbles; massive rocks, like granite, shed more equidimensional fragments. Effects of the agent or environment are much less clear. It has been said that beach pebbles are flatter than river pebbles. This statement has been supported by observations of Landon (1930), Cailleux (1945), and Lenk-Chevitch (1959), but disputed by Gregory (1915), Wentworth (1922b), Kuenen (1964), and Grogan (1945).

Roundness of the pebbles in gravels and conglomerates is readily observed and can easily be estimated even in well-lithified rocks. To some extent roundness is a function of the character of the material of which the pebble or cobble is composed. Some rocks such as chert are prone to fracture under given conditions, whereas other materials such as quartzite are not. For the same rigor and distance traveled the chert will be less rounded than the associated quartzites or vein quartz (Sneed and Folk, 1958).

How far must a pebble be transported to become well rounded? Limestone fragments in a tumbling barrel become well rounded at about the equivalent of 7 miles of travel (Krumbein, 1941). Using Daubrée's (1879) figures of 0.001 to 0.004 loss in weight per kilometer of travel for granite cobbles, to become well rounded (that is, to lose one-third of the original weight) requires a transport of between 50 and 200 miles (84 to 333 km). The calculations are very crude, but the results are probably of the correct order of magnitude.

Field studies support these estimates. Limestone pebbles in Rapid and Battle Creeks in South Dakota became well rounded at about 11 and 23 miles, respectively (Plumley, 1948). Quartzite pebbles in the Brandywine upland gravels of Maryland were well rounded; the nearest ledges from which they could have come are about 45 miles distant. Knowing that most observed rounding is acquired in the first few miles of transport, it is evident that an angular or subangular gravel cannot have been moved more than 1 or 2 miles by a stream (Fig. 6-6).

Gravels and conglomerates have various internal fabrics. Larger clastic elements of a gravel tend to have a preferred orientation. It has long been observed that flatter stones in river gravels have an upcurrent dip (Fig. 6-7). This feature is readily seen in ancient gravels in appropriate cross section. Imbrication has been observed in both fluviatile and marine gravels. It has been said that the angle of inclination in river gravels, 15 to 30 degrees, is larger than that of marine deposits, 2 to 12 degrees (Cailleux, 1945). Other investigators report inclinations of pebbles in fluvial deposits inconsistent with these figures (White, 1952). The long axes of the pebbles are said to be oriented in the direction of current flow (Krumbein, 1939, 1940, 1942; Johansson, 1965) or transverse to the same (Lane and Carlson, 1954; Fraser, 1935, p. 986; Twenhofel, 1947, p. 121). Even ice-deposited till tends to have a preferential orientation of elongate till stones parallel to the ice flow (Richter, 1932; Krumbein, 1939; Holmes, 1941).

Orientation of pebbles in ancient conglomerates may enable us to determine both the direction of current flow and the angle of initial dip of the bed (White, 1952; Bluck,

1965). Orientation of the more elongate clasts in tillites has enabled us to reconstruct the pattern of ancient ice movement (Lindsey, 1969). Refer to Potter and Pettijohn (1963, pp. 28–36) for a full review of the subject of oriented clasts and tills.

COMPOSITION OF GRAVELS AND CONGLOMERATES

The composition of a gravel or conglomerate can be estimated by pebble counts. Because some rock types yield larger cobbles and others smaller pebbles, counting does not yield the same results as methods based on Rosiwal traverses or point-counts (Donaldson and Jackson, 1965, p. 628; Boggs, 1969). The composition of a gravel or conglomerate can be represented by grouping rock types into extrusive igneous (E), plutonic igneous (P), sedimentary (S), and metamorphic (M) classes. Commonly these are plotted as a triangular diagram with sedimentary and metamorphic taken together. As in the case of sandstones, it is desirable to distinguish between a supracrustal and plutonic provenance. Such a distinction is in some degree a measure of the extent of uplift and depth of erosion in the source area —both functions of tectonism.

The composition of a gravel or conglomerate is not an exact expression of the kind and abundance of rock types in the source area. Because of varying block-forming capacities of the several rock types and of varying resistance to abrasion of these rocks, proportions present in the gravel are not a direct reflection of the relative abundance in the source area. Under certain conditions some rocks readily yield blocks; others do not. Vein quartz and chert, for example, are common as pebbles. Granite, on the other hand, tends to disintegrate and become arkosic sand (gruss); limestone tends to dissolve and yield no gravelly detritus other than its insoluble chert. The flint gravel beaches of England are derived from friable chalk cliffs. Both granite and limestones yield blocks if conditions are such that disintegration and solution are inhibited or subordinated. Such conditions are those marked by high relief and rigorous climate with its attendant rapid erosion and accelerated frost action, or, more rarely, by glacial conditions, which even in regions of low relief are productive of gravels of mixed character rich in metastable rock fragments. As Garner (1959) has shown, aridity promotes gravel formation, and such aridity will produce gravels of mixed character. In general, the proportion of gravel increases with increasing immaturity of the associated sands, both of which appear to be functions of relief and climate and hence of tectonism.

FIG. 6-6. Effect of stream transportation on granodiorite pebbles, San Gabriel Canyon, California. The upper sample was collected near the source of the gravel and has a roundness of 0.28; the lower sample was 5.5 miles (8.9 km) downstream and shows a roundness of 0.44. (Krumbein, 1940, Pl. 1.)

FIG. 6-7. Imbricated stream gravel, Greybull River, Meeteetse, Wyoming. Current flowed from left to right. (Photograph by D. A. Lindsey.)

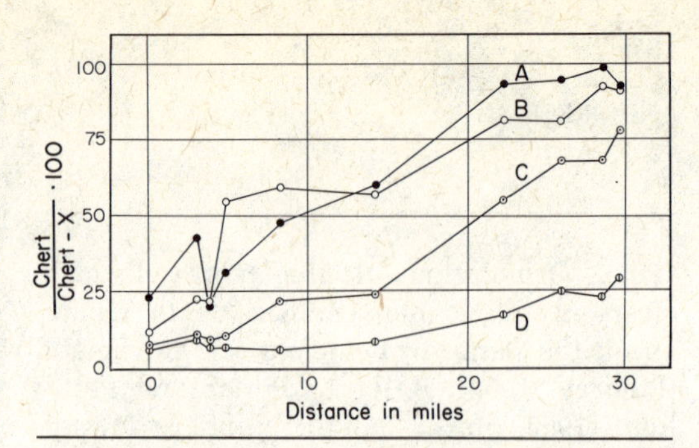

FIG. 6-8. Effect of transport on rock constituents of gravel in Rapid Creek, Black Hills, South Dakota. Ratio of chert to chert plus each component: A: sandstone; B: limestone; C: Precambrian metamorphics; D: quartz plus quartzite. (After Plumley, 1948.)

The composition of the gravel produced can be significantly modified during transport, as has been well shown in a number of field studies. Downstream changes in the composition of river gravels have long been noted. Hochenburger (Grabau, 1913, p. 247) described the disappearance of certain rock types in the River Mur and estimated the distance of travel required to destroy various rock types completely. Plumley (1948) has recorded the changing composition of the Black Hills (South Dakota) stream gravels and has shown that these materials undergo marked modifications downstream (Fig. 6-8). This rapid elimination of unstable components (granite and dolomite) and enrichment in stable species (quartzite and vein quartz) is confirmed by study of the gravels of the Dunajec River derived from the Tatra Mountains in Poland (Unrug, 1957), of the Piave in Italy (Dal Cin, 1967, Fig. 52), and of the Colorado River of Texas (Sneed and Folk, 1958). Clearly gravels, unlike sands, can become compositionally mature, that is, be reduced to the most stable constituents—vein quartz, quartzite, and chert—in a short distance of transport.

Gravels of restricted composition (*oligomictic*, Schwetzoff, 1934) are of two types, those formed by reduction of a gravel of an initially varied composition to a stable end-residue of vein quartz and quartzite, and those of a very local derivation, from a small watershed or a pocket beach, which derive their clasts from a single source rock. Gravels with a more varied composition (*petromictic*) indicate a larger gathering ground and diverse source terrane.

Intraformational conglomerates are a special class to which these rules do not apply.

STRUCTURE OF GRAVELS AND CONGLOMERATES

Gravel deposits are crudely bedded, with rather thick bedding units. Channeling is common so that gravel forms pods, lenticular bodies, and channel fills. Except for the imbrication readily seen in properly oriented exposures, gravels have no characteristic internal structure. Constituent clasts, especially those of tabular form, of most gravels lie with their longest dimensions parallel to the bedding. Exceptionally this fabric is disturbed by slumping or by the action of solifluction or frost, so that the pebbles lie in diverse positions. Cross-bedding is rare, except in interbedded sands. Large-scale grading is known in some fluvial gravels (a fining-upward sequence) and in some turbidite gravels. Some gravels show a large-scale inclined bedding with dips of the order of 20 degrees. Such a structure does not appear to be cross-bedded as usually defined. It may be caused by deposition on delta foresets or to lateral accretion on a migrating gravel bar.

BASIC DIFFERENCES BETWEEN GRAVELS AND SANDS

Although both gravels and sands are residues—washed residues—left after rock weathering, and both are mechanically transported and deposited, they differ from each other in several important ways.

Gravels (and conglomerates) are dominantly *rock* fragments, including both coarse- and fine-textured fragments, whereas sands are dominantly *mineral* particles. Rock particles (of fine-textured rocks only) are present in many sands and are important in some (lithic arenites). The proportion of rock particles is size-dependent increasing with increasing grain size (Fig. 6-9) and culminating in the coarse clastics—the gravels.

Gravels, certainly the most common gravels, tend to have a bimodal size distribution; sands are largely unimodal. This comes about from the observation that openwork gravels are rare; normal gravel has an abundance of sand between the gravel clasts. On the other hand, openwork sand is the rule. Even sands carried and deposited from muddy rivers tend to be "clean," having a pore system filled only with fluids. Even the exceptional sands—the wackes—may have been deposited with such an open pore system; the matrix they now have may be of diagenetic origin.

Gravels by and large have rounded clasts, rounding which is achieved in a comparatively short distance of travel. Associated sands, on the other hand, may be subangular or angular. Sand rounds very slowly and is not likely to become rounded in short river travel. Similarly gravels become compositionally mature very quickly because of abrasion, which selectively eliminates the less durable rocks. Selective abrasion operates very slowly on sands; many soft and cleavable minerals survive long distances of travel (Russell, 1937).

Gravels display a marked downcurrent size decline, a feature readily mapped and used to ascertain direction of transport and even distance to the source area. Sands show only slight variations in grain size over large areas; some show none at all.

Sands are commonly cross-bedded; gravels rarely show true cross-bedding but commonly display imbrication of the flatter pebbles. Imbrication is also known in sands, but it is more difficult to see and is not so strikingly developed.

Lastly, gravels are much less abundant than sands and are much more restricted in extent. They are local deposits related to a channel or beach or to an escarpment. Sands are more widespread and cover vast areas.

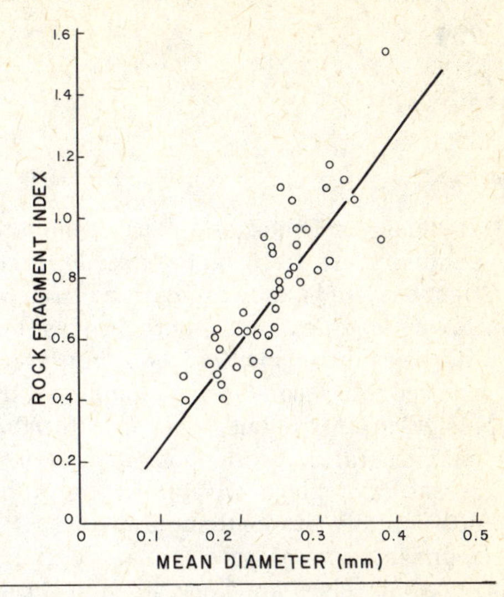

FIG. 6-9. Relations between grain size and proportion of rock particles in Clee Sandstone (Devonian) of Wales. "Rock fragment index" is percent rock fragments divided by percent quartz plus percent feldspar or Rx/Q+F. (After Allen, 1962, Jour. Sed. Petrology, v. 32, Fig. 9.)

CLASSIFICATION OF GRAVELS AND CONGLOMERATES

Conglomerates and gravels have been classified in various ways. They may be classified in a purely descriptive sense based on texture (such as boulder conglomerate or cobble conglomerate), on composition of their fragments (such as limestone conglomerate or chert conglomerate), or on cement (such as ferruginous conglomerate or calcareous conglomerate). Commonly they are classed according to the agent or to the environment responsible for their transport or accumulation, such as beach conglomerate, fluviatile conglomerate, or glacial conglomerate, or, in broader terms, marine, littoral, or continental conglomerates. Only in the case of present-day conglomerates can one be certain of the class to which a gravel deposit should be assigned. Also proposed is a classification based on the agent or process responsible for fragmentation (hence, epiclastic, cataclastic, and pyroclastic conglomerates or breccias). Here, also, assignment of ancient conglomerates to one or another of these groups requires discriminating criteria which are not infallible.

It is first important to recognize that conglomerates and breccias are not a homogeneous group (are not of a common origin). It is therefore necessary to separate the several types into geologically meaningful categories, even if the assignment of the deposit to one or another of these is at times difficult. Conglomerates and breccias belong to five major categories, not all equally abundant or important. Most common and abundant are *terrigenous gravels*. Like terrigenous sands, they are derived from breakdown of preexisting source rocks outside the basin of deposition; i.e., they are land-derived. A second group, common but of insignificant volume, are *intraformational conglomerates*, whose fragmentation processes take place within the basin of deposition and are contemporaneous with the sedimentation process. A larger and more important class are those of volcanic origin —the *volcanic conglomerates* and *breccias*, including *agglomerates*. These products of volcanic eruption correspond to sands of similar origin (tuffs). Coarse clastics produced by earth movements are *cataclastic breccias;* no significant volume of sand is associated with these materials. Included here are fault and founder breccias, the latter related to solution and sometimes called solution breccias. Glacial action and the resulting till and tillite, commonly considered with coarse clastics, could be considered cataclastic (the result of earth movement—movement of one rock over another), so that the till formed at the base of a moving ice sheet might be considered as gouge along an overthrust. It is thus a species of tectonic moraine, but, because tills are related to fluvioglacial and glaciomarine deposits and are interstratified with ordinary sediments, and because the overriding rock mass (ice) has vanished, they are more traditionally considered with terrigenous gravels.

Recent work has focused attention on a very rare type of breccias—those caused by *meteoric impact*. Fall-back breccias have been described; and although they are very common on the moon, they are miniscule in the terrestrial environment.

The dominant group, terrigenous gravels and conglomerates, can be divided into two major subgroups, one very common, the other much less so. Gravels collected by ordinary water currents have an intact or clast-supported framework; these have been designated *orthoconglomerates*—ordinary or normal conglomerates. Those deposited by subaqueous turbidity flows and slides, by solifluction, and by glacial ice or other modes of mass transport do not have an intact framework of gravel clasts but display, instead, a dominant matrix of fine-grained materials in which the larger clasts are embedded or "float." These have been called *paraconglomerates* or *diamictites*. They are, in simple words, conglomeratic mudstones or argillites. They are ill sorted or not sorted at all and exhibit a polymodal size distribution with the principal mode in the finer grades unlike ordinary conglomerates which, if bimodal, have the principal mode in the gravel range.

Normal terrigenous gravels (orthoconglomerates) may consist of a single rock type (vein quartz, for example), because all other debris has been eliminated by weathering or by long transport. Such supermature conglomerates with only a single component have been termed *oligomictic* (Schwetzoff, 1934) or *monogenetic* (Hatch and Rastall, 1971, p. 76). Others with a mixed composition, including many unstable materials (such as granite, basalt, and limestone), are referred to as *polymictic* or polygenetic. The several classes of conglomerates and breccias are shown in Table 6-2.

For other attempts to classify conglomerates and breccias, particularly genetic classifications, refer to Norton's classic paper (1917), to Field (1916), to Reynolds (1928), to the work of Fisher on volcanic breccias (1960), and to Maslov (1938).

ORTHOCONGLOMERATES

Orthoconglomerates have an intact framework of pebbles and coarse sand and are bound together by a mineral cement. They were deposited from highly turbulent waters, either high-velocity streams or the surf. They may be divided into two groups: mature orthoquartzitic conglomerates, mainly vein quartz and chert, and immature or petromictic conglomerates, composed of an assortment of metastable rock fragments.

ORTHOQUARTZITIC CONGLOMERATES

These conglomerates have a simple composition. Their pebbles are materials very resistant to wear and decomposition, such

TABLE 6-2. Classification of conglomerates and breccias

Epiclastic	Extraformational	Orthoconglomerates (matrix < 15%)	Metastable < 10%	Orthoquartzitic (oligomict) conglomerate
			Metastable > 10%	Petromict conglomerate (limestone conglomerate, granite conglomerate, and so on)
		Paraconglomerates (matrix > 15%)—diamictites	Laminated matrix	Laminated conglomeratic mudstone or argillite
			Nonlaminated matrix	Tillite (glacial)
				Tilloid or Geröllton (nonglacial)
	Intraformational	Intraformational conglomerates and breccias		
Pyroclastic	Volcanic breccias and agglomerates			
Cataclastic	Landslide and slump breccias			
	Fault and fold (*Reibungs*) breccias; "tectonic moraines"			
	Collapse and solution breccias			
Meteoric	Impact breccias			

as vein quartz, several types of quartzite and chert. These materials are a concentrate —a residuum derived from the destruction of a much larger volume of rock. Chert, for example, is concentrated from a large body of limestone containing a scattering of chert nodules. Vein quartz implies the destruction of a large volume of igneous and metamorphic rocks transected by widely separated quartz veins. Consequently orthoquartzitic gravels do not, as a rule, form very large deposits. They occur as sporadic pebbles or as pebbly layers and lenses interbedded with strongly cross-bedded sands. These pebbly layers may occur at the base of the sandstone, or they may recur at several levels within the formation. In general, orthoquartzitic conglomerates are not coarse. Pebbles several centimeters in diameter are common, but those a centimeter (more or less) in size are more typical. The pebbles, even the chert, are well worn and rounded. Material of this sort is virtually indestructible and may be reworked and redeposited through several cycles of sedimentation. The several varieties of chert, especially the fossiliferous cherts, may provide clues to the ultimate source of the gravel.

Gravels of this type are common in the geologic record. Nearly everywhere the basal Cambrian is marked by a few centimeters, in places a few meters, of such conglomerate. Basal conglomerates of this type, however, are more likely to be contaminated with locally derived subjacent metastable materials than are conglomerates which recur at higher levels in the same formation. Sporadic pebbles scattered through the Mississagi and Lorrain quartzites (Huronian) of Ontario are typical Precambrian examples. The latter formation contains the justly famous red jasper-bearing conglomerate found in nearly all rock collections. The gold-bearing conglomerates of the Witwatersrand district of South Africa are relatively thin gravel banks composed of vein quartz pebbles. Most pebbles are under 3 cm in diameter and have a compressed elliptical cross section. Scattered orthoquartzitic gravels are common in several Paleozoic formations in the central Appalachians. Portions of the Silurian Tuscarora quartzite contain sporadic vein quartz pebbles; locally they contain thin beds of well-rounded quartz gravel (Yeakel, 1962). The Devonian Chemung contains a few lenses of conglomerate with well-rounded discoidal quartz pebbles, in contrast to the poorly rounded, subspherical pebbles of the Catskill (McIver, 1961). The Mississippian Pocono is notably conglomeratic (Pelletier, 1958). The conglomerates, forming 3 to 10 percent of the formation, are vein quartz gravels cemented by quartz containing pebbles 1 to 8 cm in diameter. The thickest conglomerate beds range from under 1 m to 3 m in thickness. Similarly, those somewhat thicker and coarser quartz conglomerates characterize the Pennsylvanian Pottsville of the Anthracite Coal Basin (Meckel, 1967). Gravels of equivalent character and age occur in western New York (Olean Conglomerate) and in eastern Ohio (Sharon Conglomerate). These consist dominantly of vein quartz, well-rounded, 2 to 3 cm in diameter on the average. Except for the Chemung, all these orthoquartzitic materials appear to be river deposited.

Much younger gravels of similarly restricted composition include the Brandywine upland gravels of Maryland (Schlee, 1957) and the similar Lafayette upland gravels of western Kentucky (Potter, 1955). The Maryland gravels form a thin blanket up to 10 m thick and are composed of well-rounded chert, vein quartz, and quartzite. The Lafayette gravels, of comparable thickness, are dominantly chert. Both are late Tertiary in age and are presumed to be fluvial; the Maryland gravels are Potomac gravels, and those of the Lafayette are mainly gravels of the Tennessee River.

In summary it appears that orthoquartzitic gravels are highly mature, well-rounded accumulations of ultraresistant materials. They form deposits of no great thickness and form only a small part of the formation in which they occur; the balance is usually coarse, cross-bedding pebbly sands. They can be accumulated by rivers or waves and may be either beach or alluvial in origin. Most of those mentioned above are alluvial. They are a concentrate—a residuum winnowed out of a vast body of materials. The Paleozoic Appalachian examples are mainly vein quartz from the present Piedmont metamorphic complex.

It should be noted that, although the gravels are very mature, the associated sands are not. In the Appalachian examples cited, the associated sands are largely lithic sandstones.

PETROMICT CONGLOMERATES

The great conglomerates of the past belong to the group of petromict conglomerates. In general these are thick, wedge-shaped, basin-margin accumulations of gravel that were shed from sharply elevated highlands. These gravels form a conspicuous part of the sequence to which they belong, forming striking cliffs and ledges or prominent ridges. They may be basal or intercalated at several levels in the sequence.

These conglomerates are coarse-grained representatives of lithic and arkosic sandstones. Although their composition is varied, all are marked in that the chief constituents are metastable rocks, generally of several kinds. The most common type is a mixture of pebbles or cobbles of plutonic, eruptive, or sedimentary and metamorphic origin. In many cases, however, one type or class of pebble predominates.

Granite cobbles play the same role in coarse sediments that feldspars do in sands. Granite-bearing conglomerate is thus the coarse-grained equivalent of an arkose. It is not surprising, therefore, that granite-bearing conglomerates are commonly associated with arkoses. Because of the limited block-forming capacity of much granite, granite-bearing conglomerates are likely to be chiefly arkose with scattered pebbles of granite, or at best lenses of granitic gravel in arkose. Both granite-bearing conglomerate and arkose record rapid erosion of the crystalline basement. Accordingly the appearance of granite-bearing conglomerates denotes a major uplift.

Limestone-bearing conglomerates record unusual conditions which permitted the erosion of limestone as gravel rather than its usual removal by solution and resultant accumulation of chert gravel. This implies sharp uplift and locally high relief—best achieved along a fault scarp—such as that which prevailed during the accumulation of the Triassic limestone conglomerates of Maryland and Pennsylvania. Abundant limestone gravels are a product of glaciation, as the Pleistocene limestone outwash gravels of parts of Illinois and Wisconsin testify. These gravels were derived from ice of the Silurian Niagaran dolomites of that region.

In regions of active volcanism lavas may yield extensive gravels. In general, these gravels tend to consist of fragments of felsitic lavas, even though the bulk of the associated flows are more basic. Apparently the gravel-forming capacity and the survival ability of the former are greater than that of the latter. The principle is well illustrated by the conglomerates of the Precambrian Keweenawan of Minnesota and Michigan, which are composed mainly of felsitic fragments rather than of basaltic lavas which are vastly more abundant. Volcanic gravels may be very thick (1000 m or more) and are more or less contemporaneous with the associated lava flows. Unlike the granite-bearing gravels, they record no great uplift or deep erosion.

All these conglomerates are marked by their coarseness. The thickest beds tend to be the coarsest (Fig. 6-10). Although many boulders are a meter or more in diameter, the mean size is more apt to be 10 to 20 cm, about one-tenth of the maximum size (Fig. 6-5). Marked downcurrent decrease in size is the rule, so mapping of maximum pebble size over a region clearly reflects the paleocurrent system (Fig. 14-5). Sorting is fair to poor, because the interstices of the gravel are filled with sandy materials.

In general, the rounding of these gravels is fair to good, although exceptionally—as very near the source ledges—the rounding is very poor and the rock may best be described as a breccia. The rounding of the larger cobbles is in sharp contrast to the very poor rounding of the associated sands.

Pebbles may display the effects of solution at their points of contact; in some cases they show stylolitic interpenetration, especially in the limestone gravels (Bastin,

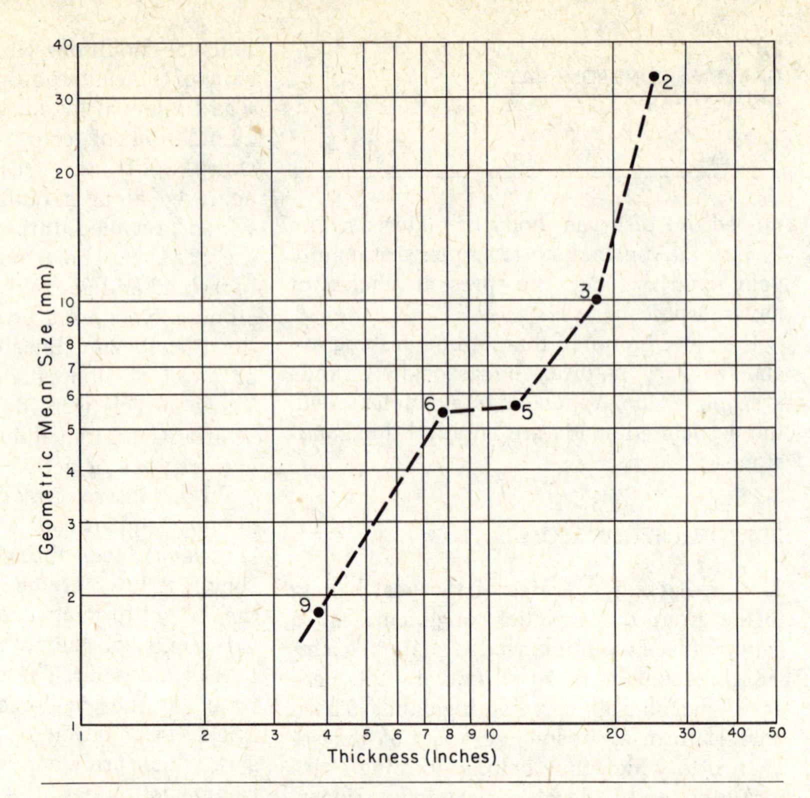

FIG. 6-10. Relation of thickness of bed to mean grain size in Pleistocene sands and gravels. Figures denote number of measurements. (Unpublished data from Kurk, 1941.)

1940). In others are conspicuous solution pits at points of contact (Kuenen, 1942). Pebbles of some conglomerates are fractured and show healed microfaults (Fig. 6-25).

The bedding of these conglomerates tends to be large-scale and lenticular and marked by a rude parallelism of the flatter clasts (or, in many cases, a well-defined clast imbrication seen best in sections parallel to the flow of the depositing current). Cross-bedding is absent except in the finest gravels and interbedded sands.

As noted, the petromict orthoconglomerates are the most common, most conspicuous, and most impressive conglomerate deposits in the geologic record. There are, therefore, many examples of this type of deposit which have been well described. Notable examples include the Archean conglomerates found in the shield areas of the world. Among these are the conglomerates of Abram Lake in northwestern Ontario (Pettijohn, 1934, 1943; Turner and Walker, 1973), which are in places 1000 m or more thick, containing granitic cobbles, some exceeding 1 m in diameter. Conglomerates of similar nature occur elsewhere in the Canadian Shield (Henderson, 1970; Boutcher, Edhorn, and Moorehouse, 1966) and in the Archean of Finland (Simonen, 1953, p. 26). The late Precambrian Keweenawan of Michigan contains numerous conglomerates, some of which are copper-bearing. The thickest of these, the Great Conglomerate, over 2000 feet (600 m) thick, is steeply tilted and forms a prominent ridge running much of the length of the Keweenaw Peninsula in Lake Superior. It consists of well-rounded coarse volcanic detritus, mainly felsitic (Irving, 1883, p. 167; White, 1952). Pebbles average 6 to 8 inches (15–20 cm); some exceed 1 foot (30 cm). These conglomerates are associated with red clastic sediments and lavas. Other Precambrian conglomerates of note include the Murky Formation of the East Arm of Great Slave Lake (Hoffman, 1969, p. 455). This formation, at least 3000 feet (914 m) thick in places, occurs in massive beds 30 to 150 feet (9 to 45.7 m) thick, with boulders up to 4 feet (1.2 m). The deposits thin rapidly downdip. The maximum size of the boulders also decreases rapidly in size, diminishing to half their initial size in about 16 miles (27 km). See Fig. 6-11.

Several conglomerates of this type of late Paleozoic age occur in the Arbuckle and Wichita mountains of Oklahoma. The Collings Ranch Conglomerate (Ham, 1954), over 2000 feet (600 m) thick, is a good example. It is a limestone boulder conglomerate with clasts up to 75 cm in diameter. The Permian Post Oak Conglomerate of the Wichita mountains (Chase, 1954) is another example. This arkosic boulder conglomerate is in places granite-bearing and is elsewhere a limestone conglomerate.

The conglomerates in the Triassic basins of the eastern United States are thick, wedge-shaped deposits marginal to the Triassic basins, commonly related to a border fault. In places these are subangular, limestone pebble conglomerates ("Potomac marble"). The conglomerates of the Newark Series (Triassic) in Connecticut have been described by Krynine (1950), those of the Deep River Basin of North Carolina by Reinemund (1955). These all seem to be extensive alluvial fan accumulations (McLaughlin, 1939). The Roxbury Conglomerate of the Boston Bay area (Mansfield, 1906; Dott, 1961) is another well-known conglomerate in the Appalachian area.

Famous conglomerates of the western United States include the San Onofre "Breccia" (Miocene) in California (Woodford, 1925), the Gila Conglomerate in Arizona, the Price River Conglomerate (Cretaceous) in the Wasatch Plateau of Utah, and the Wasatch Conglomerate (Eocene) itself. All of these conglomerates are notably thick, locally 5000 feet (1500 m) or more; all have a considerable along-strike persistence; and all wedge out rapidly downdip. Many are related to or are associated with contemporaneous uplift and faulting.

Thick, petromict conglomerates characterize molasse sedimentation in many areas. Illustrative of such deposits are the conglomerates in the molasse basin north of the Alps (Blissenbach, 1957; Füchtbauer, 1967; Gasser, 1966). They form most of the proximal molasse section (several thousand meters) and diminish northward until they form 10 percent or less of the section in a distance of 200 km. Maximal pebble size declines from over 1 m to about 10 cm in the same distance. The gravel composition shows a progressive enrichment in stable components in the downcurrent direction (Fig. 6–12).

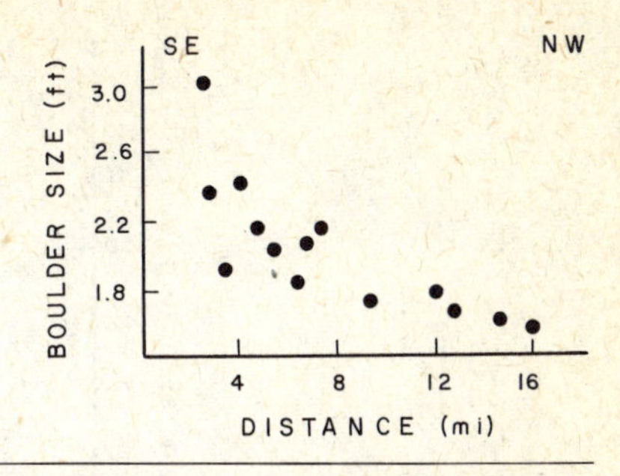

FIG. 6-11. Variation in maximum boulder size in conglomerate, Murky Formation (Precambrian), East Arm, Great Slave Lake, Northwest Territories, Canada. Average of 10 largest boulders at each point. (After Hoffman, 1969, Canad. Jour. Earth Sciences, v. 6, Fig. 9. By permission of the National Research Council of Canada.)

The conglomerates described above are largely alluvial. They vary from alluvial fans (*fanglomerates*, Lawson, 1925) to the deposits of braided and meandering rivers. Some petromict conglomerates, however, seem to have been deposited in deep water. These are the turbidite conglomerates of which the Cretaceous conglomerates of Wheeler Gorge in California (Rust, 1966; Fisher and Mattinson, 1968) and the Archean conglomerates of Minnitaki Lake, Ontario (Walker and Pettijohn, 1971) are examples. The conglomerate of Wheeler Gorge displays beds up to 95 feet (30 m) thick with rounded boulders up to 3 feet (about 1 m) in diameter. Conglomerate beds show very large-scale grading. The undersurface of the main conglomerate is marked by gigantic flute casts. These deposits are interbedded with thinly laminated shales and siltstones that were presumably deposited on a flat sea floor in deep water. Conglomerates of the Archean are likewise interbedded with dark slates and thin fine-grained siltstones. The gravels and sands are graded; cross-bedding is lacking. It is clear that coarse conglomerates need not signify shallow-water sedimentation.

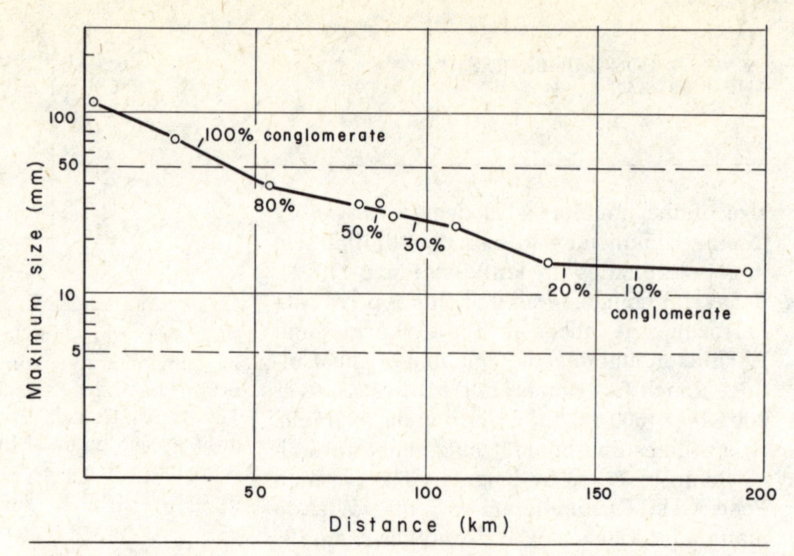

FIG. 6-12. Downstream decrease in maximal pebble size and composition of gravels in Miocene Molasse north of the Alps. (After Füchtbauer, 1967, Geol. Rundschau, v. 56, Fig. 8.)

PARACONGLOMERATES (CONGLOMERATIC MUDSTONE)

Conglomerates with more matrix than clasts are in reality mudstones with a sparse to liberal sprinkling of pebbles or cobbles. In many cases the pebbles form 10 percent or less of the rock. However, these deposits are commonly described as conglomerates rather than as mudstone.

No satisfactory terms have been proposed for this interesting though relatively uncommon class of conglomerates. It seems that most terms that can be applied to these rocks—applicable at the hand specimen or outcrop level and not implying genesis—and are either inaccurate or inconveniently long. The nomenclatural problems have been reviewed by Blackwelder (1931); Miller (1953); Folk (1954); Flint, Sanders, and Rodgers (1960a, 1960b); Harland, Herod, and Krinsley (1966, p. 228); and Schermerhorn (1966). Pettijohn has proposed the term *paraconglomerate* (1957, p. 261). As the term implies, the rock is amiss or aberrant—not deposited in the manner of an ordinary gravel. These rocks are not the products of normal aqueous transport. Some pebbly or cobbly mudstones are boulder clays, a term which, although presumably descriptive, is generally considered synonymous with *till*. The latter is a particular kind of boulder clay, namely a deposit made by glacial ice. The term *Geröllton* has been applied to nonglacial pebbly mudstones (Ackermann, 1951); the term *tilloid* had been earlier applied to a "till-like deposit of doubtful origin" (Blackwelder, 1931, p. 902). More recently the term *diamictite* has been proposed for "any nonsorted or poorly sorted terrigenous sediment that consists of sand and/or larger particles in a muddy matrix" (Flint, Sanders, and Rodgers, 1960b). These are the rocks which Folk (1954, p. 347) would call *conglomerate mudstone;* Crowell (1957) termed them *pebbly mudstone.* The term *mudstone conglomerate* is quite different, because it is applied to a conglomerate in which the fragments (not the matrix) are pelitic. The terms *till* and *tillite* are reserved for pebbly mudstones or boulder clays of glacial origin.

Of the two basic types of pebbly mudstone, one has a stratified matrix; the other has an unstratified or structureless matrix.

LAMINATED PEBBLY MUDSTONE

Laminated pebbly mudstones are very rare rocks of distinctive character. They consist of delicately laminated argillites or slates in which occur thinly scattered phenoclasts, some no larger than sand grains (Fig. 6-13) and others full-sized cobbles or even boulders. Laminations near the larger clasts are distorted and bend down beneath them as well as lap on or arch over them (Fig. 6-14).

These conglomeratic laminated argillites were obviously produced by the dropping of pebbles (*dropstones*) into still bottom waters in which the finest silts and muds were accumulating. Dropstones are normally the product of raft action—most commonly by glacial ice, although it is possible for cobble-sized fragments to be carried in the roots of floating trees or by river or shore ice. Where such stones are numerous, especially in pelitic rocks displaying laminations of varvelike character, they are almost certainly glacial. Such deposits have been called *pelodites* (Woodworth, 1912, p. 78). Some thinly laminated, fine-grained ash beds containing larger infallen stones or blocks closely resemble glacial pelites with dropstones. The abundance of volcanic fragments and especially glass serve to identify the volcanic nature of these deposits and distinguish them from true pelodites.

In general, laminated argillites with rafted blocks are closely associated with tillites and are, perhaps, the best evidence of the presence of glacial ice and of a glacial origin for associated strata. Pelodites have been described from deposits of all ages and places. Dropstones are common in many varved glaciolacustrine clays of the Pleistocene of Scandinavia and Canada. Lithified counterparts of these deposits have been reported in the Fern Creek glacial beds (Precambrian) in Michigan (Pettijohn, 1952), in the Gowganda Formation (Huronian) in Ontario (Collins, 1925, pl. 10; Ovenshine, 1965; Lindsey, 1969, pl. 4), in the Tapley Hill slates of the Adelaide Series of Australia (Caldenius, 1938), in the Eocambrian of Sweden (Kulling, 1938), and in the Itu varvite of the São Paulo district of Brazil.

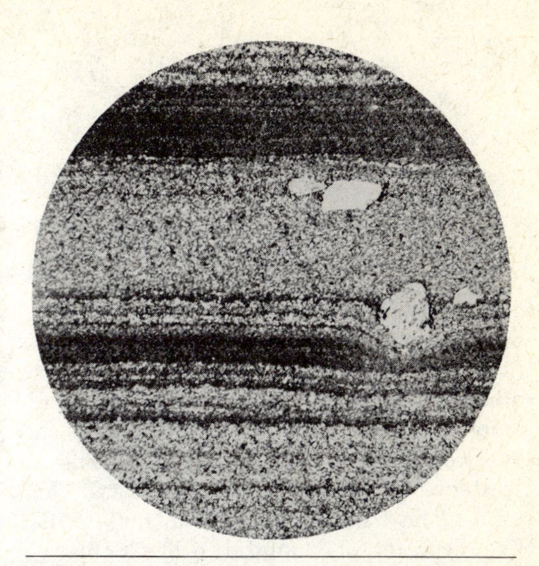

FIG. 6-13. Argillite, Cobalt Group (Precambrian), Ontario, Canada, showing coarse, ice-rafted sand grains embedded in siltstone (light) and slate (dark) laminations (see also Fig. 8-3). Ordinary light, ×90. The banding is probably seasonal, two dark winter layers and three coarser summer silts as shown. Minor sublaminations record subseasonal freshets. This argillite is closely associated with tillite (see Fig. 6-22).

FIG. 6-14. Varved argillite, Cobalt Group (Precambrian), Ontario, Canada, showing rafted granite pebble and distortion of laminations. Length of visible specimen about 8 inches (20 cm).

TILL AND TILLITE

The term *till* was apparently first applied in Scotland to "stiff, unstratified clays containing angular, subangular and rounded blocks of rock, mostly polished and striated" (Woodward, 1887; Gary, McAfee, and Wolf, 1972, p. 741). It is synonymous with *boulder clay*, which was defined as "a deposit which owes its origin more or less directly to the grinding-action of glaciers" (Geikie, 1874, p. 190). Thus till has come to be a genetic term applied to the unstratified deposits of glacial ice (Fig. 6-15).

The term *tillite* has been applied to lithified tills (Fig. 6-16). It is attributed to Penck, according to Du Toit (1954, p. 269).

The glacial origin of Pleistocene tills is no longer a debatable question, although the origin of a particular deposit may be uncertain. A good deal of uncertainty, however, prevails with respect to many so-called tillites. Difficulty arises because mechanisms other than glacial ice can produce massive, more or less structureless deposits with an overwhelming clay matrix that contains a scattering of embedded phenoclasts. The term *tilloid* has been applied to these "till-like deposits of doubtful origin" (Blackwelder, 1931, p. 902). The term has come to mean essentially a nonglacial diamictite, tillite being a diamictite of glacial origin.

Some difficulties of discrimination between tillites and tilloids have arisen because most of our knowledge and concepts about till have come from a study of Pleistocene glacial deposits. These deposits are nearly all continental, thin, and unlikely to be preserved in the geologic record. Most of the glacial record of the past appears to be marine, and the marine tills differ in important ways from those of continental origin. Harland, Herod, and Krinsley (1966) have recognized these differences and apply the term *orthotill* to those deposits formed by immediate release from transporting ice by ablation and melting and apply the term *paratill* to those deposits formed by ice rafting in a marine or lacustrine environment. The problem is complicated because some ice-laid, subaqueous till is indistinguishable from tills deposited on land. True paratills, however, are marine sediments with a smaller or larger amount of ice-rafted material, primarily *dropstones.* The ice-laid subaqueous till is an orthotill even though it is submarine. Only if rafting is involved is the term paratill applicable. The nomenclature difficulties arise, of course, because the various terms applied are defined in terms of the process by which the deposit was formed—a feature not seen in the outcrop.

Textures Because of the conspicuous role played by the boulders and cobbles which they contain, tillites would generally be termed conglomerates by the field geologist, even though they are dominantly fine-grained materials (mostly clay or its lithified equivalent). In tills analyzed by Krumbein (1933), this fine-grained matrix constitutes four-fifths to nine-tenths of the total and only one-fifth to one-twentieth of the whole falls in the gravel range. In most tills, silt and clay form one-half to two-thirds of the deposit. The median size in most tills falls between 3 and 10 microns. These tills are obviously clay tills (Fig. 6-17).

The sorting of till is notably poor. All of Krumbein's 48 analyzed tills had 12 or more grades containing 1 percent or more of the material. No very large amount of material was present in any one size class. The largest modal class in the tills analyzed contained only 20 percent by weight of the whole sample; obviously the modal class is inconspicuous. Although some of the tills

FIG. 6-15. Pleistocene till, Bull Lake, Wind River Range, Wyoming. (Photograph by C. D. Holmes.)

FIG. 6-16. Sturtian Tillite, Adelaide Series (?Precambrian), near Adelaide, Australia. (Photograph by R. T. Chamberlin.)

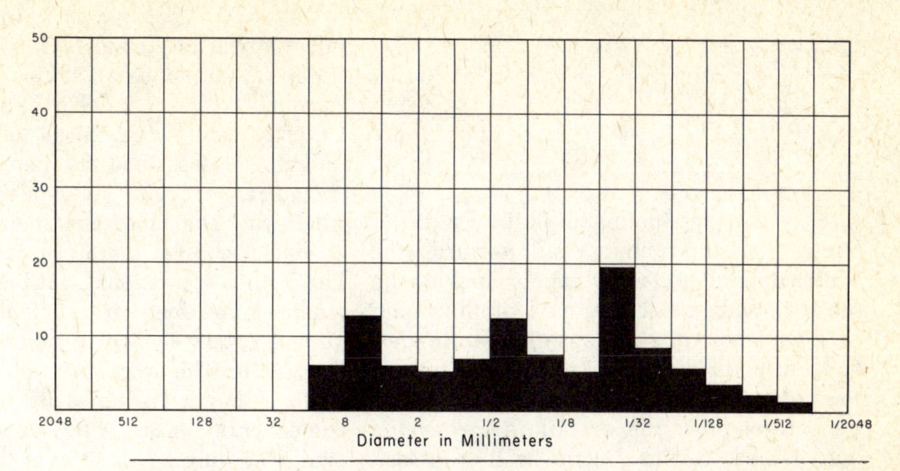

FIG. 6-17. Histogram of Wisconsin till (Pleistocene), Cary, Illinois. Composite of 10 samples.

FIG. 6-18. Till stones (Pleistocene), Illinois. Note striations, faceting, and snubbed edges and corners. Largest cobble about 5 inches (13 cm) in diameter.

analyzed appear to have only a single modal class, most have two or more such classes.

The size distribution of some tills, especially water-laid tills, may be modified in various ways. After subjection to water currents or wave action, there may be selective loss of the fines. In an extreme case, these may all be removed leaving a lag deposit predominantly boulders or cobbles. In some cases, also, glaciers may pass over sandy deposits and acquire a significant load of sand so that the resulting till is very sandy.

Although composed predominantly of fine materials, till and tillite may contain very large boulders. Boulders 1 m in diameter are rather rare but not unknown. A few glacial erratics, whose weight exceeds 1000 tons are known. Because much of a till is 1 micron or smaller and the largest 1 m or more, the size range is a millionfold or more. Till is thus the poorest sorted of all sediments.

Von Engelen (1930) maintained that glacial cobbles and pebbles tended to have a unique form which was diagnostic of glacial action (Fig. 6-18). The characteristic form is that of a striated and faceted flatiron. In detail the features of such a cobble are: (1) the roughly triangular shape in plan with the facet of the largest and flattest area down, (2) the pointed but scour-snubbed nose at the apex of the narrowest angle of the bottom facet, (3) an only slightly scoured or hackly backside above the base line of the bottom triangle, (4) a tendency to hump form of the top side of the flatiron with (5) lateral facets running off toward the snubbed point, (6) chipping or nicking on the underside or at the apex of the point, (7) a tendency for striations on lateral facets to be directed diagonally downward toward the point, and (8) indication that the variations from the norm or failure to develop one of these features in the well-processed pebble are attributable to a particular, and still obvious, configuration of the original fragment or the nature, rock structure, and composition of the specimen.

Von Engelen's description of the end-form toward which glacial action tends to shape rock fragments was supported by quantitative data collected by Wentworth (1936a), who analyzed 626 glacial cobbles. Three hundred, selected for perfection of glacial action, were studied in detail. The cobbles were classified according to their general shape. The dominant shape was tabular, in part because of the original tabular shape of the fragments and of the sustained abrasion on the two initially opposed principal faces. Study of the marginal profiles (as observed when the pebble lies on its most stable face, that is, the profile normal to the shortest axis) shows that the pentagonal outline is indeed the most common. Approximately two-thirds of the margins might be described as pentagonal, quadrangular, triangular, polygonal, trapezoid, or reniform. The most characteristic shape of a glacial cobble, therefore, is parallel tabular with a pentagonal outline. The flatiron shape noted by Von Engelen is verified by actual count. Typical profiles are shown in Fig. 6-19. The average cobble was a slab with a length 1.4 times its width and 2.25 times its thickness. A few cobbles were four times as long as they were thick and were over twice as long as wide. Of the 300 cobbles examined, 128 have well-rounded margins, 116 moderately rounded margins, and 56 were rough and broken. By far the majority, therefore, showed definite evidence of superficial shaping of the margins by abrasion. The characteristic "pushoff" ends and edges of the cobbles, termed *snub scars* by Wentworth, were strongly marked on 43 cobbles, clearly present on 107, faintly seen on 42. Though not seen on 108, these lee-end pressure spalls are sufficiently abundant to be considered a characteristic feature of glacial cobbles.

Typical till stone is striated, although such striations are more common on certain rock types than others and as a whole are by no means as abundant as is commonly sup-

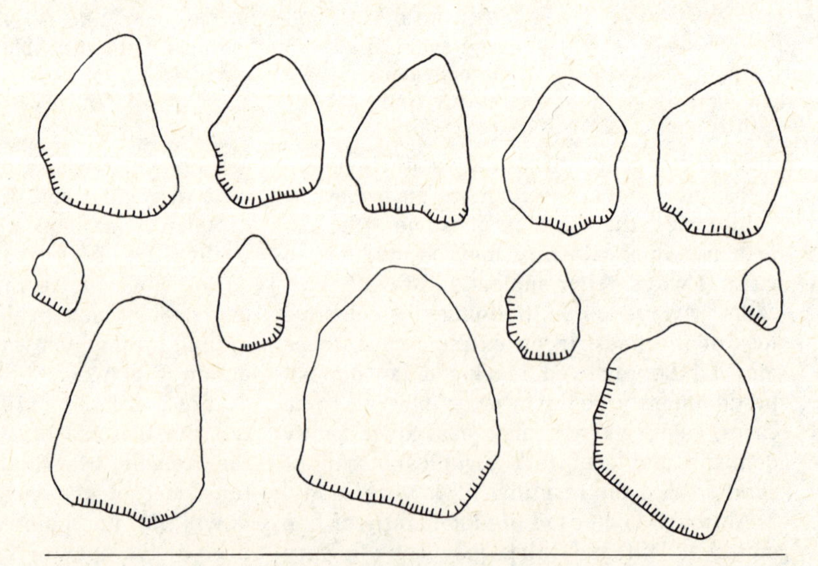

FIG. 6-19. Profiles of the pentagonal and other "flat-iron" glacial cobbles, largest about 18 cm long. Stoss end at top of cut; margin showing snub scars hachured. (From Wentworth, 1936.)

posed. Wentworth's study of striations (1936a) showed that in his material only 10 percent of the limestone and only 1 percent of all other rocks were found to show striations (Fig. 6-20). Wentworth found these striations to be subparallel to one another and generally parallel to the long axis of the till stone.

Because of the lithified character of most tillites, especially those of Precambrian age, it is difficult to remove cobbles intact from the matrix. Consequently it is difficult to determine the true shape of cobbles or to see any striations. The common failure to find striated stones in many ancient tillites is not surprising, in view of the scarcity of striations even in Pleistocene tills. Striations are so definitive, under the proper conditions, that they are one of the most important criteria for glacial action. But their absence does not necessarily oppose a glacial origin. It should be remembered also that weak striations can be produced by river ice and other agents (Wentworth, 1936b).

Structures of till Though many tills, especially orthotills, appear structureless and devoid of bedding or any other internal organization, it has long been known that there is a weak but distinct tendency for the more elongated rock fragments to lie with their longest axis nearly parallel to the direction of the ice flow at the time the deposit was formed (Fig. 6-21). This arrangement obviously would be true only of the stones in orthotills; dropstones show a different fabric. Although it was noted in 1884 by Hugh Miller the younger, it was not until 1932 that Richter began his systematic study of till fabric. He measured and plotted the long-axis orientation. The ice movement pattern inferred from his maps was wholly consistent with known bedrock striations and morainic and other ice-constructed land forms. Since Richter's work was published, various papers have applied his techniques to problems of ice movement in the United States (Krumbein, 1939; Holmes, 1941), in Finland (Okko, 1949; Virkkala, 1951, 1960;

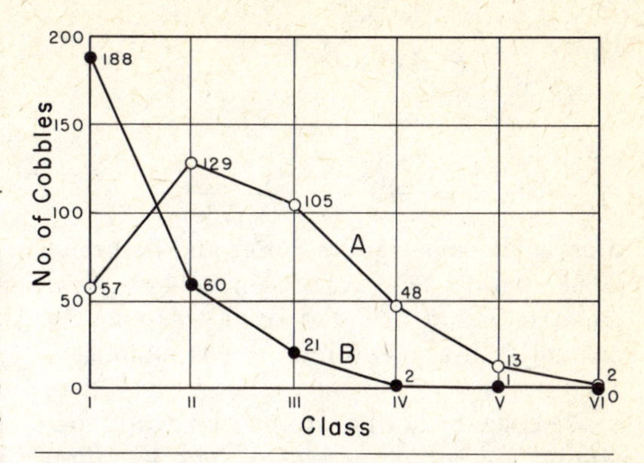

FIG. 6-20. Degree of shaping and marking of 626 pebbles and cobbles from Wisconsin (Pleistocene) drift near Baraboo, Wisconsin. A: limestone; B: all other rocks. Class I, no striations or facets; II, faint striations; III, clear striations on one side; IV, clear striations on two sides or grid on one side; V, grid on two sides; VI, grid on several sides. (After Wentworth, 1936.)

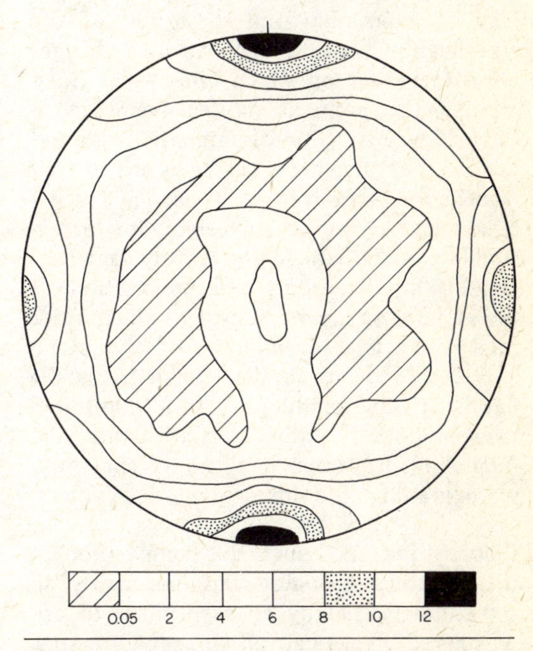

FIG. 6-21. Composite contoured diagram of long-axis orientation of till stones, based on 10 localities and 1180 till stones. Each individual pattern was superposed by placing its recorded ice-flow direction in north–south position. The direction from which the ice approached is indicated by tick mark at top of diagram. (From Holmes, 1941.)

Kauranne, 1960), and in Great Britain (West and Donner, 1956). Till also has a microfabric related to ice motion (Dreimanis, 1959). The whole subject of fabric has been thoroughly reviewed by Potter and Pettijohn (1963, pp. 31–35). And efforts have been made to study orientation of tillstones in ancient tillites (Pettijohn, 1962; Lindsey, 1969).

Whereas orthotills are massive and devoid of any structure except the orientation fabric noted above, paratills show bedding in varying degrees. Outsize clasts, that is, those which exceed the thickness of the stratification units (beds or laminations), could not have been carried laterally into place contemporaneously with the sediment. These *dropstones* are randomly oriented or show a vertical axial symmetry. They display a penetration and distortion of the stratification beneath, with an unconformable covering on top (Ovenshine, 1965). Such relations are, perhaps, the decisive evidence for rafting. Distortion of laminations around a stone may, however, be the result of clay compaction; therefore some asymmetry between underlying and overlying distortion should also be looked for (Hardy and Legget, 1960). Disruption and penetration of the underlying layers is strong indication of rafting. Vertical orientation of the stone axis and tendency for flat stones not to be parallel to the bedding are good additional evidence of raft action. Unlike tilloids, tills (including dropstone tills) rarely show upward grading into finer-textured sediment.

Composition Although the composition of till and tillite is highly variable, nearly all are characterized by an assortment of unweathered blocks and till stones in a matrix or paste of unweathered materials—"rock flour." Till stones, although composed most commonly of the underlying basement, are also in part materials foreign to the area. All types of rocks—sedimentary, metasedimentary, and both plutonic and effusive igneous rocks—may be present. In Pleistocene till, careful pebble counts and plotting the distribution and quantity of the till stones, or of a particular rock type, make possible the determination of the flow pattern of the ice and also yield significant information on the nature of the concealed bedrock and even, in some cases, the location of hidden ore deposits (Potter and Pettijohn, 1963, pp. 195–198; see also Fig. 14-7).

The matrix of till, if unweathered, is usually a dark bluish gray; if oxidized it is buff. The usual tillite matrix is dark gray to greenish black. It closely resembles graywacke (which it perhaps is) and under the microscope appears to consist of fresh, angular grains of quartz, feldspar, and rock fragments set in a "paste" of fine grain (Fig. 6-22). In tillites themselves, the latter appears to be richly chloritic and micaceous and is probably the product of low-grade metamorphism of the original clay component of till. Triturated carbonate is common in tills of limestone regions and renders such deposits strongly calcareous.

The bulk composition of till or tillite matrix is much like that of graywackes and related rocks (Table 6-3). Normally these materials, constituting an unweathered rock flour, are rich in alumina, iron, alkaline earths, and alkali metals. Tills of limestone regions are notably calcareous and hence have a high CaO and CO_2 content, as well as MgO if the source rocks were dolomitic.

Dropstone tills and tillites will differ from orthotills because of the effects of washing and removal of fines. The matrix may be more sandy, and consequently lower in Al_2O_3, iron, and K_2O and higher in SiO_2.

Stratigraphic aspects The thickness of individual till sheets or tillites is highly variable, from a meter or two to hundreds of meters or more. Because glaciations are commonly multiple, more than one till or tillite beds are likely to be present in a given section. These glacial beds preserved in the past were products of continental rather than mountain glaciers; hence tillites are apt to be found in rocks of the same age over wide areas. A glacial epoch is a regional and not a local phenomenon. Tillites, therefore, are not apt to be local as would be slump, landslide or other deposits of a related sort.

Many, but by no means all, tills are found to rest on a striated pavement. This is true only of the first or lowest till or tillite in a sequence. Higher and younger till may be interbedded with glaciolacustrine and fluvioglacial beds. It is also more apt to be true

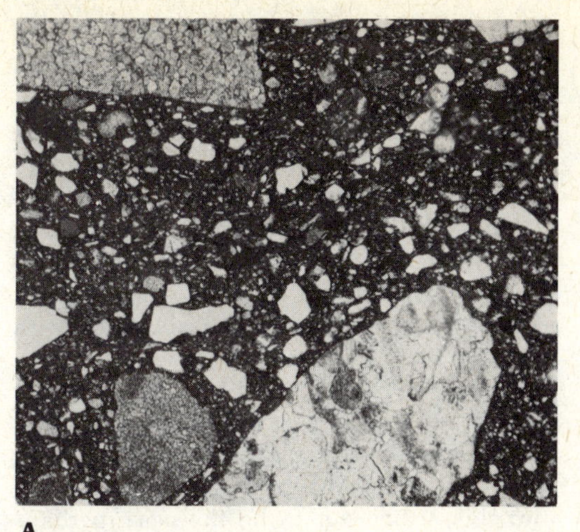

A

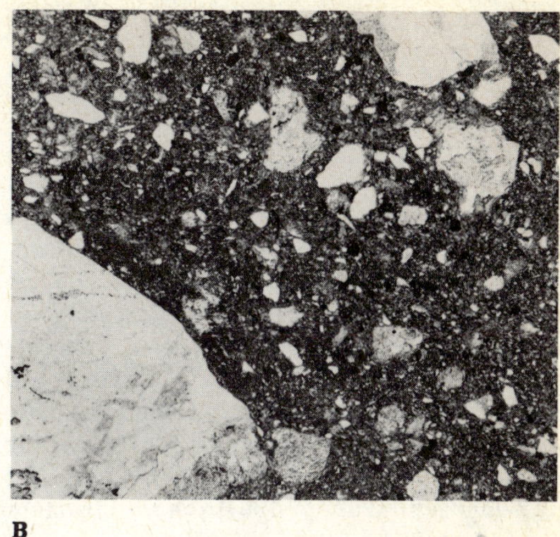

B

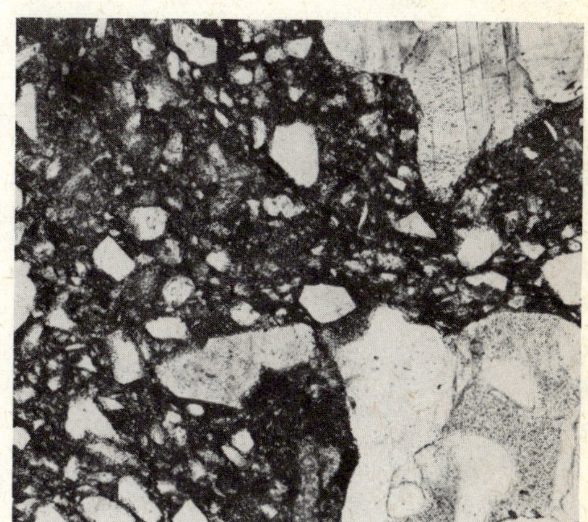

C

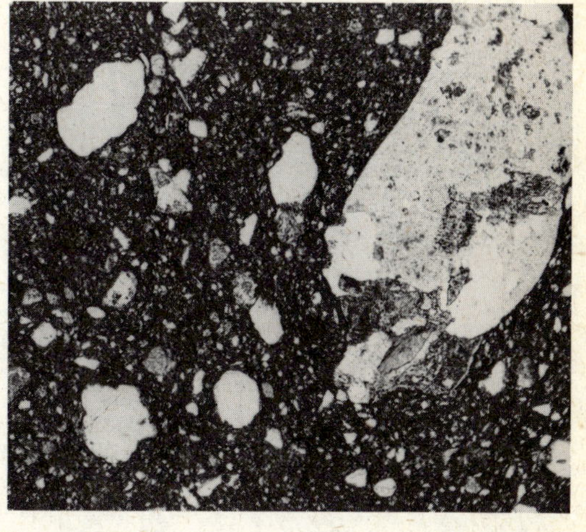

D

FIG. 6-22. Till and tillite. A: Wisconsin (Pleistocene) till, Illinois. Ordinary light, ×30. Large rock fragments are fossiliferous and dolomitic limestone; smaller fragments are mainly clear angular quartz, some feldspar, and many smaller limestone grains, together with a few pieces of shale (black). Matrix is a paste of calcareous silt and clay. B: Gowganda Tillite, Cobalt Group, (Precambrian), Bruce Mines, Ontario, Canada. Ordinary light, ×30. Largest fragments are granite; small ones are quartz (clear), and feldspar (clouded with inclusions). Matrix is an intimate pastelike intergrowth which when resolved under high power is found to be rich in chlorite and sericite. C: Dwyka Tillite (Permian), South Africa. Ordinary light, ×80. Rock fragments are granite, composed of microcline (dusty and traversed by cleavage cracks) and quartz (clear blebs); sand grains are clear quartz and dusty feldspars. Matrix is a fine paste of indeterminate character. D: Sturtian Tillite, Adelaide Series (?Precambrian), Australia. Ordinary light, ×20. Largest fragments are pieces of granite or feldspar (inclusion-filled) and quartz (clear).

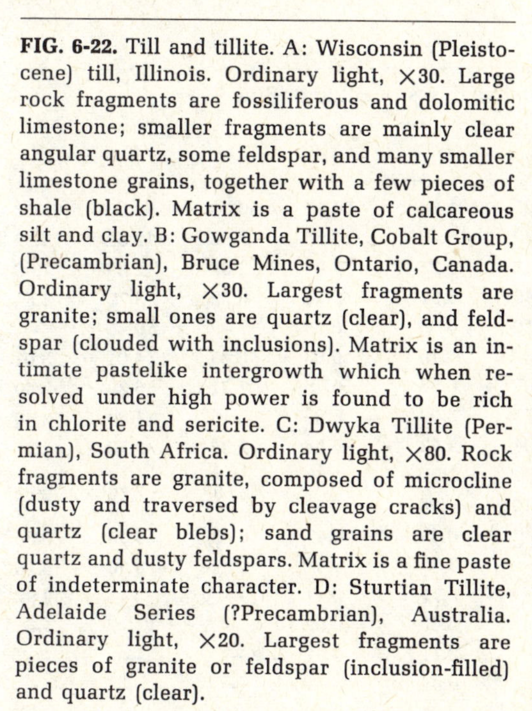

of continental tills. Marine tills, especially dropstone tills, are not so apt to rest on glacial pavement.

An almost universal and a most significant associate of till or tillite is varved clay or its lithified equivalent (varvite or pelodite). These materials show exceptionally even laminations, which represent seasonal deposition of clay (perhaps now an argillite or slate) in quiet fresh-water lakes (Fig. 6-13). Abundant rafted angular blocks, both small and large, which are embedded in these strata constitute the strongest evidence of glaciation, excepting only perhaps the striated stones of the till itself. Conversely the absence of such rafted cobbles casts doubt on many ancient deposits attributed to glacial action. Even in thin section, coarse, angular grains of quartz will be observed in the finest siltstone or slate laminations, which is indicative of rafting on a microscopic scale (Fig. 6-13). As noted by Lindsey (1969, p. 1694), varve structure is widely known in a glacial lacustrine beds but is not generally reported in marine sediments. It seems likely that, although dropstones in a finely laminated matrix may be marine, those with a true varve structure are of fresh-water origin.

Fluvioglacial beds associated with tills and tillites are similar in all important respects to nonglacial alluvial gravels. Such deposits are commonly associated with tillites.

Because Pleistocene tills are commonly associated with loess (an eolian silt), it may be supposed that the lithified equivalent would be found in ancient glacial sequences. But this seems not to be the case, perhaps

TABLE 6-3. Chemical analyses of till and tillite

Constituent	A	B	C	D	E	F
SiO_2	61.98	61.68	64.59	80.34	59.32	61.57
Al_2O_3	17.20	16.48	14.66	7.66	12.34	14.57
Fe_2O_3	1.42	1.73	2.89	1.39	2.29	1.80
FeO	4.49	5.61	3.55	0.72	3.72	5.54
MnO	0.10	0.09	0.08	—	0.45	—
MgO	3.27	3.18	3.83	0.83	4.01	3.01
CaO	1.00	0.53	0.46	1.51	5.20	2.40
Na_2O	5.27	3.99	1.60	2.13	1.80	2.46
K_2O	2.04	2.62	5.86	1.94	2.52	2.27
H_2O+	2.70	2.95	1.66	2.63[b]	7.76[b]	2.88
H_2O-	0.10	0.08	0.08			
TiO_2	0.60	0.49	0.47	0.42	0.55	0.50
P_2O_5	—	0.20	0.22	trace	0.30	0.18
CO_2	—	0.12	0.31	—	—	2.87
C	—	0.00	0.07	—	—	—
S	—	0.11	0.01	—	—	—
BaO	—	0.04	—	—	—	—
		99.97				
Less O		0.08				
Total	100.15	99.89[a]	100.35	99.59	100.27	99.69

[a] Includes SO_3, 0.01; F, 0.05.
[b] Loss on ignition.

A. Gowganda Tillite, Huronian, Ontario. M. F. Conner, analyst (Collins, 1925, p. 66).
B. Varved argillite (pelodite), Gowganda Formation, Ontario. Lucille N. Tarrant, analyst (Pettijohn and Bastron, 1959, Table 1).
C. Fern Creek Tillite (Precambrian), Dickinson County, Michigan. B. Brunn, analyst.
D. Ipanema Tillite (Permo-Carboniferous), São Paulo, Brazil. M. Fontoura, analyst (Leinz, 1937).
E. Barra Bonita Tillite (Permo-Carboniferous), Paranã, Brazil. M. Fontoura, analyst (Leinz, 1937).
F. Dwyka Tillite (Permian), South Africa (Univ. Toronto Studies, geol. ser., no. 12, 1921, pp. 63–67).

because most of the "fossil" glacial beds are marine rather than continental.

Distribution in time and space—examples Tillites belong to no particular time or place. Coleman (1926) has summarized known and problematic occurrences of glacial beds. Such deposits are, however, less common than once supposed; more and more presumed tillites are now thought to be nonglacial tilloids rather than true till. Crowell and Frakes (1970) have summarized what is known about Phanerozoic glaciations.

Four well-known periods of widespread glaciation include the middle Precambrian (circa 2150–2500 my), the Eocambrian, the Permocarboniferous of the Southern Hemisphere, and the Pleistocene.

The Middle Precambrian (Aphebian of the Geological Survey of Canada) glacial beds have long been known. The best described of these are the tillites of the Gowganda Formation (Upper Huronian) of eastern Ontario and adjoining Quebec. Early described by Wilson (1913), they have been more recently studied by Ovenshine (1965) and Lindsey (1969). The Gowganda extends over an area of several thousand square miles, has yielded striated stones, rests on a glacial pavement in several places (Schenk, 1965), and is associated with varved argillites containing rafted blocks (Figs. 6-13, 6-14, 6-22, and 6-23). Rocks of similar character and presumed to be of the same age occur in the Hurwitz Group in the Northwest Territories of Canada (Bell, 1970, p. 163). The Lower Huronian Ramsey Lake Conglomerate and the Bruce Conglomerate are paraconglomerates also believed to be glacial (Frarey and Roscoe, 1970, p. 150). These deposits are similar to the Fern Creek Tillite of northern Michigan (Pettijohn, 1952), considered by some to be correlative (Young, 1966, p. 209), and to the presumed tillites of the Reany Creek Formation (Puffett, 1969) of the same region.

Perhaps better known and more widespread are the glacial deposits of the Eocambrian. These seem to be worldwide. They are best known perhaps in northern Norway

FIG. 6-23. Gowganda Tillite, Cobalt group (Precambrian), near Green Bay, north shore, Lake Huron, Ontario, Canada. For thin-section appearance see Fig. 6-22, B.

(Spjeldnaes, 1965; Reading and Walker, 1966). Tillites of the same age have been reported from Spitzbergen and eastern Greenland; from Australia, the Sturtian glacial beds of the Adelaide Series (Howchin, 1908; David, 1907; Browne, 1940; Mawson, 1949); from the Wasatch Range of Utah (Blackwelder, 1932), and from other places. The glacial origin of some of these deposits, the Wasatch "tillites" in particular, has been challenged (Condie, 1967). Those of Norway and Australia seem to be firmly established as glacial, presumably marine rather than continental glacial beds. Critical evidence to Eocambrian glaciation has been carefully reviewed by Harland (1965).

The most famous of all ancient glacial deposits are undoubtedly those of Permo-Carboniferous age in the Southern Hemisphere. Included here are the Dwyka tillites of South Africa (Du Toit, 1921; Hälbich, 1962; Crowell and Frakes, 1972), the glacial and the periglacial beds of the Congo Basin (Hübner, 1965), the Carboniferous tillites of Brazil and other parts of South America (Leinz, 1937; Frakes and Crowell, 1969), the glacial beds of Australia and Tasmania (Wanless, 1960; Crowell and Frakes, 1971), the Talchir glacial deposits of India (Smith, 1963), the Lafonian Tillite of the Falkland Islands (Frakes and Crowell, 1967), and the glacial beds of Antarctica (Frakes and Crowell, 1968). These deposits have been critically reexamined and seem to be truly glacial beds of correlative age. They constitute one of the best documented and extensive ancient glaciations.

The Pleistocene glacial epoch left a clear

and widespread record in North America and Europe. The deposits of this epoch are readily studied in continental exposures, which show them to be a product of multiple glaciations and, for the most part, thin (a few tens of meters) and unconsolidated. The marine record of this glacial epoch is far less clear. It has been described in part by Miller (1953). Refer to the monographic summary of the Pleistocene glacial epoch by Flint (1971) for further information.

In addition to the well-established glacial records listed above, various other deposits, of other ages, are thought to be glacial. Restudy of some of these casts serious doubt on their glacial origin. An example is the so-called Squantum "Tillite" of the Boston Bay area (Sayles, 1914) which, although it resembles a tillite in many respects, lacks many critical features and is probably nonglacial in origin (Dott, 1961). Likewise, deposits generally thought to be glacial occur in the West Congo geosyncline, but recent study has led to doubt about their glacial origin (Schermerhorn and Stanton, 1963). There seem to be, however, other truly glacial deposits, other than those belonging to the four major epochs. Certain Ordovician deposits in northern Africa are included here (Arbey, 1968).

Origin and geologic significance Although much has been written on till, the mechanism of its formation has not been wholly understood. The relation between the composition of till and the subjacent bedrock has been studied here (Holmes, 1952) and abroad (Lundquist, 1935). Size distribution of tills has been partially explained by Krumbein (1933) and Krumbein and Tisdel (1940). Mechanisms of glacial deposition, including the processes involved, have been reviewed by Harland, Herod, and Krinsley (1966). Those relevant to marine glacial deposits were analyzed by Carey and Ahmad (1961). The subject is also reviewed at length in a symposium volume (Goldthwait, 1971) and by Flint (1971, p. 171). It is important to note that, although our knowledge of glacial action and glacial deposits (based on Pleistocene and recent glaciers) is largely that of continental glaciation, the record of the past seems to be largely one of marine glaciation. Failure to recognize this fact has led to confusion and differences in interpretation. One needs to distinguish between dropstone tillites (paratills) and ice-deposited till (orthotill).

The geologic significance of tillite, however, is self-evident. Conceivably related only to glaciers of Alpine nature, the tillites of the past, especially those of considerable areal extent, record periods of intense and widespread refrigeration. The cause of such drastic climatic changes and the reason for the occurrence of many ancient tillites in very low latitudes are still unknown. Apparently from earliest times the earth has been subjected to glaciations as widespread as it has in the recent past and at the present. The fact that such glaciations were widespread leads to hope that such events could be used for correlation of Precambrian strata which lack fossils or other means of correlation by usual methods.

TILLOID ("GERÖLLTON"): NONGLACIAL PEBBLY MUDSTONES

Every chaotic deposit with large blocks embedded in a clayey matrix is not a tillite. Great caution should be exercised when discriminating between true tillite and other materials which resemble it—the *tilloids*. Many mudflow, landslide, and solifluction deposits and some volcanic tuffs and breccias closely resemble till. The discrimination between tillites and tilloids is often difficult and has led to differences of opinion on specific deposits and to considerable controversy. A good summary of the problem and a summary of useful criteria for the recognition of tills and tilloids is given by Schermerhorn and Stanton (1963, especially Table 2), by Crowell (1957), and by Harland, Herod, and Krinsley (1966).

Definitions As stated before (p. 170), the term *diamictite* or *paraconglomerate* has been applied to those rocks with scattered phenoclasts embedded in a very fine-grained matrix. *Till* (and tillite) is a diamictite of glacial origin and *tilloid* has come to be applied to nonglacial diamictites. Ackermann (1951) assigned the term *Geröllton* to these deposits.

Textures, structures and composition of tilloids Because tilloids are products of several unlike processes, they show a diversity of textures and structures and differ also in the character of the associated deposits and in their overall geometry and dimensions.

Larger, more extensive, and more impressive (and perhaps the most common) tilloids are attributable to subaqueous gravity flow. Subaerial tilloids related to landslide, solifluction, and so forth are local or restricted in character and, because they are subaerial, are prone to erode and are hence unlikely to be preserved. Except to students of Quaternary geology, they are unimportant. On the other hand, deposits of subaqueous mass flow are common and extensive, and many have been misinterpreted and considered glacial.

Nonglacial and conglomeratic mudstones or tilloids vary from a chaotic assemblage of coarse materials set in a mudstone matrix to a mudstone with sparsely distributed cobbles (Fig. 6-24). The matrix may be either subordinate to the larger clasts, or it may be dominant. Many pebbly mudstones described by Crowell (1957) were 80 percent matrix and 20 percent scattered clasts. The largest clasts may be boulder size.

Close inspection shows many of the boulder beds, some up to 50 m or more in thickness, to be a chaotic structureless deposit. In other cases, however, vestigial bedding, albeit indistinct, is present. Especially significant are contorted masses of sandstone or shale, rocks similar to those below the pebbly mudstone. These display hook-shaped and twisted bedding, designated "slump overfolds" by Crowell (1957). More resistant pebbles are commonly impressed into the included shale or sandstone blocks showing the latter to have been soft at the time of their incorporation in mudstone. In addition to these contorted "intraclasts," pebbly mudstones contain many foreign pebbles and cobbles, in many cases a polymict assortment and in many cases also very well rounded. Presumably the subaqueous flow engulfed well-worn polymict gravels of an ordinary sort. Consequently these xenoclasts can be of any rock type, including granite.

FIG. 6-24. Pebbly mudstone (Upper Cretaceous), Pigeon Point, California. Note roundness of clasts.

Stratigraphic associations Tilloids caused by subaqueous mud flows are associated with marine sediments, commonly deep-water turbidite sandstones and shales. They occur as thicker or thinner beds and lenses intercalated with such deposits. Lacking, however, are the delicately stratified argillites with rafted blocks diagnostic of the glacial beds.

Distribution in time and space Excellent descriptions of ancient, nonglacial pebbly mudstones have been published by Crowell (1957), Schermerhorn and Stanton (1963), McBride (1966), and Dott (1961).

Massive marine mudstones containing scattered pebbles occur in several places in California, especially in Upper Jurassic and Lower Cretaceous strata. These have been described and illustrated in Crowell's classic paper on pebbly mudstones (1957). The so-called Squantum "Tillite" of the Boston Bay area is another excellent example of this class of rocks. Long-considered a glacial deposit (Sayles, 1914), it has been reinterpreted by Dott (1961) as a subaqueous mud flow. The Squantum beds contain several conglomeratic beds, some 10 m or more thick, which are 50 or more percent mudstone and contain large angular or bent intraclasts of laminated argillite. Associated strata included graded gravels and graywackes.

The boulder beds of the Haymond Formation (Pennsylvanian) of the Marathon region of Texas are justly famous. Blocks over 100 feet (30.5 m) in diameter have been reported. These bouldery mudstones have long been an enigma (King, 1958). They were interpreted by Baker (1932) as glacial. A recent restudy (McBride, 1966) considers them to be the product of submarine slumping, mudflows, and turbidity currents. Larger blocks of enormous size are considered too large to have been transported by turbidity currents; it is inferred that they were sloughed from submarine thrust fault scarps (King, 1958, p. 1734).

The Eocambrian of the West Congo geosyncline contains pebbly mudstones, over a large area, which have been thought by some to be glacial. They are now explained by Schermerhorn and Stanton (1963) as the products of submarine mudflows and turbidity currents despite their widespread nature.

Various other tilloids have been reported, but few have been adequately studied. Possible subaqueous mud flow deposits include the Cow Head Breccia of Newfoundland (Kindle and Whittington, 1958), the so-called Wasatch "Tillite" (Condie, 1967), the Levis Conglomerate of Quebec (Osbourne, 1956, p. 183), the Tertiary Gunnison "Tillite" of Colorado (Van Houten, 1957), and the pebbly and bouldery Jurassic or Cretaceous mudstones from Cape Blanco, Oregon (Dott, 1961, Fig. 2, Pl. 1). Devonian carbonate "debris flow" conglomerate have been described by Cook et al. (1972).

Origin of tilloids Most tilloids, certainly the more extensive ones, are the products of subaqueous gravity movement, although some rather small and localized deposits may be subaerial. As pointed out by Dott (1963), these movements are of several kinds: subaqueous sliding or slumping, subaqueous plastic mass flow, and viscous fluid flow. The first two lead to formation of tilloid, the last leads to generation of a turbidity current. In the first two, some stratification may be preserved, even though it is commonly much distorted. If velocity and turbulence increase sufficiently, all cohesion is lost and materials are thrown into suspension and flow as a turbidity current. Most subaqueous unsorted pebbly mudstones represent arrested flows which just surpassed their liquid limits—the critical percentage of pore water beyond which the material can no longer behave as a plastic solid. Tilloids are relatively rare in the geologic record, because, once liquid limit is reached, viscosity diminishes rapidly. Thus if plastic flow is exceeded, formation of a turbidity current is inevitable with its accompanying sorting and grading of clasts.

Inducement of mass failure and resulting subaqueous gravity movement can result from overloading, oversteepening of depositional slopes, earthquakes, and hydraulic pressures caused by tides, seepages, or storms.

Subaqueous slump and subaqueous mudstreams have been reported from Lake Zug in Switzerland and from certain Norwegian fjords (Ackermann, 1951), although not much is known about the character and structure of the resulting deposits. Those of the fjords traveled a considerable distance. The extraordinary extent and vigor of the presumed Grand Banks flow (Heezen and Ewing, 1952) make it clear that these agents are fully capable of producing the conglomeratic mudstones (tilloids) and make unnecessary an appeal to any other agent. Their occurrence, in some cases interbedded with fine-grained rocks containing deep-water Foraminifera, as in the Santa Paula Creek section (Crowell, 1957, p 998), lend further support to this interpretation.

Some conglomerates, less appropriately called conglomeratic mudstones, have also been ascribed to subaqueous slumping and turbidity flows. These deposits are not too different from the more poorly sorted immature terrestrial gravels, such as the fanglomerates. But the associations of these beds and other considerations require an extraordinary mode of transport and deposition. The conglomerates are intercalated with marine shales, they are commonly graded, and the associated sands lack cross-bedding and other evidences of shallow-water deposition, all of which would be expected of near-shore or terrestrial gravels. Such conglomerates are common in Upper Cretaceous beds of the San Joaquin Valley in California (Briggs, 1953, p. 440); in the Pliocene of the Ventura Basin, where they are believed to have accumulated in no less than 4000 to 5000 feet (1220 to 1525 m) of water (Natland and Kuenen, 1951); and in the Cerro Torro Formation (Cretaceous) of the Patagonian Andes (Scott, 1966). The Cretaceous conglomerates of Wheeler Gorge in California and the Archean conglomerates of Minnitaki Lake, Ontario, are probably of this type (see p. 169).

DIAGENESIS OF CONGLOMERATES

Conglomerates, like all sedimentary rocks, undergo modifications after deposition. Most dramatic of these is lithification or induration usually brought about by the introduction of a cement that binds the constituent clasts together. The cementation of ordinary gravels differs in no important way from the cementation of sands. (Refer to the chapter on sandstones for a full discussion of cements.) It is of interest to note that in some gravels weakly cemented by carbonate, the cementation process begins on the undersides of the pebbles or cobbles. If these are removed from the rock fabric, they display a thin carbonate plaster on the underside in which some interstitial sand grains are embedded.

Openwork gravels are cemented by a drusy crystalline cement, commonly dogtooth spar. In some the carbonate forms a layered crust on the pebbles, a relationship particularly characteristic of caliche-cemented gravels. Not uncommonly openwork gravels have a mud coating on the pebbles (Cary, 1951). This coating can effectively reduce permeability.

Conglomeratic mudstones become lithified, not through the introduction of a cement, but rather by a diagenetic or low-

FIG. 6-25. Faulted pebbles, age and locality unknown. (Photograph by C. Weber.)

grade metamorphic recrystallization of the mud and conversion of this material to a dense argillite. (For discussion of these alterations refer to the section on the diagenesis of shales and mudstones.)

Other diagenetic changes in conglomerates involve intrastratal solution. The solution is concentrated at the contacts between pebbles. In many cases, in the limestone conglomerates in particular, there is mutual stylolitic interpenetration between pebbles (Bastin, 1940). Such stylolitic boundaries are also known in quartzitic conglomerates, especially those with jasper and chert pebbles. At times solution at pebble contacts leads to pits where one, presumably less soluble, pebble appears to be impressed into another (Kuenen, 1942).

Mechanical pressures lead to deformation of pebbles, even in conglomerates not notably deformed. Some pebbles, even those of quartz or quartzite, display fractures or microfaults. These appear as minute step-like displacements of the pebble surface. They are usually healed fractures, so the pebble remains a coherent body (Fig. 6-25).

INTRAFORMATIONAL CONGLOMERATES

An intraformational conglomerate or breccia is a rudaceous deposit formed by penecontemporaneous fragmentation and redeposition of the stratum in question (Walcott, 1894; Field, 1916). Such fragmentation and redeposition are but a minor interlude in the deposition of the formation and in some cases may be wholly subaqueous. Debris is always of very local origin, has undergone very little or no transportation, and is only slightly worn. Although many breccias, such as reef talus breccia, could in a sense be considered intraformational, they are not so designated. Intraformational breccias are generally confined to a single sedimentation unit, usually a thin bed from a few centimeters up to a meter thick. They may be widespread and traceable for several kilometers or more, but they are probably generally very restricted in extent.

The term *intraclast* was introduced by Folk (1959, p. 4) to describe "fragments of penecontemporaneous, usually weakly consolidated carbonate sediment that have been eroded from adjoining parts of the sea bottom and redeposited to form a new sediment." Although generally applied to limy marine sediments, shale fragments in sandstone are very common and of intraformational origin and are therefore properly called intraclasts. One can say, therefore, that pebbles of any intraformational conglomerate, regardless of composition or place of origin, are intraclasts.

Intraclasts are the result of several different processes; most commonly they are formed by shoaling and temporary withdrawal of the waters, followed by desiccation and mud cracking. Subsequent flooding of mud-cracked layers disturb the fragments, which may be slightly shifted and deposited together in a thin, persistent conglomerate of flat pebbles. In a few cases the fragmentation has been attributed to subaqueous gliding or slump (Potter, 1957), and the brecciated zone can be traced into sharply folded and much contorted strata. The desiccation conglomerates, on the other hand, show no such relation to contemporaneous deformation and are associated, instead, with mud-cracked zones and other strand-line features.

Two types of intraformational conglomerates are most common. One is characteristic of some limestones and dolomites, especially those which are sandy and oolitic. The fragments or intraclasts of such conglomerates or breccias are generally small pieces of limestone (or dolomite) embedded in a limestone or sandy limestone (or dolomite) matrix (Fig. 6-26). Such breccias, the intraclastic rudites of Folk (1959), contain flat, discoid, well-rounded pebbles formed by the erosion of a semiconsolidated carbonate sediment. The fragments may consist of any kind of limestone or dolomite, but in Pettijohn's experience they are most commonly fine-grained or micritic. In coarser conglomerates the fragments are embedded in a matrix of carbonate sand, commonly oolitic and also commonly containing rounded quartz grains. Such conglomerates are probably the product of desiccation and induration of lime muds but may also be the result of subaqueous fragmentation and transport by turbidity flows. The former, perhaps the most common, are associated with mud-cracked beds and with other strand-line features, especially stromatolites, and are apt to have a sandy matrix. The latter are widespread and unrelated to the strand line; they have a mud matrix and are associated with graded beds or other turbidity current features. The flat pebbles of some limestone conglomerates stand on edge and are packed together to form a so-called *edgewise conglomerate*. Such structure is apparently the result of the tabular form of the fragments and the more than normal agitation by waves and currents.

A second very common type of intraformational conglomerate is a shale pebble conglomerate or breccia, in which the intraclasts are thin or tabular pieces of shale embedded in a sandy matrix. Such flat-pebble conglomerates are common in formations composed of alternating beds of shale and sandstone. Shale-pebble conglomerates generally are found at the base of the sandy beds. The fragments may be a few scattered chips or flakes of shale or even an appreciable number of pieces, several centimeters across, confined to the lower 5 or 10 cm of the sandstone bed. If these shale pebbles occur in the redbed sequences, as they commonly do, they are probably desiccation fragments. If they are found in interbedded graywacke–shale sequences, the shale pieces may be a product of subaqueous fragmentation (the "rip-up clasts" produced by turbidity flows).

Intraformational conglomerates may be sometimes confused with true reibungsbreccias of tectonic origin or with some rare *inter*formational conglomerates consisting dominantly or solely of limestone fragments.

In summary, it may be said that intraformational conglomerates, although common, are not indicative of any great break in sedimentation. They are characterized by thinness, flat-pebble form, edgewise arrangement in some cases, restricted composition of the intraclasts (shale or limestone only), association of some with mud-cracked beds, or, in less common types, relation to subaqueous folds and graded beds. Those generated by subaqueous slump and turbidity flows are significant, because the deposit so formed, even though very thin, may be widespread and be a precise time marker—the event producing such a breccia is that of a few hours duration.

So common are intraformational conglomerates and breccias that only a few can be cited as examples. They abound in the lowest Paleozoic limestones and dolomites of the Appalachian region. Well known are the desiccation features and associated flat-pebble limestone conglomerates of the Bellefonte region in Pennsylvania (Walcott, 1894); conglomerates of a similar character and origin occur in the Conococheague (Upper Cambrian) of Maryland. At the top of the Gros Ventre and below the base of the Gallitan Formation (Cambrian) of central Wyoming are excellent examples of flat-pebble conglomerates (Fig. 6-26). The flat-pebble conglomerates of the Muav Limestone (Cambrian) of the Grand Canyon area form thin beds said to be traceable for more than 60 miles (97 km) normal to the presumed shoreline (McKee, 1945, pp. 26, 65). They have been attributed to subaqueous rather than subaerial fragmentation.

Sandstones with shale clasts are exceedingly common, especially in the sandstones of the fining-upward alluvial sequences. Red shale pebbles are found in the sandstones of the Juniata (Ordovician) of central Pennsylvania and in the sandstones of the Catskill (Devonian) and the Mauch Chunk (Mississippian) of the same region. All are alluvial facies. They occur in similar sandstones in other places too numerous to mention.

The Precambrian "slate breccia," which is a thin but widespread marker bed in the Iron River–Crystal Falls district of Michigan, consists of small, sparsely distributed slate fragments in a dense pyritic mudstone matrix. It has a thickness in most places of less than 10 feet (3 m). It has been attributed to a catastrophic subaqueous slump or turbidity flow (James et al., 1968, pp. 41, 71).

FIG. 6-26. Intraformational conglomerate, Gallitan Formation (Cambrian), Teton Mountains, Wyoming. Length of specimen about 5 inches (13 cm). Consists of flat pebbles of laminated calcilutite in a matrix of lime mud.

BRECCIAS

The term *breccia* is generic and applies to a rock of any kind which is composed of an aggregate of angular fragments. Their mode of aggregation is ignored except by the introduction of a supplementary word or phrase. The usefulness of the term inheres in that comprehensiveness which is characteristic of the terminology of an immature science. It is a term so overburdened with meaning that it has become totally uninformative (see Fig. 6-27). In general the term has been used in an imprecise manner —as a "wastebasket" term applied to a very diverse group of rocks. The classification of breccias has been reviewed by Norton (1917) and Reynolds (1928).

In this book we differentiate between nonsedimentary breccias, those rudaceous rocks composed of angular pieces which are formed by post-depositional processes, and synsedimentary breccias, whose fragments are related to and contemporaneous with the sedimentational process. We would apply the term *sharpstone conglomerate* to the latter and restrict the term *breccia* to the former.

Nonsedimentary breccias (breccias in the

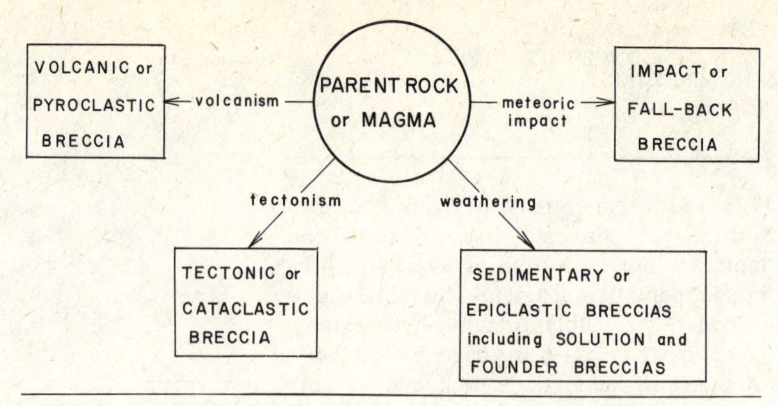

FIG. 6-27. Origin of breccias.

restricted sense) are a diverse group in both their manner of origin and their physical attributes. They arise in three principal ways and can be classed, accordingly, as cataclastic breccias, pyroclastic breccias, and impact breccias. Cataclastic (autoclastic) breccias are those whose fragmentation is caused by earth movements (one rock mass moving over another). The stresses involved may be gravitational or tectonic. In many cases the movement is minimal, with little or no lateral component. Pyroclastic breccias are those formed by explosive volcanism. Impact breccias (fallback breccias) are the result of meteoric impact and shattering.

CATACLASTIC BRECCIAS

In cataclastic rocks (*autoclastic* of Grabau, 1904), fragmentation was accomplished by movement of large masses of rock past one another. Boundary materials are literally crushed. The movement that occurs along a fault surface gives rise to fault breccias and to gouge. If the fault is an overthrust sheet, the deposit created is a thin blanket along the sole of such an overthrust. Like till at the base of an overriding ice sheet, it is a moraine—a tectonic moraine rather than a glacial moraine.

Normally considered here are *fault breccias, fold breccias* (reibungsbreccias), and *crush conglomerates.* Fault breccias are distinguished by cross-cutting relations and by the presence of gouge. Blackened, slickensided blocks, and shaly matter marked by similar evidences of shearing usually serve to identify these materials. Tectonic moraines or boulders and similar materials incorporated in the base or sole of an overthrust mass are more readily mistaken for normal rudaceous beds, inasmuch as they have a local concordant relation to the associated strata.

More common are fold breccias, or reibungsbreccias, which are the result of sharp folding of thin-bedded brittle layers between which are incompetent plastic beds. Interbedded chert and shale are likely to form a reibungsbreccia on sharp folding. Such breccias are local, are confined to sharply folded strata, and are likely to pass into unbroken beds.

Crush conglomerates are produced by deformation of brittle, closely jointed rocks. Rotation of joint blocks, and granulation and crushing, may produce a rock that closely simulates a normal conglomerate. The lozenge shape of the fragments, the similarity in composition of the fragments and matrix, and the restricted composition of both, generally to one rock type, are distinguishing features of these rocks. Crush conglomerates are most likely to be confused with the "recomposed" basal conglomerates which may have been deformed following their deposition.

LANDSLIDE AND SLUMP BRECCIAS

Some breccias are formed by earth movements brought about by simple gravitational stress. These landslide and slump breccias are in part subaerial and in part subaqueous. They are generally very local and are quantitatively unimportant. Subaqueous slumps generate slump folds and slump breccias.

(These have been discussed in Chapter 4 on structures.) Slumps and landslides, either subaerial or subaqueous, pass by addition of water into mudflows; in the subaqueous environment these pass into mudstreams and turbidity currents. The resulting deposit is a tilloid—a species of conglomeratic mudstone which has been discussed above.

Landslides are related to slope, structure, and lithology, particularly certain species of clay minerals that upon wetting provide lubricant for movement of superjacent earth masses. The geology of landslides and related phenomena has been summarized by Sharpe (1938).

Although slide breccias are minor deposits, they have been identified in the older geologic record—in the Cambrian of the Grand Canyon, for example (Sharp, 1940).

COLLAPSE (FOUNDER) AND SOLUTION BRECCIAS

Some breccias are related to simple downward earth movements or foundering as a result of removal of the underlying materials by solution. These are particularly related to the solution of salt beds and foundering of the overlying strata, especially limestones. Such breccias have a widespread and fixed position in the stratigraphy, and are commonly replaced by salt beds at depth.

Such breccia beds vary from a few centimeters to 10 or more meters in thickness. They consist of highly angular fragments, generally tabular pieces, and blocks of limestone of variable size. The basal contact of a breccia bed is sharp; the upper contact is less so. Breccias generally have a fine-grained matrix making distinction between fragments and matrix obscure; in other cases the breccia is cemented by a coarse drusy calcite or by a comb-structured tufa. Some solution breccias have floored cavities filled with fine sediment. The criteria for distinguishing between solution breccias and limestone breccias of other origins have been summarized by Blount and Moore (1969).

Solution breccias have been identified and described in the Mississippian of southwestern Montana (Middleton, 1961), in the Windsor Group (Mississippian) of Nova Scotia (Clifton, 1967), in the Upper Glen Rose (Cretaceous) of central Texas, in the St. Louis Limestone (Mississippian) near Alton, Illinois (Morse, 1916), and in the Silurian limestones of the Mackinac area of Michigan (Landes and Ehlers, 1945).

Solution of limestone beneath a cover of sandstone and shale leads to the formation of a basinal structure, a sinkhole structure into which the overlying strata have settled. Downward movement into the growing sink leads to deformation, brecciation, and slickensiding. These filled sinks, some tens of meters across and of comparable depth, have been described from northern Illinois (Bretz, 1940) and Missouri (Bretz, 1950).

PYROCLASTIC BRECCIAS

Coarse pyroclastic deposits include *volcanic breccia*, a deposit consisting of blocks of previously deposited materials, as distinguished from *agglomerate*, formed primarily of bombs or lava solidified in flight (Wentworth and Williams, 1932). (Volcanic breccias belong to the volcaniclastic sediments, which are described in Chapter 9.)

IMPACT OR FALLBACK BRECCIAS

In recent years, perhaps because of lunar explorations, geologists have focused attention on meteoric impact. Although a few terrestrial impact craters had long been known, only recently have the associated "fallback" and "base surge" materials been recognized and studied. In most cases these materials fill or are closely associated with craters. Particular attention has been given to the effects of shock metamorphism (French and Short, 1968), especially the formation of coesite and the effects of shock on quartz (Chao, 1967).

Although meteoric impact craters and associated materials are very prone to destruction by surface processes and hence are unlikely to be preserved, some ancient deposits have been interpreted as being of impact origin. Most notable is the Onaping "Tuff" (Precambrian), which earlier was thought to be a pyroclastic deposit of the ignimbrite type (Williams, 1957). It has recently been considered to be fallback material (French, 1967, 1968) related to meteoric impact. Shatter cones in basement quartzites and features attributable to shock metamorphism in the breccia itself lend support to this view.

PSEUDOCONGLOMERATES AND PSEUDOBRECCIAS

A few rocks simulate conglomerates and may be misinterpreted by the novice. Such rocks, if metamorphosed, might confuse even an experienced worker.

Diabases and related rocks weathered *in situ* yield large, rounded *boulders of exfoliation* which, if still in place and surrounded by clayey weathering products, may bear a superficial resemblance to a boulder conglomerate. Close scrutiny will dispel the illusion. Beneath the concentric exfoliation shells of such a boulder is a hard unweathered core.

A sandstone packed with many rounded concretionary bodies might also bear a likeness to a conglomerate. The composition of these bodies, commonly calcareous, and the fact that in many cases the laminations of the host rock pass through them reveal their secondary, nondetrital origin.

Some limestones consist in part on in their entirety of "algal balls"—pebble-sized concentrically structured *oncolites* (see p. 127). These bodies resemble well-rounded pebbles. Some, in fact, seem to have formed around pebbles. Unlike concretions, they are a primary component of the rock. That they are nondetrital is readily detected when they are broken and reveal the growth structure.

As pointed out on page 186, shearing of close-jointed brittle rocks rounds off the joint blocks and produces a gougelike matrix and a "tectonic conglomerate," which may easily be mistaken for a normal sedimentary conglomerate.

Certain diagenetic processes produce a pseudobreccia in some limestones. These have been described by Bathurst (1959), Royer (1938), and Wallace (1913).

References

Ackermann, E., 1951, Geröllton: Geol. Rundschau, v. 39, pp. 237–239.

Arbey, F., 1968, Structures et dépôts glaciaires dans l'Ordovicien terminal des chaines d'Ougarta (Sahara algérien): Acad. Sci., C. R. Sér. D, v. 266, pp. 76–78.

Baird, D. M., 1960, Observations on the nature and origin of the Cow Head Breccias of Newfoundland: Geol. Surv. Canada, paper 60-3, 26 pp.

Baker, C. L., 1932, Erratics and arkoses in the Middle Pennsylvanian Haymond Formation of the Marathon area, Trans-Pecos, Texas: Jour. Geol., v. 40, pp. 577–607.

Barrell, J., 1925, Marine and terrestrial conglomerates: Bull. Geol. Soc. America, v. 36, pp. 279–341.

Bastin, E. S., 1940, Discussion: a note on pressure stylolites: Jour. Geol., v. 48, pp. 214–216.

Bathurst, R. G. C., 1959, Diagenesis in Mississippian calcilutites and pseudobreccias: Jour. Sed. Petrology, v. 29, pp. 365–376.

Bell, R. T., 1970, The Hurwitz Group, a prototype for deposition on metastable cratons, *in* Basins and geosynclines of the Canadian Shield (Baer, A. J., ed.): Geol. Surv. Canada, paper 70-40, pp. 159–168.

Blackwelder, E., 1931, Pleistocene glaciation in the Sierra Nevada and Basin Ranges: Bull. Geol. Soc. Amer., v. 42, pp. 865–922.

———, 1932, An ancient glacial formation in Utah: Jour. Geol., v. 40, pp. 289–304.

Blenk, M., 1960, Ein Beitrag zur morphometrischen Schotteranalyse: Zeitschr. Geomorph., v. 4, pp. 202–242.

Blissenbach, E., 1952, Relation of surface angle distribution to particle size distribution on alluvial fans: Jour. Sed. Petrology, v. 22, pp. 25–27.

———, 1954, Geology of alluvial fans in semiarid regions: Bull. Geol. Soc. Amer., v. 65, pp. 175–190.

———, 1957, Die jungtertiäre Grobschotterschüttung in Osten des bayerischen Molassetroges: Beih. Geol. Jahrb., v. 26, pp. 9–48.

Blount, D. N., and Moore, C. H., Jr., 1969, Depositional and non-depositional carbonate breccias, Chiantla Quadrangle, Guatemala: Bull. Geol. Soc. Amer., v. 80, pp. 429–442.

Bluck, B. J., 1964, Sedimentation of an alluvial fan in southern Nevada: Jour. Sed. Petrology, v. 34, pp. 395–400.

———, 1965, The sedimentary history of some Triassic conglomerates in the Vale of Glamorgan, South Wales: Sedimentology, v. 4, pp. 225–245.

———, 1967, Sedimentation of beach gravels: examples from South Wales: Jour. Sed. Petrology, v. 37, pp. 128–156.

Boggs, S., Jr., 1969, Relationship of size and composition in pebble counts: Jour. Sed. Petrology, v. 39, pp. 1243–1247.

Boutcher, S. M. A., Edhorn, A. S., and Moorehouse, W. W., 1966, Archean conglomerates and lithic sandstones of Lake Temiskaming, Ontario: Proc. Geol. Assoc. Canada, v. 17, pp. 21–42.

Bradley, W. C., Fahnestock, R. K., and Rowehamp, E. T., 1972, Coarse sediment transport by flood flows on Knik River, Alaska: Bull. Geol. Soc. Amer., v. 83, pp. 1261–1284.

Bretz, J H., 1940, Solution cavities in the Joliet Limestone of northeastern Illinois: Jour. Geol., v. 48, pp. 337–384.

———, 1950, Origin of filled sink-structures and circle deposits of Missouri: Bull. Geol. Soc. Amer., v. 61, pp. 789–834.

Briggs, L. I., Jr., 1953, Upper Cretaceous sandstones of Diablo Range, California: Univ. California Publ. Geol. Sci., v. 29, pp. 417–452.

Browne, W. R., 1940, Late Proterozoic (?) glaciation in Australia: 17th Int. Geol. Congr., v. 6, pp. 57–63.

Bull, W. B., 1964, Alluvial fans and near-surface subsidence in western Fresno County, California: U.S. Geol. Surv. Prof. Paper 437-A, 71 pp.

Cailleux, A., 1945, Distinction des galets marins et fluviatiles: Bull. Soc. Géol. France, ser. 5, v. 15, pp. 375–404.

Caldenius, C., 1938, Carboniferous varves, measured at Paterson, N.S.W.: Geol. Fören. Stockholm Förh., v. 6, pp. 349–364.

Carey, S. W., and Ahmad, N., 1961, Glacial marine sedimentation, *in* The geology of the Arctic (Raasch, G. O., ed.): Toronto, Toronto Univ. Press, v. 2, pp. 865–894.

Carozzi, A., 1956, An intraformational conglomerate by mixed sedimentation in the Upper Cretaceous of the Roc-de-Chère, autochthonous chains of High Savoy, France: Jour. Sed. Petrology, v. 26, pp. 253–257.

Cary, A. S., 1951, Origin and significance of openwork gravel: Trans. Amer. Soc. Civ. Eng., v. 116, pp. 1296–1308.

Cayeux, L., 1929, Les roches sédimentaires de France: Roches siliceuses: Paris, Imprimerie Nationale, 774 pp.

Chao, E. C. T., 1967, Shock effects in certain rock-forming minerals: Science, v. 156, pp. 192–202.

Chase, G. W., 1954, Permian conglomerate around Wichita Mountains, Oklahoma: Bull. Amer. Assoc. Petrol. Geol., v. 38, pp. 2028–2035.

Clifton, H. E., 1963, The Pembroke Breccia of Nova Scotia: Ph.D. dissertation, Johns Hopkins Univ., 209 pp.

———, 1967, Solution-collapse and cavity filling in Windsor Group, Nova Scotia, Canada: Bull. Geol. Soc. Amer., v. 78, pp. 819–832.

Coleman, A. P., 1908, The Lower Huronian ice age: Jour. Geol., v. 16, pp. 149–158.

———, 1926, Ice ages, recent and ancient: New York, Macmillan, Inc., 296 pp.

Collins, W. H., 1925, North shore of Lake Huron: Geol. Surv. Canada Mem. 143, 160 pp.

Condie, K. C., 1967, Petrology of the Late Precambrian tillite (?) association in northern Utah: Bull. Geol. Soc. Amer., v. 78, pp. 1317–1344.

Conkling, H., Eckis, R., and Gross, P. J. K., 1934, Ground water storage capacity of valley fill: Bull. California Div. Water Resources 45.

Cook, H. E., McDaniel, P. N., Mountjoy, E. W., and Pray, L. C., 1972, Allochthonous carbonate debris flows at Devonian bank ("reef") margins, Alberta, Canada: Bull. Canadian Petrol. Geol., v. 20, pp. 439–497.

Crowell, J. C., 1957, Origin of pebbly mudstones: Bull. Geol. Soc. Amer., v. 68, pp. 993–1010.

Crowell, J. C., and Frakes, L. A., 1970, Phanerozoic glaciation and the cause of ice ages: Amer. Jour. Sci., v. 268, pp. 193–224.

———, 1971, Late Paleozoic glaciation of Australia: Jour. Geol. Soc. Australia, v. 17, pp. 115–155.

———, 1972, Late Paleozoic glaciation. Part V, Karroo Basin, South Africa: Bull. Geol. Soc. Amer., v. 83, pp. 2887–2917.

Dal Cin, R., 1967, Le ghiaie de Piave: Mem. Mus. Tridentino Sci. Nat., v. 16, pp. 3–112.

Daubrée, A., 1879, Études synthetiques de géologie expérimentale: Paris, Dunod, 828 pp.

David, T. W. E., 1907, Glaciation in Lower Cambrian, possibly in Pre-Cambrian time: C. R. Congr. Geol. Int. 10, Mexico, 1906, v. 1, pp. 271–274.

Denny, C. S., 1965, Alluvial fans in the Death Valley region, California and Nevada: U.S. Geol. Surv. Prof. Paper 466, 62 pp.

Donaldson, J. A., and Jackson, G. D., 1965, Archaean sedimentary rocks of North Spirit Lake area, northwestern Ontario: Canad. Jour. Earth Sci., v. 2, pp. 622–647.

Dott, R. H., Jr., 1961, Squantum "tillite," Massachusetts—evidence of glaciation or subaqueous movements?: Bull. Geol. Soc. Amer., v. 72, pp. 1289–1306.

———, 1963, Dynamics of subaqueous gravity depositional processes: Bull. Amer. Assoc. Petrol. Geol., v. 47, pp. 104–128.

Dreimanis, A., 1959, Rapid microscopic fabric

studies in drill-cores and hand specimens of till and tillite: Jour. Sed. Petrology, v. 29, pp. 459–463.

Du Toit, A. I., 1921, The Carboniferous glaciation of South Africa: Trans. Geol. Soc. South Africa, v. 24, pp. 188–277.

———, 1954, The geology of South Africa, 3rd ed.: Edinburgh, Oliver and Boyd, 625 pp.

Emery, K. O., 1955, Grain size of marine beach gravels: Jour. Geol., v. 63, pp. 39–49.

Erdmann, E., 1879, Bidrag till Känn edomen on rull stenars bildande: Geol. Fören. Stockholm Förh., v. 4, p. 407.

Fernald, F. A., 1929, Roundstone, a new geologic term: Science, v. 70, p. 240.

Field, R. M., 1916, A preliminary paper on the origin and classification of intraformational conglomerates and breccias: Ottawa Naturalist, v. 30, pp. 29–36, 47–52, 58–66.

Fisher, R. V., 1960, Classification of volcanic breccias: Bull. Geol. Soc. Amer., v. 71, pp. 973–982.

Fisher, R. V., and Mattinson, J. M., 1968, Wheeler Gorge turbidite-conglomerate series, California, inverse grading: Jour. Sed. Petrology, v. 38, pp. 1013–1023.

Flint, R. F., 1971, Glacial and Quaternary geology: New York, Wiley, 892 pp.

Flint, R. F., Sanders, J. E., and Rodgers, J., 1960a, Symmictite: a name for nonsorted terrigenous sedimentary rocks that contain a wide range of particle sizes: Bull. Geol. Soc. Amer., v. 71, pp. 507–510.

———, 1960b, Diamictite, a substitute term for symmictite: Bull. Geol. Soc. Amer., v. 71, p. 1809.

Folk, R. L., 1954, The distinction between grain size and mineral composition in sedimentary rock nomenclature: Jour. Geol., v. 62, pp. 344–359.

———, 1959, Practical petrographic classification of limestones: Bull. Amer. Assoc. Petrol. Geol., v. 43, pp. 1–38.

Frakes, L. A., and Crowell, J. C., 1967, Facies and paleogeography of Late Paleozoic Lafonian diamictite, Falkland Islands: Bull. Geol. Soc. Amer., v. 78, pp. 37–58.

———, 1968, Late Paleozoic glacial geography of Antarctica: Earth and Planetary Sci. Letters, v. 4, pp. 253–256.

———, 1969, Late Paleozoic glaciation: I. South America: Bull. Geol. Soc. Amer., v. 80, pp. 1007–1042.

Frarey, M. J., and Roscoe, S. M., 1970, The Huronian Supergroup north of Lake Huron, *in* Basins and geosynclines of the Canadian Shield (Baer, A. J., ed.): Geol. Surv. Canada, 70–40, pp. 143–158.

Fraser, H. J., 1935, Experimental study of the porosity and permeability of clastic sediments: Jour. Geol., v. 43, pp. 910–1010.

French, B. M., 1967, Sudbury structure, Ontario: some petrographic evidence for origin by meteoric impact: Science, v. 156, pp. 1094–1098.

———, 1968, Sudbury structure, Ontario: some petrographic evidence for an origin by meteoric impact, *in* Shock metamorphism of natural materials (French, B. M., and Short, N. M., eds.): Baltimore, Mono Book, pp. 383–412.

French, B. M., and Short, N. M., eds., 1968, Shock metamorphism of natural materials: Baltimore, Mono Book, 644 pp.

Füchtbauer, H., 1967, Die Sandsteine in der Molasse nördlich der Alpen: Geol. Rundschau, v. 56, pp. 266–300.

Garner, H. F., 1959, Stratigraphic-sedimentary significance of contemporary climate and relief in four regions of the Andes Mountains: Bull. Geol. Soc. Amer., v. 70, pp. 1327–1368.

Gary, Margaret, McAfee, Robert, Jr., and Wolf, C. L., eds., 1972, Glossary of geology: Washington, Amer. Geol. Inst., 805 pp.

Gasser, U., 1966, Sedimentologische Untersuchungen in der äusseren Zone der subalpinen Molasse des Entlebuchs (Kt. Luzern): Eclogae Géol. Helvetiae, v. 59, pp. 724–772.

Geikie, J., 1874, The great ice age, 2nd ed.: Englewood Cliffs, N.J., Prentice-Hall, 545 pp.

Göbler, K., and Klaus-Joachim, R., 1968, Entstehung und Merkmale de Olisthostrome: Geol. Rundschau., v. 57, pp. 484–514.

Goldthwait, R. P., 1971, Till, a symposium: Columbus, O., Ohio State Univ. Press, 402 pp.

Grabau, A. W., 1904, On the classification of sedimentary rocks: Amer. Geol., v. 33, pp. 228–247.

———, 1913, Principles of stratigraphy: New York, Dover, 1185 pp.

Gregory, H. E., 1915, The formation and distribution of fluviatile and marine gravels: Amer. Jour. Sci., ser. 4, v. 39, pp. 487-508.

Grogan, R. M., 1945, Shape variation of some Lake Superior beach pebbles: Jour. Sed. Petrology, v. 15, pp. 3–10.

Gubler, Y., Bugnicourt, D., Faber, J., Kubler, B., and Nyssen, R., 1966, Essai de nomenclature et caractèrisation des principales structures sédimentaires: Paris: Editions Technip, 291 pp.

Hälbich, I. W., 1962, On the morphology of the Dwyka Series in the vicinity of Loeriesfontein, Cape Province: Univ. Stellenbosch, v. 37, ser. A, no. 2.

Ham, W. E., 1954, Collings Ranch Conglomerate, Late Pennsylvanian, in the Arbuckle Mountains, Oklahoma: Bull. Amer. Assoc. Petrol. Geol., v. 38, p. 2035.

Hardy, R. M., and Legget, R. F., 1960, Boulder in varved clay at Steep Rock Lake, Ontario, Canada: Bull. Geol. Soc. Amer., v. 71, pp. 93–94.

Harland, W. B., 1965, Critical evidence for a great infra-Cambrian glaciation: Geol. Rundschau, v. 54, pp. 45–61.

Harland, W. B., Herod, K. N., and Krinsley, D. H., 1966, The definition and identification of tills and tillites: Earth-Sci. Rev., v. 2, pp. 225–256.

Harrison, P. W., 1957, A clay-till fabric: its character and origin: Jour. Geol., v. 65, pp. 275–308.

Hatch, F. H., and Rastall, R. H. (rev. Greensmith, J. T.), 1971, Petrology of the sedimentary rocks: New York, Hafner, 502 pp.

Heezen, B. C., and Ewing, M., 1952, Turbidity currents and submarine slumps and the 1929 Grand Banks earthquake: Amer. Jour. Sci., v. 250, pp. 849–873.

Henderson, J. B., 1970, Petrology and origin of the sediments of the Yellowknife Supergroup (Archean), Yellowknife, District of Mackenzie: Ph.D. dissertation, Johns Hopkins Univ., 263 pp.

Higgins, C. G., 1956, Formation of small ventifacts: Jour. Geol., v. 64, pp. 506–516.

Hoffman, P. F., 1965, Proterozoic paleocurrents and depositional history of East Arm fold belt, Great Slave Lake, Northwest Territories: Canad. Jour. Earth Sci., v. 6, pp. 441–462.

Holmes, C. D., 1941, Till fabric: Bull. Geol. Soc. Amer., v. 51, pp 1299–1354.

———, 1952, Drift dispersion in west-central New York: Bull. Geol. Soc. Amer., v. 63, pp. 993–1010.

Hough, J. L., 1932, Suggestion regarding the origin of rock bottom areas in Massachusetts Bay: Jour. Sed. Petrology, v. 2, pp. 131–132.

Howchin, W., 1908, Glacial beds of Cambrian age in South Australia: Quart. Jour. Geol. Soc. London, v. 64, pp. 234–259.

Hübner, H., 1965, Permokarbonische glazigene und periglaziale Ablagerungen aus dem zentralen Teil des Kongobeckens: Acta Universitatis Stockholmiensis, Stockholm Contributions in Geology, v. 13, no. 5, pp. 41–61.

Humbert, F. L., 1968, Selection and wear of pebbles on gravel beaches: Ph.D. dissertation, Univ. Groningen, 144 pp.

Irving, R. D., 1883, The copper-bearing rocks of Lake Superior: U.S. Geol. Surv. Monogr. 5, 464 pp.

James, H. L., Dutton, C. E., Pettijohn, F. J., and Wier, K. L., 1968, Geology and ore deposits of the Iron River–Crystal Falls District, Iron County, Michigan: U.S. Geol. Surv. Prof. Paper 570, 134 pp.

Johansson, C. E., 1965, Structural studies of sedimentary deposits: Geol. Fören. Stockholm Förh., v. 87, pp. 3–61.

Kauranne, L. K., 1960, A statistical study of stone orientation in glacial till: Bull. Comm. Geol. Finlande 188, pp. 87–97.

Kerr, P. F., Bodine, M. W., Jr., Kelley, D. R., and Keys, W. S., 1957, Collapse features, Temple Mountain uranium area, Utah: Bull. Geol. Soc. Amer., v. 68, pp. 933–982.

Kindle, C. H., and Whittington, H. B., 1958, Stratigraphy of the Cow Head region, western Newfoundland: Bull. Geol. Soc. Amer., v. 69, pp. 315–342.

King, P. B., 1958, Problems of boulder beds of Haymond Formation, Marathon Basin, Texas: Bull. Amer. Assoc. Petrol. Geol., v. 42, pp. 1731–1735.

Krumbein, W. C., 1933, Textural and lithologic variations in glacial till: Jour. Geol., v. 41, pp. 382–408.

———, 1939, Preferred orientation of pebbles in sedimentary deposits: Jour. Geol., v. 47, pp. 673–706.

———, 1940, Flood gravel of San Gabriel Canyon, California: Bull. Geol. Soc. Amer., v. 51, pp. 639–676.

———, 1941, The effects of abrasion on the size, shape, and roundness of rock fragments: Jour. Geol., v. 49, pp. 482–520.

———, 1942, Flood deposits of Arroyo Seco, Los Angeles County, California: Bull. Geol. Soc. Amer., v. 53, pp. 1355–1402.

Krumbein, W. C., and Griffith, J. S., 1938, Beach environment in Little Sister Bay, Wisconsin: Bull. Geol. Soc. Amer., v. 49, pp. 629–652.

Krumbein, W. C., and Tisdel, F. W., 1940, Size distribution of source rocks of sediments: Amer. Jour. Sci., v. 238, pp. 296–305.

Krynine, P. D., 1948, The megascopic study and field classification of sedimentary rocks: Jour. Geol., v. 56, pp. 130–165.

———, 1950, Petrology, stratigraphy, and origin of the Triassic sedimentary rocks of Connecticut: Bull. Connecticut State Geol. Nat. Hist. Surv. 73, 247 pp.

Kuenen, Ph. H., 1942, Pitted pebbles: Leidsche Geol. Meded., v. 13, pp. 189–201.

———, 1964, Experimental abrasion. 6. Surf action: Sedimentology, v. 3, pp. 29–43.

Kulling, O., 1938, Notes on varved boulder-bearing mudstone in Eocambrian glacials in the mountains of northern Sweden: Geol. Fören. Stockholm Förh., v. 60, pp. 303–306.

Kurk, E. H., 1941, The problem of sampling heterogeneous sediments: M.S. thesis, Univ. Chicago.

Landes, K. K., and Ehlers, G. M., 1945, Geology of the Mackinac Straits area, Ch. 3, Mackinac breccia: Michigan Geol. Surv. Publ. 44, ser. 37, pp. 123–153.

Landim, P. M. B., and Frakes, L. A., 1968, Distinction between tills and other diamictons based on textural characteristics: Jour. Sed. Petrology, v. 38, pp. 1213–1223.

Landon, R. E., 1930, An analysis of beach pebble abrasion and transportation: Jour. Geol., v. 38, pp. 437–446.

Lane, E. W., and Carlson, E. J., 1954, Some observations on the effect of particle size on movement of coarse sediments: Trans. Amer. Geophys. Union, v. 35, pp. 453–462.

Lane, E. W., and others, 1947, Report of the sub-committee on sediment terminology: Trans. Amer. Geophys. Union, v. 28, pp. 936–938.

Lawson, A. C., 1925, The petrographic designation of alluvial fan formations: Univ. Calif. Publ., Dept. Geol. Sci., v. 7, pp. 325–334.

Leinz, V., 1937, Estudos sobre a glaciacão permo-carbonifera do sul do Brasil: Brazil Serv. Fomento Prod. Min. Boletim 21, 55 pp.

Lenk-Chevitch, P., 1959, Beach and stream pebbles: Jour. Geol., v. 67, pp. 103–108.

Lindsey, D. A., 1969, Glacial sedimentology of the Precambrian Gowganda Formation, Ontario, Canada: Bull. Geol. Soc. Amer., v. 80, pp. 1685–1702.

Lundquist, G., 1935, Blockundersokningar, Historik och methodik: Sveriges Geol. Undersökn., ser. 3, no. 390, 45 pp.

McBride, E. F., 1966, Sedimentary petrology and history of the Haymond Formation (Pennsylvanian), Marathon Basin, Texas: Univ. Texas Bur. Econ. Geol., Rept. Inv. 57, 101 pp.

McIver, N. L., 1961, Upper Devonian marine sediments in the central Appalachians: Ph.D. thesis, Johns Hopkins Univ., 530 pp.

McKee, E. D., 1945, Cambrian history of the Grand Canyon region. Part I: Stratigraphy and ecology of the Grand Canyon Cambrian: Carnegie Inst. Washington Pub. 563, pp. 3–168.

McLaughlin, D. B., 1939, A great alluvial fan in the Triassic of Pennsylvania: Mich. Acad. Sci. Papers, v. 24, pp. 59–74.

Mansfield, G. R., 1906, The origin and structure of the Roxbury Conglomerate: Bull. Harvard Mus. Comp. Zool., v. 49, geol. ser., pp. 91–271.

———, 1907, The characteristics of various types of conglomerates: Jour. Geol., v. 15, pp. 550–555.

Maslov, V. P., 1938, Classification of breccias: Bull. Soc. Nat. Moscow, n.s., no. 46, sect. geol. 16, pp. 313–321.

Mawson, D., 1949, The Late Precambrian ice-age and glacial record of the Bibliando dome: Jour. Proc. Roy. Soc. New South Wales, v. 82, pp. 150–174.

Meckel, L. D., 1967, Origin of Pottsville conglomerates (Pennsylvanian) in the central Appalachians: Bull. Geol. Soc. Amer., v. 78, pp. 223–258.

Middleton, G. V., 1961, Evaporite solution breccias from the Mississippian of southwest Montana: Jour. Sed. Petrology, v. 31, pp. 189–195.

Miller, B. M., 1936, Cambrian stratigraphy of northwestern Wyoming: Jour. Geol., v. 44, pp. 113–144.

Miller, D. J., 1953, Late Cenozoic marine glacial sediments and marine terraces of Middleton Island, Alaska: Jour. Geol., v. 61, pp. 17–40.

Miller, H., 1884, On boulder glaciation: Proc. Roy. Phys. Soc. Edinburgh, v. 8, pp. 156–189.

Morse, W. C., 1916, The origin of the coarse breccia of the St. Louis Limestone: Science, n.s., v. 43, pp. 399–400.

Natland, M. L., and Kuenen, Ph. H., 1951, Sedimentary history of the Ventura Basin, California, and the action of turbidity currents: Soc. Econ. Paleont. Min. Spec. Publ. 2, pp. 76–107.

Nawara, K., 1964, Recent transport and sedimentation of gravels in the Dunajec and some tributaries: Prace Muzeum Ziemi, nr. 6, pp. 3–111.

Nordin, C. F., Jr., and Curtis, W. F., 1962, Formation and deposition of clay balls, Rio Puerco, New Mexico: U.S. Geol. Surv. Prof. Paper 450-B, pp. 37–40.

Norton, W. H., 1917, A classification of breccias: Jour. Geol., v. 25, pp. 160–194.

Okko, V., 1949, Glacial drift in Iceland: its origin and morphology: Bull. Comm. Geol. Finlande 170, 133 pp.

Osbourne, F. F., 1956, Geology near Quebec City: Nat. Canadien, v. 83, pp. 157–224.

Ovenshine, A. T., 1965, Sedimentary structures in portions of the Gowganda Formation, north Shore of Lake Huron: Ph.D. dissertation, Univ. California at Los Angeles.

Pelletier, B. R., 1958, Pocono Paleocurrents in Pennsylvania and Maryland: Bull. Geol. Soc. Amer., v. 69, pp. 1033–1064.

Pettijohn, F. J., 1934, The conglomerate of Abram Lake, Ontario, and its extensions: Bull. Geol. Soc. Amer., v. 45, pp. 475–506.

———, 1943, Archean sedimentation: Bull. Geol. Soc. Amer., v. 54, pp. 925–972.

———, 1952, Precambrian tillite, Menominee district, Michigan: Bull. Geol. Soc. Amer., v. 63, p. 1289.

———, 1957, Sedimentary rocks (2nd ed.): New York, Harper, 718 pp.

———, 1962, Dimensional fabric and ice flow, Precambrian (Huronian) glaciation: Science, v. 135, p. 442.

Pettijohn, F. J., and Bastron, H., 1959, Chemical composition of argillites of the Cobalt Series (Precambrian) and the problem of soda-rich sediments: Bull. Geol. Soc. Amer., v. 70, pp. 593–600.

Plumley, W. J., 1948, Black Hills terrace gravels: A study in sediment transport: Jour. Geol., v. 56, pp. 526–577.

Potter, P. E., 1955, The petrology and origin of the Lafayette Gravel. Part I. Mineralogy and petrology: Jour. Geol., v. 63, pp. 1–38.

———, 1957, Breccia and small-scale Lower Pennsylvanian overthrusting in southern Illinois: Bull. Amer. Assoc. Petrol. Geol., v. 41, pp. 2695–2709.

Potter, P. E., and Pettijohn, F. J., 1963, Paleocurrents and basin analysis: New York, Springer, 296 pp.

Puffett, W. P., 1969, The Reany Creek Formation, Marquette County, Michigan: Bull. U.S. Geol. Surv. no. 1274-F, 25 pp.

Reading, H. G., and Walker, F. G., 1966, Sedimentation of Eocambrian tillites and associated sediments in Finmark, northern Norway: Palaeogeog., Palaeoclim., Palaeoecol., v. 2, pp. 177–212.

Reinemund, J. A., 1955, Geology of the Deep River coal field of North Carolina: U.S. Geol. Surv. Prof. Paper 246, 159 pp.

Reynolds, S. H., 1928, Breccias: Geol. Mag., v. 65, pp. 97–107.

Richter, K., 1932, Die Bewegungsrichtung des Inlandeises, rekonstruiert aus Kritzen and Langsachsen der Geschiebe: Zeitschr. Geschiebeforschung, v. 8, pp. 62–66.

Royer, Louis, 1938, Les causes possibles de l'aspect bréchoïde de certaines roches: Soc. Géol. France, Ser. 5, v. 8, pp. 37–41.

Russell, R. D., 1937, Mineral composition of Mississippi River sands: Bull. Geol. Soc. Amer., v. 48, pp. 1307–1348.

Rust, B. R., 1966, Late Cretaceous paleogeography near Wheeler Gorge, Ventura County, California: Bull. Amer. Assoc. Petrol. Geol., v. 50, pp. 1389–1398.

Sayles, R. W., 1914, The Squantum tillite: Bull. Harvard Mus. Comp. Zool., v. 66, geol. ser., v. 10, pp. 141–175.

Schenk, P. E., 1965, Precambrian glaciated surface beneath the Gowganda Formation, Lake Timagami, Ontario: Science, v. 149, pp. 176–177.

Schermerhorn, L. J. G., 1966, Terminology of mixed coarse-fine sediments: Jour. Sed. Petrology, v. 36, pp. 831–835.

Schermerhorn, L. J. G., and Stanton, W. I., 1963, Tilloids in the West Congo geosyncline: Quart. Jour. Geol. Soc. London, v. 119, pp. 201–241.

Schlee, J., 1957, Upland gravels of southern Maryland: Bull. Geol. Soc. Amer., v. 68, pp. 1371–1410.

Schwarzbach, M., 1964, The recognition of ancient glaciations *in* Problems in palaeoclimatology (Nairn, A. E. M., ed.): New York, Wiley-Interscience, pp. 77–79.

Schwetzoff, M. S., 1934, Petrography of sedimentary rocks: Moscow (review in 1935, Jour. Sed. Petrology, v. 5, p. 106) (in Russian).

Scott, K. M., 1966, Sedimentology and dispersal pattern of a Cretaceous flysch sequence, Patagonian Andes, southern Chile: Bull. Amer. Assoc. Petrol. Geol., v. 50, pp. 72–107.

Sedimentary Petrology Seminar, 1965, Gravel fabric in Wolf Run: Sedimentology, v. 4, pp. 273–283.

Sharp, R. P., 1940, A Cambrian slide breccia, Grand Canyon, Arizona: Amer. Jour. Sci., v. 238, pp. 668–672.

Sharpe, C. F. S., 1938, Landslides and related phenomena, A study of mass-movements of soil and rock: New York: Columbia Univ. Press, 137 pp.

Simonen, A., 1953, Stratigraphy and sedimentation of the Svecofennidic, Early Archean supracrustal rocks in southwestern Finland: Bull. Comm. Geol. Finlande 160, 64 pp.

Smith, A. J., 1963, Evidence for a Talchir (Lower Gondwana) glaciation: striated pavement and boulder bed at Irai, central India: Jour. Sed. Petrology, v. 33, pp. 739–750.

Sneed, E. D., and Folk, R. L., 1958, Pebbles in the lower Colorado River, Texas; a study in particle morphogenesis: Jour. Geol., v. 66, pp. 114–150.

Spjeldnaes, N., 1965, The Eocambrian glaciation in Norway: Geol. Rundschau, v. 54, pp. 24–45.

Sternberg, H., 1875, Untersuchungen über längen-und Querprofil geschiebeführende Flusse: Zeitschr. Bauwesen, v. 25, pp. 483–506.

Teruggi, M. E., Mazzoni, M. M., and Spalletti, L. A., 1971, Sedimentologia de las gravas del Rio Sarmiento (Provincia de la Rioja): Rev. Mus. La Plata, new ser., geol. sect., v. 7, pp. 77–146.

Turner, C. C., and Walker, R. G., 1973, Sedimentology, stratigraphy, and crustal evolution of the Archean greenstone belt near

Sioux Lookout, Ontario: Canadian Jour. Earth Sci., v. 10, pp. 817–845.

Twenhofel, W. H., 1947, The environmental significance of conglomerates: Jour. Sed. Petrology, v. 17, pp. 119–128.

Udden, J. A., 1914, Mechanical composition of clastic sediments: Bull. Geol. Soc. Amer., v. 25, pp. 655–744.

Unrug, R., 1957, Recent transport and sedimentation of gravels in the Dunajec valley (western Carpathians): Acta Geol. Polonica, v. 7, pp. 217–257 (Polish with English summary).

Van Houten, F. M., 1957, Appraisal of Ridgway and Gunnison "tillites," southwestern Colorado: Bull. Geol. Soc. Amer., v. 68, pp. 383–388.

Vassoevich, N. B., 1953, O nekotorykh flishevykh texturah (Znakakh) (On some flysch textures): Trudy Lvovs. Geol. Obsh. Univ. Ivan Franko; geol. ser., no. 3, pp. 17–85.

Virkkala, K., 1951, Glacial geology of the Soumussalmi area, east of Finland: Bull. Comm. Geol. Finlande 155, pp. 1–66.

———, 1960, On the striations and glacier movements in the Tampere region, southern Finland: Bull. Comm. Geol. Finlande 188, pp. 161–176.

Von Engelen, O. D., 1930, Type form of faceted and striated glacial pebbles: Amer. Jour. Sci., ser. 5, v. 19, pp. 9–16.

Walcott, C. D., 1894, Paleozoic intraformational conglomerates: Bull. Geol. Soc. Amer., v. 5, pp. 191–198.

Walker, R. G., and Pettijohn, F. J., 1971, Archaean sedimentation: analysis of the Minnitaki Basin, northwestern Ontario, Canada: Bull. Geol. Soc. Amer., v. 82, pp. 2099–2130.

Wallace, R. C., 1913, Pseudobrecciation in Ordovician limestones in Manitoba: Jour. Geol., v. 21, pp. 402–421.

Wanless, H. R., 1960, Evidences of multiple Late Paleozoic glaciation in Australia: Intl. Geol. Congr., Rept. 21st Sess. Norden, pt. 12, pp. 104–110.

Wentworth, C. K., 1922a, A scale of grade and class terms for clastic sediments: Jour. Geol., v. 30, pp. 377–392.

———, 1922b, The shapes of beach pebbles: U.S. Geol. Surv. Prof. Paper 131-C, pp. 74–83.

———, 1935, The terminology of coarse sediments (with notes by P. G. H. Boswell): Bull. Nat. Res. Coun. 80, pp. 225–246.

———, 1936a, An analysis of the shapes of glacial cobbles: Jour. Sed. Petrology, v. 6, pp. 85–96.

———, 1936b, The shapes of glacial and ice jam cobbles: Jour. Sed. Petrology, v. 6, pp. 97–108.

Wentworth, C. K., and Williams, H., 1932, The classification and terminology of the pyroclastic rocks: Bull. Nat. Res. Council, v. 89, pp. 19–53.

West, R. C., and Donner, J. J., 1956, The glaciation of East Anglia and the East Midlands: a differention based on stone orientation measurement of tills: Quart. Jour. Geol. Soc. London, v. 112, pp. 69–91.

White, W. S., 1952, Imbrication and initial dip in a Keweenawan conglomerate bed: Jour. Sed. Petrology, v. 22, pp. 189–199.

Willard, B., 1930, Conglomerite, a new rock term: Science, v. 71, p. 438.

Williams, Howel, 1957, Glowing avalanche deposits of the Sudbury Basin: Ann. Rept. Ontario Dept. Mines, v. 65, pp. 57–89.

Willman, H. B., 1942, Geology and mineral resources of the Marseilles, Ottawa, and Streater Quadrangles: Bull. Illinois Geol. Survey, no. 66, p. 344.

Wilson, M. E., 1913, The Cobalt Series: its character and origin: Jour. Geol., v. 21, pp. 121–141.

Woodford, A. O., 1925, The San Onofre breccia: Univ. California Publ. Dept. Geol. Sci., v. 17, pp. 159–280.

Woodward, H. B., 1887, Geology of England and Wales: with notes on the physical features of the county (2nd ed.): London, G. Philip, 670 pp.

Woodworth, J. B., 1912, Geological expedition to Brazil and Chile, 1908–1909: Bull. Harvard Mus. Comp. Zool., v. 56, pp. 1–137.

Yeakel, L. S., Jr., 1962, Tuscarora, Juniata, and Bald Eagle paleocurrents and paleogeography in the central Appalachians: Bull. Geol. Soc. Amer., v. 73, pp. 1515–1540.

Young, G. M., 1966, Huronian stratigraphy of the McGregor Bay area, Ontario—relevance to the paleogeography of the Lake Superior region: Canad. Jour. Earth Sci., v. 3, pp. 203–210.

7

SAND AND SANDSTONES

INTRODUCTION

Sandstones constitute an important class of sediments. Excluding carbonate and volcanic sands, they form roughly one-fourth of the total sedimentary record. Besides their volumetric importance, many sands and sandstones are economic resources—as abrasives; as raw materials in the chemical, glass, and metallurgical industries; as construction materials, both building stones and as ingredients of plaster and concrete; as molding sand, paper filler, and so forth. Sands constitute important reservoirs for natural gas, oil, and artesian water. Some placer sands are a source of ore minerals and gems. Sand erosion and deposition are of engineering importance on beaches, in rivers, and in dune areas.

Sandstones, more than any other sediment, contribute to our understanding of geologic history. Their composition is a clue to provenance, their directional structures are a guide to paleocurrents, and their geometry, as well as internal structures, shed light on depositional environments.

Sands may be divided into three major groups: terrigenous, carbonate, and pyroclastic. *Terrigenous* sands are those produced by weathering and breakdown of preexisting rocks. They are transported, sorted, and modified by moving fluids—both air and water—and derived from sources external to the basin of deposition.

Carbonate sands are for the most part marine, and are primarily skeletal grains, oolites, locally derived detrital carbonate *intraclasts* (intraformational particles). These constituents are produced within the basin of deposition and are not the debris formed by breakdown of preexisting rocks. An exception are those sands rich in carbonate particles shed by very rapid erosion of thick carbonate sequences of orogenic chains. Such carbonate sands are in fact terrigenous sands derived from preexisting limestones and dolomites.

Pyroclastic sands are those produced by volcanic explosions. They may be deposited in diverse environments—in air or in water. The term *volcaniclastic* is also applied to some sands, rich in volcanic debris, which may be either truly pyroclastic or terrigenous (if derived from an older volcanic terrane).

The distinction between the several types of sands becomes blurred where materials of several different origins are deposited together. Pyroclastic materials may be mixed with either carbonate or terrigenous sands; carbonate sands may be mixed in all proportions with normal terrigenous sands.

In this chapter we deal only with the truly terrigenous sands. Carbonate sands, upon lithification, are generally classed as limestones, although they are in fact also a species of sandstone. They are dealt with in thc chapter on carbonate rocks. Pyroclastic sands are commonly considered to be igneous, but they are also quite properly regarded as sedimentary. Because of their special origin they are considered separately in Chapter 9.

MODERN SANDS

Our understanding of sandstones would be enhanced if we had a thorough knowledge of modern sands—their manner of formation and their accumulation.

Where is sand found in the present-day world? Excluding carbonate and pyroclastic sands, it occurs principally in rivers and on beaches and to a lesser extent as dunes and in shallow near-shore seas. Alluvial sands include those found in fans, in river channels, and on floodplains as well as on the deltas of lakes and seas. Most fluvial sands are associated with stream channels, although some escape and form overbank deposits (sand splays) on the floodplain. Shoreline sands include not only beaches but associated offshore bars, barriers, tidal deltas, and in some cases tidal flats. Eolian sands include coastal dunes and the more extensive dune fields of some desert basins. Marine sands are largely shelf sands, but some sand is carried over the shelf edge by turbidity currents to accumulate on the continental rise and in isolated sediment ponds in subsea regions.

In summary, there is no geomorphic area of the earth where sand is not found. Deep-ocean basins, although the most extensive geomorphic feature of the earth, are almost devoid of sand—containing only a scattering of wind-blown grains and thin turbidite sands in the proximity of the continents. In short, sand is a continental sediment; it originates on the continent and for the most part comes to rest on the continent.

It is noteworthy that the most common sites of sand accumulation in the modern world are linear (beaches and rivers); yet most sands of the past form extensive stratiform deposits. This difference between the essentially linear loci of sand deposition today and the extensive sheetlike accumulations of the past suggest that the latter are the result of displacement of loci of sand deposition through time, by lateral migration of streams or by transgression or regression of the shoreline. The broad expanse of sand on some shelf areas seems to be an exception to this. These sands, however, may be relict sands, perhaps fluvial deposits inherited from a low stand of the sea in glacial times (Emery, 1966, p. 12). More extensive dune fields of the present world seem to be underrepresented in the geologic record. But if Kuenen (1959a) is right in believing that the rounding of quartz sand is the result of eolian action, much sand in the geologic record must have had a desert eolian episode at some time in its history. Kuenen estimates that 2×10^6 km^2 of desert is needed to keep the world's average roundness constant—to offset the new, sharp-cornered sand added each year.

Not all environments of sand accumulation are equally represented in the geologic record. In the Paleozoic of the central Appalachians, where sand forms some 23 percent of the total section (Colton, 1970, p. 11), it is estimated that one-half or more is alluvial, about one-fourth marine turbidite, and the remainder littoral or shallow marine. None is identified as eolian.

Present-day sands, with few exceptions, are moderately to well-sorted materials and, except where derived from older supermature sands, are generally rather poorly rounded. A great deal of effort has been expended in an attempt to relate textural attributes of sand to the environment and/or agent of deposition. Studies have been made, for example, to discriminate between beach, dune, and river sands by some size parameter or ratio of such parameters (Friedman, 1961, 1967; Moiola and Weiser, 1968). Udden (1914) was one of the first to utilize size distribution as a whole or certain aspects of it, such as the "sorting index," to identify the environment of sediment deposition. More recent work has been summarized by Pettijohn, Potter, and Siever (1972, p. 86). In general, these efforts have met with limited success. They seem to work for sands of some areas but break down in others (Klovan and Solohub, 1968; Schlee, Uchupi, and Trumbull, 1964). In most cases they are dependent on techniques readily applied only to modern uncemented sands and are inapplicable to ancient quartzites.

The composition, both mineralogical and chemical, of modern sands varies widely. It seems to be more dependent on size and lithology of the source area than on climate, environment, or agent of deposition. The components of most sands are a mixture of quartz, feldspar, and rock fragments. What are the proportions in modern sands? Unfortunately data are hard to come by, because most mineralogical studies of present-day sands ignore the dominant light-mineral fraction and concentrate on the rare heavy-mineral constituents. A compilation of data on over 400 samples of modern North American sands (Pettijohn, Potter, and Siever, 1972, Table 2-1) shows that the feldspar content averages 15.3 percent, a figure identical with that of 434 sandstone samples of the Russian Platform (Ronov, Mikhailovskaya, and Solodkava, 1963, Table 2). The feldspar content ranges from 1 to 77 percent. River sands contain 22 percent, and beach and dune sands have about 10 percent each. The higher concentration in river sands may be attributable to the inclusion of an undue proportion of glacial materials; perhaps the lower content in the beaches and dunes is the result of incorporation of more mature sand from associated Coastal Plain formations, inasmuch as the sample is weighted with sands from the Atlantic and Gulf Coast areas.

Data on rock particle content are even less satisfactory than those for feldspar. Rock particles are widely distributed in modern sands, but total quantity and kind are seldom specified. Most rock fragments, moreover, are opaque to transmitted light unless thin sectioned and hence have generally been ignored. What is needed is *in situ* induration of modern sands and preparation of thin sections for study in the same manner as is done with sandstones of the geologic past. Such data as we do have show that sands of present-day rivers contain an abundance of rock fragments. The sands of the Ohio River, for example, contain on the average 31 percent (Friberg, 1970). An incomplete compilation (Pettijohn, Potter, and Siever, 1972, Table 2-4) shows an average of 20 percent for rivers. Twenty to 30 percent is perhaps a fair estimate for rivers as a whole. Unlike feldspars, rock particle content is strongly size-dependent. Coarser sands are notably higher in rock fragments than associated finer grades (Allen, 1962, p. 673; Shiki, 1959).

The average river sand, therefore, contains about 22 percent feldspar, 20 percent rock fragments, and, by difference, 58 percent quartz. As noted above, alluvial sands form the largest single category of sands; the average modern sand is, therefore, an immature (or at least submature) sand, both compositionally and texturally. Representative modal analyses of modern sands are given in Table 7-1.

The chemical composition of modern sands reflect, as might be expected, their mineralogical composition. Representative analyses are given in Table 7-2. Because they lack cement, the bulk composition of modern terrigenous sands shows a somewhat lesser silica content and a markedly lesser CO_2 and CaO content than their ancient counterparts.

What conclusions can we draw from this brief look at modern sands? They differ from ancient sands in some important ways. Unlike the graywackes of the older geologic past, they are essentially matrix-free, a difference that suggests that matrix is a diagenetic or post-depositional product. Modern sands are generally submature or immature in composition. The only exceptions are those sands immediately derived from older supermature sandstones. Pure quartzites (orthoquartzites) are common in

TABLE 7-1. Modal analyses of modern sands

Constituent	A	B	C	D	E
Quartz	86	15	64	17.6	64
K-Feldspar	5	5	7	10.3	5[c]
Plagioclase		28	4	7.5	2[d]
Rock fragments	8	36	19	57.2	29[e]
Other light minerals	1	—	4	3.1[a]	—
Heavy minerals	4.7	6	(4)[b]	6.8	—
Pyroxene, amphibole, mica	—	8	P	6.8	—

[a] "Matrix."
[b] Not included in total.
[c] "Acid" feldspar.
[d] "Basic" feldspar.
[e] Include 2% chert, 17% "quartz aggregates" and 10% other rock particles.

A. Ohio River sand near Cairo, Illinois (Friberg, 1970, Ph.D. thesis, Indiana Univ.). A sublitharenite sand.
B. Columbia River sediment (sand and coarse silt), Bonneville, Washington (Whetten, 1966, p. 1057). A lithic arenite; precursor of graywacke.
C. Mississippi River near Gulf of Mexico (Russell, 1937, Table 1, Sample 1083¾). A lithic arenite.
D. Sand from small streams, western Chihuahua, Mexico. Average of seven samples (Webb and Potter, 1969, Table 3). Volcaniclastic lithic arenite.
E. Sand of Rhine in Ruhr region (Koldewijn, 1955, Fig. 2). Values read from figure. A lithic arenite.

the older record but none seem to be forming today. As noted also, modern sand accumulations are mostly linear in distribution; extensive sheet accumulations which mark the past are generally absent. Finally, the texture and composition of modern sands bear little relation to the environment of deposition but are instead more largely products of source rock composition. Current action does improve sorting, but discrimination among dune, beach, and river sands on the basis of size parameters is uncertain at best. Provenance is the main factor in determining sand composition in the present-day world.

PROPERTIES OF SANDSTONES

THE FABRIC OF SANDSTONES

A sand consists primarily of a *framework*, which is the detrital fraction, and *voids*, which make up the pore system or empty spaces between the framework elements. The voids or pores in older sandstones may be partially or completely filled. The study of a sand or sandstone therefore centers on the framework, on its composition and microgeometry, and on the nature and volume of the pores and pore-filling materials.

The framework is, by definition, formed

TABLE 7-2. Chemical analyses of modern sands

Constituent	A	B	C	D	E	F
SiO_2	67.7	86.1	73.50	99.72	31.10	76.70
TiO_2	0.8	0.74	0.34	—	trace	—
Al_2O_3	13.6	5.45	13.3	0.27	4.06	13.48
Fe_2O_3		1.20	1.55			
	5.1			0.08	0.79	0.92
FeO		1.41	0.56			
MnO	0.1	0.06	0.04	—	—	—
MgO	1.9	0.61	0.56	0.02	1.20	0.74
CaO	3.4	0.98	1.13	0.06	30.55	1.40
Na_2O	2.8	0.55	2.34	0.09	1.51	1.75
K_2O	2.1	0.96	4.01	0.01	1.07	3.41
H_2O+	nd	1.10				
			1.80	—	3.25	—
H_2O-	nd	0.27				
P_2O_5	—	0.14	0.02	—	trace	trace
CO_2	nd	0.58	0.12	—	25.20	trace
Ignition loss	—	—	—	—	—	—
S	—	0.02	—	—	—	—
Total	97.5	100.17	99.27	100.25	98.73	

A. Average Columbia River sediment—sand and coarse silt—(Whetten, 1966, Science, v. 152, Table 2, pp. 1057–1058). A lithic arenite precursor of graywacke.
B. Ohio River sand, Cairo, Illinois (Friberg, 1970, Ph.D. thesis, Indiana Univ.). A sublithic arenite.
C. Modern rivers, western Chihuahua, Mexico (Webb and Potter, 1969, Tables 5, 6). A volcanic arenite.
D. Beach sands (average of 22), eastern Gulf of Mexico (Hsu, 1960, Table 9). Orthoquartzitic sand, second cycle.
E. Dune sand, Saint-Cast, (Côtes-du-Nord) France (Cayeux, 1929, p. 73). Rich in shell detritus, quartz with small quantities of feldspar and biotite.
F. Fluvial sand, River Loire, France (Cayeux, 1929 p. 53). Contains detrital quartz, feldspar, chert, and a little glauconite.

of sand-sized materials, 1/16 to 2 mm in diameter. Normally these are packed together in such a way that each grain is in contact with its neighbors and the whole framework is a mechanically stable structure in the earth's gravitational field. Unlike the grains of igneous and metamorphic rocks, which are in continuous contact with each other, the grains of a sand are in tangential contact only. The concentration of stress at these contacts may lead to solution and deposition of material dissolved elsewhere, resulting in an increase in area of grain contact and decrease in pore space. The end-product of such action is a rock with grains in continuous contact and zero porosity. These post-depositional changes in sand fabric are discussed in the section on sandstone diagenesis.

Some sands do not display simple framework–void geometry. Instead of a detrital framework with a pore system partially or wholly filled by a precipitated cement, the rock exhibits a continuum of sizes from the sand range downward into silt- and clay-sized materials. These are the *wackes* of which graywacke is the best-known example. There is no clear break in size between the coarser sand fraction and the finer materials. The sand-sized material is embedded in a *matrix*. The arbitrary line one draws between grains and matrix, is, however, an important distinction for interpretative purposes. The line has been placed at 1/16 mm (0.0625 mm) or at 0.05 mm or even as low as 0.03 mm. What proportion of matrix must a sand have to be designated a wacke? Some have chosen 15 percent; others have placed it lower (Dott, 1964, Fig. 3; Williams, Turner, and Gilbert, 1954, p. 290). The origin of this matrix is somewhat uncertain; it may be a primary depositional feature, or it may be formed by a post-depositional diagenetic process. Inasmuch as most present-day sands are free of matrix, it seems more probable that interstitial fines are a secondary or diagenetic product.

The framework of the normal sand or sandstone can be described in terms of its geometry and composition. Geometry has to do with the attributes of grains or framework elements—size, sorting, shape, roundness and surface textures—and especially with their packing and orientation. Size distribution can be defined by statistical measures of size and uniformity of size. These characteristics are related to the specific hydraulic regimen that governed the deposition of the sand and to the size of the materials available to the depositing currents. A rough measure of sorting is the ratio of the largest to the smallest grain. In well-sorted sands this ratio is under 10; in poorly sorted sands it may exceed 100. Sand grains display a variety of shapes and degrees of roundness. The interpretation of the size distribution and these other geometrical properties has been discussed elsewhere (Chapter 3).

Sands tend to assume a close packing. Nonspherical grains tend to lie with their longer axes parallel to the surface of deposition; in some cases they display an imbrication. In most cases grains in sections cut parallel to the bedding show a weak orientation aligned with the flow direction of the depositing current. In rare cases the orientation is random because of post-deposition disturbance, especially by organisms (bioturbation).

Voids form 30 to 35 percent of the normal sandstone. As a result of matrix formation or of precipitation of a mineral cement, void volume is reduced. In the "average" sandstone porosity is nearer to 15 percent; in extreme cases it is near zero. Cement may be deposited in crystallographic continuity with detrital grains (as quartz on quartz, or calcite on calcite), or it may be deposited on the detrital grains as drusy coatings, or as a microcrystalline mosaic in the voids. Exceptionally the carbonate cements may be coarsely crystalline and envelop one or more of the detrital grains as inclusions within these crystalline units or "sand crystals" (Fig. 12-3). Some cements, especially carbonate cements, show encroachment on the framework and partially replace portions of it. The nature of cementing constituents, their fabrics and relation to framework grains, and problems of their origin are discussed in more detail elsewhere (p. 239).

Sands vary greatly in *maturity*. The ultimate end-product of formative processes is a deposit in which grains are of only one mineral (the most stable mineral—quartz), are of one size (perfect sorting), and are completely rounded. No such sand exists, but some approach this ideal very closely. The concept of maturity—both textural and compositional—is important, and an effort should always be made to assess the maturity of a sand (Folk, 1951).

THE STRUCTURES OF SANDSTONES

Sandstones display a variety of sedimentary structures, which are best seen in exposed outcrops. The *internal* structures of individual beds is of great importance. Commonly sandstones display cross-bedding, the scale of which is a function of both coarseness of the sand and thickness of the accumulation unit. Many sands display smaller-scale ripple bedding. Other sandstones have a graded internal structure. As noted elsewhere, graded bedding and cross-bedding (certainly large-scale cross-bedding) are mutually exclusive features and are indicative of two quite different sandstone facies. Graded bedding is indicative of deposition below wave base and is characteristic mainly of deep-water sandstones. Some, perhaps most, sandstones display neither; some may even appear to lack any internal structure. X-ray examination, however, shows that most of these seemingly massive sandstones do possess internal lamination of some kind (Hamblin, 1962). Sedimentary structures have been described in greater detail in Chapter 4.

THE MINERALOGY OF SANDSTONES

Interpretation of the history of a sandstone depends on a thorough knowledge of its mineral composition. A single list of minerals present is not, however, enough. We need, instead, several lists—lists based on the assignment of minerals to meaningful genetic categories, such as primary detrital, precipitated cement, and post-depositional alteration products. Such categorization involves interpretation and judgment based on observable details (largely textural) and on grain-to-grain relations between component minerals. Some mineral species, quartz for example, may appear in several categories. It may be both a detrital component and a cementing constituent.

The list of primary detrital minerals that can occur in a sand is long. If the source rock was subject to incomplete weathering and the transportation brief, almost any known mineral can occur in sands that are of sand size in the parent rock. For detailed descriptions of common detrital minerals, refer to standard works by Milner (1962), Krumbein and Pettijohn (1938), Tickell (1965), Duplaix (1948), and Russell (1942).

Although the list of possible minerals is long, in practice relatively few species are encountered, and in most thin sections the number of species is even more restricted. Quartz is dominant in most sandstones; in many it forms over 90 percent of the detrital fraction. Feldspars, although common, play a subordinate role in contrast to their importance in the igneous rocks. In addition to quartz and feldspar, mica is the only other constituent in the parent rock likely to form an appreciable part of the detritus in normal sandstone. Rock particles occur in some sandstones and are abundant in a few.

Quartz, opal, and chalcedony Quartz, the most ubiquitous mineral of sands, is the chief constituent of most sandstones. The average sandstone is composed of about two-thirds quartz. Primarily a detrital mineral, it is also authigenic, in most cases as an overgrowth on the detrital grains; as such it is an important cementing constituent.

Detrital quartz of most sandstones is under 1.0 mm in diameter and is generally less than 0.6 mm (Dake, 1921). Grains larger than 1.0 mm are apt to be composite—*polycrystalline*—as opposed to the more common *monocrystalline* grains. The size of the quartz grains is determined mainly by the size of the quartz grains in the source rock. Quartz-bearing plutonic rocks, principally granites, are the ultimate source of nearly all detrital quartz. Dake (1921) has shown that these rocks seldom yield grains over 1.0 mm. Grains larger than this in the same source rock are so badly fractured and broken that only 9 percent of the quartz in rocks examined by Dake would exceed 1.0 mm. Twenty percent would exceed 0.6 mm.

The shapes of quartz grains vary greatly, but in the main they are subspherical. Most show a slight elongation, which tends to be greatest in the direction of the crystallographic *c*-axis. Wayland (1939) attributed this to vectorial abrasion caused by slight differences in hardness between the *c*- direction and the *a*-direction in quartz. Ingerson and Ramisch (1942), however, observed that quartz grains of igneous and metamorphic rocks, even those of granites, tended to be elongated and that such elongation was commonly parallel to the *c*-axis, an expression of the prismatic habit of quartz. The end-shape of sedimentary quartz, therefore, is an expression of its initial shape. Bloss (1957) and Moss (1966) showed experimentally that quartz has both weak prismatic and rhombohedral cleavage, so that fracturing produced grains elongated either parallel to the *c*-axis or at a fixed angle to it. The elongation has been used as a criterion of provenance; the quartz of metamorphic rocks is more elongated than that of igneous origin. This seems to be confirmed by data of Bokman (1952), who found a mean elongation ratio (long axis–short axis) of 1.43 for granites and 1.75 for schists.

Most detrital quartz grains contain inclusions. These, usually small, are commonly scattered at random within the grain, although they tend to be arranged in planes. Close attention has been given to inclusions as a guide to the provenance of quartz. Mackie (1896) classified detrital quartz into four groups on the basis of their inclusions: acicular, regular, irregular, and inclusionless. Acicular inclusions are, for example, rutile needles. Regular inclusions are mineral inclusions of euhedral form. And those of irregular form are fluid lacunae with or without gas bubbles. Mackie's observations led to the view that acicular and irregular inclusions characterized the quartz of igneous rocks; that of schists and gneisses had regular inclusions. More extended observations by Keller and Littlefield (1950) tended to support these views, although no one type of inclusion was diagnostic of either igneous or metamorphic quartz.

The extinction of quartz varies from sharp to wavy. Quartz which has been subjected to considerable pressure shows "strain shadows" or "undulatory extinction" observable under crossed nicols. In general, therefore, quartz of metamorphic rocks was presumed to show a marked wavy extinction, whereas that of igneous rocks was not. Observations show, however, that the larger quartz of some granitic rocks is more strained than the small annealed quartz of some metamorphic rocks. It has been concluded that undulatory extinction is an unreliable guide to the provenance of quartz (Blatt and Christie, 1963).

On the basis of one or another of these varietal characteristics of quartz, attempts have been made to classify detrital quartz into groups indicative of provenance. Sorby (1880), as well as Mackie (1896), was among the first to study sedimentary quartz; a more recent study is that of Krynine (1940, 1946b) where quartz is classified as *igneous* (including plutonic, volcanic, and hydrothermal quartz), *metamorphic* (including pressure and injection quartz), and *sedimentary* (which may be authigenic overgrowth and vein or vug fillings). Krynine's classification is difficult to apply, partly because the attributes used to discriminate overlap class limits, and partly because of inadequate knowledge of the quartz in source rocks. Although there may be differences in the statistical average for several source rock types, it is often impossible to assign specific grains to one or another class.

Even if one cannot discriminate between igneous and metamorphic quartz in all cases, one can frequently recognize volcanic quartz —the quartz derived from effusive volcanic rocks, notably the quartz porphyries. Such quartz is essentially strain-free and displays sharp extinction; it may exhibit embayed or rounded outlines caused by resorption into the magma and in some cases may show straight borders of hexagonal dipyramidal forms. Volcanic quartz is apt to be associated with felsic rock particles and perhaps with grains of zoned feldspars. In some exceptional cases, volcanic quartz is an important constituent of both modern (Webb and Potter, 1969) and older sands (Todd and Folk, 1957, p. 2550).

More important is *polycrystalline* quartz

—quartz grains which are composite and are formed of two or more crystal units. In many sands polycrystalline quartz is as abundant as monocrystalline quartz. Composite quartz grains may be microcrystalline chert, fine-grained quartzites, or more coarsely crystalline quartz of either igneous or metamorphic origin. Recent investigations of more coarsely crystalline quartz (Blatt and Christie, 1963; Blatt, 1967; Conolly, 1965; Voll, 1960) have shown that the larger the grain, the more apt it is to be polycrystalline. Also, the ratio of polycrystalline to monocrystalline quartz is lowest in more mature sands, perhaps because the polycrystalline forms are less stable. Polycrystalline forms vary also in the size and fabric of their constituent crystal elements. Some are polygonal, that is, with straight simple boundaries which tend to meet at 120°; whereas in other cases the boundaries are intricately sutured. The former is thought to be caused by "static annealing" and the latter to "cold working" (Voll, 1960, p. 536).

The high stability of quartz explains, of course, its abundance in sands. The average igneous rock contains 12.0 (Clarke, 1924, p. 33) to 20.4 (Leith and Mead, 1915, p. 74) percent quartz, whereas the average sandstone derived from the igneous rocks contains 67 to 70 percent quartz (Leith and Mead, 1915, p. 76). Such enrichment implies a high degree of mechanical and chemical stability. Quartz is not wholly insoluble and under certain conditions, as in some soils, shows marked corrosion and rounding (Crook, 1968; Cleary and Conolly, 1971). *Within* a sandstone, however, it is not only stable but tends to grow, as secondary outgrowths prove. It is notably more stable than chert, which is commonly subject to intrastratal solution (Sloss and Feray, 1948).

Quartz is mechanically very durable. Daubrée (1879, p. 256) estimated, from experimental studies, that quartz lost only 1 part in 10,000 per kilometer of travel. Abrasion mill studies by Thiel (1940) and Kuenen (1957; 1959b, p. 14; 1960) tend to confirm the slowness with which quartz is abraded.

Quartz also plays an important role as the cement of sandstones. It is the dominant cement in most sandstones, especially Paleozoic and older sandstones. Cementing quartz in many cases appears as a secondary outgrowth on detrital quartz grains. These overgrowths were described by Sorby (1880). Shortly thereafter, Irving and Van Hise (1892) recorded many examples of such enlargement of quartz from North American formations. This phenomenon is widespread and may be universal in all sandstones in which crystalline quartz serves as a cement. In the least-cemented sandstones, quartz grains may be broken apart readily and examined under the binocular microscope. The quartz overgrowths (Fig. 7-1) restore the fundamental form and symmetry of the quartz crystal. Even in the hand specimen such overgrowth can be detected by the sparkle seen in bright sunlight and attributable to light reflection from numerous newly formed crystal facets. In thin section, it can be seen that these overgrowths are separated from the detrital core by a thin line of impurities—a line made very distinct if the detrital grain was stained with iron oxide before enlargement (Fig. 7-20). In many better-welded quartzites, however, the distinction between core and overgrowth is less clear and in some cases cannot be made. If the thin section is examined by cathodoluminescence, the trace element content of the quartz evokes a fluorescence which makes possible differentiation between the core and overgrowth (Sippel, 1968).

Opal and chalcedony also occur commonly in sandstones as constituents of detrital chert grains and also, less commonly, as a cement. Opal is never seen in very old sandstones, because it devitrifies and crystallizes into chalcedony. The chalcedonic cement appears as a coating on detrital grains and is commonly fibrous, with its fibers normal to the grain surface.

Feldspar Feldspar, although the most abundant mineral of igneous rocks, plays a role subordinate to quartz in the sandstones. In some modern sands, however, it is more abundant than quartz. The average feldspar content of 404 samples of modern North American sands is 15.3 percent (Pettijohn, Potter, and Siever, 1972, p. 35) which is the same as that of 435 samples of sandstones, Precambrian to Quaternary, of the

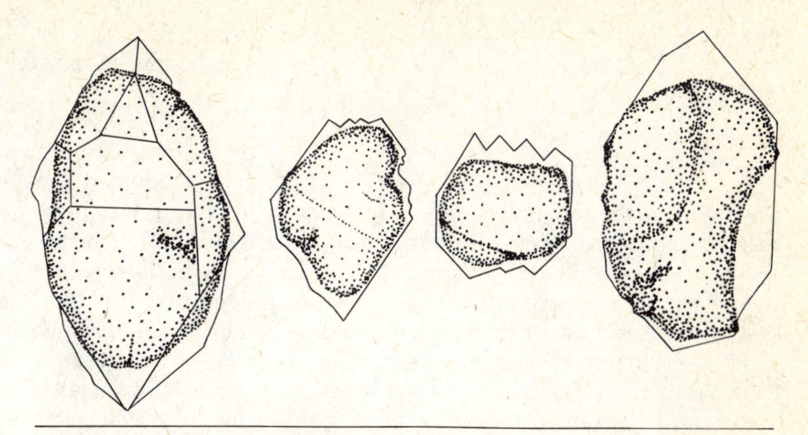

FIG. 7-1. Quartz grain enlarged by secondary growth. The shaded outline represents the boundary of the sand grain; the solid lines, the reconstructed crystal after secondary growth. (After Irving and Van Hise, 1884, Bull. U.S. Geol. Surv., 8, Pl. II.)

Russian Platform (Ronov, Mikhailovskaya, and Solodkova, 1963, Table 2). The unweighted mean of 98 North American sandstones is 10.2 percent.

Identification of feldspar and discrimination between the several types are troublesome, especially when modal analyses require a decision for every grain. Staining is the only satisfactory answer (Russell, 1935; Laniz, Stevens, and Norman, 1964).

Feldspars in sandstones include K-feldspar, especially microcline, and plagioclase, generally a somewhat albitic subspecies. Zoned feldspar is not common and is more apt to come from volcanic sources, as are feldspar euhedra or broken euhedra. Detrital feldspar may be clear or partially to completely clouded with alteration products. Some effort has been made to relate the variety of feldspar to potential source rocks (Rimsaite, 1967). Oscillatory zoning suggests volcanic or hypabyssal origin (Pittman, 1963). Feldspar of acid volcanics is more apt to be sanidine; that of plutonic rocks orthoclase or microcline. Pyroclastic feldspars are euhedra or broken euhedra in some cases with an envelope of glass or devitrified glass.

Obviously feldspar is less stable than quartz, so that weathering residues derived from plutonic rocks are impoverished in feldspar and enriched in quartz. This impoverishment, the result of decomposition in the weathering profile, may be enhanced by destruction of the feldspar during abrasive transport.

Although feldspar is supposed to be more susceptible to mechanical wear than quartz, data on this subject are contradictory. Mackie (1896) noted a decrease from 42 to 21 percent in the feldspar content of the sands of the Findhorn River in Scotland in some 30 to 40 miles of travel; he attributed this to abrasion. Plumley (1948) found a similar downstream decrease in the coarse-sand fraction of Battle Creek in the Black Hills of South Dakota. On the other hand, he found the feldspar of the Cheyenne River sands to decrease only slightly (from 29 to 24 percent) in 150 miles (240 km) and, as Russell (1937) noted, the feldspar content of the Mississippi River sands went from 25 percent near Cairo, Illinois, to 20 percent at the Gulf, an almost negligible decrease in over 1100 miles (1770 km) of travel. These limited data suggest that feldspar may be reduced rather rapidly in turbulent, gravel-carrying, high-gradient streams but lost only slowly in larger rivers. Perhaps vigorous beach action is capable of reducing materially the feldspar content of a sand.

That feldspar is chemically unstable is easily demonstrated. Goldich's (1938) data on soil profiles show that instability varied with species, lime-rich varieties being less stable than alkali feldspars, especially microcline. The stability of feldspars in a sandstone is not the same as that within a soil. In many sandstones, detrital feldspars acquire secondary overgrowths. Because the latter are nearly pure K- or Na-feldspar, it seems likely that these species are stable at the low-pressure, low-temperature, aqueous environments that prevail in the pore system of a sandstone.

The significance of detrital feldspar has been the subject of a protracted discussion. It is at the heart of the "arkose problem." The survival of feldspar appears to be a function of both *intensity* of the decay process and in *time* or duration of the action. Where relief is high and erosion rapid, feldspar will escape complete decomposition and appear in the sands; where relief is low and erosion retarded, feldspar will be destroyed. The role of climate seems secondary, but the destruction of feldspar will certainly be retarded in extreme arid or very cold climates. This question, and the evidence bearing on it, is discussed more fully in the section on arkose (p. 218).

Feldspar may, like quartz, display secondary enlargement, and thus to a small degree it, too, may serve as cement. In sandstones the differentiation between detrital nucleus and secondary rim is readily made (Fig. 7-29). The nucleus is usually rounded and is commonly kaolinized or otherwise altered so that the contrast between the clouded core and limpid overgrowth is marked. In its growth, marginal material tends to develop crystal faces and assume a regular crystal form, usually of simple rhombic outline. As a rule, authigenic feldspar, unlike secondary quartz, forms only a small fraction of the whole rock.

The nucleus of most grains seems to be triclinic feldspar (microcline). Marginal growths appear to be untwinned potash feldspar. Overgrowths are known also in detrital plagioclase. Secondary overgrowth on such nuclei is pure soda feldspar (albite). Although the overgrowth is added in crystallographic continuity with the core, it generally shows a very small difference in extinction angle, indicating a slight difference in composition (Fig. 7-29). Authigenic feldspar is nearly always pure alkali feldspar. It is known from petrological and experimental evidence that mixed feldspars, that is, both soda-lime and potash-soda feldspars, are formed only at higher temperatures. Sedimentary, low-temperature feldspar is the purest feldspar known (Baskin, 1956).

Secondary growth takes place following deposition of the nucleus, as the new feldspar is molded around adjacent interfering quartz grains. In some cases, however, evidence for several periods of secondary growth is present (Fig. 7-2). The first-deposited rim itself may be worn prior to the last outgrowth of feldspar (Goldich, 1934; Stewart, 1937).

The circumstances leading to the formation of authigenic feldspars are not fully understood. Most workers today concede a diagenetic origin and reject a metamorphic or hydrothermal origin (Goldich, 1934), although there are proven cases of relationships between secondary feldspar and igneous intrusions (Heald, 1956a). It has been said that marine waters are necessary for the formation of secondary feldspar and that such feldspar is, therefore, a criterion of marine origin (Crowley, 1939).

Rock fragments Coarser-grained rocks, both igneous and metamorphic, do not occur as detrital grains in medium-grained clastic sediments. These sediments are derived primarily from disintegration of plutonic rocks. On the other hand, fragments of fine-grained rocks may appear in the sands; in some (lithic arenites) they are the dominant constituent, exceeding even quartz. The average modern sand, based on a sample of 85, contains 20 percent rock particles (Pettijohn, Potter, and Siever, 1972, Table 2-4). The sands of the Ohio River average 31 percent (based on 187 samples, Friberg, 1970). In 13 Paleozoic sandstone formations in the central Appalachians, the rock particle content ranges from 0 to 33 percent and averages 13.0 percent (Pettijohn, Potter, and Siever, 1972, Table 12-2). Late Mesozoic sandstones of the Sacramento Valley have 20 to 75 percent rock detritus (Dickinson, 1969).

The number of rock species varies but may be large. Some 19 species were recognized in the Kulm Graywacke of the Harz Mountains of Germany (Mattiat, 1960).

Rock particles are strongly size-dependent. They are more abundant in the coarser sand fraction, but they can be found in the finest sands. As grain size decreases, identification becomes more difficult and more subjective

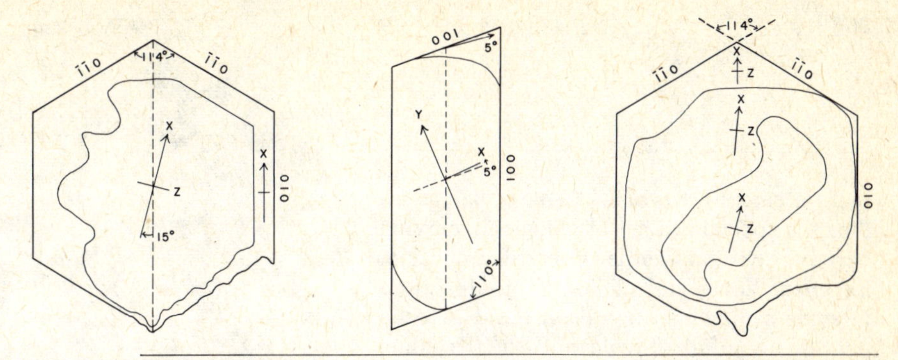

FIG. 7-2. Optical relations in enlarged feldspar grains. Left, (001) section; center, (010) section; right, (001) section of grain with two zones of secondary growth. (After Goldich, 1934.)

(Boggs, 1968). Identification of small rock grains presents difficulties. Dickinson (1970, p. 700) suggests use of operational criteria, mainly textural, to classify rock particles. He recognizes (1) volcanic rock fragments which have the textures of igneous aphanites, (2) clastic rock fragments with fragmental textures, (3) tectonite fragments with schistose or semischistose fabrics, (4) microgranular particles, that is, those with roughly equant, well-sized grains, and (5) carbonate rock fragments. These categories can be further subdivided, the volcanic broken down into grains that are felsic, microlitic, lathwork and vitric; the clastic may be silty-sandy or argillaceous, or volcaniclastic; the tectonic are either metasedimentary or metavolcanic; the microgranular are subdivided, with difficulty, into hypabyssal, hornfelsic, and sedimentary types. Refer to Dickinson's paper for the criteria for making the above discriminations.

Two identification problems deserve special mention. The distinction between felsites and chert is important and may be troublesome. Felsites may be marked by microphenocrysts, relict glass shards (as in welded tuffs), slight internal relief (attributable to differences in refractive indices of quartz and feldspar) and, if the thin section was stained for feldspars, a yellow or red stain. Chert may contain relict spicules, radiolarians or diatoms, or oolitic fabric. The problem of discriminating between felsites and chert has also been discussed by Wolf (1971).

The other identification problem is that of distinguishing between terrigenous particles of micritic carbonate and intraclastic and other carbonate materials of intrabasinal origin. The latter are common in the calcarenites. Terrigenous carbonates are commonly coarsely dolomitic; they are more apt, also, to be associated with other terrigenous rock particles.

The *durability* of rock fragments is highly variable. Chert and felsitic fragments tend to survive abrasive action; schist fragments, on the other hand, are very susceptible to destruction (Cameron and Blatt, 1971). Hence a short transport is implied for those sands containing an abundance of such rock particles.

Micas Detrital micas are found in sandstones, especially lithic sandstones, graywackes, and arkoses. Because of their flaky shape and despite their higher density and large size, they tend to collect with finer sands and silt. They are not ordinarily found with clean-washed sands.

Both biotite and muscovite are common; biotite in many cases has been altered to chlorite or, in rare cases, to glauconite (Galliher, 1935).

Micas occur in well-defined plates to very finely comminuted flakes and shreds. Where in contact with adjacent more rigid grains, larger plates tend to be bent or deformed because of compaction of the sand. The flakes are generally oriented parallel to the bedding and hence to each other. They are more abundant in finer-grained layers and in some rocks are concentrated on some bedding surfaces to which they impart a sheen and perhaps enhance the original bedding plane fissility.

The mica plates may be well rounded, a

feature said by Krynine (1940, p. 82) to indicate a specialized set of sluggishly moving currents with a gentle to-and-fro motion. Excellent hexagonal plates of biotite found in some sediments have been thought to be part of an infall of volcanic ash (Krynine, 1940, p. 22).

Detrital micas were derived from mica-bearing granites and gneisses and especially from mica schists. They are particularly abundant in the phyllarenites. Mica has been cited as a criterion of continental or littoral sedimentation (Lahee, 1941, p. 39), but perhaps this is only because the alluvium of large rivers and their deltaic deposits are mainly lithic sandstones of a mixed sedimentary and metamorphic provenance. Micas are known from turbidite sediments and are more abundant in distal than in proximal sections of submarine fans (Lovell, 1969, p. 950).

Heavy minerals Among the minerals of the parent rock surviving destruction are the so-called "heavy minerals." These minor accessory minerals of sandstones are marked by a higher than average specific gravity (greater than that of bromoform, 2.85). Rarely exceeding 1 percent and more commonly forming less than 0.1 percent of the rock, they are derived from the minor accessory minerals of the parent rock; more exceptionally, they are surviving remnants of the rather abundant but unstable mafic components of the source rocks. Zircon is an example of the stable mineral accessories, and hornblende is representative of the more abundant but unstable mafic components of the source rock. The number and kind of heavy minerals vary between wide limits—from 2 or 3 species to over 20 and from a few hundredths of 1 percent to those exceptional concentrations found in some placers. In most sandstones they form less than 1 percent.

If heavy minerals are newly derived from crystalline rocks, they are little worn. Cleaved fragments and more or less euhedral crystals characterize the assemblage (Fig. 7-3). If, however, the "heavies" are derived from earlier sediments, the less stable species tend to be absent and the more stable survivors show notable rounding (Fig. 7-4).

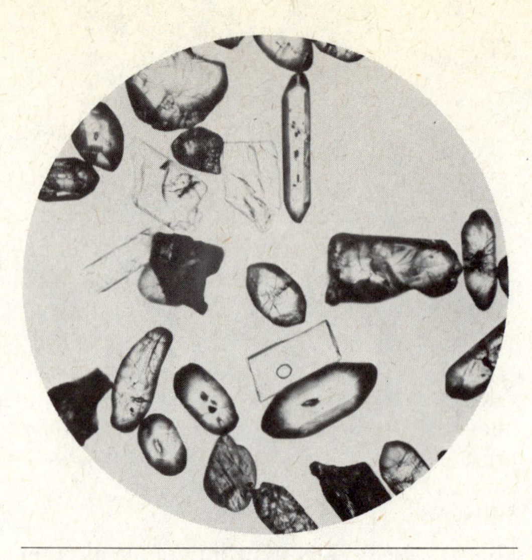

FIG. 7-3. Heavy minerals, ordinary light, ×120. Principally euhedral to slightly worn zircon.

So rare are heavy minerals that, except for a grain or two, they are seldom seen in thin section. In order to investigate them, they must be concentrated and isolated from the light minerals with which they are associated. Methods for achieving this separation are discussed in the standard manuals on sedimentary petrography (Krumbein and Pettijohn, 1938; Milner, 1962, vol. 1; Tickell, 1965, Müller, 1964; Carver, 1971, p. 427).

Study of heavy-mineral residue has proved useful in some cases for stratigraphic correlation, because, theoretically, each stratigraphic unit differs in some degree from every other in character and abundance of its suite of minor accessory minerals (Boswell, 1933, pp. 29, 60; Milner, 1962, vol. 2, p. 372). That this is commonly true has been verified many times and is the basis for "petrographic correlation." Such correlation depends for its success not only on the recognition of a distinctive association of minerals but also on its peculiar varieties and on changing proportions of the constituent minerals with time. Such differences are secured by progressive denudation of a varied terrane. Each new rock mass unroofed contributes new species or varieties to the accumulating sediments or changes the proportions of species already present. Correlation is complicated by reworking of older sediments so that the new deposit has many species in common with the sediment from which it was derived.

The problem of interpreting heavy-mineral suites is further complicated by the selective solution which the assemblage undergoes after deposition. The actual assemblage, therefore, is a function of both source rock composition and mineral stability (and hence capacity for survival both in the soil profile and in the sediment itself). Questions of stability and intrastratal solution have been debated at length (Boswell, 1933, p. 37; Pettijohn, 1941; van Andel, 1952, 1959; Weyl, 1950).

The heavy-mineral suite has, then, proved a useful guide to the type of source rock from which the sand was derived (Boswell, 1933; Feo-Codecido, 1956). Some minerals are diagnostic of certain source rocks. Others, like quartz, are more nearly ubiquitous and occur in nearly all possible parent materials. In this case, varietal features such as inclusions or color serve as a guide to the source rock. Krynine's work on tourmaline (1946a), in which he recognizes 13 varieties or subtypes, and that of Vitanage on zircons (1957) illustrate the use of varietal features. (The question of heavy minerals and provenance is further discussed in Chapter 13.)

The areal distribution of a particular heavy-mineral suite is defined as *sedimentary petrologic province.* Mapping such provinces contributes greatly to our understanding of paleocurrents and paleogeography. Refer to Potter and Pettijohn (1963, p. 193) for an extended discussion of this topic and to Füchtbauer's work (1964) on the sandstones of the German Molasse as an example of this use of heavy minerals. (The topic is further treated in Chapter 14.)

Calcite, dolomite, and siderite Carbonates occur most commonly in sandstones as a cement, although they also may occur as detrital carbonate rock particles, and, in the case of calcite, as a constituent of fossils and skeletal detritus.

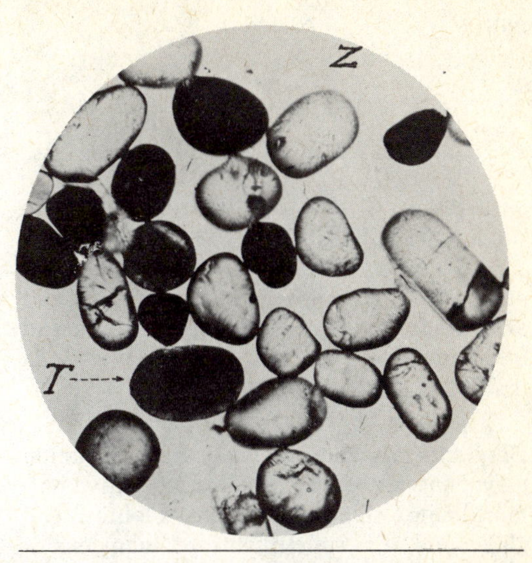

FIG. 7-4. Heavy minerals from St. Peter Sandstone (Ordovician). Ordinary light, ×60. Many well-rounded zircons (Z) and tourmaline (T). (Thiel, Bull. Geol. Soc. Amer., v. 46, 1935, Pl. 48.)

Calcite is a very common cement, as common as quartz in the Mesozoic and Cenozoic sandstones (Tallman, 1949). In partially cemented sandstones calcite appears under the microscope as a fringelike deposit on the borders of sand grains. In most carbonate-cemented sands, however, calcite forms a crystalline mosaic between the grains. Each pore is filled with a single crystal or at most with two or three such crystals (Fig. 7-30). In a few sandstones, crystals of calcite are large—1 cm or more in diameter. Such sandstones are said to show *luster-mottling.* Incomplete cementation of such sandstones results in the formation of *sand crystals,* which are large calcite euhedra, usually scalenohedra, that are loaded with inclusions of detrital sand (Fig. 12-3). (For a further discussion of these crystals and of calcitic concretions in sandstone, refer to Chapter 12.)

In some sandstones, quartz grains are widely separated or "float" in a field of carbonate (Fig. 7-30C). Such sandstones have been interpreted as an original mixture of clastic quartz and *clastic* carbonate, the latter recrystallized and now without any trace of its clastic origin.

Dolomite is also known as a cement of some sandstones. In a few sandstones it is a detrital constituent both as dolomitic rock particles and as large detrital dolomite rhombs (Sabins, 1962).

Siderite cement characterizes some sandstones; but, because of its instability and easy oxidation, it is seen only in cores and other fresh samples and seldom or never in natural outcroppings.

Clay minerals and other silicates Common clay minerals, such as kaolinite and montmorillonite, chlorites, various zeolites, and glauconite, are constituents of many sandstones. Some are primary detrital materials; most are diagenetic.

Kaolinite is found in the pores of some sandstones as a well-crystallized "blocky" cement. Clearly this mineral is a precipitate from solution (Glass, Potter, and Siever, 1956, p. 752).

Sericite and *chlorite* are the most important constituents of the matrix of the wacke class of sandstones—materials considered by some to be crystallized mud deposited concurrently with associated sand and as authigenic products derived by breakdown of unstable rock particles by others. The matrix problem is discussed more fully on page 227.

Various *zeolites* are common, particularly in volcaniclastic sands and other sands which contain volcanic admixtures (especially glass). Zeolites most commonly found in sandstones are analcime, laumontite, heulandite, clinoptilolite, and mordenite. They are diagenetic and are largely derived from volcanic materials, including glass (Hay, 1966).

Glauconite is a minor constituent of some sandstones and a major constituent of a few—the greensands. It occurs mainly as a framework grain of sand-sized granules. These are usually of subspherical form, polylobate in outline, seen under the microscope to be made of yellow-green to grass-green microcrystalline material. The mineralogy of glauconite has been investigated by Gruner (1935), Burst (1958), and Foster (1969). Burst in particular has shown that the materials designated "glauconite" vary considerably in composition and crystal structure. Burst identifies four subspecies, some of which are low in potassium. The iron content of glauconite depends on the iron concentration in the environment of formation; the potassium content is also variable, low potash indicating immaturity or degeneration (Foster, 1969).

Collophane A few sandstones are phosphatic. The phosphatic matter, a complex amorphous carbonate fluorapatite—collophane—occurs as a few scattered phosphatic skeletal fragments and/or as phosphatic nodules or granules, rarely ooids. In some sandstones associated with phosphorites, collophane is abundant both as framework grains and as a cement (Bushinsky, 1935; Cressman and Swanson, 1964, p. 310), which forms a drusy coating on the quartz grains or as a microcrystalline pore filling.

CHEMICAL COMPOSITION OF SANDSTONES

The composition of a sandstone may be expressed in terms of its bulk chemical composition. Such bulk chemical analyses are very useful. Sands (and other sediments) are, in a sense, products of large-scale chemical and mechanical fractionation processes, which, although somewhat imperfect, often lead to surprisingly good results. The processes, if long continued, separate elements into more or less chemically homogeneous end-products. To understand fully the geochemical processes and the evolution of various types of sediments or differentiates, chemical analyses are needed. Such data provide a norm or "bench mark" for the study of higher-grade metamorphic products and for a study of gains or losses if the process is not isochemical or for ascertaining the origin of the end-product where original sedimentary textures and structures are no longer recognizable. Analyses are useful also in the case of finer sands or those with a fine-grained matrix of which modal analyses are difficult to make. We need chemical data, especially averages, to study the mass balance and flow of materials in the overall evolution of the earth.

The chemical composition of a sand (as well as of most other rocks) is ordinarily reported by the analyst in terms of "oxides." The oxygen content itself is not actually determined, so the practice of reporting the constituents as oxides is based on the assumption that the elements determined combine with oxygen in stoichometric proportions (an assumption not always valid). If iron sulfides are present, for example, it is obviously incorrect to report the iron as FeO and the sulfur as SO_3. In most sediments, fortunately, sulfides are rather rare, and this exception and others like it are generally unimportant.

Chemical analyses vary a great deal in reliability and completeness. To evaluate and to use analyses require considerable judgment and some knowledge of how analyses are made. Washington (1930, pp. 3–12) has written an excellent discussion of the problem of completeness of chemical analyses as well as the method of evaluating them. Many analyses are woefully incomplete, and even major constituents may not be separately determined. Some analysts, for example, report "loss on ignition." This loss may include free and combined water, carbon dioxide, sulfide sulfur, and carbon or organic matter. Titania, an important constituent, may not be determined. If unreported, it is combined in the figure for alumina (Al_2O_3), which therefore makes that figure too high. In many sediments, the alkalies Na_2O and K_2O are not separately determined. The lesser constituents, MnO, P_2O_5, BaO, SO_3, S, and even CO_2, are commonly omitted. Such incomplete analyses are a decided handicap in the study of sediments.

The oxides reported by the analyst are not usually present as such but are bound up with other oxides to form mineral species. Hence the chemical analysis of a sand is properly understood only if one has some knowledge of the mineral composition of the sand. It should be recalled, also, that the bulk chemical analysis of a sandstone does not distinguish between the framework constituents and the cement. For this reason, analyses of sandstones are not comparable with those of modern sands of the kinds from which the sandstones were formed. Another point to be remembered is that the mineral composition, and hence the chemical composition of clastic sediments, is size-dependent. With progressive decrease in grain size comes a decline in quartz content and a concomitant increase in clay-mineral content. Thus there is a decline in SiO_2 and a rise in Al_2O_3 and K_2O content. (This is well shown in Table 8-1.)

Minerals of sandstones, unlike those of igneous rocks, are not an equilibrium assemblage, and so it is not possible to calculate a "normative" composition from the bulk chemical analysis as in analyses of the igneous rocks.

Table 7-3 shows the variation in average chemical composition of the common terrigenous sands. Included in this table is the "average igneous rock" so that one may see the extent to which the sands have been fractionated. It is clear that quartz arenites (orthoquartzites) are the most differentiated and enriched in silica and the most impoverished in the other constituents. Less mature sands may contain undecomposed feldspar and other minerals. It is self-evident, therefore, that sands will show a wide range of chemical composition depending on their maturity. Sands, moreover, show a much greater diversity in their composition than do shales. This is because sands are coarse unmodified residue from source rocks, whereas shales are the fine end-products of decomposition processes. But, whereas the primary chemical nature of a sand is determined by the completeness of the weathering process which gave rise to the sand and to the thoroughness of mechanical fractionation (washing) attendant on transportation and deposition, final composition is modified in various ways by processes of diagenesis, particularly by the introduction of a pore cement.

Chemical characteristics of each of the major sandstone classes and representative analyses are given in the various parts of this chapter.

In summary, the bulk chemical composition of a sandstone is dependent on, and markedly altered by, cementation. The composition of a particular species or class of sandstones is further governed by the manner in which the class itself is defined.

CLASSIFICATION OF SANDSTONES

The problems of classification of sedimentary rocks were first seriously tackled by ceptions forms the base for nearly all classificatory systems. All workers distinguish between cementing minerals and those constituting the framework and set the detrital components as defining parameters of various sandstone classes. Inspection of any sandstone (or modern terrigenous sand) shows that its principal components are quartz, feldspar, and rock fragments—the latter constituting sand-sized particles of fine-grained igneous (such as the felsites), sedimentary (such as chert and micritic limestone), and metamorphic (such as slate) origin. Other detrital components are rare and only exceptionally are a major constituent (glauconite in greensand and magnetite in black sands, for example). The framework composition, therefore, can be expressed in terms of those three constituents and the proportions of each shown graphically by the use of an equilateral triangular plot (Fig. 7-5). Principal sandstone classes or families can be defined by the proportions of these major constituents. The triangle can be, therefore, appropriately

---

Grabau (1904). Interest in the problem was renewed by Krynine (1948), Pettijohn (1948, 1954), Shrock (1946, 1948), and Rodgers (1950). In recent years there has been a very large literature on the subject, especially on classification of sandstones. The status of sandstone classification has been reviewed by Klein (1963), McBride (1963), Okada (1971), and most recently by Pettijohn, Potter, and Siever (1972, p. 149). Refer to these works both for a historical review of the subject and for basic philosophies underlying sandstone classification and nomenclature.

For the most part, it turns out that sandstones are best described—and hence classified—on the basis of their texture and mineralogical composition. Composition has proved most meaningful and with few ex-

TABLE 7-3. Mean composition of principal sandstone classes

Constituent	Orthoquartzite[a]	Lithic arenite[b]	Graywacke[c]	Arkose[d]
SiO_2	95.4	66.1	66.7	77.1
TiO_2	0.2	0.3	0.6	0.3
Al_2O_3	1.1	8.1	13.5	8.7
Fe_2O_3	0.4	3.8	1.6	1.5
FeO	0.2	1.4	3.5	0.7
MnO	—	0.1	0.1	0.2
MgO	0.1	2.4	2.1	0.5
CaO	1.6	6.2	2.5	2.7
Na_2O	0.1	0.9	2.9	1.5
K_2O	0.2	1.3	2.0	2.8
H_2O+	0.3	3.6	2.4	0.9
H_2O-	—	0.7	0.6	—
P_2O_5	—	0.1	0.2	0.1
CO_2	1.1[e]	5.0	1.2	3.0
SO_3	—	—	0.3	—
S	—	—	0.1	—
C	—	—	0.1	—
Total	100.7	100.0	100.4	100.0

Source: From Pettijohn (1963), Table 12.

[a] Computed from 26 published analyses.

[b] Computed from 20 analyses.

[c] Based on total of 61 analyses including 28 New Zealand graywackes (Paleozoic and Mesozoic) from Reed (1957, p. 16).

[d] Computed from 32 published analyses.

[e] Estimated from CaO.

subdivided in whatever detail deemed necessary.

The above system handles normal sands, those with a well-defined framework–pore system. It runs into difficulty with those sands which contain an appreciable matrix—the wackes. Hence for classficatory purposes many have proposed to divide sands into two groups—those with and those without matrix—and then to subdivide each of these groups. This division of all terrigenous sands into these two groups has been criticized by Dickinson (1970, p. 697), not only because of difficulty of framing an operational definition of *matrix* but also because of the polygenetic origin of matrix and the resulting interpretative problems. But despite these difficulties, the scheme proposed by Pettijohn (1954) and modified by Dott (1964), shown in Fig. 7-6, will be used in this book.

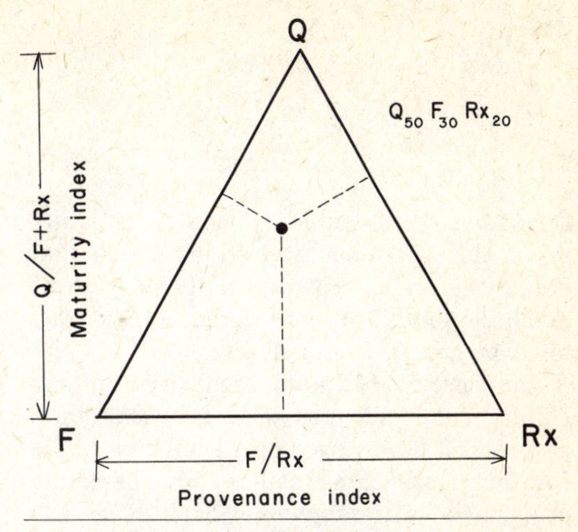

FIG. 7-5. Three-component sand represented by a ternary diagram.

Although there is a general consensus that this approach to sandstone classification is both the most workable and the most meaningful, there are many variations in use. The positioning of the boundaries and the naming of the several sandstone families have varied somewhat (see for example,

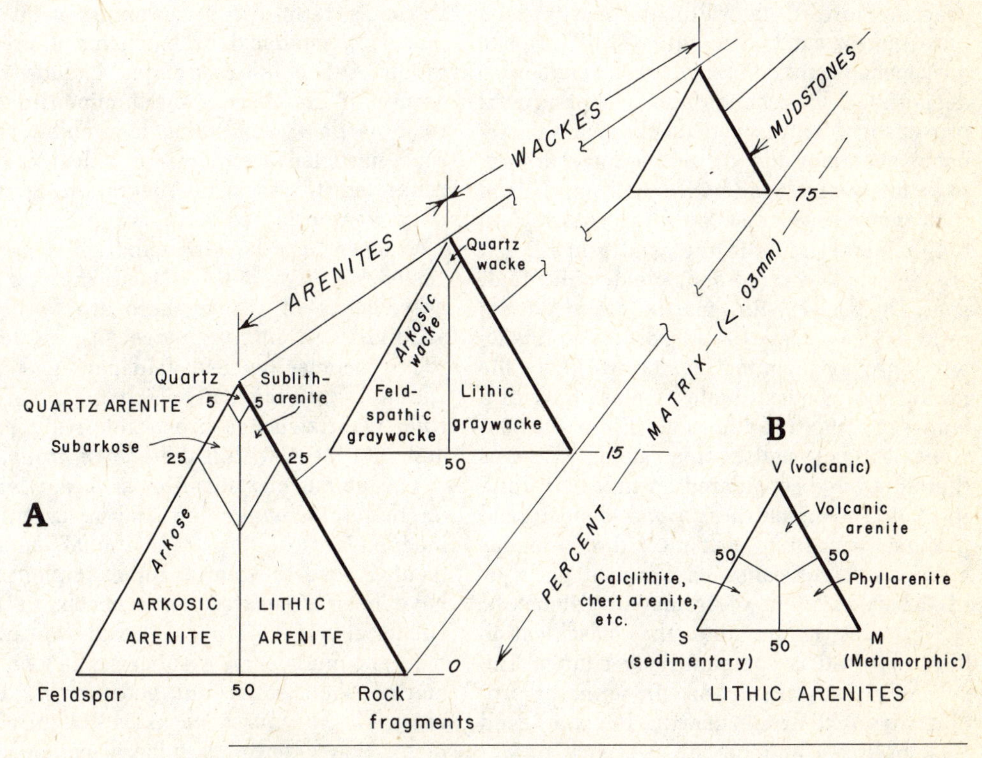

FIG. 7-6. A: classification of terrigeneous sandstones (modified from Dott, 1964, Jour. Sed. Petrology, v. 34, Fig. 3); B: subdivision of lithic arenites (after Folk, 1968, p. 124).

Dickinson, 1970, Table 1). Some ambiguity exists also about such questions as whether polycrystalline quartz, especially such forms as chert, should be considered a rock fragment or grouped with quartz.

The most confusing divergence in nomenclature has been engendered by the term *graywacke*. It was used by Krynine and Folk for that class of sandstone now generally designated *lithic sandstone*. Such usage has been generally abandoned, and the term is now used by most in its more classical sense, i.e., for those dark gray sandstones with a significant matrix content.

It is worth noting that the choice of mineral parameters, which form the basis for the classifications usually used and the one here adopted, is not only sufficient to describe and categorize most sandstones; they also have genetic significance. The ratio $Q/(F + Rx)$ is a rough measure of compositional maturity. The ratio measures the progress toward the ultimate end-type—a pure quartz sand. The ratio F/Rx reflects provenance and distinguishes between a deep-seated provenance and a supracrustal provenance. Supracrustal rocks, whether igneous, metamorphic, or sedimentary, are apt to be fine-grained and hence yield sand-sized *rock* particles. Coarse plutonic rocks yield only mineral grains in the sand range. Most are feldspar-bearing and yield only feldspar. The $(Q+F+Rx)$/matrix (grain–matrix) ratio is less easy to interpret. Sediments with an overwhelming matrix are likely the products of a quasi-liquid or mass flow of a mud–sand mixture; normal dilute suspensions deposit matrix-free sands. It was therefore once considered an *index of fluidity*. But if some matrix is a post-depositional product (perhaps diagenetic), the ratio has a different significance—a measure, perhaps, of the degradation of framework elements.

Fig. 7-6A shows that the classification here followed is comparatively simple. The several classes are defined in terms of proportions of detrital quartz, feldspar, and rock particles and on the presence or absence of an interstitial matrix. Those with 15 or more percent matrix constitute the *wackes;* those with less are the "ordinary" (ortho) sandstones. In this class—the matrix-free or matrix-poor sandstones—three main families are defined: (1) those in which quartz forms 95 or more percent of the framework fraction, the *quartz arenites* (orthoquartzites); (2) those containing 25 or more percent feldspar, which exceeds rock particles, the *arkoses;* and (3) those characterized by over 25 percent of rock particles, the *lithic sandstones* (lithic arenite). At times it is appropriate to define and name subclasses, classes transitional between major families. Included here are *subarkose* and *sublithic sandstones* or *arenites.* The lithic arenite class may be itself subdivided in a meaningful manner on the basis of the kind of rock fragments present, as suggested in Fig. 7-6B. The most common variety is one in which rock particles of low-grade pelitic metamorphic rocks, such as slate, phyllite, and mica schist, prevail. Sandstones with such phylloid fragments have been designated *phyllarenites* (Folk, 1968, p. 131). The term *calclithite* has been proposed (Folk, 1968, p. 141) for those terrigenous sands containing a large quantity of detrital limestone and dolomite particles, to distinguish this type of sand from a *calcarenite,* Grabau's term for a carbonate sand produced by chemical or biochemical precipitation. Other lithic arenites include *chert arenite,* if the chert is the dominant detrital rock particle, and *volcanic arenite,* if the rock particles are those of volcanic rocks generated by rock disintegration by ordinary weathering and erosion.

Wackes may be also subdivided as suggested in Fig. 7-6A. Dominant are the *graywackes* of which there are two main subdivisions: *lithic graywackes,* in which rock particles exceed feldspar, and *feldspathic graywacke,* in which the reverse is true. *Quartzwackes* are a relatively minor and rare class within the wacke group.

The above classification is based largely on mineral composition and is essentially independent of the depositional environment. A quartz arenite, for example, might have been deposited on a beach, or as a subaerial dune, or by a stream. An arkose may accumulate on a subaerial fan or as a marine shelf sand. But because the character of the source rocks is the principal factor that determines mineral composition, this classification more closely reflects source area composition (provenance) than anything else. It may thus be indirectly related to tectonics, as Krynine once contended (1945, Fig. 2).

A complete classification of sandstones should recognize textural attributes and also the characters of cements. These aspects can be accommodated by adjectival modifiers, thus: a well-sorted, calcareous subarkose or a poorly sorted, siliceous phyllarenite.

A large number of names has been used to characterize the diverse types of sandstones found in nature. Some of these, such as grit, gannister, flagstone, and brownstone, arose from common language and express some particular attribute or use. Many of these have no petrographic value or status. Other terms have been coined and were intended to give a more precise designation of a petrographic type, such as calcareous phyllarenite. Unfortunately, terminology tends to overbreed; we now have become burdened with a multiplicity of names. We shall here try to keep matters within bound. Refer to the several glossaries for definitions of rare types and especially the earlier work of Allen (1936) and the more recent compilation of Pettijohn, Potter, and Siever (1972, p. 161), whose glossaries are devoted exclusively to sandstone nomenclature.

SANDSTONE PETROGRAPHY

The petrography of a sandstone is dependent in large measure on the composition of the source rock or rocks. This is especially true of immature sandstones. Because quartz is by far the dominant constituent of sands, the ultimate source of most sands must be quartz-bearing plutonic rocks—granites, quartz monzonites, and related gneisses. A large class of sands—*arkoses*—are the product of disintegration, without much decomposition, of these rocks. Those sands rich in rock particles—*lithic arenites*—are derived more from supracrustal rather than plutonic source rocks. Their quartz is derived from preexisting sandstones, their rock particles from fine-grained sedimentary, metamorphic, and effusive igneous rocks. The effect of provenance is greatly diminished in mature sands, especially orthoquartzites or *quartz arenites*. All sands evolve toward this end-type, so it becomes increasingly difficult to determine either the immediate or the ultimate source of these sands.

In addition to the major classes of sandstones, outlined above, we have the *wackes* (*graywacke* in most cases), which also constitute a significant class of sandstones. We shall consider these separately.

The petrography of sandstones should be pursued by careful study and analysis of a well-chosen set of hand specimens and thin sections. The beginner would do well also to consult some of the larger works on this subject, including the classic monographs of Cayeux (1929) and Hadding (1929) and more recent works (Pettijohn, Potter, and Siever, 1972). In addition there are other comprehensive petrographic studies of sandstones of a particular place or a specific formation such as the classic paper of Krynine on the Devonian Third Bradford Sand of Pennsylvania (1940). An annotated list of such papers has been published (Pettijohn, Potter, and Siever, 1972, pp. 245–252), and various examples are cited in this section of the chapter.

FELDSPATHIC SANDSTONES AND ARKOSE

Definitions The term *feldspathic sandstone* refers only to sandstones in which feldspar is an important detrital constituent, usually abundant enough to be seen with the unaided eye. The term *arkose*, on the other hand, is a special class of feldspathic sandstones. It is an old term whose derivation is uncertain; it has been attributed to Brongniart (Oriel, 1949) who wrote one of the first papers on arkose and its geological significance (Brongniart, 1826). The meaning of arkose as originally defined has changed very little. It is a sandstone, generally coarse and angular, moderately well sorted in most cases, composed principally of quartz and feldspar, and presumably derived from a granite or rock of somewhat granitic composition. Quartz is generally the dominant mineral, although in some arkoses feldspar exceeds quartz. Other constituents are very subordinate.

Arkose There has been no general agreement on how little feldspar a sandstone can contain and still be designated arkose. Allen (1936) placed the lower limit at 25 percent; Krynine once suggested 30 percent (1940, p. 50) but later (1948) proposed 25 percent. Pettijohn (1949, p. 227) accepted the 25 percent lower limit and proposed to restrict the term *feldspathic sandstone* to those sandstones with 10 to 25 percent (of their detrital fraction). The term *subarkose* is now generally applied to this group of sandstones. Arkose has been redefined (Pettijohn, 1954) as a sandstone characterized by 25 or more percent of labile constituents (feldspar and rock fragments) of which feldspar forms one-half or more. By this definition arkose might contain as little as 12.5 percent feldspar.

The definitions of arkose given above all fail to discriminate between true arkoses and the feldspathic graywackes—some with over 25 percent feldspar. The graywackes by definition contain a substantial matrix, whereas normal arkose has a precipitated mineral cement, often calcite. But some arkoses also contain interstitial clay. How can we discriminate between these and feldspathic graywacke? In general, arkose is indeed derived by disintegration of granitic rocks, and it is rich in K-feldspar; on the other hand the characteristic feldspar of graywackes is Na-feldspar. Graywackes, unlike arokses, are also rich in a variety of rock particles of diverse origins. In general, too, graywacke matrix is chloritic, whereas the interstitial clay of *arkosic wacke* tends to be kaolinitic and commonly red because of ferric pigments. These compositional differences are generally associated with differing modes of occurrence and differing internal organization or structure. The relations between these several types of sands are shown diagrammatically in Fig. 7-7.

Fabric and composition Arkose is typically a coarse-grained rock consisting mainly of quartz and feldspar (Fig. 7-9). It is pink or reddish, colors imparted to the rock by the feldspar, or in some cases, by a red clay matrix. Some arkoses are derived from granitic or gneissic rocks containing gray or white feldspar and hence are themselves gray to white unless red ferruginous materials are present.

The dominant mineral of arkose is generally quartz, although exceptionally it is feldspar. Because of coarseness of grain much of the quartz is polycrystalline; some composite granules consist of both feldspar and quartz. The grains are generally poorly rounded. With few exceptions the feldspar is microcline. It varies from wholly fresh to markedly weathered (kaolinized) to a mixture of both. In arkoses cemented by calcite, feldspars show varying degrees of replacement or mere corrosion of grain boundaries to complete replacement. In other arkoses, feldspars show regeneration, that is, overgrowths of clear, untwinned feldspar on detrital cores. Large detrital micas, both muscovite and biotite (and chloritized biotite), characterize arkoses. Micas tend to lie parallel to the bedding and hence

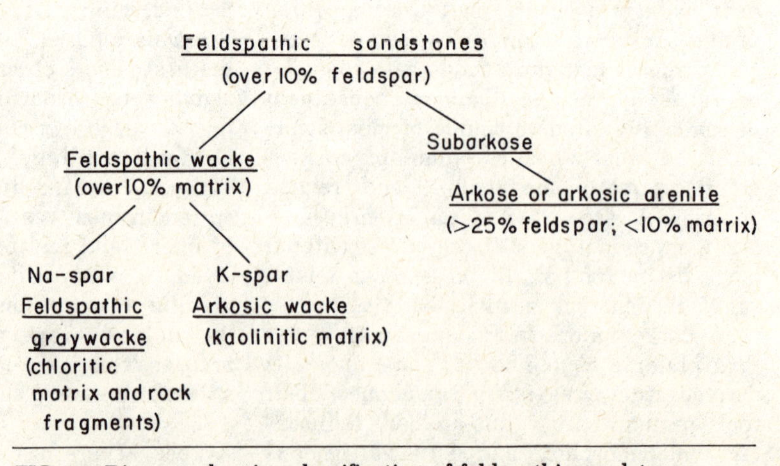

FIG. 7-7. Diagram showing classification of feldspathic sandstones.

TABLE 7-4. Mineral composition of subarkose and arkose

Constituent	A	B	C	D	E[a]	F	G	H	I
Quartz	60	57	60	71	35	37.7	57	51	53.1
Microcline	34	27	13	25	59[b]	0.7	24	30	18.5
Plagioclase	—	1				45.4	6	11	0.4
Micas	—	—	trace[c]	—	—	4.2	3	1	6.9
Clay	—	—	5	—	—	12.0	9	7	17.0
Carbonate	—	—	—	—	2	—	trace[c]	trace[c]	—
Other	6[d]	14	8	4	4[e]	—	1	—	4.1

[a] Normative or calculated values.
[b] Modal feldspar given by Mackie (1905) as 60.
[c] Present but under 1 percent.
[d] Chlorite.
[e] Iron oxide (hematite and kaolin).

A. Sparagmite (Precambrian), Norway (Barth, 1938, p. 60).
B. Jotnian (Precambrian), Finland (Simonen and Kuovo, 1955, Table 5).
C. Subarkose, Lamotte Sandstone (Cambrian), Missouri (Ojakangas, 1963, p. 863).
D. Subarkose, Potsdam Sandstone (Cambrian), New York (Wiesnet, 1961, Jrl. Sed. Petrol., v. 31, p. 9).
E. Lower Old Red (Devonian), Scotland (Mackie, 1905, p. 58).
F. Arkose (Permian), Auvergne, France (Huckenholtz, 1963, p. 917).
G. Pale arkose (Triassic), Connecticut (Krynine, 1950, p. 85).
H. Red arkose (Triassic), Connecticut (Krynine, 1950, p. 85).
I. Arkose (Oligocene), Auvergne, France (Huckenholtz, 1963, p. 917).

to one another. They are commonly bent or deformed by pressure of adjacent grains. Biotite may be chloritized or altered by oxidation. Arkosic sandstones of mixed provenance may contain rock particles and pass by degrees into coarse lithic arenites.

Calcite is a common cement in younger arkoses. Some older arkosic sandstones show secondary overgrowths on both feldspar and quartz. Such enlargements, if carried to completion, produce a strongly lithified rock superficially resembling a granite gneiss for which, especially in small outcrops, it may be mistaken. Some arkoses lack a precipitated mineral cement and have instead a kaolinitic clay matrix, very commonly red because of ferric oxide pigmentation. These are the *redstones* of Krynine (1950, p. 103).

Representative modal analyses of arkoses are given in Table 7-4; representative chemical analyses are shown in Table 7-5.

The bulk chemical composition of arkose reflects its mineral composition. That consisting largely of quartz and feldspar shows mainly SiO_2, Al_2O_3, and K_2O. If cemented by calcite, CaO and CO_2 become important constituents. Typical arkose differs from quartz arenites (orthoquartzite) in its lower SiO_2 and higher Al_2O_3 and K_2O content and differs from graywacke in being rich in K_2O and poor in Na_2O. See Fig. 7-8. Unlike ar-

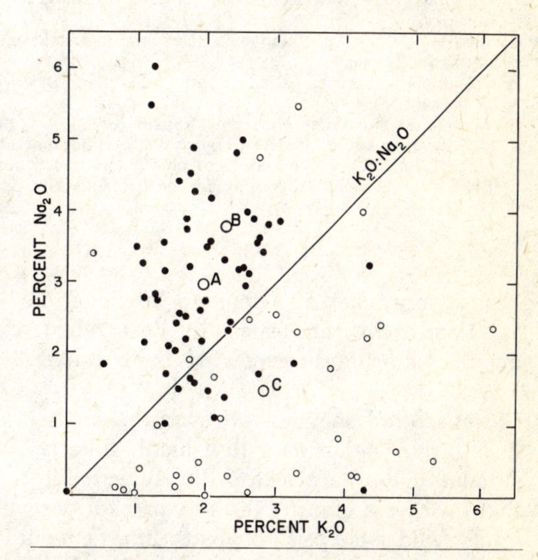

FIG. 7-8. Na_2O/K_2O ratio in arkoses and graywackes; Solid black circles, graywackes; open circles, arkoses. A: average graywacke; B: composite New Zealand graywacke (Reed, 1957, p. 16); C: average arkose (Pettijohn, 1963, Fig. 2; data from same source).

TABLE 7-5. Representative chemical analyses of arkose and subarkose

Constituent	A	B	C	D	E	F	G	H	I	J	K
SiO_2	79.30	75.80	80.89	73.32	59.24	69.94	92.13	87.02	85.74	72.21	76.6
TiO_2	.22	.15	.40	—	—	—	—	—	.38[c]	.22	0.6
Al_2O_3	9.94	11.74	7.57	11.31	6.65	13.15	4.42	2.86	6.84[d]	10.69	12.4
Fe_2O_3	1.00	.59	2.90	3.54	2.02 }		{ .37	.49	.79[e]	.80	0.7
						2.48					
FeO	.72	1.31	1.30	.72	.31 }		{ .33	.28	—	.72	0.2
MnO	.02	.05	—	—	.50[a]	.70	.24[a]	—	—	.22	—
MgO	.56	.54	.04	.24	.12	trace	.14	.20	1.11	1.47	0.3
CaO	.38	1.41	.04	.75	16.04	3.09	1.27	3.41	.49	3.85	0.4
Na_2O	2.21	2.40	.63	2.34	.19	5.43	.11	.00	1.16	2.30	0.3
K_2O	4.32	4.51	4.75	6.16	2.30	3.30	.72	1.98	2.19	3.32	3.8
H_2O+	.55	.86 }			{ 1.26	—	—	—	—	1.46 }	
			1.11	.30							2.7
H_2O-	.41	.03 }			{ —	—	—	—	—	.08 }	
P_2O_5	.05	.60	—	—	—	—	—	—	.01	.10	0.2
CO_2	—	trace	—	.92	12.16	—	none	—	—	2.66	—
Ignition loss	—	—	—	—	—	1.01	.42	3.35	1.12	—	—
Total	99.68	99.99	99.63	99.60	100.79	99.10	100.15	99.65[b]	99.83	100.10[f]	100.6

[a] Reported as MnO_2.
[b] Includes 0.06 percent S.
[c] Contains ZrO_2 and V_2O_5.
[d] Contains MnO_2.
[e] Total iron.
[f] Sum given in original as 99.90.

Source: After Pettijohn (1963), Table 8, with modifications.

A. Jotnian (Precambrian), Köyliö, Muurunmäki, Finland. H. B. Wiik, analyst (Simonen and Kouvo, 1955, p. 63). 44 percent normative feldspar.
B. Torridonian (Precambrian), Kinlock, Skye. M. H. Kerr, analyst (W. Q. Kennedy, 1951, p. 258). 53 percent normative feldspar.
C. Sparagmite (Lower Cambrian), Engerdalen, Norway (Barth, 1938, p. 58). 33.5 percent normative feldspar.
D. Lower Old Red Sandstone (Devonian), Foyers, Loch Ness, Scotland (Mackie, 1905, p. 58). 52 percent normative feldspar.
E. Calcareous arkose, Old Red Sandstone (Devonian), Red Crags, Fochabers-on-Spey, Scotland (Mackie, 1905, p. 58). 16 percent normative feldspar and 28 percent normative calcite.
F. Portland Stone, Newark Series (Triassic), Portland, Connecticut (Merrill, 1891, p. 420). 74 percent normative feldspar.
G. Subarkose, Rosebrae Sandstone (Devonian), Rosebrae, Elgin, Scotland (Mackie, 1905, p. 59). About 12 percent normative feldspar.
H. Calcareous subarkose (Cambrian or Ordovician), Bastard Township, Ontario, Canada (Keith, 1940, p 21). About 12 percent feldspar and 7 percent calcite.
I. Whitehorse subarkose (Permian), Kansas (Swineford, 1955, Kansas State Geol. Surv. Bull. 111, p. 122).
J. Molasse Arkose (Oligocene [Zugertypus]), Unterägeri, Mt. Zug, Switzerland. F. de Quervain, analyst (Niggli et al., 1930, Beitr. Geol. Schweiz. Geotech. Ser., no. 114, p. 262).
K. Arkose (Oligocene), Auvergne, France (Huckenholtz, 1963, p. 917). 19 percent feldspar.

kose, graywackes also tend to be richer in MgO and total iron, especially FeO, reflecting the chloritic character of their matrix.

Varieties and manner of occurrence Arkose occurs either as a thin blanketlike residuum at the base of a sedimentary series that overlies a granitic terrane, or as a very thick wedge-shaped deposit interbedded with much coarse granite-bearing conglomerate and a lesser quantity of red siltstone and shale.

Basal arkose, because of its thin and discontinuous character, is seldom of large volume. It may pass rapidly upward into more normal feldspar-poor sands. A well-documented example is the lowest part of the Cambrian Lamotte Sandstone of the Ozark area in Missouri, where this formation rests on Precambrian granite (Ojakangas, 1963). Basal arkose is a slightly reworked feldspathic residuum. Encroachment of the sea on a land area of granitic rocks results in reworking of the arkosic mantle rock. Reworking and removal of the more completely decayed and finer portions leaves a feldspathic residue which, on consolidation, may be called an arkose or subarkose depending on the feldspar content. Such material is restricted to the base of the formation or to intercalated wedges of *granite wash* near the base or shed from buried hills of granite. In some cases the residuum is so little reworked and so little decomposed that on lithification the deposit looks very much like the granite itself. It is then termed a *recomposed* or *reconstituted* granite. Such rocks can readily be misidentified in the field, especially in small outcrops or in drill cuttings or cores. It may be difficult in the latter case to know whether the granitic "basement" has been reached or whether the drill bit has only penetrated a tongue of granite wash. Even in outcrop, especially in some Precambrian terranes where rocks are welded by metamorphism, controversy has arisen regarding whether the meta-arkose is in truth an arkosic sediment, a granite, or a granitized sediment. Such contacts between granite and overlying residual arkose may be gradational and, if truly sedimentary, have been described as a *gradational unconformity*. A well-known example is the contact between Archean arkoses and granite on Saganaga Lake on the Ontario-Minnesota boundary, described by Grant (Winchell et al., 1899, p. 322) and Clements (1903, p. 270).

The criteria for discriminating between true granite and its recomposed equivalent are numerous but commonly difficult to apply. True granite may exhibit a faint gneissic foliation which is lost on complete disintegration and the slightest reworking. True granite is also cut by aplites and other complementary dikes. Recomposed granite, on extended search of the outcrop, usually will contain a few granite fragments or pebbles and some faint bedding. Under the microscope, recomposed granite exhibits an unusual range of sizes of grain. This characteristic of recomposed granite is in contrast to the even-grained nature of granite or to the porphyritic textures of some intrusives. Arkose that has undergone little or no transportation—*residual arkose*—is essentially unsorted and usually has a clay-rich matrix, commonly stained red, which may constitute 20 percent or more of the rock (Fig. 7-10). The term *redstone* has been applied to these arkosic wackes (Krynine, 1950, p. 103; Hubert, 1960, p. 65). The recomposed rock has a greater percentage of quartz than is normal for granite. A slight rounding of the feldspars (not to be mistaken for resorbed phenocrysts) is also probable. Well cuttings offer fewer criteria, although slight rounding and abundance of quartz characteristic of sediments are probably the most useful.

FIG. 7-9. Fountain Arkose (Pennsylvanian), Colorado. Crossed nicols, ×20. A coarse reddish sandstone consisting of angular to poorly rounded quartz (clear) and feldspar (clouded and showing multiple twinning).

Arkose also occurs as deposits related to granite uplifts. The arkose related to these uplifted and denuded granite plutons forms a thick wedge-shaped deposit, usually very coarse-grained and commonly conglomeratic. Well-described examples of arkose of this type include the New Haven, Portland, and other arkoses of the Newark Series (Triassic) in Connecticut and other eastern states (Krynine, 1950); the arkosic beds of the Lyons and Fountain formations (Pennsylvanian) of the Front Range of Colorado (Hubert, 1960); the Old Red Sandstone of Scotland (Mackie, 1899); and parts of the Tertiary Molasse of southern Germany and Switzerland (Gasser, 1968; Table 10). See Fig. 7-9. The arkoses within some mobile belts are anomalous, in that the feldspar is sodic rather than potassic. This is true of some Archean arkoses (Walker and Pettijohn, 1971) and of the well-known Swauk Arkose (Paleocene) in Washington (Foster, 1960, p. 105). For a more extensive tabulation of arkosic deposits of all ages and types, refer to Barton's classic paper (1916).

Shield areas are predominantly granitic in character and are therefore potentially generators of a great volume of arkosic sand. Arkoses apparently related to such areas include the late Precambrian Jotnian sandstones of Finland (Simonen and Kuovo, 1955, p. 60) and of Sweden (Gorbatscher and Klint, 1961); the Sparagmites of Norway and Sweden (Hadding, 1929, p. 151); the early Cretaceous (pre-Aptian) sands in western Venezuela derived from the Guayana Shield; the Precambrian Kazan Formation of the Dubawnt Group in the Northwest Territories (Donaldson, 1967); and the lower part of the Huronian Lorrain Quartzites of Ontario (Hadley, 1968).

FIG. 7-10. Sugarloaf Arkose, Newark Series (Triassic), Mt. Tom, Massachusetts. Crossed nicols, ×22. A coarse red arkose ("redstone") consisting of a very poorly sorted mixture of angular quartz and feldspar together with a little mica set in a red, ferruginous clayey matrix. (From Pettijohn, Potter and Siever, 1972, Sand and sandstone, Fig. 6-3. By permission of Springer-Verlag.)

Origin and geologic significance Field relations and mineral composition clearly show the close relation between arkose and granitic provenance. Arkosic sands are therefore restricted to local basins or to areas deriving all their detritus from an uplifted and denuded granitic block or to an area adjacent to a granitic shield. Large drainage basins are petrographically diverse and do not yield arkosic sands. Although the sands of large rivers may be somewhat feldspathic, they are not arkoses but are instead lithic sands. Arkoses are therefore restricted in time and place in the geologic record.

But quartz-rich sands, the quartz arenites, are also of granitic provenance. Indeed, virtually all the quartz of sands is ultimately derived from quartz-bearing plutonic rocks —"granite" in a very broad and loose sense. Why then are some sands highly feldspathic and others not?

The significance of detrital feldspar has been the subject of considerable controversy. The presence of a large quantity of feldspar in some sandstones (arkoses) has led to the prevalent theory that certain special climatic conditions, which inhibited the decomposition of feldspar, were required to ensure its survival and accumulation in the sediment (Mackie, 1899). Accordingly, very arid climate (implying absence of water and hence arrested chemical decay) or very cold climate (and hence much retarded chemical action) was postulated. Sufficient evidence has now accumulated to make necessary a modification of the rigorous climate theory. Krynine (1935) observed arkose formation under humid tropical conditions, with average temperatures of 80°F (26°C) and annual rainfall of 120 inches (300 cm). Not only is feldspar seen to accumulate in sediments under such conditions, but critical study of ancient feldspathic deposits yields abundant evidence that many of these were not the products of rigorous climate. Reed (1928), for example, noted that Eocene sandstones of California, which contain nearly 50 percent feldspar, yield a flora which could only have lived under warm humid conditions. The Catahoula Sandstone of Texas, also of (?) Eocene age, contains a tropical coastal flora, although it contains nearly 50 percent of feldspar (Goldman, 1915). As noted by Barton (1916), arkosic sandstones produced under humid conditions contain a high proportion of weathered or partially weathered feldspars. The mixture of brilliantly fresh with somewhat clouded and altered feldspar in the same sediment may be explained as the product of torrential erosion of a deeply dissected highland area underlain by feldspar-bearing rocks under warm climatic conditions. A rigorous-climate arkose should contain little or no weathered feldspar.

If feldspar content is independent of climate, what then is its significance? Weathering of feldspar requires not only a suitable climate but also a proper length of time. *Intensity* of the decay process is climate controlled, but *duration* of time through which the processes of decomposition operate is determined by relief. Regions of high relief undergo rapid erosion, so feldspar escapes destruction and is contributed to the basin of deposition. If relief is low, the rate of erosion is slow, and if climate is favorable, the feldspar will be wholly decomposed. The presence or absence of feldspar, therefore, is the result of the balance struck between rate of decomposition and rate of erosion. Arkose is thus an index of both climate rigor and tectonic activity. Whether it is indicative of one rather than the other will have to be decided by criteria other than merely the presence or absence of feldspar.

LITHIC SANDSTONES AND SUBGRAYWACKES

Definitions Sandstones in which rock fragments exceed feldspar particles are called *lithic sandstones* (Pettijohn, 1954), lithic arenites (Williams, Turner, and Gilbert, 1954, p. 304), or litharenite (McBride, 1963) in the same way in which tuffs with an abundance of rock fragments are called lithic tuffs. The proportion of rock particles in a sandstone varies widely. Those containing 25 or more percent may properly be called lithic sandstones. Those with 10 to 25 percent form a transitional class which has been designated *sublitharenite* (McBride, 1963) or protoquartzite (a term suggested by Krynine; Payne et al., 1952).

Krynine has applied the term *graywacke* to both graywackes of the classical Harz Mountain type and to the lithic sandstones. In fact, as first defined by Krynine (1940), the term could only be applied to lithic sandstones; Krynine (1945) later redefined the term and recognized two types, high-rank (with feldspars) and low-rank (feldspar-poor) graywackes. Low-rank graywacke thus defined is the same as lithic sandstone. This usage was also followed for a time by Folk (1954, Fig. 2) but has now been abandoned and the term *litharenite* applied to this class of rocks (Folk, Andrews, and Lewis, 1970, Fig. 8).

The term *subgraywacke* was first used for rocks transitional between quartz arenites and graywackes (Pettijohn, 1949, p. 255). As first defined, it had less than 10 percent feldspar and over 20 percent matrix. It was redefined (Pettijohn, 1954, Fig. 1) as a sandstone with less than 15 percent matrix but with 25 percent labile grains of which rock particles exceed feldspar. Thus defined it is essentially a lithic arenite. Subgraywacke (and lithic arenite generally) superficially resemble graywacke, especially in color and rock particle content.

General description and varieties Lithic arenites are most commonly light gray "salt-and-pepper" sandstones with a substantial content of rock particles, especially of sedimentary and low-rank metamorphic origin (Fig. 7-11). Quartz is generally subangular to rounded; mica flakes are especially abundant; feldspar is sparse or absent. The sand as a whole is moderately well sorted and bound by a chemically precipitated cement, in most cases quartz or calcite. Detrital matrix is scarce or absent, although a pseudomatrix generated by squashed shale particles or authigenically precipitated clay may be present (Fig. 7-12).

Rock particles of lithic arenites are not only abundant but also are diverse. A given lithic arenite may contain a dozen or more species. In some, particles of *volcanic* or flow rocks are abundant; in others, low-grade *metamorphic* and *sedimentary* rock particles are common. Those rich in volcanic detritus are volcanic or volcaniclastic arenites and are not to be confused with pyroclastic arenites which are direct and immediate products of volcanic eruptions. In general, particles of acid volcanic rocks tend to be more prominent; those of basic volcanic flows tend to alter to matrix. Most lithic arenites have an abundance of low-grade, pelitic metamorphic rocks such as slate, phyllite, and sericitic schists. These have been called *schist arenites* (Krynine,

1937, p. 427). The term *phyllarenite* (Folk, 1968, p. 131) has also been used, in part perhaps because of the abundance of mica, both in contained rock particles and as detrital flakes present in these sands. Many lithic sands are also rich in rock particles of sedimentary origin. Two varieties deserve special mention. One are *chert arenites* (Fig. 7-11). Chert forms, for example, 20 to 90 percent of the Cutbank Sandstone (Cretaceous) of Montana (Sloss and Feray, 1948, p. 6). Extremely chert-rich Jurassic sands in Montana also have been described (Suttner, 1969, Fig. 11). Care should be taken to discriminate between detrital chert and particles of devitrified rhyolitic rocks—not always an easy distinction. Chert-rich sands probably signify a very local origin either from a cherty limestone terrane in a humid climate or from bedded chert formations.

Detrital carbonates are common in some lithic arenites. Sandstones of this type have been called *calclithites* (Folk, 1968, p. 141) to distinguish them from *calcarenite*, a term customarily applied to carbonate sands of which the constituent grains were skeletal particles, oolites, and other materials formed *within* the basin of deposition (Fig. 7-13). Detrital carbonate sands of terrigenous origin are common but are relatively rare in the geologic record. Abundant detrital dolomite is found in Cretaceous sandstones in the western interior of the United States (Sabins, 1962, p. 1185). Both detrital dolomite and limestone grains are abundant in some Molasse sandstones north of the Alps (Füchtbauer, 1967b, Fig. 3). See Fig. 7-14. Because reduction of most limestone terranes is normally by solution and the only clastic residues are clay and chert, calclithites must require rapid erosion and high relief, whereas chert arenites imply low relief and removal of limestones by solution. Calclithites are the sand-sized analogue of limestone gravels and conglomerates. In addition to chert and micritic limestone or dolomite, lithic arenites contain finer-grained pelitic rock particles such as shale, mudstone, and siltstone. Because, these are relatively weak grains, their presence suggests short transport or even intraformational origin. Moreover, they are subject to deformation by overburden pressure, and the shale grains in particular become squashed and squeezed into the interstices so that they resemble a matrix-filled pore (Allen, 1962, p. 669). They may not then be recognizable as rock particles. True

FIG. 7-11. Viking or Cardium "B" Sandstone (Cretaceous), Alberta, Canada. Crossed nicols, ×53. A lithic sandstone consisting of a moderately well-sorted mixture of subangular to rounded quartz and chert, the latter forming nearly one-third of the framework of sand. (From Pettijohn, Potter, and Siever, 1972, Sand and sandstone, Fig. 6-9. By permission of Springer-Verlag.)

FIG. 7-12. Pottsville Formation (Pennsylvanian), Pottsville, Pennsylvania. Crossed nicols, ×20. A coarse, ill-sorted lithic arenite composed of quartz and rock particles, the latter in part weak shale and siltstones forming a pseudomatrix.

matrix would be present in all pores, whereas the pseudomatrix described would occupy some pores; others would be filled with precipitated mineral cement. Moreover, the rock particles are not all alike—hence the presumed pore fillings would vary from pore to pore.

As might be expected from the foregoing, lithic arenites are a diverse group of sands and show wide variations in their compositional makeup. This is reflected both in modal analyses (Table 7-6) and in their bulk chemical composition (Table 7-7). Modal analyses fail to tell the whole story unless the "rock particle" category is broken down into species present. Recognition of rock species in very small particles is somewhat difficult and requires considerable experience. The characteristic fabric of rocks is best preserved in larger grains, but that of finer-grained rocks is still visible, even in the fine size grade (Boggs, 1968). Quartz is, of course, the main constituent of most lithic arenites. It may be of volcanic origin in the volcanic sands (Webb and Potter, 1969). That of the phyllarenites is likely to be sedimentary quartz, derived from sandy sediments associated with sedimentary rock particles. Hence it is likely to show roundness inherited from earlier cycles and perhaps even abraded rims of secondary quartz. The quartz of true phyllarenites, those with low-rank metamorphic rock particles, is apt to be more markedly undulatory and polycrystalline than the quartz of volcanic or normal sedimentary arenites. These arenites also contain a high proportion of detrital mica (whose flakes lie parallel to the bedding), are in many cases concentrated on certain bedding planes to which they impart a sheen, and are likely to show bending or deformation caused by some compaction of the sand.

Lithic arenites are generally cemented with calcite—especially those of Mesozoic or Cenozoic in age—or with quartz. Little or no matrix is present though some display a pseudomatrix derived from squashed shale particles. Some also contain precipitated clay minerals or even zeolites.

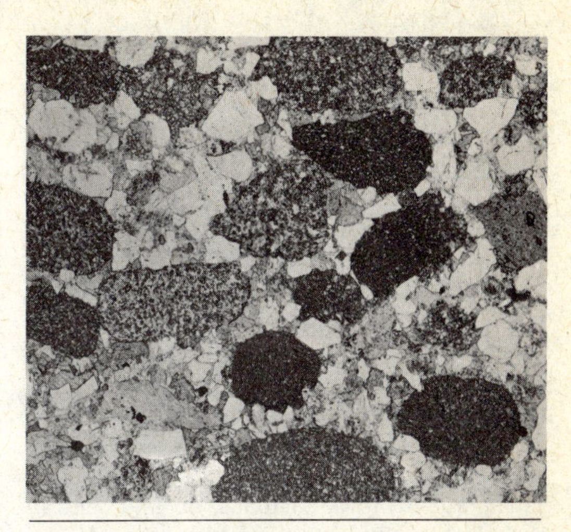

FIG. 7-13. Calclithite, Tochatwi Formation, Christie Bay Group (Lower Proterozoic), Great Slave Lake, Northwest Territories, Canada. Ordinary light, ×20. Larger polycrystalline carbonate rock particles and smaller, less rounded quartz grains.

FIG. 7-14. Molasse Sandstone (Lower Miocene) [Aquitanian] 20 mi (32 km) east-northeast of Bregenz, Bodensee, Germany. Crossed nicols, ×20. Note abundance of carbonate rock particles.

The chemical compositions reflect the varied composition of both framework fractions and infilling cements. The analyses can only correctly be understood if one knows also the mineral composition. High CO_2 and CaO may mean a calcite cement, it may mean shell fragments, or it may indicate detrital limestone particles. High SiO_2 may reflect abundance of detrital quartz, it may be attributable to detrital chert, or to a silica cement, or to all three. Most lithic arenites, unlike graywackes, are low in Na_2O and MgO and high in K_2O. Exceptions occur; detrital dolomite will greatly increase MgO.

Occurrence and significance Lithic sandstones are perhaps the most abundant of all sandstones (see p. 248). Many, if not most, Paleozoic sandstones of the central Appalachians are lithic sandstones. Included here are the Ordovician Oswego (Krynine and Tuttle, 1941), Fig. 7-15; the Devonian Third Bradford Sand described in Krynine's classic paper (1940); the Mississippian Pocono (Pelletier, 1958) and Mauch Chunk (Hoque, 1968; Meckel, 1967) formations; and the Pennsylvanian Pottsville Formation (Meckel, 1967). These are quartz-rich, feld-

FIG. 7-15. Oswego Sandstone (Ordovician), Bald Eagle Mountain, near State College, Pennsylvania. Crossed nicols, ×15. The formation consists of 50 percent or more quartz, and 30 to 40 percent rock particles. Original outlines in quartz are difficult to see; the secondary quartz results in interlocking sutured grain boundaries. Rock particles are mainly siltstone, fine-grained sandstone, fine-grained quartzite, and phyllitic low-grade metamorphic grains. Feldspar is very rare—only 1 or 2 percent.

TABLE 7-6. Mineral composition of lithic sandstones and protoquartzites (sublitharenites)

Constituent	A	B	C	D	E	F	G	H
Quartz	50	60–65	78	65.4 (quartz + feldspar)	32.0	71	30.9	27
Feldspar	3–5	1.5	3		2.2	8	10.0	2
Rock fragments	40	30	15	10.6	43.0[b]	22[c]	33.0[d]	46[e]
Mica	—	2–3	—	—	0.2	trace	0.5	—
"Clay" or matrix	10	5–10	4	6.8	6.9	2	5.5	5
Silica cement	—[a]	—	—	11.9	trace	—	—	—
Calcite cement	—	2–18	—	8.5	13.0	—	19.2	20

[a] 5 to 10 percent, author's observation.
[b] Includes 28.0 percent chert.
[c] Includes 5 percent chert.
[d] Includes 15.0 percent chert.
[e] Includes 3 percent chert, 12 percent limestone, and 17 percent dolomite.

Source: From Pettijohn (1963), Table 3, with additions.

A. Oswego Sandstone (Ordovician), Pennsylvania (Krynine and Tuttle, 1941).
B. Bradford Sand (Devonian), Pennsylvania (Krynine, 1940, Table 3).
C. Deese Formation (Pennsylvanian), Oklahoma (Jacobsen, 1959, Table 4).
D. Salt Wash Member of Morrison Formation (Jurassic), Colorado Plateau. Mean of 22 thin sections (Griffiths, 1956, p. 25).
E. "Calcareous graywacke" (Cretaceous), Torok, Alaska. Average of three samples (Krynine, *in* Payne et al., 1952).
F. Basal Claiborne Sand (Eocene), Texas (Todd and Folk, 1957).
G. "Frio" Sandstone (Oligocene), Seeligson Field, Jim Wells and Kleberg counties, Texas. Average of 22 samples (Nanz, 1954, p. 112).
H. Molassesandstein (Tertiary), Germany (Füchtbauer, 1964, p. 256).

spar-poor, and quartz cemented. Rock particles from low-rank metamorphic and sedimentary sources are abundant.

Many Jurassic and Cretaceous sandstones in the Rocky Mountain area are lithic arenites. Included here are chert-rich arenites in Montana (Sloss and Feray, 1948; Suttner, 1969) and Alberta (Lerbekmo, 1963). The Chico Formation (Cretaceous) of California is also a lithic arenite (Williams, Turner, and Gilbert, 1954, p. 306).

Most of the Tertiary sandstones of the Molasse Basin of southern Germany and Switzerland are lithic arenites (Fig. 7-14), many containing detrital grains of micritic carbonate (Füchtbauer, 1964, Table 9B;

TABLE 7-7. Representative chemical analyses of lithic sandstones (subgraywackes) and protoquartzites (sublitharenites)

Constituent	A	B	C	D	E	F	G	H
SiO_2	84.01	65.0	56.80	51.52	92.91	47.75	40.35	74.45
TiO_2	.05	—	.10	.32	—	.20	.30	.50
Al_2O_3	2.57	9.57	8.48	5.77	3.78	6.41	7.43	10.83
Fe_2O_3	.17	1.59	1.67	2.43	trace	2.39	3.27	4.62
FeO	.26	1.08	—	—	.91	—	—	—
MnO	.04	—	—	.14	—	—	—	—
MgO	.67	.4	1.24	.95	trace	4.48	10.28	1.30
CaO	5.41	10.1	15.25	16.96	.31	18.75	12.00	.35
Na_2O	.17	2.14	1.31	1.32	.34	1.20	.54	1.07
K_2O	.86	1.43	1.46	1.90	.61	1.02	.93	1.51
H_2O+	.54	.82	.50	2.25	1.19 (H₂O+ and H₂O− combined)	1.32	6.75	4.95
H_2O-	.19	.23	—	2.54		—	—	—
P_2O_5	.04	—	trace	.10	—	.10	—	trace
CO_2	4.65	6.9	12.95	13.34	—	17.78	17.80	trace
SO_3	—	.04	—	.52	—	—	—	—
S	.02	.16	—	—	—	—	—	—
Total	99.73[a]	99.54[b]	99.76	100.06	100.05	101.40[c]	99.65	99.58

[a] Includes Cl, 0.02; F, 0.01; BaO, 0.05.
[b] Includes C, 0.06; Cu, 0.002; V, 0.017; Zn < 0.03; Cr, 0.003.
[c] Sum given as 99.40 in original.

Source: From Pettijohn (1963), Table 4.

A. Protoquartzite, Salt Wash Member, Morrison Formation (Jurassic). Composite of 96 unmineralized samples, Colorado Plateau, V. C. Smith, analyst (unpublished U.S. Geol. Surv. analysis). See Table 7-6D.
B. Calcareous subgraywacke (lithic arenite), "Frio," (Oligocene), Seeligson Field, Jim Wells and Kleberg counties, Texas. Composite of 10 samples (Nanz, 1954, p. 114). See Table 7-6G.
C. Calcareous subgraywacke, Molasse, (Aquitanien), Lausanne, Switzerland (Cayeux, 1929, p. 161).
D. Calcareous subgraywacke, Molasse, Gränichen, Burghalde, Mt. Aargau, Switzerland. J. Jakob, analyst (Niggli et al., 1930, Beitr. Geol. Schweiz. Geotech. Ser., no 114, p. 262).
E. Protoquartzite, Berea Sandstone (Mississippian), Berea, Ohio. L. G. Eakins, analyst (Bull. U.S. Geol. Surv. 60, p. 159).
F. (?) Calcareous subgraywacke, Molasse (Burdigalien), Voreppe, Isère, France (Cayeux, 1929, p. 163).
G. Coal measure sandstone (? calcareous subgraywacke) (Carboniferous), Westphalian coal basin, France-Belgium (Hornu and Wasmes) (Cayeux, 1929, p. 227).
H. Coal measure sandstone (? subgraywacke) (Carboniferous), Westphalian coal basin, France-Belgium (Hornu and Wasmes) (Cayeux, 1929, p. 227).

Gasser, 1968). Not all lithic arenites come from orogenic belts. Many of the Tertiary sands of the Gulf Coast are lithic arenites. Some well-described examples include the Oligocene "Frio" (Nanz, 1954), the Miocene Oakville (Folk, 1968a, p. 141), and the Wilcox Formation (Williams, Turner, and Gilbert, 1954, p. 305).

Protoquartzite or sublitharenites are equally abundant. Examples include parts of the Tuscarora Quartzite (Silurian) of the central Appalachians (Yeakel, 1962), and the Anvil Rock Sandstone (Pennsylvanian) of the Illinois Basin (Hopkins, 1958).

The sands of most present-day large rivers are probably lithic sands. An average of 187 samples of Ohio River sands shows 62 percent quartz, 31 percent rock fragments, and 6 percent feldspar (Friberg, 1970). This is certainly a typical lithic arenite.

Ancient lithic arenites and protoquartzites abound and are perhaps the most common type of sandstone. On the basis of a sample of 121 sandstones of diverse age and distribution, Pettijohn (1963) estimated that lithic arenites formed 26 percent of all sands, whereas arkose constituted only 15 percent. Lithic arenites are most common in Cretaceous and Tertiary sequences. Of 718 sandstones of Cretaceous, Paleocene, and Eocene age in western Venezuela, 400 or about 56 percent were lithic or sublithic sandstones as here defined (van Andel, 1958). Only 79 or 11 percent were arkose or subarkose. Lithic arenites play somewhat the same role that graywacke does in the older flysch sequences, but they occur outside the geosyncline as well as within it. The typical molasse sandstone is probably a lithic arenite.

Lithic arenites are immature sands in that most of their rock particles are either mechanically weak or are chemically unstable. Just how large volumes of sand are generated from fine-grained rocks is not clearly understood. Disintegration without decomposition of coarse-grained rocks yields sand, but one might expect the finer-grained rocks to disintegrate into silt-sized particles. Mechanically unstable rock components appear to disintegrate readily in areas of intense abrasion, as in the surf zone. Chemically unstable rock particles break down during diagenesis to produce matrix materials. Perhaps lithic arenites are antecedents of the graywackes of older geologic sequences. To produce an arkose requires a restricted provenance, likely to be found only in a small drainage basin. Lithic arenites generally reflect a wider provenance, that of a large drainage basin which is most likely to have a diverse bedrock lithology. Lithic arenites, therefore, would be the deposits of large rivers, either alluvial or deltaic, and most conspicuous in but not restricted to miogeosynclines.

GRAYWACKES AND RELATED ROCKS

Because of the possible link between lithic sandstones and graywackes, it is appropriate to next consider the sandstones to which the term *graywacke* has been applied. The term itself has engendered a good deal of controversy not only with respect to the origin of the rocks so named but also with the definition of the term itself. A large number of papers deal with these questions; a good summary of the nomenclatural problem was given by Dott (1964) and has more recently been reviewed by Okada (1971, p. 514) and Pettijohn, Potter, and Siever (1972, p. 197). The term is old and was apparently first applied to the sandstones of the Upper Devonian–Lower Carboniferous strata of the Harz Mountains in Germany. These rocks have been carefully restudied in recent years (Helmbold, 1952; Mattiat, 1960). Their distinctive features are a dark gray appearance, marked induration, abundance of both feldspar and rock particles, and a lack of a normal pore-filling cement in place of which is a matrix consisting of a fine-grained intergrowth of sericite and chlorite together with some silt-sized quartz and feldspar. These rocks, unlike arkoses, are notably rich in FeO, MgO, and Na_2O. Although there has been some difficulty in precisely defining graywackes, there exists a fairly homogeneous group of rocks, which resemble the classic graywackes of the Harz in all essential respects and which by tradition have been designated graywacke. The dark, fine-grained matrix is an essential aspect of these rocks.

As noted on page 219, the term has been defined more loosely to include what are now generally designated as lithic arenites. Whereas the latter superficially resemble graywackes, they differ in lacking the matrix and in being porous or in having a precipitated mineral cement. This overextension of the term *graywacke* has now been generally abandoned.

The matrix is the heart of the graywacke problem, both of definition and of interpretation. How much matrix and what is its upper size limit? As Okada (1971) notes, the percent of matrix which distinguishes wackes (of which graywacke is the most important class) from the "clean" sandstones has been variously set from 5 to 25 percent. Dott (1964) and Gilbert (Williams, Turner, and Gilbert, 1954, p. 290) accepted 10 percent. Most workers have chosen 15 percent, the figure adopted here. The upper size limit has also varied; Okada (1971) notes that most authors adopt 20 microns. Following Pettijohn, Potter, and Siever (1972, p. 159), the figure here used is 30 microns.

FIG. 7-16. Kulm Graywacke (Carboniferous), Harz Mountains, Germany. Crossed nicols, ×22. The type or "classical" graywacke consisting mainly of angular, ill-sorted quartz, feldspar, and rock fragments in fine-grained chloritic matrix.

Fabric and composition In general, graywackes are dark gray to black, tough rocks, commonly without internal stratification or parting, and they are generally, but not everywhere, graded. They may be mistaken in a small outcrop or a hand specimen for a basic igneous rock. Under the microscope graywacke has the appearance of a microbreccia made of sharp, angular, sliverlike quartz, together with angular feldspar and rock particles, embedded in a paste which in some examples equals or exceeds the volume of the larger detrital grains (Fig. 7-16). This paste or matrix is a microcrystalline aggregate of quartz, feldspar, chlorite, and sericite and is, in places, replaced by patchy carbonate. Under crossed nicols, the distinction between matrix and fine-grained rock particles almost disappears; they are indeed difficult to distinguish. It has been contended that, because of this difficulty, the volume of matrix in graywackes has been greatly overestimated (de Booy, 1966). In some graywackes there may be an alignment of the folia of sericite and chlorite, an alignment which is an incipient schistosity. But in most the matrix fabric is chaotic.

The composition of graywackes is somewhat variable. Quartz is generally the most abundant constituent. In most graywackes it forms less than one-half of the sand fraction, in many no more than one-fourth or even in some one-tenth (Table 7-8). Quartz is almost everywhere very angular, generally with strongly undulatory extinction. Some exceptional graywackes have volcanic quartz, and, with an increasing proportion of volcanic rock particles, strongly zoned feldspars, and broken crystal euhedra, these graywackes (Fig. 7-17) pass insensibly into water-laid tuffs and tuffaceous sandstones. Feldspar is generally present in graywackes; in some it equals or exceeds quartz in volume (Fig. 7-18). Generally it is plagioclase, usually very sodic, near albite, which explains the high Na_2O value in the bulk chemical analyses. Perhaps the feldspar was once more calcic and the exsolved calcium now appears in the rock as replacement patches of calcite. K-feldspar is wholly absent in many graywackes, an observation difficult to explain. It may be attributable to provenance,

FIG. 7-17. Kuskokwim Graywacke (?Cretaceous), Kuskokwim region, Alaska. Crossed nicols, ×22. A graywacke with a low quartz content, an abundance of feldspar much of which is subhedral, and many aphanitic flow rocks. Very poorly sorted. This is a graywacke with considerable volcanic input.

FIG. 7-18. Franciscan Graywacke (?Upper Jurassic) near San Francisco, California. Crossed nicols, ×22. A typical feldspathic graywacke, chiefly quartz, feldspar, and rock fragments in a fine-grained chloritic and sericitic matrix. Note poor rounding and lack of assortment of the material.

TABLE 7-8. Representative modal analyses of graywackes

Constituent	A	B	C	D	E	F	G	H	I	J
Quartz	4	24	56	33	9	trace	26	33	27	30
Feldspar	10	32	37	15	43	30	5	21	19	12
Rock fragments	50	19	7	3	10	13	26	7	30	16
"Matrix"	32	[a]	[b]	45	25	45	43	33	21	39
Mica and chlorite	—	16	—	—	4	—	—	6	—	3
Miscellaneous	2	8	—	—	4[c]	10[c]	—	—	3	—

[a] Not separately reported; 38 percent of rock is "clay and silt."
[b] Not separately reported.
[c] Hornblende and pyroxene.

A. Lithic graywacke (Devonian), Australia. Average of 5 analyses (Crook, 1955, p. 100).
B. Tanner feldspathic graywacke (Devonian-Mississippian), Harz Mountains, Germany (Helmbold, 1952, p. 256).
C. Franciscan (? Jurassic) feldspathic graywacke, California. Average of 17 analyses (Taliaferro, 1943, p. 135).
D. Feldspathic graywacke (Precambrian), Ontario. Average of 3 analyses (Pettijohn, 1943, p. 946).
E. Purari graywacke (Cretaceous), Papua. Average of 4 analyses (Edwards, 1950b, p. 164).
F. Tuffaceous Aure graywacke (Miocene), Papua. Average of 2 analyses (Edwards, 1950a, p. 129).
G. Lithic Martinsburg graywacke (Ordovician), Pennsylvania (McBride, 1962, p. 62).
H. Lower Mesozoic graywacke, Porirua district, New Zealand (Webby, B. D., 1959, New Zealand Jour. Geol. Geophys., v. 2, p. 472).
I. Harz Kulm graywacke (Mattiat, 1960).
J. Burwash Formation (Archean), Yellowknife, Northwest Territories, Canada (Henderson, 1972, Table 1).

that is, to derivation from a quartz dioritic or sodic granite source, or it may be attributable to some kind of diagenetic or low-grade metamorphic action (Gluskoter, 1964, p. 341).

Detrital micas, both muscovite and biotite (and chloritized biotite), are common but not abundant constituents.

Rock particles are both abundant and diverse. Mattiat (1960), for example, identified 19 different types in the Kulm Graywacke of the Harz Mountains. These formed, on the average, 26 percent of the sand fraction. Included in the rock particle category were acid to basic effusive volcanics, several types of low-grade metamorphic rocks (including quartzite, mica, sericite, and chlorite schist), and a sprinkling of sedimentary rock particles (including several types of sandstones and slates).

Graywacke can be grouped into two major classes, *feldspathic graywackes* and *lithic graywackes*, depending on the relative abundance of feldspar and rock fragments, respectively. As noted above, graywackes pass into *volcanic graywackes* marked by volcanic quartz, zoned feldspar, broken phenocrysts, and increasing abundance of volcanic rock particles. Many volcanic wackes are very poor in quartz (Fig. 7-17). These rocks grade into water-laid tuffs. The term *quartz wacke* has been used for those rocks exceptionally rich in quartz. These relatively rare rocks have been attributed to a sedimentary provenance.

With few exceptions graywackes have a fairly well-defined bulk chemical composition (see Table 7-9). As can be seen from the table, the graywackes are rich in Al_2O_3, FeO, MgO, and Na_2O. The high Na_2O reflects the albitic nature of the feldspar; the high MgO and high FeO of these rocks are a result of the iron-rich chlorite in the matrix. Graywackes differ from arkose in the dominance of Na_2O over K_2O, of MgO over CaO, and of FeO over Fe_2O_3. Graywackes have a bulk composition not greatly different from that of granodiorite. Sands normally differ from their source rocks by enrichment in SiO_2 and depletion in Al_2O_3, iron, and the alkalies and alkaline earths. The chemical similarity of graywackes and granodiorites suggests, therefore, a greatly suppressed weathering and sorting. The present-day Columbia River sands (Whetten, 1966) have a bulk chemical composition very similar to that of a graywacke. Their mineral composition indicates a mixed granitic and volcanic provenance.

The matrix problem As Cummins (1962) has noted, the matrix is "the essence of the graywacke problem." The matrix has been variously explained. Inasmuch as it has the composition of a slate, it was thought to be the result of recrystallization under low-grade metamorphic conditions of original detrital mud. If so, the problem becomes one of explaining simultaneous sedimentation of sand and mud. Normal aqueous currents sort so that sand and mud are separately accumulated. Even sands of dominantly mud-carrying rivers are relatively "clean" and mud-free. Woodland (1938), one of the first to consider the problem, thought that electrolytes of sea water would flocculate the clays so that the sand and mud would come down together. But few, if any, present-day nearshore shallow-marine sands bear witness to such action.

The matrix of graywackes has also been explained by their manner of transport and deposition. Kuenen and Migliorini (1950, p. 123), for example, note that ". . . all the exceptional and puzzling features of a typical series of graded graywackes can be readily accounted for by the activity of turbidity currents of high density." Such currents of suspended fine mud are capable of transporting sands down submarine slopes into deep waters where, upon ponding, would deposit their load of both sand and fine mud, producing a graded graywacke with a very muddy matrix. Deposition of such sands in deep marine basins would explain our failure to find the modern counterpart of the ancient graywackes. But such deposits do not appear to be present in the deep sea, and the sands that do occur in this region, which have been attributed to turbidity current action, are generally matrix-free (Hollister and Heezen, 1964). Theoretical considerations (Kuenen, 1966) also suggest that no more than 10 percent of the deposit is apt to be mud matrix.

Other suggested explanations for the matrix include post-depositional infiltration of muds by movement of interstitial waters introduced from overlying or underlying beds (Emery, 1964; Klein, 1963, p. 571).

The matrix is, in any case, clearly a recrystallized product, and that some of it involves reactions with detrital sand was noted long ago. The original water-worn boundaries of the quartz have wholly disappeared, the existing boundary being a kind of *chevaux-de-frise* of green chlorite crystals projecting into clear quartz, an observation early made by Greenly (1897, p. 256) and confirmed by others (Krynine, 1940, p. 22). Irving and Van Hise (1892, p. 334) attributed the matrix to a "... micaceous alteration of the feldspar." More recently Cummins (1962) attributed all, or almost all, the matrix to diagenesis. He noted that most Tertiary and Recent turbidites do not have an abundant matrix as is found in comparable Paleozoic and older turbidites. The latter have presumably undergone deep burial and incipient metamorphism which led to alteration of the labile constituents, mainly unstable rock fragments and feldspar

TABLE 7-9. Representative chemical analyses of graywackes

Constituent	A	B	C	D	E	F	G	H	I	J
SiO_2	60.51	66.24	76.84	69.11	68.85	74.43	70.6	71.1	68.84	65.05
TiO_2	.87	0.64	—	.60	.74	.83	0.7	.5	.25	.46
Al_2O_3	15.36	15.28	11.76	11.38	12.05	11.32	13.5	13.9	14.54	13.89
Fe_2O_3	.76	0.70	.55	1.41	2.72	.81	2.2	trace	.62	.74
FeO	7.63	4.53	2.88	4.64	2.03	3.88	1.6	2.7	2.47	2.60
MnO	.16	0.06	trace	.17	.05	.04	0.1	.05	nil	.11
MgO	3.39	2.74	1.39	2.06	2.96	1.30	1.6	1.3	1.94	1.22
CaO	2.14	1.70	.70	1.15	.50	1.17	1.3	1.8	2.23	5.62
Na_2O	2.50	3.12	2.57	3.20	4.87	1.63	2.9	3.7	3.88	3.13
K_2O	1.69	1.91	1.62	1.76	1.81	1.74	1.6	2.3	2.68	1.41
H_2O+	3.38	2.49	1.87[a]	4.13	2.30	2.15	2.8	1.9	1.60	2.30
H_2O-	.15	0.08	—	.05	.77	.20	0.3	.26	.35	.28
P_2O_5	.27	0.12	—	.03	.06	.18	0.2	.10	.15	.08
ZrO_2	—	—	—	—	—	—	—	—	.05	—
CO_2	1.01	0.38	—	—	.08	.48	0.6	.12	.14	2.83
SO_3	—	—	—	—	—	—	—	—	.15	—
S	.42	—	—	—	.08	.12	—	trace	—	.05
BaO	—	—	—	—	trace	—	—	—	.04	—
C	—	—	—	—	.07	.17	—	.09	—	—
Total	100.24	99.99	100.18	99.69	99.94	100.45	100.0	99.8	99.93	99.77

[a] Loss on ignition.

A. Archean, Manitou Lake, Ontario. B. Brunn, analyst (Pettijohn, 1949, p. 250).
B. Average of three Burwash Formation graywackes (Archean), Yellowknife Bay, Northwest Territories, Canada. L. Seymour, analyst (Henderson, 1972, p. 890).
C. Tyler Slate (Animikean), Hurley, Wisconsin. H. N. Stokes, analyst (Diller, 1898, p. 87).
D. (?) Ordovician, Rensselaer, near Spencertown, New York. H. B. Wiik, analyst (Balk, 1953, p. 824).
E. Tanner Graywacke (Upper Devonian–Lower Carboniferous), Scharzfeld, Germany. R. Helmbold, analyst (Helmbold, 1952, p. 256).
F. Carboniferous graywacke from Stanley Shale, near Mena, Arkansas. B. Brunn, analyst (Pettijohn, 1957, p. 319).
G. Average of six Kulm graywackes from Oberharz, Germany (Mattiat, 1960, Table 14).
H. Lower Mesozoic composite sample prepared by using equal parts of 20 graywackes exposed along shoreline between Palmer Head and Hue-Te-Taka, Wellington, New Zealand. J. A. Richie, analyst (Reed, 1957, p. 16).
I. Franciscan (? Jurassic) Quarry, Oakland Paving Co., Piedmont, California. J. W. Howson, analyst (Davis, 1918, p. 22).
J. Eocene, near Solduc, Olympic Mountains, Washington. B. Brunn, analyst (Pettijohn, 1949, p. 250).

(Fig. 7-19). Modern sands, chemically similar graywacke (Hawkins and Whetten, 1969). water pressure of 1 kilobar, led to matrix formation and the production of a synthetic graywacke (Hawkins and Whetton, 1969). These experiments, coupled with Brenchley's observation (1969) that the matrix of some Ordovician volcanic graywackes, normally forming 40 to 60 percent of the whole rock, was absent where a calcite cement was present, indicate a diagenetic origin of the matrix. The early calcite cement apparently inhibited formation of a matrix.

We may conclude that matrix can be produced in more than one way. As Dickinson (1970) points out, we may have *protomatrix, orthomatrix, epimatrix,* and *pseudomatrix;* the first is trapped detrital clay, the second is recrystallized material, the third is the product of diagenetic alteration of sand-sized grains, and the fourth is a result of deformation and squashing of soft pelitic fragments. Although all are possible, the matrix of most older graywackes seems to be epimatrix, the result of deep burial and low-grade metamorphism or high-grade diagenesis—a process termed *graywackization* by Kuenen (1966, p. 296).

The soda problem The high Na_2O content of graywackes has been explained in various ways (Engel and Engel, 1953, pp. 1086–1091).

The Na_2O seems to be primarily in the feldspar, which is near albite in composition. The Tanner Graywacke (Helmbold, 1952), for example, contains 3.5 percent of Na_2O which, if contained in the albite, would imply 30 percent of this mineral, an estimate consistent with the observed 30 to 40 percent feldspar, 85 to 90 percent of which has a composition of $An_{3\text{-}10}$.

The feldspar is a detrital component. Is its albitic nature original or is it diagenetic? The close association of eugeosynclinal graywackes with soda-rich greenstones (spilites) suggests that the problem is closely related to the origin of the spilites (Turner and Verhoogen, 1960, p. 270). There is some evidence of post-depositional albitization. Graywackes contain irregular patches of replacive calcite together with vugs and veins of the same mineral. Perhaps albitization of the feldspar provided the CaO. Moreover, if the albite were an original constituent why are not the associated and interbedded slates also soda-rich? The pelitic strata interbedded with graywackes in some Mesozoic graywackes of New Zealand have a normal Na_2O–K_2O ratio in that K_2O much exceeds Na_2O, whereas the reverse prevails in the cogenetic graywackes (Reed, 1957, p. 28). On the other hand, the plutonic rocks

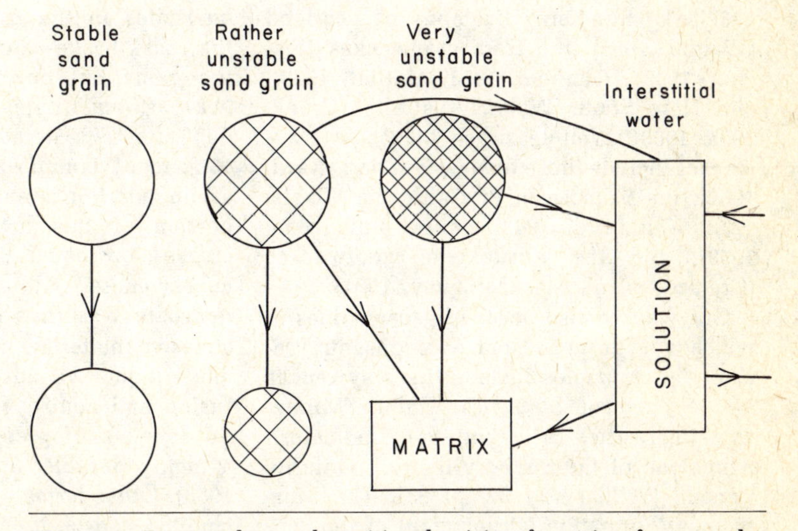

FIG. 7-19. Presumed post-depositional origin of matrix of graywacke. (Modified from Cummins, 1962, Liverpool and Manchester Geol. Jour., v. 3, Fig. 4.)

of many eugeosynclinal tracts are notably soda-rich, and some associated pelitic strata have a higher than normal Na_2O content (Walker and Pettijohn, 1971, Table 3). As noted elsewhere, the absence of K-feldspar is a noteworthy feature of graywackes. Its absence has been attributed to post-depositional solution and removal (Gluskoter, 1964, p. 341).

The Na problem remains unresolved.

Occurrence and geologic significance The bulk of the graywackes of the geologic record are old, mostly Paleozoic or older, and most are part of a flysch-like sequence. They are commonly associated with "greenstones" (Tyrrell, 1933) and are the most typical sandstones of eugeosynclinal fold belts. Eugeosynclinal graywackes are apt to be feldspar-rich and grade into volcanic wackes. Precambrian examples include the Archean sequences of the Canadian Shield (Pettijohn, 1943; Donaldson and Jackson, 1965; Walker and Pettijohn, 1971), of the Fennoscandian Shield (Simonen and Kuovo, 1951), and of the South African older Precambrian (Anhaeusser et al., 1969, p. 2184). Well-known Paleozoic graywackes include those from Wales (Woodland, 1938; Okada, 1967), the Southern Highlands of Scotland (Walton, 1955), the Harz graywackes (Fischer, 1933; Helmbold, 1952; Mattiat, 1960), and New South Wales, Australia (Crook, 1955, 1960). Well-described Mesozoic graywackes include those of New Zealand (Reed, 1957), the Franciscan of California (Davis, 1918; Taliaferro, 1943; Bailey and Irwin, 1959), and the Kuskokwim graywackes (Cretaceous) of Alaska (Loney, 1964).

Graywackes also occur in geosynclines in which volcanism is sparse or absent. Precambrian examples include the graywackes of the Thomson Slate near Duluth, Minnesota (Schwartz, 1942) and the Chelmsford Sandstone of the Sudbury district, Ontario, Canada (Williams, 1957, p. 83). The Martinsburg Formation (Ordovician) of the central Appalachians (McBride, 1962), the graywackes of the Normanskill Shale (Ordovician) of New York (Weber and Middleton, 1961), and the Haymond Formation (Pennsylvanian) of Texas (McBride, 1966) are good Paleozoic examples. Younger graywackes include those of Cretaceous age in Japan (Shiki, 1962; Okada, 1960, 1961) and many of the graded sandstones of the Apennines of Italy (Sestini, 1970). All of these examples of miogeosynclinal graywackes are characterized by their abundance of rock particles rather than feldspar.

In summary, graywackes (as here defined) form one-fifth to one-fourth of all sandstones (see p. 248), are most common in the Paleozoic and earlier orogenic belts. They are absent in undeformed sequences deposited on the stable cratonic or shield areas. They are generally marine and, in the examples cited above, are believed to be turbidite sands of the flysch facies. Not all flysch sands, however, nor all turbidite sands, are graywackes.

QUARTZ ARENITES (ORTHOQUARTZITES)

Definitions All terrigenous sands are quartz-rich; there are few indeed in which quartz is not the most abundant constituent. All therefore can be called quartzose sandstones. But in some, quartz is almost the only component. These sands were termed *orthoquartzites* by Tieje (1921, p. 655), a term popularized and more rigorously defined by Krynine (1945), who applied it to sands made up entirely of quartz grains cemented by silica. These rocks were indeed quartzites in the traditional sense in that the rocks broke across, rather than around, the grains. But they are sedimentary quartzites—*orthoquartzites* rather than *metaquartzites*. As noted by Krynine (1948, p. 152), many contain some carbonate cement as an additional end-member; as the proportion of this nonsiliceous material increases, orthoquartzites begin to show less cohesiveness. Although the term *orthoquartzite* was thus extended to these less cohesive materials and even to friable or modern quartz sands, it has led to some confusion and controversy, because it negates the long-standing concept of "quartzite" as a highly durable, nonfriable rock (Mathur, 1958). Other terms have been suggested, including *quartz arenite* (Williams, Turner, and Gilbert, 1954, p. 294) and *quartzarenite* (McBride, 1963). The term *orthoquartzite* has been widely used for many years, but recent literature suggests that the term is gradually being replaced by *quartz arenite*.

Fabric and composition As here defined, quartz arenites are sands, the detrital fraction of which is 95 or more percent quartz. They commonly are cemented by quartz deposited in optical continuity with detrital quartz and thus are truly orthoquartzites (Fig. 7-20). Others, however, are cemented by calcite or are friable and not cemented at all. The quartz-cemented varieties, which are true quartzites, are erosion resistant and consequently cap high hills or, if upturned, form prominent ridges. Many quartz arenites are relatively thin, blanketlike deposits; others are thick (1000 m, more or less), especially those of Precambrian age. Ripple marking and cross-bedding are prominent in many.

The quartz of the orthoquartzites is largely monocrystalline, polycrystalline grains being less stable and largely eliminated (Blatt, 1967, p. 422), and, for similar reasons, display sharp rather than undulatory extinction (Blatt and Christie, 1963, p. 571). The quartz is very well sorted and highly rounded. Many of these sands, therefore, are the most texturally and compositionally mature known.

Other constituents are rare, but a few well-rounded grains of chert or equally durable quartzite particles may be present. Heavy minerals are very scarce and consist mainly of very well-rounded tourmaline and zircon with perhaps some rutile and ilmenite (or leucoxene derived therefrom).

Silica is the usual cement. It is generally quartz, deposited in optical and crystallographic continuity with the rounded detrital quartz grains. In less than fully cemented sands, secondary quartz exhibits crystal facets which reflect light so that the rock sparkles in bright sunlight. Regenerated quartz crystals (Fig. 7-20) bound the pores of the sandstone; in more friable sands, the grains show well-formed pyramidal terminations (Fig. 7-1). Inspection shows that these coincide approximately with the long dimension of the detrital

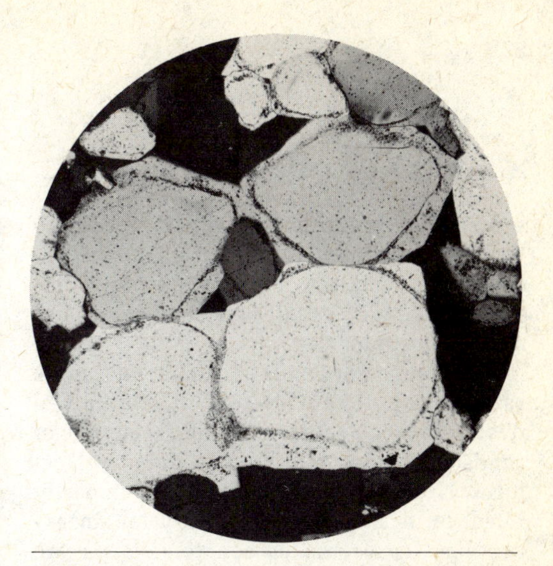

FIG. 7-20. Cambrian sandstone, Iron Mountain, Michigan. Crossed nicols, ×40. Dusty borders mark outlines of original quartz grains. A quartzite formed by cementation of quartz in crystallographic continuity with detrital quartz grains.

core. The end-result of the enlargement process is a reduction of pore space, with the intergrowths of adjacent grains growing together to form an interlocking quartz mosaic within which a thin line of inclusions reveals the rounded outlines of the original sand grains (Fig. 7-20). In some quartz arenites this "dust ring" is faint or absent. It is most prominent where the detrital grains have an iron oxide coating.

Many orthoquartzites show such pressure-solution features as well-developed stylolitic seams (Heald, 1955) and microstylolitic grain contacts, especially involving chert grains. This pressure solution helps transform a quartz sand into a quartzite, a process which has been described in detail by Skolnick (1965) and duplicated by subjecting loose quartz sand to high pressure and an elevated temperature in the presence of appropriate pore solutions (Maxwell, 1960; Ernst and Blatt, 1964). Appearances are, however, very deceiving, and orthoquartzites, if examined by luminescence microscopy, show that many cases of presumed pressure solution are nothing more than the end-product of quartz enlargement (Sippel, 1968).

Some quartz arenites are cemented with other forms of silica, such as opal or chalcedony which form coatings on the grains (Fig. 7-24). Chalcedony cements may show a microfibrous fabric, with fibers normal to the grain surfaces. In general, opaline cements are confined to very young sandstones. In older sands it is devitrified and now chalcedonic.

Commonly carbonates, calcite and more rarely dolomite, are present in addition to quartz; in many cases, especially the younger sandstones, they may be the dominant cement. Normally each pore is filled, or partly filled, by a single carbonate crystal. In exceptional cases the carbonate crystals are large and envelop a number of detrital grains. These form the luster-mottled sandstones (Fig. 7-21). As noted elsewhere (p. 240), the carbonate cements may encroach on and corrode the detrital quartz grains.

With the appearance of feldspar the quartz arenites pass by degrees into subarkose, or, if rock fragments become abundant enough, into protoquartzites (sublitharenites). The transition is characterized also by the appearance of more polycrystalline quartz and a less perfect rounding of quartz grains.

Orthoquartzites and related sandstones are sparingly fossiliferous. Unless richly calcareous, well-preserved carbonate shells are rare. Scattered aggregates of calcite may be present; these have been interpreted as dissolved and reprecipitated shelly matter. The effects of partial solution of shells, notably their thinning, and the complete cementation of the sand *within* the shell by calcite, have been noted by Krynine (1940, Fig. 8).

As shown in Table 7-10, quartz arenites and related sandstones are exceedingly high in SiO_2, in some cases containing much less than 1 percent of other oxides. Such sandstones constitute some of the largest and purest concentrations of silica known. They form a commercial source of silica and, if low enough in iron, may be suitable for the manufacture of glass. Those sandstones with appreciable carbonate cement will contain significant amounts of CaO and CO_2, and in some cases MgO, with corresponding reduction in the quantity of SiO_2. The finer-grained varieties, and those which verge on subarkose or protoquartzite, will have appreciable amounts of Al_2O_3 and K_2O.

FIG. 7-21. Dakota Formation (Cretaceous), Lincoln, Kansas. Crossed nicols, ×84. A calcareous quartz arenite showing "luster mottling" in the hand specimen, composed of rounded quartz grains markedly etched and corroded by the carbonate cement which binds them together. The cement is coarsely crystalline and single crystals enclose many detrital quartz grains.

Distribution in time and space Orthoquartzites are known from many places and are of many ages; but most seem to be late Precambrian and early Paleozoic. The oldest Precambrian Archean seems not to contain much orthoquartzite. Well-known and well-described North American Precambrian examples include the Sioux Quartzite of Minnesota, Iowa, and South Dakota (Rothrock, 1944, p. 147); the Baraboo and Waterloo quartzites of Wisconsin (Brett, 1955), the Sturgeon Quartzite of the Menominee district and the Mesnard Quartzite of the Marquette district of Michigan; the Palms Quartzite of the Gogebic district of Wisconsin and Michigan; the Upper Lorrain Quartzite of the north shore of Lake Huron, Ontario (Hadley, 1968); and the Odjick Formation in the Coronation Gulf area in the Northwest Territories (Hoffman, Fraser, and McGlynn, 1970). These formations are thick, up to 1000 m or more and are commonly cross-bedded and ripple marked. Interbedded pelitic sediments are rare or absent. Other examples are the Keweenawan Sibley Sandstone of the Thunder Bay area of Lake Superior and the Hinckley Sandstone (Thiel and Dutton, 1935) of Minnesota.

The Athabaska Formation of northern Saskatchewan (Fahrig, 1961), covering about 40,000 miles^2 (104,000 km^2), and the Thelon Formation of like extent and character of the Northwest Territories (Donaldson, 1967) are quartz arenites of latest Precambrian age.

The older Paleozoic of North America abounds in quartz arenites (Figs. 7-20, 7-22, 7-23). Unlike the Precambrian examples, these are less well cemented, a great deal thinner but fairly widespread. In the Upper Mississippi Valley are Cambro-Ordovician examples—the Cambrian Dresbach, Franconia, and Jordan sandstones (Graham, 1930) and the Ordovician New Richmond and St. Peter sandstones (Dake, 1921; Thiel, 1935). In the central Appalachians the Cambrian Gatesburg, the Chickies Quartzite, and the Devonian Oriskany are good quartz

TABLE 7-10. Representative chemical analyses of orthoquartzites (quartz arenites)

Constituent	A	B	C	D	E	F	G	H	I	J
SiO_2	98.87	95.32	97.58	97.36	98.91	83.79	99.54	99.40	97.30	93.13
TiO_2	—	—	—	.05	.05	—	.03	.02	.28	—
Al_2O_3	.41	2.85	.31	.73	.62	.48	.35	.20	1.40	3.86
Fe_2O_3	.08	.05	1.20	.63	.09	.063	.09	.01	.30	.11
FeO	.11	—	—	.14	—	—	—	—	—	.54
MnO	—	—	—	.01	—	—	—	trace	.003	—
MgO	.04	.04	.10	.01		.05	.06	.01	.03	.25
					.02					
CaO	—	trace	.14	.04		8.81	.19	<.01	<.05	.19
Na_2O	.08		.10	.08	.01	—	—	.08	<.05	—
		.30								
K_2O	.15		.03	.19	.02	—	—	trace	.20	—
H_2O+		—	—	.54	—	—	.25	.04	—	—
	.17									
H_2O-		—	—	.14	—	—	—	.01	—	—
P_2O_5	—	—	—	.02	—	—	—	none	—	—
ZrO_2	—	—	—	—	—	—	—	<.01	.06	—
CO_2	—	—	—	—	—	6.93[b]	—	—	—	—
Ignition loss	—	1.44	.03	—	.27	—	—	.28	—	1.43
Total	99.91	100.00	99.62[a]	99.94	99.99	100.13[c]	100.51	100.05[d]	99.57	99.51

[a] Including SO_3 0.13.
[b] Calculated.
[c] Includes organic matter 0.006.
[d] Includes Cr_2O_3 0.00008; BaO and SrO none; NiO less than 0.0001; CuO less than 0.00027; CoO less than 0.0002.

Source: After Pettijohn (1963), Table 2.

A. Mesnard Quartzite (Precambrian), Marquette County, Michigan. R. D. Hall, analyst (Leith and Van Hise, 1911, p. 256).
B. Lorrain Quartzite (Precambrian), Plummer Township, Ontario, Canada. M. F. Conner, analyst (Collins, 1925, p. 68).
C. Sioux Quartzite (Precambrian), Sioux Falls, South Dakota (Rothrock, 1944, p. 151).
D. Lauhavuori Sandstone (? Cambrian), Tiiliharju, Finland. Pentti Ojanperä, analyst (Simonen and Kuovo, 1955, p. 79). Computed mineral composition: quartz 70–75, feldspar 0.1–1.4, rock fragments 0.1–5.6, silica cement 18–20.
E. St. Peter (Ordovician), Mendota, Minnesota. A. William, analyst (Thiel, 1935, p. 601).
F. Simpson (Ordovician), Cool Creek, Oklahoma (Buttram, 1913, p. 50).
G. Tuscarora (Silurian), Hyndman, Pennsylvania (Fettke, 1918, p. 263).
H. Oriskany (Devonian), Berkeley Springs Quarry, Pennsylvania Glass Sand Corp. Sharp-Schurtz Co., analysts. Analysis supplied courtesy of Pennsylvania Glass Sand Corp.
I. Mansfield (basal Pennsylvanian), Crawford County, Indiana. M. E. Coller, R. K. Leininger, and R. F. Blakely, analysts. Computed mineral composition: quartz 95.3, orthoclase 1.2, kaolin 3.0, ilmenite 0.3 (Murray and Patton, 1953, Indiana Dept. Conserv. Geol. Surv. Prog. Rept. 5, p. 28).
J. Berea (Mississippian), Berea, Ohio. N. W. Lord, analyst (Cushing, Leverett, and Van Horn, 1931, p. 110). A protoquartzite.

arenites. The Tulip Creek (Ordovician) of the Arbuckle Mountains of Oklahoma (Fig. 7-23) and the Blakeley and Crystal Mountain sandstones of the Ouachitas are likewise orthoquartzites. In the western United States quartz arenites are represented by the Flathead Quartzite (Cambrian) of Wyoming, the Eureka Quartzite (Ordovician) of Nevada, and the Swan Peak Quartzite of Idaho (Ketner, 1966). Late Paleozoic examples include the Tensleep Sandstone of Wyoming and the correlative Casper Formation and the Weber Quartzite of Utah.

The Dakota Sandstone (Cretaceous) (Fig. 7-21) of the Great Plains and some relatively thin Tertiary sands, such as the Cohansey of New Jersey (Carter, 1972), are also quartz arenites.

European examples of quartz arenites include the Hardeberga of Sweden (Hadding, 1929, p. 77), the Malvern Quartzite (Cambrian) of England, and the sandstone of Lauhavuori, Finland (Simonen and Kuovo, 1955, p. 75). A number of Cretaceous sandstones, including the Quadersandstein of the Harz (Rinne, 1923, p. 280), are good quartz arenites. The Fontainebleau Sandstone (Tertiary) of France (Cayeux, 1929, p. 154) is also a good example.

It is clear from the above review that quartz arenites are very common (they form perhaps one-third of all sandstones), that they are mostly late Precambrian or early Paleozoic in age (the thickest is Precambrian), and that they are abundant in the stable cratonic regions. Many are closely associated with limestones and dolomite; most are free of shaly interbeds. Although orthoquartzite is the characteristic sandstone of the stable shield areas, there is some spillover into marginal miogeosynclinal belts. The presumption is, however, that the sands were formed in the stable region. Orthoquartzites are almost unknown in the eugeosynclinal facies, but there are a few exceptions. Ketner (1966), for example, has described an Ordovician orthoquartzite (Valmy Formation) in the eugeosynclinal province of Nevada where it is associated with greenstones and bedded cherts.

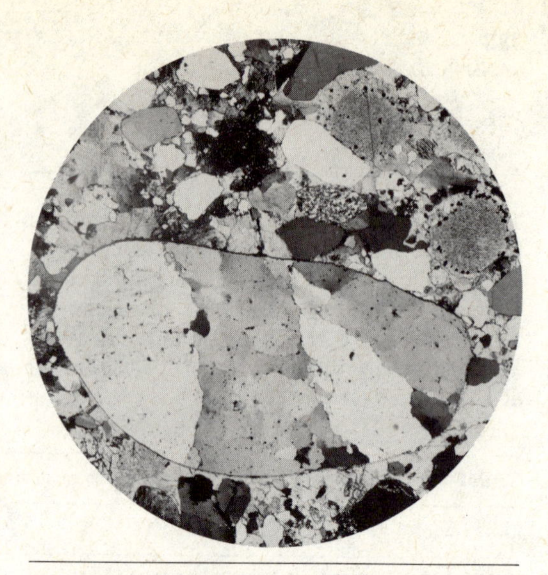

FIG. 7-22. Conglomeratic quartz arenite from Coldwater Shale (Devonian), Michigan. Crossed nicols, ×14. Vein quartz pebble (largest fragment), a good example of polycrystalline quartz, quartz, carbonate detritus ("dusty"), and chert (microcrystalline aggregates) cemented by carbonate. Authigenic pyrite granules in ringlike arrangement around detrital carbonate.

FIG. 7-23. Tulip Creek Sandstone (Ordovician), Oklahoma. Crossed nicols, ×55. A supermature quartz sand slightly cemented by carbonate; cementation is slight and porosity high.

Origin and geologic significance The high quartz content and the excellent sorting and rounding displayed by the orthoquartzitic sandstones are indicative of a high degree of textural and mineralogical maturity. These rocks are obviously the end-product of protracted and profound weathering, sorting, and abrasion. In order that there be sufficient time to achieve these results, it is imperative either that the source area and site of deposition be tectonically very stable or that the sand go through several cycles of sedimentation. The prevailing view, based both on experimental study of sand abrasion (Kuenen, 1959b) and on the study of the sands of large rivers, such as the Mississippi (Russell and Taylor, 1937; Russell, 1937, p. 1346), is that it is improbable that quartz arenites of the type found in the geologic record could ever be produced by river action, no matter how prolonged.

Kuenen (1960, p. 109) concluded that the rounding of sand is achievable only in the eolian environment and that highly rounded sands must, at some time in their history, have had an eolian episode which is not necessarily that of the last depositional environment. This conclusion implies a multicycle origin for the quartz arenites—an origin difficult to establish on internal evidence from the sand itself. Conclusive evidence such as worn overgrowths on the quartz are known but are seldom seen.

Despite Kuenen's conclusions, some field evidence supports the view that repeated washing and winnowing of sand in the surf zone will give rise to a supermature sand. Folk's observation (1960, p. 55) of interbedding of highly rounded with markedly less rounded sand in the Tuscarora Quartzite (Silurian) of West Virginia suggests that the differences noted were a product of the local environment of deposition which in the case of the most mature sands was thought to be the beach. Such local winnowing or washing seems to be capable of "cleaning up" of otherwise less mature sands.

Unlike the geologic past, most modern sands are not quartz arenites. The few apparent exceptions (Mizutani and Suwa, 1966) are those sands derived in the present cycle from nearby quartz arenites. Such is the case with some beach sands of the Gulf Coast of northwestern Florida where the sand is over 99 percent silica (Burchard, 1907, p. 382). The lack of present-day quartz arenites, their general scarcity in the younger geologic times (except possibly in the Cretaceous), and their abundance in the Cambrian and general absence in the Archean require an explanation. One might conclude that orthoquartzites mark periods of great stability—base leveling and prolonged weathering. If so, the earth is now, and has been since the Cretaceous, in a period of instability and high relief and is characterized, therefore, by immature sands. The relationship of tectonic stability and sandstone petrography is explored further at the end of this chapter.

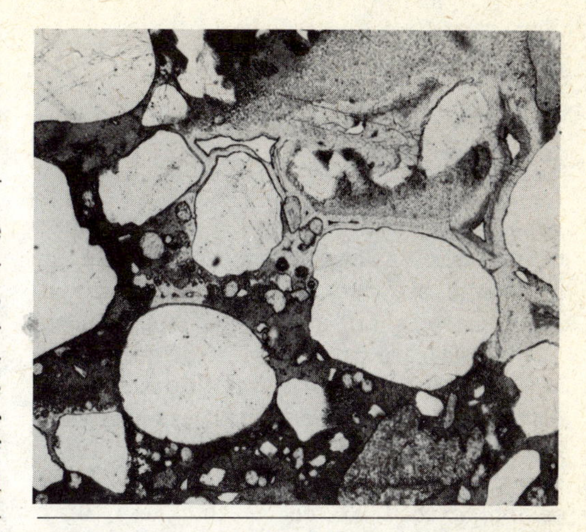

FIG. 7-24. Opal-cemented sandstone, Ogallala "Quartzite" (Pliocene), Kansas. Ordinary light, ×20. A coarse sandstone composed of quartz together with a little feldspar cemented by opal.

MISCELLANEOUS SANDSTONES

A number of sandstones do not fall into the major groups reviewed in the foregoing portion of this chapter. Even excluding volcanic and carbonate sands, there remain a few aberrant types. Among these are the greensands.

Greensand The term *greensand* has been applied to glauconite-rich sands (Fig. 7-25). Under the hand lens the best greensands appear to be composed entirely of glauconite, less than 1 percent being grains of quartz. More commonly quartz is the dominant constitutent forming one-half or more of the whole. Sands composed mainly of glauconite are dark to light green; mixed sands have a salt-and-pepper appearance. Greensands are prominent in beds of Cretaceous and Eocene age in the Coastal Plain of the eastern United States, especially in New Jersey and Delaware (Ashley, 1918; Mansfield, 1920). Although individual beds seldom exceed 25 feet in thickness, they are areally extensive and potentially an important source of potash. The chemical composition of several representative greensands is given in Table 7-11. As might be expected, the composition of greensands varies greatly depending on the proportions of glauconite to other detrital materials and on the kind and volume of cement present.

The origin and significance of greensands are tied up with the glauconite problem. Our knowledge of the geology and distribution of glauconite has been summarized by Hadding (1932), Cloud (1955), and others (Galliher, 1936, 1939; Goldman, 1919; Takahashi, 1939; Schneider, 1927). According to Cloud, glauconite is formed only in marine waters of normal salinity; it requires slightly reducing conditions (weakly oxidizing, according to Chilingar, 1955); its formation is facilitated by the presence of organic matter; it is mainly characteristic of waters 10 to 400 fathoms in depth; it is formed only in areas of slow sedimentation; and it is chiefly formed from micaceous minerals or bottom muds rich in iron. As noted by Hadding (1932) and others, the site of formation of glauconite and its place of accumulation may not be the same as some glauconite is reworked or transported.

Galliher (1936, 1939) concluded that glauconite was derived from biotite by a process of submarine weathering. He observed a series of transition grains that demonstrate this transformation. He also noted, in Monterey Bay, California, that biotite-rich sands near shore grade horizontally into mixed glauconite-mica silty sands offshore and that they thence progress to glauconite muds at a depth of 100 fathoms. Gruner (1935) has shown that the ionic arrangement of the unit cells of glauconite and biotite was very similar, if not identical, so that the transformation of biotite to glauconite requires no great change. Although Galliher's observations have been confirmed in other places (Edwards, 1945), some glauconite seems not to have formed from mica (Allen, 1937). As summarized by Takahashi (1939), glauconite originates from a number of mother materials, such as fecal pellets; clayey substances filling cavities in foraminifers, radiolarians, and tests of other marine organisms; or such silicate mineral substances as volcanic glass, feldspar, mica, or pyroxene. Organic matter seems to facilitate the process of glauconitization. Parent materials undergo a loss of alumina, silica, and alkalies (except potash) and gain ferric iron and potash. Seawater seems essential to the process.

FIG. 7-25. Greensand, Ft. Augustus Formation (Cretaceous), Alberta, Canada. Ordinary light, ×55. A sandstone rich in glauconite (large round pellets) together with subangular quartz grains cemented by carbonate. This formation also contains some detrital chert. (After Pettijohn, Potter, and Siever, 1972, Sand and sandstone Fig. 6-30. By permission of Springer-Verlag.)

Although glauconitic sands range from Precambrian to the present-day in age, they seem particularly common in Cambrian rocks. Likewise, the Cretaceous seems to have been a period of glauconitization. This period continued into the Paleocene and Eocene in many places, as in the Atlantic Coastal Plain, for example.

Phosphatic sandstones The phosphatic materials, commonly designated *collophane*, a poorly organized carbonate fluorapatite, may form cement of the sand, or they may constitute part or all of the framework fraction.

Many sands contain a little phosphatic debris, as phosphatic skeletal materials, more rarely as phosphatic nodules or granules. The glauconitic sands (Table 7-11) are apt to be rich in phosphatic matter. In some sands the phosphate forms a cement, a drusy coating on quartz grains, or a microcrystalline pore filling (Bushinsky, 1935; Cressman and Swanson, 1964, p. 307). Some phosphorites (see Fig. 11-31) are themselves sands consisting of phosphatic granules or ooids which may be mixed in all proportions with detrital quartz.

Calcarenaceous sandstones Sandstones, especially quartz arenites, grade into calcarenitic limestones. The transitional types are a mixture of detrital quartz and detrital car-

TABLE 7-11. Representative chemical analyses of greensands and other sandstone types

Constituent	A	B	C	D	E	F
SiO_2	57.40	50.74	75.95	45.43	48.85	49.81
TiO_2	.29	—	.20	.11	trace	—
Al_2O_3	6.89	1.93	2.91	.03	11.82	5.17
Fe_2O_3	11.98	17.36	10.29	2.92	1.83	29.17
FeO	3.04	3.34	—	—	1.22	.35
MnO	.03	—	—	.02	—	—
MgO	2.41	3.76	1.37	.61	.45	.95
CaO	1.78	2.86	.10	26.21	12.85	2.43
Na_2O	1.11	1.53	.35	.34	.47	.84
K_2O	4.85	6.68	2.99	.16	.64	.48
H_2O+	5.36	9.08	5.40	2.78	2.75	6.56
H_2O-	4.46	—	—	—	—	3.85
P_2O_5	.22	1.79	—	16.05	10.70	.42
CO_2	—	.88	—	3.12	3.40	—
SO_3	.45	—	—	.86	—	—
F	—	—	—	1.87	2.86	—
Total	100.29[a]	99.95	99.56	101.25[b]	97.84	100.03
Less O				— .79	−1.20	
				100.46[c]	96.64	

[a] Includes BaO, 0.02.
[b] Includes C, 0.45; FeS_2, 0.29.
[c] Given as 99.01 in original.

Source: After Pettijohn (1963), Table 9.

A. "Greensand" (Middle Eocene), Pahi Peninsula, New Zealand (Ferrar, 1934, p. 47).
B. "Greensand marl" (Upper Cretaceous), New Jersey. R. K. Bailey, analyst (Mansfield, 1922, p. 124).
C. Greensand, opal-cemented, Thanetien, Angre, Belgium (Cayeux, 1929, p. 130).
D. Phosphate sandstone, "upper phosphorite stratum" (Cenomanian), Kursk, Shchigry, USSR (Bushinsky, 1935, p. 90). About 38 percent quartz, 45 percent phosphorite, 5 percent glauconite.
E. Phosphatic sandstone, St. Pôt, Boulonnais, France (Cayeux, 1929, p. 191).
F. Ferruginous sandstone, "carstone," Hunstanton, Norfolk, England (Phillips, 1881, Quart. Jrl. Geol. Soc., v. 37, p. 18). Consists of quartz grains cemented by brown iron ore, with a very little feldspar and mica.

bonate debris—skeletal and oolitic (Fig. 7-26). The term *calcarenaceous* sandstone was suggested for such sandstones to avoid confusion with *calcareous* sandstone, a term applied to ordinary sandstones with a carbonate cement. Care should be taken also to discriminate between calcarenites and calcarenaceous sandstones and *calclithites,* sandstones which are lithic sandstones with terrigenous carbonate debris.

Modern sands of some areas consist of a mixture of biochemical or chemical carbonate detritus and ordinary quartz sand. On the east coast of Florida, siliceous components are carried southward by shore drift and mixed with locally produced shelly debris. In northern areas quartz is dominant; farther south the carbonate exceeds quartz (Martens, 1935).

Ancient examples are common. Well-described calcarenaceous sands include the Loyalhanna of Pennsylvania (Adams, 1970, p. 83) and the Greenbrier of West Virginia (Rittenhouse, 1949), both of Mississippian age. The Loyalhanna consists of about 40 to 45 percent quartz, 25 percent detrital carbonate, and 30 percent carbonate cement. The chemical composition of this rock is given in Table 7-11. The Cow Creek Formation (Cretaceous), near Fredericksburg, Texas (Fig. 7-26), is probably of beach origin and is likewise a calcarenaceous sand.

FIG. 7-26. Cow Creek "Limestone" (Cretaceous) near Fredericksburg, Texas. Crossed nicols, ×22. Consists of subrounded detrital quartz, skeletal debris (fibrous structure), and some micritic carbonate grains cemented by calcite. (From Pettijohn, Potter, and Siever, 1972, Sand and sandstone, Fig. 6-34. By permission of Springer-Verlag.)

Placer sands Placer sands constitute a minor sand type, mineralogically very interesting; and although of small volume they are often of great economic value for titania, zirconium, and thorium. Placers are local accumulations produced by rivers and beach action. They are of small extent and slight thickness, rarely being more than 1 or 2 m thick.

Examples of modern beach placers are numerous. Those of Oregon (Griggs, 1945) have local concentrations of chromite; those of Florida have been exploited for ilmenite, zircon, and rutile (Martens, 1928). "Fossil" placers are also known and exploited as, for example, are the rutile-bearing sands of the Cohansey (Pliocene) sands of New Jersey (Markewicz, 1969) and the Upper Cretaceous titaniferous magnetite sands of Montana (Stebinger, 1914, p. 329).

A very unusual placer type of sandstone is that from Rand Basin (Precambrian) of South Africa which contains thin seams of detrital pyrite (Ramdohr, 1958, p. 16). See Fig. 7-27.

Itacolumite Itacolumite is a peculiar quartz schist named for Itacolumi, a mountain in Brazil. The name is attributed to Humboldt by Holmes (1928, p. 126). Itacolumite is composed of interlocking quartz grains with some mica; it exhibits some flexibility and hence is often referred to as "flexible sandstone" (Cayeux, 1929, p. 196; Ginsburg and Lucas, 1949). It is perhaps a metamorphic rock rather than a sandstone. It also occurs in North Carolina (Foushee, 1954).

DIAGENESIS OF SANDSTONE

Very shortly after deposition, a sand begins a process of aging, which profoundly modifies its character. It is no longer a loose granular material; it is instead transformed into a dense lithified rock. The aging processes are complex and not fully understood. In part these are purely mechanical: grain fracturing, bending and deformation of detrital micas, and squashing of weaker pelitic grains. But mainly they are chemical, involving solution, reprecipitation, decomposition, and intergranular reactions. Redistribution of materials such as solution of quartz at one place and its precipitation at another leads to cementation and pore space reduction. The less-stable framework grains are degraded and lose their identity as such and are transformed into a microcrystalline matrix that may interact with other more stable grains. The net result of pressure solution, devitrification (of opaline and glassy materials), and decomposition of unstable, detrital components is an alteration of the rock fabric, a dramatic loss of porosity, a blurring of original textures, and a transformation of the rock to a more nearly equilibrium mineral assemblage.

The changes affect various sands differently. Pure quartz sands undergo only solution transfer of quartz and conversion to orthoquartzite. Immature sands, especially lithic arenites with a high proportion of unstable rock particles, are converted to graywackes. Volcaniclastic sands undergo perhaps the most drastic alterations—alterations that are not unlike those of retrograde metamorphism.

Larger-scale features such as load casts, slump folds, and the like caused by "soft-sediment deformation" are not usually considered diagenetic. Diagenesis involves mineralogic and textural changes and not production of structures. Exceptionally, however, such structures as concretions and stylolites are considered diagenetic. In this book, larger diagenetic segregations—concretions and nodules—are separately considered (Chapter 12).

We shall first consider those diagenetic features brought about by chemical processes.

CEMENTATION

The chemically precipitated material which forms the cement in many sandstones is an important constituent of such rocks. If voids are completely filled, the cement constitutes from one-fourth to one-third of the whole rock. A cemented sandstone stratum 100 m thick, for example, contains within it enough cementing material to form a layer 25 to 30 m thick if these materials were segregated. Cementation, moreover, is the last step in the formation of the sandstone, and our knowledge is incomplete and unsatisfactory unless the origin and manner of emplacement of the cement are fully understood.

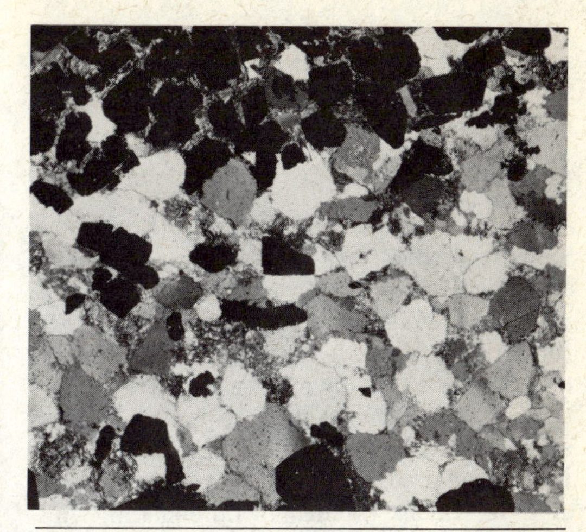

FIG. 7-27. Precambrian pyritiferous quartzite, Rand Basin, South Africa. Crossed nicols, ×22. Detrital pyrite interlaminated with quartz sand.

The introduction of a cement obviously affects both the porosity and the permeability of the rock and is therefore of considerable interest in the study of the movement of fluids through the rock and in the estimation of the total volume of such fluids. Cementation, if carried to completion, produces a "tight" sand capable of neither holding nor transmitting such fluids as groundwater, petroleum, or natural gas.

Many species of minerals are known to play the role of cement. Some, however, are relatively rare and therefore quantitatively unimportant. The most common cementing material is silica, generally in the form of quartz. Quartz is usually deposited as an overgrowth on detrital quartz grains (Figs. 7-1 and 7-20). Under some relatively uncommon conditions, silica is deposited as opal or chalcedony instead of quartz (Fig. 7-24). Sandstones with opaline cement are chiefly of late geologic age. Factors leading to the formation of opal rather than quartz are not fully understood, although they seem to be related to ion concentration (Millot, Lucas, and Wrey, 1963). The opal-cemented Ogallala of Kansas (Frye and Swineford, 1946) has been interpreted as an opal replacement of a calcite-cemented sandstone. Ash beds associated with this formation are thought to be the source of the replacement silica. The close association of opal-cemented sands and ash beds seems rather common. This is the case in the Gueydan (Catahoula) Formation (Tertiary) in Texas (McBride, Lindemann, and Freeman, 1968, p. 46).

Various carbonate minerals, especially calcite, are also common cementing agents. Dolomite is less common, and siderite is comparatively rare. Siderite is not so rare, perhaps, as commonly believed; it is seldom seen in outcrop materials, simply because it is very unstable in the presence of the atmosphere. Many ferruginous- or iron-oxide-cemented sandstones may have been in fact sideritic sandstones. Examination of some of the spotted sandstones shows that each spot, a small area cemented by limonite, is derived by oxidation of siderite, some of which is preserved in the center of the limonite-cemented area (Fig. 7-28).

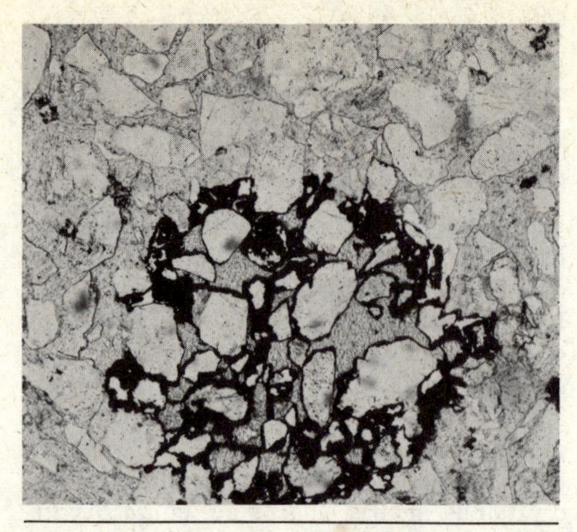

FIG. 7-28. Concretion from Dakota Sandstone (Cretaceous), Ellsworth County, Kansas. Ordinary light, ×80. A "spotted" sandstone, each spot of which is a small area cemented by limonite (black) derived by oxidation of siderite cement. Some unaltered siderite is still present in center of one of the limonite-cemented areas.

Iron oxide and very rarely iron sulfide occur as cements. Among the silicates which play the role of cement are feldspars (Fig. 7-29), at best a very minor constituent, kaolinite and other clay minerals, and zeolites. Although clays may be trapped at time of deposition, some kaolinite is a truly precipitated coarsely crystalline pore filling (Donaldson, 1967; Carrigy and Mellon, 1964). Zeolites are notably common in volcanic sandstones or those with volcanic glass (Hay, 1966; Weeks and Eargle, 1963). Barite and anhydrite are minor cementing agents and are only locally important.

Tallman's study of sandstone cements (1949) confirmed earlier impressions that silica was a more common cement in older formations, whereas silica and carbonate cements were about equally abundant in Mesozoic and younger sandstones. The significance of this observation is not clear. Perhaps carbonate cements of older sandstones have been replaced by silica; perhaps they have been removed by leaching.

The relation of the cement to the detrital framework of the sand is of considerable interest and importance. If the mineral composition of cement is the same as that of the detrital grains, the end-product resulting from secondary overgrowth on the mineral grains is an interlocking crystalline aggregate—a quartzite if the rock is largely quartz. If cement is mineralogically unlike that of the detrital grains it may show various textural relations (Fig. 7-30). Calcite, for example, in partially cemented sandstone may be deposited as a drusy coating on the grains, or as a crystalline mosaic between the grains, or in rare cases as large poikiloblastic grains containing many detrital grains (Fuhrmann, 1968). See Fig. 7-30B. Opaline and chalcedonic cements may form an agatelike coating on framework grains (Fig. 7-24) or a botryoidal deposit with fan-shaped radial fibers. Other cements such as kaolin may appear as a blocky polycrystalline pore filling.

In some cases, especially with the carbonate cements, there is a reaction between the cement and the framework grains. The cement appears to corrode the detrital grains, as shown by an irregular and embayed contact between cement and grains (Fig. 7-21). In places this is carried so far that all that remains of the original grain are a few small oriented residuals that extinguish together. Chert, feldspar, and even

A

B

FIG. 7-29. Enlarged feldspar in Croixan Sandstone (Cambrian), Wisconsin. A: Ordinary light, ×160. Note simple rhombic outline of reconstructed feldspar, cleavage crossing both secondary outgrowth and nucleus, and faint outline of detrital core. B: Crossed nicols, ×160. Note sharp differentiation between detrital core and secondary rim, because of difference in composition, which results in slightly different extinction positions for two parts of the reconstructed crystal.

quartz are susceptible to such corrosion and replacement by carbonate.

Normally cement fills or partially fills the pores in the sand (Fig. 7-30A). In a few cases carbonate cement constitutes an undue volume of the whole sandstone equaling or even exceeding the detrital quartz, which then appears to be "floating" (Fig. 7-30C). This has been attributed by some to recrystallization of detrital carbonate presumed to have been deposited with detrital quartz and by Waldschmidt (1941) to the growth of the cement and the forceful wedging apart of the grains. It could also

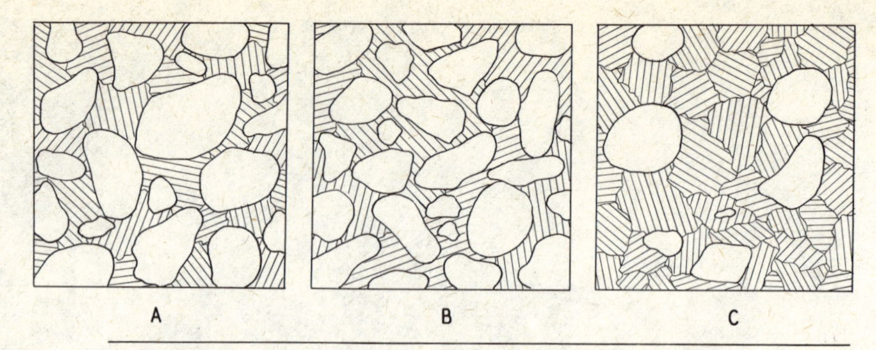

FIG. 7-30. Schematic diagram showing the relation of carbonate cement to detrital grains in sandstone; carbonate ruled, detrital quartz blank. A: normal calcareous sandstone; B: "luster-mottled" sandstone; C: arenaceous limestone.

be caused partially by corrosion and replacement of detrital grains by carbonate as noted above.

Some sandstones have more than one species of cementing mineral. In such cases it is important to determine the paragenesis or relative ages of the several cementing minerals. Waldschmidt (1941) believed a definite order of precipitation could be established for the Rocky Mountain sandstones studied by him. Calcite followed quartz; if there were three cementing minerals, dolomite followed quartz and was in turn followed by calcite. If four cementing minerals were present, the above three were followed by anhydrite. In some sandstones containing three species of cement, the sequence was quartz, dolomite, and anhydrite. Heald (1950) also observed that carbonates were later than quartz in Paleozoic sandstones in West Virginia, but Gilbert (1949) found that dolomite and, in some cases, calcite were earlier than the quartz in Tertiary sandstones of California.

In general, the order of precipitation of several cementing minerals was established on the principle that those minerals first deposited will be better formed or more euhedral and will be attached to the walls of the pores. The last-precipitated minerals must occupy the remaining unfilled space and be molded about the early formed crystals. But can it be assumed that the minerals were formed in open space? Hadding (1929, p. 17), Cayeux (1929, p. 138), and Swineford (1947, p. 86) have described sandstones first cemented by calcite which was later partly or completely replaced by quartz. Evidence of such replacement lies in scattered inclusions of calcite in secondary quartz. Possibly, therefore, euhedralism is no certain guide to relative age.

The problems of *how* and *when* sands become cemented and the *source* of the cementing material are still unresolved. It was long supposed that meteoric or artesian circulation carried cementing materials into the sandstone and there deposited their silica or carbonate. It is known that groundwaters do carry these materials in solution, that artesian flow does occur, and that materials are precipitated from solution. Van Hise, 1904, p. 866) reviewed the problem at length. He supposed that silica (or other cementing material) was dissolved in the zone of weathering and redeposited in the sands as cement. Van Hise pointed out, however, that the silica content of groundwater was very low. The silica content of waters from igneous terranes ranges from about 10 to 70 ppm (White, Hem, and Waring, 1963, Table 1). Using a figure of 20 ppm, Van Hise estimated that to cement 1 cubic mile of sand (porosity 26 percent) would require 130,000 cubic miles of average groundwater. Because some sandstones, especially those of deeper structural basins, are filled with salt waters, many have supposed that no meteoric water ever circulated through them; thus another source for their cement must be sought.

R. H. Johnson (1920) proposed, therefore, that the silica be derived from connate water. Connate water is essentially trapped seawater; because seawater contains even less silica on the average (about 4 ppm) than groundwater, it is clear that the connate waters of sandstone would provide only a negligible and wholly inadequate quantity of silica. So he suggested that the connate waters of shales may be the source of silica. Shales were originally more porous than the sands and undergo a much greater compaction. Entrapped fluids must escape, and interbedded sandstones may indeed be the channelways of escape. Under elevated temperatures incident to burial, shale waters may contain a higher than normal content of silica, which might be deposited in sandstones. Interstitial waters of modern deep-sea clays are in places supersaturated in silica (up to 80 ppm), but this high silica content seems to be the result of solution of diatom tests (Siever, Beck, and Berner, 1965). A more plausible source of silica, however, is the post-depositional transformation of montmorillonite and/or the mixed layer illite-montmorillonite of shales to illite, a transformation which releases silica (Towe, 1962). This transformation is promoted by deep burial, and the general prevalence of illite in older shales suggests that this mechanism operated on a wide scale. Füchtbauer's (1967a) observation that quartz cementation in a Dogger sandstone increases toward the shale margin of the bed supports the concept of derivation of silica-supersaturated waters from shales. But many cemented sandstones occur without any associated shales. How did these become cemented?

The difficulties of ascribing cements either to artesian or to the waters expelled from shales have led some workers to look *within* the formation for a source of silica. Intrastratal origin has been postulated by Waldschmidt (1941), Gilbert (1949), and others. Waldschmidt concluded that solution of silica at points of grain contact and precipitation of the same in voids were responsible for silica-cemented sandstones. This concept, as noted by Waldschmidt, is essentially Riecke's principle applied to nonmetamorphic rocks. As evidence that the process was operative, Waldschmidt pointed to interlocking boundaries of quartz grains. Concave–convex contacts between quartz grains (analogous to the pitted pebbles of some conglomerates, Kuenen, 1942) and sutured (microstylolitic) boundaries between others are apparent evidences in support of Waldschmidt's conclusions (Sloss and Feray, 1948; Thomson, 1959; Trurnit, 1968). If this concept is correct, one might expect such solution effects and resulting cementation to show some relation to depth of burial and resulting increase in both temperature and pressure, which would promote the process. A number of investigators (Taylor, 1950; Maxwell, 1964; Füchtbauer, 1967a) have shown or postulated such a correlation—a correlation grossly expressed by a porosity decline with depth. Taylor's work is of particular interest; she made direct observations on the nature and number of grain contacts with increasing depth and concluded that sandstones do undergo a "condensation" process whereby individual grain contacts progress from tangential, through concave–convex, to sutured so that the number of apparent contacts per grain in thin sections increases from 1.6 in uncompacted sands to 2.5 at 2885 feet to 5.2 at 8343 feet in depth (Fig. 7-31). Taylor believed these changes to be brought about by intrastratal solution and also by solid flow of the quartz grains. In the latter case, pore volume would be reduced and little chemical precipitation needed. Evidences of pressure, such as bent micas and fractured quartz grains, are common; but solid flow itself is difficult to prove and concave–convex contacts may be a solution effect. On the other hand, relationships between pressure solution and overburden in Pennsylvanian sandstones in the eastern United States was investigated by Siever (1959), with somewhat inconclusive results. Experiments by Fairbairn (1950), and by Maxwell and others (Maxwell and Verrall, 1954; Borg and Maxwell, 1956; Maxwell, 1960; Ernst and Blatt, 1964; Heald and Renton, 1966), demonstrate that silica solution and redeposition in a closed system are a reality and that sand cementation can be achieved by such internal transfers.

Petrographic evidence of solution at grain contacts is not always clear. In many cases little or no grain penetration is observed. In most cases of sutured quartz boundaries, the outlines of the original quartz grains are difficult or impossible to see. Moreover, Sippel (1968) has recently shown that original detrital grains in some of these sandstones are clearly seen with cathodolumi-

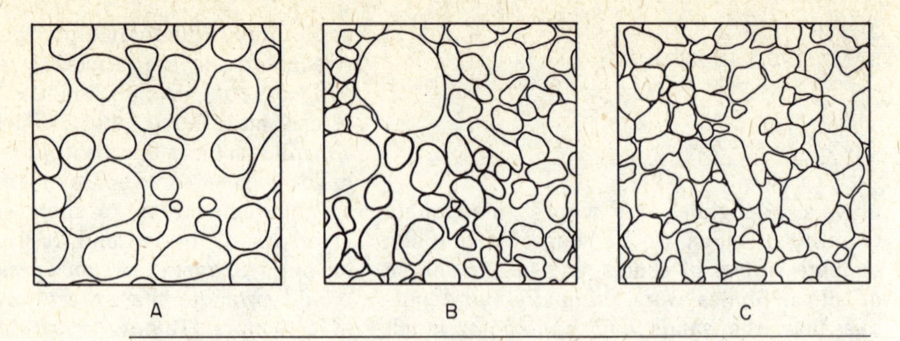

FIG. 7-31. Camera lucida sketches showing packing of grains and condensation. A: Arbuckle calcareous oolite, fewer than 0.5 contacts per grain; B: Lake Superior Sandstone (Cambrian), 1.5 contacts per grain; C: Montebello Sandstone (Devonian), 2.6 contacts per grain.

nescence microscopy and that no penetration of boundaries is present. Secondary quartz is almost certainly not derived by solution at grain contacts. Pye (1944, p. 102) and also Goldstein (1948, p. 114) thought that some or most silica cement might have derived from solution of the fines which the sand may have once contained. It is a well-known observation that, in a given solution, finer particles may dissolve at the same time the larger ones are growing. Conceivably, therefore, larger grains in a sandstone might grow at the expense of smaller ones, but, because fines are destroyed in the process, proof of the validity of this concept is difficult to establish. Heald (1955; 1956a) has suggested that silica was derived by intrastratal solution along stylolitic seams. Stylolites are more common in sandstones than is commonly realized, and solution along such seams might provide some of the silica required for cementation.

Krynine (1941, p. 112), who deemed both intrastratal solution and artesian flow to be inadequate mechanisms of cementation, concluded, from a study of the Devonian Oriskany and other sandstones of the central Appalachians, that ". . . probably close to 95 percent of the 'secondary' silica in the Oriskany (and in many if not most other quartzites and cherts) is really of primary penecontemporaneous origin." Krynine thought that precipitation of silica cement "takes place at the bottom of the sea immediately following the deposition of the sand grains." He does not present evidence to support this conclusion, and the absence of present-day cementation of marine sands contemporaneous with their deposition makes this hypothesis untenable.

Hydrothermal waters, notably those of some hot springs, are especially rich in silica containing 500 or more ppm. That such waters have played a role in sand cementation on a large scale is very unlikely, although some cementation, particularly precipitation of secondary feldspar near igneous bodies, has been reported (Heald, 1956b).

Carbonate cements pose problems similar to those of silica. The principal questions are the *source* of the carbonate and the *time* of its emplacement. Artesian waters contain dissolved carbonates and could presumably provide cement for carbonate-cemented sandstones.

Some modern sands are cemented *in situ*. Beach rock is a good example. These are, however, generally carbonate sands and were, perhaps, self-cemented. Aragonite-cemented sandstone attributed to submarine cementation has been recovered from the outer continental shelf near Delaware Bay (Allen et al., 1969). Seawater trapped in pores of marine sands may be carbonate-supersaturated, but the quantity, if precipitated, is inadequate to cement the sand. Much carbonate cement is later than quartz and is, therefore, not contemporaneous with the sedimentation process or even related to the environment of deposition.

A more probable source of carbonate is the enclosed shelly materials. Such shells may dissolve and reprecipitate as cement. Cementation in the proximity of such shelly material or within a bivalve (Krynine, 1940, Fig. 8) demonstrates, on a small scale, the process of redistribution of carbonate. Pressure solution of carbonate grains within sandstone or in nearby formations, either limestones or sandstones, probably with pore-water transport over limited distances, may be the most likely explanation for late carbonate cement.

Many unresolved questions remain. What governs the type—aragonite, calcite, dolomite, siderite—of carbonate cement? A sandstone with two or more cementing minerals is a further problem; none of the theories of sandstone cementation has adequately dealt with this problem. Distribution of cement requires further analysis. Why is a sand "tight" here and porous there? Is it related to original cementation or to subsequent decementation? Answers to these questions involve a study of the relations between quantity and kind of cement in sandstones and gross geological factors, such as structure and paleocurrents. Warner (1965) showed that cement in the Duchesne River Formation (Eocene) of the Uinta Mountains area was most abundant in the "upcurrent" direction, as shown by cross-bedding, and the kind of cement, calcite, or quartz, is related to rocks of the source region, with calcite more abundant where the source rocks were mostly limestones. There appeared to be also some relation between abundance of cement and structure. Clearly we need more *mapping* of sandstone cements.

DECEMENTATION

If void-filling fluids and solid grains of a sandstone do not form a closed system, that is, if fluids move out as well as in or if ions can diffuse in and out, materials precipitated in voids might also be dissolved out of them. In other words, leaching of the cement or *decementation* might take place. Such occurs in the case of calcareous sandstones in the weathering profile. But does it occur on a large scale at depth?

Partial replacement of quartz and other detrital grains by carbonate is evidence that some silica was removed. If the cement is quartz and there is evidence that it formed by replacement of a carbonate matrix, there is implied removal of much carbonate. That carbonate removal has occurred in the Oriskany Sandstone was a conclusion reached by Krynine (1941), who observed small quantities of carbonate in the recesses of some pores and believed it to be residual following leaching. If it were only a partial pore filling, it would show euhedral outlines on free margins; apparently these were lacking. Perhaps many highly friable and loosely cemented sandstones, such as the St. Peter Sandstone (Ordovician) of the Upper Mississippi Valley, were once carbonate cemented and have lost their cement by downdip artesian flow. The etched surfaces of the quartz grains of this and other friable sandstones may be the record of attack on the quartz grains of the now-vanished cement.

In conclusion, there seems to be no reason why there should not be extensive carbonate removal from calcareous sandstones, because there is extensive internal solution in many limestones. There is, in fact, some evidence that such has indeed been the case. And as with limestones, such solution can go on under phreatic conditions far below the water table. As in the limestones also, the process may be reversed and voids filled once more by precipitated materials.

INTRASTRATAL SOLUTION

In addition to evidences of the solution of cementing constituents, especially carbonates, sandstones display other evidences of intrastratal solution.

Most conspicuous are stylolitic seams. Stylolites, most commonly thought of as a feature of limestones, also occur in sandstones and quartzites. They occur not only as microstylolites between framework grains, as noted above, but also as conspicuous bedding plane features (Heald, 1955; Stockdale, 1936; Conybeare, 1949).

Stylolites of sandstones display all the usual characteristics. They appear as a surface marked by the interlocking or mutual interpenetration of the two sides. Toothlike projections of one side fit into sockets on the other. The relief varies from a few millimeters to several centimeters. The stylolitic surface itself is marked by a thin deposit of relatively insoluble material. In sandstones this may be a parting of coaly matter; in quartzites the seams are marked by iron oxide.

Further evidence of intrastratal solution is documented by selective loss of some heavy-mineral species. The intrastratal solution of detrital heavy minerals has been demonstrated by Bramlette (1941), who found the mineral assemblage within a calcareous concretion in the Modello Sandstone of California to differ markedly from that in the sand outside the concretionary body (Table 7-12). The problem of intrastratal solution of heavy minerals in sandstones has been discussed in many papers, particularly those of Pettijohn (1941), van Andel (1952; 1959), and Weyl (1950). It is of particular importance where zonation of a sedimentary sequence is made on the presumption of no intrastratal solution. The zones may be stability zones rather than valid stratigraphic levels (Pettijohn, 1941).

MATRIX

As has been noted elsewhere, some sandstones, the graywackes in particular (see p. 227), have a matrix of silt- and clay-sized materials in place of a precipitated mineral cement. The origin and significance of this matrix have been discussed by Cummins (1962) and Kuenen (1966), who conclude that it is probably diagenetic, formed by a process of "graywackisation"—essentially a breakdown of the unstable detrital particles. This process is most effective in those sands which have a high proportion of volcanic debris, such as aphanitic rocks of intermediate to basic composition and volcanic glass. The older, and once more deeply buried, sands generally contain much more matrix than do their younger counterparts, an observation made by Cummins and considered strong evidence of matrix formation by a process of aging. The matrix problem has been discussed in detail in the section on graywackes.

FRACTURED AND DEFORMED GRAINS

Sandstones apparently do not compact in the manner characteristic of shales. Some observations, however, suggest some mechanical or purely physical readjustment to pressure. Large detrital micas are commonly bent or wrapped around more resistant quartz grains. In a few sandstones, quartz grains display extensive fracturing. Fractures traverse individual grains; two or more may radiate from the point of contact with adjacent grains. In a few, a fractured part may rotate slightly with respect to the parent grain.

In general, under normal sedimentary conditions, even in deeply buried sands—sands from wells reaching 30,000 feet (9,120 m)—quartz grains retain their integrity and show few effects of superincumbent pressure.

PETROGENESIS OF SANDSTONES

We have reviewed the properties of sandstones, their classification, and the salient facts about the principal families and some of the problems relating to their formation. We need now to look at sandstones from a larger point of view. We need to know that geologic factors control the production

TABLE 7-12. Percentage of heavy-mineral grains in calcareous concretions and their matrix

	Zircon	Garnet	Titanite	Epidote-zoisite	Hornblende
Hambre sandstone	12	3	10	37	5
Hambre concretion	5	3	6	17	44
Modelo sandstone	20	15	22	2	—
Modelo concretion	12	5	10	53	rare

Source: After Bramlette (1941), Table 1.

of sand and determine its petrography. Or to turn the question around, what do sandstones tell us about the past—about the nature of the source rocks, the relief and climate of the source area, and the agent of transport and the environment of deposition?

Sand is formed by a complex group of processes, and the nature of the sand depends on the relative importance of the several processes (Fig. 7-32). Sands may be generated by weathering processes (both chemical and mechanical), by volcanic action, by earth movements (and even by meteoric impact), and by chemical and biochemical action. A given sand may be produced by only a single process; others are complex and contain materials of several origins. In some cases the end-product, although texturally a sand, is classed as a limestone or a tuff rather than as ordinarily conceived sandstone. We are concerned here with normal epiclastic sands; volcanic and carbonate sands are discussed in other chapters.

As noted in Chapter 1, sandstones form a significant part of the sedimentary record. Estimates of their relative abundance (Table 2-2) indicate that sandstones form one-fourth to one-third of the total sedimentary section. If we accept Poldervaart's estimate (1955) of the total volume of sediment on the continent to be 176×10^6 km^3 and assume that one-fourth is sand, then the total volume of sandstone is 44×10^6 km^3, a volume which corresponds to 120×10^{15} metric tons.

What is the composition of this sand? The average mineral composition can be estimated by recasting the chemical analysis. The result is quartz 59 percent, feldspar 22 percent, kaolin 6 percent, chlorite 4 percent, calcite 6 percent, and iron oxide 2 percent. But such a calculated composition does not distinguish between grains and cement, nor does it permit an estimate of rock particles. If all of the kaolin and chlorite and one-third of the feldspar are considered to be present in rock particles, the average sandstone on a cement-free basis (omitting calcite and iron oxide) would be quartz 65 percent, feldspar 15 percent, and rock particles 18 percent.

What is the relative importance of the several sandstone clans or families? Several estimates have been published (Krynine, 1948; Tallman, 1949; Middleton, 1960; Pettijohn, 1963). These vary somewhat because of differences in the way in which the several classes were defined and because of differences in size and nature of the sample on which the estimates were based. If graywackes are derived from lithic arenites as a result of degradation of framework grains (which would therefore include them with lithic sands), the proportions are quartz arenite about 35 percent, arkose 15 to 20 percent, and lithic sandstones 45 to 50 per-

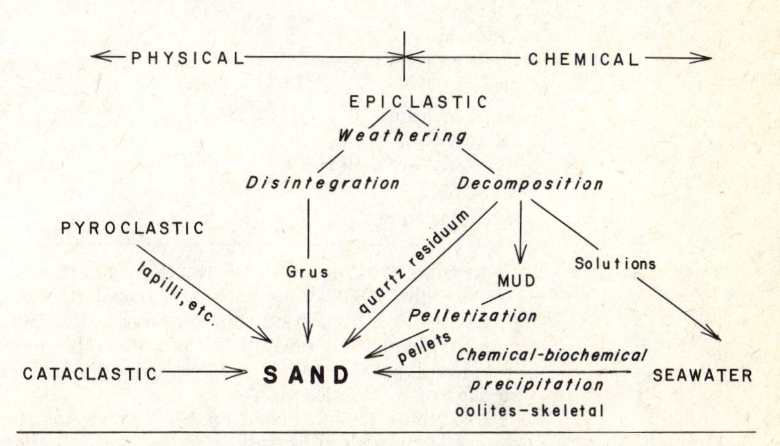

FIG. 7-32. Diagram showing origin of sand. (After Pettijohn, Potter, and Siever, 1972, Sand and sandstone, Fig. 8-1. By permission of Springer-Verlag.)

deposited? About the source land, its nature, climate, and relief? And by what agent was sand transported and in what kind of an environment was it deposited? What principles can we enumerate that govern sand production and sandstone petrography?

cent. (See Table 7-13.) Lithic sandstones are certainly dominant. They are very much like the sand of large present-day rivers, such as the Ohio (Friberg, 1970). Arkoses and orthoquartzites are, in a sense, unusual sands requiring special conditions for their formation and preservation, restricted provenance in the first case and exceptional tectonic stability in the second.

Estimates of relative and absolute abundance of sand, its average chemical and mineralogical composition, and relative abundance of the several types of sandstones are of interest, perhaps, to geochemists concerned with sedimentary cycling and mass balance on a global scale—the study of production of sediment, its transfer from land to sea and back again, and its destruction and incorporation into the subcrust. Interesting as this point of view is, most geologists seek answers to questions of more immediate and specific nature. What does *this* sandstone tell me about the paleogeography at the time it was

Geologic factors that control production of sand and that determine the kind of sand produced have been mentioned at various places in this chapter (Fig. 7-32). Briefly summarized for epiclastic sands these are (1) the *source rock,* (2) the *climate* in both the source area and at the site of deposition, (3) the *environment* and/or *agent* of transport and deposition, and (4) the *tectonics* in both the source area and at the place of deposition. Relations between these geologic factors and the textures, structures, and composition of sandstones are very complex and only partially understood. Comprehensive theories of sandstone petrogenesis which attempt to deal with the problem are few and often incomplete; most are debatable in one or another respect.

Certainly the source rock, or rocks, play a vital role in determining the character of a sand. Sands supplied by small watersheds of differing bedrock differ from one another. Studies of modern rivers, the Rhine in particular, have shown the effect of local bed-

TABLE 7-13. Relative abundance of sandstone classes

Class	Krynine[a] (1948)	Pettijohn[b] (1960)	Middleton[c] (1960)	Tallman[d] (1949)
Orthoquartzite	22.5	34	34	45
Arkose[e]	32.5	15	16	17
Graywacke ("high rank graywacke")	10.0	26	24	17
Lithic arenite ("low rank graywacke" or "subgraywacke")	35.0	20	26	21
Miscellaneous		5		

[a] Krynine (1948), basis of estimate not stated.

[b] Pettijohn (1963), new estimate, based on 121 sandstones in Johns Hopkins collection for which thin sections are available. From point of view of time involved the Precambrian is underrepresented; from point of view of area involved the Maryland-Pennsylvania area is oversampled (about 15 percent of total collection).

[c] Middleton (1960), based on 167 sandstones for which chemical analyses appear in published literature.

[d] Tallman (1949), based on sample of 275 Cambrian to Tertiary sandstones from all parts of United States.

[e] Includes subarkose.

rock on the composition of the sands in the river (Hahn, 1969; Koldewijn, 1955). Sands derived from a rhyolitic volcanic terrane (Webb and Potter, 1969) will differ materially from those produced in a region underlain by deep-seated granites and gneisses (Hayes, 1962). The first has a paucity of quartz (all of which is volcanic) and feldspar and a superabundance of rhyolitic rock fragments, whereas the latter is primarily quartz and feldspar with few or no rock fragments. Studies of modern sedimentary basins show that the mineralogy of sands reflects closely the composition of the source area. This is especially true of heavy-mineral suites in sands, as illustrated by the studies of the Gulf of Mexico (Davies and Moore, 1970) and of the North Sea (Baak, 1936).

Yet sands derived from a given source area differ in composition from the source rock itself. Some minerals are more susceptible to weathering than others and are selectively lost. The loss is governed by the nature of the mineral itself (Gruner, 1950). Order of loss has been empirically determined (Goldich, 1938). But it also depends on the nature and intensity of the weathering process itself and on the duration of time through which the action is continued. Climate governs weathering; tectonics controls relief. Under conditions of low relief and a warm humid climate only stable species survive; with high relief erosion is accelerated and weathering is interrupted midcourse so that unstable species escape destruction and appear in sands. Observational data to support these deductive general principles are scant indeed. Krynine's observations on the production of highly feldspathic sandstones in tropical Mexico seem to support these concepts (see p. 218).

The effects of the agent of transport (wind, waves, rivers, ice) and the environment of deposition (beach, delta, dune, etc.) on the petrography of sand are relatively minor. Despite a large effort to relate textural parameters, grain rounding, and the like to either the agent or to the environment, the results are largely negative. This is in part because many grain properties—size, shape, roundness, and composition—are only slightly modified by abrasion during transport, and, more important, many are inherited from a previous cycle of sedimentation and do not indicate the last environment of deposition. Despite impressive experimental and field evidence that support this view, some contrary geologic evidence shows that the environment may leave an imprint on the petrography of the sandstone. Folk (1960) interprets the well-rounded orthoquartzitic phase of the Silurian Tuscarora of West Virginia as a product of "cleaning up" of a subangular, less mature sand in a beach environment. Likewise the Lower Carboniferous Kellerwald Quartzite, a pure quartz sand in the Variscan geosyncline west of the Rhine, is interpreted as a shelf edge bar—a product of winnowing and reworking graywacke material (Meischner, 1971, p. 19).

The notion that the primary control of petrography of a sandstone is tectonic arises from the observation that the maturity of the sands on cratons, or sands derived from cratons, is markedly greater than that of those sands deposited in geosynclines, particularly sands derived from "tectonic lands." The typical sandstone of the eugeosyncline is graywacke; that of the craton is orthoquartzite. Sands in a miogeosyncline of cratonic or foreland derivation are generally orthoquartzites; those from interior tectonic land are lithic sandstones or graywackes. It was presumed that tectonic stability governed petrographic character. In the central Appalachians, for example, sandstones with a westerly or cratonic source —the Weverton and Antietam (Schwab, 1970)—are orthoquartzites and protoquartzites; those with a southeasterly source—sandstones of the Martinsburg (McBride, 1962), the Juniata, the Bald Eagle (Yeakel, 1962), the Pocono (Pelletier, 1958), and the Mauch Chunk and Pottsville (Meckel, 1967) —are lithic arenites or graywackes. The southeast-derived Tuscarora (Yeakel, 1962) is a subgraywacke or protoquartzite in most places but is locally modified to an orthoquartzite. Likewise, sands of the Coronation Gulf geosyncline (Hoffman, Fraser, and McGlynn, 1970) derived from a stable cratonic area to the east of the trough (Hornby Channel, Kluziai of Great Slave Lake and the Western River, Odjick, and Burnside lands and those shed from interior regions of the geosyncline—the "tectonic lands."

River of the Epworth-Goulburn areas) are subarkoses and orthoquartzites. Sands derived from the opposite side of the geosyncline and somewhat younger (the Recluse of the Epworth area and the sands of the Pethei Group in Great Slave Lake) are graywackes.

Krynine (1942) was one of the first to formulate a complete theory of sandstone petrogenesis in which tectonics was considered to be the overriding factor. He related each sandstone type to a particular stage of the diastrophic cycle (Fig. 16-8), a view which, with minor modifications, was supported by others (Dapples, 1947; Dapples, Krumbein, and Sloss, 1948; Pettijohn, 1943). Van Andel (1958) has reassessed these concepts and attempted to sort out the geologic controls which determine the petrography of some Cretaceous, Paleocene, and Eocene sandstones of western Venezuela. He believed that source materials rather than tectonism are the basic controls over sandstone composition and that environment determines only texture and textural maturity. He states (p. 762) that "there is no systematic tectofacies control of texture and textural maturity" and adds that "mineralogical maturity does not reflect the tectofacies of the depositional basin"; hence, only insofar as the relief of the source area is controlled by tectonics does the petrography of a sandstone reflect the latter.

To a considerable degree most workers have supposed that basin subsidence and tectonic uplift are coupled together—an assumption not valid when the source area is far removed from the basin of deposition, as is the case for most large river systems.

We can only say in conclusion that in general the petrography of a sandstone is a valuable clue to provenance, and indirectly to the climate and relief of the source area, and is generally insensitive to the environment of deposition. Relations between petrographic character and tectonism are not fully understood, but there do seem to be significant differences in many geosynclines between the sands derived from stable fore-

References

Adams, R. W., 1970, Loyalhanna Limestone—cross-bedding and provenance, *in* Studies of Appalachian geology—central and southern (Fisher, G. W., Pettijohn, F. J., Reed, J. C., Jr., and Weaver, K. N., eds.): New York, Wiley-Interscience, pp. 83–100.

Allen, J. R. L., 1962, Petrology, origin and deposition of the higher Old Red Sandstone of Shropshire, England: Jour. Sed. Petrology, v. 32, pp. 657–697.

Allen, R. C., Gavish, Eliezer, Friedman, G. M., and Sanders, J. E., 1969, Aragonite-cemented sandstone from outer continental shelf off Delaware Bay: submarine lithification mechanism yields product resembling beachrock: Jour. Sed. Petrology, v. 39, pp. 136–149.

Allen, V. T., 1936, Terminology of medium-grained sediments, *in* Rept. Comm. Sedimentation: Nat. Res. Coun., 1935–1936, pp. 18–47.

———, 1937, A study of Missouri glauconite: Amer. Mineral., v. 22, pp. 842–846.

van Andel, Tj. H., 1952, Zur Frage der Schwermineralverwitterung in Sedimenten: Erdöl und Kohle, v. 5, pp. 100–104.

———, 1958, Origin and classification of Cretaceous, Paleocene, and Eocene sandstones of western Venezuela: Bull. Amer. Assoc. Petrol. Geol., v. 42, pp. 734–763.

———, 1959, Reflections on the interpretation of heavy mineral analyses: Jour. Sed. Petrology, v. 29, pp. 153–163.

Anderson, D. W., and Picard, M. D., 1971, Quartz extinction in siltstone: Bull. Geol. Soc. Amer., v. 82, pp. 181–186.

Andresen, M. J., 1961, Geology and petrology of the Trivoli Sandstone in the Illinois Basin: Illinois Geol. Surv. Circ. 316, 31 pp.

Anhaeusser, C. R., Mason, R., Viljoen, M. J., and Viljoen, R. P., 1969, A reappraisal of some aspects of Precambrian Shield geology: Bull. Geol. Soc. Amer., v. 80, pp. 2175–2200.

Ashley, G. H., 1918, Notes on the greensand deposits of the eastern United States: Bull. U.S. Geol. Surv. 660-B, pp. 27–49.

Baak, J. A., 1936, Regional petrology of the southern North Sea; Wageningen, H. Veenman u. Zonen, 127 pp.

Bailey, E. H., and Irwin, W. P., 1959, K-feldspar content of Jurassic and Cretaceous graywackes of northern Coast Ranges and Sacramento Valley, California: Bull. Amer. Assoc. Petrol. Geol., v. 43, pp. 2797–2809.

Balk, R., 1953, The structure of graywacke areas and Taconic Range, east of Troy, New York: Bull. Geol. Soc. Amer., v. 64, pp. 811–864.

Barth, T. F. W, 1938, Progressive metamorphism of sparagmite rocks of southern Norway: Norsk Geol. Tidsskr., v. 18, pp. 54–65.
Barton, D. C., 1916, The geological significance and genetic classification of arkose deposits: Jour. Geol., v. 24, pp. 417–449.
Baskin, Y., 1956, A study of authigenic feldspars: Jour. Geol., v. 64, pp. 132–155.
Blatt, H., 1959, Effect of size and genetic quartz type on sphericity and form of beach sediments, northern New Jersey: Jour. Sed. Petrology, v. 29, pp. 197–206.
———, 1966, Diagenesis of sandstones: processes and problems: Symp. 12th Ann. Conf. Wyoming Geol. Assoc., pp. 63–65.
———, 1967, Original characteristics of clastic quartz grains: Jour. Sed. Petrology, v. 37, pp. 401–424.
Blatt, H., and Christie, J. M., 1963, Undulatory extinction in quartz of igneous and metamorphic rocks and its significance in provenance studies of sedimentary rocks: Jour. Sed. Petrology, v. 33, pp. 559–579.
Bloss, F. D., 1957, Anistropy of fracture in quartz: Amer. Jour. Sci., v. 255, pp. 214–225.
Boggs, S., Jr., 1968, Experimental study of rock particles: Jour. Sed. Petrology, v. 38, pp. 1326–1339.
Bokman, J., 1952, Clastic quartz particles as indices of provenance: Jour. Sed. Petrology, v. 22, pp. 17–24.
de Booy, T., 1966, Petrology of detritus in sediments, a valuable tool: Proc. Konink. Nederl. Akad. van Wetensch., v. 69, ser. B., pp. 277–282.
Borg, I. Y., and Maxwell, J. C., 1956, Interpretation of fabrics of experimentally deformed sands: Amer. Jour. Sci., v. 254, pp. 71–81.
Boswell, P. G. H., 1933, On the mineralogy of the sedimentary rocks: London, Murby, 393 pp.
Bramlette, M. N., 1941, The stability of minerals in sandstone: Jour. Sed. Petrology, v. 11, pp. 32–36.
Brenchley, P. J., 1969, Origin of matrix in Ordovician greywackes, Berwyn Hills, North Wales: Jour. Sed. Petrology, v. 39, pp. 1297–1301.
Brett, G. W., 1955, Cross-bedding in the Baraboo Quartzite of Wisconsin: Jour. Geol., v. 63, pp. 143–148.
Brongniart, A., 1826, De l'arkose, caractères minéralogiques et histoire géonostique de cette roche: Ann. Sci. Nat., v. 8, pp. 113–163.
Burchard, E. F., 1907, Notes on various glass sands mainly undeveloped: Bull. U.S. Geol. Surv. 315, pp. 377–382.
Burst, F. F., 1958, "Glauconite" pellets; their mineral nature and applications to stratigraphic interpretation: Bull. Amer. Assoc. Petrol. Geol., v. 42, pp. 310–327.
Bushinsky, G. I., 1935, Structure and origin of phosphorites of the U.S.S.R.: Jour. Sed. Petrology, v. 5, pp. 81–92.
Buttram, Frank, 1913, The glass sands of Oklahoma: Oklahoma Geol. Surv. Bull. 10, 91 pp.
Cadigan, R. A., 1967, Petrology of Morrison Formation in the Colorado Plateau region: U.S. Geol. Surv. Prof. Paper 556, 113 pp.
Cameron, K. L., and Blatt, H., 1971, Durabilities of sand size fluvial transport, Elk Creek, Black Hills, South Dakota: Jour. Sed. Petrology, v. 41, pp. 565–576.
Carrigy, M. A., and Mellon, G. B., 1964, Authigenic clay mineral cements in Cretaceous and Tertiary sandstones of Alberta: Jour. Sed. Petrology, v. 34, pp. 461–472.
Carter, C. H., 1972, Miocene-Pliocene beach and tidal flat sedimentation, southern New Jersey: Ph.D. dissertation, The Johns Hopkins Univ., pp. 186.
Carver, R. E., ed., 1971, Procedures in sedimentary petrology: New York, Wiley-Interscience, 653 pp.
Cary, A. S., 1951, Origin and significance of openwork gravel: Trans. Amer. Soc. Civil Eng., v. 116, pp. 1296–1308.
Casshyap, S. M., 1969, Petrology of the Bruce and Gowganda formations and its bearing on the evaluation of Huronian sedimentation in the Espanola-Willisville area, Ontario (Canada): Paleogeogr. Paleoclimat. Paleoecol., v. 6, pp. 5–36.
Cayeux, L., 1929, Les roches sédimentaires de France: roches siliceuses: Paris, Imprimerie Nationale, 774 pp.
Chilingar, G. V., 1955, Joint occurrence of glauconite and chlorite in sedimentary rocks—a review: Bull. Amer. Assoc. Petrol. Geol., v. 40, pp. 493–498.
Clarke, F. W., 1924, The data of geochemistry: Bull. U.S. Geol. Surv. 770, 841 pp.
Cleary, W. J., and Conolly, J. R., 1971, Distribution and genesis of quartz in a Piedmont–Coastal Plain environment: Bull. Geol. Soc. Amer., v. 82, pp. 2755–2766.
Clements, J. M., 1903, The Vermilion iron-bearing district of Minnesota: U.S. Geol. Surv. Monogr. 45, 463 pp.
Cloud, P. E., Jr., 1955, Physical limits of glauconite formation: Bull. Amer. Assoc. Petrol. Geol., v. 39, pp. 484–492.
Collins, W. H., 1925, The north shore of Lake Huron: Geol. Surv. Canada, Mem. 143, 160 pp.
Colton, G. W., 1970, The Appalachian Basin—its depositional sequences and their geologic relationships, *in* Studies of Appalachian geol-

ogy—central and southern (Fisher, G. W., Pettijohn, F. J., Reed, J. C., Jr., and Weaver, K. N., eds.): New York, Wiley-Interscience, pp. 5–47.

Conolly, J. R., 1965, The occurrence of polycrystallinity and undulatory extinction in quartz in sandstones: Jour. Sed. Petrology, v. 35, pp. 116–135.

Conybeare, C. E. B., 1949, Stylolites in pre-Cambrian quartzite: Jour. Geol., v. 57, pp. 83–85.

Cressman, E. R., and Swanson, R. W., 1964, Stratigraphy and petrology of the Permian rocks of southwestern Montana: U.S. Geol. Surv. Prof. Paper 313-C, pp. 275–569.

Crook, K. A. W., 1955, Petrology of graywacke suite sediments from Turon River, Coolamigal Creek district, N.S.W.: Proc. Roy. Soc. New South Wales, v. 88, pp. 97–105.

———, 1960, Petrology of Tamworth Group Lower and Middle Devonian, Tamworth, Nundle district, New South Wales: Jour. Sed. Petrology, v. 30, pp. 353–369.

———, 1968, Weathering and rounding of quartz sand grains: Sedimentology, v. 11, pp. 171–182.

Crowley, A. J., 1939, Possible criterion for distinguishing marine and nonmarine sediments: Bull. Amer. Assoc. Petrol. Geol., v. 23, pp. 1716–1720.

Cummins, W. A., 1962, The greywacke problem: Liverpool and Manchester Geol. Jour., v. 3, pp. 51–72.

Cushing, H. P., Leverett, F., and Van Horn, F. R., 1931, Geology and mineral resources of the Cleveland district, Ohio: U.S. Geol. Surv. Bull. 818, 138 pp.

Dake, C. L., 1921, The problem of the St. Peter Sandstone: Bull. Missouri School Mines and Metall., v. 6, 228 pp.

Dapples, E. C., 1947, Sandstone types and their associated depositional environments: Jour. Sed. Petrology, v. 17, pp. 91–100.

———, 1972, Some concepts of cementation and lithification of sandstones: Bull. Amer. Assoc. Petrol. Geol., v. 56, pp. 3–25.

Dapples, E. C., Krumbein, W. C., and Sloss, L. L., 1948, Tectonic control of lithologic associations: Bull. Amer. Assoc. Petroleum Geol., v. 32, pp. 1924–1947.

Daubrée, A., 1879, Études synthétiques de géologie expérimentale, 2 vols., Paris: Dunod, 828 pp.

Davies, D. K., and Moore, W. R., 1970, Dispersal of Mississippi sediment in the Gulf of Mexico: Jour. Sed. Petrology, v. 40, pp. 339–353.

Davis, E. F., 1918, The Franciscan sandstone: Bull. Univ. California Univ. Publ., Dept. Geol., v. 11, pp. 6–16.

Dickinson, W. R., 1968, Singatoka dune sands, Viti Lebu (Fiji): Sed. Geol., v. 2, pp. 115–124.

———, 1969, Evolution of calc-alkaline rocks in the geosynclinal system of California and Oregon: Proc. Andesite Conference, Bull. Oregon Dept. Geol. Min. Ind. 65, pp. 151–156.

———, 1970, Interpreting detrital modes of graywacke and arkose: Jour. Sed. Petrology, v. 40, pp. 695–707.

Diller, J. S., 1898, The educational series of rock specimens, etc.: U.S. Geol. Surv. Bull. 150, 400 pp.

Donaldson, J. A., 1967, Two Proterozoic clastic sequences: a sedimentological comparison: Proc. Geol. Assoc. Canada, v. 18, pp. 33–54.

Donaldson, J. A., and Jackson, G. D., 1965, Archean sedimentary rocks of North Spirit Lake area, northwestern Ontario: Canad. Jour. Earth Sci., v. 2, pp. 622–647.

Dott, Robert H., Jr., 1964, Wacke, graywacke and matrix—what approach to immature sandstone classification?: Jour. Sed. Petrology, v. 34, pp. 625–632.

Duplaix, S., 1948, Détermination microscopique des minéraux des sables, 80 pp. Paris-Liège: Librarie Polytech. Ch. Beranger.

Edwards, A. B., 1945, The glauconitic sandstone of the Tertiary of East Gippsland, Victoria: Proc. Roy. Soc. Victoria, no. 5, v. 57, pp. 153–167.

———, 1950a, The petrology of the Miocene sediments of the Aure Trough, Papua: Proc. Roy. Soc. Victoria, v. 60, pp. 123–148.

Edwards, A. B., 1950b, The petrology of the Cretaceous greywackes of the Purari Valley, Papua: Proc. Royal Soc. Victoria, n.s., v. 60, pp. 163–171.

Ehrenberg, H., 1928, Sedimentpetrographische Untersuchungen an Nebengesteinen der Aachener Steinkohlenvorkommen: Preuss. Geol. Landesanst. Jahrb., v. 49, pp. 33–58.

Emery, K. O., 1964, Turbidites—Precambrian to present, *in* Studies on oceanography: Tokyo, Univ. Tokyo Press, 568 pp.

———, 1966, Geologic background, *in* The Atlantic Continental Shelf and slope of the United States: U.S. Geol. Surv., Prof. Paper 529-A, pp. 1–23.

Engel, A. E. J., and Engel, C. G., 1953, Grenville Series in the northwest Adirondack Mountains: Bull. Geol. Soc. Amer., v. 64, pp. 1013–1097.

Ernst, W. G., and Blatt, H., 1964, Experimental study of quartz overgrowths and synthetic quartzites: Jour. Geol., v. 72, pp. 461–470.

Fahrig, W. F., 1961, The geology of the Athabaska Formation: Bull. Geol. Surv. Canada 68, 41 pp.

Fairbairn, H. W., 1950, Synthetic quartzite: Amer. Mineral., v. 35, pp. 735–748.

Feo-Codecido, G., 1956, Heavy-mineral techniques and their application to Venezuelan stratigraphy: Bull. Amer. Assoc. Petrol. Geol., v. 40, pp. 948–1000.

Ferrar, H. T., 1934, The geology of the Dargaville-Rodney Subdivision: New Zealand Geol. Surv. Bull. 34, 78 pp.

Fischer, G., 1933, Die Petrographie der Grauwacken: Preuss. Geol. Landesanst. Jahrb., v. 54, pp. 320–343.

Folk, R. L., 1951, Stages of textural maturity in sedimentary rocks: Jour. Sed. Petrology, v. 21, pp. 127–130.

———, 1954, The distinction between grain size and mineral composition in sedimentary-rock nomenclature: Jour. Geol., v. 62, pp. 344–359.

———, 1960, Petrography and origin of the Tuscarora, Rose Hill, and Keefer formations, Lower and Middle Silurian, of eastern West Virginia: Jour. Sed. Petrology, v. 30, pp. 1–58.

———, 1968, Petrology of sedimentary rocks: Austin, Tex., Hemphills, 170 pp.

Folk, R. L., Andrews, P. B., and Lewis, D. W., 1970, Detrital sedimentary rock classification and nomenclature for use in New Zealand: New Zealand Jour. Geol. Geophys., v. 13, pp. 937–968.

Foster, M. D., 1969, Studies of celadonite and glauconite: U.S. Geol. Surv. Prof. Paper 614-F, 17 pp.

Foster, R. J., 1960, Tertiary geology of a portion of the central Cascade Mountains, Washington: Bull. Geol. Soc. Amer., v. 71, pp. 99–125.

Foushee, E. D., 1954, A report on the flexible sandstone or itacolumite of Stokes County, North Carolina: Compass, v. 31, pp. 78–80.

Friberg, J. F., 1970, Mineralogy and provenance of the Recent alluvial sands of the Ohio River Basin: Ph.D. dissertation, Indiana Univ.

Friedman, G. M., 1961, Distinction between dune, beach, and river sands from their textural characteristics: Jour. Sed. Petrology, v. 31, pp. 514–529.

———, 1967, Dynamic processes and statistical parameters compared for size frequency distribution of beach and river sands: Jour. Sed. Petrology, v. 37, pp. 327–354.

Frye, J. C., and Swineford, A., 1946, Silicified rock in the Ogallala Formation: Bull. State Geol. Surv. Kansas, no. 64, pt. 2, pp. 37–76.

Füchtbauer, H., 1964, Sedimentpetrographische Untersuchungen an der älteren Molasse nördlich der Alpen: Eclogae Géol. Helvetiae, v. 57, pp. 157–298.

———, 1967a, Influence of different types of diagenesis on sandstone porosity: Proc. 7th World Petrol. Congr., pp. 353–369.

———, 1967b, Die Sandsteine in der Molasse nördlich der Alpen: Geol. Rundschau, v. 56, pp. 266–300.

Fuhrmann, W., 1968, "Sandkristalle" und Kugelsandsteine. Ihre Rolle bei der Diagenese von Sanden: Der Aufschluss, v. 5, pp. 105–111.

Galliher, E. W., 1936, Glauconite genesis: Bull. Geol. Soc. Amer., v. 46, pp. 1351–1356.

———, 1939, Biotite-glauconite transformation and associated minerals, *in* Recent marine sediments (Trask, P. D., ed.) Tulsa, Okla., Amer. Assoc. Petrol. Geol., pp. 513–515.

Gasser, U., 1968, Die innere Zone der subalpinen Molasse des Entlebuchs (Kt. Luzern). Geologie und Sedimentologie: Eclogae Géol. Helvetiae, v. 61, pp. 229–319.

Gilbert, C. M., 1949, Cementation in some California Tertiary reservoir sands: Jour. Geol., v. 57, pp. 1–17.

Ginsburg, L., and Lucas, G., 1949, Présence de quartzites élastiques dans les grès armoricains metamorphiques de Berrien (Finistère): Acad. Sci. Paris, C. R., v. 228, pp. 1657–1658.

Glass, H. D., Potter, P. E., and Siever, R., 1956, Clay mineralogy of some basal Pennsylvanian sandstones, clay, and shales: Bull. Amer. Assoc. Petrol. Geol., v. 40, pp. 750–754.

Glover, J. E., 1963, Studies in the diagenesis of some Western Australian sedimentary rocks: Jour. Roy. Soc. Western Australia, v. 46, pp. 33–56.

Gluskoter, H. J., 1964, Orthoclase distribution and authigenesis in the Franciscan Formation of a portion of western Marin County, California: Jour. Sed. Petrology, v. 34, pp. 335–343.

Goldich, S. S., 1934, Authigenic feldspar in sandstone of southeastern Minnesota: Jour. Sed. Petrology, v. 4, pp. 89–95.

———, 1938, A study in rock weathering: Jour. Geol., v. 46, pp. 17–58.

Goldman, M. I., 1915, Petrographic evidence on the origin of the Catahoula Sandstone of Texas: Amer. Jour. Sci., ser. 4, v. 39, pp. 261–287.

———, 1919, General character, mode of occurrence and origin of glauconite: Jour. Wash. Acad. Sci., v. 9, pp. 501–502.

Goldstein, A., Jr., 1948, Cementation of the Dakota Sandstone of the Colorado Front Range: Jour. Sed. Petrology, v. 18, pp. 108–125.

Gorbatscher, R., and Klint, O., 1961, The Jotnian Mälar Sandstone of the Stockholm region: Bull. Geol. Inst. Univ. Uppsala, v. 40, pp. 51–68.

Grabau, A. W., 1904, On the classification of sedimentary rocks: Amer. Geol., v. 33, pp. 228–247.

Graham, W. A. P., 1930, A textural and petrographic study of the Cambrian sandstones of Minnesota: Jour. Geol., v. 38, pp. 696–716.

Greenly, E., 1897, Incipient metamorphism in the Harlech Grits: Trans. Edinburgh Geol. Soc., v. 7, pp. 254–258.

Greensmith, J. T., 1957, Lithology, with particular reference to cementation, etc.: Jour. Sed. Petrology, v. 27, p. 405.

Griffiths, J. C., 1956, Petrographical investigations of the Salt Wash sediments: U.S. Atomic Energy Comm. Tech. Rept. RME-3122 (Pts. I and II), 84 pp.

Griggs, A. B., 1945, Chromite-bearing sands of the southern part of the coast of Oregon: Bull. U.S. Geol. Surv. 945-E, pp. 113–150.

Gruner, J. W., 1935, The structural relationship of glauconite and mica: Amer. Mineral., v. 20, pp. 699–714.

———, 1950, An attempt to arrange silicates in the order of reaction energies at relatively low temperatures: Amer. Mineral., v. 35, pp. 137–148.

Hadding, A., 1929, The pre-Quaternary sedimentary rocks of Sweden, III. The Paleozoic and Mesozoic sandstones of Sweden: Lunds Univ. Årsskr., N. F., Avd. 2, v. 25, 287 pp.

———, 1932, The pre-Quaternary rocks of Sweden, IV. Glauconite and glauconitic rocks: Medd. Lunds Geol. Min. Inst., no. 51, 175 pp.

Hadley, D. G., 1968, The sedimentology of the Huronian Lorrain Formation, Ontario and Quebec, Canada: Ph.D. dissertation, Johns Hopkins Univ., 301 pp.

Hahn, C., 1969, Mineralogisch-Sedimentpetrographische Untersuchungen an den Flussbettsanden im Einzugsbereich des Alpenrheins: Eclogae Géol. Helvetiae, v. 62, pp. 227–278.

Hamblin, W. K., 1962, X-ray radiography in the study of structures in homogeneous sediments: Jour. Sed. Petrology, v. 32, pp. 201–210.

Harms, J. C., 1969, Hydraulic significance of some sand ripples: Bull. Geol. Soc. Amer., v. 80, pp. 363–396.

Hawkins, J. W., Jr., and Whetten, J. T., 1969, Graywacke matrix minerals: hydrothermal reactions with Columbia River sediments: Science, v. 166, pp. 868–870.

Hay, R. L., 1966, Zeolites and zeolite reactions in sedimentary rocks: Geol. Soc. Amer. Spec. Paper 85, 130 pp.

Hayes, J. R., 1962, Quartz and feldspar content in South Platte, Platte, and Missouri river sands: Jour. Sed. Petrology, v. 32, pp. 793–800.

Heald, M. T., 1950, Authigenesis in West Virginia sandstones: Jour. Geol., v. 58, pp. 624–633.

———, 1955, Stylolites in sandstone: Jour. Geol., v. 63, pp. 101–114.

———, 1956a, Cementation of Simpson and St. Peter sandstones in parts of Oklahoma, Arkansas, and Missouri: Jour. Geol., v. 64, pp. 16–30.

———, 1956b, Cementation of Triassic arkoses in Connecticut and Massachusetts: Bull. Geol. Soc. Amer., v. 67, pp. 1133–1154.

Heald, M. T., and Renton, J. J., 1966, Experimental study of sandstone cementation: Jour. Sed. Petrology, v. 36, pp. 977–991.

Helmbold, R., 1952, Beitrag zur Petrographie der Tanner Grauwacken: Heidelberger Beitr. Min. Petrog., v. 3, pp. 253 288.

Henderson, J. B., 1972, Sedimentology of Archean turbidites at Yellowknife, Northwest Territories: Canad. Jour. Earth Sci., v. 9, pp. 882–902.

Henningsen, D., 1961, Untersuchungen über Stoffbestand und Paläeogeographie der Giessener Grauwacke: Geol. Rundschau, v. 51, pp. 600–626.

Hoffman, P. F., Fraser, J. A., and McGlynn, J. C., 1970, The Coronation Gulf Geosyncline of Aphebian age, District of Mackenzie: Geol. Surv. Canada Paper 70-40, pp. 201–212.

Hollister, C. D., and Heezen, B. C., 1964, Modern graywacke-type sands: Science, v. 146, pp. 1573–1574.

Holmes, A., 1928, The nomenclature of petrology, 2nd ed.: London, Murby, 284 pp.

Hopkins, M. E., 1958, Geology and petrology of the Anvil Rock Sandstone of southern Illinois: Illinois Geol. Surv. Circ. 256, 48 pp.

Hoppe, W., 1927, Beiträge zur Geologie und Petrographie des Buntsandsteins im Odenwald II: Notizbl. Vereinst. Erdkunde, Hessischen Geol. Landesanstalt, ser. 5, v. 10, pp. 54–103.

Hoque, M. ul, 1968, Sedimentologic and paleocurrent study of the Mauch Chunk sandstones (Mississippian) of south-central and western Pennsylvania: Bull. Amer. Assoc. Petrol. Geol., v. 52, pp. 246–263.

Hsu, K. J., 1960, Texture and mineralogy of the Recent sands of the Gulf Coast: Jour. Sed. Petrology, v. 30, pp. 380–403.

Hubert, J. F., 1960, Petrology of the Fountain and Lyons formations, Front Range, Colorado: Colorado School Mine Quart., v. 55, 242 pp.

Huckenholtz, H. G., 1963, Mineral composition and texture in graywackes from the Harz Mountains (Germany) and in arkoses from the Auvergne (France): Jour. Sed. Petrology, v. 33, pp. 914–918.

Hunter, R. E., 1967, The petrography of some Illinois Pleistocene and Recent sands: Sed. Geol., v. 1, pp. 57–75.

Ingerson, E., and Ramisch, J. L., 1942, Origin of

shapes of quartz sand grains: Amer. Mineral., v. 27, pp. 595–606.

Irving, R. D., and Van Hise, C. R., 1884. On secondary enlargements of mineral fragments in certain rocks: U.S. Geol. Surv. Bull., 8, 56 pp.

Jacobsen, Lynn, 1959, Petrology of Pennsylvanian sandstones and conglomerates of the Ardmore Basin: Oklahoma Geol. Surv. Bull. 79, 144 pp.

Johnson, R. H., 1920, The cementation process in sandstone: Bull. Amer. Assoc. Petrol. Geol., v. 4, pp. 33–35.

Keith, M. L., 1949, Sandstone as a source of silica sands in southeastern Ontario: Ontario Dept. Mines Ann. Rept., v. 55, pt. 5, 36 pp.

Keller, W. D., and Littlefield, R. F., 1950, Inclusions in quartz of igneous and metamorphic rocks: Jour. Sed. Petrology, v. 20, pp. 74–84.

Kennedy, W. Q., 1951, Sedimentary differentiation as a factor in the Moine-Torridonian correlation: Geol. Mag., v. 88, pp. 257–261.

Ketner, K. B., 1966, Comparison of Ordovician eugeosynclinal and miogeosynclinal quartzites of the Cordilleran geosyncline, *in* Geological Survey Research 1966, U.S. Geol. Surv. Prof. Paper 550-C, pp. C54–C60.

Klein, G. deV., 1963, Analysis and review of sandstone classification in the North American geological literature: Bull. Geol. Soc. Amer., v. 74, pp. 555–576.

Klovan, J. E., and Solohub, J. T., 1968, Grain-size parameters: a critical evaluation of their significance: Geol. Soc. Amer., Prog. with Abstr., Ann. Mtg. Mexico City, pp. 161–162.

Koldewijn, B. W., 1955, Provenance, transport, and deposition of Rhine sediments. II. An examination of the light fraction: Geol. Mijnb. (n. s.), v. 17, pp. 37–45.

Krumbein, W. C., and Pettijohn, F. J., 1938, Manual of sedimentary petrography: New York, Plenum, 549 pp.

Krynine, P. D., 1935, Arkose deposits in the humid topics: a study of sedimentation in southern Mexico: Amer. Jour. Sci., ser. 5, v. 29, pp. 353–363.

———, 1937, Petrography and genesis of the Siwalik series: Amer. Jour. Sci., ser. 5, v. 34, pp. 422–446.

———, 1940, Petrology and genesis of the Third Bradford Sand: Bull. Pennsylvania State Coll. Min. Ind. Exp. Sta. 29, pp. 13–20.

———, 1941, Petrographic studies of variations in cementing material in the Oriskany Sand: Proc. 10th Pennsylvania Min. Ind. Conf., Bull. Pennsylvania State Coll. 33, pp. 108–116.

———, 1942, Differential sedimentation and its products during one complete geosynclinal cycle: Santiago, Chile, An. Congr. Panamer. Ing. Minas Geol., pt. 1, v. 2, pp. 536[b]561.

———, 1945, Sediments and the search for oil: Producers Monthly, v. 9, no. 3, pp. 12–22.

———, 1946a, The tourmaline group in sediments: Jour. Geol., v. 54, pp. 65–87.

———, 1946b, Microscopic morphology of quartz types: An. 2nd Congr. Panamer. Ing. Minas Geol., v. 3, pp. 35–49.

———, 1948, The megascopic study and field classification of sedimentary rocks: Jour. Geol., v. 56, pp. 130–165.

———, 1950, Petrology, stratigraphy and origin of the Triassic sedimentary rocks of Connecticut: Bull. Connecticut State Geol. Nat. Hist. Surv. 73, 247 pp.

Krynine, P. D., and Tuttle, O. F., 1941, Petrology of the Ordovician-Silurian boundary in central Pennsylvania (abstr.): Bull. Geol. Soc. Amer., v. 52, pp. 1917–1918.

Kuenen, Ph. H., 1942, Pitted pebbles: Leidsche Geol. Meded., v. 13, pp. 189–201.

———, 1957, Some experiments on fluviatile rounding: Proc. Konink. Nederl. Akad. van Wetensch., ser. B, v. 61, no. 1, pp. 47–53.

———, 1959a, Sand—its origin, transportation, and accumulation: Geol. Soc. South Africa, Annexure, v. 62, 33 pp.

———, 1959b, Experimental abrasion. 3. Fluviatile action of sand: Amer. Jour. Sci., v. 257, pp. 172–190.

———, 1960, Experimental abrasion. 4. Eolian action: Jour. Geol., v. 68, pp. 427–449.

———, 1966, Matrix of turbidites: experimental approach: Sedimentology, v. 7, pp. 267–297.

Kuenen, Ph. H., and Migliorini, C. I., 1950, Turbidity currents as a cause of graded bedding: Jour. Geol., v. 58, pp. 91–127.

Lahee, F. H., 1941, Field geology, 5th ed.: New York, McGraw-Hill, 883 pp.

Laniz, R. V., Stevens, R. E., and Norman, M. N., 1964, Staining of plagioclase feldspar and other minerals: U.S. Geol. Surv. Prof. Paper 501-B, pp. B152–B153.

Leith, C. K., and Mead, W. J., 1915, Metamorphic geology: New York, Holt, Rinehart, and Winston, 337 pp.

Leith, C. K., and Van Hise, C. R., 1911, The geology of the Lake Superior region: U.S. Geol. Surv. Mono. 52, 641 pp.

Lerbekmo, J. F., 1963, Petrology of the Belly River Formation, southern Alberta foothills: Sedimentology, v. 2, pp. 54–86.

Loney, R. A., 1964, Stratigraphy and petrography of the Pybus-Gambier area, Admiralty Island, Alaska: Bull. U.S. Geol. Surv. 1178, 103 pp.

Lovell, J. P. B., 1969, Tyee Formation: a study

of proximality in turbidites: Jour. Sed. Petrology, v. 39, pp. 935–953.

McBride, E. F., 1962, Flysch and associated beds of the Martinsburg Formation (Ordovician), central Appalachians: Jour. Sed. Petrology, v. 32, pp. 39–91.

———, 1963, Classification of common sandstones: Jour. Sed. Petrology, v. 33, pp. 664–669.

———, 1966, Sedimentary petrology and history of the Haymond Formation (Pennsylvanian), Marathon Basin, Texas: Univ. Texas Bur. Econ. Geol. Rept. Inv. 57, 101 pp.

McBride, E. F., Lindemann, W. L., and Freeman, P. S., 1968, Lithology and petrology of the Gueydan (Catahoula) Formation in south Texas: Univ. Texas Bur. Econ. Geol. Rept. Inv. 63, 122 pp.

McEwen, M. C., Fessenden, F. W., and Rogers, J. J. W., 1959, Texture and composition of some weathered granites and slightly transported arkosic sands: Jour. Sed. Petrology, v. 29, pp. 477–492.

Mackie, W., 1896, The sands and sandstones of eastern Moray: Trans. Edinburgh Geol. Soc., v. 7, pp. 148–172.

———, 1899, The felspars present in sedimentary rocks as indications of the conditions of contemporaneous climate: Trans. Edinburgh Geol. Soc., v. 7, pp. 443–468.

———, 1905, Seventy chemical analyses of rocks: Trans. Edinburgh Geol. Soc., v. 8, pp. 33–60.

Mansfield, G. R., 1920, The physical and chemical character of New Jersey greensand: Econ. Geol., v. 15, pp. 547–566.

———, 1922, Potash in the greensands of New Jersey: U.S. Geol. Surv. Bull. 727, 146 pp.

Marchese, H. G., and Garrasino, C. A., 1969, Clasificacion descriptiva de areniscas: Rev. Asoc. Geol. Argentina, v. 24, no. 3, pp. 281–286.

Markewicz, F. J., 1969, Ilmenite deposits of the New Jersey Coastal Plain, *in* Geology of selected areas in New Jersey and eastern Pennsylvania and guidebook of excursions (Subitzky, S., ed.): New Brunswick, N.J., Rutgers Univ. Press, pp. 363–382.

Martens, J. H. C., 1928, Beach deposits of ilmenite, zircon and rutile in Florida: Florida State Geol. Surv. 19th Ann. Rept., pp. 124–154.

———, 1935, Beach sands between Charleston, South Carolina, and Miami, Florida: Bull. Geol. Soc. Amer., v. 46, pp. 1563–1596.

Mathur, S. M., 1958, On the term "Orthoquartzite": Eclogae Geol. Helvetiae, v. 51, pp. 695–696.

Matisto, A., 1968, Die Meta-Arkose von Mauri bei Tampere: Bull. Comm. Geol. Finlande, no. 235, p. 4–20.

Mattiat, B., 1960, Beitrag zur Petrographie der Oberharzer Kulmgrauwacke: Beitr. Min. Petrogr., v. 7, pp. 242–280.

Maxwell, J. C., 1960, Experiments on compaction and cementation of sand, *in* Rock deformation (Griggs, D., and Handin, J., eds.): Geol. Soc. Amer. Mem. 79, pp. 105–132.

———, 1964, Influence of depth, temperature, and geologic age on porosity of quartzose sandstone: Bull. Amer. Assoc. Petrol. Geol., v. 48, pp. 697–709.

Maxwell, J. C., and Verrall, P., 1954, Low porosity may limit oil in deep sands: World Oil, v. 138, no. 5, pp. 106–113; no. 6, pp. 102–104.

Meckel, L. D., 1967, Origin of Pottsville conglomerates (Pennsylvanian) in the central Appalachians: Bull. Geol. Soc. Amer., v. 78, pp. 223–258.

Meischner, D., 1971, Clastic sedimentation in the Variscan Geosyncline east of the River Rhine, *in* Sedimentology of parts of central Europe (Müller, G., ed.): Guidebook 8th Int. Sed. Congr., Heidelberg, pp. 9–43.

Mellon, G. B., 1964, Discriminatory analysis of calcite- and silicate-cemented phases of the Mountain Park Sandstone: Jour. Geol., v. 72, pp. 786–809.

Merrill, G. P., 1891, Stones for building and decoration, 3rd ed.: New York, Wiley, 551 pp.

Middleton, G. V., 1960, Chemical composition of sandstones: Bull. Geol. Soc. Amer., v. 71, pp. 1011–1026.

Millot, G., Lucas, J., and Wrey, R., 1963, Research on evolution of clay minerals and argillaceous and siliceous neoformation: Clays and clay minerals, 10th Conf.: New York, Pergamon, pp. 399–412.

Milner, H. B., 1962, Sedimentary petrography, v. 1, Methods in sedimentary petrography, 643 pp.; v. 2, Principles and applications, 715 pp.: New York, Macmillan, Inc.

Mizutani, S., and Suwa, K., 1966, Orthoquartzitic sand from the Libyan Desert, Egypt: Jour. Earth Sci., Nagoya Univ., v. 14, pp. 137–150.

Moiola, R. J., and Weiser, D., 1968, Textural parameters: an evaluation: Jour. Sed. Petrology, v. 38, pp. 45–53.

Moss, A. J., 1966, Origin, shaping, and significance of quartz sand grains: Jour. Geol. Soc. Australia, v. 13, pp. 97–136.

Müller, G., 1964, Methoden der Sedimentuntersuchungen: Stuttgart: E. Schweizerbart'sche Verlagsbuchhandlung, 303 pp.

Nanz, R. H., Jr., 1954, Genesis of Oligocene sandstone reservoir, Seeligson Field, Jim Wells and Kleberg counties, Texas: Bull. Amer. Assoc. Petrol. Geol., v. 38, pp. 96–117.

Ojakangas, R. W., 1963, Petrology and sedimentation of the Upper Cambrian Lamotte Sandstone in Missouri: Jour. Sed. Petrology, v. 33, pp. 860–873.

Okada, H., 1960, Sandstones of the Cretaceous Mifuné Group, Kyushu, Japan: Kyushu Univ. Mem. Fac. Sci., ser. D., Geology, v. 10, pp. 1–40.

———, 1961, Cretaceous sandstones of Goshonoura Island, Kyushu, Japan: Kyushu Univ. Mem. Fac. Sci., ser. D., Geology, v. 11, pp. 1–48.

———, 1967, Composition and cementation of some Lower Paleozoic grits in Wales: Kyushu Univ. Mem. Fac. Sci., ser. D., Geology, v. 18, pp. 261–276.

———, 1971, Classification of sandstone: analysis and proposal: Jour. Geol., v. 79, pp. 509–525.

Ondrick, C. W., and Griffiths, J. C., 1969, Frequency distribution of elements in Rensselaer Graywacke, Troy, New York: Bull. Geol. Soc. Amer., v. 80, pp. 509–518.

Oriel, S. S., 1949, Definitions of arkose: Amer. Jour. Sci., v. 247, pp. 824–829.

Payne, T. G., et al., 1952, The Arctic slope of Alaska: U.S. Geol. Surv., Oil and Gas Invest. Map, O. M. 126, sheet 2.

Pelletier, B. R., 1958, Pocono paleocurrents in Pennsylvania and Maryland: Bull. Geol. Soc. Amer., v. 69, pp. 1033–1064.

Pettijohn, F. J., 1941, Persistence of heavy minerals and geologic age: Jour. Geol., v. 49, pp. 610–625.

———, 1943, Archean sedimentation: Bull. Geol. Soc. Amer., v. 54, pp. 925–972.

———, 1948, A preface to the classification of sedimentary rocks: Jour. Geol., v. 56, pp. 112–118.

———, 1949, Sedimentary rocks, 1st ed.: New York, Harper & Row, 526 pp.

———, 1954, Classification of sandstones: Jour. Geol., v. 62, pp. 360–365.

———, 1963, Chemical composition of sandstones—excluding carbonate and volcanic sands, *in* Data of geochemistry, 6th ed.: U.S. Geol. Surv. Prof. Paper 440-S, 19 pp.

Pettijohn, F. J., Potter, P. E., and Siever, R., 1972, Sand and sandstone: New York, Springer, 618 pp.

Pittman, E. D., 1963, Use of zoned plagioclase as an indicator of provenance: Jour. Sed. Petrology, v. 33, pp. 380–386.

van der Plas, L., 1966, The identification of the detrital feldspars: Amsterdam, Elsevier, 305 pp.

Plumley, W. J., 1948, Black Hills terrace gravels: a study in sediment transport: Jour. Geol., v. 56, pp. 526–577.

Poldervaart, A., 1955, Chemistry of the earth's crust, *in* Crust of the earth—a symposium (Poldervaart, A., ed.): Geol. Soc. Amer. Spec. Paper 62, pp. 119–144.

Potter, P. E., 1963, Late Paleozoic sandstones of the Illinois Basin: Illinois Geol. Surv., Rept. Inv. 217, 92 pp.

Potter, P. E., and Pettijohn, F. J., 1963, Paleocurrents and basin analysis: New York, Springer, 296 pp.

Pye, W. D., 1944, Petrology of the Bethel Sandstone of south-central Illinois: Bull. Amer. Assoc. Petrol. Geol., v. 28, pp. 63–122.

Ramdohr, Paul, 1958, New observations on the ores of the Witwatersrand in South Africa and their genetic significance: Trans. Geol. Soc. South Africa, v. 61, annexure, 50 pp. (Engl. translation of paper originally publ. in German, 1955.)

Ramez, M. R. H., and Mosalamy, F. H., 1969, The deformed nature of various size fractions in some clastic sands: Jour. Sed. Petrology, v. 39, pp. 1182–1187.

Reed, J. J., 1957, Petrology of the lower Mesozoic rocks of the Wellington District: Bull. New Zealand Geol. Surv. (n.s.), 57, 60 pp.

Reed, R. D., 1928, The occurrence of feldspar in California sandstones: Bull. Amer. Assoc. Petrol. Geol., v. 12, pp. 1023–1024.

Rickard, M. J., 1964, Metamorphic tourmaline overgrowths in the Oak Hill Series of southern Quebec: Canad. Mineral., v. 8, pp. 86–91.

Rimsaite, J., 1967, Optical heterogeneity of feldspars observed in diverse Canadian rocks: Schweiz. Min. Petrog. Mitt., v. 47, pp. 61–76.

Rinne, F., 1923, Gesteinskunde, Leipzig: Dr. Max Jänecke, 374 pp.

Rittenhouse, G., 1944, Sources of modern sands in the middle Rio Grande Valley: Jour. Geol., v. 52, pp. 145–183.

———, 1949, Petrology and paleogeography of Greenbrier Formation: Bull. Amer. Assoc. Petrol. Geol., v. 33, pp. 1704–1730.

Rodgers, J., 1950, The nomenclature and classification of sedimentary rocks: Amer. Jour. Sci., v. 248, pp. 297–311.

Ronov, A. B., Mikhailovskaya, M. S., and Solodkova, I. I., 1963, Evolution of the chemical and mineralogical composition of arenaceous rocks, *in* Chemistry of the earth's crust, (Vinogradov, A. P., ed.): Israel Program Sci. Trans., 1966, v. 1, pp. 212–262.

Rothrock, E. P., 1944, A geology of South Dakota: Bull. South Dakota Geol. Surv., no. 15, 255 pp.

Russell, R. D., 1935, Frequency percentage determinations of detrital quartz and feldspar: Jour. Sed. Petrology, v. 5, pp. 109–114.

———, 1937, Mineral composition of Mississippi River sands: Bull. Geol. Soc. Amer., v. 48, pp. 1307–1348.

———, 1942, Tables for the determination of detrital minerals: Rept. Committee Sedimentation 1940–1941, Div. Geol. Geog., Nat. Res. Coun., pp. 6–8.

Russell, R. D., and Taylor, R. E., 1937, Roundness and shape of Mississippi River sands: Jour. Geol., v. 45, pp. 225–267.

Sabins, F. F., Jr., 1962, Grains of detrital, secondary, and primary dolomite from Cretaceous strata of the Western Interior: Bull. Geol. Soc. Amer., v. 73, pp. 1183–1196.

Schlee, J., Uchupi, E., and Trumbull, J. V. A., 1964, Statistical parameters of Cape Cod beach and eolian sands, *in* Geological Survey research: U.S. Geol. Surv. Prof. Paper 501-D, pp. 118–122.

Schneider, H., 1927, A study of glauconite: Jour. Geol., v. 35, pp. 299–310.

Schwab, F. L., 1970, Origin of the Antietam formation (Late Precambrian-Lower Cambrian), central Virginia: Jour. Sed. Petrology, v. 40, pp. 354–366.

Schwartz, G. M., 1942, Correlation and metamorphism of the Thomson Formation, Minnesota: Bull. Geol. Soc. Amer., v. 52, pp. 1001–1020.

Sestini, G., 1970, Flysch facies and turbidite sedimentology: Sediment. Geol., v. 4, pp. 559–597.

Sheppard, R. A., 1971, Clinoptilolite of possible economic value in sedimentary deposits of the coterminous U.S.: Bull. U.S. Geol. Surv., no. 1332-B, pp. B1–B15.

Shiki, T., 1959, Studies on sandstones in the Maizuru Zone, southwest Japan. I. Importance of relations between mineral composition and grain size: Mem. College Sci., Univ. Kyoto, v. 25, pp. 239–246.

———, 1962, Studies on sandstones in the Maizuru Zone, southwest Japan. III. Graywacke and arkose sandstones in and out of the Maizuru Zone: Mem. College Sci., Univ. Kyoto, v. 29, pp. 291–324.

Shrock, R. R., 1946, Classification of sedimentary rocks: Bull. Geol. Soc. Amer., v. 57, p. 1231.

———, 1948, Classification of sedimentary rocks: Jour. Geol., v. 56, pp. 118–120.

Siever, R., 1959, Petrology and geochemistry of silica cementation in some Pennsylvanian sandstones, *in* Silica in sediments (Ireland, H. A., ed.): Soc. Econ. Paleont. Min. Spec. Publ. 7, pp. 55–79.

Siever, R., Beck, K. C., and Berner, R. A., 1965, Composition of interstitial waters of modern sediments: Jour. Geol., v. 73, pp. 39–73.

Simonen, A., and Kuovo, O., 1951, Archean varved schists north of Tampere in Finland: Soc. Geol. Finlande, Comptes Rendus, v. 24, pp. 93–117.

———, 1955, Sandstones in Finland: Bull. Comm. Geol. Finlande 168, pp. 57–87.

Sippel, R. F., 1968, Sandstone petrology, evidence from luminescence petrography: Jour. Sed. Petrology, v. 38, pp. 530–554.

Skolnick, H., 1965, The quartzite problem: Jour. Sed. Petrology, v. 35, pp. 12–21.

Sloss, L. L., and Feray, D. E., 1948, Microstylolites in sandstone: Jour. Sed. Petrology, v. 18, pp. 3–13.

Smith, E. R., 1946, Sand: Indiana Acad. Sci., v. 55, pp. 121–143.

Sorby, H. C., 1880, On the structure and origin of non-calcareous stratified rocks: Proc. Geol. Soc. London, v. 36, pp. 62–64.

Stauffer, P. H., 1967, Grain-flow deposits and their implications, Santa Ynez Mountains, California: Jour. Sed. Petrology, v. 37, pp. 487–508.

Stebinger, E., 1914, Titaniferous magnetite beds on the Blackfeet Indian Reservation, Montana, *in* Contributions to economic geology: Bull. U.S. Geol. Surv. 540, pp. 329–337.

Stewart, D., Jr., 1937, An occurrence of authigenic feldspar: Amer. Mineral., v. 22, pp. 1000–1003.

Stockdale, P. B., 1936, Rare stylolites: Amer. Jour. Sci., ser. 5, v. 32, pp. 229–233.

Suttner, L. J., 1969, Stratigraphic and petrographic analysis of Upper Jurassic–Lower Cretaceous Morrison and Kootenai formations, southwest Montana: Bull. Amer. Assoc. Petrol. Geol., v. 53, pp. 1391–1410.

Swineford, A., 1947, Cemented sandstones of the Dakota and Kiowa formations in Kansas: Bull. State Geol. Surv. Kansas 70, pt. 4, pp. 53–104.

Takahashi, J., 1939, Synopsis of glauconitization, *in* Recent marine sediments (Trask, P. D., ed.): Okla., Amer. Assoc. Petrol. Geol., pp. 503–512.

Taliaferro, N. L., 1943, Franciscan-Knoxville problem: Bull. Amer. Assoc. Petrol. Geol., v. 27, pp. 109–219.

Tallman, S. L., 1949, Sandstone types, their abundance and cementing agents: Jour. Geol., v. 57, pp. 582–591.

Taylor, J. M., 1950, Pore-space reduction in sandstones: Bull. Amer. Assoc. Petrol. Geol., v. 34, pp. 701–716.

Thiel, G. A., 1935, Sedimentary and petrographic analysis of the St. Peter Sandstone: Bull. Geol. Soc. Amer., v. 46, pp. 559–614.

———, 1940, The relative resistance to abrasion of mineral grains of sand size: Jour. Sed. Petrology, v. 10, pp. 103–124.

Thiel, G. A., and Dutton, C. E., 1935, The architectural, structural, and monumental stones of Minnesota: Bull. Minnesota Geol. Surv., no. 25, 160 pp.

Thomson, A., 1959, Pressure solution and porosity, *in* Silica in sediments (Ireland, H. A., ed.): Soc. Econ. Paleont. Min. Spec. Publ. 7, pp. 92–111.

Tickell, F. G., 1965, The techniques of sedimentary mineralogy: Amsterdam, Elsevier, 220 pp.

Tieje, A. J., 1921, Suggestions as to the description and naming of sedimentary rocks: Jour. Geol., v. 29, pp. 650–666.

Todd, T. W., and Folk, R. L., 1957, Basal Claiborne of Texas, record of Appalachian tectonism during Eocene: Bull. Amer. Assoc. Petrol. Geol., v. 41, pp. 2545–2566.

Towe, K. M., 1962, Clay mineral diagenesis as a possible source of silica cement in sedimentary rocks: Jour. Sed. Petrology, v. 32, pp. 26–28.

Trurnit, P., 1968, Pressure solution phenomena in detrital rocks: Sed. Geol., v. 2, pp. 89–114.

Turner, F. J., and Verhoogen, J., 1960, Igneous and metamorphic petrology: New York, McGraw-Hill, 694 pp.

Tyrrell, G. W., 1933, Greenstones and greywackes: reunion intern, pour l'étude du Precambrien 1931, Comptes Rendus, pp. 24–26.

Udden, J. A., 1914, Mechanical composition of clastic sediments: Bull. Geol. Soc. Amer., v. 25, pp. 655–744.

Van Hise, C. R., 1904, Treatise on metamorphism: U.S. Geol. Surv. Monogr. 47, 1286 pp.

Vitanage, P. W., 1957, Studies of zircon types in Ceylon Pre-Cambrian complex: Jour. Geol., v. 65, pp. 117–138.

Voll, G., 1960, New work on petrofabrics: Liverpool and Manchester Geol. Jour., v. 2, pt. 3, pp. 503–567.

Waldschmidt, W. A., 1941, Cementing materials in sandstones and their influence on the migration of oil: Bull. Amer. Assoc. Petrol. Geol., v. 25, pp. 1839–1879.

Walker, R. G., and Pettijohn, F. J., 1971, Archaean sedimentation: Analysis of the Minnitaki Basin, northwestern Ontario, Canada: Bull. Geol. Soc. Amer., v. 82, pp. 2099–2130.

Walton, E. K., 1955, Silurian greywackes of Peebleshire: Proc. Roy. Soc. Edinburgh, v. 65, 1952–1955, pp. 327–357.

Warner, M. M., 1965, Cementation as a clue to structure, drainage patterns, permeability, and other factors: Jour. Sed. Petrology, v. 35, pp. 797–804.

Washington, H. S., 1930, The chemical analysis of rocks, 4th ed.: New York, Wiley, 296 pp.

Wayland, R. G., 1939, Optical orientation in elongate clastic quartz: Amer. Jour. Sci., v. 237, pp. 99–109.

Webb, W. M., and Potter, P. E., 1969, Petrology and chemical composition of modern detritus derived from a rhyolitic terrain, western Chihuahua: Bol. Soc. Geol. Mexicana, v. 32, no. 1, pp. 45–61.

Weber, J. N., and Middleton, G. V., 1961, Geochemistry of turbidites of the Normanskill and Charny formations: Geochim. Cosmochim. Acta, v. 22, pp. 200–288.

Weeks, A. D., and Eargle, D. H., 1963, Relation of the diagenetic alteration and soil-forming processes to the uranium deposits of the southeast Texas Coastal Plain, *in* Clays and clay minerals, 10th Conf.: New York, Pergamon, Macmillan, Inc., pp. 23–41.

Wermund, E. G., 1964, Geologic significance of fluvio-detrital glauconite: Jour. Geol., v. 72, pp. 470–476.

Weyl, R., 1950, Schwermineralverwitterung und ihr Einfluss auf die Mineralführung klastischer Sedimente: Erdöl und Kohle, v. 3, no. 5, pp. 209–211.

Whetten, J. T., 1966, Sediments from the lower Columbia River and origin of graywacke: Science, v. 152, pp. 1057–1058.

Whetten, J. T., Kelley, J. C., and Hanson, L., 1969, Characteristics of Columbia River sediment and sediment transport: Jour. Sed. Petrology, v. 39, pp. 1149–1166.

White, D. E., Hem, J. G., and Waring, G. A., 1963, Chemical composition of subsurface waters, *in* Data of geochemistry: U.S. Geol. Surv. Prof. Paper, 440-F, 67 pp.

Wiesnet, D. R., 1961, Composition, grain size, roundness and sphericity of the Potsdam Sandstone (Cambrian) in northeastern New York: Jour. Sed. Petrology, v. 31, pp. 5–14.

Williams, H., 1957, Glowing avalanche deposits of the Sudbury Basin: Ontario Dept. Mines, Ann. Rpt., v. 65, pt. 2, pp. 57–89.

Williams, H., Turner, F. J., and Gilbert, C. M., 1954, Petrography: San Francisco, Freeman, 406 pp.

Winchell, N. H., Grant, U. S., Todd, J. E., Upham, W., and Winchell, H. V., 1899, Geology of Minnesota: Geol. Nat. Hist. Surv. Minnesota, Final Rept., v. IV, 630 pp.

Wolf, K. H., 1971, Textural and compositional transitional stages between various lithic grain types: Jour. Sed. Petrology, v. 41, pp. 328–332.

Woodland, A. W., 1938, Petrological studies in the Harlech Grit series of Merionethshire II: Geol. Mag., v. 74, pp. 440–454.

Wurster, P., 1964, Geologie des Schilfsandstein: Hamburg, Mitt. Geol. Staatsinst., v. 33, 140 pp.

Yeakel, L. S., Jr., 1962, Tuscarora, Juniata, and Bald Eagle paleocurrents and paleogeography in central Appalachians: Bull. Geol. Soc. Amer., v. 73, pp. 1515–1540.

8

SHALES, ARGILLITES, AND SILTSTONES

INTRODUCTION

Of common sediments, shales are the most abundant. They form about one-half of the geologic column, being estimated at 44 percent by Schuchert (1931, p. 12), 46 percent by Leith and Mead (1915, p. 60), and 56 percent by Kuenen (1941). Shale forms about 32 percent of all Paleozoic and later sediments on the North American craton (an estimate based on data by Sloss, 1968); it forms 44 percent of the geosynclinal sequence at Jackson, Wyoming (Schwab, 1969). Blatt (1970) estimates 69 percent of the continental sediment of the world to be shale. On the basis of certain geochemical considerations, shales should form 80 percent of all the sediment produced through geologic time (Clarke, 1924, p. 24).

Despite their abundance they are not so well exposed as are the more resistant limestones and sandstones. And because of their fine texture and complex composition they are not so well understood as the other sedimentary materials. Their fine grain makes thin-section study difficult. Many of their constituents are not readily resolved under the microscope, so they cannot be identified by usual optical means. Recourse must be to analyses of gross chemical composition or to special techniques such as x-ray and differential thermal analyses. Even these methods fail to provide all the relevant data needed for petrographic analysis. For these reasons, description, classification, and interpretation of shales and argillites are inadequate and incomplete at present.

The argillaceous deposits are, nonetheless, of considerable economic worth. Many clays and some shales are raw materials for the manufacture of brick, building tiles, roofing tiles, drain tile, pottery, and other ceramic wares. Clay shale is mixed with limestone in the proper ratio, fired, and ground to form portland cement. Some high-purity clays serve as filler for paper. Slate, a metamorphic derivative of shale, can be split to form roofing shingles, electrical panels, and blackboards. Some shales yield a distillate on heating which can be refined into motor fuel and other products.

Interest in argillaceous sediment has been greatly stimulated by our better understanding of the clay minerals. The latter have become better known as a result of x-ray diffractometry and other techniques. Major contributions have been published in the proceedings of the several clay-mineral conferences, beginning about 1951, and in the larger monographic works such as those of Grim (1968), Millot (1949, 1964), and Carroll (1970). In general, earlier literature dealt with clay minerals as such, and problems of argillaceous sediments themselves received less attention. Their economic aspects have been covered by Ries (1927) and Grim (1962); other aspects were treated by Boswell (1961). Refer to these works and to the special papers cited in this chapter.

DEFINITIONS AND TERMINOLOGY

The terminology applied to clays and shales was reviewed at some length by Twenhofel

(1937) and more recently by Clark (1954) and Tomadin (1964).

A *clay* has been defined as a natural plastic earth (though some clays are non-plastic) composed of hydrous aluminum silicates (the "clay minerals") and of fine grain (a clay is a sediment with grains less than 0.002 or 1/256 mm in diameter). The definition based on grain size is least satisfactory, because most commercial clays are not clays by this definition. The definition based on mineral composition errs in that the clay minerals may constitute as little as one-third or even less of the whole rock. Twenhofel (1937) says that small particles should be dominantly clay minerals and that a clay should have an excess of particles (over 50 percent) of clay size. Clay minerals could, by this definition, form as little as one-fourth of the total.

Clark (1954) would define *shale* as a detrital rock whose particles have a diameter less than 1/16 mm. Thus defined, the term is all-inclusive, embracing both siltstone as well as shale as usually defined. Most workers, however, divide the finer materials into two classes: silt and clay (forming siltstone and claystone, respectively). Udden (1914) placed the division between silt and clay at 1/256 mm. Krumbein and Sloss (1951, p. 14), however, considered 1/100 mm more desirable because coarser sediment has the field characteristics of sandstone (interstitial cement and induration, ripple cross-lamination), whereas finer sediments have the usual characteristics of shale (slaking, plasticity when wet, and the like).

Other writers have used other parameters in their classification and nomenclature of finer-grained sediments. *Claystone* is indurated clay. If it possesses bedding fissility, it may be called shale. However, some writers (Shrock, 1948; Flawn, 1953) would use the term *claystone* for a rock *less* indurated than a shale.

Shale is a laminated or fissile rock. The term is restricted to buried or ancient deposits. To those claystones which are neither fissile nor laminated but are blocky or massive, the term *mudstone* may be applied.

In a more restricted manner, Ingram (1953) defined *claystone* as a massive rock in which clay predominates over silt, and reserved *siltstone* for massive rocks in which silt exceeds clay. To those rocks in which the proportions of clay and silt are not known or specified, Ingram applied the term *mudstone*. The terms *clay shale, silt shale,* and *mud shale* were proposed for the correlative *fissile* rocks.

Twenhofel (1937) would extend the term *mudstone* to include the whole family of argillaceous rocks. Most writers, including Pettijohn, tend to restrict the term to those rocks having the grain size and composition of a shale but lacking its laminations and/or its fissility (Picard, 1953).

Silt is the material between 1/16 and 1/256 mm in diameter or a sediment in which 50 percent or more of the particles fall in this range. *Siltstone* is indurated silt. As noted by Krumbein and Sloss (1951, p. 14), most rocks to which the term is applied in the field are coarse silts (over 1/100 mm) and, unlike shales, are commonly bonded by chemical cements; they may be cross-bedded on a small scale and may show convolute bedding, injection, and so forth.

The term *argillite* is used in various ways. Twenhofel applies the term to a rock derived from a siltstone or shale that has undergone a somewhat higher degree of induration than is usually present in those rocks. It is thus intermediate in character between a shale and a slate. Grout (1932, p. 365) uses the term *argillite* for a clay or shale hardened by recrystallization and applied the term *slate* to a similar rock if it possesses secondary cleavage. Flawn (1953) would use it in much the same sense as Twenhofel and would use the term *meta-argillite* for more completely recrystallized rocks. Both terms, however, would be restricted to rocks without cleavage or parting.

The terminology here utilized is shown in Fig. 8-1.

TEXTURES AND STRUCTURES

GRAIN SIZE AND FABRIC

Particle size distribution or "mechanical composition" of clays and shales has been extensively investigated. Size analyses of such materials, however, are subject to marked limitation. Because of their fine-grained character, particle size of clays is

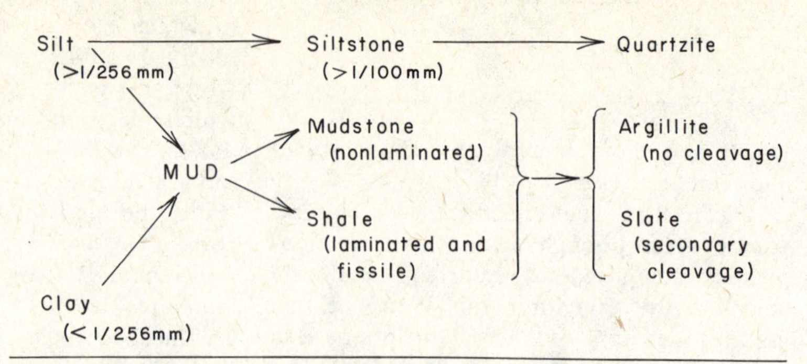

FIG. 8-1. Terminology of the argillaceous sediments. Note that a slate, as well as an argillite, can develop from either a shale or a mudtsone.

usually determined by methods based on differential settling velocities. These velocities are notably affected by the shape and specific gravity of particles as well as by their size. Analytical results, therefore, are misleading in that the size values computed from fall velocities are based on the premise that the particles are spheres of quartz (Krumbein and Pettijohn, 1938, p. 96). The analyzed sample, moreover, is fully dispersed prior to the start of the analysis. Such dispersal, achieved by physical or chemical agents, probably destroys, or at least profoundly modifies, original size distribution. Many clays, especially those which accumulate in marine waters, were in a state of partial or complete flocculation at the time of deposition. Grading curves as determined by analysis may be quite unlike those of the original sediment. Some now homogeneous muds were once aggregated into pellets and perhaps transported and deposited as such. The usual size analysis of such a mud reveals little about its depositional history (Harrison, 1971, p. 72).

A more serious limitation is encountered in older shales because of the effect of diagenesis on size distribution. Because of the fine state of division of materials and the resulting large total surface area of the grains, as well as the instability of some clay minerals, these materials are prone to diagenetic change. Such reorganization must alter greatly size distribution. For these reasons, therefore, size analyses of clays and shales must be interpreted with great caution.

The most significant result of size analyses, or even casual inspection of thin sections, is the disclosure that most shales—certainly the more common types—contain a very large proportion of silt. The marine Perry Farm Shale (Pennsylvanian) of Missouri, although a plastic rock, contains 74 percent by weight of fine sand and silt and but 14 percent clay-sized material and balance carbonate (Keller and Ting, 1950). A similar shale in Illinois contained 68 percent silt (Krumbein, 1938). Krynine estimated (1948, p. 154) the average shale to be about 50 percent silt. Recent *mineralogical* studies of shale suggest that shale is about two parts silt and one part clay, a proportion more nearly consistent with the observations of Keller and of Ting and Krumbein. If so, their composition is about that of the average material forming the Mississippi Delta (Table 8-1).

A feature of some clays is their pellet structure (Grim and Allen, 1938; Allen and Nichols, 1943; Harrison, 1971). Pellets are small rounded aggregates of clay minerals and fine quartz scattered through a matrix of the same materials. The pellets may be separated from the matrix by a shell of organic material. In size, pellets are 0.1 to 0.3 mm in diameter and in a few cases several millimeters in length. They have been ascribed to the action of water currents; in other cases they may be fecal pellets (Moore, 1939; Harrison, 1971).

Some argillaceous rocks of residual origin display *relict* textures inherited from the parent materials from which they were derived. Examples are *saprolites* derived from various coarse igneous and metamorphic rocks. In these, the "ghosts" of the original minerals are well enough preserved so that the original gneissic foliation, porphyroblasts, and the like can be seen. Another example of relict texture is found in bentonites and related materials formed by *in situ* alteration of volcanic ash (p. 308). Other textures, not relict, include oolitic and pisolitic forms developed in some bauxitic and diaspore clays. Also known are pseudomorphic replacements of shells by montmorillonite and diagenetic recrystallization textures such as "metacrysts" of illite mica in a fine-grained illitic groundmass. Most shales, however, show none of these features; they are either structureless or laminated.

Laminated shales characteristically show a fabric produced by the orientation of platy micaceous constituents parallel to the bedding. Under the microscope such tendency to parallelism can readily be seen. Although many individual crystals do not lie exactly parallel to the bedding, sections cut perpendicular to the bedding show a mass extinction effect very much as if the slide were cut from a single crystal. Platy minerals have their slow ray vibrating parallel to their cleavage and hence show parallel extinction—thus the aggregate extinction effect.

In some clays and shales, however, clay minerals show a random orientation (Keller, 1946). This may in some cases be the result of authigenic crystal growth in place. In other cases it is caused by disruption of the original fabric by burrowing and mud-eating bottom-dwelling organisms.

Newly deposited muds have an extremely high water content and a very high porosity. The initial porosities may be as much as 70 to 80 percent (Trask, 1931). Because the average shale has a porosity of only 13 percent, the initial deposit has been greatly compacted and dewatered. That the porosity reduction takes place by compaction rather than by infilling of pores (as in the case of sandstones) is shown by the progressive modification of the fabric, which tends to bring the clay platelets into greater parallelism with one another and with the bedding (Oertal and Curtis, 1972).

FISSILITY

Many shales display primary fissility, the tendency of the rock to split or separate along relatively smooth surfaces parallel to the bedding. This property is related to the orientation of the constituent micaceous minerals. Some shales are markedly fissile; others weakly so.

Alling (1945) and Ingram (1953) attempted to establish a scale of fissility (Table 8-2)

TABLE 8-1. Composite Mississippi Delta sediment[a]

Size grade	Diameter (mm)	Percent	
Coarse sand	1.0–0.5	trace	
Medium sand	0.5–0.25	trace	
Fine sand	0.25–0.125	6	29
Very fine sand	0.125–0.0625	23	
Silt	0.0625–0.0312	30	60
	0.0312–0.0156	16	
	0.0156–0.0078	7	
	0.0078–0.0039	7	
Clay	Under 0.0039		11
Total			100

[a] Based on 300 surface and borehole samples.
Source: After Russell and Russell (1939).

TABLE 8-2. Fissility scales

Alling (1945)	Ingram (1953)	McKee and Weir (1953)
Massive	Massive	Massive; Blocky
Platy-flaggy		
Heavy-bedded	Flaky	Slabby; Flaggy
Thin-bedded		
Fissile	Flaggy	Shaly, platy; Papery

and to relate fissility to the composition. As noted by both investigators, an increasing content of siliceous or calcareous matter decreases fissility of a shale (Fig. 8-2). Rubey (1931) also noted that fissility of shale bore an inverse relationship to its calcium carbonate content. Shales rich in organic matter, on the other hand, seem to be exceptionally fissile, as the black shales demonstrate. Bioturbated shales, however, are nonfissile, as are the more silty mudstones.

Rubey has noted that fissility is not everywhere parallel to the bedding, that it is more pronounced in older beds of any given section, and that those rocks which have the steepest dip and the most pronounced aggregate orientation show the most pronounced fissility. Possibly fissility is in part a secondarily induced structure brought about by rotation or growth of micaceous minerals by pressure. Such is certainly the case in slates, where rock cleavage is commonly at high angles to the bedding.

More puzzling, perhaps, are argillites, which although finely laminated and which have the mineral composition of a slate, are without fissility, either parallel to the bedding or otherwise.

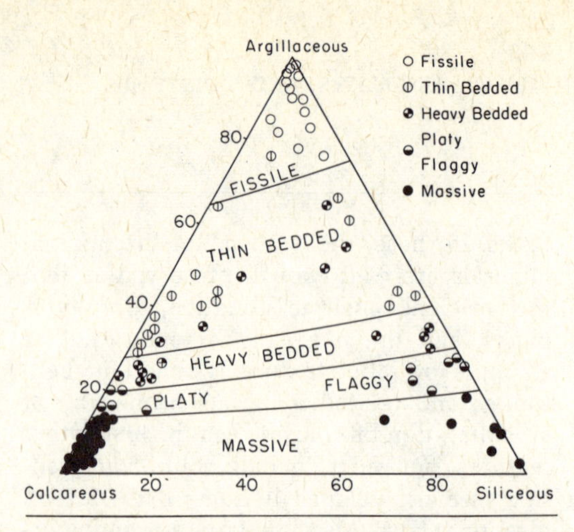

FIG. 8-2. Relationship of composition of sediments to shaliness and ease of splitting. (From Alling, 1945.)

FIG. 8-3. Argillite Cobalt Group (Upper Huronian), Ontario, Canada. Ordinary light, ×75. Laminations of silt, mainly angular quartz and feldspar, interbedded with layers of finer detritus (dark) of similar composition but much richer in chlorite.

LAMINATIONS

Laminations of shales range from 0.05 to 1.0 mm in thickness with most laminations in the range of 0.1 to 0.4 mm and appear to be of three kinds: alternations of coarse and fine particles, such as silt and clay; alternations of light and dark layers distinguished only by organic content, which is responsible for their color; and alternations of calcium carbonate and silt (Figs. 8-3 and 8-4). These alternations of various materials seem to be the result of differential settling rates of the several constituents or differing rates of supply of these materials to the basin of deposition.

Laminations may be caused by storms or floods or other more or less capricious or accidental causes. Or they may be attributable to fluctuations in supply that are seasonally controlled (Bradley, 1929, 1931; Rubey, 1931). If the very thinnest laminations are persistent and show no effects of scour, it seems likely that storms, or bottom currents which arise from them, can hardly be responsible. Because many laminations are of the right order of thickness (as shown by the rates of sedimentation estimated for times past, or observed at present) and have a structure similar to the year-laminations now forming, it seems probable that many are of an annual nature (varves) and depend on the yearly climatic cycle (Figs. 8-5 and 8-6). This cycle affects temperature, salinity, and silt content of waters, as well as seasonal production of plankton.

The absence of laminations, which is rather common, is in some ways more remarkable than the laminations themselves. Extremely uniform sedimentation over a very long period of time may produce a structureless sediment; a more probable cause of such sediment is the reworking and ingestion of the mud by scavenging benthonic organisms (Dapples, 1942; Moore and Scruton, 1957). In the latter case there are usually vestiges of original laminations.

FIG. 8-4. Green River oil shale (Eocene), Colorado. Ordinary light, ×75. Consists of organic-rich laminations (black), and laminations containing less organic matter (gray) and much carbonate (white). The latter forms one-third to one-half of the total. A very little detrital quartz and authigenic analcite is also present.

CONCRETIONS AND OTHER STRUCTURES

Shales and siltstones commonly contain concretionary bodies. Calcareous concretions, somewhat flattened parallel to the bedding and with bedding planes passing through them, are common in many shales. They are more prominent in the silt or siltstone interbeds. Black shales tend to contain cone-in-cone layers and, in rare cases, chert nodules and layers. Not uncommon in many shales are septarian nodules and clay ironstone concretions. (These are all described in Chapter 12.)

MINERAL COMPOSITION OF SHALES AND ARGILLITES

The composition of transported clays and shales is complex and varied, because these materials consist of products of abrasion (mainly silt), end-products of weathering (residual clays), and chemical and biochemical additions (Fig. 8-7). These chemical additions either are materials precipitated from solution and deposited concurrently with accumulating clays, such as calcium carbonate; or they are materials added by reaction or exchange with the surrounding medium (usually sea water), such as potas-

FIG. 8-5. Varved clay (Pleistocene), Baraboo, Wisconsin. Length of specimen about 7 cm. Five varves or year layers are present. Darker bands are winter clay; laminated thicker layers are summer silt.

FIG. 8-6. Varved argillite, Cobalt Group (Precambrian), Wells Township, Ontario, Canada. Note scattered, small, ice-rafted pebbles.

sium or magnesium. The several varieties or subclasses of shales are mainly dependent on the relative importance of the several contributing sources; hence both mineralogy and chemical composition fluctuate within wide limits. The kinds and proportion of mechanically derived silts are dependent on the relief and climate of the source area. If such materials are absent or rare, mudstones are enriched in the residual materials, and under appropriate conditions they are enriched in precipitates such as calcite, aragonite, siderite, chamosite, silica, and in some cases organic matter.

The fineness of grain makes identification of the mineral constituents of shale difficult. Under the microscope only larger grains (over 0.01 mm) can be identified with any certainty. These prove to be the same as the grains found in silts and finer sandstones. The residue is a paste which cannot be resolved. The finer fraction can be separated and placed on the x-ray diffractometer so that its constituents can be identified and their proportions crudely estimated. Knowing what minerals are present, it is possible to calculate the probable mineral composition of a clay or shale from its chemical analysis (Imbrie and Poldervaart, 1959; Nicholls, 1962; Miesch, 1962). Such calculations show that the coarser fraction is largely quartz and feldspar and that the finer fraction is richer in clay minerals, clay micas, chlorite, and hydroxides of iron.

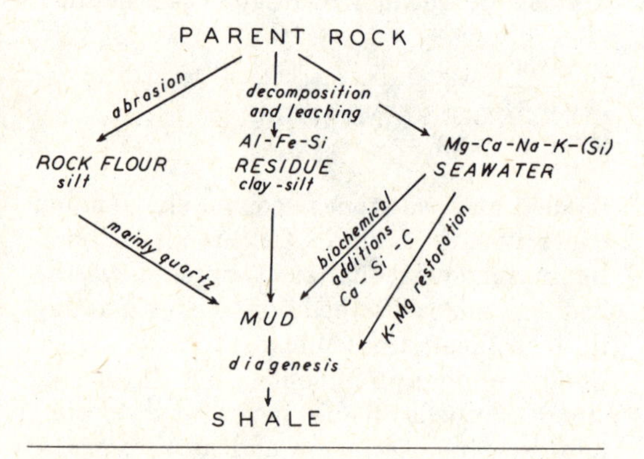

FIG. 8-7. Provenance of shale.

THE CLAY MINERALS

Composition and structure When silicates of primary crystalline rocks are decomposed by weathering, they yield, among other things, a group of minerals known as *clay minerals.* These minerals are hydrated silicates of aluminum commonly with some replacements by iron and magnesium. They are of fine grain, generally less than 5 microns in size, commonly even smaller, and in some cases as small as a millimicron. These minerals occur not only in residual clays formed by the decomposition *in situ* of parent materials, but they are also transported and deposited as sediment. They form an important part of such clays and shales and give these rocks their distinctive properties. They also occur mixed with carbonates in argillaceous limestones, and mixed with sand-sized debris in some sandstones.

Because of their very fine size, clay minerals are difficult to identify. Positive identification can rarely be made from the thin section alone. Special techniques for the isolation of clay minerals and study by chemical, optical, x-ray, and other means are required for certain identification (Carroll, 1970).

Common clay minerals of shales are phyllosilicates; that is, they have a sheet structure somewhat like that of the micas. They consist of two types of layers (Fig. 8-8). One is a silica tetrahedral layer consisting of SiO_4 groups linked together to form a hexagonal network of the composition Si_4O_{10} repeated indefinitely. The other type of layer is the alumina or aluminum hydroxide unit, consisting of two sheets of close-packed oxygens or hydroxyls between which octahedrally coordinated aluminum atoms are embedded in such a position that they are equidistant from six oxygens or hydroxyls. Actually only two-thirds of the aluminum positions are occupied in this layer, which is the gibbsite structure.

Clay minerals belong to two groups. In the *kaolinite group*, the mineral is characterized by a two-layer (1 : 1 layer) lattice consisting of one octahedral or gibbsite layer linked to one silica tetrahedral layer. This lattice does not expand with varying water content, and no replacements by iron or magnesium in the gibbsite layer are known. The other group of clay minerals is characterized by a three-layer (2 : 1) lattice. In this type of lattice an alumina octahedral layer is sandwiched between two silica tetrahedral layers. Several important clay minerals belong to the three-layer group. In *montmorillonite* these three-layer units are loosely held together in the *c*-direction with water and cations between them. The amount of water varies so that the *c*-dimension ranges from 9.6 to 21.4 angstrom units. The mineral is said to have an expanding lattice. Three-unit layers may also be held together by potassium which, because of favorable ionic diameter and coordination capacity, binds the structure together so tightly that expansion is impossible. The clay mica thus formed is *illite*. The *chlorite group* also has a three-layer structure characterized by insertion of a brucite layer, $Mg(OH)_2$, between the three-layer units. Many compositional varieties are possible in each structural group. Although many of these have been given specific names as varieties based on composition, one can generally consider each of the groups to display a wide and indefinite range of compositions. Clay minerals are classed primarily on the basis of their structure (Fig. 8-8).

The principal clay mineral groups, therefore, are the kaolinite group, the montmorillonite group, the illite or muscovite group, and the chlorite group. The chief member of the kaolinite group is *kaolinite* which has the composition $(OH)_8Al_4Si_4O_{10}$. *Anauxite*, similar to kaolinite except for a $SiO_2 : Al_2O_3$ molecular ratio of about 3 instead of 2, is much less common. *Dickite* and *nacrite*, similar to kaolinite in composition but with slightly different crystal forms, are also members of the group. They rarely occur, however, in sediments.

The montmorillonite group, named for the chief member of the group, *montmorillonite*, has the composition of

$$(OH)_4Al_4Si_8O_{10} \cdot nH_2O$$

Magnesium generally replaces some of the aluminum in the lattice. *Beidellite*, which has a $SiO_2 : Al_2O_3$ molecular ratio of 3, and *nontronite*, in which ferric iron replaces the aluminum, are placed in the montmorillonite group.

The illite or clay mica group includes *illite* which has the general formula

$$(OH)_4K_y(Al_4 \cdot Fe_4 \cdot Mg_4 \cdot Mg_6)(Si_{8-y} \cdot Al_y)O_{20},$$

with y varying from 1 to 1.5. Illite is related to white micas but differs, perhaps, in that it contains less potash and more water than do the micas proper. Apart from the variety illite, the group contains *glauconite* in the strict sense (Burst, 1958).

The chlorite group consists of magnesium-rich minerals widely present in shales and in which ferrous ions are prominent.

Many "mixed-layer" clay minerals are also known. The structure of this group is the result of ordered or random stacking of the basic clay mineral units one upon the other in the *c*-direction. Some, for example, alternate two-layer with three-layer type of arrangements. Such mixed layer types are commonly referred to as kaolinite-illite, chlorite-illite, and so forth, instead of devising new names for each mixture.

In addition to the principal groups listed above, some clay minerals, of less common occurrence and of somewhat different crystal structure, include *halloysite,*

$$(OH)_{16}Al_4Si_4O_6$$

together with the less hydrated *metahalloysite,* $(OH)_8Al_4Si_4O_{10}$, and *allophane,* a noncrystalline mutual solution of silica, alumina, and water in varying proportions. Also found in some clays are *vermiculites* and *palygorskites* (sepiolite and attapulgite). None of these minor clay minerals appears to be found in shales.

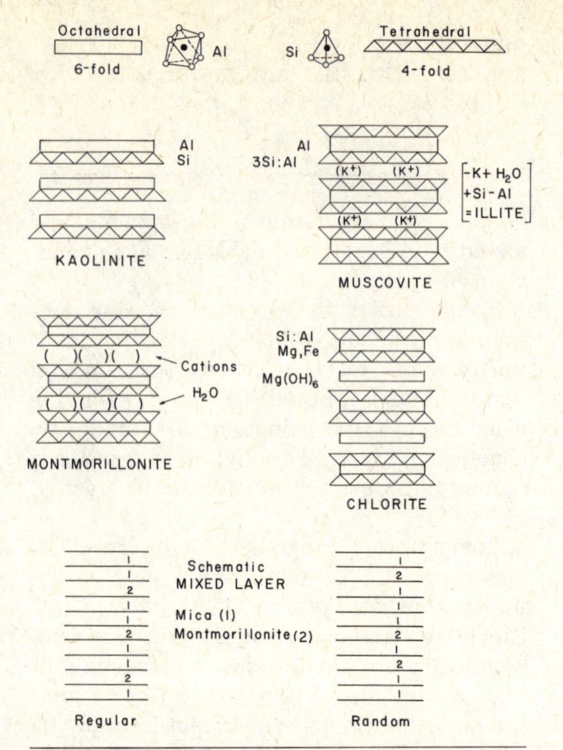

FIG. 8-8. Terminology and structure of clay minerals. (From Pettijohn, Potter, and Siever, 1965, Geology of sand and sandstone Fig. 2-2.)

Base exchange Base exchange is the exchange of ions in solution for those of a solid. It follows, therefore, that upon contact with a solid, the solution will undergo a change reciprocal to that of the solid. Clay minerals show this property in varying, though marked, degrees. In general montmorillonite shows a large base exchange capacity, whereas kaolinite has only slight exchange ability. Illite is intermediate in behavior.

The exchangeable ion can be displaced only by other ions. It does not of itself move freely out into the liquid. The manner in which the exchangeable ions are held and the exact mechanism by means of which base exchange takes place is not fully understood. Rival concepts have arisen; refer to papers by Kelley (1939, 1942) and Grim (1953, pp. 128–155; 1968).

Occurrence and origin of clay minerals Although the mineralogy of clay is now becoming rather well understood, the geology of this material is as yet far from clear. The occurrence, origin, and transformation of clay minerals are not fully known. Clay minerals are generated in large part by weathering of aluminous silicates, but they may be formed also in the environment of sedimentation (early diagenetic) or in the post-burial environment (late diagenetic). Some also are a product of hydrothermal rather than weathering or diagenetic processes. Some may be formed by precipitation from ionic solutions, others by the crystallization of a gel. Keller (1970) has reviewed the processes leading to the formation of clay minerals.

In general, clay minerals seem to form by alteration of some preexisting mineral. This is especially true of those formed in the soil profile. Kaolinite, montmorillonite, and illite all appear to arise from the weathering of many kinds of rocks under various climatic conditions. Any aluminum silicate parent material can yield kaolin by weathering provided that leaching removes K, Na, Ca, Mg, and ferrous iron and that hydrogen is added (Keller, 1970, p. 797). Kaolin is readily formed from granites, whereas gabbro tends to yield minerals of the montmorillonite group. The K- and Na-silicates formed by the hydrolysis of the alkali feldspars are highly soluble and leached away, whereas Ca, Mg, and Fe tend to combine with silica and form montmorillonite. The kaolinization is promoted in an "acid" realm such as characterizes freshwater environments. Montmorillonite commonly originates from calcic-mafic rocks, including volcanic ash, in environments conducive to the retention of the bivalent metals and silicic acid. Such conditions are promoted by alkaline conditions. The conditions for formation of illite in the soil profile are less clear. Adequate potassium is necessary for its formation. In-place alteration of feldspars to illite has been well established.

Various clay minerals, whatever their parent material and geochemical environment of formation, can be transported and deposited in an environment very different from that in which they formed. Some petrologists believe that clay minerals are very susceptible to change and readjust to the new environment. Further changes may take place upon burial with consequent rise in temperature and pressure; still others occur when the rock enters the metamorphic realm. Upon erosion and exposure of the rock to the atmosphere, still further alterations occur. These changes and their effects on clay mineralogy are discussed in the section on diagenesis.

OTHER MINERALS OF SHALES

As noted, most shales have a large silt fraction, most of which is detrital quartz; feldspar forms a lesser part. These minerals have been described in the chapter on sandstones. Nondetrital constituents of shales include biochemical carbonates. These have been described in the chapter on the limestones. Various iron-bearing minerals, including glauconite, volcanic glass, biogenic silica, and phosphatic components which may be mingled with ordinary constituents of shales are described elsewhere.

Of special interest is the work on the stability fields of iron-bearing minerals in terms of the oxidation–reduction potential (*Eh*) of the environment of deposition (Krumbein and Garrels, 1952; James, 1954). Iron sulfide (mainly pyrite), iron carbonate (siderites), iron silicates (iron-rich chlorites, chamosite, glauconite), and iron oxide (hematite) constitute a mineral series correlated with increasing oxidation potential. The presence of these minerals, even in small amounts, if they are truly contemporaneous with sedimentation, is indicative of the oxidation state of the environment of deposition. These minerals may, however, form in the diagenetic environment, perhaps after burial; hence it is important to establish the time of their formation.

The *p*H of shale (determined in an aqueous suspension of the ground rock) is believed to be the same as that of the waters of the depositional environment (Shukri, 1942; Millot, 1964, p. 386). Freshwater shales are said to have a mean *p*H of 4.7, whereas the mean *p*H of shales deposited in marine, lagoonal, or lime-depositing lakes is about 7.8.

THE AVERAGE SHALE

The mineral composition of the average shale as determined by various investigators is summarized in Table 8-3. Differences

TABLE 8-3. Average mineral composition of shale

Constituent	Clarke (1924)	Leith and Mead (1915)	Yaalon (1962a)	Shaw and Weaver (1965)
Quartz	22.3	32	20	36.8
Feldspar	30.0	18	8	4.5
Clay minerals	25.0	34[a]	50	66.9
Iron oxides	5.6	5	3	< 0.5
Carbonates	5.7	8	7	3.6
Other minerals	11.4	1	3	< 2.0
Organic matter	—	1	—	1

[a] Kaolinite and clay minerals 10, sericite and paragonite 18, chlorite and serpentine 6.

between earlier estimates and more recent ones are a result of our better understanding of the clay minerals. Much of the chemical constituiton attributed to feldspars, iron oxides, and other constituents must now be assigned to clay minerals. The percentage of clay-mineral content is notably higher in new analyses because of this and also because, in part, silty shales were excluded from the sample. The high quartz content nevertheless substantiates the view that the average shale contains a great deal of silt (about 40 percent or more).

Because modal analyses of shales are very difficult to make, very few have been published. A most thorough effort to derive a quantitative estimate of shale composition was made by Shaw and Weaver (1965), who used an x-ray diffraction-absorption technique. Analysis of some 300 samples of Paleozoic and younger shales showed a quartz content ranging from less than 10 to near 80 percent and averaged about 34 percent; feldspar varied from near 0 to nearly 30 percent but averaged only 3.6 percent. In the shales analyzed, carbonates were generally absent, averaging 2.7 percent, although some contained over 50 percent. Clay-mineral content (obtained by difference) averaged 64 percent but ranged from under 50 percent to nearly 90 percent. See Table 8-5 for comparison with mineral composition derived by recasting bulk chemical composition. The Paleozoic shales of Illinois are generally similar in character (Grim, Bradley, and White, 1957).

Because the mineral composition of the silt fraction differs from that of the clay fraction of a shale, clearly mineralogy and chemical composition are texturally dependent. The relation between grain size and composition is illustrated by chemical analyses of artificially separated silt and clay fractions by Grout (1925). See Table 8-4. If finer fractions consist of the same minerals as observed in coarser fractions (but in different proportions), it is possible to calculate the probable mineral composition of each fraction. The results of such calculations are given in Table 8-5; as can be seen, the finer fractions are poorer in quartz and richer in clay minerals (kaolinite, sericite, paragonite and iron oxides). These mineralogical differences are closely correlated, of course, with the differences in chemical composition of these same materials.

CHEMICAL COMPOSITION

Chemical analysis remains one of the chief sources of information about the composition of shales.

TABLE 8-4. Relation of chemical composition to size of grain[a]

Constituent	Fine sand	Silt	Coarse clay	Fine clay
SiO_2	71.15	61.29	48.07	40.61
TiO_2	0.50	0.85	0.89	0.79
Al_2O_3	10.16	13.30	18.83	18.97
Iron oxides	3.72	3.94	6.91	7.42
MgO	1.66	3.31	3.56	3.19
CaO	3.65	5.11	4.96	6.24
Na_2O	0.86	1.32	1.17	1.19
K_2O	2.20	2.33	2.57	2.62
Ignition	5.08	7.05	10.91	12.51

[a] Based on average of 12 clays: 1 residual clay, 1 Ordovician shale, 2 Cretaceous clays, and the remainder (8) of glacial or Recent origin. "Fine clay" is under 1 micron, "coarse clay" is 1 to 5 microns, and "silt" is 5 to 50 microns. See Table 8-5 for recalculated mineral composition of the same materials.

Source: After Grout (1925), Table 13.

Silica is the dominant constituent of all clays and shales. It is present as a part of the clay-mineral complex; as undecomposed detrital silicates; and as free silica, both detrital quartz and biochemically precipitated silica (opal of radiolarians, diatoms, spicules). Alumina is an essential constituent of the clay-mineral complex as well as a component of unweathered detrital silicates—primarily feldspars. Exceptionally high alumina suggests free aluminum hydrate (diaspore) or bauxitic materials. Iron in shales is present as an oxide pigment, as part of the chloritic matter present, and exceptionally as pyrite or marcasite, siderite, or iron silicate. The state of oxidation of the iron greatly affects the color of the shale (Fig. 8-9). Magnesia occurs in the chlorite complex or as a component of dolomite. Lime occurs chiefly as the carbonate, although in some shales it is present in amounts larger than necessary to form carbonate and must, therefore, be contained in unweathered silicates or in other cases in the form of gypsum. Alkalies occur, in part, in unweathered detrital silicates (especially feldspars). Potassium is adsorbed by the clay-mineral present and is a constituent of illite or clay mica; it may also be contained in glauconite. Minor constituents are titania (as rutile), manganese, phosphorus, and organic matter.

Interpretation of chemical analyses of shales is fraught with considerable difficulty, because, as pointed out elsewhere (p. 270), chemical composition is dependent on grain size, on maturity of the sediment, and on restoration by chemical or biochemical processes of many of the constituents removed by weathering during the production of the residual soils from which sediment was derived.

The effect of grain size is well illustrated by comparison of the analysis of the summer silt fraction of the same varve with the associated winter clay (Table 8-6A and B). Here maturity and post-depositional histories of the materials are identical so that composition differences are related wholly to differences in grain size. As can be seen, the coarser fraction is richer in silica, whereas fine materials are richer in alumina,

TABLE 8-5. Calculated mineral composition of size fractions of clays and shales

Constituent	Size fraction of "clay"[a]		
	Silt[b]	Coarse clay[b]	Fine clay[b]
Kaolinite and clay minerals	7.5	17.0	23.2
Sericite and paragonite	16.6	21.2	22.1
Quartz	36.7	19.3	13.1
Chlorite and serpentine	8.2	10.3	7.3
Limonite, hematite, and pyrite	3.0	5.5	8.0
Calcite and dolomite	10.5	7.5	5.7
Feldspars	12.6	7.2	7.3
Zeolites	3.0	7.5	6.9
Titanite and rutile	1.7	2.0	1.7
Carbonaceous matter	0.2	0.9	0.6
Moisture	0.9	1.3	4.1
Total	100.9	99.7	100.0

[a] Includes 1 residual clay, 1 Ordovician shale, 2 Cretaceous clays, and the remainder (8) of glacial or Recent origin.

[b] "Fine clay" is under 1 micron; "coarse clay" is 1 to 5 microns; and silt is 5 to 50 microns.

Source: After Grout (1925).

iron, potash, and water. These differences reflect, no doubt, the enrichment of the silt in detrital quartz and of the finer fraction in the clay minerals—potassium-bearing hydrous aluminum silicates and iron-rich magnesium-bearing chlorites.

The composition of the average shale (Table 8-7A) differs materially in character from a typical residual clay. The differences are in part attributable to grain size. Inspection of Grout's analyses (Table 8-4) of the several size fractions present in clays suggests that the average shale is about two parts silt and one part clay. Such a mixture would have the composition approximating that of the average shale. Close comparison of the analyses of the average shale with those of residual clays (Table 8-6) shows differences other than those produced by the addition of silt. Residual clays are extremely low in alkalies and alkaline earths, whereas the average shale contains a notable content of these elements. In other words, the average shale is not just a transported residual clay. Not only is it a mixture of clay and silt, but in the clay fraction some of the constituents removed by weathering have been partially restored during or after the sedimentation process. Potassium, in particular, and also to some extent magnesium are apparently taken up by the clay-mineral complex and built into authigenic sericites and chlorites. Lime is not normally so added but may be fixed by biochemical action. Sodium alone is not restored.

The chemical composition of some shales is notably aberrant, that is, departs markedly from the common or "average" shale. Nonsilty shales with abnormally high silica (siliceous shales) are likely to be diatomaceous or to contain silica from volcanic ash. Those notably low in silica and abnormally high in alumina are either exceptionally fine-grained or are kaolinitic or, in unusual cases, bauxitic. Exceptionally iron-rich shales or slates probably contain iron sulfide (pyrite) if black, or are siderite-rich, or contain appreciable quantities of iron silicate. The iron-silicate-bearing shales grade into chamositic and related mudstones. The red shales are often described as "ferruginous," although in fact they may not contain any abnormal quantity of iron (see p. 275); they are characterized by ferric rather than by ferrous compounds of iron. Shales abnormally high in lime and magnesia are most likely calcareous and contain calcite or dolomite. A high content of CO_2 will substantiate such a conclusion, as will a high content of acid soluble material. If lime exceeds that required by CO_2, some unweathered silicates or sulfates are probably present. In the latter case, the high lime content is associated with a high content of SO_3.

Potash nearly always exceeds soda in shales and slates. The few shales in which the reverse is true must be those which contain products of abrasive action rather than clay minerals. Some glacial clays and silts are thus characterized (Table 8-6). Shales which have important volcanic admixtures may also have a higher-than-normal soda content; soda may exceed potash (Table 8-10A and B).

Highly carbonaceous shales have accumulated slowly under anaerobic conditions. Such shales are apt to be rich in sulfide sulfur. Some are also phosphate-rich.

Aside from the light which chemical composition sheds on the probable mineralogical makeup of shale (see p. 271) and its relation to grain size distribution, it has been thought to bear some relation to the environment of deposition. Millot (1949) in particular thought that the shales of freshwater, brackish, and marine origin differed in their bulk chemical composition. Freshwater shales were lower in both K_2O and MgO than either marine or lagoonal shales. These differences were presumed, because kaolinite was the characteristic clay mineral of acid freshwater, whereas clay minerals of the more alkaline marine waters was montmorillonite and its diagenetic derivative illite. These views have been questioned (Greensmith, 1958). Many investigators, however, believe that some differences exist, especially in minor element composition. The boron and radioactivity contents of marine shales seem to be markedly different from that of nonmarine shales (Potter, Shimp, and Witters, 1963; Couch, 1971). But the reliability of trace elements as environmental indicators of shales has been challenged (Cody, 1971).

Aside from variations in chemical composition related to texture and those brought about by addition of a chemical, biochemical, or volcanic contaminant, some have been attributed to differing residual character. Vogt (1927, p. 489) regarded clay sediments as washed residual products of weathering and presumed that alumina, as the least mobile oxide, will tend to be enriched in the most mature weathered residuum. Such enrichment, however, may be obscured by variations in the amount of quartz silt present in the shale, so it is necessary to recalculate the chemical analysis excluding SiO_2 and TiO_2. Three oxides, MgO, CaO, and Na_2O, *decrease* gradually and regularly with increasing residual character; three oxides, K_2O, SiO_2, and TiO_2, *increase* with increasing residual character. The amount of iron remains constant over a large range of composition (Kennedy, 1951; Nanz, 1953, p. 61). As suggested above, glacial and volcanic shales may be abnormally rich in soda; the most mature weathering products are enriched in alumina. The alumina–soda ratio, therefore, is perhaps a good index of maturity.

A number of efforts have been made to determine the chemical composition of the average shale (Table 8-7). Clarke (1924, p. 34) was among the earlier workers who computed such an average. Such averages are useful for geochemical mass balance calculations; they are most useful as bench

TABLE 8-6. Chemical composition of varved sediments and pellodites

Constituent	A	B	C	D	E
SiO_2	59.20	50.33	52.00	62.74	66.87
TiO_2	1.20	1.13	—	—	0.47
Al_2O_3	16.14	19.17	16.11	16.94	15.36
Fe_2O_3	4.36	6.50		5.07	2.81
			4.69		
FeO	3.24	2.52		1.59	1.89
MnO	0.09	0.13	—	—	0.05
MgO	3.14	3.77	4.10	3.05	2.40
CaO	2.52	1.43	8.26	1.39	0.34
Na_2O	3.82	1.78	2.76		1.21
				6.07	
K_2O	1.97	4.03	1.74		6.60
H_2O+	1.16	4.87	—	3.20[a]	1.35
H_2O-	1.15	3.74	9.64[a]	0.36	none
P_2O_5	0.17	0.14	—	—	0.23
CO_2	—	—	—[b]	—	0.28
SO_3	—	—	0.09	—	—
C	1.94	0.41	—	—	0.04
Total	100.10	99.95	99.39	100.41	99.93

[a] Loss on ignition.
[b] Included in ignition loss.

A. Summer silt, Late Glacial varved sediment, Leppakosi, Finland. L. Lokka, analyst (Eskola, 1932).
B. Winter clay, same as A.
C. Varved clay, north end Lake Timiskaming (Miller, 1905. p. 27).
D. Argillite, Cobalt Series (Precambrian), Cobalt District. Ontario (Miller, 1905, p. 42).
E. Argillite, Fern Creek Formation (Precambrian), Dickinson County, Michigan. B. Brunn, analyst.

marks with which to compare shale analyses so that one may recognize significant departures from the norm.

COLOR

More emphasis is placed on the color of shale than on that of most sedimentary rocks. Field geologists in particular are likely to describe a shale or slate in terms of its color such as a black shale, red slate, and so forth.

The color of shales and slates generally results from a pigmentation of some kind. In a crude way, the darker the shale the higher the content of organic matter (Trask, 1937, Fig. 2). Black shales especially are rich in carbonaceous materials (Table 8-9). As shown by Tomlinson (1916), red slates contain no more iron than do black, gray, or green ones (Fig. 8-9). Color differences reflect only the state of oxidation of the iron (MacCarthy, 1926; Grim, 1951). More recent data on the iron-oxide content of associated red and nonred shales of several

TABLE 8-7. Chemical composition of average shale, average Mississippi Delta sediment, and related materials

Constituent	A	B	C	D	E	F
SiO_2	58.10	55.43	60.15	60.64	56.30	69.96
TiO_2	0.65	0.46	0.76	0.73	0.77	0.59
Al_2O_3	15.40	13.84	16.45	17.32	17.24	10.52
Fe_2O_3	4.02	4.00	4.04	2.25	3.83	3.47 (Fe₂O₃ + FeO)
FeO	2.45	1.74	2.90	3.66	5.09	
MnO	—	trace	trace	—	0.10	0.06
MgO	2.44	2.67	2.32	2.60	2.54	1.41
CaO	3.11	5.96	1.41	1.54	1.00	2.17
Na_2O	1.30	1.80	1.01	1.19	1.23	1.51
K_2O	3.24	2.67	3.60	3.69	3.79	2.30
H_2O+	5.00 (H_2O+ and H_2O-)	3.45	3.82	3.51	3.31	1.96
H_2O-		2.11	0.89	0.62	0.38	3.78
P_2O_5	0.17	0.20	0.15	—	0.14	0.18
CO_2	2.63	4.62	1.46	1.47	0.84	1.40
SO_3	0.64	0.78	0.58	—	0.28	0.03
Cl	—	—	—	—	—	0.30
Organic	0.80[a]	0.69[a]	0.88[a]	—	1.18[a]	0.66
Misc.	—	0.06[c]	0.04[c]	0.38[b]	1.98[b]	0.32
Total	99.95	100.48	100.46	99.60	100.00	100.62

[a] Carbon.
[b] FeS_2.
[c] BaO.

A. Average shale (Clarke, 1924, p. 24). Based on cols. B and C.
B. Composite sample of 27 Mesozoic and Cenozoic shales. H. N. Stokes, analyst (Clarke, 1924, p. 552).
C. Composite sample of 51 Paleozoic shales. H. N. Stokes, analyst (Clarke, 1924, p. 552).
D. Unweighted average of 36 analyses of slate (29 Paleozoic, 1 Mesozoic, 6 early Paleozoic or Precambrian) (Eckel, 1904).
E. Unweighted average of 33 analyses of Precambrian slates (Nanz, 1953).
F. Composite analysis of 235 samples of Mississippi delta. G. Steiger, analyst (Clarke, 1924, p. 509).

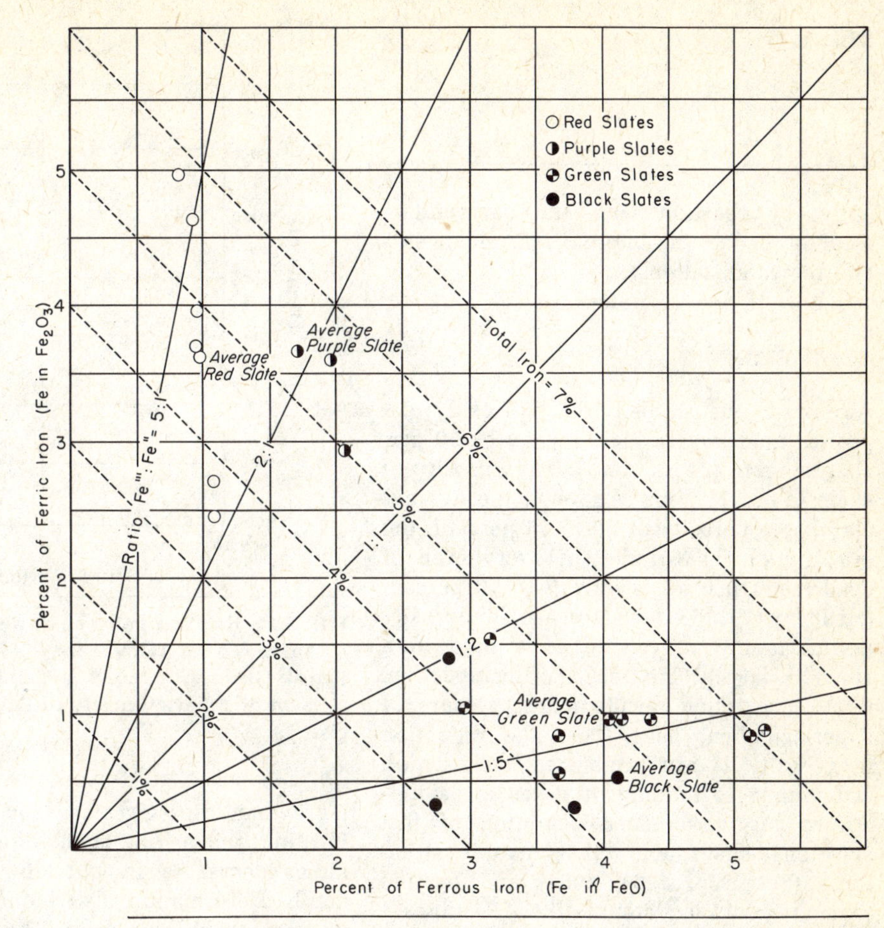

FIG. 8-9. Relation of ferrous ferric iron content to color of slates. Note that the red slates contain no more *total* iron than the black slates. (Redrawn from Tomlinson, 1916.)

ages showed that, although red shales do not owe their color simply to a greater total iron content or to a greater ferric iron content than normal rocks, there is more of both in most red shales (Van Houten, 1961). Most have an excess of Fe_2O_3 over FeO. The data seem to suggest that color difference is not just a question of reduction of iron in the nonred shales but also of removal of some reduced iron.

Red shales are so colored because of the presence of finely divided ferric oxide (hematite). In green and black shales the iron is largely in the ferrous state, but, exceptionally, ferric iron is present in such minerals as glauconite. Siderite-bearing shales tend to be gray to bluish on wholly fresh surfaces, such as cores, but because of the instability of siderite they become brown or buff in the outcrop even after very brief exposure.

The origin and significance of red color remain a troublesome problem. Van Houten (1961) thought the ferric oxide pigment to be detrital, derived from red upland soils formed in a tropical to subtropical climate. The essential condition for preservation of the red color is an oxidizing environment at the place of deposition. Such conditions are largely related to local geographic factors rather than to climatic ones. In general red beds are nonmarine; many are alluvial; many also are associated with evaporites, suggesting a dry climate.

DIAGENESIS OF SHALES

Shales are subject to many post-depositional modifications, some physical and some chemical. The first includes compaction, involving spore-space reduction and improved clay mineral orientation; the second involves mineral transformation either early, by reaction with the surrounding medium, or late, by internal rearrangements and recrystallization.

COMPACTION

As noted, the porosity of a freshly deposited mud is very high. It may be 50 percent or more (Trask, 1931). The porosity of shale is notably less. Although the average clay has a porosity of 30 to 35 percent, the porosity of shale under an overburden of 6000 feet (1800 m) is only 9 to 10 percent (Hedberg, 1936). This decrease in porosity accompanying conversion of a mud to a shale is largely the result of *compaction.* Such compaction results from pressure of superincumbent beds. That the reduction in porosity is attributable to such action rather than to infilling of pores is demonstrated by the crushed condition of fish skeletons, shells, and other organic structures (Ferguson, 1963, 1964) and by the crumpling of sandstone dikes (Dzulynski and Radomski, 1957) and the fillings of shrinkage cracks (Shelton, 1962). The compaction factor, that is, the ratio of original thickness to present thickness, shown by these observations ranges from 2.6 for sand crack fillings to 6.0 for deformed shells. Such thickness reductions can lead to formation of compaction dips where shale deposits overlie noncompactable "highs" such as reefs (Conybeare, 1967).

Athy (1930) and others (Hedberg, 1936; Jones, 1944) have demonstrated that porosity is a function of the thickness of the overlying strata (Fig. 8-10). Rubey (1931) has attempted to express the relations between porosity and depth below surface in algebraic terms:

$$P = 100\,C/(B + C + D)$$

where P is porosity, B is thickness of rocks eroded away, C is a constant, and D is depth below surface. The porosity–depth relations are complicated by two other factors, texture and deformation. In general, fine-grained rocks tend to be compacted

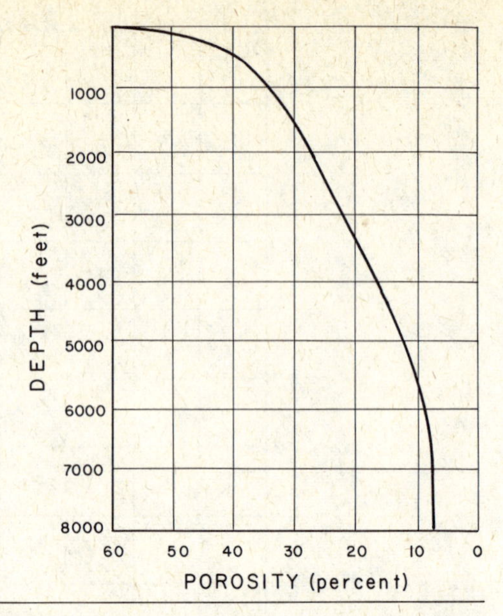

FIG. 8-10. Relation between porosity and depth of burial. (After Conybeare, 1967, Bull. Canad. Petrol. Geol., v. 15, no. 3 p. 332, Fig. 1. By permission of Alberta Society of Petroleum Geologists.)

more than coarse-grained rocks; therefore, if other things are equal, they will show larger decreases in porosity for a given depth. Deformation also induces a decline in porosity as was shown by Rubey, who expressed the relations between deformation and porosity as:

$$P_u = 100 - \cos d\,(100 - P_p)$$

where P_u is porosity of untilted rock, d is present angle of dip, and P_p is present porosity.

Compaction begins at the moment of deposition and continues over a long interval of time. As shown by studies of carbonate concretions in shale, compaction continues during the growth of the concretion (Raiswell, 1971) and is accompanied by a progressive improvement in the orientation of clay platelets (Oertal and Curtis, 1972).

DIAGENESIS

Chemical and mineralogical changes are presumed by some to begin almost as soon as the mud is deposited. Other changes take place with aging of the deposit and still others when temperatures and pressures become high enough to induce more drastic recrystallization. The relative importance and extent of these changes, especially those presumed to be early, have been subject to much debate and uncertainty. Some petrologists believe that clay minerals generated in the weathering profile and carried by streams are very susceptible to change once they reach the marine environment (Grim and Johns, 1954; Millot, 1949). Some go so far as to conclude that an equilibrium assemblage is established in newly deposited materials (Zen, 1959; Peterson, 1962; McNamara, 1966). Grim thought that kaolinitic clays, on reaching the marine environment, were altered to illite and chlorite and that montmorillonite was converted to chlorite. Hence the clay minerals are environment-sensitive and can be used to ascertain the nature of ancient, now-vanished environments. This view has been vigorously challenged. Studies of the clay-mineral content of a very large number of oceanic sediments seem to show that clay-mineral species present in streams draining adjacent land areas are the same as those in contiguous deposits (Griffin, 1962; Biscaye, 1965). Any of the clay minerals can occur in any of the major depositional environments (Keller, 1970, Table 9). Examination of ancient shales is even more convincing. Virtually all major clay minerals occur in any given shale. Clay mineralogy seems not to be correlated with the fresh, brackish, or marine facies of the Molasse (Lemcke, von Engelhardt, and Füchtbauer, 1953, p. 57). Likewise, clay mineralogy of the shales of the marine and nonmarine parts of the same cyclothem is much the same (Potter and Glass, 1958, Table 13). Some students of shales deny any correlation between clay minerals and environment and attribute the differences between one shale and the next as a record of variation in character of the source region at the time of accumulation. In other words, clay minerals are considered inert and behave as do any other detrital mineral (Weaver, 1958). The only exception, perhaps, is glauconite, a product of marine diagenesis. The only modification admitted by the proponents of this view is secondary cation adsorption (base exchange) in marine environments which does not, however, alter the basic clay-mineral lattice.

Examples of variations in clay-mineral character in one and the same deposit—either ancient or modern—can be explained by processes other than conversion of one mineral to another. In a study of Cretaceous and Tertiary deposits of the Upper Mississippi Embayment, Pryor and Glass (1961) noted that kaolin was dominant in those deposits of fluviatile origin, whereas montmorillonite dominated the outer neritic facies. They excluded diagenesis as an explanation and ascribed the differences to differential settling related to differences in grain sizes of the several clay-mineral species, to differential flocculation, or to different floccule size of the two species. Essentially similar conclusions were reached by van Andel and Postma (1954, p. 78) in their study of the Gulf of Paria.

It has been clearly shown that the clay mineralogy of a sediment may be altered on burial. Montmorillonite and kaolinite tend to disappear with increasing depth; illitic and chloritic materials take their places. The sequence of changes with depth of burial (and associated rise in pressure and temperature) has been studied in detail from shale samples collected from deep borings (Weaver and Beck, 1971). Other changes are related to aging of the material. In general, for example, older Paleozoic shales are largely illitic; Mesozoic and younger shales are more apt to be montmorillonitic. Kaolinite is also relatively less abundant in ancient sediments than in younger ones. These differences might, in part, be attributable to differing atmospheric and biospheric conditions in the earlier times (Weaver, 1967).

With increase in temperature and pressure, shales show progressive mineralogical changes which can be described as metamorphic. In general, transformations are mineralogical, and bulk chemical composition remains essentially unchanged except for loss of water and carbon dioxide (Eckel, 1904; Nanz, 1953; Shaw, 1956). The changes are, therefore, largely isochemical. There is a tendency for some reduction of iron during metamorphism. Compare, for example, the averages for Paleozoic shales with those for Paleozoic slates (Table 8-7).

Some clay minerals themselves can be

weathered. Extraction of potassium from illite, for example, is possible under certain climatic conditions. Kaolin may be altered to bauxite. These changes are reversible, however, and in the proper environment the degraded illite may take up potassium (Grim, Dietz, and Bradley, 1949, p. 1806); the bauxite may even be converted to kaolinite (Gordon and Tracey, 1952).

CLASSIFICATION AND PETROLOGY OF SHALES

CLASSIFICATION

Shales, unlike sandstones and limestones, have never been classified in a comprehensive and rational manner. Current nomenclature is only for convenience and tradition and does not necessarily express differences of first-order significance. Most commonly our nomenclature recognizes only color and fissility; we have, therefore, such designations as red mudstone, black shale, green argillite.

Carozzi (1960, pp. 124–184) classified clays and shales mainly on the basis of texture. He recognized those with (1) a residual igneous texture, (2) a residual pyroclastic texture, (3) a massive to slightly laminated texture (Fuller's earth, underclays), (4) a laminated character (the common shales), (5) diagenetic oolitic and pisolitic textures (diaspore clays and bauxite), (6) textures indicating "diagenetic migration" (such as replacement of shells by montmorillonite), and (7) a diagenetic recrystallization texture. This classification leads to insight into the origin of the clay or shale, a laudable goal, but tends to lump together all common shales and to separate out only rather rare and exotic types.

Efforts have been made to identify the types of shale associated with principal sandstone groups (Krynine, 1948; Nanz, 1953). Krynine believed the shales associated with graywacke were chloritic, those with arkoses, kaolinitic, those with orthoquartzites, illitic. Nanz's effort was to identify these types by their bulk chemical composition. Neither of these efforts have proved very successful, perhaps because variations in texture and concomitant variations in composition are greater than differences caused by genesis. Moreover, shales are much less well differentiated chemically than are sandstones and are, except for unusual dilution by chemical or biochemical contaminants, a much more homogeneous group of rocks.

We cannot be sure with our present knowledge just what the significant attributes really are. We do not know with certainty the significance of clay-mineral species, or the extent to which chemical and mineralogical composition is influenced by source rocks or climate in the source region, or the relations, if any, between composition and chemistry of the medium in which the mud accumulated, or the nature and effects of diagenetic modification. Until we have adequate data on these questions, the problem of a rational classification of shales based on composition and texture—analogous to that of the sandstones—will have to remain in abeyance.

The trouble with shales arises from our inability to make the kind of analysis that we do with sandstones. A list of minerals present is not enough; nor even is a list with proportions of each constituent adequate. We need to know the *origin* of the minerals. In a sandstone we can distinguish, in thin section, between authigenic and detrital, between those minerals precipitated from solution and those produced by *in situ* alteration. The geometry of the fabric of a sandstone enables us to sort out minerals and place them in genetically significant categories and thus read the history of the rock. In shales this can be done only very poorly or not at all. Hence our knowledge of shales, compared with that of sandstones or limestones, is primitive and incomplete.

To be meaningful a classification must be based on genetically significant properties. Can we define any such propertes of shales? As indicated, perhaps the alumina–soda ratio is a maturity index. The ferrous–ferric iron ratio is perhaps an oxidation index, an index, however, subject to modification during metamorphism. The presence of relict structures in residual deposits and the presence or absence of laminations in transported shales and mudstones are genetically meaningful textural attributes.

Because some of these parameters are measurable only by chemical analysis, and because our list is probably incomplete, we cannot at this time form the basis of a workable field classification. We have, therefore, no classification. For sake of presentation, we treat separately those deposits which are residual or formed in place and those which have been transported. Among the latter we recognize those marked by exceptional textural or compositional attributes such as carbonaceous and siliceous shales and loess.

RESIDUAL CLAYS

Residual clays form in place and are, in fact, a soil or a product of the soil-forming processes. Although some such clays have important commercial uses, they are seldom encountered in older strata except rarely as paleosols or "fossil" soils at an unconformity surface. Because most clays and shales contain materials of residual origin, it is important to know what these materials are in order to recognize the presence of such materials and to correctly interpret their significance.

Residual soils (*regolith* of Merrill; *saprolite* of Becker; *sathrolite* of Sederholm) are products of weathering, formed *in situ*. The character of these deposits is dependent on climate, drainage, and parent-rock materials. In mature soils (normal or zonal types) climate is the more important factor. In immature soils (intrazonal or azonal) effects of drainage and the nature of the parent rock are readily discernible. That essentially similar products are formed from diverse source rocks is well shown by comparison of analyses A, B, C, and D of Table 8-8 which are residual clays formed from basalt, limestone, gneiss, and granite, respectively.

In general, in humid regions residual materials are enriched in hydroxides of aluminum and ferric iron (pedalfers) and impoverished in lime, magnesia, and alkalies. Under extreme conditions even silica is removed (analysis E of Table 8-8), so that the end-product will consist of little more than alumina and iron oxide. These alumina and iron-rich residual materials are the *laterites* (Harrison, 1934). Laterization requires both the high rainfall and the high temperatures of tropical regions. Laterites, both ferruginous and bauxitic, are marked by concretionary structures, pisolites, and larger bolsterlike bodies.

In the arid regions the soils are enriched in lime (pedocals). A *caliche* or *duricrust* (Woolnough, 1928) may be formed. A duricrust is a peculiar type of deposit which is formed on a peneplain in a climate that leads to a sharply defined alteration of saturation and desiccation. It is an armorlike deposit produced by upward capillary migration of groundwaters during the arid period. From these solutions are precipitated ferruginous (ferricrete), siliceous (silcrete), or calcareous (calcrete or caliche) materials.

A detailed treatment of soils, their properties and origin, is clearly beyond the scope of this book; refer to the standard works of pedologists (Glinka, 1927; Byers et al., 1938; Robinson, 1951). For a discussion more oriented toward the geological aspects of soils, refer to Goldich (1938), Reiche (1950), and especially Hunt (1972).

Examination of an unconformable contact should include observations on the undisturbed residuum which may be present on such a surface and on the freshness or degree of alteration of the rocks below the unconformity. The characters of such a residuum will be, in part, those of a soil but also in part those brought about by changes since burial. These changes are diagenetic and are a reversal of soil-forming processes. "Fossil" soils may not be present at all unconformities because of their removal before deposition of the younger beds. A "fossil" regolith is characterized by its ill-sorted nature. The textures and structures of parent rocks have been obliterated largely without being replaced by textures and structures that characterize a normal sediment. The regolith may grade downward into unaltered source rock and may also grade upward into overlying sedimentary rocks. Such graded unconformities commonly occur where arkose overlies granite. The arkosic residuum may be indurated by later cementation so that it constitutes a recomposed granite (p. 217).

Because the residuum is mainly the product of chemical decay, it should be marked by certain chemical features. It contains a notably higher content of ferric oxide and alumina than does the parent rock and is deficient in fugitive oxides, such as soda. Moreover, if the residuum is a mature soil, it will show a soil profile or zonal arrangement, as do present-day soils. The profile may be incomplete because of removal of the upper zones by the advancing sea; or incomplete because of the climatic conditions under which it formed. Most fossil soils are azonal or intrazonal. The original profile in any case will be more or less altered by the authigenesis which has taken place since burial.

"Fossil" soils have been given cursory field and laboratory study. An exception, perhaps, are the soils developed on older Pleistocene glacial deposits. These form important time markers and so have been given close scrutiny (Kay and Pearce, 1920; Leighton, 1930). Pre-Pleistocene soils are seldom mentioned, less often described. The criteria for recognizing such materials need to be worked out, because such criteria are also the means for recognizing an unconformity. The student must exercise caution in studying unconformity surfaces to be sure that any rock alterations are not pres-

TABLE 8-8. Chemical composition of residual clays

Constituent	A	B	C	D	E	F	G
SiO_2	40.7	55.42	55.07	66.01	44.80	9.28	48.00
TiO_2	7.3	trace	1.03	0.10	2.44	3.78	1.00
Al_2O_3	30.9	22.17	26.14	21.21	38.84	69.76	34.56
Fe_2O_3	8.7[a]	8.30	3.72	2.11	0.36[a]	1.14[a]	1.54[a]
FeO	—	trace	2.53	0.57	—	—	—
MnO	—	—	0.03	—	—	—	—
MgO	—	1.45	0.33	0.05	0.10	0.15	0.35
CaO	1.0	0.15	0.16	none	0.05	0.40	0.23
Na_2O	0.4	0.17	0.05	none	0.30	0.40	0.51
K_2O	0.3	2.32	0.14	0.85	0.23	0.95	0.59
H_2O+	11.0[b]	7.76	9.75	7.55	13.62[c]	13.37[c]	12.30
H_2O-	—	2.10	0.64	1.24	—	—	1.30
P_2O_5	—	trace	0.11	—	—	—	—
SO_3	—	—	trace	—	—	—	—
CO_2	—	—	0.36	0.08	—	—	—
BaO	—	—	0.01	—	—	—	—
S	—	—	0.04	—	—	—	—
Total	100.3	99.84	100.11	99.77	100.74	99.23	100.38

[a] Total iron.
[b] Ignition loss.
[c] Total water.

A. Residual clay from basalt, Spokane County, Washington (Scheid, 1945).
B. Residual clay from dolomite, Morrisville, Alabama. W. F. Hillebrand, analyst (Russell, 1889).
C. Residual clay from Morton gneiss, Redwood Falls, Minnesota. S. S. Goldich, analyst (Goldich, 1938).
D. Weathered Hong Kong granite ("completely decomposed"), Gindrinker's Bay, Hong Kong area. A. Willman, analyst (Brock, 1943).
E. White flint clay, Phelps County, Missouri. R. T. Rolufs, analyst (Allen, 1935).
F. Diaspore clay, same location as "D." R. T. Rolufs, analyst (Allen, 1935).
G. Plastic Ione clay, Jones Butte, California (Allen, 1929, p. 379).

ent-day products. Such surfaces may be zones of accelerated groundwater flow and leaching. In recent years great interest has been shown in paleosols (Yaalon, 1971).

There are but few well-described examples of fossil soils. One of the oldest on record occurs at the contact between the Huronian and the Archean basement on which it rests in the area north of Lake Huron (Frarey and Roscoe, 1970, p. 152). This particular regolith is not enriched in ferric iron, an observation taken to mean by some that the atmosphere was anaerobic rather than oxidizing as at present. There is presumed to be an evolutionary sequence in the character of very ancient paleosols which reflects the changing nature of both the atmosphere and biosphere (Yaalon, 1962b). An even older regolith, and one of the first to be described, is that on the sub-Bothnian unconformity in Finland (Sederholm, 1931). Other examples of fossil soils include that at the base of the Cambrian in Arizona (Sharp, 1940), the Pennsylvanian underclays beneath coals interpreted as paleosols by Grim and Allen (1938), and the weathering of the Lewisian Gneiss beneath the Torridonian of Scotland (Williams, 1968).

COMMON SHALES

As has been noted, common shales are a mixture of clay-sized materials, mainly clay minerals, and silt, mainly quartz. The clay-mineral content varies from 40 percent or less to 100 percent. In those shales with a very high clay-mineral content, clay particles are very small, mostly less than 0.002 mm. They show uniform aggregate orientation parallel to the lamination of the shale. In these shales other minerals tend also to be very fine-grained. Shales with a lesser clay-mineral content are generally coarser or silty. Maximum grain size may be as much as 0.01 mm. Such shales contain larger flakes of muscovite, biotite, and chlorite. These tend to be present in distinct books. The larger flakes tend, also, to show a uniform orientation parallel to the bedding, although some are bent around detrital quartz grains and a few are actually perpendicular to the laminations.

The clay-mineral fraction is usually complex. Pennsylvanian shales in Illinois contain a mixture of illite, kaolinite, and chlorite, with illite dominant (Grim, Bradley, and White, 1957). Many contain illite-montmorillonite mixed-layer assemblages. Older Paleozoic shales have little or no kaolinite. Essentially similar observations were made by Weaver (1958, 1967). Mesozoic and younger shales are apt to contain montmorillonite as, for example, those in the Tertiary Molasse of Germany (Lemcke, von Engelhardt, and Füchtbauer, 1953). Even younger shales, however, contain illite, kaolinite, chlorite, and various mixed-layer subspecies (Fig. 17-1).

The silt fraction is mainly quartz, generally poorly rounded or angular, Feldspar can be identified in some siltier shales. Carbonate, generally calcite, is a common accessory mineral in many shales, mostly as fossil fragments (in some cases recognizable) but more often as irregular grains and aggregates up to 0.06 mm in size. Siderite is a rare constituent in other shales. Pyrite, as either aggregates or as minute grains, commonly euhedral, is a minor constituent in many shales. Shales are, in general, poor in heavy minerals, although it is possible to recover a significant suite of such minerals from them. It has been observed that unstable species dissolved in sandy beds may be preserved in contemporaneously deposited shales (Blatt and Sutherland, 1969).

In general, the silt fraction is scattered through the shale; that is, silt shows no tendency to concentrate in distinct layers. The shale is, therefore, a heterogeneous mixture of clay minerals and other materials of all sizes. Exceptionally the silt forms thin discontinuous laminations. An additional cause of lamination is concentration of organic matter. Shales with appreciable amounts of such material tend to split preferentially along bedding surfaces on which such material is concentrated. The organic matter is present as a pigment or as minute discrete yellow bodies.

As noted, the color of the shale tends to reflect the quantity of organic matter present and, most particularly, the oxidation state of the iron.

RED SHALES AND MUDSTONES

Red shales and red mudstones are the dominant rock types in the so-called red-bed facies. They are the most deeply colored; associated sandstones are less red, commonly pale pink or even green or gray. Many are very silty and might properly be called siltstones. But they are, in contrast to associated sandstones, quite fine-grained. Many are without laminations, are without fissility, and are best described as massive mudstones. They are notably unfossiliferous. Some contain small irregular calcareous concretionary bodies, probably a "fossil" form of caliche.

Most of the red mudstones probably are terrestrial—perhaps alluvial—because they occur in fining-upward cycles. Notable Appalachian examples include the red mudstones of the Catskill (Devonian) and Mauch Chunk (Mississippian) formations. Similar red deposits occur in the terrestrial Newark Series (Triassic) and to a lesser extent in the Juniata (Ordovician) and Bloomsburg (Silurian) formations of the same region. Red mudstones and red shales are known from similar deposits the world over.

Red shales and mudstones are not particularly iron-rich despite their red color (Fig. 8-9). They owe their color primarily to their accumulation in an oxidizing rather than reducing environment (Grim, 1951; Van Houten, 1961). They are, therefore, more often than not, terrestrial and perhaps commonly deposited under an arid or semi-arid climate. Were the depositional site or climate otherwise, organic matter would be present and tend to reduce the ferric iron to a nonred ferrous state.

BLACK (CARBONACEOUS) SHALES

Black shales are fissile, and many split readily into thin, slightly flexible sheets of large size. They are exceptionally rich in organic matter. They also tend to be rich in iron sulfide, usually pyrite, which replaces fossils, forms nodules, or occurs as finely disseminated grains. Black shale rarely contains any fossils, or at best, has a sparse, depauperate, and restricted fauna. Except for phosphatic forms present, organisms are present only as graphitic or carbonaceous films or as pyrite replacements. Concretionary carbonate layers or nodules, commonly showing cone-in-cone structure and septarian veins, are abundant locally in some black shales.

The faunas of black shales are remarkably uniform. Littoral and benthonic forms, of the usual warm-water facies, are missing. Brachiopods are represented only by phosphatic, inarticulated shells, such as those of *Lingula* or *Discina*, which are ubiquitous and hardy types capable of survival under adverse conditions. The mollusks are all thin-shelled and as a rule depauperate. Among them are byssiferous pelecypods (*Posidonia*) which are as likely to attach themselves to floating as to fixed objects. The noncalcareous black shale fauna is very unlike that of the limestones and calcareous shales that are found in the same area (Ruedemann, 1934, p. 13). Conodonts, rare fish remains, and spores and spore cases complete the list of organic remains found in black shales. The black shales of the Ordovician and Silurian, however, are marked, in addition, by graptolitic remains preserved as carbon films.

Black shales contain an unusual amount of carbonaceous matter, in part distillable, and sulfide sulfur (Table 8-9). Whereas the average shale has about 1 percent carbon, carbonaceous shales contain from 3 to 15 percent. Some are also noted for their unusual concentration of certain trace elements, notably V, U, Ni, and Cu. Because of the unusual metal content, some black shales have been exploited, such as the Kupferschiefer shales at Mansfeld in Germany (Gregory, 1930) and the Nonesuch Shale (Precambrian) in Michigan (White and Wright, 1954). The oxidation of the iron sulfides produces a swelling and disintegration of the rock in outcrops and covers the surface with a white efflorescence.

Components of the black shales are coarse and fine detrital materials, which form the bulk of the rock, pyritic matter, and a carbonaceous fraction which gives the shale its distinct character. A considerable part, perhaps as much as one-third, is silt differing in no respect from that of ordinary shales; another third is clay material, mainly illite. The pyritic fraction varies greatly in abundance, forming in extreme cases as much as 38 percent, as in the Wauseca "graphitic slate" (Precambrian) in Iron County, Michigan. The pyrite in this example is extremely fine-grained, and individual crystals are invisible to the naked eye. In Paleozoic or younger black shales, carbonaceous materials may occur in discrete bodies—spores and spore cases (Hough, 1934). These appear in transmitted light as yellow to amber-colored objects of discoidal form averaging 0.2 mm in diameter. The organic matter also occurs as irregular shreds and flakes and as very fine material that cannot be resolved (Fig. 8-4).

The paper-thin bedding so characteristic of many black shales has been ascribed to

TABLE 8-9. Chemical composition carbonaceous shales

Constituent	A	B	C	D	E	F
SiO_2	51.03	60.65	36.67	58.03	63.09	29.36
TiO_2	—	0.62	0.39	0.64	0.99	0.41
Al_2O_3	13.47	11.62	6.90	15.00	18.58	8.18
Fe_2O_3	8.06	0.36	—[c]	3.67	2.17	4.84
FeO	—	—	2.35[c]	5.82	2.73	—
MnO	—	0.04	0.002	0.09	0.22	0.01
MgO	1.15	1.90	0.65	1.64	2.67	0.32
CaO	0.78	1.44	0.13	0.26	1.11	16.85
Na_2O	0.41	0.60	0.26	3.52	4.54	1.27
K_2O	3.16	3.10	1.81	3.60	0.54	1.36
H_2O+		3.77	1.25	3.46		
(H_2O+ and H_2O-)	0.81				2.69	5.09[e]
H_2O-		1.19	0.55	0.84		
P_2O_5	0.31	0.18	0.20	0.16	0.12	0.21
CO_2	—	1.65	—	0.03	—	10.80
SO_3	—	—	2.60	—	—	—
S	7.29	3.20[a]	—	0.04	trace	5.33
C	13.11	9.20	7.28	3.27	—	16.01[f]
FeS_2	—	—	38.70	—	—	—
V_2O_5	—	—	0.15	—	—	—
Total	102.90[b]	99.52	99.91	100.24[d]	99.45	100.04

[a] FeS_2.
[b] Less O = S; total becomes 100.17.
[c] Direct determination not possible because of organic matter; iron in excess of pyrite reported as FeO.
[d] Includes 0.14 SO_3.
[e] "Ignition" loss at 900°C less CO_2, S, and organic matter.
[f] "Organic matter."

A. Black shale (Devonian), Dry Gap, Walker County, Georgia. L. G. Eakins, analyst (Diller, 1898). Includes 3.32 hydrocarbons.
B. Ohio shale (Devonian), Logan County, Ohio. Downs Schaaf, analyst (Lamborn, Austin and Schaaf, 1938, p. 20).
C. "Graphitic slate," Wauseca Member, Dunn Creek slates (Precambrian), tenth level Buck Mine, Iron River district, Michigan. C. M. Warshaw, analyst (James, 1951, p. 255).
D. Black slate, Dunn Creek slates (Precambrian), Crystal Falls district, Michigan. R. H. Nanz, Jr., analyst (Nanz, 1953).
E. Nonesuch Shale (Keweenawan), Michigan (Lane, 1911, p. 118).
F. Posidonienschiefer, Germany (von Gaertner, 1955, p. 455).

the colloidal nature of the materials, which have shrunk and have been compressed perhaps to one-fifth or less of their original thickness.

Black shales occur in many places and formed at many times during the geologic past. The most notable North American examples are the older Precambrian "graphitic slates" of Michigan (James, 1951), the younger Precambrian Nonesuch shale of the same region (White and Wright, 1954), the Ordovician black shales of the Taconic region (Ruedemann, 1934), the Chattanooga and other Devonian black shales of the eastern interior United States (Rich, 1951; Bates and Strahl, 1957), and the black shales of the Pennsylvanian coal measures (Payton and Thomas, 1959). Well-known European examples include the Permian Mansfeld Kuperschiefer (Gregory, 1930) and the Posidonia shales (von Gaertner, 1955) of Germany. As pointed out by James (1954, p. 279), black shales and related products of strongly reducing environments seem to appear in the geosynclinal cycle just after the sediments formed on the aereated shelf and just before those derived from the rising geanticlinal ridge. The rise of the latter, prior to emergence within the geosyncline, produces the semi-isolation required for black shale deposition.

The origin of black shales has been much debated. Certainly they were deposited under anaerobic conditions. How such conditions were achieved is less certain. The stagnated water body may have been deep and sealed off from the atmosphere by a density stratification of waters produced by a layer of relatively fresh water overlying more saline waters as in the present Black Sea (Androussow, 1897) and in certain Norwegian fjords (Ström, 1936). Some writers contend that black shales were deep-water marine (geosynclinal) sediments; others have postulated comparatively shallow waters, either lagoonal or marine. Ulrich (1911), Schuchert (1915), Grabau and O'Connell (1917), Ruedemann (1934), Twenhofel (1939) and Rich (1951), in particular, have reviewed the black shale problem.

Siliceous shales have an abnormally high content of silica. Whereas the average shale has 58 percent of silica, siliceous shales may contain as much as 85 percent (Table 8-10). Other constituents, notably ferrous iron or carbonate, are absent or very minor. Calculations, assuming silicate minerals with the highest probable silica ratios, show at least 70 percent of the rock is uncombined silica. Such highly siliceous shales, as might be expected, are hard, durable rocks that resist disintegration.

The siliceous character seems not to be attributable to large quantities of detrital

TABLE 8-10. Chemical composition of siliceous shales and related materials

Constituent	A	B	C
SiO_2	84.14	73.71	84.54
TiO_2	0.22	0.50	0.35
Al_2O_3	5.79	7.25	4.14
Fe_2O_3	1.21	2.63	1.48
FeO	—	0.44	0.51
MnO	—	—	—
MgO	0.41	1.47	0.52
CaO	0.13	1.72	1.25[a]
Na_2O	0.99	1.19	0.46
K_2O	0.50	1.00	0.64
H_2O+	5.56 (H₂O+ and H₂O−)	6.94	3.11
H_2O-		2.88	3.25
P_2O_5	—	0.24	0.28
CO_2	none	trace	—[a]
SO_3	—	0.16	0.18
S	—	—	—
C	1.21	0.00	0.12
Total	100.03	100.13	100.74[b]

[a] CO_2 and a corresponding proportion of CaO to combine as $CaCO_3$, calculated out of the composition.

[b] $CaCO_3$ calculated as 5.16 not included in total.

A. Siliceous shale, Mowry Formation (Cretaceous), Black Hills, South Dakota. J. G. Fairchild, analyst (Rubey, 1929).

B. Diatomaceous shale, Modelo Formation (Miocene), California. J. G. Fairchild, analyst (Hoots, 1931, p. 108).

C. Cherty shale, Modelo Formation (Miocene), Mulholland Highway, Santa Monica Mountains, California. J. G. Fairchild, analyst (Bramlette, 1946, p. 13).

quartz but rather to be derived from amorphous silica, such as opal, or from volcanic ash. Rubey (1929) estimated the siliceous Mowry Shales (Cretaceous) to be one-third cryptocrystalline quartz derived from chemically precipitated opaline silica. More or less devitrified rhyolitic ash also forms a major part of the shale and accounts for its peculiar chemical composition (Table 8-10A).

Rubey concluded that the Mowry Shale owed its highly siliceous character to the volcanic ash with which it is closely associated. Chemical decomposition of slowly accumulated, very fine-grained, highly siliceous volcanic glass, in the presence of decaying organic matter, provided the silica, which was precipitated concurrently with the accumulation of the normal shale constituents.

Goldstein and Hendricks (1953) reached similar conclusions regarding the siliceous shales of the Stanley, Jackfork, and Atoka (Carboniferous) formations of Arkansas and Oklahoma. The silica of these shales was thought to be derived from submarine weathering of volcanic ash. The restricted faunas of these shales (linguloid and orbiculoid brachiopods, conodonts, radiolarians, and sponge spicules) suggest waters rich in silica and poor in lime. Fine laminations suggest quiet waters and a slow rate of deposition.

Bramlette (1946) concluded that the siliceous shales of the Monterey Formation (Miocene) of California, called porcelanous shale and porcelanites or cherty shales by him, were produced by the addition of biochemical silica to normal shale at the time of formation of the deposit. Infalls of volcanic ash may have been a source of the silica utilized by the abundant microorganisms.

CALCAREOUS SHALES AND MARLS

The lime carbonate content of most shales is small. The average shale has about 2.63 percent CO_2, equivalent to about 6 percent calcite. As the carbonate content is increased, the rock becomes less fissile and effervesces in acid and may properly be termed a calcareous shale (Table 8-11A).

Marls proper are semifriable mixtures of clay materials and lime carbonate. The better-indurated rocks of like composition are *marlstones* or *marlite* and are more correctly an earthy or impure limestone rather than a shale. Marl has been defined as a rock with 35 to 65 percent carbonate and a complementary content of clay (Fig. 10-41).

The carbonate of a calcareous shale may consist of finely precipitated materials or small particles of organically fixed carbonate (microfossil tests, pulverized skeletal materials, and the like).

Calcareous shales are very common and hence have attracted less attention than more unusual types. An exception is the rather thorough study of Campbell and Oliver (1968) of the Devonian shales associated with the Leduc Reefs of Alberta, Canada. They found a complete range in composition from nearly pure limestone to calcareous shale. The clay-mineral fraction was largely illite, forming 13 to 39 percent of the shale. Quartz made up 3 to 26 percent. The carbonate fraction was dolomite and calcite in amounts from 1 to 36 percent dolomite and 16 to 76 percent calcite. Most of the constituents were thought to be detrital, with carbonates derived from organic carbonate buildups within the basin and noncarbonates derived from sources exterior to the basin.

MISCELLANEOUS SHALES

The unusual shale types described above (red shales, black shales, siliceous shales, and calcareous shales) all have gross physical properties which enable the field geologist to recognize them at sight or by simple test with an acid bottle. There are, however, some shales of rather unusual composition which cannot be so detected and which require chemical analyses for their recognition. These are shales unusually rich in alumina, iron, soda, or potash (Table 8-11).

High alumina shales are uncommon. The average shale contains 15.4 percent Al_2O_3 (Table 8-11F). Those with a high silt content will be lower, whereas those with a higher-than-average clay content will be higher. Unlike residual clays, few shales or slates contain more than 20 percent alumina (Fig. 8-11). Probably fewer than 5 percent of all shales contain over 22 percent. Some residual clays, on the other hand, are very high in alumina (Table 8-8). But because most shales are a mixture of both residual clay and detrital silt (mainly quartz), the alumina content is much depressed below that in the residual deposit. The finer fraction of any given shale is more largely residual in character and hence shows an alumina content higher than the shale as a whole. Kaolinitic clays, and hence kaolinitic shales, tend to be richer in alumina than

TABLE 8-11. Chemical analyses of shales (excluding carbonaceous and siliceous shales)

Constituent	A	B	C	D	E	F
SiO_2	25.05	51.38	49.85	56.73	58.82	58.10
TiO_2	—	1.22	1.45	0.88	0.73	0.65
Al_2O_3	8.28	23.89	13.88	19.27	16.46	15.40
Fe_2O_3	0.27	2.05	3.75	5.57	1.10	4.02
FeO	2.41	5.01	14.10	1.89	7.20	2.45
MnO	4.11	0.02	0.24	—	0.09	—
MgO	2.61	2.71	3.32	1.93	4.92	2.44
CaO	27.87	0.24	0.20	0.01	0.76	3.11
Na_2O	—	0.59	0.10	0.49	4.03	1.30
K_2O	—	7.08	2.74	8.85	1.60	3.24
H_2O+	2.86	4.66	4.90	3.77[a]	3.73	}
						5.00
H_2O-	1.44	0.21	0.14	0.38	0.11	}
P_2O_5	0.08	0.01	0.09	—	0.17	0.17
CO_2	24.20	0.14	4.09	0.00	0.01	2.63
SO_3	—	none	—	—	0.02	0.64
S	—	none	1.51	—	0.05	—
C	—	0.16	0.69	—	—	0.80
Total	99.18	99.52	101.05	99.77	99.87[b]	99.95
Less O for S			−0.76		−0.06	
			100.29		99.81	

[a] "Loss on ignition."
[b] Includes 0.03 Cl, 0.04 F.

A. Cretaceous shale, Mt. Diablo, California. Analyst, W. H. Melville (Clarke, 1924, p. 552). A very calcareous shale, also with abnormal MnO content.

B. Slate from the Tyler Formation (Precambrian), about 1 mile west of Montreal, Wisconsin. R. Nanz and B. Brunn, analysts (Nanz, 1953). A high alumina rock, also potassic.

C. Dunn Creek Slate (Precambrian), Homer Mine, Iron River, Michigan. C. Warshae, analyst (James et al., 1968, Table 4). An iron-rich slate.

D. Average of six analyses of Cartersville Slate (Cambrian), Georgia (Shearer, 1918, p. 134). A very high potassic slate.

E. Varved argillite (Huronian), Olive Township, Ontario, Canada. M. Balazs, analyst (Pettijohn and Bastron, 1959). Includes 0.03 Cl, 0.04 F. A soda-rich argillite.

F. Clarke's average shale for comparison (Clarke, 1924, p. 34).

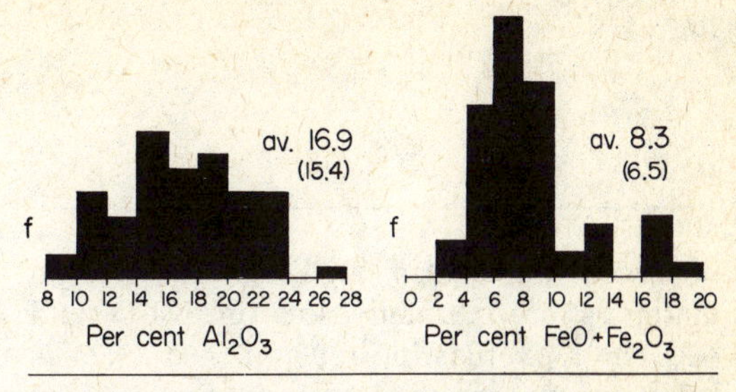

FIG. 8-11. Distribution of alumina and the iron oxides in 60 Paleozoic and Precambrian slates. The average content of these oxides is also given; numbers in parentheses indicate the content of these oxides in Clarke's average shale.

those with three-layer clay-mineral species. A high alumina shale, therefore, is either an exceptonally fine-grained shale, or one which is kaolinitic. Such shales are, as noted, uncommon.

The average shale contains about 6.47 percent iron oxides: 4.02 percent Fe_2O_3 and 2.45 percent FeO (Table 8-7). The average late Precambrian slate contains 8.9 percent iron oxides; the average Paleozoic slate contains 5.9 percent (Nanz, 1953). In slates, unlike shales, ferrous oxide generally exceeds ferric oxide. It is clear that the normal pelitic sediment contains 6 to 8 percent iron oxide. As is apparent from Fig. 8-11, an iron oxide content in excess of 12 percent is unusual; certainly if more than 15 percent is present, the rock is a *ferriferous shale* or slate.

Such iron-bearing rocks grade into true ironstones—rocks with at least 15 percent iron (about 20 percent iron oxides). The high iron content of a shale or mudstone is indicative of the presence of iron sulfide, iron carbonate, iron silicate, or iron oxide. Inspection of the analyses should disclose the nature of the iron-bearing mineral (Table 8-11C). These shales are evidently mixed sediment produced by cosedimentation of fine argillaceous sediment and an iron-bearing mineral. The petrology of the ironstones proper is covered elsewhere (page 407).

Clay ironstone is an old term for a mudstone rich in siderite, generally containing more siderite than argillaceous matter. It occurs as thin beds and nodules in the Coal Measures of both the United States and Great Britain. *Chamositic* or *chloritic mudstones,* mixtures of chamosite or chlorite and clay, are relatively rare rocks associated with sideritic mudstones and bedded siderites in some iron-bearing districts such as the Cleveland area of Great Britain (Taylor, 1949, pp. 27, 29–30) and the Iron County area of Michigan (James, 1954, Table 8, analysis 7).

The potash content of the average shale is very nearly the same as that of the average igneous rock. Yet the weathering of the latter tends to form potash-poor residual clays. Apparently potash removed by weathering is somehow restored to shales. The average shale, according to Clarke (1924, p. 34), contains 3.24 percent potash. The average late Precambrian or Paleozoic slate contains about 3.6 percent. As can be seen from Fig 8-12, a potash content of 2 to 5 percent is normal for most shales. Only about 1 shale in 20 has 5 or more percent. Such shales are *potassic shales* (Table 8-11B and D).

Illustrative of such shale is the Decorah Shale (Ordovician) of Minnesota (Schmitt, 1924) once considered as a possible potash reserve. This shale contains nearly 6 percent K_2O, which on a carbonate-free basis is about 8 percent K_2O. The potash content of the finer fraction (under 1/32 mm) of the Glenwood Shale of the same age is nearly 11 percent. Potassic shales are also known from Georgia (Shearer, 1918, p. 134). The Cartersville Slate (Cambrian) contains over 8 percent K_2O (Table 8-11D). It also was once looked on as a source of potash. According to Gruner and Thiel (1937), potash in Minnesota shales was believed to be present as orthoclase thought by them to be of authigenic origin. These conclusions do not explain the high potash content. Perhaps a related problem is the feldspathization of a Cambrian sandstone described by Berg (1952), who ascribed the feldspar of this rock to reaction between clay minerals and sea water. Weiss (1954) attributed the high potash content of several shale seams to "feldspathization of a potassic bentonite."

Sodic shales are those in which the Na_2O content not only exceeds the 1.3 percent of the average shale but also greatly exceeds the K_2O percentage. It has been long observed that shales, argillites, and related materials are normally low in Na_2O and high in K_2O (Bastin, 1909, p. 463). These relationships are well seen in Fig. 8-13 in which is shown the Na_2O–K_2O content of various Paleozoic and Precambrian slates. Shown in this figure also are several examples of abnormally sodic argillites and slates (see also Table 8-11E). The soda in the Huronian argillites seems to be present as albitic feldspar. Various explanations have been offered to account for the abnormal soda content. If the argillite is glacial, as in the Huronian, the glacial silt may be sodic if the source terrane is rich in sodic rocks; it seems to be in this case (Pettijohn and Bastron, 1959), but the low content of lime in these same sediments is incompatible with this hypothesis, because the presumed source rocks are lime-bearing. Possibly, therefore, soda has been introduced—a process of albitization. Possibly also, deposition in a sodic lake would form soda-rich shales. A somewhat higher-than-normal Na_2O content in the pelitic rocks of the Carolina Slate Belt has been attributed to the inclusion of significant quantities of unweathered rhyodacitic ash (Sundelius, 1970, p. 359).

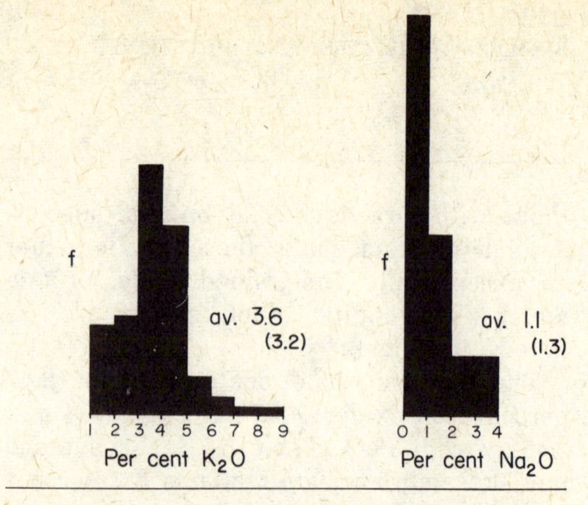

FIG. 8-12. Distribution of potash and soda in 66 Paleozoic and Precambrian slates. The average content of these oxides is also given; numbers in the parentheses indicate the content of these oxides in Clarke's average shale.

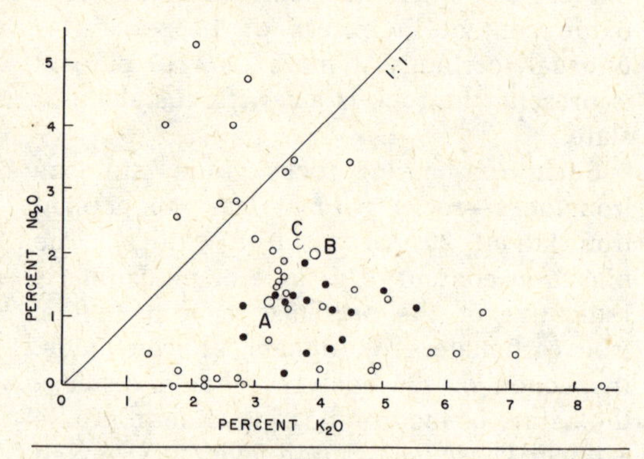

FIG. 8-13. Na_2O/K_2O content of argillites and shales. Black circles: slates (after Dale, 1914, p. 51); open circles: Precambrian slates and argillites (Nanz, 1953, pp. 53–54). A: average shale (Clarke, 1934, p. 34); B: average Norwegian glacial clay (Goldschmidt, 1954, p. 53); C: composite New Zealand slate or argillite (Reed, 1957, p. 28). (After Pettijohn and Bastron, 1959, Bull. Geol. Soc. Amer., v. 70, Fig. 2.)

GEOLOGIC OCCURRENCE OF MUDS AND SHALES

IN THE MODERN WORLD

Finer-grained products of weathering are of great volume, much exceeding coarser residues, and are very widely distributed today. The principal load of the great rivers of the world is mud. Where are the mud traps in which this material comes to rest?

Much weathered residuum, of course, has not yet moved. This is the regolith or mantle rock—that material derived by *in situ* disintegration and decomposition of the underlying bedrock. Despite the widespread character of regolithic deposits and their economic importance, they receive little attention from the field geologist because they form only a minute fraction of the older geologic record. "Fossil soils," although of rare occurrence and small volume, may be geologically significant as indicators of an unconformity and perhaps as a guide to past climate or even atmospheric composition.

Our main concern, therefore, is for transported muds and the deposits they form. In general muds tend to come to rest in relatively quiet waters—waters beyond the reach of waves and subject only to the weakest currents. The most extensive deposits are in oceans, especially in those areas bordering continents—the continental rises. These muds lie in deep water, perhaps a thousand meters or more, and form abyssal plains. They are clearly terrigenous, that is, land-derived. They were identified as such by Murray and Renard of the *Challenger* expedition (1891), who classed them as red muds (100,000 mi^2), blue and gray muds (14,500,000 mi^2), and green muds (1,000,000 mi^2). Red muds are found adjacent to the mouths of large tropical rivers such as the Amazon; blue and gray muds, which are the most extensive, are most typical of the terrigenous muds; green muds are glauconitic. As shown by Griffin (1962) and Biscaye (1965), the clay-mineral character of these deposits reflects the character of the clays carried by streams on adjacent land areas. These terrigenous sediments are believed to have been transported seaward by turbidity currents. Some geophysical work has shown the volume of sediment, largely mud, off the east coast of North America, to be comparable in thickness and extent with that which filled the Appalachian geosyncline during the Paleozoic (Drake, Ewing, and Sutton, 1960).

Muds also collect in shallow marine waters, estuaries, lagoons, and tidal flats—wherever turbulence is minimal. They form a significant part of large deltas, such as the Mississippi Delta.

Muds are also deposited in various continental environments, most notably the floodplains of large rivers, and, to a lesser extent, in fresh and saline lakes. Of special interest, perhaps, are the varved clays of Pleistocene proglacial lakes.

Other continental accumulations include clay-tills or boulder clays of the Late Glacial epoch and the wind-blown silt, loess, derived from river floodplains. Loess forms a blanket of variable thickness mantling the topography on the lee side of large river valleys, especially in the midwestern United States.

IN THE ANCIENT WORLD

Where were the great shale accumulations of the geologic past? In general, shale seems to form a lesser part of some stratigraphic sequences than of others. It forms, for example only 5 percent of a 2,700 foot (821 m) section of Cambro-Ordovician rocks in southeastern Missouri, but 17 percent of the 2,600 foot (790 m) Paleozoic section in the Bighorn Mountains of Wyoming. On the other hand, it forms 61 percent of the much thicker section (9,000 feet; 2736 m) of Carboniferous in the Anadarko Basin of Oklahoma (Bokman, 1954). Schwab (1969) reports 43 percent shale in a 34,000 foot (10,330 m) section at Jackson, Wyoming. Kuenen (1941) estimated 56 percent of a 7,000 m (23,100 foot) Tertiary section in the Dutch East Indies to be shale. It is clear from these examples that the impressive shales of the geologic past are found in thick geosynclinal sequences and not in thinner deposits on the stable craton. Not only is the percentage of shale greater in the geosyncline, but the actual thickness of the shale formations is greater. The Stanley Shale (Carboniferous) of Arkansas, for example, is over 6,000 feet (1,824 m) thick; the Martinsburg (Ordovician) of the central Appalachians may exceed 9,000 feet (2,736 m) (Drake and Epstein, 1967). The Michigamme Slate (Precambrian) is of the order of 6,000 feet (1824 m) in Iron County, Michigan (James et al., 1968). All of these formations belong to the flysch facies—aptly described by Jones (1938) as the "argillaceous facies."

Shales are common also in much of the molasse facies, although this facies is characterized by coarse sandy and conglomeratic materials. The Mauch Chunk Shale (Mississippian) of Pennsylvania, for example, is over 3,000 feet (912 m) thick.

The big shale deposits of the geologic past seem to have accumulated in geosynclinal tracts, during the deposition both of the flysch and the molasse. These are the common shales for the most part, somewhat siltier and commonly red in the molasse facies.

Shales did also form in other situations. Black shales were deposited in somewhat "starved"—euxinic basins—in some cases in the geosyncline itself prior to the deposition of the flysch. Shales are an important part of the accumulation in interior basins, in some cases lacustrine, in others alluvial, as in the Newark-type basins of the eastern United States. Nonetheless it remains true that the greatest mud traps of all were geosynclines.

SILTSTONES AND LOESS

Although silt seems to be very abundant in nature, forming, for example, 60 percent of the material deposited in the Mississippi Delta (Table 8-1), siltstone appears not to be as common a rock as either sandstone or shale. Most silt "disappears" when it is incorporated in the shales. Shales *normally* contain one-third to one-half silt, and some contain more. In some geologic sections, however, siltstones are common as thin flags or as thin seams interbedded with more abundant shales or slates (Fig. 8-14). Siltstones rarely form beds of any considerable thickness and almost never constitute a "formation."

Silt has been defined as material 1/16 to 1/256 mm in diameter (also 0.05 to 0.005 mm and 0.1 to 0.01 mm). Siltstone is consolidated silt. The rocks to which the term *siltstone* is applied by the field geologist are usually composed of two-thirds or more of silt particles, generally greater than 0.01 mm in diameter. These rocks tend to be flaggy; that is, they form hard durable layers, generally thin, which weather in relief in the outcrop. They may show small-scale cross-bedding, and various primary current structures, notably groove and striation casts, rib-and-furrow, and primary current lineation. When water-saturated, silts tend to be "quick"; siltstones, therefore, may show evidence of intrastratal flowage such as convolute bedding. Associated clays, on the other hand, are tenaceous and not prone to flow.

Silt particles are angular rather than round like sand grains. Siltstones are cemented with a mineral cement and in part simply bonded by recrystallization of the component clay materials.

In composition silt and siltstone are intermediate in character between sandstone and shale. They are richer in silica and poorer in alumina, potash, and water than a shale, but not so rich in silica as are more mature sands (Tables 8-4 and 8-12). Siltstones rarely, if ever, consist of pure quartz silt (unless, perhaps, *ganister* is such a rock). Most siltstones contain an abundance of mica, or micaceous clay minerals and chlorite. Feldspar may be present, but rock particles are virtually absent.

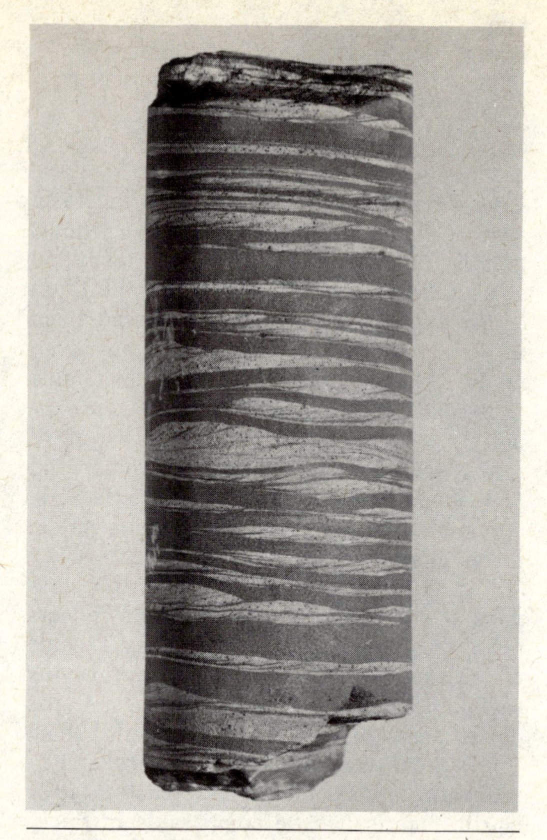

FIG. 8-14. Shale and siltstone (Upper Pennsylvanian), Cumberland County, Kentucky. Diameter of core 2 inches (5 cm). Note bedding typical of zone of intermittent turbulence. Silt layers (light) form plano-convex lenses ("starved" ripples); shale (dark) laminations are more continuous. Note small-scale cross-lamination in silt.

The siltstone flags of the Devonian (especially the so-called "Portage" Group or facies) of New York State are among the best-known examples of these rocks (Fig. 8-15).

LOESS

A silt of very special character is *loess* (German, *löss*). Loess is an unconsolidated porous silt, commonly buff in color (locally gray, yellow, brown, or red), characterized by its lack of stratification and remarkable ability to stand in a vertical slope. It commonly shows a crude columnar structure. It is generally highly calcareous and effervesces in weak acid.

Loess is essentially a silt. Udden's analyses of loess from the Upper Mississippi Valley show the 1/16 to 1/32 mm (0.06 to 0.03 mm) grade to be the modal class and contain about 60 percent of the whole size distribution (Udden, 1898; 1914, p. 741). The material is well sorted; very little sand or clay is present. The Chinese loess is said to have an average grain size of about 0.01 (Barbour, 1927); the Dutch loess consists mainly of materials 0.01 to 0.05 mm in size (Doeglas, 1949); western European and Kansan loess is mostly in the 1/16 to 1/32 mm range (Swineford and Frye, 1955).

The Muscatine, Iowa, loess has quartz as its chief constituent (Diller, 1898, p. 65). Other components are orthoclase, plagioclase, hornblende, occasional biotite, and some carbonate and clay colored by iron oxide. Loess from St. Charles, Missouri (Oefelein, 1934), contains quartz and feldspars (in a quartz–feldspar ratio ranging from 72–28 to 57–43) and a clay mineral (beidellite). Heavy minerals form 0.05 to 0.20 percent and are mainly green and brown hornblende, garnet, tourmaline, zircon, and epidote. Loess from the Lower Mississippi Valley is similar; heavy minerals include hornblende, zircon, garnet, epidote, and opaques (Doeglas, 1949). The essential similarity of the heavy minerals of both the Mississippi Valley and the Dutch loesses to those of the associated glacial deposits was pointed out by Doeglas (1949, 1952). Swineford and Frye (1955), however, have shown that despite close similarity of gross properties, loess from different localities shows a wide range in mineral composition. They attribute differences to variations in source materials.

The chemical composition of several loesses is given in Table 8-12. As can be seen, loess is an oxidized, immature silt. Most loesses are notably calcareous, although as can be seen from the table there are exceptions. The calculated calcium carbonate content of 17 samples of loess from western Europe and Kansas ranged from 0.77 to 40.69 percent (Swineford and Frye, 1955).

FIG. 8-15. Devonian siltstone, Pennsylvania. Crossed nicols, ×33. Mainly angular quartz and some detrital micas in a chloritic and sericitic matrix. Typical flagstone.

Loess occurs primarily as a thin blanket (generally under 100 feet, 30 m thick) of Pleistocene age in central Europe (especially in the Netherlands and Germany), in the Mississippi Valley, in the Pacific Northwest, in portions of China, and in Argentina and New Zealand. In the Upper Mississippi Valley several loessal deposits are closely associated with Pleistocene glacial beds. The most recent is closely related in position and thickness to larger streams. Loess thins rapidly and regularly eastward from these streams (Krumbein, 1937; Smith, 1942; Simonson and Hutton, 1954). No lithified pre-Pleistocene loess has been positively identified in the geologic record.

The origin of loess has been debated for over 50 years. The prevailing view is that it is an eolian silt derived mainly from river floodplains. Much of the world's loess seems to be associated with Pleistocene glaciers. Some authors, however, have attributed loess to a soil-forming process, eluviation (Russell, 1944). Refer to recent literature on the "loess problem." (Thwaites, 1944; Doeglas, 1949, 1952) and especially to Smalley's review of the controversy (1971).

ORIGIN OF SILT

What is the origin of silt-sized quartz that constitutes the bulk of modern dusts (Laprade, 1957) and siltstones and is a substantial fraction of common shales? This question has been briefly reviewed elsewhere (Chapter 3). Rogers, Kreuger, and Krog (1963) presumed silt to originate by chipping off of particles from larger quartz grains. This view was also once advanced by Smalley and Vita-Finzi (1968), who thought the process to be most effective during wind action in deserts. But Kuenen's experiments on eolian action (1969) failed to substantiate this hypothesis. Vita-Finzi and Smalley (1970) and Smalley (1971) later concluded that glacial grinding was responsible for most of the silt in the geologic record. Inasmuch as the deltas of large rivers are largely silt, it seems that neither wind nor glacial action is necessary for silt production. Although some of the river-borne silt may be recycled, most of it must have been produced during soil formation.

References

Allen, V. T., 1929, The Ione formation of California: Univ. Calif. Publ., Dept. Geol. Sci., v. 18, pp. 347–448.

———, 1935, Mineral composition and origin of Missouri flint and diaspore clays: Missouri Geol. Surv. and Water Resources, 58th Biennial Rept., Appendix IV, 24 pp.

TABLE 8-12. Chemical composition of silt, siltstone, and loess

Constituent	A	B	C	D	E	F
SiO_2	64.61	60.69	74.46	59.30	59.20	59.19
TiO_2	0.40	0.52	0.14	0.60	1.20	1.45
Al_2O_3	10.64	7.95	12.26	11.45	16.14	14.61
Fe_2O_3	2.61	2.61	3.25	2.32	4.36	1.51
FeO	0.51	0.67	0.12	1.55	3.24	11.28
MnO	0.05	0.12	0.02	—	0.09	0.10
MgO	3.69	4.56	1.12	2.29	3.14	2.94
CaO	5.41	8.96	1.69	9.78	2.52	0.09
Na_2O	1.35	1.17	1.43	1.80	3.82	0.12
K_2O	2.06	1.08	1.83	2.17	1.97	2.38
H_2O+	2.05	1.14	2.70	0.96	1.16	4.69
H_2O-					1.15	0.07
P_2O_5	0.06	0.13	0.09	0.20	0.17	0.01
CO_2	6.31	9.63	0.49	7.41	—	1.25
SO_3	0.11	0.12	0.06	—	—	0.08[a]
C (organic)	0.13	0.19	0.12	—	1.94	0.25
Cl	0.07	0.08	0.05	—	—	—
Total	100.06	99.62	99.83	99.83	100.10	100.02

Note: For A–D, the single value is bracketed to cover both H_2O+ and H_2O-.

[a] Sulfide sulfur.

A. Loess, near Galena, Illinois. R. B. Riggs, analyst (Clarke, 1924, p. 514).
B. Loess, Vicksburg, Mississippi. R. B. Riggs, analyst (Clarke, 1924, p. 514).
C. Loess, Kansas City, Missouri. R. B. Riggs, analyst (Clarke, 1924, p. 514).
D. Loess, Kansu, China (Barbour, 1927, p. 283).
E. Summer silt, Late Glacial varved sediment, Leppakosi, Finland. L. Lokka, analyst (Eskola, 1932).
F. Siltstone, Dunn Creek Slate (Precambrian), drill core, Homer Mine, Iron River, Michigan. C. Warshaw, analyst (James et al., 1968, Table 4). An unusually iron-rich siltstone.

Allen, V. T., and Nichols, R. L., 1943, Clay-pellet conglomerates of Hobart Butte, Lane County, Oregon: Jour. Sed. Petrology, v. 15, pp. 25–33.

Alling, H. L., 1945, Use of microlithologies as illustrated by some New York sedimentary rocks: Bull. Geol. Soc. Amer., v. 56, pp. 737–756.

van Andel, Tj., and Postma, H., 1954, Recent sediments of the Gulf of Paria: Verhandl. Konink. Neder. Akad. van Wetensch., Afd. Natuurkunde, E. R., v. 20, no. 5, 245 pp.

Androussow, N., 1897, La Mer noire: 7th Int. Geol. Congr., Guide des Excursions, v. 29.

Athy, L. F., 1930, Density, porosity, and compaction of sedimentary rocks: Bull. Amer. Assoc. Petrol. Geol., v. 14, pp. 1–35.

Barbour, G. B., 1927, The loess of China: Smithsonian Inst. Ann. Rept. 1926, pp. 279–296.

Bastin, E. S., 1909, Chemical composition as a criterion in identifying metamorphosed sediments: Jour. Geol., v. 17, pp. 445–472.

Bates, F. T., and Strahl, E. O., 1957, Mineralogy, petrography and radioactivity of representative samples of Chattanooga Shale: Bull. Geol. Soc. Amer., v. 68, pp. 1305–1314.

Berg, R. R., 1952, Feldspathized sandstone: Jour. Sed. Petrology, v. 22, pp. 221–223.

Biscaye, P., 1965, Mineralogy and sedimentation of recent deep-sea clay in the Atlantic Ocean and adjacent seas and oceans: Bull. Geol. Soc. Amer., v. 76, pp. 803–832.

Blatt, H., 1970, Determination of mean sediment thickness in the crust: a sedimentologic method: Bull. Geol. Soc. Amer., v. 81, pp. 255–262.

Blatt, H., and Sutherland, B., 1969, Intrastratal solution and non-opaque heavy minerals in shales: Jour. Sed. Petrology, v. 39, pp. 591–600.

Bokman, J., 1954, Relative abundance of common sediments in Anadarko Basin, Oklahoma: Bull. Amer. Assoc. Petrol. Geol., v. 38, pp. 648–654.

Boswell, P. G. H., 1961, Muddy sediments: Cambridge, Heffer, 140 pp.

Bradley, W. H., 1929, The varves and climate of the Green River epoch: U.S. Geol. Surv. Prof. Paper 158-E, pp. 87–110.

———, 1931, Non-glacial marine varves: Amer. Jour. Sci., ser. 5, v. 22, pp. 318–330.

Bramlette, M. N., 1946, The Monterey Formation of California and the origin of its siliceous rocks: U.S. Geol. Surv. Prof. Paper 212, 57 pp.

Brock, R. W., 1943, Weathering of igneous rocks near Hong Kong: Bull. Geol. Soc. Amer., v. 54, pp. 717–738.

Burst, J. F., 1958, "Glauconite" pellets: their mineral nature and applications to stratigraphic interpretations: Bull. Amer. Assoc. Petrol. Geol., v. 42, pp. 310–327.

———, 1969, Diagenesis of Gulf Coast clayey sediments and its possible relation to petroleum migration: Bull. Amer. Assoc. Petrol. Geol., v. 53, pp. 73–93.

Byers, H. G., Kellogg, C. E., Anderson, M. S., and Thorp, James, 1938, Formation of soil, *in* Soils and men, U.S. Dept. Agric. Yearbook, pp. 948–978.

Campbell, F. A., and Oliver, T. A., 1968, Mineralogic and chemical composition of Ireton and Duvernay formations, central Alberta: Bull. Canadian Petrol. Geol., v. 16, pp. 40–63.

Carozzi, A. V., 1960, Microscopic sedimentary petrography: New York, Wiley, 485 pp.

Carroll, D., 1970, Clay minerals: a guide to their x-ray identification: Geol. Soc. Amer. Spec. Paper 126, 80 pp.

Clark, T. H., 1954, Shale: a study in nomenclature: Trans. Roy. Soc. Canada, v. 48, ser. 3, sec. 4, pp. 1–7.

Clarke, F. W., 1924, Data of geochemistry: Bull. U.S. Geol. Surv., no. 770, 841 pp.

Cody, R. D., 1971, Adsorption and the reliability of trace elements as environment indicators of shales: Jour. Sed. Petrology, v. 41, pp. 461–471.

Conybeare, C. E. B., 1967, Influence of compaction on stratigraphic analysis: Bull. Canad. Petrol. Geol., v. 15, pp. 331–345.

Couch, E. L., 1971, Calculation of paleosalinities from boron and clay mineral data: Bull. Amer. Assoc. Petrol. Geol., v. 55, pp. 1829–1837.

Dapples, E. C., 1942, The effect of macro-organisms upon near-shore marine sediments: Jour. Sed. Petrology, v. 12, pp. 118–126.

Diller, J. S., 1898, The educational series of rock specimens: Bull. U.S. Geol. Surv., no. 150, 400 pp.

Doeglas, D. J., 1949, Loess, an eolian product: Jour. Sed. Petrology, v. 19, pp. 112–117.

———, 1952, Loess, an eolian product: Jour. Sed. Petrology, v. 22, pp. 50–52.

Drake, A. A., Jr., and Epstein, J. B., 1967, The Martinsburg Formation (Middle and Upper Ordovician) in the Delaware Valley, Pennsylvania–New Jersey: Bull. U.S. Geol. Surv., no. 1244-H, pp. H1–H16.

Drake, C. L., Ewing, M., and Sutton, G. H., 1960, Continental margins and geosynclines: the east coast of North America north of Cape Hatteras, *in* Physics and chemistry of the earth, v. 3: New York, Pergamon, pp. 110–198.

Dzulynski, S., and Radomski, A., 1957, Clastic

dikes in Carpathian Flysch: Ann. Soc. Geol. Pologne, v. 26, pp. 225–264.

Eckel, E. C., 1904, On the chemical composition of American shales and roofing slates: Jour. Geol., v. 12, pp. 25–29.

Eskola, Pentti, 1932, Conditions during the earliest geologic times: Ann. Acad. Sci. Fennicae, ser. A., v. 36, pp. 5–74.

Ferguson, L., 1963, Estimation of the compaction factor of a shale from distorted brachiopod shells: Jour. Sed. Petrology, v. 33, pp. 796–798.

———, 1964, A comparison of two techniques for measuring shale compaction: Jour. Sed. Petrology, v. 34, pp. 694–695.

Flawn, P. T., 1953, Petrographic classification of argillaceous sedimentary and low-grade metamorphic rocks in subsurface: Bull. Amer. Assoc. Petrol. Geol., v. 37, pp. 560–565.

Frarey, M. J., and Roscoe, S. M., 1970, The Huronian Supergroup north of Lake Huron, *in* Basins and geosynclines of the Canadian Shield: Geol. Surv. Canada Paper 70-40, pp. 143–158.

von Gaertner, H. R., 1955, Petrographische Untersuchungen am nordwestdeutschen Posidonienschiefer: Geol. Rundschau, v. 43, pp. 447–463.

Glinka, K. D., 1927, The great soil groups of the world, and their development (Marbut, C. F., trans.): Ann Arbor, Mich., Edwards, 235 pp.

Goldich, S. S., 1938, A study in rock-weathering: Jour. Geol., v. 50, pp. 225–275.

Goldstein, A., Jr., and Hendricks, T. A., 1953, Siliceous sediments of Ouachita facies in Oklahoma: Bull. Geol. Soc. Amer., v. 64, pp. 421–442.

Gordon, M., and Tracey, J. I., 1952, Origin of the Arkansas bauxite deposits, *in* Problems of clay and laterite genesis: Amer. Inst. Min. Eng., pp. 12–34.

Grabau, A. W., and O'Connell, M., 1917, Were the graptolitic shales, as a rule, deep- or shallow-water deposits?: Bull. Geol. Soc. Amer., v. 28, pp. 2–5, 959.

Greensmith, J. T., 1958, Preliminary observations on chemical data from some British Upper Carboniferous shales: Jour. Sed. Petrology, v. 28, pp. 209–210.

Gregory, J. W., 1930, The copper-shale (Kupferschiefer) of Mansfeld: Trans. Inst. Min. Metall., v. 40, pp. 1–55.

Griffin, G. M., 1962, Regional clay-mineral facies—products of weathering intensity and current distribution in the northeastern Gulf of Mexico: Bull. Geol. Soc. Amer., v. 73, pp. 737–768.

Grim, R. E., 1942, Modern concepts of clay materials: Jour. Geol., v. 50, pp. 225–275.

———, 1951, The depositional environment of red and green shales: Jour. Sed. Petrology, v. 21, pp. 226–232.

———, 1953, Clay mineralogy, 1st ed.: New York, McGraw-Hill, 384 pp.

———, 1962, Applied clay mineralogy: New York, McGraw-Hill, 422 pp.

———, 1968, Clay mineralogy, 2nd ed., New York: McGraw-Hill, 596 pp.

Grim, R. E., and Allen, V. T., 1938, Petrology of the Pennsylvanian underclays of Illinois: Bull. Geol. Soc. Amer., v. 49, pp. 1485–1513.

Grim, R. E., Bradley, W. F., and White, W. A., 1957, Petrology of the Paleozoic shales of Illinois: Illinois Geol. Surv. Rept. Inv. 203, 35 pp.

Grim, R. E., Dietz, R. S., and Bradley, W. F., 1949, Clay mineral composition of some sediments from the Pacific Ocean off the California coast and the Gulf of California: Bull. Geol. Soc. Amer., v. 60, pp. 1785–1808.

Grim, R. E., and Johns, W. D., 1954, Clay-mineral investigation of sediments in the northern Gulf of Mexico: Proc. 2nd Nat. Conf. Clay and Clay Minerals, Nat. Acad. Sci.–Nat. Res. Council Pub. no. 327, pp. 81–103.

Grout, F. F., 1919, Clays and shales of Minnesota: Bull. U.S. Geol. Surv. 678, 259 pp.

———, 1925, Relation of texture and composition of clays: Bull. Geol. Soc. Amer., v. 36, pp. 393–416.

———, 1932, Petrography and petrology: New York, McGraw-Hill, 522 pp.

Gruner, J. W., and Thiel, G. A., 1937, The occurrence of fine grained authigenic feldspar in shales and silts: Amer. Mineral., v. 22, pp. 842–846.

Harrison, J. B., 1934, The katamorphism of igneous rocks under humid tropical conditions: Harpenden, Eng., Imperial Bur. Soil Sci., 79 pp.

Harrison, S. C., 1971, The sediments and sedimentary processes of the Holocene tidal flat complex, Delmarva Peninsula, Virginia: Ph.D. dissertation, The Johns Hopkins Univ., 202 pp.

Hedberg, H. D., 1926, The effect of gravitational compaction on the structure of sedimentary rock: Bull. Amer. Assoc. Petrol. Geol., v. 10, pp. 1035–1072.

———, 1936, Gravitational compaction of clays and shales: Amer. Jour. Sci., ser. 5, v. 31, pp. 241–281.

Holmes, A., 1937, The age of the earth: London, Nelson, 196 pp.

Hoofs, H. W., 1931, Geology of the eastern part of the Santa Monica Mountains, Los Angeles

County, Calif.: U.S. Geol. Survey Prof. Paper 165, pp. 83–134.
Hough, J. L., 1934, Redeposition of microscopic Devonian plant fossils: Jour. Geol., v. 42, pp. 646–648.
Hunt, C. B., 1972, Geology of soils: their evolution, classification, and uses: San Francisco, Freeman, 344 pp.
Imbrie, J., and Poldervaart, A., 1959, Mineral compositions calculated from chemical analyses of sedimentary rocks: Jour. Sed. Petrology, v. 29, pp. 588–595.
Ingram, R. L., 1953, Fissility of mudrocks: Bull. Geol. Soc. Amer., v. 64, pp. 869–878.
James, H. L., 1951, Iron formation and associated rocks in the Iron River District, Michigan: Bull. Geol. Soc. Amer., v. 62, pp. 251–266.
———, 1954, Sedimentary facies of iron-formation: Econ. Geol., v. 49, pp. 236–293.
James, H. L., Dutton, C. E., Pettijohn, F. J., and Wier, K. L., 1968, Geology and ore deposits of the Iron River–Crystal Falls District, Iron County, Michigan: U.S. Geol. Surv. Prof. Paper 570, 184 pp.
Jones, O. T., 1938, On the evolution of a geosyncline: Proc. Geol. Soc. London, v. 94, pp. lx–cx.
———, 1944, The compaction of muddy sediments: Quart. Jour. Geol. Soc. London, v. 100, pp. 137–160.
Kay, G. F., and Pearce, J. N., 1920, The origin of gumbotil: Jour. Geol., v. 28, pp. 89–125.
Keller, W. D., 1946, Evidence of texture on the origin of the Cheltenham fire clay of Missouri and associated shales: Jour. Sed. Petrology, v. 16, pp. 63–91.
———, 1970, Environmental aspects of clay minerals: Jour. Sed. Petrology, v. 40, pp. 788–813.
Keller, W. D., and Ting, C. P., 1950, The petrology of a specimen of the Perry Farm Shale: Jour. Sed. Petrology, v. 20, pp. 123–132.
Kelley, W. P., 1939, Base exchange in relation to sediments, *in* Recent marine sediments (Trask, P. D., ed.): Tulsa, Okla., Amer. Assoc. Petrol. Geol., pp. 454–465.
———, 1942, Modern clay researches in relation to agriculture: Jour. Geol., v. 50, pp. 307–315.
Kennedy, W. Q., 1951, Sedimentary differentiation as a factor in the Moine-Torridonian correlation: Geol. Mag., v. 88, pp. 257–266.
Krumbein, W. C., 1937, Sediments and exponential curves: Jour. Geol., v. 45, pp. 577–601.
———, 1938, Size frequency distributions of sediments and the normal phi curve: Jour. Sed. Petrology, v. 8, pp. 84–90.
Krumbein, W. C., and Garrels, R. M., 1952, Origin and classification of chemical sediments in terms of pH and oxidation–reduction potentials: Jour. Geol., v. 60, pp. 1–33.
Krumbein, W. C., and Pettijohn, F. J., 1938, Manual of sedimentary petrography: New York, Plenum, 549 pp.
Krumbein, W. C., and Sloss, L. L., 1951, Stratigraphy and sedimentation, 1st ed.: San Francisco, Freeman, 497 pp.
Krynine, P. D., 1948, The megascopic study and field classification of sedimentary rocks: Jour. Geol., v. 56, pp. 130–165.
Kuenen, Ph. H., 1941, Geochemical calculations concerning the total mass of sediments in the earth: Amer. Jour. Sci., v. 239, pp. 161–190.
———, 1969, Origin of quartz silt: Jour. Sed. Petrology, v. 39, pp. 1631–1633.
Lamborn, R. E., Austin, C. R., and Schaaf, D., 1938, Shales and surface clays of Ohio: Ohio Geol. Survey, ser. 4, Bull. 39, 281 pp.
Lane, A. C., 1911, The Keweenaw series of Michigan: Michigan Geol. Surv., Publ. 6 (g.s. 4), 2 vols., 983 pp.
Laprade, K. E., 1957, Dust storm sediments of Lubbock area, Texas: Bull. Amer. Assoc. Petrol. Geol., v. 41, pp. 709–726.
Leighton, M. M., 1930, Weathered zones of the drift-sheets of Illinois: Jour. Geol., v. 38, pp. 28–53.
Leith, C. K., and Mead, W. J., 1915, Metamorphic geology: New York, Holt, Rinehart and Winston, 337 pp.
Lemcke, K., von Engelhardt, W., and Füchtbauer, H., 1953, Geologische und sedimentpetrographische Untersuchungen im Westteil der ungefalten Molasse des suddeutschen Alpenvorlandes: Beitr. Geol. Jahrb., v. 11, 108 pp.
MacCarthy, G. R., 1926, Colors produced by iron in minerals and the sediments: Amer. Jour. Sci., ser. 5, v. 12, pp. 17–36.
McKee, E. D., and Weir, G. W., 1953, Terminology of stratification and cross-stratification: Bull. Geol. Soc. America, v. 64, pp. 381–390.
McNamara, M. J., 1966, The paragenesis of Swedish glacial clays: Geol. Fören. Stockholm Förh., v. 87, pp. 441–454.
Miesch, A. T., 1962, Computing mineral composition of sedimentary rocks from chemical analyses: Jour. Sed. Petrology, v. 32, pp. 217–225.
Miller, W. G., 1905, The cobalt-nickel arsenides and silver deposits of Temiskaming: Ontario Bur. Mines Ann. Rept., v. 14, pt. 2, 66 pp.
Millot, G., 1949, Relations entre la constitution et la genèse des roches sédimentaires argileuses: Géol. Appliq. Prosp. Min., v. 2, pp. 1–352.

———, 1964, Géologie de argiles: Paris, Masson, 499 pp.

———, 1970, Geology of clays (trans.): New York, Springer, 429 pp.

Milne, I. H., and Earley, J. W., 1958, Effect of source and environment on clay minerals: Bull. Amer. Assoc. Petrol. Geol., v. 42, pp. 328–338.

Moore, D. G., 1939, Faecal pellets in relation to marine deposits, *in* Recent marine sediments (Trask, P. D., ed.): Tulsa, Okla., Amer. Assoc. Petrol. Geol., pp. 516–524.

Moore, D. G., and Scruton, P. C., 1957, Minor internal structures of some Recent unconsolidated sediments: Bull. Amer. Assoc. Petrol. Geol., v. 41, pp. 2723–2751.

Murray, J., and Renard, A. F., 1891, Report on deep-sea deposits based on the specimens collected during the voyage of H.M.S. *Challenger* in the years 1872 to 1876: *Challenger* Repts., pp. 378–391.

Nanz, R. H., 1953, Chemical composition of pre-Cambrian slates with notes on the geochemical evolution of lutites: Jour. Geol., v. 61, pp. 51–64.

Nicholls, G. D., 1962, A scheme for re-calculating the chemical analyses of argillaceous rocks for comparative purposes: Amer. Mineral., v. 47, pp. 34–46.

Oefelein, R. T., 1934, A mineralogical study of loess near St. Charles, Missouri: Jour. Sed. Petrology, v. 4, pp. 36–44.

Oertal, G., and Curtis, C. D., 1972, Clay-ironstone concretion preserving fabrics due to progressive compaction: Bull. Geol. Soc. Amer., v. 83, pp. 2597–2606.

Payton, C. E., and Thomas, L. A., 1959, The petrology of some Pennsylvanian black "shales": Jour. Sed. Petrology, v. 29, pp. 172–177.

Peterson, M. N. A., 1962, The mineralogy and petrology of Upper Mississippian carbonate rocks of the Cumberland Plateau in Tennessee: Jour. Geol., v. 70, pp. 1–31.

Pettijohn, F. J., and Bastron, H., 1959, Chemical composition of argillites of the Cobalt Series (Precambrian) and the problems of soda-rich sediments: Bull. Geol. Soc. Amer., v. 70, pp. 593–599.

Pettijohn, F. J., Potter, P. E., and Siever, Raymond, 1965, Geology of sand and sandstone: Bloomington, Ind., Indiana Univ., 205 pp.

Picard, M. D., 1953, Marlstone—a misnomer as used in Uinta Basin, Utah: Bull. Amer. Assoc. Petrol. Geol., v. 37, pp. 1075–1077.

Potter, P. E., and Glass, H. D., 1958, Petrology and sedimentation of the Pennsylvanian sediments in southern Illinois: a vertical profile: Illinois Geol. Surv. Rept. Inv. 204, 60 pp.

Potter, P. E., Shimp, N. F., and Witters, J., 1963, Trace elements in marine and fresh-water argillaceous sediments; Geochim. Cosmochim. Acta, v. 27, pp. 669–694.

Pryor, W. A., and Glass, H. D., 1961, Cretaceous-Tertiary clay mineralogy of the Upper Mississippi Embayment: Jour. Sed. Petrology, v. 31, pp. 38–51.

Raiswell, R., 1971, The growth of Cambrian and Liassic concretions: Sedimentology, v. 17, pp. 147–171.

Reiche, P., 1950, A survey of weathering processes and products: Univ. New Mexico Publ. Geol. 3, 95 pp.

Rich, J. L., 1951, The probable fondo origin of Marcellus–Ohio–New Albany–Chattanooga bituminous shales: Bull. Amer. Assoc. Petrol. Geol., v. 35, pp. 2017–2040.

Ries, H., 1927, Clays, origin, properties, and uses, 3rd ed.: New York, Wiley, 613 pp.

Robinson, G. W., 1951, Soils, their origin, constitution and classification, 3rd ed.: London, Murby, 573 pp.

Rogers, J. J. W., Kreuger, W. C., and Krog, M., 1963, Sizes of naturally abraded materials: Jour. Sed. Petrology, v. 33, pp. 628–632.

Rubey, W. W., 1929, Origin of the siliceous Mowry Shale of the Black Hills region: U.S. Geol. Surv. Prof. Paper 154-D, pp. 153–170.

———, 1931, Lithologic studies of fine-grained Upper Cretaceous sedimentary rocks of the Black Hills region: U.S. Geol. Surv. Prof. Paper 165-A, 54 pp.

Ruedemann, R., 1934, Paleozoic plankton of North America: Geol. Soc. Amer. Mem. 2, 140 pp.

Russell, I. C., 1889, Subaerial decay of rocks and origin of the red color of certain formations: U.S. Geol. Surv. Bull. 52, 63 pp.

Russell, R. J., 1944, Lower Mississippi Valley loess: Bull. Geol. Soc. Amer., v. 55, pp. 1–40.

Russell, R. J., and Russell, R. D., 1939, Mississippi River delta sedimentation, *in* Recent marine sediments (Trask, P. D., ed.): Tulsa, Okla., Amer. Assoc. Petrol. Geol., pp. 153–177.

Scheid, V. E., 1945, Preliminary report on Excelsior high-alumina clay deposit, Spokane County, Washington: Unpublished report, U.S. Geol. Surv., 66 pp.

Schmitt, H. A., 1924, Possible potash production from Minnesota shale: Econ. Geol., v. 19, pp. 72–83.

Schuchert, C., 1915, The conditions of black shale deposition as illustrated by the Kupferschiefer and Lias of Germany: Trans. Amer. Phil. Soc., v. 54, pp. 259–269.

———, 1931, Geochronology or the age of the earth on the basis of sediments and life, *in*

The age of the earth: Bull. Nat. Res. Coun. 80, pp. 10–64.

Schwab, F. L., 1969, Geosynclines: what contribution to the crust?: Jour. Sed. Petrology, v. 39, pp. 150–158.

Sederholm, J. J., 1931, On the sub-Bothnian unconformity and on Archaean rocks formed by secular weathering: Bull. Comm. Geol. Finlande, no. 95, 81 pp.

Sharp, R. P., 1940, Eo-Archean and eo-Algonkian erosion surfaces: Bull. Geol. Soc. Amer., v. 51, pp. 1235–1270.

Shaw, D. B., and Weaver, C. E., 1965, The mineralogical composition of shales: Jour. Sed. Petrology, v. 35, pp. 213–222.

Shaw, D. M., 1956, Geochemistry of pelitic rocks. III: Major elements and general geochemistry: Bull. Geol. Soc. Amer., v. 67, pp. 919–934.

Shearer, H. K., 1918, The slate deposits of Georgia: Bull. Georgia Geol. Surv., no. 34, 192 pp.

Shelton, J. W., 1962, Shale compaction in a section of Cretaceous Dakota Sandstone, northwestern North Dakota: Jour. Sed. Petrology, v. 32, pp. 873–877.

Shrock, R. R., 1948, A classification of sedimentary rocks: Jour. Geol., v. 56, pp. 118–129.

Shukri, M. N., 1942, The use of pH-values in determining the environment of deposition of some Liassic clays and shales: Bull. Fac. Sci. Fouad I Univ., v. 24, pp. 61–65.

Simonson, R. W., and Hutton, C. E., 1954, Distribution curves for loess: Amer. Jour. Sci., v. 252, pp. 99–105.

Sloss, L. L., 1968, Sedimentary volumes on the North American craton: Geol. Soc. Amer., Program with abstracts, 1968 Ann. Mtg., Mexico City, p. 281.

Smalley, I. J., 1966, The properties of glacial loess and the formation of loess deposits: Jour. Sed. Petrology, v. 36, pp. 669–676.

———, 1971, "In-situ" theories of loess formation and the significance of the calcium-carbonate content of loess: Earth Sci. Rev., v. 7, pp. 67–85.

Smalley, I. J., and Vita-Finzi, C., 1968, The formation of fine particles in sandy deserts and the nature of "desert" loess: Jour. Sed. Petrology, v. 38, pp. 766–774.

Smith, G. D., 1942, Illinois loess-variations in its properties and distribution: Bull. Illinois Agric. Exp. Sta. 490, pp. 139–184.

Ström, K. M., 1936, Land-locked waters; hydrography and bottom deposits in badly-ventilated Norwegian fjords with remarks upon sedimentation under anaerobic conditions: Skrifte Norske Videnskaps. Akad. Oslo, Mat. Natur. Kl., v. 1, no. 7, pp. 1–85.

Sundelius, H. W., 1970, The Carolina Slate Belt, *in* Studies of Appalachian geology (Fisher, G. W., Pettijohn, F. J., Reed, J. C., Jr., and Weaver, K. N., eds.): New York, Wiley-Interscience, pp. 351–367.

Swineford, A., and Frye, J. C., 1955, Petrographic comparison of some loess samples from western Europe with Kansas loess: Jour. Sed. Petrology, v. 25, pp. 3–23.

Tank, R., 1969, Clay mineral composition of the Tipton Shale member of the Green River Formation (Eocene) of Wyoming: Jour. Sed. Petrology, v. 39, pp. 1593–1595.

Taylor, J. H., 1949, Petrology of the Northampton sand ironstone formation: Mem. Geol. Surv. Great Britain, 111 pp.

Thwaites, F. T., 1944, Review of R. J. Russell's article on loess: Jour. Sed. Petrology, v. 14, pp. 246–248.

Tomadin, L., 1964, Orientament attuali sulla sistematica delle rocce argillose: Ann. Mus. Geol. Bologna, ser. 2, v. 32, pp. 531–543.

Tomlinson, C. W., 1916, The origin of red beds: Jour. Geol., v. 24, pp. 153–179.

Trask, P. D., 1931, Compaction of sediments: Bull. Amer. Assoc. Petrol. Geol., v. 15, pp. 271–276.

———, 1937, Studies of source beds in Oklahoma and Kansas: Bull. Amer. Assoc. Petrol. Geol., v. 21, pp. 1377–1402.

Twenhofel, W. H., 1937, Terminology of the fine-grained mechanical sediments: Exhibit F —report of Committee on Sedimentation 1936–1937: Nat. Res. Coun., Div. Geol. Geog., pp. 81–104.

———, 1939, Environments of origin of black shales: Bull. Amer. Assoc. Petrol. Geol., v. 23, pp. 1178–1198.

Udden, J. A., 1898, Mechanical composition of wind deposits: Augustana Library Publ. 1.

———, 1914, The mechanical composition of clastic sediments: Bull. Geol. Soc. Amer., v. 25, pp. 655–744.

Ulrich, E. O., 1911, Revision of the Paleozoic system: Bull. Geol. Soc. Amer., v. 22, p. 358.

Van Houten, F. B., 1961, Climatic significance of red beds, *in* Descriptive paleoclimatology: New York, Wiley-Interscience, pp. 89–139.

Vita-Finzi, C., and Smalley, I. J., 1970, Origin of quartz silt: comments on a note by Ph. H. Kuenen: Jour. Sed. Petrology, v. 40, pp. 1367–1368.

Vogt, T., 1927, Geology and petrology of the Sulitelma district: Norges Geol. Undersökelse, no. 121, pp. 449–560.

Weaver, C. E., 1958, Geological interpretation

of argillaceous sediments: Bull. Amer. Assoc. Petrol. Geol., v. 42, pp. 254–271.

———, 1967, Potassium, illite, and the ocean: Geochim. Cosmochim. Acta, v. 31, pp. 2181–2196.

Weaver, C. E., and Beck, K. C., 1971, Clay water diagenesis during burial: how mud becomes gneiss: Geol. Soc. Amer. Spec. Paper 134, 96 pp.

Weiss, M. P., 1954, Feldspathized shales from Minnesota: Jour. Sed. Petrology, v. 24, pp. 270–274.

White, W. S., and Wright, J. C., 1954, The White Pine copper deposit, Ontonagan County, Michigan: Econ. Geol., v. 49, pp. 675–716.

Williams, G. E., 1968, Torridonian weathering, and its bearing on Torridonian paleoclimate and source: Scottish Jour. Geol., v. 4, pp. 164–184.

Woolnough, W. G., 1928, Origin of white clays and bauxite, and chemical criteria of peneplanation: Econ. Geol., v. 23, pp. 887–894.

Yaalon, D. H., 1962a, Mineral composition of the average shale: Clay Min. Bull., v. 5, pp. 31–36.

———, 1962b, Weathering and soil development through geologic time: Bull. Res. Coun. Israel, sect. G., v. 11G, Proc. Israel Geol. Soc., 4th Congr. Israel Assoc. Adv. Sci., 1961.

———, ed., 1971, Paleopedology—origin, nature and dating of paleosols: Internat. Soc. Soil Sci. and Israel Univ. Press, 350 pp.

Zen, E-an, 1959, Clay mineral–carbonate relations in sedimentary rocks: Amer. Jour. Sci., v. 257, pp. 29–43.

9

VOLCANICLASTIC SEDIMENTS

INTRODUCTION AND DEFINITIONS

The term *pyroclastic* is the adjective applied to rocks produced by explosive or aerial ejection of material from a volcanic vent. Such materials may accumulate on land or under the sea.

The term *volcaniclastic* has been used to encompass the entire spectrum of fragmental volcanic rocks formed by any mechanism or origin, emplaced in any physiographic environment (on land, under water, or under ice), or mixed with any nonvolcanic fragment types in any proportion (Fisher, 1961, 1966a). It is used here for those rocks with a preponderance of fragments of volcanic origin. If the materials are dominantly nonvolcanic, the usual sedimentary nomenclature is modified by an appropriate adjective. The hybrid rock may be called a tuffaceous sand or sandstone, a tuffaceous clay or shale, and the like. If epiclastic material is subordinate, the mixtures may be designated sandy tuff or clayey tuff, and so forth.

Volcaniclastic materials include pyroclastic debris but also include deposits derived from volcanic source rocks by ordinary processes of weathering. The stream sands in Chihuahua, Mexico, for example, that are derived by disintegration of middle Tertiary rhyolitic lavas are volcaniclastic but are not pyroclastic sands even though they are composed of one-half to two-thirds rhyolitic debris (Webb and Potter, 1969). They would be, if lithified, a species of lithic arenite.

Most confusion is apt to arise when pyroclastic materials, newly deposited on land, are reworked and redeposited by rivers or by surf. Such deposits are truly volcaniclastic but are they pyroclastic? Such redeposited tephra are generally considered pyroclastic rather than epiclastic, because fragments are of pyroclastic origin. Thus there may be "primary" (unreworked) as well as "secondary" (reworked) pyroclastic deposits (Fisher, 1966a). However, objection has been raised to the designation *tuff* for these reworked deposits (Hay, 1952). We prefer to call them *redeposited tuffs*, if it can be in fact determined that they are such; where this distinction cannot be made they should be called tuff, for, as Fisher notes, the components are indeed pyroclastic and they record contemporary volcanism.

Volcaniclastic sediments form very large accumulations. According to Sapper (1928), total volcanic output over the world for the last four centuries is estimated to have been 83 miles^3 (320 km^3) of pyroclastic materials and 15.5 miles^3 (50 km^3) of lava. The volume of ignimbrites in the Great Basin is probably more than 50,000 miles^3 or 238,500 km^3 (Mackin, 1960). In some places thousands of meters of stratified tuffs are interbedded with normal sediments or mixed with these sediments in all proportions. Subaqueous pyroclastic flows form almost one-half of the 10,000 foot (3,050 m) Ohanapecosh Formation (Eocene) of the Mount Rainier region (Fiske, 1963). Single flows range from 10 to 200 feet (3 to 70 m). The extent and importance of volcaniclastic materials has been emphasized by Ross (1955).

Recognition of volcanic materials may be one of the most difficult tasks facing the sedimentary petrographer. Volcanic debris is especially susceptible to diagenetic alteration and rapidly loses its distinctive character. If the rock is further altered by metamorphism, its original character may be greatly obscured. Refer to the excellent papers by Pirsson (1915) and Ross (1928) for criteria for the identification of volcanic tuffs, both fresh and altered.

TEXTURES AND STRUCTURES OF VOLCANICLASTIC SEDIMENTS

If the sediments are in fact epiclastic but derived from older volcanic source rocks, they are described in the same textural terms as are any other ordinary clastic sediments. If, however, the materials are truly pyroclastic—ejectementa from a volcanic vent—they are collectively designated *tephra* (Thorarinsson, 1954), and a set of special terms is applied (Wentworth and Williams, 1932) to the several sizes of material (Table 9-1). The term *block* is used for large fragments (over 32 mm) broken from a piece of previously consolidated lava; the term *bomb* is an equally large object formed by solidification of lava while in flight. The former is sharp-edged; the latter is of rounded contour, subspherical, oblate or irregular in form and, in some cases, with a checked, fissured surface ("bread-crust" bombs). An aggregate of blocks is a *volcanic breccia* (Fig. 9-1); an aggregate of bombs is an *agglomerate*. If the fragments are of erupting lava, they are termed *essential;* if they are debris of earlier lavas and pyroclasts of the same cone they are *accessory;* if the fragments are of other rocks they are called *accidental*. *Lapilli* is a term given to the materials 4 to 32 mm in diameter. A deposit of such material is a *lapilli tuff*. *Coarse ash* is ¼ to 4 mm; *fine ash* is under ¼ mm. Corresponding deposits are *coarse tuff* and *fine tuff*. These size limits and terminology have been modified somewhat by other investigators (Blyth, 1940; Fisher, 1958, 1961, 1966a). But on the whole the several usages are not greatly different from that of Wentworth and Williams.

An important textural aspect is the sorting of the deposit. Volcaniclastic materials exhibit all possible degrees of sorting. Some are very well sorted and finely laminated; others are chaotic and unsorted and contain debris ranging from the finest ash to great blocks of either cognate or noncognate (ac-

TABLE 9-1. Grade size limits and size terms of pyroclastic debris

Size (mm)	Wentworth and Williams, 1932	Twenhofel, 1950, p. 319	Fisher, 1961	
256	Blocks[a] = (volcanic breccia)	Bombs	Coarse	Blocks and bombs
128 64	Bombs[b] = (agglomerate)	Lapilli	Fine	
32 16 8	Lapilli = (Lapilli tuff)		Lapilli	
4 2 0.5	Coarse-ash = (coarse tuff)	Coarse ash		
0.250			Coarse ash	
0.125 0.0625	Fine ash (fine tuff)	Fine ash		
			Fine ash	

[a] Fragments of preexisting volcanic rock.
[b] Lava solidified in flight.

cidental) rocks. The size distribution of these several types of pyroclastic materials shows striking contrasts (Moore, 1934; Walker, 1971). See Fig. 9-2. Materials transported by air and water currents have grading characteristics similar to those of ordinary epiclastic sediments (Table 9-2). Those transported as "glowing clouds" (*nuées ardentes*) or as ash flows show a size distribution similar to that of crushed materials—essentially a Rosin's Law distribution (Krumbein and Tisdel, 1940)—not greatly different from that displayed by colluvium, mud flows, and other mass-transported materials.

As Correns and Leinz (1933) have pointed out, many primary pyroclastic deposits are *porphyritic* (Table 9-3). Relatively large

FIG. 9-1. Volcanic breccia, Upper Minoan pumice flow deposit, Santorin. (Photograph by H.-U. Schmincke.)

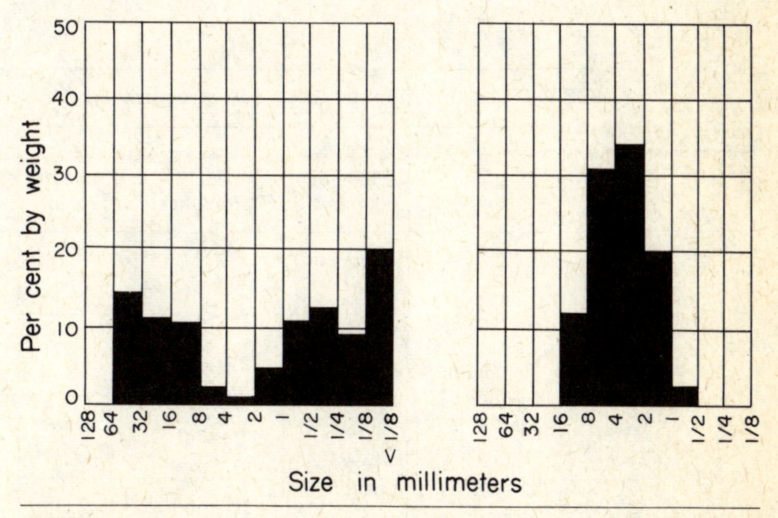

FIG. 9-2. Mechanical composition of tuffs, Crater Lake area, Oregon. Right, normal air fall deposit; left, *nuée ardente* deposit. (Data from Moore, 1934.)

crystals and rock fragments lie in a matrix of much finer debris (Fig. 9-3). Some of the larger rock fragments, bombs or blocks, are hurled into areas where finer ash is accumulating. These large pieces resemble the dropstones of ice-rafted origin seen in glaciolacustrine and glaciomarine deposits. Like glacial dropstones, they distort the laminations of the host deposit. This disparity in grain size also reflects a hiatus in composition; the larger fragments are chiefly rock particles or crystal debris, whereas the finer materials are mainly glass (Table 9-4). If the glass is present as pumice, however, it may be more abundant in the coarser grades and, as noted below, may constitute the coarser part of a reverse graded bed.

In terms of poor sorting, angularity of constituents, abundance of matrix, and normal graded structure of many tuff beds, these deposits resemble graywackes and in many cases are distinguished only with difficulty from these immature sediments.

Those materials reworked by waves and currents or by rivers tend to show the best stratification and sorting (Table 9-3). Some airborne materials are likewise well sorted and may show graded bedding, as do some water-laid tuffs. Some subaqueous ash-flow deposits show *reversed* grading caused by

TABLE 9-2. Comparison of pyroclasts and hydroclasts and mode of transport

Composition	Water and sediment (hydrosol)		Gases and volcanic ejectamenta (aerosol)	
	Concentrated	Dilute	Concentrated	Dilute
Type	Mudflows and turbidity currents	Normal water currents	*Nueés ardentes*	Normal air currents
Product	Tilloids and Graywackes	Orthoarenites and conglomerates	"Ignimbrites" or welded tuffs	Bedded tuffs and ash

Source: After Pettijohn (1950).

TABLE 9-3. Classification, textures, and structures of the pyroclastic arenites

Type	I. Transport	I. Place of deposit	II. Transport	II. Place of deposit	Texture	Structure
A 1	Volcanic explosions	Terrestrial	—	—	Porphyritic	Without bedding
A 2		Lacustrine	—	—		Bedded with parallel structure
B 1		Terrestrial	Fluvial	Lacustrine	Uniform grained	
B 2			By the surf	Littoral		Without bedding

Source: After Correns and Leinz (1933).

separation of large porous pumice fragments from the smaller more dense rocks and mineral fragments (Fiske, 1963, p. 400). Coarse pyroclastics are the most poorly sorted or lack sorting altogether. They are also the most poorly bedded or lack bedding altogether—particularly the products of *nuée ardente* type of eruptions. They weather readily and accordingly present a rusty weathered appearance in outcrop.

As noted, certain textural and compositional attributes distinguish pyroclastic sediments, both primary and secondary, from ordinary epiclastic deposits. Are there distinctive structures which mark a pyroclastic origin? As might be expected, secondary or reworked pyroclastic materials will exhibit both the textures and structures that characterize the agent or environment of reworking. True pyroclastic sediments show a spectrum of structures related to their particular mode of transport and deposition. Those deposited as fallout from the atmosphere mantle the surface on which they are deposited, never show cross-bedding, and show normal grading. Those deposited from a high-velocity incandescent pyroclastic flow tend to be massive, or only crudely bedded, and have a variable thickness related to the subjacent topography. They may become *welded* and, in some cases, form a *reconstituted lava,* which moves slowly as a viscous body and rotates included clasts, stretches and pulls apart enclosed pumice fragments, and develops tension cracks and other structures related to this movement (Schmincke and Swanson, 1967). Those deposited from lower temperature, steam-saturated, base-surge eruptions more resemble the air-fall deposits in having good stratification and grading but differ in exhibiting antidune and cross-bedding structures (Crowe and Fisher, 1973). Subaqueous eruptions tend to show structures akin to those of turbidites (Fiske, 1963). Volcanic mudflows, either subaqueous or subaerial, show little internal structure; were it not for their composition, they would be difficult to distinguish from ordinary mudflows or other mass-transported materials. Isolated oversized blocks found in some stratified tuffs with their related down-

FIG. 9-3. Stratified pumice showing embedded volcanic block. Note distortion of stratification ("bomb sag"). (Photograph by H.-U. Schmincke.)

TABLE 9-4. Composition of tuff with "porphyritic" texture

Mechanical composition		Mineralogical composition (percent)			
Size classes (mm)	Percent	Mineral component	Coarse fraction	Fine fraction	Total
>0.5	21.8	Quartz	40	31	34
0.5 to 0.2	12.8	Sanidine	35	15	21
0.2 to 0.06	51.2	Plagioclase	—	3	3
<0.06	14.2	Biotite and hornblende	3	3	3
		Glass	22	48	40

Source: After Correns and Leinz (1933).

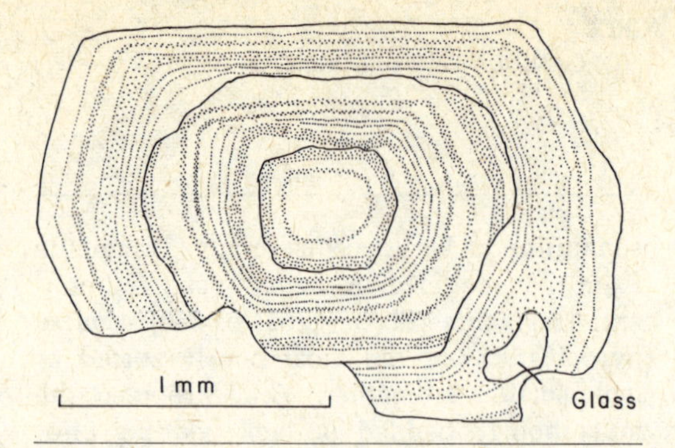

FIG. 9-4. Volcanic feldspar shows zoning. Quaternary ash, New Zealand. Note embayment. (After Ewart, 1963, *Jour. Petrology*, v. 4, Fig. 12. By permission of Oxford University Press.)

bending of laminations in the tuff are rather distinctive features (Fig. 9-3).

COMPOSITION

Volcanic debris is of three sorts: rock fragments, crystals and crystal fragments of intratelluric origin, and glass. Glass is the most important of these and the most distinctive earmark of pyroclastic origin.

Rock fragments include both cognate and noncognate (accidental) blocks. Cognate blocks are part of the solidified lava which have been broken up by later eruptions; noncognate blocks are fragments of the wall rock through which the volcanic vents are blasted. Blocks of volcanic origin are most prominent in the volcanic breccias which accumulate nearest the centers of eruption. Rock particles are chiefly acid aphanitic igneous rocks (often mistaken for chert under the microscope), although basic tuffs contain andesitic and basaltic rock particles. Both rock particles and the matrix of the tuff may contain an abundance of microlites.

Crystal debris is more widespread. The crystals are those which commonly form phenocrysts in lava, such as volcanic quartz with its characteristic resorbed outlines and, less commonly, dipyramidal form, and feldspars, commonly showing oscillatory zoning (Fig. 9-4). Especially significant is sanidine, the high-temperature feldspar. Less common but very significant are crystals or broken crystals of amphiboles, pyroxenes, and olivine—all exceedingly rare or unknown in ordinary sediments. Crystals may be intact but more generally they are broken euhedra. In many cases they are surrounded by a thin envelope or pellicle of glass. Biotite plates are especially likely to be widely distributed.

Generally, however, the crystals are associated with a larger quantity of volcanic *glass*, which may be very fine-grained, and, if reworked or redeposited, it may be mingled with normal sedimentary materials in all proportions. Fresh glass is characterized by its colorless to pale yellow appearance in thin section, by its low index (mostly 1.50 to 1.52), by its isotropic character, and principally by its shape. The glass fragments of acidic lavas exhibit curious, curved, spiculelike forms, termed *shards* (Fig. 9-9); those of more basic lavas tend to have a droplet form (Heiken, 1972). The approximate silica content of the glass can be determined by use of its index of refraction (George, 1924; Huber and Reinhart, 1966); hence in younger sequences the several ash falls may be distinguished from one another by the index of the glass of which they are composed (not an infallible guide, however; see Swineford, Frye, and Leonard, 1955). See Fig. 9-5. In time the glass undergoes alteration and becomes devitrified.

Some silicic, air-fall tuffs contain *accretionary lapilli*. These pelletlike objects are small ovoid to subspherical bodies (2 to 10 mm) with concentric structure (Fig. 9-6). They are presumed to have formed by raindrops falling into a cloud of ash or to have formed on the ground by the rolling of lapilli nuclei over fresh ash surfaces (Wentworth and Williams, 1932; Moore and Peck, 1962). These accretionary lapilli should not be confused with *peperites*, which are globular (but not water-worn) bodies, 0.5 to 10.0 cm in size, consisting of obsidian glass containing bubble cavities and in some cases microlites of pyroxene embedded in an interstitial matrix of finely granular calcite darkened by clay minerals.

As might be expected, tuffs have a chemical composition similar to that of the igneous rocks of the same family (Table 9-5). The composition of intermediate to basic tuffs is quite unlike that of normal sediments. The composition of the more acid pyroclastics—which are by far the most abundant—is, however, not unlike that of some immature sediments. And in those cases where diagenesis or metamorphism have obscured or erased distinctive textures and have altered mineralogy, it may be difficult to recognize their volcanic origin. In such cases one may have to rely on some chemical "anomalies" disclosed by bulk chemical analysis. A high Na_2O–K_2O ratio (greater than 1.0) in a slate, for example, suggests volcanic contamination or origin, for in a normal slate K_2O greatly exceeds

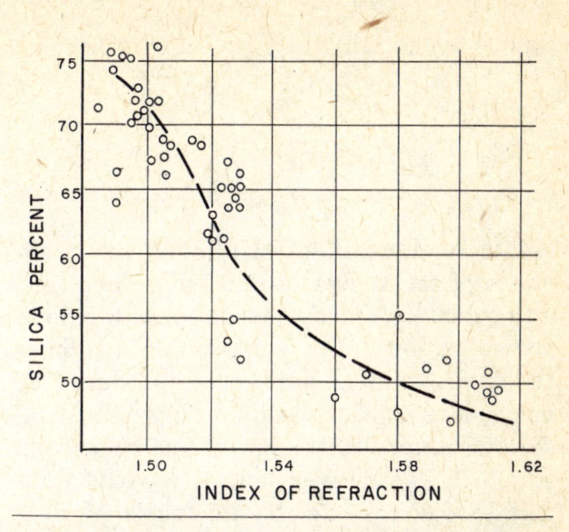

FIG. 9-5. Relation of index of refraction of volcanic glass and silica content. (After George, 1924.)

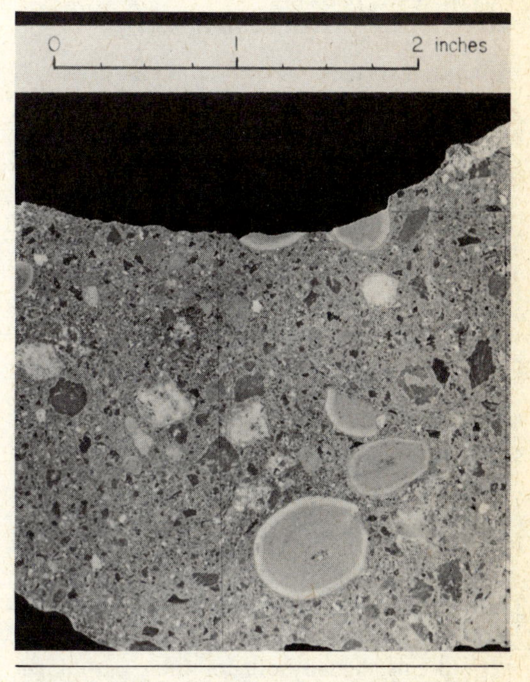

FIG. 9-6. Accretionary lapilli tuff, Ohanapecosh Formation (Eocene), Mt. Rainier National Park, Washington. (Photograph by R. S. Fiske.)

TABLE 9-5. Representative chemical analyses of volcaniclastic sediments

	Pyroclastic deposits			Volcanic arenites	
Constituent	A	B	C	D	E
SiO_2	70.40	53.63	48.67	73.50	61.69
TiO_2	0.21	0.96	1.99	0.34	1.03
Al_2O_3	13.65	19.59	14.15	13.3	13.89
Fe_2O_3	1.18	5.70	9.07	1.55	3.82
FeO	1.81	n.d.	0.83	0.56	2.20
MnO	0.04	—	—	0.04	0.11
MgO	0.07	3.35	6.36	0.56	2.20
CaO	1.58	3.53	6.16	1.13	3.10
Na_2O	3.76	3.64	1.61	2.34	2.20
K_2O	3.90	1.62	0.96	4.01	1.88
H_2O+	4.03	7.91	9.39	1.80	1.81
H_2O-		—	—		—
P_2O_5	0.06	—	0.36	0.02	0.25
CO_2	—	—	—	0.12	—
Total	100.69	99.93	99.5	99.3	94.0[a]

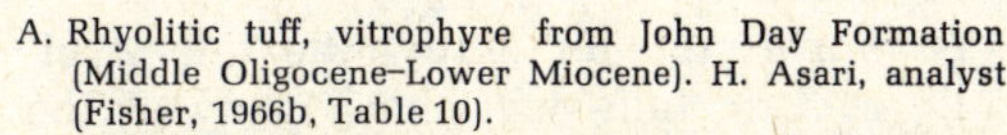

A. Rhyolitic tuff, vitrophyre from John Day Formation (Middle Oligocene–Lower Miocene). H. Asari, analyst (Fisher, 1966b, Table 10).
B. Andesitic tuff, Saleijer Island, Celebes. A Wichmann, analyst (Washington, 1917, pp. 846–847).
C. Basaltic tuff, Szentgyorghegy, Zala, Hungary. K. Eniszt, analyst (Washington, 1917, pp. 894–895).
D. Stream sand (rhyolite arenite), western Chihuahua, Mexico. R. K. Leinger and M. E. Coller, analysts (Webb and Potter, 1969, Tables 5, 6).
E. Stream sand (mixed arenite), Columbia River, Oregon. A. Stelmach, analyst (Whetten, 1966, Table 2, CC29).

[a] Loss on ignition 5.99.

Na_2O. The use of chemical criteria for identifying sediments has been reviewed by Bastin (1909). Significant features of ordinary sediments include the dominance of magnesia over lime and of potash over soda, an excess of alumina (excess over that necessary to satisfy the 1-to-1 ratio in which it is combined with lime and alkalies in common rock-forming silicates), and high silica. To these should be added the dominance of ferric over ferrous iron. There are, unfortunately, many exceptions to these generalizations. The composition of those sediments, such as graywacke (produced by mechanical disintegration only and unsorted), is much like that of the rock from which it was derived, and it may be difficult to discriminate, on the basis of bulk chemical composition, from some volcanic sediments.

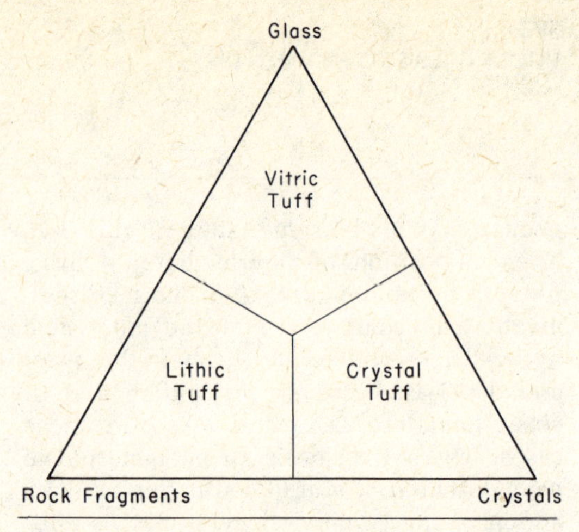

FIG. 9-7. Nomenclature and classification of tuffs.

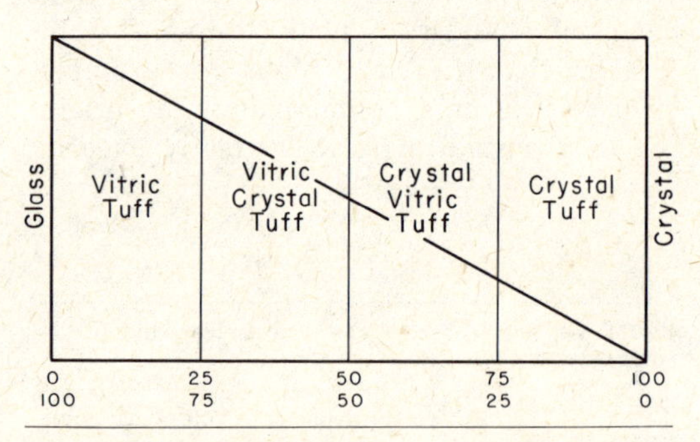

FIG. 9-8. Nomenclature and classification of vitric and crystal tuffs.

PETROGRAPHY OF VOLCANICLASTIC SEDIMENTS

CLASSIFICATION

Volcaniclastic sediments of coarsest grain, volcanic breccias and agglomerates, are largely rock fragments (blocks) or bombs (scoria). Those of medium grain, the tuffs, consist of glass, mineral grains (crystal or crystal fragments), and rock particles. They may therefore be classed according to the proportions of these several components (Fig. 9-7). Further subdivision of the various possible mixtures may follow the scheme in Fig. 9-8. Tuffs may be further designated as essential, accessory, or accidental, according to the source of materials of which they are composed.

Tuffs and other pyroclastic sediments also may be classified by their magmatic affinities, as, for example, rhyolite ash, andesite tuff, or basaltic agglomerate. They should, perhaps, be named both for their parentage and their components: andesitic crystal tuff, rhyolitic vitric tuff, and so forth. We have already indicated their textural classification. As noted, the terms *tuff, breccia, lapilli tuff,* and the like imply specific size grades (Table 9-1).

PETROGRAPHY

Vitric tuffs Vitric tuffs are characterized by abundance or dominance of glassy materials, shattered pumice, forming shards: lunate or sickle-shaped pieces of glass, also cusplike shapes, also some pumice fragments with many vesicular cavities (Fig. 9-9). Some pumice fragments, however, have been drawn out into tubelike forms. In welded tuffs these various forms are also present but show the effects of pressure and are "collapsed" and distorted or molded about any crystals or crystal fragments that the tuff may contain. These textural features of glassy particles or "bogen structure" have been termed *vitriclastic* (Pirsson, 1915, p. 198).

A thin section of a rhyolitic tuff—typically a vitric tuff—will show many of the glass shard forms described. These are, of course, isotropic unless they are devitrified, in which case they show a fine aggregate polarization. Interstitial materials may be more minute fragments of glass which in ordinary light appear as an excessively fine microgranular substance of a brownish color. In some, the matrix is chalcedonic material, locally radially fibrous, which has a low index and very low birefringence. The brown color observed is an optical effect caused by refraction and internal reflection such that blue rays are absorbed and red-orange transmitted.

Even vitric tuffs are apt to contain a few crystals or crystal fragments, especially of quartz and feldspar but perhaps also augite, hornblende, or biotite. Their characteristics are described in the next section.

Crystal tuffs Crystals of minerals, the kind depending largely on the nature of the magma, either perfect in form or more or less fragmental, are found in nearly all tuffs. When they become dominant or a striking feature of the deposit, the term *crystal tuff* is applied (Fig. 9-10).

The crystals are, as it were, premature phenocrysts in the lava ejected at the time of explosive eruption. A few, perhaps, might come from disrupted or shattered portions of the rocky walls through which the vent was drilled. True intratelluric crystals may be inclusion-filled—in some cases euhedral, even zoned in the case of the feldspars, in other cases (quartz, for example), rounded and embayed by corrosion or resorption in the magma. In some cases the crystals stand with their longer axes at right angles to the laminations of the tuff as if they had fallen into it from above.

The matrix material is largely comminuted glass, often nearly submicroscopic in size or the devitrification products of such glass.

Lithic tuffs Lithic tuffs are marked by dominance of volcanic rock particles but include also some rock fragments produced by shattering of the walls of the volcanic vent (Fig. 9-11). Lava particles derived from previously solidified magma tend to be of one type and tend to be marked by microlites and other features characteristic of aphanitic-flow rocks. As in other tuffs, the matrix is largely glassy ash, usually devitrified in older deposits.

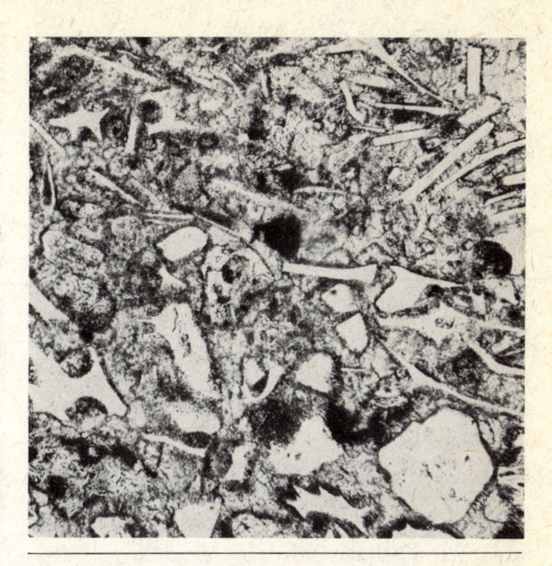

FIG. 9-9. Vitric tuff, Wasatch tuffaceous sandstone (Eocene), Jackson Hole, Wyoming. Ordinary light, ×80. Quartz and feldspar sand grains (clear) and numerous glass fragments (shards) in a carbonate matrix. Note bedding plane separating finer tuff with few mineral grains from coarser bed with many mineral grains. Mineral grains concentrated near the base of the bed. A redeposited tuff.

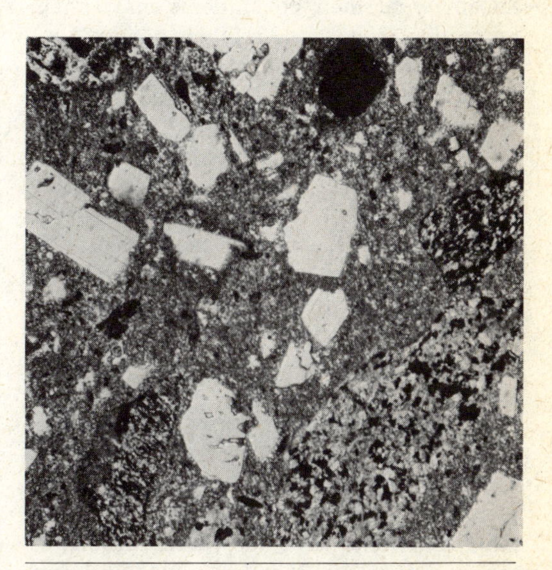

FIG. 9-10. Crystal tuff, Laacher See, Eifel, Germary. Ordinary light, ×22. Composed largely of whole and broken crystal euhedra, mainly untwinned feldspar (sanidine), a sodic pyroxene with a few rock fragments in a fine-grained groundmass.

Tuffaceous sedimentaries Where tuffs have fallen into water, or have been quickly eroded and redeposited, they may contain varying quantities of ordinary sedimentary materials and in some cases grade into ordinary sediments, either clastic or calcareous (see, for example, Ross, Miser, and Stephenson, 1929). Where pyroclastic materials are relatively fresh, tuffaceous origin is fairly clear, because vitriclastic textures are preserved. Where volcanic debris has been weathered, these tell-tale textures are destroyed, and volcanic debris becomes difficult to recognize. Crystal or broken euhedra, or volcanic quartz with its resorption forms, may still be recognizable. But in some cases the origin is obscure and only suspect but not provable.

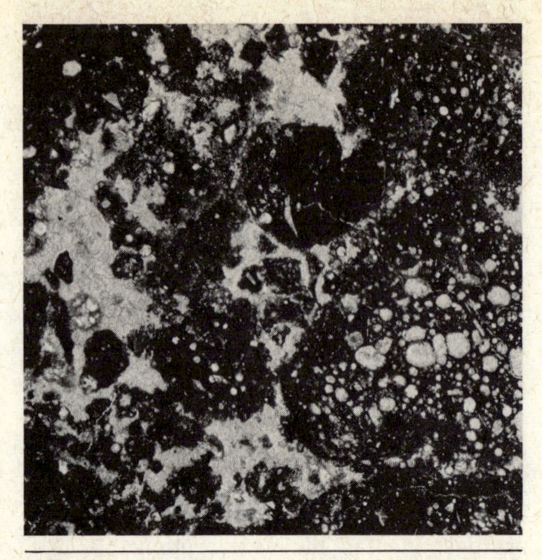

FIG. 9-11. Lithic tuff, Mt. Vesuvius, Italy. Ordinary light, ×22. Composed of vesicular lapilli (vesicles containing zeolites) in matrix probably much altered glass plus some carbonate.

ALTERATION OF TUFFS

It is the exception rather than the rule to find tuffs in the condition they were when freshly deposited. Their porosity and the instability of their constituents make them especially prone to alteration.

Weathering and consolidation proceed rapidly. Weathering is accelerated by ready access of air and moisture and the relatively large surface area, a consequence of fineness of grain. Feldspathic tuffs are readily kaolinized and converted to soft earthy masses. One of the earliest changes in vitric tuff is the release of silica and the deposition of hydrated silica—opal and chalcedony—which may convert these felsic tuffs to a dense flinty rock very much resembling chert. Vestiges of vitriclastic texture, together with the form and composition of associated crystals, and bulk chemical composition, distinguish the silicified tuffs from chert, porcellanite, and related materials. In other cases carbonates are introduced so that the whole rock becomes very calcareous. Less commonly the tuff is converted to microcrystalline sericitic material mingled with granules of quartz. Basaltic tuffs, so-called "palagonites," are altered to secondary silica, zeolites, chlorites, carbonate, and limonite. Alteration may destroy the vitriclastic texture and leave the origin of the rock in doubt.

Except in a "sealed" environment, glass is seldom seen in rocks older than Tertiary. Glass undergoes devitrification, a process that leads to the formation of clay minerals, zeolites, and silica. The result is that, between crossed nicols, the whole rock appears to be composed of a mosaic of feebly polarizing particles. Careful study in ordinary light may disclose "ghosts" of shards, threads, cusps, or vesicles of the original glass (Fig. 9-12). Montmorillonite, and also halloysite, form colloform vermicular masses which may precipitate in cavities in volcanic sands (Lerbekmo, 1957). Or glass may devitrify to a microcrystalline aggregate resembling chert. Complete alteration of ash produces bentonite.

Bentonite is a rock composed of a crystalline, claylike mineral formed by devitrification and accompanying chemical alteration of a glassy igneous material, usually a tuff or volcanic ash. It may contain grains that were originally phenocrysts in the volcanic glass. The claylike mineral has a micaceous habit, high birefringence, and a texture inherited from the original ash. It is usually montmorillonite, less commonly beidellite (Ross and Shannon, 1926). Bentonite has the property of swelling into a

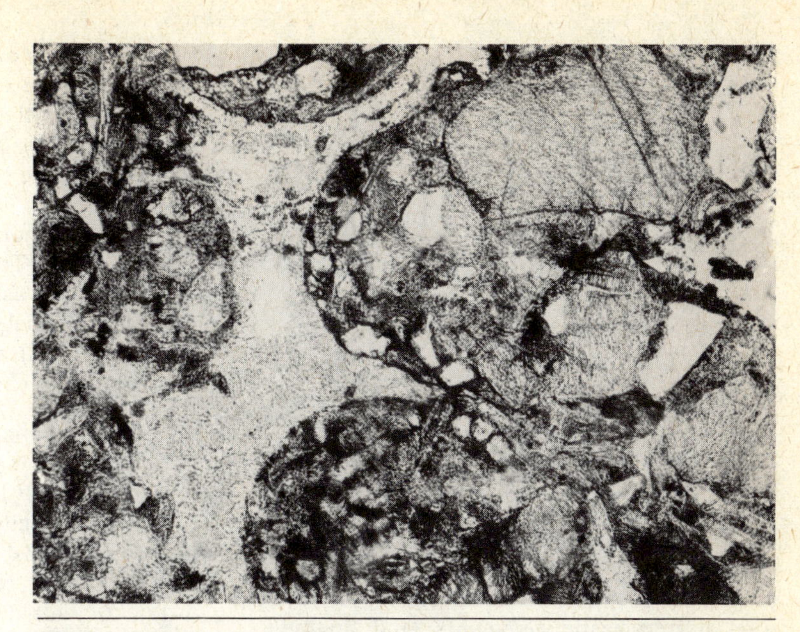

FIG. 9-12. Hattan Tuff Member, Stanley Shale (Carboniferous), near Mena, Arkansas. Ordinary light, ×80. A much altered tuffaceous sandstone, associated and interbedded with shales and graywackes. Angular quartz (clear), much sericitized feldspar (clouded), and devitrified glass (large shard) in chloritized and sericitized matrix.

gelatinous mass or disintegrating into a granular to fluffy aggregate. Bentonites, unlike ordinary shales and clays, commonly has a waxy translucence in thin chips. The most telling evidence, however, of their volcanic origin are relict shard structures and the presence of minerals and crystal euhedra characteristic of volcanic rocks.

Basic tuffs are most susceptible to change during diagenesis. The chief product formed is chlorite, which forms along with epidote within volcanic rock fragments. It commonly forms a pale-green fibrous cement in cavities between detrital grains. Cementation and replacement of some detrital grains by calcite is also very common.

Zeolites are of particular interest. A survey of the known occurrences of sedimentary zeolites shows that most, if not all, of these minerals were formed during diagenesis of volcanic materials, especially rhyolitic glass (Deffeyes, 1959). Zeolites precipitate in void spaces and form coatings on framework grains of volcaniclastic sands (Hay, 1966; Kaley and Hanson, 1955; Gilbert and McAndrews, 1948). The most common zeolites are analcime and clinoptilolite, which generally occur in aggregates too fine-grained for optical study; hence they are most readily identified by x-ray techniques. The process of zeolite alteration of pyroclastics has been studied by Bramlette and Posnjak (1933) as well as by Deffeyes (1959).

The effects of metamorphism are thorough recrystallization and, at higher grades, appearance of new minerals such as biotite, green hornblende, and actinolite. Original phenocrystic feldspars are replaced with aggregates of new feldspar and quartz and some biotite. The particular mineral assemblage depends somewhat on the original composition of the tuff as well as the grade of metamorphism. In general, sericite is the most characteristic mineral of metamorphosed felsic tuffs and is accompanied by quartz, biotite, chlorite, epidote and clinozoisite; mafic tuffs are largely actinolite and chlorite with more or less biotite, clinozoisite, and epidote.

Where metamorphism is not too profound, primary characteristics of tuffs may persist. Remains of vitriclastic textures are decisive; a secondary line of evidence is the nature of the included crystals or crystal fragments. Such evidence is strong if accompanied by a well-defined stratification; otherwise the parent rock may have been a lava. Coarser pyroclastics still show megascopic evidence—blocks and pieces of vesicular and/or porphyritic lavas. Bulk chemical composition may also serve to corroborate the volcanic nature of the original sediment.

ORIGIN AND NATURE OF VOLCANICLASTIC DEPOSITS

We will first discuss true pyroclastic deposits—those of primary origin, the products of an eruption—then secondary or reworked pyroclastic sediments, and finally those volcaniclastic sediments produced by the breakdown of ancient volcanic rocks.

PRIMARY PYROCLASTIC SEDIMENTS

Pyroclastic materials are the chief product of volcanism in island arcs and along aggressive continental margins. In the ancient record, they are principally confined to mobile belts. Most are felsic in composition and tend to be rhyolitic. Lavas rich in silica have a greater viscosity and higher gas content than those poor in silica and are apt, therefore, to be related to explosive rather than effusive volcanism (the latter is more characteristic of plateau basalts). Although coarse tephra may accumulate locally to form cinder cones, finer materials may be transported high in the atmosphere over great distances to form widespread layers of ash that may serve as useful time-equivalent marker beds in many stratigraphic sections. Such debris blown into the atmosphere may fall either on land or into water. Such a fallout deposit is an *ash fall.* In other eruptions, a hot incandescent, turbulent mixture of debris and gas may be erupted from a vent and move rapidly downslope in much the same manner as a submarine turbidity flow. Such high-velocity, destructive flows deposit materials some distance from their source. These deposits constitute an *ash-flow* deposit. They may be released from either a subaerial or a subaqueous vent. The latter gives rise to a submarine ash flow. *Base-surge* eruptions are steam-laden clouds of debris which are ejected laterally from the base of the eruptive column. In behavior they resemble *nuées ardentes,* but the resulting deposits more closely resemble those of ash falls.

Mudflows are secondary phenomena related to eruptions. Torrential rains falling on newly deposited ash produce a mudflow which moves downslope without appreciable sorting of entrained materials. The resulting deposit is chaotic and unsorted.

Ash falls In ash falls the ejected materials are carried downwind from the eruptive center. They tend to be sorted out according to their fall velocity; hence the resulting accumulation diminishes both in thickness and grain size away from the point of origin (Fig. 9-13). As noted by Scheidegger and Potter (1968), both size and thickness show a decline in a regular fashion downcurrent. The fallout pattern itself is related to wind direction, velocity, and turbulence and to the height attained by the eruption cloud. The dispersal pattern is perhaps the most reliable guide to paleowind directions (Eaton, 1964).

Typical ash falls are marked by good to very good sorting and well-defined bedding (Fig. 9-14). Unlike ash flows, which are restricted to volcanic centers, ash falls may cover wide areas, upwards of 300,000 to 400,000 miles^2 or 7.8 $\times 10^5$ to 10.4 $\times$ 10^5 km^2 (Ross, 1955). The distal margin of an ash fall may be only a bentonitic clay seam a few centimeters thick. Such bentonitic seams are time markers of great interest to the stratigrapher (Kay, 1935; Swineford, Frye, and Leonard, 1955; Dennison and Textoris, 1970).

Ash flows Ash-flow deposits, also known as ignimbrites, are a product of the *nuée ardente* type of eruption made famous by the Mt. Pelée disaster in 1902 in Martinique in the West Indies. This type of eruption is a swift-moving, hot density current, traveling downslope, commonly following larger topographic lows. The high velocity, perhaps approaching 100 mi/hr (160 km/hr), and high temperature, some 550° to 950°C, make it a very destructive agent. Such flows may extend as much as 20 to 60 miles (32 to 97 km) from their source. They are somewhat irregular in thickness, the upper boundary being approximately level; the lower contact

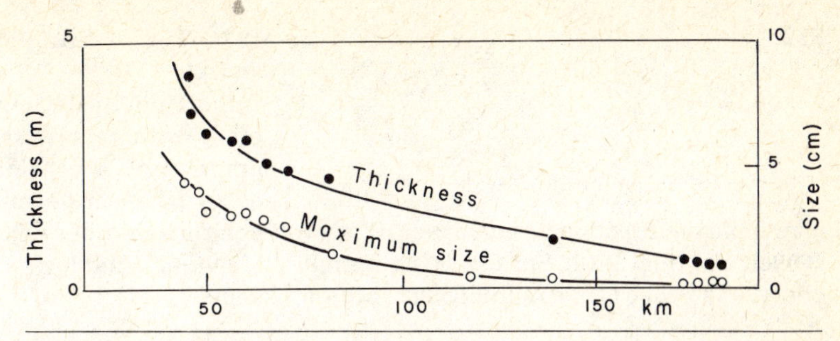

FIG. 9-13. Decline in grain size and thickness of ash fall away from source. (Modified from Katsui, 1963, *Jour. Fac. Sci. Hokkaido Univ.*, ser. 4, v. 11, Figs. 4 and 5.)

FIG. 9-14. Well-stratified airfall ash, Laacher See, Eifel, Germany. (Photograph by H.-U. Schmincke.)

is controlled by the underlying topography. Single units, representing one flow, may be as much as 100 m thick, but most vary from 15 to 30 m. Internal stratification is obscure, but an overall zoning, resulting from differential cooling between the lower and upper parts of the unit, may be apparent. Rapid accumulation of the incandescent tuff retards dissipation of the heat with the result that glass shards and pumice fragments are partially or totally fused or collapsed. The result is a welded tuff (Fig. 9-15). In general the lower part of the deposit is more completely welded; the upper part is apt to be less so and hence is more porous. The greater the degree of welding, the denser the end-product. Many so-called rhyolites turn out to be, on close inspection, welded tuffs. Field evidence for an ash flow are flattened and stretched pumice fragments, many of which display tension cracks, pull-aparts, imbrication, and rotation, all of which can be used to infer direction of transport. In some cases movement during or after welding (rheoignimbrites of Cook, 1966), resulting in deformation and laminar flowage features (Schmincke and Swanson, 1967), can also be used to determine the direction of transport. Additionally, some diminution in both thickness and class size may indicate direction of movement (Fisher, 1966c). See Fig. 9-16. Sorting is noticeably poorer than that of air-fall deposits (Moore, 1934, Figs. 2 and 3; Walker, 1971, Figs. 1 and 2).

Individual ash flows may follow one another without breaks between, whereas in other cases a considerable time lapse may occur between flows. In the latter case soil horizons or other deposits, such as ash falls, alluvium, mudflows, and lavas, may intervene.

FIG. 9-15. Welded tuff, San Miguel Mountains, Colorado. Ordinary light, ×22. Large plagioclase euhedra with some clinopyroxene and biotite embedded in glass showing compressed shards.

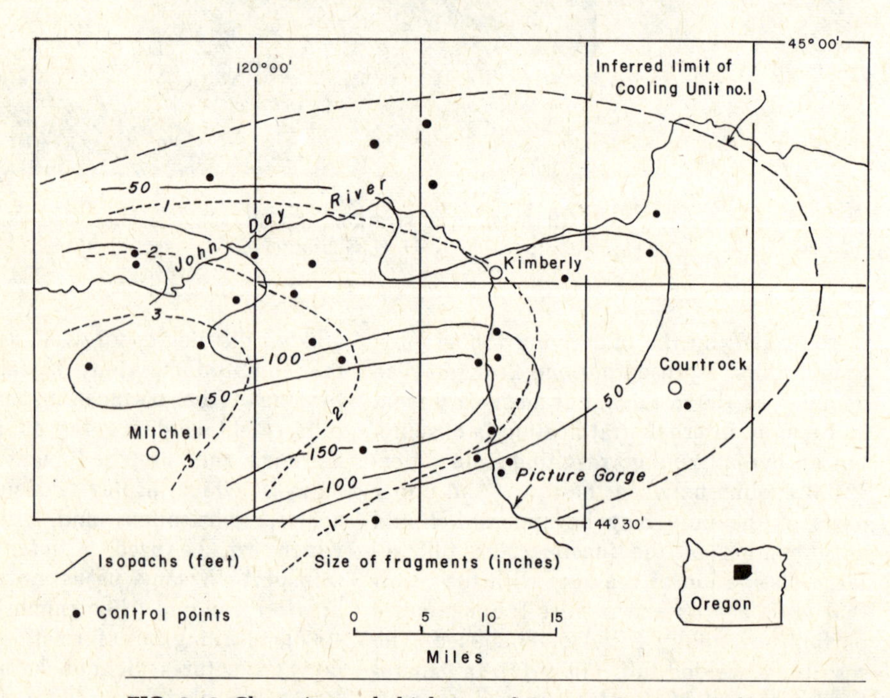

FIG. 9-16. Clast size and thickness of Picture Gorge ignimbrite (Oligocene-Miocene) in John Day Basin, Oregon. (Modified from Fisher, 1966, Univ. Calif. Publ. Geol., v. 67, Figs. 1 and 4.)

Good descriptions of ash-flow deposits have been published by Gilbert (1938), Fisher (1966b), and Schmincke and Swanson (1967). The whole subject has been reviewed and summarized by Smith (1960), Ross and Smith (1961), and Cook (1966).

Fiske (1963) and Fiske and Matsuda (1964) have described submarine ash-flow deposits. Debris of deep subaqueous ash flows is rapidly chilled and settles to the bottom where it may form a dense turbidity flow. Such deposits are much better sorted than are subaerial ash flows. As subaqueous turbidity flows, they may travel long distances. Crystal and lithic fragments tend to be deposited first; fine ash travels far; pumice fragments tend to float to the surface and become dispersed by waves and currents or, if deposited, form the upper (and coarser) part of the accumulate.

Base-surge deposit Base-surge deposits (Waters and Fisher, 1970) are deposits typical of maar volcanoes. They are produced by high-velocity, horizontally directed steam-laden eruptive clouds. They differ markedly, however, from the deposits of the incandescent *nuée ardente* eruptions in that they show thin and continuous stratification, may have low-angle cross-stratification, and, in still other cases, display repeated graded sequences. Air-fall blocks may be embedded in these beds; such blocks create "bedding sag" in underlying stratified materials. Base-surge deposits most resemble air-fall deposits, but cross-stratification and undulatory wavelike forms (antidunes) serve to identify them. Characteristic bed forms have been well described by Fisher and Waters (1970) and Crowe and Fisher (1973).

Mudflows Torrential rains that accompany or follow eruptions may, upon falling on newly deposited ash and cinders, generate large mudflows or *lahars* (Mullineaux and Crandell, 1962). These move downslope and give rise to chaotic deposits marked by no assortment or stratification, although some show a crude zonation (Schmincke, 1967). They contain large meter-sized and smaller blocks embedded in a tuffaceous matrix. That such deposits are mudflows and not ash flows is shown by lack of welding, absence of collapsed pumice, and heterogeneity of rock fragments. Volcanic mudflows may also enter the sea or be formed by subaqueous slump. Such flows pass by degrees into turbidity currents, in which case they show all the usual features of turbidites.

SECONDARY (REWORKED) PYROCLASTIC SEDIMENTS

Vegetative cover is effectively blanketed or destroyed by volcanic eruptions. Newly deposited tephra, resting in many places on steep slopes, is therefore especially susceptible to rapid erosion. Loss of vegetative cover is usually accompanied by landsliding. Such action plus torrential rains can deliver an enormous volume of debris to streams, giving rise to debris flows that choke headwaters and spread quickly downstream. These materials may assume the character of mudflows or eventually become a part of the stream alluvium and may ultimately be carried out to sea where they are spread by submarine currents. Some may eventually be swept off the shelf and move seaward as a turbidity current.

Many of these reworked deposits will show good assortment, stratification, and other structures characteristic of the agent involved. Reworked tephra may be diluted by ordinary clastic debris. The subaqueous deposits may even be mingled with calcareous shelly materials. The essential volcanic nature of these deposits depends on the recognition of vitriclastic textures, the presence of crystal debris of such minerals as hornblende and olivine, and chemical anomalies in their bulk composition.

EPICLASTIC VOLCANIC SEDIMENTS

Ancient volcanic terranes undergo disintegration and erosion, and the debris derived from such areas will be rich in volcanic detritus. Webb and Potter (1969), for example, noted that streams draining a rhyolitic plateau carried sands consisting mainly of rhyolitic rock particles with a lesser quantity of volcanic quartz and feldspar. About two-thirds of the grains in the sands from The Dalles of the Columbia River in Washington are of volcanic origin (Whetten, 1966). How does one distinguish lithic sandstones of volcanic provenance from pyroclastic sands? Because these sands are derived from somewhat older volcanic rocks, they are apt to be nearly free of volcanic glass; the unstable glass has long since been devitrified and converted to clay-sized debris. Such deposits will lack the usual vitriclastic texture that an altered tuff would display. They almost surely will be diluted with materials derived from nonvolcanic sources.

References

Bastin, E. S., 1909, Chemical composition as a criterion in identifying metamorphosed sediments: Jour. Geol., v. 17, pp. 445–472.

Blyth, F. H., 1940, The nomenclature of clastic deposits: Bull. Volcanologique, v. 6, ser. 2, pp. 145–156.

Bramlette, M. N., and Posnjak, E., 1933, Zeolite alteration of pyroclastics: Amer. Mineral., v. 18, pp. 167–171.

Cook, E. F., 1966, Paleovolcanology: Earth Sci. Rev., v. 1, pp. 155–174.

Correns, C. W., and Leinz, Viktor, 1933, Tuffige Sedimente des Tobasees (Nordsumatra) als Beispiele für die sediment-petrographische Bedeutung von Struktur und Textur: Centralbl. Min. Geol., Ab. 4, pp. 382–390.

Crowe, B. M., and Fisher, R. V., 1973, Sedimentary structures in base-surge deposits with special reference to cross-bedding, Ubehebe Craters, Death Valley, California: Bull. Geol. Soc. Amer., v. 84, pp. 663–682.

Deffeyes, K. S., 1959, Zeolites in sedimentary rocks: Jour. Sed. Petrology, v. 29, pp. 602–609.

Dennison, J. M., and Textoris, D. A., 1970, Devonian Tioga Tuff in northeastern United States: Bull. Volcanologique, v. 34, pp. 289–294.

Eaton, G. P., 1964, Windborne volcanic ash, a possible index to polar wandering: Jour. Geol., v. 72, pp. 1–35.

Fisher, R. V., 1958, Definition of volcanic breccia: Bull. Geol. Soc. Amer., v. 69, pp. 1071–1073.

———, 1961, Proposed classification of volcaniclastic sediments and rocks: Bull. Geol. Soc. Amer., v. 72, pp. 1409–1414.

———, 1966a, Rocks composed of volcanic fragments and their classification: Earth Sci. Rev., v. 1, pp. 287–298.

———, 1966b, Geology of a Miocene ignimbrite layer, John Day Formation, eastern Oregon: Univ. California Publ. Geol. Sci., v. 67, pp. 1–58.

———, 1966c, Mechanism of deposition from pyroclastic flows: Amer. Jour. Sci., v. 264, pp. 350–363.

Fisher, R. V., and Waters, A. C., 1970, Base surge bed forms in maar volcanoes: Amer. Jour. Sci., v. 268, pp. 157–180.

Fiske, R. S., 1963, Subaqueous pyroclastic flows in the Ohanapecosh Formation, Washington: Bull. Geol. Soc. Amer., v. 74, pp. 391–406.

Fiske, R. S., and Matsuda, T., 1964, Submarine equivalents of ash flows in the Tokiwa Formation, Japan: Amer. Jour. Sci., v. 262, pp. 76–106.

George, W. O., 1924, Relation of the physical properties of the natural glasses to their chemical composition: Jour. Geol., v. 32, pp. 353–372.

Gilbert, C. M., 1938, Welded tuff in eastern California: Bull. Geol. Soc. Amer., v. 49, pp. 1829–1862.

Gilbert, C. M., and McAndrews, M. G., 1948, Authigenic heulandite in sandstone, Santa Cruz County, California: Jour. Sed. Petrology, v. 18, pp. 91–99.

Hay, R. L., 1952, The terminology of fine-grained detrital volcanic rocks: Jour. Sed. Petrology, v. 22, pp. 119–120.

———, 1966, Zeolites and zeolite reactions in sedimentary rocks: Geol. Soc. Amer. Spec. Paper 85, 130 pp.

Heiken, Grant, 1972, Morphology and petrography of volcanic ashes: Bull. Geol. Soc. Amer., v. 83, pp. 1961–1988.

Huber, N. K., and Reinhart, C. D., 1966, Some relationships between refractive index of fused glass beads and the affinity of volcanic rock suites: Bull. Geol. Soc. Amer., v. 77, pp. 101–110.

Kaley, M. E., and Hanson, R. F., 1955, Laumontite and leonhardite in Miocene sandstone from a well in San Joaquin Valley, California: Amer. Mineral., v. 40, pp. 923–925.

Kay, G. M., 1935, Distribution of Ordovician altered volcanic materials and related clays: Bull. Geol. Soc. Amer., v. 46, pp. 225–244.

Krumbein, W. C., and Tisdel, F. W., 1940, Size distribution of source rocks of sediments: Amer. Jour. Sci., v. 238, pp. 296–305.

Lerbekmo, J. F., 1957, Authigenic montmorillonoid cement in andesitic sandstone: Jour. Sed. Petrology, v. 27, pp. 298–305.

Mackin, J. H., 1960, Structural significance of Tertiary volcanic rocks in southwestern Utah: Amer. Jour. Sci., v. 258, pp. 81–131.

Moore, B. N., 1934, Deposits of possible nuée ardente origin in the Crater Lake region, Oregon: Jour. Geol., v. 42, pp. 358–375.

Moore, J. G., and Peck, D. L., 1962, Accretionary

lapilli in volcanic rocks of western continental United States: Jour. Geol., v. 70, pp. 182–194.

Mullineaux, D. R., and Crandell, D. R., 1962, Recent lahars from Mt. St. Helens: Bull. Geol. Soc. Amer., v. 73, pp. 855–870.

Pettijohn, F. J., 1950, Turbidity currents and graywackes—a discussion: Jour. Geol., v. 58, pp. 169–171.

Pirsson, L. V., 1915, The microscopical characters of volcanic tuffs—a study for students: Amer. Jour. Sci., ser. 4, v. 40, pp. 191–211.

Ross, C. S., 1928, Altered volcanic materials and their recognition: Bull. Amer. Assoc. Petrol. Geol., v. 12, pp. 143–164.

———, 1955, Provenance of pyroclastic materials: Bull. Geol. Soc. Amer., v. 66, pp. 427–434.

Ross, C. S., Miser, H. D., and Stephenson, L. W., 1929, Water-laid volcanic rocks of early Upper Cretaceous age in southwestern Arkansas, southeastern Oklahoma, and northeastern Texas: U.S. Geol. Surv. Prof. Paper 154-F, pp. 175–202.

Ross, C. S., and Shannon, E. V., 1926, The minerals of bentonite and related clays and their physical properties: Amer. Ceramic Soc. Jour., v. 9, pp. 77–96.

Ross, C. S., and Smith, R. L., 1961, Ash-flow tuffs—their origin, geologic relations, and identifications: U.S. Geol. Surv. Prof. Paper 366, 81 pp.

Sapper, K., 1928, Die Tätigsten Vulcangebiete der Gegenwart: Zeitschr. Vulk., v. 11, pp. 181–187.

Scheidegger, A. E., and Potter, P. E., 1968, Textural studies of graded bedding: Sedimentology, v. 11, pp. 163–170.

Schmincke, H.-U., 1967, Graded lahars in the type sections of the Ellensburg Formation, south central Washington: Jour. Sed. Petrology, v. 37, pp. 438–448.

Schmincke, H.-U., and Swanson, D. A., 1967, Laminar viscous flowage structures in ash-flow tuffs from Gran Canaria, Canary Islands: Jour. Geol., v. 75, pp. 647–664.

Smith, R. L., 1960, Ash flows: Bull. Geol. Soc. Amer., v. 71, pp. 795–842.

Swineford, Ada, Frye, J. C., and Leonard, A. B., 1955, Petrography of the Late Tertiary volcanic ash falls in the central Great Plains: Jour. Sed. Petrology, v. 25, pp. 243–261.

Thorarinsson, S., 1954, The tephra-fall from Hekla on March 29, 1947. The eruption of Hekla, 1947–1948: Mus. Nat. Hist. Soc. Sci. Islandica, Reykjavik, 68 pp.

Twenhofel, W. H., 1950, Principles of sedimentation (2nd ed.): New York, Wiley, 673 pp.

Walker, G. P. L., 1971, Grain-size characteristics of pyroclastic deposits: Jour. Geol., v. 79, pp. 696–714.

Washington, H. S., 1917, Chemical analyses of igneous rocks: U.S. Geol. Surv. Prof. Paper 99, 1201 pp.

Waters, A. C., and Fisher, R. V., 1970, Maar volcanoes: Proc. 2nd Columbia River Basalt Symp., Cheney: Eastern Washington State Coll. Press, pp. 157–170.

Webb, W. M., and Potter, P. E., 1969, Petrology and geochemistry of modern detritus from a rhyolitic terrain, Western Chihuahua, Mexico: Bol. Soc. Geol. Mexicana, v. 32, pp. 45–61.

Wentworth, C. K., and Williams, H., 1932, The classification and terminology of the pyroclastic rocks: Rept. Comm. Sedimentation, Bull. Nat. Res. Coun. no. 80, pp. 10–53.

Whetten, J. T., 1966, Sediments from the lower Columbia River and origin of graywacke: Science, v. 152, pp. 1057–1058.

10

LIMESTONES AND DOLOMITES

INTRODUCTION

DEFINITIONS

To the lime manufacturer, *limestone* is a general term for that class of rocks which contain at least 80 percent of the carbonates of calcium and magnesium and which, when calcined, gives a product that slakes upon the addition of water. Although this is the literal meaning of limestone, geologists now use it to embrace a larger group of rocks. The term *carbonatite* has also been used for these rocks (Kay, 1951, p. 5), but this usage has been abandoned because this term is applied also to certain nonsedimentary rocks (Heinrich, 1967).

In general, the term *limestone* is applied only to those rocks in which the carbonate fraction exceeds the noncarbonate constituents. If, for example, sand-sized detrital quartz is present in excess of 50 percent, the term calcareous sandstone would be more appropriate. Likewise, rocks in which shaly matter exceeds the carbonate fraction are calcareous shales rather than limestones.

Normally the term *limestone* is used for those rocks in which the carbonate fraction is composed primarily of calcite or aragonite, and the term *dolomite* is reserved for those rocks which are composed primarily of mineral dolomite, even though dolomite is, of course, a lime-bearing rock. There are, moreover, rocks containing both calcite and dolomite. The nomenclature of these rocks of intermediate composition is further elaborated in the section on dolomite.

Unfortunately, in many ways the term *limestone* is a wastebasket term for a diverse group of rocks. Limestones are polygenetic (Fig. 10-1). Some are fragmental or detrital and are mechanically transported and deposited; others are chemical or biochemical precipitates and are formed in place. The first type exhibits hydrodynamic textures and structures no different from any other rock transported and deposited by waves and currents. The second type has a very different fabric. Detrital limestones display *current bedding*; those formed *in situ* show *growth bedding*. Both types may be profoundly modified by various post-depositional changes, both mechanical and chemical, so that the original characters are obscured or erased. These epigenetic or diagenetic rocks show their own distinctive fabrics.

That all these rocks of such diverse origins are designated limestone obscures rather than elucidates their history. It is, no doubt, for this reason that Grabau (1904) would abandon the term *limestone* and designate the detrital carbonates as calcarenites, and so forth, and the carbonate precipitates as calcigranulites (calcigranulytes) and the like. But the term has persisted and will, no doubt, continue in use as a trade or commercial term and perhaps as a field term. Established usage is difficult to change, although abandonment of the term would result in closer observations and better understanding of lime-bearing strata.

OCCURRENCE OF LIMESTONES

Limestone is a very common sedimentary rock. Estimates based on field measurements show that limestone (and dolomite) form one-fifth to one-fourth of the stratigraphic record (see p. 20).

Limestones (and dolomites) are of all ages, even earliest Precambrian (Archean), although they are much less abundant in the earlier record than in younger rocks. Most notable of the earliest limestones are the Steeprock Limestone of Canada (Jolliffe, 1955) and the limestones of the Bulawayan Series of South Africa (Schopf et al., 1971). These are over 2.6 billion years old. Thick, widespread limestones and dolomites are common in the later Precambrian (Proterozoic) rocks and are very abundant in the early Paleozoic record, especially in North America. In general, earlier carbonate deposits are more largely dolomite, the Ca–Mg ratio seemingly showing a progressive increase with decreasing age in North America (Daly, 1909).

Limestones (and dolomites) are thin and most widespread on stable cratonic areas and covered almost the entire North American craton during the Ordovician. They are considerably thicker in marginal miogeosynclinal tracts, forming sequences over 15,000 feet (5,000 m) thick in some cases (Sharp, 1942). They tend to be absent in eugeosynclinal troughs but in some cases are represented by thin turbidite or allodapic limestones.

ECONOMIC VALUE OF LIMESTONES

Limestones are of interest in part because of their economic worth (Lamar, 1961). Some, such as the popular "Bedford stone" of Indiana, are widely used in the United States as building stone. Crushed limestone is used as concrete aggregate, as road metal, and, if pulverized, as an agricultural dressing. It is calcined to form lime, is a major constituent in the manufacture of portland cement, is used as a flux in the reduction of iron ores, and is utilized as a paper filler. The economic value of limestones and their role as storage reservoirs for both petroleum and groundwater have provided incentives for their study (Geze, 1965). One-third to one-half of the world's oil is produced from limestone or dolomite reservoirs. The yield of some limestone wells is phenomenal —over 100,000 barrels per day. Limestone regions are characterized by karst drainage, a product of solution. The groundwaters flow through an intricate network of channelways and issue as very large springs. Such a water system is especially susceptible to contamination.

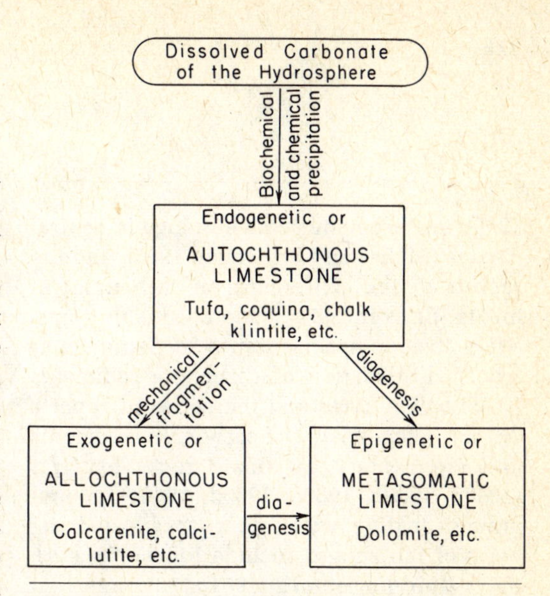

FIG. 10-1. Genetic classification of limestones.

GENERAL REFERENCES

Because limestones form so much of the sedimentary record, the sedimentary petrologist must be familiar with their attributes and occurrence, and an understanding of their significance and origin is essential to reading earth history.

It is very difficult to summarize our present knowledge into a single chapter. The serious student of the subject should turn to the larger works. Among these is the monographic work of Cayeux (1935), now available in English (Carozzi, 1970). The publications of Hadding (1933, 1941, 1957), and more recently of Bathurst (1971), and of Chilingar, Bissell, and Fairbridge (1967) are good general works on limestones. More recent publications include the symposia volumes on regional aspects of carbonate deposition (LeBlanc and Breeding, 1957), on the classification of carbonates (Ham, 1962), on dolomitization and limestone diagenesis (Pray and Murray, 1965), on depositional environments in carbonate rocks (Friedman,

1969), on carbonate sedimentology in central Europe (Müller and Friedman, 1968), and the reports of the conference on carbonate cements (Bricker, 1971). Other valuable references include special issues of *Sedimentology* on lithification of carbonate sediments (Füchtbauer, 1969) and the series of reports on the geochemistry of carbonate sediments and sedimentary carbonate rocks by Graf (1960a, 1960b, 1960c, 1960d, 1960e). Monographs dealing with other specialized aspects of the subject include Johnson's work on organic limestones (1951, 1961) and that of Horowitz and Potter (1971) on the petrology of skeletal detritus. Numerous papers on present-day carbonate deposits have been published. The bibliography on carbonates (Potter, 1968) is a good guide to the vast literature on both Recent and ancient carbonates.

MODERN CARBONATE SEDIMENTS

Calcareous sediments are widely distributed in the modern world. Good recent summaries of the vast literature on this subject are given by Rodgers (1957) and by Bathurst (1971, pp. 93–216); an earlier and important monograph on modern carbonates is that of Pia (1933). It is our intent here to enumerate and classify the several environments in which carbonate sediments accumulate and to indicate the nature of these deposits. If the actualistic approach is valid, we might expect these environments and sediments to be represented in the geologic record.

OCCURRENCE AND DISTRIBUTION

In general, there are five groups of modern carbonate deposits (not all of which are of equal importance): shallow-water marine deposits, deep-water marine carbonates, carbonates of evaporitic basins, carbonates of freshwater lakes and springs, and eolian carbonates. Because most deposits of the geologic past belong to the first category, these have received special attention. Nevertheless, deep-sea carbonates are the most widespread in the modern world, and a considerable number of ancient carbonates are now recognized as of deep-water origin. Evaporitic carbonates, those deposited in freshwater lakes and springs, and carbonate dunes, although of small volume, are also worthy of study.

Shallow-water carbonates Although shallow-water marine carbonates were very extensive in the geologic past, they are today found in only a few places. The best known and largest occurrences include those of the Florida-Bahama region. The Great Bahama Bank is a barely submerged plateau off the coast of Florida some 700 km long and up to 300 km in width. Most of it is covered by waters less than 10 m in depth. Emergent parts constitute an archipelago, the largest of which is Andros Island. Sediments of this region have been described by Illing (1954); Newell and Rigby (1957); Newell, Purdy, and Imbrie (1960); Purdy (1963a, 1963b); and Imbrie and Purdy (1962). In brief, the deposits are largely calcareous sands, both skeletal and oolitic, with a lesser volume of fine carbonate muds and reef rock. The sands cover large areas of the banks and in places are in motion, forming large subaqueous dune fields (Ball, 1967). Muds are confined to very shallow waters and to tidal flats on the west or lee side of Andros Island; reefs form narrow restricted bodies mainly on the windward (east) edge of the Bahama Platform. Some carbonate sediments escape from the platform and are redeposited in associated deep-water basins, such as the Tongue of the Ocean and the Columbus Basin (Bornhold and Pilkey, 1971). Carbonate sands also accumulate as subaerial dunes on the emergent parts of the banks.

Carbonate deposits of the Florida Platform are similar (Ginsburg, 1956, 1964) and consist of reef carbonates, back-reef sands, and lagoonal muds. The sands are in part skeletal, in part oolitic. Of special interest are the mud banks in Florida Bay, which seem to be localized by "meadows" of sea grass (*Thalassia*) that act as a baffle to reduce turbulence and thus induce mud deposition.

Other areas of extensive shallow-water carbonate sedimentation include the Campeche Bank, adjacent to Yucatán, and the shelf off British Honduras. The Campeche Bank has been described by Logan and co-workers (1969), and the sediments of this area have been investigated by Hoskins (1963), Folk and Robles (1964), and Harding (1964). The Honduran shelf carbonates are less well known but are presently under active study (Matthews, 1966; Ginsburg, James, and Marszalek, 1973).

Deposits of carbonate sands and muds and some reefs are found on the Trucial Coast area of the Persian Gulf (Houbolt, 1957). The near-shore lagoons and the tidal flats of this region contain hypersaline waters; very saline environments are present in the adjacent, and general emergent sabkha salt flats (Shearman, 1963; Kinsman, 1964). As a result, evaporitic deposits are closely associated with carbonates.

Other areas of shallow-water carbonate sedimentation generally less well known include the coasts of Western Australia (Logan et al., 1970) and Queensland. The Shark Bay area on the west coast is in part hypersaline and is particularly well known for its stromatolitic facies (Logan, 1961; Logan et al., 1970). The Queensland shelf is comparable in area to the Bahama Platform and is the site of the justly famous thousand-mile-long Great Barrier Reef (Maxwell and Swinchatt, 1970). Fringing barrier reefs and atolls throughout the Pacific and other tropical waters have been described in the literature. Although these deposits are miniscule, they provide much information on reef formation, structure, and evolution (Kuenen, 1933). Those of Funafuti (Cullis, 1904), Guam (Schlanger, 1964), and Bikini (Emery, Tracey, and Ladd, 1954) are perhaps among the best known.

In summary, shallow-water marine carbonates have accumulated in several distinct subenvironments: reef, tidal flat, *Thalassia* meadows, open bank or shelf, and subaerial dune. For the most part, mud accumulates in the tidal flat and *Thalassia* meadow environment; the sands are found on the open shelf and in dunes.

Deep-sea carbonates Present deep-water carbonates belong to two classes: turbidite or basinal deposits and pelagic deep-sea deposits. The first, although less extensive, is more common in the geologic record; pelagic carbonates, the most widespread lime deposits in the modern world, are poorly represented, if at all, in the ancient record.

Turbidite or basinal carbonates have been only recently identified as such, either in the present seas or in the geologic record. The deposits form in basins some thousands of meters deep. They were introduced as graded beds which alternate with indigenous sediment, either pelagic carbonate or fine terrigenous sediments. Their shallow-water origin is shown by the contained organic detritus such as fragments of calcareous red algae (*Halimeda*). Sediments of this type are associated with and derived from reefs and banks and transferred to the deep-water environment by turbidity flows. Such sediments have been reported from the Tongue of the Ocean in the Bahamas, the Columbus Basin in the southern Bahamas (Bornhold and Pilkey, 1971), and in the deeper part of the Gulf of Mexico off the Campeche Bank (Davies, 1968). Deposits of this type appear to be common in the ancient geologic record (see p. 378).

The most widespread deep-sea carbonates are pelagic pteropod and globigerina oozes, the latter much more widespread. Globigerina ooze covers 125×10^6 km^2; pteropod ooze covers 2×10^6 km^2 (Kuenen, 1950, p. 362). The average depth of accumulation of these deposits is 3600 and 2000 m, respectively. Globigerina ooze consists mainly of planktonic foraminiferal tests, the most important of which is *Globigerina*. The carbonate content of the ooze ranges from 30 to over 90 percent, averages 65 percent. These deposits are most abundant in lower latitudes, and their distribution seems to correlate with the salinity of the surface waters; deposits are most abundant where surface salinity is highest (Trask, 1937). These deposits are absent, however, where the ocean floor is deepest. Apparently calcareous foraminiferal tests tend to dissolve in deeper, colder waters. The depth below which solution dominates over accumulation is about 4000 m. The carbonate content falls off rapidly below this depth; carbonates are virtually absent below 6000 m. Although a few carbonate sediments, such as chalk, resemble these modern oozes, very few pelagic calcareous deposits in the geologic record are presumed to have formed in waters to the surface where, by evaporation, lime-rich caliche is formed. Because it forms only in regions of limited rainfall, it is an important climatic index.

abyssal depths. This may in part be attributable to the fact that no pelagic deposits could form prior to the evolution of carbonate-secreting plankton until Cretaceous time.

Freshwater carbonates In some present-day freshwater lakes friable carbonate earths, designated *marl*, are forming (Davis, 1900). Marl-depositing lakes are common in North American areas covered by calcareous glacial drift (Blatchley, 1900; Thiel, 1933). Marl beds also underlie many freshwater swamp peats, recording an earlier lacustrine stage. Many such marls are somewhat argillaceous and hence usable in the manufacture of portland cement.

Lime deposits are also forming today by evaporation of some spring and river waters. *Tufa*, a spongy, porous material that forms a thin, surficial deposit about springs and seeps and exceptionally in rivers, is seldom extensive and is restricted mainly to Recent or Quaternary deposits. *Travertine*, a dense, banded deposit, especially common in limestone caverns, forms relatively small deposits and like tufa, is primarily Quaternary or Recent in age.

Evaporitic carbonates Of some minor accumulations of calcium carbonate associated with arid climates, *caliche* (and *calcrete*) is the most widespread. This impure lime-rich deposit is found in the soils of semi-arid regions. Capillary action draws lime-bearing

Eolian carbonates Small deposits of carbonate sand—debris from offshore reefs—accumulate on beaches and in dunes associated with these beaches. Calcareous eolianites occur in Bermuda (Sayles, 1931; Mackenzie, 1964) and on many of the islands of the Bahama Platform. These dune deposits are self-cementing and become lithified early, as do also some of the beach sands; the latter is the familiar *beach rock* (Ginsburg, 1953; Stoddart and Cann, 1965).

ENVIRONMENTS OF CARBONATE DEPOSITION

This brief review of modern carbonate sediments has shown that carbonate accumulates in many differing environments (Fig. 10-2). Of greatest interest to the geologist are shallow-water marine carbonates and the satellitic deposits to which they contribute, namely the eolianites and, more importantly, the basinal turbidites. The shallow-water environment is itself complex and includes tidal and supratidal flats, more extensive shelf and bank areas, marginal reefs, and back-reef lagoons. Each environment leaves its own imprint on the textures and structures of the accumulating carbonate. In regions of marked aridity, the salinity of shallower and somewhat restricted lagoonal and tidal areas may lead to evaporitic deposits closely associated with carbonates.

Of deep-water environments, turbidite basins adjacent to carbonate platforms are

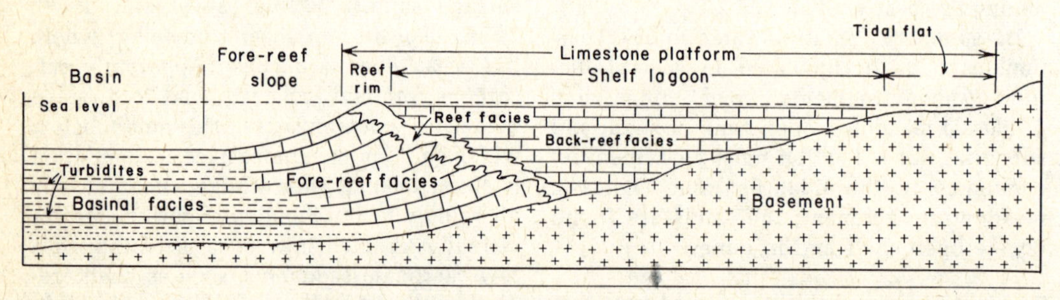

FIG. 10-2. Schematic cross section shows principal environments of marine carbonate deposition. (Based on diagram by Playford, 1972, Ann. Soc. Geol. Belgique, v. 95, Fig. 2.)

the most important. The abyssal environment with its pelagic deposits, although widespread today, leaves little or no geologic record.

Lacustrine and spring environments play a minor role in the record. On the other hand, the soil environment in which caliche is formed is widespread in older strata. Accumulations of this type appear in the uppermost part of alluvial cycles—in the Old Red Sandstone of Wales, for example (Allen, 1965). Similar lime accumulations are known in the alluvial deposits of the Devonian Catskill Formation (Allen and Friend, 1968), the Mississippian Mauch Chunk Formation in Pennsylvania, and in the Triassic Newark deposits near York, Pennsylvania (Cloos and Pettijohn, 1973, p. 528).

TABLE 10-1. Percentage distribution of constituents of modern lime-bearing sediments

Constituent	A	B	C
Organic constituents			
Algae, calcareous	22.8	25.1	18.0
Mollusks	15.8	17.5	12.2
Foraminiferans	11.7	9.0	17.3
Coral	9.0	9.3	8.2
Spicules, total	3.6	4.3	2.1
Worm tubes	1.8	1.4	3.0
Crustaceans	1.2	1.4	0.7
Bryozoans	0.3	0.4	trace
Other constituents			
Silt	13.2	13.9	11.7
Clay (with aragonite needles)	10.2	7.8	14.8
Minerals (mostly quartz)	2.8	3.9	0.5
$CaCO_3$ (unidentifiable forms)	5.5	5.3	6.0
Oolites	0.8	0.4	1.6
Pellets (? fecal)	1.3	trace	3.8
Aggregates	0.2	—	0.8
Total	100.2	99.7	100.7

Source: After Thorp (1936).
A. Average of 50 samples from Florida and 24 samples from the Bahamas.
B. Average of 50 Florida samples.
C. Average of 24 Bahama samples.

PETROLOGY OF MODERN CARBONATE SEDIMENTS

The sediments of marine shelf-bank deposits consist of reef rock, sands (of several types), and muds. The sands are by far the most extensive and of the greatest volume. Locally these are oolitic. In the Bahamas, especially near the south end of the Tongue of the Ocean, and elsewhere, these sands accumulate as very large bars—veritable subaqueous dunes—which may be awash at lowest tides (Ball, 1967). In other places, sands are of mixed composition, with skeletal sands very common. Many constituent sand grains are microcrystalline aggregates which may be micritized or degraded oolites and skeletal grains or may be fecal pellets of carbonate mud. The proportions of the several constituents vary and define subfacies of the sand class (Purdy, 1963b; Newell, Purdy, and Imbrie, 1959). In some places, where the sands are no longer in active motion, cementation of grain clusters bound together by micritic carbonate form "grapestone" aggregates (Illing, 1954). This facies is quite extensive in some areas.

The relative abundance of organic constituents in sands and proportions of other materials are given in Table 10-1. These analyses and others (Ginsburg, 1956, for example) show that only a few types of organisms make large contributions. Noteworthy of present time is the importance of lime-secreting algae (Thorp, 1936). Notable also is the relatively subordinate position of the corals, even in so-called coral reefs (Table 10-2). The proportions shown by analyses are not necessarily a true reflection of the relative importance of the several forms as producers of lime, inasmuch as some are more resistant and survive as sand grains; others disintegrate rapidly and contribute only as silt- and mud-sized particles (Chave, 1960). Nonetheless, the algae seem to dominate as producers of both sand and mud.

Sands, especially in areas of active movement, are well sorted, stratified, and cross-bedded (Imbrie and Buchanan, 1965). Where sands move onto the beach they may give rise to beach rock (Ginsburg, 1953; Emery and Cox, 1956; Taylor and Illing, 1969); where they are blown inland they form dunes with large-scale cross-bedding. These dunes tend to be self-cementing and become lithified very rapidly (Sayles, 1931; Mackenzie, 1964).

The carbonate muds of the Bahamas are

less extensive, as a rule, than the carbonate sands. They are most abundant in the very shallow waters and on the tidal flats west of Andros Island. They form 50 percent or more of the deposits of Florida Bay (Ginsburg, 1956). In intertidal areas, muds show thin laminations apparently related to algal mat development. Where the mat is exposed, and only infrequently flooded, mud cracking occurs. Subaqueous muds are commonly pelleted and thus form a species of sand. The origin of these aragonitic muds has been the subject of many investigations. Although there is some uncertainty as to whether sea water is saturated with calcium carbonate, there can be no doubt that very shallow marine waters, partially isolated, may be saturated and from such waters carbonate may be precipitated as aragonite. Under turbulent conditions oolites may form; in tranquil water aragonite is precipitated as minute acicular crystals. Aragonite needles found in the muds of shallow lagoons in the Bahamas have been regarded as evidence of direct precipitation of calcium carbonate (Cloud, 1962). Recent work, however, has shown that aragonite needles may have been formed in algal tissues and released upon disintegration of organic matter (Lowenstam, 1955). They have the same size, morphology, and isotopic composition (Lowenstam and Epstein, 1957) as those in the muds. (The isotopic composition of precipitated carbonate is not the same.) Recent studies of the productivity of algae in the Florida region demonstrate the adequacy of this mechanism of mud formation (Stockman, Ginsburg, and Shinn, 1967). In the Trucial Coast area, however, the evidence for inorganic precipitation is reasonably good (Kinsman, 1966). It is possible also, that some mud is the finest product of trituration of skeletal materials (Chave, 1960).

Reef rock is a product of a community of organisms. As seen in Table 10-2, corals occupy a relatively subordinate position even in so-called "coral reefs." The order of abundance of the organisms that contribute to the Funafuti Reef was found to be encrusting coralline algae (*Lithothamnium*), other algae (*Halimeda*), foraminiferal tests, and corals (Finckh, 1904, p. 133). Essential to reef development are forms that construct a wave-resistant framework. Other forms contribute detritus which becomes incorporated in this framework and is overgrown by it. The reefs are islands of intense vital activity. They grow primarily from calcareous material deposited by the inhabiting organisms (mainly benthonic). Almost immediately after initiation, reefs rise above the surrounding bottoms. With entrance into the zone of wave-generated turbulence, they shed debris which forms flanking beds on the seaward side and thus broadens the reef platform. The growing core may then expand and overgrow the flanking detrital apron. Some reef debris is transferred great distances and into deep water by turbidity currents. Reefs thus give rise to much more extensive deposits than are represented by the reef structure itself. Some reef debris is thrown upon the reef to become an emergent island eventually stabilized by vegetation.

All of the above sediment types, and the environments responsible for them, are represented in the ancient carbonate record. Calcarenites or carbonate sands are very abundant. So also are calcilutites—the lithified carbonate muds—relatively more abun-

TABLE 10-2. Quantitative biologic composition of modern reef sediments

Constituents	A	B	C	D
Algae, calcareous	48.5	25.1	18.0	42.5
Mollusks	17.8	17.5	12.2	15.2
Corals, madreporarian	16.6	9.3	8.2	34.6
Foraminifera	6.3	9.0	17.3	4.1
Total	89.2	60.9	55.7	96.4
Constituent ratios				
Algae–coral	2.92	2.70	2.20	1.23
Algae–mollusk	2.72	1.43	1.47	2.79
Algae–foraminiferan	7.70	2.80	1.04	10.03
Mollusk–foraminiferan	2.82	1.94	0.71	3.71
Mollusk–coral	1.07	1.88	1.49	0.44
Coral–foraminiferan	2.64	1.01	0.47	8.44

Source: After Thorp (1936).

A. Pearl and Hermes Reefs.
B. Southeastern Florida.
C. The Bahamas.
D. Murray Island, Australia (after Vaughn, 1917).

dant, perhaps, than carbonate sands in Recent deposits. Reef rock is also represented, and, as at the present time, it forms only a small part of the total volume of carbonate rock. Many carbonate buildups of the geologic past, however, are not coral-algal reefs but appear to be, instead, carbonate mud mounds.

MINERALOGICAL AND CHEMICAL COMPOSITION

CARBONATE MINERALS

Inasmuch as limestones are carbonate rocks, the essential minerals are carbonates calcite, aragonite, and dolomite. Some limestones contain also a little ankerite and siderite. Of these, calcite and dolomite are the most abundant; less so are aragonite, ankerite, and siderite. All except aragonite are hexagonal, uniaxial, and negative and display excellent rhombohedral $(10\bar{1}1)$ cleavage. Dolomite, siderite, and ankerite have curved cleavage faces, a feature normally observed only in very large crystals. All have very high birefringence and variable relief.

Calcite and dolomite are the most difficult to distinguish from one another. The normal crystal habit of calcite is scalenohedral; seldom does it assume the rhombic form characteristic of dolomite. Except for certain vein and geode fillings, and the rare "sand crystals," calcite rarely shows any euhedral outlines but forms, instead, an anhedral mosaic. In oolites and in some precipitated cements it may have a radial or fibrous habit.

The calcite of limestones is both an original (or primary) constituent and a diagenetic (or secondary) mineral. Rock-building organisms use both calcite and aragonite in their skeletal structures (see pp. 331–332). Some skeletal materials are aragonitic, some are exclusively calcitic, and some are partly aragonite and partly calcite (Lowenstam, 1963). The calcites of limestones, with few exceptions, are quite pure $CaCO_3$ and are relatively free of iron and magnesium. On the other hand, hard parts of some invertebrate skeletal structures contain a significant proportion of $MgCO_3$ in solid solution—the so-called "high magnesian calcites." Some contain up to 18 percent of $MgCO_3$ (Chave, 1954a). These calcites are metastable, and those from Mesozoic and older rocks are now virtually free of magnesium (Chave, 1954b). Only rarely does metastable magnesian calcite persist over long geologic time.

Several generations of calcite are apt to be present in many limestones—some as a primary constituent of shelly debris, some as a recrystallization product of aragonite, some as a precipitated cement. These can be distinguished from one another by their textural geometry. They may also differ slightly in their composition; two generations of cement may differ slightly in their iron content. These compositional differences can be detected by microprobe analyses or more readily by use of stains (Lindholm and Finkelman, 1973).

Dolomite is intimately associated with calcite in some limestones. It is commonly difficult to distinguish between these two minerals. The chief differences are summarized in Table 10-3. Commonly its rhombohedral habit will alert the petrographer to the presence of dolomite, but this difference is not an infallible criterion. Staining must be resorted to for proper discrimination (Friedman, 1971). Dolomite is not commonly primary; it is in most cases a replacement of calcite or aragonite. No shell structures are originally dolomitic; they are a product of post-depositional replacement. Scattered dolomite rhombs that transect primary structures of the rock may be attributable to the magnesium released by breakdown of metastable high-magnesian calcites. Where the rock as a whole is a mosaic of dolomite crystals, introduction of magnesium is implied. Some dolomite is quite iron-rich (ferrandolomite); this fact makes possible the staining techniques and also the generally buff-weathering character of dolomitic rocks in the field. In many cases the ferrous iron content of dolomite is shown by staining to be zonally distributed in rhombs.

Aragonite is a polymorph of calcium carbonate. But it is orthorhombic and differs therefore in its optical and other physical

TABLE 10-3. Chief differences between calcite and dolomite

	Calcite	Dolomite
Crystal habit (in rocks)	Anhedral; rarely rhombohedral	Rhombohedral; may be zoned
Refractive indices	O 1.658 E 1.486	O 1.680 E 1.501
On cleavage face	1.566	1.588
Birefringence	0.172	0.179
Specific gravity	2.71	2.87
Solubility in acid	Readily soluble in cold dilute acid	Very slowly soluble in cold dilute acid
Staining[a]	Readily takes a silver chromate stain	Not stained by silver chromate
Weathering	Does not become buff or pink	Weathers buff or pink because of small amount $FeCO_3$ present

[a] See Friedman (1971) for use of stains.

properties. It is the chief constituent of pelecypod and gastropod shells and of some corals. It is also the form taken by chemically precipitated carbonate. Present-day carbonate muds consist mainly of minute aragonite needles; modern calcareous ooids are also aragonitic, with the aragonite deposited as tangentially oriented crystals in concentric layers. Because aragonite is unstable, it is found only in recent materials. Even the aragonite of modern shells may change into calcite in the course of only a few years (Lowenstam, 1954, p. 288). The visible effect of this transformation is loss of internal structure and formation of an anhedral crystalline mosaic. Aragonitic ooids undergo a similar recrystallization and may be converted to a mosaic of micritic calcite with loss of much or all of the original internal structure. Aragonite, unlike calcite, does not take magnesium carbonate into solid solution, and hence aragonitic shells are largely magnesium free (Table 10-5). They seldom contain as much as 1 percent $MgCO_3$ (Chave, 1954a). They may, on the other hand, contain appreciable strontium.

Siderite is a rare and generally very minor constituent of some limestones. As noted, ferrous iron is commonly present in the mineral dolomite, but in a few cases, as in those limestones associated with sideritic ironstones, it occurs as scattered sideritic rhombs. Slight oxidation results in breakdown of siderite, which is then readily detected by heavy iron-oxide stains along the cleavage and grain boundaries.

SILICA AND SILICATES

Although some limestones are very pure and consist almost wholly of one or more carbonate minerals, others have a smaller or larger content of other minerals. *Silica* in various forms is a common minor constituent of many limestones. Most generally it is present as chalcedony. It may be disseminated throughout the rock; more commonly it is segregated into large nodules—the chert or flint nodules of many limestones and dolomites. If the chalcedonic silica is fine-grained and disseminated, it is difficult to detect in thin section. It may occur also as small spherulites (Fig. 10-3) or as a filling between the dolomite rhombs of some dolomitized rocks. In the latter case solution of the dolomite leaves a spongy porous residue of "dolocastic chert" (Ireland, 1936; McQueen, 1931). Such material is most conspicuous in the acid-insoluble residues of such rocks. Silica also occurs as small quartz euhedra which transect primary structures and are therefore authigenic. Many limestones and dolomites, especially calcarenites, contain detrital quartz grains. In many cases these are overgrown with secondary quartz and form a large regenerated quartz crystal (Fig. 10-4).

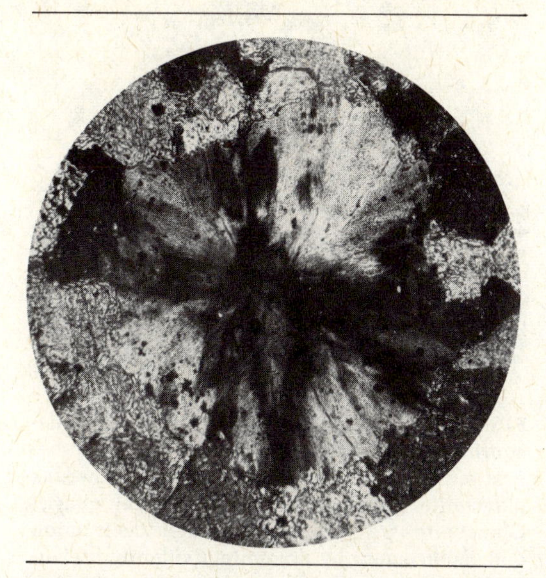

FIG. 10-3. Spherulite of chalcedony in Niagaran Dolomite (Silurian), Thornton, Illinois. Crossed nicols, ×160. Note irregular outline of spherulite, lack of concentric structure, and radial arrangement of fibers of chalcedony. The latter is shown by the crude pseudo-uniaxial cross which appears when the nicols are crossed.

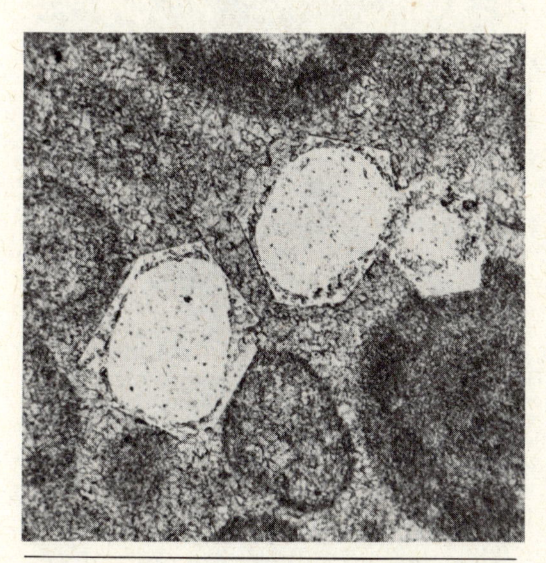

FIG. 10-4. Arbuckle Limestone (Cambrian), Carter County, Oklahoma. Ordinary light, ×80. Originally a calcareous oolite. Quartz grains have been enlarged by the addition of quartz which has replaced both ooids and matrix. Note carbonate inclusions in the overgrowth at boundary of original detrital grains.

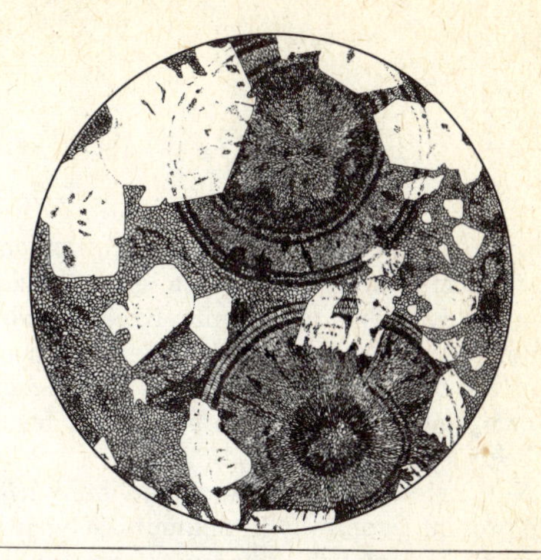

FIG. 10-5. Feldspathized oolite, Siyeh Limestone (Precambrian), Glacier National Park, Montana. (Drawing based on Hatch, Rastall, and Black, 1938, Fig. 58.)

Feldspar, like some quartz, occurs as authigenic euhedra as a very minor constituent in many limestones and dolomites, although in rare cases feldspar may form as much as 40 percent of the whole rock (Fig. 10-5).

Clay minerals are one of the most common contaminants of the carbonate rocks (Robbins and Keller, 1952). Clay is not conspicuous in thin section because of its fine grain and disseminated state, but it is readily apparent in the acid-insoluble residues prepared from many limestones. The nature of the clay mineral is determined best by x-ray diffraction. In Illinois limestones and dolomites illite is the most common; kaolinite is found in some, but no montmorillonite has been reported (Grim, Lamar, and Bradley, 1937). The predominance of illite in carbonate rocks was confirmed by Weaver (1959, p. 172), although other clay minerals are also commonly present.

MINOR CONSTITUENTS

Other minor constituents of limestones include glauconite, collophane, and pyrite. Glauconite occurs as larger rounded granules and locally may be an abundant component (Fig. 10-6). Collophane occurs primarily as phosphatic skeletal debris: shells of linguloid brachiopods, fish bones, and similar materials. Although a subordinate constituent, it too may be locally important and the limestone properly described as phosphatic. Pyrite is nearly ubiquitous as small scattered grains, which upon oxidation are converted to limonite. Exceptionally a limestone is rich enough in pyrite to warrant the description pyritic. The pyrite may occur principally along the borders of fossil debris.

Not uncommonly gypsum and anhydrite are present, especially in dolomites. Gypsum occurs as isolated euhedra, a few millimeters in length. In places it is present only as a crystal mold; in other places it is replaced by calcite. Gypsum also occurs as isolated subspherical to ovoid nodules from 1 mm to 3 cm in diameter (Fig. 10-7). They may be sparsely distributed or crowded together. Commonly they are strung out in layers parallel to the bedding. These nodules are made up of a swirl of tiny grains. In some dolomites the sulfates are anhydrite rather than gypsum.

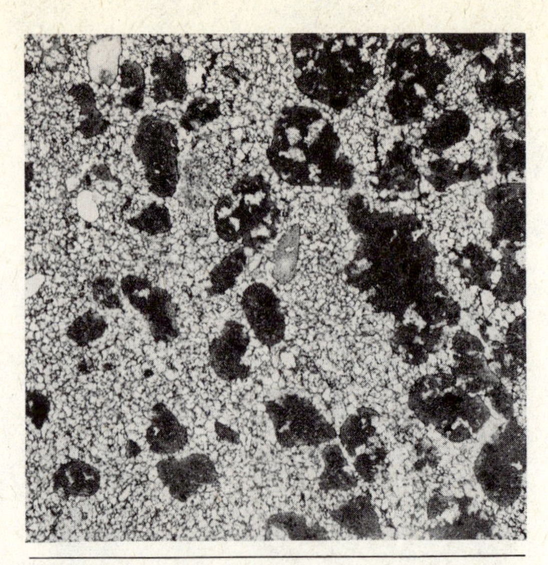

FIG. 10-6. Bonneterre Dolomite (Cambrian), Missouri. Ordinary light, ×15. Fine-grained anhedral mosaic of dolomite in which are numerous glauconite grains and a little detrital quartz. Glauconite granules show partial destruction and replacement by dolomite. Evidently a glauconitic, arenaceous calcarenite which has been profoundly dolomitized.

FIG. 10-7. Gypsum nodules strung out parallel to bedding in sandy dolomite, Bellerophon Formation (Upper Permian), Italy. Scale in centimeters. (From Bosellini and Hardie, 1973, Sedimentology. v. 20, no. 1, Fig. 11a. By permission of Blackwell Scientific Publication.)

CHEMICAL COMPOSITION

The chemical composition of limestones, as might be expected, reflects rather closely their mineral composition. Because limestones are primarily calcite, the content of both CaO and CO_2 is extremely high (Table 10-4), forming in some cases more than 95 percent of the whole. Other constituents which commonly become important include MgO, which, if it exceeds 1 or 2 percent, probably indicates the presence of the mineral dolomite. Rodgers (1954) would restrict the term *magnesian limestone* to those rocks containing several percent MgO but without the mineral dolomite. Except in calcite deposited by living organisms, no more than 2 percent $MgCO_3$ (less than 1 percent MgO) in solid solution is possible. The solid solution is unstable, so that fossil calcite invariably contains less than 1 or 2 percent $MgCO_3$ (Chave, 1952).

The magnesium content is a function of both the magnesium content of skeletal debris in the limestone and of post-depositional additions. Normally magnesium is added by the dolomitization process, although some ancient rocks contain less magnesium than that demanded by organic debris, suggesting that the magnesium released by the breakdown of the high-magnesium calcite has been lost (Chave, 1954b,

TABLE 10-4. Chemical analyses of representative limestones

Constituent	A	B	C	D	E	F	G	H
SiO_2	5.19	0.70	7.41	2.55	1.15	13.80	2.38	0.09
TiO_2	0.06	—	0.14	0.02	—	—	—	—
Al_2O_3	0.81	0.68	1.55	0.23	0.45	7.00	1.57[e]	
Fe_2O_3		0.08	0.70	0.02	—	4.55	0.56	
	0.54							0.11
FeO		—	1.20	0.18	0.26	—	—	
MnO	0.05	—	0.15	0.04	—	0.29	—	—
MgO	7.90	0.59	2.70	7.07	0.56	1.32	0.59	0.35
CaO	42.61	54.54	45.44	45.65	53.80	38.35	52.48	55.37
Na_2O	0.05	0.16	0.15	0.01		2.61	—	—
					0.07			
K_2O	0.33	none	0.25	0.03		0.86	—	0.04
H_2O+	0.56[a]	—	0.38	0.05	0.69	—	—	
								0.32
H_2O-	0.21	—	0.30	0.18	0.23	—	—	
P_2O_5	0.04	—	0.16	0.04	—	0.25	none	—
CO_2	41.58	42.90	39.27	43.60	42.69	31.31	41.85[f]	43.11
SO_3	0.05	0.06	0.02	0.03	none	—	—	0.44
Cl	0.02	—	—	—	—	—	—	—
S	0.09	0.25	0.25[b]	0.30[b]	—	none	none	—
BaO	—	—	none	0.01[d]	—	—	—	—
SrO	—	—	none	0.01	—	—	—	—
Li_2O	trace	—	—	—	—	—	—	—
Organic	—	trace	0.09[c]	0.04[c]	—	—	—	0.17
Total	100.09	99.96	100.16	100.04	99.90	100.34	98.84	100.10

[a] Includes organic matter.
[b] Calculated as pyrites.
[c] Organic carbon.
[d] Constituent does not exceed figure given.
[e] Includes TiO_2.
[f] Calculated from MgO and CaO ("ignition loss," 40°–1000°C, given as 40.95).

A. Composite analysis of 345 limestones. H. N. Stokes, analyst (Clarke, 1924, p. 564).
B. "Indiana limestone," Salem (Mississippian). A. W. Epperson, analyst (Loughlin, 1929, p. 150).
C. Crystalline, crinoidal limestone, Brassfield (Silurian), Ohio. Downs Schaaf, analyst (Stout, 1941, p. 77).
D. Dolomitic limestone, Monroe Formation (Devonian), Ohio. Downs Schaaf, analyst (Stout, 1941, p. 132).
E. Lithographic limestone, Solenhofen, Bavaria. G. Steiger, analyst (Clarke, 1924), p. 564).
F. Argillaceous limestone (natural cement rock), Lower Freeport Limestone (Pennsylvanian), Columbiana County, Ohio. P. J. Demarest, analyst (Stout and Lamborn, 1924, p. 195).
G. Chalk, Fort Hays (Cretaceous), Ellis County, Kansas (Runnels and Dobins, 1949).
H. Travertine, Mammoth Hot Springs, Yellowstone, Wyoming. F. A. Gooch, analyst (Clarke, 1924, p. 204).

p. 598). Although the average limestone contains 7.9 percent MgO, equivalent to 16.5 percent $MgCO_3$, most limestones contain either much less magnesia or much more (Fig. 10-8). Most limestones have less than 4 percent or over 40 percent $MgCO_3$ equivalent; rocks with intermediate composition are uncommon. Clarke's average limestone (Clarke, 1924, p. 30) is therefore an average carbonate rock—not a limestone—and includes dolomites as well as limestones. These chemical data reflect the simple fact

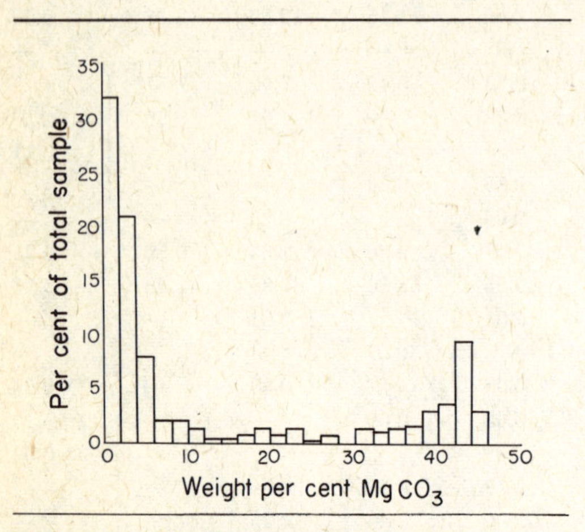

FIG. 10-8. Distribution of magnesium in 317 quarried Paleozoic limestones from Illinois. (Based on data by Krey and Lamar, 1925.)

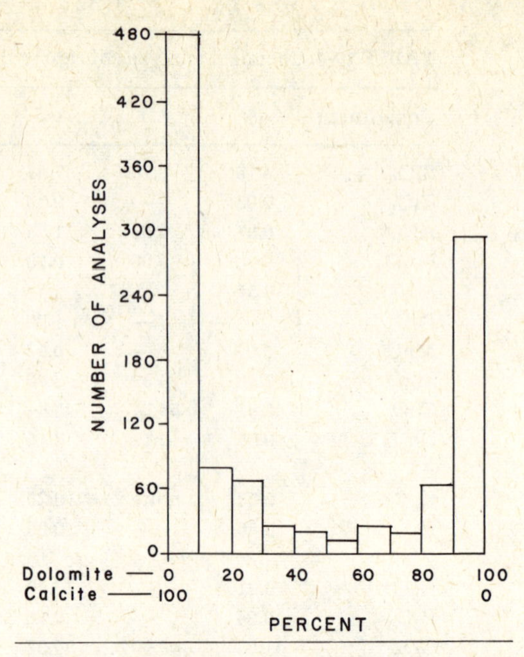

FIG. 10-9. Computed percentages of calcite and dolomite for 1148 analyses of carbonate rocks from North America (Steidtmann, 1917). (Graf, 1960, Fig. 4.)

that normal carbonate rocks are dominantly calcitic or dominantly dolomitic—calcite–dolomite mixtures are rare (Fig. 10-9). The magnesium content of modern sediments, on the other hand, displays no such bimodal distribution (Fig. 10-10).

Silica, if present in significant amounts, indicates the presence of much noncarbonate detritus such as clay, silt, or sand or the presence of siliceous spicules or chert (Table 10-4C and D). If alumina is high also, silica is probably a constituent of contaminating shaly matter. With the advent of such argillaceous materials, potassium (K_2O) and combined water (H_2O+) are also higher than normal (Table 10-4 col. F). Exceptionally a limestone may be notably rich in a minor constituent such as phosphorus, iron oxide, or sulfide sulfur. The presence of significant sulfate sulfur suggests the presence of anhydrite or gypsum. These are most apt to appear in dolomites.

Inasmuch as many limestones are composed of skeletal structures or debris derived therefrom, the composition of such rocks is an expression of the bulk composition of skeletal elements. The composition of shell and other hard parts varies with the nature of the organism and the conditions under which it lived. The chemical composition of the inorganic constituents has been investigated by Clarke and

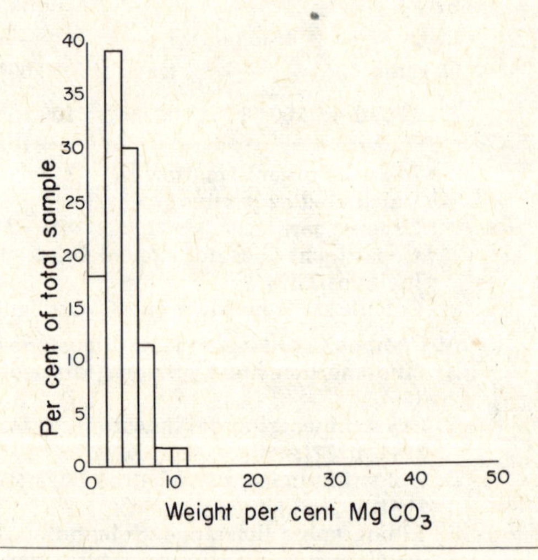

FIG. 10-10. Distribution of magnesium in modern calcareous sediments. (After Chave, 1954b.)

Wheeler (1922) and Chave (1954a). As seen in Table 10-5, even though each phylogenetic group shows a range in composition, there are differences between groups. Calcareous algae, for example, are notably richer in $MgCO_3$ than are mollusks; the carapaces of crustaceans are notably phosphatic; and so forth. Chave (1952, 1954a, 1954b) has made a special study of the magnesium content of carbonates of marine invertebrates. He found magnesium to decrease with increasing level of organization of the organism, those of higher phyla being less rich in MgO. The content of magnesium was also a function of the temperature of formation and was generally higher in shells living in warmer waters. The most important factor was shell mineralogy. Aragonitic shells were poor in magnesia; calcitic shells were commonly rich in this constituent.

It is clear from inspection of Chave's data that simple accumulation of skeletal materials will not explain the distribution of $MgCO_3$ in older limestones. High-magnesium-bearing rocks must be either inorganic precipitates of dolomite or skeletal accumulations which have been enriched in this substance. The most common type of limestone, that is, one with less than 4 percent $MgCO_3$, must be either an inorganic precipitate of calcite (or aragonite) or an accumulation of organic detritus of only those organisms which secrete magnesium-poor skeletons. Or the material, if produced by other organisms, has lost its magnesia. The last possibility seems proved in those cases where organic debris is of a type known to be magnesium-rich at present—such as crinoidal limestone (Table 10-4C, for example, with crinoids, Table 10-5). These relationships have an important bearing on the dolomite problem (see p. 369).

ISOTOPIC COMPOSITION

In recent years there has been a great deal of interest in the isotopic composition of limestones (Graf, 1960d). The ratio ^{18}O–^{16}O in marine calcium carbonate is unlike that

TABLE 10-5. Inorganic constituents of marine invertebrates

Class[a]	Number of Analyses	$CaCO_3$	$MgCO_3$	$Ca_3P_2O_8$	SiO_2	$(Al, Fe)_2O_3$	$CaSO_4$
Foraminiferans	7	77.0–90.1	1.8–11.2 avg. 8.2	T[b]	T–15.3	T–4.0	—
Calcareous sponges	4	71.1–85.0	4.6–14.1	?–10.0	T–7.8	1–5.7	—
Corals							
Madreporaria	30	97.6–99.7	0.1–0.8	0–T	0–1.2	0–0.6	—
Alcyonaria	22	73.0–98.9	0.3–15.7 avg. 11.1	T–8.3	0–1.7	T–1.0	0.5–5.4
Echinoderms							
Crinoids	24	83.1–91.5	7.9–13.7 avg. 10.8	T–1.1	T–2.0	0.1–1.4	—
Echinoids	14	77.9–91.7	6.0–13.8	T–1.8	T–9.9	0.1–5.2	T–2.6
Bryozoans	13	63.3–96.9	0.2–11.1	T–2.9	0.2–16.7	0.1–2.2	—
Brachiopods							
Calcareous	5	88.6–98.6	0.5–8.6	T–0.6	0.1–0.5	T–0.5	—
Phosphatic	4	?–8.3	1.7–6.7	74.7–91.7	0.5–0.9	0.3–1.2	?–8.4
Mollusks							
Pelecypods	11	98.6–99.8	0–1.0	T–0.4	0–0.4	T–0.5	—
Gastropods	20	96.6–99.9	0–1.8	T–0.8	0–2.2	T–1.9	—
Cephalopods	3	93.8–99.5	0.2–6.0	T	0–0.2	T–0.1	—
Crustaceans	13	28.6–82.6	3.6–16.0	6.6–49.6	0–1.1	T–8.8	—
Algae, calcareous	16	73.6–88.1	10.9–25.2 avg. 17.4	T–0.4	T–3.5	T–1.6	—

[a] Clarke and Wheeler also give analyses of hydroids, annelids, starfishes, ophiurans, scaphopods, and amphineurans—all of which make minor contributions to the sedimentary rocks.
[b] T stands for "trace," or less than 0.1 percent.

Source: From Clarke and Wheeler (1922).

of freshwater and hence can be used to discriminate between marine cements and meteoric cements. The ratio in marine waters is a function of temperature and hence conceivably can be used as a "geologic thermometer" (Epstein et al., 1951, 1953). It is also presumably modified by living matter so that the carbonate precipitated by organisms is different from that deposited inorganically (Lowenstam and Epstein, 1957).

Carbon isotopes, especially ^{14}C, the radioisotope, makes it possible to determine absolute age in years of those carbonates under 20,000 years old. This becomes important in studying rates of deposition, ooid growth, and so forth. The ^{13}C–^{12}C ratio seems to be different in freshwater and marine carbonates.

TEXTURES AND STRUCTURES OF LIMESTONES

INTRODUCTION

Inasmuch as the constituents of some limestones are emplaced in the rock fabric as solid particles by waves and currents, they differ not at all in their textures and structures from ordinary "clastic" sediments. They have the same rock fabric, namely a framework–cement relationship. And they show similar hydrodynamic or current structures such as ripple marking and cross-bedding. Other carbonate rocks form more or less *in situ,* often in a currentless environment. These deposits show no sorting, nor any other evidence of current activity, and are without bedding or at best have a poorly developed stratification. Exceptionally, though, they may have a well-defined algal or "growth" bedding. A brief, well-illustrated treatment of the textures of carbonate rocks is that of Bosellini (1964). The subject is also covered in the larger works of Bathurst (1971, pp. 77–92) and others.

Limestones, whether current-deposited or formed *in situ,* are made up of large complex grains or "allochems"; "micrite," a very fine-grained carbonate which commonly is a matrix for larger elements; and "spar," a coarsely crystalline calcite which in many limestones is a cement binding allochems together. Before we can discuss the two major groups of limestone we should take a closer look at their components.

ALLOCHEMS ("FRAMEWORK" ELEMENTS)

In the case of a normal sandstone, framework elements are debris derived from the wasting away of a source land. In carbonate sands, however, grains are produced chemically or biochemically *within* the basin of accumulation. They are *intrabasinal* in origin. To these grains Folk (1959, 1962) has applied the term *allochems.* These are not ordinary chemical precipitates as the chemist thinks of them but are instead complex grains that have a higher organization. The four principal allochem types include oolites, skeletal structures and debris, intraclasts, and pellets. The distinction between intraclasts and pellets is less clear. Moreover, some fossil and even oolitic material may be altered (micritized) so that it, too, is difficult to distinguish from the other two. In such ambiguous cases, grains may be termed *peloids* (McKee and Gutschick, 1969, p. 101).

Oolites Oolites and the problems of oolite formation have been covered in some detail in Chapter 3 (p. 83); see also Bathurst (1971, pp. 77–84). We are concerned here only with carbonate ooids, either aragonite or calcite. Aragonitic ooids, of course, are confined largely to Recent deposits; they have inverted to calcite in older rocks. Care should be taken to discriminate between concentrically structured ooids and rarer spherulites with radial structure. Some ooids display both radial and concentric structure. Spherulites show only a radial habit. Ooids are transported bodies; spherulites form in *situ* and, if close-packed, tend to assume a polyhedral form.

Fossils Skeletal materials, or fossils, are exceedingly abundant and may occur scattered through any type of limestone and, of course, in other sedimentary rocks as well. They may be winnowed out, may be transported, and may form a current-deposited accumulation in which they are the dominant framework constituent. Different organisms secrete calcareous skeletal structures, and it is of some importance to discriminate between them. This discrimination is relatively easy if skeletal materials are unbroken but is more difficult if they consist only of small bits and pieces. Several recent publications are devoted to the task of identification of this material in thin section (Horowitz and Potter, 1971; Majewske, 1969; Bathurst, 1971, pp. 1–76). The "petrology" of fossils is summarized below.

Calcareous algae deposit either aragonite or calcite. Certain genera (*Halimeda*) are aragonitic; others (*Lithothamnium*) are calcitic. The calcite is generally a high-magnesian calcite (Table 10-5).

Algae contribute large and important quantities of lime carbonate to present-day reefs and therefore to clastic debris derived from such reefs (Table 10-2). Thorp (1936) reports 20 to 50 percent algal material in modern reef-rock and 20 to 25 percent identifiable algal detritus in modern lime-bearing sediments of the Florida-Bahama region. Algae also made important contributions to ancient limestone deposits (Wolf, 1962). Some algae, however, are not believed to be agents of lime deposition but are, instead, agents for entrapment of fine carbonate particles. These algae are thus sediment-binding—not sediment precipitating. Many of the so-called algal structures—the stromatolites—found in Precambrian to Recent rocks are believed to be produced by blue-green algae of the sediment-binding type. (These algal structures are described in Chapter 4.)

Lime-secreting algae give rise to crustose or nodular growths or to erect and commonly branching forms. Calcification varies from weak to substantial. Erect and articulated forms disintegrate into fragments which may be redeposited as framework components of carbonate sands. The microstructure of the lime-depositing algae varies somewhat from one group to another. Some display a layered, rectangular microstructure which bears a superficial resemblance to that seen in some stromatoporoids and corals. The cell-like structure, however, is on a finer scale, commonly less than 0.1 mm. Others show an internal fine tubelike structure. The tubes, commonly less than 0.1 mm in diameter, may be intertwined but are generally perpendicular to the walls of the fragment.

Although both calcareous and agglutinated *foraminifera* are found in sediments, calcareous forms are more commonly seen in thin section. They commonly are multichambered and whole and hence are readily recognized under the microscope, although in some cases the chambers are detached. They are generally under 1 mm in maximum diameter. In some genera the tests are composed of calcite; in others they are aragonitic. In older rocks the aragonite has been converted to calcite. The foraminifers are important rock builders and their remains may constitute the bulk of the rock, as in the case of certain fusilinid and nummulitic limestones.

The interior of a foraminiferal test may be filled with crystalline calcite, in some cases with a radial arrangement of fibers normal to the walls, so that the fibers present a black cross under crossed nicols. Glauconite is also deposited within the foraminiferal test; by removal of the latter the filling is left in the form of an interior mold or "cast." According to some writers this is the origin of glauconite granules in many sediments.

Spicules of siliceous *sponges* are a constituent of some sediments. They are most prominent in some Paleozoic and Mesozoic cherts where they appear as clear, slender, curved structures of chalcedonic silica. They show a great variety of rayed shapes, both ornamented and unornamented, and range from less than 1 mm to over 1 cm in length but usually under 1 mm in transverse section. Some sponges have calcareous spicules which were deposited as single crystals of high-magnesium calcite.

Many *corals* consist of little fibers set perpendicular to the walls and septal surfaces and in some cases of granules of aragonite. Most Paleozoic corals were calcitic, and calcite occurs also in some modern forms, notably those inhabiting the deep sea. The modern reef-building forms are chiefly aragonite. In general the aragonitic forms have been later converted to calcite with consequent loss of original structural details.

The hard parts of *echinoderms* are most singular in that each plate or skeletal element is a single crystal of calcite. Larger ones clearly show the calcite cleavage to the unaided eye, and the limestone composed primarily of such remains accordingly has a marked crystalline appearance. Crinoidal limestones are commonly described as "crystalline limestones" and "encrinites" and in the building trade pass as "marbles." In most cases the ossicles and plates have been cemented with clear calcite in crystallographic and optical continuity with crinoid fragments. The original fragment is distinguished by a dusty area showing the usual circular or elliptical (in oblique section) outline with internal canals. The original fragment, however, is traversed by cleavage cracks that pass uninterrupted into a secondary cement. The fragment and adjoining cement extinguish as a unit between crossed nicols (Fig. 10-30). Usually echinoderms disarticulate rapidly so that one sees only scattered debris whose shapes are highly variable because of original growth or the plane sectioned. Their crystallographic unity is their most diagnostic feature.

Bryozoans are common in many limestones. Their chambered cell-like structure (zooecia) makes identification under the microscope easy, because their cell pattern is considerably coarser than that of calcareous algae. They may be either aragonite or calcite and consist of tangentially set fibers. Individual cells may be filled either with clear calcite or micritic mud or both.

Brachiopods, except phosphatic forms, are chiefly calcite. Their shells are built up of bundles of prisms, the prisms of each bundle being parallel and having a quadrangular cross section. The prisms do not have straight extinction. The bundles are arranged obliquely to the shell, and adjacent bundles abut upon one another and in part are interlocking. Brachiopods occur as disarticulated valves, 1 to 10 cm in maximum dimension, or as fragments, or in some cases entire with a sedimentary or sparry filling, not uncommonly both with geopetal fabric.

Mollusk shells are chiefly aragonite and hence appear as a mosaic of calcite in older rocks. Some exceptions, however, are noteworthy. Some pelecypod shells have an inner layer of aragonite that is protected by an outer layer of calcite. In some genera (notably *Ostrea*, *Pecten*, and *Inoceramus*) the whole shell consists of calcite arranged in two layers. The outer and principal layer has a prismatic structure in which the prisms, unlike those of brachiopods, are perpendicular to the shell surface. The inner pearly layer has a fine lamellar structure.

Similarly, most *gastropods* have an aragonite shell, but a few exhibit a two-layer structure consisting of an inner aragonitic layer covered by an outer calcitic layer. *Cephalopod* shells are aragonitic, though the aptychi of some ammonites were calcite. The guard of belemnites is calcite, with the calcite fibers set radially about an axis.

Of the *arthropods* only trilobite and ostracod debris is likely to be seen in thin section. The first yields small to large curved to hook-shaped fragments with fine calcite prisms normal to the surface of the fragments. Ostracods are smaller, many being under 1 mm, and display an ovoid bivalve form commonly with a sparry filling. Disarticulated valves and fragments are also common.

Intraclasts The term *intraclast* was introduced by Folk (1959) to designate fragments of penecontemporaneous, generally weakly cemented carbonate sediment that has been broken up and redeposited as a clast in a new framework (Fig. 10-27). As noted in Chapter 6 (p. 184), intraclasts of shale are common in some ordinary sandstones. We are concerned here, however, only with carbonate intraclasts, most of which seem to have been produced by erosion of a layer of semiconsolidated carbonate sediment. Hence many of them are flat; the largest are slablike, and elongated parallel to their bedding. They may show internal laminations parallel to their flat sides. They have been somewhat abraded and rounded.

They range from fine-sand size to the larger slablike pieces of intraformational (and perhaps "edgewise") limestone conglomerates (Fig. 6-26). Most are composed of microcrystalline carbonate ("lithographic" limestone). If these are small and well-rounded, they are difficult to distinguish from pellets.

The term *grapestone* (Illing, 1954) was applied to complex carbonate grains seemingly made of pellets and other grains weakly to firmly cemented by micritic aragonite found on the Bahama banks. These composite grains are not intraclasts in a strict sense because they are aggregates or built-up grains—not broken pieces of a previously cemented layer. Beales (1958) finds such grains to be a common constituent of some Paleozoic limestones of Canada and interprets them as the result of accretion, not attrition. They are indeed intraformational but not intraclastic.

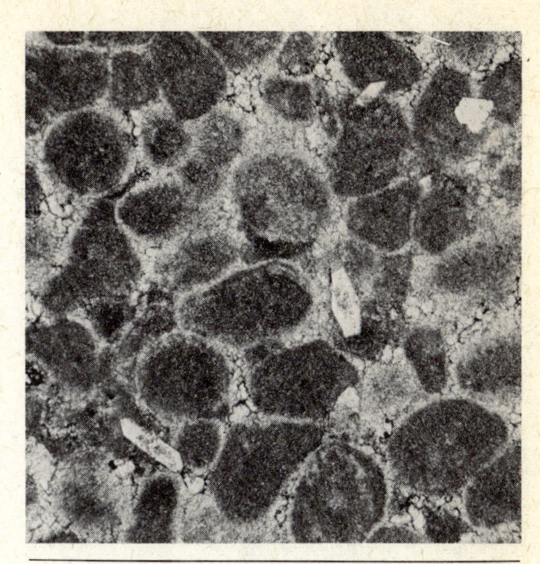

FIG. 10-11. Pellet limestone, Trenton Limestone (Ordovician), Trenton Falls, New York. Ordinary light, ×58.

Pellets Pellets are small spherical to ellipsoidal ovoid bodies or aggregates of microcrystalline calcite, devoid of any internal structure (Fig. 10-11). In any given rock they tend to be all of the same size and shape. They range from 0.03 to 0.15 mm in size; most are between 0.05 and 0.10 mm. They closely resemble the fecal pellets (p. 88) of modern sediments (Moore, 1939; Dapples, 1942). They are distinguished from oolites by their lack of radial or concentric structure, from fossils by their lack of internal structures, and from intraclasts by their uniformity of size and shape. They may be confused with very small intraclasts.

Some diagenetic processes lead to formation of small micritic patches set in a coarser sparry cement, the *grumeleuse* fabric of Cayeux (1935, p. 271; see also Bonet, 1952, Fig. 22). These residual clots of micritic limestone superficially resemble pellets. They differ in being less uniform in size and shape and have vague, ill-defined boundaries.

Pellets are difficult to see in hand specimens, and even under the binocular microscope. Many limestones described in the field as pellet limestones are instead intraclastic limestones; true pellet limestone consisting largely of pellets looks homogeneous and would be designated *calcilutite* or *micritic* in the field.

MICRITE

Many limestones are very fine grained (aphanitic). Grabau long ago (1904) applied the term *calcilutite* to these carbonate rocks with grains 50 microns or less in diameter. These were, however, deposits in which grains were emplaced by currents; those formed by precipitation were called *calcipulverites*. This distinction is very difficult to make in rocks of such fine grain.

This ambiguity can be avoided by use of the term *micrite*, a term used by Folk (1959) as a contraction of "microcrystalline calcite." Folk defined it as "clay-sized carbonate," that is, materials 1 to 4 microns in size. Material 5 to 10 (or even up to 50 microns) was termed *microspar*. But as Bathurst (1971, p. 89) notes, the break between the clay and silt grade, 4 microns, of the Wentworth-Udden scale, which was designed for terrigenous sediments, has little or no relevance to carbonates. The actual size range of most fine-grained ("lithographic") limestones crosses this boundary so that the distinction between micrite and microspar is not particularly meaningful. There is a tendency to call all microcrystalline carbonate micrite, even if it is aragonite rather than calcite. If dolomite, it may be called dolomicrite. The term is here used as originally defined, that is, for microcrystalline calcium carbonate but without the size and mineralogical restrictions placed on the term by Folk.

In the hand specimen micrite is a dull, ultrafine-grained material ranging from white through gray to nearly black. Under the microscope the constituent grains are more or less equant and irregularly rounded.

The origin of micrite is far from clear. It may be formed by inversion of aragonite to calcite, with the original aragonitic ooze either a chemical precipitate or a product formed by the release of minute argonitic needles by degradation of algal tissue in which they were precipitated. Some modern carbonate oozes are biochemical, rich in coccoliths (Fischer, Honjo, and Garrison, 1967). Some also may be the finest product of shell attrition and thus mechanical in origin. Some clearly are a diagenetic product formed by a grain diminution process—micritization. Partially to wholly micritized ooids, fossils, and like materials are common. The problem is explored further in the section on calcilutites.

SPARRY CALCITE ("SPAR")

Many limestones contain coarsely crystalline calcite. This material, often designated "spar," is a clear, coarsely crystallized material showing well-defined grain boundaries and cleavage traces.

Its occurrence is highly varied. Most commonly it is a pore-filling cement. In the absence of a micritic matrix, it fills the pores of the framework whose elements are ooids, fossils, pellets, and the like. The size of spar crystals depends on the coarseness of the framework net and the intervening pores; in most limestones it averages 0.02 to 0.10 mm. Larger crystals, 1.0 mm or more, are also present in larger voids—voids not intergranular in nature. These larger voids may be in part sediment-filled or in part spar-filled, in which case the sediment forms a floor with the spar occupying the upper part of the cavity and thus creating a *geopeptal fabric* (Fig. 10-12). Spar is also formed as a result of a crystallization of finer-grained, micritic carbonate.

A hybrid fabric involving both micritic and sparry calcite is the *grumose texture* (French, *structure grumeleuse*) described by Cayeux (1935, p. 272). This texture is characterized by clots or clusters of clots of micritic carbonate surrounded by a matrix of sparry calcite (Fig. 10-13). The micritic clots are small (0.05 mm) and have fuzzy rather than sharp boundaries. They are ascribed to incomplete conversion of a homogeneous carbonate mud to coarse sparry

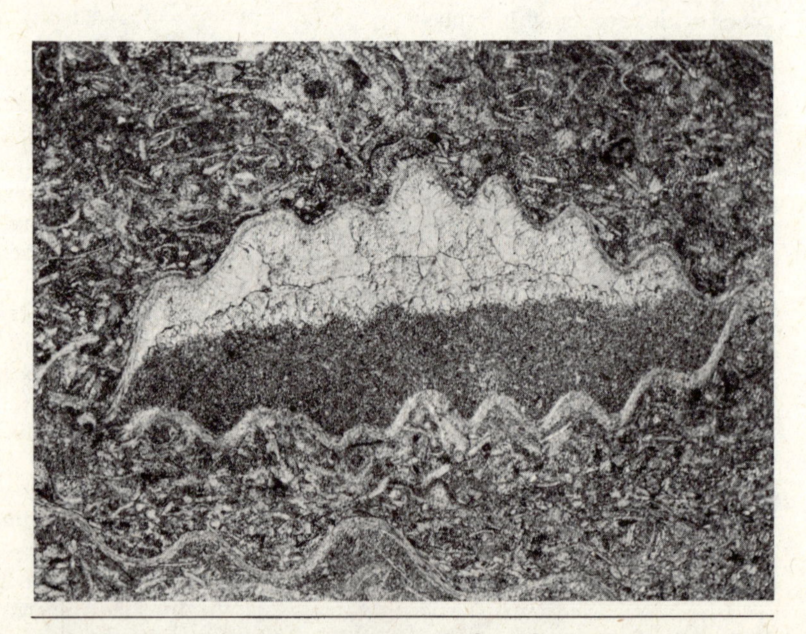

FIG. 10-12. Geopetal fabric, *Zygospira* beds, Lebanon Limestone, Oregonia, Warren County, Ohio. Ordinary light, ×22.

calcite—the clots are relicts of the original material (Cayeux, 1935, p. 272; Folk, 1959, p. 7). The clots, however, bear some resemblance to pellets; it may be that the grumose fabric is caused by some diagenetic recrystallization or other process that blurs the boundaries of the pellets (Bathurst, 1971, p. 513). Bonet (1952, p. 172), however, thought he could trace the recrystallization of a carbonate mud beginning with scattered patches of coarse spar to a network formed by partial coalescence of such patches to the grumose stage with its small islandlike residuals of the original micrite. The problem of distinguishing between these several interpretations of the grumose texture has been reviewed by Bathurst (1959a).

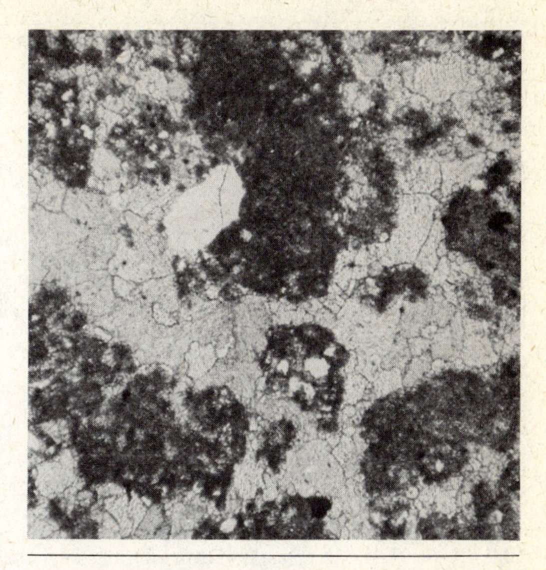

FIG. 10-13. Grumose limestone, El Abra Formation (Cretaceous), Canyon of El Abra, Mexico. Ordinary light, ×32. The dark small islands are remains of the original cryptocrystalline structure; clear portions are recrystallized. (From Bonet, 1952, Bol. Asoc. Mexicana Geol. Petrol., v. 4, Fig. 22.)

NONCARBONATE COMPONENTS

Noncarbonate components of limestones and dolomites include chalcedony as spherulites and filling of intercrystalline voids in dolomite, glauconite as granules and as partial replacement of fossil structures, phosphatic shells and pellets, pyrite spherulites and grains, and crystal euhedra of quartz, of feldspar, and of anhydrite or gypsum crystals. In many cases sulfates are present as calcitized pseudomorphs. Replacement patches and nodular masses of anhydrite and gypsum are also seen in some limestones.

POROSITY OF CARBONATE ROCKS

Most sedimentary carbonate rocks have very little porosity, but those that are porous have great economic significance. About one-half of the world's known reserves of oil and gas are contained in porous limestone and dolomite. The carbonate sediment as deposited may have had a large porosity which has since been greatly reduced or obliterated by post-depositional processes. Some porosity, however, may have been formed by other post-depositional processes. It is our task to consider the nature of porosity of carbonate rocks. The process of pore elimination, or creation, by diagenetic processes is covered in the section on the diagenesis of the carbonate rocks.

One of the most comprehensive reviews of the subject of carbonate porosity is that of Choquette and Pray (1970; see also Bosellini, 1964). Basic porosity falls into two classes: that related to fabric or textures of the rock and that independent of it (Fig. 10-14). In the first group are pores present at the time of sediment accumulation and those formed later but fabric controlled. Initial porosities are of several types. A carbonate sand, like any other sand with a grain-supported framework, has a large (30 to 40 percent) initial porosity. Some shell accumulations, those with a "potato-chip" fabric, may have as much as 80 percent porosity; this is *interparticle* porosity. In limestones in particular are other initial pores formed at time of deposition. Relatively large openings exist in reefs, for example. In such a *growth-framework* porosity, openings are commonly partially infilled by reef detritus and later by incrustations of crystalline carbonate. Body cavities, as within crinoid calyxes, corals, bryozoans, or other intact skeletal structures, are *intraparticle* pores, which also may be the sites of internal sedimentation and crystal filling.

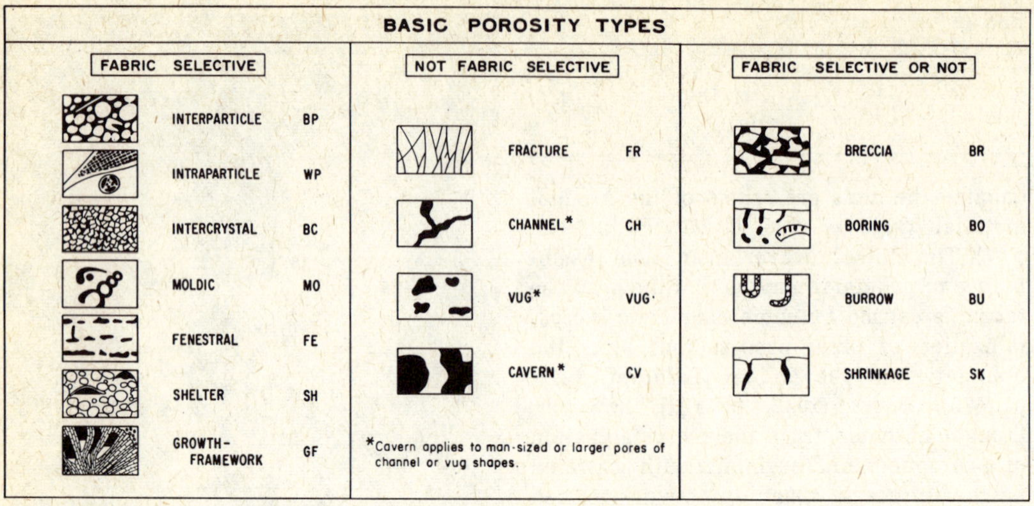

BASIC POROSITY TYPES					
FABRIC SELECTIVE		NOT FABRIC SELECTIVE		FABRIC SELECTIVE OR NOT	
INTERPARTICLE	BP	FRACTURE	FR	BRECCIA	BR
INTRAPARTICLE	WP	CHANNEL*	CH	BORING	BO
INTERCRYSTAL	BC	VUG*	VUG	BURROW	BU
MOLDIC	MO	CAVERN*	CV	SHRINKAGE	SK
FENESTRAL	FE				
SHELTER	SH				
GROWTH-FRAMEWORK	GF				

*Cavern applies to man-sized or larger pores of channel or vug shapes.

FIG. 10-14. Geologic classification of pores in carbonate rocks. (From Choquette and Pray, 1970, Bull. Amer. Assoc. Petrol. Geol., vol. 54, no. 2, Fig. 2.)

Even single valves of mollusks or brachiopods, if oriented convex upward, create cavities—the *shelter* pores of Choquette and Pray. Other cavities are formed by solution of shells of destruction of other original components of the rock, creating *moldic* porosity and, in other cases, *fenestral* porosity. Coarse dolomites may show *intercrystal* porosity caused primarily by solution of nonreplaced calcite (Murray, 1960).

The second major porosity group includes pores unrelated to primary fabric and includes *fracture* porosity, irregular *channel* porosity, *vugs*, and *cavern* porosity. The last three are solution features which differ from one another only in scale.

Some porosity, such as that caused by brecciation, by organic boring and burrowing, and by shrinkage, may or may not be related to rock fabric.

The actual pores may have had a complex history (Fig. 10-15). A skeletal element, for example, may be removed by solution, leaving a mold, which may be enlarged by furthe solution and converted to an irregular vug. Either the mold or the enlarged product may be partially to completely filled at any stage. The history of pore filling, whatever the origin of the pore, is complex and is the subject of the section on limestone diagenesis.

The term *fenestral* (Tebbutt, Conley, and Boyd, 1965) has been applied to carbonate rocks having primary or penecontemporaneous voids in the rock framework larger than the grain-supported interstices. These may be open spaces in the rock, or they may be partially or completely filled by secondarily introduced sediment or sparry cement or both. The distinguishing characteristic of fenestrae is that the spaces have no apparent support in the framework of the primary grains forming the sediment. Features meeting these qualifications include calcite-flecked "birdseye" limestones and the *Stromatactis* of certain reef cores.

"Birdseye" limestone is characterized by irregular flecks and masses of clear calcite thought by Ham (1954) to be fillings of irregular spaces left following decomposition of gelatinous algal material, particularly blue-green algae of the family *Spongiostroma*. Shinn (1968), however, considered that the more nearly spherical openings were caused by trapped gas bubbles and that the less regular planar forms were caused by shrinkage. These features were thought by him to have formed in the supratidal environment.

Another structure of somewhat problematic origin but perhaps related to algae, found in the autochthonous limestones, is the *Stromatactis* of American (Lowenstam, 1950, p. 439) and Belgian (Lecompte, 1937) authors and the "tufa" or "fibrous calcite" of British literature (Black, 1952). This structure is a subhorizontal feature characteristic of some reef knolls. It consists of layers or irregular masses of coarsely crystalline calcite in an otherwise finer-grained limestone (Fig. 10-16). The layers are 1 to 5 mm thick and up to 10 or more cm long and are oriented more or less parallel to the bedding of the limestone. They have a smooth lower floor and a highly irregular upper surface. The lower part consists of finer-grained carbonate sediment; the upper part is drusy and coarsely crystalline calcite. *Stromatactis* has been variously explained as a result of subaerial evaporation ("tufa") or diagenetic recrystallization of a fine carbonate. Neither concept adequately explains the structure. Bathurst (1959b) considers this structure a result of chemical deposition from solution on the walls of a post-depositional cavity in the primary-reef sediment. The cavity is formed early, perhaps by decay of organic tissue, followed by mechanical deposition of internal sediment and finally chemical precipitation of the drusy or fibrous calcite in the upper part of the cavity (Wolf, 1962; Lees, 1964, p. 522; Philcox, 1965). Lees (1964,

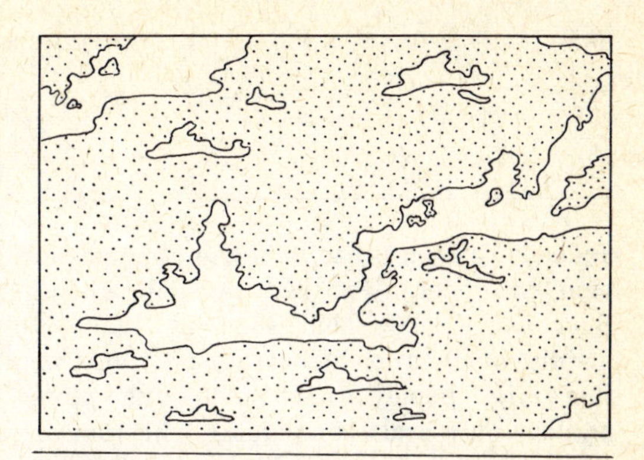

FIG. 10-16. Characteristic forms of *Stromatactis*. Small forms up to 1 cm; large forms 10 to 20 cm. (After Lees, 1964, Trans. Roy. Phil. Soc. London, ser. B, v. 247, Fig. 22.)

p. 522, Fig. 28), in particular, thought the original cavity to be a collapse structure brought about by decay of organic materials in an original mud. Heckel (1972) also thought the cavity to be a kind of collapse structure caused, however, by a differential thixotrophy of fine carbonate mud. In either case the mud released settles to the bottom of the cavity to form a level floor. The structures observed by Lowenstam (1950, p. 439) were ascribed to an "organic origin" and considered to be "frame builders." They form as much as 80 percent of the core area of some Niagaran (Silurian) reefs in the Great Lakes region. These have been more recently attributed to solution of fistuliporid bryozoans (Textoris and Carozzi, 1964, p. 412).

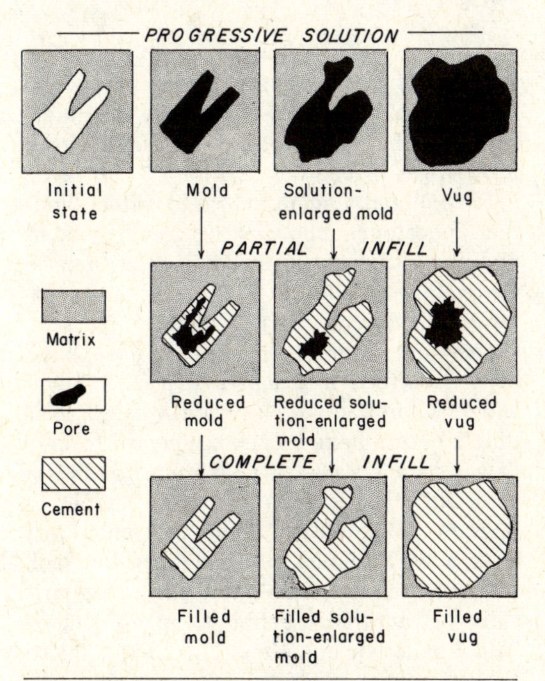

FIG. 10-15. States in evolution of one basic type of pore, a mold. Starting material is a crinoid columnal (top left). It, and the matrix adjoining it, then may be dissolved in varying degrees. Depending on extent of solution (top row), resulting pore is classed as a mold, solution-enlarged mold, or vug. Filling by cement can occur after each solution stage. (From Choquette and Pray, 1970, Bull. Amer. Assoc. Petrol. Geol., v. 54, Fig. 4.)

BEDDING AND OTHER STRUCTURES OF LIMESTONES

Hydrodynamic structures The limestones that consist solely of sand- or silt-sized grains of one or another of the types described above are clearly current-transported and sorted materials. It is not surprising, therefore, that they display larger structures which characterize such deposits. Lime sands of this type are commonly cross-bedded, both on a large and a small scale (Fig. 10-17). Carbonate eolianites may show large-scale cross-bedding. The thickness of the cross-bedded units in the Bermuda eolianites ranges from 1 foot to about 75 feet (0.3 to 25 m) with inclinations mainly between 30 and 35 degrees (Mackenzie, 1964, p. 55). The cross-bedding of modern sub-

FIG. 10-17. Cross-bedded limestone, Loyalhanna Limestone (Mississippian), near Somerset, Pennsylvania.

aqueous carbonate sands of the Bahama region has been described by Imbrie and Buchanan (1965). That seen by them is mainly small-scale, but carbonate sands of this region display a subaqueous dune morphology on a scale comparable with that of subaerial dune fields (Potter and Pettijohn, 1963, Pl. 10B and 11); hence one might expect larger-scale cross-bedding to be present in these deposits. Many marine limestones are strongly cross-bedded. The thickness of the cross-bedded sets is variable, ranging up to 2 m in the Mississippian Salem Limestone of Indiana (Sedimentation Seminar, 1966) and up to nearly 5 m in the Loyalhanna Limestone of Pennsylvania (Adams, 1970). In general, the azimuths of the cross-bedding in marine limestones show a bimodal distribution—an expression of reversing tidal currents (Hamblin, 1969). Small-scale ripple laminations are common in the finer-grained limestones; large-scale rippling is also characteristic of some limestones (Bucher, 1917; Pettijohn and Potter, 1964, Pl. 90).

Some carbonates are transported and deposited by turbidity currents. These show characteristic graded bedding (Meischner, 1964; Eder, 1970), and also a host of sole markings, especially flute casts (Potter and Pettijohn, 1963, Fig. 5-3).

Growth fabrics and structures Limestones formed *in situ* lack the usual framework-cement fabric and current bedding and other structures. They tend to be generally fine-grained, reflecting their accumulation in a nearly currentless environment. They seldom contain oolites or intraclasts, but fossils or skeletal materials may be abundant. Such a rock then has a bimodal size distribution—of larger skeletal fragments embedded in a lime mud or calcilutite. Pellets may also be present in a fine-grained matrix. These bimodal carbonates are the "wackestones" of Dunham (1962) in contrast to the grain-supported, mud-free "grainstones" of hydrodynamic rather than *in situ* origin.

Bedding may be crudely defined—by layers richer in skeletal debris and by orientation of much skeletal debris parallel to the bedding. Such bedding may be interrupted by borings; bioturbation may destroy all vestiges of bedding.

Exceptionally some *in situ* or autochthonous limestones show very good "growth bedding," the most striking of which is *stromatolitic bedding* (Fig. 10-18). Stromatolites display a great diversity of size, shape, and internal structure. These have been described in Chapter 4, p. 127. Some show subhorizontal layering—"algal bedding"; others have a convex-upward form—"algal structures"—varying from thimble to soup-bowl dimensions and from simple to digitate (Fig. 4-31). Some (thrombolites) have a well-defined external form but lack internal laminations. A few are concentrically structured, pebble-sized objects—"algal balls" or oncolites (Fig. 10-19). These should not be confused with caliche pisolites (Dunham, 1969). Stromatolites are particularly common in Precambrian and early Paleozoic limestones.

Nodular bedding A structure particularly characteristic of fine-grained, somewhat argillaceous limestone, is *nodular bedding*. Limestone with this type of bedding (German, *Knollenkalk*) stands in contrast to that

FIG. 10-18. Algal limestone, ×3. Polished transverse section of algal reef rock showing alternate algal and inorganic layers. The irregular light-colored layers consist of molds of *Chlorellopsis coloniata* and typical spongy algal deposit. The black and finely banded layers are of inorganic origin and owe their dark color to disseminated pyrite. (U.S. Geological Survey photograph. Bradley, 1929, Pl. 45.)

showing thin, platy or tabular bedding (German, *Plattenkalk*). Argillaceous limestones are apt to be nodular. Such limestones consist of an alternation of wavy-bedded to nodular layers separated by thinner more argillaceous partings or seams. The limestone layers, with little or no argillaceous content, pinch and swell or separate into lenticular nodular bodies lying parallel to the bedding. The nodular masses range from 1 to several centimeters in thickness and several to 10 or more centimeters in length. In places their boundaries are indistinct. Clay seams or partings are richer in insoluble materials and are more continuous. In some limestones these partings are somewhat dolomitic.

Nodular bedding of this type has been attributed to a "compaction flow" of interbedded calcareous shale (McCrossan, 1958), and it is thus a type of boudinage (Voigt, 1962). Others have attributed the shaly partings to intrastratal solution and fusion of several stylolitic seams (Trurnit, 1968, p. 81), with the argillaceous seams the insoluble residue left following solution of considerable limestone and not a primary bed. All agree that the nodular bedding is not pri-

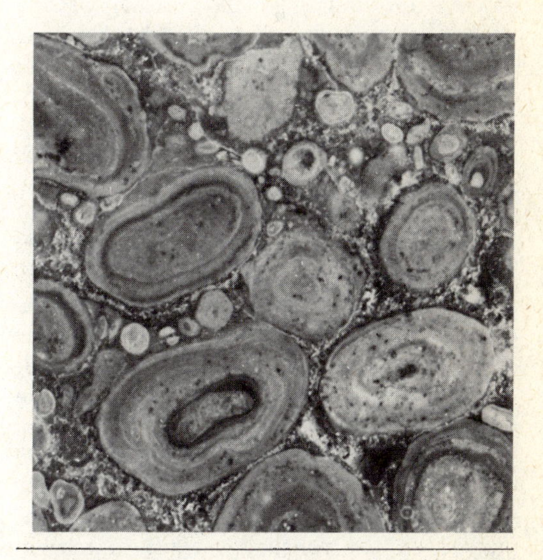

FIG. 10-19. Algal pisolite ("algal structure"), ×5. Polished specimen showing details of internal structure. These bodies consist of layers of spongy algal deposit and thin dense inorganic layers. Between the algal pebbles, or oncolites, are small oolitic grains mixed with fine limy sand. (U.S. Geological Survey photograph. Bradley, 1929, Pl. 47.)

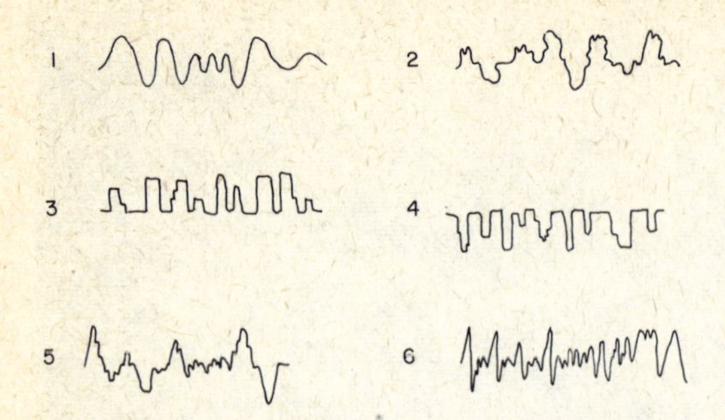

FIG. 10-20. Cross-sectional view of a stylolitic seam in limestone. Six types of stylolitic patterns. (After Park and Schot, 1968, Recent developments in carbonate sedimentology in central Europe, Fig. 1, p. 68. By permission of Springer-Verlag.)

mary, but whether it is caused by flowage, by solution, or by some other process is not clear. Wilson (1969, p. 17) thought that nodular limestones were deposited in shallow waters and were the result of "differential compaction" of clay lenses erratically and randomly distributed in the limy deposit. Matter (1967) believed some "lumpy limestones" were formed by desiccation modified perhaps by burrowing organisms. On the other hand, Garrison and Fischer (1969, p. 26) described nodular limestones presumed to be bathyl. They accept Hollmann's nterpretation (1962, 1964) of the nodules as solution relicts of carbonate beds formed and subsequently destroyed on the sea floor by a process termed *subsolution*.

Stylolites A stylolitic seam is a surface marked by interlocking or mutual interpenetration of two sides (Fig. 10-20). The toothlike projections of one side fit into sockets of like dimension on the other. In cross section the stylolitic surface resembles a suture (German, *Drucksuturen*). Stylolitic seams are exceedingly common in certain types of rocks. They are very abundant in carbonate rocks and are found in many limestones, dolomites, and bedded siderites. They are present in metamorphosed carbonates, especially marbles, and are elegantly displayed on the polished surface of marble and limestone panels used in many public buildings. They are known also to occur in some noncarbonate rocks, especially the sandstones and quartzites (see p. 245). They occur also in gypsum and probably in anhydrite and salt. They are most conspicuous and most common in relatively pure or homogeneous rocks.

Stylolites vary greatly in size, from microstylolites with amplitudes of a fraction of a millimeter to those 10 to 20 cm. The width of the teeth or columns and the corresponding sockets into which they fit varies with the height or amplitude of the structure. Larger columns have a proportionally greater width. Larger and better developed columns (Fig. 10-21) are commonly striated, with striations parallel to the column axis and indicating movement of the column into the socket in which it occurs. Very rarely the column is slightly curved.

The geometry of a stylolitic seam is quite variable, not only with respect to size but also with respect to morphologic details. These range from somewhat conical interlocking teeth, to more regular rectilinear columns and sockets, to multiple or compound stylolites, to smooth pressure-solution surfaces. Close study of these variations has led to various systems of classification (Park and Schot, 1968; Trurnit, 1968, Fig. 4).

The stylolitic seam proper is traceable for varying distances—a few centimeters to several meters. The seams commonly overlap and may even split into two subparallel seams. The stylolitic surface itself is marked by a thin deposit of relatively insoluble material which is a very minor constituent of

FIG. 10-21. Portion of stylolitic column from Salem Limestone (Mississippian), Indiana. Shows grooves and striations. Length of specimen about 10 cm.

the rock in which the stylolite occurs. The residue on the seams in carbonate rocks is largely clay, in part carbonaceous and in part ferruginous. Particles of quartz silt or fine sand tend to collect along the seam, as do idiomorphic quartz crystals (Brown, 1959).

Stylolites are generally parallel to the bedding, although there are stylolites which are transverse, even perpendicular, to the bedding plane (Heald, 1955, Rigby, 1953; Stockdale, 1943). In general, columns are normal to the stylolitic surface; in a few cases columns are vertical even though the seam is inclined.

Most significant are the relations of stylolite to other structures. The stylolitic penetration of fossils has been described many times (Stockdale, 1922; Dunnington, 1954), as has the penetration of oolites (Bastin, 1951). Of special interest, too, is the relation of stylolites to veins (Dunnington, 1954; Conybeare, 1949; James, 1951, p. 263; p. 1.3, Fig. 4). The stylolites both cut and are cut by veinlets. Oblique veins that are cut by a stylolitic surface show an *apparent* displacement by the stylolite proper (Conybeare, 1950).

Microstylolites are of an order of magnitude smaller than common stylolitic seams. They are commonly intergranular and mark contacts between adjacent grains—ooids, fossils, pebbles. If numerous, they may form an irregular network.

Although some uncertainties are connected with the origin of stylolites, there is no doubt that they are a pressure-solution phenomenon and were formed in consolidated rock despite some dissenting views (for example, Shaub, 1939, Prokopovich, 1952). Stockdale (1922) has presented an extensive review of this subject; short and more recent summaries are those of Dunnington (1954) and Manten (1966). Pettijohn agrees with both these writers that geometrical relations between stylolites and such features as fossils, oolites, bedding, and veins demand removal of considerable rock material. Inasmuch as stylolites transect structures that were indurated (shells and the like) and, because they may be transverse to the bedding and cut post-consolidation features such as veins, there can be no doubt that they were formed by post-consolidation solution. Concentration of idiomorphic quartz crystals, which themselves are post-indurational, along stylolitic surfaces further supports this view (Brown, 1959). That pressure is involved is shown by evidences of movement between columns and sockets—striated columns—and the orientation of the stylolite. The stylolitic surface is horizontal (normal to the gravitational field) or, in rare cases, near vertical and parallel to the axial plane of the fold. The pressure involved in most cases, however, is gravitational; hence the stylolitic columns are vertical, even though the seam itself may follow an inclined bedding plane or even a cleavage plane. These relations establish the stylolite as not only post-consolidation but as post-deformation in age. Although Park and Schot (1968) consider stylolites as a pressure-solution phenomena, they thought that stylolitization occurred before the rock was completely cemented and "was still relatively plastic."

Though a pressure-solution and post-consolidation origin seems well established, the mechanism by which solutions, acting under directed pressure, make a stylolite is far from clear. Stockdale (1926) believed that zones of differentially soluble limestone distributed laterally along both sides of a parting subject to solution would soon produce an undulatory surface; pressure would become greatest on the crests and troughs of these undulations and least on the sloping sides; "Since pressure is important in increasing the solubility of certain solids in liquids, the rock opposite to the top and bottom of adjacent undulations will succumb to a greater rate of solution, producing (a) a deepening of the interpenetrating parts, (b) a further decrease in pressure (and consequent dissolving of the rock) on the sides of the undulation, and (c) a possible final development of vertical columns with a decided concentration of pressure and accompanying solution at the ends." Although admitting that stylolites can be produced only by differential solution and directed pressure, Dunnington (1954) does not consider Stockdale's mechanism satisfactory.

The manner in which the process of solution and removal of dissolved materials was maintained was explored by Weyl (1959). The whole problem is summarized by Bathurst (1971, p. 459).

Stylolites abound in carbonate rocks and are a record, therefore, of extensive intrastratal solution. In a few cases one can estimate the volume of rock dissolved in the production of a stylolite seam. Such is possible by comparison of concentration of quartz silt grains in the host rock with their concentration along a stylolitic seam, and also by the amount of apparent displacement of an oblique vein by a cross-cutting stylolite (Fig. 10-22). Stockdale (1926) counted the number of stylolitic seams in limestone and from this count and the amplitude of seams estimated the total quantity of material removed by solution from such surfaces. He assumed that the height of the longest stylolitic column was a measure of the thickness lost by the seam under observation. His assumption seems to be justified in that the thickness of the clay capping on columns seems to be consistent with the quantity of insolubles known to occur in the rock presumed to have been dissolved. Stockdale estimated that 5 to 40 percent of the original thickness of a formation was lost by stylolitic solution.

As pointed out by Dunnington (1954), solution along the stylolitic surface is almost certainly concomitant with precipitation from the same solutions elsewhere in the rock. Transfer of materials from stylolite to pores or other openings in the rock is thus a kind of large-scale application of Riecke's principle of solution at simultaneous points of pressure and reprecipitation in adjacent places of lower pressure. As noted by Dunnington (1967) and others, porosity and permeability are lower in those limestones where stylolites are most abundant, an observation that seems to support the concept of solution transfer and reprecipitation. It may well be that intergranular solution along grain contacts is responsible for much of the cement in some limestones. As noted by Ramsden (1952) and Dunnington (1954), such stylolitic solution and pore filling would result in wholesale expulsion of pore fluids. And if, as suggested by Dunnington, there is some critical pressure required before stylolitic solution begins, cementation should be restricted to rocks subjected to some minimum load. Such cementation, however, would be difficult to distinguish from that caused by downward movement of meteoric solutions such as those that lead to "self-cementation" in eolianite limestones.

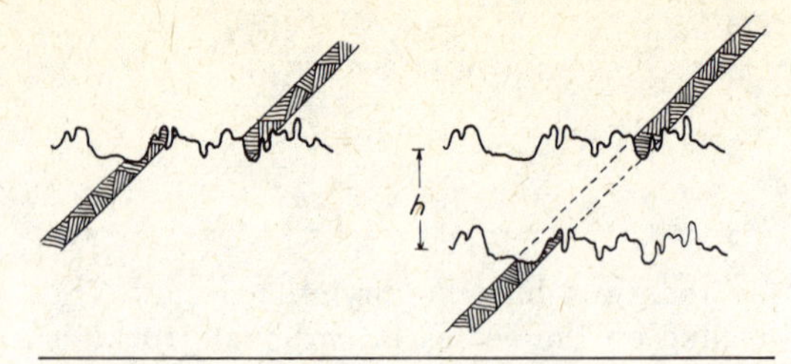

FIG. 10-22. Apparent offset of vein by stylolite can be used to estimate thickness of layer dissolved during formation of stylolite. (After Conybeare, 1949.)

More difficult to recognize and to evaluate are the smooth, unlimited pressure-solution planes (Trurnit, 1968, p. 81). These may arise through fusion of several closely spaced stylolitic seams, and the end-result may be a thin clayey seam or parting in the limestone—a seam difficult to distinguish from an original shaly parting. Some less regular, more argillaceous partings in wavy-bedded and nodular limestones may be the end-product of this action.

CLASSIFICATION OF LIMESTONES

Because of the polygenetic nature of limestones, they are difficult to classify. Defining parameters should be easily observable and genetically significant; but because of the polygenetic origin of limestones it is difficult to find one set of parameters applicable to all limestones. In this respect, the classification of limestones is more involved or complex than that of sandstones which form a much more homogeneous class of rocks.

Grabau (1904) was one of the first to recognize the problems of limestone classification; he proposed an appropriate rational scheme and nomenclature, some of which we use today (*calcarenite*, for example, is one of Grabau's terms). Little progress was made, however, until Folk's milestone paper was published in 1959. Following Folk's lead and because of a renewed interest in carbonate rocks, a large number of papers on the subject have been published. These include papers by Wolf (1960, 1961), Bramkamp and Powers (1958), Mamet (1961), and Memoir 1 of the American Association of Petroleum Geologists (Ham, 1962) which embodies the materials presented at a symposium on carbonate rock classification. Papers by Ham and Pray, Feray and others, Plumley and others, Dunham, Powers, Nelson and others, Folk, Leighton and Pendexter, and Imbrie and Purdy all touch on one or another aspect of the classification problem. Refer to the bibliography and also to those which have appeared more recently (Rich, 1964; Sander, 1967; Todd, 1966; Müller-Jungbluth and Toschek, 1969).

The classification of carbonate rocks, like that of all other rocks, has a genetic basis. In part, carbonates are rocks whose constituents are mechanically emplaced in the rock fabric, that is, are deposited by waves and currents. These include lime gravels, lime sands, and some lime muds. Unlike their noncarbonate counterparts, these materials, although characterized by textures and structures similar to those of usual clastic sediments, are not the waste products of a landmass undergoing erosion. Instead, the debris of which they are composed is derived from within the basin in which they are accumulating. They are, therefore, in a sense intraformational—certainly intrabasinal—deposits. Although much of the detritus is of clastic origin (broken pieces of reef rock, fragmented and worn shell debris), some of it is unbroken and is biochemical or chemical in origin. Many lime sands contain entire foraminiferal tests; others consist in part or wholly of oolites. All of these materials, as well as fragmental constituents, are current-sorted and deposited and have the textures and structures of a normal clastic rock. For this reason these mechanically deposited limestones are treated as a separate group, despite the fact that in a strict sense they are not wholly clastic. Although one should discriminate carefully between the several types of materials found in these calcarenites, a nomenclature discriminating between clastic, detrital, fragmental, and mechanical seems unnecessary. Because the material has been transported and redeposited, these limestones have been called *allochthonous* (without roots).

A second major group of limestones has been formed *in situ* by an accumulation of organically or chemically precipitated materials. These rocks have not been subjected to current transport and redeposition. They are, therefore, not sorted or bedded as are current-laid materials; but they may show a crude stratification in places and, in some cases, have well-defined stromatolitic or growth bedding. In part these limestones were consolidated during the depositional process and may have formed wave-resistant masses or reefs. In other cases the accumulating lime carbonate was not so consolidated. These limestones which grow in place have been termed accretionary or *autochthonous* limestones. Biochemical deposits of moundlike form are *bioherms,* including reefs; similar deposits of a more layered form or aspect are *biostromes* (Cumings and Shrock, 1928, p. 599). Not all autochthonous limestones are biochemical accumulations. Some rather minor deposits, such as travertine, caliche, and calcareous tufa, are also formed in place. These are purely chemical precipitates from supersaturated solutions and are products of localized precipitation in springs and lakes or in the soil profile. Moreover, even within autochthonous limestones rich in skeletal materials may be pockets of mechanically deposited carbonate as well as important additions directly precipitated from sea water without any evident biological cause.

Commonly the distinction between these two classes of limestone is easy. But this is not always so. Criteria for distinguishing between them are related to sorting, to current structures, to the state of articulation of skeletal materials, and to the content and size of the noncarbonate detritus (Table 10-6). Distinction can be made with great difficulty, if it can be made at all, in the case of the calcareous muds.

Any given limestone *formation* may be a composite of these two fundamental types.

The Maastrichian Limestone (Andrée, 1915), for example, consists of regularly repeated crinoidal limestone, coral limestone, and bryozoan limestone. The first is transported; the other two are sedentary. In the Conococheague (Cambrian) of Maryland are alternating beds of detrital limestone (calcarenites and calcilutites) and stromatolitic or algal limestones formed *in situ*.

Any of the above limestones, both those transported and those formed in place, may be greatly altered by recrystallization or by replacement which obscures or erases original textures and structures and may greatly alter rock composition. Such rocks, including dolomite, constitute a third major class of carbonate rocks. Dolomite is, of course, not the only replacing material. Limestones may be replaced by silica, by phosphate, by hematite, and so forth. Such replacement rocks are distinguished with some difficulty from original or primary precipitates of these same materials. (They are, therefore, considered in detail in Chapter 11).

Allochthonous or transported carbonates exhibit a framework–cement relationship analogous to that found in ordinary sandstones. These are indeed carbonate sands, the calcarenites of Grabau. Their classification can be handled in a manner similar to that applied to sandstones (Fig. 10-23). This treatment involves a classification based on the nature and proportions of the framework elements and on the nature and proportions of the cement and/or matrix materials.

A cursory examination of a few *calcarenites* (carbonate "sandstones") shows them to consist of sand-sized grains of skeletal materials, ooids (or pisolites), and "lithic" particles (fine-grained aggregates). These may be built up from fine carbonate mud by organisms (fecal pellets), or they may be broken-up, weakly lithified clasts of carbonate muds (intraclasts). But particles of similar nature may be attributable also to a process of degradation of skeletal or oolitic materials. Such "micritized" grains are difficult to distinguish from those built up by aggrading processes. Folk presumed that it was possible to identify four basic components (intraclasts, pellets, skeletal elements, and ooids) and based his classification on these distinctions. Collectively he termed these *allochems*. The ambiguities which arise when the distinctions cannot be made can be handled by calling all the small rounded micritic grains *peloids* (McKee and Gutschick, 1969). Reduced to three components (ooids, skeletal, and peloids), the composition of any carbonate sand can be represented by a point in a triangle, and a classification of calcarenites can be made by appropriate subdivisions of the triangle (Fig. 10-23).

Framework components may be cemented by a precipitated calcite cement ("spar"), or they may be embedded in a fine-grained carbonate matrix (the *micrite* of Folk). These alternatives are analogous to the ordinary sandstones cemented with a precipitated mineral cement and those sandstones

TABLE 10-6. Criteria for distinguishing between autochthonous and allochthonous limestones

Autochthonous limestone	Allochthonous limestone
1. Associated with shales	1. Associated with orthoquartzites
2. Grades into calcareous shales and mudstones	2. Grades into and is interbedded with orthoquartzites (calcareous sandstones)
3. Interstices between fossils filled with lime muds	3. Interstices filled with clear calcite cement
4. Bryozoan-encrusted fossils	4. Contains rolled fossils
5. Unsorted as to size	5. Sorted as to size
6. Fossils articulated	6. Fossils disarticulated
7. Reef structures	7. Cross-bedded

of allochems and to distinguish between sparry and micritic cements. To do so we follow Folk's terminology. But distinctions between frameworks that are grain-supported, those that are related to organic growth and concurrent cementation, and those that refer to unsorted lime muds with scattered skeletal elements can better be made using Dunham's terminology.

PETROGRAPHY OF LIMESTONES (AND DOLOMITES)

The five major classes of carbonate rocks differ from one another in their fabric, and the differences relate to processes of deposition and diagenesis. One class, the grainstones, are marked by a framework–pore system, a product of current action, of medium and coarse debris mechanically deposited and emplaced in the fabric of the rock. These materials have been winnowed out and segregated. Included here are calcarenites and calcirudites—the carbonate sands and gravels.

A second group, the *boundstones*, are those composed of a coarse biogenic framework bound together during accumulation by encrusting algae, the whole being marked by large internal voids in which internal sedimentation is common. These are reef-core deposits—the biohermites.

The third group, the *carbonate mudstones*, is perhaps the most common. These are marked by a greatly predominant fine-grained matrix with many to few larger components (the "wackestones") or in some cases no larger constituents ("lithographic" limestones). The wackestones are probably the normal biomodal indigeneous sediments of many differing environments. Some have a thin tabular bedding (*Plattenkalk*); others are nodular (*Knollenkalk*).

The *allodapic limestones* form a fourth class. These are resedimented carbonates—materials moved from their place of origin to deeper waters by turbidity currents. These are basinal carbonates. In one sense they are not a separate class, in that the lower portion of a bed may be a coarse grainstone and the uppermost part a mudstone. But because of their graded structure, they are treated separately.

Dolomites constitute a fifth class of carbonates. They are the most common diagenetic carbonates. They may be replacements of any of the other carbonate types, especially the reef-core boundstones, but they may also be primary sediments and hence they fall into two classes: the fine-grained primary dolomites and the coarse-grained or saccharoidal replacement dolomites.

Some carbonates, such as travertine and caliche, do not fit neatly into these five groups. They are considered separately.

GRAINSTONES (CALCARENITES AND CALCIRUDITES)

Grainstone is a term introduced by Dunham (1962) for rocks which have a grain-supported framework. The intergranular pore system is generally, but not always, filled with a clear crystalline cement. The term *calcarenite* (Grabau, 1904) is a general term to describe these mechanically deposited grainstones of sand size (1/16 to 2 mm in diameter). If the fragments are over 2 mm in diameter, the term *calcirudite* may be applied to the rock.

The carbonate detritus is of subaqueous origin and consists mainly of fossil materials (both entire and broken) of pebbles and granules of calcilutite (intraclasts of Folk), and of oolites. Some detritus is therefore biofragmental, some strictly clastic, and in part both biochemical (skeletal) and chemical carbonate (oolites). All of it, however, is current-transported and -sorted and mechanically deposited so that the accumulation has the structure of a detrital sediment (Fig. 10-24).

The distinction between materials of biological, clastic, and even chemical origin is arbitrary. Some of the fossil debris is coated with one or more layers of precipitated carbonate; many of the oolites themselves have nuclei of skeletal carbonate or even quartz. Moreover, calcarenites vary greatly in the importance of contributing materials. Some are nearly all oolites (oosparites); others are wholly fossil debris (biosparites). Still others consist mainly of granules of calcilutitic material (intrasparites and pel-

characterized by a fine-grained, silt-clay matrix (the graywackes). In some cases the volume of the matrix exceeds that of the framework elements or allochems, and the latter appear to "float" in a micritic paste. Utilizing these concepts, Folk set up his classification (Table 10-7) based on the kind and proportions of allochems (vertical axis) and the spar–matrix ratio (horizontal axis). Folk recognized two other carbonate families—autochthonous biohermal deposits and diagenetic or epigenetic dolomites. The former was not subdivided; the latter were broken into two major classes, those with relict fabrics of the rocks from which they were derived and those which had lost all traces of their origin.

Folk developed a nomenclature to identify each subclass in his classification scheme. For the most part these are self-explanatory, although some have been shortened for brevity's sake. Each term is a composite. The prefix indicates the nature of the allochem, the stem indicates the nature of the cement or matrix, and the suffix indicates the texture or grain size. For example, *biosparrudite*, indicates a skeletal or shelly allochem, a sparry or precipitated crystalline calcitic cement, and a coarse-grained texture—in other words, a coquina. A sand-sized equivalent would be biosparearenite, which can be shortened to *biosparite* or, in other words, a biocalcarenite or skeletal sand with a crystalline calcite cement (Table 10-7). Perusal of recent literature shows that Folk's nomenclature has received wide acceptance and therefore fills a real need. Many other classifications are either slight modifications of Folk's scheme or utilizations of the same basic principles. Dunham (1962) placed greater emphasis on depositional texture and recognized five fundamental types (Table 10-8). His terminology has also found its way into the literature.

In this book we tend not to adhere rigidly to any single scheme. It does seem desirable to discriminate between the several types

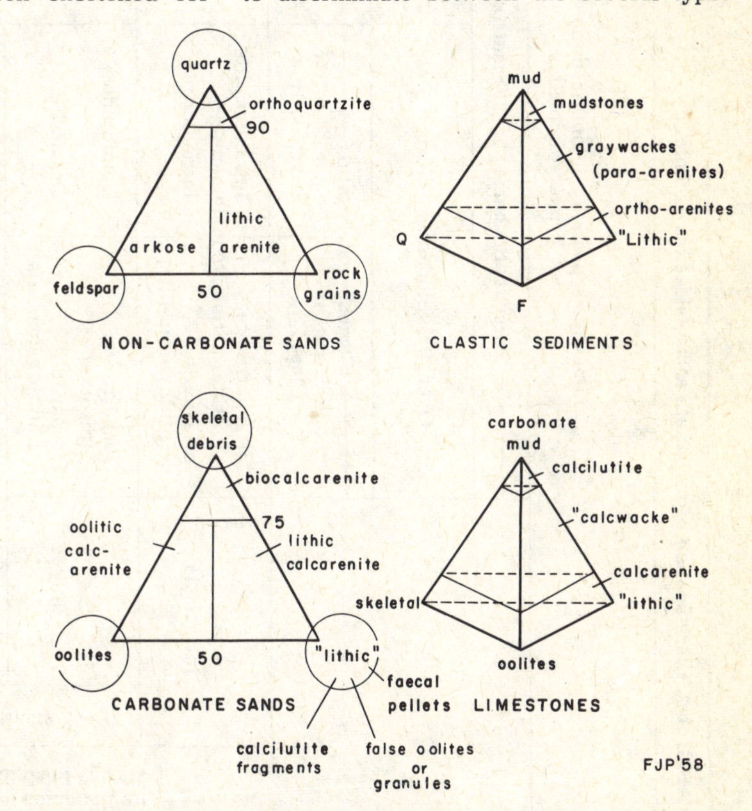

FIG. 10-23. Relation between classification of calcarenites and calcilutites and their siliceous clastic analogues.

TABLE 10-7. Classification of carbonate rocks (siderites excluded)

Allochem composition (varying kinds of "clasts")			Limestones, partially dolomitized limestones and primary dolomites — Percent allochems ("clast"–matrix ratio): Over 10% allochems ("clasts"), Cement–matrix ratio: spar > matrix Class I	Over 10% allochems ("clasts"), Cement–matrix ratio: spar < matrix Class II	< 10% allochems (1–10% / < 1%): Microcrystalline rocks Class III	Non-"clastic" limestone class IV	Replacement dolomites class V: Allochem ghosts	Replacement dolomites class V: No allochem ghosts
> 25% intraclasts			Intrasparrudite (intraformational conglomerate)[a] Intrasparite (lithic calcarenite)	Intramicrudite Intramicrite —	1–10%: — < 1%: Micrite and dolomicrite (calcilutite)	Biohermite (klintite)	Intra-clastic dolomite	Coarse, medium, fine crystalline dolomites
< 25% intraclasts	> 25% oolites		Oosparrudite (pisolite) Oosparite (oolitic calcarenite)	Oomicrudite — Oomicrite —	1–10%: — < 1%: Micrite and dolomicrite (calcilutite)	Biohermite (klintite)	Oolitic dolomite	Coarse, medium, fine crystalline dolomites
< 25% intraclasts	< 25% oolites: fossil–pellet ratio	3 : 1	Biosparrudite (coquina)	Biomicrudite (coquinoid limestone)	1–10%: — < 1%: Micrite and dolomicrite (calcilutite)	Biohermite (klintite)	Biogenic dolomite	Coarse, medium, fine crystalline dolomites
< 25% intraclasts	< 25% oolites: fossil–pellet ratio	3 : 1 to 1 : 3	Biosparite (biocalcarenite)	Biomicrite (fossiliferous calcilutite)	1–10%: — < 1%: Micrite and dolomicrite (calcilutite)	Biohermite (klintite)	Biogenic dolomite	Coarse, medium, fine crystalline dolomites
< 25% intraclasts	< 25% oolites: fossil–pellet ratio	1 : 3	Pelsparite (pellet calcarenite)	Pelmicrite (pelletiferous calcilutite)	1–10%: — < 1%: Micrite and dolomicrite (calcilutite)	Biohermite (klintite)	Pellet dolomite	Coarse, medium, fine crystalline dolomites

[a] Pettijohn's common terms in parentheses.

Source: Modified from Folk (1959), Table 1. By permission Amer. Assoc. Petrol. Geol.

TABLE 10-8. Classification of limestones according to depositional texture

Depositional texture recognizable					Depositional texture not recognizable
Original components not bound together during deposition				Original components bound together during deposition	
Contains mud (fine silt and clay size particles)			Lacks mud		
Mud-supported		Grain-supported			
Less than 10 percent grains	More than 10 percent grains				
Mudstone	Wackestone	Packstone	Grainstone	Boundstone	Crystalline carbonate (subdivide according to physical or diagenetic texture)

Source: After Dunham (1962), Table 1. By permission Amer. Assoc. Petrol. Geol.

characterized by a fine-grained, silt-clay matrix (the graywackes). In some cases the volume of the matrix exceeds that of the framework elements or allochems, and the latter appear to "float" in a micritic paste. Utilizing these concepts, Folk set up his classification (Table 10-7) based on the kind and proportions of allochems (vertical axis) and the spar–matrix ratio (horizontal axis). Folk recognized two other carbonate families—autochthonous biohermal deposits and diagenetic or epigenetic dolomites. The former was not subdivided; the latter were broken into two major classes, those with relict fabrics of the rocks from which they were derived and those which had lost all traces of their origin.

Folk developed a nomenclature to identify each subclass in his classification scheme. For the most part these are self-explanatory, although some have been shortened for brevity's sake. Each term is a composite. The prefix indicates the nature of the allochem, the stem indicates the nature of the cement or matrix, and the suffix indicates the texture or grain size. For example, *biosparrudite,* indicates a skeletal or shelly allochem, a sparry or precipitated crystalline calcitic cement, and a coarse-grained texture—in other words, a coquina. A sand-sized equivalent would be biospararenite, which can be shortened to *biosparite* or, in other words, a biocalcarenite or skeletal sand with a crystalline calcite cement (Table 10-7). Perusal of recent literature shows that Folk's nomenclature has received wide acceptance and therefore fills a real need. Many other classifications are either slight modifications of Folk's scheme or utilizations of the same basic principles. Dunham (1962) placed greater emphasis on depositional texture and recognized five fundamental types (Table 10-8). His terminology has also found its way into the literature.

In this book we tend not to adhere rigidly to any single scheme. It does seem desirable to discriminate between the several types

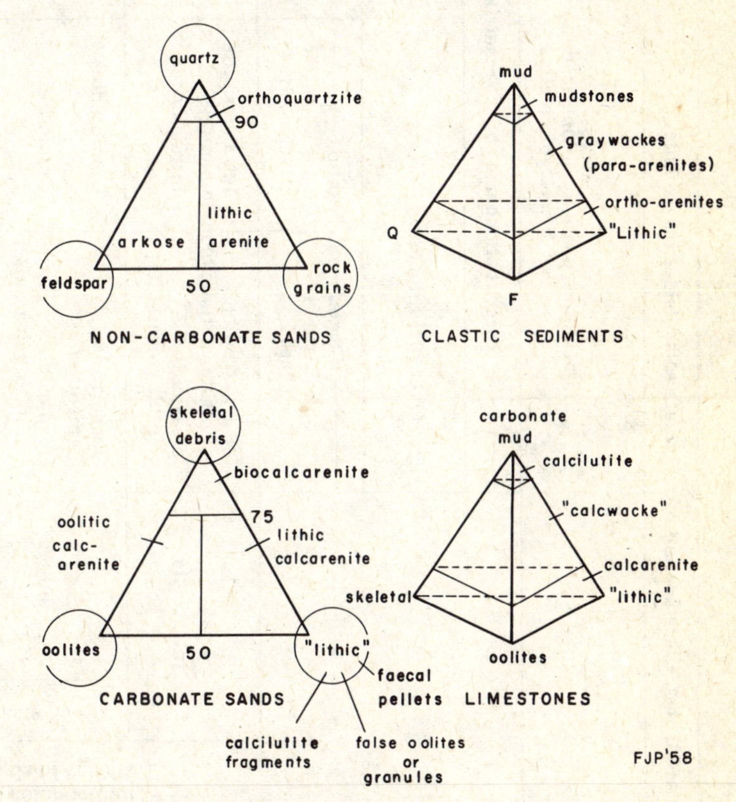

FIG. 10-23. Relation between classification of calcarenites and calcilutites and their siliceous clastic analogues.

TABLE 10-7. Classification of carbonate rocks (siderites excluded)

Allochem composition (varying kinds of "clasts")	Limestones, partially dolomitized limestones and primary dolomites: Percent allochems ("clast"–matrix ratio) — Over 10% allochems ("clasts"): Cement–matrix ratio, spar > matrix, Class I	Over 10% allochems ("clasts"): Cement–matrix ratio, spar < matrix, Class II	< 10% allochems, 1–10%: Microcrystalline rocks, Class III	< 10% allochems, < 1%: Microcrystalline rocks, Class III	Non-"clastic" limestone class IV	Replacement dolomites class V: Allochem ghosts	Replacement dolomites class V: No allochem ghosts
> 25% intraclasts	Intrasparrudite (intraformational conglomerate)[a] Intrasparite (lithic calcarenite)	Intramicrudite Intramicrite —		Micrite and dolomicrite (calcilutite)	Biohermite (klintite)	Intra-clastic dolomite	Coarse, medium, fine crystalline dolomites
< 25% intraclasts; > 25% oolites	Oosparrudite (pisolite) Oosparite (oolitic calcarenite)	Oomicrudite — Oomicrite —				Oolitic dolomite	
< 25% intraclasts; < 25% oolites; fossil–pellet ratio 3 : 1	Biosparrudite (coquina)	Biomicrudite (coquinoid limestone)				Biogenic dolomite	
< 25% intraclasts; < 25% oolites; fossil–pellet ratio 3 : 1 to 1 : 3	Biosparite (biocalcarenite)	Biomicrite (fossiliferous calcilutite)					
< 25% intraclasts; < 25% oolites; fossil–pellet ratio 1 : 3	Pelsparite (pellet calcarenite)	Pelmicrite (pelletiferous calcilutite)				Pellet dolomite	

[a] Pettijohn's common terms in parentheses.

Source: Modified from Folk (1959), Table 1. By permission Amer. Assoc. Petrol. Geol.

TABLE 10-8. Classification of limestones according to depositional texture

<table>
<tr><td colspan="4">Depositional texture recognizable</td><td></td><td>Depositional texture not recognizable</td></tr>
<tr><td colspan="4">Original components not bound together during deposition</td><td>Original components bound together during deposition</td><td rowspan="5">Crystalline carbonate (subdivide according to physical or diagenetic texture)</td></tr>
<tr><td colspan="3">Contains mud (fine silt and clay size particles)</td><td>Lacks mud</td><td rowspan="4">Boundstone</td></tr>
<tr><td colspan="2">Mud-supported</td><td colspan="2">Grain-supported</td></tr>
<tr><td>Less than 10 percent grains</td><td>More than 10 percent grains</td><td rowspan="2">Packstone</td><td rowspan="2">Grainstone</td></tr>
<tr><td>Mudstone</td><td>Wackestone</td></tr>
</table>

Source: After Dunham (1962), Table 1. By permission Amer. Assoc. Petrol. Geol.

of allochems and to distinguish between sparry and micritic cements. To do so we follow Folk's terminology. But distinctions between frameworks that are grain-supported, those that are related to organic growth and concurrent cementation, and those that refer to unsorted lime muds with scattered skeletal elements can better be made using Dunham's terminology.

PETROGRAPHY OF LIMESTONES (AND DOLOMITES)

The five major classes of carbonate rocks differ from one another in their fabric, and the differences relate to processes of deposition and diagenesis. One class, the *grainstones,* are marked by a framework–pore system, a product of current action, of medium and coarse debris mechanically deposited and emplaced in the fabric of the rock. These materials have been winnowed out and segregated. Included here are calcarenites and calcirudites—the carbonate sands and gravels.

A second group, the *boundstones,* are those composed of a coarse biogenic framework bound together during accumulation by encrusting algae, the whole being marked by large internal voids in which internal sedimentation is common. These are reef-core deposits—the biohermites.

The third group, the *carbonate mudstones,* is perhaps the most common. These are marked by a greatly predominant fine-grained matrix with many to few larger components (the "wackestones") or in some cases no larger constituents ("lithographic" limestones). The wackestones are probably the normal biomodal indigeneous sediments of many differing environments. Some have a thin tabular bedding (*Plattenkalk*); others are nodular (*Knollenkalk*).

The *allodapic limestones* form a fourth class. These are resedimented carbonates—materials moved from their place of origin to deeper waters by turbidity currents. These are basinal carbonates. In one sense they are not a separate class, in that the lower portion of a bed may be a coarse grainstone and the uppermost part a mudstone. But because of their graded structure, they are treated separately.

Dolomites constitute a fifth class of carbonates. They are the most common diagenetic carbonates. They may be replacements of any of the other carbonate types, especially the reef-core boundstones, but they may also be primary sediments and hence they fall into two classes: the fine-grained primary dolomites and the coarse-grained or saccharoidal replacement dolomites.

Some carbonates, such as travertine and caliche, do not fit neatly into these five groups. They are considered separately.

GRAINSTONES (CALCARENITES AND CALCIRUDITES)

Grainstone is a term introduced by Dunham (1962) for rocks which have a grain-supported framework. The intergranular pore system is generally, but not always, filled with a clear crystalline cement. The term *calcarenite* (Grabau, 1904) is a general term to describe these mechanically deposited grainstones of sand size (1/16 to 2 mm in diameter). If the fragments are over 2 mm in diameter, the term *calcirudite* may be applied to the rock.

The carbonate detritus is of subaqueous origin and consists mainly of fossil materials (both entire and broken) of pebbles and granules of calcilutite (intraclasts of Folk), and of oolites. Some detritus is therefore biofragmental, some strictly clastic, and in part both biochemical (skeletal) and chemical carbonate (oolites). All of it, however, is current-transported and -sorted and mechanically deposited so that the accumulation has the structure of a detrital sediment (Fig. 10-24).

The distinction between materials of biological, clastic, and even chemical origin is arbitrary. Some of the fossil debris is coated with one or more layers of precipitated carbonate; many of the oolites themselves have nuclei of skeletal carbonate or even quartz. Moreover, calcarenites vary greatly in the importance of contributing materials. Some are nearly all oolites (oosparites); others are wholly fossil debris (biosparites). Still others consist mainly of granules of calcilutitic material (intrasparites and pel-

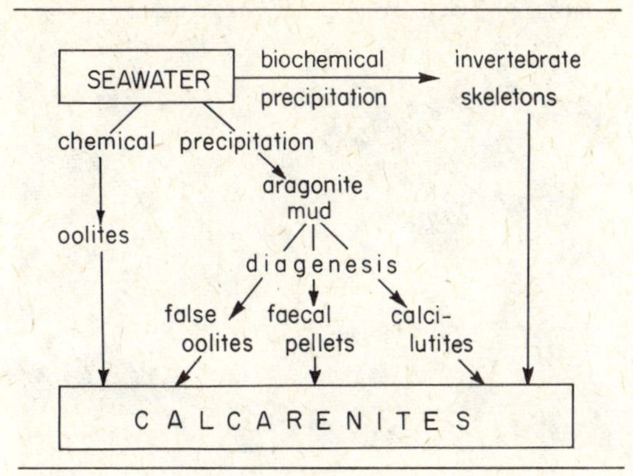

FIG. 10-24. Origin of calcarenitic grainstones.

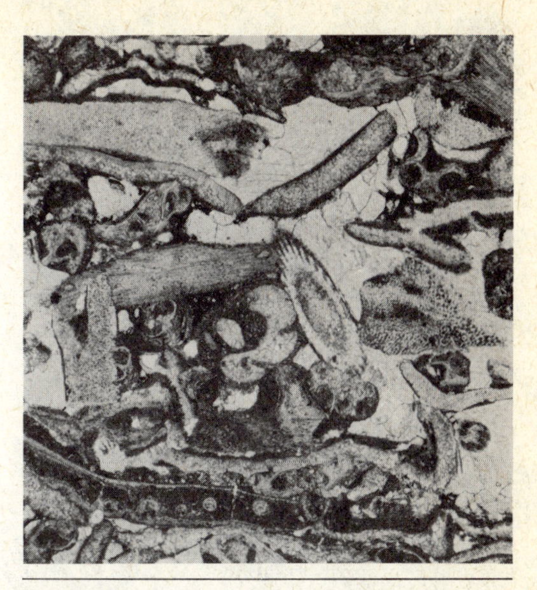

FIG. 10-26. Salem ("Bedford") Limestone (Mississippian), Indiana. Ordinary light, ×30. A microcoquina or biosparite, cemented with clear calcite. Consists of sorted fossil debris, including bryozoan fragments and a few entire foraminifera (*Endothyra*), together with carbonate detritus of unknown origin. Note dense, dark micritic rims on some detritus.

sparites) and contain only a few scattered ooids or recognizable fossil materials (Figs. 10-25, 10-26, 10-27).

Those detrital limestones consisting wholly or nearly so of sorted fossil debris are coquinas (biosparrudites) of one kind or another. The term *coquina* is most commonly applied to the more or less cemented coarse shelly debris (Fig. 10-28). These mechanically deposited materials should not be

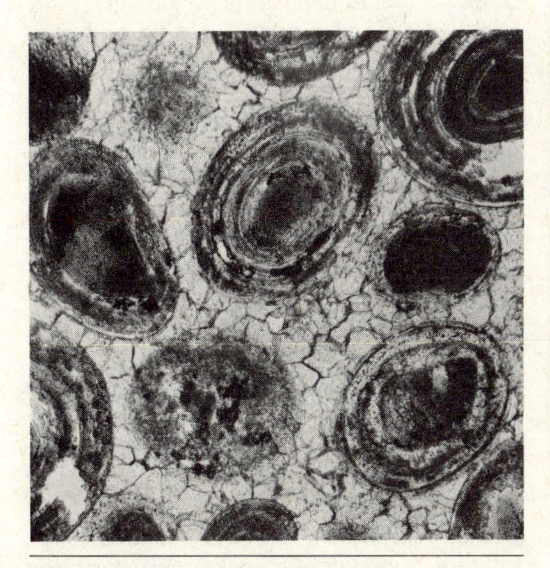

FIG. 10-25. Aragonitic, concentrically structured ooids in a coarsely crystalline calcite matrix; Ooids in part degraded by micritization; Miami Oolite (Pleistocene), Florida. Ordinary light, ×58.

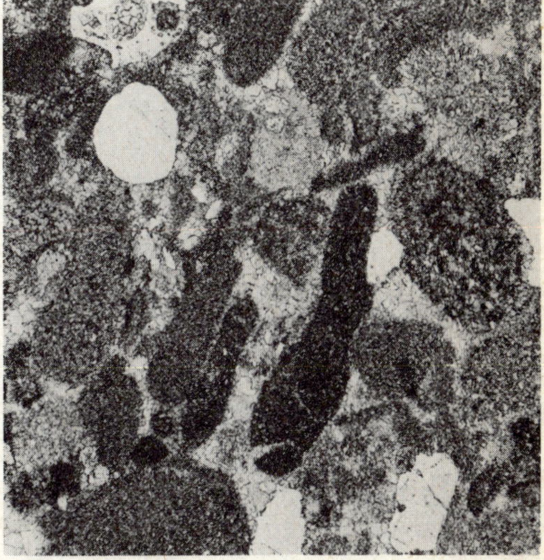

FIG. 10-27. Altyn Limestone, Belt Series (Precambrian), Glacier National Park, Montana. Ordinary light, ×30. A calcarenite that consists mainly of rounded carbonate detritus associated with a little detrital quartz, feldspar, particles of granite and chert.

confused with rudistid and oyster bank materials—materials formed and cemented in place at the time of deposition (boundstones), to which the term *coquinoid* rather than *coquina* has been applied (Fig. 10-29). For finer shell detritus, the term *microcoquina* (biosparite) is used. These rocks are apt to have a low grain to cement ratio, indicating a very high original porosity because of the "potato-chip" or "corn-flake" fabric of original materials (Dunham, 1962).

A special variety of coquina is *encrinite*, a variety consisting almost wholly of crinoidal debris. This has also been called *organogenic conglomerate* (Hadding, 1933, p. 48). Encrinites consist of disks and plates that are detached from one another and which are commonly worn and sorted into beds that differ in the size of constituent fragments. They may be strongly cross-bedded or in some cases even graded. Well-preserved crowns and long stems are not found in these rocks; they occur in the marlstones. The latter, which are also crinoid-bearing rocks (but not encrinites) were formed in more tranquil waters than were crinoid coquinas, so that, after the death of the animals, the remains were entombed more or less intact in the bottom mud. Marls with scattered but well-preserved articulate crinoid remains and encrinites that are built up of sorted and washed crinoid debris are indices of two contrasting facies of lime deposition. Hadding (1933, p. 49) attributed removal of mud and concentration of crinoid debris to bottom currents upon uplift of the sea bottom and shoaling of the waters. The cementation of the encrinites takes place by deposition of calcite in crystallographic continuity with the framework grains, thus converting the whole rock into a coarse, interlocking crystalline mosaic—a sedimentary "marble" (Fig. 10-30).

FIG. 10-28. Coquina, or biorudite (Recent), Florida. About natural size. Note fragmental character of shell debris, absence of matrix, and uniformity of size.

FIG. 10-29. Coquinoid limestone (biomicrudite). About natural size. Note that many fossils are entire, that the matrix is fine-grained, the lack of assortment, and the diversity of the fossil types represented.

The calcarenites in which oolites are the chief ingredient are the oosparites (if pisolitic, the oosparrudites) of Folk. The oolitic texture is almost certainly a primary feature characteristic of shallow, strongly agitated waters. The uniformity of size of oolites, the association of oolites and detrital quartz grains (Fig. 10-31), the cross-bedded character of many oolitic rocks, and the clear crystalline carbonate cement (and the absence of fine interstitial carbonate mud) are supporting evidence for this interpretation. Observations on the calcareous oolites of the Bahamas has led Illing (1954, p. 43) to the conclusion that these oolites were precipitated where "fresh oceanic waters sweeping on the shallow banks have been sufficiently warmed to become saturated with calcium carbonate." According to Illing, "oolitic accretion depends fundamentally on the movement of sand grains under the impetus of marine currents."

In addition to organic detritus and oolites, many calcarenites contain small, sand-sized grains which are fine-grained and generally without any internal structure, organic or otherwise (Fig. 10-11). Such rounded nonskeletal grains have been called pellets, granules, false oolites, pseudo-oolites, and so forth. In some calcarenites these grains greatly exceed either recognizable organic debris or oolites proper. They seem to have originated in several ways. Most were probably deposited as carbonate mud which by one process or another has been built up into sand-sized grains. Conceived in this manner, these calcarenites are sands somewhat analogous to lithic arenites, which contain shale, slate, and argillite sand grains (Fig. 10-32). They differ, however, in that the reconstitution process in the one case is penecontemporaneous, whereas in the other it is mainly epigenetic and in some cases metamorphic (although, of course, some sandstones also contain mudstone clasts of intraformational or penecontemporaneous origin). In other cases, micritic grains may be diagenetic and formed by degradation of either skeletal or oolitic materials.

The upbuilding process, whereby carbonate muds become aggregated into sand-sized grains, may be the result of several different causes. In some cases subaqueous erosion of previously consolidated calcilutites is responsible. There is no fundamental differ-

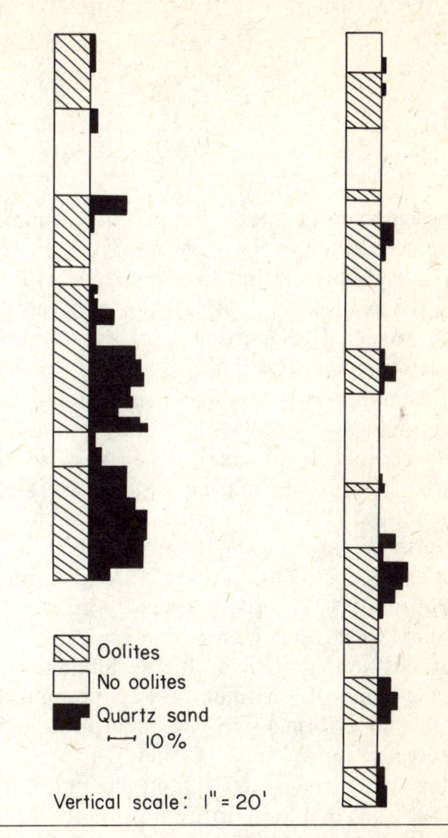

FIG. 10-31. Co-occurrence of carbonate oolites and detrital quartz grains in Greenbrier (Mississippian) Limestone, West Virginia. (After Rittenhouse, 1949.)

FIG. 10-30. Crystalline limestone (biosparite), Fernvale Limestone (Ordovician), Carter County, Oklahoma. Ordinary light, ×30. Consists almost wholly of crystalline crinoid debris cemented by calcite. Cement deposited in crystallographic continuity with detritus. Note cleavage which passes without interruption from dusty detrital grain into clear cement. Properly called a calcarenite, variety microcoquina (encrinite).

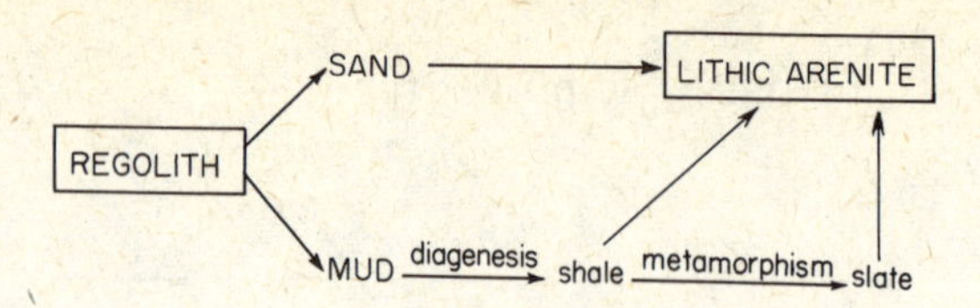

ence between the larger flat pebbles of many intraformational conglomerates and the smaller pebbles found in some calcarenites, and the sand-sized debris of the calcarenites themselves. These are the intrasparrudites and intrasparites of Folk.

On the other hand, these composite grains may indeed be pellets or fecal pellets, a term applied to the small ovoid bodies produced by various marine organisms. Such present-day fecal pellets are very common in both calcareous and noncalcareous sediments (Moore, 1939; Eardley, 1938, pp. 1401–1408; Illing, 1954). Many are soft and, unless lithified, lose their identity on burial. Those that become hard prior to burial might be washed out of the muddy environment in which they formed and be incorporated in carbonate sands. But, as Folk (1959, p. 25) notes, most rocks designated "pellet limestones" in the field turn out to be intraclastic limestones. True pellet limestones are so fine-grained that they are almost without exception mistaken for micrites. Some pellets turn out to be micritized oolites or fossils which have been rendered microcrystalline and nearly structureless by a process of grain diminution. These can usually be identified by shape, size, or relict structures in others or by finding partially micritized grains.

The normal cement of calcarenites is sparry calcite, except where it has been dolomiflzed. It may form 50 or more percent of the whole rock. In some calcarenites carbonate cement appears both as a comblike fringe on the detrital grains and as a clear coarsely crystalline mosaic between the grains. The carbonate detritus, which is crystallographically a unit (such as echinoderm debris), is invariably surrounded by cementing carbonate in crystallographic and optical continuity. This is Bathurst's syntaxial rim cement (1958). The original fragment is seen as a ghostlike body distinguished only by an abundance of dusty inclusions (Fig. 10-30). Some have considered the clear sparry calcite cement as a diagenetic product replacing an earlier fine-grained micrite (Bonet, 1952, p. 174). In the

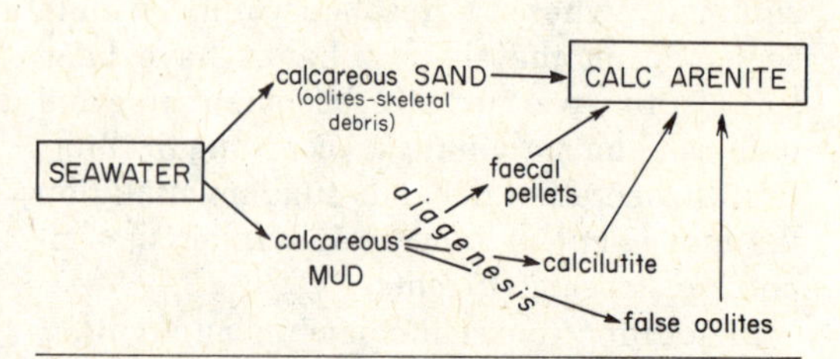

FIG. 10-32. Diagram to show similarity in provenance of calcarenites and lithic arenites.

Fredonia Oolite (Mississippian) of Illinois, Graf and Lamar (1950) considered the sparry calcite to be a second-generation cement emplaced after the removal, by solution, of an earlier finely crystalline brown calcite. Incomplete removal of the latter left hourglass-shaped residuals in the recesses of the intergranular pores particularly at the contact between two grains. This interpretation, however, was considered and rejected for the calcarenites associated with the Scurry Reef (Permo-Carboniferous) of Texas by Bergenback and Terriere (1953). It seems improbable that well-washed and cross-stratified carbonate sands and carbonate mud would be deposited concurrently in the same place. The calcite cement, therefore, is a true pore-filling cement and not a recrystallization product.

One feature common in many calcarenites are the micritic rims seen on many of framework grains. Such micritic envelopes could be either a precipitated micritic coating or a micritic alteration of the original grain. The single-layer coating—the superficial oolites of Carozzi (1960, p. 238)—generally has a uniform thickness and follows closely the contour of the grain. On the other hand, micritic alteration is uneven and penetrates the grain to varying depths. It is present on some grains not at all; others are wholly converted to micrite. The inner boundary of the micritic zone may be fuzzy—not sharp like that of the precipitated coating.

Calcarenites may be closely associated

with quartz arenites. Not only are they interbedded with one another, but they grade into each other also (Fig. 10-33). Clastic quartz sand and carbonate sand are mingled in all proportions. Even formations that are commonly termed *limestone* may contain a significant volume of detrital quartz (Winchell, 1924; Decker and Merritt, 1928; Rittenhouse, 1949; Henbest, 1945; Adams, 1970). See Fig. 10-34. With increasing proportions of detrital quartz the calcarenite becomes an arenaceous limestone. The larger conspicuous quartz grains are commonly very well rounded. In some cases the quartz forms the nuclei of calcareous oolites; and in some of these euhedral quartz overgrowths are seen to replace the host ooid (Henbest, 1968). See Fig. 10-35.

In the field, calcarenites resemble noncarbonate sands. They are commonly current-bedded with cross-bedding similar in scale and kind to that found in ordinary sandstones. The cross-bedding of marine calcarenites averages less than 1 foot (18 cm), but beds up to several meters in thickness are known (Sedimentation Seminar, 1966; Adams, 1970). The cross-bedding is notably bimodal; the two modes are about 180° apart, indicating tidal control (Sedimentation Seminar, 1966; Knewtson and Hubert, 1969; Hrabar, Cressman, and Potter, 1971). Eolian calcarenites, on the other hand, display unimodal cross-bedding, much of it

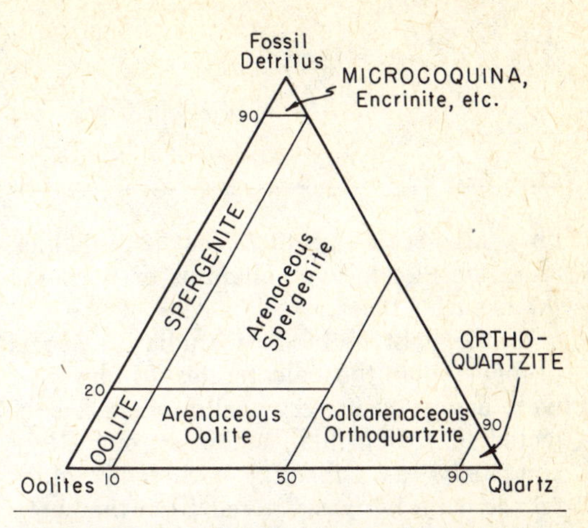

FIG. 10-34. Classification of the calcarenites and related rocks based on the composition of the framework fraction.

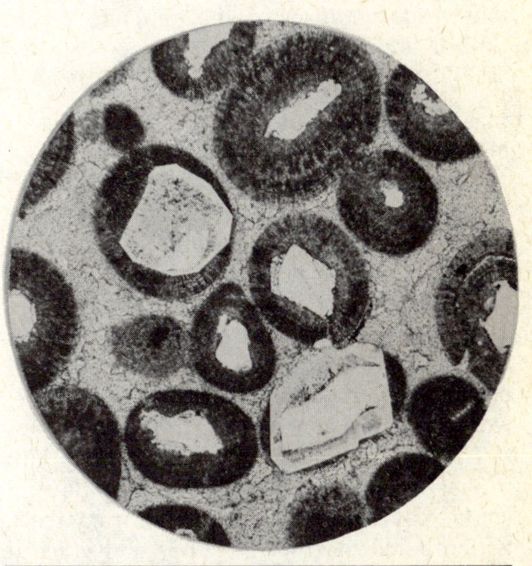

FIG. 10-35. Hale Formation, Morrow Group (Pennsylvanian), Fayetteville, Arkansas. Ordinary light, ×40. A calcareous oolite (oosparite) cemented by clear calcite. Detrital quartz nuclei of ooids have been enlarged by outgrowth which has replaced ooid. Note minute carbonate inclusions of quartz in which faintly seen is the structure of the replaced ooid.

FIG. 10-33. St. Louis Limestone (Mississippian), Beardstown Quadrangle, Illinois. Crossed nicols, ×25. Large rounded quartz, quartzite, and chert grains, together with small silt-sized particles of quartz in a carbonate matrix; matrix dominant. Hand specimen shows limestone pebbles.

on a large scale—up to 75 feet (23 m) thick as in the Pleistocene eolianites of Bermuda (Mackenzie, 1964).

In the light of modern studies it seems probable that the calcarenites of the past were deposited in very shallow waters subjected to strong tidal action or, much less commonly, as subaerial carbonate dunes whose material was derived from the shore zone to which the carbonate detritus from shallow-shelf seas migrates. The petrography and internal structure of carbonate sand banks in the Bahamas resemble in all important respects those of the geologic record (Imbrie and Buchanan, 1965).

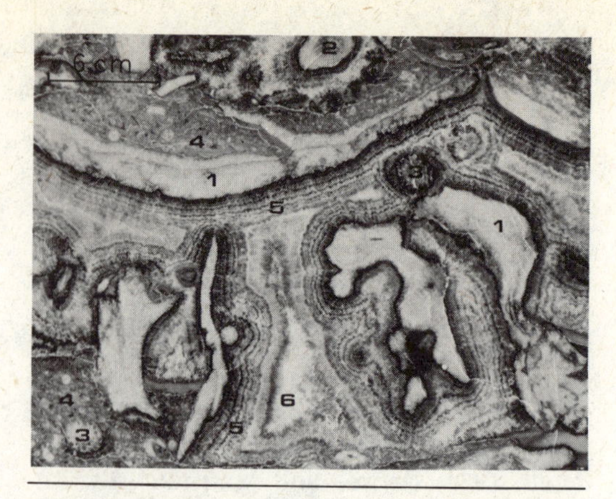

FIG. 10-36. Reef rock—boundstone or biohermite —(Devonian), central Germany. Fibrous calcite in primary voids. 1: Alveolites; 2: solitary rugose coral; 3: tabulate coral; 4: detritus; 5: fibrous calcite; 6: drusy calcite. (From Krebs, 1971, Fig. 11, p. 63.)

BOUNDSTONES (REEF ROCK)

Boundstone is a term introduced by Dunham (1962) for rocks bound together during deposition. The term is more or less equivalent to "reefoid carbonate," "biohermal carbonate," and "klintite," all of which are rocks formed by encrusting and sediment-binding organisms and which constitute the framework structure of wave-resistant reefs.

The rock itself tends to be a loosely knit reticulating network with large voids and coated cavities (Fig. 10-36). The framework which binds the rock together and gives it its rigidity and wave resistance varies with place and age of the deposit. Corals are, and have been, important framework builders in the past—hence the designation "coral reefs." But as seen today, they play a lesser role than do some other forms, notably the encrusting algae such as *Lithothamnium* and related forms (Adey and Macintyre, 1973). In other places stromatoporoids, even fenestrate bryozoans (Pray, 1958), and rudistids produce the boundstones of reeflike deposits.

One-half the volume of the rock may be voids and coated cavities. The cavities have a drusy lining of calcite crystals. Larger cavities are coated with explanate masses of thinly laminated onyxlike calcite which resembles the flowstone of caves. A few of the larger pockets have been filled with a laminated sediment—shale in some cases, laminated micrite in others. Many cavities display such internal sedimentation in the lower part of the void with drusy spar in the upper half, thus forming a *geopetal* fabric. Many cavities appear to be original voids in the reef; others show by their shape that they are openings left by solution of fossils; still others are irregular openings of uncertain origin. The *Stromatactis* of many reefs, especially the Silurian reefs of the Great Lakes area, are such structures of unknown origin. Some of the reef rock has the appearance of a rude breccia. Reef rock seems to be especially susceptible to dolomitization, and many fossils reefs are wholly dolomite.

In general, the central part of the reef core is devoid of bedding or related structures except for a *Stromatactis* framework. The core may be exceptionally fossiliferous, but in other cases extensive dolomitization and solution seem to have destroyed all organic forms. In general, the fossils mainly are internal and external molds, casts, and impressions.

Reefs and reef rock have been described from many places. Among the most intensively studied and best known are the Niagaran (Silurian) reefs of the Great Lakes region (Cumings and Shrock, 1928; Shrock, 1939; Lowenstam, 1950), the Silurian reefs of Gotland (Hadding, 1941), the Lower Carboniferous "reef knolls" of Ireland (Lees,

1964) and England (Parkinson, 1957), the Mississippian (Pray, 1958) and Pennsylvanian bioherms of New Mexico (Plumley and Graves, 1953), the Pleistocene reefs of Guam (Schlanger, 1964), the Devonian reef complexes of the Rhenish Geosyncline (Krebs, 1971; Franke, 1971), and the Devonian reefs of Alberta, Canada (Klovan, 1964).

CARBONATE MUDSTONES (CALCILUTITE AND CALCIGRANULITES)

Carbonate mudstones consist largely of a very fine-grained carbonate. These rocks have been called *calcilutites* and *calcipulverites* by Grabau (1904), the former a current-deposited mud (the mud being the finest product of attrition) and the latter a fine precipitated product. This distinction, although an important one, is not readily made, and the term *calcilutite* is commonly used for all fine-grained limestones without respect to their origin. Folk (1959) applied the term *micrite* to this class of rocks; although he restricted it to microcrystalline calcite under 4 microns in size, it has been applied to somewhat coarser materials and has also applied to modern aragonitic muds.

Some micritic limestones on close examination are found to contain some larger constituents, such as pellets, skeletal elements, and spherulites (rarely true oolites and intraclasts). To these Folk would apply such terms as *biomicrite* and *pelmicrite* if these constituents formed over 10 percent of the rock. If these were sparse (under 10 percent), the rock is described as a fossiliferous micrite or the like (Table 10-8). Rocks lacking any such larger components are simply micrite. Micrites with 10 or more percent of skeletal and other grains are Dunham's (1962) *wackestones* (Figs. 10-37, 10-38, 10-39), unless they display a grain-supported framework, in which case they are designated *packstones*. Many micritic limestones lack any larger components and are dense and homogeneous. These are the so-called "lithographic limestones" (mudstones of Dunham); under the microscope a surprising number show a well-defined pellet structure. Those micritic limestones which contain irregular patches, tubules, or lenses of sparry calcite—thus having a fenestral fabric—were designated *dismicrite* by Folk. The origin of these fenestrae have been discussed elsewhere in this chapter. The bed-

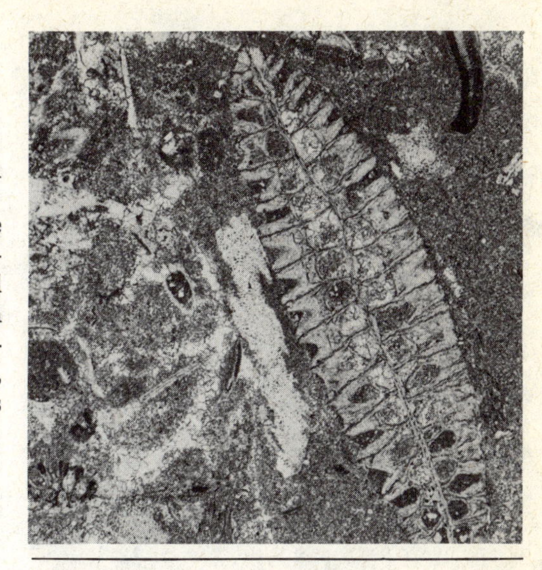

FIG. 10-37. Limestone (biomicrite, wackestone) bed, Decorah Shale (Ordovician), St Paul, Minnesota. Ordinary light, ×30. Shell debris—brachiopods and bryozoans—in a lime mud matrix.

FIG. 10-38. Foraminiferal limestone (biomicrite), (?) Eocene from the Pyramids, Ghizeh, Egypt Ordinary light, ×36. Several entire (?) *Nummulites* tests, embedded in a lime mud which contains much comminuted fossil debris.

ding of calcilutites varies from platy (*Plattenkalk*) to nodular (*Knollenkalk*).

Many fine-grained calcilutites are so homogeneous and devoid of structure that they have indeed been used for lithographic purposes. Some, however, display delicate laminations. The term *laminite* has been applied to these limestones. Although some laminations may be current-generated, most are attributed to accumulation under the influence of algal mats. These form, in a sense, a species of stromatolite (see p. 127). Presumably the sediment surface was covered by a thin filamentous mat of blue-green algae. An influx of sediment covered the mat with a thin (1 mm or less) coating of sediment. The mat promptly reestablished itself and in due time received another coating of sediment. Sedimentary particles tend to adhere to the mat or be trapped by it. In addition to such sediment-binding action, some carbonate may have been precipitated by algae. Ancient carbonate laminites preserve the morphology of the mats even after organic matter has become degraded and has disappeared.

Algal laminations tend to be discontinuous, are commonly irregular, showing small to larger convex archings (Fig. 4-30). They may be interrupted by downward-wedging cracks—probably mud cracks. Laminites are perhaps indicative of hypersaline waters or other environments in which snails and other grazing forms are absent. Extensive grazing destroys the mat and prevents the formation of laminites (Garrett, 1970).

Thin-section study of the micritic limestones has contributed little. Except as noted above, where one can see distinct pellets, the microcrystalline carbonate paste is generally difficult to resolve in the standard thin section. Ultrathin sections are more helpful, but most progress is apt to be made by use of the electron microscope (Loreau, 1972; Fischer, Honjo, and Garrison, 1967; Lobo and Osborne, 1973). The latter reveals a variety of grain shapes and grain fabrics, so that it may well become possible to determine the origin of the grains, to unravel diagenetic history, and perhaps to determine

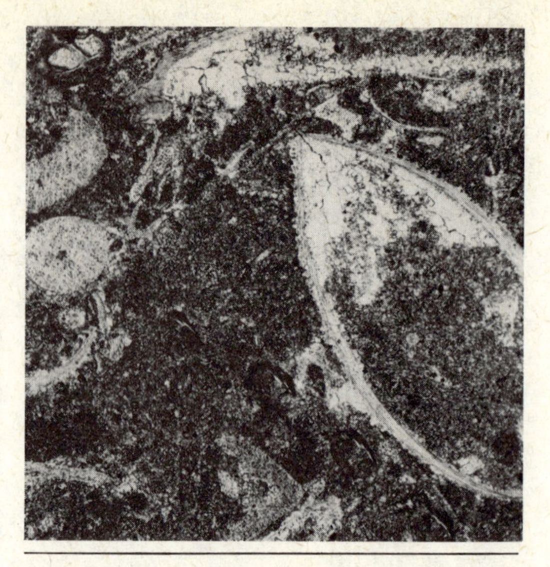

FIG. 10-39. Coral limestone (biomicrite), Wenlock (Silurian), Shropshire, England. Ordinary light, ×30. Entire brachiopod shell and debris from echinoderms embedded in lime mud.

the manner and environment of deposition. Some calcilutites, particularly those of Jurassic age or younger, contain an abundance of coccoliths and hence are clearly of biologic origin. Indeed many younger fine-grained limestones may be accumulations of coccoliths or "algal dust" (Wood, 1941). Examination under the scanning electron microscope convinced Lobo and Osborne that much, if not most, of an Ordovician micrite in Ohio was comminuted skeletal material.

The origin and significance of the several types of fine-grained, dense, and structureless limestone is a moot question. As Grabau (1904) recognized, some may be the finest products of marine attrition just as many calcarenites are formed by such action. But others may be chemical or biochemical precipitates. It is not even clear how present-day carbonate muds originate. The fine impalpable aragonitic muds of the Bahama region ("drewite") has been attributed to bacterial action (Drew, 1914; Bavendamm, 1931, 1932). They have been also ascribed to direct chemical precipitation of calcium carbonate (Cloud, 1962) and to release of aragonitic needles from algal tissue on disintegration of the latter (Lowenstam, 1955; Stockman, Ginsburg, and Shinn, 1967). The oxygen isotopic composition of inorganic carbonate and that biochemically precipitated are different, and isotopic analyses of the Bahamian carbonate muds seems to indicate an algal origin (Lowenstam and Epstein, 1957). In the Persian Gulf, however, aragonitic muds are purely chemical precipitates (Kinsman, 1964).

Chalk is also a fine-grained carbonate rock which differs from the usual micritic limestone in being friable and porous rather than dense and well indurated. Normally it is white and consists almost wholly of calcium carbonate (Table 10-4G) as calcite. The carbonate content varies from 90 to 98 percent in the French chalk; the Kansas chalk, 88 to 98 percent, averages 94 percent (Runnels and Dubins, 1949). Under the microscope, chalk consists of the tests of microorganisms composed of clear calcite set in a structureless matrix of fine-grained carbonate. In the Kansas chalk microfossils form 17 to 34 percent of the rock; the balance is the matrix. *Globigerina*, *Textularia*, and other foraminiferians are the most conspicuous. Also present in chalk are spikes and cells of planktonic algae known as *rhabdoliths* and *coccoliths*, together with a few sponge spicules and radiolarian tests.

The best-known chalks seem to be of Cretaceous age. The most famous is that exposed on both sides of the English Channel, the type locality of the Cretaceous (Latin, *creta*, chalk) system. In North America chalk occurs extensively in Cretaceous beds in Alabama, Mississippi, and Tennessee (Selma Chalk) and in Nebraska and adjoining states (Niobrara Chalk). The Fort Hays Chalk of Kansas has been described in some detail (Runnels and Dubins, 1949).

Chalk is a friable carbonate. Although it may contain chert nodules, in some cases in rhythmically spaced layers (Richardson, 1919), it is an almost unaltered deposit. That solutions moving through such a porous and easily alterable material affect so little change is indeed remarkable. It has been suggested that, unlike other lime-carbonate deposits, chalk was precipitated as calcite instead of aragonite, and, because of the greater stability of this substance, no reorganization took place as it would have if the original precipitated carbonate was aragonite. It failed, therefore, to become a dense hard rock. There are, however, notable exceptions, and "hard grounds" of well-cemented chalk are known (Voigt, 1959).

ALLODAPIC (GRADED) LIMESTONES

The term *allodapic* was applied to certain limestones believed to have been deposited by turbidity currents in relatively deep water that derived their load from shallow-water reef areas (Meischner, 1964).

Allodapic limestones are rhythmically interstratified regular tabular limestone beds and pelitic layers. Individual beds are traceable over long distances without appreciable change in thickness. They lack any features indicative of shallow-water deposition. They exhibit graded bedding with a sharp lower contact and an indistinct top; the base of the bed is in places marked by flute and groove casts (Kuenen and ten Haaf, 1956; Tucker, 1969; Ketner, 1970). The characteristic turbidite cycle is shown by some beds, that is, a coarse lower-graded interval followed by plane-laminated bedding and in a few cases by current-ripple lamination and convolute bedding (Fig. 10-53).

Allodapic limestones, like their siliceous counterparts, are a "petrologic complex" and are not a well-defined rock species, but because of their unique character they are here given separate status. In many, the lower part of each bed is coarse. Sorting is poor; the material is largely poorly rounded, coarse fossil debris. The fossils are mainly shallow-water marine benthonic forms, mostly reef dwellers with an admixture of pelagic fossils; pelitic intercalations contain only pelagic fossils—both nectonic and planktonic, benthos being absent. The lower interval is generally a packstone (grain-supported detritus with a mud matrix); the upper interval is a fine-grained calcilutite or micrite. The more distal portions of a bed may be largely fine-grained.

TUFA, TRAVERTINE, AND MARL

Tufa and travertine are limestones formed by evaporation of spring and river waters. *Tufa* is a spongy porous material that forms a thin, surficial deposit about springs and seeps and exceptionally in rivers. The carbonate of lime is deposited on growing plants and therefore commonly bears the imprint of leaves and stems. It has a reticulate or explanate structure; much of it is weak and semifriable (Fig. 10-40). Tufa is seldom extensive and is restricted mainly to Recent or Quaternary deposits. Tufa of a dense and more durable nature is found today associated with slightly hypersaline lakes, such as Pyramid Lake in Nevada. It is found also as somewhat more widespread deposits along the shores of the now-extinct Lake Lahontan (Russell, 1885, pp. 189–222) and Lake Bonneville (Gilbert, 1890, pp. 167–169) of the Great Basin region. The Lahontan tufa takes several forms: *lithoid tufa*, a comblike coating on various materials, *thinolitic tufa*, interlaced crystalline calcite pseudomorphic after gaylussite, and *dendritic tufa*, which forms spherical to hemispherical bolsterlike bodies with a coarse radial structure. These tufa deposits form sizable mounds or domes in many places along ancient lake shores.

Travertine is a more dense, banded deposit especially common in limestone caverns where it forms the well-known flowstone and dripstone, including stalactites and stalagmites (Allison, 1923). Like tufa it forms relatively small deposits of no great geologic importance and is primarily Quaternary or Recent in age.

Marl, a term of varied meaning, has been applied to glauconitic greensands but is used most commonly to designate certain friable carbonate earths accumulated in Recent or present-day freshwater lakes (Davis, 1900). Certain plants, notably the stonewort (*Chara*), are able to obtain carbon dioxide for photosynthesis from CO_2 in solution. Calcium carbonate, therefore, is precipitated as a crust on the leaves and stems of the plant. This deposit is sloughed off from time to time and accumulates on the bottom of the lake. Freshwater limestone as thus formed will be characterized by a pseudobrecciation. Small angular chips or flakes of calcium carbonate are embedded in a mud of similar composition. To the material thus produced may be added carbonate precipitated by microorganisms (Williams and McCoy, 1934) and that secreted by freshwater mollusks. Freshwater marls are somewhat argillaceous and are commonly used as an ingredient in the manufacture of portland cement (Fig. 10-41).

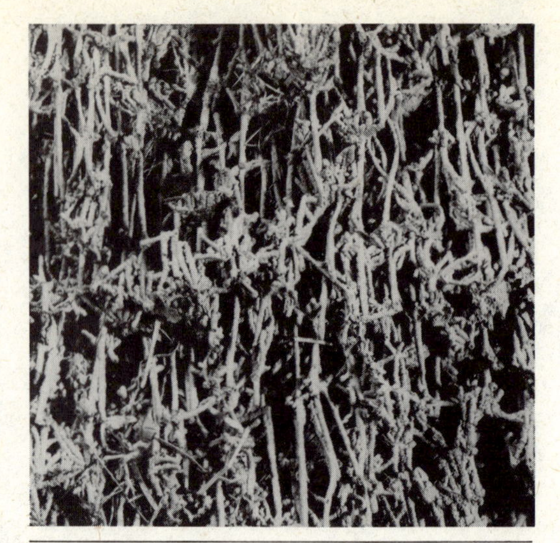

FIG. 10-40. Calcareous tufa (Recent). About natural size. Calcareous material deposited on plant stems. Note high porosity and spongy character.

Freshwater limestones in the geologic record are characterized by a paucity of marine fauna; they may contain *Spirorbis*, various ostracods, and occasional mollusks. Many are algal and show algal structures. Some are nodular. The nodular masses may show calcite-filled syneresis cracks. Under the microscope the rock has a clotted appearance, being made up of rounded clotlike masses of dense fine-grained carbonate held together with clearer, coarser-textured carbonate (Fig. 10-42).

Marlstone (or marlite) is a term applied to better-indurated rocks of about the same composition as marl. They are less fissile than shales and, like mudstone, have a blocky subconchoidal fracture. The term *marlstone* has been extended to include other calcareous rocks—a practice which has been criticized (Picard, 1953).

CALICHE AND OTHER EVAPORITIC CARBONATES

Caliche is a lime-rich deposit formed in the soils of certain semi-arid regions. Capillary action draws the lime-bearing waters to the surface where, by evaporation, lime-rich caliche is formed. Old caliches are likely to be well indurated (*calcrete*). Some are marked by large, concentrically banded, bolsterlike masses. In some regions caliche forms the cap rock of mesas and buttes (Price, 1933; Bretz and Horberg, 1949). The several stages in caliche development are well documented by Gile and co-workers (1966).

When examined under the microscope, these caliches show pisolitic bodies, concentric rings of calcite filling shrinkage cracks, corrosion and varying degrees of replacement of detrital quartz and feldspar, and concentrically banded structures resembling those produced by algae. These pseudo-oncolites and pisolitic bodies have been thought by some to be algal but are now generally considered to be pedogenic and related to pisolites of bauxitic and lateritic soils (Swineford, Leonard, and Frye, 1958). Caliches are primarily calcium carbonate as calcite; magnesium carbonate is rare (Goudie, 1972). Opal is present in some caliches (Emery, 1945).

Caliche is also found in the older geologic record as small calcareous nodules ("race") in the upper parts of fining-upwards alluvial cycles deposited under arid climatic conditions. In some cases these nodules coalesce and form a limy bed as in the Old Red Sandstone of North Wales (Allen, 1965). Under the microscope these caliches display a pelletoidal structure, containing silt particles, surrounded by coarsely crystalline calcite free of the detrital silt. Some of the larger pellets are rudely banded. Similar caliche nodules are known from the upper red mudstone members of similar alluvial cycles of the Catskill (Devonian) and Mauch Chunk (Mississippian) in Pennsylvania. Caliche pisolites have also been described from the Capitan Reef (Permian) of the Guadalupe Mountains of New Mexico which were formerly considered to be algal (Dunham, 1969).

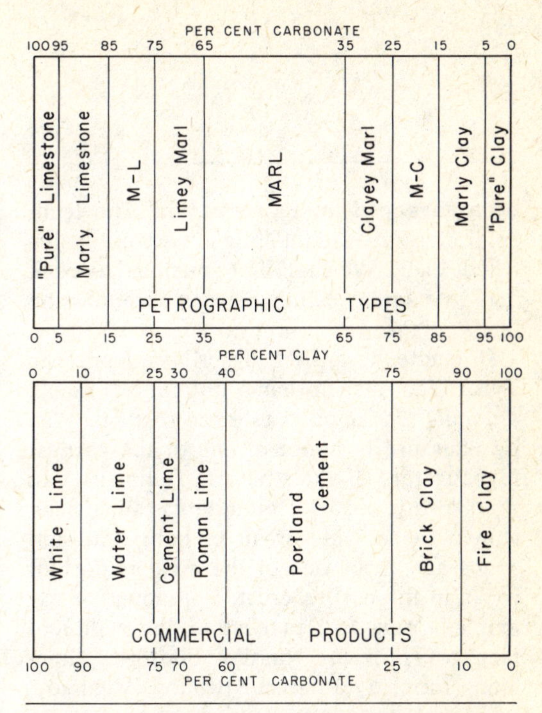

FIG. 10-41. Classification of clay-lime carbonate mixtures; top, petrographic classification; bottom, classification based on commercial use. (After Barth, Correns, and Eskola, 1939.)

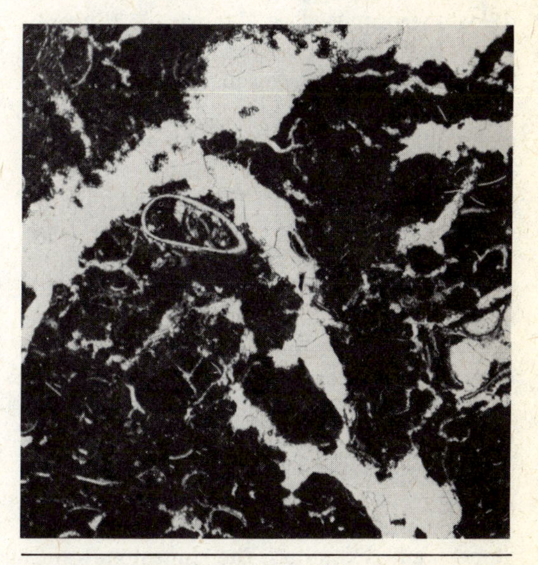

FIG. 10-42. Greenbush Limestone (Pennsylvanian), Illinois. Ordinary light, ×35. An exceedingly fine aggregate of carbonate (black) in which are embedded a few shell fragments (clear lemniscate areas) cut by irregular cracks and openings filled with vein calcaite and microcrystalline silica (clear). A freshwater limestone.

DOLOMITE

As noted by Van Tuyl (1916a), dolomite and the problem of its origin have long occupied the minds of geologists; many theories have been advanced for its formation. The technical literature on dolomite is very extensive (Clee, 1950). We shall first consider the rock and in a later section discuss the question of its origin.

Dolomite was first described from the Tyrol. The specific name dolomite (French, *dolomie*) was applied to these rocks in 1792 by Saussure in honor of one of the earliest investigators of their nature, Dolimeu (Van Tuyl, 1916a, p. 257). Dolomite has since been shown to be widespread in both time and space and to be one of the most important rocks in the earth's crust. Dolomite, for example, forms 10.2 percent of the sedimentary cover of the Russian Platform; limestone forms 17.6 percent (Ronov, Migdisov, and Barskaya, 1969).

Nomenclature *Dolomites* are those varieties of limestone containing more than 50 percent carbonate, of which more than one-half is dolomite. Because it is also used as a mineral name, its abandonment as a rock term has been advocated and the term *dolostone* proposed as a substitute (Shrock, 1948). Despite the possible ambiguity arising from the usage of the same term for both a rock and a mineral, the term has persisted and will probably continue to be used for both.

Rocks intermediate in composition between limestone and dolomite have been variously named. In general, those in which calcite exceeds dolomite are called dolomitic limestone; those in which dolomite exceed calcite are limy, calcitic, calciferous, or calc- dolomites. The precise line of division between these mixed rocks and the more homogeneous end-members has not been agreed upon. The predominance of calcite over dolomite or vice versa is taken as the basis of the division between limestone and dolomite. The limestone end-member is called a *high-calcium limestone* if it contains only a little magnesium. If it contains appreciable magnesium, it is a *magnesian limestone*, and, if dolomite is conspicuous, it is a *dolomitic limestone* (Fig. 10-43). If the content of magnesium is small,

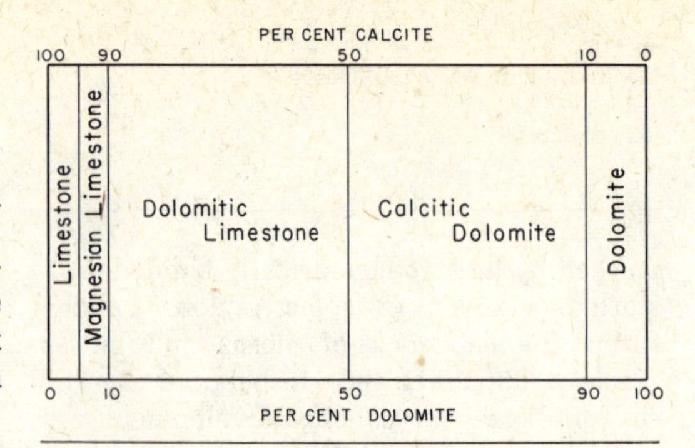

FIG. 10-43. Classification of calcite–dolomite mixtures.

this element is present as $MgCO_3$ in solid solution in the calcite, and the mineral dolomite is absent. Calcite, however, ordinarily can contain only 1 or 2 percent of $MgCO_3$ in solid solution. Only in some present-day skeletal materials is it present in larger quantities. In ancient rocks, therefore, a magnesian limestone will contain an appreciable, although small, amount of dolomite.

Rocks consisting of mixtures of calcite and dolomite are in fact far less common than the end-members consisting primarily of calcite or dolomite. As Fig. 10-8 and Table 10-9 show, carbonate rocks with more than 10 percent but less than 90 percent dolomite (less than 90 but more than 10 percent calcite) are comparatively rare, a fact early noted by Steidtmann (1917) and confirmed by Goldich and Parmalee (1947), Chave (1954b), and others. If the boundary between the dolomite end-member and the

TABLE 10-9. Relative abundance of limestones and dolomites[a]

Percent dolomite[b]	Number of samples
0–10	48
10–50	8
50–90	5
90–100	97
Total	158

[a] In the Cedar Valley, Wapsipinicon, Gower, and Hopkinton formations (Silurian and Devonian) of Iowa.
[b] Petrographically determined.
Source: After Scobey (1940).

Even traces of iron lead to a buff color upon weathering, a characteristic caused by oxidation of the iron which serves to distinguish between limestone and dolomite in the field.

mixed calcite–dolomite rock is placed at 90 percent dolomite and that between limestone and the mixed rocks is placed at 10 percent dolomite, the bulk of the carbonate rocks is either limestones or dolomites. A suggested nomenclature is summarized in Table 10-10. A more precise classification, based on mineral composition as calculated from chemical analysis, has been developed by Martinet and Sougy (1961). See also Chilingar (1957), Rodgers (1954), Cayeux (1935, p. 339) and Fairbridge (1957, p. 128) for various compositional classifications.

Composition In chemical composition dolomites resemble limestones except that MgO is a large and important constituent (Table 10-11. The magnesium content of the sedimentary carbonates appears to show a secular trend (Fig. 17-2). Older carbonates are higher in magnesia than those of later geologic times (Daly, 1909, p. 165); that is, dolomites are more common in the older record.

Some dolomites are closely associated with evaporites and contain anhydrite or gypsum. Cherty varieties are notably siliceous as, of course, are arenaceous dolomites. Ferrous carbonate occurs in solid solution in dolomite but not calcite. Iron-rich varieties are referred to as ferrandolomites.

Textures and structures Some observers (Steidtmann, 1917, pp. 445, 449; Folk, 1973, p. 129) have presumed that dolomites fall into two classes separated by a crystal-size boundary of 10 to 20 microns: an extremely fine-grained crystalline dolomicrite which occurs characteristically in thin, commonly laminated beds which show ripple marks, mud cracks, and erosional features and lacks the ghosts of fossils, oolites, and the like; and a more coarsely crystalline or saccharoidal dolomite with abundant evidence of replacement. The first is presumed to be "primary"; the second, an epigenetic or diagenetic replacement of limestone. Because the latter is perhaps the most common, we will describe it first.

Dolomites closely resemble limestones in their grosser aspects, but close inspection shows many differences. In dolomites which are the product of dolomitization of a limestone, earlier textures and structures tend to be obscured or even obliterated. The clastic texture of a calcarenite, for example, may become very obscure. The clastic character is suggested by a ghostlike pattern faintly outlining the original framework grains. Dolomitized oolites and fossils appear as faint shadows outlining the original shapes but are largely without any internal detail.

TABLE 10-10. Nomenclature of sedimentary calcitic and dolomitic carbonates

Type	Percent dolomite	Approx. MgO equivalent percent	Approx. $MgCO_3$ equivalent percent
Limestone			
High calcium	0–10[a]	0–1.1	0–2.3
Magnesian		1.1–2.1	2.3–4.4
Dolomitic limestone	10–50	2.1–10.8	4.4–22.7
Calcitic dolomite	50–90	10.8–19.5	22.7–41.0
Dolomite	90–100	19.5–21.6	41.0–45.4

[a] Dolomite not present in high-calcium limestone; magnesium carbonate is in solid solution in calcite.

In other cases, only scattered grains of detrital quartz bear witness to the original clastic nature of the rock. Although dolomites are generally unfossiliferous, in some examples fossils remains are visible to the unaided eye as internal and external molds. The preservation of detail is poor and such cavities tend to be lined with drusy dolomite.

Dolomitization involves large-scale recrystallization. The end-product has a granoblastic texture. Complete recrystallization produces a medium to coarsely crystalline mosaic in which, however, many of the dolomite crystals tend to show a euhedral form (Fig. 10-44). This is the idiotopic fabric

TABLE 10-11. Chemical composition of dolomites

Constituent	A	B	C	D	E	F
SiO_2	—	2.55	7.96	3.24	24.92	0.73[a]
TiO_2	—	0.02	0.12	—	0.18	—
Al_2O_3	—	0.23	1.97	0.17	1.82	0.20
Fe_2O_3	—	0.02	0.14	0.17	0.66	—
FeO	—	0.18	0.56	0.06	0.40	1.03[b]
MnO	—	0.04	0.07	—	0.11	—
MgO	21.9	7.07	19.46	20.84	14.70	20.48
CaO	30.4	45.65	26.72	29.58	22.32	30.97
Na_2O	—	0.01	0.42	—	0.03	—
K_2O	—	0.03	0.12	—	0.04	—
H_2O+	—	0.05	0.33	0.30 (H₂O+ and H₂O− combined)	0.42	—
H_2O-	—	0.18	0.30		0.36	—
P_2O_5	—	0.04	0.91	—	0.01	0.05
CO_2	47.7	43.60	41.13	45.54	33.82	47.51[f]
SO_3	—	0.03	—	—	0.01[c]	—
S	—	0.30[d]	0.19	—	0.16[d]	—
BaO	—	0.01[c]	none	—	none	—
SrO	—	0.01[c]	none	—	none	—
Organic	—	0.04[e]	—	—	0.08[e]	—
Total	100.0	100.06	100.40	99.90	100.04	100.97

[a] "Residue (mostly silica)."
[b] Calculated from reported iron.
[c] Constituent does not exceed figure given.
[d] Calculated as pyrite.
[e] Organic carbon.
[f] Calculated from ferrous iron, magnesia, and lime.

A. Theoretical composition of pure dolomite.
B. Dolomitic limestone, Monroe Formation (Devonian). Downs Schaaf, analyst (Stout, 1941, p. 564).
C. Niagaran Dolomite (Silurian), Joliet, Illinois. D. F. Higgins, analyst (Fisher, 1925, p. 34). MgO dolomite equivalent is 89.5 percent.
D. "Knox" Dolomite (Cambro-Ordovician), Morrisville, Alabama. W. F. Hillebrand, analyst (Russell, 1889, p. 25). MgO dolomite equivalent is 96.5 percent.
E. Cherty dolomite, Niagaran Group (Silurian), Highland County, Ohio. Downs Schaaf, analyst (Stout, 1941, p. 82).
F. Randville Dolomite (Precambrian), Dickinson County, Michigan. E. E. Brewster, analyst (Bayley, 1904, p. 215). MgO dolomite equivalent is 94.5 percent

of Friedman (1965). Incomplete dolomitization produces a scattering of dolomite euhedra in an unaltered calcitic matrix (porphyroid texture of Friedman). See Fig. 10-45. These dolomite rhombs are commonly zoned with a central portion clouded with inclusions and a relatively clear peripheral part (Fig. 10-46). In a few cases the crystals show alternately clear and clouded zones. In other rhombs the central part is calcite and the exterior dolomite. Dolomite rhombs in some arenaceous dolomites may enclose small grains of quartz. The dolomite is automorphic against calcite. The rhombs transect primary structures such as fossils, oolites, and glauconite grains (Fig. 10-6); dolomite, therefore, is a replacement product.

Incomplete dolomitization results in a mottled rock marked by a patchy distribution of dolomite. The mottled appearance of the rock is best seen in a weathered or etched surface. Mottled dolomite has been described many times (Van Tuyl, 1916a, p. 342, pls. 21, 22; Birse, 1928; Griffin, 1942; Beales, 1953; Sando, 1953, p. 40) and is probably rather common. The dolomite areas in such rocks are very irregular in shape and in some cases form an anastomosing network (Fig. 10-47). Few patches of dolomite are wholly isolated. The mottled pattern has

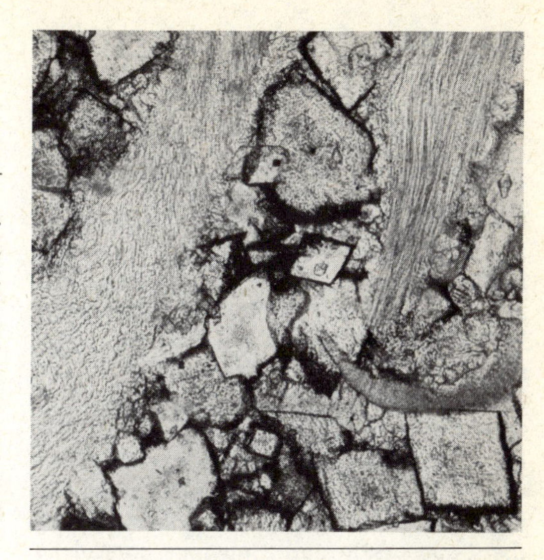

FIG. 10-45. Dolomitic limestone, Rochester, New York. Ordinary light, ×80. A coquinoid limestone (biomicrite or biomicrudite), composed chiefly of brachiopod shells (*Leptocoelia*), now fibrous calcite. The calcareous mud between the brachiopod shells has been converted to coarse dolomite rhombs.

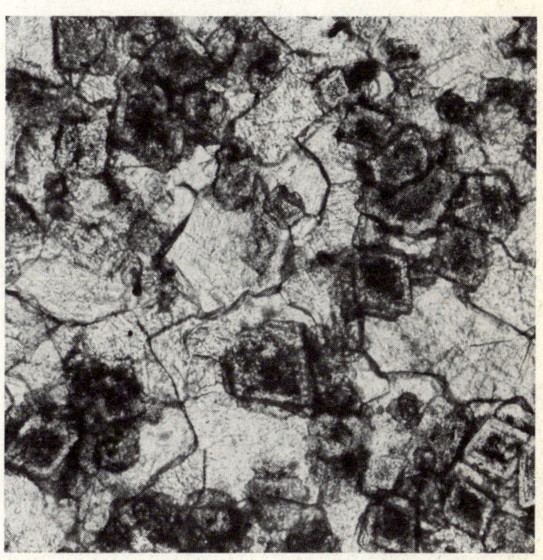

FIG. 10-46. Dolomitic limestone, Minnesota. Ordinary light, ×80. Zoned porphyroblasts of dolomite scattered through an anhedral mosaic of calcite.

FIG. 10-44. Glauconitic dolomite (Cambrian), Colorado. Ordinary light, ×31. A coarsely crystalline dolomite (gray rhombs) with glauconitic granules (black) and some scattered quartz (white). Not well shown are overgrowths on quartz.

been interpreted as a product of arrested dolomitization produced by migration of a magnesium-bearing solution through the rock (Griffin, 1942; Beales, 1953). If mottling is indeed an expression of incomplete dolomitization, its pattern may reflect control by some preexisting structure in the limestone being replaced (Osmond, 1956). This control has been thought by some to be organic—perhaps "algal" tubes (Birse, 1928) or more likely "burrows." In some cases there appears to be no obvious control by bedding or other structures, and the isolation of some dolomite patches is difficult to explain. Perhaps the mottling is the product of diagenetic differentiation resulting from unmixing of the high-magnesium calcites of organic origin. The exsolved magnesian carbonate might form dolomite either as isolated rhombs or as segregated patches forming the mottles. The quantity of dolomite, some 0 to 30 percent in the Ordovician Platteville Limestone in the upper Mississippi Valley (Griffin, 1942) and the 10 to 40 percent in the Devonian Palliser of Alberta (Beales, 1953) is approximately that to be expected by reorganization of algal carbonate which is notably rich in magnesia. The incompleteness of dolomitization and its uniformity over vast areas and great thickness further suggest an internal source for magnesium.

Relations between dolomitic carbonate and precipitated silica (chert) are contradictory. In some rocks dolomite appears to be the earliest and to have been replaced by later silica, as chert pseudomorphs of dolomites demonstrate (Fig. 11-7). Evidence for the converse is less clear. *Dolocastic chert,* a common constituent of insoluble residues of some carbonate rocks, is a spongy chert in which are many rhomb-shaped cavities that were left by solution of the dolomite. Were the original dolomite rhombs metacrysts in the chert? If so, they must have replaced the chert. They might have been formed by the replacement of a calcite matrix in which were embedded numerous dolomite crystals.

Dolomites also occur, as noted earlier, as very fine-grained rocks which lack most of the features attributed to replacement. They contain no fossils or, at best, a restricted and somewhat different fauna than associated limestones (Strakhov, 1958). They tend to be thinly laminated, with partings of more bituminous materials. Not uncommonly they contain selenite crystals or nodules of gypsum or anhydrite. They are commonly associated with evaporites and display such primary sedimentary structures as ripple cross-bedding and mud cracks.

FIG. 10-47. Mottled dolomite, Palliser Limestone (Devonian), Banff National Park, Alberta, Canada. About half natural size. (Photograph by C. Weber.)

Occurrence and associations Dolomite occurs in rocks of all ages, although, as noted by Daly (1909), Strakhov (1958) and others, it is more common in the Paleozoic and older rocks. It is closely associated with limestone with which it is commonly interbedded. Beds of dolomite a foot or so thick or several tens of feet thick may be interbedded with limestone beds of a like order of thickness (Sarin, 1962). In some cases, however, the dolomite–calcite boundary crosses stratigraphic planes (Deininger, 1964). In some cases the distribution of dolomite seems to be controlled by structures such as faults (Ham, 1951) or folds (Landes, 1946). It is not rare for dolomite to grade laterally into limestone; such facies changes may be abrupt. Although not much is yet known about the paleogeography of dolomite, it appears that it is more commonly a near-shore facies, whereas limestone is the product of offshore and perhaps more normal marine waters—the near-shore area is presumed to be more saline (Fig.

10-48). There are cases, however, where the pattern is reversed, with limestone in peripheral regions and dolomite in the center of the basin (Strakhov, 1958). Dolomite may characterize back-reef areas, and limestones form in the reef or more seaward (Cloud and Barnes, 1948, p. 92). Dolomites of the Lower Silurian seem to be geographically distinct from coexisting limestones (Amsden, 1955), as do those of the Cambro-Ordovician of east Tennessee and southwest Virginia (Dunbar and Rodgers, 1957, Fig. 113; Harris, 1973).

In many places dolomite shows a close association with evaporites, particularly gypsum and anhydrite. In some cases the only evaporitic materials are scattered crystals of gypsum or anhydrite in dolomite. In other cases, dolomite is interbedded with sulfates. These dolomites tend to be fine-grained rather than saccharoidal and are commonly thinly laminated and brownish in color because of the presence of bituminous matter.

Where is dolomite being deposited today? Nowhere does dolomite seem to be forming on a large scale; by some this has been taken as evidence that dolomite is a diagenetic facies—not an original sediment—or that conditions in the past were markedly different from those of today. However, in recent years dolomite has been found, albeit in rather small volume, in a number of somewhat restricted environments. Despite several early references to modern dolomite, it was not until Alderman and Skinner (1957) reported dolomitic marl forming in the Coorang Lagoon of South Australia that much attention was given to the subject. Dolomite has since been found in supratidal crusts in Florida Bay (Shinn, 1964) and the Bahamas (Shinn, Ginsburg, and Lloyd, 1965), in the sabkha flats of the Persian Gulf (Illing, Wells, and Taylor, 1965), in the Netherlands Antilles (Deffeyes, Lucia, and Weyl, 1965) and in certain Pacific atolls (Berner, 1965). Other occurrences are also known. Among these are the dolomites of the shore flats of Lake Balkash (Strakhov, 1958).

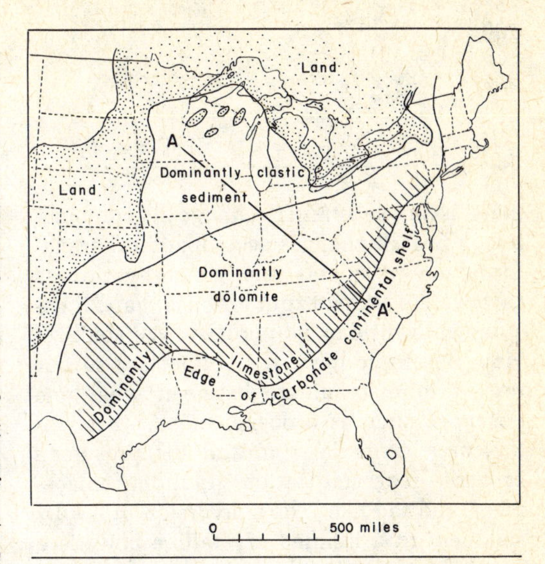

FIG. 10-48. Geography of dolomite in eastern United States. (After Harris, Fig. 1, 1973.)

These modern occurrences, although small in volume, seem to support the inferences drawn from the fine-grained and laminated dolomites of the geologic record regarding the conditions of dolomite formation. They seem to be tropical, many in arid or semi-arid environments, commonly in a tidal lagoon or flat. They seem to involve hypersaline brines enriched in magnesium. The close association of older dolomites with gypsum, their algal laminations and structures, and their mud cracking and other evidences of exposure or of hypersaline brines are consistent with the observations on modern dolomite development.

DIAGENESIS OF CARBONATE ROCKS

More than any other common sedimentary rock, perhaps, limestone is prone to alteration, both pre- and post-consolidation. Most profound are those changes in texture and composition which lead to the formation of dolomite (p. 369) and to replacement by silica, phosphate, and so forth (pp. 402, 425, and 434). In some cases these replacements involve introduction of foreign materials into the rock; in others the changes involve only rearrangements of materials already present—a process termed diagenetic differentiation.

But even in ordinary limestones, where

there is no substantial change in composition, a host of changes accompany the transformation of a soft and generally porous carbonate sediment into a dense, hard limestone with little or no porosity. We shall review this topic here. It is important in our review to discriminate between what one *observes*, such as a coarse mosaic, a stylolite, or a syntaxial rim, and the presumed *process*—recrystallization, solution, replacement, and the like. The conversion of a limy sediment to a lithified crystalline body is a species of metamorphism, the resulting textures are not dissimilar to those found in the more conventional metamorphic rocks. The student would do well, therefore, to become familiar with metamorphic textures (Spry, 1969), and certainly all students of limestone fabrics should be familiar with the monographic study of Bruno Sander (1951).

Questions of diagenesis and dolomitization of limestones have received a great deal of study. Attention is called to the special issues of *Sedimentology* devoted to the lithification of carbonate sediments (Füchtbauer, 1969), the symposium on dolomitization and limestone diagenesis (Pray and Murray, 1965), the Bermuda conference on carbonate cements (Bricker, 1971), as well as the early but still readable classic work of Van Tuyl (1916a) on dolomites and dolomitization. In addition are innumerable papers on these subjects, some of which are cited in the following section. Diagenetic fabrics of carbonate rocks and the presumed processes responsible for them are shown schematically in Fig. 10-49.

Although we have concentrated here on the effects of diagenesis—pore filling, syntaxial rims, micritic envelopes and the like—and have suggested the processes whereby these features have arisen, it should be kept in mind that the diagenetic changes which take place in a carbonate deposit are complex. Changes may be selective, such selectivity in part related to the original mineralogy of a skeletal element or ooid (Land, 1967). Moreover, the processes are complex, because they take place over a period of time, some at the time of deposition and

DIAGENETIC FABRICS IN AVONIAN LIMESTONES

PROCESS	FABRIC
CEMENTATION AND CAVITY FILLING	GRANULAR CEMENT
	DRUSY MOSAIC
	DRUSY FIBROUS CALCITE
	SYNTAXIAL CEMENT RIM
GRAIN GROWTH	COARSE MOSAIC
	SYNTAXIAL REPLACEMENT RIM
	REPLACEMENT FIBROUS CALCITE
GRAIN DIMINUTION	
	GRANULAR MOSAIC

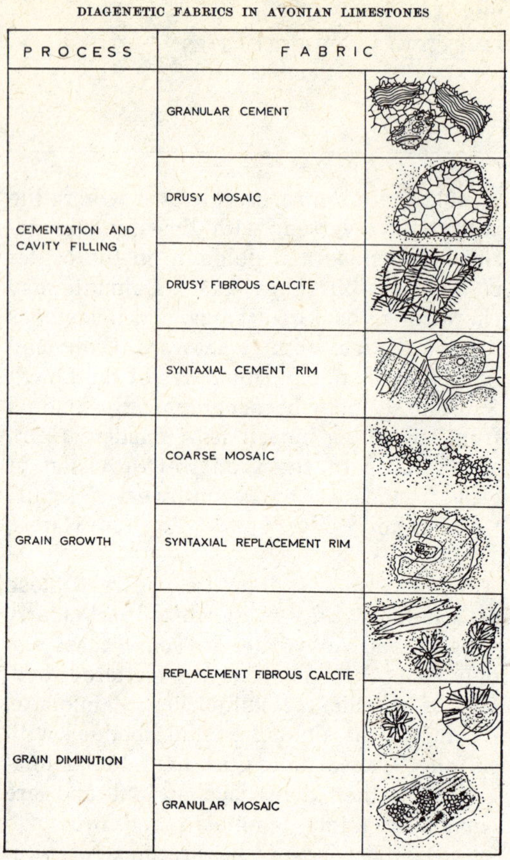

FIG. 10-49. Diagenetic fabrics in limestone. (From Orme and Brown, 1963, Proc. Yorkshire Geol. Soc., v. 34.)

some in the post-uplift stage. The diagenetic history can be very difficult to resolve (Friedman, 1964).

CEMENTS AND CAVITY FILLINGS

Carbonate cements form a significant part of the binding materials that hold limestones together. Until recently our knowledge concerning the lithification of carbonate sediments was based on microscopic analysis of the cement fabrics in ancient rocks. To a large extent it still is, but studies of active sedimentation of modern and Pleistocene carbonate sediments and the descriptive chemistry of carbonate cements and the waters associated with them have greatly added to our understanding of the cementation process. A detailed study of carbonate cementation, the topic of a symposium, appears in Bricker (1971).

A newly deposited carbonate sediment contains a variety of pores and openings, both large and small (see p. 335). Total porosity of some carbonate sediments may be as much as 75 to 85 percent (Dunham, 1962, pl. II). This intergranular porosity is lost by precipitation of a cement in the interstices of the framework in the carbonate sands and reduced by compaction and pressure solution at grain contacts in carbonate muds (Wolfe, 1968). The cement may appear as drusy coatings on the grains, a fibrous coating in some cases, radial into the void, coarser in the centers of the voids, and with planar boundaries between crystal units. The early-formed cement—as in beach rock—may be aragonitic or, in some cases a Mg-calcite, but in the ancient rocks the cement is calcite.

In some limestones the cement appears as *syntaxial rims* or outgrowths on some framework grains—those which are single crystals as are the skeletal elements of the echinoderms. This secondary overgrowth is a clear calcite distinguished from the original fragment, which is clouded with inclusions. The crystalline mosaic produced by syntaxial growth is marked by planar intergranular boundaries.

Larger cavities (such as body cavities of some skeletal structures and larger voids in reef rock) are in part mechanically filled. Fine carbonate mud or silt sifts into the cavity and forms a *floored cavity*. The rest of the cavity is then filled with a precipitated calcite, first coating the walls, then filling the remaining void space. Such composite fillings have a *geopetal* fabric (Sander, 1951, p. 56). Geopetal fabrics are very common in many limestones and are a guide to the "way up" in steeply tilted and overturned rocks (Shrock, 1949, pp. 281–284). Calcite is the usual void filling material, but in some cases dolomite is also present. It has been observed that if calcite and dolomite are both present, dolomite is earlier than calcite (Freeman, 1973).

There has been considerable discussion of the *time* at which the cementation took place (for example, Purdy, 1968). Some, certainly, are synsedimentary, as in the case of *beach rock* (Bricker, 1971, pp. 1–46). Beach rock, formed by simple pore filling, is forming today in the intertidal zone mainly in tropical regions. In general, the cement consists of micritic grain coatings but in other cases fibrous coatings, commonly aragonitic, are present.

Even reef rock becomes cemented while still in the marine environment (Krebs, 1969; Ginsburg, Schroeder, and Shinn, 1971, p. 54; Land, 1971, p. 59), as do other marine sediments (Purser, 1969; Bricker, 1971, pp. 47–49). These submarine cements seem to be fibrous aragonite or high-magnesium calcite, especially in grainstones and pelletal limestones. Cementation also takes place after uplift by circulating meteoric waters. The cements so formed tend to be a stable low-magnesium calcite and one having more negative oxygen and carbonate isotope values than those found in marine cements (Bricker, 1971, p. 121).

SOLUTION

Carbonate rocks are especially susceptible to solution. Such solution is commonly made evident by removal of a shell or other skeletal element or, in rarer cases, by removal of ooids, in either case leaving a void whose shape is a clue to its origin. Selective removal of ooids gives rise to so-called "oolicastic porosity." In other cases one sees irregular channelways, of both small and large scale. The latter are the well-known limestone caverns. Solution porosity may, at some later time, be reduced by refilling of the cavities by precipitated carbonates. Even caverns undergo a late stage of filling by flowstones—stalactites and stalagmites.

A more pervasive though less spectacular product of solution—pressure solution, to be more precise—are stylolitic seams; their character and origin have been discussed elsewhere (p. 340). As noted, they may record large-scale removal of the host rock, reducing stratigraphic thickness in some cases by as much as 40 percent (Stockdale, 1926). Many carbonate rocks display an abundance of microstylolites—small-scale true stylolites best seen in thin sections. These mark grain boundaries between one grain or fossil and another, or even between the grains or fossils and their matrix.

GRAIN GROWTH (AGGRADING NEOMORPHISM)

Many limestones show clear evidence of increase in grain size and crystallinity with aging of the deposit. To this process Folk (1965, p. 23) has given the name *aggrading neomorphism* (as opposed to degrading neomorphism or grain diminution). In some cases grain enlargement involves a single element in the rock; in other cases it is pervasive and involves the whole rock. An example of the first case, and a dramatic example of increase in the grain size, is the *syntaxial replacement rim*—an outgrowth or overgrowth on a crystal element, such as a crinoid ossicle, that extends into the matrix in which that element is embedded. The secondary rim, enlarging the original fragment, can only grow by replacing the surrounding matrix (Evamy and Shearman, 1965). Another example of neomorphism involving new crystal growth is the radially structured spherulites found in some fine-grained limestones. More common is the appearance of a coarse crystalline mosaic which invades, in an irregular way, shell debris, oolites, and matrix alike. It obliterates earlier textural features of the rock. This mosaic is characterized by curved, consertal, or implicate boundaries between grains.

In general the distinction between a precipitated void-filling calcite and neomorphic calcite is relatively clear, but in some cases this is not so (Stauffer, 1962, pp. 361–362; Folk, 1965, pp. 42–45). Mention has been made elsewhere of grumose limestone (p. 335). Are the micritic "islands" modified pellets in a sparry cement or are they relicts of incomplete recrystallization of a fine-grained limestone? Even the cement of some calcarenites has been interpreted as neomorphic, a recrystallization product of a micritic groundmass (p. 352). Whereas the sparry calcite of a calcarenite conceivably could be either a cement or neomorphic, that found in a non-grain-supported rock, a wackestone, is probably neomorphic. That spar filling the upper part of a floored cavity and that filling fractures is surely cement—not neomorphic (unless perchance the vein is a replacement vein). Bathurst (1971, p. 294) has pointed out that calcite cements show enfacial junctions much more commonly than does neomorphic calcite. An *enfacial junction* is a triple junction (meeting of three intercrystalline boundaries) in which one of the three angles is 180°. In the pattern of crystal growth of sparry cement, orientation of crystals normal to the surface and increasing crystal size away from the surface are features lacking in neomorphic calcite.

The mechanisms underlying this transformation of fine-grained carbonate to a coarse carbonate are not clear (Bathurst, 1958). The problem has been discussed at some length by Bathurst (1971, pp. 493–503) and Folk (1965, pp. 21–23). In some cases the neomorphic mosaic is a product of inversion of aragonite to calcite as in some ooids and some shells. In other cases the mosaic is coalescive recrystallization of a carbonate mud or micrite. The process involves the growth of some crystals at the expense of others. If such crystals are widely separated, the initial stage is a pseudoporphyritic fabric (porphyroid neomorphism of Folk, 1965, p. 22). If the centers of initial growth are more closely spaced, enlarging crystals soon impinge on one another and the porphyrotopic fabric will not develop, but instead we have a coalescive neomorphic fabric (Folk, 1965, p. 22).

GRAIN DIMINUTION (MICRITIZATION OR DEGRADING NEOMORPHISM)

In general, neomorphism leads to the coarsening of the rock fabric—to a general increase in crystal size similar to that which accompanies the conversion of a limestone to a marble. But in a few special cases diagenesis leads to grain diminution or the conversion of larger crystal elements to a mosaic of much smaller grains. The term *micritization* has been employed to designate this process (Bathurst, 1966).

The effects of this process are most strikingly shown in the micritic rim seen on some ooids and skeletal grains even in Recent sediments. The micritic rim could readily be mistaken for an accretionary micritic envelope. The micritic alteration zone, unlike the accretionary layer, has an irregular contact with the core of the grain. An accretionary layer, on the other hand, is apt to rest in sharp contact with the core grain and to be of more uniform thickness; it may even show two or more laminations. Bathurst (1966) attributes micritic alteration to boring algae, whose emptied bores are apt to be filled with micritic carbonate. Ancient carbonate rocks also show micritic envelopes which probably had a similar origin.

Wardlaw (1962) has described the conversion of a calcarenite to a calcilutite by recrystallization marked by grain diminution. He attributes this primarily to strain recrystallization—a feature common in metamorphic rocks—even though some of the limestones displaying this kind of alteration were only slightly deformed.

COMPACTION

Carbonate sands, by analogy with their siliceous counterparts, might be expected to show little or no compaction effects, whereas carbonate muds, like ordinary muds, might be compactable. Most investigators, however, believe that lime muds compact very little because of early cementation (Bathurst, 1971, p. 439; Pray, 1960; Zankl, 1969). Evidence supporting this view is diverse, but perhaps the most persuasive is the unbroken, uncrushed delicate shells embedded in many calcilutites. Although newly deposited lime mud undergoes early water loss, it apparently retains a high primary porosity.

There is, however, some contrary evidence on the compaction question. Terzaghi (1940) called attention to the oversteepened dip of the reef-flanking beds, which she attributed to greater compaction of these beds as compared with the reef core. Her experimental compaction of modern carbonate mud seemed to support this view (see also Robertson, 1965, p. 170; Fruth, Orme, and Donath, 1966). Biggs (1957, p. 23) noted that delicate fossils were preserved in chert nodules, whereas their counterparts in the host limestone were shattered, an observation he attributed to chertification prior to compaction of the lime mud. The question of compaction, therefore, is still open; perhaps those lime muds which do not undergo early cementation do compact, whereas others do not.

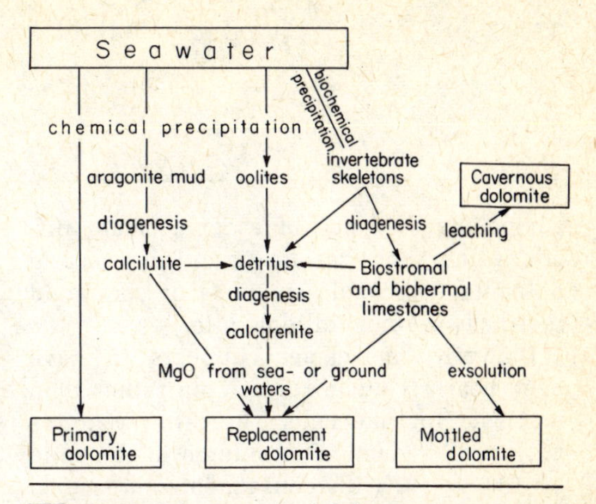

FIG. 10-50. Provenance of dolomite.

LIMESTONE REPLACEMENT AND THE DOLOMITE PROBLEM

The replacement of limestones by silica, by various iron-bearing minerals, by phosphate, and by other substances is well known. Such replacements are generally minor; only exceptionally will the limestone be wholly replaced. A much more common and widespread replacement is that which forms *dolomite*.

The origin of dolomite has been much debated, and the literature on the subject is extensive (Skeats, 1905, 1918; Steidtmann, 1911, 1917; Van Tuyl, 1916a, 1916b; Linck, 1937; Canal, 1947; Weynschenk, 1951; Fairbridge, 1957; Strakhov, 1958; Teodorovich, 1961; Murray and Pray, 1965, and many others). It is probable that dolomite is polygenetic (Fig. 10-50). Most dolomites are clearly replaced limestones. Evidence for replacement includes the automorphic boundaries of dolomite against calcite, even against clastic quartz and glauconite; the inclusion of clastic quartz in dolomite euhedra; the transection of oolites, fossil structures, and the like by dolomite euhedra; the shadowy palimpsest patterns of original calcite, bioclastic, or oolitic textures; the structural control of the distribution of dolomite; and the transgression of stratigraphic planes by the calcite–dolomite boundary. Most compelling evidence of a replacement origin is the fact that no organisms secrete dolomite; yet whole coquinas or coquinoid reef beds are now all dolomite.

Dolomite replacement appears to have been nearly volume for volume rather than molecule for molecule. The latter requires an overall reduction in volume in the ratio of 100 to 88, with consequent increase in porosity. Steidtmann (1917) has shown that there is no significant difference in porosity between limestone and dolomite, although Landes (1946) has described a case in which only the dolomitized portion of a formation is porous (and productive of petroleum). The porosity of such dolomites has been ascribed to selective removal of calcitic constituents of an incompletely dolomitized rock (Hall and Sardeson, 1895, p. 198).

Although a replacement origin is clearly established for many, perhaps most, dolomites, the time of replacement is not clear. It can take place in the environment of deposition and before burial. The dolomitized old reef in the Funafuti Atoll is an example (Judd, 1904). Replacement might take place after burial but before uplift, or it might take place after both burial and uplift. The principal argument for early replacement is the stratigraphic persistence of many dolomite beds. It is difficult to believe that a thin bed, extending over many square kilometers, would be dolomitized by circulating waters while the overlying and underlying beds remain unaffected. On the other hand, later dolomitization can be proved if the distribution of dolomite is confined to the vicinity of faults or is otherwise structurally controlled. A further argument for later dolomitization is Daly's observation that the older the rock, the more magnesium-rich it appears to be (Daly, 1909). The older the rock, the greater the probability that it would come in contact with magnesium-bearing waters and become dolomitized. However, Daly and some others (Steidtmann, 1911, p. 427) took the view that the composition of the early seas was different from that of later times, and hence earlier rocks were more dolomitic than younger ones. No doubt dolomites were formed by both early and late replacement.

What were the source of the magnesium and the nature of the fluids needed for replacement? Where dolomitization is only partial, it may be that the magnesium was supplied by unmixing of the high-magnesium calcites of organic origin. Some algal limestones contain $MgCO_3$ (up to 24 percent) in solid solution in the calcite. Exsolution of this material will not in itself be sufficient to convert a limestone to a dolomite, but it would be enough to form a scattering of dolomite crystals in a limestone or, perhaps, to lead to the formation of a mottled dolomite. Two generations of dolomite are therefore implied in those cases where mottled dolomites have been converted to solid dolomite.

The addition of magnesium from external sources is implied in most dolomites. Epigenetic replacement related to faults and other structures must be the product of magnesium-bearing connate waters or circulating meteoric waters. Presumably earlier, synsedimentary dolomitization is a result of the reaction of lime carbonate sediment and magnesium-bearing sea water. But if so, why is not every limy sediment which is in contact with sea water converted to dolomite? Experimental studies and our observations on modern dolomite formation suggest that dolomitizing fluids were much richer in magnesium than ordinary sea water. It may be that the waters of basins that are partially isolated under conditions of aridity are enriched in magnesium by the continual inflow of normal sea water and by the precipitation of calcium carbonate and sulfate. Enrichment in this manner would form a heavy brine that moved down through porous sediments below and dolomitized those sediments and reefs with which it came in contact. This is the "reflux" hypothesis applied first to the formation of evaporites by King (1947) and later to the formation of dolomite by Adams and Rhodes (1960). It explains back-reef dolomites and fore-reef limestones. The close association of many dolomites with evaporitic minerals and deposits lends support to this hypothesis of brine formation and dolomitization.

Not all dolomites, however, show unequivocal evidence of replacement. Many are fine-grained, are thinly laminated, are mud-cracked, show ripple cross-lamination, and contain a restricted fossil assemblage different from that of associated limestones (Strakhov, 1958). In some cases these dolomites are in sharp contact with limestones; lime materials may even fill cracks in the dolomite mud, and the dolomite may contain intraclasts of limestone. Limy stromatolites may be surrounded by thinly laminated dolomite and themselves show no evidence of dolomite replacement. Are these dolomites primary, in the sense that constituent grains were dolomite when emplaced in the rock fabric? Various workers have supported the concept of a primary origin: Sloss (1953) attributed an evaporitic origin to certain thinly laminated wholly unfossiliferous dolomites closely associated with bedded anhydrite. Sando (1953, p. 41) reached a similar conclusion respecting thinly laminated dolomites interbedded with limestones of the Beekmantown (Ordovician) of Maryland. Although no anhydrite was present in the Beekmantown, the dolomite was found to enclose small pebbles of underlying limestones. It is difficult to escape the conclusion that this dolomite was primary. Sarin (1962) also considered the dolomites in the lowest Beekmantown, which alternate with limestones in a cyclic fashion, to be primary. Strakhov (1958) believed dolomite could be precipitated from "salinized" marine waters. The interbedding of dolomite and limestone on a micro scale, the laminations only a fraction of a millimeter thick, has been cited as evidence that some dolomite is a primary constituent (Sander, 1951, p. 127).

Experimental studies and observations on modern dolomites suggest that the waters from which dolomite precipitated was not normal sea water. It was probably richer in magnesium (relative to calcium) and had a higher-than-normal *p*H and also a higher-than-normal temperature. These conditions seem to best be realized in shallow, hypersaline lagoons or tidal flats of warm arid regions. The high salinity suppresses the normal fauna and leads to precipitation of associated evaporitic sulfates; mud cracking and stromatolitic development are further indication of shallow waters and even exposure. Periodic freshening of the waters would lead to limestone deposition with a normal fauna.

For a more extended review of the dolomite problem refer, in addition to Van Tuyl's classic paper (1916a), to papers by Fairbridge (1957), Friedman and Sanders (1967), and Michard (1969), as well as the symposium volume edited by Murray and Pray (1965).

Dedolomitization Although replacement of limestone by dolomite is very common, the reverse process, that is, replacement of dolomite by calcite, is rare. The replacement of dolomite rhombs by calcite, termed *dedolomitization,* has been reported by several investigators (Shearman, Khouri, and Taha, 1961; Goldberg, 1967; Evamy, 1967; Schmidt, 1965, p. 149). Evamy showed excellent examples of dolomite porphyroblasts replaced by a calcite mosaic. De Groot's (1967) experiments on dedolomitization suggested that the process required solutions with a high Ca–Mg ratio, rapid flow of such waters, and temperatures under 50°C. These restrictions imply near-surface conditions.

Other replacements Limestone diagenesis in some cases involves replacement by silica in various forms and less commonly replacement by phosphates, various iron-bearing minerals, especially pyrite, and other substances. These replacements may be extensive and in some cases complete; they have been discussed mainly in the section on cherts, iron-bearing formations, and phosphorites (Chapter 11).

It is, perhaps, worth noting here that in ordinary circumstances, replacements are the result of redistribution of materials already present in the deposit and require no introduction from without. This is particularly true of silica. Silica diagenesis takes several forms such as disseminated chert, especially "dolocastic" silica, chert nodules, silicified fossils, euhedral quartz crystals with or without detrital nuclei, chalcedonic spherulites, and the like. Many aspects of silica diagnesis in limestones have been described by Wilson (1966).

means that the clay minerals present are the result of establishment of equilibrium after deposition and are not, therefore, a reliable indicator of source material.

DIAGENETIC PARAGENESIS

As noted, diagenetic changes in limestones can be varied and complex. In a specific case it may be possible to work out a time sequence of changes—a paragenesis (Friedman, 1964). The question remains, however, of whether diagenetic trends prevail in many or all limestones or whether each paragenetic sequence is in fact unique. Are the time relations between silica and dolomite, or between one generation of cement and another, or between secondary calcite and dolomite universal, or are they highly variable and reflect the particular history of the limestone involved? The truth is, perhaps, both. Aragonite is almost always replaced by calcite, not the reverse. In general, calcite and/or aragonite is replaced by dolomite. Dolomite is only rarely replaced by calcite, never by aragonite. High-magnesium calcite gives rise to low-magnesium calcite, never the opposite. As noted by Freeman (1973), the sequence of filling of void spaces is dolomite followed by calcite and not the reverse. Nevertheless, at present it is not possible to formulate a normal diagenetic trend, or even to indicate the various pathways that diagnesis might take. Such a goal requires more study of what *did* happen and perhaps a better theoretical understanding of what *should* happen.

The view that the mineral assemblage in some limestones, especially those of fine grain, is an equilibrium assemblage has been advanced by Zen (1959) and Peterson (1962). Zen, who studied sediments in the Peruvian Trench, believed that chemical equilibrium between the several clay minerals and carbonates was attained in a reasonably short time. If so, a moderately complex multicomponent system of this kind can be successfully analyzed by thermodynamic methods. Peterson carried this approach over into the study of the mineral assemblages in some Upper Mississippian limestones of Tennessee, from which he concluded that "the mineral assemblages are the result of mineral paragenesis adjusted to the original bulk composition of the sediment." This concept means that the clay minerals present are the result of establishment of equilibrium after deposition and are not, therefore, a reliable indicator of source material.

LIMESTONE FACIES

A distinction should be made between limestone petrography and limestone facies. A given limestone formation, although a stratigraphic entity, may display several facies. It may be here a biohermal, perhaps reef facies, and elsewhere an even-bedded nonreef facies. A particular facies, in turn, may consist of one or several rock types such as reef rock (boundstone), internal pockets of clay, and flanking fore-reef sediments which may be calcirudites and calcarenites, and a subaerial facies of an eolian calcarenite. It is most important to discriminate between the several facies of limestone deposition—which correspond to several paleoenvironments in which they formed and several rock types of which they are made.

Various facies no doubt correspond in a general way to the several different kinds of environments in which they accumulated. There is no general agreement at this time on a classification of carbonate environments and the facies generated in each (Friedman, 1969). In a general way there are two major categories: basin and shelf. Each of these is in turn divided into lesser categories (Wilson, 1969, p. 18). We present here a brief resumé of the more common facies. Our classification is based on the distinguishing physical aspect of the facies—something we can see—rather than the environment in which the facies is presumed to have formed. We have, however, indicated the environment most commonly thought to be responsible for the facies in question. These include the stromatolitic facies (tidal flat), the biohermal facies (shelf-edge reef), the cross-bedded facies (shelf banks and bars and eolianites), the allodapic or graded facies (basinal), the nodular facies, the chalk (pelagic) facies, and the evaporitic facies (back-reef lagoons and sabkha). We also include some freshwater facies.

STROMATOLITIC (TIDAL FLAT) FACIES

The stromatolitic facies is so named because of the presence of algal bedding and algal structures. It is a facies consisting mainly of carbonate muds. The abundance of stromatolites and algal bedding associated with mud cracks and related features make it apparent that these limestones were deposited in very shallow water and that some were formed in a tidal-flat environment and were subjected to intermittent flooding and exposure. This environment is characterized by a diversity of carbonate rock types and sedimentary structures.

The tidal flat is essentially a mud trap and in arid regions an area of hypersaline waters. Abnormally high salinity represses the activity of grazing benthos which otherwise destroy the algal mat and prevent stromatolitic growth (Garrett, 1970). Likewise the tidal flats of arid regions constitute an environment favorable for dolomite formation and even for precipitation of gypsum or anhydrite. The stromatolitic facies, therefore, is commonly associated with both dolomitic and evaporitic layers.

The New Market Limestone (Ordovician) of Maryland described by Matter (1967), an example of the stromatolitic facies, consists of six major rock types: thinly laminated dolomite, a "ribboned" rock of alternating thin dolomite and limestone beds with wavy bedding, "lumpy" limestone consisting of discontinuous beds or "lumps" of limestone surrounded by dolomite, stromatolitic limestone, and intraclastic and bioclastic limestones. These subfacies are thought to be related to the subenvironments within the tidal-flat complex. The lumpy limestones are thought to be tidal-marsh deposits, the laminated dolomites supratidal, and the ribboned and stromatolitic rocks intertidal. Because these subfacies are many times repeated, it is probable that there were numerous transgressions and regressions of the tidal zone.

Of special interest are the sedimentary structures. Mud cracks are abundant and record repeated exposure. Lumpy or nodular limestone is considered to be a desiccation feature. Stromatolites and associated "birdseye" voids are taken as evidence of deposition at or above the air-water interface.

The same rock types are found in most of the Paleozoic limestone formations of the central Appalachians and in other areas as, for example, in the Conococheague (Long, 1953) and the Beekmantown (Sando, 1953) groups of western Maryland and Pennsylvania, in the Ellenburger Group of Texas (Cloud and Barnes, 1948), and in the Precambrian Randville (Greenman, 1951) of northern Michigan. Although the deposits formed in a tidal-flat complex at a given stand of the sea are quite thin, repeated regressions and advances of the sea accompanied by continued subsidence have led to great thickness of carbonates—some 3000 m in the central Appalachians, for example. A thorough review of the carbonate shoreline and tidal-flat environment is that of Lucia (1972). Close inspection of these deposits shows that some display a recognizable pattern of cyclic sedimentation (Pelto, 1942; Root, 1965). See the section on envirormental analysis in Chapter 15 for a discussion of regressive cycles of carbonate tidal flats.

BIOHERMAL (REEF AND MOUND) FACIES

The term *bioherm* has been defined by Cumings and Shrock (1928) as any domelike, moundlike, lenslike, or otherwise circumscribed mass, built exclusively or mainly by sedentary organisms and enclosed in a normal rock of different lithologic character. By some, bioherm has been considered synonymous with reef, inasmuch as the structures to which the term was first applied were reefs (Lowenstam, 1950). If a reef is defined as a wave-resistant structure (or having potential wave resistance), then some bioherms were indeed reefs whereas others were not. We first consider true reefs.

Reefs vary in size and shape, and are designated by types of constituent organic remains. They may be composed in part or entirely of encrusting and sediment-binding algae, stromatoporoid colonies, coral colonies, even rudisted and oyster shells. They may be merely small mounds a meter or two in height or they may be impressive structures a thousand or more meters across and a hundred or more meters thick.

Only if the organisms responsible were sediment-binding and made the structure wave-resistant, so that it stood somewhat above the surrounding bottom, is the deposit a true reef. The structure will, in time, emerge and be in part subaerial.

The Silurian reefs of Indiana (Cumings and Shrock, 1928), Illinois (Lowenstam, 1950; Ingels, 1963), and Wisconsin (Shrock, 1939) are perhaps typical fossil reefs. A recent restudy of these reefs has shown a more complex history than heretofore thought (Textoris and Carozzi, 1964). These structures consist of a massive central mound of uneven-textured, fossiliferous dolomite (the reef core), surrounded by a relatively narrow zone (the reef flank) consisting of well-bedded, granular, porous, and sparsely fossiliferous strata which lap against and grade into the core and which commonly show steep dips away from it (Fig. 5-12). The distal flanking beds grade imperceptibly into the horizontal and relatively unfossiliferous rock of the inter-reef region. Upon erosion the massive core commonly remains as a prominent knob or hill known as a *klint*.

The rock composing such klints is a boundstone or *klintite* (also biohermite). It is a loosely knit, reticulating network of dense dolomite, which is hard and tough, and which because of its rigid framework gives the reef mound its strength and resistance to erosion. The reef core may contain lenses or irregular masses of dolomitized coquina which, though well cemented, is still very porous. One-half of the volume of the rock may be voids and coated cavities. Cavities have a drusy lining of calcite crystals. Larger cavities are coated with explanate masses of thinly laminated onyxlike calcite which resembles flowstone of caves. A few of the larger pockets have been filled with a laminated or massive shale. Some cavities appear to be original voids in the reef; others show by their shape that they are openings left by the solution of fossils; still others are irregular solution cavities not related to organic forms. Some of the reef rock has the appearance of a rude breccia.

In general, the central part of the reef core is devoid of bedding or related structures except for a *Stromatactis* framework. The peripheral portion, on the other hand, shows a crude stratification which grades radially outward into the well-defined bedding of the inclined flanking strata.

The core rock may be exceptionally fossiliferous, but, in other cases, extensive dolomitization and solution seem to have destroyed all organic forms. In general, the fossils are mainly internal and external molds, casts, and impressions.

There is, in some of the larger reefs, a slight downward flexure under part of the core, as though the underlying strata had been bent downward somewhat from the weight of overlying rocks.

The flanking strata form a narrow zone that is concentric with the core. In some few cases, however, the arrangement is asymmetrical, perhaps reflecting asymmetrical growth of the reef in response to a prevailing wind or current direction (Ingels, 1963). The flanking strata of well-bedded porous dolomite dip away from the core at angles up to 50 degrees or more. These inclined beds pass into the core on one side and grade outward into the inter-reef limestone of the surrounding area. The high dips of some of the flanking beds are attributed to steepening by slumping and compaction. The material of the flanking strata is well-bedded, even-textured, porous, or cavernous granular dolomite with few fossils. This material has the appearance of a poorly cemented calcareous sand. It is, in fact, the detritus formed by the grinding up of the biohermal core material. In places, large fragments of the rough porous, nonfossiliferous reef rock are embedded in the flanking beds. If these fragments are numerous enough, the term *reef breccia* may be applied. The porosity of the core and dip of the flanking beds made the reef a potential trap for oil and gas and indeed many fossil reefs are oil-producing (Lowenstam and DuBois, 1946).

The reefs were islands of intense vital activity. They grew primarily from the calcareous material deposited by the inhabiting organisms (mainly benthonic). Almost immediately after the initiation the reefs rose above the surrounding bottoms. With entrance into the zone of wave-generated turbulence, they shed debris which formed flanking beds; this deposition broadened the reef platform. The growing core commonly expanded and overgrew the flanking detrital apron. The surrounding bottoms were never heavily populated and seem to have been built up slowly by fine calcareous muds and silts derived by erosion of the growing reef and perhaps also by carbonate precipitated directly from sea water, except perhaps in the proximity of the reef where reef-induced turbulence inhibited deposition of these fine materials, with the result that the inter-reef strata thin in the immediate vicinity of the reef (Lowenstam, 1950, Fig. 6). In some cases coarse debris shed from the reef moved away as a slump or a turbidity current. Such currents may travel several tens of kilometers and formed a well-defined debris fan (Eder, 1970).

The dip of the flanking strata diminishes away from the reef core and ultimately is reversed, in the case of very large reefs. The gentle reverse dip and the peripheral sag which it bounds are interpreted as the response of the strata to the sinking of the heavy reef structure and compensating upbulging of surrounding strata (Shrock, 1939).

The high porosity of the reef rock itself favored dolomitization, which greatly modified the reef rock and in some cases destroyed most of the organic structures characterizing the reef. Today such reef structures appear as irregular bodies of massive dolomite that interrupt the uniform bedding of the regional stratigraphic sequence.

Because reef-building organisms seem to be limited by depth and temperature, reefs will form only under conditions which permit growth of the reef community. Reefs seem to occur in belts which are mainly depth-controlled. Shifting of the strandline related to transgression or regression of the sea will result in migration of the reef belt (Link, 1950). Not uncommonly the lagoons between the reef belt and the shoreline proper become hypersaline and the site of salt deposition.

Fossil reefs are widespread in time and space. The best known Silurian reefs are those of the Great Lakes region described above and those of Gotland (Manten, 1962, 1971). Those of the Devonian in Alberta are oil-bearing and have been extensively drilled and studied (Klovan, 1964). Well known also are the Paleozoic reefs of Belgium (Lecompte, 1938), the Devonian reefs of the eastern Rhenish Schiefergebirge (Krebs, 1971) and those of the Harz in Germany (Franke, 1971) (Fig. 10-51), the Mississippian bryozoan reefs of the southwestern United States (Pray, 1958), and the Lower Carboniferous Waulsortian reef knolls in Ireland (Lees, 1964) and England (Parkinson, 1957). The Permian reef complex of the Guadalupe Mountains of Texas and New Mexico (Newell et al., 1953) is a classic example of reef and associated sedimentation. Lower Cretaceous rudisted reefs have been described from south Texas (Griffith, Pitcher, and Rice, 1969). Tertiary reefs of Guam have been studied in great detail (Schlanger, 1964). The literature of ancient and modern reefs is too vast to list here. For summaries of the topic of reefs in the geologic past see Hadding (1941) and Twenhofel (1950). For bibliographies see Pugh (1950) and Milliman (1965).

Not all biohermal bodies of limestone or dolomite are reefs. Some are localized facies that persisted through time but never stood much above the surrounding bottoms, such as the crinoidal bioherms of the Borden of Indiana (Stockdale, 1931, p. 717; Carozzi and Soderman, 1962). They never were emergent wave-resistant bodies. The distinction between such accumulations and true reefs may be difficult; the subject has been discussed by Lowenstam (1950, pp. 431–435), Ginsburg and Lowenstam (1958), and Braithwaite (1973). The major difficulty is that many bioherms, some of which have been called reefs, have a core without any conspicuous framework builders. The core of some bioherms is a very fine-grained lime mud—a calcilutite—instead of a boundstone formed by encrusting algae, corals, and the like. How was such a mud buildup localized, and was it in any sense wave-resistant and hence a true reef? Pray (1958) described biohermal mounds up to 350 feet (106 m) thick of Mississippian age in New Mexico, which consist of a seemingly unfossiliferous calcilutite, but which on closer inspection is found to consist of a meshwork of fenestrate bryozoans filled by lime mud. These bryozoans may have acted as a current barrier and sediment trap. Pray thought that these bioherms stood up above the surrounding bottom but called them a "subreef" rather than a true reef. Somewhat similar bioherms are the Waulsortian reef knolls of the Carboniferous of England (Parkinson, 1957) and Ireland (Lees, 1964). A modern analogue perhaps are the mud banks localized by sea grass (*Thalassia*) or related forms (Ginsburg and Lowenstam, 1958). Even the classic Niagaran "reefs" of Indiana may have begun as mud mounds stabilized by bryozoans and perhaps later evolved into a wave-resistant stromatoporoid structure (Textoris and Carozzi, 1964).

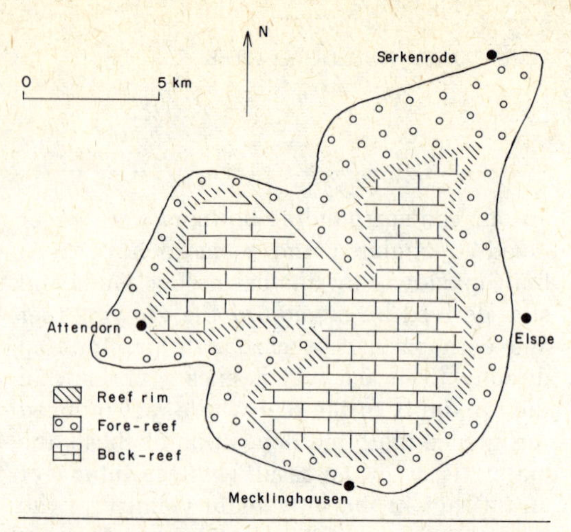

FIG. 10-51. Reconstruction of the original size of the Attendorn reef complex, Germany, before the Variscan orogeny. (After Gwosdz, 1971, in Krebs, 1971, Sedimentology of parts of central Europe, Fig. 15, p. 66. By permission of Springer-Verlag.)

CROSS-BEDDED (WINNOWED SHELF) FACIES

A common facies of limestone is that marked by an abundance of cross-bedded calcarenite. These limestones are essentially carbonate sand bodies, in part submarine and in part subaerial. The sands are calcarenites of various sorts, mainly skeletal and oolitic. They are marked by a high original porosity, now largely filled with sparry calcite.

The petrography of these calcarenites has been described in some detail. The Fredonia oolite member of the St. Geneviève (Mississippian) of Illinois is a oolitic calcarenite (Graf and Lamar, 1950; Connolly, 1948); the Salem Limestone (Mississippian) of Indiana is a skeletal calcarenite (Loughlin, 1929, p. 131; Pinsak, 1957; Sedimentation Seminar, 1966, p. 98); the Lexington Limestone (Ordovician) of Kentucky is a skeletal calcarenite (Hrabar, Cressman, and Potter, 1971); the Loyalhanna Limestone (Mississippian) of Pennsylvania and western Maryland is a calcarenite with a large admixture of detrital quartz (Adams, 1970). In these cases and innumerable others, the deposit consists of well-sorted, winnowed sands, in part skeletal debris, in other cases *in situ* precipitated oolites.

The most striking and distinctive feature of this facies is its large-scale cross-bedding. The thickness of the cross-bedded sets is variable—up to 2 m in the Salem Limestone and nearly 5 m in the Loyalhanna. The average in the Salem is about 20 cm. Both planar and wedge cross-bedding are present in this facies. The most striking characteristic of cross-bedding is its bimodality (Fig. 10-52). Two opposed modes are generally present; but one dominates (Sedimentation Seminar, 1966; Hrabar, Cressman, and Potter, 1971). These have been interpreted as recording tidal ebb and flow.

Limestone bodies which belong to this facies seem to have a lenslike form and to have been a bank or shoal on a shallow marine shelf—perhaps a shelf-edge deposit. The bodies are of limited thickness, 50 to 100 m, and of limited extent. The cross-bedded facies of the Salem Limestone, for example, more or less coincides with the quarried outcrop belt in southern Indiana; it passes downdip into a calcilutite facies (Pinsak, 1957).

This facies has been termed a "winnowed platform edge" facies (Wilson, 1969, p. 18) and is believed to form as subaqueous bars and dunelike accumulations such as today occur in the Florida and Bahama carbonate province, especially near the south end of the Tongue of the Ocean. The morphology and composition of these sand bodies have been described in detail by Ball (1967). Their internal sedimentary structures, mainly cross-bedding, in the Bahamian sands have been described by Imbrie and Buchanan

(1965). A Pleistocene deposit of this genre is the Miami Oolite (Tanner, 1959, Fig. 1).

ALLODAPIC (BASINAL) LIMESTONE FACIES

The allodapic facies is the carbonate equivalent of the turbidite facies (see Chapter 15, p. 560). It has most of the characteristics of that facies, namely rhythmic interbedding of detrital limestones and pelitic layers; the limestone beds are traceable over long distances without change in thickness and commonly have a sharp base, internal coarse- to fine-upward grading, and an indistinct upper contact. Not uncommonly the base of the bed shows flutes and groove casts. The limestone beds may display partial to complete Bouma cycles (Fig. 10-53).

The detritus forming the limestones is largely skeletal, mainly of a shallow benthonic fauna; that of the intervening pelitic

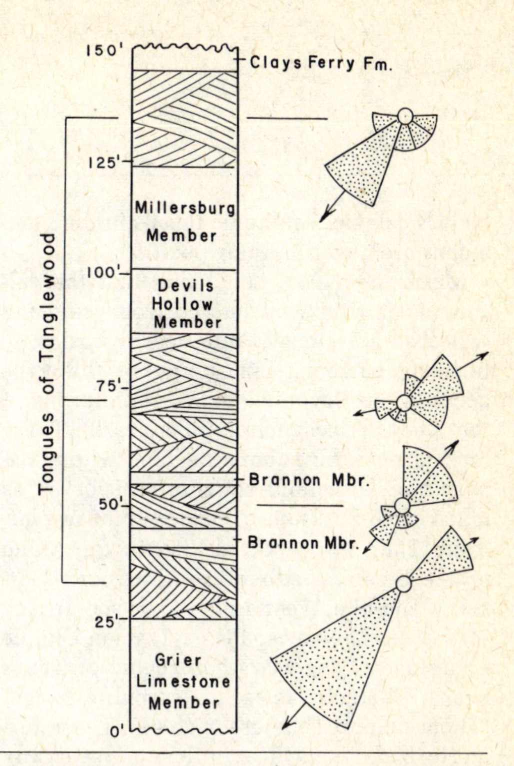

FIG. 10-52. Cross-bedding directions in an outcrop section in the Tanglewood Limestone Member, Kentucky. (From Hrabar, Cressman, and Potter, 1971, Brigham Young Univ. Studies, v. 18, Fig. 9.)

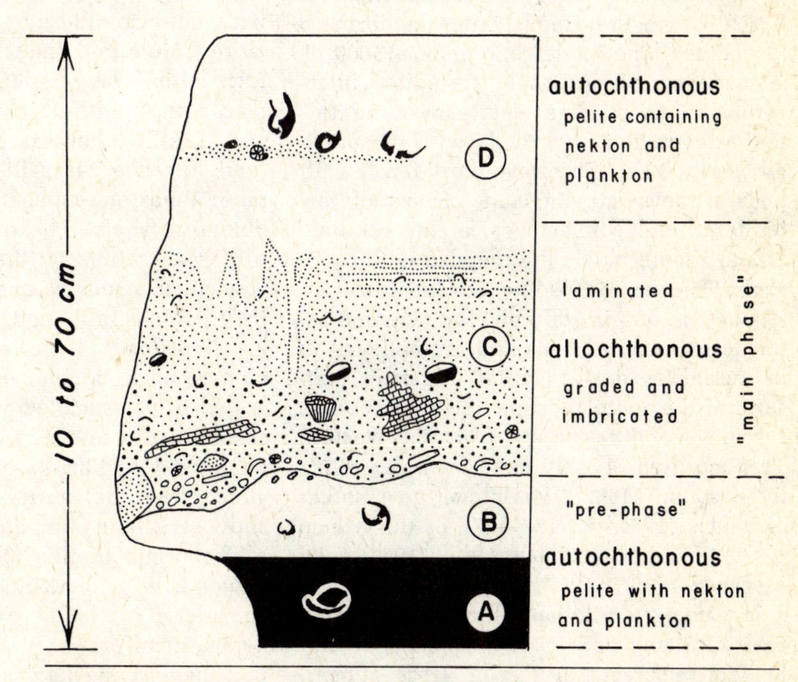

FIG. 10-53. Ideal zonation of a "Barbecker Kalk" bed. The thickness varies from 10 to 70 cm. (After Eder, 1970, Verh. Geol. Bundesanst., v. 4, Fig. 2.)

beds is pelagic. Benthonic fossils in the limestones are never in living position.

Meischner (1964) concluded that the calcareous detritus was derived from reefs and was shed episodically into nearby basins by turbidity currents. Limestones of this type occur in the Devonian and Carboniferous of the Rhenish geosyncline (Eder, 1970). These limestones form submarine fans whose paleocurrent directions show a transport of as much as 60 km from the source reef or platform. They are reported also in the Malm of the Swabian Alb of southwestern Germany and the West Hellenic flysch trough of Greece. The Lower Pennsylvanian Dimple Limestone of the Marathon region of Texas has a basinal facies of turbidite origin (Thomson and Thomasson, 1969). These turbidite beds are calcarenitic and generally graded; they are chiefly less than 30 cm thick, though a few exceed 1 m. The interbedded mudstones and spicular radiolarian cherts are pelagic in origin. Extraordinarily thick allodapic limestones have been reported by Scholle (1971) in the Monte Antola flysch (Upper Cretaceous) from the northern Apennines. The total section, some 5000 m, is made up of carbonate turbidites interbedded with black shales. Limestone turbidites average 85 cm thick, with the thickest nearly 30 m. The Devonian Marble Cliff beds of Cornwall, England, consist of turbidite crinoidal limestones, a few centimeters to a meter thick, interbedded with black shale (Tucker, 1969). The Eocene calcarenites of Tuscany in Italy display repeated sequences of graded beds and are believed to be turbidites derived from a neritic platform to the south (Sestini, 1964). Carbonate turbidites of Jurassic age from the western Trentino area of Italy have been described by Bosellini (1967). Meischner gives other examples (1964, pp. 177–187), as do Wilson (1969), Kuenen and ten Haaf (1956), and Mackenzie (1970).

In summary, allodapic limestones constitute thick and widespread deposits, resemble noncalcareous flysch in gross appearance and organization, and form turbidite fans that radiate from specific reefs or carbonate platforms. They are an exception to the general rule that carbonates are a product of shallow-water sedimentation, although the material of which they are composed was generated in a shallow-water environment.

The characteristics of limestones thought to be basinal have been described in some detail by Wilson (1969). He thought that allodapic or limestone turbidites were more common in deep water near steep slopes but that nongraded and nonturbiditic basinal limestones are also common. These are generally dark, fine-grained lime mudstones and siltstones. They may be somewhat siliceous and intercalated with beds of chert.

NODULAR LIMESTONE FACIES

Nodular limestones (*Knollenkalk*) constitute a distinct facies. But unlike most other carbonate facies, the origin of the nodular aspect is not at all clear. It is not known whether the nodular character is an original primary feature or whether it is in part or wholly diagenetic. Nor can the nodular facies be identified with a specific environment.

This facies is dominated by nodular and irregularly bedded limestones intercalated with more argillaceous (and in some cases more dolomitic) materials. The limestones form thin wavy-bedded layers which in places pass into thin disconnected lenses and in still other cases into very irregular (and in some cases ill-defined) nodules of purer limestone embedded in calcareous or dolomitic shale. The nodules do not seem to be concretionary; they lack the radial or concentric structure of such bodies. Nor are they pebbles, although in some cases nodular limestones have a superficial resemblance to a conglomerate. They are not sorted nor associated with current-scoured surfaces, nor are they found with current-winnowed sediments. Unlike normal conglomerates, the matrix is clayey and may show swirl-like laminations around nodules.

The origin of this facies has been much debated. In general, nodular limestones are considered to be of a secondary or post-depositional origin. It is contended by some that they are mechanically formed by a kind of flowage of semiconsolidated sediment. They are thus a species of boudinage. Mc-

Crossan (1958) thought that original thin limy beds sandwiched between more clayey and hence more plastic beds were pulled apart by laterally moving plastic beds when subjected to unconfined compaction. A somewhat similar view was expressed by Voigt (1962). Wilson (1969, p. 17) ascribed the nodules to "differential compaction of the clay and lime mudstones," such compaction being controlled by clay mud lenses "erratically and randomly distributed through carbonate muds." Others ascribe a diagenetic origin to nodular limestones, believing that they formed by an unmixing process from a once homogeneous lime-clay mixture (Hallam, 1964, 1967). Garrison and Fischer (1969, p. 26) reject this view and believe that Hollmann's hypothesis (1962, 1964) best explains these features, according to which the nodules are solution relicts of once continuous beds partially destroyed on the sea floor by a process termed subsolution. Nodules were further modified by solution after burial as abundant stylolitic contacts between nodules attest. Wanless (1973) attributed this facies in the Grand Canyon Cambrian to post-depositional overburden compaction combined with intrastratal solution along stylolitic surfaces and flowage.

It is difficult to relate this facies to any specific environment; if it is post-depositional in origin, no such relation should exist unless the original sediment was of a particular composition or structure that would facilitate later nodule formation. McDaniel (Wilson, 1969, p. 17) presumed that nodular argillaceous limestones were indicative of a basinal environment. Garrison and Fischer (1969, p. 26) thought the nodular Adnet beds (Jurassic) of the Austrian Alps were formed in very deep waters. Wilson (1969, p. 17), on the other hand, thought the structure was characteristic of a shallow neritic environment but states that he had never seen sedimentary boudinage in tidal carbonates. Matter (1967) describes a similar structure in such carbonates of Paleozoic age. Root (1965, Fig. 2) also reports a "ribbon rock"—a rock marked by lenses and nodulelike limestones and separated by somewhat argillaceous and dolomitic partings—in a tidal-flat sequence of Cambrian age. Wanless (1973) assigned the Grand Canyon Cambrian nodular limestones to a platform carbonate environment.

Clearly the nodular limestone facies is not the product of a particular environment; it seems more probably a diagenetic rather than a depositional facies.

CHALK (PELAGIC) FACIES

As noted, chalk is a soft friable micritic deposit, generally very pure, consisting almost entirely of organically derived calcitic particles mainly coccoliths, rhabdoliths and calcispheres, all presumably the remains of planktonic organisms or "nannofossils" and mostly 5 microns or less in size. Foraminifera may also be present, but they generally form no more than 10 percent of the chalk.

The environment of deposition of the chalk facies has been much debated. In general, it seems probable that the water was moderately deep, that the bottom was below the zone of wave action, perhaps 200 m or more. Coccoliths form today in waters 60 to 100 m deep, so the depth of chalk accumulation must have been somewhat greater. Some have considered chalk comparable to the modern Globigerina ooze and hence thought it to be a deep-sea deposit. The molluscan fauna of some chalk beds indicates otherwise. The pelagic origin is not in doubt; the choice is between waters of moderate depth far removed from sources of clastic material or the deep sea. The weight of evidence is for the former.

EVAPORITIC CARBONATE FACIES

The several limestone facies reviewed above are perhaps the most common—certainly the best known. Each of them is, however, a facies complex, there are subfacies in each. The tidal complex, for example, consists of tidal channel, levee, tidal marsh, tidal delta, and other subenvironments, each of which no doubt generates a subfacies. In addition, there probably are other environments of lime deposition, perhaps less extensive, which give rise to limestone facies other than those reviewed here. Some such restricted facies include the tufa and marl facies and perhaps the caliche facies. The latter is perhaps the soil analogue of the dolomite and gypsiferous facies of the saline tidal flat. In both cases, capillary rise and evaporation of mineral-bearing solutions lead to precipitation in interstices of the soil in the one case and in tidal sediment in the other.

In addition to these evaporitic carbonates, there are others presumed to form in penesaline back-reef lagoons. Included in this group is dolomite interbedded with anhydrite. According to Sloss (1953, p. 145), the penesaline evaporitic carbonates include dense, finely laminated dolomite with anhydrite blebs and stringers, a finely crystalline carbonate matrix with isolated larger dolomite rhombs, saccharoidal dolomites made up of well-sorted rhombs, perhaps cross-bedded, and oolitic carbonates. The saccharoidal dolomite is believed to be a product of winnowing of the fine matrix carbonate with the scattered rhombs.

We have mentioned the freshwater marls and other freshwater limestones. These constitute a specialized carbonate facies of limited extent and importance.

RELATIONS BETWEEN FACIES

The various limestone facies recognized have been presented as though they bore no relation to one another. Such is probably not the case. We have suggested a close spatial and genetic relation between the reef and the allodapic facies. There is also a presumed relation between the reef and the back-reef lagoonal facies, with the reef the barrier which separates the open marine environment from the shallow and often hypersaline lagoon and tidal flat. In some cases the barrier is not a reef but is, instead, a wave-built shoal of carbonate sands (the "winnowed shelf edge" facies). Some efforts have been made to explore the relations between the several carbonate environments and to construct an appropriate model, based in part on studies of present-day areas of carbonate deposition and in part on stratigraphic relations between the several facies (Fig. 10-2). A good analysis of this sort is that of Irwin (1965; see also Link, 1950, and Harris, 1973). It is of interest that the more porous barrier facies—either a reef or a cross-bedded calcarenite facies—is, in many cases, the reservoir rock of significant oil fields. Clearly the study of limestone facies and their aerial distribution is of prime economic importance.

References

Adams, J. E., and Rhodes, M. L., 1960, Dolomitization by seepage refluxion: Amer. Assoc. Petrol. Geol., v. 44, pp. 1912–1920.

Adams, R. W., 1970, Loyalhanna Limestone—cross-bedding and provenance, *in* Studies of Appalachian geology: central and southern (Fisher, G. W., Pettijohn, F. J., Reed, J. C., Jr., and Weaver, K. N., eds.): New York Wiley-Interscience, pp. 83–100.

Adey, W. H., and Macintyre, I. G., 1973, Crustose coralline algae: a re-evaluation in the geological sciences: Bull. Geol. Soc. Amer., v. 84, pp. 883–904.

Alderman, A. R., and Skinner, H. C. W., 1957, Dolomite sedimentation in the southeast of South Australia: Amer. Jour. Sci., v. 255, pp. 561–567.

Allen, J. R. L., 1965, The sedimentation and palaeogeography of the Old Red Sandstone of Anglesey, North Wales: Proc. Yorkshire Geol. Soc., v. 35, pp. 140–185.

Allen, J. R. L., and Friend, P. F., 1968, Deposition of the Catskill facies, Appalachian region with notes on some other Old Red Sandstone basins, *in* Late Paleozoic and Mesozoic continental sedimentation, northeastern North America (Klein, G. deV., ed.): Geol. Soc. Amer. Spec. Paper 60, pp. 21–74.

Allison, V. C., 1923, The growth of stalagmites and stalactites: Jour. Geol., v. 31, pp. 106–125.

Amsden, T. W., 1955, Lithofacies map of Lower Silurian deposits in central and eastern United States and Canada: Bull. Amer. Assoc. Petrol. Geol., v. 39, pp. 60–74.

Andrée, K., 1915, Ursachen und Arten der Schichtung: Geol. Rundschau, v. 6, pp. 351–397.

Asquith, G. B., 1967, The marine dolomitization of the Mifflin member Platteville limestone in southwest Wisconsin: Jour. Sed. Petrology, v. 37, pp. 311–326.

Ball, M. M., 1967, Carbonate sand bodies of Florida and the Bahamas: Jour. Sed. Petrology, v. 37, pp. 556–591.

Barth, T., Correns, C. W., Eskola, P., 1939, Die Entstehung der Gesteine: New York, Springer, 422 pp.

Bastin, E. S., 1951, A note on stylolites in oolitic limestone: Jour. Geol., v. 59, pp. 509–510.
Bathurst, R. G. C., 1958, Diagenetic fabrics in some British Dinantian limestones: Liverpool and Manchester Geol. Jour., v. 2, pp. 11–36.
———, 1959a, Diagenesis in Mississippian calcilutites and pseudobreccias: Jour. Sed. Petrology, v. 29, pp. 365–376.
———, 1959b, The cavernous structure of some Mississippian *Stromatactis* reefs in Lancashire, England: Jour. Geol., v. 67, pp. 506–521.
———, 1966, Boring algae micrite envelopes and lithification of molluscan biosparites: Geol. Jour., v. 5, pp. 15–32.
———, 1971, Carbonate sediments and their diagenesis: Developments in sedimentology no. 12, Amsterdam, Elsevier, 620 pp.
Bavendamm, W., 1931, The possible role of micro-organisms in the precipitation of calcium carbonate in tropical seas: Science, v. 73, pp. 597–598.
———, 1932, Die mikrobiologische Kalkfällung in der tropischen See: Arch. Mikrobiol, v. 3, pp. 205–276.
Bayley, W. S., 1904, The Menominee iron-bearing district of Michigan: U.S. Geol. Surv. Mono. 46, 513 pp.
Beales, F. W., 1953, Dolomitic mottling in Palliser (Devonian) Limestone, Banff and Jasper National Parks, Alberta: Bull. Amer. Assoc. Petrol. Geol., v. 37, pp. 2281–2293.
———, 1958, Ancient sediments of Bahaman type: Bull. Amer. Assoc. Petrol. Geol., v. 42, pp. 1845–1880.
Bergenback, R. E., and Terriere, R. T., 1953, Petrography of Scurry Reef, Texas: Bull. Amer. Assoc. Petrol. Geol., v. 37, pp. 1014–1029.
Berner, R. A., 1965, Dolomitization of the mid-Pacific atolls: Science, v. 147, pp. 1297–1299.
Biggs, D. L., 1957, Petrography and origin of Illinois nodular cherts: Illinois Geol. Surv. Circ. 245, 25 pp.
Birse, D. J., 1928, Dolomitization processes in the Paleozoic horizons of Manitoba: Proc. Trans. Roy. Soc. Canada, ser. 3, soc. IV, v. 22, pp. 215–222.
Black, W. W., 1952, The origin of the supposed tufa bands in Carboniferous reef limestones: Geol. Mag., v. 89, pp. 195–200.
Blatchley, W. S., and Ashley, G. H., 1901, The lakes of northern Indiana and their marl deposits: Ann. Rept. Dept. Geol. and Nat. Res. Indiana, v. 25, pp. 31–321.
Bonet, F., 1952, La facies Urgoniana del Cretácico Medio de la region de Tampico: Bol. Asoc. Mexicana Geol. Petrol., v. 4, pp. 153–262.
Bornhold, B. D., and Pilkey, O. H., 1971, Bioclastic turbidite sedimentation in Columbus Basin, Bahama: Bull. Geol. Soc. Amer., v. 82, pp. 1341–1354.
Bosellini, A., 1964, Sul significato genetico e ambientale di alcuni tipi di rocce calcaree in base alle-piu recenti classificazioni: Mem. Mus. Storia Nat. Venezia Tridentina Ann. 27–28, 1964–1965, v. 15, pp. 5–58.
———, 1967, Turbiditi carbonatiche nel Giurassico delle Giudicarie e loro significato geologico: Ann. Univ. Ferrara, n.s., sez. 9, Sci. Geol. Paleont., v. 4, pp. 101–115.
Bosellini, A., and Hardie, L. A., 1973, Depositional theme of a marginal marine evaporite: Sedimentology, v. 20, pp. 5–28.
Bradley, W. H., 1929, Algae reefs and oolites of the Green River formation: U.S. Geol. Surv. Prof. Paper 154-G, pp. 203–223.
Braithwaite, C. J. R., 1973, Reefs: Just a problem of semantics?: Bull. Amer. Assoc. Petrol. Geol., v. 57, pp. 1100–1116.
Bramkamp, R. A., and Powers, R. W., 1958, Classification of Arabian carbonate rocks: Bull. Geol. Soc. Amer., v. 69, pp. 1305–1318.
Bretz, J H., and Horberg, L., 1949, Caliche in southeastern New Mexico: Jour. Geol., v. 57, pp. 491–511.
Bricker, O. P., ed., 1971, Carbonate cements: Studies in geology no. 19, Baltimore, Johns Hopkins Univ. Press, 376 pp.
Brown, W. W. M., 1959, The origin of stylolites in the light of a petrofabric study: Jour. Sed. Petrology, v. 29, pp. 254–259.
Bucher, W. H., 1917, Large current-ripples as indicators of paleogeography: Proc. Nat. Acad. Sci., v. 3, pp. 285–291.
Canal, P., 1947, Observations sur les caractères pétrographiques de calcaires dolomitique et de dolomies: Comptes Rendus Soc. Geol. France, pp. 161–162.
Carozzi, A. V., 1960, Microscopic sedimentary petrography: New York, Wiley, 485 pp.
———, 1970, Carbonate rocks (transl. Cayeux, Les roches sédimentaires): New York, Hafner, 506 pp.
Carozzi, A. V., and Soderman, J. G. W., 1962, Petrography of Mississippian (Borden) crinoidal limestones at Stobo, Indiana: Jour. Sed. Petrology, v. 32, pp. 397–414.
Cayeux, L., 1935, Les roches sédimentaires de France: roches carbonatees: Paris, Masson, 447 pp.
Chave, K. E., 1952, A solid solution between calcite and dolomite: Jour. Geol., v. 60, pp. 190–192.
———, 1954a, Aspects of the biogeochemistry of magnesium. 1. Calcareous marine organisms: Jour. Geol., v. 62, pp. 266–283.
———, 1954b, Aspects of the biogeochemistry

of magnesium. 2. Calcareous sediments and rocks: Jour. Geol., v. 62, pp. 587–599.

———, 1960, Carbonate skeletons to limestone: problems: Trans. New York Acad. Sci., v. 23, pp. 14–24.

Chilingar, G. V., 1957, Classification of limestones and dolomites on basis of Ca/Mg ratio: Jour. Sed. Petrology, v. 27, pp. 187–189.

Chilingar, G. V., Bissell, H. J., and Fairbridge, R. W., 1967, Carbonate rocks. Part A: Origin, occurrence and classification, 471 pp., Part B: Physical and chemical aspects, 413 pp.: Amsterdam, Elsevier.

Choquette, P. W., and Pray, L. C., 1970, Geologic nomenclature and classification of porosity in sedimentary carbonates: Bull. Amer. Assoc. Petrol. Geol., v. 54, pp. 207–250.

Clarke, F. W., 1924, Data of geochemistry: Bull. U.S. Geol. Surv. 770, 841 pp.

Clarke, F. W., and Wheeler, W. C., 1922, The inorganic constituents of marine invertebrates: U.S. Geol. Surv. Prof. Paper 124, 62 pp.

Clee, V. E., 1950, Bibliography on dolomite: Supp. Rept. Comm. Sed. 1949, Nat. Res. Coun., 91 pp.

Cloos, E., and Pettijohn, F. J., 1973, Southern border of Triassic Basin, west of York, Pennsylvania: fault or overlap?: Bull. Geol. Soc. Amer., v. 84, pp. 523, 536.

Cloud, P. E., 1962, Environment of calcium carbonate deposition west of Andros Island, Bahamas: U.S. Geol. Surv. Prof. Paper 350, 138 pp.

Cloud, P. E., and Barnes, V. E., 1948, The Ellenburger Group of central Texas: Univ. Texas Bur. Econ. Geol. Publ. 4621, 473 pp.

Connolly, F. T., 1948, The geology of the Passport oil pool, Clay County, Illinois: M.S. dissert., Univ. Cincinnati.

Conybeare, C. E. B., 1949, Stylolites in pre-Cambrian quartzite: Jour. Geol., v. 57, pp. 83–85.

———, 1950, Microstylolites in pre-Cambrian quartzite: a reply: Jour. Geol., v. 58, pp. 652–654.

Crowley, D. J., 1973, Middle Silurian patch reefs, etc.: Bull. Amer. Assoc. Petro. Geol., v. 57, pp. 283–300.

Cullis, C. G., 1904, The mineralogical changes observed in the cores of the Funafuti borings, *in* The atoll of Funafuti (Bonney, T. G., ed.): Proc. Roy. Soc. London, pp. 392–420.

Cumings, E. R., and Shrock, R. R., 1928, Niagaran coral reefs of Indiana and adjacent states and their stratigraphic relations: Bull. Geol. Soc. Amer., v. 39, pp. 579–620.

Daly, R. A., 1909, First calcareous fossils and the evolution of the limestones: Bull. Geol. Soc. Amer., v. 20, pp. 153–170.

Dapples, E. C., 1942, The effect of macro-organisms upon near-shore marine sediments: Jour. Sed. Petrology, v. 12, pp. 118–126.

Davies, D. K., 1968, Carbonate turbidites, Gulf of Mexico: Jour. Sed. Petrology, v. 38, pp. 1100–1109.

Davis, C. A., 1900, A contribution to the natural history of marl: Jour. Geol., v. 8, pp. 485–497.

Decker, C. E., and Merritt, C. A., 1928, Physical characteristics of the Arbuckle Limestone: Oklahoma Geol. Surv. Circ. 15, p. 49.

Deffeyes, K. S., Lucia, F. J., and Weyl, P. K., 1965, Dolomitization of Recent and Plio-Pleistocene sediments in marine evaporite waters on Bonaire, Netherlands Antilles, *in* Dolomitization and limestone diagenesis (Pray, L. C., and Murray, R. C., eds.): Soc. Econ. Paleont. Min. Spec. Publ. 13, pp. 71–88.

De Groot, K, 1967, Experimental dedolomitization: Jour. Sed. Petrology, v. 37, pp. 1216–1220.

Deininger, R. W., 1964, Limestone–dolomite transition in the Ordovician Platteville Formation in Wisconsin: Jour. Sed. Petrology, v. 34, pp. 281–288.

Donaldson, A. C., 1960, Interpretation of depositional environments of Lower Ordovician carbonates in central Appalachians: Proc. West Virginia Acad. Sci., v. 31, pp. 153–161.

Drew, G. H., 1914, On the precipitation of calcium carbonate in the sea by marine bacteria and on the action of denitrifying bacteria in tropical and temperate seas: Carnegie Inst. Washington Publ. 182, pp. 7–45.

Dunbar, C. O., and Rodgers, J., 1957, Principles of stratigraphy: New York, Wiley, 356 pp.

Dunham, R. J., 1962, Classification of carbonate rocks according to depositional texture, *in* Classification of carbonate rocks (Ham, W. E., ed.): Tulsa, Okla., Amer. Assoc. Petrol. Geol., pp. 108–121.

———, 1969, Vadose pisolite in the Capitan Reef (Permian), New Mexico and Texas, *in* Depositional environments in carbonate rocks (Friedman, G. M., ed.): Soc. Econ. Paleont. Min. Spec. Publ. 14, pp. 182–191.

Dunnington, H. V., 1954, Stylolite development post-dates rock induration: Jour. Sed. Petrology, v. 24, pp. 27–49.

———, 1967, Aspects of diagenesis and shape change in stylolitic limestone reservoirs: Proc. 7th World Petrol. Congr., Mexico, v. 2, pp. 339–352.

Eardley, A. J., 1938, Sediments of Great Salt Lake, Utah: Bull. Amer. Assoc. Petrol. Geol., v. 22, pp. 1305–1411.

Eder, F. W., 1970, Genese Riff-naher Detritus-

Kalk bei Balve im Rheinischen Schiefergebirge (Garbecker Kalk): Verh. Geol. Bundesanst., v. 4, pp. 551–569.

Emery, K. O., 1945, Mineralogy of caliche from San Diego County, Calif.: Bull. Southern California Acad. Sci., v. 44, pp. 130–135.

Emery, K. O., and Cox, D. C., 1956, Beachrock in the Hawaiian Islands: Pacific Sci., v. 10, pp. 382–402.

Emery, K. O., Tracey, J. I., and Ladd, H. S., 1954, Geology of Bikini and nearby atolls: U.S. Geol. Surv. Prof. Paper 260-A, pp. 1–265.

Epstein, S., Buchsbaum, R., Lowenstam, H. A., and Urey, H. C., 1951, Carbonate-water isotopic temperature scale: Bull. Geol. Soc. Amer., v. 62, pp. 417–426.

———, 1953, Revised carbonate-water isotopic temperature scale: Bull. Geol. Soc. Amer., v. 64, pp. 1315–1326.

Evamy, B. D., 1967, Dedolomitization and the development of rhombohedral pores in limestone: Jour. Sed. Petrology, v. 37, pp. 1204–1215.

Evamy, B. D., and Shearman, D. J., 1965, The development of overgrowths from echinoderm fragments: Sedimentology, v. 5, pp. 211–234.

Fairbridge, R. W., 1957, The dolomite question, *in* Regional aspects of carbonate deposition (LeBlanc, R. J., and Breeding, J. G., eds.): Soc. Econ. Paleont. Min. Spec. Publ. 5, pp. 125–178.

Feray, D. E., Heuer, E., and Hewatt, W. G., 1962, Biological, genetic, and utilitarian aspects of limestone classification, *in* Classification of carbonate rocks—a symposium (Ham, W. E., ed.): Amer. Assoc. Petrol. Geol. Mem. 1, pp. 20–32.

Finckh, A. E., 1904, Biology of reef-forming organisms at Funafuti Atoll, *in* The atoll of Funafuti (Bonney, T. G., ed.): Royal Society of London, p. 133.

Fischer, A. G., Honjo, S., and Garrison, R. E., 1967, Electron micrographs of limestones and their nannofossils: Princeton Monogr. Geol. Paleont., no. 1, 141 pp.

Fisher, D. J., 1925, Geology and mineral resources of the Joliet quadrangle: Illinois Geol. Surv. Bull. no. 51, 160 pp.

Folk, R. L., 1959, Practical petrographic classification of limestones: Bull. Amer. Assoc. Petrol. Geol., v. 43, pp. 1–38.

———, 1962, Spectral subdivision of limestone types, *in* Classification of carbonate rocks—a symposium (Ham, W. E., ed.): Amer. Assoc. Petrol. Geol. Mem. 1, pp. 62–84.

———, 1965, Some aspects of recrystallization in ancient limestones, *in* Dolomitization and limestone diagenesis (Pray, L. C., and Murray, R. C., eds.): Soc. Econ. Paleont. Min. Spec. Publ. 13, pp. 14–48.

———, 1973, Carbonate petrography in the post-Sorbian age, *in* Evolving concepts in sedimentology (Ginsburg, R. N., ed.): Baltimore, Johns Hopkins Univ. Press, pp. 118–158.

Folk, R. L., and Robles, Rogelio, 1964, Carbonate sands of Isla Perez, Alacran Reef complex, Yucatán: Jour. Geol., v. 72, pp. 255–292.

Franke, W., 1971, Structure and development of the Iberg/Winterberg reef (Devonian to Lower Carboniferous), Harz, West Germany, *in* Sedimentology of parts of central Europe (Müller, G., ed.): Frankfurt, Waldemar Kramer, pp. 83–90.

Freeman, T., 1973, Temporal dolomite-calcite sequence and its environmental implications (abstr.): Bull. Amer. Assoc. Petrol. Geol., v. 57, p. 780.

Friedman, G. M., 1964, Early diagenesis and lithification in carbonate sediments: Jour. Sed. Petrology, v. 34, pp. 777–813.

———, 1965, Terminology of crystallization textures and fabrics in sedimentary rocks: Jour. Sed. Petrology, v. 35, pp. 643–655.

———, ed., 1969, Depositional environments in carbonate rocks: Soc. Econ. Paleont. Min. Spec. Publ. 14, 209 pp.

———, 1971, Staining, *in* Procedures in sedimentary petrology (Carver, R. E., ed.): New York, Wiley, pp. 511–530.

Friedman, G. M., Amiel, A. J., Braun, M., and Miller, D. S., 1973, Generation of carbonate particles and laminites in algal mats—example from sea-marginal hypersaline pool, Gulf of Aqaba, Red Sea: Bull. Amer. Assoc. Petrol. Geol., v. 57, pp. 541–557.

Friedman, G. M., Geblein, C. D., and Sanders, J. E., 1971, Micritic envelopes of carbonate grains are not exclusively of photosynthetic algal origin: Sedimentology, v. 16, pp. 89–96.

Friedman, G. M., and Sanders, J. E., 1967, Origin and occurrence of dolostones, *in* Carbonate rocks (Chilingar, G., Bissell, H., and Fairbridge, R., eds.): Amsterdam, Elsevier, pp. 267–348.

Fruth, L. S., Jr., Orme, G. R., and Donath, F. A., 1966, Experimental compaction effects in carbonate sediments: Jour. Sed. Petrology, v. 36, pp. 747–754

Füchtbauer, H., ed., 1969, Lithification of carbonate sediments: Sedimentology, spec. issue, v. 12, pp. 117–322.

Garrett, P., 1970, Phanerozoic stromatolites: noncompetitive ecologic restriction by grazing and burrowing animals: Science, v. 169, pp. 171–173.

Garrison, R. E., and Fischer, A. G., 1969, Deep-water limestones and radiolarites of the

Alpine Jurassic, *in* Depositional environments in carbonate rocks: Soc. Econ. Paleont. Min. Spec. Publ. 14, pp. 20–55.

Geze, B., 1965, Les conditions hydrogéologiques des roches calcaires: Bur. Rech. Geol. Min., no. 7, pp. 9–40.

Gilbert, G. K., 1890, Lake Bonneville: U.S. Geol. Surv. Mono. 1, 438 pp.

Gile, L. H., Peterson, F. F., and Grossman, R. B., 1966, Morphological and genetic sequence of carbonate accumulation in desert soils: Soil Sci., v. 101, pp. 347–360.

Ginsburg, R. N., 1953, Beach rock in south Florida: Jour. Sed. Petrology, v. 23, pp. 85–92.

———, 1956, Environmental relationships of grain size and constituent particles in some south Florida carbonate sediments: Bull. Amer. Assoc. Petrol. Geol., v. 40, pp. 2384–2427.

———, 1964, South Florida carbonate sediments, Guidebook for Field Trip 1, Geol. Soc. America convention, 1964: New York, Geol. Soc. Amer., 72 pp.

Ginsburg, R. N., James, N. P., and Marszalek, D. S., 1973, Sedimentation and diagenesis in deep forereef, British Honduras barrier and atoll reefs (abstr.): Bull. Amer. Assoc. Petrol. Geol., v. 57, p. 781.

Ginsburg, R. N., and Lowenstam, H. A., 1958, The influence of marine bottom communities on the depositional environment of sediments: Jour. Geol., v. 66, pp. 310–318.

Ginsburg, R. N., Schroeder, J. H., and Shinn, E. A. 1971, Recent synsedimentary cementation in subtidal Bermuda reefs, *in* Carbonate cements: Studies in geol. no. 19 (Bricker, O. P., ed.), Baltimore, Johns Hopkins Univ. Press, pp. 54–58.

Goldberg, M., 1967, Supratidal dolomitization and dedolomitization in Jurassic rocks of Hamakhtesh Hagatan, Israel: Jour. Sed. Petrology, v. 37, pp. 760–773.

Goldich, S. S., and Parmalee, E. B., 1947, Physical and chemical properties of Ellenburger rocks, Llano County, Texas: Bull. Amer. Assoc. Petrol. Geol., v. 31, pp. 1982–2020.

Goudie, A., 1972, The chemistry of world calcrete deposits: Jour. Geol., v. 80, pp. 449–463.

Grabau, A. W., 1904, On the classification of sedimentary rocks: Amer. Geol., v. 33, pp. 228–247.

Graf, D. L., 1960a, Geochemistry of carbonate sediments and sedimentary carbonate rocks. Part I. Carbonate mineralogy, carbonate sediments: Illinois Geol. Surv. Circ. 297, 39 pp.

———, 1960b, Geochemistry of carbonate sediments and sedimentary rocks. Part II. Sedimentary carbonate rocks: Illinois Geol. Surv. Circ. 298, 43 pp.

———, 1960c, Geochemistry of carbonate sediments and sedimentary carbonate rocks. Part III. Minor element distribution: Illinois Geol. Surv. Circ. 301, 71 pp.

———, 1960d, Geochemistry of carbonate sediments and sedimentary carbonate rocks. Part IV-A. Isotopic composition, chemical analyses: Illinois Geol. Surv. Circ. 308, 42 pp.

———, 1960e, Geochemistry of carbonate sediments and sedimentary carbonate rocks. Part IV-B. Bibliography: Illinois Geol. Surv. Circ. 309, 55 pp.

Graf, D. L., and Lamar, J. E., 1950, Petrology of the Fredonia oolite in southern Illinois: Bull. Amer. Assoc. Petrol. Geol., v. 34, pp. 2318–2336.

Greenman, N., 1951, Origin of Randville dolomite: Ph.D. thesis, Univ. of Chicago.

Griffin, R. H., 1942, Dolomitic mottling in the Platteville limestone: Jour. Sed. Petrology, v. 12, pp. 67–76.

Griffith, L. S., Pitcher, M. G., and Rice, G. W., 1969, Quantitative environmental analysis of a Lower Cretaceous reef complex, *in* Depositional environments in carbonate rocks (Friedman, G. M., ed.): Soc. Econ. Paleont. Min. Spec. Publ. 14, pp. 120–138.

Grim, R. E., Lamar, J. E., and Bradley, W. F., 1937, The clay minerals in Illinois limestones: Jour. Geol., v. 45, pp. 829–843.

Hadding, A., 1933, The pre-Quaternary sedimentary rocks of Sweden. V. On the organic remains of the limestones: Lunds Univ. Arrskr. N.F. Avd. 2, v. 29, no. 4, 93 pp.

———, 1941, The pre-Quaternary sedimentary rocks of Sweden. VI. Reef limestones: Medd. Lunds Geol. Min. Inst., Lunds Univ. Arsskr. N.F. Avd. 2, v. 37, no. 10, 137 pp.

———, 1957, The Pre-Quaternary sedimentary rocks of Sweden. VII. Cambrian and Ordovician limestones: Lunds Univ. Arsskr. N.F. Avd. 2, v. 54, no. 5, 262 pp.

———, 1958, Origin of the lithographic limestones: Kungl. Fysiogr. Sällsk. Lund Förh., v. 28, no. 4, pp. 21–32.

Hall, C. W., and Sardeson, F. W., 1895, The Magnesian series of the northwestern states: Bull. Geol. Soc. Amer., v. 6, pp. 167–198.

Hallam, A, 1964, Origin of the limestone-shale rhythm in the Blue Lias of England: a composite theory: Jour. Geol., v. 72, pp. 157–169.

———, 1967, Sedimentology and palaeogeographic significance of certain red limestones and associated beds in the Lias of the Alpine region: Scottish Jour. Geol., v. 3 (2), pp. 195–220.

Ham, W. E., 1951, Dolomite in the Arbuckle Limestone, Arbuckle Mountains, Oklahoma (abstr.): Bull. Geol. Soc. Amer., v. 62, pp. 1446–1447.

———, 1954, Algal origin of the "birdseye" limestone in the McLish Formation: Proc. Oklahoma Acad. Sci., v. 33, pp. 200–203.

———, ed., 1962, Classification of carbonate rocks: Amer. Assoc. Petrol. Geol. Mem. 1, 279 pp.

Ham, W. E., and Pray, L. C., 1962, Modern concepts and classification of carbonate rocks, *in* Classification of carbonate rocks—a symposium (Ham, W. E., ed.): Amer. Assoc. Petrol. Geol. Mem. 1, pp. 2–19.

Hamblin, W. K., 1969, Marine paleocurrent directions in limestones of the Kansas City Group (Upper Pennsylvanian) in eastern Kansas: Bull. Geol. Surv. Kansas 194, pt. 2, pp. 1–25.

Harbaugh, J. W., 1959, Small scale cross-lamination in limestones: Jour. Sed. Petrology, v. 29, pp. 30–37.

Harding, J. L., 1964, Petrology and petrography of the Campeche lithic suite, Yucatán shelf, Mexico: Texas A & M Univ., Dept. Ocean. Meteorol. Tech. Rept. 64-11T, 139 pp.

Harris, L. D., 1973, Dolomitization model for Upper Cambrian and Lower Ordovician carbonate rocks in the eastern United States: Jour. Res., v. 1, pp. 63–78.

Hatch, F. H., Rastall, R. H., and Black, M., 1938, The petrology of the sedimentary rocks, 3rd ed.: London, Murby, 383 pp.

Heald, M. T., 1955, Stylolites in sandstones: Jour. Geol., v. 63, pp. 101–114.

Heckel, P. H., 1972, Possible inorganic origin for *Stromatactis* in calcilutite mounds in the Tully Limestone, Devonian of New York: Jour. Sed. Petrology, v. 42, pp. 7–18.

Heinrich, E. W., 1967, Carbonatites: nonsilicate igneous rocks: Earth Sci. Rev., v. 3, pp. 203–210.

Henbest, L. G., 1945, Unusual nuclei in oolites from the Morrow Group near Fayetteville, Arkansas: Jour. Sed. Petrology, v. 15, pp. 20–24.

———, 1968, Diagenesis in oolitic limestones of Morrow (early Pennsylvanian) age in northwestern Arkansas and adjacent Oklahoma: U.S. Geol. Surv. Prof. Paper 594-11, 22 pp.

Hollmann, R., 1962, Ueber subsolution und die "Knollenkalke" des calcare Ammonitico Russo Superiore im Monte Baldo: Neues Jahrb. Geol. Paleont., Mh., v. 4, pp. 163–179.

———, 1964, Subsolutions—Fragmente (zur Biostratonomie der Ammonoiden im Malm des Monte Baldo/Norditalian): Neues Jahrb. Geol. Paleont. Abh., v. 119, pp. 22–82.

Horowitz, A. S., and Potter, P. E., 1971, Introductory petrography of fossils: New York, Springer, 302 pp.

Hoskins, C. M., 1963, Recent carbonate sedimentation on Alacran Reef, Yucatán, Mexico: Nat. Acad. Sci., Nat. Res. Coun. Publ. 1089, pp. 1–160.

Houbolt, J. J. H. C., 1957, Surface sediments of the Persian Gulf near the Qatar Peninsula: Thesis, Univ. Utrecht, 113 pp.

Hrabar, S. V., Cressman, E. R., and Potter, P. E., 1971, Cross-bedding of the Tanglewood Limestone Member of the Lexington Limestone (Ordovician) of the Blue Grass Region of Kentucky: Brigham Young Univ. Geol. Stud., v. 18, no. 1, pp. 99–114.

Illing, L. V., 1954, Bahaman calcareous sands: Bull. Amer. Assoc. Petrol. Geol., v. 38, pp. 1–95.

Illing, L. V., Wells, A. J., and Taylor, J. C. M., 1965, Penecontemporary dolomite in the Persian Gulf, *in* Dolomitization and limestone diagenesis (Pray, L. C., and Murray, R. C., eds.): Soc. Econ. Paleont. Min. Spec. Publ. 13, pp. 89–111.

Imbrie, J., and Buchanan, H., 1965, Sedimentary structures in modern carbonate sands of the Bahamas, *in* Primary sedimentary structures and their hydrodynamic interpretation (Middleton, G. V., ed.): Soc. Econ. Paleont. Min. Spec. Publ. 12, pp. 149–172.

Imbrie, J., and Purdy, E. G., 1962, Classification of modern Bahaman carbonate sediments, *in* Classification of carbonate rocks—a Symposium (Ham, W. E., ed.): Amer. Assoc. Petrol. Geol. Mem. 1, pp. 253–272.

Ingels, J. J. C., 1963, Geometry, paleontology, and petrography of Thornton Reef complex, Silurian of northeastern Illinois: Bull. Amer. Assoc. Petrol. Geol., v. 47, pp. 405–440.

Ireland, H. A., 1936, Use of insoluble residues for correlation in Oklahoma: Bull. Amer. Assoc. Petrol. Geol., v. 20, pp. 1086–1121.

Irwin, M. L., 1965, General theory of epeiric clear water sedimentation: Bull. Amer. Assoc. Petrol. Geol., v. 49, pp. 445–459.

James, H. L., 1951, Iron formation and associated rocks in the Iron River district, Michigan: Bull. Geol. Soc. Amer., v. 62, pp. 251–266.

Johnson, J. H., 1951, An introduction to the study of organic limestones: Colorado Sch. Mines Quart., v. 46, 185 pp.

———, 1961, Limestone-building algae and algal limestones: Golden, Colo., Colorado Sch. Mines, 297 pp.

Jolliffe, A. W., 1955, Geology and iron ores of Steep Rock Lake: Econ. Geol., v. 50, pp. 373–398.

Judd, J. W., 1904, The atoll of Funafuti: Rept. Coral Reef Comm., Roy. Soc. London, pp. 364–365.

Kay, M., 1951, North American geosynclines: Geol. Soc. Amer. Mem. 48, 143 pp.

Ketner, K. B., 1970, Limestone turbidite of Kinderhook age and its tectonic significance, Elko County, Nevada: U.S. Geol. Surv. Prof. Paper 700-D, pp. D18–D22.

King, P. B., 1942, Permian of west Texas and southeast New Mexico: Bull. Amer. Assoc. Petrol. Geol., v. 26, pp. 535–763.

King, R. H, 1947, Sedimentation in Permian Castile Sea: Bull. Amer. Assoc. Petrol. Geol., v. 31, pp. 470–477.

Kinsman, D. J. J., 1964, The Recent carbonate sediments near Halat el Bahrain, Trucial Coast, Persian Gulf, *in* Deltaic and shallow marine deposits (van Straaten, L. M. J. U., ed.): Amsterdam, Elsevier, pp. 185–192.

Klovan, J. E., 1964, Facies analysis of the Redwater Reef complex, Alberta, Canada: Bull. Canad. Petrol. Geol., v. 12, pp. 1–100.

Knewtson, S. L., and Hubert, J. F., 1969, Dispersal patterns and diagenesis of oolitic calcarenites in the Ste. Geneviève Limestone (Mississippian), Missouri: Jour. Sed. Petrology, v. 39, pp. 954–68.

Krebs, W., 1969, Early void-filling cementation in Devonian fore-reef limestones (Germany): Sedimentology, v. 12, pp. 279–299.

———, 1971, Devonian reef limestones in the eastern Rhenish Schiefergebirge, *in* Sedimentology of parts of central Europe (Müller, G., ed.): Frankfurt, Verlag Waldemar Kramer, pp. 45–82.

Krey, F., and Lamar, J. E., 1925, Limestone resources of Illinois: Bull. Illinois Geol. Surv., 46, 392 pp.

Kuenen, Ph. H., 1933, The *Snellius* expedition. 5. Geological results. 2. Geology of coral reefs: Utrecht, Kemink., 125 pp.

———, 1950, Marine geology: New York, Wiley, 568 pp.

Kuenen, Ph. H., and ten Haaf, E., 1956, Graded bedding in limestone: Proc. Konink. Nederl. Akad. van Wetensch., ser. B., v. 59, no. 4, pp. 314–317.

Lamar, J. E., 1961, Uses of limestone and dolomite: Illinois Geol. Surv. Circ. 321, 41 pp.

Land, Lynton S., 1967, Diagenesis of skeletal carbonate: Jour. Sed. Petrology, v. 37, pp. 914–930.

———, 1971, Submarine lithification of Jamaican reefs, *in* Carbonate cements: Stud. in geol. no. 16 (Bricker, O. P., ed.): Baltimore, Johns Hopkins Univ. Press, pp. 59–62.

Landes, K. K., 1946, Porosity through dolomitization: Bull. Amer. Assoc. Petrol. Geol., v. 30, pp. 305–318.

Loughlin, G. F., 1929, Indiana oolitic limestone: Bull. U.S. Geol. Surv. 811, pp. 113–202.

LeBlanc, R. J., and Breeding, J. G., eds., 1957, Regional aspects of carbonate deposition: Soc. Econ. Paleont, Min. Spec. Publ. 5, 178 pp.

Lecompte, M., 1937, Contribution à la connaissance des recifs du Devonien de l'Ardenne. Sur la presence de structures conservées dans des efflorescences crystalline du type "stromatactis": Bull. Mus. Roy. Hist. Nat. Belgique, v. 13, pp. 1–14.

———, 1938, Quelques types de recifs Siluriens et Devoniens de l'Amerique du Nord. Essai de comparaison avec les recifs coralliens actuels: Bull. Mus. Roy. Hist. Nat. Belgique, 14, 51 pp.

———, 1958, Les recifs Paleozoiques en Belgique: Geol. Rundschau, v. 47, pp. 384–401.

Lees, A., 1964, The structure and origin of the Waulsortian (Lower Carboniferous) "reefs" of west-central Eire: Trans. Roy. Phil. Soc. London, ser. B, v. 247, pp. 483–531.

Leighton, M. W., and Pendexter, C., 1962, Carbonate rock types, *in* Classification of carbonate rocks—a symposium (Ham, W. E., ed.): Amer. Assoc. Petrol. Geol. Mem. 1, pp. 33–61.

Linck, G., 1937, Bildung des Dolomits und Dolomitisierung: Chemie der Erde, v. 11, pp. 278–386.

Lindholm, R. C., and Finkelman, R. B., 1972, Calcite staining: semiquantitative determination of ferrous iron: Jour. Sed. Petrology, v. 42, pp. 239–242.

Link, T. A., 1950, Theory of transgressive and regressive reef (bioherm) development and origin of oil: Bull. Amer. Assoc. Petrol. Geol., v. 34, pp. 263–294.

Lobo, C. F., and Osborne, R. H., 1973, The American Upper Ordovician standard. XVIII: Investigation of micrite in typical Cincinnatian limestones by means of scanning electron microscopy: Jour. Sed. Petrology, v. 43, pp. 478–483.

Logan, B. W., 1961, *Cryptozoon* and associate stromatolites from the Recent, Shark Bay, Western Australia: Jour. Geol., v. 69, pp. 517–533.

Logan, B. W., Davies, G. R., Read, J. F., and Cebulski, D. E., 1970, Carbonate sedimentation and environments, Shark Bay, Western Australia: Amer. Assoc. Petrol. Geol. Mem. 13, 223 pp.

Logan, B. W., Harding, J. L., Ahr, W. M., Williams, J. D., and Snead, R. G., 1969, Carbonate sediments and reefs, Yucatán shelf, Mexico: Amer. Assoc. Petrol. Geol. Mem. 11, 198 pp.

Long, M. B., 1953, Origin of the Conococheague Limestone: Ph.D. thesis, Johns Hopkins Univ.

Loreau, J.-P., 1972, Pétrographie de calcaires fins au microscope électronique: introduction à une classification des "micrites": Comptes Rendus Acad. Sci. Paris, v. 274, pp. 810–813.

Lowenstam, H. A., 1950, Niagaran reefs of the Great Lakes area: Jour. Geol., v. 58, pp. 430–487.

———, 1954, Factors affecting the aragonite/calcite ratios in carbonate-secreting marine organisms: Jour. Geol., v. 62, pp. 284–322.

———, 1955, Aragonite needles secreted by algae and some sedimentary implications: Jour. Sed. Petrology, v. 25, pp. 270–272.

———, 1963, Biologic problems relating to the composition and diagenesis of sediments: The Earth Sciences, Problems in Current Research, Rice Univ., Semicentennial Publ., pp. 137–195.

Lowenstam, H. A., and DuBois, E. P., 1946, Marine pool, Madison County: Illinois Geol. Surv. Rept. Inv. 114, 30 pp.

Lowenstam, H. A., and Epstein, S., 1957, On the origin of the sedimentary aragonite needles of the Great Bahama Bank: Jour. Geol., v. 65, pp. 364–375.

Lucia, F. J., 1972, Recognition of evaporite-carbonate shoreline sedimentation, *in* Recognition of ancient sedimentary environments (Rigby, J. K., and Hamblin, W. K., eds.): Soc. Econ. Paleont. Min. Spec. Publ. 16, pp. 160–191.

Mackenzie, F. T., 1964, Bermuda Pleistocene eolianites and paleowinds: Sedimentology, v. 3, pp. 52–64.

MacKenzie, W. S., 1970, Allochthonous reef-debris limestone turbidites Powell Creek, Northwest Territories: Bull. Canad. Petrol. Geol., v. 18, pp. 474–492.

McCrossan, G. R., 1958, Sedimentary "boudinage" structures in the Upper Devonian Ireton Formation of Alberta: Jour. Sed. Petrology, v. 28, pp. 316–320.

McKee, E. D., and Gutschick, R. C., 1969, History of Redwall Limestone of northern Arizona: Geol. Soc. Amer. Mem. 114, 726 pp.

McQueen, H. S., 1931, Insoluble residues as a guide to stratigraphic study: Missouri Bur. Geol., 56th Bien. Rept. App.

Majewske, O. P., 1969, Recognition of invertebrate fossil fragments in rocks and thin sections: Leiden, Brill, 101 pp.

Mamet, B., 1961, Reflexions sur la classification des calcaires: Bull. Soc. Geol. Belge, v. 70, pp. 48–64.

Manten, A. A., 1962, Some Middle Silurian reefs of Gotland: Sedimentology, v. 1, pp. 211–234.

———, 1966, Note on formation of stylolites: Geol. en Mijnb., v. 45, pp. 269–274.

———, 1971, Silurian reefs of Gotland: Developments in sedimentology 13, Amsterdam, Elsevier, 539 pp.

Martinet, B., and Sougy, J., 1961, Utilisation pratique des classifications chimiques des roches carbonatées: Ann. Fac. Sci. Univ. Dakar, Ann., 1961, v. 6, pp. 81–92.

Matter, A., 1967, Tidal flat deposits in the Ordovician of western Maryland: Jour. Sed. Petrology, v. 37, pp. 601–609.

Matthews, R. K., 1966, Genesis of Recent lime mud in southern British Honduras: Jour. Sed. Petrology, v. 36, pp. 428–454.

Maxwell, W. G. H., and Swinchatt, J. P., 1970, Great Barrier Reef: Regional variation in a terrigenous-carbonate province: Bull. Geol. Soc. Amer., v. 81, pp. 691–724.

Meischner, K.-D., 1964, Allodapische Kalke, Turbidite in Riff-nahen Sedimentations-Becken, *in* Turbidites (Bouma, A. H., and Brouwer, A., eds.): Amsterdam, Elsevier, pp. 156–191.

———, 1971, Clastic sedimentation in the Variscan Geosyncline east of the Rhine, *in* Sedimentology of parts of central Europe (Müller, G., ed.), Heidelberg: VIII Int. Sediment. Congr. 1971, Frankfurt, Verlag Waldemar Kramer, pp. 9–43.

Michard, A., 1969, Les dolomies, une revue: Bull. Serv. Carte Géol. Alsace-Lorraine 22, pp. 1–92.

Milliman, J. D., 1965, An annotated bibliography on recent papers on corals and coral reefs: Nat. Acad. Sci., Nat. Res. Coun., Pacific Sci. Board, Atoll Res. Bull. 111, 58 pp.

Moore, H. B., 1939, Faecal pellets in relation to marine deposits, *in* Recent marine sediments: Tulsa, Okla., Amer. Assoc. Petrol. Geol., pp. 516–524.

Müller, G., and Friedman, G. M., eds., 1968, Recent developments in carbonate sedimentology in central Europe: New York, Springer, 255 pp.

Müller-Jungbluth, W. U., and Toschek, P. H., 1969, Karbonat sedimentologische Arbeitsgrundlagen: Alpenkundliche Studies no. 4, 32 pp.

Murray, R. C., 1960, Origin of porosity in carbonate rocks: Jour. Sed. Petrology, v. 30, pp. 59–84.

Murray, R. C., and Pray, L. C., 1965, Dolomitization and limestone diagenesis: Soc. Econ. Paleont. Min. Spec. Publ. 13, 180 pp.

Nelson, H. F., Brown, C. W., and Brineman, J. H., 1962, Skeletal limestone classification, *in* Classification of carbonate rocks—a symposium (Ham, W. E., ed.): Amer. Assoc. Petrol. Geol. Mem. 1, pp. 224–252.

Newell, N. D., Purdy, E. G., and Imbrie, J., 1960, Bahamian oolitic sand: Jour. Geol., v. 68, pp. 481–497.

Newell, N. D., and Rigby, J. K., 1957, Geological studies on the Great Bahama Bank, *in* Regional aspects of carbonate deposition (LeBlanc, R. J., and Breeding, J. A., eds.): Soc. Econ. Paleont. Min. Spec. Publ. 5, pp. 15–72.

Newell, N. D., Rigby, J. K., Fischer, A. G., Whiteman, A. J., Hickox, J. E., and Bradley, J. S., 1953, The Permian reef complex of the Guadalupe Mountains, Texas and New Mexico: San Francisco, Freeman, 236 pp.

Orme, G. R., and Brown, W. W., 1963, Diagenetic fabrics in the Avonian limestone of Derbyshire and North Wales: Proc. Yorkshire Geol. Soc., v. 34, pp. 51–66.

Osmond, J. C., 1956, Mottled carbonate rocks in the Middle Devonian of eastern Nevada: Jour. Sed. Petrology, v. 26, pp. 32–41.

Park, W. C., and Schot, E. H., 1968, Stylolitization in carbonate rocks, *in* Carbonate sedimentology in central Europe (Müller, G., and Friedman, G. M., eds.), pp. 66–74.

Parkinson, D., 1957, Lower Carboniferous reefs of northern England: Bull. Amer. Assoc. Petrol. Geol., v. 41, pp. 511–537.

Pelto, C. R., 1942, Petrology of the Gatesburg Formation of central Pennsylvania: M.S. thesis, Pennsylvania State Univ., 60 pp.

Peterson, M. N. A., 1962, The mineralogy and petrology of Upper Mississippian carbonate rocks of the Cumberland Plateau in Tennessee: Jour. Geol., v. 70, pp. 1–31.

Pettijohn, F. J., and Potter, P. E., 1964, Atlas and glossary of primary sedimentary structures: New York, Springer, 370 pp.

Philcox, M. E., 1965, Sedimentation of Upper Devonian *Stromatactis* bioherm, Alberta, Canada (abstr.): Geol. Soc. Amer. Spec. Paper 82, p. 150.

Pia, J., 1933, Die rezenten Kalksteine: Zeitschr. Krist. Min. Petrol., Erganbundesband, Abt. B, pp. 1–420.

Picard, M. D., 1953, Marlstone—a misnomer as used in the Uinta Basin: Bull. Amer. Assoc. Petrol. Geol., v. 37, pp. 1075–1077.

Pinsak, A. P., 1957, Subsurface stratigraphy of the Salem limestone and associated formations in Indiana: Indiana Geol. Surv. Bull. 11, 62 pp.

Plumley, W. J., and Graves, R. W., Jr., 1953, Virgilian reefs of the Sacramento Mountains, New Mexico: Jour. Geol., v. 61, pp. 1–16.

Plumley, W. J., Risley, G. A., Graves, R. W., Jr., and Kaley, M. E., 1962, Energy index for limestone interpretation and classification, *in* Classification of carbonate rocks—a symposium (Ham, W. E., ed.): Amer. Assoc. Petrol. Geol. Mem. 1, pp. 85–107.

Potter, P. E., 1968, A selective, annotated bibliography on carbonate rocks: Bull. Canad. Petrol. Geol., v. 16, pp. 87–103.

Potter, P. E., and Pettijohn, F. J., 1963, Paleocurrents and basin analysis: New York, Springer, 296 pp.

Potter, P. E., and Scheidegger, A. E., 1966, Bed thickness and grain size: graded beds: Sedimentology, v. 7, pp. 233–240.

Powers, R. W., 1962, Arabian Upper Jurassic carbonate reservoir rocks, *in* Classification of carbonate rocks—a symposium (Ham, W. E., ed.): Amer. Assoc. Petrol. Geol. Mem. 1, pp. 122–192.

Pray, L. C., 1958, Fenestrate-bryozoan core facies, Mississippian bioherms, southwestern United States: Jour. Sed. Petrology, v. 28, pp. 261–273.

———, 1960, Compaction in calcilutites (abstr.): Bull. Geol. Soc. Amer., v. 71, p. 1946.

Pray, L. C., and Murray, R. C., eds., 1965, Dolomitization and limestone diagenesis—a symposium: Soc. Econ. Paleont. Min. Spec. Publ. 13, 180 pp.

Price, W. A., 1933, Reynosa problem of south Texas and origin of caliche: Bull. Amer. Assoc. Petrol. Geol., v. 17, pp. 488–522.

Prokopovich, N., 1952, The origin of stylolites: Jour. Sed. Petrology, v. 22, pp. 212–220.

Pugh, W. E., 1950, Bibliography of organic reefs, bioherms, and biostromes: Tulsa, Okla., Seismograph Service Corp., 139 pp.

Purdy, E. G., 1963a, Recent calcium carbonate facies of the Great Bahama Bank. 1. Petrography and reaction groups: Jour. Geol., v. 71, pp. 334–355.

———, 1963b, Recent calcium carbonate facies of the Great Bahama Bank. 2. Sedimentary facies: Jour. Geol., v. 71, pp. 472–497.

———, 1968, Carbonate diagenesis: an environmental survey: Geologica Romana, v. 7, pp. 183–228.

Purser, B. H., 1969, Syn-sedimentary marine lithification of Middle Jurassic limestones in the Paris Basin: Sedimentology, v. 12, pp. 205–230.

Ramsden, R. M., 1952, Stylolites and oil migration: Bull. Amer. Assoc. Petrol. Geol., v. 36, pp. 2185–2186.

Rich, M., 1964, Petrographic classification and method of description of carbonate rocks of the Bird Spring Group in southern Nevada: Jour. Sed. Petrology, v. 34, pp. 365–378.

Richardson, W. A., 1919, The origin of Cretaceous flint: Geol. Mag., v. 56, pp. 535–547.

Rigby, J. K., 1953, Some transverse stylolites: Jour. Sed. Petrology, v. 23, pp. 265–271.

Rittenhouse, G., 1949, Petrology and paleogeography of Greenbrier Formation: Bull. Amer. Assoc. Petrol. Geol., v. 33, pp. 1704–1730.

Robbins, C., and Keller, W. D., 1952, Clay and other non-carbonate minerals in some limestones: Jour. Sed. Petrology, v. 22, pp. 146–152.

Robertson, E. C., 1965, Experimental consolidation of carbonate mud (abstr.), *in* Dolomitization and limestone diagenesis (Pray, L. C., and Murray, R. C., eds.): Soc. Econ. Paleont. Min. Spec. Publ. 13, p. 170.

Rodgers, J., 1954, Terminology of limestones and related rocks: an interim report: Jour. Sed. Petrology, v. 24, pp. 225–234.

———, 1957, The distribution of marine carbonate sediments, *in* Regional aspects of carbonate deposition (LeBlanc, R. J., and Breeding, J. G., eds.): Soc. Econ. Paleont. Min. Spec. Publ. 5, pp. 2–14.

Ronov, A. B., Migdisov, A. A., and Barskaya, N. V., 1969, Tectonic cycles and regularities in the development of sedimentary rocks and paleogeographic environments of sedimentation of the Russian Platform (an approach to a quantitative study): Sedimentology, v. 13, pp. 179–212.

Root, S. I., 1965, Cyclicity of the Conococheague Formation: Proc. Pennsylvania Acad. Sci., v. 38, pp. 157–160.

Runnels, R. T., and Dubins, I. M., 1949, Chemical and petrographic studies of the Fort Hays Chalk in Kansas: Bull. Geol. Surv. Kansas 82, pt. I, pp. 1–36.

Russell, I. C., 1885, Geological history of Lake Lahontan, a Quaternary lake of northwestern Nevada: U.S. Geol. Surv. Monogr. 11, 288 pp.

———, 1889, Subaerial decay of rocks and origin of the red color of certain formations: U.S. Geol. Surv. Bull. 52, 63 pp.

Sander, B., 1951, Contributions to the study of depositional fabrics (Knopf, E. B., trans.): Tulsa, Okla., Amer. Assoc. Petrol. Geol., 207 pp.

Sander N. J., 1967, Classification of carbonate rocks of marine origin: Bull. Amer. Assoc. Petrol. Geol., v. 51, pp. 325–336.

Sando, W. J., 1953, Beekmantown group (Lower Ordovician) of Maryland: Geol. Soc. Amer. Mem. 68, 161 pp.

Sarin, D. D., 1962, Cyclic sedimentation of primary dolomite and limestone: Jour. Sed. Petrology, v. 32, pp. 451–471.

Sayles, R. W., 1931, Bermuda during the Ice Age: Proc. Amer. Acad. Arts Sci., v. 66, pp. 382–467.

Schlanger, S. O., 1964, Petrology of the limestones of Guam: U.S. Geol. Surv. Prof. Paper 403-O, 52 pp.

Schmidt, V., 1965, Facies, diagenesis and related reservoir properties in the Gigas Beds (Upper Jurassic), northwestern Germany, *in* Dolomitization and limestone diagenesis (Pray, L. C., and Murray, R. C., eds.): Soc. Econ. Paleont. Min. Spec. Publ. 13, pp. 124–168.

Scholle, P. A., 1971, Sedimentology of fine-grained deep-water carbonate turbidites, Monte Antola flysch (Upper Cretaceous), northern Apennines, Italy: Bull. Geol. Soc. Amer., v. 82, pp. 629–658.

Schopf, J. W., Oehler, D. Z., Horodyski, R. J., and Kvenvolden, K. A., 1971, Biogenicity and significance of the oldest known stromatolites: Jour. Paleont., v. 45, pp. 477–485.

Scobey, E. H., 1940, Sedimentary studies of the Wapsipinicon Formation in Iowa: Jour. Sed. Petrology, v. 10, pp. 33–44.

Sedimentation Seminar, 1966, Cross-bedding in the Salem limestone of central Indiana: Sedimentology, v. 6, pp. 95–114.

Sestini, G., 1964, Paleocorrenti eoceniche nell' area tosco-umbra: Boll. Soc. Geol. Italiana, v. 83, pp. 1–54.

Sharp, R. P., 1942, Stratigraphy and structure of the southern Ruby Mountains, Nevada: Bull. Geol. Soc. Amer., v. 53, pp. 647–690.

Shaub, B., 1939, The origin of stylolites: Jour. Sed. Petrology, v. 9, pp. 47–61.

Shearman, D. J., 1963, Recent anhydrite, gypsum, dolomite and halite from the coastal flats of the Arabian shore of the Persian Gulf: Proc. Geol. Soc. London, no. 1607, pp. 63–65.

Shearman, D. J., Khouri, J., and Taha, S., 1961, On the replacement of dolomite by calcite in some Mesozoic limestones from the French Jura: Proc. Geol. Assoc., v. 72, pp. 1–12.

Shinn, E. A., 1964, Recent dolomite, Sugarloaf Key, Florida, *in* South Florida sediments (Ginsburg, R. N., ed.): Geol. Soc. Amer. Guidebook, Field Trip 1, Ann. Mtg., Florida, pp. 26–33.

———, 1968, Practical significance of birdseye structures in carbonate rocks: Jour. Sed. Petrology, v. 38, pp. 215–223.

Shinn, E. A., Ginsburg, R. N., and Lloyd, R. M., 1965, Recent supratidal dolomite from Andros Island, Bahamas, *in* Dolomitization and limestone diagenesis (Pray, L. C., and Murray, R. C., eds.): Soc. Econ. Paleont. Min. Spec. Publ. 13, pp. 112–123.

Shrock, R. R., 1939, Wisconsin Silurian bioherms (organic reefs): Bull. Geol. Soc. Amer., v. 50, pp. 529–562.

———, 1948, Classification of sedimentary rocks: Jour. Geol., v. 56, pp. 118–129.

———, 1949, Sequence in layered rocks: New York, McGraw-Hill, 507 pp.

Siever, R., and Glass, H. D., 1957, Mineralogy of some Pennsylvanian carbonate rocks of Illinois: Jour. Sed. Petrology, v. 27, pp. 56–63.

Skeats, E. W., 1905, The chemical and mineralogical evidences as to the origin of the dolomites of southern Tyrol: Quart. Jour. Geol. Soc. London, v. 61, pp. 97–114.

———, 1918, The formation of dolomite and its bearing on the coral reef problem: Amer: Jour. Sci., ser. 4, v. 45, pp. 185–200.

Sloss, L. L., 1953, The significance of evaporites: Jour. Sed. Petrology, v. 23, pp. 143–161.

Spry, A., 1969, Metamorphic textures: New York, Pergamon, 350 pp.

Stauffer, K. W., 1962, Quantitative petrographic study of Paleozoic carbonate rocks, Caballo Mountains, New Mexico: Jour. Sed. Petrology, v. 32, pp. 357–396.

Steidtmann, E., 1911, The evolution of limestone and dolomite: Jour. Geol., v. 19, pp. 323–345, 393–428.

———, 1917, Origin of dolomite as disclosed by stains and other methods: Bull. Geol. Soc. Amer. v. 28, pp. 431–450.

Stockdale, P. B., 1922, Stylolites: their nature and origin: Indiana Univ. Stud., v. 11, pp. 1–97.

———, 1926, Stratigraphic significance of solution in rocks: Jour. Geol., v. 34, pp. 399–414.

———, 1931, Bioherms in the Borden Group of Indiana: Bull. Geol. Soc. Amer., v. 42, pp. 707–718.

———, 1943, Stylolites: primary or secondary?: Jour. Sed. Petrology, v. 13, pp. 3–12.

Stockman, K. W., Ginsburg, R. N., and Shinn, E. A., 1967, The production of lime mud by algae in south Florida: Jour. Sed. Petrology, v. 37, pp. 633–648.

Stoddart, D. R., and Cann, J. R., 1965, Nature and origin of beach rock: Jour. Sed. Petrology, v. 35, pp. 243–247.

Stout, W. E., 1941, Dolomites and limestones of western Ohio: Ohio Geol. Surv., 4th ser., Bull. 42, 468 pp.

Stout, W. E., and Lamborn, R. E., 1924, Geology of Columbiana County: Ohio Geol. Surv., 4th ser., Bull. 28, 408 pp.

Strakhov, N. M., 1958, Facts and hypotheses concerning the genesis of dolomite rocks: Izv. Akad. Nauk. SSSR, ser. geol., no. 6, pp. 1–18 (AGI trans. 1960).

———, 1970, Principles of lithogenesis (trans.): New York, Plenum, v. 3, 577 pp.

Swineford, A., Leonard, A. B., and Frye, J. C., 1958, Petrology of the Pliocene pisolitic limestone of the Great Plains: Bull. Geol. Surv. Kansas 130, pt. 2, pp. 98–116.

Tanner, W. F., 1959, The importance of modes in cross-bedding data: Jour. Sed. Petrology, v. 29, pp. 221–226.

Taylor, J. C. M., and Illing, L. V., 1969, Holocene intertidal calcium carbonate cementation, Qatar, Persian Gulf: Sedimentology, v. 12, pp. 69–107.

Tebbutt, G. E., Conley, C. D., and Boyd, D. W., 1965 Lithogenesis of a distinctive carbonate rock fabric, *in* Contributions to geology (Parker, R. B., ed.): Laramie, Wyoming, Univ. Wyoming, pp. 1–13.

Teodorovich, G. I., 1961, On the origin of sedimentary dolomite: Int. Geol. Rev., v. 3, no. 5, pp. 373–384.

Terzaghi, R. D., 1940, Compaction of lime mud as a cause of secondary structure: Jour. Sed. Petrology, v. 10, p. 78–90.

Textoris, D. A., and Carozzi, A. V., 1964, Petrography and evolution of Niagaran (Silurian) reefs, Indiana: Bull. Amer. Assoc. Petrol. Geol., v. 48, pp. 397–426.

Thiel, G. A., 1933, A correlation of marl beds with types of glacial deposits: Jour. Geol., v. 38, pp. 717–728.

Thomson, A., and Thomasson, M. R., 1969, Shallow to deep water facies development in the Dimple Limestone (Lower Pennsylvanian), Marathon Region, Texas, *in* Depositional environments in carbonate rocks (Friedman, G. M., ed.), Soc. Econ. Paleont. Min. Spec. Publ. 14, pp. 57–77.

Thorp, E. M., 1936, The sediments of the Pearl and Hermes reefs: Jour. Sed. Petrology, v. 6, pp. 109–118.

Todd, T. W., 1966, Petrogenetic classification of carbonate rocks: Jour. Sed. Petrology, v. 36, pp. 317–340.

Trask, P. D., 1937, Relation of salinity to the calcium carbonate content of marine sediments: U.S. Geol. Surv. Prof. Paper 186-N, pp. 273–299.

Trurnit, P., 1968, Analysis of pressure-solution contacts and classification of pressure-solution phenomena, *in* Carbonate sedimentology in central Europe (Müller, G., and Friedman, G. M., eds.): New York, Springer, pp. 74–84.

Tucker, M. E., 1969, Crinoidal turbidites from the Devonian of Cornwall and their paleogeographic significance: Sedimentology, v. 13, pp. 281–290.

Twenhofel, W. H., 1950, Coral and other organic reefs in geologic column: Bull. Amer. Assoc. Petrol. Geol., v. 34, pp. 182–202.

Van Tuyl, F. M., 1916a, The origin of dolomite: Ann. Rept., Iowa Geol. Surv., v. 25, pp. 251–422.

———, 1916b, New points on the origin of dolomite: Amer. Jour. Sci., ser. 4, v. 42, pp. 249–260.

Vaughn, T. W., 1917, Chemical and organic deposits of the sea: Bull. Geol. Soc. America, v. 28, pp. 933–944.

Voigt, E., 1959, Die ökologische Bedeutung der Hartgründe ("hardgrounds") in der oberen Kreide: Palaeont. Zeitsch., v. 33, pp. 129–147.

———, 1962, Früh diagenetische Deformation der turonen Plänerkalke bei Halle/Westf. als Folge einer Grossgleitung unter besonderer Berückschichtung des Phacoid-Problems: Hamburg, Geol. Staatsinst. Mitt., v. 31, pp. 146–275.

Wanless, H. R., 1973, Microstylolites, bedding, and dolomitization (abstr.): Abstracts of Papers, Ann. Meeting Soc. Econ. Paleont. Min., Anaheim, Bull. Amer. Assoc. Petrol. Geol., v. 57, p. 811.

Wardlaw, N. C., 1962, Aspects of diagenesis in some Irish Carboniferous limestones: Jour. Sed. Petrology, v. 32, pp. 776–780.

Weaver, C. E., 1958, Geologic interpretation of argillaceous sediments. Part I. Origin and significance of clay minerals in sedimentary rocks: Bull. Amer. Assoc. Petrol. Geol., v. 42, pp. 254–271.

———, 1959, The clay petrology of sediments: Proc. 6th Nat. Conf. Clays and Clay Minerals, New York, Pergamon, pp. 154–187.

Weyl, P. K., 1959, Pressure solution and the force of crystallization—a phenomenological theory: Jour. Geophys. Res., v. 64, pp. 2001–2025.

Weynschenk, R., 1951, The problem of dolomite formation considered in the light of research on dolomites in the Sonnwend-mountains (Tirol): Jour. Sed. Petrology, v. 21, pp. 28–31.

Williams, F. T., and McCoy, E., 1934, On the role of microorganisms in the precipitation of calcium carbonate in the deposits of fresh water lakes: Jour. Sed. Petrology, v. 4, pp. 113–126.

Wilson, J. L., 1969, Microfacies and sedimentary structures in "deeper water" lime mudstones: Soc. Econ. Paleont. Min. Spec. Publ. 14, pp. 4–16.

Wilson, R. C. L., 1966, Silica diagenesis in Upper Jurassic limestones of southern England: Jour. Sed. Petrology, v. 36, pp. 1036–1049.

Winchell, A. N., 1924, Petrographic studies of limestone alterations at Bingham: Trans. Amer. Inst. Min. Metall. Eng., v. 70, pp. 884–902.

Wolf, K. H., 1960, Simplified limestone classification: Bull. Amer. Assoc. Petrol. Geol., v. 44, pp. 1414–1416.

———, 1961, An introduction to the classification of limestones: Neues Jahrb., pp. 236–250.

———, 1962, The importance of calcareous algae in limestone genesis and sedimentation: Neues Jahrb., pp. 245–261.

Wolfe, M. J., 1968, Lithification of a carbonate mud: Devonian chalk in northern Ireland: Sed. Geol., v. 2, no. 4, pp. 263–290.

Wood, A., 1941, Algal dust and the finer-grained varieties of Carboniferous limestone: Geol. Mag., v. 78, pp. 192–200.

Zankl, H., 1969, Structural and textural evidence of early lithification in fine-grained carbonate rocks: Sedimentology, v. 12, pp. 241–256.

Zen, E., 1959, Clay mineral–carbonate relations in sedimentary rocks: Amer. Jour. Sci., v. 257, pp. 29–43.

11

NONCLASTIC SEDIMENTS (EXCLUDING LIMESTONES)

INTRODUCTION

The nonclastic sediments here described fall into three main categories: primary precipitates, secondary or diagenetic segregations and metasomatites, and certain organic accumulations (the bioliths of Grabau, 1904).

Primary precipitates include evaporites, which are deposits formed by evaporation of saline solutions, principally brines derived from sea water. Evaporites include rock sulfates (mainly anhydrite), chlorides (chiefly rock salt), some carbonates (principally dolomite), and various rare nitrates.

A second group of nonclastic sediments includes cherts, iron-bearing sediments, and phosphates. These generally occur in two forms—as nodules or as bedded deposits. The origins of these deposits are not wholly resolved. Bedded deposits are generally regarded as precipitates resulting from reactions between dissolved materials and the finest suspended clay sediment, or as a result of change in acidity or shift in the oxidation–reduction potential. Some may be the product of a diagenetic reorganization of biochemically precipitated skeletal detritus. Nodular deposits are in part the result of rearrangement *within* the sediment which produces segregations of the minor constituents of the sediment. (This type of diagenetic segregation is the subject of Chapter 12). We have to deal here also with *metasomatites* (Berkey, 1924), which are a group of rocks produced by the chemical alteration of existing rocks. For the most part they are limestones that have been dolomitized, silicified, phosphatized, and so forth. In these rocks, changes are on such a large scale that introduction of material from *without* must have taken place. The source of the introduced materials may be sea water (as it is for many dolomites) or circulating waters, either meteoric or magmatic. Metasomatism is a process that forms new rocks, chemically unlike the parent rock, that retain in some degree the textures and structures of the original rock. The changes, therefore, are a species of metamorphism produced at relatively low temperature and pressure.

Theoretically the differences between metasomatites and diagenetic segregations and the primary precipitates are clear and unequivocal. The actual assignment of a particular deposit to one or another of these groups, however, is often difficult. Moreover, many of these nonclastic sediments are polygenetic and may form in more than one way. Chert, for example, may form as a segregation (chert nodules), as a replacement (silicified limestone), as a modified biolith (silicified diatomite), or perhaps as a precipitate, either of silica or of a sodium silicate precursor which is converted to chert. Because of such complexities of ori-

gin and of an economy of space which is thereby gained, nonclastic sediments in this group are treated according to their chemical composition. Deposits of silica, iron, and phosphorus, moreover, have certain characteristics in common, and their geologic occurrence and formation have certain common factors.

The *bioliths,* or organic accumulates, do not form a homogeneous group but fall into two unlike subgroups, *acaustobioliths* and *caustobioliths,* which are noncombustible and combustible rocks, respectively (Grabau, 1904). See Fig. 11-1. Of the former, caliliths form an important part of the limestone group and therefore have been placed with these rocks and described elsewhere. Siliciliths (mainly diatomite and radiolarite) have been grouped with the cherts to which they are closely related. Phosphatic bone beds belong with the phosphorites. Combustible bioliths, including coal and the bitumens, although relatively rare rocks, are of great economic and scientific interest. They have accordingly been given special treatment.

It should be pointed out here that many of the nonclastic sediments described in this chapter are the end-products of nature's geochemical fractionation processes which in some cases lead to deposits of remarkable purity. They are, therefore, the main source of many materials useful to man. From the evaporites, for example, we recover salts of various elements: Na, Cl, Br, I, B, N, and S. The greatest accumulations of Si, P, and Fe are found in the second group of nonclastics, which also yield some of the rarer elements such as V and U. And caustobioliths are the most important concentrations of carbon (the carbonates excepted) in the world and are the primary source of energy ("fossil" fuels).

CHERT AND OTHER SILICEOUS SEDIMENTS

DEFINITIONS AND CLASSIFICATION

Tarr (1938) prepared a very extended summary of the proper names of the nonclastic siliceous sediments; for the most part, the

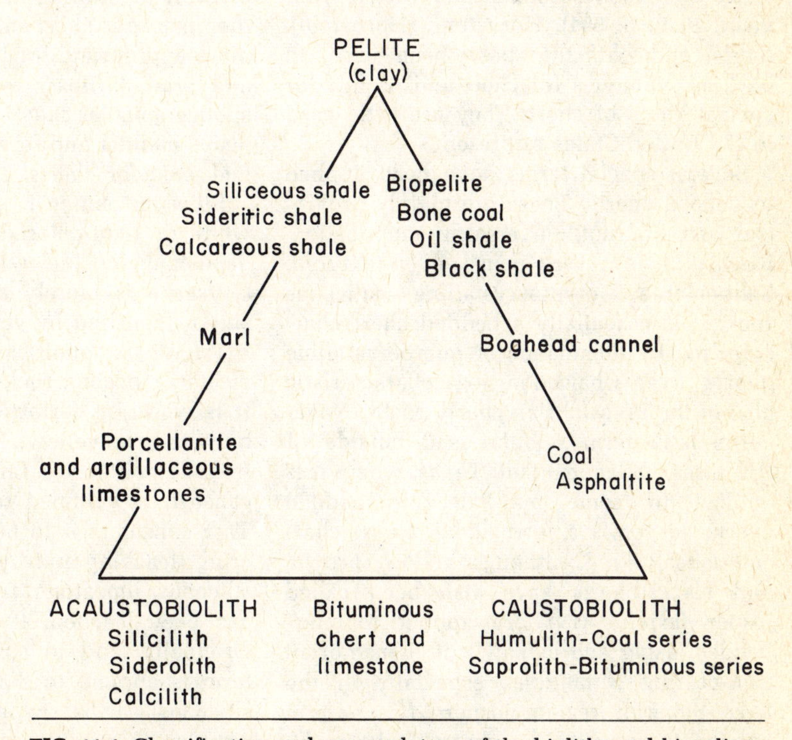

FIG. 11-1. Classification and nomenclature of the bioliths and biopelites.

usage followed here is that recommended by Tarr.

Chert and *flint* are the most common chemical siliceous sediments. Much current confusion concerns the origin of these terms, their exact meaning, and the differences, if any, between them. Chert is a dense rock composed of one or several forms of silica —opal, chalcedony (microcrystalline fibrous quartz), or microcrystalline quartz. It has a tough, splintery to conchoidal fracture. It may be white or variously colored gray, green, blue, pink, red, yellow, brown, and black. Flint (*Feuerstein*) is a term widely used both as a synonym for chert and as a variety of chert. Tarr states that it is identical with chert, and the term, therefore, should be dropped or reserved for artifacts to which it is most often applied. Although the term antedates the term *chert*, usage favors the latter as the proper designation of the materials to which both the terms have been applied.

Silexite is the French term for chert, especially the black carbonaceous variety (Cayeux, 1929, p. 554). *Hornstone* (*Hornstein*), *lydite*, and *phthanite* have been used, in part, as synonyms for chert and in part for special kinds of chert. They are little used in the United States at present.

Several special terms have been applied to bedded cherts. These are mainly provincial and are not universal in application. *Novaculite* is a very dense, even-textured, light-colored, cryptocrystalline siliceous rock. It is essentially a bedded chert characterized by dominance of microcrystalline quartz over chalcedony—a characteristic also of the Precambrian cherty rocks. Novaculite is a term seldom used outside of Arkansas, Oklahoma, and Texas, where it is applied to some mid-Paleozoic bedded cherts. *Jasper* is a ferruginous chert, characteristically red, although yellow, brown, and black cherts have also been called jasper. *Jaspilite* is a term applied to interbedded jasper and hematite of Precambrian iron-bearing formations, especially in the Lake Superior region. *Jasperoid* is a term applied to some of the cherty beds in the mining districts of Oklahoma, Missouri, and Kansas (Siebenthal, 1915, p. 15; Lovering, 1972).

Porcelanite (porcellanite) is a term applied to dense, hard rocks having the texture and fracture of unglazed porcelain. Most commonly, though not always, such rocks are cherts with an abundance of included materials. These impure cherts are in part argillaceous and in part calcareous, or more rarely sideritic. Argillaceous porcelanites grade into siliceous shales; calcareous porcelanites grade into the siliceous limestones. Some porcelanites, however, are silicified tuffs.

Not all siliceous sediments are dense, tough rocks. Some are friable and porous— these are the siliceous earths, including *diatomaceous* and *radiolarian earths*. They are composed of the opaline tests of diatoms and latticelike skeletal frameworks of radiolarians, respectively. They are usually white or cream-colored; more rarely they are buff, red, or brown. These rocks are homogeneous, porous, and friable and have a dry earthy feel and appearance. The terms *diatomite* and *radiolarite* have also been applied to these deposits, although they have also been used for these materials which have become dense and chertlike as a result of infilling of the pores. The terms *diatomaceous chert* and *radiolarian chert* are more appropriate for the lithified diatomites and radiolarites, respectively. Siliceous sponge spicules may accumulate in an analogous manner and form *spiculites*, or if lithified, *spicular cherts*.

Siliceous sinter is also porous material, white or light-colored and light in weight, deposited by the waters of hot springs. *Geyserite* is merely a variety of siliceous sinter deposited by geysers.

Tripoli is another very porous, lightweight, siliceous rock (mainly chalcedony). It is white or variously colored gray, pink, buff, red, or yellow. It has a harsh, rough feel. Tripoli forms fairly large masses; because it is confined to the earth's surface, Tarr considers it to be a product of weathering (leaching and hydration) of chert and siliceous limestone from which carbonate has been leached. It includes *rottenstone*. Originally, and to some extent today, the term is applied to diatomaceous and other siliceous earths, which tripoli closely resembles.

Another product of weathering is *silcrete*, a term applied to a siliceous duricrust or soil formation and presumed to be the product of silica deposited from waters drawn upward by capillary action in an arid or semi-arid region. It is analogous to caliche (also called calcrete). The term is widely used by South African geologists (Cressman, 1962, p. 1).

COMPOSITION OF CHERTS

Mineralogic composition The principal constitutents of nondetrital siliceous sediments are opal, chalcedony, and quartz. Other minerals present, usually those found in adjacent sedimentary rocks, must be considered impurities.

The crystal chemistry of silica minerals is more complex than previously thought. What is known has been summarized by Cressman (1962, p. 5) and Deer, Howie, and Zussman (1963, pp. 178–230). Opal, $SiO_2 \cdot nH_2O$, is primarily amorphous silica with some water. It is marked by its low density, about 2.1, and its solubility in KOH. It has a low refractive index which varies from 1.38 to 1.46, depending on the water content (Taliaferro, 1935). Bramlette (1946, p. 17) measured and index of 1.440 ± 0.002 in the opal of siliceous organisms. This corresponds to a water content of about 9 percent. Opal occurs in many cherts and is the dominant constituent of some. It is found, however, only in Mesozoic and Cenozoic cherts and is presumed to have been converted to chalcedony and quartz in older rocks. The conversion—a chemical dehydration—gives rise to certain shrinkage effects, notably concentrically banded spheroids (Taliaferro, 1934). The shrinkage features of other cherts, however, may be the result of conversion of magadiite—a precursor of chert—to chert (Surdam, Eugster, and Mariner, 1972).

Chalcedony, the dominant constituent of most cherts, is a natural microscopic fibrous silica with atomic arrangement of quartz. Sosman (1927) did not regard chalcedony as a variety of quartz because of its distinctive properties. It is dominantly fibrous, normally shows negative elongation ("length fast"), indices of 1.533 and 1.540, birefringence of 0.009 to 0.011, density of 2.55 to 2.63, absence of inversion at 573°C, and heat capacity unlike that of quartz. Although chalcedony consists dominantly of microfibrous silica, some investigators have explained the anomalous properties by supposing that various amounts of submicroscopic amorphous silica or opal were present. Washburn and Navais (1922) concluded that "flint and chalcedony consists of colloidal quartz. In the purer form of chalcedony, the colloid is of the gel type and the individual colloid particles are microscopic or submicroscopic in size." Sosman, however, emphasized shape, not size, of the particles and notes that the particles are threadlike or fibrous. If these are short and randomly oriented, the fracture is conchoidal as in flint and chert. Electron micrographs of chalcedonic or fibrous silica show that the material has a spongy texture presumed to be caused by minute spherical water-filled cavities. There is no evidence of fibrous structure or that opal is present (Folk and Weaver, 1952).

As noted, chalcedonic fibers have a negative elongation ("length fast"), that is, with the *c*-axis (slow ray) perpendicular to the fibers. Some chalcedony, however, is length slow. Folk and Pittman (1971) have recently called attention to this fact and further noted that the rare, length-slow type occurs almost exclusively in association with evaporitic deposits, with the silica as a replacement of evaporitic minerals.

Tarr's study of thin sections (1926, 1938) convinced him that, although chalcedony predominates in chert, quartz is also present, and in some deposits it is predominant. He thought that microfibrous chalcedony would in time pass over into quartz. The sequence of changes is: original hydrous gel (opal), chalcedony, and quartz. James (1955, p. 1473) has shown that grain size of quartz in some Precambrian cherts is closely correlated with degree of metamorphism; the higher the grade, the coarser the chert.

Constituents other than the forms of silica listed above are rare in cherts. Some of the less common "impure" cherts contain calcite, dolomite, or siderite. Even some common cherts contain scattered rhombs of dolomite. A few cherts contain large detrital quartz grains. Chemical analyses high in alumina suggest clay mineral impurities, although some aluminum may substitute for silicon.

Chemical composition As might be expected, cherts are very high in silica (Table 11-1). In some silica exceeds 99 percent. Variations in composition are closely related to differences in lithology of the associated sediment (Cressman, 1962, pp. 6–10). Radiolarian cherts are commonly associated with shales; both radiolarite and diatomite are associated with pyroclastics; both, therefore, are more aluminous than spicular cherts which are more apt to be associated with sandstones or carbonate rocks. Iron is present as pyrite or magnetite in some black cherts; in red jaspers as hematite. Titanium is a minor component in most cherts and is significant only in some silcretes. Alkali

TABLE 11-1. Chemical analyses of cherts and other nondetrital siliceous sediments

Constituent	A	B	C	D	E	F	G	H
SiO_2	93.54	98.93	99.47	82.69	70.78	43.43	73.71	82.94
TiO_2	—	0.005	—	—	0.03	—	0.50	0.27
Al_2O_3	2.26	0.14	0.17	1.76	0.45	11.25[a]	7.25	0.1
Fe_2O_3	0.48	0.06	0.12	1.00	0.02	0.18	2.63	3.4
FeO	—	0.08	—	0.31	0.30	21.00	0.44	—
MnO	0.79	0.01	—	0.01	0.02	—	—	—
MgO	0.23	0.02	0.05	1.08	1.88	1.39	1.47	0.19
CaO	0.66	0.04	0.09	2.93	12.90	0.70	1.72	1.60
Na_2O	0.37	trace	0.15	0.50	0.05	1.21	1.19	0.65
K_2O	0.51	trace	0.07	2.61	0.06	3.99	1.00	1.40
H_2O+	0.72	0.17	0.12[b]	4.75[b]	0.32 }		{ 6.94 }	
						0.50		0.33
H_2O-	0.21	0.27	—	—	0.48 }		{ 2.88 }	
P_2O_5	—	trace	—	0.21	0.16	—	0.24	0.8
CO_2	—	0.02	—	2.28	12.04	15.76	?trace	0.40
SO_3	—	none	—	—	trace	—	0.16	—
Cl	—	—	—	0.15	—	—	—	—
C	—	0.18	—	—	0.33	0.08	0.00	—
Total	99.86	99.92	100.24	100.28	100.14	99.45	100.13	92.38[c]

[a] Includes P_2O_5.
[b] Loss on ignition.
[c] Includes F, 0.10; S, 0.14; V_2O_5, 0.06.

A. Franciscan chert (?Jurassic), Bagley Canyon, Mount Diablo, California (Davis, 1918, p. 268).
B. Vanport "Flint" (Carboniferous), Flint Ridge, Ohio. Downs Schaaf, analyst (Stout and Schoenlaub, 1945, p. 82).
C. Novaculite (Devonian), Rockport, Arkansas. R. N. Brackett, analyst (Clarke, 1924, p. 551).
D. Average of ten cherty rocks, Monterey Formation (Miocene), California (Bramlette, 1946, p. 49).
E. Calcareous nodular chert, Delaware Limestone (Devonian), Ohio. Downs Schaaf, analyst (Stout and Schoenlaub, 1945, p. 28).
F. Sideritic porcellanite (Precambrian), Iron County, Michigan. B. Bruun, analyst.
G. Diatomaceous "shale," Monterey Formation (Miocene), Hollywood, California. J. G. Fairchild, analyst (Hoots, 1931, p. 108).
H. Chert, Phosphoria Formation (Permian), Brazer Canyon, Utah. Tennessee Valley Authority, analyst (Smith and others, 1952, p. 12).

metals are present in trace amounts except in those cherts closely associated with volcanics.

PETROGRAPHY OF CHERTS

Typical chert, either nodular or bedded, is a hard, dense rock that shows a smooth conchoidal fracture. The color varies with impurities and ranges from white to gray to black, or in other cases gray-green, yellow, brown, or red.

Under the microscope chert is a colorless, exceedingly fine-grained microcrystalline aggregate (Fig. 11-2). In some cherts circular or elliptical clear areas represent radiolarians which, if not too damaged by recrystallization of the matrix, show spines or more rarely internal lattice characteristics of these organisms. Siliceous sponge spicules are abundant in spiculite cherts. Individual grains of chert, under the highest magnifications, commonly show a fibrous wavy extinction which is different from that of the adjacent grains. Some younger cherts contain isotropic material in which may be seen a few or many minute polarizing specks. Very possibly these specks are crystallites of quartz—the first step in the devitrification of the original amorphous silica gel. Like volcanic glass, opaline silica becomes wholly crystallized with passage of time. Older cherts (all those of Precambrian age) now consist entirely of a fine-grained mosaic of quartz. Under the electron microscope some cherts show well-defined polyhedral blocks of quartz; others are less dense and contain numerous bubblelike holes, in some cases numerous enough to give the chert a scoriaceous aspect (Folk and Weaver, 1952). See Fig. 11-3. The holes are believed to have been filled with water.

Some cherts are relatively impure and under the microscope show carbonate disseminated throughout the body of the chert. Porcelanite in particular contains much such carbonate. Those with carbonate inclusions exhibit all gradations, from chert that contains a few scattered rhombs of calcite or

FIG. 11-2. Hertfordshire "puddingstone" conglomerate (Eocene), England. Crossed nicols, ×28. Detail of two flint pebbles and enclosing matrix. Flint is exceedingly fine-grained mosaic of microcrystalline silica.

FIG. 11-3. Electron micrograph of cherty Caballos Novaculite (Devonian), Marathon Basin, Texas. Field approximately 8 microns long. Note blocky subhedral grains 3 to 5 microns in size. (Photograph courtesy of E. F. McBride.)

dolomite (Fig. 11-4) to rocks in which carbonate rhombs become so numerous that they touch one another and form a spongelike mesh whose interstices are filled with opal or with opal and chalcedony. The carbonate may be uniformly distributed throughout the chert; in other cases it may be concentrated in laminations which become accentuated on weathering. Siderite, rather than calcite or dolomite, may occur in the porcelanites associated with iron-bearing formations (Table 11-1F). Weathering of such material is accompanied by oxidation and conversion of the carbonate to a dense, hard limonite, which forms a black crust over the rock and conceals its true nature.

Rocks consisting of a mixture of clay or silty clay and a large but variable proportion of opaline silica are also porcelaneous and have been termed porcelanite (Bramlette, 1946, p. 15). These grade into siliceous shales, which consist of a mixture of the common constituents of shale with an exceptional amount of precipitated silica (see p. 284). The siliceous shales are flintlike in hardness and have a platy parting and a subconchoidal fracture. The silica of the Mowry Shale is thought to have been derived from associated fine volcanic ash. Decomposition of the glass in sea water and concurrent precipitation of the silica by diatoms or other agents are believed to be the origin of this rock (Rubey, 1929). A similar explanation has been proposed for the origin of the siliceous shales in the Stanley (Carboniferous) beds of Oklahoma (Goldstein and Hendricks, 1953).

Sandy cherts are not common, although all gradations between chert, sandy chert, and sandstone with chert cement are known (Cayeux, 1929, p. 131; Humphries, 1956; Sargent, 1923). In some cherts, large rounded sand grains are scattered sparsely through the body of the chert. (See, for example the Sweepings Bed of the Precambrian Ironwood Formation, Aldrich, 1929, Pl. IIF).

Closely related to cherts and much like them in appearance are silicified tuffs (Chapter 9). These rocks differ from true cherts

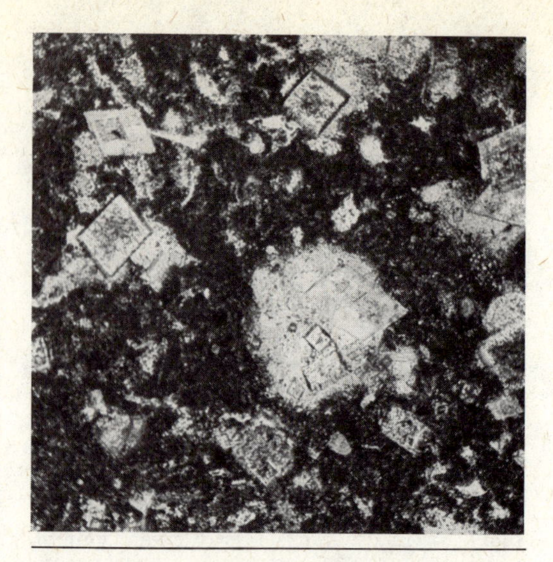

FIG. 11-4. Huntersville Chert (Devonian), Wood County, West Virginia. Ordinary light, ×80. A dense chert consisting of chalcedony with scattered rhombs of dolomite. Mottling of chert suggests texture of fossiliferous calcarenite which has become chertified.

by the inclusion of glassy materials—distinguished with some difficulty from isotropic opaline spicules and other curved structures. Devitrification of the glass may render recognition difficult.

Oolitic cherts are a special class. Although a minor variety of chert, the siliceous oolites show most clearly the evidence of a replacement origin, and they are, therefore, of the greatest interest in connection with the chert problem. Best known are the siliceous oolites from State College, Pennsylvania (Diller, 1898, p. 95; Krynine, Honess, and Myers, 1941; Choquette, 1955). Under the microscope (Fig. 11-5) this chert shows numerous oolitic bodies, about 1 mm in diameter, composed of microcrystalline quartz and concentrically structured chalcedony. Most contain a well-rounded detrital quartz nucleus which is enclosed in an overlay of secondary quartz in optical continuity with the detrital grain. The boundary between the original sand grain and the overgrowth is marked by carbonate inclusions. Interstices between the oolites are filled with finely crystalline quartz mosaic. In some oolites the quartz nucleus is displaced to one side; others, lacking a quartz nucleus, show a geopetal fabric resulting from removal in solution of the central core of the ooid, leaving the insolubles displaced to the lower side, followed by infilling with precipitated coarsely crystalline quartz (Choquette, 1955). See Fig. 11-6.

Other siliceous oolites have similar characteristics. The oolitic chert from the Shakopee Dolomite (Ordovician) at Utica, Illinois, is noteworthy for the festoonlike deposits on the interior of the ooid as well as for the chert pseudomorphs of dolomite present both in the oolites and their matrix (Fig. 11-7). These oolites clearly are replacements of a calcareous oolite, and all stages of transformation can be seen in a properly collected suite of specimens. Similarly there are cherts produced by the silification of biocalcarenites such as the cherts from the Osage limestones (Mississippian) of Indiana, which show clearly the biogenic detritus from which they were derived (Fig. 11-8).

Although generally structureless, some nodules have a dense black interior surrounded by a lighter, and in some cases

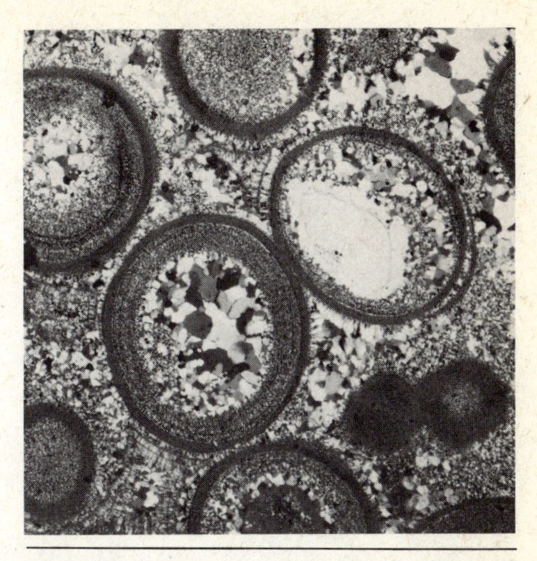

FIG. 11-5. Siliceous oolite, Mines Formation (Cambrian), State College, Pennsylvania. Crossed nicols, ×30. This is the well-known State College Oolite. The ooids commonly have detrital quartz grains, usually secondarily enlarged, as nuclei (Fig. 11-7). Silification of calcareous oolite has destroyed much of the concentric structure and converted the carbonate to a microcrystalline mosaic of quartz. Quartz micromosaic surrounds and binds ooids together.

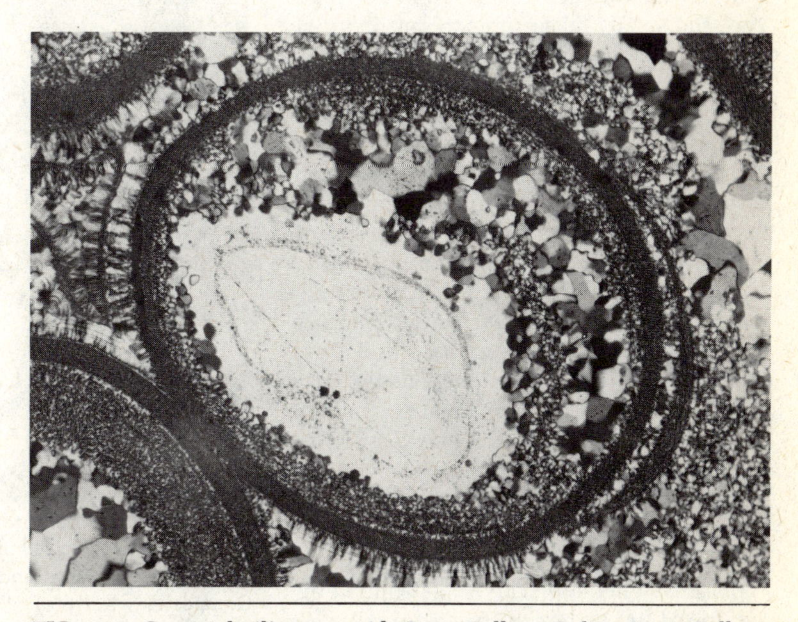

FIG. 11-6. Geopetal siliceous ooid, State College Oolite, State College, Pennsylvania. Crossed nicols, ×58. Detail of Fig. 11-5. Note secondarily enlarged quartz grain displaced to lower half ooid.

white, exterior ("cotton" rock). A few show traces of the bedding that is continuous with that of the host rock; still others are marked by a concentric structure, contraction spheroids (Taliaferro, 1934), or shrinkage cracks and reticulate patterns (Surdam, Eugster, and Mariner, 1972). A few enclose patches of the host rock. Rather commonly cherts are fossiliferous, the fossils either silicified or calcareous. In the latter case, the carbonate may be dissolved, leaving only a cavity or mold.

GEOLOGIC OCCURRENCES OF CHERTS AND RELATED ROCKS

Chert, the most extensive siliceous rock of chemical origin, occurs either as nodular segregations, mainly in a carbonate host rock, and as areally extensive bedded deposits.

Chert (or flint) *nodules* are very irregular, usually structureless dense masses of microcrystalline silica. Such nodular masses vary from more or less regular disks 2 to 5 cm in diameter to large, highly irregular, and tuberous bodies 20 to 30 cm in length. The shape of these objects is infinitely varied, but the larger ones, although of rounded contour, are marked by warty or knobby exteriors (Fig. 12-10). In most cases, nodules are elongated parallel to the bedding and are commonly concentrated along certain bedding planes. In some limestones they are numerous enough to coalesce into more or less continuous though irregular or "wavy" layers. Chert-bearing layers may be more or less rhythmically spaced (Richardson, 1919). In a given layer the chert forms an irregular two-dimensional network; if layers are closely spaced, some connections or bridging between them may occur, and the chert network becomes three-dimensional. Few limestones are so cherty; most are characterized only by scattered discrete nodules.

Chert nodules are widely distributed and are found in carbonate rocks of all ages. Well-known are the cherts of the Paleozoic limestones of the upper Mississippi Valley region (Van Tuyl, 1912; Tarr, 1917; Biggs,

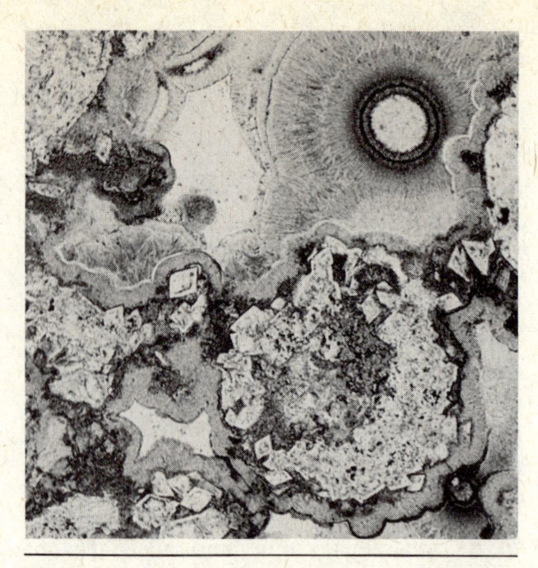

FIG. 11-7. Shakopee oolitic chert (Ordovician), Utica, Illinois. Ordinary light, ×90. Original calcareous ooids no longer show either concentric or radial structure. Under crossed nicols they reveal a micromosaic of crystalline silica. Note silicified rhomb-shaped pseudomorphs of dolomite which lie within and without the ooids. A later generation of chalcedony has been deposited between the ooids. The chalcedony has assumed a radial and concentric habit and was deposited in part as a conformal envelope around the silicified ooids. (Photograph by Chester Johnson.)

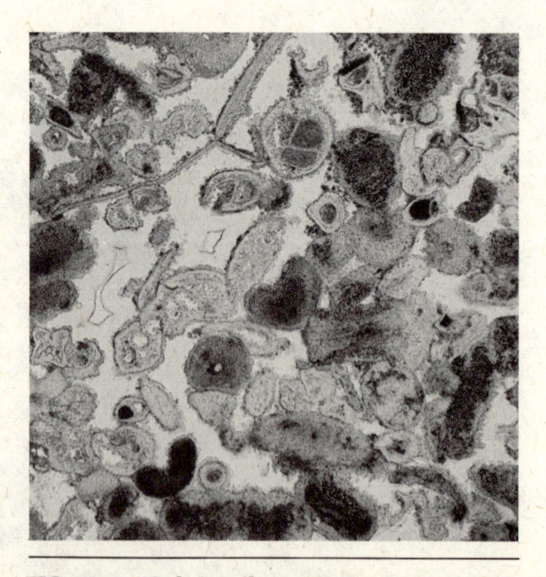

FIG. 11-8. "Oolitic" chert, (?) Osage Formation (Mississippian), Indiana. Ordinary light, ×35. Composed wholly of silica but retaining much of the structure of the original limestone from which it was formed by silicification. Original rock contained detrital carbonate, fossil debris, and calcareous ooids.

1957) and the flint nodules of the Cretaceous Chalk of England and France (Richardson, 1919). Although nodular cherts are not known in modern sediments, nodular cherts are found in Pleistocene deposits in East Africa. These nodules are irregular, lobate to spinous bodies, with reticulate surface patterns, 5 to 15 cm long, embedded in silicified claystones and zeolitized tuffs (Eugster, 1967; Hay, 1968). Similar cherts are known from other places, especially the Green River beds (Eocene) of Wyoming (Surdam, Eugster, and Mariner, 1972).

Bedded cherts are the most impressive occurrence of precipitated silica. They not only make up whole formations but are both thick and areally extensive. Thicknesses of several meters are common, and a hundred or more are not unknown.

Bedded cherts seemingly fall into three classes. The first are *cratonic* cherts, that is, those associated with shallow-water limestones and quartz arenites—the "stable shelf" association. The second are *geosynclinal* cherts or those associated with a presumed deep-water assemblage—the siliceous black shales. They are the presumed analog of deep-sea radiolarian and diatomaceous oozes. The third class of cherts are associated with evaporitic or at least hypersaline deposits (presumably ephemeral alkaline lake deposits).

Cratonic cherts are associated with limestones which contain numerous chert nodules, nodules in some forming an anastomosing, three-dimensional network. With increasing proportion of chert these rocks pass into cherts with a few limestone interbeds. Bedding is prone to be irregular and wavy. Abundant silicified calcareous fossils characteristic of normal shallow marine water—brachiopods, bryozoans, and the like—are common. Cherts of this type include such formations as the Shriver Chert (Devonian) of Maryland, the Boone Chert (Mississippian) of Missouri (Fowler et al., 1934), and perhaps the Bigfork Chert (Ordovician) of Oklahoma (Honess, 1923) and the Ft. Payne Chert (Mississippian) of Georgia (Hurst, 1953).

Geosynclinal cherts are rhythmically

FIG. 11-9. Calcareous concretion in cherty shale, San Luis Obispo County, California. (U.S.G.S. photograph. Bramlette, 1946, Pl. 6.)

layered cherts. The layers, a few centimeters thick, are separated by partings of less thickness of siliceous or dark shales or in the Precambrian iron-formations by layers of siderite (see "cherty iron carbonates," p. 413). Although apparently even-bedded (Figs. 11-9 and 11-10), the chert layers may pinch and swell in an irregular manner, lens out, or even bifurcate. In most cases the chert beds make up the bulk of the formation with the partings forming a lesser fraction of the deposit. As in nodules, the chert is cryptocrystalline, dense, and subvitreous. The bedded cherts are brittle, close-jointed rocks. Many are associated with siliceous shales, pelagic limestones, turbidites and ophiolites (Grunau, 1965).

The best-known bedded cherts of this type include the Miocene Monterey Chert of California (Bramlette, 1946), the cherts of the Franciscan (?Jurassic) of the same region (Davis, 1918), the Permian Rex Chert of Idaho (Keller, 1941; Cressman and Swanson, 1964, p. 369), the Caballos Novaculite of Texas (McBride and Thomson, 1970), and the famous Arkansas Novaculite (Honess, 1923; Goldstein and Hendricks, 1953; Park and Croneis, 1969). Similar bedded cherts are present in the Ordovician Normanskill of New York State (Ruedemann and Wilson, 1936) and in the Ordovician of the Cordilleran eugeosyncline of Nevada and Idaho (Ketner, 1969). The jaspilites and cherty iron carbonates are Precambrian bedded cherts

FIG. 11-10. Lateral uniformity of layering in rhythmically bedded chert, Monterey Formation (Miocene), California. U.S.G.S. photograph. Bramlette, 1946, Pl. 10.)

with an abnormally high content of iron. They are discussed in the next section of this chapter. Important European chert deposits occur in the Culm Measures of North Devon (Prentice, 1958) and in rocks of the same age in Germany (Hoss, 1957, 1959) and include the Mesozoic radiolarian cherts of the Alps and the Apennines (Winkler, 1925). The distribution of radiolarian cherts in space and time has been summarized by Grunau (1965).

Cherts that characterize ephemeral alkaline lakes have only recently been identified; their nature and distribution in the geologic record is less well known. These are nodular and bedded, and the chert nodules are marked by a reticulate pattern. The chert is interbedded with siliceous claystones, zeolitized tuffs (Hay, 1968), and associated in some cases also with oil shales and resedimented dolomite (Eugster and Surdam, 1971). Cherts of this type occur in Pleistocene lacustrine deposits in East Africa (Eugster, 1967, 1969) and in Green River (Eocene) beds of Wyoming and possibly elsewhere. Some Precambrian cherts may also belong to this group (Eugster and Chou, 1973).

ORIGIN OF CHERT

The literature on the chert problem is very extensive. Various theories that have been advanced to explain these materials fall into two groups. According to one concept, the silica is contemporaneous with sedimentation; according to the other view, the silica is a post-depositional replacement of the host rock, generally a limestone. There are various modifications of each of these theories (Fig. 11-11).

Athough there is no general agreement on the chert problem, majority opinion seems to incline toward a replacement origin of nodular cherts and flints found in limestone and other carbonate rocks. A few writers, notably Tarr (1917, 1926) and some others (Twenhofel, 1950, pp. 405–414; Trefethen, 1947; Fernandez, 1961; Harris, 1958) stoutly maintain the view that nodular cherts are formed by direct precipitation of masses of silica gel on the sea floor. Evidence for replacement (and therefore post-depositional origin) of the nodular chert is abundant and clear. Van Tuyl (1918) and Biggs (1957) have given a good resumé of the evidence bearing on this question. Supporting the replacement origin are (1) the occurrence of chert along fissures in limestone, (2) the very irregular shape of some chert nodules, (3) the presence of irregular patches of limestone within some nodules, (4) the association of silicified fossils and cherts in some limestones, (5) the presence of replaced fossils (Fig. 11-8), (6) the preservation of textures and structures (especially bedding) in some cherts (Bastin, 1933; White, 1947), (7) the failure of some cherts to follow definite

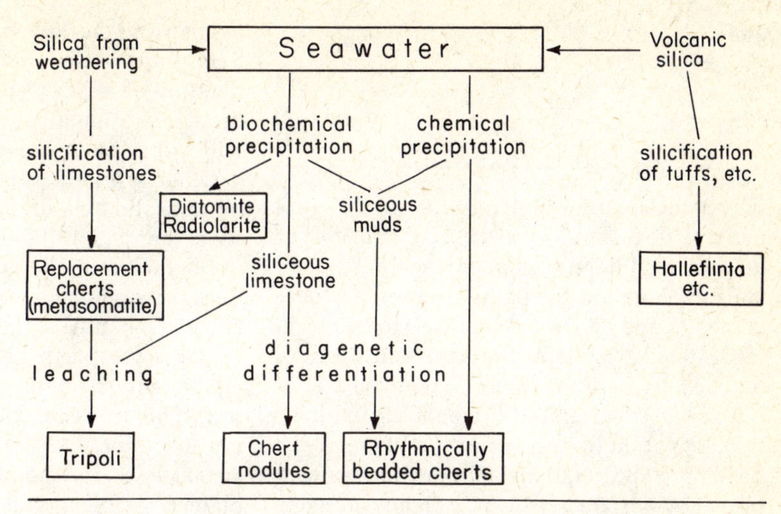

FIG. 11-11. Provenance of chert and other siliceous sediments.

zones in limestone formations, and (8) the occurrence of silicified oolites formed by replacement of calcareous ones.

The first criterion, perhaps, is not one of replacement but rather of age. Veinlets of chert place the formation of chert in the post-consolidation period of the host rock. Chert veinlets are rather rare, and Twenhofel (1950, p. 412) is quite correct in citing this fact as evidence that chert is not generally introduced by circulating waters. The observation that some chert nodules have a stylolitic contact with adjacent limestone has been cited as evidence of an epigenetic replacement origin (Bastin, 1933), although similar contacts between interbedded chert and iron carbonates have not been so interpreted (James et al., 1968, Fig. 24).

Various geochemical considerations make it seem unlikely that silica gel would be precipitated from normal marine waters. At 25°C amorphous silica is soluble to the extent of 100 to 140 ppm and, because sea water contains only about 4 ppm silica, it is unlikely that silica gel would be precipitated (Krauskopf, 1959; Siever, 1962). Seemingly the low concentration of silica in sea water is maintained by extraction by organisms—radiolarians, diatoms, and siliceous sponges—so that the higher figure required for precipitation has never been attained. It might be supposed that, prior to the appearance of these silica-precipitating organisms, the concentration of silica in sea water might rise to the point at which precipitation occurs. But, as experiments by Garrels and Mackenzie (1971, pp. 283–284) have shown, the introduction of clay minerals into sea water will remove excess silica and so keep it below saturation level. On the other hand, recent studies by Eugster, Hay, and others (Eugster, 1967, 1969; Hay, 1968) have shown that nodular (and also bedded) cherts can form by leaching of sodium silicate gels precipitated in silica-rich alkaline lakes. In general, however, the replacement origin of chert nodules in carbonate host rocks is supported by a wealth of observations, both in the field and under the microscope. Certainly the anastomosing two- and three-dimensional networks of some cherts in limestones are difficult to explain except by replacement.

The source of the silica and the mechanics of nodule formation, however, are less clear. The principal problem is, Was the replacing silica introduced from without or was it derived from within the host bed? It has been suggested that the silica was precipitated concurrently with the deposition of limestone, perhaps by radiolarians, diatoms, or sponges and then, in time, dissolved and reprecipitated as nodular masses which replace the matrix in which they occur (Hinde, 1887; Richardson, 1919; Biggs, 1957, p. 19). The silica needed for nodule formation is therefore extracted from sea water by normal, widespread biochemical processes. No special supply of silica, nor abnormal concentrations, nor special conditions of precipitation are necessary. And as Ramberg (1952, p. 222) has given theoretical reasons why solution and reprecipitation of disseminated silica *should* occur, the concept of diagenetic differentiation is the most plausible explanation for nodular cherts in limestones. Beds most susceptible to replacement would explain the concentration of nodules in certain layers; greater ease of replacement parallel to the bedding explains the elongation of nodules in the plane of the bedding. Field relations support this concept also. Lowenstam (1948, p. 36) has noted a definite relation between the type of chert and the state of preservation and distribution of siliceous sponges in the Niagaran (Silurian) dolomites of Illinois. He was led to the view that the silica of cherts was derived from sponges, a conclusion earlier reached regarding some Palestinian cherts (Lowenstam, 1942) and supported by observations of Middleton (1958).

If these concepts are correct, the question arises: At what time was the chert formed, early diagenetic or later, after consolidation and uplift? Petrographic evidence suggests that it followed dolomitization (Biggs, 1957, p. 22). It has been suggested that nodular chert is a product of silicification associated with weathering. But the stratigraphic continuity of most cherty zones and the distinctive character of cherts of various beds indicate the chert to be unrelated to the present-day surface (and in most cases unrelated to ancient surface recorded as unconformities) and to be related, instead, to sedimentation responsible for the host rock. Better preservation of fossils within the chert suggests their early incorporation before the overload crushed them (Biggs, 1957, p. 23).

The origin of even-bedded cherts associated with dark siliceous shales is a controversial subject; many theories to explain these deposits have been advanced. Some investigators, noting that chert nodules are replacements of the host limestone, consider bedded cherts to be formed by an extension of the process of replacement to the whole rock. Logan and Chase (1961), for example, believe the 3300 foot (1666 m) Coomberdal Chert of Western Australia to have formed by the silicification of a limestone. Such large-scale silicification requires introduction of silica; clearly redistribution of the silica originally present in the limestone will not explain the deposit.

Other workers consider bedded cherts to be consolidated or indurated siliceous bioliths, that is, a lithified radiolarian ooze such as is found on the present deep-sea floor. Because organisms are today forming pure siliceous deposits, the uniformitarian view requires that such deposits or their lithified equivalents be present in the geologic record. The Jurassic cherts of Borneo, which are 300 feet (152 m) thick and cover over 40,000 km^2, for example, have been so explained (Molengraaff, 1910), as have also the Carboniferous cherts of the Independence Range in Nevada (Fagan, 1962) and almost all radiolarian cherts (Grunau, 1965).

Many students of bedded cherts, particularly Precambrian cherts, reject both the concept of large-scale silicification of limestones and the theory of a biochemical origin for these deposits and ascribe them instead to the direct precipitation of silica (Davis, 1918; Sampson, 1923; Sargent, 1929; Aldrich, 1929, p. 135; Moore and Maynard, 1929; James, 1954). In the absence of modern deposits of this nature and because geochemical considerations make silica precipitation in open marine waters unlikely, some workers have looked elsewhere for an answer to the chert problem. A partial answer seems to be the view that, although chert itself is not a primary precipitate, it is formed by leaching of sodium silicate gels which are precipitated in certain ephemeral alkaline lakes. Such an origin has been proposed for the cherty beds of the Eocene Green River (Eugster and Surdam, 1971) and for the cherts of the Precambrian iron formations (Eugster and Chou, 1973).

Evidence of silicification of limestone is impressive. Logan and Chase (1961) cite the observed transition of dolomitic limestone to chert, the presence of siliceous fossils and ooliths believed to have been originally calcareous, relict carbonate inclusions in the chert, siliceous rhombs apparently after dolomite, and quartz overgrowths in some clastic members. The progressive chertification of a limestone is well illustrated by McKee's (1960) study of the 70 to 120 foot bed in the Redwall Limestone (Mississippian) in Arizona. The transformation can be traced from a limestone with a few irregular lenses and nodules of chert to a three-dimensional network of chert to a chert with a few relict pods or lenses of limestone. Cherts are abundantly fossiliferous; limestone is barren. The chert replacement is believed to be early diagenetic, before dolomitization. It is quite possible that cratonic cherts—those closely associated with shallow-water limestone and clean quartz sands—are the end-product of a silicification process. The uniformity of bedding, rhythmic nature, and absence of evidence of replacement in geosynclinal cherts make the hypothesis of a metasomatic origin of these deposits unlikely.

Inasmuch as the remains of radiolarians, diatoms, and siliceous sponge spicules are found in many cherts, and are very abundant in some, the conclusion that these organisms are responsible for the formation of the chert is not surprising. Although sea water contains only 4 ppm silica, organisms have been able to extract this material and in many places form extensive deposits of nearly pure silica Such deposits today are restricted to areas receiving little or no land-derived sediment and to waters too deep for the deposition of calcareous sediment—waters below the calcium carbonate compensation depth. That bedded cherts are lithified siliceous bioliths has been stated by many writers on the problem of bedded cherts. Some of the cherts of the Phosphoria Formation (Permian) of Montana have been ascribed to ". . . the accumulation and partial diagenetic reorganization of sponge spicules and other siliceous remains . . ." (Cressman, 1955, p. 25). Bramlette (1946) believed that the bedded Monterey Chert (Miocene) of California (Fig. 11-10) was produced by the addition of silica to diatomite. For the most part, the added silica was derived from diatom materials themselves at an early stage in the formation of the rock. The Monterey cherts, therefore, are a product of diagenetic rearrangement of the silica in a diatomaceous ooze. The rhythmic bedding is a feature inherited from unaltered diatomite and diatomaceous shales. If pure silica and average diatomaceous shale are taken in equal proportions, the resulting rock will have a composition substantially like that of associated bedded cherts. The concept of bedded chert as the equivalent of a radiolarian ooze has been invoked to explain Jurassic radiolarian cherts of Borneo (Molengraaff, 1910), the Alps (Steinmann, 1925), and elsewhere (Fagan, 1962; Grunau, 1965; Thurston, 1972).

Many writers, however, regard the radiolarians in these cherts as incidental, especially in those cherts in which their remains are sparse, and consider them accidental components unrelated to the origin of the chert. Hence many writers ascribe bedded cherts to the direct precipitation of silica. As noted by James (1954), the chert layers are, in places, involved in slump structures and intraformational breccias; the chert, therefore, must be essentially contemporaneous with the sedimentation and even prelithification.

If bedded cherts are primary precipitates of silica, it is pertinent to inquire into the condition which permits their formation. Silica is carried to the sea by river waters; and, although these contain more silica than does sea water, the concentration of silica is not very great. Correns (1950) states that concentrations of silica substantially greater than those of present-day streams cannot be flocculated by any known agent, and hence the silica of the cherts must be biochemically deposited. To avoid this difficulty, others have appealed to volcanism to supply the silica and build up the concentration of this material to the point where inorganic precipitation is possible (Davis, 1918; Taliaferro, 1933; Van Hise and Leith, 1911, p. 516). Although some cherts are associated with submarine flows (greenstones) and tuffs, many have no known association either with flows or infalls of volcanic ash. Lacking such association, volcanism cannot be appealed to for silica.

If ocean waters are much undersaturated in amorphous silica, it seems that such materials cannot precipitate in the normal marine environment. And indeed no such precipitates are known anywhere today. It must be presumed, therefore, that the seas were of a different composition in the past or that the environment of deposition was not open marine but was restricted and of a very different chemistry. Siever (1962) has suggested that biochemical precipitation has kept the silica content of the open oceans at its present low level at least since the Cambrian. We must, therefore, postulate an abnormal supply of silica or a special environment for silica precipitation. The requisite environment eluded sedimentologists until Eugster (1967) noted that chert was forming by the leaching of sodium silicate precursors such as magadiite—$NaSi_7O_{13}(OH)_3 \cdot 3H_2O$—which was precipitated from the alkaline waters of Lake Magadi, an ephemeral lake in East Africa. The conversion of magadiite to chert proceeds fairly rapidly so that earlier Pleistocene deposits of Lake Magadi contain an abundance of nodular and bedded chert. The observations of Eugster have been confirmed by Hay (1968) and others. How general is the mechanism of chert formation and how widely can it be applied to the bedded cherts of the geologic record? Cherts of the Magadi type have been reported in rocks as old as Jurassic (Surdam, Eugster, and Mariner, 1972). They are abundant in the Green River Formation (Eocene) of Wyoming (Eugster and Surdam, 1971), and Eugster and Chou (1973) have suggested that cherts of Precambrian iron-formations may have been formed in the same manner.

It does not seem probable that this mechanism of chert formation can be applied to all bedded cherts. The depth of water in the Magadi-type deposits is very slight indeed. Some of the associated beds are mud cracked, implying exposure. On the other hand, "geosynclinal" cherts are presumed to be abyssal. The radiolarian Normanskill and Deepkill (Ordovician) of New York were considered to have been deposited in waters 12,000 feet (3,660 m) deep on the basis of paleontological evidence (Reudemann and Wilson, 1936). Similar views have been expressed respecting Alpine cherts (Bailey, 1936). On the other hand, Davis (1918) and Taliaferro (1943, p. 147) supposed the Franciscan (?Jurassic) cherts of California were deposited in shallow water primarily because of their close association with coarse sandstones—sandstones, however, now interpreted as turbidites. Deposition of such coarse clastics, even at abyssal depths, has been demonstrated and makes unnecessary either the hypothesis of shallow-water origin or that of frequent and radical changes in water depth for those sections containing both bedded chert and sandstone. The common association of pyritic black shale with geosynclinal bedded cherts suggests that these were formed in a restricted basin. This basin has the attributes of a "starved" basin, that is, one so situated as to receive little or no clastic material and one too deep or otherwise unfavorable for normal benthonic organisms and with such a low *p*H that planktonic calcareous remains are taken into solution. This interpretation is supported by modern oceanic research which has disclosed chert in deep-sea deposits (Pimm, Garrison, and Boyce, 1971). Under such conditions sedimentation is exceedingly slow and is confined to the finest papery black shales and to the deposition of compounds of iron, phosphorus, and silicon.

After deposition, the silica (opaline tests of organisms) undergoes some diagenetic changes. Opal is converted to chalcedonic and microcrystalline quartz. Disseminated silica in limestones is segregated into the familiar chert nodules. That deposited in muds may be segregated into layers. Such presumed diagenetic unmixing in the explanation given by Davis (1918, p. 394) for the rhythmic bedding of California cherts. Others have explained the rhythmic bedding as a product of seasonal deposition (Sakamoto, 1950; Hough, 1958). In the alkaline lake environment, diagenetic changes involve formation of chert by leaching of the original magadiite precipitate.

In conclusion, it can be said that chert—both nodular and bedded—is a polygenetic rock and that no single mode of origin is responsible for all cherts. It seems clear, in the light of geologic evidence and what is known about modern silica accumulates, that chert is a replacement product in some cases—silicified limestones, a biochemical accumulate in a starved basin below the calcium carbonate compensation depth in other cases, and a product of ephemeral silica-rich alkaline lake environment in still other cases. No doubt the bedded cherts of the geologic past were deposited by one or another of these or closely similar processes. Cratonic cherts seem to be silicification products; geosynclinal cherts are abyssal biochemical accumulates. Alkaline lake cherts are perhaps more widespread than heretofore realized. It is not always clear to which environment a particular chert belongs, because the criteria for differentiating between them are not yet fully known. The Arkansas novaculate, for example, displays some aspects of the alkaline lake environment, as do some of the Precambrian cherts associated with the iron-formation (Folk, 1973; Eugster and Chou, 1973).

SILCRETE

Silcrete is a surficial chertlike deposit of limited thickness presumed to be a product of chemical weathering in relatively arid regions of low relief. It is the siliceous analog of caliche. It tends to be massive or at best poorly bedded. A brecciated structure is a characteristic feature; it is ascribed to resilicification of earlier formed silcrete that has been brecciated during denudation. The silica is in part chalcedonic, in part crystalline quartz. Silcretes apparently can form on many different rock types. The accessory mineral content of the silcrete is closely related to that of the parent material. In many examples derived from igneous rocks the TiO_2 content is relatively high (Cressman, 1962, Table 8).

Silcretes of Australia and South Africa are among the best known (Mountain, 1952; Williamson, 1957). An ancient silcrete, believed to be a relict duricrust formed during a late Precambrian cycle of aridity and deep weathering, has been described by James et al. (1968, pp. 72–75).

IRON-BEARING SEDIMENTS

DEFINITIONS

Iron is one of the most abundant elements in the earth's crust, and few rocks indeed are iron-free. The average shale, for example, contains 6.47 percent FeO and Fe_2O_3 (Clarke, 1924, p. 34). Hence in the broadest sense all sediments are likely to be iron-bearing. Generally, however, the term *iron-bearing* is reserved for those rocks which are much richer in iron than is usually the case. The term *ferruginous* would be more appropriate, perhaps, had it not been applied to sandstones and shales colored red by ferric oxide. Such rocks, although called ferruginous, may have no more iron than similar strata not so colored (p. 274). Commonly the terms *iron-formation* and *ironstone* are applied to the iron-rich rocks; less often these rocks are termed *ferriferous*. James (1966, p. 11) would distinguish between iron-formation, the typically laminated, *chert*-bearing deposits (mainly of Precambrian age) and ironstones, which are *noncherty* (and mainly post-Precambrian). The term *iron-formation* has been given local names such as itabirite (Brazil), banded hematite quartzite (India), and quartz-banded ore (Scandinavia). See Brandt et al. (1972). The term *taconite* is used in the Lake Superior region for unoxidized iron-formation, that is, one not oxidized as a result of the weathering process. The term *iron ore* is often applied to these rocks. In a strict sense, *ore* is something which can be mined at a profit; it is, therefore, defined by both metal content and economic considerations. In the Lake Superior region iron ore must contain at least 50 percent iron. Most such deposits are secondary and derived by alteration and enrichment of an iron-formation.

A precise definition of iron-formation or ironstone is difficult to frame, because the term embraces a mineralogically and texturally diverse group of rocks. The only restriction is that the total iron content be significantly higher than that of normal sediments. Usage in the Lake Superior region suggests that 15 percent iron be present before the term is applied (James, 1954, p. 239). This would correspond to 21.3 percent Fe_2O_3 or 19.4 percent FeO.

Other iron-rich sediments include *bog iron ore*, *clay ironstones*, and *laterite*. Clay ironstones are sideritic nodules of diagenetic origin; bog iron ores are minor accumulations in small freshwater lakes of high latitudes; laterites are residual iron-rich deposits akin to silcrete and bauxite—all weathering products.

The grouping of the iron-bearing sediments together is a general departure from the usual classification of the chemical sediments, a classification generally based on the anion as in the case of sulfates, phosphates, and carbonates. Yet the iron-bearing sediments are related, one to the other, and usage favors a unified treatment of the group as a whole.

MINERALOGY AND CLASSIFICATION

Iron-bearing sediments are iron-rich because of the presence or dominance of one or more iron-bearing minerals in abnormal amounts (Table 11-2). The central problem confronting the petrographer is to untangle the complex mineralogy of the iron-bearing sediments and to determine which minerals are formed by primary sedimentary processes, which are products of diagenesis, which are caused by metamorphism, and which are the result of weathering—either in the present cycle or during an earlier epoch. Petrographers have differed widely on their interpretations, and these differences have led to considerable controversy and confusion. We discuss here the minerals which are primary or diagenetic, although many of these are also produced by metamorphism and weathering. These minerals fall into four groups: the oxides, the carbonates, the silicates, and the sulfides. Many iron-bearing sediments are complex in that more than one type of iron-bearing mineral, as well as other minerals are present (Fig. 11-12).

Iron oxides Goethite ("limonite"), hematite, and magnetite are found as important constituents of the iron-rich sediments; lepidocrocite and maghemite have been reported only rarely.

Goethite, α-FeO(OH), is the principal constituent of many post-Precambrian deposits, especially the Jurassic deposits of Europe. The "limonite" often reported is presumed to be geothite mixed with other minerals and with additional adsorbed water. The goethite commonly occurs as ooliths consisting of successive shells of oxide or of oxide alternating with chamosite. In general, goethite is rare or absent in the older ironstones; in these, hematite seems to take its place, suggesting a progressive dehydration and transformation of goethite to hematite with age.

Hematite, α-Fe_2O_3, is a dominant mineral in many early Paleozoic ironstones, especially those of the Clinton Group (Silurian) of New York State. It is the principal mineral of the Precambrian iron-formations in which it is interbedded with chert. Hematite, like goethite, is commonly oolitic in habit. Maghemite, the strongly magnetic di-

TABLE 11-2. Iron-bearing minerals of sediments

Sulfides:	Pyrite	FeS_2
	Marcasite	FeS_2
	Hydrotroilite	$FeS \cdot nH_2O$
Oxides:	Limonite	$FeO(OH) \cdot nH_2O$
	Goethite	$HFeO_2$
	Hematite	Fe_2O_3
	Magnetite	Fe_3O_4
Silicates:	Glauconite	$KMg(Fe, Al)(SiO_3)_6 \cdot 3H_2O$
	Chamosite	$3(Fe, Mg)O \cdot (Al, Fe)_2O_3 \cdot 2SiO_2 \cdot nH_2O$
	Stilpnomelane	$2(Fe, Mg)O \cdot (Fe, Al)_2O_3 \cdot 5SiO_2 \cdot 3H_2O$
	Minnesotaite	$(OH)_2(Fe, Mg)_3Si_4O_{10}$
	Greenalite	$FeSiO_3 \cdot nH_2O$
Carbonates:	Siderite	$FeCO_3$
	Ankerite	$Ca(Mg, Fe)(CO_3)_2$
Phosphates:	Vivianite	$Fe_3(PO_4)_2 \cdot 8H_2O$

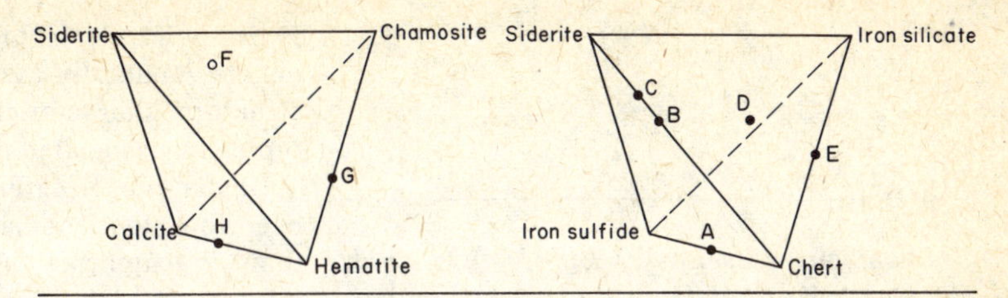

FIG. 11-12. Approximate mineral composition of iron-bearing rocks; their analyses are given in Table 11-3. Minor constituents omitted.

morph of hematite, is known though of rare occurrence. More common is martite, the hematite pseudomorph after magnetite.

Magnetite, Fe_3O_4, is abundant in Precambrian iron-formations in which it may be either diagenetic or metamorphic in origin. It is known but not common in the younger ironstones, where it occurs as both isolated crystals and as a constituent of some ooliths of unmetamorphosed iron-bearing sediments. In Precambrian iron-formations it occurs as impure layers alternating with chert.

Iron carbonates The only important iron-bearing carbonate is siderite, $FeCO_3$, which is found in abundance in iron-bearing sediments of all ages; it is also a minor component of many ordinary sediments. In younger ironstones it is closely associated with chamosite, less commonly with limonite. Many Precambrian iron-formations are dominantly interbedded siderite and chert. Iron ores are a product of alteration of these deposits—an alteration involving oxidation of siderite to hematite and removal by leaching of cherts. Most of the siderite of sedimentary rocks is very fine-grained and intimately intermixed with other materials. In some, however, siderite forms more coarsely crystalline spherulites (Fig. 11-21); in many others it forms large rhombs replacing chamositic oolites. Chemical analyses show siderite to contain, in solid solution, small but variable quantities of $MnCO_3$, $MgCO_3$, and $CaCO_3$. Siderite shows the usual characteristics of the class of carbonates to which it belongs: rhombic crystal forms, rhombic cleavage, very high birefringence (0.242), indices 1.603 and 1.875. It is nearly colorless in thin section but is commonly oxidized along borders and cleavage to the oxide and hence stained yellow to brown.

Siderite is also known from clay ironstones, which are concretionary bodies found in some shales, especially those associated with Paleozoic coal measures where it occurs also in thin layers, forming the so-called black-band ores (Tyler, 1950).

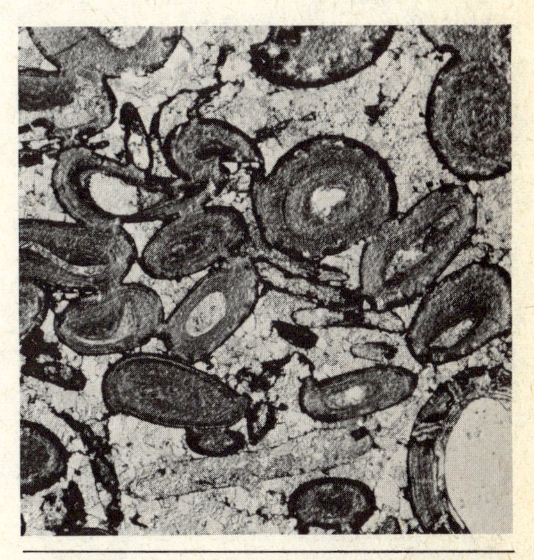

FIG. 11-13. Oolitic chamosite, Westmoreland Ironstone (Silurian), near Lairdsville, New York. Ordinary light, ×22. Chamosite (septechlorite) oolites, some with quartz sand grain nuclei, in carbonate matrix.

Because of its susceptibility to oxidation, siderite is almost everywhere altered to limonite in the outcrop.

Iron silicates The only iron silicates definitely of primary origin and importance are chamosite, glauconite, and greenalite. Thuringite, minnesotaite, and stilpnomelane are of common occurrence and hence are included here, although they are of diagenetic or perhaps low-grade metamorphic origin. Riebeckite is also a component of some of the older iron-formations, especially those of Australia.

Chamosite, essentially $3(Fe,Mg)O \cdot (Al, Fe)_2O_3 \cdot 2SiO_2 \cdot nH_2O$, is the most abundant primary iron silicate of ironstones other than those of Precambrian age. It occurs as ooliths of green material in a matrix of siderite or calcite and is commonly associated with ooliths of goethite; in some cases these consist of alternating shells of chamosite and goethite as in the Northampton Ironstone (Taylor, 1949, pp. 17–19).

The exact nature of the mineral chamosite has been the subject of considerable research. What is known is summarized by Deer, Howie, and Zussman (1963, p. 164) and James (1966, pp. 4–6). Chamosite is closely related in structure and composition to chlorites. Chamosite is itself characterized by a 7Å spacing but is readily transformed to a chlorite (thuringite) with a 14Å spacing. The chamosite ooliths of the Mesozoic ironstone of England (Hallimond, 1925; Taylor, 1949) consist of very fine flakes, pale to dark green, arranged tangentially so that the ooids show a dark extinction cross under crossed nicols. The flakes have positive elongation, indices that range from 1.62 to 1.66, and a low birefringence of 0 to 0.03. The index varies and is related to the ferrous–ferric ratio. According to Hallimond (1925), chamosite was an original precipitate, perhaps an amorphous gel which later assumed the form it now shows. Or it may have originated by progressive replacement of clay in which oolites of the same material may or may not have formed. According to Deverin (1945), chamosite in the Swiss Dogger ironstones is not an original precipitate but is a replacement of carbonate detritus, chiefly echinoderm debris.

Greenalite, essentially a hydrated ferrous silicate, is a common and abundant primary iron silicate in the Mesabi and Gunflint districts of the Lake Superior region where it was first identified and named (Leith, 1903, pp. 101–115). It was classed as a septechlorite by Deer, Howie and Zussman (1963, p. 167).

Unlike chamosite and glauconite, greenalite is important only in Precambrian rocks, but it has been reported in younger deposits. It occurs, like glauconite, in rounded to irregular, isotropic, light- to dark-green pellets generally without the concentric structure characteristic of chamosite. It is typically associated with magnetite.

Glauconite, $KMg(Fe, Al)(SiO_3)_6 \cdot 3H_2O$, has been reported from some ironstones younger than Precambrian. Although it is a valid mineral species, it shows considerable variation. It appears as bright green granules but is also known as intergranular fillings and disseminations, as replacements of fecal pellets, as fillings of openings in fossils (Takahashi, 1939), and even as coatings on heavy minerals (Grim, 1936). The granules range from 0.01 to about 0.50 mm in diameter. Granular glauconite is closely associated with detrital quartz sand (Fig. 7-25).

Burst (1958a; 1958b) divides glauconite into four classes: well-ordered, nonswelling, high-potassium mica-type lattice; disordered, nonswelling, low-potassium mica-type lattice; extremely disordered, expandable, low-potassium montmorillonite-type lattice; and mixtures of two or more clay minerals, such as kaolinite and illite. There seems to be no correlation between glauconite composition and geologic age (Foster, 1969).

Glauconite is a constituent of modern sediments in many parts of the world (Cloud, 1955). It ranges from near-shore sandy deposits (Galliher, 1935a, 1935b) to the deeper sea; most commonly it is found between 60 and 2400 feet (18 to 730 m). In some older sediments ("greensands") it accumulates as beds some tens of feet thick containing 75 or more percent glauconite (Mansfield, 1920). These are, on their iron content, technically ironstones, although they are rarely exploited for their iron.

Thuringite (bavalite), $(Si_{4.8}Al_{3.2})(Mg_{1.4}Fe_{7.4}Fe_{1.6}Al_{1.7})(OH)_{16}O_{20}$ is an abundant mineral in a Lower Silurian ironstone at Thuringia, Germany, where it forms oolites in a matrix of quartz and magnetite (Engelhardt, 1942). Thuringite is very similar in composition to chamosite from which it is thought to have formed by mild metamorphism.

Minnesotaite, $(Fe, Mg)_3Si_4O_{10}(OH)_2$, is an iron talc found in some Precambrian iron-formations. Because of its extremely fine grain size, it is difficult to recognize and has probably been overlooked. Although considered by Gruner (1946, p. 12) to be a primary constitutent, others believe it to be a low-grade metamorphic product.

Stilpnomelane, $2(Fe, Mg)O \cdot (Fe, Al)_2O_3 \cdot$

$5SiO_2 \cdot 3H_2O$, is a micalike mineral very similar in appearance to biotite. It is a common constituent in many of the iron-formations of the Lake Superior region (James, 1954). It is known elsewhere as a secondary mineral. As a rock-forming mineral it occurs as microscopic needles and plates showing strong pleochroism and absorption. It is olive green to dark brown, depending on the ferrous–ferric ratio. The indices likewise show a large range related to this ratio. It may occur in veins but most commonly is disseminated in siderite or forms thin laminations.

Riebeckite is a soda-amphibole, $Na_2Fe_3Fe_2(Si_8O_{22})(OH,F)_2$ whose fibrous variety is known as crocidolite. Although normally a constituent of igneous rocks, riebeckite is present either in massive or fibrous form in the bedded iron-formations of both South Africa and Australia. In these formations riebeckite is diagenetic, or possibly low-grade metamorphic in the South African deposits, in which the riebeckite consists of aggregates and radiating groups of minute fibers.

Iron sulfides Pyrite and marcasite are the only sulfides of importance in iron-rich sedimentary rocks. These iron sulfides may have been derived from less stable amorphous black iron sulfides such as those in some present-day unlithified deposits.

Pyrite, FeS_2, the most common sulfide, occurs as scattered isolated crystals of diagenetic origin. In a few cases it forms layers consisting of pellets, spherules, and replacements of shell fragments. In some occurrences the pyrite is exceedingly fine-grained (0.003 mm) and disseminated; in other cases it is segregated into thin layers one to several millimeters thick.

Marcasite, FeS_2, is rare or absent in ironstones. Its most common occurrence is as nodules in the coal measures.

Iron-rich sediments also contain many other minerals. Some of the ironstones are highly calcareous and contain much calcite; others are argillaceous and contain assorted clay minerals. Some, especially the oxide facies, are sandy and contain large detrital quartz grains.

Of greatest significance to the problem of genesis and distribution is the recognition of primary sedimentary facies identified by the dominant iron mineral present—sulfide, carbonate, oxide, and silicate.

Bedded iron sulfides The rocks in which iron sulfide (mainly pyrite) is an important constituent, or even the dominant one, are uncommon and insignificant in total volume. Beds of pyrite, 6 inches (15 cm) to a little over 1 foot (30 cm) thick, interbedded with fissile black shale, occur in the Wabana district of Newfoundland in Ordovician rocks (Hayes, 1915). The pyrite beds are composed of minute spherulites of pyrite together with both pyritized and unpyritized fossil fragments, all cemented by an extremely fine-grained siliceous groundmass. Pyrite forms about 65 percent of these beds.

Similarly, in some of the Precambrian black slates associated with bedded cherty iron carbonates of the Iron River–Crystal Falls district of Michigan are interbedded layers of pyrite varying from mere films to those 1 or 2 inches (2.5 to 5.0 cm) in thickness (James et al., 1968, p. 41). The most important occurrence of sedimentary pyrite in the district, however, is the Wauseca Pyritic Member, the so-called "graphitic slate" of the footwall formation, which is about 5 to 20 feet (1.5 to 6.1 m) thick. On the basis of a few samples it averages about 38 percent pyrite. The pyrite in this rock is extremely fine-grained, and the individual crystals (0.003 mm) are nearly invisible to the naked eye. The pyrite tends to be concentrated in certain layers and regularly interbedded with dark carbonaceous slate (Fig. 11-14). The iron sulfide content of these richer layers is about 75 percent. The bulk chemical composition of the rock is given in Table 11-3A.

Pyrite also occurs in some limestones. Representative of such a limestone is a 4 foot (1.2 m) bed at the base of the Greenhorn Formation (Upper Cretaceous) of Wyoming (Rubey, 1930). It consists of calcite, 45.4 percent; pyrite, 25.2 percent; gypsum (in part of secondary origin), 17.6 percent; iron oxides, 6.1 percent; organic matter, 2.6 percent, and bone phosphate, 2.0 percent. The Tully Limestone (Devonian) of New York State is notably pyritic and in places grades into a thin but widespread pyrite bed (Loomis, 1903). The Tully is closely associated with carbonaceous black shales. The association of pyrite with carbonates and with organic matter (or carbonaceous residues therefrom) is rather general. The black shales or slates have the most pyrite. This suggests that the source of the sulfur is the nitrogeneous component of organic matter; it is possible also that the association is attributable primarily to the

TABLE 11-3. Chemical composition of iron-bearing sediments

Constituent	A	B	C	D	E	F	G	H
SiO_2	36.67	28.86	24.25	31.84	61.90	8.51	12.59	8.71
TiO_2	0.39	0.20	—	0.12	none	0.36	0.27	—
Al_2O_3	6.90	1.29	1.71	2.09	0.37	6.12	5.71	3.67
Fe_2O_3	—	1.01	0.71	14.83	15.00	1.77	75.12	30.24
FeO	2.35	37.37	35.22	20.59	10.28	36.91	—	—
FeS_2	38.70	—	—	—	—	—	—	—
MnO	trace	0.90	2.11	2.35	—	0.42	0.06	—
MgO	0.65	3.64	3.16	3.80	2.33	3.75	0.42	7.84
CaO	0.13	0.74	1.78	1.49	0.28	5.54	1.49	20.64
Na_2O	0.26	—	0.04	nd	none	0.05	—	—
K_2O	1.81	—	0.20	nd	none	0.03	—	—
H_2O+	1.25		0.21		4.17	4.05	2.17[b]	—
		0.68[a]		1.80				—
H_2O-	0.55		—		2.50	10.00	0.52	—
P_2O_5	0.20	trace	0.91	0.83	none	1.30	1.63	0.75
CO_2	—	25.21	27.60	19.40	2.04	20.70	—	24.78
SO_3	2.60	—	—	none	—	—	—	0.15
S	—	—	nd	0.33	—	0.05	none	—
C	7.60	—	1.96	—	—	0.27	—	—
Total	100.21[c]	99.90	99.86	99.47	99.54[d]	99.86	99.98	96.78

[a] Ignition.
[b] Ignition.
[c] Includes 0.15 percent V_2O_5.
[d] Includes 0.64 percent iron and aluminum oxides in soluble part.

A. Sulfide iron-formation (Precambrian), 10th level, Buck Mine, Iron River, Michigan. C. M. Warshaw, analyst (James, 1954, p. 250).
B. Cherty iron carbonate, Ironwood Iron-formation (Precambrian), Michigan. W. F. Hillebrand, analyst (Irving and Van Hise, 1892, p. 192). Approximately 29 percent chert and 62 percent siderite.
C. Cherty iron carbonate (Precambrian), Iron River district, Michigan. Leonard Shapiro, analyst (James, 1951, p. 257). About 24 percent chert and 70 percent siderite.
D. Cherty iron carbonate (Precambrian), Crystal Falls, Michigan. J. G. Fairchild, analyst (Pettijohn, 1952). About 20 percent chert, 48 percent siderite, and 23 percent stilpnomelane.
E. Greenalite Iron-formation, Biwabik Formation (Precambrian), Minnesota. George Steiger, analyst (Leith, 1903, p. 108). Approximately 50.4 percent greenalite and 50 percent chert.
F. Sideritic-chamositic mudstone, Cleveland Ironstone (Jurassic), Great Britain, J. E. Stead, analyst (Hallimond, 1925, p. 51). About 34.2 percent chamosite, 34.7 siderite.
G. Chamositic hematite, Dominion bed, zone 2 (Ordovician), Wabana, Newfoundland, Canada (Hayes, 1915, p. 45). About 65 percent hematite and 24 percent chamosite.
H. Hematite "fossil ore," Clinton Formation (Silurian), Alabama. E. C. Sullivan, analyst (Burchard et al., 1910, p. 34). About 30 percent hematite, 17 percent calcite, and 36 percent dolomite.

reducing environment needed for the preservation of organic material and the bacterial reduction of the sulfates of sea water. Rubey (1930) has noted the relation of pyrite, organic matter, and carbonate in certain Cretaceous formations (Table 11-4).

There is a growing literature on the "exhalative-sedimentary" bedded sulfides such as the *Kupferschiefer*, the copper-bearing shales of Germany. Possibly the pyritic member of the Precambrian iron-formation of the Michipicoten district of Ontario is of this type (Goodwin, 1962).

Bedded siderite (cherty iron carbonates, clay ironstones) The iron carbonate, siderite, occurs intimately interbedded with chert ("cherty iron carbonate") or mixed in all proportions with clay (clay ironstones). Typical bedded cherty carbonate consists of fine-grained, light- to dark-gray siderite and dense black chert (Fig. 11-15). The constituents are rhythmically interbedded in layers a fraction of an inch to 3 or 4 inches (7–10 cm) thick. In less cherty formations the chert occurs as small nodules. The proportions are variable, but commonly the formation is one-fourth to one-third chert and one-half to two-thirds siderite. In places, formations of this character reach 300 feet (100 m) in thickness. The iron content of the rock as a whole is 20 to 30 percent (50 to 70 percent siderite). See Table 11-3B, C, and D.

The siderite layers may show paper-thin

TABLE 11-4. Content of carbonates plus organic matter, and pyrites in Black Hills Cretaceous rocks

	Carbonates plus organic matter	Pyrite
Colorado group		
Average of five samples	2.3	0.4
Pierre shale		
Average of three samples	8.7	0.8
Calcareous parts of Greenhorn and Niobrara formations	42.7	9.1

Source: After Rubey (1930).

FIG. 11-14. Iron-sulfide facies (Precambrian), Iron River, Michigan. Slightly enlarged. Pyritic layers appear light-colored; unpolished specimens are dull and nearly black. (U.S.G.S. photograph. James, 1951, Pl. 1.)

FIG. 11-15. Iron-carbonate facies, Riverton Iron Formation (Precambrian), Iron River, Michigan. Chert dark, siderite light. About half natural size. (U.S.G.S. photograph. James, 1966, Fig. 23.)

laminations of iron silicate or iron sulfide; they may be nearly pure carbonate. The latter are commonly stylolitic. The iron carbonate formation is generally free of detrital materials. Oolitic textures are absent.

Cherty iron carbonates are common in the

Lake Superior district of the United States. This formation has been generally considered to be the major primary iron-bearing sediment in this district, and most other types of iron formation were regarded as post-depositional modifications of this rock. Except for truly metamorphosed cherty carbonate rocks, it now seems clear that the other types of iron formation (silicate and oxide) are themselves primary sedimentary facies and not oxidized (weathered) or silicated (metamorphic) iron carbonate rocks (James, 1954, p. 238). Primary bedded iron carbonate is the principal type of iron-formation in the Crystals Falls–Iron River district of Michigan. It is also present in important volume in both the Gogebic and Marquette districts.

In post-Precambrian ironstones, siderite may form a groundmass for chamositic or limonitic oolites, commonly showing considerable diagenetic replacements of ooids. It may occur in fine-grained sideritic mudstone, or it may occur as matrix and partial replacements of shell fragments and calcitic ooids or as replacement of matrix calcite. It also occurs as nodular masses, "sphaerosiderite," in chamositic and kaolinitic ironstone. Only fine-grained sideritic mudstone is a primary sediment; all others are thought to be diagenetic.

Clay ironstone is the term applied to the argillaceous sideritic concretions and beds found in the coal measures of both Great Britain and the United States (Stout, 1944). Clay ironstone occurs as layers of nodules, many of which display a septarian structure and as more or less continuous thin beds. Their color is dark gray to brown and less commonly black (blackband ore). Clay ironstones are fine-grained; the clay content varies from 1 to 30 percent. Ironstone nodules or beds are found most commonly to overlie coal seams; fossils are common.

Bedded iron oxides (including bog iron) Some iron-formations consist largely of iron oxide. Of these hematite is perhaps the most common. The sedimentary hematites are best illustrated by the Silurian Clinton iron-

FIG. 11-16. Clinton oolitic hematite or "flaxseed ore" (Silurian). Ordinary light, ×30. A nitrocellulose peel of polished surface. Many ooids apparently had quartz sand grains as nuclei. Note flattening of some oolitic bodies.

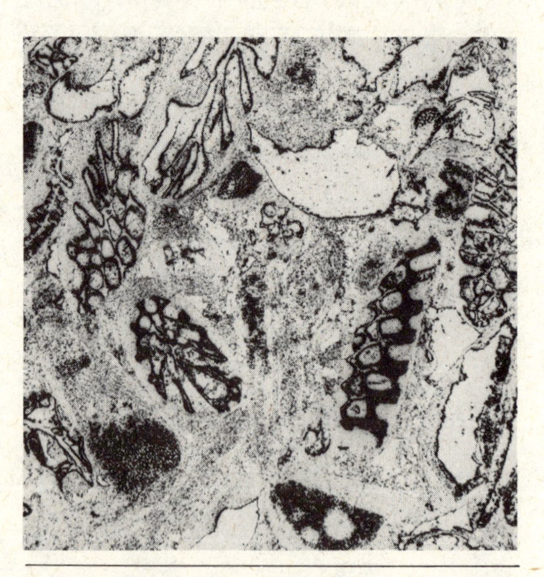

FIG. 11-17. Clinton "fossil" ore (Silurian). Ordinary light, ×30. A nitrocellulose peel of polished surface of fossil ore. The fossil structures are well preserved in hematite.

bearing strata (Smyth, 1892; Alling, 1947; Schoen, 1964; Hunter, 1970). These several beds are found throughout the Appalachian region and form important deposits of iron ore in the Birmingham district of Alabama (Burchard et al., 1910). Most of the iron-bearing rocks are oolitic hematite and "fossil ore" (Figs. 11-16 and 11-17). In places thin beds of oolitic chamosite are associated with shales. Hematitic ooids have detrital quartz and fossil grains as nuclei. The fossil grains are typically partially replaced by hematite, and their internal cavities are filled with iron minerals. All of the oolitic hematites are cemented with calcite and dolomite.

That these hematites were early, perhaps contemporaneous, replacements or original precipitates is proved by their stratigraphic continuity, their distribution (unrelated to either outcrops or unconformities), the occurrence of fragments of ore in overlying beds of limestone of the same formation, the presence of hematitic oolites in calcareous beds in which the matrix consists of dolomite or calcite, and the occurrence of calcareous fossils enveloped in concentrically deposited shells of hematite.

The Wabana ores (Ordovician) of Newfoundland are complex and contain hematite, chamosite, and, less commonly, siderite (Hayes, 1915). The hematite is oolitic (Fig. 11-18); some oolites are themselves complex and consist of alternating layers of hematite and chamosite. Chamosite occurs as pure oolites and is present also in the matrix of the oolitic rocks. The average ironstone consists of 50 to 70 percent hematite; 15 to 25 percent chamosite, 0 to 50 percent siderite, 4 to 5 percent calcium phosphate, 0 to 1 percent calcite, and 0 to 10 percent detrital quartz. The compositions of three typical iron-bearing beds (A, B, C) and a ferruginous sandstone (D) are shown in Fig. 11-19. Hayes cites much evidence to establish the primary marine origin for hematite oolites. The sorting, ripple marking, and cross-bedding of the oolitic members, the close association with marine fossils, the borings of marine organisms into the oolitic layers which are filled with mud but which contain few oolites, and the intraformational conglomerates holding pebbles of the hematite-chamosite beds—all point to the primary marine character of the deposit.

Bedded, oolitic hematites occur also in the Precambrian iron-formations of the Lake Superior region and of the Labrador trough.

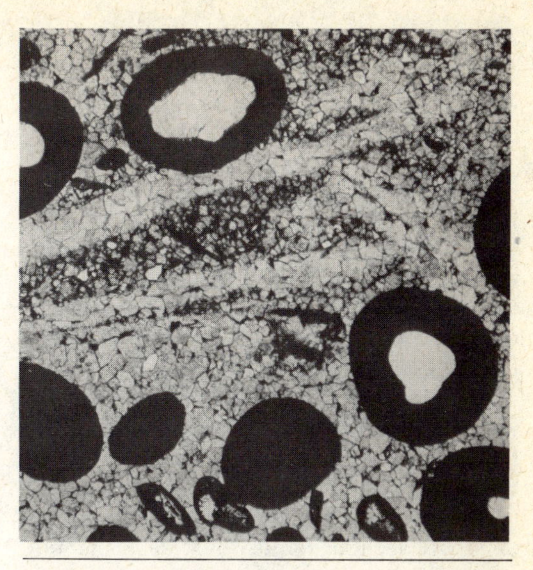

FIG. 11-18. Hematite ooids, Wabana (Ordovician), Newfoundland, Canada. Ordinary light, ×20. Thin section of hematite ooids (black), some with detrital quartz nuclei (white), in a dolomite matrix. The dolomite shows some fossil structures.

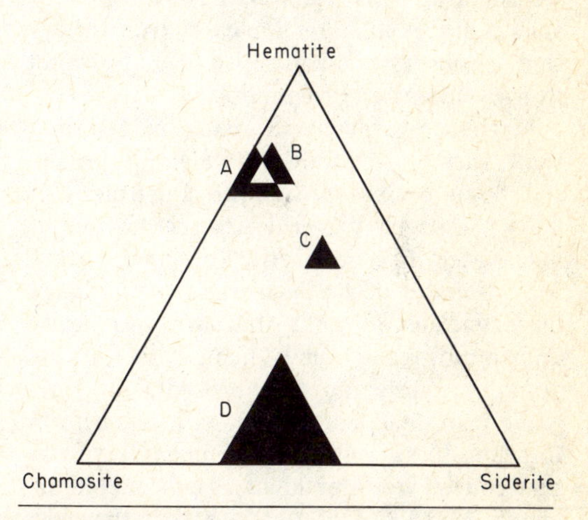

FIG. 11-19. Mineral composition of the iron-bearing sediments from the Wabana district, Newfoundland, Canada. A, B, C: iron-bearing beds; D: a ferruginous sandstone. (Data from Hayes, 1915).

The oolitic structure is seen most readily in some of the jaspers where the ooids consist mostly of silica but where the structure is defined by hematite (Fig. 11-20). Oolites consisting more largely of hematite are also present; no doubt many of the hematitic layers were once oolitic. More commonly, however, the hematitic layers appear not to be oolitic. The rhythmically interbedded jasper and steel-gray hematite forms one of the most spectacular rocks in the Lake Superior region. Jaspilite, as it is termed, occurs in the Marquette, Menominee, and Gogebic districts of Michigan and in the Vermilion district of Minnesota. It was once regarded as the product of premetamorphic weathering of a cherty iron carbonate. James (1954), however, considered it to be a primary sedimentary deposit. The preservation of exceedingly fine bedding and knife-edged contacts between hematite and jasper is not characteristic of weathered rocks. Moreover, the presence of continuous, although thin, layers of hematite in unoxidized carbonate formation further argues for the primary nature of the hematite. The close association of the oolitic structure of much hematite with quartz sand grains and the absence of such features in other facies of iron-formation are further evidence of a primary origin in a turbulent environment.

Magnetite, present in many iron-formations, may be a product of metamorphism. It is known also as a major constituent of black sands—a placerlike concentration of magnetite. The magnetite of many iron-formations is of neither origin. Abundance of magnetite in rocks that are essentially unmetamorphosed, as indicated by the fine grain of cherts and the presence of low-grade iron silicates, should serve as criteria to separate synsedimentary magnetite from that caused by later higher-grade metamorphism. Magnetite of nondetrital nature has been reported from modern sediments (Berz, 1922; Friedman, 1954). It is even known as a component of the teeth of chitons (Lowenstam, 1962). The magnetite of many Precambrian iron-formations occurs as impure layers interbedded with cherts. James considers the magnetite to be diagenetic (James, 1966, p. 3), as does Dimroth (1968a, p. 264). Magnetite-banded rocks are typically, almost invariably, associated with the iron silicate facies of iron formation, and every gradation can be found between the two.

"Limonite" (goethite) occurs in oolitic form in some ironstones, mainly those of Mesozoic and Cenozoic age—an exception

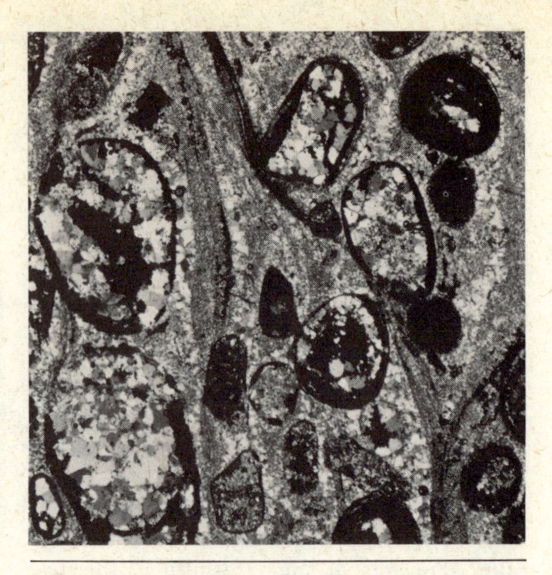

FIG. 11-20. Oolitic ironstone (Precambrian), Mesabi Range, Minnesota. Ordinary light, ×22. Ooids consist mainly of silica but with structure defined by hematite.

is the Ordovician Mayville deposit in Wisconsin (Hawley and Beavan, 1934). This facies consists of oolitic or pelletal grains of limonite, many with cores of clastic material, in a matrix of chamosite or calcite. Some consist of mixtures of chamosite and limonite ooids, with chamosite either partly altered to limonite or containing limonite in alternating layers with chamosite. Siderite, present in quantities of 10 percent or more, seems to be of diagenetic origin.

Bog ore is principally an earthy mixture of yellow to dark brown ferric hydroxides (Harder, 1919; James, 1966, p. 42). Such ores are deposited along the borders of certain lakes and in bogs. Lake ores consist mainly of flat, disklike, or irregular concretionary masses of ferric hydroxide, 1 or 2 feet thick, or as a layer of soft to hard porous, yellow, bedded limonite or as a limonitic cement in sand. Bog ores proper occurs in a bog or marsh as a spongy mass mixed with peat. Some are tubular, others pisolitic, and still others nodular or concretionary bodies. A few form solid bodies, commonly rather thin and filled with sand and clay impurities. Bog iron ores are abundant in glaciated regions of North America and Europe. Iron dissolved from glacial drift is precipitated either chemically or biologically. Although they are rather small deposits, they are of interest because they are among the few iron-bearing sediments known to be forming at present.

Bedded iron silicates (except glauconite) The silicate facies is that in which the primary silicates chamosite, greenalite, and glauconite are dominant, plus those in which minnesotaite, stilpnomelane, and riebeckite (probably diagenetic or metamorphic in origin) are present.

The chamositic ironstone facies is the most characteristic silicate facies of the post-Precambrian ironstones. The ironstones are closely associated with limonitic ironstones in some districts; in others they grade into or are associated with sideritic rocks.

Chamositic ironstones are typically oolitic chamosite. Chamosite ooliths are, for example, found in the Clinton ironstones (Silurian) of the Appalachian region. Chamositic ooids are largely an iron-rich septechlorite with a 7Å spacing (Schoen, 1964). The ooids, like those of hematite, contain shell debris as nuclei; they are concentric-structured and display an extinction cross. They are associated with fossil debris and cemented by calcite. In addition to the Clinton, other North American occurrences include the Wabana ironstones (Ordovician) of Newfoundland (Hayes, 1915) and the thin iron-bearing beds of the Montebello Formation (Devonian) of Pennsylvania (Sheppard and Hunter, 1960). See Figs. 11-21 and 11-22. Chamositic ironstones are common in the Jurassic ironstone of Great Britain, of France, and of Switzerland.

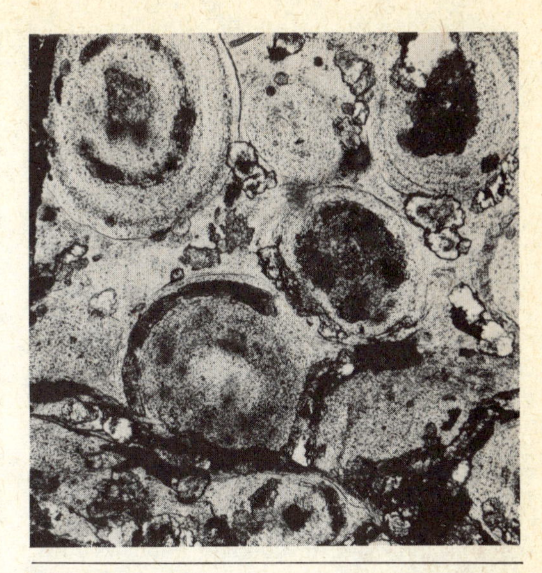

FIG. 11-21. Oolitic chamosite, Lorraine (Jurassic) ore, France. Ordinary light, ×55. Mainly chamosite (ooids) with a few small spherulites of siderite (clear granules with high relief).

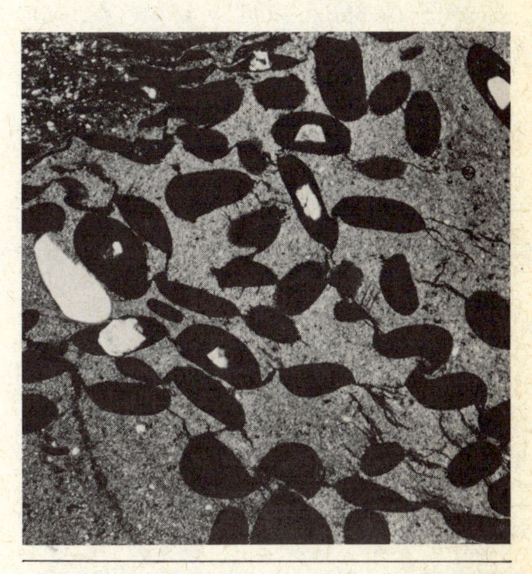

FIG. 11-22. Chamosite ooids, deformed and oxidized, Montebello Formation (Devonian), Harrisburg, Pennsylvania. Ordinary light, ×22.

In the Precambrian iron-formations, silicate is not chamosite but is instead greenalite or one or another of the diagenetic silicates: minnesotaite, stilpnomelane, or riebeckite. James (1954) recognized two major types of silicate iron-formation. In one, the silicate mineral is present as granules or as oolites; in the other the silicate is nongranular and is typically thinly bedded or laminated. Greenalite, like glauconite, is characteristically granular. Although some granules have a microcrystalline and structureless interior, others show an obscure mottling in thin section (Fig. 11-23). Typical silicate granules are about 0.5 mm in diameter and are rudely elliptical in section. They may occur in a matrix of chert or of other silicate materials. The second type of silicate iron-formation—the nongranular—is thinly bedded or laminated. Characteristically these laminations are only a millimeter or fraction thereof in thickness (Fig. 11-24). The variations in the FeO–Fe_2O_3 ratio in those composed of stilpnomelane, as indicated by refractive indices, are good evidence that the present composition of layers is essentially that of the primary mineral.

The composition of the iron-silicate facies varies within wide limits, chiefly because of the varying proportions of carbonate and iron oxide present in addition to silicates which are themselves of variable composition (Table 11-3). Nongranular iron silicate rocks seem to have a high proportion of clastic materials. In younger deposits these are aptly described as chloritic mudstones, chamositic mudstones, and so on.

Silicate iron-formations are a very common type of iron-bearing sediment. Glauconitic sandstones and limestones are common in Cambrian and younger deposits. Some of these, such as the New Jersey greensands (Cretaceous), are 50 percent or more glauconite and contain 20 or more percent FeO and Fe_2O_3 (Table 7-11, A, B, and C) and hence are in a sense ironstone. Chamositic-bearing beds are common in the Jurassic ironstones of Great Britain, France, and Switzerland. Chamosite is a minor constituent of the Silurian Clinton and the Ordovician Wabana. It seems not to be present in Lake Superior formations where greenalite, stilpnomelane, and minnesotaite are the common iron silicates. Greenalite is presumed to be primary; the others may be diagenetic or metamorphic in origin. The stilpnomelane of the Brockman Iron-formation of Australia appears to have formed from volcanic ash.

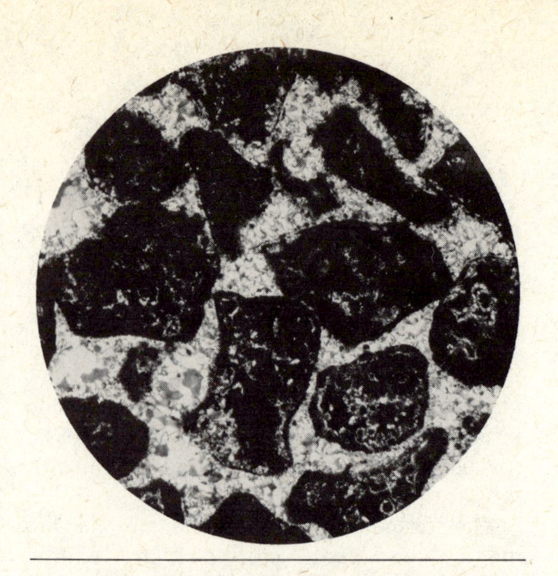

FIG. 11-23. Greenalite granules, Gunflint Formation (Precambrian), Ontario. Crossed nicols, ×30. Greenalite granules in chert matrix. Greenalite shows peculiar mottling caused by recrystallization of chert. (Photograph by H. L. James. James, 1955, p. 292, Fig. 27.)

FIG. 11-24. Siderite-stilpnomelane ironstone, Riverton Iron-formation (Precambrian), Crystal Falls, Michigan. About natural size. Fine-grained siderite (light) with interbedded laminations of stilpnomelane (dark).

OCCURRENCE AND DISTRIBUTION OF IRON-BEARING SEDIMENTS

Iron-bearing sediments are widespread in space and time. The thickest and most impressive deposits are of Precambrian age. Many local iron-formations of Archean (2500 my or older) age are interbedded with ellipsoidal greenstones and are associated also with thick sections of graywacke and slate. These are the so-called "Keewatin-type" iron formations (Dunbar and McCall, 1965; Goodwin, 1962).

The most spectacular deposits are those of Proterozoic age, deposited mainly between 1700 and 2500 my ago. The best known of these include formations in the Belcher Islands, Richmond Gulf, Cape Smith, and Labrador trough areas in Canada (Dimroth et al., 1970; Dimroth, 1968b), and the Lake Superior district (Van Hise and Leith, 1911; James, 1951, James et al., 1968), mainly in the United States. If not deposited in a single continuous basin, they were formed in a series of basins during one major episode of iron sedimentation which encircles what is now known as the Superior Province of the Canadian shield. Two other Proterozoic basins of iron deposition are also well known. These are the Hammersley Basin of Western Australia and the Transvaal System basin of South Africa (Trendall, 1968). The first is a basin in excess of 300 miles (480 km) in length and contains an iron-formation (Brockman) over 3000 feet (915 m) thick. The second is an even larger basin containing one, possibly two, iron-formations of great thickness. A good brief comparison of these two basins with those of the Animikie (Lake Superior) basin has been published by Trendall (1968). Other well-known Precambrian iron-formations occur in the Krivoy-Rog area of Russia, in peninsular India, and in Brazil.

Very probably there has been a period of iron sedimentation somewhere in every geologic period of the Phanerozoic era. In North America, for example, ironstones are known from the Cambrian Bliss Sandstone of New Mexico (Kelley, 1951), the Ordovician Mayville of Wisconsin (Hawley and Beavan, 1934), the Wabana of Newfoundland (Hayes, 1915, 1929), the Silurian Clinton of the Appalachians (Alling, 1947; Schoen, 1964; Hunter, 1970), the Devonian Boyle Limestone of Bath County, Kentucky (Bucher, 1918, pp. 598, 600), the Carboniferous clay ironstones and black-band ores of the Appalachian Coal Basin (Stout, 1944), the Cretaceous oolitic ironstone of Alberta (Mellon, 1962), and the Cretaceous greensands especially of New Jersey (Mansfield, 1922) and Texas (Eckel, 1938).

Very well known are the Minette (Jurassic) ores of France (Cayeux, 1909, 1922), the equivalent Dogger beds of Switzerland (Deverin, 1945), and the Northampton ironstones of Great Britain (Hallimond, 1925; Taylor, 1949). The list of both Precambrian and Phanerozoic iron-bearing deposits could be greatly extended. James (1966) has published a very complete and annotated tabulation.

What is the geologic framework in which iron-bearing sediments are found and what light does the geologic milieu shed on the problem of iron sedimentation? In a very general way, iron-bearing sediments can be grouped into two classes: cherty iron-bearing sediments—the iron-formations, which are mainly Precambrian—and the noncherty deposits—the ironstones, which are largely Phanerozoic. They can also be looked at in another way. One group, which might be called the *oolitic facies,* is one in which iron minerals occur in oolitic or granule form and the primary constituents are mainly ferric oxides or ferro-ferric silicates. These materials are interbedded or intermixed with mature quartz sands and are commonly associated with stromatolitic cherts, dolomites, and, in the Phanerozoic, marine limestones. Cross-bedding and ripple marking are not rare. Siderite, if present, is apt to be diagenetic and to replace silicates or other carbonates. Many iron-formations of the Lake Superior region are of this type. The Gunflint Iron-formation of the Animikie district of Ontario rests on a thin conglomerate overlying the Archean basement; the Biwabik Iron-formation of Minnesota rests on a thin quartzite (Pokegama), which in turn lies on the Archean floor. The Ironwood Iron-formation of the Penokee district of Wisconsin rests on the supermature Palms Quartzite, and the lowest chert member contains large rounded quartz sand grains and oolites. As noted by several workers, the texture of many of these iron-formations are similar to those of calcarenites with oolites, peloids, and intraclasts and with structures indicative of a turbulent, shallow-water environment (Dimroth, 1968a; Dimroth and Chauvel, 1973; Gross, 1972). Paleozoic and younger ironstones are likewise generally oolitic with oolites that have detrital quartz cores, are cross-bedded and are associated with calcarenites. The greensands are a mingling of quartz and glauconitic sands. The fossil content of the Phanerozoic beds indicate a shallow-marine environment. There is no reason to believe that the Precambrian oolitic facies was different.

A second group of iron-bearing sediments form the *laminated facies,* a facies in which iron minerals form thin laminations, mainly of magnetite and iron silicates with siderite, and in some cases pyrite interbedded with dark slates and pyritic shales. These are the "slaty" iron-formations of some reports (largely silicate-magnetite rocks) and the cherty iron carbonates. The iron-formations of the Cuyuna and the Iron River–Crystal

Falls districts and most of the Archean "Algoma" ironstones (Dunbar and McCall, 1965) are of this type (Fig. 11-25). The iron-formation of the Transvaal and Hammersley basins (Trendall, 1968) seem also to belong here. Oolitic and granule structures are generally absent, as is the ferric oxide facies. The mineralogy, associated sediments, and sedimentary structures suggest a strongly reducing environment, a starved basin, essentially currentless except for breccias and graywackes emplaced by slump or by turbidity currents—such as the "slate breccia" and other widespread and locally thick breccias (olistostromes) in the Iron River district (James et al., 1968, p. 71)—and other evidence of tectonic instability such as graywacke dikes.

This division into the oolitic or shallow-water (cratonic) facies and the laminated or deep-water (basinal) facies does not preclude the presence of both in the same district caused by a significant change in water depth. Like the limestones, we have both carbonates of shoal-water or platform facies and those of deep-water basinal facies.

ORIGIN OF IRON-BEARING SEDIMENTS

The origin of iron-bearing sediments has been one of the most debated topics in sedimentary petrology (Fig. 11-26). There is, as yet, no acceptable model for the deposition of ironstones or iron-formations. Yet all the main classes of iron-bearing minerals are being deposited today, and the processes leading to their deposition may provide clues to the origin of the larger deposits of the past. Iron sulfides are found today in certain black muds deposited under strongly reducing conditions. Siderite is a component of some bog iron deposits of lakes and swamps. Glauconite is apparently forming today in Monterey Bay, California (Galliher, 1935a, 1935b). Iron oxides are accumulating in modern lakes and bogs, in tropical lateritic soils (which in places are pisolitic), and as ooliths on the floor of the North Sea (Pratje, 1930). None of these iron-bearing accumulations are on a scale to match that of the geologic record, but neither are modern deposits of chert or of evaporites on a scale comparable with similar materials of the past. Yet these deposits show that it is possible to transport iron and to precipitate it as a sulfide, carbonate, silicate, or oxide as in the past.

Although it is widely believed that iron-rich sediments are chemically precipitated rocks (albeit with biological intervention perhaps in the case of the sulfides and oxides), there is strong disagreement concerning their depositional chemistry, their paleogeographic significance, and their source of iron. Two prevailing views on the source problem have been aptly termed *from below* and *from above* (Oftedahl, 1958). According to the first concept, iron (and silica of the cherty deposits) was derived from a volcanic source, either as exhalations—vapors and volcanic waters—or by volcanic infall and submarine eruptions. This view, early formulated by Van Hise and Leith (1911) and reiterated by Aldrich (1929) and others (Davis, 1918), has found more recent support in the writings of Tanton (1950), Guild (1953, 1957), Goodwin (1956, 1961, 1962), Oftedahl (1958), Harder (1963), Trendall (1965), and LaBerge (1966a, 1966b). The other concept, perhaps more actualistic, presumes that surface processes were adequate to extract, transfer, and concentrate iron derived from adjacent land areas, a view early expressed by Gruner (1922) and Moore and Maynard (1929), and in more recent years championed by Sakamoto (1950), James (1951, 1954, 1966), White (1954), Alexandrov (1955), Hough (1958), Huber (1959), Lepp and Goldich (1964), and Govett (1966). Recently Borchert (1960) proposed a "from within" theory involving an intramarine redistribution of the iron by a diagenetic segregation process—a view akin to that expressed by Cayeux (1909, 1911), who regarded ironstones as products of diagenetic modification of an original calcareous deposit.

The appeal to volcanism as a source of iron (and silica) arises from the presumed inadequacies of ordinary processes to supply and transport iron in sufficient quantities. The evidence for a volcanic source is the presumed close association in time and space of iron sedimentation and volcanism. There are undeniable examples of interbedding of iron-formations and volcanic rocks and even in some cases a relict shard fabric

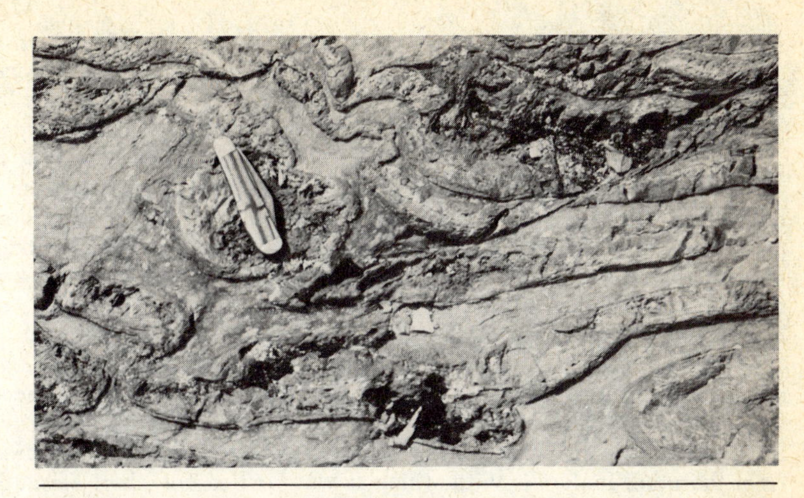

FIG. 11-25. Outcrop of "Algoma-type" iron-formation (Archean). Minnitaki Lake, Ontario, Canada. Chert interbedded with layers of magnetite-bearing iron silicate.

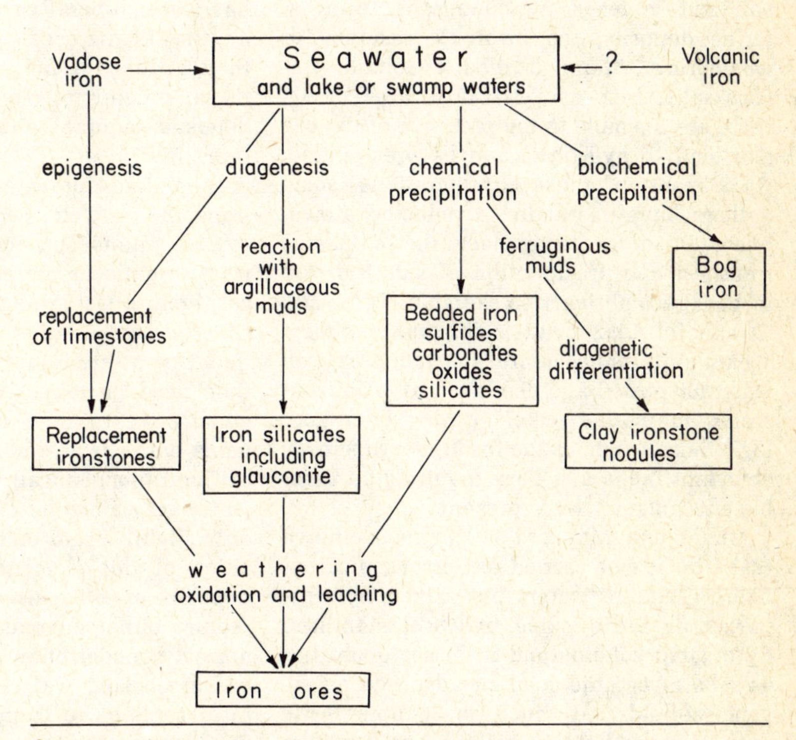

FIG. 11-26. Provenance of iron-bearing sediments.

in some stilpnomelane beds, suggesting derivation of this mineral from an infall of iron-rich volcanic ash (LaBerge, 1966a). Most of the later iron-formations, and many, if not most, of the Precambrian deposits are not associated with volcanic rocks of contemporary age. If surficial processes can be shown to be adequate to supply iron (and silica), there is then no compelling reason to ascribe them to volcanism.

As noted by Gruner (1922), Gill (1927), and James (1954), the iron and silica content of stream waters in tropical or subtropical regions is high enough to account for iron-formation deposition. As noted by Gruner (1922), the Amazon River, with an iron content of only three parts per million, is competent to transport 1,940,000 million metric tons of iron in 176,000 years, an amount about equal to that contained in the Biwabik Formation of Minnesota.

The transportation of iron has been considered a problem. Some means is required to remove iron from the soil, to transport it to the site of deposition, and to separate it from the other materials with which it was once associated—both the clastic detritus and the calcareous deposits. Ferric compounds are notably insoluble; iron oxides and hydroxides tend to accumulate in the soil and under some conditions form lateritic deposits, the insoluble residues of weathering. Although readily soluble ferrous salts, such as $Fe(HCO_3)_2$, are known, these are unstable in the presence of oxygen and tend to hydrolyze and be precipitated. As a result of these considerations, some authors have postulated a reducing atmosphere in order to explain the extensive transport and precipitation of sideritic iron in the Precambrian (for example, Tyler and Twenhofel, 1952, p. 134). This extreme view seems quite unnecessary, inasmuch as iron was transported and deposited to form ironstones in marine waters during later geologic times when, as the fossil record shows, abundant fauna and flora requiring an aerobic environment were present.

Moore and Maynard (1929) have shown that iron is not carried as ferrous bicarbonate in surface waters but rather is transported as ferric oxide hydrosol stabilized by organic colloids and in lesser quantities as salts of organic acids or adsorbed by organic colloids. As much as 36 ppm ferric oxide can be held in colloidal solution by 16 ppm organic matter. Very probably the bulk of iron today is transported in the form of colloidal iron oxide, and much of it is inextricably tied up with the clay-mineral fraction. The iron thus transported can be released under reducing conditions and can enter into the formation of such minerals as glauconite and chamosite (Carroll, 1958). This concept of transport and release is thus a facet of the diagenetic redistribution (from within) process (Borchert, 1960; Strakov, 1959)—a process deemed quantitatively inadequate by James (1966, p. 47) to account for iron-formations.

James (1966, p. 48) notes that at present the iron leached from the adjacent land is transported by bicarbonate *groundwaters* of low *Eh* and *p*H to lakes or bogs where it is precipitated to form a sideritic or oxide deposit. He invokes this mechanism, operating on a somewhat larger scale, to carry iron to somewhat restricted basins of the past, and there to precipitate according to *Eh* and *p*H conditions that prevailed. Groundwater transport has been presumed to supply the silica for chert formation in the Lake Magadi basin of central Africa (Eugster, 1969) and has been postulated even for the iron-depositing basins of the Precambrian (Eugster and Chou, 1973). This mechanism might have been operative during the deposition of the Phanerozoic ironstone, but James thought that the far greater dimensions and the abundance of chert in the Precambrian deposits required another mode of formation.

It is doubtful whether there is an iron transport problem, because, as James (1954, p. 276) points out, the accumulation of an iron-formation requires only that the chemical factors governing its precipitation be exactly right; neither the concentration nor quantity of iron in the waters at a given time need be excessive. The enormous accumulation of lime carbonate on the Bahama banks is a product of localized chemical (and biochemical) precipitation from sea water in which the calcium content differs only slightly from that in areas of nondeposition of this element. Nor do widespread deposits of siliceous ooze on the sea floor require either an unusual source or an abnormal concentration of silica. Iron is present in oceanic waters and is able to enter into reactions to form glauconite. It did so in Cretaceous time and formed greensand beds 40 feet (12 m) thick in New Jersey. There is no reason why iron silicates, whether greenalite or chamosite, could not accumulate in the past as glauconite now does in beds 40 or 400 feet thick if geologic stability is maintained long enough.

The form in which the iron is *precipitated* —even as ferrous carbonate—is a function of the local environment of accumulation, not the manner in which the iron is *trans-*

ported. Local reducing conditions, despite an oxidizing atmosphere, are common today, and even ferrous sulfide is known to be forming in these environments. The compound of iron precipitated is primarily a function of the oxidation–reduction potential (*Eh*). Experiments by Casteño and Garrels (1950; see also Huber, 1958; Krumbein and Garrels, 1952) and geological observations show that the stability of several iron-bearing minerals is more closely determined by the oxidation–reduction potential than it is by the acidity–alkalinity (*pH*) of the medium with which it is in contact (Fig. 11-27). At the lowest potential only iron sulfide will form; at a slightly higher potential ferrous carbonate is stable. Under fully oxidizing conditions ferric hydroxides are formed. Magnetite seems to require a lower potential than other oxides. The stability of various iron silicates is less clear. Apparently they form over a considerable range of *Eh* values, inasmuch as they show a varying $FeO–Fe_2O_3$ ratio within themselves. According to Teodorovich (Chilingar, 1955), iron chlorites form in a neutral zone, whereas glauconite requires a weakly oxidizing environment.

The naturally occurring assemblage (mineral facies) supports the experimental and theoretical considerations. Sulfides occur alone or in association with siderite; silicates are associated with siderite and magnetite; oxides occur alone or are with silicates (as in the case of magnetite). Oxides (and some silicates, such as glauconite) exhibit features, such as oolites or granules, and are associated with quartz sand grains, which indicate turbulence—a condition more likely to be realized in an aerobic rather than in an anaerobic environment. The absence of these features or associations in the sulfide and carbonate facies clearly shows that the oxide and silicate facies are not merely diagenetic modifications of sulfides and carbonates; it shows also that sulfides and carbonates were formed under very different environmental conditions. The relation between the several ironstone facies and the physicochemical

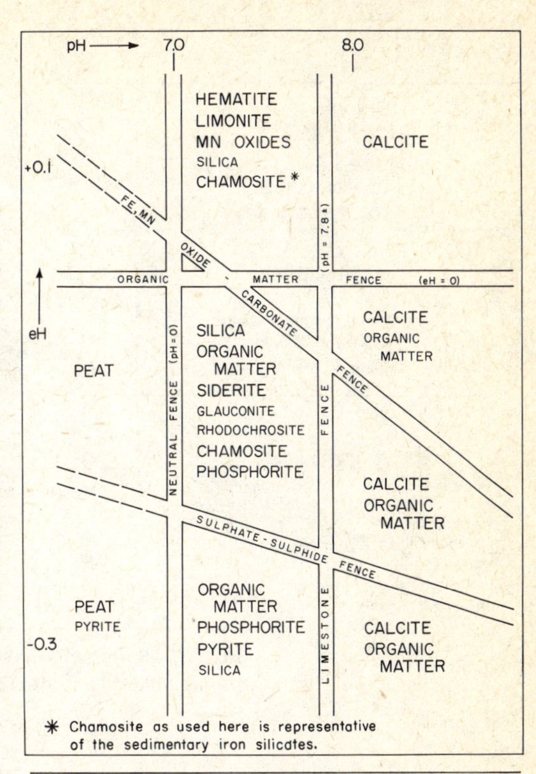

FIG. 11-27. Chemical classification of nonclastic sediments (evaporites excepted) based on *Eh* and *pH*. (After Krumbein and Garrels, 1952.)

conditions have been pointed out by several investigators (Borchert, 1952; Strakhov, 1959; James, 1966, p. 15). Although no complete array of major facies grading laterally into one another as indicated in Fig. 11-28 is known, partial arrays, that is, time-equivalent gradations from one facies to another are known in many areas: oxide to chamosite in the Clinton (Hunter, 1970), hematite to magnetite, to silicate and carbonate in the Wabush Lake area (Gastil and Knowles, 1960).

The role of organisms in the formation of ironstones and cherts is difficult to evaluate. Although bacteria are known to precipitate iron, there is no good evidence that they have played a major role in the formation of iron-bearing strata. Harder (1919) attributed many ferric hydroxide deposits to bacteria. Although some bog ores may be bacterial in origin and some iron sulfides are a product of sulfate-reducing bacteria (Androussow, 1897; Galliher, 1933), bacteria do not precipitate iron carbonate or iron silicates. And because the primary constitu-

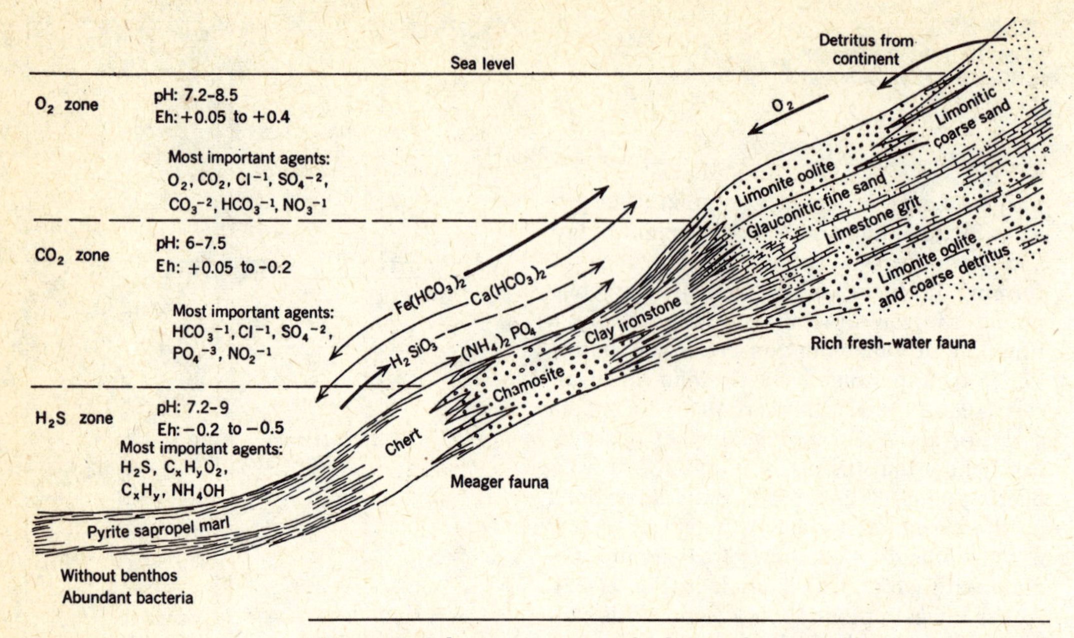

FIG. 11-28. Schematic section of relations between ironstone facies and physicochemical conditions. (From James, 1966, after Borchert, 1960.)

ents of the more important iron-bearing sediments are carbonates and silicates, the role of bacteria in the formation of iron-rich sediments may be rather minor.

As noted in the preceding section, radiolarians and other organisms do precipitate silica. It is not clear what role (if any) they have had in the precipitation of the chert associated with iron-formation. No unquestioned radiolarian remains have been identified from these rocks, although they do contain what appear to be primitive algal forms. The chert stromatolites are presumed to be of organic origin but whatever role algae play in silica precipitation is not clear.

The principal problem of iron deposition seems to be to define those conditions which *permit* the deposition of iron compounds but inhibit the deposition of lime and retard or prevent the introduction of clastic sediments into the basin of sedimentation. The absence or paucity of clastics is understandable, if the relief of land areas is extremely low or if they were deposited elsewhere, that is, if the basin were a "starved" basin. The absence of lime is perhaps related to a low *p*H. Iron sedimentation has been in part shown to occur in near-shore areas where the waters are brackish or at least fresher than those of the open sea (Hunter, 1970, pp. 117–118).

The exact nature of the basin of deposition has been much debated. Many ironstones were formed under near-marine if not fully marine conditions, as contained fossils indicate. It has been supposed by some that Precambrian banded iron-formations were deposited in a wholly freshwater lake (Hough, 1958), a conclusion apparently reached because it provided a mechanism to explain the rhythmic layering of various iron minerals and chert. The alternation was attributed to seasonal overturn in the lake waters, in which the silica was precipitated from the lower, more acid waters during the summer and the iron was precipitated when the lake waters overturned and became oxidized in the winter. Sakamoto (1950) likewise attributed the banding to seasonal changes in water chemistry, with the iron precipitating during the dry season and the silica forming during the wet season. Most workers, however, opt for a marine environment, albeit somewhat restricted or isolated from the open sea by a barrier (James, 1954, p. 243). The origin of the rhythmic spacing of the chert layers is rather inadequately explained. Such rhythmicity, however, is seen also in the cherts of the English Chalk and in the bedded cherts explained as a result of diagenetic segregation (Davis, 1918, p. 394).

DIAGENESIS OF IRON-BEARING SEDIMENTS

Most troublesome to recognize and interpret are the post-depositional changes which the iron-bearing sediment undergoes. Diagenetic reorganization, metamorphism, and weathering all produce marked changes in mineralogy and texture.

Evidences of replacement—siderite replacing fossils, hematite fossil ore, and similar observations—prove that iron migrates and readily replaces calcium carbonate. The importance of these replacements was early recognized by Cayeux (1909, 1911, 1922) and more recently described by Deverin (1945). Sideritization is common in the Northampton ironstone (Taylor, 1949, pp. 81–82). Cayeux' studies in particular led to the conclusion that ironstones were essentially metasomatic replacements, and to the rejection of the direct precipitation theory. Chamosite ooids were explained as replacements of crinoidal and other shelly debris. Deverin postulated formation of ooids at one place and their deposition in another. According to Cayeux, the original or primary calcite was altered to siderite, which in turn was replaced by chlorite (chamosite). Subsequent oxidation led to conversion to limonite or even to magnetite and hematite. This sequence of replacements is not inflexible. Deverin, for example, points out that siderite has usually replaced chamosite and that in some rocks hematite has directly replaced the calcite. James (1954, pp. 274–276) has noted that diagenetic reactions are generally in the direction required by a lower *Eh*, such as siderite replacing glauconite. Interstitial waters in which such reactions take place are more reducing than those above the sediment–water interface. Weathering changes go in the reverse direction required by a higher *Eh*.

Hallimond (1925) and most other students of ironstones believe that the theory of a metasomatic origin of various facies of ironstone must be abandoned. Hallimond noted the presence of unaltered calcite shells in a siderite matrix. In other cases, sideritized shells and limestone pebbles, seen in a groundmass of clear calcite, were interpreted to mean that siderization has taken place on the sea floor. The stratigraphic persistence of the iron-bearing beds, the varying facies of iron-formation, the preservation of fine sedimentary features, and the presence of each of the facies as fragments in intraformational breccias and clastic dikes make the subsequent replacement origin of iron-formations seem unlikely (James, 1951, p. 263).

The evidences of replacement (sideritized fossils and the like) cited above are therefore to be considered evidences of later reorganization—early diagenetic in most cases and metamorphic in a few. Siderite in particular seems to be prone to solution and redeposition, because siderite spherulites and siderite metacrysts (Fig. 11-21) appear in various silicate formations and as replacements of fossils in calcareous ironstones.

Weathering and metamorphism may profoundly alter iron-bearing strata. This is especially true of formations in the Lake Superior district where primary iron minerals are oxidized and converted to limonite or hematite and where silica (primarily chert) has been removed by "wholesale" leaching, which converted the rock to ore. Metamorphism has converted the original rock to a complex grunerite-magnetite rock. It is not within the province of this book to discuss either the formation of oxide ore bodies or the metamorphic derivatives of these strata. Refer to original works dealing with these subjects (Leith, Lund, and Leith, 1935; James, 1955; James et al., 1968; French, 1968).

GLAUCONITE

Glauconite-bearing sediments are closely related to silicate iron-formations, although few such sediments are rich enough in glauconite to be designated iron-formation, and fewer still give rise to an iron ore. Glauconite, however, is more widespread than are other iron-bearing silicates and hence is more often encountered. It, therefore, deserves special mention.

As noted, glauconite is a dioctahedral microcrystalline mineral rich in iron and potash. Several structural types have been described by Burst (1958a, 1958b). Roughly one-half of the glauconite molecule is silica, one-fourth iron oxides, one-tenth alumina and magnesia, and one-sixth potash and water. It forms granules, rudely elliptical in section, averaging ½ mm in diameter. These have no well-defined internal structure; they have, however, a polylobate outline and in some cases a system of internal shrinkage cracks. The granules are commonly shiny, green to greenish black to the naked eye, and light greenish yellow to grass green in this section (yellow-brown if oxidized). Exceptionally glauconite forms an envelope or pellicle around grains of collophane, quartz, feldspar, micas, or even heavy minerals (Grim, 1936).

Glauconite granules occur in many quartz-rich sandstones (Fig. 7-25); although found in some feldspathic sandstones, they do not seem to be characteristic of graywackes. They also occur in calcarenites and dolomite derivatives (Fig. 10-6). Granular glauconite, therefore, seems to be deposited in a somewhat current-agitated environment. Glauconite is said to occur finely divided and disseminated in some shales. Our knowledge of the geology and distribution of glauconite has been summarized by Hadding (1932), Cloud (1955), and others (Goldman, 1919; Schneider, 1927; Galliher, 1935a, 1935b; Takahashi, 1939). The mineralogy of glauconite has been investigated by Gruner (1935), Burst (1958a, 1958b), Hower (1961), and Bentor and Kastner (1965); the geochemistry has been summarized by James (1966, pp. 6–7, 23, Table 15).

Greensands and "greensand marls" are the most important glauconite-bearing sediments. The term *greensand* has been applied to unconsolidated glauconite-rich sands. Under the hand lens the best greensands appear to be composed entirely of glauconite, less than 1 percent grains of quartz sand. More commonly quartz is the dominant constituent forming one-half or more of the whole sediment. Sands composed mainly of glauconite are dark to light green; mixed sands have a salt-and-pepper appearance; where oxidized they become red or brown.

Greensands are prominent in Cretaceous and Eocene beds in the Coastal Plain of the eastern United States, especially in New Jersey and Delaware (Ashley, 1918; Mansfield, 1920). Although individual beds seldom exceed 25 feet (7.6 m) in thickness, they are areally extensive and potentially an important source of potash (and iron and phosphorus). Greensands are strongly developed in the Lower Cretaceous (Rastall, 1919) and in the Cambrian sandstones (Hadding, 1932, p. 54) of Europe. The Cambrian of the Upper Mississippi Valley is also glauconite-bearing, as is the Cambrian of Missouri (Allen, 1937), of the Arbuckle Mountains of Oklahoma, and of the Grand Canyon. Precambrian glauconite, although uncommon, has been reported from the Belt Series of Canada (Mudge, 1972) and the Vindhyan sandstones of peninsular India (Auden, 1933, p. 212).

Glauconite appears to be forming in marine environments at present. It has been recovered by dredging in waters of 200 to 300 fathoms or 366 to 550 m (Murray and Renard, 1891, p. 378) and in comparatively shallow waters of 5 to 60 fathoms or 9 to 110 m (Galliher, 1935a, 1935b). It has been dredged from the Atlantic Shelf and slope from Cape Hatteras to Florida in depths of 30 to over 800 m (Ehlmann, Hulings, and Glover, 1963).

Much has been written on the origin of glauconite. The geological aspects of the problem have been reviewed by Cloud (1955), who concluded that it is formed only in marine waters of normal salinity; it requires slightly reducing conditions (weakly oxidizing according to Chilingar, 1956); its formation is facilitated by organic matter; it is mainly characteristic of waters 10 to 400 fathoms (18 to 730 m) in depth; it is formed only in areas of slow sedimentation; and it is chiefly formed from micaceous minerals or bottom muds rich in iron. It has been observed to form by alteration of mud fillings of foraminiferal tests (Ehlmann, Hulings, and Glover, 1963). As noted by Hadding (1932, pp. 82, 145) and others, the place of formation and the place of accumulation may not be the same. Some glauconite is reworked or transported.

Galliher (1935a, 1935b) concluded that glauconite was derived from biotite by a process of submarine weathering. He observed a series of transition grains that demonstrate this transformation. He noted that biotite-rich sands near shore grade horizontally into mixed glauconite-mica silty sands offshore and that these in turn passed progressively into glauconite muds at a depth of 100 fathoms (183 m). Gruner (1935) has shown that the ionic arrangement and the structure of the unit cells of glauconite and biotite were very similar, if not identical, so that the transformation of biotite to glauconite requires no great change.

Although Galliher's observations have been confirmed in other places (Edwards, 1945), many glauconites seem not to have formed from mica (Allen, 1937). Takahashi (1939) says, "glauconitization is one of the processes of submarine metamorphism that gives rise to the mineral glauconite. The phenomenon is known only in marine sediments that are formed under anerobic or reducing conditions. It is usually associated with the presence of iron sulfide, though subsequent reworking may cause the glauconite to be concentrated in sandy deposits without the presence of iron sulfide." In summary Takahashi says, "glauconite seems to be formed under marine conditions by a process of hydration of silica and subsequent absorption of bases and loss of alumina. Glauconite may originate from a number of mother materials, such as faecal pellets, clayey substances filling cavities of foraminifera, radiolaria, and tests of other marine organisms, or from silicate mineral substances, such as volcanic glass, feldspar, mica, or pyroxene. The presence of organic matter seems to facilitate the formation of glauconite. In salt water . . . the mother substances during glauconitization lose alumina, silica, and alkalies except potash, and gain ferric iron and potash. Sea water, therefore, seems essential. . . ."

It is of interest that chamosite and glauconite generally seem to be mutually exclusive. Hunter (1970) noted that the chamosite-bearing ironstones of the Clinton gave way seaward to glauconitic sediments. The rare joint occurrence of these two minerals has been attributed to redeposition of glauconite or in other cases to formation of these two minerals at different times and under differing oxidation–reduction states (Chilingar, 1956)—the glauconite under weakly oxidizing conditions and the chamosite under more reducing conditions.

PHOSPHORITES AND OTHER PHOSPHATIC SEDIMENTS

INTRODUCTION

Nearly all sediments contain phosphorus in small amounts; some contain exceptional amounts. The latter are described as phosphatic, such as phosphatic limestone or phosphatic shale. The average shale, for example, contains 0.17 percent P_2O_5, the average limestone, 0.04 percent (Clarke and Washington, 1924). Most of the phosphate is present in organic skeletal structures, especially those of certain phosphatic brachiopods, crustaceans, and bones and teeth of vertebrates (Clarke and Wheeler, 1922). The phosphatic content of sediments is quite variable and in some is abnormally high; some limestone, for example, may contain several percent P_2O_5.

Rocks containing more than 19.5 percent P_2O_5 (about 50 percent "apatite") are defined as *phosphorites*; if they contain more than 7.8 percent P_2O_5 (about 20 percent "apatite"), they are described as *phosphatic* (Cressman and Swanson, 1964, p. 282). Most writers, however, would probably designate rocks as phosphatic if the P_2O_5 content were one or two orders of magnitude greater than normal. Phosphorites may be comparatively free from contaminating minerals. In some, however, phosphatic minerals form a cement for an assemblage of detrital minerals, or the phosphatic constituent is mixed with other materials.

Whereas the term *phosphorite* has been used for those sediments in which a phosphate mineral is the principal constituent, other terms such as *rock phosphate, bedded phosphate,* and the like have also been applied to these rocks. Distinctions are also made between those rocks which were initially highly phosphatic and those which have been *phosphatized* much later. Some limestones, for example, have been altered by phosphate-bearing solutions and converted to a phosphate rock just as some limestones have been altered by silica-bearing waters and become silicified. So also a distinction is commonly made between sediments containing phosphatic nodules, the *nodular phosphates,* and the *bedded phosphates,* a distinction analogous to that between nodular and bedded cherts. *Residual phosphate* is a term applied to a surface accumulation of insoluble phosphatic materials left as a residuum from the solution of limestone in which it was once dispersed. It is analogous to a residual chert accumulation, and, like the chert gravels which are a residuum reworked by streams, we also have *pebble phosphates* formed in the same way. A special type of phosphatic accumulation, which has no chert analogue, is *guano,* which is generally a bird manure found on some desert islands of the eastern Pacific Ocean and the West Indies. *Bone beds* or *bone phosphate* are another organic residue which, as the name suggests, are an accumulation of vertebrate (generally fish) bones of sufficient volume to constitute a bed and hence a phosphate deposit.

In summary, the terminology reviewed suggests the usual genetically based classification:

Organic remains: bone beds
Organic excrement: guano
Metasomatic phosphate: phosphatized limestone
Residual phosphate
Resedimented phosphate: pebble phosphates
Phosphatic nodules: nodular phosphates
Bedded phosphate: phosphorite, rock phosphate and the like

Most of these deposits, except the last one, are small, local accumulations (Fig. 11-29).

MINERALOGY AND CHEMICAL COMPOSITION

Various studies have shown the composition of the phosphorites to be complex (Lacroix, 1910; Schaller, 1912; Rogers, 1922, 1944; Hendricks, Jefferson, and Mosely, 1932; Bushinsky, 1935; McConnell, 1950). McConnell (1950) lists 38 phosphate minerals known in rock phosphates, many of which are rare. The mineral components of phosphorites are difficult to study because of their submicroscopic crystals and admixtures of fine-grained impurities. Crystal-structure studies show that isomorphic replacements are common, which explains why the mineral composition of phosphorites is so poorly defined and understood. The term *collophane* (Rogers, 1922) applied to this mineral complex is probably not a true mineral species but may be no more than a convenient name for a group of closely related minerals.

The most common components are the phosphates of calcium, especially the several varieties of apatite:

Fluor-apatite, $Ca_{10}F_2(PO_4)_6$
Chlor-apatite, $Ca_{10}Cl_2(PO_4)_6$
Hydroxy-apatite, $Ca_{10}(OH)_2(PO_4)_6$
Oxy-apatite (voelckerite), $Ca_{10}(PO_4)_6$
Carbonate-apatite, $Ca_{10}CO_3(PO_4)_6$

The general formula may be written $Ca_{10}(PO_4, CO_3)_6(F, Cl, OH)_2$. The structure of apatite favors a wide variety of minor substitutions (McConnell, 1938). For example, VO_4, As_2O_4, SO_2, SO_4, and CO_3 may be substituted for equivalent amounts of PO_4; the F position may be partly or completely occupied by F, Cl, or OH; and minor amounts of Mg, Mn, Sr, Pb, Na, U, Ce, and Y and other rare earths may substitute for Ca. These substitutions explain why the phosphates have unusual amounts of such elements as vanadium and uranium and explain also the keen interest in their occurrence in the atomic age.

Apatite is the primary mineral, but a number of others are common in deposits formed during weathering of phosphate rock and guano (McKelvey, 1967, p. 4).

The carbonate-fluorapatite of most marine phosphorites consists of microcrystalline particles, 1 or 2 microns in size, is weakly birefringent, and shows a "pinpoint" extinction not unlike that of fine-grained chert. The material varies from amber to black. Normally it consists of pellets and oolitic bodies which in some cases are surrounded by banded zones of crystalline fibers. Many phosphatic skeletal grains appear to be composed of a light yellow to brown isotropic material. Others show a

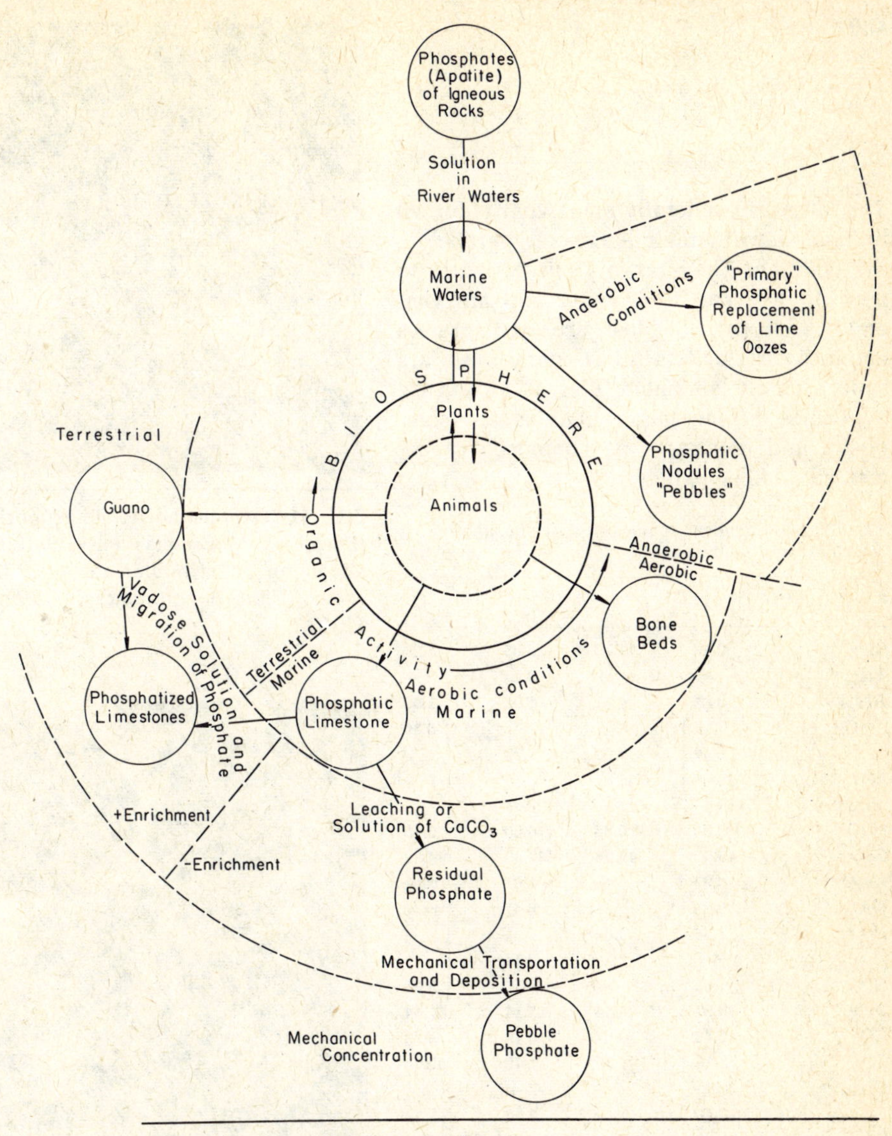

FIG. 11-29. The phosphorus cycle.

weak irregular birefringence caused perhaps by orientation of minute crystals.

Chemical analyses of phosphorite show the material to be mainly hydrous tricalcium phosphate with varying amounts of calcium carbonate and fluoride. Because of admixtures of nonphosphatic materials, such as calcite, dolomite, and chalcedony as cement and also detrital contaminants such as quartz and clay, analyses are highly variable (Table 11-5). Hydrocarbon commonly is included in phosphorite and is responsible for the fetid odor when phosphorite is struck with a hammer.

PETROLOGY

Most of the bedded primary phosphates are black. Even the so-called "brown" phosphates of Tennessee have black phases. Secondary concentration by meteoric waters, on the other hand, are white, yellow, or, more rarely, brown.

In bedded phosphates, part of the phosphate is in the form of interstitial cement, and in a few layers most of the phosphate is in phosphatic brachiopods and fish scales. The great bulk of the phosphatic material is conglomerated into pellets and nodules.

The latter are generally somewhat elliptical in section, with their long axes parallel to the bedding. They range from 0.05 mm to more than 3 cm in diameter and are generally well sorted. Most are structureless granules or "pellets," but many also are concentrically laminated (Figs. 11-30 and 11-31). Some of the larger nodules are compound and appear to be composed of cemented smaller pellets.

TABLE 11-5. Chemical composition of phosphatic sediments and nodules

Constituent	A	B	C	D
SiO_2	0.46	11.70	36.65	1.21
TiO_2	—	—	0.16	—
Al_2O_3	0.97	4.11	1.02	1.30
Fe_2O_3	0.40 }	3.75	1.77	{ 8.36
FeO	— }			{ —
MnO	—	—	0.04	—
MgO	0.35	0.84	0.50	0.10
CaO	48.91	40.96	29.43	40.38
Na_2O	0.97	—	1.01	trace
K_2O	0.34	—	0.47	trace
H_2O+	1.34	3.65[b] }	3.14[c]	{ —
H_2O-	1.02	— }		{ 3.02
P_2O_5	33.61	23.54	17.14	21.44
CO_2	2.42	10.64	5.19	9.20
SO_3	2.16	1.39	1.48	none
Cl	trace	—	—	trace
F	0.40	—	2.08	1.52
S	0.40	—	0.05[d]	—
Organic matter	nd	—	0.58[e]	0.37
Total	95.97[a]	100.18	99.25	100.14[f]

[a] Includes 2.62 percent "insoluble."
[b] Ignition.
[c] Includes "organic" substances.
[d] FeS_2.
[e] Carbon.
[f] Includes 13.24 percent "insoluble in HCl."

A. Phosphate bed, Phosphoria Formation (Permian), Cokeville, Wyoming. George Steiger, analyst (Clarke, 1924, p. 534).
B. Phosphate nodule from sea floor, 1900 fathoms. C. Klement, analyst. (Murray and Renard, 1891, p. 383). Recalculated from separate analyses of soluble and insoluble portions.
C. Upper phosphorite stratum (Cenomanian), Briansk, Russia (Bushinsky, 1935, p. 90).
D. Oolitic phosphate, Modelo Formation (Miocene), California. J. G. Fairchild, analyst (Hoots, 1931, p. 106).

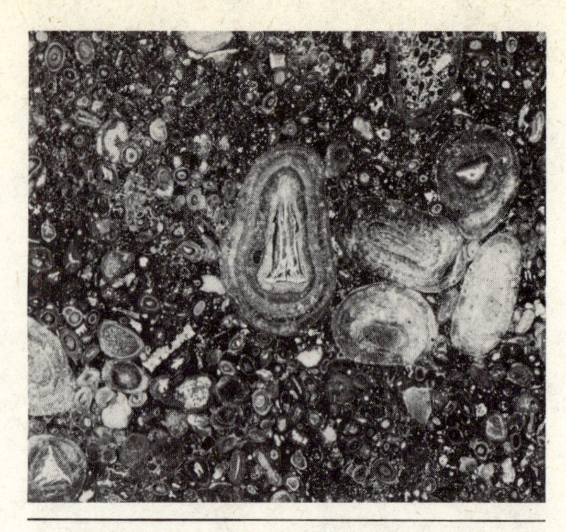

FIG. 11-30. Phosphorite, Phosphoria Formation (Permian), McDouglas Pass, Salt River Range, Wyoming. ×3. Nodular and pisolitic phosphate rock. (U.S.G.S. photograph. Mansfield, 1927, Pl. 63.)

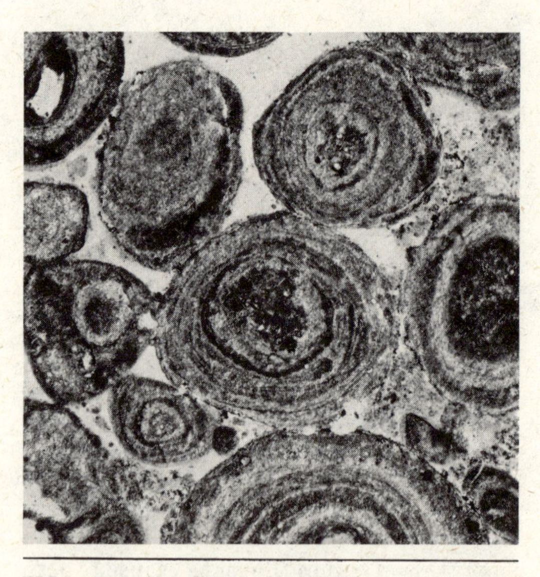

FIG. 11-31. Oolitic phosphorite, Phosphoria Formation (Permian), Montana. Ordinary light, ×22. Phosphatic ooids with chert matrix.

The phosphatic rocks are generally well cemented with carbonate-fluorapatite, argillaceous matter, chert, calcite, or dolomite. In the Phosphoria Formation (Permian) of Montana, phosphorite occurs in layers, ranging from a millimeter or two to several meters thick (McKelvey, Swanson, and Sheldon, 1953). Most of them are a few millimeters thick and are interbedded with less phosphatic mudstones or carbonate rocks.

Phosphatic nodules or "pebbles" are found not only in phosphate deposits themselves but are widely scattered in some limestones (Pettijohn, 1926) and especially in the Cretaceous Chalk (Fisher, 1873) and other sediments (Adams, Groot, and Hiller, 1961). They also occur on the present sea floor (Murray and Renard, 1891, p. 396; Dietz, Emery, and Shepard, 1942). These objects vary from small granules to pebble-like objects several centimeters in diameter. They are typically black, are of irregular form, and have a hard shiny surface. Larger nodules contain much foreign matter, including sand grains, mica flakes, shell debris, and sponge spicules. The black color is most intense near the outer rim.

Metasomatic phosphates—those formed by replacement of limestones by phosphate-bearing solutions—are apt to show relict textures of the rocks replaced in a manner similar to that shown by silicified limestones.

ASSOCIATIONS AND GEOLOGIC OCCURRENCE

Marine phosphorites commonly are associated with glauconite or greensand. This is true of nodules now forming (Murray and Renard, 1891, p. 391; Dietz, Emery, and Shepard, 1942, p. 820) and of the Cretaceous nodule layers of the English Chalk (Fisher, 1873) and of the Russian deposits (Bushinsky, 1935). It is also true of the Tennessee phosphates (Hayes, 1896) and the Cambrian nodule layers of southern New Brunswick (Matthew, 1893, p. 109). This association has been pointed out by Cayeux (1905) and Goldman (1922).

Phosphate beds have been said to occur at major and minor unconformities; the phosphatic and glauconitic materials have even been cited as criteria for such surfaces (Grabau, 1919; Goldman, 1922). Pettijohn (1926) interpreted zones rich in phosphatic granules or "pebbles" as the residuum on a "corrosion surface" or diastemic plane caused by submarine solution, and it is probable that unconformities related to phosphatic and glauconite materials are submarine unconformities—surfaces of nondeposition rather than erosion or exposure. The extreme slowness of sedimentation is shown by the phosphorites of Morocco, where a few meters of accumulation record, without breaks, the whole of time from the Cenomanian (Upper Cretaceous) to the Lutetian (Eocene).

Bones, shark teeth, fish scales, and the remains of *Lingula* and other phosphatic brachiopods and of trilobites—all of which are highly phosphatic—are common in some phosphate accumulations. In a few places bone beds form significant deposits—the so-called bone beds.

Phosphorite occurs in rocks of nearly all ages and is probably more common than has been supposed. One of the best known and most extensive are the bedded phosphates of the Phosphoria Formation (Permian) of Utah, Idaho, Wyoming, and Montana and adjoining parts of Colorado and Nevada (Mansfield, 1918, 1927; McKelvey et al., 1956, 1959). The Phosphoria underlies an area of at least 135,000 miles^2 (225,650 km^2). It contains 20 or more phosphate-rich beds having an aggregate thickness of 73 feet (22 m); the thickest bed is 7 feet (2.1 m). The phosphate beds are in part oolitic, pelletoidal, and arenaceous with chalcedonic cement and in part carbonaceous, pyritic, phosphatic mudstones. Individual beds can contain 30 percent or more P_2O_5. These are associated with black shales and bedded cherts. Other well-known and extensive phosphorites include those of Cretaceous and Eocene age in western and northern Africa. These and other occurrences have been briefly summarized by McKelvey (1967) and Twenhofel and Blackwelder (1932, pp. 546–561) and were the subject of a session at the 19th International Geological Congress in Tunis (1953).

Phosphatic deposits are seemingly rare in Precambrian rocks presumably because of the general lack of animals with phosphatic skeletal materials. Abiotic precipitation was thought not to form large accumulations (Geijer, 1962). Phosphatic nodules are, however, known in the Precambrian Torridonian of Scotland (Downie, 1962).

ORIGIN

Various theories have been promulgated to explain phosphatic deposits (Fig. 11-29). Most of these are inadequate and are of historic interest only. The phosphate beds have been considered, for example, as accumulations of coprolites of fish and higher animals. The deposit, therefore, would be a kind of submarine guano. Nondeposition caused by failure of land-derived sediment is invoked to explain the concentration of coprolitic materials (Fisher, 1873). Other writers have supposed that unfavorable conditions for the formation of carbonate of lime permitted the accumulation of phosphatic hard parts of organisms during a period of nondeposition (Miller, 1896). Hayes and Ulrich (1903) thought that the black Devonian phosphates of Tennessee were formed by the mechanical reworking by the sea of residual concentrations of known Ordovician phosphates of the same area. Fisher (1873) explained the coprolitic beds of the Gault of England as phosphatized sponges.

None of the above theories explains all of the observed features of areally extensive bedded phosphorites. Phosphate is known to be extracted from sea water by living organisms and concentrated within their hard parts. Abnormal concentrations of such bones, teeth, fish scales, or phosphatic invertebrate hard parts has produced phosphate-rich beds. However, the relative role of biogenic versus abiotic precipitation of phosphorus is not yet clear. The close association of phosphorites and organic matter, and the abundance of phosphatic skeletal structures, suggest a cause-and-effect relation. The general absence of Precambrian phosphorites and their appearance in the Phanerozoic with the first appearance of skeletal-secreting organisms further suggest a biotic causation. On the other hand, phosphorites and organic materials, like associated cherts, may be only products of the same environment. Phosphatic granules and pellets have been ascribed to a fecal origin. But as noted by McKelvey and coworkers, their wide range in size, good sorting in individual layers, regional trends in average size, and compound nature of the larger ones, as well as the oolitic structure of many, do not favor an organic origin. It has been suggested that ammonium phosphate generated by decay of pelagic organisms may be the agent of precipitation of the phosphate (Blackwelder, 1916, p. 293).

Fossils prove the phosphorites to be marine. The black color and commonly associated hydrocarbon materials indicate anaerobic conditions. The absence of fossils of sessile and bottom-living types and the presence of depauperized forms, as well as pyrite and associated black shale, further confirm this interpretation. The scarcity of carbonate of lime either in skeletal form or as precipitated carbonate and the presence of much chert suggest a *pH* slightly less than normal.

Kazakov (1937) observed that phosphorites form mainly in the border zone between shallow-water platform sediments and deep-water geosynclinal accumulation. Platform phosphorites are nodular in habit and are associated with glauconite and arenaceous materials. Geosynclinal phosphorites are bedded, platy, rich in P_2O_5, and, as noted by McKelvey and others, associated with black shales and cherts. This general facies concept seems to be substantiated by the studies of the stratigraphy of the Permian Phosphoria and associated formations (Fig. 11-32).

Although phosphorus, like iron, is present in sea water only in small quantities, it will be precipitated in important quantities locally if conditions are favorable. At a particular *Eh* and *p*H either will accumulate relatively free of the deposits of calcium. Similarities of many microstructures (granules and oolites) and gross lithologic associations (bedded cherts and black shales), along with the co-occurrence of iron silicates (glauconite) with some phosphatic deposits and abnormally high concentrations of P_2O_5 of some iron silicates and iron carbonates, especially basinal iron-formations (Table 11-3F and G), suggest that the environment of phosphorite and ironstone deposition have many factors in common.

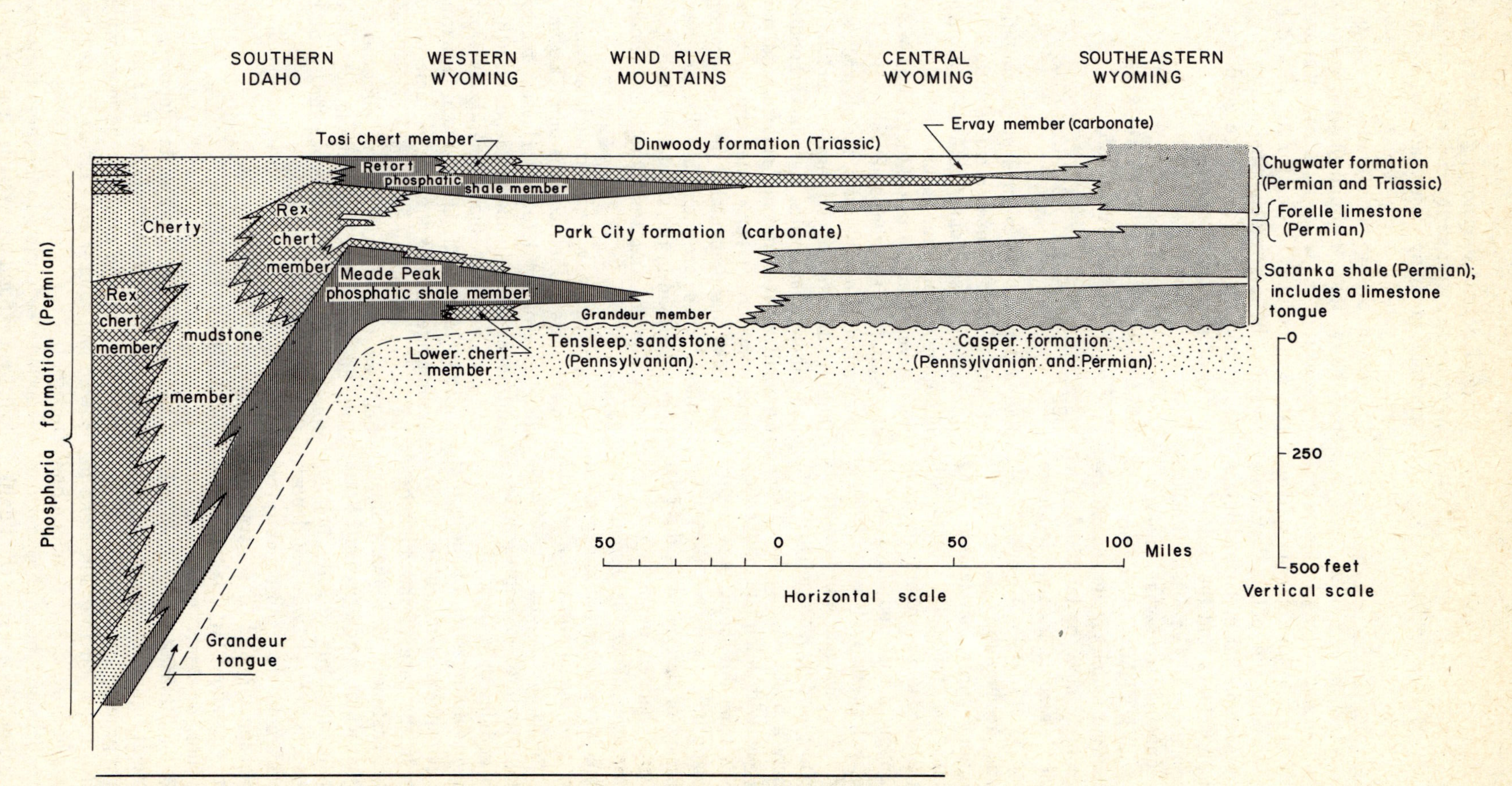

FIG. 11-32. Stratigraphic relations of the Phosphoria, Park City, and Chugwater formations in Idaho and Wyoming. (After McKelvey and others, 1959, Fig. 1.)

Each displays two contrasting facies: a cratonic or platform facies (nodular phosphates with glauconite and the oolitic ironstone facies) and a geosynclinal or basinal facies (the bedded phosphates and the bedded or nonoolitic ironstones). The basinal facies seems to require a somewhat anaerobic milieu, one with a slightly lower than normal *p*H and one in which clastic sedimentation is exceedingly slow or inhibited. The platform facies is aerobic, is somewhat more turbulent, and is apt to be arenaceous although clastic accumulation is small. If these conditions are maintained long enough, circulation of oceanic waters will supply all the needed phosphorus or iron. It will also supply the needed silica to produce the bedded cherts which are characteristic of both Precambrian iron-formations and some phosphorites. The unusual conditions, namely a basin with impeded convection and yet with adequate connection with an inflow from the ocean, plus greatly restricted clastic influx and crustal stability for a long period of time, explain the relative scarcity of bedded deposits of both iron and phosphorus. When these conditions are not fully achieved, the iron is deposited in a scattering of glauconite granules, and the phosphorus is precipitated in invertebrate skeletal structures or is present as a few isolated nodules or pellets.

There remains the question of the actual process of phosphate sedimentation. We may ask, as in the case of the dolomites, cherts, and ironstones: Was the primary component that which is now present? Were the phosphatic materials directly precipitated from sea water? Or was the deposit originally calcareous and subsequently replaced by phosphate? And, if so, was the replacement synsedimentary, taking place on the sea floor, or was it epigenetic and took place after lithification and uplift? In general, current opinion favors the first view, namely primary precipitation of the phosphate. No transition from limestone to phosphate has been seen. Nor does one find shelly beds which have been phosphatized. The pellets and ooliths are somewhat unlike those of normal limestones; they are somewhat oblate, with shrinkage cracks, without structure in many cases, and with asymmetric layers in some. These do not look like replaced carbonate ooids. There is a possibility that the phosphorus was extracted from sea water by a biological agent (algae, according to Charles, 1953), and released to be later precipitated abiotically as oolites and granules. Evidence on this matter is at present inconclusive.

SALINES AND OTHER EVAPORITES (EXCLUDING CARBONATES)

DEFINITIONS AND CLASSIFICATIONS

Saline deposits are formed by precipitation of salts from concentrated solutions or brines. Because concentration is brought about by evaporation, saline deposits have been termed *evaporites*. Most common are the sulfates, gypsum and anhydrite; less common are the chlorides, mainly rock salt (halite); and relatively rare are polyhalite and other potash salts. Certain carbonates, especially travertine and caliche, and perhaps some oolitic carbonates and dolomites may also be precipitated by evaporation. Although these are chemical salts, they have been described in the chapter on limestones and dolomites and are, therefore, excluded here.

Evaporites may be classified, as above, on the basis of their composition. They may also be grouped into two classes: marine and nonmarine. Nonmarine deposits include sodic deposits of such minerals as trona, $Na_2CO_3 \cdot NaHCO_3 \cdot 2H_2O$, not found in marine assemblages but as the salt of some inland lakes. Because many nonmarine deposits are believed to be derived from airborne sea salt, the composition of some is not unlike those of marine origin; hence they contain both gypsum and halite.

MODERN EVAPORITES

Although modern evaporite deposits are thin and cover relatively small areas, they shed light on the origin of the more widespread and thicker deposits of the geologic record.

Because evaporites form where the water lost by evaporation exceeds that supplied by rainfall, they are restricted to arid regions. Many are forming in inland basins on the lee side of mountains that cross the paths of the prevailing winds. These are the nonmarine evaporitic basins some, of which contain permanent brine lakes such as the Great Salt Lake of Utah (Eardley, 1938) and the Dead Sea (Neev and Emery, 1967), whereas others have ephemeral lakes or *playas* such as those of Saline Valley (Hardie, 1968), Death Valley, Deep Springs Lake (Jones, 1965), and Searles Lake (Smith and Haines, 1964; Eugster and Smith, 1965), all in California. Some, such as those of Saline Valley, are little more than efflorescent crusts (up to 1 m thick) and salt-rich muds and sands. Some, such as Searles Lake, are sodic and trona-depositing. The salts in these basins are in large part precipitated in the playa from brines drawn upward to the surface by capillary action.

A marginal marine environment analogous to the continental playa which has received increasing attention in recent years is the *sabkha.* The sabkha is a coastal salt flat whose type example is found in the Persian Gulf region (Kinsman, 1969). For the most part salt deposition on these flats is the result of evaporation of brines drawn to the surface by capillary action and precipitated in the interstices of the sediment or as an efflorescent crust.

True marine evaporites form by partial or complete isolation of the evaporating body of water from the open sea under conditions of marked aridity. The sites which thus qualify are marginal salt pans, marine salinas, lagoons, and relict seas (Grabau, 1920). The Runn of Cutch in northwest India is an oft-cited example of a modern marginal salt pan. Here some 7,000 square miles (18,130 km) are flooded annually and in time are evaporated to form a salt crust several feet thick. There have been almost no modern studies of the Runn of Cutch, and the earlier interpretations have been challenged (Platt, 1962). Marine salinas receive sea water by percolation through a continuous permeable barrier. The Lake of Larnace on Cyprus (Bellamy, 1900) is an example. The deposits of salt pans and salinas do not reach any great thickness. Those of relict seas and lagoons may be much larger bodies. A modern example of a salt lagoon is the Gulf of Karaboghaz on the east side of the Caspian Sea and separated from it by two narrow sand spits, between which is a shallow channel no more than a few hundred meters wide. Approximately 130 $\times$ 10^6 tons of salt are carried into the gulf each year, and gypsum halite, and various sulfates of magnesium and sodium are deposited (Urasov and Polyakov, 1956). The deposits of the Great Bitter Lake of Suez (Grabau, 1920, pp. 139–142) are a recent example of a marine lagoonal deposit wherein up to 20 m of salts were deposited over an area of 80 km^2.

COMPOSITION OF EVAPORITES

More than 80 mineral species (excluding clastic materials) have been reported from marine evaporitic deposits. Most of these are chlorides, sulfates, carbonates, and borates (Stewart, 1963, pp. 6–7). Only about 12 rank as major constituents (Table 11-6). Many are of secondary origin. In nonmarine evaporites other species, such as trona ($Na_2CO_3 \cdot NaHCO_3 \cdot 2H_2O$), mirabilite ($Na_2SO_4 \cdot 10H_2O$), and glauberite ($Na_2Ca(SO_4)_2$) may be major constituents.

TABLE 11-6. Major mineral constituents of marine evaporites

Chlorides	
Halite	$NaCl$
Sylvite	KCl
Carnallite	$KMgCl_3 \cdot 6H_2O$
Sulfates	
Anhydrite	$CaSO_4$
Langbeinite	$K_2Mg_2(SO_4)_3$
Polyhalite	$K_2Ca_2Mg(SO_4)_6 \cdot H_2O$
Kieserite	$MgSO_4 \cdot H_2O$
Gypsum	$CaSO_4 \cdot 2H_2O$
Kainite	$KMg(SO_4)Cl \cdot 3H_2O$
Carbonates	
Calcite	$CaCO_3$
Magnesite	$MgCO_3$
Dolomite	$CaMg(CO_3)_2$

Source: Stewart, 1963, U.S. Geol. Survey Prof. Paper 440-Y, Table 2.

Because of the high solubility of most of these minerals, they are seldom seen in outcrop except in the most arid regions. Moreover, deposits of these very soluble minerals are relatively rare. Only the sulfates, gypsum and anhydrite, are apt to be encountered by the field geologist. Both may occur in a relatively pure state as a chemical deposit of considerable thickness and extent. Anhydrite is known also as a minor authigenic constituent of sandstone; gypsum forms authigenic crystals or crystal aggregates in some clays and shales. Both occur as nodular bodies or as isolated crystals in some dolomites, although in many cases they appear only as calcitic pseudomorphs in these rocks.

Inasmuch as anhydrite is converted to gypsum by hydration, the sulfate seen in outcrop or near-surface situations is apt to be gypsum ($CaSO_4 \cdot 2H_2O$). It is a colorless, monoclinic mineral with 010 cleavage, low specific gravity (2.32), low hardness (1.5 to 2.0), low indices (1.530, 1.523, and 1.520), and low birefringence (0.010). Anhydrite ($CaSO_4$) is also a colorless mineral, is orthorhombic with pinacoidal cleavage (001, 010, and 100), with a somewhat higher density (2.93) and higher refractive indices (1.614, 1.575, 1.570) and birefringence (0.044) (Larsen and Berman, 1934, pp. 101, 107). The hemihydrate ($CaSO_4 \cdot \frac{1}{2}H_2O$) is also reported.

Other salt minerals are so water-soluble that they neither crop out in humid regions nor appear in ordinary thin sections. Rocks suspected of being halite-bearing must be sectioned by special processes. Other chlorides and sulfates occur in the potash-bearing salt beds. These minerals are so rare, however, that there is no need to review their properties here. Refer to special papers dealing with this group of minerals (Clarke, 1924, pp. 222–229; Stewart, 1963, pp. 6–11; Braitsch, 1962).

Neither halite, nor gypsum or anhydrite, is found in nature in a state of absolute purity. Study of the water-insoluble residues of salt beds sheds much light on their minor constituents. Salt from the salt domes of Louisiana (Taylor, 1937) contains 5 to 10 percent water-insoluble material. Of the insolubles, 99 percent is anhydrite in the form of cleavage fragments and euhedra. The carbonates are common as crystal euhedra, especially dolomite and calcite. Rarer constituents identified include pyrite, quartz crystals, limonite, hematite, hauerite, sulfur, celestite, marcasite, barite, kaolinite, gypsum, danburite, and boracite.

The bulk chemical composition of the evaporitic sulfate rocks is given in Table 11-7.

PETROGRAPHY OF EVAPORITES

Rock gypsum varies from coarsely crystalline to fine crystalline granular; the latter is the most common. The manner of occurrence of gypsum is quite variable and is in part related to the primary processes of deposition and in part also to post-depositional diagenesis. In some cases, gypsum is thinly laminated and may be interlaminated with dolomite. In Sicilian Upper Miocene deposits, many laminations show both normal and reversed grading; in places cross-lamination is present, all of which are hydrodynamic features (Hardie and Eugster, 1971). In places the laminated gypsum has apparently been broken up and redeposited as an intraformational conglomerate.

Gypsum also occurs in some deposits as scattered nodules, generally in a carbonate matrix (Fig. 10-7). The nodules, 1 or 2 cm in size, may be scattered strung out along the bedding, or may coalesce to form irregular "wavy"-bedded gypsum layers or be so close-packed as to form a massive bed, the partings between the nodules forming a netlike or "chicken-wire" pattern (Bosellini and Hardie, 1973).

Gypsum also occurs in large crystals, of the selenite variety. In Sicilian deposits these form spectacular crystals showing "swallow tail" twinning, are up to 1 m in length, and tend to be oriented more or less perpendicular to the bedding according to Mattura's rule (Ogniben, 1957). Hardie and Eugster consider these crystals to be primary, although Ogniben believed them to be a replacement of an original sulfate, probably anhydrite. In some deposits gypsum may have a pseudoporphyritic or pseudo-ophitic texture. A pseudoporphyritic texture is characterized by large prismatic crystals embedded in a finely crystalline groundmass of the same or different material; the pseudo-ophitic texture is marked by large platy crystals which enclose small, well-formed euhedra. Larger selenite crystals probably are of later origin than the matrix in which they are found. They are of a porphyroblastic character rather than being phenocrysts (Holliday, 1970).

Gypsum also occurs in veins both in rock gypsum and in associated beds. These veins commonly show a cross-fiber structure and in rare cases a cone-in-cone structure. Large gypsum euhedra and rosettes also occur in some muds and shales. These occurrences are probably authigenic and are formed in muds after deposition (Masson, 1955).

Gypsum commonly appears to be formed by hydration of anhydrite. In some cases this process involves an increase in volume of 30 to 50 percent. The resulting swelling produces notable effects, such as the enterolithic folding of thin anhydrite layers enclosed in rock salt or other beds. Local crumpling and intense folding of hydrate layers may take place without much effect on the enclosing strata (Fig. 11-33). Kirkland and Anderson (1970), however, interpret the folding as tectonic and not related to volume change.

Anhydrite, like rock gypsum, occurs in beds which in some places are thick and extensive. It may be delicately laminated. In the Castile Formation (Permian) in the Delaware Basin of Texas and adjoining New Mexico these laminations average 1.6 mm

TABLE 11-7. Chemical composition of evaporitic sulfates

Constituent	A	B	C	D	E	F
SiO_2	2.20	0.40	0.10	—	trace	—
Al_2O_3	0.20 (Al_2O_3 + Fe_2O_3)	2.97	0.12	0.14 (Al_2O_3 + Fe_2O_3)	trace	0.03
Fe_2O_3		0.77	—			
MgO	2.11	1.53	0.33	0.24	—	0.02
CaO	36.76	30.76	32.44	38.46	40.61	42.64
Na_2O	—	—	—	0.07	—	—
K_2O	—	—	—	0.19	—	—
SO_3	36.11	43.70	45.45	39.53	56.82	51.52
CO_2	6.43	2.80	0.85	7.73	—	5.15
H_2O	16.27	17.53	20.80	12.69	1.87	—
Organic	—	—	—	—	0.46	0.05[b]
Total	100.00	100.54	100.09	99.54[a]	99.76	99.86[c]

[a] "Insoluble," 0.45.
[b] Soluble in chloroform.
[c] Includes 0.28 insoluble.

A. Gypsum (Silurian), Caledonia, Ontario, Canada (Caley, 1940, p. 113). Gypsum calculated, 77.67%.
B. Gypsum (Silurian), New York. G. E. Willcomb, analyst (Stone et al., 1920, p. 214). Gypsum calculated, 94.26%.
C. Gypsum (Triassic), east of Cascade, Black Hills, South Dakota. George Steiger, analyst (Stone et al., 1920, p. 248).
D. Gypsum (?Jurassic), Nephi, Utah. E. T. Allen, analyst (Clarke, 1924, p. 232). Calculated composition: calcite 17.5%, gypsum 60.5%, and anhydrite 19.3%.
E. Anhydrite rock (?Permian), Gypsum, Colorado. J. G. Fairchild, analyst (Clarke, 1915, p. 357).
F. Anhydrite-carbonate rock, Caesar Grande Well no. 1, Eddy County, New Mexico. E. T. Erickson, analyst (Wells, 1937, p. 120). Calculated composition: anhydrite, 88.0%; calcite, 11.7%.

in thickness and are separated from one another by a thin, brown, bitumen-rich film. Individual laminations have been correlated from one drill core to another up to as much as 70 miles (113 km) apart (Anderson et al., 1972). They have been interpreted as annual layers or varves (Udden, 1924; Anderson and Kirkland, 1960, 1966). A similar interpretation has been made of the laminations in other evaporites (Richter-Bernburg, 1964). Anhydrite beds are commonly finely granular, although fibrous and coarsely crystalline masses are also known. In some places crystals of gypsum are scattered throughout anhydrite, thereby giving the rock a porphyritic appearance (Fig. 11-34).

Because of their susceptibility to solution and reprecipitation, evaporites have undergone many large-scale and often complex secondary changes so that it is difficult to know what the primary mineralogy and textures were (Murray, 1964). This is especially true of gypsum and anhydrite. Which was the primary sulfate mineral? Some geologic evidence suggests that much gypsum is derived from anhydrite. Supporting this view is the observation that gypsum beds in outcrop grade into anhydrite at depth (West, 1964), that gypsum occurs as veins and penetrates anhydrite in an irregular manner, and that small patches of anhydrite occur in gypsum. The porphyroblastic occurrence of selenite crystals in anhydrite further supports the view that gypsum is secondary (Holliday, 1970). These observations have led to the notion that the original precipitate was anhydrite. It has been suggested, however, that if the original precipitate were gypsum, it might, after formation, sink into a lower, denser bottom brine and there be converted to anhydrite before burial (Hollingsworth, 1948). But the matter is not so simple. Pseudomorphous structures show that gypsum was once abundant in the sulfate zones of the Salado Formation (Schaller and Henderson, 1932). Similar pseudomorphs show that some of the thick sulfate zones in the English Zechstein originally consisted of gypsum but are now entirely replaced, mainly by anhydrite (Stewart, 1949, 1953). Similar pseudomorphs of anhydrite after early gypsum are also known from the German Zechstein (Borchert and Baier, 1953). It is noteworthy that the only sulfate forming in modern evaporites, with one possible exception, is gypsum (Hardie, 1967, p. 189).

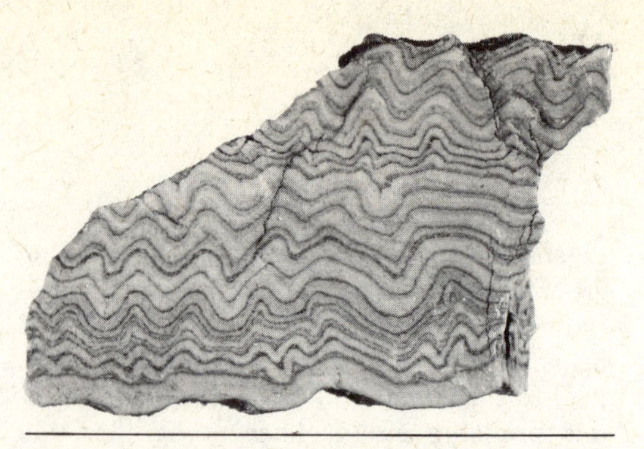

FIG. 11-33. Laminated gypsum showing enterolithic folding, Castile Formation (Permian), Texas. Laminations are considered to be annual.

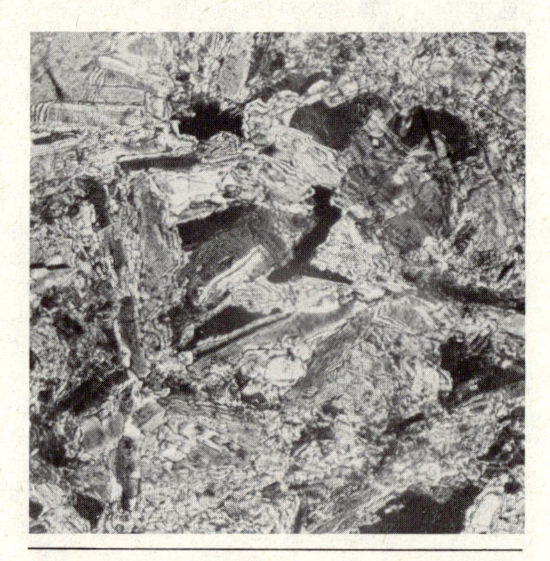

FIG. 11-34. Gypsum from Chugwater Formation, Little Sheep Mountain, Wyoming. Crossed nicols, ×22.

Considerable experimental work has been done on the gypsum–anhydrite equilibrium. The most recent work, in which the earlier studies have been summarized, is that of Hardie (1967). Hardie showed that, although anhydrite could be synthesized (by dehydration of gypsum) at one atmosphere pressure and under geologically reasonable conditions of temperature and activity of H_2O

in a geologically reasonable time, primary precipitation of anhydrite could not be achieved. Gypsum, if held in the stability field of anhydrite, will be converted to anhydrite. Hardie's work showed that a considerably higher temperature than heretofore suggested was required for anhydrite formation. These observations suggest that gypsum would be the first-formed sulfate on evaporation of natural waters.

The problem of metasomatic changes in evaporites is not confined to relations between gypsum and anhydrite. Included here are replacement of carbonate rocks by anhydrite and of sulfate rocks by halite and polyhalite. Many complex replacement sequences have been worked out and have been summarized by Stewart (1963, p. 41).

Halite, or rock salt, is a massive, coarsely crystalline material without joints; in some deposits it is laminated. Layers of salt several centimeters thick are separated by partings of anhydrite or dolomite. In some cases layers of dark salt are interbedded with white salt; in other cases cloudy salt is interbedded with clear salt. The dark salt is filled with anhydrite inclusions; the cloudy salt is marked by an abundance of liquid inclusions. The salt crystals may be well-defined cubes; in other cases the crystals are hopper-shaped. According to Dellwig (1955), the hopper-shaped crystals were formed at the surface of the brine, whereas the cubic habit is a product of growth at the depositional interface between the salt accumulation and its supernatant brine.

Study of fluid inclusions and enclosed bubbles in the Salina salt of Michigan indicates a temperature of crystallization of 32°C to 48.4°C (Dellwig, 1955). Because these determinations were made only on inclusions in crystals presumed to have formed at the surface of the brine, they are indicative only of the temperature attained by such a surface layer, which may be appreciably higher than that of the whole body of the brine.

Salt is a rock which is prone to flow at relatively low temperatures and pressures. Salt from a deeply buried stratum may rise as a piercement plug or salt "dome" (Fig. 5-14). Adjacent strata are intruded by the salt and may be ruptured and sheared. Salt domes are common on the Gulf Coast of Texas and Louisiana and are known also in Germany, Russia, and Iran. The salt intrusion is circular, most commonly 0.5 to 2 miles (0.8 to 3.2 km) in diameter, and has a nearly vertical axis. The top tends to be flat or domical, is surmounted by a "cap rock," which may exceed 100 feet (30 m) in thickness, and is composed of limestone, gypsum, and anhydrite (Brown, 1931; Goldman, 1952). The salt within domes displays an intricate system of large- and small-scale flow folds, lineations, and similar characteristics produced by the upward flow of salt (Balk, 1953).

Although several theories have been advanced to explain salt domes, they are now generally regarded as intrusive bodies of salt. They are therefore, a tectonic structure, and extended treatment of them is out of place here. Refer to the extensive literature on these interesting and economically important structures (Moore, 1926).

Evaporites are associated with common sediments such as shales and dolomites. An association with red beds is common but by no means universal. Many evaporites are interbedded with, or rest upon, carbonate rocks. Dolomite is more abundant than limestone. In some sections it is notably brown, probably because of bituminous matter, and is thinly laminated. The rock actually may be fetid (*Stinkstein*), although megascopic fossils are very rare indeed. Anhydrite is common in dolomites and locally forms a large part of the rock. Very probably these dolomites are chemical precipitates (Sloss, 1953). In many places the anhydrite is overlooked, because it has been replaced and now occurs as calcite pseudomorphs of anhydrite crystals or as calcitized nodules in limestone. This calcitization may occur on a large scale.

Shales also are associated with anhydrite and salt and may be interbedded with them. Cross-cutting veins of gypsum are common.

The absence of megascopic fossils from evaporite and associated rocks is nearly universal, although there are striking exceptions. The absence is not surprising, however, in view of the high salinity of the waters from which the deposits crystallized. Note has been made of the association of bitumen and evaporites. Bitumen is present in associated dolomites, may form laminations in the anhydrite itself, and may be found as inclusions in salt crystals. It probably owes its origin to planktonic organisms swept into the highly saline gulf or lagoon from the open ocean.

OCCURRENCE OF EVAPORITES

Evaporites are known from all continents, with the possible exception of Antarctica, and from all geologic systems including even the Precambrian. Their distribution in time and place has been briefly summarized by Kozary, Dunlap, and Humphrey (1968), for the world, and in more detail for the United States by Krumbein (1951), and, most recently, for marine evaporites by Stewart (1963). A bibliography of over 700 titles on evaporites has been compiled by Cramer (1969).

In brief, one-fourth of the continental areas of the world are underlain by evaporites; in 60 percent of the cases the evaporitic sequence contains chlorides, virtually all are in intracratonic basins, and very few indeed are in geosynclines. Most occur in the Northern Hemisphere, and many are associated with petroleum-bearing sediments. Most are Cambrian or later in age. Certain areas are more prone to evaporitic sedimentation. In the Persian Gulf region, for example, evaporites of late Precambrian, Cambrian, Jurassic, and Tertiary age are known (Stöcklin, 1968). These are among the most famous of the world's evaporites. Other well-known deposits include those of the Zechstein province (Permian) of northwestern Europe, those of the Silurian salt basins of New York and Michigan, and the vast Permian salt basin of west Texas and New Mexico. For further information on the evaporitic deposits of the world refer to the proceedings of the international salt symposium (Mattox, 1968) and the monographic works of Lötze (1957).

The New York Salina (Silurian) evaporites underlie about 10,000 miles^2 (25,900 km^2) in western New York, Pennsylvania, eastern Ohio, and northern West Virginia (Alling, 1928). Single beds 40 to 80 feet (12.2 to 24.4 m) thick are known. Seven salt beds which alternate with shales and total 250 feet (76 m) in thickness are reported from one section between depths of 1900 and 3120 feet (580 and 952 m). Salt and gypsum of Salina age also occur in the Michigan Basin and adjoining parts of Ontario (Dellwig, 1955; Dellwig and Evans, 1969). The maximum aggregate thickness of the salt beds in Michigan is in excess of 1600 feet (488 m).

The most spectacular of the evaporite deposits is the Castile Formation (Permian), which underlies an area roughly 200 miles (322 km) in diameter in Texas, New Mexico, and adjoining parts of Mexico. More than 95 percent of the formation is composed of salts deposited by the evaporation of a brine (Kroenlein, 1939). The Castile and Salado formations have a maximum thickness of about 4000 feet (1220 m), of which 1200 to 1500 feet (366 to 457 m) are laminated anhydrite. It thus is one of the thickest known evaporite deposits in the world. The lower portion is chiefly delicately laminated anhydrite (Fig. 11-33); the laminations average 1.6 mm in thickness. A thin brown bitumen-rich film separates the laminations from one another. Udden (1924) interprets the laminations as annual layers or varves and has estimated that the precipitation of the laminated anhydrite took place over 306,000 years. Individual laminations have been correlated over distances up to 70 miles (113 km) (Anderson et al., 1972). Near the outcrop the anhydrite is converted to gypsum. The upper part of the sequence is chiefly halite, although there are some potash-bearing salts, chiefly polyhalite, $Ca_2K_2Mg(SO_4)_4 \cdot 2H_2O$, as well as some intercalated gypsum beds. The ratio of halite to gypsum is about 1 : 1 instead of the 30 : 1 ratio that would be the case if the deposit were formed by simple isolation and evaporation of sea water.

The Permian evaporitic basin of northwestern Europe underlies much of the North Sea and adjacent land areas including the northeastern part of England, most of Denmark and the Netherlands, the north German plain, and its extension eastward into Poland

and beyond. Permian rocks within this basin, generally concealed by younger beds, are especially characterized by evaporites particularly in the upper division or Zechstein. The Permian sequence is overlain by the Bundsandstein and the Muschelkalk; evaporites are present in the latter as well as in the overlying Jurassic (Brunstrom and Walmsley, 1969). Zechstein evaporites underlie at least 250,000 km^2 and in the Stassfurt region of Germany; they exceed 1000 m in thickness. Gypsum and anhydrite are thick and widespread and extend to within a few miles of the limits of the Zechstein; the more soluble evaporites are found basinward, particularly in the Stassfurt area where they are exploited for potash. Salt movement and diapiric intrusions occur where halite is thickest. The Zechstein sequence consists of a series of evaporitic cycles (four in Germany) which start with carbonates followed by anhydrite, which in turn is overlain by halite (and potash salts in some places). The stratigraphy, petrology, and geochemistry have been described in many papers (Lötze, 1957; Richter-Bernburg, 1955; Kühn, 1968; Stewart, 1949, 1951, 1953).

ORIGIN OF EVAPORITES

All salt deposits are formed by the evaporation of a brine. The ultimate source of such brines is generally sea water. The brine may be formed directly from sea water by evaporation in a semi-isolated or wholly isolated arm of the sea in an arid region. Brines may also form in interior basins of arid regions into which waters flow that derived their salt either from connate waters of marine sediments ("fossil" sea water or trapped brine), from dissolution of older salt beds, or, as now thought by some, by airborne salt particles derived from ocean spray transported into the continental interior by the atmosphere (Broecker and Walton, 1959).

The evaporation of sea water and the course of crystallization that follows have been studied experimentally. Sea water contains about 3.5 percent by weight of dissolved solids, about four-fifths of which is sodium chloride. The average composition of these materials is given in Table 11-8.

The experiments of Usiglio (1849) show that, when the original volume of sea water is reduced by evaporation to about one-half, a little iron oxide and some $CaCO_3$ are precipitated. When the volume is about one-fifth that of original sea water, gypsum is formed. Upon reduction to approximately one-tenth the original volume, NaCl begins to crystallize. Further reduction in volume leads to the appearance of sulfates and chlorides of magnesium and finally to NaBr and KCl.

Although the order observed by Usiglio agrees in a general way with the sequence found in some salt deposits (Schmalz, 1969), many exceptions are known. Also, many minerals known from salt beds did not appear in experimentally formed residues. The crystallization of a brine is very complex and depends not only on the solubility of the salts involved but also on the concentration of the several salts present and the temperature. For a further discussion of phase relations refer to Stewart's (1963) review of the crystallization of chloride brines and to Eugster and Smith's (1965) study of crystallization of brine derived from carbonate-rich waters.

If all the salt in a 1000 foot (305 m) column of sea water were precipitated, it would form a deposit 15 feet (4.6 m) thick, of which 0.5 feet (0.15 m) would be calcium sulfate, 11.8 feet (3.6 m) would be NaCl, and the remaining 2.6 feet (0.8 m) would be potassium- and magnesium-bearing salts. Salt

TABLE 11-8. Composition of oceanic salts

Salt	Percent
NaCl	77.76
$MgCl_2$	10.88
$MgSO_4$	4.74
$CaSO_4$	3.60
K_2SO_4	2.46
$CaCO_3$	0.35
$MgBr_2$	0.22
Total	100.00

Source: From Clarke (1924, p. 23).

deposits over 300 feet (100 m) thick must, therefore, require evaporation of a very great volume of water. Deep water, however, is not implied. Indeed, deep water is generally presumed to be improbable because of the association of many salt beds with stromatolitic structures, mud cracking, and the like. As far as is known, all modern evaporites are of very shallow-water origin. Schmalz (1969), however, has presented cogent arguments for a deep-water origin of some ancient evaporites.

If precipitation were carried to completion, the salts deposited should be in an order approximately that given by Usiglio and roughly in the proportions in which they are present in sea water. Moreover, once a salt has begun to precipitate, its crystallization should continue until the end-stage is reached, unless it reacts with the residual liquid to form a different solid phase. Inasmuch as many evaporite deposits show marked exceptions to the above requirements, simple evaporation of sea water did not occur, and either the parent brine was not formed from sea water or the evaporation took place under special conditions that will explain these anomalies.

An extended discussion of salt deposits of various derivations is given by Grabau (1920) and Lötze (1957). A briefer and more modern review is given by Stewart (1963). At present two schools of thought consider the origin of evaporites in the geologic record. One presumes the deposits to have been precipitated from a substantial, more or less permanent, standing body of brine—a brine lake or a hypersaline arm of the sea. The other presumes that salts were deposited from subsurface brines brought to the surface of a playa or a sabkha flat by capillary action and that a water body, if any, was at best extremely shallow and ephemeral.

In order to explain the thick deposits of salts, Ochsenius (1877) put forward his "bar theory." This concept assumes a lagoonal area cut off from oceanic circulation by a permanent bar, except for a narrow channel through which sea water can enter, offsetting the losses caused by evaporation and through which there is a little or no outflow. As more salt is constantly carried into the lagoon, salinity will increase until salt deposition takes place. As long as the channel remains shallow enough to prevent escape of the dense bottom brines, this process will continue. The model for this concept is the Gulf of Karaboghaz on the east side of the Caspian Sea.

Several modifications of the barred basin concept are needed to explain the stratigraphy of various salt accumulations. The multiple-basin hypothesis (Branson, 1915), for example, postulates a succession of connected basins. The waters flow through successive basins and become progressively more saline. In a second or third basin, perhaps, halite might be precipitated without a subjacent deposit of anhydrite or gypsum. Because this concept requires a complex arrangement of basins and concentrations, it seems rather improbable. Fractional crystallization can be accomplished in a simpler manner. King (1947), for example, advanced an ingenious explanation for the thick anhydrite deposits of the Permian Castile Formation of Texas and New Mexico. He postulates deposition in a semi-isolated sea into which normal sea water flowed through a somewhat restricted channel The concentrated brine of the Castile sea tended to sink to the bottom and in part return, by a sort of reflux action, through a permeable barrier, to the sea. The salinity achieved was sufficient to precipitate calcium sulfate but not sodium chloride. The initial deposit would be gypsum, but King thought that the anhydrite stage would be reached in about 1/1000th part of Castile time (assuming a temperature of 30°C). An evaporation of 114 inches (290 cm) is estimated to be necessary in order to produce the observed annual increment of anhydrite. Calculation shows that the rate of influx to reflux was about 10 to 1. Such a balance would lead to anhydrite precipitation with relatively little halite and would permit the more soluble salts to be carried out of the basin and back to the sea. Modern oceanographic studies of circulation patterns in somewhat restricted lagoons or embayments provide support for the influx-reflux concept (Scruton, 1953).

In recent years the sabkha model has been called on to explain many evaporite deposits in the geologic record. Bosellini and Hardie (1973), for example, ascribe the evaporitic facies of the Bellerophon (Upper Permian) formation of northern Italy to a sabkhalike environment. Some 46 regressive sabkha cycles, averaging 3 m thick, consist of an earthly dolomite passing upward into gypsiferous dolomite which in turn is overlain by a layered nodular gypsum capped by massive gypsum with a "chicken-wire" structure. Similar cyclic sequences are reported in the Baibl Formation (Triassic) of the Dolomites, in the Upper Burano Formation of the central Appennines and in the Nordenskioldbreen Formation (Carboniferous) of Spitzbergen (Hardie, pers. commun., 1972). The evaporitic deposits of Devonian age in southern Alberta and Saskatchewan, Canada, likewise display sabkha cycles and have, therefore, been attributed to deposition on sabkha flats subject to repeated exposure (Fuller and Porter, 1969) Even some halite deposits have been so explained (Smith, 1971).

Similarly the Solfifera Series (Upper Miocene) of Sicily is believed to have been deposited in a very shallow strandline lagoon and sabkha environment (Hardie and Eugster, 1971).

Some deposits of gypsum form by reaction of sulfate-bearing waters on limestones. The oxidation of pyrite, especially the pyrite of black shales, has led to the formation of large quantities of acid sulfate waters which react with limestone and convert large parts of it to gypsum. These deposits are not, of course, evaporites and are generally of small volume and local in character.

Gypsum is found in some places as *gypsite*, which is gypsum earth. This material is an efflorescent deposit, found only in arid regions which occurs over the ledge outcrop of gypsum or gypsum-bearing stratum. Locally it is a source of gypsum.

CARBONACEOUS SEDIMENTS

INTRODUCTION

Organic compounds in living matter undergo oxidation and are converted to CO_2 and water. This oxidation or "slow combustion" does not cease on the death of the organism but is replaced by bacteriological decay or direct oxidation. Depending on where they accumulate and the amount of oxygen available, organic residues undergo only partial or incomplete oxidation, known as humification or putrefaction. Such incompletely oxidized products are methane (CH_4), carbon monoxide (CO), hydrogen (H_2), and so forth. These gaseous products escape, but some of the partially degraded solid organic residues are buried and preserved in the lithosphere where further changes take place.

Organic matter that escapes scavengers and undergoes burial becomes a constituent of the accumulating sediment. The percentage of organic matter thus incorporated varies widely. According to Trask (1939), few typically marine sediments contain more than 10 percent and equally few contain less than 0.5 percent organic matter. The average quantity for near-shore sediments is about 2.5 percent and for those of the open ocean about 1.0 percent. Because some 50 to 60 percent of the organic matter is carbon, the average organic carbon content of near-shore marine sediments is thus about 1.5 percent. Ancient sediments also contain carbonaceous residues. In general, clay shales contain about twice as much as sandy shales or siltstones; these in turn contain about twice as much as fine-grained sandstones. The average shale contains 2.1 percent organic matter; average limestones contain about 0.29; and the average sandstone has about 0.05 percent (Degens, 1965, p. 202). A study based on 25,000 samples (Trask and Hammar, 1934) showed that the organic content of few formations exceeded 4 percent and that the general average was 1.5 percent (equivalent to 1 percent carbon). But even with this small carbon content, the total quantity of organic carbon trapped in sediments is very large. If we take 1 percent as the average and if the total mass of the sediments is 480×10^{15} metric tons (Poldervaart, 1955), the quantity of entombed organic carbon is 4.8×10^{15} tons. This is 500 times greater than that locked up as coal (6×10^{12} metric tons).

Carbonaceous residues largely free of ordinary detrital sediment accumulate in some places. Such residues include peat in the modern world and the "fossil" equivalent in the ancient record—coal. Although these and related materials have great value to man, they form a very small part of the record and, as noted above, only a small part of the total "fossil" carbon. A sedimentary layer 1000 m thick contains enough organic matter to form a layer 20 m thick. But for each 1000 m of sediment deposited only 5 cm is coal.

The ultimate source of the organic matter which eventually gives rise to carbonaceous residues in sediments is the carbon dioxide of the atmosphere. This substance is converted, by photosynthesis in green plants, to complex organic compounds which, as noted, are in part broken down by metabolic processes (respiration), in part by putrefaction, and in part by rapid combustion by fire (Fig. 11-35). Those that escape complete degradation become entombed in sediments or, in rare cases, form a sedimentary deposit free of mineral matter (coal and bitumen). Obviously this entrapment of organic carbon in the lithosphere would lead to a gradual depletion of atmospheric and hydrospheric CO_2 were not additional CO_2 supplied to the system. These additions are largely volcanic, with the CO_2 derived from the earth's mantle by a degassing process (Rubey, 1951).

NATURE AND FORM OF ORGANIC RESIDUES

Resistance to decomposition of organic materials varies (Tomkeieff, 1954, p. 14). Most susceptible to decomposition are proteins, sugars, starches, and other food materials. Cellulose and fats decompose less readily; amber, the chiton of certain brachiopods and other forms, resin, and waxes remain even in rocks of Cambrian age. The degradation of plant tissues is highly selective and is not well understood (Barghoorn, 1952).

Organic residues at present accumulate in the forms of humus, peat, and sapropel (Polynov, 1937).

Humus is the accumulation of organic residues in the uppermost part of the lithosphere, mainly in the soil. For the most part, this form of accumulation undergoes oxidation. In a normal soil, therefore, humus consists of newly added organic matter and a large number of compounds representing various stages of decay. Some of the intermediate products, the so-called humic acids, are very active. These acids are, in fact, a colloid complex capable of adsorbing cations from solution. If sesquioxides and dissolved humus are prepared in certain proportions, mixed sols are formed. If the proportions are altered, coagels of humus with the hydrated oxides of iron and alumina are formed.

Peat-forming conditions are nearly everywhere associated with freshwater swamps. These organic residues may form 70 to 90 percent of the total accumulation. Mineral components are nil. Peat deposition takes place where there is rapid growth and reproduction of plants, excessive development of organic compounds that are difficult to decompose, and development of such conditions in the medium that the life activity of microorganisms is reduced to a minimum or completely extinguished. Plants vary in their peat-forming ability. Involved in peat formation are rushes, sedges, horsetails (*Equisetum*), various woody plants—notably certain species of pine (*Pinus*), birch (*Betula*), black alder (*Alnus*), and spruce (*Picea*)—and most important of all, the peat mosses (*Sphagnum* and *Hypnum*).

Peat-moss accumulations may be meters thick and cover many square kilometers. They are hydroscopic and absorb up to 14 times their own volume of water. Peat mosses fulfil the conditions for peat accumulation because they grow rapidly and consist largely of cellulose and waxy substances with little protein. Organic acids formed on putrefaction cannot obtain bases for their neutralization; hence the resulting acidity inhibits microorganisms.

In the modern world, peat accumulation is most extensive in the northern latitudes and is associated with extensive freshwater swamps. Those in Canada are known as *muskegs* (Radforth, 1969), and there are some 500,000 $miles^2$ (1,295,000 km^2) of such swamps in Canada alone. Peat is also found forming today in coastal mangrove swamps and in swamps flooded with brackish or marine waters (Spackman, Riegel, and Dolsen, 1969).

Sapropel is a silt rich in, or composed wholly of, organic compounds which collect at the bottoms of various water basins:

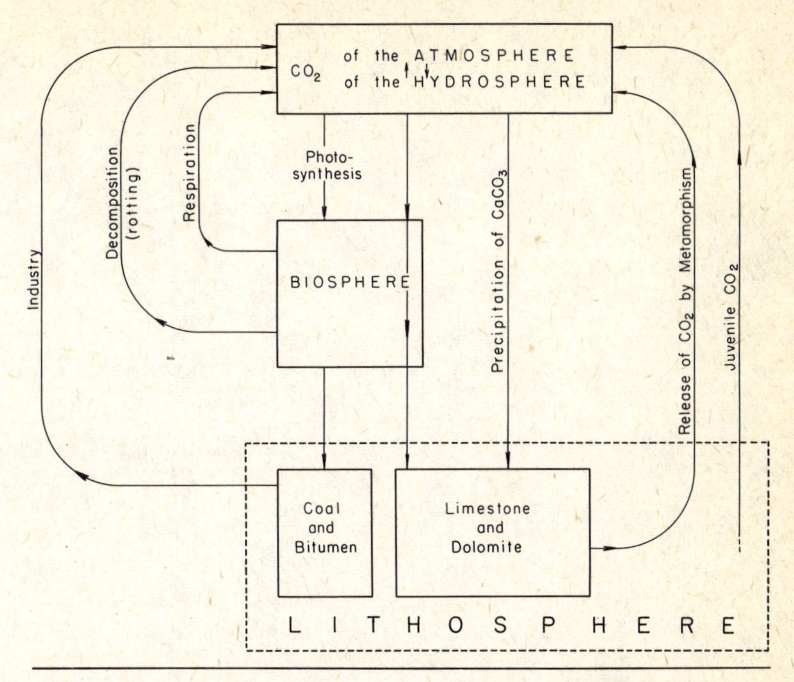

FIG. 11-35. The carbon dioxide cycle. (After Goldschmidt, 1933.)

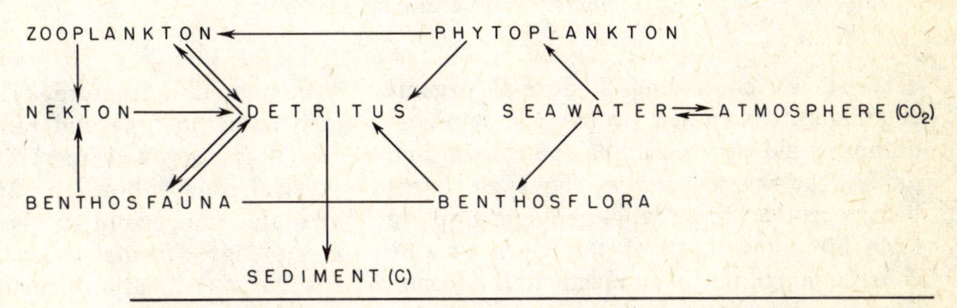

FIG. 11-36. Relations of carbon, organisms, and marine sediment. Nekton, active swimming surface forms; plankton, passive surface forms; benthos, bottom-living forms.

lakes, lagoons, estuaries, and the like. The remains of phyto- and zooplankton are richer in fatty and protein substances than is peat. Hence decomposition (putrefaction) in the presence of little oxygen takes place. Various types of hydrocarbons form, as do reduced forms of iron (ferrous carbonate and ferrous sulfide). The progressive accumulation of sapropel is governed largely by rapid multiplication of the organisms responsible for it (Fig. 11-36). Free organic acids are lacking, because the environment is essentially neutral. Such accumulations may not occur or may take place with considerable mixture of mineral matter.

CLASSIFICATION OF CARBONACEOUS SEDIMENTS

Organic residues which accumulate as sediments are, thus, of two major types—peaty or humic residues which upon burial lead to various forms of coal (humic or woody coals), and sapropelic residues which form various sapropelic coals (cannel and boghead coals). Where residues are mixed with mineral matter, impure coals of various types are formed. Impure humic coal is bone coal. Sapropelites such as oil shale are shales rich in sapropelic residues (Fig. 11-37). Black shales are carbonaceous shales richer than average in carbonaceous matter. Even ordi-

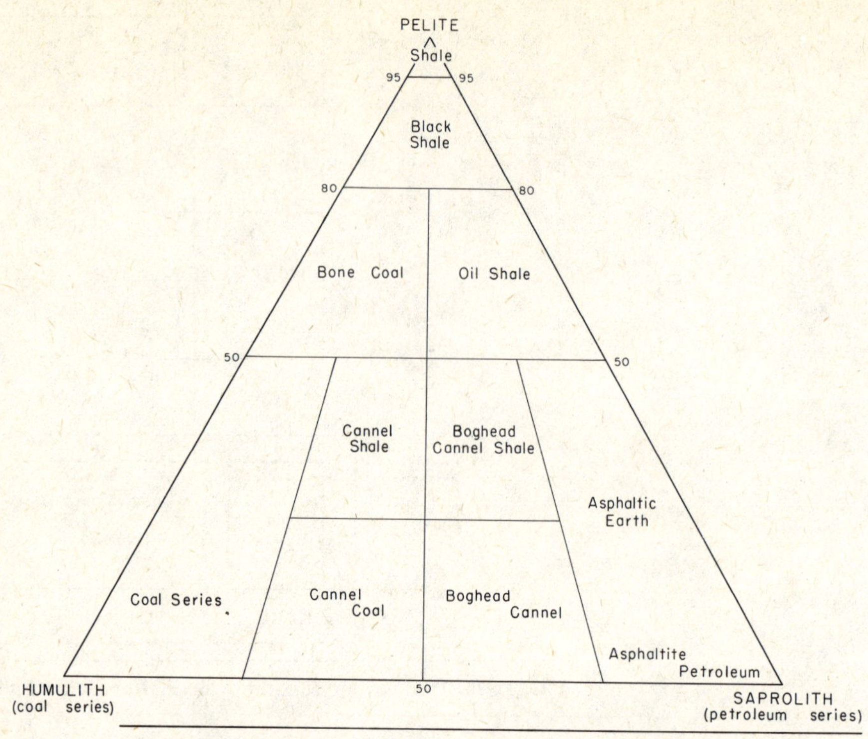

FIG. 11-37. Classification and nomenclature of the carbonaceous sediments. (Source unknown).

nary shales containing a normal organic component can, under proper geologic conditions, yield organic fluids (petroleum and natural gas) which migrate from such source beds to collect in suitable geologic traps to form oil pools. Loss of the more volatile constituents of petroleum leads to the formation of solid asphaltic residues (asphaltite, uintaite, etc.).

THE COAL SERIES

Because of the great importance of coal as an energy source in the western world, a great deal has been written on its properties, occurrence, origin, mining, and uses. Among the more comprehensive works on coal are those of Moore (1940), Raistrick and Marshall (1948), and Tomkeieff (1954).

General characteristics and classification

Coal is a combustible, opaque (except in very thin slices), noncrystalline solid which varies from light brown to black. It is dull to brilliant in luster and has a low specific gravity (1.0 to 1.8). The hardness varies from 0.5 to 2.5. It is brittle and has a hackly to conchoidal fracture. These properties vary with the type and rank of coal.

Coal is classified according to rank and physical constitution. The first classification antedates the second and is based primarily on the degree of metamorphism of the coal; it is the classification primarily used in the coal trade. The second classification is biopetrologic and is based in large part on the microscopic study of coal.

Ranks of coal depend on the degree of coalification. They are more or less arbitrary, although carefully selected chemical criteria are used to supplement gross features which characterize each rank (Fig. 11-38). Commonly recognized ranks of coal are: (1) brown coal or *lignite*, (2) subbituminous, (3) *bituminous*, (4) semibituminous, (5) semianthracite, and (6) *anthracite*. In addition certain other coal types, not part of the usual graded series, are commonly recognized. These include cannel and bone coal.

Brown coal or *lignite* is a low-rank coal which is brown, brownish black, but rarely black. It commonly retains structures of the original wood. It is high in moisture, low in heat value, and checks badly on drying. It

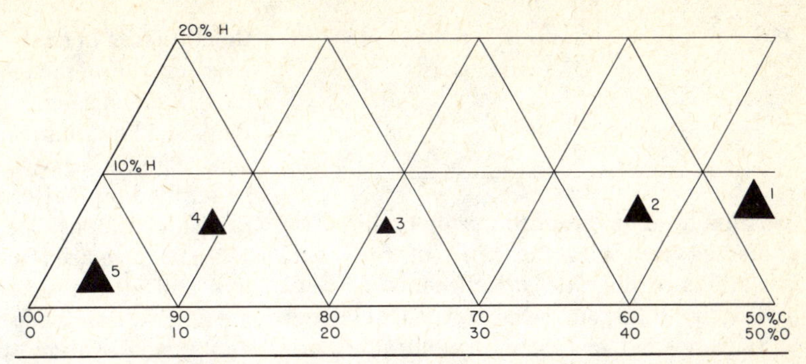

FIG. 11-38. The coal series. (1) wood; (2) peat; (3) lignite; (4) bituminous coal; (5) anthracite coal. (Data from Clarke, 1924, p. 773.) ,

FIG. 11-39. Banded bituminous coal, No. 6 bed, Franklin County, Illinois. About natural size. Note alternations of vitreous bright coal (vitrain) and thinly laminated coal (clarain). (U.S. Bureau of Mines photograph. Sprunk, 1942, p. 413.)

FIG. 11-40. Nonbanded canned coal. About natural size. Note blocky structure, conchoidal fracture, and absence of stratification. (U.S. Bureau of Mines photograph. Sprunk, 1942, p. 414.)

burns readily with a smoky flame. Most lignite is Cretaceous or younger in age.

Bituminous coal is a higher rank coal, that is, contains a higher percentage of carbon and less water. It also burns readily and does not disintegrate so easily on exposure. Most bituminous coals show a fine banding caused by alteration of laminations of bright and dull coal (Fig. 11-39).

Anthracite coal is marked by its bright, almost submetallic luster and its conchoidal fracture. It is high in fixed carbon and low in volatile hydrocarbons. It ignites less readily than the lower ranks of coal and burns with a short flame which produces much heat and little smoke. It is relatively high in ash.

Subbituminous, semibituminous, and semianthracitic coals are transitional coal types.

Cannel coal is a nonbanded dull, black coal of bituminous rank with a conchoidal fracture (Fig. 11-40). It burns readily with a long smoky flame. It is high in volatile constituents. *Bone coal* is a very impure coal, high (33 percent or more) in ash. The term *graphitoid* has been applied to the coals metamorphosed beyond the anthracite stage.

Chemical composition of coal Coal is composed chiefly of carbon, hydrogen, and oxygen. Nitrogen and sulfur are lesser constituents. Mineral matter (ash) is a variable constituent. The proportions of major constituents vary with the rank of the coal. Variations of the chief constituents, carbon, hydrogen, nitrogen, and oxygen, starting with wood and ending with anthracite, are given in Table 11-9. This table may be restated to show the proportions of hydrogen, nitrogen, and oxygen to 100 parts of carbon (Table 11-10). The steady decrease in hydrogen and oxygen during the transitions from peat to the various ranks of coal is more clearly seen. Furthermore, the proportional decrease in oxygen is greater than for hydrogen. In cellulose ($C_6H_{10}O_5$) these two elements exist in exactly the proportions in which they are found in water (1 to 8). In wood, hydrogen is in excess of that ratio (1 to 7), and the excess steadily increases until in anthracite it is roughly 1 to 1. The changing composition (based on Table 11-9) is shown in Fig. 11-38.

Constituents in coal Two schools of thought exist on the nomenclature and classification of the constituents of coal: American and British. The American nomenclature (Thiessen, 1920) is botanical and genetic and can be applied only after microscopic study. The British or Stopes classification (Stopes, 1919, 1935) is primarily megascopic and descriptive and is hence applicable to hand specimens. Each of these systems has its proponents; some attempts have been made to reconcile or to combine the rival schemes and to formulate an acceptable international standard (for a review of the subject see Cady, 1939, 1942; Dapples, 1942; Marshall, 1942). The result is the Stopes-Heerlen classification (Table 11-11).

Any specimen of common bituminous coal reveals, to the unaided eye, three or four varieties of coal, each with individual and characteristic features. These components, which are more or less segregated into bands, are the ingredients of coal and are designated vitrain, clarain, durain, and fusain (Fig. 11-39).

Vitrain forms thin, horizontal laminations up to 20 mm thick. It is a brilliant, vitreous

TABLE 11-9. Average composition of wood, peat, and coals

	Carbon	Hydrogen	Nitrogen	Oxygen
Wood	49.64	6.23	0.92	43.20
Peat	55.44	6.28	1.72	36.56
Lignite	72.95	5.24	1.31	20.50
Bituminous coal	84.24	5.55	1.52	8.69
Anthracite	93.50	2.81	0.97	2.72

Source: After Clarke (1924, p. 773).

TABLE 11-10. Relative proportions of constituents of wood, peat, and coals

	Carbon	Hydrogen	Nitrogen	Oxygen
Wood	100	12.5	1.8	87.0
Peat	100	11.3	3.5	64.9
Lignite	100	7.2	1.8	28.1
Bituminous coal	100	6.6	1.8	10.3
Anthracite	100	3.0	1.3	2.9

Source: After Clarke (1924, p. 773).

or glassy-looking, jetlike coal which alternates with broader bands of other kinds of coal. Vitrain has a conchoidal fracture, is brittle, and is clean to the touch. To the naked eye it commonly appears homogeneous and structureless (eu-vitrain). In other cases it exhibits "striations" attributed to plant structures (pro-vitrain).

Clarain is a coal whose thin and thicker layers or bands, unlike vitrain, are laminated. It has a smooth fracture, is marked by a glossy or shiny luster, and is especially distinguished by its silky appearance that arises from the sublaminations. The silky luster is distinctly different from the smooth brilliance of vitrain in the same coal.

Durain is a dull coal characterized by its lack of luster, matte or earthy appearance, and black or lead-gray color. It occurs in solid bands which possess a close, firm texture. Internal stratification is generally absent.

Fusain or "mineral charcoal" is readily identified because of its resemblance to common charcoal. If not mineralized, it is very friable and highly porous. The pores are in some cases filled with calcite or with pyrite. In thin sections it is opaque and highly cellular (Fig. 11-41). In the hand specimen fusain forms irregular wedges lying on bedding planes at various angles.

FIG. 11-41. Fusain, Arley Seam, Atherton, North Manchester, England. Ordinary light, ×80.

Microscopic examination has led to refinement and subdivision of these mega-

TABLE 11-11. Classification of the ingredients of coal

Rock Types (termination, *ain*)	Macerals (termination, *inite*)		Notes
Vitrain	Vitrinite	Collinite	May be subdivided if desired
		Telinite	May be subdivided according to botanical origin if desired
	Semifusinite		Intermediate between vitrinite and fusinite
Fusain	Fusinite		Opaque cellular tissue
	Micrinite		Opaque "residuum"
	Exinite	Sporinite	Spore material
		Cutinite	Cuticle material
Clarain	—		Dominantly vitrinite with some exinite and other macerals
Durain	—		Dominantly micrinite with exinite

Source: Heerlen Committee, after Marshall (1942).

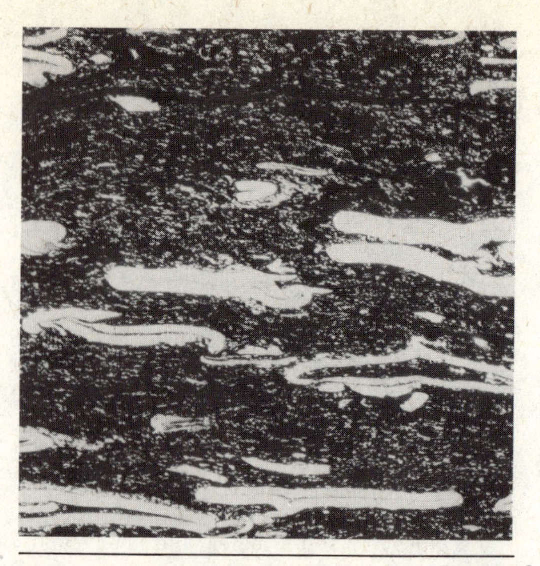

FIG. 11-42. Durain, Low Main Coal, Nottingham, England. Ordinary light, ×80. Chiefly opaque attritus with numerous collapsed megaspores.

scopically defined terms (Table 11-11). The coal types are considered rock species ("lithotypes") made up of *macerals* (analogous to the *minerals* of the noncarbonaceous rocks). A maceral is an organic unit, that is, a single fragment of plant debris or material derived therefrom. The basic macerals of vitrain are *vitrinite*, consisting of collinite and telinite. *Collinite* is a structureless jellified plant residue, whereas *telinite* is a translucent golden gel that retains some cell structure. The macerals of clarain and durain belong to the *leptinite* or *exinite* group, which are primarily the smaller and more resistant debris of plants. Included here are *alginite* (of algal derivation), *sporinite* (spore cases), *resinite* (resin), and *cutinite* (macerated fragments of cuticle). The spores are yellow, transparent bodies (in most cases collapsed) of both small and large spore bodies (microspores and megaspores, Fig. 11-42). Resinite occurs as small isolated bodies of a translucent, reddish nature. Fusain consists of *inertinite* macerals, which include *micrinite, fusinite,* and *sclerotinite*. Micrinite, an opaque residuum, is the dominant constituent of durain as well as a constituent of fusain. Fusinite is the opaque-walled carbonized cell structure so characteristic of fusain. Sclerotinite arises from the sclerotia of fungi.

The several types of bituminous coal consist of these macerals in varying proportions (Table 11-11). The Stopes-Heerlen classification of coal types (lithotypes) and their macerals grew out of the study of the bituminous coals. They are less readily adopted to the study of brown coals—especially soft brown coals—and efforts continue to develop a workable system for these coals (Murchison, 1969a).

Thiessen and some other American students of coal petrology consider coal to consist of two fundamental constituents, namely *anthraxylon* and *attritus*. The proportions of these two determine the type of coal. Bright (glance) coal is composed primarily of anthraxylon, whereas dull (matte or split) coal is chiefly attritus. According to Theissen, the bright layers were derived chiefly from woody parts of plants (anthraxylon). Thicker bands represent branches or trunks of trees; thinner ones are smaller branches or twigs. Several types of anthraxylon, moreover, are determined by the type of plant from which it formed. Anthraxylon is clearly translucent. Attritus, on the other hand, is applied to all material not identified as anthraxylon. Attritus, unlike anthraxylon and fusain, which are masses of integrated plant tissue preserved as units and showing cell structure, is macerated and degraded plant materials. It is both opaque and transluscent.

Common coal types according to this system are bright coal, semisplint, splint, cannel, and boghead.

Under the microscope it appears that vitrain is primarily anthraxylon; clarain is composed of transluscent attritus and thin shreds of anthraxylon. Durain, on the other hand, appears to be made up mainly of opaque attritus.

Cannel and boghead coals are clean, compact, blocky coals of massive structure and of uniform, fine-grained texture. They are normally gray to black, have a greasy luster, and have a prominent conchoidal fracture. Cannel coal is devoid of bedding. Under the microscope cannel coal appears to be made up almost wholly of microdebris which consists of spores, resins, woody fragments, and opaque attritus. Boghead coal contains large masses of oil algae. Cannel and boghead are sapropelic rather than humic coals.

Occurrence of coal Coal is a relatively rare, although widely distributed, type of rock. It occurs in rocks ranging in age from early Proterozoic to Tertiary, but it did not become common until the development of woody plants in the Devonian. Pre-Devonian coals are probably all sapropelic and perhaps algal. One of the oldest, if not the oldest, coal is that found in the Michigamme Slate (Precambrian) in Iron County, Michigan. This coal, in excess of 1700 million years in age, occurs interbedded with the slate, is anthracitic (about 80 percent fixed carbon), and is high in ash. It is probably an algal coal (Tyler, Barghoorn, and Barrett, 1957).

The first extensive coal deposits are of Carboniferous age. Even in the Coal Measures (Carboniferous), however, coal rarely forms more than 1 or 2 percent of the whole section. Individual beds of coal range from a mere film to seams over 400 feet (122 m) in thickness. The average thickness is on the order of 1 or 2 feet (0.3 to 0.6 m). Beds over 10 feet (3.0 m) are generally rare. The Pittsburgh seam, one of the most widespread and productive seams in the central Appalachian fields in the United States, maintains a thickness of 6 to 10 feet (1.8 to 3.0 m) over a wide area. It is an exceptional deposit and underlies 22,000 miles^2 (56,980 km^2) in western Pennsylvania, West Virginia, and Ohio and is of workable thickness beneath 6,000 miles^2 (15,540 km^2). In any given coal-bearing section, numerous seams, both workable and nonworkable, are likely to be present. In western Illinois are 18 coal seams; in part of Ohio and southwestern Pennsylvania over 40 seams are present; in Nova Scotia more than 70 beds have been recorded; and in the Westphalian field in northwestern Germany more than 90 beds, aggregating 274 feet (84 m), are reported. In the Jharia (Gondwana) coal field of India are 24 seams thicker than 4 feet (1.2 m) that have an aggregate thickness of 240 feet (73 m) in a section some 2000 feet (610 m) in thickness (Fox, 1930). Coal forms more than 10 percent of the whole section. The thickest coal, the Kargali seams, is 100 feet (30.5 m) thick. The Indian coal measures are atypical in that coal forms a higher percentage of the whole section, in the great thickness of the seams, and in the high ash content of the coals themselves.

Coal occurs interbedded with ordinary sedimentary rocks. In many places these strata are arranged in a cyclical manner forming cyclothems (Udden, 1912; Weller, 1930; Wanless, 1950). In many places the coal rests on a clay of special character known as *underclay* or seat earth. Some of these clays are so low in iron and alkalies that they constitute a fireclay.

Origin of coal The presence of organic structures in coal and the well-established coal series, beginning with wood and ending with anthracite, leave no room for doubting the plant origin of coal (Jeffrey, 1915; White and Thiessen, 1913). There remains, however, to consider mode of accumulation of vegetable matter and its transformation into coal.

Two views relate to accumulation. Most students of coal suppose the organic matter to have accumulated in place, thereby giving rise to an autochthonous coal (Jeffrey, 1915; White and Thiessen, 1913). Some investigators, particularly the students of the Indian Gondwana coals, have supposed the organic debris to have been transported from its place of growth to another place, where it accumulated to form an allochthonous coal seam (Fox, 1930). Coal undoubtedly has formed both ways, but the prevailing view favors accumulation in place in large freshwater swamps for most coals. Uniformity of thickness, wide extent, and absence of admixtures of inorganic detritus favor this conclusion. Some of the cannel coals, however, may represent accumulations of organic detritus.

The process of conversion of woody, peaty residues to coal is termed *coalification*. It is in part biological and in part metamorphic. Normally vegetable matter is oxidized to water and carbon dioxide. If plant material accumulates under water, however, oxygen is rapidly depleted and decomposition is only partial. Plant residues are incompletely oxidized. Such incomplete destruction lead to the accumulation of an organic deposit—peat. Although some of the decomposition and change may be the result of bacterial action, normally the toxic conditions of the peat bog suppress or extinguish the activity of such microorganisms.

The chemistry of the coalification processes involved in the change of peat to various ranks of coal is not fully understood (Murchison, 1969b). Burial with consequent rise in pressure and temperature promotes the changes. Only in the sharply folded beds is the anthracite rank reached.

OIL SHALES, PETROLEUM, AND NATURAL GAS

The subject of petroleum and natural gas has given rise to a vast literature; an adequate digest of this material is impossible here. Refer to the many standard works on this subject such as Levorsen's *Geology of Petroleum* (Levorsen, 1967). It is appropriate, however, to point out how the study of the sedimentary rocks bears on the problem of the origin of the fluid hydrocarbons contained within them.

Most students of the subject now agree that petroleum and natural gas are generated from organic matter trapped in sediments at the time of their deposition. In general it is agreed that sapropelic rather than peaty materials were the parent substance. The process of conversion has been termed *bitumenization*. Many rocks, however, contain no petroleum as such but yield oil on destructive distillation. These are the kerogen-bearing oil shales.

The term *oil shale* has been applied to any rock from which substantial quantities of oil can be extracted by heating. Normally these rocks are fine-textured and laminated. They include true shales, such as the Scottish Lower Carboniferous oil shales (Greensmith, 1962), and marlstones and dolomitic limestones, such as those of the Green River Formation (Eocene) of Wyoming, Colorado, and Utah (Donnell, Culbertson, and Cashion, 1967). The yield varies widely; the Green River oil shales, for example, yield 125 liters per metric ton. The material which gives rise to crude oil on distillation has been termed *kerogen*. It is an organic product consisting of a mixture of large hydrocarbon molecules. Although petroleum and natural gas are thought to be generated in marine sediments, oil shales are more largely nonmarine, generally lacustrine in origin, but marine oil shales are also known.

Crude oil is believed to have formed from organic debris, perhaps mainly unicellular algae or diatoms, enclosed in marine argillaceous and calcareous sediment. Study of the origin of petroleum and natural gas involves not only study of organic matter in sediments and its transformation into petroleum but also study of source beds and the ultimate reservoir rocks. Unfortunately there is no general agreement on what are the source beds. Exhaustive studies have been made of the source bed problem and the establishment of chemical and other criteria for their recognition (Trask and Patnode, 1936). Although the reservoir rocks are better known because they are more readily identified, there has been little systematic work on their petrology. Krynine (1945) has advanced the view that reservoir rocks are specialized, somewhat atypical deposits. To date this concept has not been supported by adequate data, although some attempt has been made to evaluate the idea by statistical scrutiny (Emery, 1955). Certainly, however, there are several different petrologic types of oil pools markedly different from one another. The only common denominator is adequate porosity for the storage of oil and gas, whether the reservoir rock is a limestone or sandstone.

It has been suggested that if petrologic characters of both the source and reservoir rocks were adequately known, it should be possible to predict where in time and space such characteristics would have their optimum development, so that oil prospecting could be put on a rational basis. Were the petroliferous zone closely delineated, the search for structural and other traps could then be carried out with the usual geological and geophysical tools with a minimum of expense and a maximum chance for success. In essence the problem is to frame a geological and petrological definition of the habitat of oil within a given basin. This involves essentially recognition of the depositional environment in which petroleum is formed and entrapped. (The problem of environmental recognition is the subject of Chapter 15.)

References

Adams, S. K., Groot, S. S., and Hiller, N. N., Jr., 1961, Phosphatic pebbles from the Brightseat

Formation of Maryland: Jour. Sed. Petrology, v. 31, pp. 546–552.

Aldrich, H. R., 1929, The geology of the Gogebic Iron Range of Wisconsin: Bull. Wisconsin Geol. Nat. Hist. Surv. 71, 279 pp.

Alexandrov, E. A., 1955, Contribution to studies of origin of Precambrian banded iron ores: Econ. Geol., v. 50, pp. 459–468.

Allen, V. T., 1937, A study of Missouri glauconite: Amer. Mineral., v. 22, pp. 842–846.

Alling, H. L., 1928, The geology and origin of the Silurian salt of New York: Bull. New York State Mus. 275, 139 pp.

———, 1947, Diagenesis of the Clinton hematite ores of New York: Bull. Geol. Soc. Amer., v. 58, pp. 991–1018.

Anderson, R. Y., Dean, W. E., Jr., Kirkland, D. W., and Snider, H. I., 1972, Permian Castile varved evaporite sequence, west Texas and New Mexico: Bull. Geol. Soc. Amer., v. 83, pp. 59–86.

Anderson, R. Y., and Kirkland, D. W., 1960, Origin, varves, and cycles of Jurassic Todilto Formation, New Mexico: Bull. Amer. Assoc. Petrol. Geol., v. 44, pp. 37–52.

———, 1966, Intrabasin varve correlation: Bull. Geol. Soc. Amer., v. 77, pp. 241–256.

Androussow, N., 1897, La Mer noire: 7th Int. Géol. Congr. Guide Excursions, no. 27.

Ashley, G. H., 1918, Notes on the greensand deposits of the eastern United States: Bull. U.S. Geol. Surv. 660-B, pp. 27–49.

Auden, J. B., 1933, Vindhyan sedimentation in the Son valley, Mirzapur District: Mem. Geol. Surv. India, v. 62, no. 2, 250 pp.

Bailey, E. B., 1936, Sedimentation in relation to tectonics: Bull. Geol. Soc. Amer., v. 47, pp. 1713–1726.

Balk, Robert, 1953, Salt structure of Jefferson Island salt dome, Iberia and Vermilion parishes, Louisiana: Bull. Amer. Assoc. Petrol. Geol., v. 37, pp. 2455–2474.

Barghoorn, E. S., 1952, Degradation of plant tissues in organic sediments: Jour. Sed. Petrology, v. 22, pp. 34–41.

Bastin, E. S., 1933, Relation of cherts to stylolites at Carthage, Missouri: Jour. Geol., v. 41, pp. 371–381.

Bellamy, C. V., 1900, A description of the salt lake of Lanarca: Quart. Jour. Geol. Soc. London, v. 56, pp. 745–758.

Bentor, Y. K., and Kastner, M., 1965, Notes on the mineralogy and origin of glauconite: Jour. Sed. Petrology, v. 35, pp. 155–166.

Berkey, C. P., 1924, The new petrology: Bull. New York State Mus. 251, pp. 105–118.

Berz, K. C., 1922, Ueber Magneteisen in marinen Ablagerungen: Centralbl. Min. Geol., pp. 569–577.

Biggs, D. L., 1957, Petrography and origin of Illinois nodular cherts: Illinois Geol. Surv. Circ. 245, pp. 1–25.

Blackwelder, E., 1916, The geologic role of phosphorus: Amer. Jour. Sci., ser. 4, v. 42, p. 293.

Borchert, H., 1952, Die Bildungsbedingungen mariner Eisenerzlagerstatten: Chem. Erde, v. 16, pp. 49–74.

———, 1960, Genesis of marine sedimentary iron ore: Trans. Inst. Min. Metall., 640, v. 79, pp. 261–279.

Borchert, H., and Baier, E., 1953, Zur Metamorphose ozeaner Gipsablagerungen: Neues Jahrb. Min. Abb., v. 86, pp. 103–154.

Bosellini, A., and Hardie, L. A., 1973, Depositional theme of a marginal marine evaporite: Sedimentology, v. 20, pp. 5–28.

Braitsch, O., 1962, Entstehung und Stoffbestand der Salzlagerstätten: New York, Springer, 232 pp.

Bramlette, M. N., 1946, The Monterey Formation of California and the origin of its siliceous rocks: U.S. Geol. Surv. Prof. Paper 212, 57 pp.

Brandt, R. T., Gross, G. A., Gruss, H., Semenenko N. P., Dorr, J. V. N., 1972, Problems of nomenclature for banded ferruginous-cherty sedimentary rocks and their metamorphic equivalents: Econ. Geol., v. 67, pp. 682–684.

Branson, E. B., 1915, Origin of thick salt and gypsum deposits: Bull. Geol. Soc. Amer., v. 26, pp. 231–242.

Broecker, W. S., and Walton, A. F., 1959, Reevaluation of the salt chronology of several Great Basin Lakes: Bull. Geol. Soc. Amer., v. 70, pp. 601–618.

Brown, L. S., 1931, Cap rock petrography: Bull. Amer. Assoc. Petrol. Geol., v. 15, pp. 509–530.

Brunstrom, R. G. W., and Walmsley, P. J., 1969, Permian evaporites in the North Sea Basin: Bull. Amer. Assoc. Petrol. Geol., v. 53, pp. 870–883.

Bucher, W. H., 1918, On oolites and spherulites: Jour. Geol., v. 26, pp. 593–609.

Burchard, E. F., Butts, C., and Eckel, E. C., 1910, Iron ores, fuels, and fluxes of the Birmingham district, Alabama: Bull. U. S. Geol. Surv. 400, 204 pp.

Burst, J. F., 1958a, Mineral heterogeneity in "glauconite" pellets: Amer. Mineral., v. 43, pp. 481–497.

———, 1958b, "Glauconite" pellets; their mineral nature and application to stratigraphic interpretation: Bull. Amer. Assoc. Petrol. Geol., v. 42, pp. 310–327.

Bushinsky, G. I., 1935, Structure and origin of the phosphorites of the USSR: Jour. Sed. Petrology, v. 5, pp. 81–92.

Cady, G. H., 1939, Nomenclature of the megascopic description of Illinois coals: Econ. Geol., v. 34, pp. 475–494.

———, 1942, Modern concepts of the physical constitution of coal: Jour. Geol., v. 50, pp. 337–356.

Caley, J. F., 1940, Paleozoic geology of the Toronto-Hamilton area, Ontario: Geol. Surv. Canada Mem. 224, 284 pp.

Carroll, D., 1958, Role of clay minerals in the transportation of iron: Geochim. Cosmochim. Acta, v. 14, pp. 1–27.

Casteño, J. R., and Garrels, R. M., 1950, Experiments on the deposition of iron with special reference to the Clinton iron ore deposits: Econ. Geol., v. 45, pp. 755–770.

Cayeux, L., 1905, Genèse des gîsements de phosphates de chaux sédimentaires: Bull. Soc. Géol. France, ser. 4, v. 5, pp. 750–753.

———, 1909, Les minerais de fer oolithique de France. I. Minerais de fer primaires: Paris, Imprimerie Nationale, 344 pp.

———, 1911, Les minerais de fer oolithiques primaires de France: Revue de Metall., v. 8, pp. 117–126.

———, 1922, Les minerais de fer oolithique de France. II. Minerais de fer secondaires: Paris, Imprimerie Nationale, 1052 pp.

———, 1929, Les roches sédimentaires de France: roches siliceuses: Mem. Carte Geol. France. Paris, Imprimerie Nationale, 774 pp.

Charles, G., 1953, Sur l'origine des gîsements de phosphates de chaux sédimentaires: 19th Int. Geol. Congr. Algiers, Comptes Rendus, v. 11, pp. 163–184.

Chilingar, G. V., 1955, Review of Soviet literature on petroleum source-rocks: Bull. Amer. Assoc. Petrol. Geol., v. 39, pp. 764–768.

———, 1956, Joint occurrence of glauconite and chlorite in sedimentary rocks: a review: Bull. Amer. Assoc. Petrol. Geol., v. 40, pp. 394–398.

Choquette, P. W., 1955, A petrographic study of the "State College" siliceous oolite: Jour. Geol., v. 63, pp. 337–347.

Clarke, F. W., 1915, Analyses of rocks and minerals from the laboratory of the United States Geological Survey, 1880–1914: U.S. Geol. Surv. Bull. no. 591, 376 pp.

———, 1924, Data of geochemistry: Bull. U.S. Geol. Surv. 770, 841 pp.

Clarke, F. W., and Washington, H. S., 1924, The composition of the earth's crust: U.S. Geol. Surv. Prof. Paper 127, 117 pp.

Clarke, F. W., and Wheeler, W. C., 1922, The inorganic constituents of marine invertebrates, U.S. Geol. Surv. Prof. Paper 124, 62 pp.

Cloud, P. E., 1955, Physical limits of glauconite formation: Bull. Amer. Assoc. Petrol. Geol., v. 39, pp. 484–492.

Correns, C. W., 1950, The geochemistry of diagenesis: Geochim. Cosmochim. Acta, v. 1, pp. 49–54.

Cramer, H. R., 1969, Evaporites—a selected bibliography: Bull. Amer. Assoc. Petrol. Geol., v. 53, pp. 982–1011.

Cressman, E. R., 1955, Physical stratigraphy of the Phosphoria Formation in part of southwestern Montana: U.S. Geol. Surv. Prof Paper 1027-A, pp. 1–31.

———, 1962, Nondetrital siliceous sediments: U.S. Geol. Surv. Prof. Paper 440-T, 22 pp.

Cressman, E. R., and Swanson, R. W., 1964, Stratigraphy and petrology of the Permian rocks of southwestern Montana: U.S. Geol. Surv. Prof. Paper 313-C, pp. 275–569.

Dapples, E. C., 1942, Physical constitution of coal as related to coal description and classification: Jour. Geol., v. 50, pp. 437–450.

Davis, E. F., 1918, The radiolarian cherts of the Franciscan group: Bull. Univ. California Publ. Dept. Geol., v. 11, pp. 235–432.

Deer, W. A., Howie, R. A., and Zussman, S., 1963, Rock-forming minerals, v. 4, Framework silicates: New York, Wiley, pp. 178–230 (see especially the chapter on silica minerals).

Degens, E. T., 1965, Geochemistry of sediments: Englewood Cliffs, N.J., Prentice-Hall, 342 pp.

Dellwig, L. F., 1955, Origin of the Salina salt of Michigan: Jour. Sed. Petrology, v. 25, pp. 83–110.

Dellwig, L. F., and Evans, R., 1969, Depositional processes in Salina salt of Michigan, Ohio, and New York: Bull. Amer. Assoc. Petrol. Geol., v. 53, pp. 949–956.

Deverin, L., 1945, Étude pétrographique des minerais de fer oolithiques du Dogger des Alpes suisses: Beitr. Geol. Schweiz., Lieferung 13, v. 2, 115 pp.

Dietz, R. S., Emery, K. O., and Shepard, F. P., 1942, Phosphorite deposits on the sea floor off southern California: Bull. Geol. Soc. Amer., v. 53, pp. 815–845.

Diller, J. S., 1898, The educational series of rock specimens collected and distributed by the United States Geological Survey: Bull. U.S. Geol. Surv. 150, 400 pp.

Dimroth, E., 1968a, Sedimentary textures, diagenesis, and sedimentary environment of certain Precambrian ironstones: Neues Jahrb. Geol. Palaeont. Abb., v. 130, pp. 247–274.

———, 1968b, The evolution of the central segment of the Labrador Geosyncline. Part I. Stratigraphy, facies, and paleogeography: Neues Jahrb. Geol. Palaeont. Abb., v. 132, pp. 22–54.

Dimroth, E., Baragar, N. R. A., Bergeron, R., and Jackson, G. D., 1970, The filling of the circum-Ungava Geosyncline, *in* Precambrian basins and geosynclines of the Canadian Shield

(Baer, A. J., ed.): Geol. Surv. Canada, Paper 70-40, pp. 45–142.

Dimroth, E., and Chauvel, J.-J., 1973, Petrography of the Sokoman Iron-formation in part of the central Labrador Trough, Quebec, Canada: Bull. Geol. Soc. Amer., v. 84, pp. 111–134.

Donnell, J. R., Culbertson, W. C., and Cashion, W. B., 1967, Oil shale in the Green River Formation: Proc. 7th World Petrol. Congr., Mexico, v. 3, pp. 699–702.

Downie, C., 1962, So-called spores from the Torridonian: Proc. Geol. Soc. London, no. 1600, pp. 127–128.

Dunbar, G. J., and McCall, G. J. H., 1965, Archean turbidites and banded ironstones of the Mt. Belches area (Western Australia): Sed. Geol., v. 5, pp. 93–133.

Eardley, A. J., 1938, Sediments of Great Salt Lake, Utah: Bull. Amer. Assoc. Petrol. Geol., v. 22, pp. 1305–1411.

Eckel, E., B., 1938, The brown ores of eastern Texas: Bull. U.S. Geol. Surv. 902, 151 pp.

Edwards, A. B., 1945, The glauconitic sandstone of East Gippsland, Victoria: Proc. Roy. Soc. Victoria, n.s., v. 57, pp. 153–167.

Ehlmann, A. J., Hulings, N. C., and Glover, E. B., 1963, Stages of glauconite formation in modern foraminiferal sediments: Jour. Sed. Petrology, v. 33, pp. 87–96.

Emery, J. R., 1955, The application of discriminant function to a problem in petroleum geology (abstr.): Jour. Sed. Petrology, v. 25, p. 131.

von Engelhardt, W., 1942, Die Strukturen von Thuringit, Bavalit, und Chamosit und ihre Stellung in der Chloritgruppe: Zeitschr. Krist., v. 104, pp. 142–159.

Eugster, H. P., 1967, Hydrous sodium silicates from Lake Magadi, Kenya; precursors of bedded chert: Science, v. 157, pp. 1177–1180.

———, 1969, Inorganic bedded cherts from the Magadi area, Kenya: Contr. Min. Petrology, v. 22, pp. 1–31.

Eugster, H. P., and Chou, I-M., 1973, The depositional environments of Precambrian banded iron formations: Econ. Geol., v. 68, pp. 1144–1168.

Eugster, H. P., and Smith, G. I., 1965, Mineral equilibria in the Searles Lake evaporites, California: Jour. Petrology, v. 6, pp. 473–522.

Eugster, H. P., and Surdam, R. C., 1971, Bedded cherts in the Green River Formation: Geol. Soc. Amer., abstr. with programs (Ann. Mtg.), v. 3, pp. 559–560.

Fagan, J. S., 1962, Carboniferous cherts, turbidites, and volcanic rocks in northern Independence Range, Nevada: Bull. Geol. Soc. Amer., v. 73, pp. 595–612.

Fernandez, D., 1961, Sull'origine delle seki stratificate e in noduli nel calcare: Bull. Soc. Geol. Italiana, v. 80, pp. 3–5.

Fisher, O., 1873, On the phosphatic nodules of the Cretaceous rocks of Cambridgeshire: Quart. Jour. Geol. Soc. London, v. 29, pp. 52–63.

Folk, R. L., 1973, Evidence for peritidal deposition of Devonian Caballos Novaculite, Marathon Basin, Texas: Bull. Amer. Assoc. Petrol. Geol., v. 57, pp. 702–725.

Folk, R. L., and Pittman, J. S., 1971, Length-slow chalcedony; a new testament for vanished evaporites: Jour. Sed. Petrology, v. 41, pp. 1045–1058.

Folk, R. L., and Weaver, C. E., 1952, A study of the texture and composition of chert: Amer. Jour. Sci., v. 250, pp. 498–510.

Foster, M. D., 1969, Studies of celadonite and glauconite: U.S. Geol. Surv. Prof. Paper 614-F, 17 pp.

Fowler, G. M., Lyden, J. P., Gregory, F. E., and Agar, W. M., 1934, Chertification in the Tri-state (Oklahoma-Kansas-Missouri) mining district: Am. Inst. Mining Metall. Eng. Tech. Publ. 532, 50 pp.

Fox, C. S., 1930, The Jharia coalfield: Geol. Surv. India Mem. 56, 225 pp.

French, B. M., 1968, Progressive contact metamorphism of the Biwabik iron-formation, Mesabi Range, Minnesota: Minnesota Geol. Surv., Bull. 45, 103 pp.

Friedman, S. A., 1954, Low temperature authigenic magnetite: Econ. Geol., v. 49, p. 101.

Fuller, J. G. C. M., and Porter, J. W., 1969, Evaporite formations with petroleum reservoirs in Devonian and Mississippian of Alberta, Saskatchewan, and North Dakota: Bull. Amer. Assoc. Petrol. Geol., v. 53, pp. 909–926.

Galliher, E. W., 1933, The sulfur cycle in sediments: Jour. Sed. Petrology, v. 3, pp. 51–63.

———, 1935a, Glauconite genesis: Bull. Geol. Soc. Amer., v. 46, pp. 1351–1356.

———, 1935b, Geology of glauconite: Bull. Amer. Assoc. Petrol. Geol., v. 19, pp. 1569–1601.

Garrels, R. M., and Mackenzie, F. T., 1971, Evolution of sedimentary rocks: New York, Norton, 397 pp.

Gastil, G., and Knowles, D. M., 1960, Geology of the Wabush Lake area, southwestern Labrador and eastern Quebec, Canada: Bull. Geol. Soc. Amer., v. 71, pp. 1243–1254.

Geijer, P., 1962, Some aspects of phosphorus in Precambrian sedimentation: Arkiv Min. Geol., v. 3, pp. 165–186.

Gill, J. E., 1927, Origin of the Gunflint iron-

bearing formation: Econ. Geol., v. 22, pp. 726–727.
Goldman, M. I., 1919, General character, mode of occurrence and origin of glauconite: Jour. Washington Acad. Sci., v. 9, pp. 501–502.
———, 1922, Basal glauconite and phosphate beds: Science, n.s., v. 56, pp. 171–173.
———, 1952, Deformation, metamorphism, and mineralization in gypsum-anhydrite cap rock: Geol. Soc. Amer. Mem. 50, 169 pp.
Goldstein, A., Jr., and Hendricks, T. A., 1953, Siliceous sediments of Ouachita facies in Oklahoma: Bull. Geol. Soc. Amer., v. 64, pp. 421–442.
Goodwin, A. M., 1956, Facies relations in the Gunflint Iron-formation: Econ. Geol., v. 51, pp. 565–595.
———, 1961, Genetic aspects of Michipicoten Iron-formation: Trans. Canad. Mining Inst., v. 64, pp. 32–36.
———, 1962, Structure, stratigraphy, and origin of iron-formations, Michipicoten area, Algoma District, Ontario: Bull. Geol. Soc. Amer., v. 73, pp. 561–586.
Govett, G. J. S., 1966, Origin of banded iron formations: Bull. Geol. Soc. Amer., v. 77, pp. 1191–1212.
Grabau, A. W., 1904, On the classification of sedimentary rocks: Amer. Geol., v. 33, pp. 228–247.
———, 1919, Prevailing stratigraphic relationships of the bedded phosphate deposits of Europe, North Africa, and North America (abstr.): Bull. Geol. Soc. Amer., v. 30, p. 104.
———, 1920, Geology of the nonmetallic mineral deposits other than silicates. V. 1. Principles of salt deposition: New York, McGraw-Hill, 435 pp.
Greensmith, J. T., 1962, Rhythmic deposition in the Carboniferous oil-shale group of Scotland: Jour. Geol., v. 70, pp. 355–364.
Grim, R. E., 1936, The Eocene sediments of Mississippi: Bull. Mississippi Geol. Surv. 30, 240 pp.
Gross, G. A., 1972, Primary features in cherty iron-formations: Sed. Geol., v. 7, pp. 241–261.
Grunau, H. R., 1965, Radiolarian cherts and associated rocks in space and time: Eclogae Géol. Helvetiae, v. 58, pp. 157–208.
Gruner, J. W., 1922, Organic matter and the origin of the Biwabik iron-bearing formation of the Mesabi Range: Econ. Geol., v. 17, pp. 407–460.
———, 1935, The structural relationship of glauconite and mica: Amer. Mineral., v. 20, pp. 699–713.
———, 1946, Geology and mineralogy of the Mesabi Range: St. Paul, Iron Range Res. Rehab. Comm., 127 pp.
Guild, P. W., 1953, Iron deposits of the Congonhas District, Minas Gerais, Brazil: Econ. Geol., v. 48, pp. 639–676.
———, 1957, Geology and mineral resources of the Congonhas district, Minas Gerais, Brazil: U.S. Geol. Surv. Prof. Paper 290, 90 pp.
Hadding, A., 1932, Glauconite and glauconitic rocks. Part 4 of the Pre-Quaternary sedimentary rocks of Sweden: Lunds Univ. Arsskr., n.f., ard. 2, v. 28, 175 pp.
Hallimond, A. F., 1925, Iron ores: bedded ores of England and Wales: Great Britain Geol. Surv. Spec. Rpts., Min. Res., v. 29, 139 pp.
Harder, E. C., 1919, Iron-depositing bacteria and their geological relations: U.S. Geol. Surv. Prof. Paper 113, 89 pp.
Harder, H., 1963, Zur Diskussion über die Entstehung der Quarzbändererze (Itabirite): Neues Jahrb. Min. Bh., v. 12, pp. 303–314.
Hardie, L. A., 1967, Gypsum-anhydrite equilibrium at one atmosphere pressure: Amer. Mineral., v. 52, pp. 171–200.
———, 1968, The origin of the Recent non-marine evaporite deposit of Saline Valley, California: Geochim. Cosmochim. Acta, v. 32, pp. 1279–1301.
Hardie, L. A., and Eugster, H. P., 1971, The depositional environment of marine evaporites; a case for shallow, clastic accumulation: Sedimentology, v. 16, pp. 187–220.
Harris, L. D., 1958, Syngenetic chert in the Middle Ordovician Hardy Creek Limestone of southwest Virginia: Jour. Sed. Petrology, v. 28, pp. 205–208.
Hawley, J. E., and Beavan, A. P., 1934, Mineralogy and genesis of the Mayville iron ore of Wisconsin: Amer. Mineral., v. 19, pp. 493–
Hay, R. L., 1968, Chert and its sodium silicate precursors in sodium-carbonate lakes of East Africa: Contr. Min. Petrology, v. 17, pp. 255–274.
Hayes, A. O., 1915, The Wabana ores of Newfoundland: Geol. Surv. Canada, Mem. 66, 163 pp.
———, 1929, Further studies of the origin of the Wabana iron ore of Newfoundland: Econ. Geol., v. 24, pp. 1–4.
Hayes, C. W., 1896, The Tennessee phosphates: U.S. Geol. Surv. Ann. Rept., pt. 2, pp. 519–550.
Hayes, C. W., and Ulrich, E. O., 1903, Columbia, Tennessee: U.S. Geol. Surv. Atlas, folio no. 95, 6 pp.
Hendricks, S. B., Jefferson, M. E., and Mosely, V. M., 1932, The crystal structure of some natural and synthetic apatite-like substances: Zeitschr. Krist., v. 81, pp. 353–369.
Hinde, G. J., 1887, On the organic origin of the chert in strata of North Wales and Yorkshire: Geol. Mag., v. 24, pp. 435–446.
Holliday, D. W., 1970, The petrology of sec-

ondary gypsum rocks: a review: Jour. Sed. Petrology, v. 40, pp. 734–744.

Hollingsworth, S. E., 1948, Evaporites: Proc. Yorkshire Geol. Soc., v. 27, pp. 192–198.

Honess, C. W., 1923, Geology of the southern Ouachita Mountains of Oklahoma: Oklahoma Geol. Surv. 32, pt. 1, 278 pp.

Hoots, H. W., 1931, Geology of the eastern part of the Santa Monica Mountains, Los Angeles County, California: U.S. Geol. Surv. Prof. Paper 165-c, pp. 83–134.

Hoss, H., 1957, Untersuchungen über die Petrographie kulmischer Kieselschiefer: Beitr. Min. Pet., v. 6, pp. 59–88.

Hough, J. L., 1958, Fresh-water environment of deposition of Precambrian banded iron formations: Jour. Sed. Petrology, v. 28, pp. 414–430.

Hower, J., 1961, Some factors concerning the nature and origin of glauconite: Amer. Mineral., v. 46, pp. 313–334.

Huber, N. K., 1958, The environmental control of sedimentary iron minerals: Econ. Geol., v. 53, pp. 123–140.

———, 1959, Some aspects of the Ironwood Iron-formation of Michigan and Wisconsin: Econ. Geol., v. 54, pp. 82–118.

Humphries, D. W., 1956, Chert: its age and origin in the Hythe Beds of the western Weald: Proc. Geol. Assoc. London, v. 67, pp. 296–313.

Hunter, R. E., 1970, Facies of iron sedimentation in the Clinton Group, *in* Studies of Appalachian geology (Fisher, G. W., Pettijohn, F. J., Reed, J. C., Jr., and Weaver, K. N., eds.), New York: Wiley-Interscience, pp. 101–124.

Hurst, V. J., 1953, Chertification in the Fort Payne Formation, Georgia: Bull. Georgia Geol. Surv. 60, pp. 215–238.

Irving, R. D., and Van Hise, C. R., 1892, The Penokee iron-bearing series of Michigan and Wisconsin: U.S. Geol. Surv. Mono. 19, 534 pp.

James, H. L., 1951, Iron formation and associated rocks in the Iron River district, Michigan: Bull. Geol. Soc. Amer., v. 62, pp. 251–266.

———, 1954, Sedimentary facies of iron formation: Econ. Geol., v. 49, pp. 235–293.

———, 1955, Zones of regional metamorphism in the Precambrian of northern Michigan: Bull. Geol. Soc. Amer., v. 66, pp. 1455–1488.

———, 1966, Chemistry of the iron-rich sedimentary rocks: U.S. Geol. Surv. Prof. Paper 440-W, 61 pp.

James, H. L., Dutton, C. E., Pettijohn, F. J., and Wier, K. L., 1968, Geology and ore deposits of the Iron River–Crystal Falls District, Iron County, Michigan: U.S. Geol. Surv. Prof. Paper 570, 134 pp.

Jeffrey, E. C., 1915, The mode of origin of coal: Jour. Geol., v. 23, pp. 218–230.

Jones, B. F., 1965, The hydrology and mineralogy of Deep Springs Lake, Inyo County, California: U.S. Geol. Surv. Prof. Paper 502-A, 56 pp.

Kazakov, A. V., 1937, The phosphorite facies and the genesis of phosphorites, *in* Geological investigations of agricultural ores, Trans. Sci. Inst. Fertilizers and Insecto-Fungicides no. 142: 17th Sess. Int. Geol. Congr., Leningrad, pp. 95–113.

Keller, W. D., 1941, Petrography and origin of the Rex Chert: Bull. Geol. Soc. Amer., v. 52, pp. 1279–1298.

Kelley, V. C., 1951, Oolitic iron deposits of New Mexico: Bull. Amer. Assoc. Petrol. Geol., v. 35, pp. 2199–2228.

Ketner, K. B., 1969, Ordovician bedded chert, argillite, and shale of the Cordilleran eugeosyncline in Nevada and Idaho: U.S. Geol. Surv. Prof. Paper 650-B, pp. B23–B34.

King, R. H., 1947, Sedimentation in the Permian Castile sea: Bull. Geol. Soc. Amer., v. 31, pp. 470–477.

Kinsman, D. J. J., 1969, Modes of formation, sedimentary associations, and diagnostic features of shallow-water and supratidal evaporites: Bull. Amer. Assoc. Petrol. Geol., v. 53, pp. 830–840.

Kirkland, D. W., and Anderson, R. Y., 1970, Microfolding in the Castile and Todilto evaporites, Texas and New Mexico: Bull. Geol. Soc. Amer., v. 81, pp. 3259–3282.

Kozary, M. T., Dunlap, J. C., and Humphrey, W. E., 1968, Incidence of saline deposits in geologic time, *in* Saline deposits (Mattox, R. B., ed.): Geol. Soc. Amer. Spec. Paper 88, pp. 43–57.

Krauskopf, K. B., 1959, The geochemistry of silica in sedimentary environments, *in* Silica in sediments (Ireland, H. A., ed.): Soc. Econ. Paleont. Min. Spec. Publ. 7, pp. 4–19.

Kroenlein, G. A., 1939, Salt, potash, and anhydrite in the Castile Formation of southeastern New Mexico: Bull. Amer. Assoc. Petrol. Geol., v. 23, pp. 1682–1693.

Krumbein, W. C., 1951, Occurrence and lithologic associations of evaporites in the United States: Jour. Sed. Petrology, v. 21, pp. 63–81.

Krumbein, W. C., and Garrels, R. M., 1952, Origin and classification of chemical sediments in terms of *p*H and oxidation–reduction potentials: Jour. Geol., v. 60, pp. 1–33.

Krynine, P. D., 1945, Sediments and the search for oil: Producers Monthly, v. 9, pp. 12–22.

Krynine, P. D., Honess, A. P., and Myers, W. M., 1941, Siliceous oolites and chemical sedimentation (abstr.): Bull. Geol. Soc. Amer., v. 52, pp. 1916–1917.

Kühn, Robert, 1968, Geochemistry of the German potash deposits, *in* Saline deposits (Mat tox, R. B., ed.): Geol. Soc. Amer. Spec. Paper 88, pp. 427–504.

LaBerge, G. L., 1966a, Altered pyroclastic rocks in iron-formation in the Hammersley Range, Western Australia: Econ. Geol., v. 61, pp. 147–161.

———, 1966b, Altered pyroclastic rocks in South African iron-formation: Econ. Geol., v. 61, pp. 572–581.

Lacroix, A., 1910, Sur la constitution mineralogique des phosphates français: Comptes Rendus, Soc. Geol. France, v. 150, p. 1213.

Larsen, E. S., and Berman, H., 1934, The microscopic determination of the non-opaque minerals, 2nd ed.: Bull. U.S. Geol. Surv. 848, 266

Leith, C. K., 1903, Mesabi iron-bearing district of Minnesota: U.S. Geol. Surv. Mem. 43, 316 pp.

Leith, C. K., Lund, R. J., and Leith, A., 1935, Pre-Cambrian rocks of the Lake Superior region: U.S. Geol. Surv. Prof. Paper 184, 34 pp.

Lepp, Henry, and Goldich, S. S., 1964, Origin of the Precambrian iron formations: Econ. Geol., v. 59, pp. 1025–1060.

Levorsen, A. I., 1967, Geology of petroleum, 2nd ed.: San Francisco, Freeman, 924 pp.

Logan, B., and Chase, R. L., 1961, The stratigraphy of the Moora Group, Western Australia: Jour. Roy. Soc. Western Australia, v. 44, pp. 14–31.

Loomis, F. B., 1903, The dwarf fauna of the pyrite layers of the Tully limestone of western New York: Bull. New York State Mus., 69, pp. 892–920.

Lotze, F. W., 1957, Steinsalz und Kalisalz, 2nd ed. Pt. I: Berlin, Gebrüder Borntraeger, 465 pp.

Lovering, T. G., 1972, Jasperoid in the United States—its characteristics, origin, and economic significance: U.S. Geol. Surv. Prof. Paper 710, 164 pp.

Lowenstam, H. A., 1942, Geology of the eastern Nazareth Mountains, Palestine. I. Cretaceous stratigraphy: Jour. Geol., v. 50, p. 813.

———, 1948, Biostratigraphic studies of the Niagaran inter-reef formations in northeastern Illinois: Illinois State Mus. Sci. Papers, v. 4, 146 pp.

———, 1962, Magnetite in denticle capping in Recent chitons (Polyplacophora) Bull. Geol. Soc. Amer., v. 73, pp. 435–438.

McBride, E. F., and Thomson, Alan, 1970, The Caballos Novaculite: Geol. Soc. Amer. Spec. Paper 122, 129 pp.

McConnell, D., 1938, A structural investigation of the isomorphism of the apatite group: Amer. Miner., v. 23, pp. 1–19.

———, 1950, The petrography of rock phosphates: Jour. Geol., v. 58, pp. 16–23.

McKee, E. F., 1960, Spatial relations of fossils and bedded cherts in the Redwall limestone, Arizona: U.S. Geol. Surv. Prof. Paper 400-B, pp. 461–463.

McKelvey, V. E., 1967, Phosphate deposits: Bull. U.S. Geol. Surv., 1252-D, 22 pp.

McKelvey, V. E., Swanson, R. W., and Sheldon, R. P., 1953, The Permian phosphorites of western United States: 19th Int. Geol. Congr., Algiers, Comptes Rendus 11, pp. 45–64.

McKelvey, V. E., and others, 1956, Summary description of Phosphoria, Park City, and Shedhorn formations in western phosphate field: Bull. Amer. Assoc. Petrol. Geol., v. 40, pp. 2826–2863.

———, 1959, The Phosphoria, Park City, and Shedhorn formations in the western phosphate field: U.S. Geol. Surv. Prof. Paper 313-A, 47 pp.

Mansfield, G. R., 1918, Origin of the western phosphates of the United States: Amer. Jour. Sci., v. 46, pp. 591–598.

———, 1920, The physical and chemical character of New Jersey greensand: Econ. Geol., v. 15, pp. 547–566.

———, 1922, Potash in the greensands of New Jersey: Bull. U.S. Geol. Surv. 727, 146 pp.

———, 1927, Geography, geology, and mineral resources of part of southeastern Idaho: U.S. Geol. Surv. Prof. Paper 152, pp. 75–78, 208–

Marshall, C. E., 1942, Modern conceptions of the physical constitution of coal and related research in Great Britain: Jour. Geol., v. 50, pp. 385–405.

Masson, P. H., 1955, An occurrence of gypsum in southwest Texas: Jour. Sed. Petrology, v. 25, pp. 72–77.

Matthew, W. D., 1893, On phosphate nodules from the Cambrian of southern New Brunswick: Trans. New York Acad. Sci., v. 12, pp. 108–120.

Mattox, R. B., ed., 1968, Saline deposits: Geol. Soc. Amer. Spec. Paper 88, 701 pp.

Mellon, G. B., 1962, Petrology of Upper Cretaceous oolitic iron-rich rocks from northern Alberta, Econ. Geol., v. 57, pp. 921–940.

Middleton, G. V., 1958, Diagenesis of lowermost Devonian at Hagersville, Ontario, Canada: Proc. Geol. Assoc. Canada, v. 10, pp. 95–108.

Miller, A. M., 1896, The association of the gastropod genus *Cydora* with phosphate of lime deposits: Amer. Geol., v. 17, pp. 74–76.

Molengraaff, G. A. F., 1910, On oceanic deep-sea deposits of central Borneo: Proc. Konink. Akad. Wetensch. Amsterdam, v. 12, pp. 141–147.

Moore, E. S., 1940, Coal: its properties, analysis, classification, geology, extraction, uses and distribution, 2nd ed.: New York, Wiley, 473 pp.

Moore, E. S., and Maynard, J. E., 1929, Solution, transportation, and precipitation of iron and silica: Econ. Geol., v. 24, pp. 272–303, 365–402, 506–527.

Moore, R. C., ed., 1926, Geology of salt dome oil fields: Tulsa, Amer. Assoc. Petrol. Geol., 797 pp.

Mountain, E. D., 1952, The origin of silcretes: South African Jour. Sci., v. 48, pp. 201–204.

Mudge, M. R., 1972, Pre-Quaternary rocks in the Sun River Canyon area, northwestern Montana: U.S. Surv. Prof. Paper 663-A, 142 pp.

Murchison, D., 1969a, Report of the international commission for coal petrology: the work of the nomenclature subcommission from 1963–1966, Comptes Rendus 6th Int. Congr. Strat. Geol. Carb., Sheffield (1967), v. 1. pp. lii–lviii.

———, 1969b, Some recent advances in coal petrology: Comptes Rendus 6th Int. Congr. Strat. Geol. Carb., Sheffield, v. 1, pp. 351–364.

Murray, J., and Renard, A. F., 1891, Report on deep-sea deposits based on the specimens collected during the voyage of H.M.S. *Challenger* in the years 1872–1876: *Challenger* Reports, p. 396.

Murray, R. C., 1964, Origin and diagenesis of gypsum and anhydrite: Jour. Sed. Petrology, v. 34, pp. 512–523.

Neev, D., and Emery, K. O., 1967, The Dead Sea: Bull. Israel Geol. Surv. 41, 147 pp.

Ochsenius, C., 1877, Die Bildung der Steinsalzlager und ihrer Mutterlaugensalze: Halle, Pfeffer, 172 pp.

Oftedahl, C., 1958, A theory of exhalative-sedimentary ores: Geol. Fören. Stockholm Förh. v. 80, pp. 1–19.

Ogniben, L., 1957, Secondary gypsum of the Sulphur Series, Sicily, and the so-called integration: Jour. Sed. Petrology, v. 27, pp. 64–79.

Park, D. E., Jr., and Croneis, Carey, 1969, Origin of the Caballos and Arkansas novaculite formations: Bull. Amer. Assoc. Petrol. Geol., v. 53, pp. 94–111.

Pettijohn, F. J., 1926, Intraformational phosphate pebbles of the Twin City Ordovician: Jour. Geol., v. 34, pp. 361–373.

———, 1952, Geology of the northern Crystal Falls area, Iron County, Michigan: U.S. Geol. Surv. Circ. 153, 17 pp.

Pimm, A. C., Garrison, R. E., and Boyce, R. E., 1971, Sedimentology synthesis: lithology, chemistry and physical properties of sediments in the northwestern Pacific Ocean, *in* Initial reports of the Deep-Sea Drilling Project, v. 6, pp. 1131–1252.

Platt, L. B., 1962, The Rann of Cutch: Jour. Sed. Petrology, v. 32, pp. 92–98.

Poldervaart, A., 1955, Chemistry of the earth's crust, *in* Crust of the earth—a symposium (Poldervaart, A., ed.): Geol. Soc. Amer. Spec. Paper 62, pp. 119–144.

Polynov, B. B., 1937, The cycle of weathering (Muir, A., trans.): London, Murby, 220 pp.

Pratje, O., 1930, Rezente marine Eisen-Ooide aus der Nordsee: Centralbl. Min. Geol. Palaeont. Abt. B., pp. 289–294.

Prentice, J. E., 1958, The radiolarian cherts of North Devonshire, England: Eclogae Géol. Helvetiae, v. 51, pp. 706–711.

Radforth, N. W., 1969, Environmental and structural differentials in peatland development, *in* Environments of coal deposition (Dapples, E. C., and Hopkins, M. E., eds.): Geol. Soc. Amer. Spec. Paper 114, pp. 87–104.

Raistrick, A., and Marshall, C. E., 1948, The nature and origin of coal and coal seams: London, England Univ. Press, 282 pp.

Ramberg, Hans, 1952, The origin of metamorphic and metasomatic rocks: Chicago, Univ. Chicago Press, 317 pp.

Rastall, R. H., 1919, The mineral composition of the Lower Greensand strata of eastern England: Geol. Mag., v. 56, pp. 211–226, 265–272.

Richardson, A. W., 1919, Origin of Cretaceous flint: Geol. Mag., v. 58, pp. 114–124.

Richter-Bernburg, G., 1955, Über salinare Sedimentation: Deutsche Geol. Gesell. Zeitschr., v. 105, pp. 843–854.

———, 1964, Solar cycle and other climatic periods in varvitic evaporites, *in* Problems in palaeoclimatology: New York, Wiley-Interscience, pp. 510–519.

Rogers, A. F., 1922, Collophane, a much neglected mineral: Amer. Jour. Sci., ser. 5, v. 3, pp. 269–276.

———, 1944, Pellet phosphorite from Carmel Valley, Monterey County, California: California Jour. Mines Geol., v. 40, pp. 411–421.

Rubey, W. W., 1929, Origin of the siliceous Mowry Shale of the Black Hills region: U.S. Geol. Surv. Prof. Paper 154, pp. 153–170.

———, 1930, Lithologic studies of fine-grained Upper Cretaceous sediments of the Black Hills region: U.S. Geol. Surv. Prof. Paper 165A, 54 pp.

———, 1951, The geologic history of sea water: an attempt to state the problem: Bull. Geol. Soc. Amer., v. 62, pp. 1111–1148.

Ruedemann, R., and Wilson, T. Y., 1936, Eastern New York Ordovician cherts: Bull. Geol. Soc. Amer., v. 47, pp. 1535–1586.

Sakamoto, T., 1950, The origin of the Precambrian banded iron ores: Amer. Jour. Sci., v. 248, pp. 449–474.

Sampson, E., 1923, The ferrigenous chert formations of Notre Dame Bay, Newfoundland: Jour. Geol., v. 31, pp. 571–598.

Sargent, H. S., 1923, The massive chert formations of North Flintshire: Geol. Mag., v. 60, pp. 168–183.

———, 1929, Further studies of chert: Geol. Mag., v. 66, pp. 272–303.

Schaller, W. T., 1912, Mineralogical notes, ser. 2: Bull. U.S. Geol. Surv. 509, pp. 89–100.

Schaller, W. T., and Henderson, E. P., 1932, Mineralogy of drill cores from the potash field of New Mexico and Texas: Bull. U.S. Geol. Surv. 833, 124 pp.

Schmalz, R. F., 1969, Deep-water evaporite deposition: a genetic model: Bull. Amer. Assoc. Petrol. Geol., v. 53, pp. 798–823.

Schneider, H., 1927, A study of glauconite: Jour. Geol., v. 35, pp. 299–310.

Schoen, Robert, 1964, Clay minerals of the Silurian Clinton ironstones, New York State: Jour. Sed. Petrology, v. 34, pp. 855–863.

Scruton, L. P., 1953, Deposition of evaporites: Bull. Geol. Soc. Amer., v. 37, pp. 2498–2512.

Sheppard, R. A., and Hunter, R. E., 1960, Chamosite oolites in the Devonian of Pennsylvania: Jour. Sed. Petrology, v. 30, pp. 585–588.

Siebenthal, C. E., 1915, Origin of the zinc and lead deposits of the Joplin region: Bull. U.S. Geol. Survey 606, 283 pp.

Siever, Raymond, 1962, Silica solubility, 0°–200°C and diagenesis of siliceous sediments: Jour. Geol., v. 61, pp. 127–150.

Sloss, L. L., 1953, The significance of evaporites: Jour. Sed. Petrology, v. 23, pp. 143–161.

Smith, D. B., 1971, Possible displacive halite in the Permian Upper Evaporite Group of northeast Yorkshire: Sedimentology, v. 17, pp. 221–232.

Smith, G. I., and Haines, D. V., 1964, Character and distribution of nonclastic minerals in the Searles Lake evaporite deposit, California: Bull. U.S. Geol. Surv. 1181-P, 58 pp.

Smith, L. E., Hosford, G. E., Sears, R. S., Sprouse, D. P., and Stewart, M. D., 1952, Stratigraphic sections of the Phosphoria Formation in Utah: Circ. U.S. Geol. Survey No. 211, 48 pp.

Smyth, C. H., Jr., 1892, On the Clinton iron ore: Amer. Jour. Sci., ser. 3, v. 43, pp. 487–496.

Sosman, R. B., 1927, The properties of silica: New York, Chemical Catalogue Co., 856 pp.

Spackman, W., Riegel, W. L., and Dolsen, C. P., 1969, Geological and biological interactions in the swamp-marsh complex of southern Florida, *in* Environments of coal deposition (Dapples, E. C., and Hopkins, M. E., eds.): Geol. Soc. Amer. Spec. Paper 114, pp. 1–35.

Sprunk, G. C., 1942, Influence of physical constitution of coal on its chemical hydrogenation, and carbonization properties: Jour. Geol., v. 50, pp. 411–436.

Steinmann, Gustav, 1925, Gibt es fossile Tiefseeablagerungen von erdgeschichtlicher Bedeutung?: Geol. Rundschau, v. 16, pp. 435–468.

Stewart, F. H., 1949, The petrology of the evaporites of the Eskdale No. 2 Boring, East Yorkshire: Mineral. Mag., v. 28, pp. 621–675; v. 29, pp. 445–475, 557–572.

———, 1951, The petrology of the evaporites of the Eskdale no. 2 boring, east Yorkshire: Mineral. Mag., v. 29, pt. 2, pp. 445–475; pt. 3, pp. 557–572.

———, 1953, Early gypsum in the Permian evaporites of northeastern England: Proc. Geol. Assoc., v. 64, pp. 33–39.

———, 1963, Marine evaporites, *in* Data of geochemistry (Fleischer, M., ed.): U.S. Geol. Surv. Prof. Paper 440-Y, 54 pp.

Stöcklin, J., 1968, Salt deposits of the Middle East, *in* Saline deposits (Mattox, R. B., ed.): Geol. Soc. Amer. Spec. Paper 88, pp. 157–181.

Stone, R. W., and others, 1920, Gypsum deposits of the United States: U.S. Geol. Surv. Bull. no. 697, 326 pp.

Stopes, M. C., 1919, On the four visible ingredients in banded bituminous coal: Proc. Roy. Soc. London, v. B, 90, pp. 69–87.

———, 1935, On the petrology of banded bituminous coal: London, Fuel, v. 14, pp. 4–13.

Stout, Wilbur, 1944, The iron-bearing formations of Ohio: Bull. Geol. Surv. Ohio, ser. 4, no. 45, 230 pp.

Stout, W., and Schoenlaub, R. A., 1945, The occurrence of flint in Ohio: Bull. Ohio Geol. Surv., ser. 4, no. 46, 110 pp.

Strakhov, N. M., 1959, Schéma de la diagenèse des depôts marins: Eclogae Geol. Helvetiae, v. 51, pp. 761–767.

Surdam, R. C., Eugster, H. P., and Mariner, R. H., 1972, Magadi-type cherts in Jurassic and Eocene to Pleistocene rocks: Bull. Geol. Soc. Amer., v. 83, pp. 2261–2266.

Takahashi, J., 1939, Synopsis of glauconitization, *in* Recent Marine Sediments (Trask, P. D., ed.) Tulsa, Okla., Amer. Assoc. Petrol. Geol., pp. 503–512.

Taliaferro, N. L., 1933, The relation of volcanism to diatomaceous and associated siliceous sediments: Bull. Univ. California Publ. Dept. Geol. Sci., v. 23, pp. 1–55.

———, 1934, Contraction phenomena in cherts: Bull. Geol. Soc. Amer., v. 45, pp. 196–197 ff.

———, 1935, Some properties of opal: Amer. Jour. Sci., ser. 5, v. 30, pp. 450–474.

———, 1943, Franciscan-Knoxville problem: Bull. Amer. Assoc. Petrol. Geol., v. 27, pp. 109–219.

Tanton, T. L., 1950, The origin of iron range rocks: Trans. Roy. Soc. Canada, ser. 3, v. 44, pp. 1–19.

Tarr, W. A., 1917, Origin of chert in the Burlington Limestone: Amer. Jour. Sci., ser. 4, v. 44, pp. 409–452.

———, 1926, The origin of flint and chert: Univ. Missouri Studies, v. 1, no. 2, pp. 1–54.

———, 1938, Terminology of the chemical siliceous sediments: Rept. Comm. Sed. 1937–1938, Nat. Res. Coun., pp. 8–27 (mimeographed).

Taylor, J. H., 1949, Petrology of the Northampton sand ironstone formation: Great Britain Geol. Surv. Mem., 111 pp.

Taylor, R. E., 1937, Water-insoluble residues in rock salt of Louisiana salt plugs: Bull. Amer. Assoc. Petrol. Geol., v. 21, pp. 1268–1310.

Thiessen, R., 1920, Compilation and composition of bituminous coals: Jour. Geol., v. 28, pp. 185–209.

Thurston, D. R., 1972, Studies on bedded cherts: Contr. Min Petrology, v. 36, pp. 329–334.

Tomkeieff, S. I., 1954, Coals and bitumens and related fossil carbonaceous substances; nomenclature and classification: London, Pergamon, 122 pp.

Trask, P. D., 1939, Organic content of Recent marine sediments, *in* Recent marine sediments (Trask, P. D., ed.): Tulsa, Okla., Amer. Assoc. Petrol. Geol., pp. 428–453.

Trask, P. D., and Hammar, H. E., 1934, Organic content of sediments: Amer. Petrol. Inst., 15th Ann. Mtg., Dallas, 14 pp. (reprint).

Trask, P. D., and Patnode, H. W., 1936, Means of recognizing source beds, *in* Drilling and production practice: Chicago, Amer. Petrol. Inst., pp. 368–384.

Trefethen, J. M., 1947, Some features of the cherts in the vicinity of Columbia, Missouri: Amer. Jour. Sci., v. 245, pp. 56–58.

Trendall, A. F., 1965, Origin of Precambrian iron-formations (discussion): Econ. Geol., v. 60, pp. 1065–1070.

———, 1968, Three great basins of Precambrian banded iron-formation deposition: Bull. Geol. Soc. Amer., v. 79, pp. 1527–1544.

Twenhofel, W. H., 1950, Principles of sedimentation: New York, McGraw-Hill, 673 pp.

Twenhofel, W. H., and Blackwelder, E., 1932, Phosphatic sediments, *in* Treatise on sedimentation (Twenhofel, W. H., ed.): Baltimore, Williams and Wilkins, pp. 546–561.

Tyler, S. A., 1950, Sedimentary iron deposits, *in* Applied sedimentation (Trask, P. D., ed.): New York, Wiley, pp. 506–523.

Tyler, S. A., Barghoorn, E. S., and Barrett, L. P., 1957, Anthracitic coal from Precambrian Upper Huronian black shale of the Iron River district, northern Michigan: Bull. Geol. Soc. Amer., v. 68, pp. 1293–1304.

Tyler, S. A., and Twenhofel, W. H., 1952, Sedimentation and stratigraphy of the Huronian of Upper Michigan: Amer. Jour. Sci., v. 250, pp. 1–27, 118–151.

Udden, J. A., 1912, Geology and mineral resources of the Peoria Quadrangle: Bull. U.S. Geol. Surv. 506, pp. 45–70.

———, 1924, Laminated anhydrite in Texas: Bull. Geol. Soc. Amer., v. 35, pp. 347–354.

Urasov, G. G., and Polyakov, V. D., 1956, Salts of Kara-Boghaz-Gol: Priroda, no. 9, p. 61 (in Russian).

Usiglio, J., 1849, Analyse de l'eau de la Mediterranée sur la côte de France: Ann. Chim. Phys., ser. 3, v. 27, pp. 92, 172 (see Clarke, 1924, pp. 219–220).

Van Hise, C. R., and Leith, C. K., 1911, The geology of the Lake Superior region: U.S. Geol. Surv. Monogr. 52, 641 pp.

Van Tuyl, F. M., 1912, A study of the cherts of the Osage series of the Mississippian system: Proc. Iowa Acad. Sci., v. 19, pp. 173–174.

———, 1918, The origin of chert: Amer. Jour. Sci., ser. 4, v. 45, pp. 449–456.

Wanless, H. R., 1950, Late Paleozoic cycles of sedimentation in the United States: 18th Int. Geol. Congr., Rept., pt. IV, pp. 17–28.

Washburn, E. W., and Navais, L., 1922, The relation of chalcedony to other forms of silica: Proc. Nat. Acad. Sci., v. 8, pp. 1–5.

Weller, J. M., 1930, Cyclical sedimentation of the Pennsylvanian Period and its significance: Jour. Geol., v. 38, pp. 97–135.

Wells, R. C., 1937, Analyses of rocks and minerals from the laboratory of the United States Geological Survey, 1914–36: U.S. Geol. Surv. Bull. no. 878, 134 pp.

West, I. M., 1964, Evaporite diagenesis in the Lower Purbeck Beds of Dorset: Proc. Yorkshire Geol. Soc., v. 34, pt. 3, pp. 315–330.

White, D. A., 1954, Stratigraphy and structure of the Mesabi Range, Minnesota: Bull. Minnesota Geol. Surv. 38, 92 pp.

White, D. E., 1947, Diagenetic origin of chert lenses in limestones at Soyatal, state of Querétaro, Mexico: Amer. Jour. Sci., v. 245, pp. 49–55.

White, D., and Thiessen, R., 1913, The origin of coal: Bull. U.S. Bur. Mines 38, 390 pp.

Williamson, W. O., 1957, Silicified sedimentary rocks in Australia: Amer. Jour. Sci., v. 255, pp. 23–42.

Winkler, A., 1925, Über die Bildung mesozoischer Hornsteine—ein Beitrag zur sedimentpetrographie der julischen Alpen: Tschermaks Min. Pet. Mitt., v. 38, pp. 424–455.

12

CONCRETIONS, NODULES, AND OTHER DIAGENETIC SEGREGATIONS

INTRODUCTION

Chemical action penecontemporaneous with or subsequent to sedimentation is responsible for various sedimentary structures. In this chapter we are concerned with those objects which are formed by precipitation or segregation of mineral matter and include such bodies as nodules, concretions, geodes, and septaria. Although these are usually of small volume and a minor feature, they are very common. In general they seem to be segregations of the rarer constituents of the host rock in which they occur.

Until recently concretions have been looked on as curiosities and for the most part have not been given serious study. Recent work, especially that of Raiswell (1971), has shown that concretions shed a good deal of light on the nature and sequence of diagenetic processes that affect sediment. The concretion contains a "frozen" record of the condition of the sediment at the time of deposition; it also carries a record of the consolidation process. The study of concretions, therefore, has much greater significance than was previously supposed.

DEFINITIONS AND CLASSIFICATION

So many objects of diverse character have been collectively called *concretions* that this term has only a very general meaning. It has been used interchangeably with the term *nodule*, likewise without precise meaning. All definitions, however, specify a composition different from the host rock, and all also indicate formation by precipitation from an aqueous solution within the host rock, a specification necessary to distinguish a concretion or nodule from a pebble. Both pebbles and concretions are thus genetically defined.

Recognizing the need for better definitions and terminology, Todd (1903) proposed four terms: *accretion, intercretion, excretion,* and *incretion,* based on mode of growth. Accretions grow from the center outward in a regular manner; intercretions are septaria which form by irregular and interstitial growth; excretions grow from the exterior inward; and incretions are cylindrical forms with a hollow core. Todd's proposal seems not to have been generally followed.

Usage has recognized difference in form and internal structure, and special names such as *geodes, septaria,* and *cone-in-cone* have been applied to concretionary bodies that display such differences. Most often, also, the term *nodule* is applied to less regular and structureless bodies, such as flint and chert nodules of many carbonate rocks. These are not called concretions—a term reserved for more regular forms many of which, unlike the chert nodules, have a nucleus. Rather obviously we are in need of better terminology.

Diagenetic segregations may be classified on the basis of time of formation, form, internal structure, and composition.

Various concretionary bodies may be *synsedimentary*, as are the manganese nodules of the present deep sea; *penecontemporaneous*, that is, early diagenetic before compaction of the enclosing sediment; or *epigenetic*, following consolidation of the host rock.

They may also be classified on their external form, either nodular or stratiform. Most diagenetic segregations have formed about a center, often about a foreign object, a leaf or shell, as a nucleus. But some, such as cone-in-cone, are secreted or precipitated along a bedding plane and thus are tabular in form.

Segregations may also be classified on the basis of their internal structure.

If certain pseudoconcretionary forms, such as armored mud balls, certain "lime mud balls" (Croneis and Grubbs, 1939), and certain algal structures or oncolites (Fig. 10-18) are excluded, then various secondary-mineral aggregations fall into four fundamentally different classes—classes which differ in their internal structure and organization. The first class, which includes such objects as chert and flint nodules, are rather irregular in form and have no regular internal structure. They seem to be a product of post-depositional replacement of the host rock, although some geologists have postulated a primary origin for these bodies. A second group are crystal aggregates, which are *coarsely* crystalline bodies. Some show a radial arrangement of acicular crystals (spherulites); others display a radial symmetry (rosettes) or a less regular intergrowth of coarse crystals (sand crystals). Some include and some exclude the matrix in which they grew. A third class of concretionary bodies are normally subspherical, commonly very oblate to composite forms resulting from growth about several centers, generally formed by orderly precipitation of mineral matter in the *pores* of a sediment adjacent to a nucleus. The bedding of the host rock appears to pass through these objects. They are termed simply concretions. The fourth and last class of mineral segregations are those related to the filling of a cavity or fracture or set of fractures. They are not all alike and do not form a homogeneous group. Included here are simple veins, septarian vein networks, geodes, and so forth.

Segregations may be classified according to their composition, a classification analogous to that of nondetrital sediments. Hence we have siliceous segregations (chert, flint), carbonates (calcite, aragonite, siderite), phosphates, iron oxides, sulfates (gypsum, barite), and sulfides (marcasite, pyrite).

ORIGIN

It is significant that the various structures designated *concretions* are mainly segregations of minor constituents of the rock in which they are found. They are the silica of the carbonate host rock, the lime carbonate of the shale or sandstone, the iron sulfide of the black shale, the phosphorus, manganese, barium, or other minor elements in the host rock. Introduction of foreign matter does not seem necessary for precipitation of these materials. As noted by Ramberg (1952, p. 222), because of surface energy differences, free energy will be less if these constituents occur in clusters rather than finely divided and disseminated; hence aggregates will form in time.

An important problem of origin is the *manner* of emplacement of the concretionary body. Segregations may make room for themselves by replacing the host rock, as do chert nodules; by deposition in preexisting openings or fractures, as in geodes and veins; or by deposition in available pores around loci or centers in the host rock, as are calcareous concretions of sand and silt —the *Kugelsandstein* and the sand crystals of calcite, gypsum, and barite. Finally, they make way for themselves by thrusting aside the enclosing matrix. It is incumbent, therefore, upon the observer to marshal and apply criteria needed to distinguish the several modes of emplacement. These are discussed below.

A related problem is that of the *time* of emplacement and the criteria for deciding this question. Some writers have concluded that segregations are *syngenetic*, that is, formed essentially on the sea floor. This concept has been applied especially to chert nodules (Tarr and Twenhofel, 1932; Trefethen, 1947) and certainly applies to the barite and manganese nodules now forming on the sea floor. In some concretions, enclosed fossils, unlike those of the matrix, are not crushed but are instead full-bodied and suggest, therefore, early formation of the enclosing concretionary body while the sediments are still soft and unconsolidated. These are *diagenetic.* In others, the passage of bedding planes through the concretionary structure without distortion at the concretion–matrix interface suggest a late post-consolidation origin for such structures. These are *epigenetic.* A post-depositional origin for others is suggested by deformation of the enclosing rock as a result of the growth of the concretionary body (Daly, 1900), although care must be taken to distinguish such deformation from that caused by compaction of the matrix about a resistant body.

What determines the size, number, and spacing of concretions? And their orientation and shapes? In many cases, as in calcareous concretions in shale, the locus of deposition is a bit of organic matter such as a leaf or a fish. Many concretions are renowned for their excellent fossils, as, for example, those with plant and insect fossils from Mazon Creek, Illinois. As noted by Weeks (1953), the ammonia generated by decomposing matter may have locally altered the pH so that deposition of calcium carbonate took place which was elsewhere inhibited. The location of chert nodules, barite rosettes, and the like probably cannot be so explained. The rhythmic spacing of flint nodules in Cretaceous chalk has been noted (Richardson, 1919b). In some sequences distribution is controlled by permeability; the more permeable beds contain concretions, the less permeable do not. Little is known about orientation of nonspherical concretions other than they tend to be extended parallel to the bedding. A regional orientation of long axes of carbonate concretions in Devonian strata of New York state is reported by Colton (1967). Such orientation may be a response to differential permeability related to a grain fabric controlled by a regional current system or to growth around plant fragments which were oriented by such a current system.

Many unexplained observations remain. Some segregations replace their matrix; others do not. In general those of carbonate rocks are replacement bodies; those of shales are not. Some concretions include much of the host rock, especially the carbonate concretions of sandstones and large crystal aggregates of calcite, barite, and gypsum in sand. On the other hand, concretions in shales seem to exclude most of the matrix. Some concretions are microcrystalline; others are formed of large single crystals or crystal clusters or radial aggregates. Some are homogeneous, others radial. Clearly the factors underlying these observations are not yet well understood.

Many special problems of origin are discussed in the sections dealing with the several types of concretions, notably geodes, septaria, and cone-in-cone.

CARBONATE CONCRETIONS AND NODULES

CONCRETIONS OF SAND AND SILTSTONES

Kugelsandstein These objects are the product of localized precipitation of mineral matter—mostly carbonate—in the pores of the sediment about a nucleus or center (Fuhrmann, 1968). They are generally spherical, spheroidal, or disk-shaped, although diverse shapes arise by fusion of two or more simple forms. They vary in size from small objects 1 cm in diameter to great spheroidal bodies as much as 9 m in diameter, such as those in the Cretaceous Dakota Sandstone near Minneapolis, Kansas (Fig. 12-1), and in the Fox Hills Sandstone of Colorado (Mathias, 1931). The size seems to be in part determined by the permeability of the rock; those in sandstones are larger (and more spherical) than those of siltstones.

These concretions, like sand crystals, contain a great deal of host rock material. The bedding planes of the host rock pass through them which indicates that these bodies were formed after deposition of the enclosing sediment. They are thus a phenomenon of localized cementation. Such cementation may have been relatively early; the heavy-mineral assemblage of the concre-

FIG. 12-1. Large concretions in Dakota Sandstone (Cretaceous), "Rock City," southwest of Minneapolis, Ottawa County, Kansas. (Photograph by Jim Enyeart, courtesy of Kansas State Geologic Survey.)

tionary body may differ from that of the host rock by being richer in less stable species (Bramlette, 1941).

Imatra stones Of common occurrence and described many times in the literature are the calcareous concretions of certain Late Glacial clays (Fig. 12-2). These objects, variously called *imatra stones* or *marlekor,* occur in the silt beds interstratified with the varved lake clays of Finland, Sweden, and Norway, where they were observed early, and in similar deposits in North America, especially in the glaciolacustrine sediments of the Connecticut Valley (Tarr, 1935), in

FIG. 12-2. Calcareous concretions, imatra stones, about one-half size.

the Ottawa region (Kindle, 1923), and in the Espanola region on the north shore of Lake Huron (Quirke, 1917, p. 56, Pl. V).

The most complete description of these structures is that given by Tarr (1935). Concretions occur primarily in silt layers; they are flattened parallel to the bedding, are generally disk-shaped, and are characterized by the bedding laminations passing through them. Although generally small, 2 to 3 cm in diameter, and of simple form, some are composite and larger, 10 cm or more, and commonly of bizarre form. These complex forms are the result mainly of coalescence of simple types producing dumbbell-shaped pairs and, in part, also the result of warty outgrowths and other irregular overgrowths which defy description. In some cases these complex forms display a strange bilateral symmetry.

Imatra stones are highly calcareous (Table 12-1), although all contain a large amount of silt. Those of the Connecticut Valley are about 50 percent carbonate and 50 percent clayey silt. Some from the Abitibi region of Quebec have 55 to 65 percent carbonate, mainly calcite (Warkentin, 1967), but with a few percent dolomite. The percentage of carbonate is approximately that required to fill the pores of a freshly deposited silt. Tarr's analyses show the concretions to have a surprising content of MnO, presumably as a carbonate.

It is clear, therefore, that the form, structure, and composition of these concretions are the result of localized and perhaps early precipitation of calcium carbonate in the pores of silt beds. Tarr attempted to elaborate in some detail the mechanism of the process.

Quirke (1917, p. 58) thought that many of them were formed and "rolled about by water action, while they were still in a plastic condition."

Sand crystals Sand crystals are large (5 to 10 cm and even larger) euhedral calcitic crystals or clusters of such euhedra of scalenohedral habit. These crystals or crystal aggregates incorporated much sand during their crystallization as dogtooth spar. They occur in friable sands and hence can be extracted intact from their matrix (Fig. 12-3). Best known are the sand crystals of the Fontainebleau Sandstone (Tertiary) of France (Cayeux, 1929, p. 154) and those of Miocene paleo-river channels the Arikaree Group in the Badlands of South Dakota (Barbour, 1901). The formation of sand crystals is related to luster-mottling seen in some sandstones and to the cementation of sandstone, a topic discussed elsewhere. Sand crystals and *Kugelsandstein* have been recently discussed by Fuhrmann (1968).

CALCAREOUS CONCRETIONS OF SHALE

Hundreds of examples of carbonate concretion-bearing shales might be cited the world over. They are especially common in thinly plated, black shales. They are generally spherical to oval and are apt to be slightly flattened parallel to the bedding. In size they range from a few centimeters to approximately 8 m in the longest dimension. Many

TABLE 12-1. Chemical analyses of carbonate concretions

Constituent	A	B	C	D	E
SiO_2	9.08	23.92	14.18	4.72	13.1
Al_2O_3	1.87	6.82	4.09	2.45	35.1
Fe_2O_3	5.03[a]	1.12	1.47	1.64	4.7[a]
FeO	—	2.57	35.73	41.68	—
MgO	13.80	2.26	2.75	4.34	—
CaO	27.29	31.35	4.44	6.13	22.9
Na_2O	—	1.16	0.55	0.22	—
K_2O	—	1.84	0.60	0.27	—
H_2O+	—	2.04	2.39	2.11	—
H_2O-	0.20	0.64	—	0.38	—
TiO_2	0.12	0.62	0.49	0.10	—
P_2O_5	trace	0.36	0.52	1.77	—
MnO	0.50[b]	0.93	1.04	0.60	—
CO_2	38.83	24.24	31.00	33.37	18.0
C	2.25	—	0.87	—	—
S	0.67	—	—	0.35	—
Total	99.65	99.87	100.13	100.15[c]	93.8

[a] Total iron.
[b] Given as MnO_2.
[c] Total includes SO_3, 0.07; BaO, 0.01; SrO, 0.05; F, 0.03.

A. Concretion from Ohio Shale (Devonian), Ohio. D. J. Demorest, analyst (Stout, 1944, p. 14).
B. Imatra stone, a calcareous concretion in Late Glacial clay, east Finland. L. Lokka, analyst (Eskola, 1932).
C. Sideritic concretion ("clay ironstone"), Upper Carboniferous, Cantabrian Mountains, Spain (Nederlof, 1959, p. 634).
D. Iron carbonate nodules, Middle Lias, England. C. O. Harvey, analyst (Taylor, 1950, p. 21).
E. Septarian nodule, partial analysis of interior matrix from Lias Shale, England (Richardson, 1919b).

FIG. 12-3. Sand crystals from paleo-river channel (Miocene) sandstone, Badlands, South Dakota. Scalenohedral calcite loaded with included sand grains. Length of specimen about 6 inches (15 cm).

contain fossil nuclei, notably fish or ammonites. The bedding of the shale bends or "flows" around the concretion, both above and below.

The concretions are usually composed of calcium carbonate, largely calcite; others are sideritic clay-ironstones, and some are decidedly manganiferous (Table 12-1). Some carbonate concretions have a special internal structure. Those with a network of veins are the septarian nodules. Those with a cone-in-cone structure form another class. Both septarian and cone-in-cone are described below.

Well-described examples include the fish-bearing carbonate concretions of dark Cretaceous shales of the Magdalena Valley of Colombia (Weeks, 1953; 1957), the carbonate concretions of the Devonian Ohio Shale of Ohio (Daly, 1900; Clifton, 1957) which in places have also yielded fish fossils, the concretions—some notably manganiferous—of the Pierre Shale (Cretaceous) of South Dakota (Gries and Rothrock, 1941), and the Cambrian concretions of South Wales and the Liassic concretions of Dorset, both described by Raiswell (1971).

The age of these concretions relative to time of deposition has been much debated. Proposed dating criteria are ambiguous. Sedimentary laminations are present in many concretions. Those close to the major axis are horizontal and continuous with laminations in the surrounding shale, whereas those near the margins of the concretion converge toward the major axis (Fig. 12-4). The continuity of laminations shows the concretion to have grown after deposition of the enclosing shales, whereas converging laminations indicate continued growth following partial compaction of the enclosing shale. Clearly they belong to the diagenetic period and are not epigenetic. Weeks' (1953) observation on the whole-bodied, uncrushed nature of enclosed fossil fish shows that growth preceded compaction. The deformation of shale laminations around the concretion is, therefore, a compaction phenomenon and not a result of forceful growth.

The observations of Raiswell (1971, p. 164) that the carbonate content of the central part of the concretion is higher than that in the margin (Fig. 12-5) is readily explained if it is presumed that the carbonate was deposited in the pores of the sediment and hence the carbonate content of the concretion is related to the porosity of the host rock sediment. If so, analytical data suggest continued growth of the concretion during

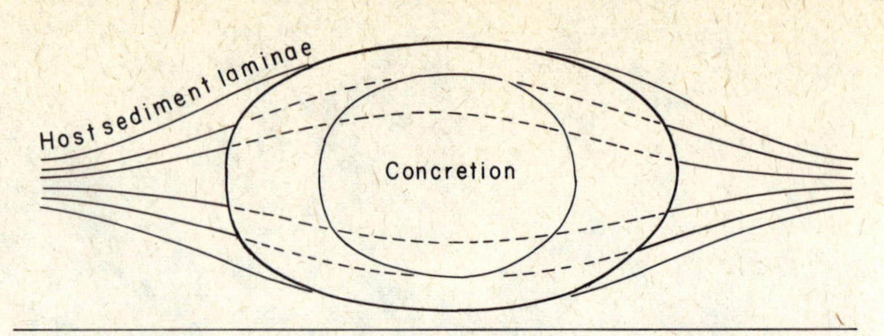

FIG. 12-4. Relations between calcareous nodule and bedding laminations. (1) initial cementation of uncompacted sediment; (2) sediment compacts around cemented concretion and is itself cemented; (3) the laminations of the host sediment are further compacted. (From Raiswell, 1971, Sedimentology, v. 17, Fig. 3.)

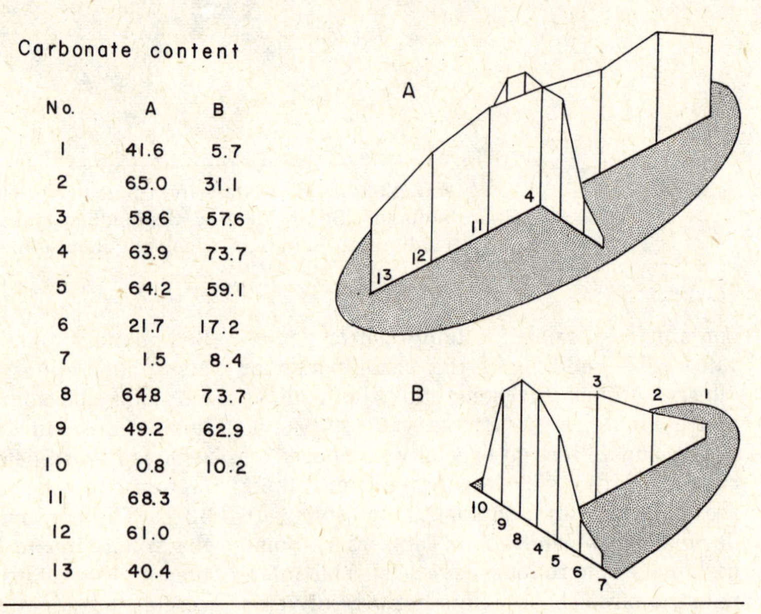
Carbonate content

No.	A	B
1	41.6	5.7
2	65.0	31.1
3	58.6	57.6
4	63.9	73.7
5	64.2	59.1
6	21.7	17.2
7	1.5	8.4
8	64.8	73.7
9	49.2	62.9
10	0.8	10.2
11	68.3	
12	61.0	
13	40.4	

FIG. 12-5. Carbonate content within concretions. Values of the carbonate content are proportional to vertical lines. (From Raiswell, 1971, Sedimentology, v. 17, Fig. 5.)

the compaction process—a conclusion supported by convergent laminations (see also Oertal and Curtis, 1972).

Based on this premise the initial porosities of the host sediment were high—85 percent in some cases. Such high porosities are recorded in the top 5 m of Recent fine-grained argillaceous sediments. The concretions, therefore, formed early, close to the sediment surface. Concretions showing a lesser porosity in the central region presumably formed at a later diagenetic stage, perhaps concurrently with the more peripheral parts of other concretions whose growth began earlier. The early growth of concretions at such shallow depths makes it seem likely that at times subaqueous erosion may wash out these bodies, an event that appears actually to have happened in some cases.

Problems of concretionary growth presented by calcareous nodules in shale also apply in a similar or modified way to those in siltstone and sandstone. The shape of the single concretionary body is a reflection of the permeability anisotropy of the host rock. Sandstone has an almost isotropic permeability; hence, calcareous concretions of this rock are almost spheroidal, whereas those of shales are flattened, with the vertical axis no more than half, on the average, of the horizontal axis. The circularity of the bedding plane sections of a concretion attest to the uniform permeability along a bedding plane; the exception is the elongated and oriented concretions of Colton (1967). Within the same sediment, earlier concretions tend to be more spheroidal than later ones (Raiswell, 1971, p. 166).

The question also arises, Did the concretions grow in a closed system, deriving all their material from the enclosing host sediment? Or were they formed in an open system involving an exchange between pore water and superjacent sea water? A high initial porosity indicating growth just below the sediment–water interface suggests the possibility of interchange of waters. A low initial porosity suggests the reverse. Mineralogy may also define the growth environment. Berner (1964) presumed that development of pyrite on the margins of concretions implies renewal of sulfate ions, possible only by diffusion from sea water. Siderite, on the other hand, requires CO_2 pressure higher than that of normal sea water (Curtis, 1967) and hence implies a closed system. The Sr^{2+} content has also been used to discriminate between open and closed systems (Hoefs, 1970). For further discussion of these and related problems of concretionary growth, refer to Raiswell's extended discussion (1971) and to papers on the subject by Berner (1964, 1968a) and Seibold (1962).

Isotopic analysis, particularly the ^{13}C–^{12}C ratio, of concretions presumed to have formed in a closed system makes it possible to interpret ancient sedimentary environments. By applying this technique to siderite nodules of Pennsylvanian age, Weber, Williams, and Keith (1964) were able to separate marine, brackish water, and freshwater shales from one another.

SEPTARIAN NODULES

Septaria are large (10 to 100 cm), distinctly oblate, nodules characterized by a series of radiating cracks that widen toward the center and die out near the margin and that is crossed by a series of cracks concentric with the margin (Fig. 12-6). Uniformity of

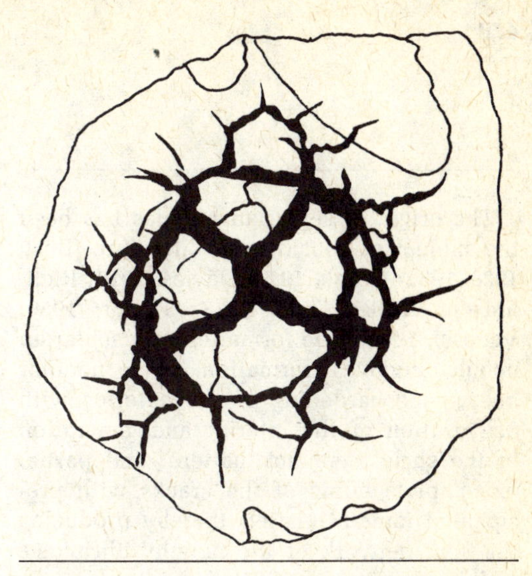

FIG. 12-6. Septarian structure. Length of specimen about 5 inches (12 cm).

pattern is the exception, and great irregularity generally prevails. The crack system, in section, more commonly appears polygonal, although near the margins the radial arrangement persists. Whereas the width of cracks in general is independent of position in the nodule, those near the margin do wedge out toward the perimeter; marginal cracks rarely extend to the outside. The cracks are invariably filled with a crystalline deposit, most commonly calcite. Many septaria, released from their shale matrix, weather and are so eroded that the interior vein system may be seen ("turtle backs"). The vein system may even be wholly freed of all adhering material. Such structures have been termed *melikaria* (Fig. 12-7). However, some are said to form in shales with no relation to nodule development (Burt, 1928).

Septarian nodules, except for vein fillings, are normally impure argillaceous carbonate bodies enclosed in shale. In some the carbonate is rich in ferrous iron, and the nodule is a true clay ironstone (Table 12-1 C and D). According to Richardson (1919a), the central part is more aluminous than the margins, an observation confirmed by Vanossi (1964, Fig. 1). The vein filling is mainly calcite, but pyrite, sphalerite, and barite are commonly present (Taylor, 1950).

A peculiarity of some septarian nodules is the shell of cone-in-cone which jackets some of these bodies.

The origin of septarian nodules has been the subject of considerable literature (Burt, 1928, 1932; Crook, 1913; Davies, 1913; Richardson, 1919a; Todd, 1913; Taylor, 1950; Vanossi, 1964). The formation of a septarian nodule involves formation of a nodular body, case hardening of the exterior with dehydration of the interior and generation of the shrinkage-crack pattern, and partial or complete filling of the cracks with precipitated mineral matter, thereby producing the vein network of the nodule. Shrinkage and production of synaresis cracks imply a gel-like character of the original body, an aspect of septarian development not well understood. Raiswell (1971, p. 156) believed septarian structures were characteristic of highly porous water-laden sediment and should, therefore, be limited to early concretions. On the other hand, the cone-in-cone structure associated with some concretions was a somewhat later product formed after some compaction of the sediment had taken place. The case hardening, dehydration, and shrinkage cracking are irreversible chemical desiccation. The mechanism of vein filling does not require elaboration.

FIG. 12-7. Melikaria. The boxwork consists of vein fillings which probably have weathered out of a septarian nodule. Length of specimen about 5 inches (12 cm).

FIG. 12-8. Cone-in-cone specimens exhibit transverse ridges and longitudinary striations. Note concave surface of cone on the right. Length of largest cone about 3 inches (7.6 cm).

CONE-IN-CONE

Calcareous cone-in-cone layers are minor features of some shales (shales with "beef"). They occur both as distinct layers in shales and as jackets enveloping some large calcareous concretions in dark shales. Layers range from 2 to 15 cm thick and are traceable for 1 or more meters in outcrop. They are particularly characteristic of some black shales. Cone-in-cone structure has been described from many places in the United States and Europe in beds ranging in age from Precambrian to Tertiary. Some 44 occurrences have been described in some detail by Woodland (1964).

The structure takes its name from its internal organization. It is characterized by an abundance of right circular cones (Fig. 12-8) which stand with the cone axis perpendicular to the cone-in-cone layer or to the surface of the concretions. Apical angles vary widely but fall most commonly between 30 and 60 degrees. The diameter of the circular base, therefore, varies from near-equality to about one-third the height of the cone. In some cases the cones have flaring bases. The sides of the cones are usually ribbed or grooved; many are marked also by annular depressions and ribs which are most pronounced near the base of the cone and which become finer and more obscure near its apex. The apexes of the cones in a layer may all be directed downward, in other cases upward. On concretions they are directed downward on the upper surface and upward on the lower surface (Gilman and Metzger, 1967; Woodland, 1964).

Internally the cones are fibrous calcite (Fig. 12-9), although cones in siderite (Hendricks, 1937) and gypsum (Tarr, 1932, p. 722) have been reported. The fibers tend to be parallel to the axis of the cone, although Woodland (1964, p. 277) has shown a considerable divergence from this position.

Although cone-in-cone is generally carbonate, it contains a considerable insoluble residue. Acid-soluble carbonate forms 70 to 90 percent of the whole; insoluble residue is largely clay. The carbonate, generally calcite, is in places ankeritic or sideritic or exceptionally a manganiferous siderite (Hendricks, 1937). See Table 12-1.

The origin of cone-in-cone has engendered more discussion, perhaps, than any other concretionary structure except perhaps that of chert nodules. The subject has been reviewed by Tarr (1932) and most extensively by Woodland (1964). The two-fold problem includes the origin of the fibrous carbonate layer and the origin of the conical structure. The fibrous carbonate layers in which much cone-in-cone is found are not original carbonate beds. This conclusion is well supported by the observation that cone-in-cone seams may separate the upper and lower carapaces of trilobites or the upper and lower impressions of fish remains (Brown, 1954). The calcite fibers set generally perpendicular to the layer or to the surface of a concretion are comparable to those of fibrous vein quartz, fibrous gypsum veins, and fibrous ice-crystals in frost-heaved soils. The force of growing crystals has separated the two walls of the vein or the two sides of the cone-in-cone layer or seam. The fibrous habit, therefore, is an expression of crystal growth in a stress field.

The origin of the conical structure is less obvious. The similarity in form to impact cones of chert, limestone, and other materials led to the view that cones were caused by shear. The source of the stresses involved was obscure; they were believed by some to be built up by the growth of the crystals themselves (Richardson, 1923), by

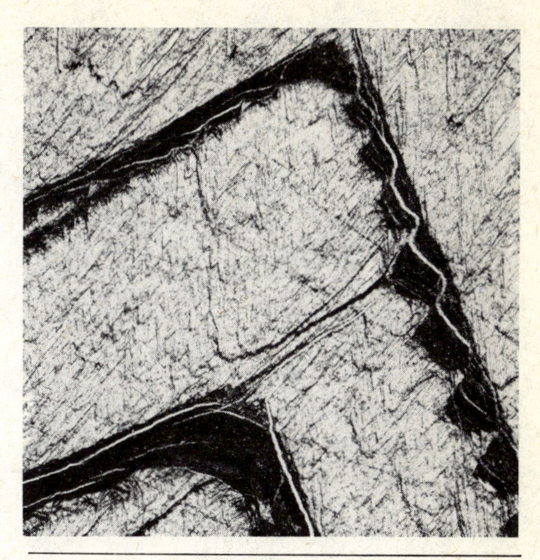

FIG. 12-9. Cone-in-cone, ordinary light, ×30. Mainly calcite; dark material is clay.

others to the weight of the superincumbent beds (Tarr, 1932, p. 729) or the expansive pressure of the concretion where cone-in cone envelops such structures. Woodland (1964, p. 292) attributes the conical structures to none of these causes. He thought the structure to be the result of a flamboyant pattern of growth. This pattern is expressed in his *c*-axis fabric diagrams (1964, Fig. 86). Although the fibrous habit is thought to indicate growth in a stress field, the conical structure, despite its resemblance to shear cones, is not a product of such shear but rather an expression of a crystal growth habit—a view not unlike that of Usdowski (1963).

CONCRETIONARY LAYERS AND VEINS

The diagenetic segregations which have been described are all more or less equidimensional globular bodies formed about a center or nucleus. Less regular forms are largely caused by coalescence or merger of two or more regular forms. Some calcareous stratiform bodies, however, are believed to be the result of diagenetic action. Evidence for such an origin is their abrupt termination within the enclosing shale.

Clearly cross-cutting veins are a secondary feature. Many are believed to be filled with materials which have migrated to a fracture from enclosing strata. Most common fillings are quartz (Adams, 1920) and calcite (Grout, 1946).

CALICHE NODULES AND LOESS CONCRETIONS

Certain other calcareous concretionary bodies differ markedly from those described above. Smaller and formed in clays and soils above the water table, these include caliche nodules of some ancient and modern soils and small carbonate concretions of loess.

Caliche itself is in a sense a species of limestone formed primarily in soils in regions of lime accumulation—regions marked by high temperatures and moderate to low seasonal rainfall. We are here concerned not with mature caliche deposits or calcrete but with those in which calcite is present only as scattered nodules or as crusts on pebbles in gravels. Such concretionary bodies are known in continental mudstones, primarily the topmost member of typical fluvial (point-bar) sequences deposited under suitable climatic conditions. They are common in the Old Red Sandstone of Wales (Allen, 1965), in similar deposits of the Catskill (Devonian) in Pennsylvania, in the Mauch Chunk (Mississippian) of eastern Pennsylvania, and no doubt in all places where a similar facies and climate prevailed.

Characteristics of these caliche nodules have been described by Allen (1965, p. 164), Flack et al. (1969, p. 444), and especially Nagtegaal (1969). They are generally small (1 to 2 cm on the average), exhibit diverse shapes, and consist mainly of calcite. Some have little or no internal structure; others are crudely concentric or even radial.

Perhaps related to these are the small carbonate concretions (*Loessmännchen* or *Loesspüppen*) found in loess. These, 1 to 2 cm in size, are simple to irregular nodules marked by an outer dense shell and a cracked clay-filled interior. Many are irregular and of complex form. Their origin is not well understood (Todd, 1913), but, like septaria, there seems to have been a case-hardened shell and shrinkage of a clay-filled interior.

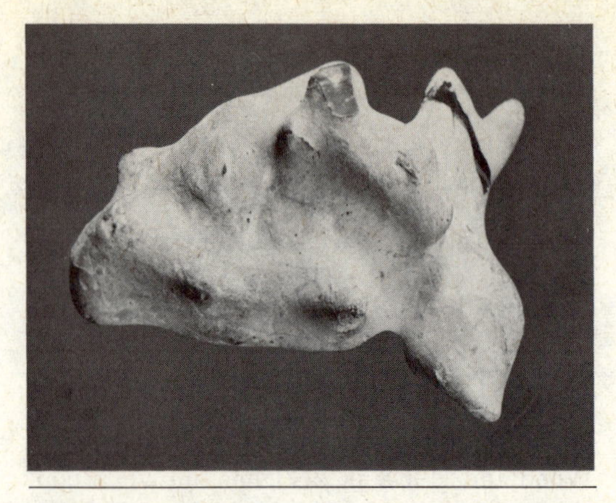

FIG. 12-10. Chert nodule. Dense black "flint" with veneer of white "cotton" rock. Length of nodule about 5 inches (12 cm).

SILICEOUS NODULES AND CONCRETIONS

In some rocks, carbonate rocks in particular, silica is segregated into nodular or other bodies. Most common are chert nodules.

CHERT NODULES

Chert nodules vary from more or less regular disks a few centimeters in diameter to large, highly irregular, and tuberous bodies 25 or more centimeters long. Their shape is infinitely varied, but the larger ones, although of rounded contour, are marked by warty or knobby exteriors (Fig. 12-10). In most cases the nodules are sharply bounded. They tend to be concentrated along certain bedding planes and tend also to be flattened or elongated parallel to the bedding. In some cases they are numerous enough to coalesce into more or less continuous layers. Chert-bearing layers in many cases seem to be rhythmically spaced (Richardson, 1919b). In a given layer the chert forms an irregular two-dimensional network; if layers are closely spaced, some connections or "bridging" between layers may occur, and the chert network becomes three-dimensional. Few limestones are so cherty; most are characterized only by scattered discrete nodules.

The nodules are primarily microcrystalline quartz and, to a lesser extent, chalcedony. Hence they are almost pure silica (Table 12-2). Some carbonate is present—perhaps a replacement relict, but some is euhedral.

Although generally structureless, some nodules have a dense black interior surrounded by a lighter, in some cases white, exterior ("cotton rock"). A few show traces of the bedding that is continuous with that of the host rock; still others are marked by a concentric structure ("contraction spheroids" of Taliaferro, 1934), and some enclose patches of the host rock. Rather com-

TABLE 12-2. Chemical analyses of cherts and other concretionary bodies

Constituent	A	B	C	D	E	F	G
SiO_2	99.47	70.78	98.36	4.46	0.73	45.13	1.80[e]
Al_2O_3	0.06	0.45	0.12	1.10	0.57[c]	0.88	—
Fe_2O_3	0.00	0.02	0.00	7.02	0.27	0.96	—
FeO	0.17	0.30	0.08	0.47	—	—	—
Fe	—	—	—	—	—	—	45.80
MnO	—	0.02	—	0.08	trace	0.02	—
MgO	—	1.88	0.01	0.30	none	none	—
CaO	0.06	12.90	0.16	44.22	none	none	—
Na_2O	0.01	0.05	0.04	0.16	—	—	—
K_2O	0.03	0.06	0.04	0.04	—	—	—
H_2O+	0.19	0.48	0.84	0.35	0.25[d]	0.31[d]	—
H_2O-	0.02	0.32	0.11	1.65			
TiO_2	—	0.03	—	0.07	—	—	—
P_2O_5	—	0.16	0.02	33.92	none	trace	—
CO_2	none	12.04	0.07	0.90	0.16	none	—
SO_3	—	—	—	—	33.15	17.87	—
BaO	—	—	—	—	64.85	34.25	—
C	0.03	0.33	0.02	0.49	—	—	—
S	—	—	—	0.04	—	—	51.10
Total	100.04	100.09[a]	99.87	99.52[b]	99.55	99.42	98.70

[a] Includes $ZrO_2 < 0.01$; FeS_2 0.27; $SO_3 < 0.01$.
[b] Includes F, 2.25; less O for F, 0.95.
[c] R_2O_3.
[d] Loss on ignition.
[e] Insoluble.

A. Oolitic chert, Mines Dolomite (Cambrian), Pennsylvania (Maxwell, 1953).
B. Chert nodule, Delaware Limestone (Devonian), Ohio. Downs Schaaf, analyst (Stout and Schoenlaab, 1945).
C. Flint nodule, Dover Chalk (Cretaceous), England (Maxwell, 1953).
D. Spherulitic phosphate nodules, Colorado Shale (Cretaceous), Bearpaw Mountains, Montana. L. M. Kehl, analyst (Pecora, Hearn, and Milton, 1962, Table 12-1).
E. Barite nodule from shale, Oklahoma. W. L. Howard, analyst (Ham and Merritt, 1944, p. 26).
F. Barite nodule from sandstone, Oklahoma (after Shead, 1923; Ham, and Merritt, 1944, p. 32).
G. Pyrite nodules, marine clay, Victoria, Australia. F. D. Drews, analyst (Edwards and Baker, 1951, p. 44).

monly cherts are fossiliferous, with the fossils either calcareous or silicified, usually the latter. Commonly the fossils are represented only by a cavity; the calcareous material has been removed by solution. Some chert is oolitic, a structure inherited from the host rock as a result of replacement (Choquette, 1955).

Chert nodules are widely distributed. They occur mainly in carbonate rocks, especially limestones and dolomites, but they are also found in bedded siderites and in a few black shales, especially those of Precambrian age (James et al., 1968, Fig. 26). Nodular cherts are found in carbonate rocks of all ages. They are abundant in Paleozoic limestones and dolomites of the Upper Mississippi Valley. Biggs (1957) has given a thorough description of nodular cherts of Illinois, ranging from Cambrian through Mississippian in age. Those of the Niagaran (Silurian) of the Great Lakes region (Lowenstam, 1942) and those of Mississippian carbonates (Van Tuyl, 1912) are especially well known. Famous also are the flint nodules of the Cretaceous Chalk of England and France (Richardson, 1919b).

Although there is no general agreement on the chert problem, the majority opinion now seems to incline toward an epigenetic formation of the nodular cherts and flints found in limestone and other carbonate rocks. A few writers, notably Tarr (1917, 1926), Tarr and Twenhofel (1932), and some others (Trefethen, 1947; Fernandez, 1961), stoutly maintain that nodular cherts are formed by direct precipitation of masses of silica gel on the sea floor. Evidence of replacement (and therefore of post-depositional origin) of the nodular cherts is abundant and clear. Van Tuyl (1918) and Biggs (1957) have given good resumés of the evidence bearing on this question. Certainly many nodules show definite evidence of a replacement origin, but field and petrographic evidence is inconclusive for many other occurrences. The principal evidence against origin of these nodules by primary precipitation may be, as Cressman (1962, p. 19) points out, recent experimental evidence indicating that sea water is undersaturated with silica. Only in water bodies of special character is chert, or some precursor of chert, likely to be precipitated (Eugster, 1967). (The chert problem is reviewed in more detail in Chapter 11.)

GEODES

Significant features of geodes are their subspherical shape, their hollow interior, their outer chalcedonic layer and, the inner drusy lining of inward-projecting crystals. Geodes are characteristic of certain limestone beds but are rarely found in shales; they seem to prefer argillaceous limestones or dolomites (Hayes, 1964, Fig. 3) to pure limestones. Their distribution, therefore, reflects a stratigraphic control.

Geodes are hollow, globular bodies, varying from a couple of centimeters to nearly a meter in diameter (most are 10 to 20 cm). They tend to be slightly flattened with the equatorial plane parallel to the bedding. Some appear to have been crushed or collapsed. Geodes are marked by a thin outermost layer of dense chalcedony (Diller, 1898, p. 112). This layer may be incomplete in some geodes and be lost in others by erosion of the exterior.

Most geodes are more or less filled. Hence on the inner surface of the chalcedonic layer are younger, usually crystalline materials. Most commonly these are inward-projecting quartz crystals. In addition to quartz, many contain scalenohedral calcite and rhombohedral dolomite. Less common are many minor and rarer constituents which include aragonite, ankerite, magnetite, hematite, pyrite, millerite, chalcopyrite, sphalerite, kaolin, and bitumen (Van Tuyl, 1916). No constant order of succession holds for all geodes, although the outer shell is invariably chalcedony which in most cases is followed directly by quartz. In some geodes is a second generation of chalcedony. Metallic sulfides, if present, are most generally the last minerals to be deposited.

Most singular and most significant feature of geodes is the incontrovertible evidence that they have grown by expansion. Notable is bending of laminations of the host rock around the geode. But most dramatic is the "exploding bomb" structure. As Bassler

(1908) has clearly shown, many geodes originate in cavities of fossils; upon growth the fossil swells, then bursts. With further growth, fragments of the fossil adhering to the geode become widely separated. Ultimately the fragments appear to have dissolved or been absorbed by the growing geode and are lost.

The origin of the geode involves first the origin of the cavity, and second the filling of the cavity. It is unlikely that subspherical cavities of the size of geodes existed in the original sediment. Robertson (1944, 1951) supposed that geodes were syngenetically deposited on the sea floor as colloidal masses of hydrated silica and that later "the loss of water from the gel produced the chalcedony shell." It is not clear why, if this is their origin, chert nodules, also presumed by some to have been similarly syngenetically precipitated silica gel, were not all converted to geodes. Hayes (1964) presumed that the precursors of geodes were not silica gel but rather were calcareous concretions of like form and size, a view earlier expressed by Van Tuyl (1916). The bodies were diagenetically converted to geodes, first by recrystallization in their central areas, silification of their exterior, followed by solution of the central portion and subsequent precipitation of the crystalline filling. If, as Bassler's collection clearly shows, an initial cavity is a necessary prerequisite for the formation of a geode, unfilled space within a crinoid calyx, a bivalve, or any similar opening will suffice. Nothing is present in the initial cavity except fluid, presumably connate salt solution. Inasmuch as the outer wall of the true geode is chalcedony, the initial deposit must have been a layer of gelatinous silica. The formation of this layer isolates the salt solution; should, in the course of time, the outside waters freshen, the gel layer might function as a semipermeable membrane, and osmosis might build up internal pressure. This outward directed pressure might lead to expansion. If this occurs prior to consolidation, surrounding lime mud would be pushed aside; if after consolidation, space might be gained by solution of limestone at the silica-limestone interface. Expansion would continue until the cell volume is much increased and the salt concentration of the containing fluids reduced to such value that expansive forces become negligible. Ultimately the silica dehydrates and crystallizes. Shrinkage and cracking follow, permitting access of mineral-bearing waters and deposition of the drusy lining over the primary chalcedonic layer. The mechanism here hypothesized to explain geodes is essentially that proposed by Tanton (1944) for the growth of *conchilites,* which are peculiar small, bowl-shaped objects of limonite or goethite growing in an inverted position on mineralized bedrock of a Canadian lake. The semipermeable membrane in these objects is gelatinous ferric hydroxide instead of silica.

SILICEOUS CONCRETIONS IN BEDDED SULFATES

Concretions in evaporite deposits are apparently rare. Various siliceous bodies, including euhedral quartz, concentrically banded chalcedony, and quartz spherulites have been reported from gypsum and anhydrite deposits in Manitoba (Brownell, 1942). They are believed to be diagenetic.

PHOSPHATIC CONCRETIONS AND NODULES

Phosphatic matter occurs both as structureless nodules and as radially structured or spherulitic concretionary forms.

Phosphatic nodules or "pebbles" are found not only in phosphate deposits themselves but are also widely scattered in some limestones (Pettijohn, 1926) and especially in the Cretaceous Chalk (Fisher, 1873). They also occur on the present sea bottom (Murray and Renard, 1891, p. 396; Dietz, Emery, and Shepard, 1942). These nodules vary from small granules to pebble-like objects several centimeters in diameter. Typically they are black or brown, are irregular in shape, and have a dense, hard, shiny surface. Larger nodules contain much foreign matter, including sand grains, mica flakes, shell debris, and sponge spicules. The black color is most intense near the outer rim. The principal constituent, however, is collophane; phosphatic spherulites, described below, are well-crystallized dahlite (McConnell, 1935).

Spherulitic phosphate nodules have also been described (Pecora, Hearn, and Milton, 1962). Best known are those in the Colorado Shale (Cretaceous) of Montana. Some are of very regular spherical form, one to several centimeters in diameter. Others are composites formed of several coalescing spherical bodies. All have a pronounced radial structure; many display a central void which in some is filled with vein materials. The composition of these spherulites is given in Table 12-2D. They are believed to have formed as homogeneous phosphatic concretions which later recrystallized to the spherulitic habit.

IRON-OXIDE NODULES AND CONCRETIONS

Ferric oxides and hydroxides are common in sediments and in places are segregated into concretionary bodies. Of particular interest are the voidal or hollow iron-oxide bodies most commonly found in sands and sandstones and in some clays (Fig. 12-11). They are especially common in Pleistocene sands but are also reported from Tertiary and even Cretaceous sands.

Like other concretions, they vary in size and shape. But unlike most concretions many are large and tubelike (Willcox, 1914), and all have a central cavity or void. The rim or shell is a hard, dense limonitic layer. Interiors are partially filled with a ferruginous sand or small pebbles which rattle when the concretions are shaken (*Klapperstein*). In a few, the core is compact and consists of clay which may be slightly separated from the walls. Not all iron-oxide concretions, however, are hollow. Those described by Emery (1950) in beach ridge sands are, instead, regular iron-oxide cemented spheres.

Hollow ferruginous concretions have been explained by "intergranular secretion of limonite" (Smith, 1948), by partial replacement of limestone pebbles by iron compounds (Shaw, 1917), by oxidation of "claystone pebbles" (Leroy, 1949), or by oxidation of sideritic concretions (Bates, 1938; Todd, 1903). The last explanation seems to account for some of these objects, inasmuch as oxidation of jointed siderite beds turns each joint block into a hollow or partially filled limonitic "box" (Taylor, 1949, pp. 38, 90). The clay core of some ironstone concretions appears to be in part unoxidized sideritic clay. These structures are, therefore, a product of weathering and should not be found below the water table. Presumed evidence of expansion seen in thin section of the limonitic rim is only a record of corrosion and partial replacement of the quartz by siderite—a common phenomenon in siderite-cemented sands.

It is probable that iron-oxide aggregates are of several kinds and that those found in clays are different in character from those formed in sands. Hence no one explanation is appropriate for all.

PYRITE AND MARCASITE NODULES AND CRYSTAL AGGREGATES

Both large and small crystalline aggregates of pyrite and marcasite occur in some clays and shales, in some coals, and even in limestone. These bodies have a diversity of sizes and shapes from millimeter-sized spherulites and single crystals to larger crystal aggregates and nodules. Iron sulfides in general seem to exclude most of the host rock material in which they are embedded. In general the sulfide in marine sedimentary rocks is pyrite, that in coal is marcasite (Newhouse, 1927). This observation led to the view that pyrite is a product of a neutral or alkaline environment (in most cases marine) and that marcasite is formed in an acid or freshwater environment (Edwards and Baker, 1951, p. 45). The exceptional marcasite in a marine formation indicates, therefore, acidification subsequent to deposition during lithification. Replacement of fossil wood by marcasite is common (Schwartz, 1927; Edwards and Baker, 1951, p. 38). Black amorphous iron sulfide seems to be present in some modern muds. Apparently after burial this material is segregated and crystallized as scattered pyrite cubes, in part

FIG. 12-11. Hollow iron-oxide concretions. Essentially a limonite-cemented shell of sand. Length of largest about 4 inches (10 cm).

replacing the matrix, as small spherulites, as larger crystal aggregates, and as a replacement of fossil wood and shells.

The source of the sulfur is uncertain. It could be a product of decomposition of organic matter or of bacterial reduction of sulfates, or be introduced by ground or magmatic waters. For reasons stated above, it is more likely an indigenous rather than introduced constituent and hence diagenetically segregated. Mathias (1928) considered the pyrite concretions of the dark Pennsylvanian shales of north-central Missouri to be of syngenetic origin. Newhouse believed the sulfides to have crystallized from colloidal gels deposited at the time the enclosing rocks were formed.

Inasmuch as sulfides may be carried by ground- and by magmatic waters, the presence of sulfide crystals is not sufficient proof of their diagenetic origin, especially in metasedimentary rocks. The close correlation between sulfide content and content of organic or carbonaceous matter in many sediments strongly suggests a diagenetic or sedimentary origin for the sulfide proper. A close correlation with veins and fractures, on the other hand, would suggest introduction from without.

BARITE NODULES AND ROSETTES

Somewhat analogous to calcite sand crystals are rosettes of barite. Barite rosettes are clusters of crystals of barite—of tabular form—somewhat symmetrically disposed, which are most characteristic of some sandstones, particularly Permian red sandstones in Oklahoma (Nichols, 1906; Shead, 1923; Tarr, 1933; Ham and Merritt, 1944). Like sand crystals, the included host rock material equals or exceeds that of the crystallized mineral itself (Table 12-2F).

Similar rosettes have been reported from many places. Pogue (1910) has reviewed this literature and has added a description of his own on rosettes from the Libyan Desert which were presumably formed in the Nubian Sandstone of that area.

Barite nodules with a spherulitic form and structure occur in some shales (Ham and Merritt, 1944, p. 25; Pl. 2; Hanna, 1936; Martens, 1925). Unlike the barite of sandstone very little of the host rock is included

(Table 12-2E). They are also known from the present-day sea floor (Revelle and Emery, 1951).

GYPSUM CRYSTALS AND CRYSTAL CLUSTERS

Many sediments contain large crystals or clusters of crystals of gypsum. Some of these grow in a friable sand matrix in a fashion similar to calcite and barite; like these minerals they include a large volume of sand. A well-known example are the large crystal aggregates from the Laguna Madre, Texas (Masson, 1955). Well-crystallized selenite occurs in some shales; one of the best-known examples is the Pennsylvanian shale of Ellsworth, Ohio. These crystals, several centimeters long, have grown at the expense of the shale, as shown by the large quantity of shaly matter included within them.

PSEUDOCONCRETIONS

A number of objects resemble concretions in some manner but are not diagenetic products or true concretions. Included here are armored mud balls, algal balls or oncolites, and lake balls.

ARMORED MUD BALLS

Armored mud balls are mechanically accreted primary structures. They are large, subspherical balls of clay that are coated or armored with fine gravel. The morphology and origin of armored mud balls have been thoroughly investigated by Bell (1940), who records their occurrence in the Las Posas Barranca, Ventura County, California. They are found in high-gradient streams, especially those subject to torrential flow (Glazek and Radwanski, 1962; Baluk and Radwanski, 1962). But they are also known from beaches (Kugler and Saunders, 1959) and are especially common in Pleistocene glacial outwash (Leney and Leney, 1957) where they are referred to as "till balls." They have also been reported from older strata such as the Pannonian (uppermost Tertiary) of the Graz Basin in Austria (Kirchmayer, 1962).

The size of mud balls varies from 1 to about 50 cm. Those between 5 and 10 centimeters are the most common. Well-formed mud balls are highly spherical (Fig. 12-12); nearly 60 percent of those studied by Bell exceeded 0.90, 32 percent were over 0.95; 12 percent exceeded 0.99; and nearly 7 percent had a sphericity near 1.00. The mean sphericity of balls that had traveled ¼ mile (400 m), 1 mile (1609 m), and 2¾ miles (4420 m) were 0.784, 0.839, and 0.898, respectively. These data indicate the sphericity to vary nearly as the cube root of the distance traveled.

The larger mud balls are coated with coarser gravel than were the smaller ones. The weight of the armor varied from 17.1 percent to 44.0 percent of that of the whole ball. The total quantity of armor was found to be a power function of surface area, varying approximately as the 1.54 power of the area, or approximately as the linear function of the ball diameter. Whereas the armor consists of a poorly graded (nine size classes) mixture of sand and gravel, the interior consists of relatively pure clay and silt.

Field relations, as well as the character of the mud balls in the Las Posas Barranca, proved rather clearly that mud balls originated as clay chunks which had been released by rapid bank erosion and undercutting and which, upon rolling downstream, had acquired a gravel armor. Sand grains and pebbles were impressed into the softened interior. As soon as the surface became well covered, it was sealed against further accretion. The growth of the ball is limited more by its structural strength than either by the size of the clay chunk available or by the transporting ability of the current. The maximum size represents a balance between forces of cohesion and destructive forces of impact. Theoretical and experimental observation led to conclusions that the velocity with which a ball may move with safety in a stream is inversely proportional to the diameter. For the Las Posas mud balls the relations determined by Bell show that a 2 inch (5 cm) ball can tolerate transport of 32 feet (9.8 m) per second, whereas a ball 12 inches (30 cm) in diameter can tolerate a velocity of 5 feet

FIG. 12-12. Armored mud balls, Las Posas Barranca, California. These have rolled about 3 miles (4.8 km). Note remarkably heavy armor and the high degree of sphericity. (Soil Conservation Service photograph. Bell, 1940, p. 18.)

(1.5 m) per second. Bell noted that the clay balls are considerably larger than other debris of the bed load of the stream.

Mud balls, although a minor and uncommon constituent of a sediment, furnish a means of estimating maximum velocities of ancient streams, nature of the bed material, approximate distance to source materials, and nature of the water body which formed them.

ONCOLITES

Oncolites are calcareous accretionary bodies of algal origin; they are spheroidal stromatolites (Fig. 10-19). They characterize many limestones. Recent marine oncolites have been described by Ginsburg (1960); a good fossil example is that from the Upper Jurassic age of the Holy Cross Mountains of Poland (Kutek and Radwanski, 1965).

Oncolites are nearly spheroidal in shape; the largest ones are 5 to 10 cm in diameter (macrooncolites); more commonly they are only 1 or 2 cm. Below this size they pass into algal pisolites. Their shape depends in part on whether or not they have a foreign core; the shape of the core determines in some degree the shape of the oncolite. Most characteristic is their internal quasiconcentric structure. Laminations vary somewhat in thickness and regularity. Microunconformities exist which express interruptions in growth.

The formation is attributed to entrapment of detritus by an enveloping mat of blue-green algae. Growth takes place on the sides and upper surface. If the oncolite is rolled over, growth ceases on the underside and starts again on the new upper surface; for this reason perfect concentricity is not attained.

LAKE BALLS

Spherical lake balls of a natural felt, found in some modern lakes, consist entirely of vegetable fibers. They have been found on the shores of Lake Michigan and are known also from some Finnish lakes (Ohlson, 1961). It is uncertain whether or not they occur in the geologic record.

References

Adams, S. F., 1920, A microscopic study of vein quartz: Econ. Geol., v. 15, pp. 623–664.

Allen, J. R. L., 1965, The sedimentation and paleogeography of the Old Red Sandstone of Anglesey, north Wales: Proc. Yorkshire Geol. Soc., v. 35, pp. 139–184.

Baluk, W., and Radwanski, A., 1962, Armored mud balls in streams in the vicinity of Nowy Sacz (Polish Carpathians): Acta Geol. Polonica, v. 12, pp. 341–366.

Barbour, E. H., 1901, Sand crystals and their relations to certain concretionary forms: Bull. Geol. Soc. Amer., v. 12, pp. 165–172.

Bassler, R. S., 1908, The formation of geodes, with remarks on the silicification of fossils: Proc. U.S. Nat. Mus., v. 35, pp. 133–154.

Bates, R. L., 1938, Occurrence and origin of certain limonite concretions: Jour. Sed. Petrology, v. 8, pp. 91–99.

Bell, H. S., 1940, Armored mud balls—their origin, properties and role in sedimentation: Jour. Geol., v. 48, pp. 1–31.

Berner, R. A., 1964, An idealized model of dissolved sulphate distribution in recent sediments: Geochim. Cosmochim. Acta, v. 28, pp. 1497–1503.

———, 1968a, Rate of concretion growth: Geochim. Cosmochim. Acta, v. 32, pp. 477–483.

———, 1968b, Calcium carbonate concretion formed by decomposition of organic matter: Science, v. 159, pp. 195–197.

Biggs, D. L., 1957, Petrography and origin of Illinois nodular cherts: Illinois Geol. Surv., circ. 245, 25 pp.

Bramlette, M. N., 1941, The stability of heavy minerals in sandstone: Jour. Sed. Petrology, v. 11, pp. 32–36.

Brown, R., 1954, How does cone-in-cone material become emplaced?: Amer. Jour. Sci., v. 252, pp. 372–376.

Brownell, G. M., 1942, Quartz concretions in gypsum and anhydrite: Univ. Toronto Studies, geol. ser., no. 47, pp. 7–18.

Burt, F. A., 1928, Melikaria: vein complexes resembling septaria veins in form: Jour. Geol., v. 36, pp. 539–544.

———, 1932, Formative processes in concretions formed about fossils as nuclei: Jour. Sed. Petrology, v. 2, pp. 38–45.

Cayeux, L., 1929, Les roches sédimentaires de France, roches silicieuses: Paris, Masson, 696 pp.

Choquette, P. W., 1955, A petrographic study of the "State College" siliceous oolite: Jour. Geol., v. 63, pp. 337–347.

Clifton, H. E., 1957, Carbonate concretions of the Ohio Shale: Ohio Jour. Sci., v. 57, pp. 114–129.

Colton, G. W., 1967, Orientation of carbonate concretions in the Upper Devonian of New York: U.S. Geol. Surv. Prof. Paper 575-B, pp. 57–59.

Cressman, E. T., 1962, Nondetrital siliceous sediments: U.S. Geol. Surv. Prof. Paper 440-T, 31 pp.

Croneis, C., and Grubbs, D. M., 1939, Silurian sea balls: Jour. Geol., v. 47, pp. 598–612.

Crook, T., 1913, Septaria: a defense of the "shrinkage" view: Geol. Mag., v. 10, pp. 514–515.

Curtis, C. D., 1967, Diagenetic iron minerals in some British Carboniferous sediments: Geochim. Cosmochim. Acta, v. 31, pp. 2109–2123.

Daly, R. A., 1900, The calcareous concretions of Kettle Point, Lambton County, Ontario: Jour. Geol., v. 8, pp. 135–150.

Davies, A. M., 1913, The origin of septarian structure: Geol. Mag., v. 10, pp. 99–101.

Dietz, R. S., Emery, K. O., and Shepard, F. P., 1942, Phosphorite deposits on the sea floor off southern California: Bull. Geol. Soc. Amer., v. 53, pp. 815–848.

Diller, J. S., 1898, The educational series of rock specimens: Bull. U.S. Geol. Surv. 150, 400 pp.

Edwards, A. B., and Baker, G., 1951, Some occurrences of supergene iron sulphides in relation to their environment of deposition: Jour. Sed. Petrology, v. 21, pp. 34–46.

Emery, K. O., 1950, Ironstone concretions and beach ridges of San Diego County, California: California Jour. Mines Geol., v. 46, pp. 213–221.

Eskola, Pentti, 1932, Conditions during the earliest geological times: Ann. Acad. Sci. Fennicae, ser. A, v. 36, no. 4, 74 pp.

Eugster, H. P., 1967, Hydrous sodium silicates from Lake Magadi, Kenya: precursors of bedded chert: Science, v. 157, pp. 1177–1180.

Fernandez, D., 1961, Sull'origine delle selci stratificate e in noduli nel calcare: Bull. Soc. Geol. Italiana, v. 80, pp. 3–5.

Fisher, O., 1873, On the phosphatic nodules of the Cretaceous rocks of Cambridgeshire: Quart. Jour. Geol. Soc. London, v. 29, pp. 52–63.

Flack, K. W., Nettleton, W. D., Gile, L. H., and Cady, J. G., 1969, Pedocementation: induration by silica, carbonates, and sesquioxides in the Quaternary: Soil Sci., v. 107, pp. 442–453.

Fuhrmann, W., 1968, Sandkristalle und Kugelsandstein, Ihre Rolle bei der Diagenese von Sanden: Der Aufschluss, v. 5, pp. 105–111.

Gilman, R. A., and Metzger, W. J., 1967, Cone-in-cone concretions from western New York: Jour. Sed. Petrology, v. 37, pp. 87–95.

Ginsburg, R. N., 1960, Ancient analogues of Recent stromatolites: 21st Sess. Int. Geol. Congr., Norden, pt. 22, pp. 26–35.

Glazek, J., and Radwanski, A., 1962, Armored mud balls in the Podmachocice Ravine (Holy Cross Mountains, central Poland): Acta Geol. Polonica, v. 12, pp. 367–376.

Gries, J. P., and Rothrock, E. P., 1941, Manganese deposits of the Lower Missouri Valley in South Dakota: South Dakota Geol. Surv., Rept. Inv. 38, 96 pp.

Grout, F. F., 1946, Microscopic characters of

vein carbonates: Econ. Geol., v. 41, pp. 475–502.
Ham, W. E., and Merritt, C. A., 1944, Barite in Oklahoma: Oklahoma Geol. Surv., circ. 23, 42 pp.
Hanna, M. A., 1936, Barite concretions from the Yazoo Clay (Eocene) of Louisiana: Jour. Sed. Petrology, v. 6, pp. 28–30.
Harnly, H. J., 1898, Cone-in-cone (an impure calcite): Trans. Kansas Acad. Sci., v. 15, pp. 22.
Hayes, J. B., 1964, Geodes and concretions from the Mississippian Warsaw Formation, Keokuk region, Iowa, Illinois, Michigan: Jour. Sed. Petrology, v. 34, pp. 123–133.
Hendricks, T. A., 1937, Some unusual specimens of cone-in-cone in manganiferous siderite: Amer. Jour. Sci., ser. 5, v. 33, pp. 458–561.
Hoefs, J., 1970, Kohlenstoff- und Sauerstoff-Isotopenuntersuchungen an Karbonat-konkretionen und umgebendem Gestein: Contrib. Min. Pet., v. 27, pp. 66–79.
Illies, H., 1949, Über die erdgeschichtliche Bedeutung der Konkretionen: Zeitschr. Deutschen Geol. Gesell., v. 101, pp. 95–98.
James, H. L., Dutton, C. E., Pettijohn, F. J., and Wier, K. L., 1968, Geology and ore deposits of the Iron River–Crystal Falls District, Iron County, Michigan: U.S. Geol. Surv. Prof. Paper 570, 134 pp.
Kindle, E. M., 1923, Range and distribution of certain types of Canadian Pleistocene concretions: Bull. Geol. Soc. Amer., v. 34, pp. 614–617.
Kirchmayer, M., 1962, Gespickte Tongerölle (armored mud balls) im steirischen Becken, Steiermark/Österreich: Neues Jahrb. Geol. Palaeont. Mh., v. 10, pp. 548–554.
Kugler, H., and Saunders, P., 1959, Occurrence of armored mud balls in Trinidad, West Indies: Jour. Geol., v. 67, pp. 563–565.
Kutek, J., and Radwanski, A., 1965, Upper Jurassic onkolites of the Holy Cross Mountains (central Poland): Bull. Acad. Polonaise Sci., v. 13, no. 2, pp. 155–160.
Leney, G. W., and Leney, A. T., 1957, Armored till balls in the Pleistocene outwash of southeastern Michigan: Jour. Geol., v. 65, pp. 105–106.
Leroy, L. W., 1949, A note on voidal concretions in the El Milagro formation of western Venezuela: Jour. Sed. Petrology, v. 19, pp. 39–42.
Lippman, F., 1955, Ton, Geoden und Minerale des Barrême von Hoheneggelsen: Geol. Rundschau, v. 43, pp. 475–503.
Lowenstam, H. A., 1942, Facies relation and origin of some Niagaran cherts (abstr.): Bull. Geol. Soc. Amer., v. 53, pp. 1805–1806.
McConnell, D., 1935, Spherulitic concretions of dahlite from Ishawooa, Wyoming: Amer. Mineral., v. 20, pp. 693–698.
Martens, J. H. C., 1925, Barite and associated minerals in concretions in the Genesee Shale: Amer. Mineral., v. 10, pp. 102–104.
Masson, P. H., 1955, An occurrence of gypsum in southwest Texas: Jour. Sed. Petrology, v. 25, pp. 72–77.
Mathias, H. E., 1928, Syngenetic origin of pyrite concretions in the Pennsylvanian shales of north-central Missouri: Jour. Geol., v. 36, pp. 440–450.
———, 1931, Calcareous concretions in the Fox Hills Formation, Colorado: Amer. Jour. Sci., ser. 5, v. 22, pp. 354–359.
Maxwell, J. A., 1953, Geochemical study of chert and related deposits: Univ. Minnesota, Ph.D. thesis.
Murray, J., and Renard, A. F., 1891, Report on deep-sea deposits based on the specimens collected during the voyage of H.M.S. *Challenger* in the years 1872–1876: *Challenger* Repts., Deep-sea deposits, pp. 391–400.
Nagtegaal, P. J. C., 1969, Microtextures in Recent and fossil caliche: Leidse Geol. Meded., v. 42, pp. 131–142.
Nederlof, M. H., 1959, Structure and sedimentology of the Upper Carboniferous of the Upper Pisuerga Valleys, Cantabrian Mountains, Spain: Ph.D. thesis, Leiden.
Newhouse, W. H., 1927, Some forms of iron sulphide occurring in coal and other sedimentary rocks: Jour. Geol., v. 35, pp. 73–83.
Nichols, H. W., 1906, New forms of concretions: Field Columbian Mus. Geol. Publ., v. 3, pp. 25–54.
Oertal, G., and Curtis, C. D., 1972, Clay-ironstone concretion preserving fabrics due to progressive compaction: Bull. Geol. Soc. Amer., v. 83, pp. 2597–2606.
Ohlson, B., 1961, Observations on Recent lake balls and ancient *Corycium* inclusions in Finland: Bull. Comm. Geol. Finlande 196, pp. 377–390.
Pecora, W. T., Hearn, B. C., Jr., and Milton, C., 1962, Origin of spherulitic phosphate nodules in basal Colorado Shale, Bearpaw Mountains, Montana: U.S. Geol. Surv. Prof. Paper 450-B, pp. 30–35.
Pettijohn, F. J., 1926, Intraformational phosphate pebbles from the Twin City Ordovician: Jour. Geol., v. 34, pp. 361–373.
Pogue, J. E., 1910, On sand-barites from Kharga, Egypt: Proc. U.S. Nat. Mus., v. 38, pp. 17–24.
Quirke, T. T., 1917, Espanola district, Ontario: Geol. Surv. Canada Mem. 102, 92 pp.
Raiswell, R., 1971, The growth of Cambrian and Liassic concretions: Sedimentology, v. 17, pp. 147–171.
Ramberg, H., 1952, The origin of metamorphic

and metasomatic rocks: Chicago, Univ. Chicago Press, pp. 220–225, 232–233.

Revelle, R., and Emery, K. O., 1951, Barite concretions from the ocean floor: Bull. Geol. Soc. Amer., v. 62, pp. 707–724.

Richardson, W. A., 1919a, On the origin of septarian structure: Mineral. Mag., v. 18, pp. 327–338.

———, 1919b, The origin of Cretaceous flint: Geol. Mag., v. 56, pp. 535–547.

———, 1921, The relative age of concretions: Geol. Mag., v. 58, pp. 114–124.

———, 1923, Petrology of shales with "beef": Quart. Jour. Geol. Soc. London, v. 79, pp. 88–89.

Robertson, P., 1944, Silica gel and Warsaw geodes: Trans. Illinois Acad. Sci., v. 37, pp. 93–94.

———, 1951, Geode note: Science, v. 114, p. 215.

Ruhland, M., 1961, Quelques observations sur les ovoides des grauwackes du Culm des Vosges meridionales: Bull. Serv. Carte Geol. Alsace-Lorraine, v. 14, pp. 65–68.

Schwartz, G. M., 1927, Iron sulphide pseudomorphs of plant structures in coal: Jour. Geol., v. 35, pp. 375–399.

Seibold, E., 1962, Kalk-konkretionen und karbonatisch gebundenes Magnesium: Geochim. Cosmochim. Acta, v. 26, pp. 899–909.

Shaw, E. W., 1917, The Pliocene history of northeastern and central Mississippi: U.S. Geol. Surv. Prof. Paper 108, 138 pp.

Shead, A. C., 1923, Notes on barite in Oklahoma with chemical analyses of barite rosettes: Proc. Oklahoma Acad. Sci., v. 3, pp. 102–106.

Smith, L. L., 1948, Hollow ferruginous concretions in South Carolina: Jour. Geol., v. 56, pp. 218–225.

Stout, W. E., 1944, The iron ore bearing formations of Ohio: Geol. Surv. Ohio, Bull. 45, 230 pp.

Stout, W. W., and Schoenlaab, R. A., 1945, The occurrence of flint in Ohio: Bull. Geol. Surv. Ohio 46, 110 pp.

Taliaferro, N. L., 1934, Contraction phenomena in cherts: Bull. Geol. Soc. Amer., v. 45, pp. 189–232.

Tanton, T. L., 1944, Conchilites: Trans. Roy. Soc. Canada, ser. 3, v. 38, sec. 4, pp. 97–104.

Tarr, W. A., 1917, The origin of the chert in the Burlington Limestone: Amer. Jour. Sci., ser. 4, v. 44, pp. 409–452.

———, 1921, Syngenetic origin of concretions in shale: Bull. Geol. Soc. Amer., v. 32, pp. 373–384.

———, 1926, The origin of chert and flint: Univ. Missouri Studies, v. 1, pp. 1–54.

———, 1932, Cone-in-cone, *in* Treatise on sedimentation (Twenhofel, W. H., ed.): Baltimore: Williams and Wilkins, pp. 716–733.

———, 1933, The origin of sand barites of the Lower Permian of Oklahoma: Amer. Mineral., v. 18, pp. 260–272.

———, 1935, Concretions in the Champlain Formation of the Connecticut Valley: Bull. Geol. Soc. Amer., v. 46, pp. 1493–1534.

Tarr, W. A., and Twenhofel, W. H., 1932, Chert and flint, *in* Treatise on sedimentation (Twenhofel, W. H., ed.), Baltimore: Williams and Wilkins, pp. 519–545.

Taylor, J. H., 1949, Petrology of the Northampton sand ironstone formation: Great Britain Geol. Surv. Mem., 111 pp.

———, 1950, Baryte-bearing nodules of the Middle Lias of the English east Midlands: Mineral. Mag., v. 29, pp. 18–26.

Todd, J. E., 1903, Concretions and their geological effects: Bull. Geol. Soc. Amer., v. 14, pp. 353–368.

———, 1913, More about septarian structure: Geol. Mag., v. 10, pp. 361–364.

Trefethen, J. M., 1947, Some features of the cherts in the vicinity of Columbia, Missouri: Amer. Jour. Sci., v. 245, pp. 56–58.

Usdowski, H.-E., 1963, Die Genese der Tutenmergel oder Nagelkalke (cone-in-cone): Beitr. Min. Petrog., v. 9, pp. 95–110.

Vanossi, M., 1964, Il problema delle septarie: Atti dell'Instituto Geol. Univ. Pavia, v. 15, pp. 32–88.

Van Tuyl, F. M., 1912, A study of the cherts of the Osage series of the Mississippian system: Proc. Iowa Acad. Sci., v. 19, pp. 173–174.

———, 1916, The geodes of the Keokuk beds: Amer. Jour. Sci., ser. 4, v. 42, pp. 34–42.

———, 1918, The origin of chert: Amer. Jour. Sci., ser. 4, v. 45, pp. 449–456.

Warkentin, B. P., 1967, Carbonate content of concretions in varved sediments: Canad. Jour. Earth Sci., v. 4, p. 333.

Weber, J. N., Williams, E. G., and Keith, M. L., 1964, Paleoenvironmental significance of carbon isotopic composition of siderite nodules in some shales of Pennsylvanian age: Jour. Sed. Petrology, v. 34, pp. 814–818.

Weeks, L. G., 1953, Environment and mode of origin and facies relationships of carbonate concretions in shales: Jour. Sed. Petrology, v. 23, pp. 162–173.

———, 1957, Origin of carbonate concretions in shales, Magdalena Valley, Colombia: Bull. Geol. Soc. Amer., v. 68, pp. 95–102.

Willcox, O. W., 1914, Iron concretions of the Redbank sands: Jour. Geol., v. 14, pp. 243–

Woodland, B. G., 1964, The nature and origin of cone-in-cone structure: Fieldiana, Geol., v. 13, no. 4, pp. 187–305.

13

PROVENANCE

INTRODUCTION

We now go beyond description and classification of sedimentary rocks and the immediate questions relating to their consolidation, to the problems of *interpretation*. And to the big question: What does the sedimentary record contribute to geology—to geologic history—other than the entombment of the life record? What do we wish to know? We might ask how and under what conditions sediment originates. For clastic sediments, this is the question of *provenance*—the climate, relief, and lithology of the source area. The answer to this question is found largely in a study of the composition of gravel clasts and the detrital mineralogy of sandstones, which constitute direct evidence of the kinds of source rocks. But the composition of sediment is not precisely that of the source region, inasmuch as debris from that area has been through a geologic "sieve" so that its composition is altered by selective losses and enrichment (a question involving mineral stability by weathering in the source area), by abrasion during transit, and by alteration or solution during diagenesis. The problem then involves mineralogical analysis coupled with a knowledge of mineral stability, both mechanical and chemical.

A second question we might ask is, Where and at what distance did the source area lie, and what is its relation to the configuration and bathymetry of the depositional basin? Or, in short, what was the *paleogeography* of the epoch during the deposition of a particular sedimentary formation? To answer this question we need to ascertain the paleoslope, the sedimentary strike, the paleocurrent system operative during the depositional process, and the facies distribution. The paleocurrent system, reconstructed from primary directional structures, distribution patterns (dispersal "fans") of debris, and lateral facies variations, including variations in such scalar properties as pebble size and roundness, will assist in solving the paleogeographic riddle. This approach involves measurement and mapping of such commonplace features as cross-bedding, pebble size, and grain fabric. It involves extended field studies.

A third question we might ask is, What was the *environment of deposition*? We must first decide how we will define environment, whether by chemical and physical parameters or whether in geomorphic (or geographic) terms. Which is the more meaningful to a geologist? We then ask what criteria can we employ to discriminate between all possible environments: textures, mineralogy, or structures of the deposit; perhaps the contained fossils or, as recent work and opinion seem to show, the vertical sequence or profile? Naturally we need *all* available criteria, but our collective experience seems to indicate that the vertical sequence of lithologies and structures, knowledge of which is gained from three-dimensional studies of modern sediments, has proved the most powerful tool for interpretation of ancient environments. This mode of attack on the problem involves field measurement of stratigraphic sections—an old procedure but one revitalized by new concepts and new data derived from the Recent.

In this and the next two chapters we will review very briefly the concepts and principles involved in determination of provenance, of paleocurrent analysis and paleogeographic reconstruction, and of depositional environment.

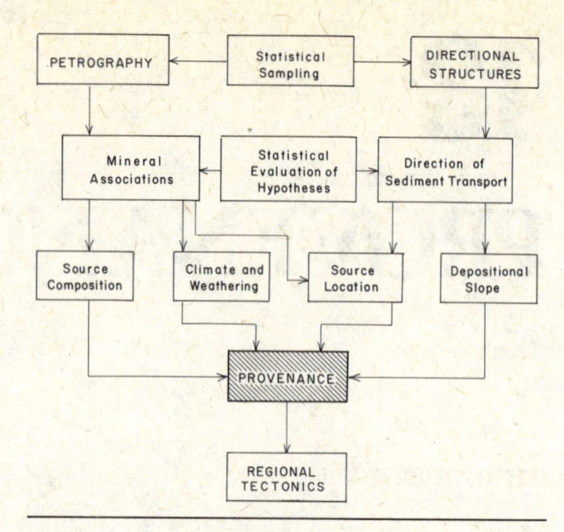

FIG. 13-1. A provenance methodology. (After Potter and Siever, 1956, Jour. Geology, v. 64, Fig. 1.)

DEFINITIONS AND CONCEPTS

Basically and fundamentally the clastic sediments are *residues*. These residues are the insoluble materials left after chemical breakdown and disintegration of some preexisting rock. The composition of the residue depends in part on the nature of the parent rock and in part on its maturity, which is a measure of the extent to which the decomposition processes were carried toward completion. The maturity is a function of the time through which the action is extended and of the intensity of that action. Time and intensity are dependent on relief and climate, respectively.

But clastic sediments are *washed* residues that have been subjected to sorting action with resultant fractionation into several size grades. These grades differ not only in size of grain but also in mineralogical and chemical composition. The finest grades are largely decomposition products—the clay minerals; the coarser grades are undecomposed residues derived from the parent or source rock.

The task of the petrologist is to examine the final fractionated residue or sediment and to determine whence it came (distance and direction of transport), to determine the kind of rock or rocks from which the residue was derived, and to deduce from the *maturity* of the residue the nature of the climate and relief of the source area (Fig. 13-1). The latter objectives constitute a study of the *provenance*, whereas the former is dealt with in the section on *paleocurrents* and paleogeography.

The term *provenance* (French, *provenir*, to originate or come forth) has been used to embrace all the factors relating to the production or "birth" of a sediment. Most often it refers to the source rocks from which the materials were derived. Each type of source rock tends to yield a distinctive suite of minerals which therefore constitute a guide to the character of that rock. But the composition of a sediment is not determined solely by the nature of the source rock; it is also a function of the climate and relief within the source region ("distributive province" of Brammall) which determines the maturity of residues derived from this region.

Erosion commonly interrupts processes of weathering in midcourse, especially in areas of high relief. Material eroded will then include rock and mineral fragments that have escaped alteration or have been only partially altered. Likewise in certain climatic regimens, processes of chemical disintegration are greatly retarded or even inhibited. The composition of the residues of weathering and the resulting sediment is, therefore, the result largely of combined effects of relief and climate on source rocks; any conclusions concerning these two factors, as well as the kind of source rocks, must be made from the chemical and mineralogical makeup of the sediments. To assay the maturity of a sediment one must know something of the mobility of the several chemical constituents of which it is composed, and especially the relative stability of various rock-making minerals. We will first discuss minerals as a guide to the nature of the source rock and then explore the question of maturity as related to oxide mobility and mineral stability.

MINERALS AND SOURCE ROCKS

The student of clastic sedimentary rocks would like to know the nature and character of the source rock or rocks from which the sediment was derived. Such knowledge might, in part, enable him to identify the source region. Fragments of a distinctive rock or an unusual mineral in sands make such identification certain and thus contribute to the understanding of the paleogeography of ancient times. In addition, events outside the basin of sedimentation influence the kind and character of sediments accumulating in the basin. Sharp uplift and erosion in the source region will release a flood of new minerals not heretofore present; volcanism outside the basin will be registered by influx of a new and distinctive mineral assemblage. It is necessary, therefore, to know and recognize mineral suites distinctive to and characteristic of contrasting sources.

It is obviously important to know if a sediment is a first-cycle sediment derived from a crystalline rock or whether it is second-cycle and derived from earlier sediments. Many properties, such as roundness, may be inherited from earlier cycles of abrasion, so interpretation of the history of a given sediment is complex. The average sandstone is estimated by Krynine (1942a) to consist of about 30 percent reworked material, about 25 percent new material of igneous derivation, and 45 percent materials from metamorphic sources.

The distinction between first-cycle and second-cycle sediments is fundamental but is a distinction difficult to make in very mature sediments. Immature sediments such as arkosic sands are apt to be first-cycle. The quartz of first-cycle sands is apt to be angular, although some may be rounded because of resorption in the magma or to corrosion in the weathering profile. Rounded overgrowths on the quartz are proof of second-cycle origin.

An important distinction is that between a plutonic and a supracrustal provenance. The *depth* from which material comes is something of a measure of the magnitude of uplift responsible for the debris. Accordingly, gneisses and plutonic igneous rocks are grouped together, as are low-rank metamorphic and sedimentary rocks. Plutonic rocks are coarse-grained and yield sand-sized quartz and feldspar; supracrustal rocks yield second-cycle quartz and rock particles of sand grade. The feldspar–rock fragment ratio of sands has been used (see the chapter on sandstones) as a provenance index and is, therefore, a measure of the relative contributions of plutonic and supracrustal zones.

Light and heavy minerals have both been used as guides to provenance. The most common mineral of sands is quartz; most efforts, therefore, have been directed to determine the relation between kinds of quartz and source rock. One of the first such efforts was that of Mackie (1896), who used the inclusions of quartz as a guide to provenance. A more comprehensive effort was that of Krynine (1940, pp. 13–20; 1946), who classified quartz as igneous (including plutonic, volcanic, and hydrothermal), metamorphic (including pressure and injection quartz), and sedimentary (which may be either authigenic overgrowths or vug or vein fillings). Criteria to recognize various quartz types utilized inclusions, extinction (undulatory or not), shape (elongation), and polycrystallinity. Krynine's classification is difficult to apply, because the attributes used overlap class limits and are difficult to assess objectively, and also because of our inadequate knowledge of the quartz in the source rocks. These questions have been critically examined by various workers and the problem reviewed by Blatt (1967a, 1967b). (The whole question has been summarized in the chapter on sandstones.)

It is clear that, despite the difficulties involved, meaningful observations can be made on detrital quartz. Even if it is not possible to distinguish with much certainty the provenance of common quartz, one can distinguish between polycrystalline quartz and common quartz, and distinguish, perhaps, between sutured and polygonized quartz (Voll, 1960, p. 536), and between plutonic quartz and volcanic quartz (Todd and Folk, 1957, p. 2550). These distinctions have value in the analysis of provenance and can, in any case, be very useful in defining mineral assemblages in mature sandstones.

Next to quartz, feldspar is the most common mineral of sands. But in general less attention has been given to it as a guide to provenance; the study of Rimsaite (1967) is an exception. In general, the feldspar of acid volcanics is apt to be sanidine; that of acid plutonic rocks is either orthoclase or microcline. Perthitic feldspar is indicative of slow cooling and hence characteristic of plutonic sources. Feldspars of pyroclastic origin tend to show euhedral forms, commonly broken, and in some cases display a thin envelope of glass, whereas plutonic feldspars are anhedral. The plagioclase in volcanic and hypabyssal rocks is characterized by oscillatory zoning; this type of zoning is rare in plutonic and metamorphic rocks (Pittman, 1963).

Rock particles are among the most informative of all detrital components, and all efforts to determine provenance should include a close study of these materials. Sandstones commonly contain rock particles, which, by reason of the definition of sand, will of necessity be fine-grained. These may be volcanic, mainly basaltic or felsic, or be sedimentary, mostly pelitic or micritic, or be fine-grained metamorphic, such as slate and phyllite. Rock particles, more than any other types of grains, carry their own evidence of provenance, although they may, in some cases, be difficult to identify (Boggs, 1968). Dickinson (1970) has proposed a simple operational definition of the several types of rock particles seen in sands. Based on texture, his classification included volcanic, clastic, tectonite, and microgranular. Volcanic particles have the texture of igneous aphanites and include altered or recrystallized volcanic rock fragments; clastic rocks have fragmental textures. Tectonite fragments display schistose or semischistose fabrics; and microgranular rock fragments are made of roughly equant, well-sorted grains. Their identity is most difficult to specify; in fact, they belong to one or another of the first three classes.

Many ambiguities can be resolved if conglomerates are present. Pebble counts, by number or by volume, are important. The several kinds and their proportions indicate both the source rocks and the distance of travel; the less resistant are rapidly eliminated with increasing distance. Because of the varying block-forming capacities of the several source rocks, however, and of the varying resistance to abrasion of these rocks, the proportions present in a gravel are not a direct reflection of the relative abundance of the several rocks in the source area. (This question has been further explored in Chapter 6.)

Heavy minerals have been found exceptionally useful as clues to the nature of the source rocks (Boswell, 1933, pp. 47–59). Some minerals are diagnostic of a particular type of source rock. Others, as in the case of quartz, are more ubiquitous and occur in nearly all possible types of parent materials. In this case, varietal features, such as inclusions, color, form, and the like, serve as a guide to the source rock type. Krynine's work on tourmaline (1946b), in which he was able to recognize 13 subspecies of this mineral, illustrates the use of varietal properties. Similarly, varieties of zircon (Tyler et al., 1940; Vitanage, 1957) have been used as guides to provenance. Mineral suites characteristic of source rock types are given in Table 13-1.

As in the case of the components of gravel, the makeup of the heavy-mineral suite may be altered not only by removal of less stable species by weathering but also by selective losses by abrasion during transit, and by losses caused by solution *after* deposition (intrastratal solution). These problems are discussed below.

Even clay minerals may have some value in provenance studies. Insofar as clay minerals are stable, the clay mineralogy of a sediment may correspond rather closely with that generated in a given source region (Weaver, 1958; Biscaye, 1965). But clay minerals are susceptible to post-burial changes so that they are a less useful guide to provenance than the more stable detrital components of sands or the pebbles of a gravel.

MINERAL STABILITY: IN THE SOIL PROFILE

Understanding the bulk chemical composition of a sediment or its mineralogical makeup requires first of all an understanding of differences in chemical and mineralogical composition of the weathered

TABLE 13-1. Detrital mineral suites characteristic of source rock types

Reworked sediments	
Barite	*Leucoxene*
Glauconite	Rutile
Quartz[a] (especially with worn overgrowths)	*Tourmaline*, rounded
Chert	*Zircon*, rounded
Quartzite fragments (orthoquartzite type)	
Low-rank metamorphic	
Slate and *phyllite* fragments	*Quartz* and *quartzite* fragments (metaquartzite type)
Biotite and muscovite	*Tourmaline* (small pale brown euhedra carbonaceous inclusions)
Feldspars generally absent	
Leucoxene	
High-rank metamorphic	
Garnet	*Quartz* (metamorphic variety)
Hornblende (blue-green variety)	Muscovite and biotite
Kyanite	Feldspar (acid plagioclase)
Sillimanite	*Epidote*
Andalusite	*Zoisite*
Staurolite	*Magnetite*
Acid igneous	
Apatite	*Zircon*, euhedra
Biotite	*Quartz* (igneous variety)
Hornblende	*Microcline*
Monazite	*Magnetite*
Muscovite	Tourmaline, small pink euhedra
Sphene	
Basic igneous	
Anatase	Leucoxene
Augite	Olivine
Brookite	*Rutile*
Hypersthene	Plagioclase, intermediate
Ilmenite and magnetite	Serpentine
Chromite	
Pegmatite	
Fluorite	*Muscovite*
Tourmaline, typically blue (indicolite)	Topaz
Garnet	*Albite*
Monazite	Microcline

[a] Italicized species are more common.

residue and the rock from which it was derived. We will consider both of these aspects of the problem.

MOBILITY OF THE OXIDES

Chemical changes that take place during weathering are determined by comparison of the composition of the weathered residue with the fresh rock from which it was derived. The changes are well illustrated by the alteration of the Morton granite gneiss (Goldich, 1938). The fresh gneiss contains about 30 percent quartz, 19 percent K-feldspar, 40 percent plagioclase, 7 percent biotite, and the remaining 3 percent hornblende, magnetite, and minor accessories. The chemical composition of unaltered gneiss is given in Table 13-2, col. A. That of altered rock is shown in col. B. It is difficult to determine by inspection of these two analyses the actual changes, because the results must, of course, total 100 percent in both analyses. The change in alumina from 14.62 to 26.14, for example, does not mean that alumina has been added but that probably alumina has increased *relative* to the other constituents, because some of these constituents actually were lost by leaching. Because alumina is rather inert and is less likely to migrate by solution than most other oxides, we may assume that it remained constant. If so, 100 grams of fresh rock contains 14.62 grams of alumina. The residual products of weathering will also contain 14.62 grams of alumina if none of it is assumed to be lost. Because this amount is 26.14 percent of the whole residue, the weight of residual materials is (14.62/26.14) 100, or 55.88 grams. This is the number of grams of altered rock which contains the same weight of alumina as 100 grams of fresh rock. This weight of residual material, therefore, will contain 55.88/100 of each item in col. B. Therefore, the residue contains 30.83 grams of silica, and so forth. By subtracting each item in col. C from the corresponding one in col. A, the gains or losses, assuming alumina constant, will be obtained (col. D). In the example given, there is an apparent gain of ferric oxide and titania, and a loss of silica, ferrous oxide, magnesia, lime, soda, and potash. Additions of water and carbon dioxide (from the atmosphere) are also evident.

It is not necessary to assume any oxide constant. If the percentage of each constituent in fresh rock is divided by its percentage in altered rock, and if the quotients (multiplied by 100) are plotted on a suitable scale, *relative* gains and losses can be seen at a glance (Leith and Mead, 1915, pp. 287–289). Fig. 13-2 illustrates the results when this procedure is applied to analyses of fresh and weathered Morton gneiss. The diagram may be interpreted as showing that 55.88 grams of altered rock contains as much alumina as 100 grams of fresh rock, but 120 grams of residue are required to contain as much silica (71.54 grams) as in the fresh rock, and so forth. It is apparent that alumina has increased relative to silica or that silica has decreased relative to alumina. The latter is the more probable conclusion. Similarly all points which plot to

TABLE 13-2. Analyses of fresh and weathered Morton gneiss

Constituent	A	B	C	D
SiO_2	71.54	55.07	30.83	−40.71
TiO_2	0.26	1.03	0.58	+ 0.32
Al_2O_3	14.62	26.14	14.62	0.00
Fe_2O_3	0.69	3.72	2.08	+ 1.39
FeO	1.64	2.53	1.43	− 0.21
MgO	0.77	0.33	0.18	− 0.59
CaO	2.08	0.16	0.09	− 1.99
Na_2O	3.84	0.05	0.03	− 3.81
K_2O	3.92	0.14	0.08	− 3.84
H_2O+	0.30	9.75	5.40	+ 5.10
H_2O-	0.02	0.64	0.36	+ 0.34
CO_2	0.14	0.36	0.20	+ 0.06
Total	99.82[a]	99.92[a]	55.88	−43.94

[a] Several minor constituents determined by Goldich omitted from totals.

Source: After Goldich (1938).

A. Fresh Morton gneiss, Cold Spring Granite Company quarry, Morton, Minnesota. S. S. Goldich, analyst.

B. Residual clay from Ramsey Park, Redwood Falls, Minnesota. S. S. Goldich, analyst.

C. Grams of each constituent present in 55.88 g of weathered material derived from 100 g of fresh rock, assuming alumina to be constant.

D. Loss in grams of each constituent in conversion of 100 g of fresh rock to 55.88 g of altered material, alumina assumed to be constant (col. A − col. C).

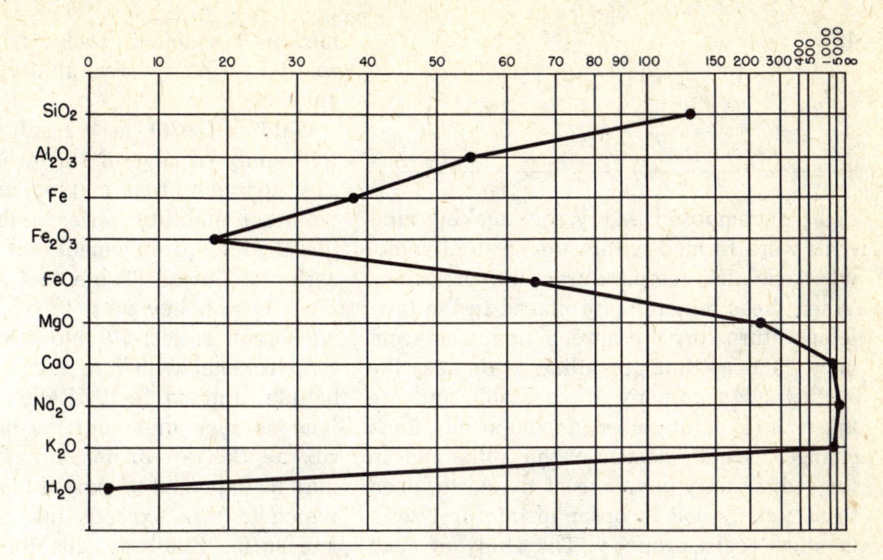

FIG. 13-2. Losses and gains of oxide constituents in the weathering of the Morton granite gneiss. (Data from Goldich, 1938.)

the left of alumina represent gains relative to that oxide, whereas all those which fall to the right represent losses. Gains or losses relative to any other oxide can be deduced similarly from this diagram. Of the original constituents in this rock, soda shows the greatest loss, whereas ferric iron displays the least. The order of loss is Na_2O, K_2O, CaO, MgO, SiO_2, Al_2O_3, and iron.

The *average* order of loss can be determined by study of the alteration of large numbers of rocks by a similar treatment (or by comparing the average igneous rock and the average sediment). Students of weathering agree that all constituents except water are lost in prolonged weathering. The order in which the oxides are lost, as interpreted by several investigators, is given in Table 13-3. The variations in the orders given in this table probably result from too limited data. Some differences might be expected in different rocks; more marked differences may be related to differences in climate (Reiche, 1950, p. 44).

It is upon differences in mobility of the several oxides that some indices of residue maturity (or of weathering potential) have been based (see p. 499).

MINERAL STABILITY

If there is a definite order or differential rate of loss of chemical oxides, it follows that rock-forming minerals in which these constituents occur should exhibit different degrees of stability. The consistency of observed and experimental findings, recorded in the literature, support this conclusion. As a group, mafic minerals are less stable than alkali feldspars; potash feldspars are more stable than soda-lime feldspars, and so forth.

The *stability* of a mineral is its resistance to alteration. We are here concerned primarily with its chemical stability or resistance to solution and decomposition rather than its mechanical stability or resistance to abrasion. Minerals are, at the time of formation, presumably in equilibrium with their environment and are therefore stable. But as they are brought into new environments, unlike those in which they formed, they are prone to go into solution

TABLE 13-3. Average order of loss of the oxides in weathering

Order	Steidtmann (1908)	Leith and Mead (1915)	Goldich (1938)
1	CaO	CaO	Na_2O
2	MgO	Na_2O	CaO
3	Na_2O	MgO	MgO
4	K_2O	K_2O	K_2O
5	SiO_2	SiO_2	SiO_2
6	Iron	Iron	Al_2O_3
7	Al_2O_3	Al_2O_3	Iron

or be decomposed. Many rock-making minerals were formed within the earth at somewhat elevated temperatures and pressures. When these minerals are placed in the low-temperature, low-pressure, aqueous environment such as that prevailing at or near the earth's surface, many are unstable and are taken into solution or decomposed. Such changes, which occur in the soils and in the sedimentary envelope of the earth, need to be understood in order to interpret sedimentary rocks properly. The study of such changes has been aptly called "mineral pathology."

Evidences of stability or instability are of several kinds. "Pathologic" features such as etched surfaces and corroded borders are indicative of instability. Conversely, outgrowths or secondary enlargements, are indicative of stability, inasmuch as the mineral appears to be growing rather than disappearing. The disappearance of minerals or their absence in soils and other residue derived from rocks in which the minerals in question are present is further evidence of instability. To a limited extent the stability of a mineral can also be experimentally determined by measuring solution losses under somewhat restricted and simplified conditions of leaching.

All minerals are not equally immune to solution and decay. Many attempts have been made to determine the *relative* stability of minerals in soils and sediments. These have been reviewed by Boswell (1933, p. 37; 1942), Milner (1940, p. 492), Pettijohn (1941), Allen (1948), Reiche (1950), and Smithson (1950).

There have been two main approaches to determining relative stabilities of detrital minerals. The first, perhaps the most direct and most important, is to study soil profiles and to note the order of disappearance of minerals as one passes from fresh to altered rock. The second approach is to compare or to determine the frequency of occurrence of the several mineral species in Recent and ancient sedimentary deposits on the premise that frequency of occurrence, adjusted for relative abundance of the several species in the source rocks, is in some way related to the survival ability of the several species.

Goldich (1938), as a result of a quantitative study of several soil profiles, arranged the common rock-making minerals in a "mineral stability series" which is nearly identical in arrangement with the reaction series of Bowen (Table 13-4). This arrangement is not, however, to be interpreted as a reaction series. Olivine does not weather to pyroxene, which in turn alters to hornblende, and so forth. Rather, in a normal igneous rock that contains olivine and pyroxene, the rate of decomposition of olivine may be expected to exceed that of pyroxene, which in turn exceeds that of hornblende, and so on. Similarly, all other things being equal, a gabbro may be expected to decompose more rapidly than a granite. Goldich dealt primarily with common rock-making minerals. Other workers have studied minor accessory minerals also and attempted to determine the stability order of these "heavy" minerals (Dryden and Dryden, 1946; Smithson, 1950; Willman, Glass, and Frye, 1966).

Sindowski and co-workers (1949) grouped the heavy minerals into several classes in order of their resistance to weathering (Table 13-5). The order was determined in part by comparison of the heavy-mineral content of younger and older terraces of the Rhine. Older terraces were impoverished in their heavy-mineral content presumably by removal by leaching or solution of less stable minerals. As noted by Sindowski, any marked differences in the heavy-mineral content of two sedimentary deposits of dif-

TABLE 13-4. Mineral-stability series in weathering

Olivine	
	Calcic plagioclase
Augite	
	Calci-alkalic plagioclase
Hornblende	Alkali-calcic plagioclase
	Alkalic plagioclase
Biotite	
Potash feldspar	
Muscovite	
Quartz	

Source: After Goldich (1938).

ferent ages may be the result of selective loss of certain mineral species *after* deposition rather than to a different provenance or differing weathering regimen *before* deposition. Loss following deposition is *intrastratal solution,* and, in general, the results of such action are similar to those of solution in the weathering profile itself.

MATURITY

The prime interest of the student of sedimentary rocks in the chemical mobility of rock-making oxides and the stability of rock-forming minerals is their bearing on the concept of sedimentary maturity. The maturity of a clastic sediment is the extent to which it approaches the ultimate end-product to which it is driven by the formative processes that operate upon it. Reiche's weathering potential index (1950, pp. 13–29) is, therefore, an index of maturity. The more mature the sediment, the less its weathering potential. This index cannot, however, be used indiscriminately. It is applicable as a maturity index only to the residues or their washed equivalents. It cannot be considered a maturity index for rocks of a mixed mechanical and chemical origin. Sandstones with infiltrated carbonate cements have a high lime and magnesia content. Restoration of materials which had been removed during weathering raises the weathering potential and obscures the primary mature nature of the sediment to which they have been added.

Vogt (Kennedy, 1951) expressed maturity as the degree of residual character. Vogt

TABLE 13-5. The stability of heavy minerals[a]

Intrastratal solution		Weathering (?)	Weathering (soil profile)	
Pettijohn (1941)	Smithson (1941)	Sindowski[b] (1949)	Goldich (1938)	Dryden (1946)
Rutile				
Zircon	Zircon	Zircon		Zircon
	Rutile	Rutile		
Tourmaline	Tourmaline	Tourmaline		Tourmaline
	Apatite			
Monazite	Monazite			Monazite
Garnet	Garnet			
Biotite			Biotite	
Apatite				
Staurolite	Staurolite	Staurolite		
Kyanite	Kyanite	Kyanite		Kyanite
Hornblende		Hornblende[c]	Hornblende	Hornblende
				Staurolite
	Ferromagnesian minerals	Garnet		Garnet
Augite		Augite	Augite	
		Apatite		
Olivine		Olivine	Olivine	

[a] The spacing is intended to draw attention to the similarities between the series. Minerals occurring in fewer than three of the five lists have been omitted.

[b] Sindowski does not arrange minerals in a continuous series but places them in groups indicated by the braces in the above table; the minerals in each group are here arranged to show the maximum possible agreement with the other lists.

[c] Given as "amphibole" in original paper.

Source: Modified from Smithson (1950).

assumed, with considerable justification, that argillaceous sediments would tend to be enriched in alumina as they become more mature. The true enrichment of alumina, however, may be obscured by a change in the amount of the independent quartz component present, with the result that clay sediments with different residual characters may contain the same amount of alumina, and vice versa. To eliminate the masking effect of silica, which fluctuates with texture (Table 8-4), Vogt recalculated the chemical analysis to the sum of 100, excluding SiO_2 (and TiO_2). In general, the three oxides MgO, CaO, and Na_2O, decrease gradually and regularly with increasing residual character, whereas K_2O, SiO_2, and TiO_2 increase. Iron tends to remain constant over a wide range in composition. Alumina is perhaps the least mobile oxide, whereas Na_2O is the oxide most readily removed and at the same time (unlike lime, magnesia, or potash) is not restored to the sediment in the ordinary cycle of sedimentation. The alumina–soda ratio of a shale, therefore, may be used as an abbreviated chemical index of maturity.

Quartz is the only chemically and physically durable mineral constituent of plutonic rocks common enough to accumulate in great volume. The mineralogical maturity of a sand, therefore, is expressed by quartz content. Because most of the quartz was originally associated with feldspars, the maturity may also be expressed by the disappearance of feldspar or by the quartz–feldspar ratio. The latter is not so appropriate for those sands derived from feldspar-poor terranes. The paucity of feldspars would lead to deceptively high quartz–feldspar ratios. Sands derived from a supracrustal complex would contain rock particles, none of which, except chert, have both chemical and mechanical stability. The ratio of chert–nonchert rock fragments would be an appropriate maturity index; of most general application would be the ratio of quartz plus chert to feldspar plus rock fragments.

It is not sufficient to assess the maturity of a sediment; more important is the significance or geological meaning of maturity. Under what conditions are highly mature sediments produced? What does immaturity mean? Because maturity is the measure of the approach of a clastic sediment to the stable end-state toward which it is driven by the formative processes operating on it, the maturity is also, therefore, a combined record of the *time* through which such processes have operated and the *intensity* of their action. If time is brief, the end-product will be immature regardless of the intensity. If intensity of action (input of energy per unit time) is low, the end-product is immature no matter how long the time. On the other hand, if intensity is high and time is long enough, the end-product will be mature.

What, in geologic terms, determines time or duration of action and its intensity? It seems probable that time or duration is determined largely by *relief*. Rapidity of erosion is a function of relief. High relief promotes a high rate of erosion, whereas low relief is associated with a retarded rate of erosion. Under conditions of rapid erosion, soil-forming processes lag behind those of transportation, and much incompletely weathered material finds its way into streams. Under conditions of retarded erosion, weathering goes to completion so that only the most stable residues appear in the sediments. In areas of high relief but with youthful topography, some relatively level interstream areas remain. Erosion of these areas yields maturely weathered detritus, whereas sharply incised canyons are sources of fresh, unweathered materials. Therefore, erosion of such an area will yield both mature and immature products of weathering. In general, mature products are fine-grained and accumulate mainly in shales; immature products are coarse-grained and appear in sandstones. Their separation, however, is never complete or perfect.

The effects of *climate* are more complex and have been of concern to students of soils for many years. Under tropical conditions, where temperature is high and moisture most abundant, weathering appears to be most rigorous, and residues formed are notably enriched in oxides of iron and alumina and are consequently relatively deficient in silica. Laterites and bauxites are the end-products of such tropical soil-forming processes. Shales formed from such parent materials would be alumina-rich, perhaps even bauxitic. In general it may be said that warmer, more humid climates lead to more complete decomposition of the source rock, whereas colder or more arid climates are marked by products of lesser maturity. The general absence of water tends to retard chemical action; coarser residues produced under arid conditions might contain many unstable minerals.

The actual composition of a sedimentary deposit, however, is the result of interaction or the combined effects of relief and climate on the source rock and the effects of abrasion and sorting of the derived residuum. In general, according to Barrell (1908, p. 183), the character of fine fluviatile or wash detritus in the region of its origin may be taken as an index of climate. The size and abundance of coarser materials, on the other hand, are a measure of the rapidity of erosion and hence a measure of topographic relief. Because relief is dependent on the balance between uplift and erosion, the character of coarser materials is therefore also an index to *tectonism*.

To what extent are the principles outlined above, deductively arrived at by Barrell (1908), Krynine (1942b, 1949) and others (Folk, 1968, pp. 84–85, 110), supported by actual observations of the sediment load of streams draining areas of differing relief and climate? In general we lack actual data on the problem. There are some exceptions. Krynine's study (1935) in the tropics of southern Mexico showed that river sands were feldspathic despite a high annual mean temperature of 80°F (26.5°C) and a high rainfall 120 inches (300 cm). Gibbs (1967, pp. 1219–1220), in his masterful study of the sediment carried by the Amazon River, confirmed Krynine's view that relief rather than climate controls the feldspar content of a sediment. Garner (1959), as a result of a study of four regions in the Andes, believed that climate rather than relief was the prime factor governing clast size. Arid climate produced coarse alluvium, implying incomplete weathering; humid climates yield mainly silts and clays. Actual studies of selected regions, designed to test the concepts linking relief and climate to sediment maturity, are very few. Webb and Potter (1969) investigated stream sands in an arid region of high relief; Robelen (1974) has examined those in a humid region of low relief. These are among the few studies attempting to sort out the effects of climate, relief, and source rocks on the character of the sediments.

It should not be forgotten that to some extent both textural and *compositional* maturity of sands and gravels are achieved by mechanical action—elimination of rock particles, polycrystalline quartz and removal of clays. This is the topic of the section below.

MINERAL STABILITY: DURING TRANSIT

It might be supposed that the residues produced by disintegration and decomposition of source rocks would undergo further alteration or change during their transport from the place of release from a source rock to the place of their ultimate deposition. Not only would the clay fraction be separated from the sand and gravel grades, but the latter two themselves would undergo further modification and fractionation. It might be supposed, for example, that processes operative during transport, which are responsible for rounding of the debris transported, would also modify the composition by selective abrasion (and sorting).

Unfortunately, careful observations on what actually does happen during transit are exceedingly few. That some changes in composition do occur during transport seems highly probable. Downstream changes in composition of river gravels have long been noted. (These studies have been summarized in Chapter 6.) In short, they show rather rapid elimination of less durable components (limestone, shale, friable sandstones) with resulting enrichment in more stable rock types (quartzite, chert, vein quartz). See Fig. 6-8. Gravels can become compositionally mature in a relatively short distance of travel.

On the other hand, evidence concerning the selective wear and elimination of minerals in the sand range is somewhat ambiguous. It might be supposed that softer and more cleavable species would be destroyed by abrasion, with a complementary enrichment in harder and more durable components. What evidence bears on this matter?

Some experiments have been made to determine the resistance of mineral grains to wear. Friese (1931) determined the durability (*Transportwiderstand*) of a considerable number of minerals. Taking crystalline hematite as 100, he assigned a numerical value to the abrasion resistance of each mineral studied (Table 13-6). Cozzens (1931) likewise determined the rate of wear of common minerals, a rate found to be a function of hardness. Apparently the durability index and Mohs scale of hardness are closely correlated, at least for minerals lower than quartz in hardness. The exact function was not worked out fully, although it appears to be of the type $y = (x/a^n)$, where y is the durability index, x is the Mohs hardness, a is the hardness of quartz (7), and n is an exponent (near 4). Erratic values were frequent and may be the result of size reduction processes other than abrasion, of the nonuniform nature of the hardness scale, and of other properties of the minerals (such as elasticity).

TABLE 13-6. Abrasion resistance of minerals (in order of increasing resistance)[a]

After Friese (1931)	After Thiel (1945)
Hematite (100)[b]	Barite
Monazite (117, av.)	Siderite
Orthoclase (150)	Fluorite
Diopside (160)	Goethite
Andalusite (220)	Enstatite
Kyanite (260)	*Kyanite*
Apatite (275)	Bronzite
Common olivine (290)	*Hematite*
Epidote (320)	*Augite*
Ilmenite (325)	*Apatite*
Garnet (378, av.)	Spodumene
Magnetite (380)	Hypersthene
Topaz (390)	Diallage
Common *augite* (420)	Rutile
Staurolite (420)	Hornblende
Cordierite (480)	Zircon
Pyrite (500)	*Epidote*
Tourmaline (817, av.)	*Garnet*
	Titanite
	Staurolite
	Microcline
	Tourmaline
	Quartz

[a] Minerals common to two lists are italicized to facilitate comparison.
[b] "Transportwiderstand"; hematite arbitrarily taken as 100.

Thiel (1940) experimentally determined relative abrasion resistance of common minerals. According to Thiel, the order of resistance is (beginning with the least resistant): (1) apatite, (2) hornblende, (3) microcline, (4) garnet, (5) tourmaline, and (6) quartz. In a later note (1945) he added several other minerals to the list. Although differing in some important particulars, the order is in the main consistent with the results of Friese (Table 13-6). Marsland and Woodruff (1937) also determined mineral resistance to abrasion. Their work was based on air-blast action and showed resistance to rounding to be, beginning with the least resistant, gypsum, calcite, apatite, magnetite, garnet, orthoclase, and quartz.

Do studies of sands in transit in nature show changes consistent with experimental data? Russell (1937, 1939) states that large streams show few or no changes in mineral composition even during prolonged transport, and the feeble changes that do occur are not the result of differential abrasion. His observations on the mineralogy of the sands of the Mississippi River between Cairo, Illinois, and the Gulf of Mexico, 1100 miles (1771 km) distant, seem to support this conclusion. There appears to be only a small loss of feldspar relative to quartz and no appreciable loss of hornblende, pyroxene, and other relatively soft and cleavable minerals. Russell's conclusions seem to be confirmed by study of the sands of the Rhine (van Andel, 1950). Observations of Mackie (1896) and Plumley (1948), on the other hand, seem to show that there is an appreciable loss of feldspar in comparatively short distances in high-gradient, gravel-carrying streams (Fig. 13-3). In Black Hills streams the decline in feldspar content can hardly have been caused by anything other than abrasion. The similarity in density and sphericity of feldspar and quartz precludes a selective sorting process, and the size fraction studied by Plumley was so coarse (1–1.4 mm class) that no progressive dilution of the feldspar by contamination was likely. There is some evidence that surf action can significantly round sands and hence should be capable of altering the mineralogy (see Chapter 7, p. 249).

There are clearly documented examples of large and significant changes in the composition of sands in transit on beaches and in streams. It can be shown that most of these are related to a large-scale selective sorting process or are the result of dilution by new sources of material. Hence, in conclusion, it can be said that the tendency to selective loss by abrasion exists; in some cases it is operative, but in other cases abrasion losses may be obscured by changes in composition from other causes. Progressive change in composition is certainly not sufficient proof of differential abrasion.

MINERAL STABILITY: INTRASTRATAL

Following deposition, sediments are subject to artesian flow and leaching. How fares their mineralogy? In general, the light minerals, primarily quartz and feldspar, remain largely unaltered, although there is some evidence of solution of feldspar in some cases (Heald and Larese, 1973). In many cases individual grains of both quartz and feldspar are secondarily enlarged, indicating their stability. Some of the rock particles may undergo breakdown and form matrix.

Although the dominant light-mineral fraction persists, the same cannot be said for minor accessory or heavy minerals. These are subject to solution either early or late in the post-depositional history of the rock. Differential solution of minor accessory or heavy minerals may obscure the interpretation of the heavy suite and thus render stratigraphic correlation based on such minerals and the question of provenance obscure.

Evidences that such solution has taken

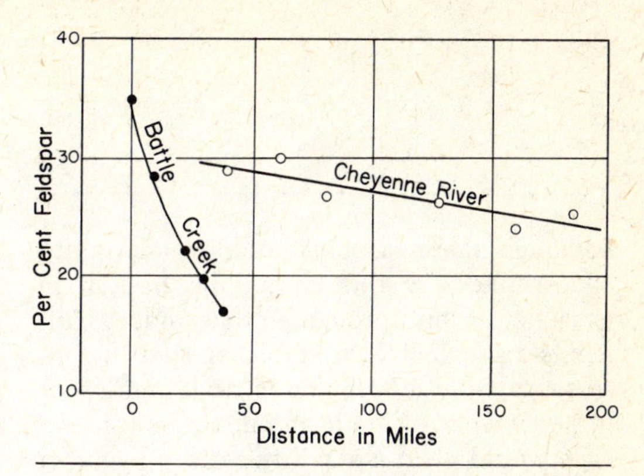

FIG. 13-3. Relation of percentage of feldspar to distance of transport in 1.0–1.414 mm size class of Black Hills streams, South Dakota. (From Plumley, 1948.)

place in a sedimentary stratum are many and varied. Etched grains constitute good evidence of solution (Bramlette, 1929). It is not always clear that etching was achieved after, rather than before, deposition, although the fragile spinelike projections or "teeth" on some of the heavy minerals, notably ampiboles and pyroxenes, could hardly have survived transport (Edelman, 1931; Edelman and Doeglas, 1931). See Fig. 13-4. In some cases it can be proved that the hacksaw or cockscomb character was formed in place after deposition (Ross, Miser, and Stephenson, 1929).

In some cases, even quartz shows some evidence of intrastratal solution. This is especially true in carbonate-cemented sands where the original rounded outline of the quartz has become irregular and embayed where it is in contact with the carbonate. In other cases, microstylolitic contacts, especially at contacts between adjacent chert grains, indicate some intrastratal solution (Sloss and Feray, 1948).

Most convincing proof of intrastratal solution is the contrast in abundance of certain heavy-mineral species within a calcareous concretion and in the matrix in which it occurs (Bramlette, 1941). See Table 7-12. Similarly, heavy minerals in cogenetic shales, also a "sealed" environment, are different from associated sandstones (Blatt and Sutherland, 1969). Students of heavy minerals long ago noted an apparent increase in complexity of the heavy-mineral suite with decreasing geologic age of the beds from which they were derived (Thoulet, 1913; Boswell, 1923). See Fig. 13-5.

Although there is considerable uncertainty concerning why this should be true, it is possible, if not probable, that older sediments have lost less stable species by intrastratal solution. If this is true, then *order of persistence* of the heavy minerals through time would yield a stability series, a subject explored by Pettijohn (1941).

Order of persistence was based on published records of minerals in sediments of all ages and places. For each species the frequency of occurrence (ratio of reported occurrences of the species in question to number of investigated formations) in Recent sediments and the average frequency of occurrence in ancient sediments were determined. The ratio of these two frequencies is taken as a measure of the survival ability of each species investigated (Table 13-7). The order of persistence thus arrived at shows a close correspondence with the stability order of common rock-making minerals as determined by Goldich (1938) from the weathering profile. Some notable exceptions may be the result of differences between the soil and subterranean environments which, although similar, are not identical. Apatite, for example, seems to be unstable in soils but is stable in ancient sediments; likewise, alkali feldspars decompose in the soil profile but may undergo growth or enlargement in sedimentary rock. Despite these and other exceptions, order of persistence is a measure of mineral stability in sedimentary rocks. Other studies seem to bear this out (Smithson, 1939, 1941, 1942).

Even in a given or particular sedimentary basin, younger sediments display a greater variety of heavy minerals. Moreover, the order of appearance of these minerals, as one proceeds from older to younger rocks,

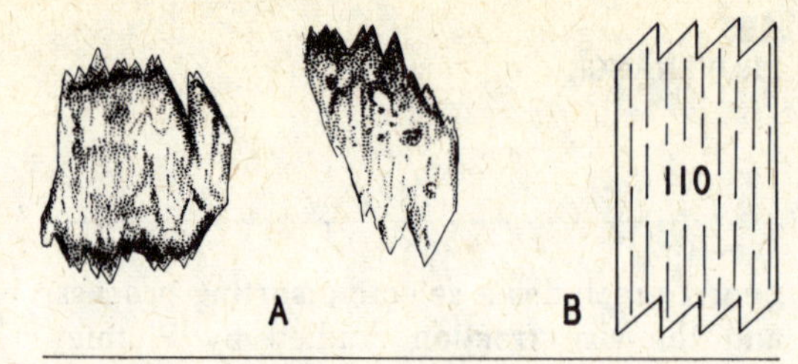

FIG. 13-4. Hacksaw terminations caused by intrastratal solution. A: detrital augite; B: schematic diagram of the hacksaw structure of a detrital augite parallel to (110). (After Edelman and Doeglas, 1931.)

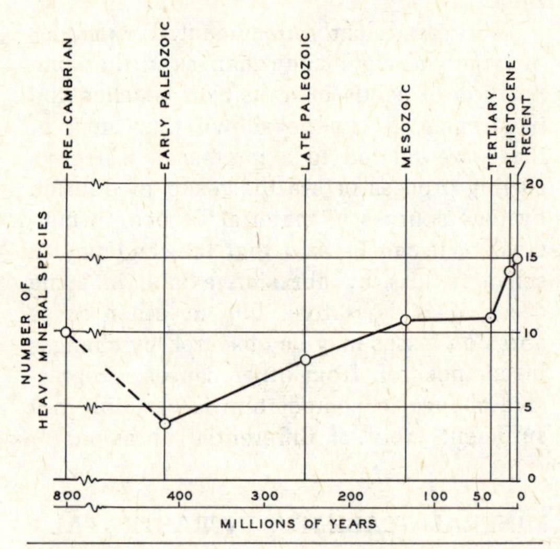

FIG. 13-5. Number of heavy-mineral species and age of deposit. The number of species in more than one-half of the reported formations is plotted against the age of the beds. (Pettijohn, 1941.)

TABLE 13-7. Order of persistence of detrital minerals

−3	Anatase[a]	10	Kyanite
−2	*Muscovite*[b]	11	Epidote
−1	Rutile	12	*Hornblende*
1	Zircon	13	Andalusite
2	Tourmaline	14	Topaz
3	Monazite	15	Sphene
4	Garnet	16	Zoisite
5	*Biotite*	17	*Augite*
6	Apatite	18	Sillimanite
7	Ilmenite	19	*Hypersthene*
8	Magnetite	20	Diopside
9	Staurolite	21	Actinolite
		22	*Olivine*

[a] A negative sign indicates the mineral to be *more* abundant in ancient than in modern sediments.

[b] Italicized species are components of Bowen's reaction series.

Source: After Pettijohn (1941).

is essentially the reverse order of their stability determined by other independent observations. The suggestion has been made, therefore, that heavy-mineral zones are stability zones and owe their existence and character to selective removal of less stable species in deeper zones by intrastratal solution. This question is explored further below.

HEAVY MINERAL ZONES

That beds of differing age, even in the same district, have differing assemblages of heavy minerals is a common observation. It was at first presumed that changing mineralogy was related to unroofing and exposure of new source rocks in the distributive province or, in other words, a record of changing provenance. If so, the heavy-mineral suite formed at a given moment in time was unique, and all sands which carried this suite, within a particular basin, must be of the same age. Heavy minerals were, therefore, a valuable adjunct in stratigraphic correlation.

But, as further studies have shown, this simplistic view may not be correct. Careful study of heavy-mineral zones in Tertiary and Mesozoic sections, in which such zones appear to be most clearly defined, reveals two things. First, the number of mineral species appears to increase as one goes from older to younger beds; second, the order of appearance of the minerals, even in widely separated and unrelated basins, is remarkably similar. If one neglects stray occurrences and counts only those minerals present in one-half or more of the samples studied, the trend appears quite clear (Fig. 13-6 and Table 13-8).

As noted above, mineral zones of several unrelated and geographically separated areas are alike, not only in the increase in number of minerals in each zone but also in the order of appearance of many of the dominant minerals; furthermore, the order of appearance is, in general, the reverse order of stability of the minerals in question.

As can be seen from Fig. 13-6 and Table 13-8, hornblende is most typical of the highest zones; the lowest zones are restricted to minerals like tourmaline, zircon, and rutile (and in some cases staurolite and garnet). Kyanite, epidote, and titanite seem more characteristic of the intermediate zones. Minerals in the lower zones, as a rule, are also present in the higher zones so that, as noted, the latter have an enlarged or enriched suite. These observations are not without exceptions. The order of appearance of the species is not always the same; in some cases, minerals of the lower zone do not persist into the higher ones (perhaps they are too much diluted by the flood of new species). And there are cases in which a mineral seems to appear prematurely, then drop out and reappear again at a higher level.

How are these observations explained? The traditional view is that increasing complexity in mineralogy is the result of progressive denudation and unroofing of new sources (Fig. 13-7, left). As erosion proceeds, deeper levels of the crust would become contributors to the basin of sedimentation. Because minerals in rocks of the deeper zone are, on the average, least stable, there might be both a normal order of succession and a succession that would correlate with stability order. This view has been championed by Krynine (1942a) and van Andel (1959).

A second hypothesis presumes a correlation between the mineral sequence and progressive uplift of the source area (Fig. 13-7, center). Under this thesis, the terrane of varied lithology would lie near base level at the initial stage and would be progressively elevated with consequent increase in gradient and accelerated erosion. During initial stages only the most stable species escape destruction in the soil profile; in the final stages even the least stable minerals would appear in the sediment (Boswell, 1923).

A third hypothesis supposes that all sediments deposited had about the same mineral suite at the time of deposition but that, because of intrastratal solution, deeper and older beds have lost all unstable species. The probability of survival is a function of depth of burial and of time. The deeper the burial and/or the older the rock, the less probable the presence of a given species

(Fig. 13-7, right; Pettijohn, 1941). In support of this view is the direct evidence of the efficacy of intrastratal solution, such as Bramlette's demonstration of the preservation of the original suite in a sealed environment—in this case in a concretion. There is, moreover, visual evidence of corrosion and removal of many species in deeper zones (Edelman and Doeglas, 1931). The preservation of the more complex suite in less permeable beds, especially shales, is further support for this view (Blatt and Sutherland, 1969).

Heavy-mineral zones, then, with some-

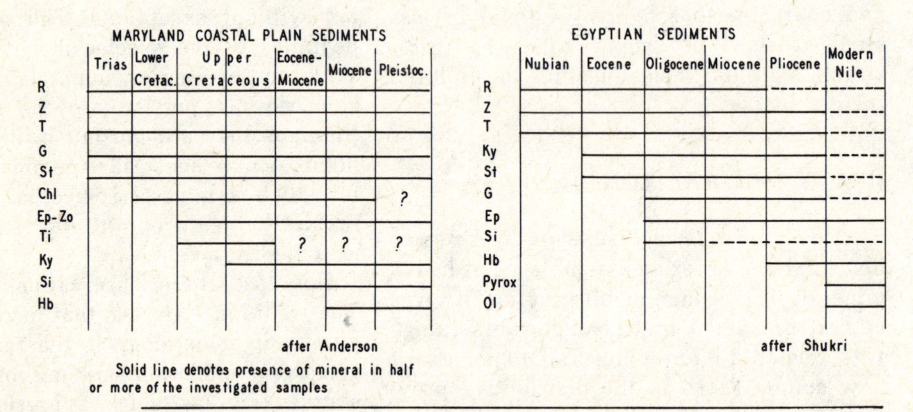

FIG. 13-6. Heavy-mineral zones. Solid lines, present in more than one-half of samples; dashed lines, present in fewer than one-half of the samples. Left, Atlantic Coastal Plain, Maryland (data from Anderson, 1948); right, Egyptian sediments (data from Shukri et al., 1954).

TABLE 13-8. Order of appearance of index species in heavy-mineral zones[a]

Stow (1938)	Cogen (1940)	Anderson (1948)	Evans, Hayman, and Majeed (1933)	Milner (1940)
Cretaceous-Tertiary, Wyoming	Tertiary, Gulf Coast	Triassic-Cretaceous-Tertiary, Atlantic Coastal Plain	Tertiary, Burma	Tertiary, Rumania
		Rutile[b]	Rutile[b]	Rutile[b]
Zircon[b]	Zircon[b]	Zircon[b]	Zircon[b]	Zircon[b]
Tourmaline[b]	Tourmaline[b]	Tourmaline[b]		Tourmaline[b]
Garnet[b]	Garnet[b]	Garnet[b]	Garnet	Garnet[b]
Staurolite	Staurolite[b]	Staurolite[b]	Staurolite	Staurolite[b]
Kyanite	Kyanite			
		Chloritoid	Chloritoid	
	Epidote	Epidote	Epidote	Epidote
	Sphene	Sphene		
		Kyanite		
Hornblende	Hornblende	Hornblende	Hornblende	Hornblende

[a] Only minerals common to three or more authors included in table.
[b] Present in lowest formation; order in table not significant.

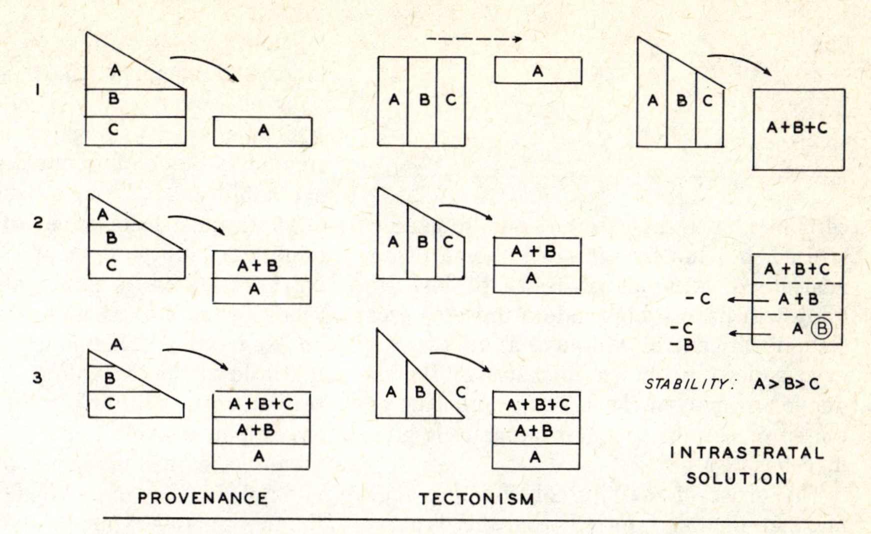

FIG. 13-7. The three hypotheses of heavy-mineral zonation.

what similar character and with a similar sequence of species might be formed in several unlike ways. Obviously if heavy-mineral zones are stability zones only, they have little or no stratigraphic significance. If they are the result of progressive denudation or progressive tectonism, the heavy suite will change with time, and the mineral zones might be useful as an aid to stratigraphic correlation. The zoning may result from simultaneous operation of several factors and be related to both stability and provenance. If so, the problem of zones and the sorting out of factors of no stratigraphic value from those of worth is much more complex than heretofore supposed. Perhaps varietal characters of the same species would assist in this task. The appearance of new and different varieties of zircon, for example, would be in no way related to intrastratal solution or differential loss by weathering in the distributive province. Only progressive denudation and unroofing of new source rocks would explain the changing character of zircon. Careful work on with tourmaline might, as in the case of zircon, assist in the solution of the mineral zone problem.

Although any of the hypotheses might be adequate for a local sequence or a single tectonic cycle, none is proved for the geologic column as a whole. Increasing complexity with decreasing age *could* be a reflection of stability only (to Pettijohn this seems most probable). It has been claimed, however, that orogenies are becoming more closely spaced and that the rate of sedimentation shows a progressive increase with passage of time. If so, mineral complexity should increase in complexity also. On the other hand, none of the older sediments, even those following notable orogenies, seems to be as rich as the Pleistocene and Holocene sands, and it is hard to escape the conclusion that their impoverishment is somehow related to long-continued intrastratal solution.

MINERAL STABILITY: THEORETICAL AND OTHER CONSIDERATIONS

Various attempts have been made to calculate or predict stability of minerals and rocks and to provide a theoretical basis for empirically determined stability orders. Reiche (1950), for example, proposed a *weathering potential index* which can be computed for a mineral or a rock from the usual chemical analysis. This index is defined as the mol percentage (percentage of any constituent divided by its molecular weight) ratio of the sum of the alkalies and alkaline earths, less combined water to the total mols present exclusive of water, or

$$\frac{100 \times \text{mols } (K_2O + Na_2O + MgO + CaO - H_2O)}{\text{mols } (SiO_2 + Al_2O_3 + Fe_2O_3 + MgO + CaO + Na_2O + K_2O)}$$

This is roughly the percentage of the four oxides empirically found to be the most fugitive. In cases of rocks containing free silica (such as granite), the amount of free silica is excluded from the calculation. Where the amount of free silica cannot be readily ascertained, as in the case of shale, total silica is used, but the results are to some degree misleading. Minerals or rocks of low stability will have a high index; those of great stability in general will have a low index or even a negative index (because of the deduction of mols of H_2O). The weathering potential of certain common minerals is given in Table 13-9.

The order of stability of common rock-making minerals inferred from the weathering potential index corresponds fairly well with that empirically determined by Goldich. The chief exception is quartz (whose index is 0) and the relative stabilities of the feldspar group as a whole as compared with the mafic suite.

The stability of minerals, according to Fersman (Baturin, 1942), is related to certain thermodynamic principles: "The greater the amount of energy evolved by an ion during its passage into the crystalline state, the more stable is the crystal obtained, the more difficult it is to reduce it to a dispersed state, to dissolve and to melt, or to divide again the atoms of the lattice into free ions; the more stable is such a system of minerals, the higher its ability to accumulate during natural processes and the less is it subject to destruction, melting, and dissolution. . . ."

TABLE 13-9. Weathering potential index

Mineral	Average	Range
Olivine	54	45–65
Augite	39	21–46
Hornblende	36	21–63
Biotite	22	7–32
Muscovite	−10.7[a]	—
Labradorite	20	18–20
Andesine	14	—
Oligoclase	15[a]	—
Albite	13[a]	—
Orthoclase	12[a]	—

[a] Average computed from averaged analysis; no data on range.

Source: After Reiche (1950).

Furthermore, according to Baturin, studies in crystal chemistry have shown that lattices characterized by a medium coordination number of average symmetry and by a smaller radius of the ions display the greatest stability.

There have been other attempts to relate mineral stability to crystal structure. Fairbairn (1943), for example, attempted to correlate the rate or ease of alteration with the *packing index.* This index was defined as the ratio of the ion volume to the volume of the unit cell. Within certain groups of minerals some correlation is apparent. Of two minerals with the same composition, the mineral that has the higher packing index is the more stable. Muir (in Polynov, 1937) has also discussed the relation between crystal structure and mineral stability. Gruner (1950) also attempted to arrange silicate minerals in the order of their stability. Gruner's arrangement was based on an "energy index" computed from electronegativities of the elements involved and coordination coefficients. In a general way, Gruner's order agrees with Goldich's stability order and other empirically determined orders.

In our discussion of mineral stability we have assumed that it was an independent entity—a property of the mineral—unalterable and unrelated to the surrounding milieu. Such is not the case. A mineral may be stable in one environment but unstable in another. Apatite, for example, appears stable in some sandstones but is unstable in the soil profile. As noted by Boswell (1942), stability is a function of the chemical properties of the pore fluid in the sediment, the temperature, and the pH, as well as the composition and crystal structure of the mineral itself. He called attention to the corroded staurolite in England's Bunter Sandstone, a formation with hard sulfate waters, and the fresh character of the staurolite in the Lower Greensand, characterized by soft pore waters. Likewise, garnet is rounded and unaltered in the Bunter but shows lunate and irregular forms in the Lower Jurassic. Consequently, all efforts to assess mineral stability, based either on empirical studies of soils or sedimentary deposits, or on theoretical considerations, are at best a statement of "average" stability. Many of the discrepancies among published stability studies arise from observations made on minerals subjected to differing environmental milieu.

READING PROVENANCE HISTORY

How, after the minerals of a source rock have passed through a complex geological sieve—eliminating many species by weathering, abrasion, or intrastratal solution—can we assess the provenance of a sediment? The goal is to unravel the "line of descent" of the sediment. Only gravels and sands lend themselves readily to this kind of analysis.

As is probably clear from the foregoing part of this chapter, reading provenance history is a difficult task. Multiple sources and multiple cycles may be involved. It may be difficult to separate the immediate source from the ultimate source of the sediment. One can follow several pathways. Even a single sand grain can throw light on the problem. As shown by Potter and Pryor (1961, Pl. 2 and p. 1250), an occasional tourmaline sand grain may show a clearly defined *rounded* overgrowth (Fig. 13-8). Such a grain implies (1) formation in an igneous or metamorphic source rock, (2) weathering and release of this grain, (3) transportation and abrasion, (4) deposition in a sand, followed by (5) authigenic or low-grade metamorphic overgrowth, (6) a second weathering and release of the grain, (7) a second episode of transport and rounding, and finally (8) redeposition in another sand. Using criteria developed by Krynine (1946b), it may even be possible to determine the kind of source rock from which the grain originally came.

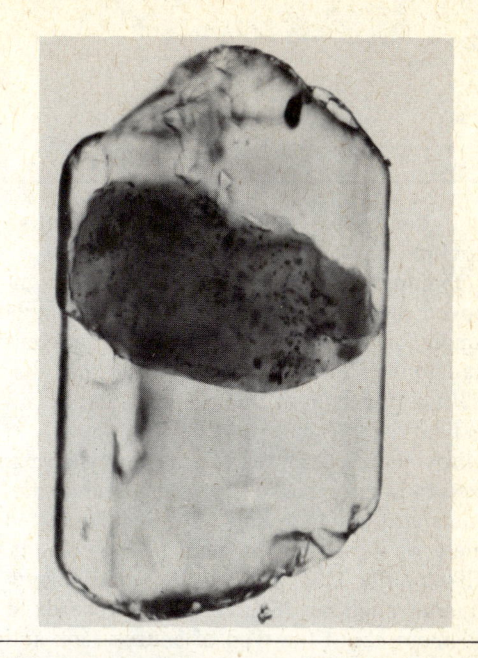

FIG. 13-8. Abraded overgrowth on detrital core of tourmaline, McNairy Sand (Cretaceous), Tennessee. Ordinary light, ×250. (After Potter and Pryor, 1961, Bull. Geol. Soc. Amer., v. 72, Pl. 2.)

A more complete provenance story can be made from a sample rather than from a single grain. What can one infer from a single sample of sand from the delta of the Mississippi River about the nature and character of the rocks in the Mississippi drainage area? A typical sample might contain 62 percent quartz, 19 percent feldspar, and 16 percent chert and other fine-grained rock particles (No. 1050 of Russell, 1937, Table 1). The quartz and feldspar are generally subangular, although a few well-rounded grains of quartz are present (Russell and Taylor, 1937). The feldspar is principally potassic feldspar and to a lesser extent oligoclase and andesine. Most grains are fresh, but some weathered feldspar is also observed. Rock particles include chert and fine-grained quartzite (4 percent) but include several varieties of volcanic rocks, some slates, and schists. A little calcite and glauconite are present. Principal heavy minerals are ilmenite, pyroxenes, and amphiboles with a smaller quantity of garnet, zircon, monazite, rutile, and titanite. From this information we might infer that the principal source was a granite or granodiorite with minor contributions from volcanic flows and metamorphic and sedimentary rocks. Such an analysis does not disclose the principal immediate source of most of the sand—the Pleistocene glacial drift—but it does identify the ultimate source, namely, the crystalline rocks of the Canadian Shield from which much of the glacial materials came. Nor is it evident that most of the Mississippi drainage area is covered by Paleozoic and younger sedimentary rocks, although the glauconite, chert, carbonate particles, and the occasional rounded quartz indicate such sources to be present. Clearly we can identify the source rocks and estimate their relative importance as contributors to the sand. But we are unable to esti-

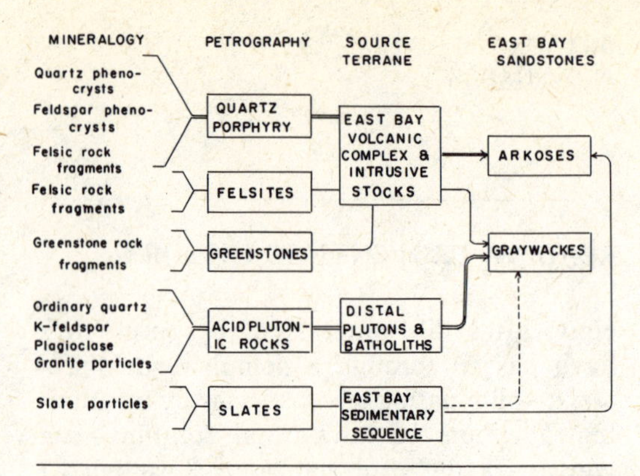

FIG. 13-9. Provenance diagram for Archean sandstones of the Minnitaki Lake area, western Ontario, Canada. (After Walker and Pettijohn, 1971, Bull. Geol. Soc. Amer., v. 82, Fig. 24.)

mate their areal extent in the distributive province.

Obviously our assessment could be improved if we had a suite instead of a single sample, especially in the case of ancient sandstones. The samples should be distributed throughout the deposit under investigation.

Provenance analyses can be best summarized in diagrammatic form. Provenance diagrams are of two kinds, one constructed solely on what can be seen in the rock itself—mainly from thin section and heavy-mineral analysis (Fig. 13-9). It records a judgment about the type of source rock and conditions in the source area from study of each of the detrital components of the sediment, whether it is a gravel or sand. A second type of provenance diagram is based only in part on laboratory study of rock samples (Fig. 13-10). It incorporates knowledge of the regional geology and stratigraphy. It shows what is possible, perhaps probable, in terms of source rock contributions.

The sedimentary petrologist need not rely solely on detrital constituents to solve the problem of provenance. He may utilize evidence provided by stratigraphy and paleocurrents.

Because much, if not most, sand is recycled or derived from older formations

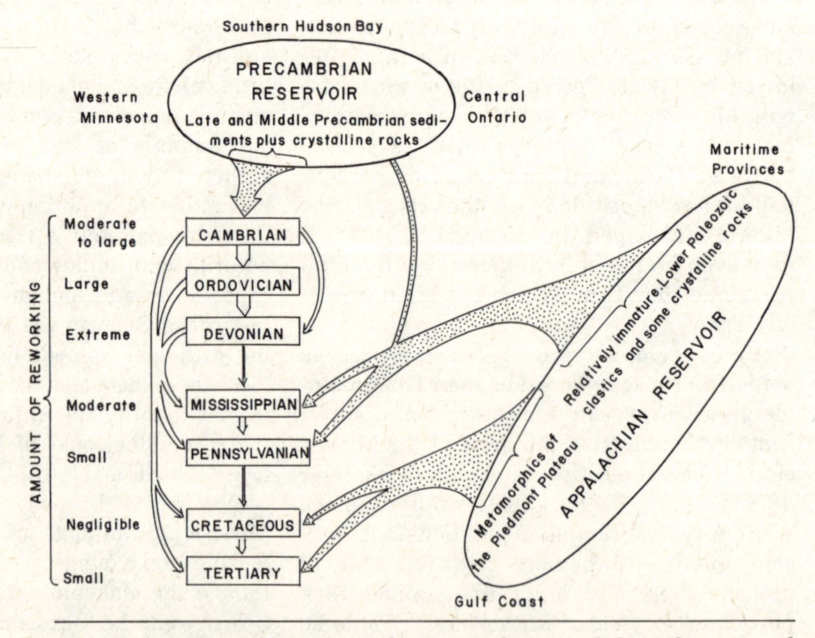

FIG. 13-10. Dispersal centers and source relations of the Paleozoic and younger sandstones of upper Mississippi Valley. (From Potter and Pryor, 1961, Bull. Geol. Soc. Amer., v. 72, Fig. 14.)

which, in turn, may be derived from still older sandstones, a knowledge of stratigraphy will contribute to provenance analysis by establishing relative ages of strata and thereby indicating what could or could not have been a source for the sandstone in question. Most important is the recognition of unconformities which record periods of erosion. Construction of a *paleogeologic* map will disclose which formations were exposed and were, therefore, potential sources. Stratigraphy may show that large potential areas —such as the craton—were effectively covered by limestones at the time the sand in question was deposited and could not have been, therefore, a source of sediment.

Paleocurrent analysis is also helpful in provenance studies, especially in formations of alluvial origin. The upcurrent direction shown by cross-bedding and other structures is in the direction of the source. It is even possible to estimate the distance to the source ledges from measurement of decrease in pebble size if conglomerates are present. (This and related topics are covered in the next chapter.)

In summary, the investigator must put together all lines of evidence—distribution and kinds of light and heavy minerals, paleocurrent information, facies relations, and distribution and kinds of rock fragments in both sands and conglomerates, knowledge of stratigraphy and of larger structural elements—in short, any and all kinds of geologic information.

Valuable insight concerning provenance analysis can be gained by referring to studies of Krynine (1940, Fig. 3), Doty and Hubert (1962, Fig. 11), Todd and Folk (1957, pp. 255–257), and Walker and Pettijohn (1971). The foregoing examples are based largely on petrology or what can be seen in the sandstone or conglomerate itself. Studies by Payne (1942, p. 1754), Suttner (1969, Fig. 15), Potter and Pryor (1961, Fig. 14), and ten Haaf (1964, p. 135) belong to the second type and utilize regional geology, tectonics, and stratigraphy to unravel provenance history.

References

Allen, V. T., 1948, Weathering and heavy minerals: Jour. Sed. Petrology, v. 18, pp. 38–42.

van Andel, Tj. H., 1950, Provenance, transport and deposition of Rhine sediments: Ph.D. dissertation, Univ. Groningen, 129 pp.

———, 1959, Reflections on the interpretation of heavy-mineral analyses: Jour. Sed. Petrology, v. 29, pp. 153–163.

Anderson, J. L., 1948, Cretaceous and Tertiary subsurface geology: Maryland Dept. Geol. Mines Water Resources, Bull. 2, p. 1–113.

Barrell, J., 1908, Relations between climate and terrestrial deposits: Jour. Geol., v. 16, pp. 159–190, 255–295, 363–384.

Baturin, V. P., 1942, On stability and formation of minerals of abyssal geospheres in the stratisphere: Comptes Rendus, Dokl. Akad. Sci. U.R.S.S., v. 37, pp. 32–34.

Biscaye, P., 1965, Mineralogy and sedimentation of Recent deep-sea clay in the Atlantic Ocean and adjacent seas and ocean: Bull. Geol. Soc. Amer., v. 76, pp. 803–832.

Blatt, H., 1967a, Provenance determinations and recycling of sediments: Jour. Sed. Petrology, v. 37, pp. 1031–1044.

———, 1967b, Original characters of clastic quartz: Jour. Sed. Petrology, v. 37, pp. 401–424.

Blatt, H., and Sutherland, B., 1969, Intrastratal solution and nonopaque heavy minerals in shales: Jour. Sed. Petrology, v. 39, pp. 591–600.

Boggs, S., Jr., 1968, Experimental study of rock particles: Jour. Sed. Petrology, v. 38, pp. 1326–1339.

Boswell, P. G. H., 1923, Some aspects of the petrology of sedimentary rocks: Proc. Liverpool Geol. Soc., v. 13, pp. 231–303.

———, 1933, On the mineralogy of sedimentary rocks: London, Murby, 393 pp.

———, 1942, The stability of minerals in sedimentary rocks: Quart. Jour. Geol. Soc. London, v. 97, pp. lvi–lxxv.

Bramlette, M. N., 1929, Natural etching of detrital garnet: Amer. Mineral., v. 14, pp. 336–337.

———, 1941, The stability of heavy minerals in sandstone: Jour. Sed. Petrology, v. 11, pp. 32–36.

Cogen, W. M., 1940, Heavy mineral zones of Louisiana and Texas Gulf Coast sediments: Bull. Amer. Assoc. Petrol. Geol., v. 24, pp. 2069–2101.

Cozzens, A. B., 1931, Rates of wear of common minerals: Washington Univ. Stud. Sci. Tech., n.s., no. 5, pp. 1—80.

Dickinson, W. R., 1970, Interpreting detrital modes of graywacke and arkose: Jour. Sed. Petrology, v. 40, pp. 695–707.

Doty, R. W., and Hubert, J. F., 1962, Petrology and paleogeography of the Warrensburg Chan nel Sandstone, western Missouri: Sedimentology, v. 1, pp. 7–39.

Dryden, A. L., and Dryden, C., 1946, Comparative rates of weathering of some common heavy minerals: Jour. Sed. Petrology, v. 16, pp. 91–96.

Edelman, C. H., 1931, Diagenetische Umwandlungerschienungen am detritischen Pyroxenen und Amphibolen: Fortschr. Min. Krist. Pet., v. 16, pp. 67–68.

Edelman, C. H., and Doeglas, D. J., 1931, Reliktstrukturen detritischer Pyroxene und Amphibole: Min. Pet. Mitt., v. 42, pp. 482–490.

Evans, P., Hayman, R. J., and Majeed, M. A., 1933, The graphic representation of heavy mineral analyses: Proc. World Petrol. Congr., 1933 (London), v. 1, pp. 251–256.

Folk, R. L., 1968, Petrology of sedimentary rocks: Austin, Tex., Hemphill's, 170 pp.

Fairbairn, H. W., 1943, Packing in ionic minerals: Bull. Geol. Soc. Amer., v. 54, pp. 1305–1374.

Friese, F. W., 1931, Untersuchungen von Mineralan auf Abnutzbarkeit bei Verfrachtung im Wasser.: Min. Pet. Mitt., v. 41, n.s., pp. 1–7.

Garner, H. F., 1959, Stratigraphic-sedimentary significance of contemporary climate and relief in four regions of the Andes Mountains: Bull. Geol. Soc. Amer., v. 70, pp. 1327–1368.

Gibbs, R. J., 1967, The geochemistry of the Amazon River System. Part I. The factors that control the salinity and the composition and concentration of the suspended solids: Bull. Geol. Soc. Amer., v. 78, pp. 1203–1232.

Goldich, S. S., 1938, A study in rock weathering: Jour. Geol., v. 46, pp. 17–58.

Gruner, J. W., 1950, An attempt to arrange silicates in the order of reaction energies at relatively low temperatures: Amer. Mineral., v. 35, pp. 137–148.

ten Haaf, E., 1964, Flysch formations of the northern Apennines, *in* Turbidities: Developments in sedimentology v. 3 (Bouma, A. H., and Brouwer, A., eds): Amsterdam, Elsevier, pp. 127–136.

Heald, M. T., and Larese, R. E., 1973, The significance of the solution of feldspar in porosity development: Jour. Sed. Petrology, v. 43, pp. 458–460.

Kennedy, W. Q., 1951, Sedimentary differentiation as a factor in Moine-Torridonian correlation: Geol. Mag., v. 88, pp. 257–266.

Krynine, P. D., 1935, Arkose deposits in the humid tropics. A study in sedimentation in southern Mexico: Amer. Jour. Sci., ser. 5, v. 29, pp. 353–363.

———, 1940, Petrology and genesis of the Third Bradford Sand: Bull. Pennsylvania State Coll. Min. Ind. Expt. Sta. 29, 134 pp.

———, 1942a, Provenance versus mineral stability as a controlling factor in the composition of sediments (abstr.): Bull. Geol. Soc. Amer., v. 53, pp. 1850–1851.

———, 1942b, Differential sedimentation and its products during one complete geosynclinal cycle: An. Primer Congr. Panamericano Ing. Minas Geol., pt. 1, v. 2, pp. 537–561.

———, 1946a, Microscopic morphology of quartz types: An. Segundo Congr. Panamericano Ing. Minas Geol., v. 3, pp. 35–49.

———, 1946b, The tourmaline group in sediments: Jour. Geol., v. 54, pp. 65–87.

———, 1948, The megascopic study and field classification of sedimentary rocks: Jour. Geol., v. 56, pp. 130–165.

———, 1949, The origin of red beds: Trans. New York Acad. Sci., ser. 2, v. 11, pp. 60–68.

Leith, C. K., and Mead, W. J., 1915, Metamorphic geology: New York, Holt, Rinehart and Winston, 337 pp.

Mackie, W., 1896, The sands and sandstones of eastern Moray: Trans. Edinburgh Geol. Soc., v. 7, pp. 148–172.

Marsland, P. S., and Woodruff, J. G., 1937, A study of the effect of wind transportation on grains of several minerals: Jour. Sed. Petrology, v. 7, pp. 18–30.

Milner, H. B., 1940, Sedimentary petrography, 3rd ed.: London, Murby, 666 pp.

Payne, T. G., 1942, Stratigraphical analysis and environmental reconstruction: Bull. Amer. Assoc. Petrol. Geol., v. 26, pp. 1697–1770.

Pettijohn, F. J., 1941, Persistence of heavy minerals and geologic age: Jour. Geol., v. 49, pp. 610–625.

Pittman, E. D., 1963, Use of zoned plagioclase as an indicator of provenance: Jour. Sed. Petrology, v. 33, pp. 380–386.

Plumley, W. J., 1948, Black Hills terrace gravels: a study in sediment transport: Jour. Geol., v. 56, pp. 526–577.

Polynov, B. B., 1937, Cycle of weathering (Muir, A., trans.): London, Murby, 220 pp.

Potter, P. E., and Pryor, W. A., 1961, Dispersal centers of Paleozoic and later clastics of the Upper Mississippi Valley and adjacent areas: Bull. Geol. Soc. Amer., v. 72, pp. 1195–1250.

Pryor, W. A., 1961, Petrography of Mesaverde sandstones in Wyoming, *in* Symposium on Late Cretaceous rocks, 16th Ann. Field Conf. Wyoming Geol. Assoc., pp. 34–46.

Reiche, P., 1950, A survey of weathering processes and products, rev. ed.: Univ. New Mexico Publ. Geol., no. 3, 94 pp.

Rimsaite, J., 1967, Optical heterogenity of feld-

spars observed in diverse Canadian rocks: Min. Pet. Mitt. Schweiz., v. 47, pp. 61–76.
Robelen, P., 1974, pers. commun.
Ross, C. S., Miser, H. D., and Stephenson, L. W., 1929, Water-laid volcanic rocks of early Upper Cretaceous age in southwestern Arkansas, southeastern Oklahoma, and northeastern Texas: U.S. Geol. Surv. Prof. Paper 154-F, pp. 175–202.
Russell, R. D., 1937, Mineral composition of Mississippi River sands: Bull. Geol. Soc. Amer., v. 48, pp. 1307–1348.
———, 1939, Effects of transportation on sedimentary particles, *in* Recent marine sediments (Trask, P. D., ed.): Tulsa, Okla., Amer. Assoc. Petrol. Geol., pp. 32–47.
Russell, R. D., and Taylor, R. E., 1937, Roundness and shape of Mississippi River sands: Jour. Geol., v. 45, pp. 225–267.
Shukri, M. N., and El-Ayouty, M. K., 1954, The mineralogy of Eocene and later sediments in the Angabia area—Cairo-Suez district: Bull. Fac. Sci., Cairo Univ., no. 32, pp. 47–61.
Sindowski, F. K. H., 1949, Results and problems of heavy-mineral analysis in Germany; a review of sedimentary-petrological papers, 1936–1948: Jour. Sed. Petrology, v. 19, pp. 3–25.
Sloss, L. L., and Feray, D. E., 1948, Microstylolites in sandstone: Jour. Sed. Petrology, v. 18, pp. 3–13.
Smithson, F., 1939, Statistical methods in sedimentary petrology: Geol. Mag., v. 76, pp. 417–427.
———, 1941, The alteration of detrital minerals in the Mesozoic rocks of Yorkshire: Geol. Mag., v. 78, pp. 97–112.
———, 1942, The Middle Jurassic rocks of Yorkshire: a petrological and palaeogeographical study: Quart. Jour. Geol. Soc. London, v. 98, pp. 27–59.
———, 1950, The mineralogy of arenaceous sediments: Sci. Prog., v. 149, pp. 17–21.
Steidtmann, E., 1908, A graphic comparison of the alteration of rocks by weathering with their alteration by hot solutions: Econ. Geol., v. 3, pp. 381–409.
Stow, M. H., 1938, Dating Cretaceous-Eocene tectonic movements in Big Horn Basin by heavy minerals: Bull. Geol. Soc. Amer., v. 49, pp. 731–762.
Suttner, L. J., 1969, Stratigraphic and petrographic analysis of Upper Jurassic–Lower Cretaceous Morrison and Kootenai formations, southwest Montana: Bull. Amer. Assoc. Petrol. Geol., v. 53, pp. 1391–1410.
Thiel, G. A., 1940, The relative resistance to abrasion of mineral grains of sand: Jour. Sed. Petrology, v. 10, pp. 103–124.
———, 1945, Mechanical effects of stream transportation on mineral grains of sand size (abstr.): Bull. Geol. Soc. Amer., v. 56, p. 1207.
Thoulet, J., 1913, Notes de lithologie sous-marine: Ann. Inst. Oceanog., v. V, fasc. 9.
Todd, T. W., and Folk, R. L., 1957, Basal Claiborne of Texas, record of Appalachian tectonism during the Eocene: Bull. Amer. Assoc. Petrol. Geol., v. 41, pp. 2545–2566.
Tyler, S. A., Marsden, R. W., Grout, F. F., and Thiel, G. A., 1940, Studies of the Lake Superior pre-Cambrian by accessory-mineral methods: Bull. Geol. Soc. Amer., v. 51, pp. 1429–1538.
Vitanage, P. W., 1957, Studies of zircon types in the Ceylon pre-Cambrian complex: Jour. Geol., v. 65, pp. 117–128.
Voll, G., 1960, New work on petrofabrics: Liverpool and Manchester Geol. Jour., v. 2, pt. 3, pp. 503–567.
Walker, R., and Pettijohn, F. J., 1971, Archaean geosynclinal basin: analysis of the Minnitaki Basin, northwestern Ontario: Bull. Geol. Soc. Amer., v. 82, pp. 2099–2129.
Weaver, C. E., 1958, Geological interpretation of argillaceous sediments: Bull. Amer. Assoc. Petrol. Geol., v. 42, pp. 254–271.
Webb, W. M., and Potter, P. E., 1969, Petrology and geochemistry of modern detritus derived from a rhyolitic terrain, western Chihuahua, Mexico: Bol. Soc. Geol. Mexicana, v. 32, pp. 45–61.
Willman, H. B., Glass, H. D., and Frye, J. C., 1966, Mineralogy of glacial tills and their weathering profiles in Illinois: Illinois Geol. Surv. circ. 400, 76 pp.

14

PALEOCURRENTS AND PALEOGEOGRAPHY

INTRODUCTION

An interest in paleocurrents—currents long vanished but which have left an imprint on rocks—dates from the early work of Sorby (1857, p. 285), who wrote: "The examination of modern seas, estuaries, and rivers, shows that there is a distinct relation between their physical geography and the currents present in them; currents so impress themselves on the deposits formed under their influence that their characters can be ascertained from those formed in the ancient periods. Therefore their physical geography can be inferred within certain limits." Today we would call this paleogeographic reconstruction.

Despite Sorby's early recognition of the relation between paleocurrents and paleogeography, little was done for nearly 100 years. Although Sorby recognized the value of measuring the orientation of current structures (1859, p. 138), he did not construct a paleocurrent map. Nor did anyone else until 1897 when Ruedemann plotted the orientation of graptolite rhabdosomes in the Utica Shales (Ordovician) of New York State. In 1911 Hyde published a map of ripple mark orientation in the Berea and Bedford formations (Mississippian) of southern Ohio. Later Rubey and Bass (1925) mapped cross-bedding in Cretaceous channel sandstone in Russell County, Kansas.

It was not until Brinkman (1933) that the objectives and methods of paleocurrent research were clearly formulated. He plainly saw that systematic mapping of cross-bedding could yield significant paleogeographic data and that this approach could also be applied to other sedimentary attributes such as grain size, fossil orientation, thickness, and the like. Following Brinkman, who studied cross-bedding in the Triassic Buntsandstein of central Europe, were a large number of similar studies: Shotten (1937) on the Lower Bunter Sandstone of England, Reiche (1938) on the Permian Coconino Sandstone of Arizona, and especially Potter and Olson (1954) on the early Pennsylvanian Caseyville and Mansfield sandstones in the Illinois Basin.

There followed a flood of papers both in Europe and North America based on cross-bedding analysis (Potter and Pettijohn, 1963, Table 5-2). At the same time there was a renewed interest in current structures referred to as "sole markings," commonly seen on the underside of many sandstone beds, especially those of the flysch facies. These markings, described by Hall as early as 1843, were little more than curiosities until the monographic studies of Vassoevich (1953) and the landmark papers by Rich (1950), Kuenen (1957), Dzulynski and Radomski (1955), and Crowell (1955), who realized their potential in paleocurrent studies.

Directional properties other than cross-bedding and sole markings have proved less helpful; they are accessory criteria. Included here are ripple marks, parting lineation, oriented fossils, and pebble and grain fabrics. Either they are uncommon features, or their measurement is too time-consuming for the results obtained, although, in the absence of other criteria of current direction, they may prove useful. An exception, perhaps, is till fabric which has proved its value in studies of glacial motion.

The direction of current flow can be deduced from other characteristics of sedimentary rocks in addition to primary current structures. Among these are pebbles or mineral grains known to be derived from a unique source rock—an approach first applied in glacial studies where boulder trains have been identified and mapped. Less well known are the identification and mapping of "dispersal fans" defined by heavy-mineral assemblages in sandstones of alluvial origin (Füchtbauer, 1954, 1958).

In addition to these compositional attributes, geologists have used scalar properties, most notably clast size (which commonly diminishes downcurrent) to determine not only the *direction* but also the *distance* of transport (Fig. 14-1). In practice the mapping of maximum pebble or cobble size in ancient conglomerates has proved most useful.

This chapter briefly summarizes the subject of paleocurrent analysis and its relevance to geologic history, especially paleogeographic analysis. For details, especially field methods and statistical or graphical summarization, refer to other works, such as *Paleocurrents and Paleogeography* (Pettijohn, 1962) and especially *Paleocurrents and Basin Analysis* (Potter and Pettijohn,

As can be seen from the foregoing, paleocurrent analysis has its greatest value in paleogeographic reconstruction. It is helpful in ascertaining sedimentary strike, delineation of paleoslope, inferring the trend of the shoreline, and also the trend and location of the margin of the depositional basin—in short, exploring the relations between the paleocurrent system and the basin architecture or geometry. It assists in delineating source lands and sedimentary basins. A lesser but important by-product of paleocurrent analysis is the establishment of a relation between paleocurrents and shape or configuration of individual sand bodies and also of carbonate reefs and distribution of debris derived from them.

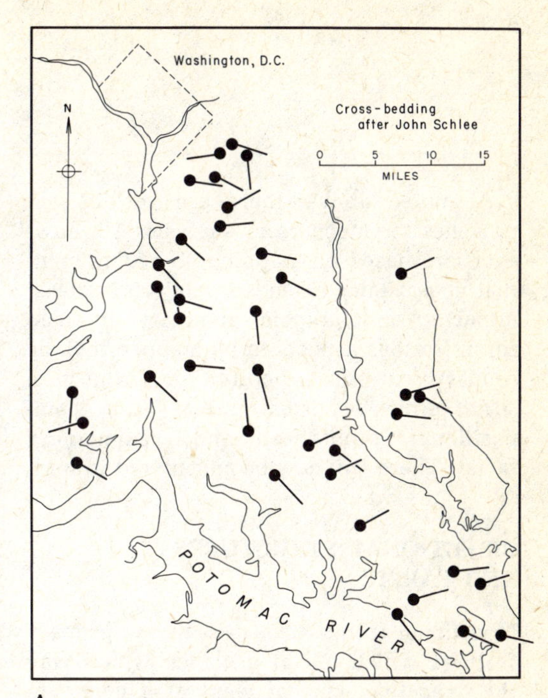

A

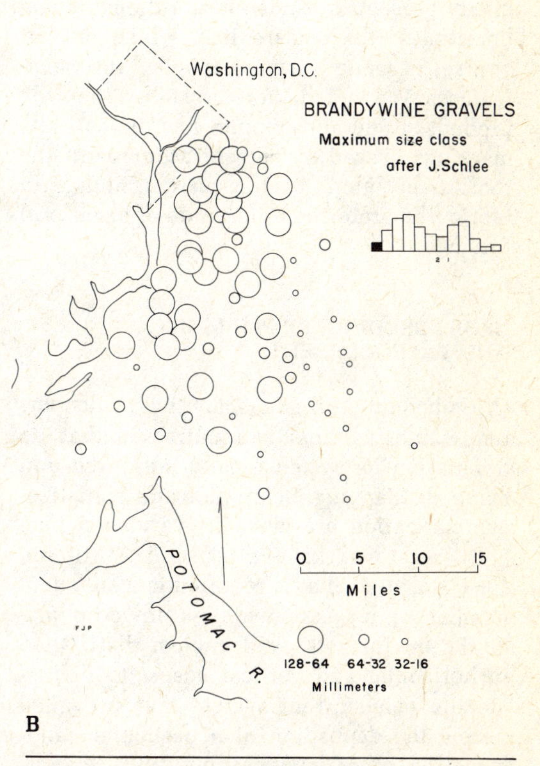

B

FIG. 14-1. Map of cross-bedding (A) and maximum pebble size (B) in Brandywine upland gravels (Pliocene-Pleistocene) near Washington, D.C. (After Schlée, 1957, Bull. Geol. Soc. Amer., v. 68, Fig. 3.)

As noted above, there are various approaches to paleocurrent analysis. All, however, are based on mapping—a requirement that necessitates extended field work, measurement, and graphic summary in map form. Various aspects to be mapped include primary current structures or properties, scalar properties (such as clast size), compositional properties, and sedimentary facies. These will now be considered in turn.

DIRECTIONAL STRUCTURES AND FABRIC

Directional properties are those primary features which tell at a glance which way the depositing current went at a given instant in geologic time. They are found in all kinds of sediments but are most commonly present in sandstones, although some limestones (the calcarenites, which are, of course, carbonate sands) display the same features. We include here cross-bedding (the product of migrating ripples and dunes), sole markings caused by current scour or to the tools swept along by the current, and grain fabric (the preferred orientation of clasts and fossils).

CROSS-BEDDING, RIPPLE MARKS, AND PALEOCURRENTS

Cross-bedding of any scale has paleocurrent value. All that is required is that its orientation be systematically measured and mapped. Mapping techniques and statistical summarization are given in various papers and larger works (Potter and Pettijohn, 1963, Ch. 10). What is required is a sufficient number of measurements of downdip azimuths on foresets, well enough distributed in horizontal and vertical space, to delineate the paleocurrent system that prevailed during the deposition of a particular sandstone or other stratigraphic unit. Planar-tabular sets are readily measured. The measurement of trough or festoon stratification is more difficult, and, if possible, the azimuth of the trough axis should be determined. On a bedding plane section, this is the bisectrix of the curved traces of foresets.

Experience has shown that azimuths of the current flow disclosed by cross-bedding are not randomly oriented. If plotted as a current rose (Fig. 4-12), these show a strong preferred orientation, with most of the azimuths falling in a single quadrant with a clearly defined mode. Moreover, in many sandstones the preferred orientation persists both vertically and laterally. Even successive sandstones in a geologic column tend to show similar patterns and a common orientation (Fig. 14-12), indicating great stability in the paleocurrent system through long periods of time (Potter and Pryor, 1961). This is particularly true of those formed by alluviating rivers and by eolian accumulation. Exceptionally some marine cross-bedding shows a bimodal distribution of current azimuths, a principal mode opposed by a lesser mode forming a "bow-tie" type of rose diagram (Fig. 10-52). This is presumed to be caused by current reversals as a result of tidal action.

Whereas the orientation of cross-bedding is a function of the *direction* of current flow, the variability is a function of other factors such as stability of the current system over the short time interval. In the case of fluvial systems the variability may be related to the sinuosity of the stream. The more complex the meander pattern, the greater the variance, which in turn may be related to the stream gradient. Variability, therefore, might be expected to increase in the downstream direction as the gradient diminishes (Brinkman, 1933, pp. 5–9; Jüngst, 1938, p. 245; Hamblin, 1958, Fig. 28).

Variability may also reflect environmental differences. In most alluvial deposits, two-thirds of the current azimuths usually lie within a 90- to 120-degree sector (Jüngst, 1938); in a marine environment there is less paleoslope control, and the variance may be larger, even approaching randomness in a few cases. Eolian deposits seem to have *a* variance not greatly different from that of rivers (Potter and Pettijohn, 1963, Table 4-2).

The relation between cross-bedding and initial depositional dip is critical to regional interpretation. This slope is the *paleoslope*. It is at right angles to depositional or sedimentary strike which is usually considered to lie subparallel to the strandline. The term *paleoslope* applies to marine or fluvial deposits; it has no meaning in an eolian situation. Clearly water ran downhill in times past, as it does now, so that fluvial cross-bedding has a simple and obvious relation to paleoslope. The mean cross-bedding azimuth is, therefore, downslope. In the proximal environment—near the basin margin—nonmarine cross-bedding may show divergences which are related to alluvial fans. The current azimuths appear to radiate from various centers—the apexes of fans (see Fig. 15-8; also Howard, 1966). The relation of marine cross-bedding to the submarine slope is less clear. Some marine cross-bedding is related to tidal and other littoral currents that may flow in diverse directions, even parallel to the shoreline and hence to the submarine contours. Studies of some ancient marine deposits show a regionally consistent paleocurrent pattern with the prevailing downcurrent azimuth normal to the shoreline and downdip (Schwarzacher, 1953, p. 327). Pettijohn is inclined to believe that this is generally true, but evidence on this point is still inconclusive.

The conclusion has been reached that eolian cross-bedding is a guide to paleowind directions. So it is, but whether such directions are those of the planetary system or are local diversions related to orographic and other features is not so clear. Some have used presumed eolian cross-bedding to reconstruct the planetary system and to infer the pole position from such reconstruction (Opdyke and Runcorn, 1960).

Observations on regional ripple mark orientation are comparatively few. Hyde (1911) long ago noted a statewide preferential orientation of oscillation (symmetrical) ripples in the Berea Sandstone (Mississippian) in Ohio. One hundred forty-nine measurements over a distance of 115 miles (185 km) show a remarkably consistent orientation. Brett (1955) found ripples in the Baraboo Quartzite (Precambrian) of Wisconsin to be essentially perpendicular to the current system shown by cross-bedding. Asymmetrical ripples might be expected to be transverse to the current flow, and the few data available bear this out. The mean ripple trend is, therefore, the depositional strike. Factors for the orientation of symmetrical ripples are poorly understood. Hyde believed that the Berea ripples were oriented parallel to the shoreline. Wave crests tend to be refracted and become parallel to the shore regardless of wind direction, and the ripples generated by them might be expected to be parallel to the shore (Kindle, 1917, p. 53; Scott, 1930, p. 53). In one study of present-day ripple marks, parallelism with the subaqueous slope (and hence general parallelism with the shoreline) has been noted (Vause, 1959). The orientation of fossil ripple marks parallel to the shoreline inferred from other considerations has also been recorded (Pelletier, 1965, p. 238).

SOLE MARKS AND PALEOCURRENTS

A great diversity of markings is found at the interface between sandstones and underlying shale. Because of disintegration of shales upon weathering, these structures appear only on the base of the sandstones, hence the appelation *sole markings*. (These were defined, described, and illustrated in Chapter 4.)

In general, sole markings described are abundant only in the flysch facies; these have been interpreted as the products of turbidity-current action. They are, therefore, presumed to be oriented parallel with the current flow which, as a density current, is presumed to behave like a heavy liquid underflow and to have moved, therefore, down the submarine slope. In some cases they have a uniform orientation over large areas (Fig. 14-2). Some recent observations seem to be inconsistent with this concept and have even led some to doubt the turbidity-current origin of structures and beds on the base of which sole markings are found (Murphy and Schlanger, 1962; Hubert, 1966). Although many groove and flute casts show an amazing uniformity in orientation from bed to bed, others show a disconcerting diversity of orientations, even on the same sole. Clearly, if one of these diversely oriented markings points downslope, the others do not. These anomalies suggest that the bottom slope has little or no effect on current flow, or that the bottom was without appreciable slope and that

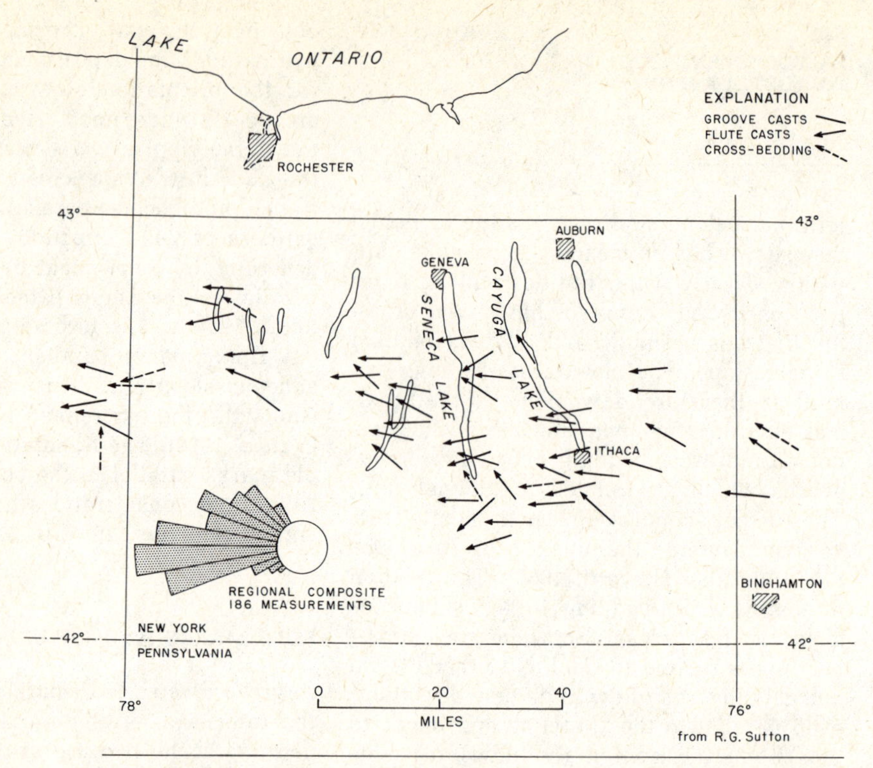

FIG. 14-2. Orientation of sole markings in "Portage" (Upper Devonian) of New York State. (After Sutton, 1959, Bull. Amer. Assoc. Petrol. Geol., v. 43, Fig. 2.)

the direction of current flow is controlled by other factors (Kuenen and Sanders, 1956, p. 662), or perhaps that these structures are not formed by turbidity currents at all.

The subject is further complicated by the observation that the direction of motion inferred from slump folds is at variance with the direction of flow indicated by sole markings. As noted by Cummins (1959, p. 177), "Slumping . . . is gravity controlled and must therefore take place down a slope." In many places slumps seemed to have moved at right angles to current movement (Cummins, 1959; Murphy and Schlanger, 1962; ten Haaf, 1959, p. 50; Richter, 1965, Fig. 51; Scott, 1966, p. 76; Hubert, 1966). These relations have been explained (Marschalko, 1961, Fig. 11) by movement of slide materials down the lateral margins of the basin which carry the coarse *Wildflysch* into the deeper part of basin where longitudinal turbidity currents prevail. As noted by Lajoie (1972, p. 584), the use of slump-fold axes as a criterion of slope can be misleading, because observed axes show a diverse orientation even in a single slump, an observation also noted by Crowell and associates (1966, p. 28).

The relation of the current system shown by sole marks to the tectonic strike has been a matter of interest for some time. Many studies (ten Haaf, 1959; Cummins, 1959, p. 177) have demonstrated that the paleocurrent system is commonly parallel to the tectonic strike. Kuenen (1957b) believed that many elongate basins were supplied primarily by large streams draining subcontinental areas and that inflow of sediment occurred primarily at one end of the basin. Others, however, have thought that smaller flanking tectonic lands were capable of producing the great volume of sediment (Hsu, 1960; Dzulynski, Ksiazkiewicz, and Kuenen, 1959) and that currents moving down marginal slopes quickly become longitudinal on reaching the axial part of the depositional trough.

Studies of modern deep basins and their margins suggest influx of turbidity currents through deep submarine canyons. Littoral drift brings material to the heads of these canyons which reach far into shelf areas, from whence it moves as a periodic turbidity flow or as a more continuous "grain flow," eventually debouching as a deep-sea fan. Current directions on the latter would presumably radiate outward from the apex (Sullwold, 1960; Potter and Pettijohn, 1963, Fig. 5-13). Such currents would have a general, somewhat transverse orientation. The question of turbidity flow, slumping, and slope control has been reviewed by Walker (1970, pp. 224–225).

For interpretation of paleocurrent systems inferred from sole markings, refer to more recent monographic summaries and symposia on the subject (Lajoie, 1970; Bouma, 1962; Bouma and Brouwer, 1964; Dzulynski and Walton, 1965).

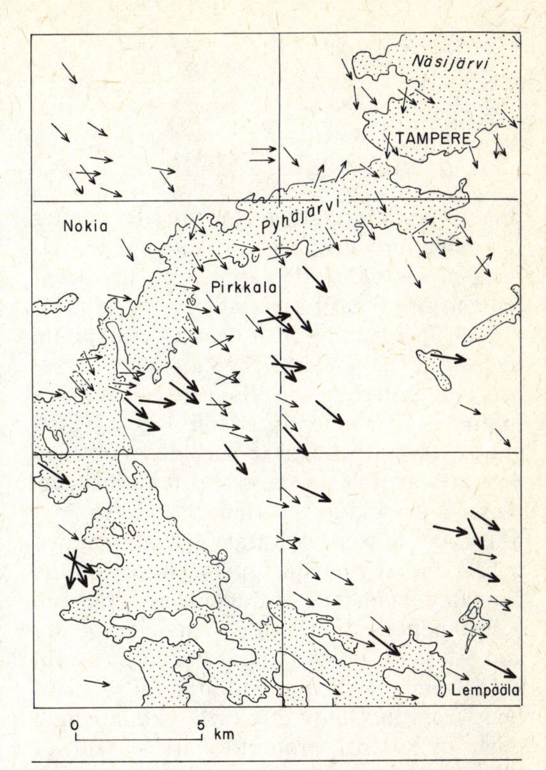

FIG. 14-3. Till fabric and striations in Tampere area of Finland. (Simplified after Virkkala, 1960, Bull. Comm. Geol. Finlande, no. 188, Fig. 2.)

FABRIC AND PALEOCURRENTS

The orientation of linear elements in a rock imposes a fabric on the rock; and if these elements are current-oriented, they constitute a criterion of current direction. Although it has long been known that flat or disk-shaped pebbles assume an imbricated pattern as a result of current flow, little use has been made of this fact for the mapping of current systems. In recent years the fabric of gravels, especially that of till stones, has been carefully investigated and used to determine current direction. The concept of sedimentary fabric and fabric symmetry and types, along with the question of relations between fabrics and depositing currents, have been summarized in Chapter 3. Our concern here is with mapping of fabrics and their interpretation.

Only in a few cases has fabric been plotted systematically on maps. Most published paleocurrent maps based on fabric have been those of till stones, mainly in Pleistocene deposits (Fig. 14-3). Included here are such studies as those of Virkkala (1960, Fig. 2) in Finland and of West and Donner (1956, Figs. 6 and 7) in southeastern England. Lindsey's (1969, Fig. 8) mapping of the orientation of elongate pebbles in the Gowganda Formation (Precambrian) of Canada is one of the few examples of a fabric analysis of a pre-Pleistocene glacial deposit. Fewer yet are studies based on sand grain orientation of tills. An exception is that of Siefert (1954, Figs. 5, 6, 7 and 8), who mapped grain fabric in tills of northern Germany. In all these studies the long dimensions of the till stones or sand grains were parallel to the ice flow.

Indeed, there are few examples of the mapping of either sand or gravel fabrics in nonglacial deposits, perhaps because it is easier to obtain the paleocurrent direction from other current features such as crossbedding.

The most extensive mapping of grain fabric appears to have been done by Kopstein (1954) in the Cambrian of the Harlech Dome in Wales. Kopstein believed the grain fabric to be a primary depositional fabric, a view not shared by other students of these rocks (Bassett and Walton, 1960, p. 99). Caution must be exercised in folded strata, inasmuch as grain orientation may be a product of deformation. It is possible, under favorable circumstances, to sort out the effects of tectonic orientation and to reconstitute the primary sedimentary fabric (Elliott, 1970). Mapping of long-axis orientation by Sestini (1964) established a regional paleocurrent pattern in some Eocene calcarenites in Italy. These deposits and the oriented grains were attributed to turbidity currents. Sestini was able to delineate the paleocurrent system and to ascertain the source area of carbonate detritus. Laming (1966) mapped pebble imbrication in fanglomerates of the New Red Sandstone of Devonshire, England.

Sedimentary fabric, if it could be readily and easily determined, might be most useful in determining the orientation of a sand body from the study of a drill core in which other directional properties are absent or are not readily seen. Obviously the orientation of the core must be known, as well as the relation of fabric to sand geometry in order that the fabric can be utilized for this purpose.

Although relations between grain or pebble orientation and current flow have been extensively studied (see p. 69), relations between grain orientation and gross geometry of the sedimentary body have seldom been observed and measured. There are a few studies of modern sediments: Curray (1956) noted that the long axes of quartz grains lie perpendicular to the trend of some Gulf Coast beaches and cheniers; and Wendler (1956, Fig. 2) found the long axes of sand grains to be essentially parallel to stream direction.

Although one of the first paleocurrent maps published was that of graptolite orientation in shale (Ruedemann, 1897), there are few *maps* of fabrics defined by oriented organic structures. But a significant number of papers deal with the subject (Chapter 3, p. 71). A conspicuous orientation of perhaps more general occurrence and use is charcoal fragment lineation. These materials are carbonized fragments of plant material—straplike leaves, stems, and other elongated plant debris. Unfortunately such materials seem, under some circumstances, to be aligned parallel to the current and in other places perpendicular to it. In general, skeletal structures which respond to current action are so scattered or sporadic that they are at best supplemental to other criteria of current flow. Exceptionally, as in the case of spindle-shaped fusilinids, they may prove to be both abundant and consistent enough to be a mappable attribute (King, 1948, pp. 50, 83).

SCALAR PROPERTIES AND PALEOCURRENTS

Theoretically any scalar property which shows a systematic downcurrent change might be used to determine direction of flow. Unlike directional properties, however, observation at a single outcrop is not sufficient. Data must be collected at no fewer than three places not on a straight line. Among the scalar properties that may exhibit systematic change are clast size and roundness. Most responsive to transport are pebbles and cobbles of rivers, which show both progressive diminution in size and increase in roundness downcurrent.

DOWNCURRENT SIZE DECLINE OF CLASTS

Despite the general observation that the size of clastic elements tends to decrease downcurrent, there are comparatively few actual field measurements of such changes. Sternberg (1875), in an oft-cited but seldom read paper, was perhaps the first to record actual sizes and study the relation of size to distance of transport and river profile. He measured the maximum and average size of pebbles in the channel of the Rhine from Basel downstream, a distance of about 160 miles (260 km). Sternberg not only observed the decline in size in the downcurent direction but concluded that the decline was proportional to the weight of the pebble in water and to the distance traveled. This rule seems to be the best expression of our knowledge of the subject to date and is in agreement with many, but not all, field observations. It applies to alluviating gravels but not to those in the channels of downcutting streams. Sternberg's law may be expressed as a negative exponential, $W = W_0 d^{-as}$, where W is the weight at any distance s, W_o is the initial weight of the pebble, and a is the coefficient of size reduction. This relation is also true if size is expressed as the diameter rather than as the weight (Fig. 3-14).

Actual field studies of other streams show that gravel does in fact decline exponentially. Such seems to be the case of the gravel of the River Mur in Austria (Hochenberger, *in* Grabau, 1913, p. 246), and the largest boulders of the Arroyo Seco in California (Krumbein, 1942, Fig. 11). Much of the earlier European data on size–distance relations are summarized by Schoklitsch, (1933, Table 25). Plumley (1948) measured the downstream decline in size of the gravels in the terraces of three streams in the Black Hills of South Dakota. Both gradient and mean size of the gravels decreased in the downstream direction (Fig. 14–4). Although the decrease was most rapid in the upper reaches of these streams and less rapid in the lower courses, as in the Rhine, the decrease does not follow Sternberg's law. Hack (1957) made similar observations on streams in Virginia and Maryland. Rapid decline in boulder size (from 1.5 m to 0.5 m in 7 miles) has been reported in flood gravels of the Arroyo Seco (Krumbein, 1942) and the Rubicon (Scott and Gravlee, 1968) of California. The coarse gravels of the Knik River of Alaska show a tenfold decrease in diameter in 16 miles (Bradley, Fahnestock, and Rowekamp, 1972, Fig. 9) of transport. The maximum size of granitic boulders in the Dunajec River of Poland decreases from about 1 m in size to under 10 cm in about 250 km of transport (Unrug, 1957, Fig. 3).

Deposits of modern alluvial fans likewise show a rapid decrease in clast size away from the apex of the fan (Bluck, 1965; Blissenbach, 1952; Denny, 1965). In general, the size decrease is more abrupt than that of rivers, excepting only the gravels deposited by flash floods. The decrease is exponential (Fig. 15–7) in many but not all cases.

Similar downcurrent decreases in both average and maximum sizes of gravel in older alluvial deposits have been reported

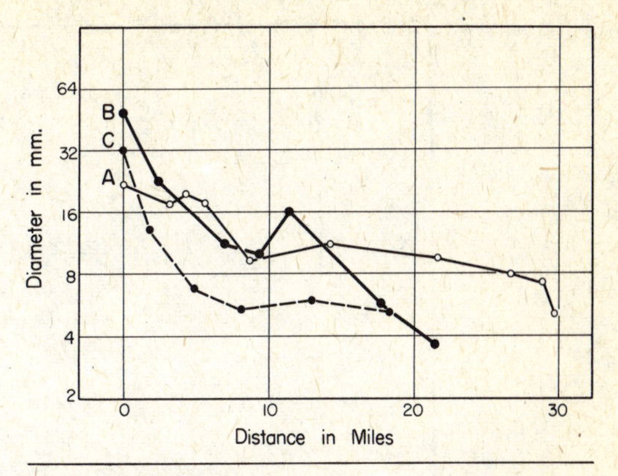

FIG. 14-4. Relation of mean size to distance of travel, Black Hills streams, South Dakota. A: Rapid Creek; B: Bear Butte Creek; C: Battle Creek. (After Plumley, 1948.)

(Schlee, 1957; Potter, 1955, Fig. 12). See Fig. 14–1. Similarly a systematic decrease in maximum pebble size in ancient Paleozoic conglomerates away from their sources has been documented by Pelletier (1958), Yeakel (1962), and Meckel (1967); in the Molasse conglomerates of Bavaria (Füchtbauer, 1967, Fig. 8); and in Precambrian fanglomerates (Hoffman, 1969, Fig. 9). See Fig. 14–5.

Downstream size changes in streams carrying primarily sand are less well known. Sands of the Mississippi River decrease notably downstream (Russell and Taylor, 1937, Table 3), as do sands of the River Tessin (Burri, 1929). In both cases the downstream decrease fluctuates widely from station to station because of accidents of sampling and, in the case of the Tessin, is marked by size increases correlated with local oversteepened portions of the stream profile.

Systematic size decrease in the downcurrent direction of littoral drift has been reported for both gravels (Krumbein and Griffith, 1938) and sands (MacCarthy, 1931; Pettijohn and Lundahl, 1943). Many airborne sediments also display a regular exponential decline in size (Thorarinsson, 1954, Table 4, Pl. 2). See Fig. 9-13.

It is apparent from the above observations that in general the size of clastic elements carried in a current decrease progressively in the direction of transport. It is apparent that the decrease in stream-carried materials is closely correlated with the gradient, with the gradient or slope a function of the third power of bed-load di-

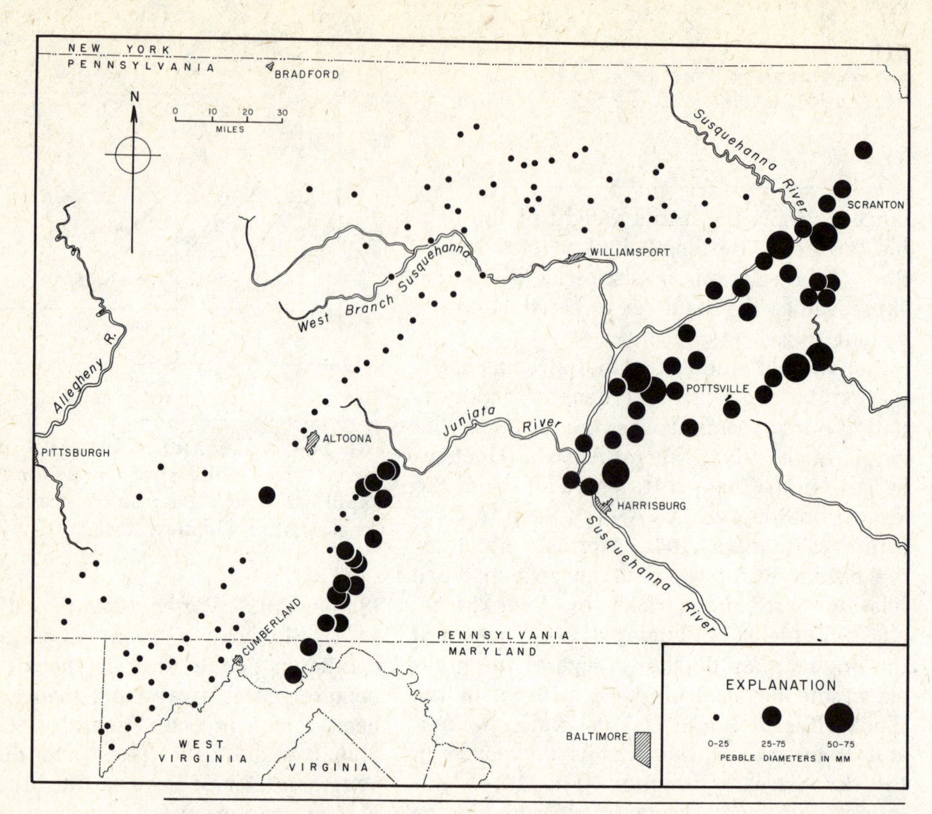

FIG. 14-5. Maximum pebble size in Pocono Formation (Mississippian) in central Appalachians. (After Pelletier, 1958, Bull. Geol. Soc. Amer., v. 69, Fig. 14.)

ameters (Schoklitsch, 1933). If the gradient diminishes exponentially, so will the size. But, as noted by Plumley, other factors, such as the mean discharge of the stream and the resistance of the materials will also govern the decrease in size. The factors that produce the decrease are not yet fully understood. Leliavsky (1955, pp. 5–10), Scheidegger (1961, pp. 168–175), and especially Humbert (1968, pp. 53–72) have reviewed the history and development of efforts to relate decrease of pebble size to distance of travel and to isolate the most important causal factors.

The downstream size decline of clastic sediments was once generally attributed to abrasion or other size reduction processes. It is by no means certain, however, that the observed diminution in size, even of river gravels, is attributable solely to abrasion. As noted by Barrell (1925, p. 327), Daubrée's experiments showed that granite fragments lose 0.001 to 0.004 of their weight per kilometer traveled in an abrasion mill, but the highest rate of wear is only 0.4 of the rate implied by Sternberg's data on the Rhine gravels, which presumably, on the average, are as resistant as granite. This is an unexpected result, because it might be supposed that conditions of laboratory wear would be more severe than those in natural streams. Barrell, therefore, thought that some conditions must exist in streams to lead to an excessive wear of gravel during its movement downstream. The wearing action postulated was "wet blasting" of the pebbles by the large amount of sand swept past them—a process also called on by Rubey (1933, p. 21) to explain the rapid disappearance of pebbles and boulders dumped into the Mississippi River by its tributaries and caving of banks of Pleistocene gravels. Kuenen (1955), however, has shown that this process is not an important factor for materials smaller than cobble size. Bradley, Fahnestock, and Rowekamp, (1972) attributed the marked size decline in the Knik River of Alaska to a selection process.

Knik gravels abraded in a circular moat were reduced in size about 8 percent in 16 miles of travel, whereas in the same distance in a natural stream they showed nearly a 90 percent decline in size. The size decline was, therefore, attributed to a sorting action.

It may be said in summary that the size of pebbles and other clasts decreases in the downcurrent direction, a decline that varies widely in different situations and is a result only in small part from abrasion and other size reduction processes. Nonetheless, such decline is a guide to the direction of transport and may be used, with some restrictions, to estimate distance of travel, provided the original size can be estimated and provided the decline is essentially that predicted from Sternberg's law. This being so, mapping of the size of clasts in a conglomerate—especially maximum size—is useful in paleocurrent and basin analysis. In general, it is very difficult to determine the mean size of the pebbles in a gravel or conglomerate at a particular site. Nothing less than a channel sample through the deposit would suffice, and, because this is generally impossible to obtain, mean size can seldom be used. Maximum pebble size, however, generally bears some relation to the mean size (Fig. 6-5). Hence, in gravel-bearing fluvial deposits, maximum pebble size can readily be measured and mapped. In general, the coarsest conglomerate beds are the thickest and in general are more prone to outcrop than the interbedded finer-grained sediments. In practice the 10 largest pebbles, collected from the most conspicuous conglomerate beds, are measured and averaged. Use of 10 measurements gives a more stable value to the maximum size. Excellent examples of this approach to paleocurrent analysis is shown in papers by Forche (1935, Fig. 8), Pickel (1937, Fig. 2), Schlee (1957), Pelletier (1958) Yeakel (1959), Meckel (1967), and others. See Figs. 14-1 and 14-5.

The log of the maximum size, determined from several localities, can be plotted against distance (corrected for "telescoping" or shortening in folded or faulted sequences). If size change follows an exponential rule, the plotted points form a straight line (Fig. 3-14). If one postulates a reasonable initial size, one can estimate the distance to the nearest point from which a clast could have come. This is the margin of the basin—the line between the area of erosion and that of sedimentation, sometimes referred to as the "fall line" (Fig. 14-6).

ROUNDNESS, SHAPE, AND PALEOCURRENTS

As noted in Chapter 3, *roundness* of gravels increases with distance of travel. It increases most rapidly at first and then changes more slowly, reaching a limiting roundness beyond which it does not change or at best increases very slowly. In most modern streams the initial rapid rounding is achieved in the first few kilometers of travel—less than 15 km for some limestone pebbles (Plumley, 1948), even in small streams. Even though vein quartz and quartzite round more slowly, they become well-rounded in some streams in less than 100 km. Hence only very near the source ledges will angular fragments be found in gravels derived from these ledges. Measurement and mapping of pebble roundness, therefore, are not apt to be helpful in determining either transport direction or distance except in the most proximal of all deposits—the alluvial fans. Theoretically, the direction of transport could be determined by mapping roundness values and drawing normals to the lines joining equal roundness values. See, for example, Laming (1966), who mapped roundness in fanglomerates of the New Red Sandstone of Devonshire.

Because the rounding of quartz sands is a very slow process, with changes in the downstream direction negligible in present-day streams, it seems unlikely to be a useful paleocurrent indicator in ancient alluvial sandstones. Although of little value in this respect, it may be helpful in making environmental discriminations, as noted elsewhere (p. 249).

Roundness, once acquired, is almost never lost, so the rounding observed in a deposit may be the cumulative effect of several cycles of transport, not just the last

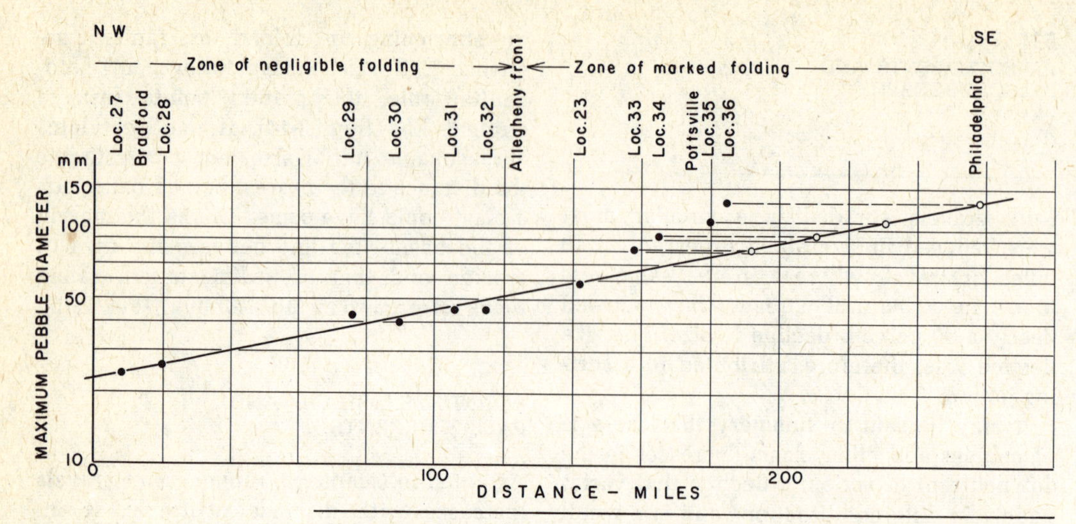

FIG. 14-6. Log of maximum pebble size and distance in Pottsville conglomerates (Pennsylvanian), Pennsylvania. Solid circles, observed data; open circles, presumed prefolding position. (After Pelletier, 1958, Bull. Geol. Soc. Amer., v. 69, Fig. 18.)

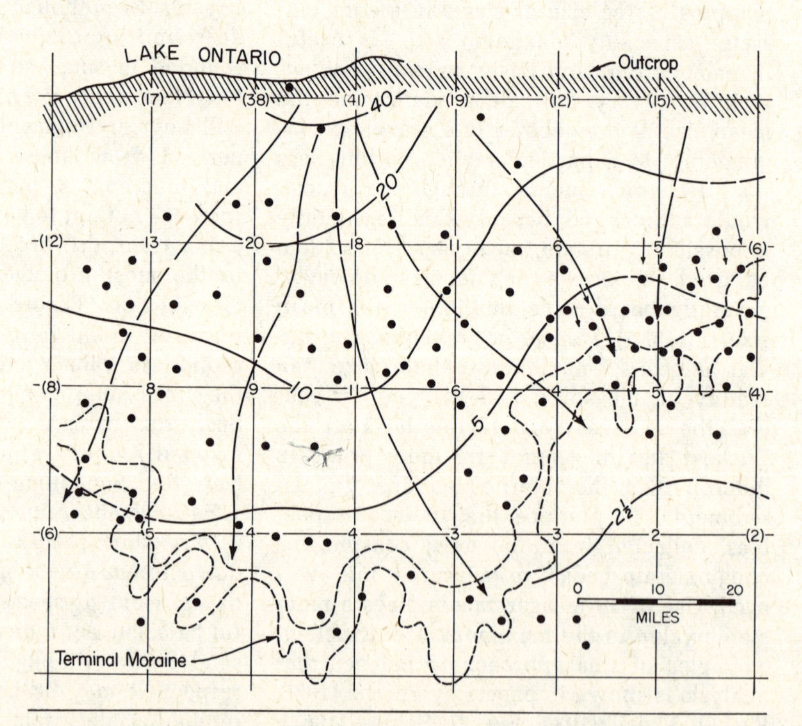

FIG. 14-7. Distribution of red-green sandstones in till, west-central New York. Generalized outcrop area of source ledges, diagonal ruling; collecting localities, black dots; numerals at intersection of grid lines denote average for the four quadrangles centered on the respective values; normals drawn to contours are the indicated directions of movement; arrows denote glacial striations and record actual movement. (Based on data from Holmes, 1952.)

one; and under marine conditions roundness bears little or no relation to the *net* distance of transport. Beach action may produce very well-rounded sands and gravels which are of very local origin. Rounding, therefore, has a very restricted value in determining either direction or distance of transport.

Although the *shape* of a clast is modified little or not at all by transport, some downcurrent changes in shape have been noted and are believed to be the result of sorting (Bluck, 1965, p. 241). Inasmuch as both roundness and sphericity are closely correlated with size, a downstream decrease in size is accompanied by a decrease of the other two properties. It is, therefore, necessary to compare clasts of the same size (and lithology) to detect any independent downcurrent changes in these properties.

COMPOSITIONAL PROPERTIES AND PALEOCURRENTS

Detrital materials released by weathering do not remain in the area in which they originate. They are dispersed and come to rest in an area whose location, size, and shape are determined by the topography (or bathymetry) and the prevailing current system. The materials move downcurrent, and, if the current system is relatively stable, the materials will tend to lodge on the lee side of the generating source. Such a lee-side accumulation forms a *dispersal fan* or *shadow*.

Dispersal fans or shadows are defined primarily by the distribution of a particular constituent. They can be identified and mapped on the basis of the *presence or absence* of the defining component. At the same time it is observed that the *concentration* of the defining component is diminished downcurrent as the area over which it is spread widens. Whether diminution is caused by selective loss by abrasion or by progressive dilution by new materials is immaterial insofar as paleocurrent analysis is concerned. Progressive downcurrent change is the significant fact. If the change follows some regular law, the distance as well as the direction in which the source ledges lie can be estimated.

Deposits of continental ice sheets illustrate these concepts especially well. The load carried by the ice is progressively modified by addition of new materials as the ice moves over varying types of bedrock. As a result, the content of any specific component diminishes rapidly away from its source; a further result is the loading of drift in any one place by a high proportion of locally derived materials. Careful analysis of the composition of the drift at several places should enable one to deduce the direction of ice movement and to estimate the distance to the contributing ledges of any specific component (Fig. 14-7).

A special case of these principles is that of boulder trains and the concentration in the drift of boulders derived from a specific source. The boulders are dispersed in the lee of the source and are most thickly concentrated in the immediate vicinity of the source ledges (Fig. 14-8). As Krumbein (1937) has shown, boulder concentration per unit area diminishes exponentially with the distance from the source. Maps published by Lundquist (1935) showing pebble counts (concentration in a small unit volume or sample) demonstrate rapid diminution away from the source; very probably also the decline is exponential. Distribution and dispersion patterns derived from a restricted or point source by moving ice have been successfully used in prospecting for otherwise concealed ore bodies (Grip, 1953; Dreimanis, 1956). For a further review of the subject of boulder trains refer to Potter and Pettijohn (1963, pp. 195–198) and Flint (1971).

Comparable dispersal fans in nonglacial deposits are more difficult to define and map. The best known are those in sandstones defined by heavy-mineral suites. Füchtbauer's (1958, Fig. 6a) study of the dispersal fans in the Tertiary Chatt sands of the foreland Molasse of the Alps is a striking example of the effective use of heavy-mineral suites to trace sediment movement (Fig. 14-9). The distribution of particular heavy-mineral assemblages defines local mineral provinces which reflect the dispersal pattern —a pattern related to a series of alluvial fans deposited along the front of the newly elevated Alpine chain. In a similar way, but on a broader scale, it is possible to trace the dispersal pattern of sands in the Gulf of Mexico, a modern marine environment (Fig. 14-10).

Again, perhaps because of the inordinate labor involved, compositional variations have been little used to map paleocurrents —a good deal more rapidly achieved by less arduous procedures. Mineralogical analyses, however, disclose other information, especially with reference to provenance.

A specialized compositional attribute is the pollen count, expressed in number of grains per gram of sediment. This has been said to be a function of the distance from the shoreline (Hoffmeister, 1954, 1960). Isobotanical contours are therefore parallel to the depositional strike and the pollen concentration is presumed to fall off systematically downdip. This approach to paleogeographic analysis might be most helpful in shales where little else exists to delineate paleocurrents or paleoslope. Little has been published on the results of such studies.

Scalar and compositional properties have both advantages and disadvantages as compared with directional properties. Many can be determined from unoriented cores, or even from drill cuttings, whereas directional properties require an outcrop or at least an

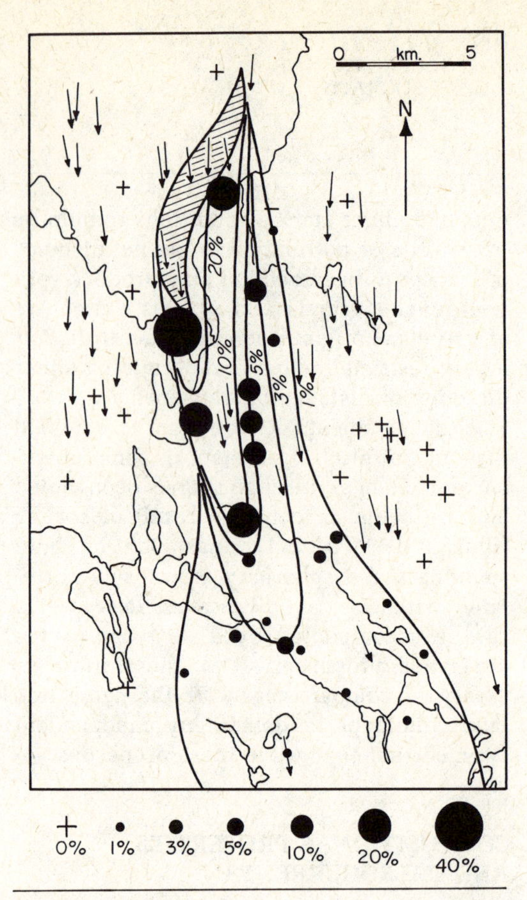

FIG. 14-8. Dispersion of boulders from the mica-slate district of Stallberg, Sweden. Transportation direction is in the direction of ice movement as shown by striations (arrows). (Boulder counts by N. H. Magnusson. From Lundquist, 1935.)

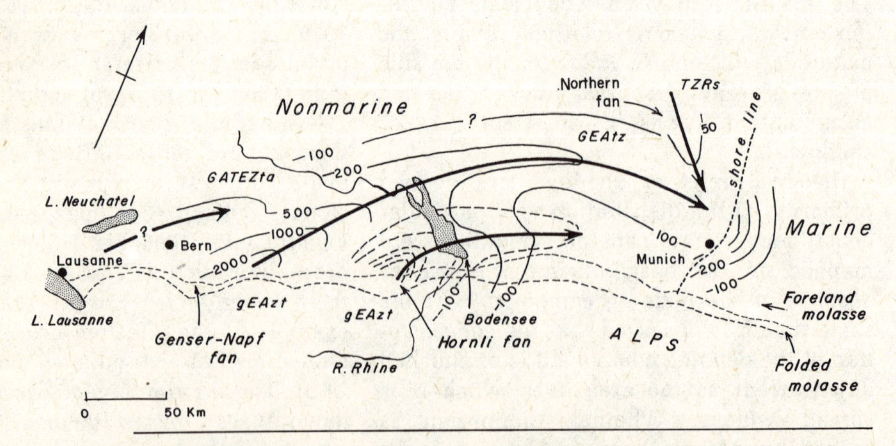

FIG. 14-9. Paleocurrent systems and mineral provinces deduced from heavy minerals in Lower Freshwater Molasse (Aquitanian) north of Alps. E: epidote; G: garnet; A: apatite; T: tourmaline; Z: zircon; S: staurolite. Capital letters denote greater abundance; lower-case letters denote lesser abundance. (After Füchtbauer, 1958, Eclogae Géol. Helvetiae, v. 57, Fig. 14e.)

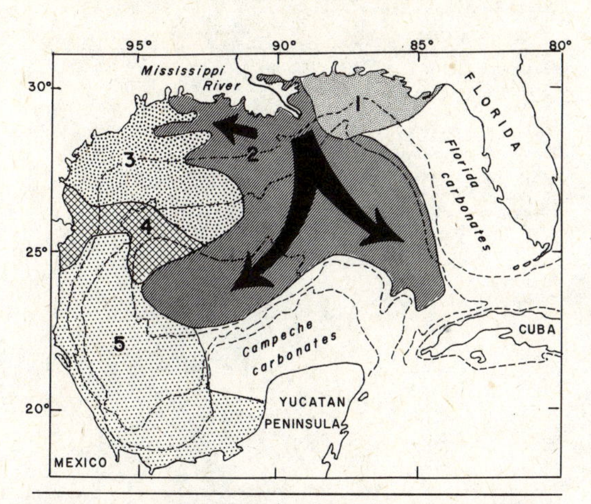

FIG. 14-10. Major heavy-mineral provinces, inferred area distribution of Mississippi sediment and principal sediment dispersal directions in Gulf of Mexico. I: eastern Gulf province; II: Mississippi province; III: central Texas province; IV: Rio Grande province; V: Mexican province. (Data from map compiled from various sources by Davies and Moore, 1970, Jour. Sed. Petrology, v. 40, Fig. 1. By permission of Society of Economic Paleontologists and Mineralogists.)

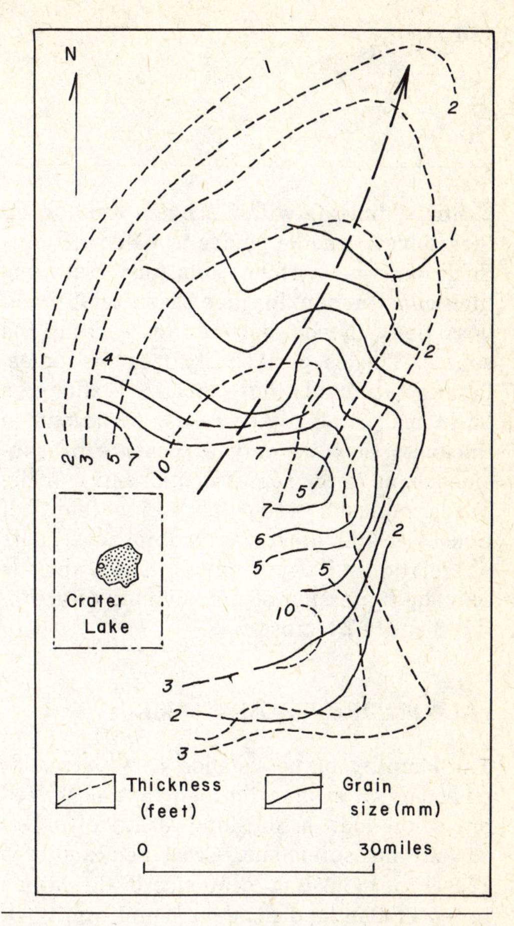

FIG. 14-11. Isopach map of decrease in thickness of ash fall away from source vent. (After Eaton, 1964, Jour. Geol., v. 72, Fig. 14.)

oriented core. On the other hand, they cannot define the current direction from a single outcrop or core; many measurements are required from many places. In contrast, a directional property such as cross-bedding tells one instantly the current direction at a given site.

BED THICKNESS, ISOPACHS, AND PALEOCURRENTS

Geologists have found it profitable to prepare maps showing the thickness of the strata under investigation. Such *isopach maps* normally show the thickness of larger stratigraphic units—a formation or sequence of formations. They are constructed by contouring the thickness of the strata involved from many measurements from both outcrops and borings. The contours are lines of equal thickness usually expressed in feet or meters.

Paleocurrent inferences that can be drawn from such maps are rather limited. On the other hand, the thickness variations of a *single* clastic layer—such as a glacial varve, an ash bed, a loess sheet, or a graded turbidite bed—may be expected to show a systematic decline in the downcurrent direction. These changes in thickness are closely correlated with decrease in grain size: Both decline exponentially. In only a few cases are field data available that demonstrate these relations, but some theoretical considerations suggest that they should be so (Potter and Scheidegger, 1966; Scheidegger and Potter, 1971). The best established case is, perhaps, that of ash falls (Thorarinsson, 1954, Table 4, Pl. 2). See Fig. 14-11 and glacial varves (DeGeer, 1940, Pl. 57). These rather specialized examples are usually only exceptionally useful to the sedimentologist in his quest for paleogeographic understanding.

In general, regional isopach patterns of

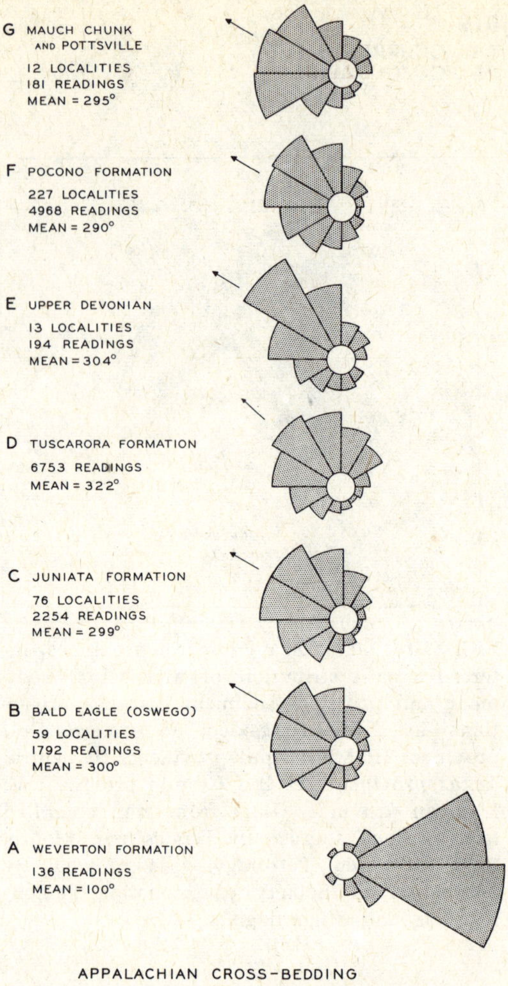

clastic sediments within a basin are not, by themselves, reliable guides to paleocurrents. Such maps have other value fully justifying their construction. In some cases, total thickness does indeed decline away from the source. This is particularly true of coarse, fanlike accumulations localized along a basin margin. In other cases, variations in thickness may be only a measure of subsidence, as is the seaward thickening of the Tertiary clastic wedge bordering the Gulf Coast. In still other cases, thickness bears no relation to depositional history but is only an expression of the beveling of gently tilted strata by erosion.

PALEOCURRENTS AND TIME

The stability or persistence of a particular paleocurrent system through time is indeed one of the most astonishing results of paleocurrent measurements. Cross-bedding in a 12,000 foot (3,660 m) sequence in the Moine series of Scotland displays a uniformity of orientation throughout which was described by Sir Edward Bailey as ". . . the most surprising single phenomenon" displayed by these strata (Wilson, Watson, and Sutton, 1953, p. 386). Similar observations have been made elsewhere. There is no significant variation in the cross-bedding of the Lorrain Quartzite, a Precambrian formation nearly 7,000 feet (2,135 m) thick (Pettijohn, 1957). Pelletier (1958, Fig. 17) has shown the mean current direction to remain constant in strata ranging from Upper Devonian (Catskill) to Pennsylvanian (Pottsville) in age of Pennsylvania and Maryland. This means an essentially stable paleoslope for a period of 150 to 200 million years. Moreover, the current directions recorded in sandstones of late Ordovician (Bald Eagle and Oswego) and early Silurian (Tuscarora) age of the same region are nearly identical to those of Late Paleozoic time (Yeakel, 1959). All these formations show a transport direction transverse to the present tectonic strike (Fig. 14-12). Perhaps the most significant reversal of slope known is that shown by the Weverton of Maryland (Whitaker, 1955) and the Antietam (Schwab, 1970), both Lower Cambrian, which seem to record west-to-east transport, a direction diametrically opposed to all other transport directions recorded in the central Appalachian area.

Potter and Pryor (1961) report a remarkably stable and persistent transport pattern in sandstones of the Upper Mississippi Valley ranging in age from early Paleozoic

G MAUCH CHUNK AND POTTSVILLE
12 LOCALITIES
181 READINGS
MEAN = 295°

F POCONO FORMATION
227 LOCALITIES
4968 READINGS
MEAN = 290°

E UPPER DEVONIAN
13 LOCALITIES
194 READINGS
MEAN = 304°

D TUSCARORA FORMATION
6753 READINGS
MEAN = 322°

C JUNIATA FORMATION
76 LOCALITIES
2254 READINGS
MEAN = 299°

B BALD EAGLE (OSWEGO)
59 LOCALITIES
1792 READINGS
MEAN = 300°

A WEVERTON FORMATION
136 READINGS
MEAN = 100°

APPALACHIAN CROSS-BEDDING

FIG. 14-12. Cross-bedding azimuthal distribution in central Appalachian area. A: Weverton (Cambrian) (after Whitaker, 1955); B: Bald Eagle and C: Juniata (Ordovician) (after Yeakel, 1959); D: Tuscarora (Silurian) (after Yeakel, 1959); E: Catskill (Devonian); F: Pocono and G: Mauch Chunk (Mississippian) (after Pelletier, 1958). (From Pettijohn, 1962, Bull. Amer. Assoc. Petrol. Geol., v. 46, Fig. 12.)

to Tertiary. The grand mean of all formations differs only little from that of present-day streams (Fig. 14-13).

In conclusion, it can be said that the stability of paleocurrent systems has been demonstrated in many cases and that such stability through time implies the existence of stable tectonic elements with their consequent slopes that govern erosion, transport, and sedimentation for long periods of time. In other areas, however, especially in mobile belts where interior sources of sediment—the somewhat ephemeral intrageosynclinal ridges or welts—exist, transport patterns may show a lesser longevity.

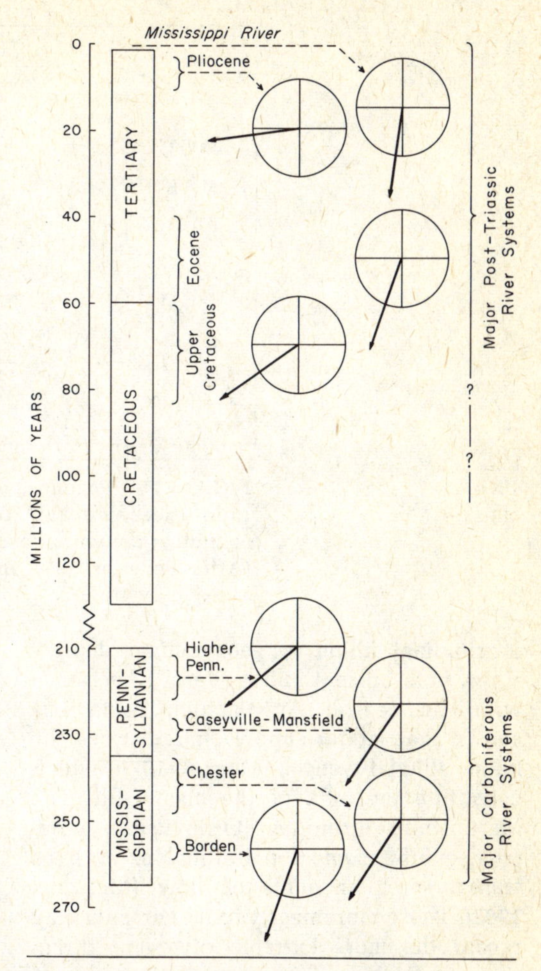

FIG. 14-13. Transport directions and time scale of post-Devonian sandstones of Illinois Basin and Upper Mississippi Embayment. (After Potter and Pryor, 1961, Bull. Geol. Soc. Amer., v. 72, Fig. 15.)

PALEOCURRENTS AND BASIN ANALYSIS

The study of a basin of sedimentation involves reconstruction of the basin, location of its margin, location of the area or region supplying the sedimentary fill, delineation of the shoreline if the basin is in part marine, and relations between structural and bathymetric axes.

The problem of locating the basin margin has been discussed in the section dealing with distance of transport and the "fall line" concept—a problem which can be solved in an approximate fashion if maximum clast size decreases downcurrent according to some regular rule—such as Sternberg's law. The upcurrent margin of the basin has been fixed in a number of cases. The fall line thus located is presumed to be parallel to the depositional or sedimentary strike, which in the central Appalachian area is parallel to the tectonic strike and is shown by both fold axes and isopachs.

The source area is presumed to be upcurrent. In marine deposits, however, the current may be littoral; the source is then indeterminate. But a widespread uniform current pattern can hardly be the result of the vagaries of littoral currents; the upcurrent direction must also be the source direction. In alluvial deposits it is theoretically possible, using particle size data only, to derive a curve that approximates the stream profile (Hack, 1957, p. 73). If so, elevations along the profile, the actual inclination of the paleoslope, and perhaps the height of the divides could be estimated, and some reconstructions of the geomorphology of the source land might be possible.

Many sedimentary basis seem to be half-grabens (Weeks, 1952, p. 2089) and filled in an asymmetrical manner. The distal margin, probably a shifting one, is difficult to locate, whereas the proximal margin—on the side from which the sediment was derived—is the fall line and is located as indicated above. The "polarity" of a sedimentary trough is one of its basic attributes.

Kuenen (1957b) has called attention to the

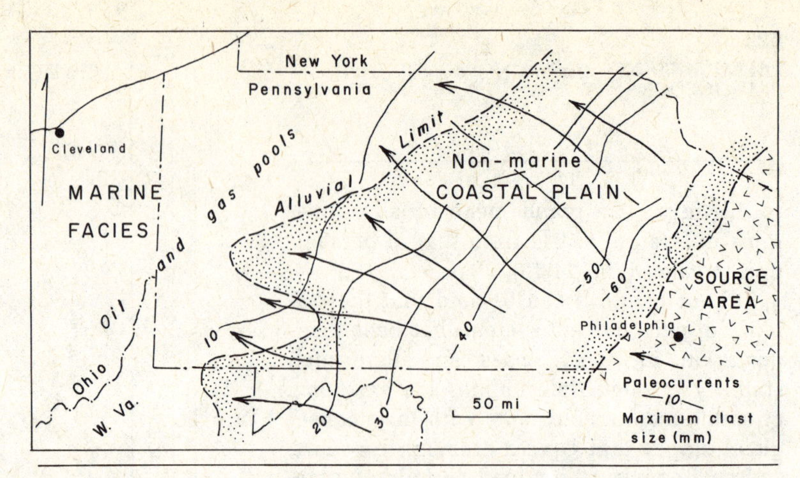

FIG. 14-14. Pocono paleogeography. Stippled area with sand–shale ratio exceeding 2. Arrows depict moving average of cross-bedding azimuths; contours based on moving average of maximum pebble size. (After Pelletier, 1958, Bull. Geol. Soc. Amer., v. 69, Fig. 16.)

longitudinal filling of geosynclinal depressions, both ancient and modern. Experience with the central Appalachian basin has shown that lateral supply characterizes the major alluvial wedges (Meckel, 1970) which constitute the bulk of the clastic fill. The Martinsburg turbidites (Ordovician), however, exhibit some longitudinal as well as transverse components of flow (McBride, 1962). These more nearly axial currents may record the more extreme divergent directions of flow—a divergence radiating from the apexes of turbidite fans which are themselves nourished by lateral supply. Or indeed it may be that transverse currents are diverted to a longitudinal flow on reaching the axial portion of the trough.

The shoreline itself is the zero contour on the paleoslope; hence, the shoreline is parallel to the depositional or sedimentary strike and the strike of the paleoslope. Its actual location is difficult to fix (Hough and Menard, 1955), but so far the most reliable criterion is fossil content of the beds. The shoreline must lie updip from the last occurrence of undoubted marine fossils. If fixed at one point, the shoreline can be extrapolated in the direction of the depositional strike. By definition it is the zero contour of the paleoslope and is normal to the paleocurrent lines. In the central Appalachian example (Fig. 14-14) it is parallel with facies boundaries and isopachs.

The axis of a sedimentary basin may be either the structural axis or the bathymetric axis. The structural axis is the line of greatest depression—hence the line of greatest thickness of accumulated fill—and can be determined from an isopach map. The bathymetric axis, on the other hand, is the line of greatest water depth. Structural and bathymetric axes may coincide or they may be far apart. In the Mississippian Pocono in the central Appalachian area, the greatest thickness was attained near or just east of the easternmost outcrop of Pocono, under the coastal plain of Pocono time. Inasmuch as the Pocono shoreline was in western Pennsylvania, the deepest water must be west of this line—somewhere in Ohio.

By bringing together data on all aspects of a clastic formation such as the Pocono, and by plotting the most essential aspects on a map, it is possible to develop a complete paleogeographic synthesis. An example of such a synthesis is shown in Fig. 14-14.

FACIES MAPS AND PALEOGEOGRAPHY

Sedimentary facies is the aspect or character of the sediment within beds of one and the same age. Moore (1949, p. 8) recommended that facies be considered to comprise any areally segregated part of a designated rock division in which physicoorganic characters differ significantly from those of another part or parts.

Because facies is an expression of areal variation in aspect, the most effective manner of showing facies is by maps. Although facies maps were made over 50 years ago, their modern development awaited the availability of subsurface data, permitting three-dimensional presentation of the stratigraphic unit. A regional facies map is one which shows broad regional trends in lithology. Such a map, for example, may show that a stratigraphic unit which is mainly shale in one area becomes a mixture of shale and limestone in another, and finally becomes dominantly limestone elsewhere. Most facies maps are what have been called *lithofacies* maps; they depict lithologic rather than biologic or other aspects of the sequence.

Data for such maps come from both measured outcropping sections and from subsurface information obtained from drilling, the latter either a sample or core log of lithologies encountered or a geophysical log translated into lithologic terms. Original data for nearly all facies maps, therefore, are thicknesses expressed as feet (or meters). These linear measurements may be converted or combined in various ways so that a great many types of maps may be constructed. The construction of facies maps and their interpretation have been given a great deal of attention. Refer to the pertinent literature on definitions and methods (Sloss, Krumbein, and Dapples, 1949; Moore, 1949; Krumbein, 1948, 1952, 1954, 1955, 1956, 1957; Krumbein and Libby, 1957) and to the many published examples.

Lithofacies maps fall into several categories. There are those which depict thickness—total thickness of the unit or the thickness of some component—such as the net sand thickness (the integrated total of the thicknesses of all the sand beds at a given site). In other cases, lithologic variations may be shown as percentages (Fig. 14-15)—the percent of sand in the sequence, for example. Such maps show the areal variation of a single component.

It is also possible to depict the variation of more than one component on a single map. Multicomponent maps are based on ratios, such as the *sand–shale ratio* or the *clastic ratio*; the latter usually involves the

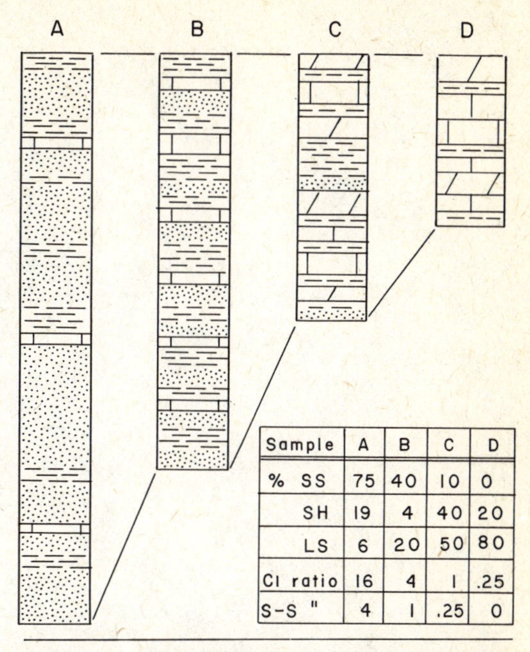

Sample	A	B	C	D
% SS	75	40	10	0
SH	19	4	40	20
LS	6	20	50	80
Cl ratio	16	4	1	.25
S–S "	4	1	.25	0

FIG. 14-15. Hypothetical cross section showing changes in content of sand, shale, and limestone. (From Krumbein, 1948, Bull. Amer. Assoc. Petrol. Geol., v. 32, Fig. 3.)

ratio of the thickness of carbonates to that of clastic beds in the section. The proportion of the three common lithologies—sand, shale, and carbonates (limestones and dolomites)—can be depicted by use of a triangular diagram (Fig. 14-16) in which the possible proportions are grouped into nine classes. Each of these classes can be shown with a proper pattern or color on a lithofacies map (Fig. 14-17). Such a map is a three-component facies map.

The ingeneous investigator can concoct a great many other multicomponent parameters which show whatever aspect is deemed significant. These include entropy maps (Pelto, 1954) which express the degree of "mixing" of the three common rock types in the section, entropy ratio maps, facies departure maps (Krumbein, 1955), and various "vertical variability" maps (showing number of sandstones, average thickness of sandstone units, and similar aspects) that present information on the composition and geometry of the component parts of the section.

It should be remembered that all these kinds of maps are derived from one linear measurement—thickness. This is primarily because thickness is about all that can be

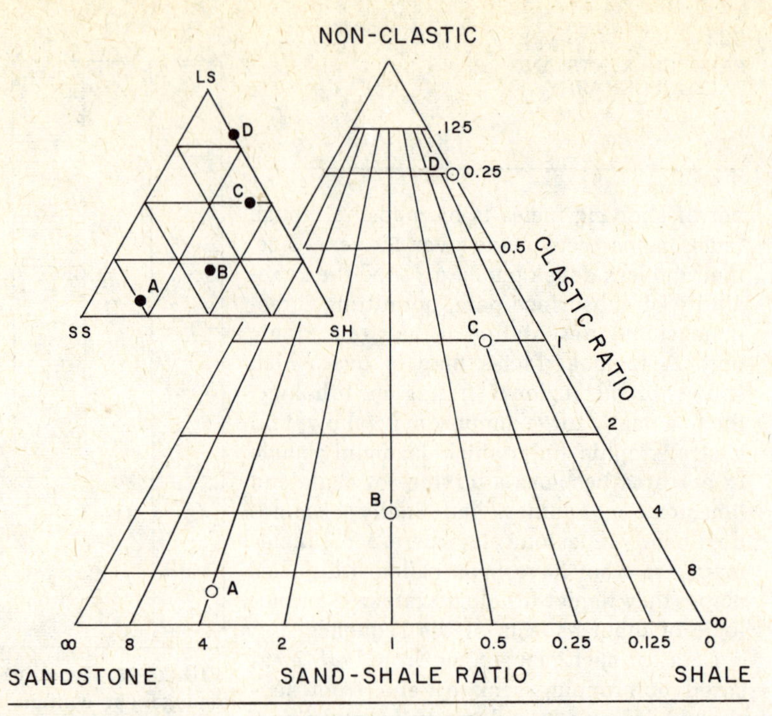

FIG. 14-16. Lithofacies triangle, showing relation of clastic ratio and sand–shale ratio to conventional 100-percent triangle (inset). Points on triangles refer to data of Fig. 14-15. (From Krumbein, 1948, Bull. Amer. Assoc. Petrol. Geol., v. 32, Fig. 2.)

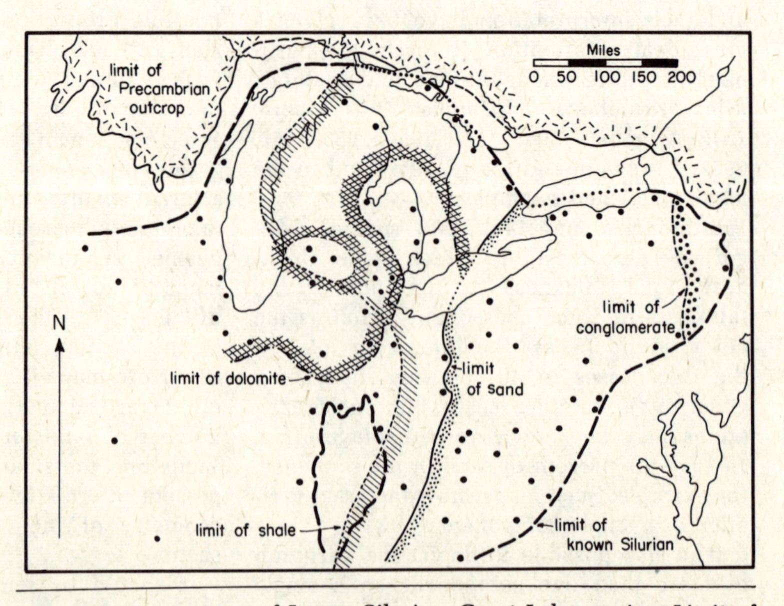

FIG. 14-17. Facies map of Lower Silurian, Great Lakes region. Limit of Conglomerate 10 percent, limit of sand 12.5 percent, limit of shale 25 percent. (After Amsden, 1955, p. 62.)

read from geophysical logs on which lithofacies mapping largely depends. Other variables, of course, including carbon ratios, boron content, or other attributes could be depicted and mapped, but such data are rarely obtainable. All maps also depend on an adequate density of control points—usually wells—and on proper contouring.

The chief drawbacks to facies maps are inadequacies of data and the problem of interpretation. The data are generally so sparsely distributed that the facies patterns delineated are necessarily very generalized. Only exceptionally is control close enough to work out the geometry of individual sand bodies and other stratigraphic details necessary for environmental analysis.

Attempts have been made to formulate principles of facies map interpretation (Krumbein, 1952; Krumbein and Sloss, 1963, p. 489). Lithofacies maps show the present areal extent of the stratigraphic unit, the thickness of the unit, and the gross lithology of the unit. Such maps are presumed to shed light on the geologic history, the environment of deposition, and the tectonic setting of the deposit in question. Relations between the data mapped and these problems are involved and often obscure. Environmental interpretation from regional facies maps is not satisfactory, although certain lithologic associations—such as a high evaporite–carbonate ratio—suggest the nature of the environment. An isopach map, if the scale is large enough and the control points closely enough spaced, may show the form or geometry of individual sand bodies so that one may recognize delta distributaries or barrier island forms.

Inasmuch as most facies maps show both facies boundaries and isopachs, the relations between the two are of interest. They may be parallel to one another or intersect in various ways (Krumbein, 1952, Fig. 2).

The relations between facies and the tectonic framework are of special importance. Positive elements tend to be source areas, and the facies might be expected to bear some relations to such positive areas. Sediments tend to diminish in grain size away from their source; sands, for example, "shale out" and pass into or intertongue with shales away from their source. Hence, one may map the sand–shale ratio or depict facies variation by plotting some other measurable aspect. Facies patterns may reflect the movement of sediment away from the source area (Fig. 14-17). The sand–shale ratio is particularly significant in some cases; the ratio 1 : 1 is at or near the transition from nonmarine to marine strata and parallel to the shoreline in some instances (Fig. 14-14). Relations between this ratio and oil accumulation have also been pointed out (Dickey and Rohn, 1955; Pelletier, 1958, p. 1059; Yeakel, 1962, p. 1536).

Although isopach and lithofacies patterns are essential to understanding the history of a sedimentary basin (and commonly are the only aspects that can be obtained from subsurface data), reconstruction of the paleocurrent system and the paleogeography can better be achieved by mapping directional properties and the environmental analysis better made by study of vertical sequences and their contained sedimentary structures.

References

Barrell, J., 1925, Marine and terrestrial conglomerates: Bull. Geol. Soc. Amer., v. 36, pp. 279–342.

Bassett, D. A., and Walton, E. K., 1960, The Hell's Mouth Grits: Cambrian greywackes in St. Tidwalls Peninsula, North Wales: Quart. Jour. Geol. Soc. London, v. 116, pp. 85–110.

Blissenbach, E., 1952, Relation of surface angle distribution to particle size distribution on alluvial fans: Jour. Sed. Petrology, v. 22, pp. 25–28.

Bluck, B. J., 1965, The sedimentary history of some Triassic conglomerates in the Vale of Glamorgan, South Wales: Sedimentology, v. 4, pp. 225–245.

Bouma, A. H., 1962, Sedimentology of some flysch deposits: Amsterdam, Elsevier, 168 pp.

Bouma, A. H., and Brouwer, A., eds., 1964, Turbidites: Developments in sedimentology, v. 3, Amsterdam, Elsevier, 264 pp.

Bradley, W. C., Fahnestock, R. K., and Rowehamp, E. T., 1972, Coarse sediment transport by flood flows on Knik River, Alaska: Bull. Geol. Soc. Amer., v. 83, pp. 1261–1284.

Brett, G. W., 1955, Cross-bedding in the Baraboo Quartzite of Wisconsin: Jour. Geol., v. 63, pp. 143–148.

Brinkman, R., 1933, Über Kreuzschichtung im deutschen Bundsandsteinbecken: Nachr.

Gesell. Wissensch. 24 Göttingen, Math. Phys. Kl. Fachtgrappe IV, no. 32.

Burri, C., 1929, Sedimentpetrographische Untersuchungen an alpinen Flüssanden: Schweiz. Min. Pet. Mitt., v. 9, pp. 205–240.

Crowell, J. C., 1955, Directional-current structures from the pre-Alpine Flysch, Switzerland: Bull. Geol. Soc. Amer., v. 66, pp. 1351–1384.

Crowell, J. C., Hope, R. A., Kahle, J. E., Ovenshine, A. T., and Sams, R. H., 1966, Deepwater sedimentary structures, Pliocene Pico Formation, Santa Paula Creek, Ventura Basin, California: California Div. Mines Geol., Spec. Rept. 89, 40 pp.

Cummins, W. A., 1959, The Lower Ludlow Grits in Wales: Liverpool and Manchester Geol. Jour., v. 2, pp. 168–179.

Curray, J. R., 1956, Dimensional grain orientation studies of Recent coastal sands: Bull. Amer. Assoc. Petrol. Geol., v. 40, pp. 2440–2456.

DeGeer, G., 1940, Geochronologia suecica, principles, atlas: Kungl. Svenska Vet. Akad. Handl., ser. 3, v. 18, 360 pp.

Denny, C. S., 1965, Alluvial fans in the Death Valley region, California and Nevada: U.S. Geol. Surv. Prof. Paper 466, 62 pp.

Dickey, P. A., and Rohn, R. E., 1955, Facies control of oil occurrence: Bull. Amer. Assoc. Petrol. Geol., v. 39, pp. 2306–2320.

Dreimanis, A., 1956, Steep Rock iron ore boulder train: Proc. Geol. Assoc. Canada, v. 8, pp. 27–70.

Dzulynski, S., Ksiazkiewicz, M., and Kuenen, Ph. H., 1959, Turbidites in flysch of the Polish Carpathian Mountains: Bull. Geol. Soc. Amer., v. 70, pp. 1089–1118.

Dzulynski, S., and Radomski, A., 1955, Origin of groove casts in the light of turbidity current hypothesis: Acta Geol. Polonica, v. 5, pp. 47–66.

Dzulynski, S., and Walton, E. K., 1965, Sedimentary features of flysch and greywackes: Amsterdam, Elsevier, 274 pp.

Elliott, D., 1970, Determination of finite strain and initial shape from deformed elliptical objects: Bull. Geol. Soc. Amer., v. 81, pp. 2221–2236.

Flint, R. F., 1971, Glacial and Quaternary geology: New York, Wiley, 897 pp.

Forche, F., 1935, Stratigraphie und Paläogeographie des Buntsandsteins im Umkreis der Vogesen: Mitt. Geol. Staats Inst. Hamburg, v. 15, pp. 15–55.

Füchtbauer, H., 1954, Transport und Sedimentation der westlichen Alpenvorlandsmolasse: Heidelberger Beitr. Min. Petrog., v. 4, pp. 26–53.

———, 1958, Die Schüttungen im Chatt und Aquitan der deutschen Alpenvorlandsmolasse: Eclogae Geol. Helvetiae, v. 51, pp. 928–941.

———, 1967, Die Sandsteine in der Molasse nördlich der Alpen: Geol. Rundschau, v. 56, pp. 266–300.

Grabau, A. W., 1913, Principles of stratigraphy: New York, Dover, 1185 pp. (reprinted 1960).

Grip, E., 1953, Tracing of glacial boulders as an aid to ore prospecting in Sweden: Econ. Geol., v. 48, pp. 715–725.

ten Haaf, E., 1959, Graded beds of northern Apennines: Ph.D. thesis, Univ. Groningen, 102 pp.

Hack, J. T., 1957, Studies of longitudinal stream profiles in Virginia and Maryland: U.S. Geol. Surv. Prof. Paper 294-B, pp. 45–97.

Hall, J., 1843, Remarks upon casts of mud furrows, wave lines, and other markings upon rocks of the New York System: Assoc. Amer. Geol. Rept., pp. 422–432.

Hamblin, W. K., 1958, Cambrian sandstones of northern Michigan: Michigan Geol. Survey Pubc. 51, 149 pp.

Hoffman, P., 1969, Proterozoic paleocurrents and depositional history of the East Arm fold belt, Great Slave Lake, Northwest Territories: Canad. Jour. Earth Sci., v. 6, pp. 441–462.

Hoffmeister, W. S., 1954, Microfossil prospecting for petroleum: U.S. Patent Office, no. 2,686,108.

———, 1960, Palynology has important role in oil exploration: World Oil, v. 150, pp. 101–104.

Hough, J. L., and Menard, H. W., Jr., eds., 1955, Finding ancient shorelines: Soc. Econ. Paleon. Min., Spec. Publ. no. 3, 129 pp.

Howard, J. D., 1966, Patterns of sediment dispersal in the Fountain Formation of Colorado: Mountain Geologist, v. 3, p. 147–153.

Hsu, J. K., 1960, Paleocurrent structures and paleogeography of the ultrahelvetic flysch basins, Switzerland: Bull. Geol. Soc. Amer., v. 71, pp. 577–610.

Hubert, J. F., 1966, Sedimentary history of Upper Ordovician geosynclinal rocks, Girvan, Scotland: Jour. Sed. Petrology, v. 36, pp. 677–699.

Humbert, F. L., 1968, Selection and wear of pebbles on gravel beaches: Ph.D. thesis, Univ. Groningen, 144 pp.

Hyde, J. E., 1911, The ripples of the Bedford and Berea formations of central Ohio, with notes on the paleogeography of that epoch: Jour. Geol., v. 19, pp. 257–269.

Jüngst, H., 1938, Paläogeographische Auswertung der Kreuzschichtung: Geol. Meere Binnengewässer, v. 2, pp. 229–277.

Kindle, E. M., 1917, Recent and fossil ripple mark: Geol. Surv. Canada, Mus. Bull. 25, pp. 1–56.

King, P. B., 1948, Geology of the southern Guadalupe Mountains, Texas: U.S. Geol. Surv. Prof. Paper 215, 183 pp.

Kopstein, F. P. H. W., 1954, Graded bedding of the Harlech Dome: Ph.D. thesis, Univ. Groningen, 97 pp.

Krumbein, W. C., 1937, Sediments and exponential curves: Jour. Geol. v. 45, pp. 577–601.

———, 1942, Flood deposits of Arroyo Seco, Los Angeles County, California: Bull. Geol. Soc. Amer., v. 53, pp. 1355–1402.

———, 1948, Lithofacies maps and regional sedimentary-stratigraphic analysis: Bull. Amer. Assoc. Petrol. Geol., v. 32, pp. 1909–1923.

———, 1952, Principles of facies map interpretation: Jour. Sed. Petrology, v. 22, pp. 200–211.

———, 1954, The tetrahedron as a facies mapping device: Jour. Sed. Petrology, v. 24, pp. 3–19.

———, 1955, Statistical analysis of facies maps: Jour. Geol., v. 63, pp. 452–470.

———, 1956, Regional and local components in facies maps: Bull. Amer. Assoc. Petrol. Geol., v. 40, pp. 2163–2194.

———, 1957, Comparison of percentage and ratio data in facies mapping: Jour. Sed. Petrology, v. 27, pp. 293–297.

Krumbein, W. C., and Griffith, J. S., 1938, Beach environment in Little Sister Bay, Wisconsin: Bull. Geol. Soc. Amer., v. 49, pp. 629–652.

Krumbein, W. C., and Libby, W. G., 1957, Application of moments to vertical variability maps of stratigraphic units: Bull. Amer. Assoc. Petrol. Geol., v. 41, pp. 197–211.

Krumbein, W. C., and Sloss, L. L., 1963, Stratigraphy and sedimentation, 2nd ed.: San Francisco, Freeman, 660 pp.

Kuenen, Ph. H., 1955, Experimental abrasion of pebbles. I. Wet sandblasting: Leidse Geol. Meded., v. 20, pp. 142–150.

———, 1957a, Sole markings of graded graywacke beds: Jour. Geol., v. 65, pp. 231–258.

———, 1957b, Longitudinal filling of oblong sedimentary basins: Verhandl. Konink. Ned. Geol., Mijn. Genootsch., geol. ser., v. 18, pp. 189–195.

Kuenen, Ph. H., and Sanders, J. E., 1956, Sedimentation phenomena in Kulm and Flözleeres graywackes, Sauerland and Oberharz, Germany: Amer. Jour. Sci., v. 254, pp. 649–671.

Lajoie, J., ed., 1970, Flysch sedimentology in North America: Geol. Assoc. Canada Spec. Paper, 7, 272 pp.

———, 1972, Slump fold axis orientations: An indication of paleoslope?: Jour. Sed. Petrol., v. 42, pp. 584–586.

Laming, D. J. C., 1966, Imbrication, paleocurrents and other sedimentary features in the Lower New Red Sandstone, Devonshire, England: Jour. Sed. Petrology, v. 36, pp. 940–959.

Leviavsky, S., 1955, An introduction to fluvial hydraulics: London, Constable, 257 pp.

Lindsey, D. A., 1969, Glacial sedimentology of the Precambrian Gowganda Formation, Ontario, Canada: Bull. Geol. Soc. Amer., v. 80, pp. 1685–1702.

Lundqvist, G., 1935, Blockundersökningar, Historik och Metodik: Sverige Geol. Undersökn., ser. 3, no. 390, 45 pp.

McBride, E. F., 1962, Flysch and associated beds of the Martinsburg Formation (Ordovician), central Appalachians: Jour. Sed. Petrology, v. 32, pp. 39–91.

MacCarthy, G. R., 1931, Coastal sands of the eastern United States: Amer. Jour. Sci., ser. 5, v. 22, pp. 35–50.

Marschalko, R., 1961, Sedimentologic investigation of marginal lithofacies in flysch of central Carpathians: Geol. Prace (Bratislava), v. 60, pp. 167–230.

Meckel, L. D., 1967, Origin of Pottsville conglomerates (Pennsylvanian) in the central Appalachians: Bull. Geol. Soc. Amer., v. 78, pp. 223–258.

———, 1970, Paleozoic alluvial deposition in central Appalachians: a summary, *in* Studies in Appalachian geology: central and southern (Fisher, G. W., et al., eds.): New York, Interscience, pp. 49–68.

Moore, R. C., 1949, Meaning of facies: Geol. Soc. Amer. Mem. 30, pp. 1–34.

Murphy, M. A., and Schlanger, S. O., 1962, Sedimentary structures in Ihlas and São Sebastião formations (Cretaceous), Reconcavo Basin, Brazil: Bull. Amer. Assoc. Petrol. Geol., v. 46, pp. 457–477.

Opdyke, N. D., and Runcorn, S. K., 1960, Wind direction in the western United States in the Late Paleozoic: Bull. Geol. Soc. Amer., v. 71, pp. 959–972.

Pelletier, B. R., 1958, Pocono paleocurrents in Pennsylvania and Maryland: Bull. Geol. Soc. Amer., v. 69, pp. 1033–1064.

———, 1965, Paleocurrents in the Triassic of northeastern British Columbia, *in* Primary sedimentary structures and their hydrodynamic interpretation (Middleton, G. V., ed.) Soc. Econ. Paleont. Min. Spec. Publ. 12, pp. 233–245.

Pelto, C. R., 1954, Mapping of multicomponent systems: Jour. Geol., v. 62, pp. 501–511.

Pettijohn, F. J., 1957, Paleocurrents of Lake Superior Precambrian quartzites: Bull. Geol. Soc. Amer., v. 68, pp. 469–480.

———, 1962, Paleocurrents and paleogeography: Bull. Amer. Assoc. Petrol. Geol., v. 46, pp. 1468–1493.

Pettijohn, F. J., and Lundahl, A. C., 1943, Shape and roundness of Lake Erie beach sands: Jour. Sed. Petrology, v. 13, pp. 69–78.

Pickel, W., 1937, Stratigraphie und Sedimentanalyse des Kulms an der Edertalsperre: Zeitschr. Deut. Geol. Gesell., v. 89, Abhandl. A, pp. 233–280.

Plumley, W. J., 1948, Black Hills terrace gravels: a study in sediment transport: Jour. Geol., v. 56, pp. 526–577.

Potter, P. E., 1955, The petrology and origin of the Lafayette Gravel: Jour. Geol., v. 63, pp. 1–38, 115–132.

Potter, P. E., and Olson, J. S., 1954, Variance components of cross-bedding direction in some basal Pennsylvanian sandstones of Eastern Interior Basin: geological considerations: Jour. Geol., v. 62, pp. 50–73.

Potter, P. E., and Pettijohn, F. J., 1963, Paleocurrents and basin analysis: New York, Springer, 296 pp.

Potter, P. E., and Pryor, W. A., 1961, Dispersal centers of Paleozoic and later clastics of the Upper Mississippi Valley and adjacent areas: Bull. Geol. Soc. Amer., v. 72, pp. 1195–1250.

Potter, P. E., and Scheidegger, A. E., 1966, Bed thickness and grain size: graded beds: Sedimentology, v 7, pp. 233–240.

Reiche, P., 1938, An analysis of cross-lamination: the Coconino Sandstone: Jour. Geol., v. 46, pp. 905–932.

Rich, J. L., 1950, Flow markings, groovings, and intrastratal crumplings as criteria for recognition of slope deposits, with illustrations from Silurian rocks of Wales: Bull. Amer. Assoc. Petrol. Geol., v. 34, pp. 717–741.

Richter, D., 1965, Sedimentstrukturen Ablagerungsart und Transportrichtung in Flysch der baskischen Pyrenäen: Geol. Mitt., v. 4, pp. 153–210.

Rubey, W. W., 1933, The size distribution of heavy minerals within a waterlaid sandstone: Jour. Sed. Petrology, v. 3, pp. 3–29.

Rubey, W. W., and Bass, N. W., 1925, The geology of Russell County, Kansas: Bull. Kansas State Geol. Surv. 10, pp, 1–86.

Ruedemann, R., 1897, Evidence of current action in the Ordovician of New York: Amer. Geol., v. 19, pp. 367–391.

Russell, R. D., and Taylor, R. E., 1937, Roundness and shape of Mississippi River sands: Jour. Geol., v. 45, pp. 225–267.

Scheidegger, A. E., 1961, Theoretical geomorphology: New York, Springer, 333 pp.

Scheidegger, A. E., and Potter, P. E., 1971, Downcurrent decline in grain thickness of single turbidite beds: a semi-quantitative analysis: Sedimentology, v. 17, pp. 41–49.

Schlee, J., 1957, Upland gravels of southern Maryland: Bull. Geol. Soc. Amer., v. 68, pp. 1371–1410.

Schoklitsch, A., 1933, Über die Verkleinerung der Geschiebe in Flussläufen: Proc. Acad. Sci. Vienna, Math. Nat. Sci. Kl., sec. IIa, v. 142, pp. 343–366.

Schwab, F. L., 1970, Origin of the Antietam Formation (Late Precambrian?–Lower Cambrian), central Virginia: Jour. Sed. Petrology, v. 40, pp. 354–366.

Schwarzacher, W., 1953, Cross-bedding and grain size in the Lower Cretaceous sands of East Anglia: Geol. Mag., v. 90, pp. 322–330.

Scott, G., 1930, Ripple marks of large size in the Fredericksburg rocks west of Fort Worth, Texas, *in* Contributions to geology, 1930, Bull. Univ. Texas 3001, pp. 53–56.

Scott, K. M., 1966, Sedimentology and dispersal pattern of a Cretaceous flysch sequence, Patagonia: Bull. Amer. Assoc. Petrol. Geol., v. 50, pp. 72–107.

Scott, K. M., and Gravlee, G. C., Jr., 1968, Flood surge on the Rubicon River, California–hydrology, hydraulics, and boulder transport: U.S. Geol. Surv. Prof. Paper 422-M, 40 pp.

Sestini, G., 1964, Paleocorrenti eoceniche nell'area tosco-umbra: Boll. Soc. Geol. Italiana, v. 83, pp. 1–54.

Shotten, F. W., 1937, The lower Bunter sandstones of north Worcestershire and east Shropshire: Geol. Mag., v. 74, pp. 534–553.

Siefert, G., 1954, Das mikroscopische Korngefüge des Geschiebemergels als Abbild der Eisbewegungen, zugleich Geschichte des Eisabbaues in Fehmarn, Ost-Wagrien und dem dänischen Wohld.: Meyniana, v. 2, pp. 126–184.

Sloss, L. L., Krumbein, W. C., and Dapples, E. C., 1949, Integrated facies analysis: Geol. Soc. Amer. Mem. 39, pp. 91–123.

Sorby, H. C., 1857, On the physical geography of the Tertiary estuary of the Isle of Wight: Edinburgh New Phil. Jour., n.s., v. 5, pp. 275–298.

———, 1859, On the structures produced by the current present during the deposition of stratified rocks: The Geologist, v. 2, pp. 137–147.

Sternberg, H., 1875, Untersuchungen über Längen-und Querprofil geschiebeführender Flüsse: Zeitschr. Bauwesen, v. 25, pp. 483–506.

Sullwold, H. H., Jr., 1960, Tarzana fan, deep submarine fan of late Miocene age, Los Angeles County, California: Bull. Amer. Assoc. Petrol. Geol., v. 44, pp. 433–457.

Thorarinsson, S., 1954, The eruption of Hekla 1947–1948. II, 3, The tephra fall from Hekla on March 29, 1947: Soc. Sci. Islandica, Reykjavik, 68 pp.

Unrug, R., 1957, Recent transport and sedimentation of gravels in the Dunajec Valley (western Carpathians): Acta Geol. Polonica, v. 7, pp. 217–257.

Vassoevich, N. B., 1953, On some flysch textures: Trans. Soc. Geol. Lwow, ser. geol., v. 3, pp. 17–85.

Vause, J. E., 1959, Underwater geology and analysis of Recent sediments off the northwest Florida coast: Jour. Sed. Petrology, v. 29, pp. 555–563.

Virkkala, K., 1960, On the striations and glacier movements in the Tampere region, southern Finland: Bull. Comm. Geol. Finlande, no. 188, pp. 161–176.

Walker, R. G., 1970, Review of the geometry and facies organization of turbidites and turbidite-bearing basins *in* Flysch sedimentology in North America (Lajoie, J., ed.): Geol. Assoc. Canada, Spec. Paper 7, pp. 219–251.

Weeks, L. G., 1952, Factors of sedimentary basin development that control oil occurrence: Bull. Amer. Assoc. Petrol. Geol., v. 36, pp. 2071–2124.

Wendler, R., 1956, Zur Frage der Quarz-Kornregelung von Psammiten: Wiss. Zeitschr. Karl Marx Univ. Leipzig, v. 5, pp. 421–426.

West, R. C., and Donner, J. J., 1956, The glaciation of East Anglia and the East Midlands. A differentiation based on stone orientation measurement of tills: Quart. Jour. Geol. Soc. London, v. 112, pp. 69–91.

Whitaker, J. C., 1955, Direction of current flow in some Lower Cambrian clastics of Maryland: Bull. Geol. Soc. Amer., v. 66, pp. 763–766.

Wilson, G., Watson, J., and Sutton, J., 1953, Current-bedding in the Moine Series of northwestern Scotland: Geol. Mag., v. 90, pp. 377–387.

Yeakel, L. S., 1959, Tuscarora, Juniata, and Bald Eagle paleocurrents and paleogeography in the central Appalachians: Ph.D. thesis, Johns Hopkins Univ., 455 pp.

———, 1962, Tuscarora, Juniata, and Bald Eagle paleocurrents and paleogeography in the central Appalachians: Bull. Geol. Soc. Amer., v. 73, pp. 1515–1540.

15

ENVIRONMENTAL ANALYSIS

INTRODUCTION

A sedimentary rock is not only a product of a specific provenance and transport history, but it is also the product of its environment of deposition. Some sedimentary rocks, such as chemical and biochemical precipitates, carry no record of their provenance or transport and reflect only their environment of deposition; others, such as clastic deposits, not only record their predepositional history, but in addition have imprinted on them some record of their depositional environment. Our task here is to analyze the problem of environmental reconstruction and to report on progress made to date in such reconstruction.

It is manifestly impossible to digest, in the small space here alloted to this topic, the vast literature generated in recent years on various modern sedimentary environments and the sediments contained therein. We cite, therefore, a few of the larger works which encompass the study of modern sediments. Likewise, the task of framing criteria for the recognition of these environments in the ancient record and the voluminous literature on this subject cannot be adequately reviewed here. Hence the treatment is restricted to a discussion of the principles involved and a brief summary of the salient features of the most common and best-known environmental models that illustrate the application of the principles.

To pursue the subject of sedimentary environments and environmental analysis further, refer to the original literature and to some of the more recent monographic works on these topics. The geology of Recent sediments has been summarized by Kukal (1970b); that of depositional sedimentary environments is covered in a recent book by Reineck and Singh (1973). Perhaps the best recent review of the environments of deposition of carbonate sediments is given by Bathurst (1971); that of sands is summarized by Pettijohn, Potter, and Siever (1972, pp. 449–515). The noncarbonate tidal-flat environment is adequately treated by Reineck (1970); the fluvial environment has been thoroughly analyzed by Allen (1965a), the desert environments by Glennie (1970); deltaic environments are the subject of a symposium edited by Morgan (1970).

An excellent primer on the subject of recognition of ancient sedimentary environments is by Selley (1970) and the symposium volume on the same topic edited by Rigby and Hamblin (1972). Clastic depositional environments have been summarized by Medeiros, Schaller, and Friedman (1971).

Environmental recognition is also briefly treated in various standard works on stratigraphy and sedimentation such as those of Twenhofel (1932, pp. 783–871), Dunbar and Rodgers (1957, pp. 3–96), Weller (1960, pp. 190–237), Krumbein and Sloss (1963, pp. 234–274), and Lombard (1972).

CONCEPT OF ENVIRONMENT

The term *environment* has been much used but rarely given a precise definition. It has often been defined in terms of one or more physical or chemical parameters, and some have assumed that environments could be defined and described in terms of these parameters alone. This concept is implied in the definition of a sedimentary environment as "the complex of physical, chemical, and biological conditions under which a sediment accumulates" (Krumbein and Sloss, 1963, p. 234). One might, for example, describe an environment as reducing or as oxidizing, defined in terms of a particular value of *Eh*, the oxidation–reduction potential. Or the environment may be characterized as saline or freshwater in terms of a certain salinity. The terms *high energy* or *low energy* have been loosely applied to environments. Popular also is the description of a deposit as the product of the "lower flow regime" or some other flow conditions. Any or all of these approaches may be of value, but most are not a geologically meaningful description of the environment.

Other terms, such as *cratonic, stable shelf,* or *geosynclinal,* used to designate the environment are too general even though they are perhaps more meaningful.

Others have defined the environment as "a spatial unit in which external physical, chemical, and biological conditions and influences affecting the development of a sediment are sufficiently constant to form a characteristic deposit" (Shepard and Moore, 1955, p. 1488). This concept is close to the geomorphic concept traditionally used by geologists and emphasized by Twenhofel (1950) and reiterated by Potter (1967, p. 340) and Selley (1970, p. 1). It is Pettijohn's view that the geomorphic concept is the most geologically meaningful approach to the problem of defining and classifying sedimentary environments. A sedimentary environment is defined by a particular set of physical and chemical variables that corresponds to a geomorphic unit of stated size and shape. Thus a dune, barrier beach, tidal flat, and delta are geologically significant entities that describe sedimentary environments. Some may be too all-inclusive and should be broken down into subenvironments when this is possible. A tidal flat, for example, could be divided into subtidal, intertidal, and supratidal, or into tidal marsh, tidal channel, tidal levees, and tidal deltas. Likewise, deltas consist of a complex of subenvironments.

In conclusion, we must discriminate between *local* environment, best defined in geomorphic terms, and *tectonic* environment, the relation of sediment accumulation to larger tectonic elements—cratons and geosynclines. In this chapter we are concerned with the former; the latter is the subject of the following chapter. Here we will first discuss the parameters of the local environment and the criteria used for their evaluation, and then, following a classification of these environments in geomorphic terms, take up the question of environmental analysis—how achieved as shown by appropriate examples.

ENVIRONMENTAL PARAMETERS

It is not clear just what the relevant environmental parameters or factors are or their relative importance, nor is it clear how they should be classified. Krumbein and Sloss (1963, p. 236) grouped them into three categories: material (nature of the medium, that is, wind, water, ice, and character of the sediment, both its texture and composition), energy (kinetic, turbulent, and thermal), and biological. Twenhofel (1950, p. 10), on the other hand, took a more geologic point of view and considered such factors as physiography (of the land areas), diastrophism, shoreline physiography, and climate. Each of these is presumed to leave an imprint on the sedimentary deposit.

We have chosen more strictly measurable parameters and have grouped them as physical, chemical, and biological. Presumably each is an independent variable and one which can be evaluated from some observable aspect of the deposit. Physical and chemical parameters are, perhaps, largely independent of one another, but biologic factors are very strongly modified or controlled by the other two and in turn modify physical or chemical factors. A sea-grass meadow, for example, may reduce current velocity and turbulence. Very obviously the biota is markedly influenced by temperature, salinity, current velocity, and many other chemical or physical factors.

How well can we ascertain the several parameters that prevailed in a now-vanished environment? And, more important, how well can we identify an environment from its parametric profile? We explore these questions below.

site poles are the air-laid materials, such as dune sand, and those which are ice-laid, namely till. No more marked textural contrasts are known. The distinction between dune sand and some beach sands, however, is far less clear.

It now appears that a whole spectrum of media involve sediment and fluid, varying from dilute suspensions in water or even air, to very turbid flows (turbidity currents), to fluidized sediment flows (slurries), to grain and debris flows (mudflows), to flowing ice with contained debris (Table 15-1). As density or viscosity increases, the lower is the velocity required to transport a given size; also, as these properties increase, the less effective is the medium as a sorting agent. Dilute suspensions moved by air or water are effectively sorted; large sizes are transported only at high velocities. Concentrated suspensions or slurries carry larger fragments at lesser velocities and deposit poorly sorted sediments. If the proportion of solid to fluid increases still more, the materials no longer flow as a Newtonian liquid but move only as a semisolid plastic body. Sorting is nil, and the largest available blocks are moved and deposited concurrently with the finest particles. Mudflows and glacial ice belong to this latter category. The problem of flow, however, is complicated: Some semisolid mixtures can, under some conditions, lose their strength rapidly and become quite fluid and flow at high velocities. Such thixotropic transformation may be triggered by a sudden shock such as an earthquake.

Deposits of dilute suspensions are the

PHYSICAL PARAMETERS

The physical milieu is best described in terms of the dynamic and static properties of the depositional medium from which the sediment is laid down. Static properties include density and viscosity of the medium (air versus water, for example) and depth of the medium above the sediment–fluid interface; dynamic factors include turbulence of the medium (quiet versus rough water), velocity of the depositing current, direction of current flow, stability of the flow pattern (in terms of both velocity and direction), and similar factors.

The nature of the depositing medium One of the factors in environmental analysis is the nature of the medium from which clastic elements of the sedimentary rock were deposited. In other words, was the material deposited from air, water, or glacial ice? Clearly identification of the depositing medium would greatly advance our efforts at reconstruction of the environment.

Because the transporting power of a current and the effectiveness of that current as a sorting agent depend on the viscosity or density (or both) of the medium involved, there should be a correlation between the textures of the deposit formed and these properties of the moving medium. At oppo-

TABLE 15.1. Transporting agents and sedimentary deposits

<table>
<tr><th colspan="5">Increasing density and/or viscosity</th></tr>
<tr><th colspan="3">Fluid</th><th>Semisolid</th><th>Solid</th></tr>
<tr><td>Air</td><td colspan="2">Water plus sediment</td><td rowspan="2">Sediment plus water</td><td rowspan="2">Ice plus sediment</td></tr>
<tr><td></td><td>Dilute suspension</td><td>Concentrated suspension</td></tr>
<tr><td colspan="2">Orthoarenites (orthoquartzite and calcarenite)</td><td>Para-arenites (graywackes)</td><td>Tilloids</td><td>Till and tillite</td></tr>
<tr><td colspan="2">Current-bedded</td><td>Graded-bedded</td><td>Nonbedded</td><td></td></tr>
</table>

common gravels and sands which have a simple grain-supported framework; deposits of concentrated slurries and mudflows are the wackes and paraconglomerates with a prominent to dominant matrix and sparsely scattered larger clasts. If the solid–fluid ratio exceeds some critical value, Newtonian flow ceases and the deposit is ungraded and unsorted. Consequently the framework–matrix ratio and the presence or lack of grading are critical criteria in attempting to identify the agent of deposition. The problem is further complicated, however, because of secondary origin of the matrix in some sediments and also because of a wholly unrelated manner of origin of certain *in situ* generated carbonate "wackes," which consist of a micritic paste with embedded skeletal elements; but other carbonate wackes are true turbidity-current, mudflow or debris-flow deposits.

Theoretically, differences in density and viscosity between air and water should lead to distinctive textural differences in the deposits of these two agents. Although various writers have attempted to formulate these differences, there seems to be no definitive textural criterion for discriminating between aqueous and eolian sands. The use of various size parameters to differentiate between them has been reviewed in Chapter 3 (p. 50). Other textural criteria such as grain rounding and surface texture have likewise been employed, also with indifferent success, to discriminate between aqueous and eolian sands. These efforts, too, have been summarized in Chapter 3 (pp. 60 and 61). Moreover, subaerial density currents and debris flows whose deposits have much the same appearance as their subaqueous counterparts can be distinguished from them only with difficulty, if at all.

Sedimentary structures are believed to be a more reliable guide to environmental analysis (McKee, 1971, p. 38); hence, attempts have been made to discriminate between aqueous and eolian structures. With few exceptions success here also has been limited.

Most weight is usually given to presumed differences in bedding of beach and dune deposits. These have been summarized by Thompson (1937). It has been said that dune sands show a much greater spread or *variance* in the direction of inclination of cross-laminations than is shown by the cross-stratification of aqueous origin (Twenhofel, 1932, p. 621). Data are somewhat inconclusive. Potter and Pettijohn (1963, Table 4-2) tabulated the variance of fluvial-delta, marine, and eolian cross-bedding. These overlap to a considerable degree. Variances of deposits thought to be eolian are comparable with those of fluvial-deltaic, whereas marine sands appear to have a greater variance. More reliance has been placed on the *scale* of cross-bedding than on the consistency of its orientation. Large-scale, tangential cross-bedding has been ascribed to wind action. There are, however, fields of subaqueous dunes remarkably similar in size and form to those of subaerial origin (Potter and Pettijohn, 1963, pls. 10B and 11); consequently their internal structures, including cross-bedding, must be presumed to be similar in kind and scale. Subaerial cross-beds are said to have a higher *inclination* than those deposited subaqueously. However, Yeakel (1959, pp. 85–91), after reviewing the literature, concluded that the angle of inclination was not a reliable means to distinguish eolian from subaqueous cross-bedding. Nor is the *kind* of cross-bedding indicative of origin. Both tabular and trough cross-bedding form under both eolian and aqueous conditions. In conclusion we can say that except, perhaps, for the "herringbone" cross-bedding indicative of tidal currents, cross-bedding is not too helpful in identifying the agent of deposition. One might add that large-scale cross-bedding produced by dune migration rules out a deep basinal origin; small-scale cross-lamination related to ripple migration, however, forms at any depth. Cross-sets over 10 feet (3 m) thick are not likely to be fluvial; they may be either shallow marine or eolian.

Sand ripples generated by wind have a larger ripple index (p. 113) than do aqueous ripples. Moreover, the coarser grains of eolian ripples are concentrated in the crests. Vertebrate tracks observed on the lee slopes of dunes always appear to go up the slope (Reiche, 1938, pp. 917–918). Unfortunately, vertebrate tracks and eolian ripples are exceedingly rare and are seldom available when needed. Some types of ripple cross-lamination may be significant. The lenticular flaser bedding in which small wave-generated ripples are enclosed by mud or contain drapes of mud covering the ripples is common in tidal and subtidal environments (Reineck and Wunderlich, 1968). Some types of ripple-drift cross-lamination ("climbing ripples") are environment sensitive (Jopling and Walker, 1968).

Deposits of glacial ice, the tills and tillites, have been described in detail elsewhere (Chapter 6, p. 171), so criteria for recognizing such glacial deposits and discriminating between these and those of debris flows and other mass transport origins need not be repeated here. Likewise, the characteristics of both subaerial and subaqueous turbidity-flow deposits have been reviewed elsewhere (p. 116).

The depth of water problem The question of water depth is one of long-standing concern to geologists. It has recently been reviewed in a special issue of *Marine Geology* (Hallam, 1967). The bathymetric position of marine sediments is defined or described first of all with reference to wave base. Those sediments above wave base are deposited in a turbulent environment and are constantly shifted about and reworked so that they tend to become texturally and mineralogically mature. Sediments deposited below wave base accumulate in a relatively quiet or still-water environment little disturbed by currents after deposition. Although environments may be described as rough water or quiet water, such designation does not necessarily correlate with absolute depth. Wave base is highly variable. In small water bodies or semi-isolated bodies of small size, the waves generated are small and their effects extend only a slight depth. On exposed coasts of the open sea, waves are larger and may sweep the bottom many meters below the surface. Although most currents and the scour they produce are wave-generated, some currents, including so-called contour currents, are generally feeble, occur at great depths (Heezen and Hollister, 1964), and may be responsible for movement of fine sands and construction of ripple marks.

The rough-water environment is characterized by sand and gravel deposits and by coquinas and calcarenites. The sands are well sorted, well rounded, ripple-marked, and presumably current bedded. Coarse clastics, however, are no certain criterion of shallow water as they were once supposed to be. Sands, and even coarse gravels, are transported and deposited *below* wave base by turbidity underflows. These deposits are marked by poor sorting, graded structure, and their intercalation with fine-grained muds and silts without much disturbance of these subjacent materials. As noted by Bailey (1930, 1936), graded bedding and current bedding are indices of two contrasting environments of sand deposition, below and above wave base, respectively. As noted, however, large-scale (over 10 cm) current bedding implies shallow water; small-scale or ripple cross-lamination is known from deep-water deposits and is, therefore, no guide to depth of water.

Criteria indicative of quiet- (and deep-) water deposition include the delicate laminations which characterize some siltstones and shales. Some paper-thin laminations may represent annual increments of sediment and may thus become a measure of the rate of sedimentation. Such laminations, moreover, may persist over wide areas and be recognized in cores some kilometers apart. Only in the absence of bottom turbulence (and in the absence of a bottom mud-ingesting fauna) can such laminations be preserved. Laminations in fine sediments are not an infallible indication of deep water. Fine-grained calcareous *laminites* are said to be of tidal-flat origin (Friedman et al., 1973).

Whereas the rough-water zone is characterized by scour-and-fill cross-bedding and the quiet-water zone is distinguished by thin and even laminations, the intermittently quiet- and rough-water environment is marked by alternately regular and irregular bedding of sand or silt and shale. The sandy beds are somewhat uneven in thickness and wavy in cross section and in places show ripple cross-lamination. The siltstone or fine sandstone beds deteriorate into a string of planoconvex segments—probably "starved" ripples (Fig. 8-14). The shale interbeds which separate coarser layers constitute a drape over the ripples. Where shale is sparse, it may be restricted to phacoid-like fillings of ripple troughs (Fig. 4-18). Wavy and flaser bedding is characteristic of some modern tidal flat deposits (Reineck and Wunderlich, 1968).

The state of preservation of many fossils, their mode of attachment, and their orientation may be further clues to the turbulence of the surrounding medium. Delicate articulated forms suggest growth and burial in a current-free environment; disarticulated broken and *sorted* fossil debris suggest strong bottom currents. The random orientation of concavo-convex shells suggest an absence of appreciable curents; conversely, a common orientation of these and other fossil structures is proof of stronger bottom currents. Preservation of delicate worm trails and similar structures on mud surfaces (represented by casts on the underside of sandstone or siltstone beds) suggest very feeble bottom currents or none at all. Grazing trails of bottom dwellers of the facies are a good indication of deep water (Seilacher, 1967). The burrowing habit of forms living in a sand environment may be a response to bottom scour.

Certain minerals are indicative of reducing conditions. Such conditions cannot be achieved in highly turbulent, aereated waters. Minerals formed in stagnant anaerobic waters include sedimentary pyrite and siderite. In certain lagoons and estuaries, however, where wave and current action is negligible, these materials may form in the muds in comparatively shallow water. Other mineralogical evidence of depth includes the chamosite-glauconite pair; the former is found in shallow tropical waters (less than 60 m), the latter in waters 30 to 2000 m (Porrenga, 1967). Phosphorites are presumed to be deposited in waters 30 to 300 m in depth (Bromley, 1967).

It is comparatively easy to distinguish between turbulent and quiet water environments, that is, between sediments deposited above wave base and those deposited below. It is difficult, however, to estimate *absolute* depth of water. Few criteria are reliable.

As Allen (1963, Fig. 13; 1967) notes, the height of larger-scale cross-strata is a function of mean water depths; the thicker the coset, the deeper the water (Fig. 4-13).

The kind of fossils present may be a guide to water depth (Hallam, 1967). Algal structures indicate deposition only in the photic zone—depths to which light has access (Bathurst, 1967). The depth of light penetration varies, of course, with the turbidity of the water, but it is never very great. In more recent sediments, especially those of Tertiary age, the depth of water can be estimated by paleoecologic data. Foraminiferal species are closely correlated with depth, and, by study of the habitats of present-day forms, the depth of older deposits containing these forms can be estimated (Phleger, 1955; Bandy, 1953; Funnell, 1967). For example, the water depth in which the Lower Pliocene Repetto Formation of southern California, including conglomerates, was deposited has been shown by foraminiferal evidence to have been 4000 to 5000 feet or 1200 to 1500 m (Natland and Kuenen, 1951, pp. 83–84). The interpretation of such faunal evidence is complicated, however, by the effects of temperature on the species range and by the mechanical transport of shallow-water forms to deep water. In older rocks, the faunal content is a less sure guide to absolute depth. Trace fossil assemblages correlate with depth (Seilacher, 1967) and thus are a guide to relative depth but no certain guide to absolute depth.

At present, carbonate dissolves in depths exceeding 3000 fathoms (5400 m). Before the Cretaceous, there were no calcareous pelagic Foraminifera. The presence of such forms in Cretaceous or younger deposits indicates a depth less than this critical value; but the absence of such fossils in the older record does not mean their removal by solution at this or greater depths.

If slump structures are prominent in the sediments, probable depth can be estimated by assuming a bottom gradient sufficient to initiate slumping. A slope of 1.5 to 3.0 degrees may be considered likely. The depth of basin or trough can be estimated if the deepest and most persistent part of the trough is well outlined by stratigraphic studies and if the location of the probable source area is known. By this method Briggs (1953, p. 436) estimated the waters of the Cretaceous San Joaquin Valley trough to have been 3000 to 7000 feet (900 to 2100 m) deep.

In some cases, of course, water depth was minimal and sediments may have been actually exposed. Evidence of such exposure are many—mud cracks, rain prints, and the like (Ginsburg, Bricker, and Wanless, 1970).

It has been said that there are no deep-water sediments anywhere on the continent. The prevalence of structures characteristic of shallow turbulent waters throughout the geologic column gives considerable support to this view. But, as noted above, increasing evidence is found for deep-water deposits, and depths of the order of several thousands of meters for some deposits now seem a certainty.

In conclusion, it is clear that criteria of absolute depth are few. In general, the problem is best handled by assessment of the appropriate environmental model. Once a stratigraphic assemblage is identified as tidal flat, or fluvial, or turbidite, reasonable limits can be placed on depth estimates and an appropriate criterion of depth applied. It is generally impossible to walk up to a particular outcrop and assign a depth without considering the context in which a particular rock or bed is situated.

Current velocity, direction, and stability Various attempts have been made to assess current strength from the size of the clastic elements in a sediment (the "clasticity index" of Carozzi, 1958, p. 133). Estimates of current velocity based on grain size are valid only for ordinary sands and gravels, not for the paraconglomerates and the other products of mass transport.

The relation between clast size and current velocity is not simple. Velocity falls off rapidly as one approaches the sediment–fluid interface, and that velocity which is responsible for traction movement is somewhat less (about 40 percent less) than that higher in the stream. The velocity at which clast movement begins—the critical tractive velocity—has been empirically determined by several investigators (Table 15-2). The relations have been expressed in the Hjulström (1935) diagram (Fig. 15-1). As shown, the size of material moved is proportional to the velocity. However, because silts and clays, which are less than 0.1 mm in diameter, are somewhat cohesive, the simplicity of the relation breaks down.

There tends to be a correlation between bed forms (and the structures which they produce) and velocity (actually the stream power of which velocity is a component).

TABLE 15-2. Grain size and current velocity

Grain diameter (mm)	Observed traction velocity (cm/sec)	Calculated suspension velocity (cm/sec)
6.08	84.5[a]	45.3[c]
4.18	61.3[a]	36.3[c]
4.08	62.2[b]	35.9[d]
3.08	53.8[a]	29.7[c]
1.38	36.0[a]	18.0[c]
0.72	—	15.2[c]
0.59	31.6[b]	14.1[d]
0.35	—	13.7[c]
0.31	24.5[b]	13.2[d]
0.20	21.5[b]	10.7[d]

[a] Observed in flume experiments (Gilbert, 1914).
[b] Observed in flume experiments, U.S. Waterways Experiment Station, Vicksburg, Mississippi, 1935 (Nevin, 1946).
[c] Calculated (Rubey, 1938).
[d] Calculated (Nevin, 1946).

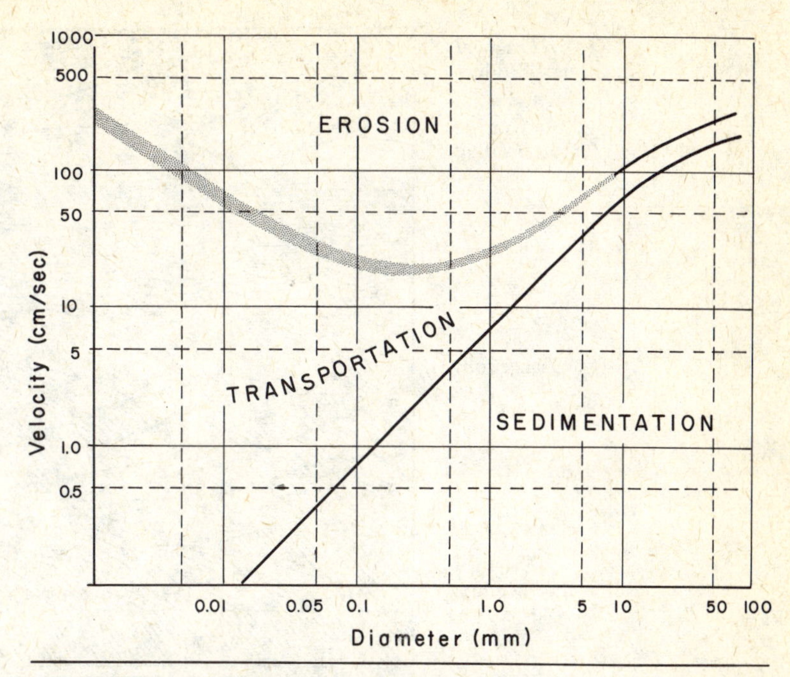

FIG. 15-1. Hjulström curves for the erosion and deposition of uniform material. (After Hjulström, 1935, Fig. 18.)

These relations were expressed diagrammatically by Andersen (1934) in his study of the sedimentary structures of glacial outwash (Fig. 15-2). As can be seen, slow-moving currents deposit even-bedded silts and clays; faster movement initiates a rippling of the sand–water interface and produces small-scale cross-lamination; at high velocities scour and fill lead to a strongly cross-bedded deposit. These concepts arrived at by Andersen from field observations and intuition have been confirmed by experiments and further field studies (Gilbert, 1914; Simons, Richardson, and Nordin, 1965). Relations between the bed forms and their internal structure, grain size of the sediment, and stream velocity are shown schematically in Fig. 15-3.

Some further indication of velocity is given by the tangency or lack of it in larger-scale cross-bedding. Experiments show that, for a given sediment mix and a constant thickness of tabular cross-bedding, an increasing velocity of flow favors a transition from angular planar foresets to tangential concave forests (Jopling, 1963, p. 119). However, a high concentration of stream load tends to induce tangency as well.

Other indices of current velocity are percussion marks found on some pebbles transported by high-velocity streams (p. 63), an abnormal percentage of broken rounds (p. 46), and rounding of micas and kyanite sand grains, possible only if the current is very gentle (p. 206).

Not only is current strength an important attribute of the environment, but so also is the persistence of current in both direction and velocity. The persistence in direction is best determined by variance of cross-bedding or other criteria of current direction; the persistence in velocity is expressed by the bed-to-bed variance in mean particle size. Cross-bedding variation in direction and its environmental significance are discussed elsewhere (p. 109).

CHEMICAL PARAMETERS

Chemical factors in the environment are primarily the oxidation–reduction potential (*Eh*), the acidity–alkalinity (pH), and salinity (concentration). The minerals precipitated and the faunal elements are closely related to these environmental factors. Another factor of interest to the geologist and of great importance in the distribution of fauna is temperature. A classification of chemical environments has been suggested

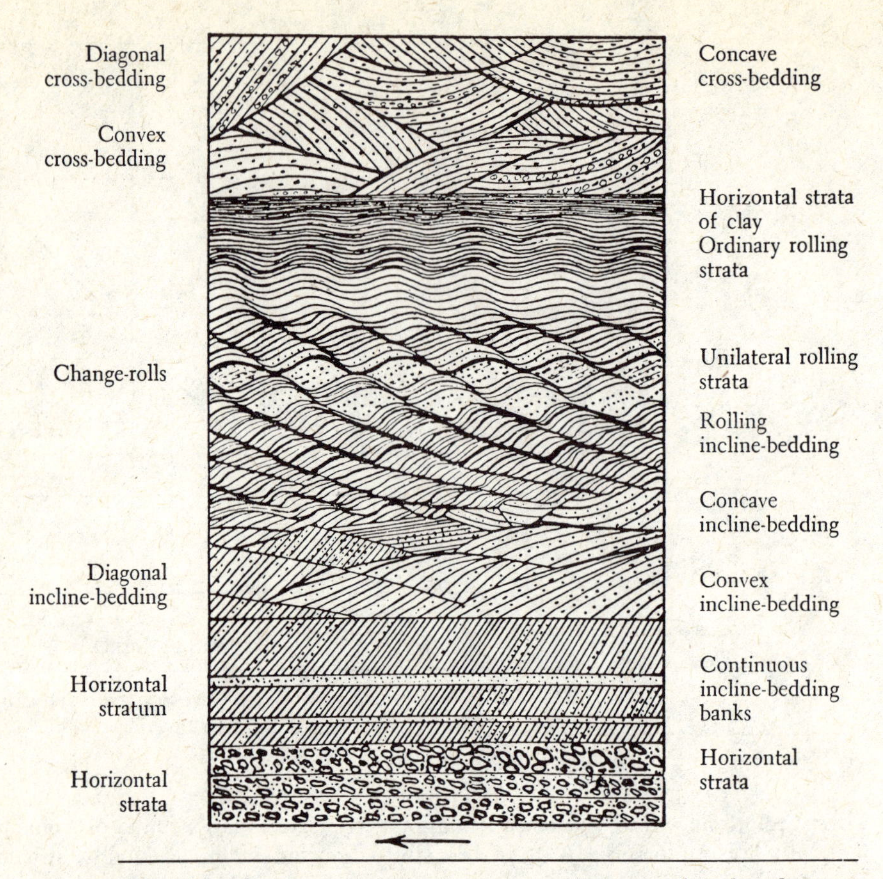

FIG. 15-2. The most important structures in aqueoglacial sediments. The arrow indicates the direction of current flow. (From Andersen, 1931, Fig. 38; the terminology used by Andersen is somewhat archaic.)

by Teodorovich (Chilingar, 1955) and by Krumbein and Garrels (1952). Although a number of factors enter into and control the precipitation of minerals, the most important seem to be the *Eh* (oxidation–reduction potential) and the *p*H (acidity–alkalinity) (Baas Becking, Kaplan, and Moore, 1960). See Fig. 11-27.

Oxidation–reduction potential (*Eh*) In a general way, sediments are deposited under either oxidizing (aerobic), or reducing (anaerobic) conditions. The measure of the oxidizing capacity of an environment is the *Eh* or oxidation–reduction potential (Mason, 1949; Zobell, 1946). Whether or not an ancient sediment was deposited under oxidizing or reducing conditions is decided mainly on the basis of the mineralogy and what is

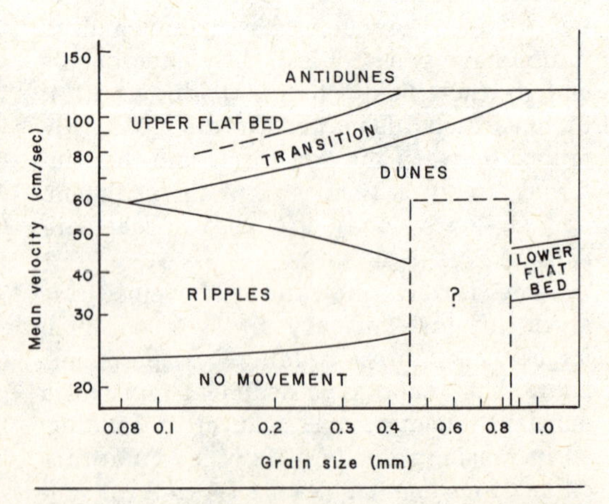

FIG. 15-3. Relation between bed forms and grain size of the sediment and velocity of flow. (After Southard, 1973, Jour. Sed. Petrology, v. 43, Fig. 1. By permission of Society of Economic Paleontologists and Mineralogists.)

known about the stability of minerals under different oxidation potentials. Iron minerals in particular are the most useful as indices of this parameter. Iron sulfides (pyrite or marcasite) signify a reducing and wholly oxygen-deficient medium; hematite indicates a fully aereated environment; siderite indicates an intermediate oxidation–reduction potential (Krumbein and Garrels, 1952).

By the use of other minerals a further subdivision, based on *Eh*, of the chemical environment is possible. Teodorovich (Chilingar, 1955) recognized six such subdivisions: strongly reducing or sulfide zone, reducing (iron carbonate and iron sulfide) zone, weakly reducing (siderite and vivianite) zone, neutral (iron-rich chlorites with both ferric and ferrous iron) zone, weakly oxidizing (glauconite) zone, and oxidizing (ferric oxide and hydroxide) zone. Each of these six environments defined on *Eh* can be subdivided into six subenvironments based on acidity or *p*H, making some 36 subdivisions, each defined by a particular suite of minerals.

Other criteria for a low oxidation potential is the absence of a normal benthonic fauna and the presence of only those forms that can tolerate toxic conditions produced by oxygen deficiency or forms which are free-swimming or are attached to floating objects. In the former category are certain phosphatic brachiopods, especially *Lingula* and *Discina*, which are ubiquitous and hardy types capable of survival under adverse conditions. Conodonts, occasional fish remains, and spores and pollen complete the fauna or flora of those shales formed in a highly reducing environment. A further indication of oxygen deficiency is an abnormally high (over 2 or 3 percent) content of organic matter. Normal microbiological and scavenger action tends to destroy the organic residues which settle to the bottom. Inhibition of such action because of oxygen deficiency leads to an increase in these materials and hence to black shales and related deposits.

As noted by Krumbein and Garrels and by Teodorovich, the oxidation–reduction surface, that is, the plane separating the oxidizing from the reducing environment, may be above, coincide, or be below the sediment–water interface (Fig. 15-4). In more strongly reducing environments it is above the mud–water interface. Inasmuch as the environment within a sediment is always reducing, how does it happen that all sediments are not reduced? Whether a sediment is reduced or not depends on whether a reducing agent is present. Such an agent is organic matter, and, as noted above, in an oxidizing environment the organic matter is mainly oxidized and destroyed by microbiological decay or normal scavenging action. In the sediment deposited in this environment, therefore, little or none remains to effect reduction of the iron. As shown by studies of modern muds (Emery and Rittenberg, 1952), diagenetic processes are reducing in character, and such organic matter as is present does in part bring about reduction of the iron.

Alkalinity–acidity (*p*H) The acidity or alkalinity of an environment is an important factor in determining whether or not certain minerals will precipitate. In a strongly acid environment, for example, carbonates will not be deposited. The deposition of calcite, therefore, is evidence of a *p*H of at least 7.8 (Fig. 11-27).

It may not be sufficient merely to discriminate between an acid and an alkaline environment. Krumbein and Garrels (1952) would define three environments based on acidity: most acid (*p*H less than 7.0), more nearly neutral (*p*H 7.0 to 7.8), and alkaline (*p*H over 7.8). Teodorovich (Chilingar, 1955) would define six environments based on *p*H: strongly alkaline—soda lakes (*p*H over 9.0), alkaline (*p*H 8.0 to 9.0), weakly alkaline (*p*H 7.2 to 8.0), neutral (*p*H 6.6 to 7.2), slightly acid (*p*H 5.5 to 6.6), and acid—swamps (*p*H 2.1 to 5.5).

Criteria for determining alkalinity or acidity of an ancient environment are mainly mineralogical. As noted, carbonates dissolve in an acid environment. Krumbein and Garrels set a *p*H of 7.8 (about that of sea water) as a "limestone fence." Calcite is freely precipitated at this or higher *p*H values; it is a minor accessory only in slightly less alkaline environments, and its precipitation is completely inhibited as the *p*H drops below 7.0. Silica, on the other hand, tends to dissolve in the alkaline environment and to be precipitated in the acid environment. Extensive deposition of chert suggests a more acid environment than that responsible for calcite. Relations between the solubility of silica and calcium carbonate, and *p*H have been reviewed by Correns (1950).

As noted by Edwards and Baker (1951), the widespread occurrence of marcasite in association with coals is in striking contrast to the occurrence of pyrite in marine clays and shales. Presumably the difference is related to the *p*H. The coal swamp is strongly acid; the marine environment is neutral or mildly alkaline. The form of iron sulfide, therefore, provides a further means of discriminating between acid and alkaline environments. Marcasite in a marine formation would be indicative of acidification subsequent to deposition.

In addition to carbonates, silica, and the forms of iron sulfide which are *p*H sensitive, it is believed that kaolin requires an acid environment for its formation, whereas an alkaline environment favors the formation of montmorillonite (Millot, 1949, p. 352).

A more direct attempt to determine the *p*H of the environment of an ancient sediment is that of Shukri (1942), who believed that the *p*H of an aqueous extract of a shale would be essentially that of the water from which the mud was originally deposited. Emery and Rittenberg (1952), however, noted a tendency for the interstitial waters of muds to have a slightly higher *p*H than the waters of the basin in which they accumulated.

Salinity The salinity of waters is essentially the weight percentage of the dissolved solids. Normal sea water contains about 3.5 percent by weight of dissolved materials. For technical reasons the salinity of sea water and other brines is expressed as *chlorinity*, approximately defined as the parts per thousand of chlorine. Normal sea water has a chlorinity of about 19.4.

The salinity of waters in ancient basins varied from fresh to supersaline. As a guide to salinity most reliance is placed on the

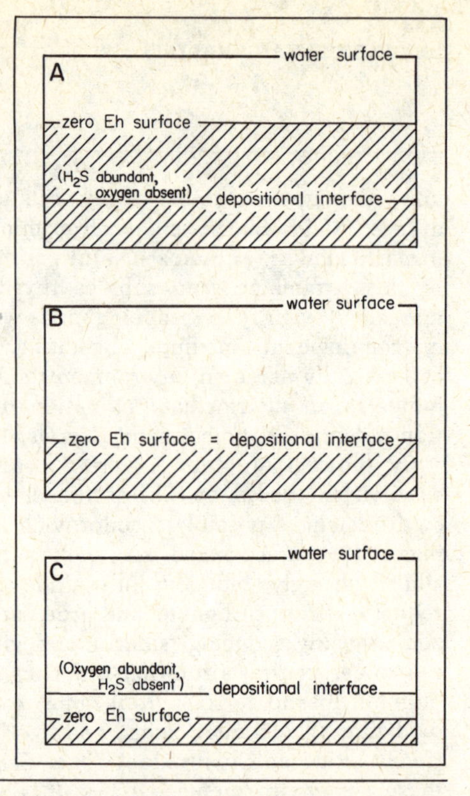

FIG. 15-4. Relation of oxidation–reduction boundary and sediment–water interface. (After Krumbein and Garrels, 1952.)

fauna. The fauna may be described as freshwater, brackish, or normal saline. Ecologic studies of present faunas, especially foraminiferal assemblages, have proved dependable guides to interpretation of the more recent past (Tertiary). Biological criteria for more ancient times are less certain.

In the case of more than normal, or hypersaline, conditions, the biota become sparse or disappear. Evidence of supersaline conditions is recorded in the salt minerals formed. The sulfates, gypsum and anhydrite, are common; greater salinity leads to the precipitation of halite. Only from the most concentrated brines are potash salts precipitated.

Most difficulty arises when waters are not saline enough to precipitate sulfates or chlorides but are appreciably more saline than normal sea water. These waters have been termed *penesaline*. Deposits formed under such conditions are generally unfossiliferous. Evaporitic carbonates are said to indicate penesaline waters (Sloss, 1953). Such carbonates are chiefly the oolitic carbonates and the finely laminated dolomites interpreted as primary.

Salt crystal molds, in the absence of salt deposits proper, indicate at least a temporary salinity greater than normal. Salt crystals can form in shallow intermittently flood areas even in regions of substantial rainfall and do not necessarily indicate prevalence of an arid condition.

To a certain extent trace elements have been used to discriminate between fresh and marine waters and thus in a sense are criteria for salinity. Boron, in particular, has been used with some success in this manner. Normally, marine shales are richer in boron than those deposited in fresh waters (Potter, Shimp, and Witters, 1963). The use of geochemical indicators in distinguishing freshwater and marine sediments is summarized by Keith and Degens (1959).

Isotopic analyses have also been used to discriminate freshwater and marine materials (see Lowenstam, 1963, for review of relation of isotopic composition of shells and its relation to both salinity and temperatures).

Temperature An important parameter of the environment of deposition is temperature. Temperature affects the solubility of many minerals and gases and has therefore an important effect on chemical precipitation. Certain salt minerals may be deposited in winter but dissolved in summer. At lower temperatures the solubility of CO_2 is greatly enhanced; hence, the solution of calcium carbonate (as the bicarbonate) is promoted in cold waters, and conversely precipitation of the same material is brought about by a rise in temperature. Temperature also affects the composition of mixed crystals or solid solutions. Shell carbonate is richer in MgO at the lower temperature (Chave, 1954); lower-temperature authigenic plagioclase is nearly pure albite, whereas that formed at higher temperature contains more lime. The effect of temperature on viscosity of water, although readily measured, seems not to be geologically important. Temperature profoundly affects the behavior of water, however, if it passes below the freezing point of that liquid. Glaciers and glacial deposits form only at materially reduced temperatures.

Criteria for determining past temperatures (or paleotemperatures) are geological, mineralogical, and ecological. Geological evidence of glacial temperatures are tillites and pellodites. Criteria for the recognition of tillite have been reviewed elsewhere (p. 172); the recognition of pellodites (glaciolacustrine varved clays) has also been summarized. Certain identification of these deposits is not easy. Tillites are closely imitated by tilloids; the only certain criterion of a glaciolacustrine deposit is bergrafted cobbles, and, because other agents of rafting are known, even this criterion is not infallible.

Mineralogical or chemical criteria of temperature are not clearly formulated and are not wholly isolated from those related to other climatic factors. The red color of many solids (and hence perhaps the red pigmentation of many sediments) seems to be correlated with latitude (and hence with temperature). The highly oxidized red pigmentation is characteristic of soils of low latitudes; soils of higher latitudes are not red. Color seems to be related to rate of oxidation of the humus in soil. Humus inhibits oxidation of the iron; its destruction permits oxidation.

Although the composition of certain solid solutions is a function of temperature, a geological thermometer based on this principle has not been successfully used to discriminate between high- and low-temperature values in the limited range of temperatures which prevail at the surface of the earth. Presumably the stability of certain hydrides is temperature controlled. Anhydrite rather than gypsum is formed at relatively higher temperatures (p. 439).

Recent efforts to measure paleotemperatures involve temperature-controlled fractionation of the isotopes of oxygen, ^{16}O and ^{18}O. Because proportions of these isotopes can be shown experimentally and theoretically to be temperature-dependent, the temperature of formation of a mineral can be theoretically determined, if the ratio has not been affected by post-depositional exchanges or replacements. Some success has been obtained in determining the temperature of the shell formation of some belem-

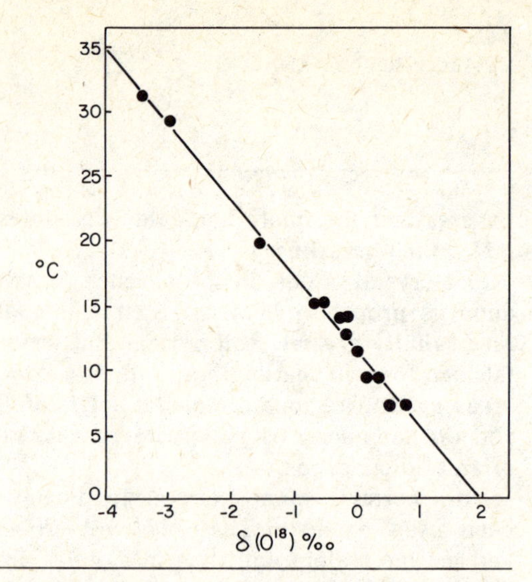

FIG. 15-5. Plot of ^{18}O concentrations versus temperature: calcium carbonate samples heat treated. Abscissa is a ratio. (After Epstein et al., 1951.)

nites and other fossil forms (Urey et al., 1951; Epstein et al., 1951). See Fig. 15-5.

Both faunal and floral evidence has been cited regarding paleotemperatures. It seems probable that, in Tertiary beds at least, valid distinctions between cold-water and warm-water faunas can be made. Foraminifera, in particular, have been useful in establishment of paleotemperatures (Bandy, 1953), although care must be taken to distinguish between surface and bottom-dwelling forms and between indigenous and transported forms (Emiliani, 1955). The determination of temperature by biological criteria is difficult in older formations, and in all cases the biota is affected by ecological factors other than temperature. These complications render interpretations difficult.

EVALUATION AND SUMMARY

In our review of the various environmental parameters and our evaluation of the criteria used for ascertaining them, it is clear that the attributes used—textural, mineralogical, structural, and biologic—do not as a rule give unambiguous answers.

Two points should be kept in mind. First, the several parameters, physical, chemical, and biological, are not wholly independent. They are to some degree interdependent. This is particularly true of the biota. The biota is very sensitive to such factors as turbulence, water depth, salinity, and temperature. The organic community, in turn, may modify turbulence (baffle effect), may alter *Eh* and *p*H (particularly the former) and lead to eutrophication which in turn drastically alters the biota. Even physical and chemical parameters may interact with one another. Salinity, if stratified, as in the case of freshwater overlying salt water, may lead to formation of reducing conditions in lower depths. High turbulence is incompatible with low *Eh*, and so forth. Moreover, sediment accumulation may modify water depth which may in turn alter current strength and direction and turbulence. There is, to use the current popular term, *feedback*. In short, because of internal or external factors, environmental parameters may change with time.

A second point to be kept in mind is that parameters do not precisely identify an environment. Except in a very restricted subenvironment, parameters vary within wide limits within an environment—a tidal flat, or lake, for example. And even though temperature, salinity, *p*H, and *Eh* are specified, one cannot be sure if one is dealing with an estuary, inland sea, lagoon, or other environmental complex. Some lakes are very salty, others fresh. Some lagoons are brackish, others briny. Some are well aerated, others reducing. Hence, even if it were possible to evaluate all the parameters, the result would not be too meaningful. We must look, therefore, elsewhere to find criteria for pinpointing the environment of deposition.

Obviously we need a new approach to environmental analysis and less reliance on textures, structures, and mineralogical composition. Even fossils may be displaced and hence misleading. The most fruitful approach is the use of the vertical profile and the geometry of sedimentary bodies. This approach to the problem of environmental analysis implies a return to the field and less dependence on laboratory analysis of samples. It requires careful measurement and study of stratigraphic sections and the mapping of sedimentary bodies. This approach is the subject of the section below.

CLASSIFICATION OF ENVIRONMENTS

A number of attempts have been made to classify domains of sedimentation. More or less traditional is that of Barrell (1906) in which sediments are classed as continental, littoral, or marine. Each of these major realms is subdivided into lesser categories. Others have sought to determine or define those realms of sedimentation which correspond to and explain the larger natural consanguineous associations of sediments, such as flysch and molasse, which recur throughout the geologic record. The geographic-geomorphic classification of Barrell as modified by Twenhofel (1950) does not do this, nor does it reveal any correlation between observable petrologic rock types and environment of deposition. More recent efforts to classify sedimentary realms according to different principles include that of Tercier (1940), for example, who believed there were more natural categories, some of which include both marine and nonmarine environments and also both deep- and shallow-water zones. Other efforts to set up realms of sedimentation include that of Weeks (1952), who developed a classification based primarily on the architecture of the basin of sedimentation, a scheme which is not, therefore, strictly a classification of sedimentary environments, although there is a presumed correlation between the nature of sedimentary deposits and the kind of basin in which they accumulate.

Clearly we have two problems; failure to state them correctly has led to much confusion. One deals with immediate *local* environments responsible for a particular bed or a sequence of beds; the other has to do with a larger assemblage of strata and the nature and tectonics of the greater basin in which this assemblage has accumulated.

The question of environmental classification has received little formal consideration (Twenhofel, 1950, p. 55; Selley, 1970, p. 2; Dunbar and Rodgers, 1957, p. 28). Most classifications adopted are modified from that of Barrell (1906), and the principles utilized in defining various classes are never made explicit. There are a few important exceptions in which an attempt was made to analyze the problem as well as to present a solution (Crosby, 1972; Krumbein and Sloss, 1963, p. 250; Laporte, 1968; Kukal, 1970b, p. 11; LeBlanc, 1972).

Even if we accept the concept of the local environment as a particular set of physical, chemical, and biological variables that correspond to a geomorphic unit of stated size and shape, we face problems of classification. We tend to mix terms involving place, medium of deposition, process or agent, and even deposited material into our nomenclature. We distinguish, for example, marine and nonmarine and thus emphasize the medium; or we talk about a coral reef or an oyster bank, a terminology reflecting the material deposited. Or we speak of the glacial or the eolian environment, referring, of course, to the agent of deposition. Even if we are careful to adhere strictly to the geomorphic concept of environments, we have difficulties. The deltaic environment is in fact an environmental complex, embracing as it does distributary channels, levees, interdistributary marshes, lagoons, and lakes, abandoned meander loops or oxbows, and various subaqueous pro-delta environments. We need, therefore, a classification of environments of sufficient generality and flexibility to encompass all second-order subenvironments. This classification should set up categories that are related in some meaningful way to the local stratigraphic column. Each major category should correlate with a particular sedimentary process which in turn is related to a particular stratigraphic motif or sequence of beds. It is perhaps not yet possible to frame such an inclusive classification, because it is not yet clear how many such models there are. The model concept is defined and analyzed further in this chapter.

We present here, therefore, a compromise classification—a compromise between the ideal classification, whose subdivisions can be correlated with observed sequences of the stratigraphic record, and other subdivisions intuitively chosen. Some of the categories relate to such cycles or motifs as shown by actual field studies; other categories have not yet been shown to be so identified (Table 15-3).

Environments are gradational; boundaries between them may not be sharp. An alluvial fan merges into a braided stream; a lake or lagoon passes into a marsh. Moreover, some environments are closely related to one another; others are not. The alluvial environment encompasses closely related alluvial fans, the floodplain of a braided or a meandering river, and the delta at the terminus of the river. There is little relation, however, between a turbidite basin and a playa lake. We have, therefore, in our classification tried to group the several environments into larger categories that express their relationships. Hence, we have the alluvial environment in which there are several closely related environments, each of which is presumed to leave a characteristic vertical sequence of beds. Second are those environments of the shore zone which can be subdivided into two somewhat coherent subgroups, namely clastic shorelines (involving tidal flat, lagoon and lagoonal marsh, and barrier beach) and carbonate shorelines (with carbonate tidal flat, hypersaline lagoon, and barrier reef). A third major category is the marine, both the shallow marine seas and the deeper basins (the latter is the recipient of either clastic or carbonate turbidites). Two other major

TABLE 15-3. Classification of sedimentary environments

Major Group	Environment	Subenvironments
Alluvial	Alluvial fan	Distributary channels, interchannel areas
	Braided river	
	Meandering river	Channel, pointbar, levee, floodplain, floodplain lakes and swamps
	Delta	Pro-delta, delta front, delta margin, delta platform
Shore zone		
Clastic	Tidal flat	Channels, levees, tidal deltas, tidal marsh
	Lagoon and estuary	
	Beach and barrier island	
Nonclastic	Tidal flat	Channels, levees, tidal deltas, tidal marsh
	Barrier reef	Reef proper, backreef, reef apron
	Lagoon (often hypersaline)	Patch reefs and the like
Marine	Shallow marine	Platform, platform margin
	Turbidite basin, noncarbonate	Subsea fan, fan valleys, levees, distal area of ponding
	Turbidite basin, carbonate	(As for noncarbonate)
Inland basin	Freshwater lake	Various subenvironments
	Brine lake, playa lake, salt flats	
	Dune field	
Glacial	Continental glacial	Subglacial, outwash aporn, kames and eskers, glaciolacustrine
	Glaciomarine	Wet base, dry base

groupings seem necessary: the inland basin which under humid conditions is a fresh-water lake and under arid conditions is a brine lake or perhaps an ephemeral playa lake and salt flat or even a dune field; and a second and unrelated category, the glacial environment including both continental glacial environments and perhaps the more important glaciomarine environment. We realize that in our classification we have used diverse parameters for defining the several categories: climate, composition, and agent as well as geomorphic entity. This somewhat illogical system, however, is geologically meaningful in that each environment leaves a recognizable *stratigraphic signature*.

We have thus defined 11 major environments, 14 if the carbonate tidal flat, barrier reef, and carbonate turbidite basins are considered separately. It is to be remembered, however, that there are many subenvironments in each of these categories. The reef, for example, has a core, a forereef apron, and perhaps, if emergent, both beach and dune environments. All of the other environments can likewise be subdivided.

For some of the most important environments we have, in recent years, developed a model and have recognized a characteristic vertical sequence of lithologies and structures—a stratigraphic signature—which identifies the environment in question. This is particularly true of the meandering river, the delta, the carbonate tidal flat, the clastic barrier beach, the carbonate reef, and the clastic and carbonate turbidite basins. For other environments, such a well-worked-out model does not exist, and the vertical sequence is largely hypothetical or at best poorly known.

FACIES MODELS AND ENVIRONMENTAL RECONSTRUCTION

INTRODUCTION

The "new" approach to environmental analysis arises from the study of Recent depositional sequences, which shows that major environments of sedimentation and their subdivisions are organized in a structured coherent manner. This approach is in reality not so new; the underlying principles involved were stated long ago by Johannes Walther (1894, p. 979), who noted that only those sedimentary deposits which are laterally adjacent can be superimposed conformably one upon the other. As a result, the vertical sequence of sedimentary facies is not random but is, instead, organized in a coherent and predictable way. In recent years, mainly as a result of three-dimensional studies of modern or Holocene sediments, disclosed by drilling and deep excavations, there has been increasing emphasis on the study of their vertical organizations of lithologies and structures—the "lithologic package." These characteristic sequences can be seen also in the ancient record and constitute the best guides to paleo-environment. We need not, therefore, rely solely on the interpretation of individual facies with their assortment of textures and structures, but rather on the vertical arrangement itself which places the individual facies and their sedimentary structures in a meaningful sequence or package. Recognition of such a pattern in the vertical profile enables us to reconstruct a detailed picture of interacting, interrelated subenvironments that cannot be obtained by simple lithologic comparison with the incomplete Recent record. One of the first clear statements of this approach is that of Visher (1965); a more recent resumé of this philosophy, based on early work of the Shell research group, is that of LeBlanc (1972). The textbook by Selley (1970) on recognizing ancient sedimentary environments also uses essentially the same approach. As noted by Visher, this approach relates the vertical profile and its ordered arrangement to a process such as delta progradation. The characteristic sequence which records such a process may be many times repeated, giving rise to a series of superposed cycles—cycles in many cases generated by the process itself and not imposed by outside events. These are the autocyclic sequences of Beerbower (1965, p. 32). Recognition of sequences of autocyclic origin is perhaps the most important way of identifying the depositional environment. One is concerned not only with the sequence of lithologies but also with the sequence of sedimentary structures seen in the vertical profile. The larger question of cycles and their recognition, classification, and origin has been summarized by Duff, Hallam, and Walton (1967).

This newer approach to environmental analysis imposes important obligations on the student of Recent sediments. It is no longer sufficient to make epidermal sedimentological studies—to collect samples from the most surficial part of the deposit for laboratory analysis. It becomes essential to determine what the vertical profile is like, what sequences are generated in the present-day environment. This implies a three- rather than two-dimensional approach to the problem.

This approach has given rise to the *facies model* concept which is no more than giving a formal name to the conceptual scheme linking the evolution of a particular sedimentary environment and the particular vertical sequence which records this evolution. Visher (1965) recognized 6 such models; Blatt, Middleton, and Murray (1972) identify 10. The total number is not unlimited. We will review here only those which are best known and which explain the largest part of the sedimentary record. Three are clastic models, namely turbidite, alluvial (point bar), and deltaic, and three are nonclastic, namely reef, tidal flat, and sabkha. In the Paleozoic of the central Appalachians, for example, nine-tenths of the clastic record can be explained in terms of deep-water turbidite deposition (Martinsburg and "Portage"), alluvial or point bar (Catskill, Pocono, Mauch Chunk, Pottsville), and prograding deltaic (Montebello). The very impressive Lower Paleozoic carbonate deposits (Conococheague, for example) are in large part tidal-flat sequences.

In our pursuit of environmental reconstruction by use of the *vertical profile*, we should not lose sight of the importance of the gross geometry of the sedimentary bodies involved. This aspect also is an outgrowth of the three-dimensional study of Holocene deposits and one which could not readily be attained by the usual two-dimensional approach to the study of such sediments. The form and configuration (geometry) of sand bodies, for example, may be a clue to the nature of their environment or deposition. The overall shape of a sedimentary facies is a function of the geomorphology of the depositional environment. Radiating and diverging distributaries of a delta and the shoestring offshore bars of other environments are illustrative of this principle. (The geometric aspect of the subject has been explored in Chapter 5.)

In this section, as elsewhere in this book, we begin with what we can see in the outcrop and tailor our discussion, insofar as possible, to those aspects of the modern world and processes which help us understand what we actually see. Emphasis, therefore, is on the rock record, because this is what we begin with and what we end with.

In summary, identification of the environment of deposition of a particular sedimentary formation depends not so much on a particular textural attribute or even a structure as it does on a characteristic sequence or package of subfacies. The particular sequence is the result of a process, such as progradation of a shoreline or meandering of a stream, which causes lateral displacement of the subenvironments in a regular manner. Such lateral migration is recorded as a sequence of subfacies, the latter being the earmark of the major environment of deposition. This concept, whose recognition has gained acceptance in the last decade, has done more to advance our interpretative ability than any other single approach or technique.

Identification of a particular paleo-environment is made possible by recognition of a characteristic motif (Walker and Harms, 1971) or ordered sequence of lithologies generally regularly repeated with, perhaps, some variations, generated by that environment. The alluvial or point-bar sequence, for example, is a fining-upward cycle; the prograding shallow-water delta generates a larger-scale coarsening-upward sequence. Our emphasis here, therefore, is on those sequences or motifs, because these are what the geologist encounters in the field, and they are the raw data upon which environmental interpretation rests. We begin our discussion of each environmental model, therefore, with a description of its characteristic stratigraphic signature and then proceed to a brief analysis of the manner in which these cycles were generated. These topics are the meeting ground of sedimentology and stratigraphy.

The identifying stratigraphic pattern may be self-evident in well-exposed sections or in drill cores. Such patterns may not be so evident in other cases because of wide departures from the norm. To guard against subjective bias in marginal cases, some kind of objective analysis of the data and construction of appropriate "tree" diagrams may be desirable (Selley, 1969; Allen, 1970, Fig. 4). Not all stratigraphic sections yield to this approach. There may be cyclic patterns not yet understood; or there may be no identifiable patterns at all. Environmental analysis has not yet attained perfection, even though great strides have indeed been made during the last two decades by use of the motif principle.

ALLUVIAL ENVIRONMENTS

Fluvial processes include both erosion and deposition. We are concerned mainly with the latter. Included here are coarse alluvial deposits formed at the edge of sharply elevated tracts at the places where streams emerge from their canyons or valleys. These *alluvial* fans merge with alluvial deposits which floor large alluvial basins—the fluvial deposits either of overloaded *braided streams* or of *meandering streams.* Large rivers eventually discharge into the sea or other large bodies of water where, under favorable conditions, they build *deltas.* Each of these alluvial environments is complex and consists of subenvironments. Deposition is marked by construction of characteristic *sequences,* which in each member has its own peculiar geometry and constellation of sedimentary structures.

Alluvial fans Alluvial fans are found adjacent to mountain ranges or high hills. They are by far the most conspicuous and largest in arid and semi-arid regions of bold relief.

Modern fans have been studied by Blissenbach (1952, 1954), Bluck (1964), Denny (1965), and Hooke (1967). The papers by the last two are notable for their detailed maps of fan surfaces and for their discussions of microrelief and materials on the fan surface, and the evolution of the fan. Ancient fan deposits have been discussed by Lawson (1925), Bluck (1965), and Bull (1972), among others. The general nature of fans and fan deposits has been summarized by Medeiros, Schaller, and Friedman (1971), LeBlanc (1972), and Bull (1972). The paper by Bull is the most complete.

Characteristic of alluvial fans is the intermittent torrential flow—under the highest energy conditions within the entire realm of sedimentation, the deposition of clastic sediment in close proximity to its source, the great range of sizes of constituent materials, and the very poor sorting.

The size of individual fans is dependent on size of drainage basin, slope, climate, and character of the rocks in the source area. Individual fans have a radius ranging from a hundred meters to several tens of kilometers. Coalescing fans form great aprons of waste several hundreds of kilometers in length. The fan surface consists of a series of diverging and anastomosing distributaries—"washes" or channels, some active and some abandoned—which are incised into a more even fan surface or "desert pavement" (Fig. 15-6). In general the upper part of the fan is more largely a pavement with one or more incised channels from 1 to 10 m deep; the lower part of the fan is characterized by coalescence of active washes. These pass downslope into braided streams or into a playa-lake environment.

Clastic debris is transported by torrential but short-lived flows. In some cases the current is so sediment-laden as to be a mudflow (or "debris flow" of Hooke). Such flows follow channels, may overtop them, and form lobate tongues. These deposits are very poorly sorted and unstratified and have sharp boundaries. More sustained water flows carry debris of all sizes; the coarser gravels are angular and poorly rounded. Under some circumstances water flows de-

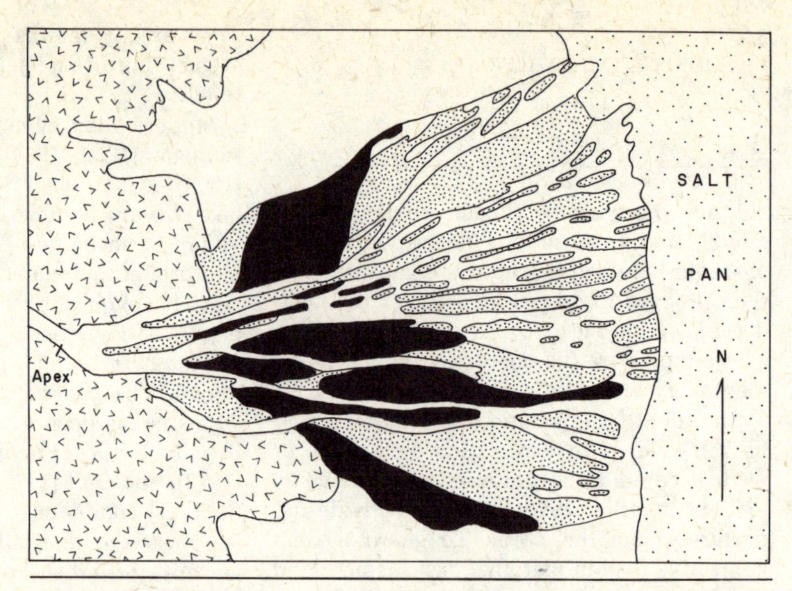

FIG. 15-6. Train Canyon fan, Death Valley, California. Note active present-day washes, abandoned washes, abandoned washes with desert-varnished gravels, and oldest fan surface covered by weathered gravels. (After Denny, 1965, Pl. 5.)

posit very coarse, openwork gravels, that is, gravels lacking fines, which Hooke (1967, p. 454) called "sieve deposits." Such deposits function as sieves permitting water to pass through. The fines collect upcurrent from such channel barriers.

Fan gravels are generally poorly sorted, are angular to poorly rounded, and consist of whatever debris the source region will provide. The water-laid gravels may be imbricated, massive, or thick-bedded. Fine-grained sediments, deposited further downslope, may be cross-bedded, massive, laminated, or thick-bedded. Debris flow or mud-flow deposits are very poorly sorted angular pebbles and cobbles set in a mudstone matrix. Viscous flows may have a rude clast fabric. The cobble size of fan materials declines rapidly downcurrent, generally exponentially (Blissenbach, 1952; Bluck, 1965, Fig. 8) but not in all cases (Denny, 1965). See Fig. 15-7. A transverse cross section of a fan would disclose beds of limited extent that are interrupted by cut-and-fill structures; longitudinal sections show greater continuity of beds.

Alluvial fan deposits ("fanglomerate" of Lawson, 1925) have been reported from the geologic record. Some of these have been tabulated (LeBlanc, 1972, p. 145) and their characteristics summarized (Bull, 1972, pp. 79–82). Among the better known ancient fan deposits are the Precambrian Keweenawan red conglomerates and sandstones of northern Michigan, the Precambrian Torridonian of Scotland (Williams, 1969), the Pennsylvanian Fountain Formation of Colorado (Hubert, 1960; Howard, 1966), and the Triassic Newark Group of the Appalachian region (Reinemund, 1955; Krynine, 1950). See Fig. 15-8.

Braided rivers The geometry and nature of deposits made by braided streams are less well known than those made by meandering streams. One of the most thorough studies of a present-day braided stream is that on the Donjek River in the Yukon by Williams and Rust (1969). Doeglas (1962) has also described the structures of braided river deposits, as have Ore (1963), Smith (1970), and Kessler (1971).

Braided streams show an interlaced network of channels of low sinuosity. They are apt to occur on steeper slopes and are formed by streams with higher discharges than those of meandering rivers; they occur on large alluvial fans in arid regions and in periglacial areas. Seemingly necessary for their development is an overload of sediment and little vegetation to stabilize banks. Channels are choked with their own detri-

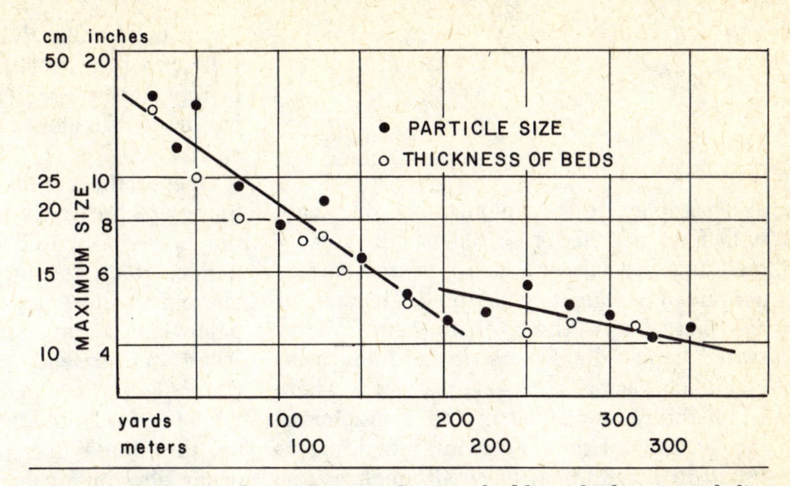

FIG. 15-7. Relations between particle size, bedding thickness, and distance on a presumed Triassic alluvial fan. (After Bluck, 1965, *Sedimentology*, v. 4, Fig. 8.)

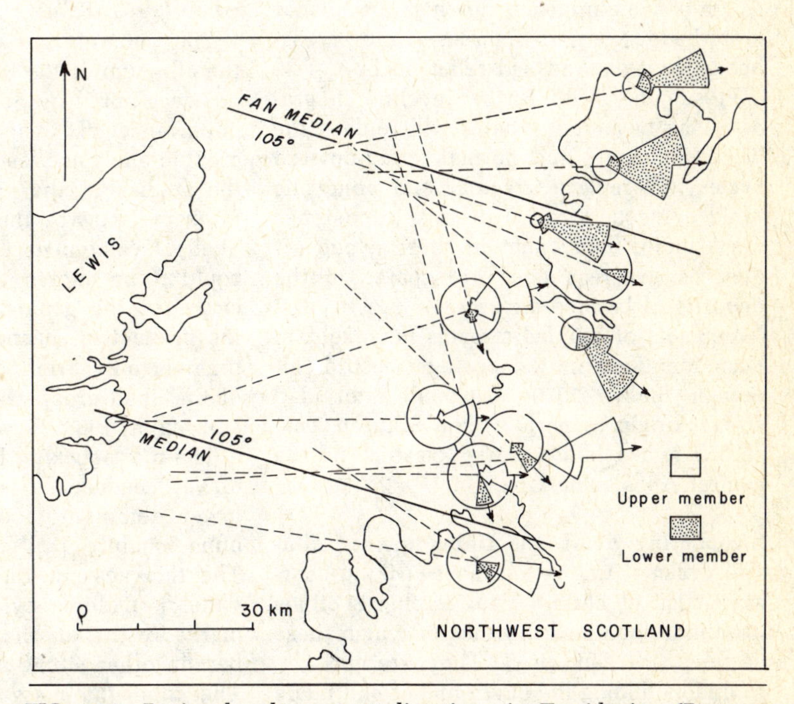

FIG. 15-8. Regional paleocurrent directions in Torridonian (Precambrian), Applecross Formation, Scotland. Radial patterns are related to alluvial fan deposition. (After Williams, 1969, Jour. Geology, v. 77, Fig. 2.)

tus which tends to accumulate as longitudinal bars in the center of channels. Repeated bar formation and channel bifurcation leads to an anastomosing network of channels and intervening bars. The alluvium of a braided river is typically sand and gravel with little overbank silt and clay. There is an absence of the normal fining-upward sequences which characterize meandering rivers. A small amount of silt may collect in abandoned channels.

Based on a study of the South Platte–Platte river system, Smith (1970) believed that braided streams show significant proximal–distal differences. The braided pattern is superposed on rivers which have a natural tendency to fractionate their bed load, with finer and better sorted particles accumulating downstream. The types of bars in a braided stream depend on grain size of the bed load. Where coarse and poorly sorted longitudinal bars form, finer sediments characterize transverse bars. There is, therefore, a change in the character of the bedding. Crude horizontal bedding is found in longitudinal bars, although coarse cross-bedded gravel accumulates on the lee side of shifting braid bars; transverse bars generate planar cross-stratification. Hence, as grain size diminishes downstream, planar cross-bedded sands increase and cross-bedded gravels and bed relief decline.

Presumably with further decline in gradient the meandering habit will replace the braided channel and normal fining-upward cycles will replace irregular and noncyclic braided sequences. Overbank mudstones are a major component in these cyclic sequences, whereas fines are sparse in the deposits of braided streams.

Deposits of braided streams have seldom been identified in the geologic record, although Smith (1970) assigned a braided stream origin to parts of the Silurian Tuscarora and related coarse clastics in the central Appalachians.

Meandering rivers and fining-upward alluvial cycles Among the most striking characteristics of certain stratigraphic sections are the fining-upward cycles of which these sections are comprised. They are the key to understanding the environment and mode of deposition of very thick stratigraphic sections. These cycles were noted by Dixon (1921) in the Old Red Sandstone of southern Wales and by Bersier (1959) in the molasse deposits of the Alpine foredeep. They are, in fact, the most characteristic feature of these deposits and appear as sequences beginning with a sharp and locally erosion-based sandstone, passing upward into finer sands and silts and terminating with a mudstone. These cycles range from several to ten or more meters in thickness. This pattern of sedimentation has been observed in deposits of many different ages and places. It is one of the most fundamental motifs and is repeated vertically many times through hundreds of meters of progressively younger strata.

One of the earliest good descriptions of fining-upward cycles is that of Allen from the Lower Old Red Sandstone of England (Allen, 1962, 1965b). Fining-upward cycles abound in the Paleozoic of the central Appalachians. They have been described by Allen and Friend (1968) in the Catskill (Devonian); they occur in the Pocono and the Mauch Chunk (both Mississippian) and in the Pottsville (Pennsylvanian). They occur also in much of the deposits of the Newark Series (Triassic) of the same region. The known occurrence of fining-upward cycles has been reviewed and tabulated by Allen (1970, Table 1).

The general fining-upward character of the alluvium of the River Klarälven in Sweden was noted by Sundborg (1956), but the importance of this observation to stratigraphic analysis was not realized. Not until the studies of the alluvium of the Brazos River of Texas by the research group of the Shell Development Company in the 1950s could it be shown that the fining-upward cycles of the geologic record were indeed the product of meandering streams. Unfortunately only brief resumés of the findings of the Shell group were published (Bernard and Major, 1963; Bernard et al., 1970).

The most striking feature of these fining-upward sequences is their division into a lower dominantly coarse member and an upper mainly pelitic member (Fig. 15-9). The thickness of the two members varies widely; in some cycles the lower unit is markedly thicker than the mudstone member; in other situations the reverse is true. The total thickness of a cycle is highly variable. In the Old Red Sandstone of Anglesey the average is about 18 feet (5.5 m); in the Mauch Chunk (Mississippian) near Pottsville, Pennsylvania, the average is 50 feet (15 m).

The lowest bed in the cycle is a very coarse sandstone, in some cases conglomeratic. Invariably some intraformational shale clasts are present. Exotic gravels, mainly quartz, may also be present. The basal contact is sharp, to some degree erosional. The conglomeratic member is overlain by a rather thick sequence of sandstone beds. These beds may be massive; in many, large-scale cross-bedding is present. The cross-bedding may be in solitary or in grouped sets. In some instances flat-bedded sandstones are also present. Sandstones become thin-bedded and finer-grained upward and pass into fine-grained sandstones or siltstones which commonly show ripple cross-lamination. These beds may alternate with the mudstones which form the upper part of the cyclic unit. Much of the pelitic member is in fact siltstone rather than shale. The siltstones are commonly red, whereas the basal sandstones of the cycle are not. The siltstones lack fissile parting and have instead a rough subconchoidal fracture. Bedding laminations are scarce or lacking. Exceptionally some fissile shales are present. Small knobbly concretions of calcareous to dolomitic composition may occur in the mudstone member of fining-upward sequences. They tend to be scattered in the rock or concentrated in thick bands. In general, the proportion of concretionary materials increases upward in the cycle. They become so abundant that they coalesce and form a thin impure limestone bed (Allen, 1965c).

As noted, these fining-upward sequences are assigned to a fluviatile origin—a meandering river—by analogy with deposits of modern streams. Coarser members resemble very closely the deposits of the channel and point bars; fine-grained members are thought to represent floodplain deposits. The calcareous component is interpreted as a caliche-like accumulation. Those cycles which have a red pelitic member and abundant calcareous matter probably formed under arid or semi-arid conditions. As can be seen in Fig. 15-10, the sequence is generated by lateral migration of the river channel across the floodplain. Point bars are constructed by lateral accretion; consequently the entire section from the base of the channel to the floodplain is deposited simultaneously. The thickness of the sequence is determined by the depth of the stream channel during flood stages and hence is related to the size of the river itself.

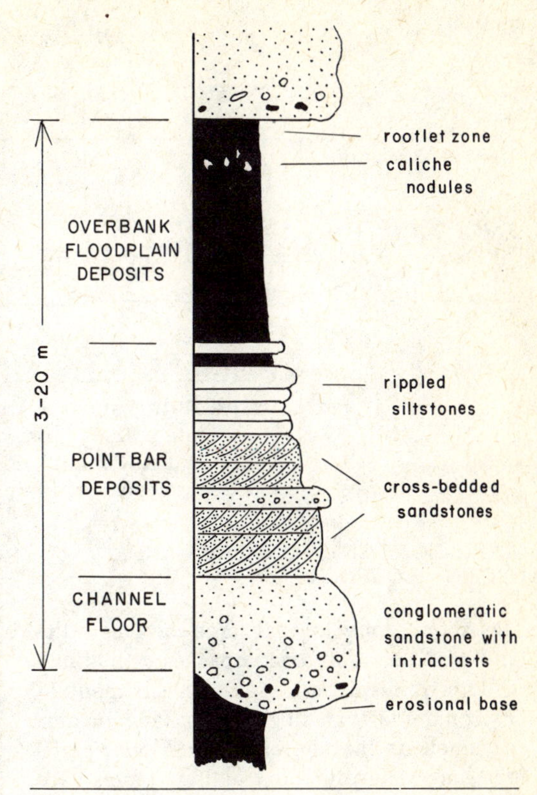

FIG. 15-9. Somewhat idealized fining-upward alluvial cycle.

Because those sections in which the point-bar facies is the dominant pattern of accumulation contain many such cycles, we must conclude that channel establishment and abandonment was repeated many times at a given site. The cycles are thus autocyclic and attributable to factors intrinsic to the fluvial regime and not to external factors such as base-level changes and tectonism. Nevertheless, their superposition one upon the other could not occur were it not for continued subsidence in the area involved.

As noted above, Allen (1970, Table 1) has tabulated the known occurrences of stratigraphic sections dominated by fining-upward cycles. Papers giving extended discussion of these cycles include those of Allen (1962, 1965b) on the Old Red Sandstone of England, Allen and Friend (1968) on the Devonian Catskill of the central Appalachians, Bersier (1958) on those of the Alpine molasse, Meckel (1970, pp. 61–65) on the cycles in the Pennsylvanian Pottsville in the Appalachians, and many others.

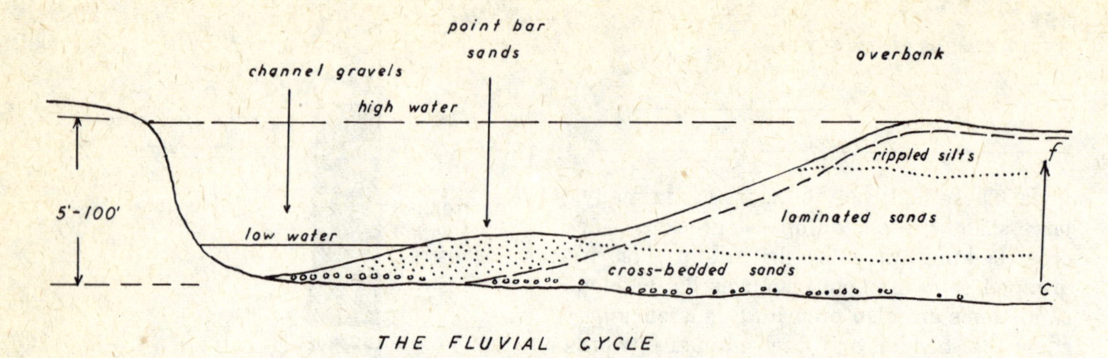

FIG. 15-10. Lateral migration of river channel and point-bar deposition generate a fining-upward cycle. (After Visher, 1965, Soc. Econ. Paleont. Min. Spec. Publ. 12, Fig. 13. By permission of Society of Economic Paleontologists and Mineralogists.)

SHORE-ZONE ENVIRONMENTS AND CYCLIC SEQUENCES

The shore zone is that area between the land and sea that encompasses something of both. It displays a greater assortment of environments than does either land or sea. Inasmuch as the shoreline is an ever-shifting line, sharply contrasting facies are intertongued one with the other often in a regular or cyclic manner.

The nature of shore-zone deposits depends on the kind and volume of sediment derived from land. Maximal sedimentation occurs at the mouths of large rivers, and the most impressive volumes of sediments occur in *deltas*. Wave and current dispersal of sediments, however, transfer the sediment influx longshore so that important accumulations occur on the seaward edge of the interdeltaic coastal plain—such as the barrier island–lagoonal complex. In situations where land supplies little or no clastic materials, because of very low relief or very arid climate, shore deposits are carbonates or even evaporites.

The ever-changing relation of land to sea and the resulting transgression and regressions—especially the latter—lead to orderly sequences such as those of the prograding delta, the prograding beach or barrier island, the carbonate tidal cycle, the sahbka cycle, and so forth.

PROGRADING DELTA AND COARSENING-UPWARD SEQUENCES

A number of authors have described rather large-scale (up to 100 meters) coarsening upward sequences. A number of such cycles may be stacked one upon the other. In brief, these sequences begin with rather fine-grained dark shales and end with coarse-grained and locally gravelly sandstones. Marine fossils are generally present in all members. These cycles record the progradation or seaward growth of a shallow-water delta. In some cases the delta cycle is topped with fluvial fining-upward cycles. Delta cycles have been observed and described in detail in the Lower Westphalian (Carboniferous) of north Devon by de Raaf, Reading, and Walker (1965). They have also been seen and carefully studied by Kaiser (1971, 1972) in the Montebello Sandstone (Middle Devonian) of south-central Pennsylvania. The same type of coarsening-upward sequence has been deposited in the present-day delta of the Brazos River of Texas. It has been described in detail by Bernard et al. (1970).

Devonian cycles of south-central Pennsylvania are fairly typical of this genre of sedimentation (Fig. 15-11). The ideal cycle begins with a considerable body of rather fissile dark shale. Fossils, if any, are thin-shelled pelecypods. A few thin (5 to 25 cm) fine-grained sandstones may be present. These form even-bedded, continuous layers. They are internally laminated; some contain displaced benthonic fossils. These flat-bedded sandstone layers are interpreted as turbidite sands derived from the steeper delta front area. Lower shales pass gradually upward into thin-bedded siltstones, rocks generally devoid of sedimentary structures. Some are intensely burrowed, are bioturbates without any original stratification preserved.

These siltstones become thicker bedded and coarser, all interbedded shales drop out, and they pass upward into sandstones which show undulatory bedding and, in places, cross-bedding. Fossils in these sandstones are thick-shelled brachiopods. The uppermost part of the sandstone member of the delta cycle may be pebbly and may display cut-and-fill structures, even channels and channel fill. These are pebbly to conglomeratic. The uppermost meter of the sandstone member is apt to be extensively riddled with vertical burrows. In some of the Montebello cycles, this member is separated from the dark shales of the next overlying cycle by a thin (10 to 30 cm) ferruginous bioturbated bed containing chamositic ooids and phosphatic granules and a rich varied marine fauna.

The character of the sandstone reflects, to some degree, shoaling waters and increasing turbulence. Not only does the grain size increase upward in the cycle, but so does the maturity of the sand. The uppermost sands are the cleanest and best rounded. Weaker rock particles have been eliminated and the sand enriched in quartz. Mapping of the cross-bedding reveals a radial dispersion pattern in keeping with the delta concept of origin on the deposit.

The coarsening-upward cycle is interpreted as a delta cycle generated by gradual progradation of a delta into shallow-marine water (Fig. 15-12). The initial water depth was about equal to the thickness of the cycle—30 to 100 m more or less. The initial shales are the pro-delta clays, deposited from suspension in quiet water below wave base. Periodically the pro-delta environment was invaded by a turbidity flow descending from the delta margin area bringing in fine sand and displaced shelly debris. The pro-delta clays give way to delta-front silts and ultimately to delta-front sands. Progressive shoaling of the water led to change in bedding thickness and textures. An active benthonic fauna ingested bottom sediments and destroyed the more delicate bedding laminations. Only in turbidites—products of near-instantaneous sedimentation—are such laminations preserved. Ultimately shoaling and

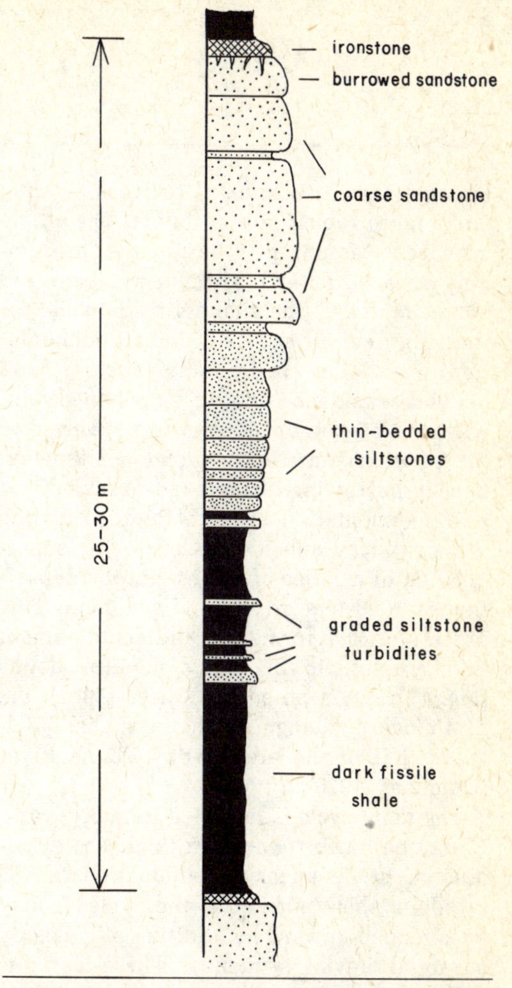

FIG. 15-11. Coarsening-upward, prograding delta cycle, Montebello Member, Mahantango Formation (Devonian), Pennsylvania. (Based on measurements by students, Johns Hopkins University.)

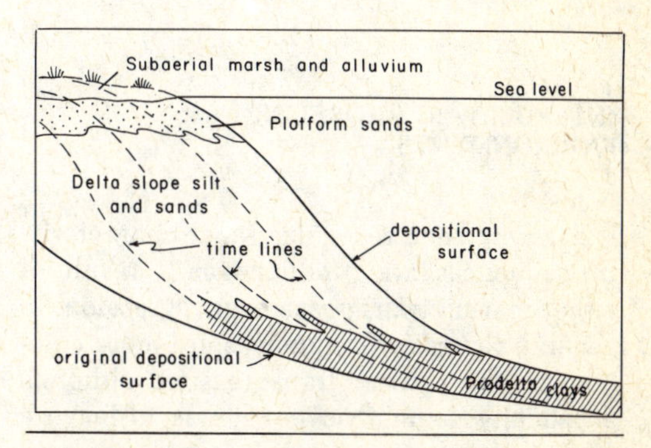

FIG. 15-12. Seaward migration of depositional environments in high constructive deltas. (From Scruton, 1960, Recent sediments of northwestern Gulf of Mexico, Fig. 9.)

stronger current action led to cross-bedding and a more robust bottom fauna. The whole sequence was abruptly terminated, presumably because of an upstream avulsion and abandonment of the channel responsible for the outgrowth of the delta lobe. Sedimentation was followed by nondeposition and gradual subsidence only to be resumed later when the stream once again was diverted to the same region. During nondeposition extensive burrowing of the sand occurred and iron sedimentation ensued. Delta switching and continued subsidence led to the superposition of a series of cycles and the deposition of 1000 feet or more of sediment. This process of lobe formation and abandonment is characteristic of many modern deltas (Fig. 15-13). As noted by Kaiser (1972), the best modern analogue of the Montebello deltas is that of the present-day Rhone River (Oomkens, 1970).

The ideal cycle is modified in many ways. It may be much reduced in thickness by reduction in thickness or elimination of the pro-delta shale portion of the cycle. It may be extended upward by addition of a strictly subaerial fluvial sequence. These variants are related to frequency and timing of delta switching and regional subsidence. Refer to the literature for a further elaboration of these and other details (de Raaf, Reading, and Walker, 1968; Kaiser, 1972). Other examples of coarsening-upward deltaic cycles have been described by Okada (1971), Okada and Fujiyama (1970), and Reading (1970).

INTERDELTAIC SHORELINE ENVIRONMENTS

It has long been recognized that characteristic sequences are produced as a result of either marine transgression or regression. In general, regressive sequences are more common in the record; transgression, although an equally common event, tends to leave a meager or no depositional record. We will consider first the environments and clastic deposits related to regression.

Regression involves encroachment of the land on the sea, a result of arrested or very slow subsidence accompanied by active sedimentation. Two models are possible. One is the deltaic model, discussed above, a model involving very rapid sedimentation. The marine environment is overwhelmed by a flood of sediment which forms a rapidly prograding delta. The other model is the interdeltaic barrier-beach complex. Although sedimentation is less aggressive, the end-result, as in the case of the delta, is an orderly succession of deposits beginning with shallow marine silts and clays, followed by the shore facies—the sands of the barrier beach—in turn followed by deposits of the tidal lagoon and marsh, and ultimately by terrestrial alluvial deposits. We will review briefly the characteristic sequences and the subenvironments which make up the interdeltaic regressive model.

The regressive sequence and the several environments represented have been extensively studied both in the present-day and the ancient record. Studies of recent alluvial and deltaic plains and the barrier-island complexes include those of the Shell research group in Texas (Bernard et al., 1970), the Dutch group in the Netherlands (van Straaten, 1954, 1961), and the German school at Wilhelmshaven (Reineck, 1970). A good recent summary of barrier-island coastlines is that of Dickinson, Berryhill, and Holmes (1972).

The interdeltaic complex occurs along linear coastlines between deltas and comprises mud flats and cheniers (abandoned beach ridges) of the chenier plains and the barrier-island, lagoon, tidal-channel complex. It also occurs along the seaward edge of a coastal plain drained by numerous small streams but devoid of any sizable deltas.

Most of the sediments are derived from land, but minor amounts come from the marine environment. Sediment brought to the sea by rivers and small streams is dispersed laterally by marine currents. Sands migrate in the shore zone and accumulate as beach ridges and barrier-beach islands. Mud is carried in suspension and in part swept out to sea and in part trapped in the lagoon-marsh areas between the barrier islands and the land area proper.

Barrier beach The barrier island separates the shallow-marine environment on one side

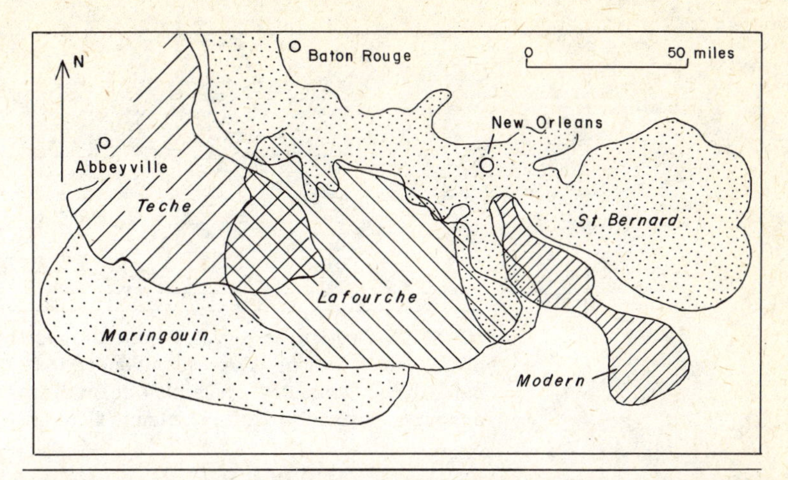

FIG. 15-13. Delta lobes of the Mississippi Delta system. (From Fisher and McGowan, 1969, Bull. Amer. Assoc. Petrol. Geol., v. 53, Fig. 6.)

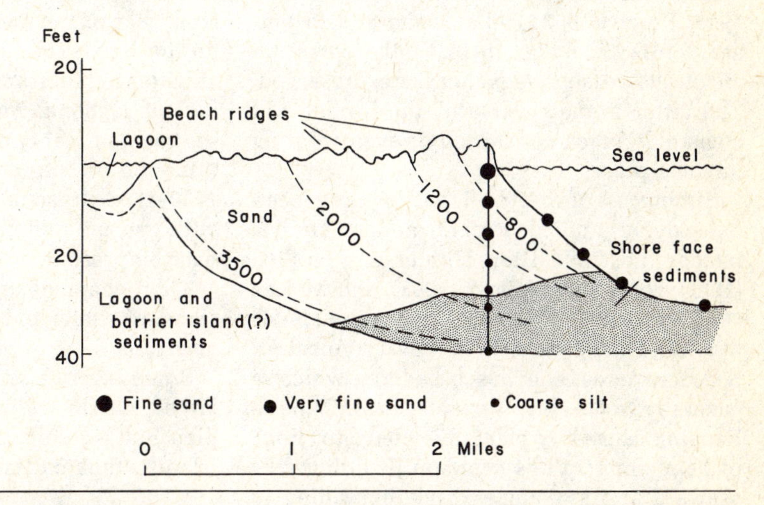

FIG. 15-14. Cross section of Galveston Island, Texas, a prograding barrier island. (After Bernard et al., 1962, Geology of Gulf Coast and central Texas, Houston Geol. Soc., Fig. 12A.)

from the lagoon and marsh behind the barrier. The seaward face of the barrier is a zone of sand accretion and therefore grows seaward. Coarser sands are deposited on the beach and upper shoreface, and finer sands are deposited in the lower shoreface area. Silt and clay accumulate seaward on the adjacent shelf bottom. As the barrier grows seaward, it encroaches on the shelf bottom and thus, like the prograding delta, generates a coarsening-upward sequence (Fig. 15-14). The emergent part of the barrier may be reworked by the wind, with resultant dune formation.

Tidal channels separate barrier islands, and at flood tide sand is swept into the lagoon through such channels. Because the current diminishes as it enters the lagoon, transported sediment builds a landward-facing *tidal delta.*

If the region as a whole is subsiding slowly, the barrier not only builds seaward but also upward. The landward, older part of the barrier subsides and is, in turn, invaded by the lagoon-marsh complex. These processes generate the typical regressive sequence (Fig. 15-15). Beach sands thus form an expanding sheet—the present barrier being the growing edge—which transgresses time planes. One of the most frequently cited ex-

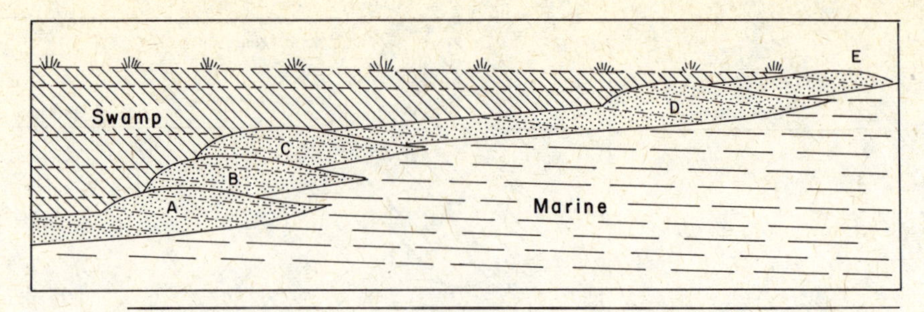

FIG. 15-15. Growth of regressive sand sheet by seaward migration of barrier beach during slow subsidence. (After Hollenshead and Pritchard, 1961, Geometry of sandstone bodies, Fig. 5. By permission of American Association of Petroleum Geologists.)

amples from the geologic record is provided by the Upper Cretaceous of the Rocky Mountains in the western United States (Spieker, 1949; Young, 1957; Hollenshead and Pritchard, 1961). See Fig. 15-16. Both regressive and transgressive sequences are preserved. The latter is the reverse of the former and commonly much condensed or even wanting in some places.

Structures of beach deposits have been extensively studied. Recent modern studies include those of Clifton, Hunter, and Phillips (1971), Hoyt and Weimer (1963), and McKee (1957, pp. 1706–1718); early work includes that of Thompson (1937). Most typical of beaches, perhaps, is beach bedding which is generally even but variable in thickness. Banding caused by placerlike concentrations of heavy minerals is common. Bedding dips seaward at a low angle, rarely exceeding 10 degrees. Scattered shell fragments are common in modern beaches. Although textural criteria (see Chapter 3) have been invoked to identify beach sands, the best clues are their position in a regressive sequence between marine shales and brackish lagoonal sands and shales and their internal structures. The Cow Creek Formation (Upper Cretaceous) is a well-described example of a "fossil" beach (Stricklin and Smith, 1973).

Lagoon and tidal marsh Behind the barrier beach is a lagoon or marsh in which finer sediments collect. This environment is subject to tidal influence and may be marked by marginal tidal flats with tidal channels. Depending on climate, location of tidal inlets, and other factors, lagoonal sediments vary in kind and volume. Generally the sediments are silts and muds with some sands near tidal inlets, and organic muck or peat (or coal in the ancient record) in marsh areas. If the waters are brackish, the fauna is different from that of the marine shales which underlie the barrier sandstone. Plant remains are abundant; bioturbation is common. In arid regions lagoonal waters are more saline and sediments notably different both in composition and in structures (McKee, 1957, pp. 1730–1738; Rusnak, 1960). Evaporites, muds with gypsum nodules, and oolitic carbonates may be present.

Tidal channels are marked by lag concentrates of shells and by intraformational mud clasts.

Some lagoons are invaded by deltas from larger streams discharging into them. In places these deltas may build out to barrier islands, thus cutting the lagoon in half.

Prograding muddy shorelines If, for whatever reason, sand is scarce, barrier islands do not form, and a mud flat and marsh lie between the land proper and the sea. If these muds contain sufficient sand, a low beach ridge forms on the seaward edge of the muddy platform by reworking and winnowing of these sediments. A new mud flat may accumulate seaward of this beach, which is then abandoned. Repeated influx of muds and reworking and beach construction produce a *chenier plain* which consists of a succession of parallel abandoned beach ridges (cheniers) separated by broad low marshy areas. The low sandy beach ridges are thus progressively younger seaward. The chenier plain and its origin and evolution have been described by Byrne, LeRoy, and Riley (1959) and by Gould and McFarlan (1959).

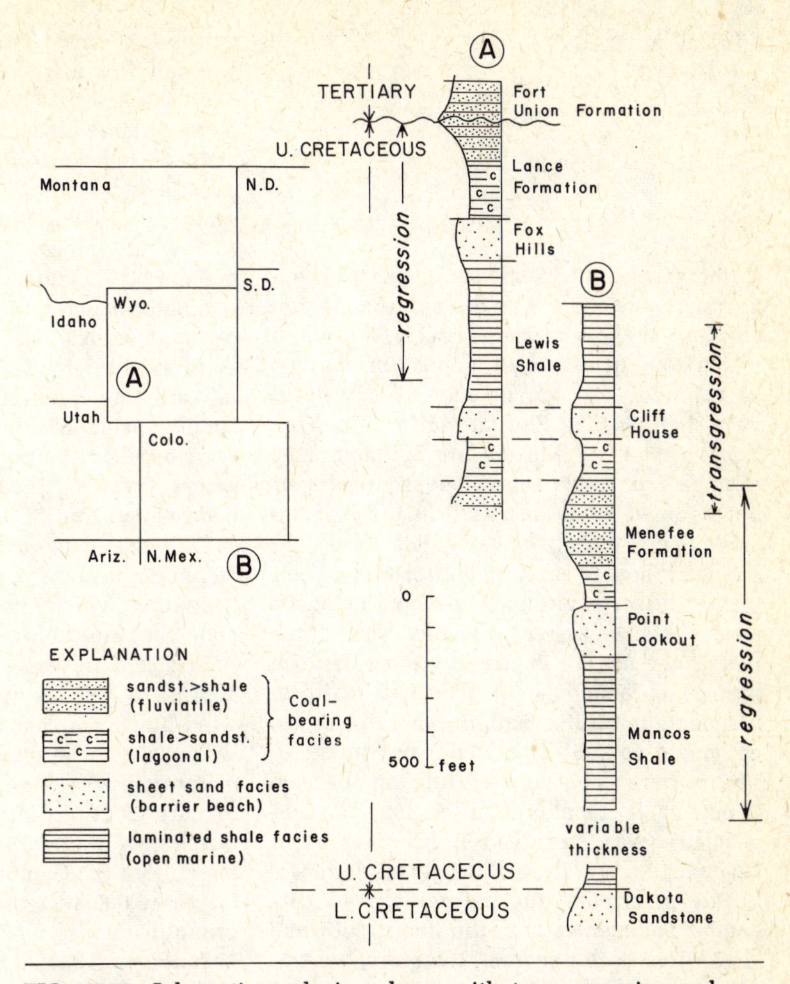

FIG. 15-16. Schematic geologic column with two regressive and one transgressive sequences, Upper Cretaceous, Wyoming. (After Selley, 1970, Ancient sedimentary environments, Fig. 6-3. By permission Cornell University Press.)

If sand is too scarce to form such beach ridges, the resulting deposits are almost wholly muds and silts, in part marine and in part nonmarine. These may alternate in a cyclic manner, as observed by Walker and Harms (1971) in the Upper Devonian of central Pennsylvania where some 25 of these sequences, termed *motifs* by Walker and Harms, have been described. They range in thickness from 4 to 45 m.

Tidal-flat and tidal sequences Some shore areas, especially estuaries, are markedly affected by tidal currents and are areas primarily of sand movement and accumulation. Such tidal action is especially marked in those estuaries where the tidal range is magnified and the tidal velocities very high. Under such conditions tidal sand bodies form. Some are lenticular; some are notably elongate. In general, grain size decreases upward, commonly from a basal lag gravel. Cross-bedding is prominent, somewhat iregular, and with a "herringbone" pattern. A current rose based on cross-bedding azimuths is of the "bow-tie" type (Reineck, 1963, Fig. 15-9; Terwindt et al., 1963, p. 256).

Klein (1971a) believes many orthoquartzites to have been deposited in a tidal environment and by tidal currents. He cites many examples of presumed tidal deposition of ancient quartzites (1972) and postulates a tidal cycle—essentially a fining-upward cycle not unlike the fining-upward alluvial cycle. From such cyclic accumulations he estimated the paleotidal range (1971b).

THE CARBONATE SHELF COMPLEX

In the absence of clastic sediments, carbonates may accumulate in the shore zone and adjacent shallow marine waters. The width of the zone of carbonate deposition may be very broad—100 km or more—as well as very shallow. The Florida platform and the Yucatán shelf of Mexico are contemporary examples of carbonate shelves. Vast carbonate shelf areas persisted in the Appalachian region during the early Paleozoic.

The carbonate shelf or platform is a complex of lesser environments including shelf-edge reefs, back-reef lagoons and patch reefs, carbonate shoals of the platform proper, carbonate tidal flats, off-platform carbonate turbidites, and, in favorable areas, carbonate eolianites. It is not our intent to review here the vast literature on the carbonate shelf complex in all its aspects. Reef geometry was discussed in Chapter 5. In Chapter 10, we discussed briefly modern carbonate sediments, the petrography of carbonate sediments, and the facies of lime deposition. In this section, therefore, we emphasize only the stratigraphic sequences generated during the upbuilding of the carbonate platform. We will consider first the carbonate tidal flat and the sequences formed in this environment, and then the sequences produced in the tidal environment under conditions of great aridity—the so-called sabkha cycle.

The carbonate tidal-flat sequence Close inspection of some limestones shows that there is commonly a recognizable pattern of cyclic sedimentation (Root, 1964; Pelto, 1942). The Conococheague Limestone (Cambrian) of western Maryland and adjacent Pennsylvania, for example, shows well-developed cycles (Fig. 15-17). In these deposits, the basic cycle consists of a lower thin- to thick-bedded flat-pebble intraformational conglomerate whose clasts form a framework with interstices filled with peloids, ooids, and fossil hash. All are cemented by a sparry cement. This unit is overlain by a calcarenite consisting of peloids, ooids, and skeletal debris. A variety of domal and digitate stromatolites are also present in this member. These are in turn overlain by a thin-bedded, laminated unit consisting of wavy-bedded and lensoid layers of micritic to peloidal limestone, commonly showing small-scale cross-laminations separated by thin, somewhat impure dolomite seams. This member is followed by a very thinly interlaminated calcite and dolomite bed marked by mud cracks. The laminations appear to be of algal origin. These laminated beds pass upward into dolomite, in places containing small gypsum nodules. Thin lenses of quartz sand occur in the cryptalgal dolomite and in places form significant but thin sandstone beds which top the cycle.

This sequence has been interpreted by Hepp (1973) as a regressive tidal-flat cycle, beginning with a subtidal intraformational conglomerate, followed by subtidal or intertidal stromatolitic beds in turn overlain by supratidal, sabkha-type deposits on which wind-blown sands may collect. The sequence is thought to be autocyclic. This pattern is many times repeated through nearly 1,000 m of carbonate section. Individual cycles are but a few meters thick. Much of the remainder of the Cambro-Ordovician sequence (some 10,000 feet or 3,000 m) is similar in origin (Donaldson, 1960; Matter, 1967; Pelto, 1942; Sarin, 1962).

The style of sedimentation displayed by the Conococheague in the central Appalachians is very common, especially in Lower Paleozoic and late Precambrian rocks. Similar lithologies occur, for example, in the Randville Dolomite (Precambrian) of the Northern Peninsula of Michigan (Greenman, 1951).

The sabkha cycle It may be that the dolomitic facies is a secondary or diagenetic rather than a primary or sedimentologic facies. Yet, as noted on the occurrence of dolomite, it is commonly a geographically distinct and mappable facies, generally a more shoreward facies than the associated limestones. And in those cases where dolomite is closely associated with evaporitic minerals or deposits, it is indeed a distinctive sedimentologic facies.

An example of the evaporitic-dolomite facies is described from the Bellerophon Formation (Upper Permian) of the Alpine region in northern Italy (Bosellini and

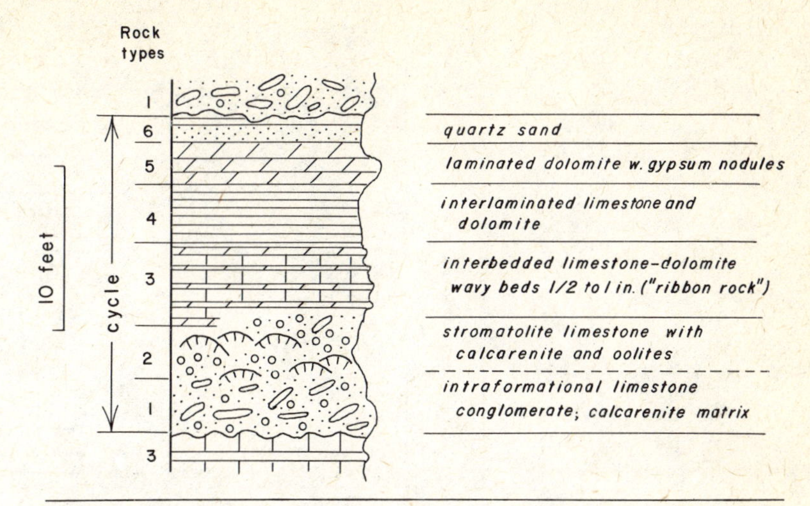

FIG. 15-17. Generalized scheme of cycles in Conococheague Limestone (Cambrian), Maryland. (After Root, S. I., *Proc. Penn. Acad. Sci.*, v. 38, Fig. 2.)

Hardie, 1973). Here dolomites and gypsum are interbedded in cyclic fashion, each cycle consisting of a lower dolomitic member and an upper gypsum member. These cycles, which average 3 m in thickness, are interpreted as the product of a prograding shallow lagoon-sabkha complex. The lowest dolomite is subtidal; the layered nodular and "chicken-wire" gypsums of the upper member were formed above the water table in a completely exposed sabkha. The ideal cycle is shown in Fig. 15-18. It is not clear whether the dolomites are primary precipitates or a synsedimentary replacement, but they constitute a distinct carbonate facies in either case. This facies is probably common in the geologic record and is reported from the Upper Triassic Raibl Formation of the southern Alps and the Upper Triassic Burano Formation of the central Apennines. It probably occurs in the Carboniferous of Spitzbergen (Bosellini and Hardie, 1973, p. 25).

The reef cycle Less well documented is the reef cycle. Lowenstam (1950, 1957) reviewed in some detail his concepts of the origin and development of the Niagaran (Silurian) reefs in the Great Lakes area. Although he did not formulate a reef cycle, he did indicate the sequence of deposits and faunas to be expected as a reef grows to maturity. He presumed an initial growth below wave base, with the nascent reef situated on a substrate of calcareous red and green shales. As the reef-to-be grows and reaches the zone of wave action, the core sheds debris which accumulates as an ever-expanding apron of waste. The core tends to expand and overgrow these flanking deposits. If the reef becomes emergent, some of the debris is thrown back onto the reef to form beaches and even dunes.

Based on this concept, and depending somewhat on where one is with respect to the reef core, one might expect a vertical sequence beginning with calcareous shales, flanking beds of debris with high initial dips, reef rock itself, and eolianites and beach rock. Essentially this sequence is, in fact, observed in some wells which penetrate the marine reef structure (Silurian) in southern Illinois (Lowenstam, 1948, Fig. 11).

The progradation of the carbonate shelf as a whole might lead to seaward migration of the reef-fringed edge so that the reef facies is overlain by the lagoonal facies with its patch reefs and perhaps even evaporites. This seems to have happened in the case of the Guadalupe reefs (Permian) in Texas and New Mexico (King, 1948; Newell et al., 1953).

MARINE ENVIRONMENTS

We have, in our discussion of the shorezone environments and sequences, touched on the contiguous shallow-water marine environments and sediments, such as clays of

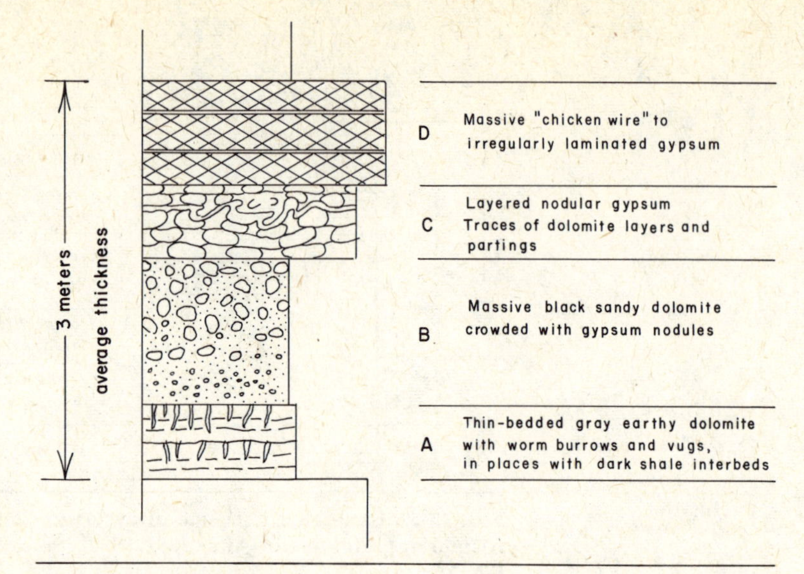

FIG. 15-18. The "modal" cycle of the Bellerophon (Permian) evaporite sequences of northern Italy. (After Bosellini and Hardie, 1973, Sedimentology, v. 20, Fig. 6.)

the pro-delta region, clays seaward of the barrier island, flanking beds of the limestone reefs and platforms. There remain to be considered the marine environment as a whole —both shallow- and deep-water marine deposits and sequences.

Shallow marine At present many so-called marine sheet sands are in fact reworked fluvial sands inundated by the post-glacial rise in sea level. Many of the ancient marine sheet sands are merely the net result of lateral accretion—the product of a prograding beach or barrier island. The sandstone formation is therefore time-transgressive (Fig. 15-15). Shale is probably the dominant product of sedimentation in the offshore shallow-marine environment.

Deep basinal marine deposits, however, contain important components of sand and even of gravel. These belong to the turbidite facies.

Turbidite ("graded-bedded") facies Certain thick sedimentary sequences are characterized by a rhythmic alteration of relatively thin sandstones with equally thin shales. These sequences may vary from nearly all sandstone with thin shaly partings to dominantly shale with scattered thin sandstones. Each sandstone layer is marked by a sharp base, possibly with an assemblage of sole marks (especially flute and groove casts), with a less sharp top. The sandstones are remarkably persistent in extent and thickness. They are commonly graded and show a characteristic internal sequence of structures—the Bouma sequence (Bouma, 1962, p. 49).

The complete five-fold Bouma sequence (Fig. 4-20) consists of a lower division (A) which may be graded or massive, overlain by a laminated division (B), in turn followed by a division (C) marked by small-scale ripple cross-lamination which may, in rare cases, also display convolute laminations. The topmost two divisions (D and E) are seldom separable in outcrop. They are primarily mudstone or shale. Not all sandstones display the complete sequence. Bottom truncation of the sequence is most common. A given sandstone bed may begin with the laminated division (B) or even with the rippled interval (C). Despite these defects in sequence, inversion of order of the several divisions is very rare indeed.

This graded sandstone facies characterizes most so-called *flysch* deposits. Its origin was, for a long time, uncertain. In 1950, however, Kuenen and Migliorini published their now classic paper in which they showed that this facies was the product of turbidity currents. Arguments to support this hypothesis were presented in detail by them and have been further elaborated by many authors since and briefly summarized in the section on graded bedding in Chapter 4. The history of the subject has been reviewed in detail by Walker (1973). In brief, deposition of the sandstone bed is believed to be the product of a turbidity underflow—a slurry of sand-laden muddy water which behaves as a heavy liquid and which is capable of flowing beneath clear water. These materials, originally deposited in shallow water, are dislodged by a slump, and the turbidity flow generated moves downslope for long distances into relatively deep water. As the current wanes, deposition progresses in such a manner to yield the observed upward sequence of structures that mark the Bouma cycle. The sandstones, in short, are the result of short-lived events that introduce these coarse materials into an alien environment, one in which the indigenous or background sediment is the finest mud. Rough estimates based on presumed rates of sedimentation of the background sediment suggest that the events responsible for the sandstones occur at infrequent intervals, perhaps once in several thousand years (Sujkowski, 1957; Kuenen, 1953, p. 28).

Biological features associated with these *turbidites,* as they have come to be called, substantiate this concept of their origin. The classic study of Natland and Kuenen (1951) showed that microfaunal forms of the pelitic interval (E) of the Pliocene turbidites in the Ventura Basin of California, indicate depths of a thousand or more meters, whereas the forams in the sandstones were those characteristic of very shallow waters. The trace fossils of turbidites are indicative of a grazing behavior and belong to Seilacher's (1967) *Nereites* facies considered by him to indicate deep, quiet water.

The brief description presented above is that of the "normal" or classic turbidite facies. Some of the other resedimented deposits cannot be described in terms of the Bouma sequence. In short, several facies of turbidites have been recognized and described. A good review of the subject has been presented by Walker and Mutti (1973).

Let us consider first those facies which are most nearly "normal." Two end-types may be defined, namely, proximal and distal. These terms are used to indicate their presumed proximity to the source from which they were derived. The *proximal* facies (Fig. 15-19) is marked by regularly bedded sandstones ranging from 10 cm to 1 meter in thickness and a sand–shale ratio of about 5 to 1. Amalgamation of sandstone beds is not uncommon. In general, these sandstones are coarse, even pebbly in some cases, and are generally graded (interval A). Laminated and rippled intervals (B and C) are rare. They are, in Bouma terms, AE turbidites. In the distal facies (Fig. 15-20) the sands are finer-grained; the sand beds range from 1 to 10 cm in thickness. The sand–shale ratio is 1 to 1 or less. They commonly begin with the laminated interval (B) or even the rippled phase (C). The proximal and distal facies grade into each other.

More aberrant resedimented deposits include massive sands, conglomeratic sands, and conglomerates. These are more difficult to recognize as belonging to the turbidite facies, but their close association with this facies and other anomalous features leave no doubt of their affinities. In some cases these coarse sediments show a well-organized internal structure—well-marked stratification and in many cases large-scale graded bedding. The superposition of one graded sandstone on another without any intervening pelitic interval has been termed *amalgamation,* a phenomenon that leads to the formation of sandstone beds an order of magnitude thicker than normal turbidite sands. Close inspection of such beds will usually reveal their composite nature. In some cases the sandstones become very coarse, even pebbly or conglomeratic, and pass by degrees into coarse conglomerates such as those exposed in the Cretaceous of Wheeler Gorge in California (Rust, 1966; Fisher and Mattinson, 1968).

In other cases the sandstones and conglomerates lack stratification and grading, and some are indeed very poorly organized. The sandstones may show the ill-defined "dish structure (Chipping, 1972, p. 589) and contain large raftlike intraclasts of shale. The conglomeratic beds are the "pebbly mudstones" of Crowell (1957) which resemble and have often been mistaken for glacial beds (see Chapter 6). These strata are perhaps the product of debris or mud flows—certainly formed by a process that leaves a chaotic internal structure.

The several facies described do not occur randomly in the stratigraphic record. It is believed that coarser deposits, both organized and chaotic, were deposited in the most proximal position—and, by analogy to present deep-sea fans, in the suprafan area where well-defined distributary channels are incised into older fan materials. The more normal proximal facies is at a greater distance downslope; the distal facies may even be beyond the obvious limits of the fan itself. If these concepts are correct, the stratigraphic record of a submarine fan prograding over a basin plain would show a regular progression of facies (Fig. 15-21).

Paleocurrent studies, based on oriented sole marks, may enable one to identify several fans in an area of complex proximal sedimentation and to identify points of sediment input. Paleocurrent directions in such an area will show a large variation; in basin areas the flow will be more uniform, not uncommonly longitudinal (Kuenen, 1957).

The turbidite facies occurs in rocks of all ages, even Archean (Walker and Pettijohn, 1971), and is a very common feature of many thick geosynclinal marine sections. It is, therefore, one of the most fundamental facies types. The most important turbidites are those of the flysch. Turbidites, although of lesser volume and importance, are also known from other environments such as the pro-delta area (see p. 552) and even in some lakes.

The turbidite facies has been intensely studied since the milestone paper of Kuenen and Migliorini in 1950. Refer to extended review papers (Walker, 1973) and special bibliographies (Kuenen and Humbert, 1964). Among the first studies of the turbidite facies in the United States were those of McBride (1962) on the Ordovician Martinsburg of the central Appalachians and that of McIver (1970) on the Upper Devonian "Portage"-type deposits of the same region. Other significant North American studies include

FIG. 15-19. Proximal turbidites (Silurian), near New Quay, Wales. (Photograph by Norman McIver.)

FIG. 15-20. Distal turbidites (Silurian) near Aberystwyth, Wales.

those of Lower Paleozoic flysch in the St. Lawrence region of Quebec (Enos, 1969; Hubert, Lajoie, and Léonard, 1970), of the Miocene Tarzana fan of southern California (Sullwold, 1960), of Cretaceous beds in the Sacramento Valley (Ojakangas, 1968), of Paleozoic and Mesozoic flyschlike sequences in the Pacific northwest and adjacent Canada (Danner, 1970), and of late Paleozoic turbidites in the Marathon Basin of Texas (McBride, 1966).

Important papers on this facies have appeared in Polish and Czech literature, especially those of Dzulynski and others (Dzulynski, 1959; Dzulynski and Smith, 1964; Dzulynski and Walton, 1965; Marschalko,

1968). The turbidites of the Apennines have been described in many papers (ten Haaf, 1959; Mutti and Ricci Lucchi, 1972; Ricci Lucchi, 1969). Alpine turbidites were early described by Crowell (1956); more recent work includes that on the Maritime Alps by Stanley (1961). There are scattered papers on the turbidite facies from other places.

In the discussion above, and in the examples cited, the deposits were normal clastic sediments; the sands were generally graywackes or subgraywackes and the lutites true shales. But, as noted in the chapter on limestones, carbonate turbidites are also known. In these the coarse fraction is generated in shallow water, on a reef or on a carbonate platform, and transported into deep water as a turbidity flow. Such carbonate turbidites resemble, in all important particulars except composition, their noncarbonate counterparts. Only a few of these deposits have been investigated. One of the best examples are the allodapic limestones in the Rhenish geosyncline of central Germany. The nature and internal structure of these beds and their downcurrent changes have been carefully studied (Meischner, 1964; Eder, 1970). (Other examples, both ancient and modern, are given in the section on allodapic limestones in Chapter 10.)

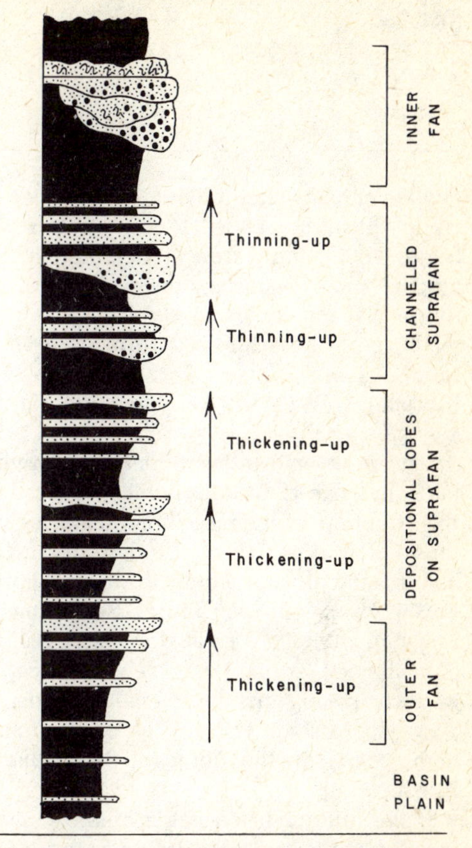

FIG. 15-21. Hypothetical vertical facies sequence produced by prograding submarine fan. (After Mutti and Ricci Lucchi, 1972, Mem. Soc. Geol. Italiana, v. 11, Fig. 14.)

INTERIOR BASIN ENVIRONMENT

Sedimentary basins may develop in the continental interior. If they form on the lee side of a high mountain chain they will be arid. Such basins commonly have a graben or half-graben structure. They have an interior drainage; that is, no water escapes to the sea.

Within such a basin are several environments. Adjoining elevated mountains shed coarse debris that forms large *alluvial fans*; finer muds are transported to a *playa flat* which is the site of an ephemeral lake or in some cases a more permanent *brine* lake. The waters in such a lake are derived from springs around the basin margin. Dissolved solids are in part supplied by the surrounding region and perhaps in part they may be airborne sea salt (Broecker and Walton, 1959). If much sand is available, *dune fields* constitute another subenvironment.

The dissolved salts find their way to the playa muds and may be precipitated in the muds or form a salt crust. The salts tend to be fractionated—calcium-magnesium carbonates are found in the peripheral areas, sulfates are precipitated basinward of the carbonates, and chlorides, if any, occur in the central area. A cross section would show this zonal arrangement and intertonguing of playa sediments and fan gravels near the playa margin and interbedding of playa muds and evaporitic materials nearer the center. Dune sand may overtop the whole sequence.

Sediments of modern arid interior basins have been described in many places. Recent studies emphasizing mineral paragenesis include those of Jones (1963), Eugster and Smith (1965), and Hardie (1968), all of whom studied small basins on the lee side of the Sierra Nevada Range in California. A good example of an ancient record of such an interior basin is that of the Green River (Eocene) of Wyoming (Eugster and Hardie, 1974).

The dune environment has been studied in many places. Studies which focused on the structure of modern dune fields include those of McKee (1966). Studies of ancient eolian sandstones include those of Shotten (1937) on the Lower Bunter Sandstone of England, of Reiche (1938) on the Coconino Sandstone (Permian) of New Mexico, of Kiersch (1950) on the Navajo Sandstone (Jurassic) of Utah, and of Bigarella and Salamuni (1961) on the Botucatu Sandstone of Brazil.

If the interior basin has a humid climate, it will have an exterior drainage and be occupied by a freshwater lake or swamp and alluvium of a more normal character.

THE GLACIAL MODEL

Glacial deposits are either continental or marine. Continental deposits are best known inasmuch as northern Europe and North America are strewn with glacial sediments of the Late Glacial epoch. They are, however, thin, subject to erosion, and with rare exceptions not apt to be preserved.

On the other hand, little is known about the glaciomarine record which is much more likely to be a permanent part of the geologic record. The nature of these deposits and the conditions of their formation have been deductively presented by Carey and Ahmad (1960). Our factual knowledge, however, is meager and is confined to bottom samples and short cores taken in the areas marginal to present or former ice sheets.

Much more is known, however, about glaciomarine deposits of the geologic past. These have been described from Norway (Reading and Walker, 1966), Australia, and other places. In general, they consist of dropstone tills interbedded with ordinary marine sediments; the latter commonly contain rafted erratics.

References

Allen, J. R. L., 1962, Petrology, origin, and deposition of the highest Lower Old Red Sandstone of Shropshire, England: Jour. Sed. Petrology, v. 32, pp. 657–697.

———, 1963, Asymmetrical ripple marks and the origin of water-laid cosets of cross-strata: Liverpool and Manchester Geol. Jour., v. 3, pp. 187–236.

———, 1965a, A review of the origin and characteristics of Recent alluvial sediments: Sedimentology, v. 5, pp. 89–191.

———, 1965b, Fining-upwards cycles in alluvial successions: Geol. Jour., v. 4, pp. 229–246.

———, 1965c, The sedimentation and paleogeography of the Old Red Sandstone of Anglesey, North Wales: Proc. Yorkshire Geol. Soc., v. 35, pp. 139–185.

———, 1967, Depth indicators of classic sequences, *in* Depth indicators in marine sedimentary environments (Hallam, A., ed.): Marine Geol., spec. issue, v. 5, nos. 5, 6, pp. 429–446.

———, 1970, Studies in fluviatile sedimentation: a comparison of fining-upwards cyclothems, with special reference to coarse-member composition and interpretation: Jour. Sed. Petrology, v. 40, pp. 298–323.

Allen, J. R. L., and Friend, P. F., 1968, Deposition of Catskill facies, Appalachian region: with notes on some other Old Red Sandstone basins: Geol. Soc. Amer. Spec. Paper 106, pp. 21–74.

Andersen, S. A., 1934, Om Aase og Terrasser inden for Sussa's Vandomraade og deres Vidnesbyrd om Isafsmeltningen Forløb: Danmarks Geol. Undersøgelse II Raekke, nr. 54 (see also Jour. Geol., 1934, v. 42, p. 551).

Baas Becking, L. G. M., Kaplan, I. R., and Moore, D., 1960, Limits of the natural environments in terms of pH and oxidation–reduction potentials: Jour. Geol., v. 68, pp. 243–284.

Bailey, E. B., 1930, New light on sedimentation and tectonics: Geol. Mag., v. 47, pp. 77–92.

———, 1936, Sedimentation in relation to tectonics: Bull. Geol. Soc. Amer., v. 47, pp. 1713–1726.

Bandy, O. L., 1953, Ecology and paleoecology of some California Foraminifera. I. The frequency distribution of Recent Foraminifera off California: Jour. Paleont., v. 27, pp. 161–182.

Barrell, J., 1906, Relative geological importance

of continental, littoral, and marine sedimentation: Jour. Geol., v. 14, pp. 316–356.

Bathurst, R. G. C., 1967, Depth indicators in sedimentary carbonates, *in* Depth indicators in marine sedimentary environments (Hallam, A., ed.): Marine Geol., spec. issue, v. 5, nos. 5, 6, pp. 447–472.

———, 1971, Carbonate sediments and their diagenesis: Amsterdam, Elsevier, 620 pp.

Beerbower, J. R., 1965, Cyclothems and cyclic depositional mechanisms in alluvial plain sediments: Bull. Kansas State Geol. Surv. 169, pt. I, pp. 31–42.

Bernard, H. A., and Major, C. F., Jr., 1963, Recent meander belt deposits of the Brazos River: an alluvial "sand" model (abstr.): Bull. Amer. Assoc. Petrol. Geol., v. 47, p. 350.

Bernard, H. A., Major, C. F., Parrott, B. S., and LeBlanc, R. J., Sr., 1970, Recent sediments of southern Texas: Bur. Econ. Geol. Univ. Texas, Guidebook 11.

Bersier, A., 1959, Séquences detritiques et divagations fluviatiles: Eclogae Géol. Helvetiae, v. 51, pp. 854–893.

Bigarella, J. J., 1972, Eolian environments: their characteristics, recognition, and importance, *in* Recognition of ancient sedimentary environments (Rigby, J. K., and Hamblin, W. K., eds.): Soc. Econ. Paleont. Min. Spec. Publ. 16, pp. 12–62.

Bigarella, J. J., and Salamuni, R., 1961, Early Mesozoic wind patterns as suggested by dune bedding in the Botucatu Sandstone of Brazil and Uruguay: Bull. Geol. Soc. Amer., v. 72, pp. 1089–1106.

Blatt, H., Middleton, G., and Murray, R., 1972, Origin of sedimentary rocks: Englewood Cliffs, N.J., Prentice-Hall, 634 pp.

Blissenbach, E., 1952, Relation of surface angle distribution to particle size distribution on alluvial fans: Jour. Sed. Petrology, v. 22, pp. 25–27.

———, 1954, Geology of alluvial fans in semiarid regions: Bull. Geol. Soc. Amer., v. 65, pp. 175–190.

Bluck, B. J., 1964, Sedimentation of an alluvial fan in southern Nevada: Jour. Sed. Petrology, v. 34, pp. 395–400.

———, 1965, The sedimentary history of some Triassic conglomerates in the Vale of Glamorgan, South Wales: Sedimentology, v. 4, pp. 225–245.

Bosellini, A., and Hardie, L. A., 1973, Depositional theme of a marginal marine evaporite: Sedimentology, v. 20, pp. 5–27.

Bouma, A. H., 1962, Sedimentology of some flysch deposits: Amsterdam, Elsevier, 168 pp.

Briggs, L. I., Jr., 1953, Upper Cretaceous sandstones of Diablo Range, California: Univ. Calif. Publ. Geol. Sci., v. 29, pp. 417–452.

Broecker, W. S., and Walton, A. F., 1959, Reevaluation of the salt chronology of several Great Basin lakes: Bull. Geol. Soc. Amer., v. 70, pp. 601–618.

Bromley, R. G., 1967, Marine phosphorites as depth indicators, *in* Depth indicators in marine sedimentary environments (Hallam, A., ed.): Marine Geol., spec. issue, v. 5, nos. 5, 6, pp. 503–510.

Bull, W. B., 1972, Recognition of alluvial-fan deposits in the stratigraphic records, *in* Recognition of ancient sedimentary environments (Rigby, J. K., and Hamblin, W. K., eds.): Soc. Econ. Paleont, Min. Spec. Publ. 16, pp. 63–83.

Byrne, J. V., LeRoy, D. O., and Riley, C. M., 1959, The Chenier Plain and its stratigraphy, southwestern Louisiana: Trans. Gulf Coast Assoc. Geol. Soc., v. 9, pp. 237–260.

Carey, S. W., and Ahmad, N., 1960, Glacial marine sedimentation, *in* Geology of the Arctic 2 (Raasch, G. O., ed.): Toronto, Univ. Toronto Press, pp. 865–894.

Carozzi, A. V., 1958, Micro-mechanisms of sedimentation in the epicontinental environment: Jour. Sed. Petrology, v. 28, pp. 133–150.

Chave, Keith, 1954, Aspects of the biogeochemistry of magnesium. 2. Calcareous sediments and rocks: Jour. Geol., v. 62, pp. 587–599.

Chilingar, G. V., 1955, Review of Soviet literature on petroleum source-rocks: Bull. Amer. Assoc. Petrol. Geol., v. 39, pp. 764–767.

Chipping, D. H., 1972, Sedimentary structures and environment of some thick sandstone beds of turbidite type: Jour. Sed. Petrol., v. 42, pp. 587–595.

Clifton, H. E., Hunter, R. E., and Phillips, R. L., 1971, Depositional structures and processes in the non-barred high-energy nearshore: Jour. Sed. Petrology, v. 41, pp. 651–670.

Correns, C. W., 1950, Zur Geochemie der Diagenese: Geochim. Cosmochim. Acta, v. 1, pp. 49–54.

Crosby, E. J., 1972, Classification of sedimentary environments, *in* Recognition of ancient sedimentary environments (Rigby, J. K., and Hamblin, W. K., eds.): Soc. Econ. Paleont. Min. Spec. Publ. 16, pp. 4–11.

Crowell, J. C., 1956, Directional-current structures from the pre-Alpine flysch, Switzerland: Bull. Geol. Soc. Amer., v. 66, pp. 1351–1384.

———, 1957, Origin of pebbly mudstones: Bull. Geol. Soc. Amer., v. 68, pp. 993–1010.

Danner, W. R., 1970, Western Cordilleran flysch sedimentation, southwestern British Columbia, Canada, and northwestern Washington and central Oregon, U.S.A., *in* Flysch sedimentology in North America (Lajoie, J., ed.): Geol. Assoc. Canada Spec. Paper 7, pp. 37–52.

Denny, C. S., 1965, Alluvial fans in the Death Valley region of California and Nevada: U.S. Geol. Surv. Prof. Paper 466, 62 pp.

Dickinson, K. A., Berryhill, H. L., Jr., and Holmes, C. W., 1972, Criteria for recognizing ancient barrier coastlines, *in* Recognition of ancient sedimentary environments (Rigby, J. K., and Hamblin, W. K., eds.): Soc. Econ. Paleont. Min. Spec. Publ. 16, pp. 192–214.

Dixon, E. E. L., 1921, Geology of the South Wales Coalfield. 13. The country around Pembroke and Tenby: Geol. Surv. Great Britain Mem., 220 pp.

Doeglas, D. J., 1962, The structure of sedimentary deposits of braided rivers: Sedimentology, v. 1, pp. 167–190.

Donaldson, A. C., 1960, Interpretation of depositional environments of Lower Ordovician carbonates in central Appalachians: Proc. West Virginia Acad. Sci., 1959 and 1960, v. 31 and 32, pp. 153–161.

Duff, P., Hallam, A., and Walton, E. K., 1967, Cyclic sedimentation: Developments in Sedimentology, v. 10, Amsterdam, Elsevier, 290 pp.

Dunbar, C. O., and Rodgers, J., 1957, Principles of stratigraphy: New York, Wiley, 356 pp.

Dzulynski, S., 1959, Turbidites in the flysch of the Polish Carpathian Mountains: Bull. Geol. Soc. Amer., v. 70, pp. 1089–1118.

Dzulynski, S., and Smith, A. J., 1964, Flysch facies: Ann. Soc. Geol. Pologne, v. 34, pp. 245–246.

Dzulynski, S., and Walton, E. K., 1965, Sedimentary features of flysch and greywackes: Amsterdam, Elsevier, 274 pp.

Eder, F. W., 1970, Genese Riff-naher Detritus-Kalke bei Balve im Rheinischen Schiefergebirge (Garbecker Kalk): Verhandl. Geol. Bundesanst., v. 4, pp. 551–569.

Edwards, A. B., and Baker, G., 1951, Some occurrences of supergene iron sulphides in relation to their environments of deposition: Jour. Sed. Petrology, v. 21, pp. 34–46.

Emery, K. O., and Rittenberg, S. C., 1952, Early diagenesis of California basin sediments in relation to origin of oil: Bull. Amer. Assoc. Petrol. Geol., v. 36, pp. 735–806.

Emiliani, C., 1955, Pleistocene temperatures: Jour. Geol., v. 63, pp. 538–578.

Enos, P., 1969, Cloridome Formation, Middle Ordovician flysch, northern Gaspé Peninsula, Quebec: Geol. Soc. Amer. Spec. Paper 117, 66 pp.

Epstein, S., Buchsbaum, R., Lowenstam, H. A., and Urey, H. C., 1951, Carbonate-water isotopic temperature scale: Bull. Geol. Soc. Amer., v. 62, pp. 417–426.

Eugster, H. P., and Hardie, L. A., 1974, Personal communication.

Eugster, H. P., and Smith, G. I., 1965, Mineral equilibria in the Searles Lake evaporites, California: Jour. Petrology, v. 6, pp. 473–522.

Fisher, R. V., and Mattinson, J. M., 1968, Wheeler Gorge turbidite-conglomerate series, California: Jour. Sed. Petrology, v. 38, pp. 1013–1023.

Friedman, G. M., and Amiel, A. J., Braun, M., and Miller, D. S., 1973, Generation of carbonate particles and laminites in algal mats, etc.: Bull. Amer. Assoc. Petrol. Geol., v. 57, pp. 541–557.

Funnell, B. M., 1967, Foraminifera and Radiolaria as depth indicators in marine environment: Marine Geol., v. 5, pp. 333–347.

Gilbert, G. K., 1914, The transportation of debris by running water: U.S. Geol. Surv. Prof. Paper 86, 263 pp.

Ginsburg, R. N., Bricker, O. P., and Wanless, H. R., 1970, Exposure index and sedimentary structures of a Bahama tidal flat (abstr.): Geol. Soc. Amer., Abstr. with Prog., 1970 Ann. Mtgs., v. 2, no. 7, pp. 744–745.

Glennie, K. W., 1970, Desert sedimentary environments: Amsterdam, Elsevier, 222 pp.

Gould, H. R., and McFarlan, E., Jr., 1959, Geologic history of the Chenier Plain, southwestern Louisiana: Trans. Gulf Coast Assoc. Geol. Soc., v. 9, pp. 261–270.

Greenman, Norman, 1951, The Randville Dolomite, nature and origin: Ph.D. thesis, Univ. Chicago.

ten Haaf, E., 1959, Graded beds of northern Apennines: Ph.D. thesis, Univ. Groningen, 102 pp.

Hallam, A., ed., 1967, Depth indicators in marine sedimentary environments: Marine Geol., spec. issue, v. 5, nos. 5, 6, pp. 329–555.

Hardie, L. A., 1968, The origin of the recent non-marine evaporite deposit of Saline Valley, Inyo County, California: Geochim. Cosmochim. Acta, v. 32, pp. 1279–1301.

Heezen, B. C., and Hollister, C., 1964, Deep-sea current evidence from abyssal sediments: Marine Geol., v. 1, pp. 141–174.

Hepp, D., 1973, Personal communication.

Hjulström, F., 1935, Studies of the morphological activity of rivers as illustrated by the River Fyris: Bull. Geol. Inst. Uppsala, v. 25, pp. 221–527.

Hollenshead, C. T., and Pritchard, R. L., 1961, Geometry of producing Mesaverde sandstones, San Juan Basin, *in* Geometry of sandstone bodies (Peterson, J. A., and Osmond,

J. C., eds.): Tulsa, Okla., Amer. Assoc. Petrol. Geol., pp. 98–118.
Hooke, R. LeB., 1967, Processes on arid-region fans: Jour. Geol., v. 75, pp. 438–460.
Howard, J. D., 1966, Patterns of sediment dispersal in the Fountain Formation of Colorado: Mountain Geol., v. 3, pp. 147–153.
Hoyt, J. H., and Weimer, R. J., 1963, Comparison of modern and ancient beaches, central Georgia coast: Bull. Amer. Assoc. Petrol. Geol., v. 47, pp. 529–531.
Hubert, C., Lajoie, J., and Léonard, M. A., 1970, Deep-sea sediments in the Lower Paleozoic Quebec Supergroup, *in* Flysch sedimentology in North America (Lajoie, J., ed.): Geol. Assoc. Canada Spec. Paper 7, pp. 103–126.
Hubert, J. F., 1960, Petrology of the Fountain and Lyons Formations, Front Range, Colorado: Colorado School of Mines Quart., v. 55, no. 1, pp. 1–242.
Jones, B. F., 1963, Hydrology and mineralogy of Deep Springs Lake, Inyo County, California: Ph.D. thesis, Johns Hopkins Univ.
Jopling, A. V., 1963, Hydraulic studies of the origin of bedding: Sedimentology, v. 2, pp. 115–121.
Jopling, A. V., and Walker, R. G., 1968, Morphology and origin of ripple-drift cross-lamination, with examples from the Pleistocene of Massachusetts: Jour. Sed. Petrology, v. 38, pp. 971–984.
Kaiser, W. R., 1971, Cyclic sedimentation in the Middle Devonian of south-central Pennsylvania (abstr.): Geol. Soc. Amer., Abstr. with Prog., 1971 Ann. Mtgs., v. 3, no. 7, pp. 616–617.
———, 1972, Delta cycles in the Middle Devonian of central Pennsylvania: Ph.D. dissertation, The Johns Hopkins Univ., Baltimore, 183 pp.
Keith, M. L., and Degens, E. T., 1959, Geochemical indicators of marine and fresh-water sediments, *in* Researches in geochemistry (Abelson, P. H., ed.): New York, Wiley, pp. 38–61.
Kessler, L. G., II., 1971, Characteristics of the braided stream depositional environment with examples from the South Canadian River, Texas: Earth Sci. Bull., v. 4, pp. 25–35.
Kiersch, G. A., 1950, Small-scale structures and other features of Navajo Sandstone, northern part of San Rafael Swell, Utah: Bull. Amer. Assoc. Petrol. Geol., v. 34, pp. 923–942.
King, P. B., 1948, Geology of the southern Guadalupe Mountains, Texas: U.S. Geol. Surv. Prof. Paper 215, 183 pp.
Klein, G. deV., 1971a, A Cambrian tidal sand body—the Eriboll Sandstone of northwest Scotland, an ancient-recent analog: Jour. Geol., v. 79, pp. 400–415.
———, 1971b, A sedimentary model for determining paleotidal range: Bull. Geol. Soc. Amer., v. 82, pp. 2585–2592.
———, 1972, Sedimentary model for determining paleotidal range, reply: Bull. Geol. Soc. Amer., v. 83, pp. 539–546.
Krumbein, W. C., and Garrels, R. M., 1952, Origin and classification of chemical sediments in terms of *p*H and oxidation–reduction potentials: Jour. Geol., v. 60, pp. 1–33.
Krumbein, W. C., and Sloss, L. L., 1963, Stratigraphy and sedimentation, 2nd ed.: San Francisco, Freeman, 660 pp.
Krynine, P. D., 1950, Petrology, stratigraphy, and origin of Triassic sedimentary rocks of Connecticut: Bull. Connecticut Geol. Nat. Hist. Surv. 73, 247 pp.
Kuenen, Ph. H., 1953, Graded bedding with observations on Lower Paleozoic rocks of Great Britain: Verh. Konink. Neder. Akad. Weten. Natur., v. 20, pp. 5–47.
———, 1957, Longitudinal filling of oblong sedimentary basins: Konink. Nederlands Geol. Mijn. Genootsch., geol. ser., v. 18, pp. 189–195.
Kuenen, Ph. H., and Humbert, F. L., 1964, Bibliography of turbidity currents and turbidites, *in* Turbidites (Bouma, A. H., and Brouwer, A., eds.): Amsterdam, Elsevier, pp. 222–246.
Kuenen, Ph. H., and Migliorini, C. I., 1950, Turbidity currents as a cause of graded bedding: Jour. Geol., v. 58, pp. 91–127.
Kukal, Z., 1970a, Sediments of alluvial fans, *in* Geology of Recent sediments: Prague, Czech. Acad. Sci., pp. 110–115.
———, 1970b, Geology of Recent sediments: London, Academic Press, 490 pp.
Laporte, L. F., 1968, Ancient environments: Englewood Cliffs, N.J., Prentice-Hall, 116 pp.
Lawson, A. C., 1925, The petrographic designation of alluvial fan formations: Univ. California Publ., Dept. Geol. Sci., v. 7, pp. 325–334.
LeBlanc, R. J., 1972, Geometry of sandstone reservoirs, *in* Underground waste management and environmental implications: Amer. Assoc. Petrol. Geol. Mem. 18, pp. 133–189.
Lombard, A., 1972, Séries sédimentaires, genèse-évolution: Paris, Masson, 426 pp.
Lowenstam, H. A., 1948, Marine Pool, Madison County, Illinois, Silurian reef producer, *in* Structure of typical American oil fields, v. 3: Tulsa, Okla., Amer. Assoc. Petrol. Geol., pp. 153–188.
———, 1950, Niagaran reefs of the Great Lakes area: Jour. Geol., v. 58, pp. 430–587.
———, 1957, Niagaran reefs in the Great Lakes area, *in* Treatise on marine ecology and

paleoecology, v. 2 (Ladd, H. S., ed.): Geol. Soc. Amer. Mem. 67, pp. 215–248.

———, 1963, Biologic problems relating to the composition and diagenesis of sediments, *in* The earth sciences (Donnelly, T. W., ed.): Houston, Rice Univ. Semicentennial, pp. 137–195.

McBride, E. F., 1962, Flysch and associated beds of the Martinsburg Formation (Ordovician), central Appalachians: Jour. Sed. Petrology, v. 32, pp. 39–91.

———, 1966, Sedimentary petrology and history of the Haymond Formation (Pennsylvanian), Marathon Basin, Texas: Bur. Econ. Geol. Univ. Texas, Rept. Inv. 57, 101 pp.

McIver, N. L., 1970, Appalachian turbidites, *in* Studies of Appalachian geology, central and southern (Fisher, G. W., Pettijohn, F. J., Reed, J. C., Jr., and Weaver, K. N., eds.): New York, Wiley-Interscience, pp. 69–82.

McKee, E. D., 1957, Primary structures in Recent sediments: Bull. Amer. Assoc. Petrol. Geol., v. 41, pp. 1704–1747.

———, 1966, Structures of dunes at White Sands National Monument, New Mexico: Sedimentology, spec. issue, v. 7, 99 pp.

———, 1971, Review of "Ancient sedimentary environments": Geotimes, v. 16, no. 12, p. 38.

Marschalko, R., 1968, Facies distribution, paleocurrents and paleotectonics of the Paleogene flysch of central west Carpathians: Slovenska Akad. Bratislava, Geol. Zbornik, Geol. Carpathica 19, pp. 69–94.

Mason, B., 1949, Oxidation and reduction in geochemistry: Jour. Geol., v. 57, pp. 62–72.

Matter, A., 1967, Tidal flat deposits in the Ordovician of western Maryland: Jour. Sed. Petrology, v. 37, pp. 601–609.

Meckel, L. D., 1970, Paleozoic alluvial deposition in the central Appalachians, *in* Studies of Appalachian geology (Fisher, G. W., Pettijohn, F. J., Reed, J. C., Jr., and Weaver, K. N., eds.): New York, Wiley-Interscience, pp. 49–68.

Medeiros, R., Schaller, H., and Friedman, G. M., 1971, Facies sedimentares: Rio de Janeiro, Dir. Doc. Tec. Patentes, 123 pp.

Meischner, K. D., 1964, Allodapische Kalke Turbidite im Riff-nahen Sedimentations-Becken, *in* Turbidites (Bouma, A. H., and Brouwer, A., eds.): Amsterdam, Elsevier, pp. 156–191.

Millot, G., 1949, Relations entre la constitution et la genèse des roches sédimentaires argileuses: Géol. Appl. Prospect. Min., v. 2, nos. 2-4, p. 352.

Morgan, J. P., ed., 1970, Deltaic sedimentation, modern and ancient: Soc. Econ. Paleont. Min. Spec. Publ. 15, 312 pp.

Mutti, E., and Ricci Lucchi, F., 1972, Le torbiditi dell'Appennino settentrionalei introduzione all'analisi di facies: Mem. Soc. Geol. Italiana, v. 11, pp. 161–199.

Natland, M. L., and Kuenen, Ph. H., 1951, Sedimentary history of the Ventura Basin, California, and the action of turbidity currents: Soc. Econ. Paleont. Min. Spec. Publ. 2, pp. 76–107.

Nevin, C. M., 1946, Competency of moving water to transport debris: Bull. Geol. Soc. America, v. 57, pp. 651–674.

Newell, N. D., Rigby, J. K., Fischer, A. G., Whiteman, A. J., Hickox, J. E., and Bradbury, J. S., 1953, The Permian reef complex of the Guadalupe Mountains region, Texas and New Mexico—a study in paleoecology: San Francisco, Freeman, 236 pp.

Ojakangas, R. W., 1968, Cretaceous sedimentation, Sacramento Valley, California: Bull. Geol. Soc. Amer., v. 79, pp. 973–1008.

Okada, H., 1971, A pattern of sedimentation in clastic sediments in geosynclines: Mem. Geol. Soc. Japan 6, pp. 75–82.

Okada, H., and Fujiyama, I., 1970, Sedimentary cycles and sedimentation of the Taishu group in the Shiohama area, central Tsushima, Kyushu: Mem. Nat. Sci. Mus. 3, pp. 9–16.

Oomkens, E., 1970, Depositional sequences and sand distribution, the post-glacial Rhone delta complex *in* Deltaic sedimentation, modern and ancient (Morgan, J. P., and Shaver, R. H., eds.): Soc. Econ. Paleont. Min. Spec. Pub. 15, pp. 198–212.

Ore, H. T., 1963, Some criteria for recognition of braided stream deposits: Univ. Wyoming Contr. Geol., v. 3, pp. 1–14.

Pelto, C. R., 1942, Petrology of the Gatesburg Formation of central Pennsylvania: M.S. thesis, Pennsylvania State Univ.

Pettijohn, F. J., Potter, P. E., and Siever, Raymond, 1972, Sand and sandstone: New York, Springer, 618 pp.

Phleger, F. B., 1955, Ecology of Foraminifera in southeastern Mississippi delta area: Bull. Assoc. Petrol. Geol., v. 39, pp. 712–752.

Porrenga, D. H., 1967, Glauconite and chamosite as depth indicators in the marine environment, *in* Depth indicators in marine sedimentary rocks (Hallam, A., ed.): Marine Geol., spec. issue, v. 5, nos. 5, 6, pp. 495–502.

Potter, P. E., 1967, Sand bodies and sedimentary environments, a review: Bull. Amer. Assoc. Petrol. Geol., v. 51, pp. 337–365.

Potter, P. E., and Pettijohn, F. J., 1963, Paleocurrents and basin analysis: New York, Springer, 296 pp.

Potter, P. E., Shimp, N. F., and Witters, J., 1963, Trace elements in marine and fresh-water

argillaceous sediments: Geochim. Cosmochim, Acta, v. 27, pp. 669–694.
de Raaf, J. F. M., Reading, H. G., and Walker, R. G., 1965, Cyclic sedimentation in the Lower Westphalian of North Devon, England: Sedimentology, v. 4, pp. 1–52.
Reading, H. G., 1970, Sedimentation in Upper Carboniferous in the Cantabrian Mountains, Spain: Proc. Geol. Assoc., v. 81, pp. 1–41.
Reading, H. G., and Walker, R. G., 1966, Sedimentation of Eocambrian tillites and associated sediments in Finnmark, northern Norway: Palaeogeogr. Palaeocl. Palaeoecol., v. 2, pp. 177–212.
Reiche, P., 1938, An analysis of cross-lamination, the Coconino Sandstone: Jour. Geol., v. 46, pp. 905–932.
Reineck, H.-E., 1963, Sedimentgefüge im Bereich der südlichen Nordsee: Senckenberg Natur. Gesell. Abb. 504, 64 pp.
———, 1970, Das Watt: Frankfurt, Waldemar Kramer, 142 pp.
Reineck, H.-E., and Singh, I. B., 1973, Depositional sedimentary environments: New York, Springer, 439 pp.
Reineck, H.-E., and Wunderlich, F., 1968, Classification and origin of flaser and lenticular bedding: Sedimentology, v. 11, pp. 99–105.
Reinemund, J. A., 1955, Geology of the Deep River Coal Field, North Carolina: U.S. Geol. Surv. Prof. Paper 246, 159 pp.
Ricci Lucchi, F., 1969, Considerazioni sulla formazione di alcune impronte da corrente: Ann. Mus. Geol. Bologna, ser. 2a, v. 36, pp. 363–415.
Rigby, J. K., and Hamblin, W. K., eds., 1972, Recognition of ancient sedimentary environments: Soc. Econ. Paleont. Min. Spec. Publ. 16, 340 pp.
Root, S. F., 1964, Cyclicity in the Conococheague Formation: Proc. Pennsylvania Acad. Sci., v. 38, pp. 157–160.
Rubey, W. W., 1938, The force required to move particles on a stream bed: U.S. Geol. Survey Prof. Paper 189-E, pp. 121–140.
Rusnak, G. E., 1960, Sediments of Laguna Madre, Texas, *in* Recent sediments, northwest Gulf of Mexico (Shepard, F. P., Phleger, F. B., and van Andel, Tj. H., eds.): Tulsa, Okla., Amer. Assoc. Petrol. Geol., pp. 153–196.
Rust, B. R., 1966, Late Cretaceous paleogeography near Wheeler Gorge, Ventura County, California: Bull. Amer. Assoc. Petrol. Geol., v. 50, pp. 1389–1398.
Sarin, D. D., 1962, Cyclic sedimentation of primary dolomite and limestone: Jour. Sed. Petrology, v. 32, pp. 451–471.
Seilacher, A., 1967, Bathymetry of trace fossils: Marine Geol., spec. issue, v. 5, nos. 5, 6, pp. 413–428.
Selley, R. C., 1969, Studies of sequence in sediments using a simple mathematical device: Quart. Jour. Geol. Soc. London, v. 125, pp. 557–581.
———, 1970, Ancient sedimentary environments: Ithaca, N.Y., Cornell Univ. Press, 237 pp.
Shepard, F. P., and Moore, D. G., 1955, Central Texas coast sedimentation—characteristics of sedimentary environment, recent history and diagenesis: Bull. Amer. Assoc. Petrol. Geol., v. 39, pp. 1463–1593.
Shotten, F. W., 1937, The Lower Bunter Sandstones of North Worcestershire and East Shropshire: Geol. Mag., v. 74, pp. 534–553.
Shukri, M. N., 1942, The use of pH values in determining the environment of deposition of some Liassic clays and shales: Bull. Fac. Sci. Fouad I Univ., v. 24, pp. 61–65.
Simons, D. B., Richardson, E. V., and Nordin, C. F., Jr., 1965, Forms generated in alluvial channels, *in* Primary sedimentary structures and their hydrodynamic interpretation (Middleton, G. V., ed.): Soc. Econ. Paleont. Min. Spec. Publ. 12, pp. 34–52.
Sloss, L. L., 1953, The significance of evaporites: Jour. Sed. Petrology, v. 23, pp. 143–161.
Smith, N. D., 1968, Cyclic sedimentation in a Silurian intertidal sequence in eastern Pennsylvania: Jour. Sed. Petrology, v. 38, pp. 1301–1304.
———, 1970, The braided stream depositional environment—comparison of the Platte River with some Silurian clastic rocks, north central Appalachians: Bull. Geol. Soc. Amer., v. 81, pp. 2993–3014.
Spieker, E. M., 1949, Sedimentary facies and associated diastrophism in the Upper Cretaceous of central and eastern Utah: Geol. Soc. Amer. Mem. 39, pp. 55–82.
Stanley, D. J., 1961, Études sédimentologiques des grès d'Annot et leurs équivalents latéraux: Inst. Franc. Pétr., Ref. 6821, Paris, Technip., 158 pp.
van Straaten, L. M. J. U., 1954, Composition and structure of Recent marine sediments in the Netherlands: Leidse Geol. Meded, v. 19, 11 pp.
———, 1961, Sedimentation in tidal flat areas: Jour. Alberta Soc. Petrol. Geol., v. 9, pp. 203–226.
Stricklin, F. L., Jr., and Smith, C. I., 1973, Environmental reconstruction of a carbonate beach complex—Cow Creek (Lower Cretaceous) Formation of central Texas: Bull. Geol. Soc. Amer., v. 84, pp. 1349–1368.
Sujkowski, Zb. L., 1957, Flysch sedimentation: Bull. Geol. Soc. Amer., v. 68, pp. 543–554.

Sullwold, H. H., Jr., 1960, Tarzana fan, deep submarine fan of late Miocene age, Los Angeles County, California: Bull. Amer. Assoc. Petrol. Geol., v. 44, pp. 433–457.

Sundborg, Åke, 1956, The River Klarälven—a study of fluvial processes: Geogr. Ann., v. 38, pp. 127–316.

Tercier, J., 1940, Dépôts marins actuels et séries géologique: Eclogae Géol. Helvetiae, v. 32, pp. 42–100.

Terwindt, J. H. J., de Jong, J. D., and van der Wilk, E., 1963, Sediment movement and sediment properties in the tidal area of the lower Rhine (Rotterdam Waterway): Konink. Nederlands Geol. Mijn. Genootsch. Verb., geol. ser., v. 21-2, pp. 243–258.

Thompson, W. O., 1937, Original structure on beaches, bars, and dunes: Bull. Geol. Soc. Amer., v. 48, pp. 723–752.

Turner, C. C., and Walker, R. G., 1973, Sedimentology, stratigraphy, and crustal evolution of the Archean greenstone belt near Sioux Lookout, Ontario: Canad. Jour. Earth Sci., v. 10, pp. 817–845.

Twenhofel, W. H., 1932, Treatise on sedimentation, 2nd ed.: Baltimore, Williams and Wilkins, 926 pp.

———, 1950, Principles of sedimentation: New York, McGraw-Hill, 673 pp.

Urey, H. C., Lowenstam, H. A., Epstein, S., and McKinney, C. R., 1951, Measurement of paleotemperatures and temperatures of the Upper Cretaceous of England, Denmark, and the southeastern United States: Bull. Geol. Soc. Amer., v. 62, pp. 399–416.

Van der Lingen, G. J., 1969, The turbidite problem: New Zealand Jour. Geol. Geophys., v. 12, pp. 7–50.

Visher, G. S., 1965, Use of vertical profile in environmental reconstruction: Amer. Assoc. Petrol. Geol., v. 49, pp. 41–61.

Walker, R. G., 1973, Mopping up the turbidite mess, *in* Evolving concepts in sedimentology: Baltimore, Johns Hopkins Univ. Press, pp. 1–37.

Walker, R. G., and Harms, J. C., 1971, The "Catskill delta"—a prograding muddy shoreline in central Pennsylvania: Jour. Geol., v. 79, pp. 381–399.

Walker, R. G., and Mutti, E., 1973, Turbidite facies and facies association, *in* Turbidites and deep water sedimentation—syllabus for a short course: Anaheim, Calif., Pacific Section Soc. Econ. Paleont. Min., pp. 119–157.

Walker, R. G., and Pettijohn, F. J., 1971, Archaean sedimentation—analysis of the Minnitaki Basin, northwestern Ontario, Canada: Bull. Geol. Soc. Amer., v. 82, pp. 2099–2130.

Walther, J., 1894, Einleitung in die Geologie als historische Wissenschaft: Jena, Fischer Verlag, 1055 pp.

Weeks, L. G., 1952, Factors of sedimentary basin development that control oil occurrence: Bull. Amer. Assoc. Petrol. Geol., v. 36, pp. 2071–2124.

Weller, J. M., 1960, Stratigraphic principles and practice: New York, Harper & Row, 725 pp.

Williams, G. E., 1969, Characteristics and origin of a Precambrian pediment: Jour. Geol., v. 77, pp. 183–207.

Williams, P. F., and Rust, B. R., 1969, The sedimentology of a braided river: Jour. Sed. Petrology, v. 39, pp. 649–679.

Yeakel, L. S., Jr., 1959, Tuscarora, Juniata, and Bald Eagle paleocurrents and paleogeography in the central Appalachians: Bull. Geol. Soc. Amer., v. 73, pp. 1515–1540.

Young, R. G., 1957, Late Cretaceous cyclic deposits, Book Cliffs, eastern Utah: Bull. Amer. Assoc. Petrol. Geol., v. 41, pp. 1760–1774.

Zobell, C. E., 1946, Studies on redox potential of marine sediments: Bull. Amer. Assoc. Petrol. Geol., v. 30, pp. 477–513.

16

SEDIMENTATION AND TECTONICS

INTRODUCTION

There is a need to look at sedimentary rocks in a larger framework: their nature and distribution relative to the larger structural elements of the continents, namely cratons and geosynclines. This approach gets us into a very controversial realm; but our approach, consistent with the philosophy underlying this volume, is that we first need to know what has been seen and recorded—that is *what is,* before we can invent hypotheses to account for the observable.

Hence we begin by describing what is known about the larger-scale distribution and character of sediments and their relation to continental structure, and we then proceed to look for some pattern from these observations—a plan of organization independent of any theories of orogenesis or continental evolution. What kinds of basins exist? What are the nature and organization of their sedimentary fill? What are the kind and relation of the sediments to the source land?

Diastrophism is the major geologic process, whose functioning produces surface irregularities and sets in motion the counteracting process of gradation. Erosion and its complementary process, sedimentation, are parts of the process of gradation, and together they tend toward the development of a planar surface. Were it not for diastrophism, the earth's surface would not be disturbed by rising and sinking movements, and sedimentation therefore would cease; thus diastrophism is the ultimate cause of sedimentation.

The influence of tectonics on sedimentation has been pointed out by a number of geologists, but that it is the most fundamental factor has been only recently appreciated. As with other generalizations, this view has been acquired by observation of the actual record. The concept of tectonic control of sedimentation was clearly stated by Jones (1938). His views grew out of a study of the Lower Paleozoic of Great Britain, especially Wales. Jones recognized two facies, the shelly or calcareous facies and the graptolitic or argillaceous facies. The former consists of glauconitic sandstones and limestones—a product of sedimentation in shallow water on a comparatively stable platform; the latter is a thick argillaceous accumulation of thin beds of graded graywackes in a vast thickness of shales—a facies to which the designation flysch is now commonly applied. Bailey (1930, 1936) pointed out relations between the structure and composition of two contrasting facies of sand deposition and the tectonic nature of their depositional sites. He placed sandstones in two groups—current- or cross-bedded and graded-bedded—the former as a product of shallow-water deposition, the latter as a deep-water deposit in an unstable trough. Tyrrell (1933) and Fischer (1933) believed graywackes to be products of geosynclinal sedimentation.

Two major facies of sedimentation, the flysch and the molasse, first designated in the Alps, were linked to the tectonic evolution of the Alpine chain. The relations of flysch and molasse to tectonic development have been extended to other Alpine-type fold belts in the world. Such usages, for example, appear in the writings of Bertrand (1897), van Waterschoot van der Gracht (1931), and Tercier (1939).

In America, the influence of tectonics on sedimentation was most strongly emphasized by Krynine (1941, 1942, 1943, 1945), Pettijohn (1943), Dapples, Krumbein, and Sloss (1948), and others (Cady, 1950; Schwab, 1969a, 1969b). Although the older concepts have been somewhat modified by current theories of plate tectonics, the interdependence of sedimentation and tectonics is fully recognized (Schwab, 1971; Dickinson, 1970, 1971).

Relations between the tectonics of the site of deposition and the character of the deposited sediments, such as noted by Jones in Wales, have also been noted elsewhere. There is a comparable difference, for example, between Paleozoic rocks of the Ouachita geosyncline of Arkansas, Oklahoma, and Texas, and rocks of the same age in the Ozark area nearby. The Ouachita facies is dominantly a clastic sequence with a great thickness of shales and interbedded turbidite sands of late Paleozoic age superposed on relatively thin early Paleozoic strata; the Ozark facies, in contrast, consists mainly of a thin carbonate sequence of early Paleozoic age followed by a thin late Paleozoic clastic sequence. Contrasts in sedimentary aspect related to the structural site of deposition have led to the generalized notion of cratonic and geosynclinal facies. The former is the *stable platform facies*; the latter is the thicker accumulation of the unstable mobile belt—the *geosynclinal facies*.

Excluding ocean basins, the major structural elements of the earth are the cratons and the mobile belts generally called *geosynclines*. A *craton*, as the term is now used, is a large area of the continent on which relatively thin, generally flat-lying sediments are deposited on a more or less stable basement, commonly but not necessarily Precambrian in age. The craton includes both the shield, where the basement is exposed, and the buried basement or platform marginal to the shield. A *mobile belt* is a linear feature consisting of one or more deep troughs and medial geanticlines which generally border a craton. The contained sedimentary fill has a large aggregate thickness and is strongly deformed by folding and thrusting.

Kay (1951, p. 92) has estimated that 82 percent by volume of the sediments of North America are geosynclinal accumulations and only 18 percent are cratonic. Gilluly, Reed, and Cady (1970), estimate the total volume of Phanerozoic sediment of North America to be about 35 $km^3 \times 10^6$, of which 14 $km^3 \times 10^6$ or a little over one-third is cratonic. In any case, it is clear that geosynclines are the major sediment traps, especially for finer silts and muds.

CRATONS

According to Aubouin (1965, p. 24), the term *craton* (or Kraton) was applied by Stille to consolidated and stable parts of the earth's crust. As usually used, however, the term is restricted to tectonically passive parts of the continent and includes both the exposed Precambrian nucleus or shield and the adjacent thinly veneered platforms consisting of essentially flat-lying late Paleozoic or younger sediments. Cratonic platforms are, therefore, large continental areas. Locally they may be very gently warped into wide, shallow, intracratonic basins, such as the Michigan and Illinois basins of North America, which are separated from one another by broad regional arches or domes such as the Cincinnati Arch and the Ozark domes in the eastern United States. The sediments overlying the concealed part of the craton are relatively thin, generally less than 5,000 feet (1,500 m) and only locally, in downwarped areas, do they attain a thickness of 10,000 feet (3,000 m) (Moss, 1936). Exceptionally the craton contains sharply downfaulted grabens, generally half-grabens, in which thick immature sediments accumulate. The Gondwana basins of India and the Keweenawan trough in North America are examples of such structures.

A craton tends to be a positive area; portions remain exposed even during times of maximum continental inundation. Much of the presently exposed Canadian Shield in North America received during the Paleozoic a thin sedimentary cover which has been subsequently stripped off. The craton is characterized by uniform and persistent paleoslopes; the latter dip very gently away from the core region toward the bordering margins. The persistence of the paleoslope is well demonstrated by paleocurrent data, especially cross-bedding in sandstones, and by their stratigraphy. Potter and Pryor (1961), for example, show that paleocurrents responsible for sandstone deposition in the Upper Mississippi Valley varied very little in direction from Cambrian to the present (Fig. 14-12). Cratonic platforms have a thin sedimentary cover, mainly of carbonates and quartz arenites; shale is subordinate. The portions of common sediments is illustrated by several stratigraphic sections (Table 16-1). Disconformities are common and widespread and become compound upslope. These relations are illustrated in Fig. 16-1.

These sediments constitute the shelly facies of Jones; they are commonly designated the platform or *cratonic facies* and consist largely of orthoquartzitic sands and limestones or dolomites. Sands are shield-derived, widespread, thin sheetlike bodies. They are clean-washed, pure quartz, well-rounded, and generally very mature. Limestones and dolomites are characterized by sandy interbeds or by numerous scattered quartz grains. Carbonates are varied in origin; some are stromatolitic and obviously algal; these form biostromes and, together with other organisms, reefs. Many limestones are calcarenites, are locally cross-bedded, and consist of oolites or sorted shell debris or nonskeletal grains or mixtures of all three. Flat-pebble conglomerates are common as are mud-cracked calcilutites. Many alternations of calcarenite, calcilutite, algal beds, and flat-pebble conglomerates characterize these carbonates. Mottled dolomite, laminated "primary" dolomite, and saccharoidal replacement dolomite are common.

Commonly this facies rests unconformably on a stable crystalline basement with only a few centimeters or a few meters of arkosic and conglomeratic material just above the unconformity. Feldspar is absent in higher beds or, if present, is restricted to areas near buried granitic uplifts or to a few scattered well-rounded grains. Most commonly the basal sandstone is followed by limestone or dolomite, although in places the carbonate rocks themselves rest directly on the underlying basement.

The cratonic facies appears to be the product of sedimentation marginal to a very low-lying stable land surface. Evidence of

TABLE 16-1. Representative orthoquartzite–carbonate sections

	A	B	C	D	E	F
Percentage sandstone	25	48	15	29	32	29
Percentage shale	7	22	5	42	17	21
Percentage limestone	68	30	80	29	51	50
Total thickness (feet)	7900	1950	2725	2450	2630	—
Number of formations	21	26	26	—	5	—
Average thickness of formation (feet)	376	75	105	—	524	—

A. Cambro-Ordovician, west-central Vermont.
B. Cambro-Ordovician, southwestern Wisconsin.
C. Cambro-Ordovician, southeastern Missouri.
D. "Lower Huronian," Marquette district, Michigan.
E. Paleozoic, Bighorn Mountains, Wyoming.
F. Average of A through E.

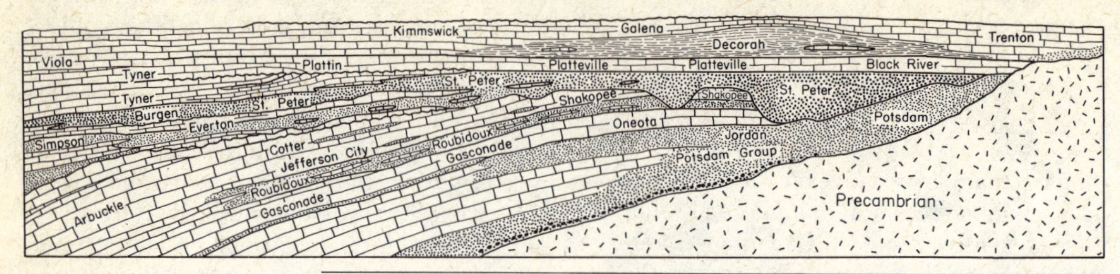

FIG. 16-1. Generalized, reconstructed, north–south section of the early Paleozoic of the Mississippi Valley. Typical quartz arenite–carbonate association. (Modified from Dake by Thiel, 1935.)

stability lies in the maturity of the sands produced and in the general paucity of clastic materials and dominance of carbonates. The slight total thickness and undeformed character of many sections further testify to the stability of the site of deposition. Although stable, the depositional surface was periodically emergent, as shown by mudcracked calcilutites and intraformational breccias—presumed to be related to desiccation. The sands, as shown by the stratigraphy (Fig. 16-1), are shield-derived. They may include some eolianites as well as shore sands.

Typical examples of this facies are the many lower Paleozoic sections in the Upper Mississippi Valley and the Great Lakes area. This region—one of extraordinary stability during most of the Paleozoic—is merely a slightly buried extension of the Canadian Shield. The deposits formed are largely limestones and dolomites, rarely more than a thousand meters thick. Sandstones become abundant only near the margins of the shield—parts presumably emergent and providing the sand.

The paucity of shale in this facies is not fully understood. If carbonates are or were (prior to dolomitization) wholly calcarenite, then calcareous and normal quartz sands would indicate deposition in a turbulent environment, and perhaps finer silts or clays from land would be bypassed into deeper, less turbulent waters. But inasmuch as many of the carbonates are calcilutites or lime muds, it is difficult to believe that argillaceous materials were bypassed. It has been suggested that clays were winnowed out in an earlier epoch and that sandstones associated with carbonates were second-cycle and derived only from sandstones of an earlier generation. Perhaps, as has also been suggested, land areas were deserts and the clay fraction was removed by the wind, leaving only a sandy dune complex to be invaded and redistributed by the advancing sea.

The broad downwarped basins within the craton contain much the same kind of sediments as do the more stable areas, except that only the sediments thicken in these basins, indicating subsidence during deposition. Some basins were at times evaporitic basins; salt and gypsum are present in one or another of them.

Sediments of the sharply downfaulted blocks within the craton contain a very different sedimentary assemblage. Coarse, immature clastic wedges accumulate near fault borders; sandstones tend to be conglomeratic and arkosic. These sediments are associated with basaltic flows and diabasic sills and dikes. The thick Keweenawan volcanics, conglomerates, and sandstones constitute an assemblage characteristic of a large intracratonic rift feature. The Gondwana basins in the Indian Peninsula are another example. The Triassic Newark basins in the crystalline Piedmont of the eastern United States are an analogous feature (Reinemund, 1955; Krynine, 1950).

The cratonic platform may at times receive less mature sand and mud from external uplifts—the tectonic lands beyond the craton proper. These deposits, such as those laid down in the eastern interior regions of the United States during the Pennsylvanian and in the High Plains during the Cretaceous, represent the distal end of great clastic wedges derived from elevated tracts outside the craton itself. Their invasion interrupted or brought to an end normal cratonic sedimentation.

GEOSYNCLINES

Because the term *geosyncline* (originally *geosynclinal*) was first applied by Dana (1873, p. 430), to the subsiding zone later to become the Appalachian Mountains, the meaning of the word has been expanded to encompass many different kinds of subsiding basins. We here agree with Aubouin (1965, p. 35) that the term thus extended has lost its meaning and that it should be either abandoned or its use restricted in such a way that it carries a precise meaning. Accordingly we use the term for those mobile belts —as contrasted to stable cratons or platforms—which exhibit a characteristic sedimentary, tectonic, and magmatic feature. Our emphasis here is on their sedimentary aspects, although these cannot be wholly divorced from tectonic and magmatic attributes.

We exclude, therefore, all downwarped and downfaulted basins within the craton; we exclude thick continental terrace-type deposits some of which have been designated geosynclines (the Gulf Coast "Geosyncline," for example). These are not mobile belts.

We shall see, however, that controversy still exists over the nature of the geosyncline and mobile belts. The main difficulty has been to identify geosynclines in the modern world, that is, to find an actualistic model.

Geosynclines, according to Aubouin (1965, p. 36), are characterized by (1) location marginal to or between cratons, (2) mobility expressed by intense folding and thrusting, (3) initial simatic igneous phase marked by ophiolitic emissions in internal (away from the craton) zones, and (4) synorogenic and post-orogenic igneous activity in internal zones.

What can we say about the sediments found within the geosynclinal belt? As one moves from the craton to the geosyncline, he notes that the thickness of the sedimentary column increases by an order of magnitude from one or two thousand meters to ten to twenty thousand meters. He notes also that one passes from an undeformed sequence of sediments into a sharply folded and thrust zone. It is apparent also that sediments undergo a change in character as one moves into the mobile belt. The cratonic record is, as noted above, one of shallow-water carbonates and thin, mature sheet sands. This is Jones's "calcareous" or "shelly" facies (1938). In the geosyncline these are replaced by a thick argillaceous facies. Both shallow- and deep-water sediments are present. Notable is the flysch and pre-flysch facies. The flysch is a thick turbidite sequence of thin, graded graywacke-like sandstones interbedded in a rhythmic fashion with indigenous shales. The pre-flysch is a varied commonly starved-basin assemblage of radiolarian cherts and black shales. Moreover, paleocurrent studies of the flysch as well as the facies patterns show these sediments to be derived mainly from a source land in the interior zone and not the craton. In the early geosynclinal stage some sediment—rare limestone turbidites and thin wedges of orthoquartzitic sand —is derived from the craton. During the later and major stages of development, abundant clastic material comes from the other side of the geosyncline. The geosyncline is thus asymmetrical and is said to have a *polarity*. The asymmetry shown by facies and paleocurrent patterns is also shown by structural and magmatic activity. Folds are overturned toward the craton; thrusts dip away from the craton. The inner parts of the mobile belt are apt to be metamorphosed and invaded by magma; the outer part is unmetamorphosed.

Detailed stratigraphic and paleocurrent studies reveal complexities. There seems to have been, in many geosynclinal belts, an internal source of sediment which apparently was a medial ridge or geanticline (Fig. 16-2). Sediments lying between this ridge and the craton differ somewhat from those lying between the medial ridge and the tectonic source land. The former are without volcanic contributions (amagmatic or miogeosynclinal); the latter contain tuffaceous materials and interbedded ophiolitic flows and sills (eugeosynclinal). These observations led Aubouin (1965, Fig. 16) to idealize the geosyncline and introduce terms to designate its parts (Fig. 16-3.)

The geosyncline was, therefore, a linear depression or trough, perhaps divided medially by an anticlinal ridge, that was filled

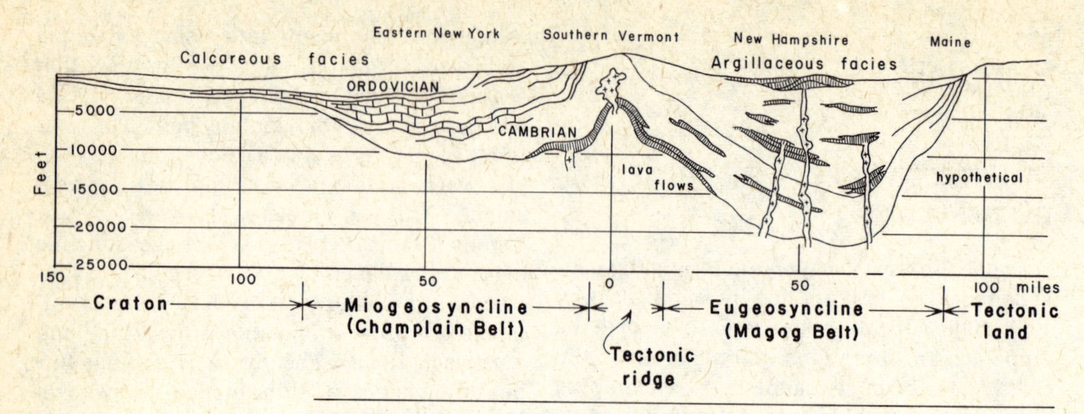

FIG. 16-2. Restored cross section of Cambrian and Ordovician in miogeosyncline and eugeosyncline, New York to Maine. (After Kay, 1951, Pl. 9.)

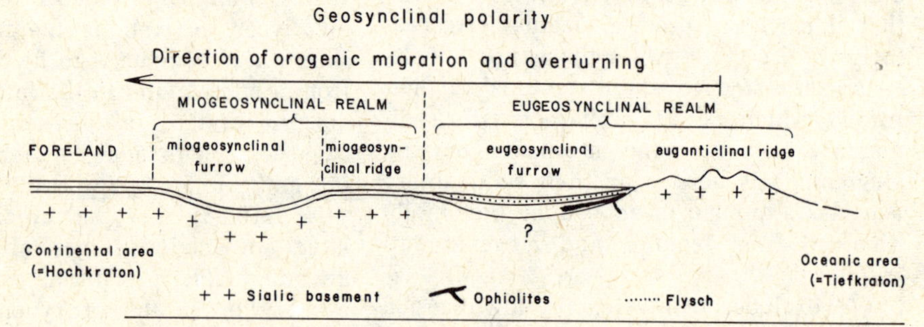

FIG. 16-3. Salient parts of the eugeosynclinal-miogeosynclinal couple. The basement of the foreland, the miogeosynclinal furrow, and the eu- and miogeosynclinal ridges is sialic, but the basement of the eugeosynclinal furrow may be simatic, hence the question mark. (After Aubouin, 1961, Fig. 1.)

mainly with clastics from the craton, the medial ridge, and most especially from a tectonic land. As noted by Aubouin and by earlier workers also, filling and ultimate deformation of the fill proceed according to a pattern. Generally there is a pre-flysch period with very slow sedimentation of radiolarites and black shales, a flysch period of rapid turbidite sedimentation, and a later molasse period during which very coarse sediments, including thick conglomerates, were deposited, generally in a subsiding molasse trough formed between the geosynclinal belt itself and the craton. Supposedly the molasse is the product of erosion of the earlier flysch, now folded and elevated into a mountain tract. Whatever interpretation one wishes to make about the geosyncline and its origin and development, it remains a fact that the flysch and the molasse facies are recurrent through geologic history and are closely associated with one another and with mobile belts.

These observations and interpretation on the nature of geosynclinal belts have raised a number of problems for which satisfactory answers are yet missing.

One such problem is the location and nature of the source of the sedimentary fill. As noted, the facies changes within a geosyncline suggest some source other than the craton. Such source or sources must supply a large volume of immature clastic sediment. Schuchert (1923) postulated a subcontinental mass, termed a *borderland,* as the source. The borderland thought to have supplied sediment to the Appalachian trough was designated Appalachia. It is presumed to have since foundered or disappeared. Foundering of similar source lands is thought to have taken place elsewhere despite isostatic considerations (Kuenen, 1959). Others have disputed the borderland concept and attributed the sediments to a volcanic island chain, even though islands supply little sediment and in many geosynclines sediments are clearly not of volcanic origin (Kay, 1951, pp. 31–33). Others have supposed the sediments to have been initially derived from the craton, deposited as a continental terrace or rise which was later uplifted and "recycled" in a reverse direction, from the uplift back into a molasse trough between the uplift and the craton (Dietz, 1963). In the case of the Appalachians, however, the craton was early covered by an armor of limestone precluding it as a source for clastics. Those who accept continental drift presume the source to have been a continental block which has since been separated by rifting and sea-floor spreading. Apparently under this view the Appalachian Paleozoic fill was derived from the African continental block.

Another problem has been the nature of the margins of the geosynclinal belt. On the one side the border is clearly that of the craton. Presumably the transition from craton to geosyncline was gradual. The other margin of the geosynclinal belt is very uncertain. The inner zone is the zone of metamorphic and magmatic rocks. In many cases, such as the Appalachians, it is not clear whether the rocks of this zone are the basement on which the geosynclinal sequence rested or whether it is a high-grade metamorphic facies of the sedimentary fill. This boundary of the geosyncline is thus lost in a magmatic-metamorphic "haze." Some—those who consider the Gulf Coast accumulation to be geosynclinal—assume that there was no other boundary, that is, the geosyncline was "open," bounded by the continent on one side and an ocean basin on the other. Such persons prefer not to use the word *geosyncline* for such accumulations but call them *geomioclines* (if they are amagmatic).

Related to this question is that of the nature of the basement on which the geosynclinal fill rests. On the cratonic margin, normal cratonic rocks (mainly granitic gneisses) of continental or sialic character pass beneath the geosynclinal sequence. Because the other margin is obscure, its basement surface is less clear. Some have supposed it to be oceanic crust rather than continental crust. The basis for this view is the flood of ophiolitic materials interbedded with sediments of some geosynclinal sequences. On the other hand, the character of the sediments themselves, particularly their quartz content, suggests a sialic source, an observation incompatible with the notion of a tectonic land of volcanic or oceanic character.

The biggest problem has been the many speculations on and search for the modern counterpart of the geosyncline. Grabau (1944, p. 126) thought "that there are no normal flooded geosynclines in the present geological age." Most geologists today think otherwise, but agreement on the matter is wholly lacking.

Dietz (1963), in his search for an actualistic model of a geosyncline, arrived at the conclusion that the continental rise was the appropriate one. This feature is indeed a thick accumulation of sediment which in comparatively stable areas is a broad feature of great length. The Gulf Coast "geosyncline" is an appropriate example. Sediments, however, are derived from the continent. They have, therefore, the wrong "polarity," inasmuch as this is the reverse of the case in the ancient geosynclinal belts.

Others have considered the oceanic trenches—those long linear deep furrows marginal to many island arcs and which also border some continents—to be modern geosynclines. Scholl and Marlow (1974), however, have shown that, in the absence of glaciation, the typical filling of a modern Pacific trench is a thin mantle (only some 500 m in the whole interval between late Mesozoic and the present). Many eugeosynclines hold an enormous volume of sediment deposited in a far shorter interval of time.

Some have considered the whole ocean basin to be a "geosyncline" whose accumulated sediments are transported to mobile belts by plate movement related to sea-floor spreading and thus by subduction incorporated into the continental margin (Fischer, 1974). This model runs into difficulty by its inability to explain the lack of pelagic carbonates in eugeosynclinal sequences. Such carbonates, if present oceans are representative, should form a very large part of the total accumulation. They do not.

More recent efforts, based on the plate-tectonic theory, lead to other models (Dickinson, 1971). In addition to the trench, two other sites are prone to become loci for sedimentary accumulation; one is the interval between the trough proper and its related island arc ("arc-trench gap"), and the other is behind the arc, that is, the continental side of the arc (Fig. 16-4). In modern arc-trench systems one finds such elongate sediment basins. These may be 75 to 275 km wide and several hundreds of kilometers in length. The Mesozoic Great Valley sequence, an inordinately thick turbidite section, is thought by Dickinson to have accumulated in an arc-trench gap basin. Under this model, sedimentary fill should come from the arc—a volcanic chain—or from medial ridges in the arc-trench gap. None can come from the oceanward side. Clastic sequences of great thickness may also accumulate in back of the arc, on the side away from the trench, in a foreland basin. The broad Late Mesozoic basin in the interior of North America between the Cordilleran fold belt on the west and the Paleozoic platform and Precambrian shield on the east, is presumed to be such a back-arc geosyncline.

From what has been said and written, the modern geosyncline is an elusive feature. Refer to the recent symposium on the problem (Dott and Shaver, 1974).

OROGENIC SEDIMENTS

FLYSCH

As noted above, there are two major facies of sediments associated with geosynclines or mobile belts. These are the *flysch* and the *molasse* facies.

The flysch is so-called from its resemblance to the Flysch of the Alps. It is the most characteristic sedimentary suite of Alpine-type orogenic belts. It was termed the argillaceous facies by Jones (1938), because of the dominance of shales, and the graywacke facies by Pettijohn, because the characteristic sandstones of this suite are graywackes.

The flysch facies is marked by its great thickness and its predominantly argillaceous nature. It is almost wholly clastic, although at some stages bedded cherts or radiolarites were deposited. The thickness is phenomenal and is measured in thousands of meters. Deposition was continuous or nearly so and hence without interruption. The bedding is well marked, uniform, and rhythmic (Fig. 16-5). The principal material is shale or silty shale regularly interbedded with graded beds of dark sandstone (usually graywackes) varying from a few centimeters to a meter or two in thickness. These strata are referred to as rhythmites—sandstones as turbidites. Graded bedding is the rule; large-scale cross-bedding is absent, although ripple bedding may be present.

The sandstones are microbreccias or grits, contain numerous shale fragments, and, as pointed out elsewhere (p. 561), are not the product of normal bottom currents. They appear to record infrequent, somewhat catastrophic, incursions of slump-generated turbidity underflows. Such flows transfer coarse sediments once deposited in comparatively shallow waters to a deep, still-water environment. In some flysch sequences beds of coarser chaotic materials often contain large erratic blocks. These are true mudflows and slump deposits.

Associated shales accumulated slowly, with accumulation periodically interrupted by the almost instantaneous deposition of sandy beds. The proportion of shale to sand varies, but in general the section is largely shale (Table 16-2). In general the mud–sand ratio is about that carried by large rivers today. The average Mississippi Delta deposit is 29 percent sand and 71 percent mud (silt plus clay). As can be seen from the table, the average of several sections shows a sand–shale ratio of about 31 : 64.

Noteworthy is the absence of carbonates, except as concretions in siltstone beds. Limestone has, however, been reported as graded beds in some sections—the presumed product of an intrusion of a carbonate-laden turbidity flow.

Some flysch sequences—such as the Great Valley sequence in California—lack volcanic materials. Others, such as the Franciscan of the same region, are closely associated with pillowed lavas, commonly altered to greenstone, an association long recognized as characteristic of Alpine belts (Tyrrell, 1933).

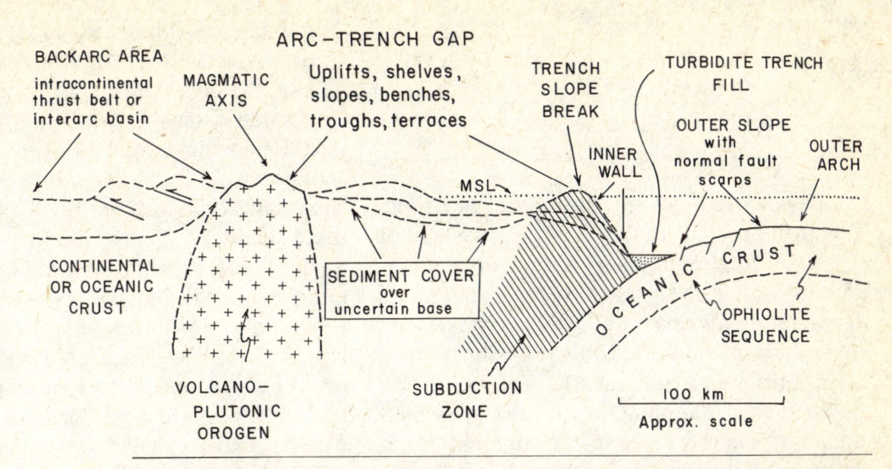

FIG. 16-4. Cross section showing main geotectonic elements of arc-trench system. Vertical exaggeration approximately ×10. (After Dickinson, 1973, Fig. 1.)

These basic flows appear to have erupted on the sea floor concurrently with sedimentation. Closely associated also are water-laid basic tuffs which grade into and resemble coarser graywackes.

Slump bedding, convolute folding, and injection of graywacke dikes and sills characterize some flysch.

The origin of the flysch facies with its rhythmic interbedding of shale and graded graywackes was long misunderstood. The coarseness of the grits and the generally unfossiliferous nature of the beds, or the presence in some cases of plant debris, led some workers to believe the sediments were

FIG. 16-5. Cretaceous turbidite flysch, Devils Slide area, California.

TABLE 16-2. Proportions of rock types in "flysch" facies

	A	B	C[b]	D	E	F
Percent sandstone	42	17	32	27	39	31
Percent shale	58[a]	66	61	73	59	64
Percent limestone	trace	17	—	trace	1–2	4
Percent conglomerate	1	—	—	—	trace	—

[a] Includes sandy shale.
[b] Includes 7 percent siltstone.

A. Upper Cretaceous (Chico), California (Trask and Hammer, 1934, p. 1366).
B. Carboniferous, Anadarko basin, Oklahoma (Bokman, 1954).
C. Carboniferous (Stanley), Arkansas–Oklahoma (Bokman, 1953, p. 155).
D. Tertiary, northern Sumatra.
E. Cretaceous, southern California (Briggs, 1953, p. 35).
F. Average of A through E.

shallow-water or even continental in origin. The evenness of the beds, the absence of cross-bedding, and the common occurrence of radiolarites led others to regard them as deep-water marine. Radiolarian cherts of some sections and graptolites of others leave no doubt about the marine origin of the flysch facies. The general absence of a benthonic fauna other than soft-bodied mud-eating forms, the thin and even-bedded sand layers, and the corresponding absence of large-scale cross-bedding or of wedging, lenticular bedding, and scour-filled channels all suggest deep water. The deep-water Foraminifera of some turbidite sequences strongly support a deep-water depositional site for the flysch facies. The turbidity-flow mechanism readily explains the grading characteristics of the coarser grits, the occurrence of such coarse clastics in a deep-water environment, and the presence of a shallow-water microfauna in the sandstones.

Inasmuch as the flysch facies is found only in excessively thick and sharply deformed sections, it is a product of geosynclinal sedimentation. It accumulated rapidly and without interruption in a comparatively deep-water (a thousand meters or more) marine environment. The sediments consist of the waste products of a high land mass, inasmuch as they are texturally and mineralogically very immature. They were derived from both low-rank metamorphic and crystalline terranes—in part, perhaps, from strata earlier deposited in the same geosyncline.

The flysch facies is very common, and good descriptions of it abound in the literature. The general features of flysch have been summarized by Sujkowski (1957), Dzulynski and Smith (1964), and Dzulynski and Walton (1965). Many of the Archean sections in the Canadian Shield are flysch-type sequences (Pettijohn, 1943; Donaldson and Jackson, 1965; Walker and Pettijohn, 1971; Henderson, 1972). The later Precambrian (Tyler, Michigamme and Thomson Slate) of the Lake Superior region are flysch sequences (Morey and Ojakangas, 1970). The Ordovician Martinsburg of the central Appalachians (McBride, 1962) and rocks of similar age in the St. Lawrence region of Quebec (Enos, 1969) are excellent flysch. The same facies is well displayed in some late Paleozoic sequences of the Ouachita Mountains of Arkansas and Oklahoma (Bokman, 1953; Cline, 1970) and the Marathon Basin of Texas (McBride, 1966). The Early Paleozoic of the Caledonian geosyncline of Great Britain, especially in Wales (Wood and Smith, 1959), and the Late Paleozoic of the Armorican geosyncline of southern England and Wales, and the Harz and central Rhine areas of Germany are excellent flysch (Fiege, 1937; Kuenen and Sanders, 1956). This suite characterizes the Mesozoic Great Valley sequence of California (Ojakangas, 1968). It is found in Alaska, the Himalayas (Valdiya, 1970), the Caucasus (Vassoevich, 1953), the Carpathians (Dzulynski, Ksiazkiewicz, and Kuenen, 1959), the Sierra Madre Orientale (Chicontopec Eocene), and the Andes of Chile (Scott, 1966). No doubt it is present in every Alpine-type chain in the world.

Not all turbidite sequences, however, are true flysch. The turbidite mechanism works even in lakes and in the pro-delta area so that one should not conclude that a graded bed is unequivocal evidence of flysch sedimentation.

MOLASSE

The other facies related to orogenic uplift is the *molasse* facies. The type molasse is a thick clastic sequence of Tertiary age found in the Swiss Plain and in the Alpine foreland of southern Germany. Rocks of similar character and relations to Alpine-type orogenies elsewhere have been also designated molasse (van Waterschoot van der Gracht, 1931, p. 998).

The molasse association consists mainly of sandstone and shales. In general it is coarser than flysch and contains notable conglomerates. In the proximal region it, like the flysch, may be several thousand meters thick. Thicker molasse belongs to the orogenic belt and is itself deformed; thinner molasse is a more distal portion of the same clastic wedge. It may cover the cratonic platform far beyond the orogenic belt.

The sands and shales of the molasse are

immature products of denudation. Sands are mainly lithic sandstones and locally protoquartzites (sublitharenites). Shales are silty and micaceous and vary from gray to red. Clay ironstone concretions, septaria, and plant fossils are common in darker shales. The more distal molasse is apt to contain coals and associated underclay; thin, nodular freshwater limestones may be present. Thin marine strata, including marine shales and limestones, may intertongue with the distal molasse.

The more proximal molasse is commonly organized into fining-upward cycles (Fig. 16-6). The sandstone member, perhaps conglomeratic, has a channeled base. It passes upward into siltstones and shales. The latter are apt to be red. These molasse cycles have been well described by Bersier (1948, 1959) and Füchtbauer (1967). In the coal-bearing molasse ("coal measures") such cycles include coal and underclay and in some cases a marine limestone. These are the "cyclothems" of Weller and others (Weller, 1930; Wanless and Weller, 1932). See Fig. 16-7.

In some molasse sequences, clastic sediments, especially mudstones, are red. The red color and the mudstone conglomerates —with mud fragments in a sand matrix— indicate oxidation and desiccation and hence deposition under subaerial conditions.

The molasse association is thus neither continental nor marine but rather both; it consists of an association of sediments formed in varying local environments including those of the beach and foreshore, the tidal lagoon, and alluvial deposits in part deltaic, in part floodplain with its backwater swamp, and in part alluvial fans. In short, the area of molasse sedimentation is a deltaic coastal plain and its inland extension to the mountain front (Fig. 14-14).

The sediment supplied to this domain is partitioned among various local environments and is differentiated texturally (and hence chemically and mineralogically) according to the energy input of the local situation. In the most proximal areas coarse fan gravels accumulate; in less proximal

FIG. 16-6. Fining-upward molasse cycles in the Catskill Formation (Devonian), near Harrisburg, Pennsylvania. Prominent ledges are coarse sandstones; recessed beds are dark red silty mudstones.

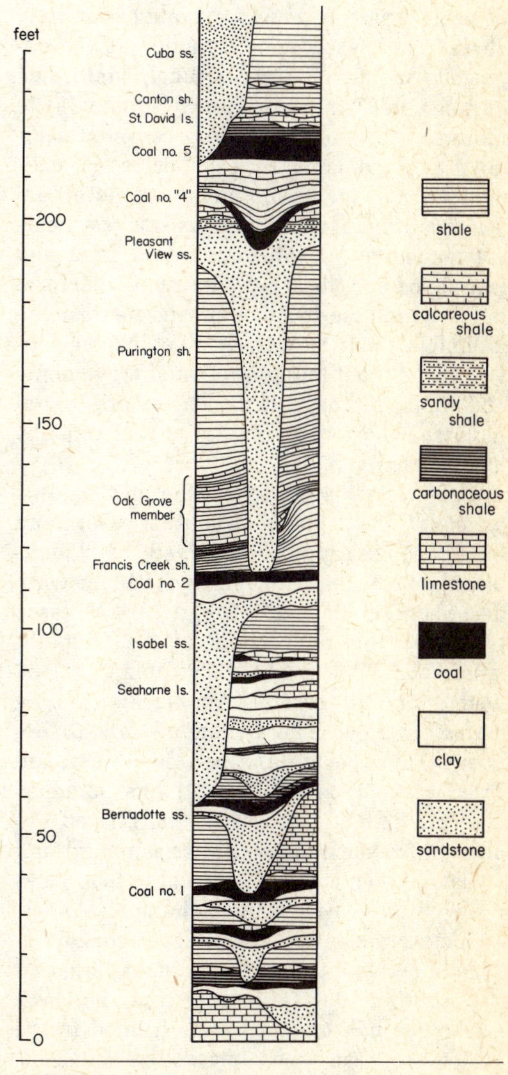

FIG. 16-7. Cyclothemic deposits. (After Wanless, 1931.)

situations the deposits are organized into fining-upward alluvial cycles. These pass more distally into shore-zone deposits of shoal-water deltas and interdeltaic deposits of lagoons and beaches. Fluvial sands are immature but may become both texturally and mineralogically more mature in the shore zone.

In general, the shore is prograding so that the molasse type sediments encroach upon and replace offshore marine sediments. The molasse in general is a vast apron of waste derived from the newly elevated mountain chain. The flysch epoch ended; newly deposited sediments were strongly deformed and elevated and laid bare by erosion. The molasse, therefore, is mainly recycled sediment and contains rock particles of sedimentary and low-grade metamorphic origin. The resulting accumulation is thickest nearest the newly elevated mountain tract and is trapped in a basin of its own—the foreland or molasse basin. Deposits nearest the mountain front may themselves become involved in continuing orogenic movements and thus be deformed. The more distal molasse which overlaps the craton remains undeformed.

The Alpine Molasse is the best-known example. The deposits of southern Germany have been carefully described by Füchtbauer (1967); those of the Swiss Plain were described by Bersier (1948, 1959). Molasse of comparable age occurs in the Carpathians. The Devonian-Mississippian clastic wedge in the central Appalachians is a molasse wedge. The Devonian Catskill deposits have been described by Allen and Friend (1968); Meckel (1970) has summarized the salient features of the related Pocono and Mauch Chunk (Mississippian) and Pottsville (Pennsylvania). Molasse sequences are well known from the Colombian Andes.

Some sedimentary assemblages closely resemble molasse in texture, composition, and organization but are not true molasse. Graben or half-graben basins formed in the craton are filled with molasselike materials. The Triassic Newark Series and the Precambrian Keweenawan clastics of the Lake Superior region are examples. These deposits differ from the molasse only in tectonic setting. They are not related to or associated with any flysch; they are not involved in Alpine-type folding, being only tilted slightly and block faulted. They resemble the molasse in that they are mainly coarse nonmarine clastics and related to sharp relief.

PREFLYSCH FACIES

A minor component of some geosynclinal accumulations is the thin sequence which in some cases precedes the flysch. This sequence, mainly black shales but including some cherts and radiolarites, records a long period of very slow sedimentation preceding the rapidly deposited influx of flysch sediments. This assemblage is a deep-water "starved basin" suite.

This association has been described in some detail by Aubouin (1965, pp. 112–129). It includes banded cherts—radiolarites and radiolarian jaspers. In some situations these are associated with the ophiolites or greenstones. In other cases a little siliceous limestone is present. Red nodular limestones, the *ammonitico rosso*, are present in the so-called generative stage of the geosyncline. The abundance of ammonites—a veritable shell bed—indicates exceedingly slow deposition. In Pettijohn's observation, black shale is the most common "starved basin" sediment. In some black shale sections chert is present both as scattered nodules, or more typically as thin layers interbedded with black shale. Closely associated with some black shales are other highly reduced sediments such as sedimentary iron sulfides and siderite.

Euxinic sediments seem commonly to occur above a carbonate-orthoquartzite assemblage and below the typical graywacke-shale of the flysch facies. Their occurrence is perhaps a record of transition between open-sea, shallow-water deposition of carbonates and craton-derived sands, and the influx of sediments from a rising island arc or other barrier. The initial rise of the barrier results in partial isolation of the locus of deposition and restricts or inhibits circulation—conditions conducive to stagnation of bottom waters and deposition of black shales. The absence of sediment either from the cratonic foreland or a barrier leads to basin starvation.

THE GEOSYNCLINAL CYCLE

Initial sediments in a mobile tract may be quite different from those which follow. Observation in various areas has led to the concept of a geosynclinal cycle. Bertrand (1897) was one of the first to formulate the concept and to recognize that several facies followed one another in a regular way. According to Bertrand, the normal sequence in a geosynclinal belt is: gneissic facies (basement of older rocks), shaly flysch (a thick argillaceous facies) deposited in the axial portion of the geosyncline which is debris derived from older rocks; coarse flysch (including *Wildflysch*), a border deposit derived by reworking of the concurrently uplifted older strata of the geosyncline; and gravels and grits (red-grit facies) deposited at the foot of the mountains after elevation of the chain. The concept was apparently largely an outgrowth of the study of the history of the Alpine geosyncline. Bertrand, however, noted that this succession of facies was repeated in other geosynclinal tracts of other periods.

Krynine (1941, 1942, 1945) attempted to generalize this concept of Bertrand and to apply it to the Appalachian geosyncline. Krynine's three tectonic stages include: peneplanation (or early geosynclinal) characterized by deposition of first-cycle orthoquartzites and carbonates on a fluctuating flat surface; a geosynclinal stage proper marked by trough deposition interrupted by marginal upwarping and shift of earlier-deposited sediments to the center of the trough after low-rank metamorphism (the graywacke suite), and postgeosynclinal stage or uplift (commonly marked by faulting) after folding and magmatic intrusion of the geosyncline marked by arkose (Fig. 16-8).

The validity of these generalizations has been questioned (Kay, 1951, p. 88). It is clear, however, that there are certain recur-

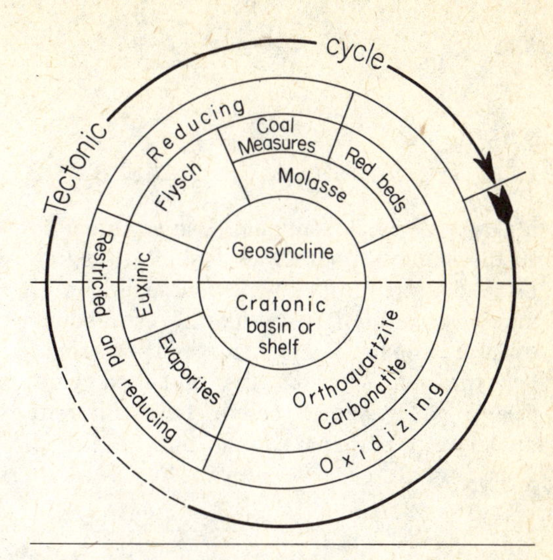

FIG. 16-8. Generalized tectonic cycle and principal sedimentary facies. (Very much modified from Krynine, 1945, Fig. 2.)

rent natural consanguineous assemblages of sedimentary rocks and that in many cases these assemblages succeed one another in a regular or systematic manner. Such sequential order repeated many times during the Precambrian and later eras gives considerable support to the concept of geosynclinal cycle. This concept is briefly summarized in Table 16-3. The concept has been more fully developed by Aubouin (1965), based mainly on his studies of the Hellenide chain. The evolution of a geosyncline, according to Aubouin, is much more complex, because it involves the history of two parallel troughs separated by a medial geanticline (Fig. 16-3). The two troughs have a somewhat different history, the internal one (eugeosyncline) being first formed and characterized by a flysch with volcanic components, the external one (miogeosyncline) being filled in a later stage with a nonvolcanic flysch. Deformation begins in the first-filled trench and proceeds during the filling of the second. Following uplift of the deposited sediments, a foredeep forms and becomes filled with a coarse molasse.

Engel and Engel (1953, p. 1038) suggest that "the recently proposed tectonic frameworks of sedimentation are so severely tailored that only a few sedimentary series can wear them well." Certainly too few sections have been properly analyzed to permit safe generalizations. Our concepts are highly colored by studies of Alpine and Appalachian geosynclinal sections. Our problem is further complicated by our inability to agree on a definition of a geosyncline and still more by our failure to identify a modern example. Some sequences, though termed geosynclinal, such as the so-called Gulf Coast "geosyncline," belong to a different model and perhaps should not be called geosynclinal at all.

GEOSYNCLINES AND PLATE TECTONICS

The concept of plate tectonics has reopened the whole question of the relations of sedimentation to tectonics—a subject of a symposium in Madison, Wisconsin, in the fall of 1972 (Dott and Shaver, 1974). According to the plate concept, a mobile belt is a zone where two crustal plates are in collision. In such a zone deep linear furrows form which become traps for sediments. It is not yet clear what the nature of these "geosynclines" is, the precise source and nature of their fill, or their sequential history, if any. Some attempt has been made to probe these questions and to formulate answers (Dickinson, 1971; Dott and Shaver, 1974). Briefly the several basins include the *trench* or furrow formed by downbending of an oceanic plate beneath a continental plate, one or more furrows in the zone between the trench and the volcanic arc generated, and the back-arc basin or trough formed between the volcanic arc and the cratonic interior of the continent.

This general model raises several questions which need further study. The source of sedimentary fill of these several basins has not been fully identified. The volcanic arc is the most obvious elevated tract susceptible to erosion, and indeed some geosynclinal sequences have a major volcanic component. Others, perhaps most, do not; the nature and volume of sediments imply a large land mass of sialic nature. A volcanic island chain is incapable of producing either the kind or volume of sediment observed. Moreover, paleocurrent data preclude the back-arc cratonic area as a source. The source land, therefore, must lie on the opposite side of the arc and consist of sialic continental rocks. Indeed, even in the Pacific coastal area of the United States, which has become something of a model for the

TABLE 16-3. Major tectonic cycles and associated sediments

Stage 1

Cratonic conditions and deposition or shield-derived orthoquartzites and of carbonates upon flooded craton or along cratonic border of geosyncline. More distal parts of geosyncline are "starved" and receive little or no sediment.

Stage 2

Mild upwarp in geosyncline and elevation of geanticlinal ridge; silled basin deposition of cherts, bedded siderites, pyrite, black shales, and phosphorites (euxinic facies).

Stage 3

Strong upward of geanticlinal ridges and island arcs with flood of clastic materials (flysch) increasing in coarseness upward; submarine extrusives and tuffs.

Stage 4

Completion of trough filling; conversion to nonmarine stage. Paralic sedimentation (molasse), at first reducing (delta and swamp) and then oxidizing (fluvial).

Stage 5

Deformation and uplift.

plate-tectonic concept, there is some evidence of a sialic source west of the present coast—an anomaly not well explained.

We can say, in conclusion, that the relation of tectonics to sedimentation remains an unsettled problem, that the plate-tectonics model raises more questions than it answers, and that far more field work is needed to resolve present ambiguities and conflicts.

References

Allen, J. R. L., and Friend, P. F., 1968, Deposition of the Catskill facies, Appalachian region: Geol. Soc. Amer. Spec. Paper 106, pp. 21–74.

Aubouin, J., 1965, Geosynclines: Developments in geotectonics, v. 1, Amsterdam, Elsevier, 335 pp.

Bailey, E. B., 1930, New light on sedimentation and tectonics: Geol. Mag., v. 67, pp. 77–92.

———, 1936, Sedimentation in relation to tectonics: Bull. Geol. Soc. Amer., v. 47, pp. 1713–1726.

Bersier, A., 1948, Les sédimentations, rhythmiques synorogéniques dans l'avant-fosse molassique alpine: 18th Intern. Geol. Congr. London, pt. 4, sec. C., pp 83–93.

———, 1959, Séquences détritiques et divagations fluviales: Eclogae Géol. Helvetiae, v. 51, pp. 854–893.

Bertrand, M., 1897, Structure des alpes françaises et recurrence de certain facies sédimentaires: Comptes Rendus, Congr. Int. Geol., 6th Sess., 1894, pp. 163–177.

Bokman, J., 1953, Lithology and petrology of the Stanley and Jackfork formations: Jour. Geol., v. 61, pp. 152–170.

Briggs, L. I., Jr., 1953, Geology of the Ortigalita Peak Quadrangle, California: California Div. Mines Bull. 167, 61 pp.

Cady, W. M., 1950, Classification of geotectonic elements: Trans. Amer. Geophys. Union, v. 31, pp. 780–785.

Cline, L. M., 1970, Sedimentary features of Late Paleozoic flysch, Ouachita Mountains, Oklahoma: Geol. Assoc. Canada Spec. Paper 7, pp. 85–101.

Dana, J. D., 1873, On some results of the earth's contraction from cooling, including a discussion of the origin of mountains and the nature of the earth's interior: Amer. Jour. Sci., v. 5, pp. 423–443; v. 6, pp. 6–14, 104–115, 161–171.

Dapples, E. C., Krumbein, W. C., and Sloss, L. L., 1948, Tectonic control of lithologic associations: Bull. Amer. Assoc. Petrol. Geol., v. 32, pp. 1936–1937.

Dickinson, W. R., 1970, Tectonic setting and sedimentary petrology of the Great Valley sequence (abstr.): Geol. Soc. Amer. Prog. 2, p. 86.

———, 1971, Clastic sedimentary sequences deposited in shelf, slope, and trough settings between magmatic arcs and associated trenches: Pacific Geol., v. 3, pp. 15–30.

Dietz, R. S., 1963, Collapsing continental rises—an actualistic concept of geosynclines and mountain building: Jour. Geol., v. 71, pp. 314–333.

Donaldson, J. A., and Jackson, G. D., 1965, Archaean sedimentary rocks of North Spirit Lake area, northwestern Ontario: Canad. Jour. Earth Sci., v. 2, pp. 622–647.

Dott, R. H., Jr., and Shaver, R. H., eds., 1974, Modern and ancient geosynclinal sedimentation: Soc. Econ. Paleont. Min. Spec. Publ. 19, 400 pp.

Dzulynski, S., and Smith, A. J., 1964, Flysch facies: Ann. Soc. Geol. Pologne, v. 34, pp. 245–266.

Dzulynski, S., and Walton, E. K., 1965, Sedimentary feature of flysch and greywackes: Amsterdam, Elsevier, 300 pp.

Dzulynski, S., Ksiazkiewicz, M., and Kuenen, Ph. H., 1959, Turbidites in flysch of the Polish Carpathian Mountains: Bull. Geol. Soc. Amer., v. 70, pp. 1089–1118.

Engel, A. E. J., and Engel, C. G., 1953, Grenville Series in the northwest Adirondack Mountains, New York: Bull. Geol. Soc. Amer., v. 64, 1013–1097.

Enos, P., 1969, Anatomy of a flysch: Jour. Sed. Petrology, v. 39, pp. 680–723.

Fiege, K., 1937, Untersuchungen über zyklische Sedimentation geosynklinaler und epikoninentaler Räume: Preuss, Geol. Landesanst. Abh. Neue Folge, v. 177, 218 pp.

Fischer, A., 1974, The odd rocks of mountain belts, *in* Modern and ancient geosynclinal sedimentation (Dott, R. H., Jr., and Shaver, R. H., eds.): Soc. Econ. Paleont. Min. Spec. Publ. 19 (in press).

Fischer, G., 1933, Die Petrographie der Grauwacken: Jahrb. Preuss. Geol. Landesanst., v. 54, pp. 320–343.

Füchtbauer, Hans, 1967, Die Sandsteine in der Molasse nördlich der Alpen: Geol. Rundschau, v. 56, pp. 266–300.

Gilluly, J., Reed, J. C., Jr., and Cady, W. M., 1970, Sedimentary volumes and their significance: Bull. Geol. Soc. Amer. v. 81, pp. 353–376.

Grabau, A. W., 1944, The world we live in: Taiwan, Geol. Soc. China, 229 pp.

Henderson, J. B., 1972, Sedimentology of Archean turbidites of Yellowknife, Northwest Territories: Canad. Jour. Earth Sci., v. 9, pp. 882–902.
Jones, O. T., 1938, On the evolution of a geosyncline: Proc. Geol. Soc. London, v. 94, pp. lx–cx.
Kay, M., 1951, North American geosynclines: Geol. Soc. Amer. Mem. 48, 143 pp.
Krynine, P. D., 1941, Differentiation of sediments during the life history of a landmass (abstr.): Bull. Geol. Soc. Amer., v. 52, p. 1915.
———, 1942, Differential sedimentation and its products during one complete geosynclinal cycle: An. Congr. Panamericano Ing. Minas Geol. Santiago, Chile, pp. 536–561.
———, 1943, Diastrophism and the evolution of sedimentary rocks: Pennsylvania Min. Ind. Tech. Paper 84-A.
———, 1945, Sediments and the search for oil: Producers Monthly, v. 9, pp. 17–22.
———, 1950, Petrology, stratigraphy, and origin of the Triassic sedimentary rocks of Connecticut: Bull. Connecticut State Geol. Nat. Hist. Surv. 73, 247 pp.
Kuenen, Ph. H., 1959, La topographie et la géologie des profondeurs océanique: Coll. Int. Cent. Nat. Rech. Sci., v. 83, pp. 157–163.
Kuenen, Ph. H., and Sanders, J. E, 1956, Sedimentation phenomena in Kulm and Flozleeres graywackes, Sauerland and Oberharz, Germany: Amer Jour. Sci., v. 254, pp. 649–671.
McBride, E. F., 1962, Flysch and associated beds of the Martinsburg Formation (Ordovician), central Appalachians: Jour. Sed. Petrology, v. 31, pp. 39–91.
———, 1966, Sedimentary petrology and history of the Haymond Formation (Pennsylvania), Marathon Basin, Texas: Bur. Econ. Geol. Univ. Texas, Rept. Inv. 57, 101 pp.
Meckel, L. D., 1970, Paleozoic alluvial deposition in the central Appalachians—a summary, *in* Studies of Appalachian geology (Fisher, G. W., Pettijohn, F. J., Reed, J. C., Jr., and Weaver, K. N., eds.): New York, Wiley-Interscience, pp. 49–68.
Morey, G. B., and Ojakangas, R. W., 1970, Sedimentology of the Middle Precambrian Thomson Formation: Minnesota Geol. Surv., Rept. Inv. 13, 32 pp.
Moss, R. G., 1936, Buried pre-Cambrian surface in the United States: Bull. Geol. Soc. Amer., v. 47, pp. 935–966.
Ojakangas, R. W., 1968, Cretaceous sedimentation, Sacramento Valley, California: Bull. Geol. Soc. Amer., v. 79, pp. 973–1008.
Pettijohn, F. J., 1943, Archean sedimentation: Bull. Geol. Soc. Amer., v. 54, pp. 925–972.
Potter, P. E., and Pryor, W. A., 1961, Dispersal centers of Paleozoic and later clastics of the upper Mississippi Valley and adjacent areas: Bull. Geol. Soc. Amer., v. 72, pp. 1195–1250.
Reinemund, J. A., 1955, Geology of the Deep River coal field of North Carolina: U.S. Geol. Surv. Prof. Paper 246, 159 pp.
Scholl, D. W., and Marlow, M. S., 1974, The sedimentary sequence in modern Pacific trenches and the apparent rarity of similar sequences in deformed circum-Pacific eugeusynclines, *in* Modern and ancient geosynclinal sedimentation (Dott, R. H., Jr., and Shaver, R. H., eds.): Soc. Econ. Paleont. Min. Spec. Publ. 19 (in press).
Schuchert, C., 1923, Sites and nature of the North American geosynclines: Bull. Geol. Soc. Amer., v. 34, pp. 151–229.
Schwab, F. L., 1969a, Cyclic geosynclinal sedimentation: a petrographic evaluation: Jour. Sed. Petrol., v. 39, pp. 1325–1343.
———, 1969b, Geosynclines: what contribution to the crust?: Jour. Sed. Petrol., v. 39, pp. 150–158.
———, 1971, Geosynclinal compositions and the new global tectonics: Jour. Sed. Petrology, v. 41, pp. 928–938.
Scott, K. M., 1966, Sedimentology and dispersal pattern of a Cretaceous flysch sequence: Bull. Amer. Assoc. Petrol. Geol., v. 50, pp. 72–107.
Sujkowski, Zb. L., 1957, Flysch sedimentation: Bull. Geol. Soc. Amer., v. 68, pp. 543–554.
Tercier, J., 1939, Dépôts marins actuels et séries géologique: Eclogae Géol. Helvetiae, v. 32, pp. 47–100.
———, 1947, Le flysch dans la sédimentation Alpine: Eclogae Géol. Helvetiae, v. 4, p. 163.
Trask, P. D., and Hammer, H. E., 1934, Preliminary study of source beds in late Mesozoic rocks on west side of Sacramento Valley, California: Bull. Amer. Assoc. Petrol. Geologists, v. 18, pp. 1346–1373.
Tyrrell, G. W., 1933, Greenstones and greywackes: Comptes Rendus Reunion Int., Étude Precambrian, pp. 24–26.
Valdiya, K. S., 1970, Simla slates; the Precambrian flysch of the lesser Himalaya, its turbidites, sedimentary structures and paleocurrents: Bull. Geol. Soc. Amer., v. 81, pp. 451–468.
Vassoevich, N. B., 1953, On some flysch textures: Trans. Soc. Geol. Lwow, ser. geol., v. 3, pp. 17–85 (in Russian).
Walker, R. G., and Pettijohn, F. J., 1971, Archaean sedimentation; analysis of the Minnitaki Basin, northwestern Ontario, Canada: Bull. Geol. Soc. Amer., v. 82, pp. 2099–2130.

Wang, C. S., 1972, Geosynclines in the new global tectonics: Bull. Geol. Soc. Amer., v. 83, pp. 2105–2110.

Wanless, H. R., and Weller, J. M., 1932, Correlation and extent of Pennsylvanian cyclothems: Bull. Geol. Soc. Amer., v. 43, pp. 1003–1016.

van Waterschoot van der Gracht, W. A. J. M., 1931, Permocarboniferous orogeny in south-central United States: Bull. Amer. Assoc. Petrol. Geol., v. 15, pp. 991–1057.

Weeks, L. G., 1952, Factors of sedimentary basin development that control oil occurrence: Bull. Amer. Assoc. Petrol. Geol., v. 36, pp. 2071–2124.

Weller, J. M., 1930, Cyclical sedimentation of the Pennsylvanian period and its significance: Jour. Geol., v. 38, pp. 97–135.

Wood, A., and Smith, A. J., 1959, The sedimentation and sedimentary history of the Aberystwyth Grits (Upper Llandoverian): Quart. Jour. Geol. Soc. London, v. 114, pp. 163–195.

17

SEDIMENTS AND EARTH HISTORY

INTRODUCTION

Perhaps because of recent lunar exploration and space probes, there is a renewed interest in the evolution of the planets and their atmospheres. It has become fashionable among cosmochemists and planetologists to construct theoretical models of planetary evolution with special emphasis on the origin and evolution of planetary atmospheres. The earth, as a planet, is believed to have undergone a long and complex geochemical evolution leading to differentiation of the crust and to formation of its hydrosphere and atmosphere. The presumed history of these fluid envelopes has been discussed by various writers (Daly, 1907; MacGregor, 1927; Rubey, 1951, 1955; Urey, 1952; Holland, 1965; Cloud, 1968; Ronov, 1968, 1972; Goody and Walker, 1972; Garrels and Mackenzie, 1971, pp. 285–296). Only a few, however, have searched the geologic past for a record of what actually, rather than what was presumed to have, happened (Eskola, 1932; Pettijohn, 1943, 1972; Rankama, 1954; Roscoe, 1973).

All writers agree that, at some time early in the earth's history, conditions at the surface of the earth were radically different from those of the present time. The oceans were presumed to have a lesser volume than now; the salinity was perhaps much less. The atmosphere may have had a higher CO_2 content and even may have been oxygen-free. The land surface was devoid of plant cover and was subject to intense and lethal ultraviolet radiation. Such differences should drastically alter processes of weathering, erosion, biological activity, and chemical sedimentation. Is there, in the sedimentary record, any evidence of a secular trend in composition or other characteristics to support these concepts? And what is the nature of the earliest sediments?

Veizer (1973) has summarized current thinking on these questions. He notes that (1) graywacke is the predominant sandstone in the Archean, that arkoses are most abundant in the early Precambrian; (2) iron-formation and manganese were deposited worldwide in greatest volume between 1800 and 3400 million years ago; (3) red beds made their appearance between 1800 and 2000 million years before the present; (4) limestones and dolomites are very rare in the Archean record; (5) sedimentary phosphates became common only since 1000 million years ago; (6) calcium sulfates and other evaporites are present only in rocks less than 600 million years old; and (7) coal is largely confined to rocks less than 350 million years old. Veizer also notes certain geochemical trends shown by analyses of shales and other common sediments, namely a secular variation in the alkalies K and Na, increasing total Fe with the age of the sediment, and increase with age in the MgO–CaO and SiO_2–Al_2O_3 ratios.

Geologists have found that the record of the past can be understood in terms of the present. This concept, variously termed the actualistic principle or the doctrine of uniformitarianism, has indeed been fruitful. There seems to be no rock or structure, even in the earliest Precambrian terranes, that cannot be found in later periods. This observation gives strong support to the view that, as far as the record goes, conditions during earliest times were not materially different from those of the present and that secular changes of large magnitude are nonexistent. This conclusion, however, has not been universally accepted. It has been pointed out that many sediments are products of biological activity, that many other sediments are modified or affected by the action of organisms, and that, inasmuch as the biota has undergone a vast and complex evolution, there should be some corresponding change in the kind and character of the sediments closely related to organic activity. The lime-secreting habit of invertebrates was not acquired until Cambrian times and clearly, therefore, limestones produced by such activity could not have formed in earlier times. Likewise, the absence of land plants in earlier times should have had a considerable effect on formation of soil, on rates of erosion, and hence on kind and rates of sedimentation. Some writers have argued that there is evidence in the geologic record of a reducing atmosphere in the earliest Precambrian and have presumed that the oxidizing atmosphere of the present is a result of the gradual release of oxygen by the photosynthetic action of green plants.

Despite the plausibility of these arguments, it is difficult to demonstrate any secular changes in the character of the sediments. The oldest strata, in excess of 3×10^9 years, consist of normal clastic rocks. Conglomerates contain well-worn cobbles and pebbles of many types of rocks (Pettijohn, 1943), showing that erosion and sedimentation went on then as now. These conglomerates "so perfectly resemble recent accumulations that, at first glance, they might be taken for such if it were not for the tilting up of the strata" (Eskola, 1932, p. 5). Tillites of Precambrian age demonstrate the presence of glaciers then as now. There certainly cannot have been a progressive cooling of the earth's surface as was once commonly believed. Many of the Precambrian sediments are mature. Some of the quartzites are accumulations of quartz sand not exceeded in roundness and silica content by any later sandstones. Such high concentrations of quartz imply thorough and complete weathering of quartz-bearing source rocks. Such mature weathering probably could not be accomplished in the absence of a plant cover on land. Despite the absence of fossils of land plants before the Devonian, it seems probable, therefore, that a plant cover existed even in the early Precambrian. The nature of the cover is not clear—perhaps it was a heavy cover of lichens and other primitive plants.

Although there is a general similarity between earlier sediments and those of later times, some recent work seems to show small or second-order differences in *average* composition of sediments of various ages. These apparent differences have been explained as related to differences in the biota of earlier times, to small but real differences in the composition of the atmosphere and hydrosphere, or to post-depositional alteration of the sediments which is progressive and nonreversible. If the differences are related to metamorphism or diagenesis, then no evolution of the earth's surface environment is implied.

EVOLUTION OF SOILS

Inasmuch as the composition of the atmosphere and the presence or absence of a plant cover might be expected to affect weathering processes and soil development, one would expect some secular change in the nature of paleosols. Yaalon (1962) has presented a theoretical analysis of the kinds of soils to be expected during earliest times. The oldest *protosoils* would be the product of anaerobic weathering of mafic rocks with little or no separation of Si, Al, or Fe. *Primitive soils,* formed under an aerobic atmosphere but without a plant cover, would form with K fixed in illite and Mg and Ca reaching the sea and there precipitated as carbonates. With a land plant cover a *rudimentary pedosphere* would develop. Intensity of leaching and weathering would increase; and kaolinite would form, as would laterites and gley soils. Finally, with the advent of flowering plants in the Cretaceous, a fully evolved pedosphere characterized by clearly differentiated soil profiles would develop. Unfortunately our knowledge of paleosols, especially those of Precambrian age, is so meager that it is impossible to say whether or not the record of soil evolution bears out Yaalon's theoretical analysis.

EVOLUTION OF SHALES

The geochemical evolution of shales and slates has been traced by Nanz (1953) and more recently and in much more detail by Vinogradov, Ronov, and associates (Vinogradov and Ronov, 1956; Ronov and Migdisov, 1971). The data for pelitic materials from the Russian Platform are given in Table 17-1. As can be seen from the table, there are no great differences in composition between the older and the younger pelitic sediments. But there are some significant minor differences. The older rocks, more largely metamorphic, show ferrous oxide to exceed ferric oxide, a difference probably a result of reduction of iron during metamorphism (compare the ferric–ferrous oxide ratio of Paleozoic shales with that of Paleozoic slates, Table 8-7). Paleozoic and younger shales are richer in $CaCO_3$, which may reflect the carbonate fixation by organisms which began in the Cambrian. Likewise, the organic carbon content shows a significant increase in later rocks. If these two components are excluded and the balance recalculated, the percentages of silica, alumina, and other constituents show no significant differences in the pelites of various ages. There seems to be a slight decrease in *total* iron and soda as one proceeds from older to younger deposits. This suggests that the provenance of older pelites

TABLE 17-1. Chemical analyses of pelitic sediments and metasediments of Russian Platform

Constituent	Archean[a]	Early Proterozoic[b]	Later Proterozoic	Paleozoic	Mesozoic and Cenozoic
SiO_2	64.15	58.42	57.65	47.94	55.61
TiO_2	0.48	0.79	0.86	0.78	0.72
Al_2O_3	15.79	16.63	17.04	14.26	14.45
Fe_2O_3	2.47	2.53	4.26	4.10	3.65
FeO	3.51	6.56	3.17	1.85	1.95
MnO	0.07	0.15	0.11	0.06	0.06
MgO	2.45	4.12	2.38	3.85	2.20
CaO	4.02	2.34	1.20	7.53	4.90
Na_2O	2.84	1.57	0.93	0.65	1.10
K_2O	2.32	3.08	4.18	3.70	2.39
H_2O	1.68	2.85	6.19	6.07	7.67
P_2O_5	0.14	0.09	0.09	0.11	0.11
CO_2	0.04	0.59	1.10	6.80	3.62
SO_3	0.08	0.20	1.05	2.18	1.40
Organic	—	0.33	0.35	0.70	0.94
Total	100.04	100.25	100.68	100.73	100.87
Number of analyses[c]	247	460	34 (1226)	401 (6734)	259 (2764)

[a] Paragneiss.
[b] Phyllites, schists, and paragneisses.
[c] Numbers in parentheses are number of specimens involved.
Source: After Ronov and Migdisov (1971).

was more largely basic volcanic than that of their younger counterparts. The appearance of sulfate in Late Proterozoic and younger shales is noteworthy and was attributed by Ronov and Migdisov (1971, p. 179) to the changeover to an oxidizing atmosphere.

These data are generally in accord with the less-well-documented conclusions of Nanz (1953). To what extent do they establish secular trends? Do they reflect real differences in the surface environment during earlier times? The differences are small. Possibly they are related to sampling. Sediments of the fold belts (geosynclines) may differ in bulk composition from those of the platforms or shelves; continental shales may differ from marine shales, and, unless the sample is adjusted to take account of these differences, it may be biased. Also, to what degree is the bulk composition of metapelites, including paragneiss, comparable with unmetamorphosed shales? Was the metamorphism isochemical? Until these questions can be answered, interpretations of the analyses must be tentative.

What about the mineral composition of shales? It has been noted by Weaver (1967) that the clay-mineral suite seems to show a progressive change with time. Early Paleozoic and Precambrian shales are largely illitic; expandable clay minerals did not become important until the Carboniferous and dominant until the late Tertiary. Kaolin is generally scarce in earlier shales and becomes relatively more important only in Pennsylvanian and later shales (Fig. 17-1). These trends were attributed by Weaver in part to the appearance of land plants which supplied much humus and acidified the soils, favoring kaolin formation. Potassium was utilized by the plants, recycled, and retained on the land, leading to a relative increase in sodium in the sea which favors formation of montmorillonite. It seems more probable, however, in the light of Weaver's own work as well as that of others, that the development of illite, at the expense of montmorillonite (and perhaps kaolinite), in older rocks is a diagenetic transformation and hence not a record of environmental change but rather a result of the "aging" of the mineral assemblage.

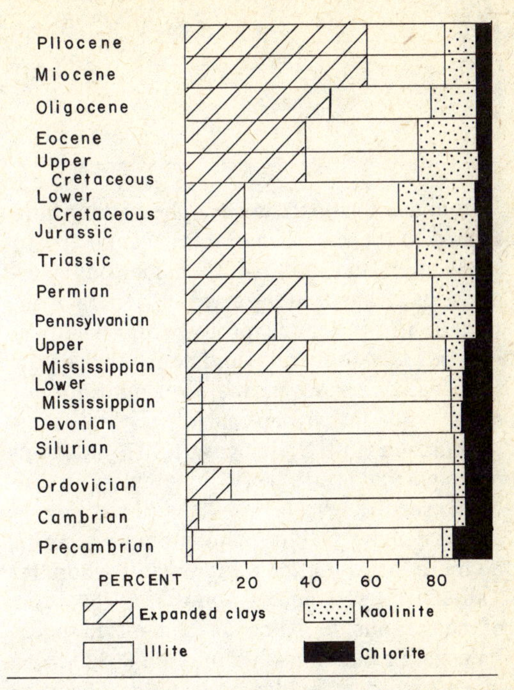

FIG. 17-1. Estimate of the variation of the clay mineral suite in North American shales with time. (After Weaver, 1967, Fig. 3, by permission of Pergamon Press Ltd.)

EVOLUTION OF THE SANDSTONES

Do older sandstones differ from those of later times? The immature character of graywackes, which are common in many Precambrian sections, has been attributed to the absence of a plant cover, and their high ferrous iron content has been ascribed to a reducing atmosphere (MacGregor, 1927). Neither conclusion is wholly justified. Younger graywackes are similar in all essential respects to those of older times (see Table 7-9).

Because the sands are much more sharply differentiated than are the shales, averages of sandstones of different ages are comparable only if the proportions of orthoquartzites, arkose, graywacke, and the like are in the same ratios as their abundance in each of the several periods being compared. We do not now have the data to make the needed comparisons. We cannot even be sure that the proportions preserved in the record are the same as those deposited.

Within each sandstone class, however, there seems to be no discernible secular trend other than that related to cementing constituents. The processes which led to fractionation of the sands, produced by weathering into the several sandstone classes or families, operated in the Precambrian not even with the exception of the Archean (Donaldson and Ojakangas, 1974), although orthoquartzites are very uncommon in the oldest record.

The cement of older sandstones seems to be more largely silica rather than carbonate. Tallman (1949) reports about a 50–50 ratio of calcareous to siliceous cement in post-Paleozoic sandstones but only an 80–20 ratio in Paleozoic and older rocks. This observation has been explained by substitution by or replacement of quartz for calcite in older rocks. Such diagenetic replacements, like the dolomitization of limestones, is largely irreversible, and its effects are cumulative.

It might be supposed that sands would become progressively enriched in quartz with the passage of time. Each cycle of weathering should reduce the feldspar content and enrich the sand in quartz. Younger sands should contain, on the average, a higher proportion of recycled quartz. Actual modal analyses of sandstones of diverse ages yield ambiguous results (Table 17-2). Data from North American sands show the Mesozoic and Tertiary sands to be notably more feldspathic than those of the Paleozoic. Sands of the Russian Platform show similar variations. The explanation is not readily apparent. In the North American case, the highest feldspar values are in Mesozoic sandstones of the Coast Ranges and Rocky Mountain areas and lowest in Paleozoic sandstones of the continental interior. However, Paleozoic sandstones in the Appalachian and Ouachita areas are also poor in feldspar although rich in rock particles. The high feldspar content of some of the Russian sandstones is attributed to tectonism, that is, rapid uplift and erosion of incompletely weathered material; the low feldspar content of others is correlated with low relief and crustal stability (Ronov, Mik-

TABLE 17-2. Feldspar content of sandstones

North America			Russian Platform		
Age	Number of formations	Feldspar (percent)	Age	Number of formations	Feldspar (percent)
Pre-Devonian	35	5.1	Precambrian	65	30.5
			Cambrian	18	16.6
			Silurian	14	9.6
			Devonian	177	8.9
Devonian-Permian	29	5.8	Carboniferous	95	4.8
Mesozoic	29	25.0	Triassic	5	61.6
			Jurassic	23	42.8
			Cretaceous	20	15.0
Tertiary	22	21.0	Tertiary	10	31.1
Pleistocene-Recent	—	15.3	Quaternary	8	22.6
Average (unweighted)		14.4	Average		15.3

Sources: North America, after Pettijohn, Potter, and Siever (1972); Russian Platform, after Ronov, Mikhailovskaya, and Solodkova (1963).

hailovskaya, and Solodkova, 1963, p. 225).

Most commonly cited differences between younger and older sandstones are the number and variety of the minor accessory minerals —the heavy minerals. As noted long ago by Boswell (1924), older sandstones have a very restricted heavy mineral suite. The number and variety of heavy minerals increase as the age of the sand decreases. Modern Pleistocene sands have the most diverse heavy-mineral assemblages (Pettijohn, 1941). See Fig. 13-5. These differences have been attributed either to increasing complexity of the terrane from which the sands were derived or to removal of less stable species from older sandstones (Boswell, 1924; Pettijohn, 1941). The older a sand, the greater is the probability that it has been leached and hence has lost the less stable heavy minerals. In a general way, the order of persistence of minerals in time is closely correlated with their stability. Not only do older sands contain a more restricted suite, but the mineral species present are the more stable ones.

A related problem is the question of the graywackes. Seemingly graywackes were a very common type of sandstone in the Precambrian and Paleozoic but are much less so in the Cenozoic. The modern equivalent is very rare or nonexistent. Cummins (1962) has attributed this observation to a process termed *graywackisation* by Kuenen (1966, p. 296). It appears that unstable minerals and rock particles were degraded and converted into a fine-grained matrix in older sands. Evidence for this is not only the secular trend in abundance of graywacke but the observation that where cemented early, as in concretions, matrix is absent, the breakdown process having been inhibited in such a "sealed environment" (Brenchley, 1969).

EVOLUTION OF THE CARBONATES

Some years ago Daly (1907, 1909) pointed out that earlier carbonate rocks appear to be richer in $MgCO_3$ than those of later times (Fig. 17-2), an observation confirmed by Vinogradov and Ronov (Ronov, 1972, Fig. 6).

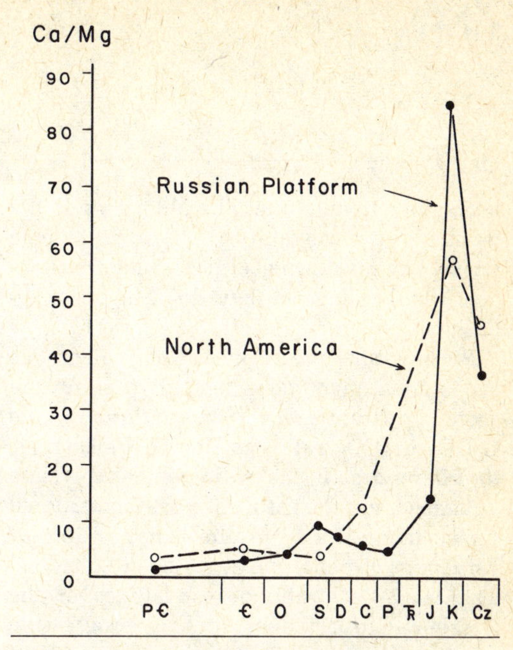

FIG. 17-2. Time variation of the Ca–Mg ratio in carbonate rocks of North America and the Russian Platform. (After Ronov, 1972, Fig. 6.)

Daly supposed that the lime and magnesia entering Precambrian (pre-Devonian) seas was quantitatively precipitated. Evidence for this is the similarity in the $CaCO_3$–$MgCO_3$ ratio in these rocks to that of streams draining the Precambrian shield of Canada. No accumulation of magnesium salts in the sea was believed to have occurred in Precambrian times. Daly explained this apparent quantitative removal of calcium and magnesium by the ammonia generated by decomposing organic matter. Later evolution of a scavenging fauna removed organic matter as rapidly as it formed. Lime-secreting forms removed calcium from sea water; magnesium, however, was left to accumulate. The composition of shales does not support the concept, inasmuch as earlier muds contain no more, perhaps less, organic carbon than those deposited after a scavenging fauna was supposed to have evolved. Daly's data may, however, be interpreted in another way. Ronov, for example, simply regarded the Ca–Mg trend as a function of compositional changes in the eroded rocks from which all sediments were derived (Ronov, 1972, p. 166). Also, circulating waters bearing magnesium are known to convert limestones to dolomite. The older a rock, the greater the probability that such change would have taken place. If so, older carbonates should show as they do, a higher magnesium content.

Presumably the cement of calcareous sandstones might be expected to show the same apparent secular variation in the $CaCO_3$–$MgCO_3$ ratio, as do the limestones, if differences in the ratio which correlate with age were the result of post-depositional replacement. Data on the composition of carbonate cements have not been collected. The MgO–CaO ratio of sandstones of the Russian Platform does show a progressive change from high values in the earliest rocks to lowest values in the youngest sandstones (Ronov, Mikhailovskaya, and Solodkova, 1963).

RELATIVE ABUNDANCE OF THE COMMON SEDIMENTS AND GEOLOGIC TIME

In Chapter 2, data on the relative abundance of sandstone, shale, and limestone were presented. The relative abundance of the several families of sandstones was presented in Chapter 7. Have these proportions remained constant through time? Some authors have noted the general absence of mature sediments, particularly orthoquartzites and limestones in Archean terranes (Pettijohn, 1943, p. 960). Ronov (1964) has shown, somewhat schematically, his general conclusions on the changing proportions of rock types and time (Fig. 17-3). Sandstones and shale form a more or less constant proportion through most of recorded time (the last 3.5 billion years). The relative importance of graywacke has declined and that of quartz arenites has increased with time. Most notable is the increase, from near zero in early Proterozoic, of evaporates and carbonate rocks, and the increase and later decline of jaspilites or banded iron-formations. Volcaniclastic sediments and lavas were presumed to be almost the only stratiform deposits in the earlier, somewhat hypothetical period of the earth's history.

There are few "hard" data to demonstrate these secular changes in proportions. As in all studies involving time, lack of a well-established Precambrian time scale has proved a major deterrent to the investigator. The well-dated portion of the record, the Paleozoic and younger eras, forms no more than one-sixth of the whole record.

RATE OF SEDIMENTATION

The rate of sedimentation shows extremely wide variations from place to place at the present time. It is virtually impossible to determine the average rate of sedimentation for the present; it is more difficult to do so for past times. Nevertheless, data presented by Barrell (1917) suggest a secular change in the rate of sedimentation. Barrell noted that if the maximum known thickness of strata deposited in each geologic period was divided by the time during which these strata were deposited, there seemed to be a progressive increase in the rate of sedimentation with decreasing age of the beds. The time allocated to each period was that assigned to the periods in question based on a few well-established dates determined by lead–uranium ratios. Barrell's observations have been substantiated by Schuchert (1931) and Holmes (1947) who, however, revised both thickness maxima and time allocations on the basis of new information (Fig. 17-4).

That the data justify the conclusion that the rate of sedimentation has in fact shown a secular change has been challenged by Gilluly (1949). Gilluly pointed out that the probability of finding the section showing the maximum thickness decreases with the age of the deposit because of concealment of the older strata beneath the younger. It is more probable, therefore, that the thickness maxima of younger beds are about correct, whereas those of the older systems are too small. Moreover, as noted by Gilluly, the maximal thickness of the Paleozoic includes much limestone generally conceded to accumulate more slowly than clastic sediments; hence, the mean rate for the Paleozoic is too low. This notion of a slower rate of accumulation for limestones, however, is not true in the thick Paleozoic section of the

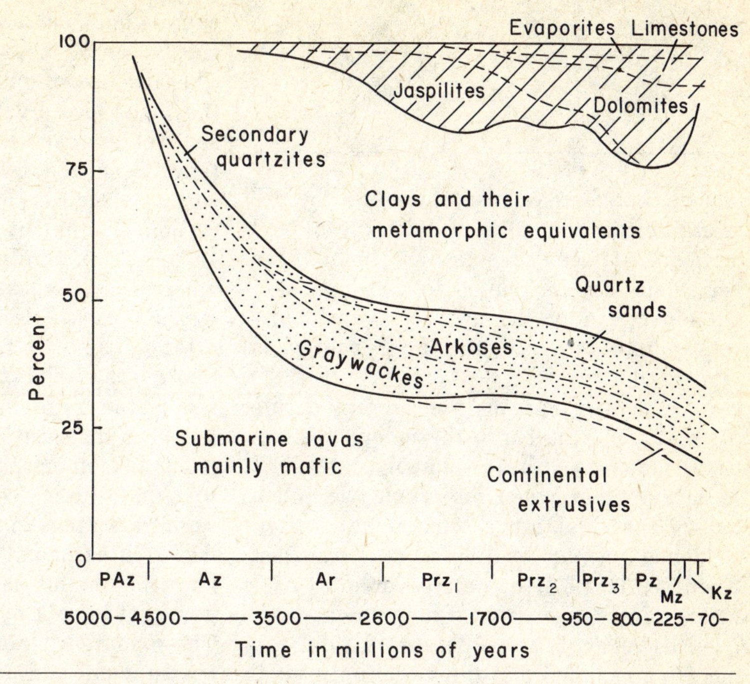

FIG. 17-3. The compositional evolution of sedimentary rocks. (After Ronov, 1964.)

central Appalachians. Here the rate of carbonate deposition was slightly *higher* than the rate for noncarbonate sediments (Colton, 1970, p. 12).

LENGTH OF THE DAY AND LUNAR TIDES

Because of tidal friction, the length of the day is presumed to be increasing, and the moon is presumed to be receding from the earth. The results would be a decrease in the number of days in a month or year and a diminution in the height of the tides. Is there any geologic record to support these conclusions?

The best evidence arises from the classical observations of Wells (1963) that some corals show daily increments of skeletal growth. Others have shown that such daily increments of growth are widespread in the organic world. Molluscan shells are particularly prone to record such growth. Pannella and his associates (1968) have summarized available data. Paleontological evidence does indeed show an increase in the number of days in a lunar month as one goes back

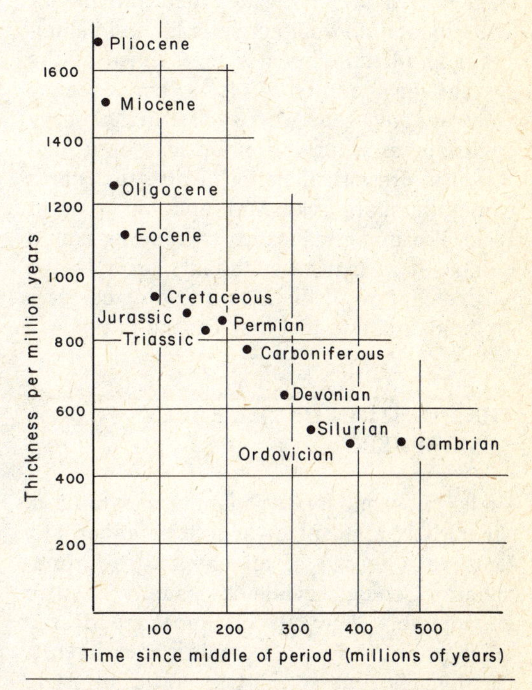

FIG. 17-4. Maximum thickness of systems and series per million years duration of the corresponding periods and epochs plotted against their median age. (Data from Holmes, 1947. Figure redrawn from Gilluly, 1949, Fig. 9.)

in time, but the data also show that the deceleration in the rate of rotation of the earth has not been constant with time. These variations may be related to the expansion or contraction of the area of the shallow seas where most of the tidal energy is dissipated.

As a consequence of the decreased rate of rotation of the earth, the moon must recede from the earth. Presumably, therefore, in earlier times when the moon was much closer, the tidal range was much greater. Evidence on this matter is very meager. Klein (1970, 1971) has endeavored to estimate tidal range from fining-upward sequences in deposits of presumed tidal origin. He concluded that the variation in tidal range since later Precambrian time is not appreciably different from variations in Holocene tidal ranges, a conclusion markedly different from the catastrophic tides postulated by Olson (1970). Merifield and Lamar (1968) postulate tidal currents having velocities 4 to 14 times greater than those of the present day and that as a consequence cross-bedding generated in the high energy environment when the moon was only one-half its present distance from the earth would be more abundant and on a larger scale. Presumably such cross-bedding would be absent in marine sediments prior to the earth's capture of the moon. Geological data are inadequate to test these speculations.

THE QUESTION OF THE EARLY ATMOSPHERE

Under most models of planetary evolution, the earliest atmosphere would be anaerobic. Is there a record of an oxygen-free atmosphere? Geologic evidences usually cited to support this view are: (1) absence of red color in the earliest Precambrian sediments, (2) high ferrous–ferric iron ratio in early Precambrian slates and graywackes, (3) great abundance of iron-formation in the Precambrian, (4) occurrence of detrital pyrite and uraninite in some Precambrian quartzites, and (5) absence of an iron-accumulation zone in some Precambrian paleosols. None of these evidences is conclusive. The absence of red color may only reflect the facies preserved. Much of the Archean sedimentary record is flyschlike. The flysch facies is not red; red coloration is more typical of terrestrial sediments. Red sediments do appear in the early Proterozoic—between 1.8 and 2.5 b.y. in the Great Slave Lake area, where they are associated with halite and gypsum casts (Hoffman, 1969, p. 453) and prior to 2.2 b.y. in the Huronian of the North Shore area. The problem of the ferrous–ferric iron ratio in graywackes and slates has been discussed elsewhere. It may be related to metamorphic reduction rather than to atmospheric composition. Iron-bearing sediments range in age from greater than 2.7 b.y. to Jurassic and younger. Even the oldest known contains jaspers, and some of the youngest have a pyritic and sideritic facies. The oxidation state is a matter of the local environment and not a function of the state of the atmosphere. Sulfide and sideritic sediments have been deposited in times supporting a large and varied biota, demonstrating a fully oxidized atmosphere. Clearly the deposition of these highly reduced sediments is no valid criterion for a reducing atmosphere. Nonetheless, the size and extent of iron sedimentation on a worldwide scale in the period 1.9 to 2.5 b.y. is a remarkable event not yet fully explained or understood. Much emphasis has been placed on the occurrence of detrital pyrite in some Precambrian quartzites, especially in the Blind River area of Ontario and in the Rand Basin of South Africa. Our knowledge of the durability of pyrite is incomplete. It is known as a detrital mineral in modern sands and has even been added to shore sands as a tracer mineral (Baker, 1956). Absence of a zone of iron enrichment in a paleosol profile is not necessarily an indication of an anerobic atmosphere. There are modern gley soils which form under conditions of poor drainage and which lack a zone of iron accumulation.

In short, none of the criteria used to indicate an oxygen-free atmosphere is conclusive. Our knowledge is as yet incomplete; the problem of atmospheric composition remains open.

The presumed absence of an ozone layer and hence lack of a shield from lethal ultraviolet radiation is likewise very speculative.

Stromatolites which characterize intertidal environments today are of very great antiquity. There is even some direct evidence of land-plant cover prior to the evolution of vascular plants (Jackson, 1973, p. 166).

SUMMARY AND CRITIQUE

Evaluation of the data which constitute evidence for presumed secular trends in the nature and character of sedimentary rocks is difficult. In all cases the validity of the sample is clearly questionable. Inasmuch as seven-eighths of the geologic record is without a valid chronology or at best with an imperfect one, the time span represented by most data is only a very small part of the whole of the recorded history of the earth's crust. Until the geologic column for the Precambrian is better established, this defect in the sample will remain. Even for later times the sample is generally small and without a good geographic spread. Most of the data are from the United States and northern Europe. In addition to these inadequacies of the sample, it may be that the sample is biased by selective preservation of certain sedimentary facies and loss of others.

Further difficulties arise following metamorphism of older rocks. How many observed or apparent differences are related to progressive, nonreversible metamorphic or diagenetic and metasomatic changes?

In conclusion, therefore, one can say that no major or significant secular trend has been well established. Some apparent trends of second-order magnitude appear to exist, but whether they are due to systematic bias of the sample, to post-depositional alterations related to age, or to real differences in the environment of deposition is not certainly known.

References

Baker, G., 1956, Sand drift at Portland, Victoria: Proc. Roy. Soc. Victoria, v. 68, pp. 151–197.

Barrell, J., 1917, Rhythms and the measurement of geologic time: Bull. Geol. Soc. Amer., v. 28, pp. 745–904.

Boswell, P. G. H., 1924, Some further considerations of the petrology of sedimentary rocks: Proc. Liverpool Geol. Soc., v. 14, pp. 1–33.

Brenchley, P. J., 1969, Origin of matrix in Ordovician greywackes, Berwyn Hills, North Wales: Jour. Sed. Petrology, v. 39, pp. 1297–1301.

Cloud, P. E., Jr., 1968, Atmospheric and hydrospheric evolution on the primitive earth: Science, v. 160, pp. 729–736.

Colton, G. W., 1970, The Appalachian basin—its depositional sequences and their geologic relationships, *in* Studies of Appalachian geology, central and southern (Fisher, G. W., Pettijohn, F. J., Reed, J. C., Jr., and Weaver, K. N., eds.) New York, Wiley-Interscience, pp. 5–47.

Cummins, W. A., 1962, The greywacke problem: Liverpool and Manchester Geol. Jour., v. 3, pp. 51–72.

Daly, R. A., 1907, The limeless ocean of pre-Cambrian time: Amer. Jour. Sci., ser. 4, v. 23, pp. 93–115.

———, 1909, First calcareous fossils and the evolution of the limestones: Bull. Geol. Soc. Amer., v. 20, pp. 153–170.

Donaldson, J. A., and Ojakangas, R. W., 1974, Orthoquartzite pebbles in Archean conglomerate, northwestern Ontario: Amer. Assoc. Petrol. Geol., Soc. Econ. Paleo. Min., Ann. Mtgs. Abst., v. 1, pp. 26–27.

Eskola, P., 1932, Conditions during the earliest geological times: Ann. Acad. Sci. Fennicae, ser. A., v. 36, pp. 5–14.

Garrels, R. M., and Mackenzie, F. T., 1971, Evolution of sedimentary rocks: New York, Norton, 397 pp.

Gilluly, J., 1949, Distribution of mountain building in geologic time: Bull. Geol. Soc. Amer., v. 60, pp. 561–590.

Goody, R. M., and Walker, J. C. G., 1972, Atmospheres: Englewood Cliffs, N.J., Prentice-Hall, 150 pp.

Herz, N., 1962, Chemical composition of Precambrian pelitic rocks, Quadrilatero Ferrifero, Minas Gerais, Brazil: U.S. Geol. Surv. Prof. Paper 450-C, pp. 75–78.

Hoffman, P., 1969, Proterozoic paleocurrents and depositional history of the East Arm fold belt, Great Slave Lake, Northwest Territories: Canad. Jour. Earth Sci., v. 6, pp. 441–462.

Holland, H. D., 1965, The history of ocean water and its effect on the chemistry of the atmosphere: Proc. Nat. Acad. Sci., v. 53, pp. 1173–1183.

Holmes, A., 1947, The construction of a geological time scale: Trans. Geol. Soc. Glasgow, v. 21, pp. 117–152.

Jackson, T. A., 1973, "Humic" matter in the bitumen of ancient sediments: variations through geologic time: Geology, v. 1, pp. 163–166.
Klein, G. deV., 1970, Paleotidal sedimentation (abstr.): Geol. Soc. Amer., Abstr. with Prog., v. 2, no. 7, p. 598.
———, 1971, A sedimentary model for determining paleotidal range: Bull. Geol. Soc. Amer., v. 82, pp. 2585–2592.
Kuenen, Ph. H., 1966, Matrix of turbidites: experimental approach: Sedimentology, v. 7, pp. 267–297.
MacGregor, A. M., 1927, The problem of the Precambrian atmosphere: South African Jour. Sci., v. 24, pp. 155–172.
Merifield, P. M., and Lamar, D. L., 1968, Sand waves and early earth-moon history: Jour. Geophys. Res., v. 73, pp. 4767–4774.
Nanz, R. H., Jr., 1953, Chemical composition of pre-Cambrian slates with notes on the geochemical evolution of lutites: Jour. Geol., v. 61, pp. 51–64.
Olson, W. S., 1970, Tidal amplitudes in geological history: Trans. New York Acad. Sci., ser. 2, v. 32, pp. 220–233.
Pannella, G., MacClintock, C., and Thompson, M. N., 1968, Paleontological evidence of variations in length of synodic month since Late Cambrian: Science, v. 162, pp. 792–796.
Pettijohn, F. J., 1941, Persistence of heavy minerals and geologic age: Jour. Geol., v. 49, pp. 610–625.
———, 1943, Archean sedimentation: Bull. Geol. Soc. Amer., v. 54, pp. 925–972.
———, 1972, The Archean of the Canadian Shield: a resumé: Geol. Soc. Amer. Mem. 135, pp. 131–149.
Pettijohn, F. J., Potter, P. E., and Siever, R., 1972, Sand and sandstone: New York, Springer, 618 pp.
Rankama, K., 1954, Geologic evidence of chemical composition of the Precambrian atmosphere (abstr.): Bull. Geol. Soc. Amer., v. 65, p. 1297.
Ronov, A. B., 1964, Common tendencies in the chemical evolution of the earth's crust, ocean and atmosphere: Geochem. Int., v. 4, pp. 713–737.
———, 1968, Probable changes in the composition of sea water during the course of geological time: Sedimentology, v. 10, pp. 25–43.
———, 1972, Evolution of rock composition and geochemical processes in the sedimentary shell of the earth: Sedimentology, v. 19, pp. 157–172.
Ronov, A. B., and Migdisov, A. A., 1971, Geochemical history of the crystalline basement and the sedimentary cover of the Russian and North American platforms: Sedimentology, v. 16, pp. 137–185.
Ronov, A. B., Mikhailovskaya, M. S., and Solodkova, I. I., 1963, Evolution of the chemical and mineralogical composition of arenaceous rocks, *in* Chemistry of the earth's crust, v. 1 (Trans. Israel Prog. Sci., 1966), pp. 212–262.
Roscoe, S. M., 1973, The Huronian Supergroup, a Paleoaphebian succession showing evidence of atmospheric evolution, *in* Huronian stratigraphy and sedimentation (Young, G. M., ed.): Geol. Assoc. Canada Spec. Paper 12, pp. 31–47.
Rubey, W. W., 1951, Geological history of sea water: an attempt to state the problem: Bull. Geol. Soc. Amer., v. 62, pp. 1111–1147.
———, 1955, Development of the hydrosphere and atmosphere with special reference to probable composition of the early atmosphere: Geol. Soc. Amer. Spec. Paper 62, pp. 631–650.
Schuchert, C., 1931, Geochronology or the age of the earth on the basis of sediments and life: Bull. Nat. Res. Coun. 80, pp. 10–64.
Tallman, S. L., 1949, Sandstone types: their abundance and cementing agents: Jour. Geol., v. 57, pp. 582–591.
Urey, H. C., 1952, On the early chemical history of the earth and the origin of life: Proc. Nat. Acad. Sci., v. 38, pp. 351–363.
Veizer, J., 1973, Sedimentation in geologic history: Recycling vs. evolution or recycling with evolution: Contr. Min. Petrology, v. 38, pp. 261–278.
Vinogradov, A. P., and Ronov, A. B., 1956, Composition of the sedimentary rocks of the Russian Platform in relation to the history of its tectonic movements: Geokhimia, v. 6, pp. 3–24 (in Russian).
Weaver, C. E., 1967, Potassium, illite, and the ocean: Geochim. Cosmochim. Acta, v. 31, pp. 2181–2196.
Wells, J. W., 1963, Coral growth and geochronology: Nature, v. 197, p. 948.
Yaalon, D. H., 1962, Weathering and soil development through geologic time (abstr.): Bull. Res. Coun. Israel, sect. G., v. 11-G, Proc. Israel Geol. Soc., 4th Congr. Israel Assoc. Adv. Sci., 1961, pp. 149–150.

INDEX OF NAMES

Numbers in italics denote complete citation.

INDEX OF SUBJECTS

75 76 9 8 7 6 5 4 3 2